1989

A PRACTICAL GUIDE
TO MOLECULAR CLONING

A PRACTICAL GUIDE TO MOLECULAR CLONING

SECOND EDITION

Bernard Perbal

Institut Curie—Section de Biologie
Centre Universitaire
Orsay Cedex, France

WILEY

A WILEY-INTERSCIENCE PUBLICATION

JOHN WILEY & SONS

New York • Chichester • Brisbane • Toronto • Singapore

Front cover illustration:

The drawing represents the construction of a shuttle phasmid vector to introduce
foreign DNA into mycobacteria. This novel strategy opens the possibility of
developing recombinant multivaccine vehicles (Jacobs *et al.* 1987) Proc. Natl.
Acad. Sci. USA 327,532

Library of Congress Cataloging in Publication Data:

Perbal, Bernard V.
    A practical guide to molecular cloning.

    "A Wiley-Interscience publication."
    Includes bibliographies and index.
    1. Molecular cloning.   I. Title.
QH442.2.P47   1988   574.87′3282   87-34648
    ISBN 0-471-85071-3
    ISBN 0-471-85070-5 (pbk.)

Printed in the United States of America

10   9   8   7   6   5   4   3   2

Dedicated to Annick
and to the Four S Team

PREFACE

Molecular cloning has become a fundamental technique in all biological fields and probably represents the most powerful tool ever developed not only for understanding the basis of life but also for modifying or creating living species. Cloning of genes has already provided a tremendous body of information for important medical problems. For instance, cancer research has been progressing notably with the isolation of activated oncogenes and the identification of many chromosomal breakage points associated with the tumoral state. The DNA sequences responsible for other severe diseases such as Duchenne distrophy or retinoblastoma have now been identified, and the recent cloning of DNA sequences encoding for factor VIII will allow treatment of hemophiles without injecting human blood compounds. This is especially important to consider now that we know of the potential existence of several infectious new human viruses such as HIV.

Another very promising application of cloning is exemplified by the introduction and expression of a functional cloned gene in defective cells so that the expression of the wild type gene compensates the cellular defect and therefore restores the normal activity. It has already been shown that treatment of anemia, such as thalassemia, is possible by this method. The low-cost mass-production of biological molecules such as interferon, insulin, and antibiotics for use in human therapy has also been made possible by the use of cloning procedures. Engineering of genes has also been applied successfully to the production of vaccines for humans. For example, the production of hepatitis B envelope protein by yeasts has led to the production of a vaccine commercialized this year in the United States. Also very promising is the use of vaccinia virus recombinants that express cloned genes of unrelated infectious agents to develop live vaccines. This approach, which is being used with recombinants expressing proteins from human immunodeficiency virus (HIV) and Hepatitis B virus, might become more powerful with the use of vaccinia recombinants also carrying genes encoding for lymphokines. Even more spectacular is the recent description of a strategy allowing construction of a multivaccine vehicle which might permit a multiple-long-term immunization following a single injection. No doubt that molecular cloning will permit a considerable improvement of our life in the near future.

Aside from medical applications, molecular cloning is also applied at a considerably expanding rate to biotechnology of animals and plants. Since the description of transgenic giant mice in 1982, human growth hormone has been expressed in transgenic rabbits, sheep, and pigs, while transgenic cows will be used for the production of foreign proteins directly excreted in milk. Transgenic plants include cereals, beetroot, tobacco plants, carrots, salads, and petunias. For example, a tobacco plant containing Tobacco Mosaic Virus (TMV) genes has been completely reconstructed in vitro and has given rise to a plant which is more resistant to TMV infection.

Other unexpected achievements permitted by gene cloning include the use of individual-specific "fingerprints" of human DNA in police investigation and expression of 2500-year-old mummy genes. The combined use of computerized programs and genetic engineering will certainly permit in the near future the creation of synthetic proteins with new activities adapted to particular needs of our society, such as elimination of hydrocarbons or heavy metals and the production of new medications.

This second edition of *A Practical Guide to Molecular Cloning* is intended to provide both the basic information needed for inexperienced people to perform cloning of any DNA fragment and eventually study its expression products in transformed (or transfected) cells, and the most sophisticated updated techniques for researchers who are already familiar with molecular biology. Therefore, this book should be very useful to students, technical staff members, and senior scientists. We have deliberately chosen to describe only a selection of techniques which have proved to be, in our hands, the most reliable and easiest to use. In a few cases, selected protocols were obtained from specialists in their fields. Because of the considerable amount of information that was added to the previous edition, we have changed its format and tried to make it even more pleasant to use. We have kept a larger printing size for the protocol sections so that the different steps in the procedures can be checked at a glance when necessary during the experimentation. A wider margin has also been set at the side of the protocols to provide space to write notes or comments. Finally, we have used a spiral binding for the regular edition to fill the need of many colleagues who prefer to open the book flat on a bench.

Aside from these changes, the content of the book has been largely reviewed and updated. A new chapter describes basic methods used in molecular biology. The aim of this chapter is to provide very basic principles of some current methods (e.g., the importance of working at an accurate pH, choosing correct buffers, determining the best conditions for electrophoresis and centrifugation, etc.). This is particularly important to beginners in the field, since many cloning experiments may fail due to a lack of knowledge in this basic matter.

Modification and specific digestion of DNA are two basic steps in molecular cloning. An updated review of commercial modifying enzymes is presented in Chapter 4 (commercial availability in both the United States and Europe being indicated). The combined use of bacterial strains expressing cloned genes and sophisticated purification methods has allowed marketing of high quality products at very reasonable prices. More and more suppliers also offer kits containing all components needed in addition to the enzyme itself. Before using such kits, make sure you can obtain individual components without having to buy the whole package again!

Because restriction endonucleases remain an essential tool in molecular cloning, we have devoted a complete chapter to describe their main properties. The number of commercial restriction endonucleases has increased considerably since the first edition was completed. We have now provided the reader with a concise table indicating commercial availability (adjusted to meet a 1988 publication date) and specific comments for all enzymes from many different sources. Several other tables provide information on thermosensitivity, salt effects, the number of cutting sites in main vectors, and an alphabetical listing of all known restriction endonucleases and isoschizomers. Special attention has been given to the search for compatibility of different restriction sites and the nature of sites generated after combining two compatible ends. This information is particularly useful when one wants to ligate DNA fragments harboring different but compatible ends without using modifying enzymes. A computerized program has been developed especially for this purpose in collaboration with C. Mugnier (CITI 2, PARIS) to provide all compatibilities between commercial enzymes listed in this manual and the resulting restriction sites generated after ligation.

Because so many different new vectors appear every month in literature, we did not feel that a complete review should be included in this book. Rather, we have presented a listing of the main commercial vectors representing the different kinds of systems being used today. Their practical advantages and limitations are also discussed.

Many of the new methods, which have been described in the past few years to obtain rapidly better yields of high quality vector and passenger DNA, have been included in this manual, with new procedures for separating DNA fragments by electrophoresis. These new techniques, which offer the ability to identify genes on intact chromosomes, open very promising roads to the understanding of genome organization and expression. We have also described new methods for propagation of recombinant DNA molecules and in situ hybridization techniques.

Labeling of DNA fragments is another essential step in molecular cloning. Because it is used on a daily basis by most researchers in various laboratories, several methods have been developed for labeling DNA with nonradioactive precursors. Their advantages and conditions for use are described in this manual. These procedures may be of great interest to laboratories involved in diagnosis. Also of considerable interest is the in vitro synthesis of labeled probes from cloned genes by SP6, T7, and T3 RNA polymerases. The methods described in this manual allow the synthesis of riboprobes with very high specific radioactivity, therefore providing increased sensitivity without high backgrounds. These procedure are suitable for in situ hybridization, as well as synthesis of probes to search genomic sequences directly on chromosomes.

When the use of purified DNA fragments is not possible, cloning the genes of interest is made possible by preparing genomic libraries in which any gene is theoretically represented or by cloning cDNA species. For the first time in literature, we give the results of a comparative test which has been performed in our laboratory with the different "lambda in vitro packaging extracts" commercially available. We hope that our experience will help our colleagues in choosing the source of extract being the most appropriate to their own needs. Performing cDNA cloning is now greatly eased by the use of specially tailored vectors which are also commercially available and proved to be of satisfactory quality for reliable, safe, and fast cloning. We describe in this manual two protocols based on the use of commercial and "house-made" vectors.

Nucleotide sequencing of DNA and RNA has been improved by the introduction of new versatile vectors, the use of reverse transcriptases and techniques allowing direct sequencing of double stranded DNA. Use of ^{35}S-labeled nucleotides and reverse transcriptases, direct sequencing of double stranded DNA, sequencing in both orientations, sequencing in microtiter plates, and other new rapid ways of sequencing are among the techniques included in this manual. Because the large amount of information obtained in sequencing often leads to use of computerized systems, we present here an updated brief review of what can be gained by using computerized programs. This chapter does not intend to review all programs, but should allow beginners to become familiar with these techniques.

Cloning of genes may be directed towards modification of living cells (see above) to improve or create new species. Mutagenesis of cloned DNA fragments is possible by several techniques which are described in this book. New techniques, such as "TAB linker mutagenesis" and "random introduction of single base mutations in defined fragments" are described in detail.

We also felt that protocols for studying DNA-protein interactions should be presented in this manual because regulatory proteins involved in differentiation processes or growth cycle regulation are often described as DNA-binding proteins. Techniques for purification of protein-DNA complexes and characterization of the DNA sequences involved in the complex are currently used by leading laboratories, and no current book describes them in detail.

Finally, in many instances, the ultimate goal of cloning remains the production of large quantities of a pure product (for research or industrial purposes). The last chapter has been completely reorganized to provide the reader with an updated review of different ways to express cloned genes in various procaryotic and eucaryotic systems. Protocols for in vitro transcription and translation of cloned sequences are described in detail. Among them are the recent techniques for the addition of cap structures to eucaryotic mRNAs and direct in vitro synthesis of capped mRNA species. We have also described in this chapter two procedures for the in vitro translation of messenger RNAs in both reticulocyte and wheat germ extracts.

Different kinds of procaryotic and eucaryotic expression vectors have become commercially available. We present a brief review of their advantages and main properties. For example, expression vectors have been designed recently to allow an easy recovery of proteins. The use of these vectors in combination with the techniques that have been described to recover intact gene products from fusion proteins should prove to be very helpful in isolating gene products corresponding to cloned sequences of various origins.

In addition to these protocols, we also describe several different techniques for the introduction of foreign DNA in animal or plant cells by means of transfection and for the characterization of polypeptides expressed in transfected or transformed cells by immunoprecipitation and immunoblotting.

The protocols described in this book have been used successfully in the past few years to characterize the cellular localization, the nature, and the function of many gene products in both procaryotes and eucaryotes.

I hope this manual will be helpful not only to all those who are about to start using molecular cloning, but also to the researchers already involved in this field. I am indebted to all the colleagues who helped me select the methods described in this manual, discussed the relative advantages and disadvantages of different techniques, and provided unpublished protocols or experimental data. I wish to thank Johann Soret for reading the manuscript, André Sentenac and collaborators who provided protocols for the study of DNA-proteins interactions, Claude Mugnier and P. Le Beux for allowing us to use the CITI2 facility programs, Anne-Lise Haenni who provided the protocol for wheat-germ extract preparation, and all the members of my laboratory for their criticism, support, and suggestions. I also wish to thank S. Kudzin for his support, stimulating discussions, and help in designing this book, and A. Besnard, C. Leray, and N. Frey for skillful typing assistance. The members of the production department at Wiley have done a great job with my original manuscript. I thank all of them sincerely.

Thanks are due to Amersham International, Beckman, Boehringer Mannheim, Pharmacia Fine Chemicals, Hoefer Instruments, International Biotechnologies Inc., Promega Biotec, New England Biolabs and United States Biochemical Corp., who allowed me to reproduce their published material. Thanks also to N. Sasavage at Bethesda Research Laboratories for allowing us to use information published in Focus and for providing documents. To all of them I dedicate the Epilogue of this manual.

Again, this is my opportunity to thank most particularly my children, Sabine, Sébastien, Séverine, and Sonia, and my wife Annick for their constant support, patience, and help during the preparation of this book.

Orsay, France
January, 1988

BERNARD PERBAL

CONTENTS

17 PREPARATION OF GENOMIC LIBRARIES 480

18 PURIFICATION AND CHARACTERIZATION OF RNA SPECIES 516

19 CLONING OF cDNA SPECIES 550

20 SEQUENCING OF DNA 578

24 METHODS FOR STUDYING DNA–PROTEIN INTERACTIONS 719

25 EXPRESSION OF CLONED DNA SEQUENCES IN VITRO OR IN PROCARYOTIC AND EUCARYOTIC CELLS 731

EPILOGUE: IF YOU WANT AVA I TO MEET BAM HI . . . 795

APPENDIX: ADDRESSES OF MAIN OFFICES FOR SUPPLIERS CITED IN THIS MANUAL 798

INDEX 801

A PRACTICAL GUIDE
TO MOLECULAR CLONING

1

LABORATORY EQUIPMENT FOR MOLECULAR CLONING

The following laboratory equipment is used in molecular cloning:

Autoclave(s).

Centrifuges for ultracentrifugation (e.g., Beckman, Sorvall, MSE).

Rotors such as Beckman 60 or 70 Ti, SW41, SW28, SW50, or equivalent. Vertical rotors are also very useful.

Centrifuges for high-speed centrifugation, such as RC5B (Sorvall) or J21 (Beckman), with rotors for large volumes (Sorvall GS3 = 6 × 500 ml, GSA = 6 × 250 ml, Beckman JS4.2 = 6 × 1000 ml or 6 × 500 ml or 6 × 250 ml with adapters) and rotors for small volumes (Sorvall SS34 or Beckman JA20), allowing runs up to 50,000 × g.

The new TL100 ultracentrifuge (Beckman) is particularly adapted to manipulation of small samples, and allows a considerable reduction of centrifugation times.

Centrifuge for microfuge tubes (1.5 ml and 1 ml). Rotors allowing horizontal centrifugation will permit pellet precipitates at the bottom of the tubes.

Centrifuge for low-speed centrifugation (used for pelleting bacteria, pelleting cells, phenol extraction, etc.).

Computer (e.g., IBM or Apple).

Sequence analysis software (e.g., IBI/Pustell).

Laminar flow hood and chemical hood.

Orbital shaker for bacterial cultures (e.g., New Brunswick), with thermostat for use anywhere; without thermostat if used in warm room.

CO_2 incubator for cell cultures.

Refrigerators and freezers (4°C, −25°C, −80°C).

Power supplies for electrophoresis. Consider buying one delivering high amperage (up to 200 mA) and high voltage (2500–3000 V) for sequencing (e.g., Bio-Rad, LKB, Pharmacia).

Power supply delivering at least 250 mA for Western blotting.

Pulse controller (such as Hoeffer PC 750) for reverse field electrophoresis (requires a 750-V DC power supply).

Apparatus for electrophoretic transfer of proteins and nucleic acids (e.g., BioRad, and Hoefer). The transphor system from Hoefer comes with a convenient power lid designed to provide the high current required for efficient and even transfer.

Camera such as Polaroid MP4 with instant type 57 (positive) and/or type 55 (positive/negative) films.

Benchtop orbital shaker for staining, destaining, and washing.

Transilluminator (long-wave UV will avoid damaging DNA).

Vacuum oven (80°C).

Microwave oven.

Gel beds for regular agarose and acrylamide gels and for sequencing gels.

Miniature electrophoresis apparatus for rapid screening of proteins (such as "mighty small" SE250 from Hoefer).

Apparatus for slot blotting (Schleicher & Schuell, BRL).

Microscope (with inverted light for examination of cell cultures).

Refractometer.

Desiccator.

Speed Vac concentrator/evaporator (Savant).

Equipment for redistillation of chemicals.

Equipment for deionization or distillation of water.

Water bath at 37°C.

Refrigerated water bath for temperatures of 10–18°C.

Heating block to keep tubes warm and dry (e.g., for phage plating).

Water bath at 80°C.

Water bath at 45°C.

Water bath with agitation.

Automatic pipettemen (e.g., Gilson or Eppendorf) for volumes of 0–20 μl, 0–100 μl, 0–200 μl, 0–1000 μl. Yellow and blue tips for pipetteman.

Cassettes for autoradiography, with intensifying screens for ^{32}P (DuPont) and films (Kodak or Cronex). Keep cassettes for ^{35}S sequencing separately.

Vacuum pump.

Liquid nitrogen tank.

pH meter and pH indicator paper.

Spectrophotometer (UV and visible).

Multicanal peristaltic pump.

Radioactivity counter.

Radioactivity monitor.

Plastic shield for protection against radiation.

Pipette acid (e.g., Drummond).

Vortexer.

Magnetic stirrer.

Balances.

Sealing-bag system and plastic bags (e.g., Sears or Krups).

Timer.

Slab gel dryers (Hoefer). Consider buying one for sequencing gels and one for protein gels.

Apparatus for electroelution, and concentration of samples to small volumes (e.g., Biotrap, Schleicher & Schuell).

Gel reader for sequencing gels. Some of them may be related directly to computers for safe analysis (IBI, Beckman).

Filtration units (e.g., Millipore).

Polypropylene bottles for centrifugation (1 liter, 500 ml, 250 ml).

Polypropylene tubes (5 ml, 15 ml, 50 ml).

Polypropylene flasks, cylinders, and beakers.

Ultraclear tubes (Beckman) for high-speed centrifugation.

Quick seal unit and tubes (Beckman).

Corex tubes (30 ml and 15 ml) with adapters.

Whatman 3MM paper.

Nitrocellulose filters.

Millipore flat forceps to manipulate filters.

BenchKote (Whatman).

Plastic disposable pipettes (10 ml, 5 ml, 1 ml).

Glass disposable pipettes (20 ml, 1 ml).

Glass plates for electrophoresis.

Combs for electrophoresis, sharkstooth combs for sequencing.

Spacers for electrophoresis, wedge spacers for sequencing.

Glass slides for minigels.

Polypropylene microfuge tubes (1.5 ml, 0.75 ml) such as Eppendorf or Beckman.

Dialyzing tubing (2.5-cm and 1-cm diameter).

Portable UV lamp.

Syringes (50 ml, 10 ml, 5 ml, 1 ml).

Hamilton syringes (5 μl, 10 μl).

Needles (18-, 21-, and 25-gauge).

Giemsa rapid stain for staining cells.

Xylene cyanol and Bromophenol blue (dyes for electrophoresis).

Gradient former and collector.

Microcaps (1–5 μl, 10 μl, 50μl, 100μl).

Adhesive tape (regular and indicating radioactivity or biohazard).

Sterilization tape (for 120°C and 180°C).

Scalpels and blades, forceps and scissors.

Plastic autoclaving bags (to eliminate biohazardous material).

Loops to seed bacterial cultures and pick colonies.

Petri dishes for eucaryotic cell cultures (special plastic) and for bacteria plating.

Isothermic containers.

All glassware to be used for RNA work should be baked at 180°C for at least 2 hours. Glassware for DNA work should be siliconized by immersion in 5% dichloromethylsilane in chloroform, followed by extensive rinsing with deionized water and drying in the oven.

2

A FEW WORDS ABOUT SAFETY

LABORATORY HOODS

Laboratory hoods (fume hoods) are open boxes connected on one side to an exhaust system which pulls the air into the open side of the hood and discharges the air outside. When the hood is operating properly and being used correctly, noxious material can be manipulated in the hood without exposing the operator and other laboratory workers to the hazards of breathing toxic compounds.

Air should move into a hood with an average inward linear velocity (face velocity) of at least 100 feet per minute (fpm), and the minimum inward velocity at any point should not be less than 70 fpm. Hoods for chemical carcinogens should have minimum average face velocities of 150 fpm with a minimum at any point of 125 fpm.

The size of the hood face opening can be adjusted by sliding windows. It is useful to have some means of indicating that there is an inward flow of air (meter, manometer, ribbons, etc.).

To ensure the proper functioning of the hood, there should be no excessive crossdrafts at the hood face. Drafts can cause reversal of the inward airflow and thus pull noxious material *out of the hood.* Drafts may be caused by open windows, air supply vents, and fans, or by people walking in front of the hood. Hoods should be located away from doors, open windows, or equipment creating appreciable air movement.

The hood will carry off only airborne material released at low velocity within the hood, such as boiling or evaporating liquids. One should work well inside the hood. At a distance of 3 feet in front of the hood, the inward air velocity drops to about 10 fpm (this is less than the velocity of random air currents in the room).

Remember that the air in the hood becomes contaminated by the material used in the hood. **DO NOT PUT YOUR HEAD INTO THE HOOD.**

BIOLOGICAL SAFETY CABINETS

Work with biological hazards must be done in a biological safety cabinet. Keep in mind that although some biological safety cabinets use laminar flow, a laminar flow hood is not necessarily synonymous with a biological safety cabinet.

A laminar flow hood provides air moving without turbulence, the direction of flow being either horizontal or vertical. The air may be cleaned by passing it through a high-efficiency particulate air (HEPA) filter. This will remove virtually all particles, including those of extremely small size, thus providing a working area free of particulate contamination. Laminar flow hoods are especially useful in preventing the spread of aerosolized material.

One can distinguish three different classes of biological safety cabinets.

Class I is similar to a chemical fume hood, providing an inward flow of air at the work opening so that aerosols cannot spread into the room. The air flow is

Table 1. Protection Provided by Various Kinds of Gloves

	Glove Material				
	Neoprene	Polyvinyl Chloride	Polyvinyl Alcohol	Latex	Polyethylene
Acids (dilute)	OK	OK	OK	OK	Bad
Alcohol	OK	OK	OK	OK	OK
Alkalis (dilute)	OK	OK	OK	OK	Bad
Chlorinated solvents	OK	OK	OK	Bad	Bad
Ketones (acetone)	OK	Bad	Bad	OK	Bad
Petroleum distillates, mineral spirits	OK	Bad	OK	Bad	Bad
Hydrocarbons	OK	Bad	OK	Bad	Bad
Epoxy resins	OK	OK	Bad	Bad	Bad

not laminar, and the working area is not protected from the air currents generated in the room. The "dirty" air is exhausted from the hood through a HEPA filter to prevent discharge of biohazardous material in the environment. Since an exhaust fan is needed for operation of the hood, which is vented to the outside, chemicals can be used inside.

Class II cabinets are usually subdivided into two groups. Group I contains hoods that recirculate 70% of the filtered air and thus cannot be used with corrosive materials, flammable solvents, or noxious chemicals, because cabinets recirculate only 30% of the filtered air. This kind of cabinet can be used with moderate amounts of chemicals and solvents.

Class III cabinets are gastight "glove boxes" that are maintained at a negative pressure with respect to the room air pressure. This kind of cabinet is used when the highest degree of personal protection is required. The hoods are ventilated internally, and HEPA filtration is provided for inlet and exhaust air.

Biological safety cabinets of all kinds should be checked periodically (at least once a year). When a cabinet must be decontaminated (for inspection or because of a spill), seal it and volatilize *para*-formaldehyde into it. Also check that the filters do not leak and that the air flow is proper.

GLOVES

It is a good policy to wear gloves for protection when working with chemicals, irritating substances, radioactive products, solvents, and so on, as long as they provide the protection needed or expected. Many different kinds of gloves are used in laboratories and the protection that they provide may vary

from a few seconds to a few hours, depending on the conditions of use.

A particular kind of glove may stop some products and be inefficient for others. Table 1 summarizes the protection given by various kinds of gloves.

ULTRAVIOLET LIGHT

Ultraviolet light (UV) may be divided into three wavelength groupings: near UV (315–400 nm), midrange UV (280–315 nm), and far UV (200–280 nm). Maximal sensitivity in humans is at about 280 nm. Exposure to direct or indirect midrange or far UV can cause acute eye irritation after a latent period of ½–24 hours. Because the retina is not sensitive to UV, eye damage may result without the subject being aware of the exposure. The latent period for feeling symptoms (sensation of sand in the eyes, tearing, sensitivity to light, difficulties in opening eyelids, etc.) is dependent on the intensity of the exposure; the more intense the exposure, the shorter the latent period. Usually discomfort disappears within 48 hours. However, it may be accompanied by a temporary reduction in visual acuity.

Skin is also sensitive to UV. Tanning provides some protection against further exposure, but remember that the midrange UV of sunlight is considered to be more important than aging in producing undesirable skin changes. UV radiation has also been implicated as a factor in causing skin cancer.

In conclusion, protect your eyes and skin from the effects of UV radiation by wearing goggles with side shields, by clothing, and by limiting exposure. UV light is used in the laboratory for several purposes. Two of them will be considered below.

Germicidal Lights

Germicidal lights are used in places where reduction of the number of microorganisms on surfaces and in the air is needed. These include biological safety cabinets, laminar flow hoods, and tissue culture and bacteriology laboratories. Germicidal UV lights resemble uncoated, clear fluorescent tubes and give off a bright bluish light. We strongly suggest the use of electrical interlocks to automatically turn off the UV when the door is opened, because the bluish light produced by the tubes may not be noticed in the daytime in a clear room. Put caution signs on doors.

Usually germicidal tubes are put on the ceiling. If properly installed they should not be hazardous to persons standing on the floor. The tubes should be cleaned periodically and replaced as soon as their intensity has dropped to ~60% of the rated initial output.

We advise the use of oil-based paint on walls and ceilings to reduce reflection of UV light. Also, be aware that a lot of plasticware and other plastic materials are progressively degraded by exposure to UV.

Detection of DNA by UV Illumination

Detection of DNA molecules having bound ethidium bromide is done by illumination with UV light. Either long- or short-wave UV can be used. However, illumination with short-wave UV light may induce damage in the DNA molecule, and in some cases will not allow further cloning or will interfere with the expression of the cloned fragment.

Three kinds of transilluminators are used to detect DNA fragments in agarose or acrylamide gels after staining with ethidium bromide. The three wavelengths used are 254 nm, 302 nm, and 365 nm. The 302-nm transilluminators are recommended for easy detection of small amounts of DNA stained with ethidium bromide without inducing damage, thus allowing further cloning of the purified fragments.

PHENOL

Phenol (carbolic acid, hydroxy benzene) has been widely used as a disinfectant and germicide. Its use as an antiseptic in surgeries in the 19th century resulted in many cases of chronic and subacute poisoning among surgeons and their assistants.

Pure phenol is a white crystalline solid with a melting point of 41°C. Upon oxidation it becomes pinkish. It is a dangerously toxic material that can produce poisoning when ingested, inhaled, or absorbed through the skin. The toxic effects include headache, dizziness, nausea, weakness, difficulty in breathing, unconsciousness, and death.

Phenol is corrosive to the skin, initially producing a white softened area, followed by severe burns. It is rapidly absorbed through the skin, and because of the local anesthetic properties of phenol, skin burns may not be felt until there has been serious damage.

First Aid

If phenol is spilled on the skin, flush off immediately with large amounts of water. **DO NOT USE ETHANOL.** Because of phenol's relatively low vapor pressure, occupational systemic poisoning usually results from skin contact with phenol or composed solutions rather than from inhalation. If eyes are contaminated, wash them with running water for about 15 minutes; call for medical help.

Redistillation of Phenol

In addition to its toxic properties, phenol is combustible. When heated above 80°C, the vapors can form an explosive mixture with air. Phenol fires can be extinguished with water or dry chemical or carbon dioxide fire extinguishers.

Distillation of phenol is performed under a fume hood with good aspiration. Set up the distillation column as shown in Figure 1.

Do not refrigerate to help condense the vapors. This would lead to phenol crystallization, pressure increase, and explosion.

Add some zinc to the liquid phenol that you heat up. Commercial phenol contains an impurity which raises the melting point to 182°C. When the vapor temperature reaches 181°C, begin to collect.

We advise keeping redistilled phenol under nitrogen to avoid oxydation and formation of quinones which may react with primary amines to form stable cross-linked products. Formation of quinones is accompanied by a yellow coloration of the phenol solution. Addition of 8-hydroxyquinoline may help to retard oxydation. Phenol solutions oxidize rapidly at 4°C but are stable in the crystalline state for several months when stored at -20°C. We keep redistilled phenol in glass bottles (e.g., 50 ml of phenol per 100-ml bottle). Before the cap is screwed on, nitrogen is blown onto the surface of the phenol for a few minutes.

When a buffer-saturated phenol solution is

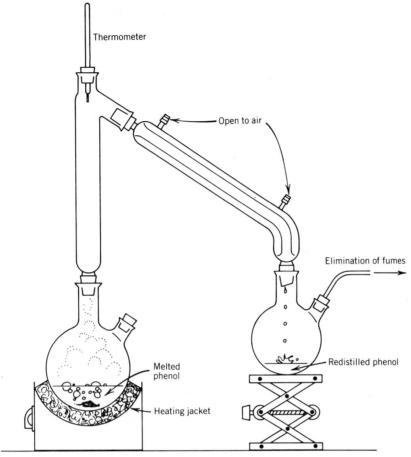

Figure 1. Setup for phenol distillation.

needed for nucleic acid extraction we proceed as described in Chapter 3.

LIQUID SCINTILLATION COCKTAILS

These solutions contain flammable solvents such as toluene, xylene, or dioxane. Since these mixtures have flash points lower than 38°C, do not store them in large glass containers.

Note that dioxane may form an explosive peroxide when it ages. Do not store for more than three months after opening a container. Dioxane is listed among suspected carcinogens.

CHLOROFORM

Chloroform is used in nucleic acid extractions in combination with isoamyl alcohol (24 parts chloroform/1 part isoamyl alcohol). Chloroform is a non-flammable, heavy, very volatile compound which is miscible with alcohol, benzene, and ether. Since it is light sensitive when pure, it is kept in brown bottles. Handle chloroform with care. Mixing chloroform with other solvents may involve serious hazard. Chloroform undergoes an endothermic reaction with acetone in basic solution to form chloretone. Do not mix chloroform and acetone.

Mixing of strong base with chloroform or other chlorinated hydrocarbons has been implicated in a number of explosions.

ISOAMYL ALCOHOL

Isoamyl alcohol, also called isopentyl alcohol, is used in combination with chloroform to help separate the phases in nucleic acid extractions. It has a disagreeable odor, and is miscible with alcohol, ether, benzene, and chloroform.

Handle isoamyl alcohol in a fume hood. Its vapors are poisonous.

ETHYL ETHER

Extraction with ether is often a last step in nucleic acid purification, following treatment with phenol and chloroform. Ether is a very volatile and highly flammable compound with vapor denser than air. Always use ether under a hood and away from any sparks. Autoignition temperature is 180–190°C.

DIETHYL PYROCARBONATE

This compound is commonly used to inactivate nucleases. It is a hazardous product because it can form urethane (a powerful carcinogen) in the presence of ammonia.

Diethyl pyrocarbonate decomposes rapidly to ethanol and carbon dioxide when added to pure water (at pH 7.0). It can be disposed of by being placed overnight in an aqueous ammonia-free solution. It must be handled with great care, because it can react with the body's endogenous ammonia and because it is irritating to eyes, mucous membranes, and skin.

METHYLMERCURY

Methylmercury is a poisonous reagent which must be handled with care. Working in a recently checked fume hood is absolutely necessary when methylmercury powder is manipulated. The use of diluted solutions of methylmercury does not require working in fume hoods. According to Junghans (*Environ. Res.* **1981**, *33* 1–31 and personal communication), methylmercury is slightly volatile (5000-fold less so than water) and there is more methylmercury (60 µg) in an 8-ounce (200-g) can of tuna than in 500 liters of air allowed to equilibrate to saturation (enclosed without ventilation) above a 50-liter volume of 5mM solution. Also, most Americans have a steady-state body burden of 1 mg of methylmercury (which is excreted), while Japanese average two to three times as much because of dietary differences, but are still 40-fold below minimum toxic levels. There is no special concern for pregnant women working with this chemical (Junghans, 1981). Health authorities in the United States permit drain disposal of methylmercury in the quantities considered here. Check for the regulations in other countries.

PLASTICWARE AND CHEMICALS

Different kinds of plastic containers (Erlenmeyer flasks, beakers, tubes, measuring cylinders, centrifugation bottles, etc.) are commonly used in many laboratories. Be aware that some plastics are severely altered by prolonged contact with chemicals. Table 2 should help you to choose containers of the appropriate material.

WORKING WITH ³²P-LABELED COMPOUNDS

Phosphorus-32 is widely used as a tracer because it is easily detected. Since the β particles emitted by ³²P have an energy of 1.71 MeV (6.1-meter range in air), labeled compounds must be handled with caution and shields must be used. Remember that when β particles hit targets electromagnetic radiation known as bremsstrahlung is produced. The yield of bremsstrahlung is directly proportional to the density of the material chosen for shielding. Therefore, a low-density material (such as plastic) must be used as first layer of the shield. A second layer of high-density material may be added, if necessary, to absorb the bremsstrahlung emitted. It is a good policy to wear gloves (two pairs are often necessary), protect your eyes, submit urine samples for regular checks, and use dosimeters. Also, it is recommended to protect benches with absorbent paper such as BenchKote and to use Geiger–Müller counters for detection of spills. Never eat, smoke, or drink while manipulating radioactive compounds. Use special tape to label containers and tubes in which radioactive materials are kept. The maximum permissible body burden of phosphorus-32 is 30µCi, but the maximum permissible burden for bone is only 6 µCi.

WORKING WITH KNOWN OR POTENTIAL CARCINOGENS

The Occupational Safety and Health Administration, an agency of the United States Government, has compiled a list of known and suspected carcinogenic compounds among more than 800 highly toxic chemicals. Category I includes *confirmed carcinogens,* as established by data obtained with humans, or with at least two mammalian species. Category II contains the compounds whose carcinogenicity has

Table 2.   Resistance of Different Plastics to Chemicals and Temperature

	Plastic Type				
	PTFE	PC	PP	HDPE	LDPE
Inorganic acids	E	G	E	E	E
Organic acids	E	G	E	E	E
Alcohols	E	G	E	E	E
Aldehydes	E	B	G	G	G
Amines	E	N	G	G	G
Bases	E	N	E	E	E
Esters	E	B	G	G	G
Ethers	E	N	E	E	E
Glycols	E	G	E	E	E
Acetone	E	N	E	E	E
Acetic acid	E	E	E	E	E
Boric acid	E	E	E	E	E
Citric acid	E	E	E	E	E
Perchloric acid	G	N	G	G	G
Trichloroacetic acid	E	B	E	G	G
Ethanol	E	E	E	E	E
Butanol	E	G	E	E	E
Methanol	E	G	E	E	E
Chloroform	E	N	G	G	B
Ethyl ether	E	N	N	N	N
Formaldehyde	E	E	G	E	E
Phenol	E	E	G	G	G
Toluene	E	B	G	G	B
Saturated urea	E	N	E	E	E
Xylene	E	N	B	G	G
Sodium hydroxide	E	N	E	E	E
Sterilization[a]					
Autoclave	Yes	Yes	Yes	120° C, 20 minutes	No
Dry	Yes	Yes	Yes	Yes	Yes
Max. temperature (°C)	250	135	135	120	80

[a]Sterilization of polycarbonate diminishes its mechanical characteristics. Do not sterilize polycarbonate parts to be used under vacuum.

Abbreviations: PTFE, polytetrafluoroethylene (Hostaflon, Algoflon, Fluon, Teflon); PC, polycarbonate (Sinvet, Makrolon, Lexan); PP, polypropylene (Kastilene, Novolen, Hostalen P, Propathene, Moplen); HDPE, high-density polyethylene (Eraclene HD, Lupolen, Hostalen, Moplen RO, Eltex); LDPE, low-density polyethylene (Eraclene LD, Lupolen, Alaton, Alkathene, Fertene, Sirtene, Hostalen); E, excellent; G, good; B, bad; N, not resistant.

been demonstrated in only one species, or is not yet confirmed by other tests. Chemicals with no known carcinogenic potential fall in category III.

Several of the chemicals routinely used in the course of molecular cloning belong to categories I and II (see Table 3). Below are some of the elementary recommendations that apply to handling of all known carcinogens or suspected carcinogens, and more generally, to handling of all highly toxic compounds.

Always wear protective clothing and gloves and clearly identify the bench area where toxic compounds are being used. It is wise to use stainless steel or plastic trays, shatterproof glass plates, and absorbent plastic-backed paper such as benchkote (Whatman). When working with a potential carcinogen check for spills and avoid using all procedures that generate aerosols. Beware of vents! Label all vessels appropriately and do not discard any known or suspected carcinogen without prior inactivation. Cleanup and disposal is efficiently achieved by one of these three methods:

1. Immersion in chromic acid followed by extensive washing down the drain with water. Do not forget to renew the chromic acid periodically.

2. Oxydation by immersion in household bleach for 24 hours at room temperature followed by extensive washing down the drain with water.

3. Inactivation by 5% KOH followed by extensive washing down the drain with water.

All solid waste can be buried in a proper manner to avoid any contamination of surrounding areas.

Table 3. Carcinogenic Potential of Some Common Chemicals Used in Cloning

Compound	Solvent	Disposal	Carcinogenic (C) or Teratogenic (T)	Toxic (T) Highly Toxic (HT)
Actinomycin	Water	Burial or chromic acid	C	HT
Colchicine	Ethanol	Burial or chromic acid	T	HT
Colcemide				
Cycloheximide	Ethanol	Burial or chromic acid	T	HT
DEP	5% KOH	See text	—	T
Ethidium bromide	Water	24 hours in bleach	C	T
Iodoacetic acid	Water	24 hours in bleach	T	T
Methotrexate	5% KOH	5% KOH	T	HT
ICR170,191	Ethanol	24 hours in bleach	C	HT
Mycostatin	Methanol	Burial or chromic acid	—	T
Spermine	Water	Burial or chromic acid	—	T

3

BASIC METHODS IN MOLECULAR BIOLOGY

GEL ELECTROPHORESIS

Introduction to Electrophoretic Theory*

Electrophoresis is a method whereby charged molecules in solution, chiefly proteins and nucleic acids, migrate in response to an electrical field. Their rate of migration or mobility through the electrical field depends on the strength of the field, on the net charge, size, and shape of the molecules, and, also, on the ionic strength, viscosity, and temperature of the medium in which the molecules are moving. As an analytical tool, electrophoresis is simple, rapid, and highly sensitive. It is used to study the properties of a single charged species, and as a separating technique.

Electrophoresis is usually carried out in gels formed in tubes, slabs, or on a flat bed. In a tube gel unit, the gel is formed in a glass tube, usually about 12 cm long and 3 to 5 mm ID. A slab gel is formed in a glass sandwich made of two flat glass plates separated by two spacer strips at the edges and clamped together to make a watertight seal. Both tube and slab gels are mounted vertically. The gel in a flat bed unit is poured on a horizontal surface and has no cover plate on it.

In most electrophoresis units, the gel is mounted between two buffer chambers containing separate electrodes so that the only electrical connection between the two chambers is through the gel.

*Courtesy of Hoefer Scientific Instruments.

Interrelation of Resistance, Voltage, Current, and Power

Two basic electrical equations are important in electrophoresis. The first in Ohm's law, $I = E/R$, which states that electrical current (I, amperes) is directly proportional to voltage (E, volts) and inversely proportional to resistance (R, ohms). The second is $P = EI$, which states that power (P, watts), a measure of the amount of heat produced, is the product of voltage and current. This can also be expressed as $P = I^2R$.

In electrophoresis, one electrical parameter, either current, voltage, or power, is always held constant. As the equations tell us, if resistance increases during the run, the consequences will differ in the following ways, depending on the constant mode selected: in constant current (velocity is directly proportional to current), the velocity is maintained but heat is generated during the course of the run; in constant power, the velocity slows but heating is kept constant.

Until recently many gel systems were designed to use either constant current or constant voltage. The important consideration, whichever mode is selected, is that diffusion and loss of sample activity caused either by heat or time be minimized.

The Net Charge of Proteins Is Determined by the pH of the Medium

Proteins are amphoteric compounds, that is, they contain both acidic and basic residues. Their net charge is determined by the pH of the medium they are in. Most of the charge of a protein comes from the pH dependent ionization of carboxyl and amino groups ($-COOH \rightleftharpoons -COO^- + H^+$; $-NH_2 + H^+ \rightleftharpoons -NH_3^+$). Each protein has its own characteristic charge properties depending on the number and kinds of amino acids carrying amino or carboxyl groups.

Nucleic acids, unlike proteins, are not amphoteric. They remain negative at any pH used for electrophoresis.

There is a pH at which there is no net charge on a protein; this pH is called the isoelectric point or pI. Each protein has its own pI. For example, the pI of human hemoglobin is 7.07, that of β-lactoglobin, 5.34. In a solution with a pH above its isoelectric point, a protein has a net negative charge and migrates toward the anode in an electrical field. Below its isoelectric point, the protein is positive and migrates toward the cathode.

A researcher can specifically design a system to separate proteins that have different charge properties. For example, when hemoglobin and β-lactoglobin are electrophoresed at pH 7.07, the hemoglobin does not migrate because it is uncharged, and the β-lactoglobin migrates toward the anode because of its net negative charge.

The pH of a solution in an electrophoresis system must be kept constant to maintain the charge, and, hence, the mobilities of the proteins. For this reason, the solutions used in electrophoresis must be buffered.

Maintaining Constant Temperature

At every stage of the electrophoretic process, temperature regulation is important.

Polymerization of the gels is an exothermic reaction. If not regulated, it will set up convections that cause irregularities in the pore size of the gel. To keep pore size uniform and reproducible the reaction should be carried out at a constant temperature. For this reason, during polymerization, gels should be surrounded by a liquid coolant held at a constant temperature.

During electrophoresis, constant temperature is essential to avoid denaturing heat labile proteins. In running slab gels, temperature has a further importance. When heat is not conducted away from the slab, even in an SDS run where denaturation is not a consideration, the samples will move faster in the center of the gel at the edges. This produces a smile effect and makes the comparison of bands difficult.

When temperature is constant, a multiple-sample separation will maintain a side-to-side flatness. All the samples will move evenly and horizontally downward.

Most systems have been developed to run at either 0°C or 25°C. In order to maintain an acceptable temperature throughout the run, the lower buffer should cover the part of the tube or slab which contains the gels. For precise regulation, the electrophoresis tank must include a system for cooling and circulating buffer in the tank. The cooled buffer will then circulate around the gels and dissipate whatever heat is generated during the run.

Support Matrix

Generally, the sample is run in a support matrix, such as paper, cellulose acetate, starch gel, agarose, or polyacrylamide gel. The matrix inhibits convections caused by heating. It provides a record of the electrophoretic run. At the end of the run, the matrix can be stained and used for scanning, autoradiography, or storage.

In addition, the most commonly used support matrices, agarose and polyacrylamide, provide the operator with a separating technique. Although agarose and polyacrylamide differ greatly in their physical and chemical structures, they both make porous gels. A porous gel may act as a sieve by retarding or, in some cases, by completely obstructing the movement of macromolecules while allowing smaller molecules to migrate freely. By preparing a gel with a restrictive pore size, the operator can take advantage of molecular size differences among proteins. For example, hemoglobin weighs 66,000 daltons and β-lactoglobin weighs 38,000 daltons. It is possible to prepare a gel with a pore size that will be more restrictive to hemoglobin than to β-lactoglobin. At a pH above the isoelectric points of both proteins, where both are anionic, the two will separate not only by charge but by molecular size.

Because the pores of an agarose gel are large, agarose is used to separate macromolecules such as nucleic acids and large proteins and proteins complexes. Polyacrylamide, which makes a small pore gel, is used to separate most proteins and small oligonucleotides, as well as small DNA fragments and RNA species.

Whichever matrix the operator chooses, it is important that the matrix be electrically neutral. This condition is essential to prevent the flow of solvent toward one or the other electrode, usually the cathode, due to the presence of immobile charged groups in the matrix. This phenomenon, called electroendosmosis, decreases the resolution of the sample.

Agarose Gels

Agarose is a highly purified polysaccharide derived from agar. Unlike agar, it is not heavily contaminated with charged material. Most preparations, however, do contain some anions such as pyruvate and sulfate, which may cause some electroendosmosis.

Agarose, which comes in powder form, dissolves when added to boiling liquid. It remains in a liquid state until the temperature is lowered to about 40°C, at which point it gels. The gel is stable; it will not dissolve again until the temperature is raised back to about 100°C.

Its pore size may be predetermined by adjusting the concentration of agarose in the gel: the higher the concentration, the smaller the pore size. Working concentrations are frequently in the range of 0.4 to 2% w/v.

Agarose gels are fragile, however. They are actually hydrocolloids, and they are held together by the formation of weak hydrogen and hydrophobic bonds. For this reason, an agarose gel should be handled with special care. If the gel bends, it will break. It should not be picked up without the use of a tray or a special tool.

Polyacrylamide Gels

Polyacrylamide gels are tougher than agarose gels. In forming the polyacrylamide gel, acrylamide monomers polymerize into long chains that are covalently linked by a cross-linker. It is the cross-linker that actually holds the structure together. The most common cross-linker is *N,N'*-methylene-bis-acrylamide or *bis* for short. Other cross-linkers whose special properties aid in solubilizing polyacrylamide are also used occasionally.

Polyacrylamide is chemically complex, and so also is the production and use of the gel. The following brief discussions indicate some of the technical requirements.

Preparing and Pouring the Mixture

Because oxygen inhibits polymerization, the monomer mixture must be deaerated. This may be done by purging the mixture with an inert gas or by evacuating the mixture with a vacuum.

When the gel is poured, either in a tube or in slab, the top of the gel forms a meniscus. If ignored, this curved top on the gel will cause a distortion of the banding pattern. To eliminate the meniscus, a thin layer of water or water-saturated butanol is carefully floated on the surface of the gel mixture before it polymerizes. After polymerization, the water layer is poured off, leaving a flat upper surface of the gel.

The layer of water or water-saturated butanol not only eliminates the meniscus but excludes oxygen which would inhibit polymerization on the gel surface.

Determining the Pore Size

The pore size in a polyacrylamide gel may be predetermined by either of two ways. One way is to adjust the total percentage of acrylamide, that is, the sum of the weights of the acrylamide monomer and the cross-linker. This is expressed as % T. For example, a 20% T gel would contain 20% w/v of acrylamide plus bis. As the %T increases, the pore size decreases.

The other way to adjust pore size is to vary the amount of cross-linker expressed as a percent of the sum of monomer and cross-linker or % C. Thus, a 20%T, 5%C gel would have 20% w/v of acrylamide plus bis, and the bis would account for 5% of the total weight of the acrylamide. It has been found that at any single % T, 5% cross-linking creates the smallest pores in a gel. Above and below 5%, the pore size increases.

When there is a wide range in the molecular weights of the material under study, the researcher may prepare a pored gradient gel. The pore size in a gradient gel is larger at the top of the gel than at the bottom; the gel becomes more restrictive as the run progresses.

Polymerization

Polymerization of a polyacrylamide gel is accomplished either by a chemical or a photochemical method. In the most common chemical method, ammonium persulfate and the quaternary amine, *N,N,N',N'*-tetramethylethylenediamine or TEMED, are used as the initiator and the catalyst, respectively.

In photochemical polymerization, riboflavin and TEMED are used. The photochemical reaction is started by shining long wavelength UV light, usually from a fluorescent light, on the gel mixture.

Since only a minute amount of riboflavin is required, photochemical polymerization is used when a low ionic strength is to be maintained in the gel. It is also used if the protein studied is sensitive to ammonium persulfate or the byproducts of chemical polymerization. Outside of these special conditions, however, the most popular electrophoretic systems use chemical polymerization.

Polymerization, it should be remembered, generates heat. If too much heat is generated, convections will form in the unpolymerized gel, resulting in inconsistencies in the gel structure. To prevent excessive heating, the concentration of initiator–catalyst chemicals should be adjusted to complete polymerization in 10 to 60 minutes.

Analysis of the Gel

After the electrophoresis run is over, the gel must be analyzed by one or more of the following procedures: staining or autoradiography followed by densitometry; liquid scintillation counting; blotting either by capillarity or by electrophoresis for DNA/RNA hybridization, or for autoradiography or immunodetection.

The most common analytical procedure is staining. Protein gels are most frequently stained with Coomassie blue, or by photographic amplification systems using silver (Giulian et al., 1983; Oakley et al., 1980; Merril et al., 1981) or other first row transition metals (Yudelson, 1983). Coomassie blue staining is only sensitive to about 1 μg of protein, whereas the photographic amplification systems are sensitive to about 10 ng of protein. Once the gel is stained it can be photographed or scanned by densitometry for a record of the position and intensity of each band.

Nucleic acids are usually stained with ethidium bromide, a fluorescent dye that glows orange when bound to nucleic acids and excited by UV light. About 5 ng of DNA can be detected with ethidium bromide. These gels are usually photographed for a record of the run.

A second common analytical procedure is autoradiography. It is used chiefly to detect radioactive samples separated on a slab gel. This procedure requires that the gel be first dried onto a sheet of paper and then placed in contact with X-ray film. The film will be exposed only where there are radioactive bands or spots. The resulting autoradiogram is usually photographed or scanned by densitometry.

A third technique, liquid scintillation spectroscopy, is used to analyze tube gels if a radioactive sample is present. Because tube gels are round, they cannot be used for autoradiography. Instead, a tube gel must be sliced and each slice counted.

The most recent highly sensitive technique, blotting, is used to transfer proteins or nucleic acids from a slab gel to a membrane such as nitrocellulose, nylon, DEAE, or CM paper. The transfer of the sample can be done by capillary or Southern blotting (Southern, 1975), or by electrophoresis (Bittner et al., 1980; Towbin et al., 1979; Burnette, 1981). Southern blotting draws the buffer and sample, usually DNA, out of an agarose gel by placing the slab in contact with blotter paper. A nitrocellulose membrane, layered between the gel and the blotter paper, binds the nucleic acids as they flow out of the gel. And since the membrane binds the DNA or RNA in the same pattern as on the original gel, it comes out a faithful copy of the original (see Chapter 15).

Electrophoretic transfers are much faster than Southern blots, taking from ½ to 2 hours to perform. The gel, which contains protein or nucleic acids, is placed next to a membrane in a gel-holding cassette and then placed in a tank filled with buffer. An electric field is applied perpendicular to the cassette and the sample moves out of the gel and onto the membrane. The result is an exact copy of the original gel (see Chapter 25).

By either method, the sample moves from the gel matrix to the surface of the membrane. Once the sample is bound to the membrane, it is detected with one of several methods. If the sample itself is radioactive, the membrane can be subjected directly to autoradiography. If it is not radioactive, radioactive probes such as complementary DNA or specific antibodies can be bound either to specific nucleic acids sequences or to specific protein antigens. The membrane is then subjected to autoradiography to detect the positions of the probes.

SDS Gel Electrophoresis: Separation of Proteins on the Basis of Molecular Weight

In SDS separations, migration is determined not by intrinsic electrical charge of polypeptides but by molecular weight (Shapiro et al., 1967). Sodium dodecylsulfate (SDS) is an anionic detergent that denatures proteins by wrapping around the polypeptide backbone. In so doing, SDS confers a negative charge to the polypeptide in proportion to its length. When treated with SDS and a reducing agent, the polypeptides become rods of negative charges with equal charge densities or charge per unit length.

The major usefulness of this system is to determine the molecular weights of polypeptides. This is done by running a gel with standard proteins of known molecular weights along with the polypeptide to be characterized. A linear relationship exists between the log of the molecular weight of a polypeptide and its R_f. By measuring R_f, that is, the ratio of the distance from the top of the gel to the polypeptide divided by the distance from the top of the gel to the dye front, a standard curve can be generated. The curve will show the R_f of the standard polypeptides and the log of their molecular weights. The R_f of the polypeptide to be characterized is determined in the same way, and the log of its molecular weight read directly from the standard curve (Figure 2).

There are two SDS systems commonly used today. The Weber and Osborn (1969) system is a continuous system and is relatively easy to set up. The Laemmli system (1970), a modification of Ornstein and Davis, is a discontinuous SDS system and is probably the most widely used electrophoretic system today. The treated peptides are stacked in a stacking gel before entering the separating gel, and hence, the resolution in a Laemmli gel is excellent.

PROTOCOL

SDS GEL ELECTROPHORESIS

Prepraration of the Separating Gel

1. Assemble a Hoeffer SE600 vertical slab gel unit (or equivalent) in the casting mode. Use 1.5-mm spacers.

2. In a 125-ml side arm vacuum flask, mix 60 ml of separating gel solution according to Table 4. Leave out the ammonium persulfate and the TEMED. Add a small magnetic stir bar.

3. Stopper the flask and apply a vacuum for several minutes while stirring on a magnetic stirrer.

4. Add the TEMED and the ammonium persulfate and gently swirl the flask to mix. Be careful not to generate bubbles.

5. Pour the solution into the sandwich to a level of about 4.0 cm from the top. Do not mouth-pipet the solutions.

6. Water layer the slabs. Use a greased 1-ml glass syringe fitted with a 2-inch 22 gauge needle. Position the needle, bevel up, at about a 45° angle so that the point is at the top of acrylamide and next to a spacer. Gently apply about 0.3 ml of water or water-saturated isobutanol. Repeat on the other side of the slab next to the other spacer. The water will layer evenly across the entire surface after a minute or two. Water layer

Table 4. Preparation of a SDS Polyacrylamide Gel

	Separating Gel 10% T, 2.7% C	Stacking Gel 4% T, 2.7% C
30% T, 2.7% C	20 ml	2.66 ml
4 × running buffer	15 ml	—
4 × stacking buffer	—	5.0 ml
10% SDS	0.6 ml	0.2 ml
H_2O	24.1 ml	12.2 ml
Ammonium persulfate	300 μl	100 μl
TEMED	20 μl	10 μl

Source: Information courtesy of Hoefer Scientific.

the second slab in the same manner. A very sharp water–gel interface will be visible when the gel is polymerized.

7. Tilt the casting stand to pour off the water layer.
8. Rinse the surfaces of the gels once with distilled water.
9. Add about 1.0 ml of running gel overlay solution.
10. Allow the gels to sit for several hours.

Preparation of the Stacking Gel

1. Pour the liquid from the surface of the gels.
2. In a 50-ml side arm vacuum flask, mix 20 ml of stacking gel solution according to Table 4. Leave out the ammonium persulfate and the TEMED. Add a magnetic stir bar.
3. Deaerate the solution as before.
4. Add the ammonium persulfate and the TEMED. Gently swirl the flask to mix.
5. Add 1–2 ml of stacking gel solution to each sandwich to rinse the surface of the gel. Rock the casting stand and pour off the liquid.
6. Fill each sandwich with stacking gel solution.
7. Insert a comb into each sandwich. Take care not to trap any bubbles below the teeth of the combs. Oxygen will inhibit polymerization and will cause a local distortion in the gel surface at the bottom of the wells.
8. Allow the gel to sit for at least a half-hour.
9. Combine equal parts of protein sample and 2X treatment buffer in a test tube.
10. Put the tube in a boiling water bath for 90 seconds.
11. Remove the sample and put it on ice until ready to use. This treated sample can be put in the freezer for future runs.

Loading and Running the Gels

1. Slowly remove the combs from the gels. Be careful to pull the comb straight up to avoid disturbing the well dividers.
2. Rinse each well with distilled water.
3. Invert the casting stand to drain the wells.
4. Right the casting stand.
5. Fill each well with tank buffer.
6. Using a Hamilton syringe, underlayer the sample in each well.

7. Put the upper buffer chamber in place. Remove lower cams and cam the sandwiches to the bottom of the upper buffer chamber. Put the upper buffer chamber in place of the heat exchanger in the lower buffer chamber.

8. Fill the lower buffer chamber with tank buffer until the sandwiches are immersed in buffer. If bubbles get trapped under the ends of the sandwiches, coax them away with a pipet.

9. Add a spinbar to the lower buffer chamber and place the chamber on a magnetic stirrer. When the lower buffer is circulated, the temperature of the buffer remains uniform. This is important, because uneven heating distorts the banding pattern of the gel.

10. Put a drop of 0.1% phenol red in the upper buffer chamber. This is the tracking dye. Alternatively, add the dye directly to the sample after it has been heat treated.

11. Fill the upper buffer chamber with tank buffer. Take care not to pour buffer into the sample wells because it will wash the sample out.

12. Put the lid on the unit and connect to the power supply. The cathode should be connected to the upper buffer chamber.

13. Set the power supply to constant current.

14. Turn the power supply on and adjust the current to 30 mA/ 1.5 mm thick gel. If you have two 1.5-mm gels you should adjust the power supply to 60 mA; if you have one 1.5-mm gel and one 0.75-mm gel, the setting should be 45 mA. The voltage should start at about 70–80 V, but will increase during the run. Keep a record of the voltage and current readings so that future runs can be compared. In this way, you will easily be able to detect current leaks or incorrectly made buffers.

15. When the dye reaches the bottom, turn the power supply off and disconnect the power cables.

Staining and Destaining the Gels

1. Disassemble the sandwiches and put the gels into stain solution.

2. Gently shake the gels for 4–8 hours.

3. Remove the gels and put them in destaining solution. Shake for 1 hour.

4. Transfer the gels to a container filled with destaining solution II.

Stock Solutions (all solutions should be filtered).

1. *Monomer solution* (30% T, 2.7% C_{Bis}).
   Acrylamide   58.4 g
   Bis   1.6 g
   H_2O to 200 ml
   Store at 4°C in the dark.

2. *4X running gel buffer* (1.5 *M* Tris–HCl pH 8.8).
   Tris   36.3 g adjust to pH 8.8 with HCl
   H_2O to 200 ml

3. *4X stacking gel buffer* (0.5 *M* Tris–HCl pH 6.8).
   Tris   3.0 g adjust to pH 6.8 with HCl
   H_2O to 50 ml

4. *10% SDS.*
   SDS   50 g
   H_2O to 500 ml

5. *Initiator* (10% ammonium persulfate).
   Ammonium   0.5 g
     persulfate
   H_2O to 5.0 ml

6. *Running gel overlay* (0.375 *M* Tris–HCl pH 8.8, 0.1% SDS).
   Tris   25 ml 4X running buffer
   SDS   1.0 ml 10% SDS
   H_2O to 100 ml

7. *2X treatment buffer* (0.125 *M* Tris–HCl pH 6.8, 4% SDS, 20% glycerol, 10% 2-mercaptoethanol).
   Tris   2.5 ml 4X stacking gel buffer
   SDS   4.0 ml 10% SDS
   Glycerol   2.0 ml
   2-Mercaptoethanol   1.0 ml
   H_2O to 10.0 ml
   Divide in aliquots and freeze.

8. *Tank Buffer* (0.025 *M* Tris pH 8.3, 0.192 *M* glycine, 0.1% SDS).
   Tris   12 g
   Glycine   57.6 g
   SDS   40 ml 10% SDS
   H_2O to 4.0 liters
   Because the pH of this solution need not be checked, it can be made up directly in large reagent bottles marked at 4.0 liters. Twelve to sixteen liters can be made up at a time. Reuse the buffer in the lower buffer chamber four or five times, but discard the upper buffer after each run.

9. *Stain stock* (1% Coomassie blue R-250).

Coomassie blue R-250	2 g
H_2O	to 200 ml

Stir and filter

10. *Stain* (0.125% Coomassie blue R-250, 50% methanol, 10% acetic acid).

Coomassie blue R-250	62.5 ml stain stock
Methanol	250 ml
Acetic acid	50 ml
H_2O	to 500 ml

11. *Destaining solution I* (50% Methanol, 10% acetic acid).

Methanol	500 ml
Acetic acid	100 ml
H_2O	to 1.0 liter

12. *Destaining solution II* (7% acetic acid, 5% methanol).

Acetic acid	700 ml
Methanol	500 ml
H_2O	to 10.0 liters

Continuous and Discontinuous Buffer Systems

There are two types of buffer systems in electrophoresis, continuous and discontinuous. A continuous system has only a single separating gel and uses the same buffer in the tanks and the gel. In a discontinuous system, a method first developed by Ornstein (1964) and Davis (1964), a nonrestrictive large pore gel, called a stacking gel, is layered on top of a separating gel. Each gel layer is made with a different buffer, and the tank buffers are different from the gel buffers.

The resolution obtainable in a discontinuous system is much greater than that of a continuous one. However, the simpler continuous system is a little easier to set up.

In a discontinuous system, a protein's mobility, a quantitative measure of the migration rate of a charged species in an electrical field, is intermediate between the mobility of the buffer ion in the stacking gel (leading ion) and the mobility of the buffer ion in the upper tank (trailing ion). When electrophoresis is started, the ions and proteins start to migrate into the stacking gel. The proteins concentrate in a very thin zone called the stack, between the leading ion and the trailing ion. The proteins continue to migrate in the stack until they reach the separating gel. This concentration of protein does not occur in a continuous system where the proteins enter the separating gel in a zone as broad as the original sample. As a result, a continuous system produces bands that are thick and poorly resolved.

An additional advantage of a discontinuous system is that the researcher can design a system to work with a particular protein. For example, the researcher can use buffers that differ only slightly in their mobilities so that a protein of interest will stack while other proteins or contaminants will not.

For a detailed discussion of the discontinuous buffer systems, see Chrambach et al. (1976).

Separation of Proteins on Nondenaturing Discontinuous Gels

The buffer system described below is nondenaturing, that is, it contains no detergents or other denaturing agents, and is designed to stack most proteins. It is a good system to use when working with a mixture of proteins such as a crude extract. A nondenaturing buffer must be used if the activity of a protein under study is to be maintained. This system can be used in slab gels as well as in tube gels. (See also Table 5).

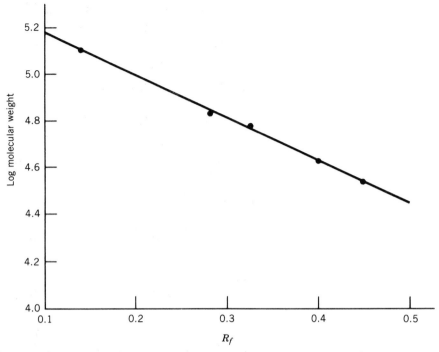

Figure 2. Standard curve for the determination of the molecular weights of polypeptides. The standard curve was generated by plotting the R_f of each standard protein against the $\log_{10}$ of its molecular weight. The molecular weight of an unknown polypeptide can then be determined by finding its R_f on the standard curve and reading the $\log_{10}$ MW from the ordinate. The antilog of this number is the molecular weight of the polypeptide (courtesy of Hoefer).

PROTOCOL

NONDENATURING DISCONTINUOUS GELS

Electrophoresis at Room Temperature

SEPARATING GEL

1. Mix the monomer, buffer, catalyst, and water in a side arm vacuum flask according to Table 5. Add a magnetic spin bar, stopper the flask, and place it on a magnetic stirrer in the dark.
2. Deaerate the gel solution by vacuum for 5 minutes.
3. Add the TEMED and gently swirl the flask to mix.
4. Pour the mixture in tubes or between plates.
5. Apply a water layer about 2 mm thick to the surface of the gel before it polymerizes. Use a greased 1.0-ml glass syringe fitted with a wide bore blunt needle. Avoid mixing the water with gel solution by observing the process at a right angle to the gel tube.
6. Illuminate the gels.

Table 5. Preparation of a Polyacrylamide Gel with Discontinuous Gel Buffer

	Separating Gel			Stacking Gel
	5% T, 5% C	10% T, 5% C	15% T, 5% C	3.125% T, 20% C
Monomer	12.5 ml	25 ml	37.5 ml	100 ml
Buffer	25 ml	25 ml	25 ml	50 ml
H_2O	50 ml	37.5 ml	25 ml	—
Catalyst	12.5 ml	12.5 ml	12.5 ml	50 ml
TEMED	150 μl	150 μl	50 μl	10 μl

Source: Information courtesy of Hoefer Scientific.

7. Pour off the water layers as soon as a sharp interface between the gel and the water layer forms. This will take about 5 minutes.

8. Carefully rinse the tops of the gels with water and drain the tubes upside down.

STACKING GEL

1. Mix the monomer, buffer, and catalyst in a 50-ml side arm vacuum flask according to Table 5.

2. Deaerate as before.

3. Add the TEMED and swirl gently.

4. Add stacking gel solution to the top of the separating gel and place comb when using a slab gel.

5. Water layer and polymerize as before.

APPLICATION OF SAMPLE AND ELECTROPHORESIS

1. Adjust the concentration of your sample to 0.2–1.0% w/v, that is, 2.0 to 10.0 mg/ml. Combine 1.0 volume of sample with 0.1 volume of sucrose–dye solution (e.g., 50 μl protein solution and 5 μl sucrose–dye solution).

2. Fill the lower buffer chamber with lower tank buffer.

3. Fill the upper buffer chamber with upper tank buffer.

4. Using a disposable microvolume pipet, load the samples.

5. Periodically note the current and voltage readings. As the upper tank buffer migrates into the gel, the voltage will rise because of an increase in the electrical resistance in the gel. A record of the voltage increase will help you standardize future runs. A difference in the starting voltage may indicate that a buffer is improperly made or that there is an electrical leak.

6. Run the gel until the tracking dye is about 1.0 cm from the bottom of the gel.

STAINING AND DESTAINING

1. When using tubes, remove gels by rimming the top and bottom of each gel with a stream of water. Use a 30-ml glass syringe

fitted with a 2-inch 22 gauge needle for this purpose. A gel should slide out of its tube after a few milliliters of water have been squirted down the inside of the tube.

2. Mark the dye front by cutting the gel.

3. Place gel in a tray and cover with fixative solution.

4. Fix the proteins in the gels by gently shaking them for 1.0 hour.

5. Add stain to the solution (2 ml stain per 50 ml fixative) and continue shaking for 2 hours.

6. Empty staining solution and cover the gel with destaining solution.

7. The gels may stay in destaining solution until they are photographed or scanned in a densitometer.

8. Determine the R_f for each band by dividing the distance the band traveled by the distance the dye front traveled. A protein should always have the same R_f regardless of how long the gel is allowed to run.

Stock Solutions

1. *Lower tank buffer* (63 mM Tris, 50 mM HCl, pH 7.47).
Tris	22.7 g
HCl	150 ml (1N)
H_2O	to 3.0 liters

2. *Upper tank buffer* (37.6 mM Tris, 40 mM glycine, pH 8.89).
Tris	4.56 g
Glycine	3.0 g
H_2O	to 1.0 liter

3. *4X separating buffer* (947 mM Tris, 0.289 N HCl, pH 8.48).
Tris	11.47g
HCl	28.92 ml (1 N)
H_2O	to 100 ml

4. *4X stacking gel buffer* (158 mM Tris, 0.256 N H_3PO_4, pH 6.90).
Tris	1.92 g
Phosphoric acid	25.6 ml (1 N)
H_2O	to 100 ml

5. *Monomer solution for separating gel* (40% T, 5% C_{Bis}).
Acrylamide	38.0 g
Bis	2.0 g
H_2O	to 100 ml

6. *Monomer solution for stacking gel* (6.25% T, 20% C_{Bis}).

Acrylamide	5.0 g
Bis	1.25 g
H_2O	to 100 ml

Store at 4°C in the dark.

7. *Catalyst* (0.06% ammonium persulfate, 0.002% riboflavin phosphate).

Ammonium persulfate	0.6 ml (10% solution)
Riboflavin phosphate	10.0 ml (0.02% solution)
H_2O	to 100 ml

8. *Sucrose-dye solution* (50% sucrose, 0.1% bromophenol blue).

Sucrose	5 g
Bromophenol blue	1.0 ml (1.0% solution)
H_2O	to 10 ml

Store aliquots in freezer.

9. *Fixative* (12.5% trichloroacetic acid).

TCA	500 g
H_2O	to 4000 ml

10. *Stain* (0.25% Coomassie blue G-250).

Coomassie blue G-250	0.5 g
H_2O	to 200 ml

Stir and filter through filter paper.

11. *Destaining solution* (7% acetic acid, 5% methanol).

Acetic acid	280 ml
Methanol	200 ml
H_2O	to 4.0 liters

Two-Dimensional Gel Electrophoresis: To Separate Complex Mixtures of Proteins into Many Components

Two-dimensional gel electrophoresis is widely used to separate complex mixtures of proteins into many more components than is possible in conventional one-dimensional electrophoresis. For example, proteins that have been resolved into perhaps 35 bands in a first-dimensional gel run, may be further resolved into hundreds in the second-dimensional run.

Each dimension separates proteins according to different properties. Conceptually, the system is simple. The first dimension is run in a tube gel which is then grafted horizontally onto the top of a polymerized slab gel. A small amount of agarose is used in the grafting process to bond the gels together. The second dimension is run on a vertical slab. The proteins that appeared as bands in the tube gel migrate out of the tube and into the slab. After the run, the proteins appear as spots in the slab.

Perhaps the most widely used two-dimensional system was developed by O'Farrell (1975). The O'Farrell system takes advantage of pI and molecular weight differences among proteins. The first dimension tube gel is electrofocused; the second dimension is an SDS slab gel.

The analysis of two-dimensional gels is more complex than that of one-dimensional gels because the components that show up as spots rather than as bands must be assigned x, y coordinates. Still, it is a remarkably sensitive and reproducible method. Anderson and Anderson (1977), for example, have used this system to analyze genetic variations in the proteins of biological fluids by comparing differences in the position of spots from a normal pattern.

Excellent resolution is achieved in the second dimension SDS gel by using a pore gradient gel, either linear of exponential, instead of a single concentration gel. A crude cell extract, for instance, with components having a wide variation of molecular weights, would show better resolution in a gradient gel than in a single concentration gel.

PROTOCOL

TWO-DIMENSIONAL GEL ELECTROPHORESIS

Preparing the Isoelectric Focusing Gels

DAY 1

1. Wrap the bottoms of 130 × 3 mm ID acid-cleaned tubes with squares of Parafilm.

2. Insert the tubes in the Hoefer TR 24 Universal tube rack (or equivalent).

3. In a side arm flask, mix the monomer solution according to Table 6, omitting the TEMED and ammonium persulfate. Add a small magnetic stir bar.

4. Allow the urea to dissolve. A few seconds in a microwave oven will speed this process, although you should not allow the solution to become warm.

5. Deaerate the solution by vacuum for 5 minutes on a magnetic stirrer.

6. Add the TEMED and ammonium persulfate according to Table 6. This starts polymerization. You must work rapidly until the gel is water layered. Gently swirl the solution to mix.

7. Using a long-tip Pasteur pipet, fill each gel tube with the solution to about 1 cm from the top. Be careful not to introduce bubbles.

Table 6. Preparation of Electrofocusing Gel

Urea[a]	5.5 g
Monomer solution	1.0 ml
10% NP-40	2.0 ml
H_2O	2.5 ml
Ampholyte 5/7	0.4 ml
Ampholyte 3.5/10	0.1 ml
TEMED	7.0 μl
Ammonium persulfate	10.0 μl

Source: Information courtesy of Hoefer Scientific.

[a]The urea should be ultrapure grade, or deionized and lyophilized.

8. Apply a water layer about 2 mm thick to the surface of the gel before it polymerizes. Use a greased 1.0-ml glass syringe fitted with a wide bore blunt cannula. Avoid mixing the water with the gel solution by observing the process at a right angle to the gel tube.

9. Allow the gel to polymerize. You will see a sharp interface between the water and the gel when the process is complete.

10. Remove the water layer. You can do this conveniently by turning the tube rack upside down and shaking it. The tubes will remain in place.

11. Apply about 20 μl sample overlay solution to the tops of the gels. Allow the gels to sit for several hours.

Running the Gels

1. Remove the liquid from the tops of the gels as before.

2. Remove the Parafilm from the bottoms of the tubes. Mount the tubes in the upper buffer chamber so that the glass tubes extend about 2 mm above the grommets in the upper buffer chamber.

3. Put the upper buffer chamber in place on the lower buffer chamber. If using a Hoefer unit, insert the central cooling core into the center of the upper buffer chamber.

4. Pour anode solution into the center of the core to fill the lower buffer chamber.

5. Pour cathode solution into the upper buffer chamber to cover the upper electrode. Coax any air bubbles out of the gel tubes with a Pasteur pipet.

6. Add 20 μl sample overlay to the tops of the gels.

7. Prerun the gels according to the following schedule:

Time	Voltage
15 min	200 V
30 min	300 V
30 min	400 V

You will notice that the current will drop with time. This is due to an increase in the resistance of the gels which accompanies the formation of the pH gradient. Take note of the current readings so that you will be able to compare future runs.

8. Turn off power.

9. Using a Pasteur pipet, gently squirt a stream of cathode solution into the tops of the tubes to flush the sample overlay solution out of the tubes.

10. Pipet the samples (in a low ionic strength buffer) onto the top of each gel.

11. Add 10 µl of sample overlay solution to the top of each gel.

12. Replace lid. Adjust the power supply to 400 constant volts and leave it on for 16 hours. Make SDS slab gels at this point.

DAY 2

1. Turn power supply up to 800 volts for 1 hour.

2. Turn power supply off.

3. Remove the gels from the tubes by rimming the top and bottom of each gel with a stream of water. Use a 30-ml glass syringe fitted with a 2-inch 22 gauge needle for this purpose. A gel should slide out of the tube after a few milliliters of water have been squirted between the gel and the tube.

4. Put the focused gels in large screw cap tubes with 10 ml treatment buffer. Shake at room temperature for 30 minutes.

5. Repeat step 4 two times.

Preparation of the SDS Separating Gel, 10% T, 2.7% C

DAY 1

1. Assemble the SE600 vertical slab gel unit in the casting mode. Use 1.5-mm spacers.

2. In a 125-ml side arm flask, mix 60 ml of separating gel solution according to Table 7. Leave out the ammonium persulfate and the TEMED. Add a small magnetic stir bar.

3. Stopper the flask and apply a vacuum for several minutes while stirring on a magnetic stirrer.

Table 7. Slab Gel for Second Dimension in Two-Dimensional Electrophoresis

	Separating Gel 10% T, 2.7% C	Stacking Gel 4% T, 2.7% C
Monomer (8)	20 ml	2.66 ml
Buffer (10)	15 ml	—
Buffer (11)	—	5.0 ml
10% SDS (12)	0.6 ml	0.2 ml
H$_2$O	24.1 ml	12.2 ml
Ammonium persulfate	132 µl	100 µl
TEMED	20 µl	10 µl

Source: Information courtesy of Hoefer Scientific.

4. Add the TEMED and ammonium persulfate and gently swirl the flask to mix. Be careful not to generate bubbles.
5. Pipet the solution into the sandwiches to a level about 1.0 cm from the tops.
6. Water layer the slabs. Use a greased 1-ml glass syringe fitted with a 2-inch 22 gauge needle. Position the needle, bevel up, at about a 45° angle such that the point is at the top of the acrylamide and next to a spacer. Gently apply about 0.3 ml of water. Repeat on the other side of the slab next to the other spacer. The water will layer evenly across the entire surface after a minute or two. Water layer the second slab in the same manner.
7. A very sharp water–gel interface will be visible when the gel is polymerized.
8. Tilt the casting stand to pour off the water layer.
9. Rinse the surfaces of the gels once with distilled water.
10. Add about 1.0 ml of running gel overlay solution.
11. Allow the gels to sit several hours or overnight.

Preparation of the Stacking Gel

DAY 2

1. Pour the liquid from the surface of the gels.
2. In a 50-ml side arm vacuum flask, mix 20 ml of stacking gel solution according to Table 7. Leave out the ammonium persulfate and the TEMED. Add a magnetic stir bar.
3. Deaerate the solution as before.
4. Add the ammonium persulfate and the TEMED. Gently swirl the flask to mix.
5. Add 1–2 ml of stacking gel solution to each sandwich to rinse the surface of the gels. Rock the casting stand and pour off the liquid.
6. Fill each sandwich to about 2 mm from the top with the gel mixture.
7. Water layer as before.
8. Allow the gels to sit for at least a half-hour.
9. Melt the agarose.

Fixing the Focused Tube Gels to the Slab Gels and Running the Gels

1. Pour off the water layer from the surface of the slab gels.
2. Put the upper buffer chamber in place and remove the cams

from the casting stand. Cam the sandwiches to the bottom of the upper buffer chamber.

3. Remove the buffer chamber sandwich assembly from the casting stand and put the assembly on the heat exchanger in the lower buffer chamber.

4. Fill the lower buffer chamber with tank buffer until the sandwiches are immersed in buffer. If bubbles get trapped under the sandwiches coax them away with a pipet.

5. Using a Pasteur pipet, fill the space between the stacking gel and the bottom of the gel trough in the upper buffer chamber with 1% agarose. Do not introduce bubbles.

6. Rinse the focused tube gels briefly in water.

7. Starting at one end of the trough, lay in the gel while at the same time running in more agarose as the gel contacts the trough. Do not introduce bubbles. Allow the agarose to gel.

8. Put a drop or two of 0.1% phenol red in the upper buffer chamber. This is the tracking dye.

9. Fill the upper buffer chamber with tank buffer.

10. Put the lid on the unit and connect to the PS 1200 power supply. The cathode should be connected to the upper buffer chamber.

11. Set the power supply to constant current.

12. Turn the power supply on and adjust the current to 30 mA/1.5 mm thick gel. If you have two 1.5-mm gels you should adjust the power supply to 60 mA. The voltage should start at about 70 to 80 volts, but will increase during the run. Keep a record of the voltage and current readings so that future runs can be compared. In this way you will easily be able to detect current leaks or incorrectly made buffers.

13. When the dye reaches the bottom of the slabs, turn the power supply off and disconnect the power cables.

Staining and Destaining the Gels

1. Disassemble the sandwiches and put the gels into stain.

2. Gently shake the gels for 4 to 8 hours.

DAY 3

1. Remove the gels and put them in destaining solution I. Shake for 1 hour.

2. Transfer the gels to the slab gel destainer filled with destaining solution II and continue destaining until the background is clear.

PROTOCOL

GRADIENT GELS

Linear Gradient, 10–20%

1. Set up the gel stand with two glass sandwiches.
2. Connect a piece of tubing to the outlet tubing connector of a gradient maker. Connect a long cannula to the other end of the tubing and put it in one of the sandwiches to the bottom center. Attach the tubing to a peristaltic pump.
3. In separate side arm vacuum flasks, mix all ingredients except the ammonium persulfate and the TEMED listed in Table 8 for the linear gradient.
4. Deaerate each flask.
5. Add the TEMED and the ammonium persulfate and gently swirl the flasks to mix.
6. Pour the 20% solution into the mixing chamber of the gradient maker. Add a small magnetic stir bar.
7. Open the stopcock between the two chambers to allow the solution to flow through the channel to, but not into, the bottom of the reservoir chamber.
8. Close the stopcock.
9. Pour the 10% solution into the reservoir.
10. Put the gradient maker on the magnetic stirrer and start to stir.
11. Turn on the pump.
12. Open the outlet stopcock to which the tubing is attached.
13. Open the stopcock between the two chambers.

Table 8. Preparation of Linear and Exponential Gradient Gels

| | Gradient Gels | | | |
| | Linear | | Exponential | |
	10%	20%	10%	20%
Monomer (8)	5.67 ml	—	11.3 ml	—
Monomer + glycerol (9)	—	11.34 ml	—	5.6 ml
Buffer (10)	4.25 ml	4.25 ml	8.5 ml	2.13 ml
SDS (12)	0.17 ml	0.17 ml	0.34 ml	85.0 µl
H_2O	6.83 ml	1.16 ml	13.67 ml	0.61 ml
Ammonium persulfate (3)	37.0 µl	37.0 µl	75.0 µl	19.0 µl
TEMED	6.0 µl	6.0 µl	11.3 µl	3.0 µl

Source: Information courtesy of Hoefer Scientific.

14. Gradually lift the cannula as the liquid level rises until all the liquid is pumped into the sandwich.

15. Water layer and continue the steps as outlined in the preparation of the 10% separating gel.

Exponential Gradient, 10–20%

1. Set up the sandwiches and gradient maker as for linear gradients.

2. Mix the ingredients according to Table 8 for the exponential gradient, omitting the ammonium persulfate and the TEMED.

3. Deaerate.

4. Add the ammonium persulfate and the TEMED. Mix.

5. Add a small magnetic stir bar and the heavy solution to the mixing chamber.

6. Fill the connecting channel with heavy solution and close the stopcock.

7. Insert the plunger in the mixing chamber until it is just over the level of the solution.

8. Close the air bleed on the plunger.

9. Fill the reservoir chamber with the light solution.

10. Place on the stirrer and turn on.

11. Proceed as with linear gradients until the reservoir chamber is empty. The volume in the mixing chamber is not delivered and must be discarded.

Stock Solutions

1. *Monomer solution* (40% T, 5% C_{Bis}).

Acrylamide	38.0 g
Bis	2.0 g
H_2O	to 100 ml

Store at 4°C in the dark.

2. *10% NP-40.*

Nonidet P-40	10 ml
H_2O	to 100 ml

3. *10% ammonium persulfate.*

Ammonium persulfate	0.5 g
H_2O	to 5.0 ml

Make fresh daily.

4. *Sample overlay solution* (9 *M* urea, 2% ampholytes).

Urea	5.5 g
Ampholytes 5/7	0.4 ml
Ampholytes 3.5/10	0.1 ml
H_2O	to 10 ml

Divide in aliquots and freeze.

5. *Anode solution* (10 m*M* phosphoric acid).

Phosphoric acid	2.0 ml (85%)
H_2O	to 3.0 liters

6. *Cathode solution* (10 m*M* histidine).

Histidine	1.55 g
H_2O	to 1.0 liter

Store in vacuo.

7. *Treatment buffer* (0.0625 *M* Tris–HCl pH 6.8, 2% SDS, 10% glycerol, 5% 2-mercaptoethanol).

Tris	6.25 ml stacking gel buffer
SDS	10.0 ml 10% SDS
Glycerol	5.0 ml
2-Mercaptoethanol	2.5 ml
H_2O	to 50 ml

8. *Monomer solution* (30% T, 2.7% C_{Bis}).

Acrylamide	58.4 g
Bis	1.6 g
H_2O	to 200 ml

Store at 4°C in the dark.

9. *Monomer solution in 75% glycerol* (30% T, 2.7% C_{Bis}).

Acrylamide	58.4 g
Bis	1.6 g
75% Glycerol	to 200 ml

Store at 4°C in the dark.

10. *Running gel buffer* (1.5 *M* Tris–HCl pH 8.8).

Tris	36.3 g adjust to pH 8.8 with HCl
H_2O	to 200 ml

11. *Stacking gel buffer* (0.5 *M* Tris–HCl pH 6.8).

Tris	3.0 g adjust to pH 6.8 with HCl
H_2O	to 50 ml

12. *10% SDS.*

SDS	50 g
H_2O	to 500 ml

13. *Running gel overlay* (0.375 *M* Tris–HCl pH 8.8, 0.1% SDS).

Tris	25 ml running gel buffer
SDS	1.0 ml 10% SDS
H_2O	to 100 ml

14. *1% agarose.*

Tris	10 ml stacking gel buffer
Agarose	0.4 g
SDS	0.4 ml 10% SDS
H_2O	to 40 ml

15. *Tank buffer* (0.025 *M* Tris pH 8.3, 0.192 *M* glycine, 0.1% SDS).

Tris	12 g
Glycine	57.6 g
SDS	40.0 ml 10% SDS
H_2O	to 4.0 liters

Because the pH of this solution need not be checked, it can be made up directly in large reagent bottles marked at 4.0 liters. Twelve to sixteen liters can be made up at a time. Reuse the buffer in the lower buffer chamber four or five times, but discard the upper buffer after each run.

16. *Stain stock* (1% Coomassie blue R-250).

Coomassie blue R-250	2 g
H_2O	to 200 ml

Stir and filter.

17. *Stain* (0.125% Coomassie blue, 50% methanol, 10% acetic acid).

Coomassie blue R-250	62.5-ml stain stock
Methanol	250 ml
Acetic acid	50 ml
H_2O	to 500 ml

18. *Destaining solution I* (50% methanol, 10% acetic acid).

Methanol	500 ml
Acetic acid	100 ml
H_2O	to 1 liter

19. *Destaining solution II* (7% acetic acid, 5% methanol).

Acetic acid	700 ml
Methanol	500 ml
H_2O	to 10.0 liters

Electrophoretic Transfer

When a protein or nucleic acid sample is separated in polyacrylamide or agarose, it is trapped in the matrix of the gel and therefore cannot be detected by the binding of probes such as radioactive complementary DNA (cDNA) or antibodies. In these cases the probes and samples are too large to diffuse through the gel to bind specifically with each other, the antibody with a specific protein antigen or the cDNA with a specific base sequence. However, if a protein or nucleic acid separation is transferred from a gel to a membrane such as nitrocellulose, it becomes extremely sensitive to detection techniques. This is because all of the sample is bound to the surface of the membrane and is available for the binding of probes or for autoradiography without quenching.

The fastest method for the transfer of proteins and nucleic acids from gels to membranes is electrophoretic transfer. To do this, the gel is simply layered next to the membrane in a gel-holding cassette and placed between two electrodes in a tank of transfer buffer. Then a voltage gradient is applied perpendicular to the gel, causing the sample to migrate off the gel and onto the membrane. The membrane is then ready for staining, autoradiography, cDNA hybridization, or immunodetection.

PROTOCOL

ELECTROPHORETIC TRANSFER

1. Immediately after completing your SDS run, remove the gel from the glass plate sandwich.
2. Place the gel in a tray containing 200 to 300 ml of transfer buffer. Incubate the gel for 10 minutes.
3. Pour about 1 inch of transfer buffer in a second tray. Fill the transphor tank with 4.0 liters of transfer buffer.
4. Place one-half of the gel cassette in the tray, followed by a foam pad, one sheet of blotter paper, and then the nitrocellulose membrane. Be careful not to trap air bubbles beneath the membrane.
5. Place the gel on the membrane, again taking care not to trap bubbles beneath.
6. Place two sheets of blotter paper over the gel. It is important to use two sheets of paper at this step in order to position the gel near the center of the cassette where the electrical field is most uniform.
7. Place the second half of the cassette on the top and hook it together.
8. Quickly, lift the cassette out of the tray of buffer and transfer it to one of the center slots in the buffer chamber. Be certain the cassette is oriented with the nitrocellulose membrane on the far side, or anode side of the gel. Tap the cassette to dislodge any bubbles that may be trapped by the cassette grid.
9. Place the power lid on the unit, taking care to engage the two banana plugs on the electrode panels. The anode will be in back.
10. Turn the power supply on and turn the voltage-adjust knob all the way clockwise.
11. Run the unit for 45 minutes.
12. Remove the cassette. Place the gel and the nitrocellulose membrane in stain solution. Shake.

13. After a few minutes, transfer the membrane to destaining solution I. Transfer the membrane to fresh destaining solution I until stain stops running off the membrane when it is picked up. Then transfer it to destaining solution II.

14. After 4 to 8 hours, transfer the gel to destaining solution I and shake for about an hour. Then transfer to destaining solution II.

Stock Solutions

1. *Transfer buffer* (0.192 M glycine, 0.025 M tris pH 8.3, 20% v/v methanol).

Tris	12 g
Glycine	57.6 g
Methanol	800 ml
H_2O	to 4 liters

   Make several batches of this buffer and store in 4-liter bottles. You can reuse this buffer three or four times.

2. *Stain solution* (50% methanol, 10% acetic acid, 0.125% Coomassie blue R-250).

Methanol	100 ml
Acetic acid	20 ml
Coomassie blue R-250	25 ml 1% solution
H_2O	to 200 ml

3. *Destaining solution I* (50% methanol, 10% acetic acid).

Methanol	250 ml
Acetic acid	50 ml
H_2O	to 500 ml

4. *Destaining solution II* (5% methanol, 10% acetic acid).

Methanol	50 ml
Acetic acid	100 ml
H_2O	to 1 liter

Separation of Low Molecular Weight Proteins by Polyacrylamide Gel Electrophoresis

It is often difficult to determine accurately the size of polypeptides whose molecular weight is lower than 14.000 daltons. We recommend that 15% polyacrylamide gels be used when samples containing polypeptides with molecular weights in the range of 30.000 to 3.000 daltons are run. Under the conditions described below, a linear relationship exists between the distance of migration in the gel and the molecular weight of the polypeptides (see Figure 3).

The gel contains 0.1% SDS and 6 M urea. It is run at 6 V/cm at room temperature, until the dye reaches the bottom of the resolving gel. We currently use 15 × 15 cm plates with 3 mm thick spacers. The quantities given below can be adapted to your own systems.

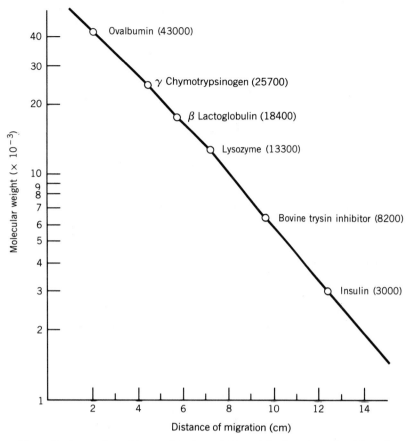

Figure 3. Separation of low molecular weight proteins in polyacrylamide gels.

PROTOCOL

POLYACRYLAMIDE GEL ELECTROPHORESIS OF LOW MOLECULAR WEIGHT PROTEINS

Preparation of the Resolving Gel

1. In 100 ml beaker, weight 18.02 g of urea and 0.69 g of sodium (monobasic) phosphate ($NaH_2PO_4 \cdot H_2O$).
2. Add 24,35 ml of acrylamide stock solution, 0.5 ml of 10% SDS, and H_2O to almost 50 ml (final volume).
3. Mix and warm up a little to help urea to dissolve.
4. Adjust pH (with NaOH) to 7.2 when the solution has reached room temperature.
5. Adjust volume to 50 ml, and transfer to a side arm flask.
6. Degas under vacuum.
7. Add 23 μl of TEMED, and 45 μl of ammonium persulfate (20%).

8. Pour between the glass plates so as to obtain a height of about 10 cm.

9. Overlay with water to allow polymerization.

Preparation of the Stacking Gel

1. In a 50 ml beaker, weight 3.60 g of urea and 0.14 g of sodium (monobasic) phosphate ($NaH_2PO_4 \cdot H_2O$).

2. Add 2.44 ml of acrylamide solution and 0.10 ml of SDS (10%).

3. Bring close to 10 ml with water.

4. Adjust pH to 7.2 with NaOH.

5. Adjust volume to 10 ml (final).

6. Degas under vacuum.

7. Aspirate the water on the top of resolving gel, and wash with a small amount of running buffer.

8. Add 5 μl of TEMED and 0.1 ml of ammonium persulfate, and immediately pour on the surface of the resolving gel.

9. Place the comb, and let polymerize at room temperature.

Treatment of Samples

1. In a microfuge tube, mix 8 volumes of sample buffer for 1 volume of sample.

2. Boil the mixture for 3 minutes before loading onto gel.

3. Flush the wells with buffer before loading.

Staining, Destaining

Cupric acid is added in the staining mixture to improve staining of small polypeptides. Destaining is performed in standard mixture.

Notes

1. Urea in this system is near its solubility limit. Do not cool the solution to avoid crystallization.

2. Do not keep the poured gel for extended periods of time.

3. It is important to rinse the wells with running buffer before loading, to obtain best results.

Buffers and Solutions

Acrylamide solution
30% acrylamide
0.8% bisacrylamide in water

Sample buffer
4.20 g urea
0.01 g $NaH_2PO_4 \cdot H_2O$
1.0 ml 10% SDS
0.1 ml 2-mercaptoethanol (14 M)
1.0 ml 0.1% bromophenol blue

Staining solution
0.1% Coomassie brillant blue
25% isopropanol
10% acetic acid
0.1% cupric acetate
in distilled water

Destaining solution	1 liter
Acetic acid 10%	100 ml acetic acid
Methanol 20%	200 ml methanol
in distilled water	700 ml H_2O

Running buffer	2 liters
0.1 M sodium phosphate pH 7.2	27.60 g sodium phosphate mono-basic ($NaH_2PO_4 \cdot H_2O$)
0.1% SDS	2.0 g SDS
	1.5 liter of H_2O
	Adjust to pH 7.2 with NaOH
	Bring to 2 liters with H_2O

Ammonium persulfate (20%)
2 g of ammonium persulfate dissolved in 10 ml of distilled water.
It is best to prepare solution just before use. May be kept a few
days at 4°C.

Coomassie brillant blue (0.1%)
10 mg of Coomassie brillant blue in 10 ml of water.

References

Anderson, L., and Anderson, N. G. (1977), *Proc. Natl. Acad. Sci. USA*, **74**, 5421.

Anker, H. S. (1970), *FEBS Lett.*, **7**, 283.

Bittner, M., Kupferer, P., and Morris, C. F. (1980), *Anal. Biochem.*, **102**, 459.

Brown, B. A., Majocha, R. E., Staton, D. M., and Marotta, C. A. (1983), *J. Neurochem.*, **40**, 299.

Burnette, W. N. (1981)., *Anal. Biochem.*, **112**, 195.

Carreira, L. H., Carlton, B. C., Bobbio, S. M., Nagao, R. T., and Meagher, R. B. (1980), *Anal. Biochem.*, **106**, 455.

Catsimpoolas, N. (ed.) (1976), *Isoelectric Focusing*, Academic Press, New York.

Catsimpoolas, N., and Drysdale, J., (eds.) (1977), *Biological and Biomedical Applications of Isoelectric Focusing*, Plenum Press, New York.

Chrambach, A., Jovin, T. M., Svendsen, P. J., and Rodbard, D. (1976), in *Methods of Protein Separation*, Vol. 2, N. Catsimpoolas, ed., Plenum Press., New York, p. 27.

Chrambach, A., and Rodbard, D. (1971), *Science*, **172**, 440.

Dahlberg, A. E., Dingman, C. W., and Peacock, A. C. (1969), *J. Mol. Biol.*, **41**, 139.

Davis, B. J. (1964), *Ann. N.Y. Acad. Sci.*, **121**, 404.

Diezel, W., Kopperschläger, G., and Hofmann, E. (1972), *Anal. Biochem.*, **48**, 617.

Dunker, A. K., and Rueckert, R. R. (1969), *J. Biol. Chem.*, **244**, 5074.

Fish, W. W., Reynolds, J. A., and Tanford, C. (1970), *J. Biol. Chem.*, **245**, 5166.

Gershoni, J. M., Davis, F. E., and Palade, G. E. (1985), *Anal. Biochem.*, **144**, 32.

Giulian, G. G., Moss, R. L., and Greaser, M. (1983), *Anal. Biochem.*, **129**, 227.

Giulian, G. G., Moss, R. L., and Greaser, M. (1984), *Anal. Biochem.*, **142**, 421.

Gordon, A. H. (1972), in *Laboratory Techniques in Biochemistry and Molecular Biology*, Vol. 1, Part I, T. S. Work and E. Work, eds., North-Holland, Amsterdam and London, p. 1.

Hansen, J. N. (1977), *Anal. Biochem.*, **76**, 34.

Laemmli, U. K. (1970), *Nature*, **227**, 680.

Manrique, A., and Lasky, M., (1981), *Electrophoresis*, **2**, 315.

Matsudaira, P. T., and Burgess, D. R. (1978), *Anal. Biochem.*, **87**, 386.

Maurer, H. R. (1971), *Disc Electrophoresis and Related Techniques of Polyacrylamide Gel Electrophoresis*, 2nd Ed., Walter de Gruyter, Berlin and New York.

Maxam, A. M., and Gilbert, W. (1977), *Proc. Natl. Acad. Sci. USA*, **74**, 560.

Merril, C. R., Goldman, D., Sedman, S. A., and Ebert, M. H. (1981), *Science*, **221**, 1436.

Oakley, B. R., Kirsch, D. R., and Morris, N. R. (1980), *Anal. Biochem.*, **105**, 361.

O'Connell, P. B. H., and Brady, C. J. (1977), *Anal. Biochem.*, **76**, 63.

O'Farrell, P. H. (1975), *J. Biol. Chem.*, **250**, 4007.

Ornstein, L. (1964), *Ann. N.Y. Acad. Sci.*, **121**, 321.

Prat, J. P., Lamy, J. N., and Weil, J. P. (1969), *Bull. Soc. Chim. Biol.*, **51**, 1367.

Richards, E. G., and Coll, J. A. (1965), *Anal. Biochem.*, **12**, 452.

Righetti, P. G., and Drysdale, J. W. (1976), in *Laboratory Techniques in Biochemistry and Molecular Biology*, Vol. 5, Part II, T.S. Work and E. Work, eds., North-Holland, Amsterdam and London, p. 341.

Sanger, F., and Coulson, R. A. (1975), *J. Mol. Biol.*, **94**, 441.

Shapiro, A. L., Vinuela, E., and Maizel, J. V. (1967), *J. Biophys. Res. Commun.*, **28**, 815.

Southern, E. M. (1975), *J. Mol. Biol.*, **98**, 503.

Towbin, H., Staehelin, T., and Gordon, J. (1979), *Proc. Natl. Acad. Sci. USA*, **76**, 4350.

Wardi, A. H., and Michos, G. A. (1972), *Anal. Biochem.*, **49**, 607.

Weber, K., and Osborn, M. (1969), *J. Biol. Chem.*, **224**, 4406.

Yudelson, J. (1983), *Kodavue kit*, Kodak Laboratories Chemicals Bulletin 74.

Zacharius, R. M., Zell, T. E., Morrison, J. H., and Woodlock, J. J. (1969), *Anal. Biochem.*, **30**, 148.

SEPARATION METHODS IN PREPARATIVE ULTRACENTRIFUGES*

Differential Centrifugation

The most common separation method is that of differential centrifugation, or pelleting (see Figure 4). In this method, the centrifuge tube is filled initially with a uniform mixture of sample solution. Through centrifugation, one obtains a separation of two fractions: a pellet containing the sedimented material, and a supernatant solution of the unsedimented material. Any particular component in the mixture may end up in the supernatant or in the pellet or it may be distributed in both fractions, depending upon its size and/or the conditions of centrifugation.

The pellet is a mixture of all of the sedimented components, and it is contaminated with whatever unsedimented particles were in the bottom of the tube initially. The only component that is in purified form is the slowest sedimenting one, but its yield is often very low.

The two fractions are recovered by decanting the supernatant solution from the pellet. The supernatant can be recentrifuged at higher speeds to obtain further purification, with formation of a new pellet and supernatant. The pellet can be recentrifuged after resuspension in a small volume of suitable solvent.

Density Gradient Centrifugation

Another method of separation is by density gradient centrifugation, a method somewhat more complicated that differential centrifugation, but one that

---

*Information courtesy of Beckman.

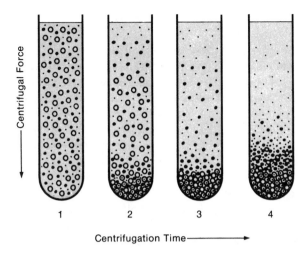

Figure 4. Differential centrifugation (courtesy of Beckman).

has compensating advantages. Not only does the density gradient method permit the complete separation of several or all of the components in a mixture, it also permits analytical measurements to be made.

The density gradient method involves a supporting column of fluid whose density increases toward the bottom of the tube. The density gradient fluid consists of a suitable low molecular weight solute in a solvent in which the sample particles can be suspended.

There are two methods of density gradient centrifugation; rate zonal and isopycnic.

Rate Zonal Technique

In the rate zonal technique, as shown in Figure 5a, a sample solution containing particles to be separated is layered on a preformed gradient column.

This sample solution creates a negative gradient at the top of the column. The sample is prevented from premature sedimentation, however, by the steep positive density gradient beneath it. Under centrifugal force, the particles will begin sedimenting through the gradient in separate zones, each zone consisting of particles characterized by their sedimentation rate.

To achieve a rate zonal separation, the density of the sample particles must be greater than the density at any specific position along the gradient column. The run must be terminated before any of the separated zones reaches the bottom of the tube. It has been demonstrated that gradients in swinging bucket rotors can support most samples if the ratio between sample concentration (% w/w) and starting

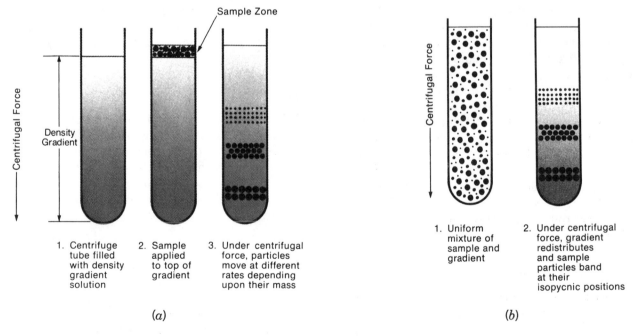

Figure 5. Density gradient centrifugation. (a) Rate-zonal centrifugation in a swinging-bucket rotor. (b) Isopycnic centrifugation using a self-generating gradient (courtesy of Beckman).

gradient concentration (% w/w) is 1:10. For example, a 6–20% w/w gradient would support a sample whose concentration is 0.6% w/w.

Isopycnic Technique

In the isopycnic technique, the density gradient column encompasses the whole range of densities of the sample particles. Each particle will sediment only to the position in the centrifuge tube at which the gradient density is equal to its own density, and there it will remain. The isopycnic technique, therefore, separates particles into zones solely on the basis of their density differences, independent of time.

In many density gradient experiments, elements of both the rate zonal and the isopycnic principles may enter into the final separations. For example, the gradient may be of such a density range that one component sediments to the bottom of the tube while another component sediments to its isopycnic position and remains there.

In the isopycnic procedure with salt solutions, it is not always convenient to form a gradient artificially and layer the sample solution on top; it is sometimes easier to start with a uniform solution of the sample and the gradient material, as noted in Figure 5b. Under the influence of centrifugal force, the gradient material redistributes in the tube so as to form the required concentration (and density) gradient. Meanwhile, sample particles, which are initially distributed throughout the tube, sediment or float to their isopycnic positions. This self-generating gradient technique often requires long hours of centrifugation. Isopycnically banding DNA, for example, takes 36 to 48 hours in a self-generating cesium chloride gradient. It is important to note that the run time cannot be shortened by increasing the rotor speed; this only results in changing the positions of the zones in the tube since the gradient material will redistribute farther down the tube under greater centrifugal force.

Caution: Sedimenting of the gradient-forming solute past its saturation concentration at the tube bottom will cause crystal precipitation. Overstressing of the rotor may result, and rotor failure may occur.

Selection of Rotors

Selection of a rotor depends on sample volume, number of sample components to be separated, particle size and density, desired run time, and desired quality of separation. Swinging bucket, fixed angle, and vertical tube rotors can be used for handling samples in tubes and bottles. For sample volumes in ex-

cess of those handled by these rotors, continuous flow and zonal rotors are available.

Swinging bucket rotors are used for pelleting, isopycnic studies, and rate zonal studies. Several components can be separated at a time. Swinging bucket rotors are best applied for rate zonal studies in which maximum resolution of sample zones are needed. Gradients of all shapes and steepness can be used. Tube caps are not required because the forces applied to the tubes and contents are axial.

Fixed angle rotors are best used for pelleting. They are also useful for rate zonal studies and isopycnic banding of DNA. Resolution in isopycnic separations is often better than with swinging bucket rotors because the reorientation of the tube increases the band volume as well as the interband volume. Run times can be reduced; however, resolution may be sacrificed in rate zonal studies. Tube caps are not required with Quick-Seal tubes, or with thick wall polyallomer and polycarbonate tubes when centrifuged partially filled.

Vertical tube rotors hold tubes parallel to the axis of rotation; therefore, zones separate across the diameter of the tube rather than down the length of the tube. These rotors should be accelerated slowly to protect the sample-to-gradient interface, and decelerated slowly to maintain the integrity of the separation during the reorientation process. Vertical tube rotors can be useful for isopycnic and rate zonal studies, and run times can sometimes be reduced. Also, these rotors hold more tubes than swinging bucket rotors. Tube caps are not required—only Quick-Seal tubes are used in vertical tube rotors.

Pelleting

Pelleting separates particles of different sedimentation coefficients, the largest particles in the sample traveling to the bottom of the tube first. Differential centrifugation is the successive pelleting of particles from previous supernatants using increasingly higher forces. The relative pelleting efficiency of each rotor is measured by its k factor, or clearing factor

$$k = \frac{\ln(r_{max}/r_{min})}{\omega^2} \times \frac{10^{13}}{3600}$$

where r_{max} is the maximum radial distance measured in the tube, r_{min} is the radial distance to the solution meniscus (of capped tubes in fixed angle rotors), and ω is the angular velocity in radians per second (2π rpm/60, or $0.10472 \times$ rpm).

This factor can be used in the following equation to estimate the time t (in hours) required for pelleting:

$$t = \frac{k}{s}$$

where s is the sedimentation coefficient of the particle of interest (because s values in seconds are such small numbers, they are generally expressed in Svedberg units, S, where 1 S is equal to 10^{-13} seconds).

$$s = \frac{dr}{dt} \times \frac{1}{\omega^2 r}$$

where $\frac{dr}{dt}$ is the sedimentation velocity.

It is usual practice to use the standard sedimentation coefficient S_{20w}, based on sedimentation in water at 20°C. Clearing factors can be calculated at speeds other than maximum rated speed by use of the following formula:

$$k = \left(\frac{\text{maximum rated speed}}{\text{actual run speed}}\right)^2 K_{\text{max speed}}$$

Run times can also be calculated from data established in prior experiments when the k factor of the previous rotor is known. For any two rotors a and b,

$$\frac{t_a}{t_b} = \frac{k_a}{k_b}$$

The high-performance rotors have small k factors and generate high centrifugal forces. These rotors can efficiently pellet small particles—small particles have a low S value and ordinarily might require a long run time.

The centrifugal force exerted at a given radius in a rotor is a function of the rotor speed. The nomogram on Figure 6 allows one to determine rotor speed, relative centrifugal field, or radial distance.

Run times can be shortened by using partially filled thick wall polyallomer and polycarbonate tubes. The short path length means less distance for particles to travel in the portion of the tubes experiencing greater centrifugal forces, and hence shortened times. The k factors for half-filled tubes can be calculated using $\ln (r_{\text{max}} / r_{\text{av}})$ in the first k factor equation above.

Isopycnic Separations (Usually Using Cesium Chloride Gradients)

A sedimentation–equilibrium, or isopycnic, experiment separates particles solely on the basis of particle buoyant density. Gradients of cesium chloride are usually selected because the density ranges achieved with this salt encompass the range of most particle densities. Each component in the sample travels through the gradient until it reaches an equilibrium position where particle density equals solution density. At equilibrium, $\rho_p - \rho_c$ is zero, and therefore particle velocity is zero according to the following equation.

$$v = \frac{d^2(\rho_p - \rho_c)}{18\,\mu} \times g$$

where

v = *sedimentation velocity (dr/dt)*

d = particle diameter

ρ_p = particle density

ρ_c = solution density

μ = viscosity of liquid media

g = standard acceleration of gravity

The gradient may be performed before the run or self-formed during centrifugation. In general, the time required for gradients to reach equilibrium in swinging bucket rotors will be longer than in fixed angle rotors. Fixed angle rotors give a shorter sedimentation path length. One way to reduce run times in swinging bucket rotors is to partially fill tubes with gradient. Refer to the appropriate rotor instruction bulletin to determine the maximum allowable speed and homogeneous CsCl solution density when using CsCl gradients.

Rate Zonal Separations (Usually Using Sucrose Gradients)

Particle separation achieved with rate zonal separations is a function of the particle sedimentation coefficient (density, size, and shape). Gradient solutes are selected so as to maximize the differences of particle size and shape, and (to some extent) density in a centrifugal field. Sucrose solutions are especially useful as gradients in rate zonal studies because of their physical characteristics (e.g., viscosity

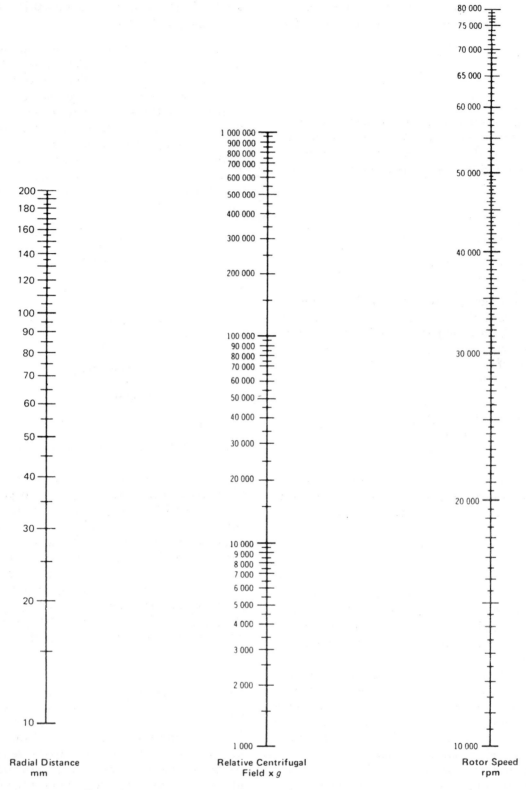

Figure 6. Nomogram for calculation of rotor speed (courtesy of Beckman).

and density) and their availability. Samples are layered on top of the gradient. Under centrifugal force, zones separate gradually, but they will travel to the bottom of the gradient unless the run is stopped at the appropriate time, or unless a component bands isopycnically.

Clearing factors of swinging bucket rotors at maximum speeds and various particle densities have been calculated in 5 to 20% (wt/wt) linear sucrose gradients at 5°C. These are called k' factors. These constants can be used to estimate the time t (in hours) required to move a zone of particles of known sedimentation coefficient and density to the bottom of the gradient:

$$t = \frac{k'}{s}$$

where s is the sedimentation coefficient in Svedberg units (s). A more accurate way to estimate run times in rate zonal studies is to use the $s\ \omega^2 t$ charts, available from the Spinco Division of Beckman Instruments, Inc. If the values of s and ω^2 are known, and gradients are either 5 to 20% or 10 to 30% (wt/wt) sucrose, the charts enable one to calculate the run time t. Also, if the value $\omega^2 t$ is known, sedimentation coefficients can be determined from zone positions.

In most cases, when banding two or three components by rate zonal separation, run times can be considerably reduced if half-filled tubes are used. Tubes are partially filled with gradient, but the sample volume is not changed. If thin wall tubes are used partially filled, suspend the sample in a 2 to 3% sucrose solution and layer it between the gradient and an upper layer of buffer (which prevents tube collapse). If thick wall tubes are used partially filled, the upper layer of buffer is not required, and the neat sample can be applied directly onto the gradient.

If swinging bucket rotors are used with preformed shallow gradients (< 5 to 20%), or if fixed angle or vertical tube rotors are used with any preformed gradient, use the number 1 setting or slowest acceleration on the control panel of your ultracentrifuge.

GEL FILTRATION WITH SEPHADEX

As a solute passes down a chromatographic bed its movement depends upon the bulk flow of the mobile phase and upon the Brownian motion of the solute molecules, which causes their diffusion both into and out of the stationary phase. The separation in gel filtration depends on the different abilities of the various sample molecules to enter pores that contain the stationary phase. Very large molecules, which never enter the stationary phase, move through the chromatographic bed fastest. Smaller molecules, which can enter the gel pores, move more slowly through the column, since they spend a proportion of their time in the stationary phase. Molecules are, therefore, eluted in order of decreasing molecular size.

Gel filtration with Sephadex has been widely used over the past 20 years for purification of enzymes, polysaccharides, nucleic acids, and many other biological macromolecules of interest.

It is essentially used in molecular cloning to remove from DNA preparations the salts, the free nucleotides, and any other type of small molecules which may interfere in subsequent manipulation. For example, removal of the ^{32}p-labeled nucleotides, that have not been incorporated in nick translation of DNA is accomplished by chromatography on G75 Sephadex columns (see Chapter 16).

Chemical and Physical Properties of Sephadex*

Sephadex is a bead-formed gel prepared by cross-linking dextran with epichlorohydrin. The large number of hydroxyl groups renders the gel extremely hydrophilic. Consequently Sephadex swells readily in water and electrolyte solutions. The G-types of Sephadex differ in their degree of cross-linking and hence in their degree of swelling and fractionation range. The degree of swelling of Sephadex is substantially independent of the presence of salts and detergents. However, the effective fractionation range will be altered if the salts or detergents change the conformation of the substances being fractionated. Table 9 lists the different G-types of Sephadex and their physical properties.

Sephadex also swells in dimethylsulfoxide and formamide. Dimethylformamide may be used with Sephadex G-10 and G-15 and mixtures of water with the lower alcohols may be used with Sephadex G-10, G-15, G-25, and G-50. It should be noted that the degree of swelling in organic solvents or their mixtures will not be the same as in water alone.

Sephadex is available in different particle size grades. The superfine grade is intended for column

_____

*Information courtesy of Pharmacia.

Table 9. Properties of Sephadex

Sephadex Type and Grade		Dry Bead Diameter, μm	Fractionalization Range (MW)		Bed Volume, ml/g Dry Sephadex
			Peptides and Globular Proteins	Dextrans	
G-10		40–120	— 700	— 700	2–3
G-15		40–120	— 1 500	— 1 500	2.5–3.5
G-25	Coarse	100–300	1 000– 5 000	100– 5 000	4–6
	Medium	50–150			
	Fine	20– 80			
	Superfine	10– 40			
G-50	Coarse	100–300	1 500– 30 000	500– 10 000	9–11
	Medium	50–150			
	Fine	20– 80			
	Superfine	10– 40			
G-75		40–120	3 000– 80 000	1 000– 50 000	12–15
	Superfine	10– 40	3 000– 70 000		
G-100		40–120	4 000–150 000	1 000–100 000	15–20
	Superfine	10– 40	4 000–100 000		
G-150		40–120	5 000–300 000	1 000–150 000	20–30
	Superfine	10– 40	5 000–150 000		18–22
G-200		40–120	5 000–600 000	1 000–200 000	30–40
	Superfine	10– 40	5 000–250 000		20–25

Source: Information courtesy of Pharmacia.

chromatography requiring very high resolution and for thin-layer chromatography. The fine grade is recommended for preparative purposes, where the extremely good resolution that can be achieved with the superfine grade is not required, but where the flow rate is of greater importance. The coarse and medium grades are intended for preparative chromatographic processes where a high flow rate at a low operating pressure is essential. In addition the coarse grade is suitable for batch procedure. We recommend the coarse grade for DNA chromatography.

Sephadex is insoluble in all solvents (unless it is chemically degraded). It is stable in water, salt solutions, organic solvents, alkaline, and weakly acidic solutions. In strong acids the glycosidic linkages in the gel matrix are hydrolyzed. However, Sephadex can be exposed to 0.1 M HCl for 1–2 hours without noticeable effects, and in 0.02 M HCl Sephadex is still unaffected after 6 months. Prolonged exposure to oxidizing agents will affect the gel and should be avoided.

Sephadex does not melt and may be sterilized in the wet state at neutral pH, or dry state by autoclaving for 30 minutes at 120°C without affecting its chromatographic properties.

ION EXCHANGE CHROMATOGRAPHY

An ion exchanger consists of an insoluble matrix to which charged groups have been covalently bound. The charged groups are associated with mobile counterions. These counterions can be reversibly exchanged with other ions of the same charge without altering the matrix.

It is possible to have both positively and negatively charged exchangers. Positively charged exchangers have negatively charged counterions (anions) available for exchange and so are termed anion exchangers. Negatively charged exchangers have positively charged counterions (cations) and are termed cation exchangers.

The matrix may be based on inorganic compounds, synthetic resins, polysaccharides, and so on. The nature of the matrix determines its physical properties such as its mechanical strength and flow characteristics, its behaviour toward biological substances and, to a certain extent, its capacity.

The first ion exchangers were synthetic resins designed for applications such as demineralization, water treatment, and recovery of ions in wastes. Such ion exchangers consist of tightly cross-linked hydrophobic polymer matrices highly substituted with ionic groups, and have very high capacities for small ions. Due to the high degree of cross-linking required to provide mechanical strength, the porosity of these matrices is low for proteins and other macromolecules. In addition the high-charge density gives very strong binding and the hydrophobic matrix tends to denature labile biological materials, so despite their excellent flow properties and capacities for small ions, these types of ion exchanger are unsuitable for use with biological samples.

The first ion exchangers designed for use with biological materials were the cellulose ion exchangers. Because of the hydrophilic nature of cellulose, they had little tendency to denature proteins. Unfortunately many cellulose ion exchangers had low capacities (otherwise the cellulose became soluble in water), and had poor flow properties because of their irregular shape.

Ion exchangers based on Sephadex and, subsequently, those based on Sepharose CL-6B and DEAE-Sephacel were the first ion exchange matrices to combine the correct spherical form with high porosity, which lead to improved flow properties and high capacities for macromolecules.

The Theory of Ion Exchange

The separation in ion exchange chromatography is obtained by reversible adsorption. Most ion exchange experiments are performed in two main stages. The first stage is sample application and adsorption. Unbound substances can be washed out from the exchanger bed using a column volume of starting buffer. In the second stage, substances are eluted from the column and separated from each other. The separation is obtained since different substances have different affinities for the ion exchanger because of differences in their charge. These affinities can be controlled by varying conditions such as ionic strength and pH. The differences in charge properties of biological compounds are often considerable, and since ion exchange chromatography is capable of separating species with very minor differences in properties, such as two proteins differing by only one amino acid, it is a very powerful separation technique indeed. This is particularly so when ion exchange is coupled with a technique such as gel filtration, which separates by a different parameter, that of size.

In ion exchange chromatography one can choose whether to bind the substances of interest, or to absorb out contaminants and allow the substance of interest to pass through the column. Generally it is more useful to adsorb the substance of interest, since this allows a greater degree of fractionation.

Ion exchange may be carried out in a column or by a batch procedure. Both methods are performed in definite stages. These stages are: equilibration of the ion exchanger, addition and binding of sample substances, change of conditions to produce selective desorption, and regeneration of the ion exchanger. These steps are illustrated in Figure 7.

Sepharose Ion Exchangers

Sepharose ion exchangers are well adapted for ion exchange of proteins, polysaccharides, and nucleic acids. For very large molecules (MW $> 4 \times 10^6$) the best choice is DEAE Sepharose or Sephacel.

These ion exchangers are based on Sepharose CL-6B, which is prepared by cross-linking agarose with 2, 3-dibromopropanol and desulfating the resulting gel by alkaline hydrolysis under reducing conditions. DEAE- or CM-groups are then attached to the gel by ether linkages to the monosaccharide units to give DEAE-Sepharose CL-6B and CM-Sepharose CL-6B.

Chemical and Physical Properties

Sepharose ion exchangers are insoluble in all solvents. They are stable in water, salt solutions, and organic solvents in the range pH 3–10. Prolonged exposure of DEAE-Sepharose CL-6B to very alkaline conditions should be avoided because of the inherent instability of the DEAE-group as a free base. Both DEAE- and CM-Sepharose CL-6B can be used in solutions of nonionic detergents such as Triton X-100. Sepharose ion exchangers can also be used with strongly dissociation solvents such as $7 M$ urea. Under oxidizing conditions, limited hydrolysis of the polysaccharide chains may occur. Sepharose is very resistant to microbial attack because of the presence of the unusual sugar 3,6-anhydro-L-galactose. However, some buffers can support growth and so a bacteriostatic reagent should be used in storage.

The cross-linked structure of Sepharose ion exchangers not only gives them increased chemical stability but also results in improved physical stability. This improves flow properties compared to normal Sepharose and prevents fluctuations in bed volume under conditions of increasing ionic

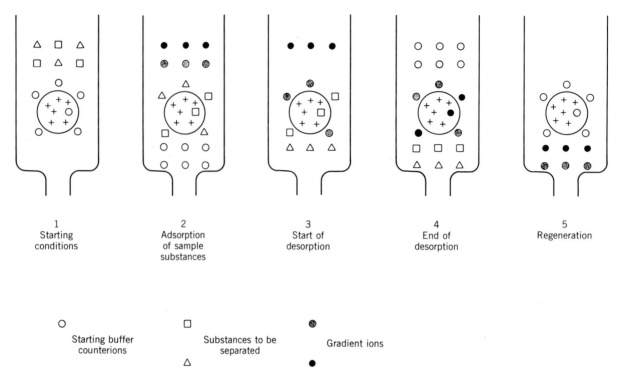

1	2	3	4	5
Starting conditions	Adsorption of sample substances	Start of desorption	End of desorption	Regeneration

○ Starting buffer counterions

□ △ Substances to be separated

⊕ Gradient ions ●

Figure 7. Chromatography on ion exchanger columns (courtesy of Pharmacia).

strength. Thus Sepharose ion exchangers can be re-equilibrated repeatedly in the column.

Sepharose ion exchangers can be used at temperatures up to 70°C and can be sterilized repeatedly in the salt form by autoclaving at pH 7 at temperatures of 110–120°C.

NACS Ion Exchangers

These resins are manufactured by BRL and can be used for analytical and preparative purification of nucleic acids and oligonucleotides. They are composed of a thin film of trialkylmethylammonium chloride which covers an inert, nonporous, noncompressible microparticulate resin. They work as anion exchangers.

The fractionation of bound nucleic acids is generally achieved by applying a linear gradient of increasing ionic strength.

Ready-to-use NACS columns are available from the manufacturer.

QUANTIFICATION OF PROTEINS

The determination of protein concentrations is often required in molecular biology (for example, to run gels, or to perform immunological screenings). Several different methods have been developed to measure the protein content of samples. The most popular are the Lowry and the Bradford assays.

Lowry Procedure

This assay relies on the formation of a protein–copper complex (biuret reaction) and the reduction of the phosphomolybdate–phosphotungstate reagent (Folin–Ciocalteu phenol reagent) by the tyrosine and tryptophan residues of the protein. This method is applicable when the protein concentration per assay ranges between 20 and 200 µg.

Since many compounds such as Tris buffer, sucrose, ammonium sulfate, and so on interact with the reagents used in this assay, it is necessary to include proper blank tubes containing the chemicals used to prepare the protein solution. The method also calls for a standard curve, which is generally established with bovine serum albumin. It is important to point out that the amino acid content of serum albumin may not be comparable to that of some other proteins. Therefore, the use of serum albumin as a standard in the Lowry procedure must be considered with caution.

Bradford Procedure

This other very popular assay for protein quantitation is based on the ability of proteins to bind Coomassie brilliant blue G-250 and form a complex whose extinction coefficient is much greater than that of the free dye.

This assay is simple, quick, and inexpensive. Alkaline protein solutions and detergents have been shown to interfere with the assay. The Coomassie brilliant blue G-250 or the Serva blue G dye exist as two forms (blue and orange) in acid solutions. The proteins bind the blue form preferentially.

PROTOCOL

LOWRY PROCEDURE

1. Prepare seven tubes containing increasing amounts of bovine serum albumin (0, 5, 10, 25, 50, 75, and 100 μg) in a total volume of 0.2 ml (see Table 10).

2. In three tubes, prepare dilutions of the samples with unknown protein concentrations. The total volume in each tube must be 0.2 ml. A blank tube with no sample must be prepared at the same time.

3. Add 1 ml of reagent C in each tube. Vortex to mix and let stand at room temperature for at least 10 minutes. It is possible to leave the tubes for longer periods of time (up to 24 hours) without problems.

4. Add 100 μl of reagent D in each tube and vortex rapidly. Let stand 30 minutes at room temperature and read absorption at 660 nm.

5. Plot the values obtained for serum albumin as a function of protein amounts (see Figure 8) and use this standard curve to determine the protein content of the samples.

Table 10. Outline of the Lowry Procedure

No. Tube	1	2	3	4	5	6	7
H_2O (μl)	200	195	190	175	150	125	100
1 mg/ml Bovin serum albumin (μl)	0	5	10	25	50	75	100
Reagent C (1 ml)	+	+	+	+	+	+	+
			Incubate 10 minutes at room temperature				
Reagent D (100 μl)	+	+	+	+	+	+	+
			Incubate 30 minutes at room temperature				
			Read optical density at 660 nm				

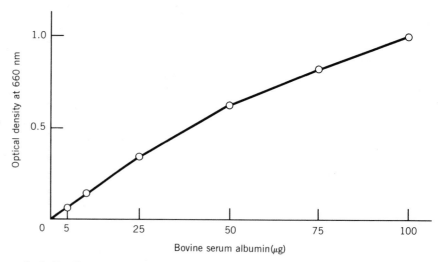

Figure 8. Calibration curve for lowry procedure. Samples containing increasing amounts of bovine serum albumin were treated as described in the text. The optical density at 660 nm was measured with a Zeiss spectrophotometer.

Reagents

Reagent A   2% Na_2CO3 in 0.1 N NaOH

Reagent B   0.5% $CuSO_4$, $5H_2O$ in 1% sodium tartrate or 1% trisodium citrate

Reagent C   Mix 50 ml of reagent A with 1 ml of reagent B; *do not keep more than one day.*

Reagent D   Commercial Folin–Ciocalteu reagent diluted with water to 1 N (approximately 1:1 dilution). You may keep the diluted reagent as long as the original stock solution. Discard this reagent if $A_{660\,nm}$ is greater than 0.2.

A modified Lowry procedure (Markwell et al., 1981) allows assay of proteins in solutions containing up to 2.5 mM EDTA, Triton X100, or 200 mM saccharose. However, this procedure is more sensitive to interference by Tris.

The procedure is as described above except that reagents are the following:

Reagent A   2% Na_2CO_3, 1% SDS, 0.16% sodium tartrate in 0.1 N NaOH

Reagent B   4% $CuSO_4$, $5H_2O$ in water

Reagent C   Reagent A mixed with reagent B at a 100:1 ratio

Reagent D   Identical of standard reagent D

PROTOCOL

BRADFORD PROCEDURE

1. Prepare a series of tubes containing 0.1 to 5.0 μg of serum albumin in a final volume of 50 μl.
2. Prepare a few tubes containing 50 μl of increasing dilutions of the sample (0.1 to 5.0 μg of protein).
3. Add 950 μl of the dye solution to all the tubes.
4. Leave the tubes at room temperature for about 5 minutes.
5. Read absorbancy at 595 nm against a blank containing 50 μl of the protein buffer and 950 μl of dye solution.
6. Use the standard curve to calculate protein concentration in the samples.

Reagent

The Serva Blue G from Serva Feinbiochemica is the purest among several other dyes from commercial source. It is chemically identical to Coomassie brilliant blue and can therefore advantageously replace it in the Bradford assay.

The dye solution is prepared by mixing 10 mg of dye powder with 10 ml of 88% phosphoric acid, and 4.7 ml of absolute ethanol. This mixture is then diluted to 100 ml with distilled water. Check that the absorption of the solution at 550 nm is 1.18. If this is not the case, adjust as required with powder or water. Since the dye solution is almost saturated, a precipitate may form with storage for long periods of time. If this happens, filter the solution through Whatmann paper no. 1 and recalibrate the dye concentration.

References

Bradford, M. M. (1976), *Anal. Biochem.*, **72**, 248.

Lowry, O. H., Rosebrough, N. J., Farr, A. L., and Randall, R. J. (1951) *J. Biol. Chem.*, **193**, 265.

Markwell, M. A. K., Haas, S. M., Tolbert, N. E., and Bieber, L. L. (1981), *Methods in Enzymology,* **72**, 296.

CHOICE OF BUFFERS

It is generally admitted that most of the biological reactions take place at a pH ranging between 6.0 and 8.0. The purpose of a buffer is to control the hydrogen ion concentration of a solution, therefore avoiding extreme changes in pH due to variations of acid–base equilibrium.

Colloquially, pH is a measure of the acidity (or alkalinity) of the solution. An acid is a substance that can give off hydrogen ions. While strong acids completely dissociate in solution, weak acids (and bases) exist in solution as an equilibrium mixture of dissociate and undissociated species. This state is usually represented as

$$HX \underset{k_2}{\overset{k_1}{\rightleftharpoons}} H^+ + X^-$$

where k_1 and k_2 are the rate constants for dissociation and association, respectively. The dissociation rate of HX is a function of its concentration in the solution. Since the k_1 and k_2 rates are identical under equilibrium, one can write

$$k_1[HX] = k_2[H^+][X^-]$$

The concentration of hydrogen ions is

$$[H^+] = K_a \frac{[HX]}{[X^-]}$$

where $K_a = \frac{k_1}{k_2}$ This concentration varies from 1 (strong acid) to 10^{-14} (strong alkali). Since it is not convenient to express acidity in terms of H^+ ions concentration, the pH symbol was introduced (by S.P.L. Sorensen) and defined as

$$pH = -\log[H^+]$$

The pH values range between 0 to 14. When it increases by one unit, the H^+ concentration decreases 10 times. Similarly, it would be possible to express alkalinity, due to OH^- ions, in pOH units defined in the same way as pH units. Since the sum of pH and pOH in aqueous solutions is constant, only the pH scale is being used.

The pK_a is defined as $-\log K_a$. One can therefore express the equilibrium reaction as follows

$$-\log[H^+] = -\log K_a - \log \frac{[HX]}{[X^-]}$$

or

$$pH = pK_a + \log \frac{[X^-]}{[HX]}$$

where $[X^-]$ and $[HX]$ represent the concentrations of the weak acid (HX) and of the corresponding anion (X^-). When the concentrations of $[HX]$ and $[X^-]$ are identical, $pH = pK_a$. Practically, the pK_a values are determined by titration.

Effective buffer systems are expected to have a range of about two pH units around the pK_a value. Some buffer systems have more than one useful pK_a. The pK_a values for common buffers range from 3 to 11, therefore allowing a buffer effect over a pH range of 2 to 12. A list of widely used buffers in molecular biology with their pK_a is reported in Table 11. Practically, however, the biological reactions occur in a pH range of 6 to 8, and many current buffers are not adapted. For example, Tris is a poor buffer below pH 7.5, phosphate buffer is not efficient above pH 7.5,

Table 11. pK$_a$ Values of Some Buffers Used in Molecular Biology

pK$_a$	Compounds
2.12 (pK$_{a_1}$)	Phosphoric acid
3.06 (pK$_{a_1}$)	Citric acid
3.75	Formic acid
4.19 (pK$_{a_1}$)	Succinic acid
4.74 (pK$_{a_2}$)	Citric acid
4.75	Acetic acid
5.40 (pK$_{a_3}$)	Citric acid
5.57 (pK$_{a_2}$)	Succinic acid
6.15	MES
6.60	ADA
6.80	Bis–Tris propane
6.80	Pipes
6.90	ACES
7.00	Imidazole
7.20	Diethylmalonic acid
7.20	MOPS
7.21 (pK$_{a_2}$)	Phosphoric acid
7.50	TES
7.55	HEPES
8.00	HEPPS
8.15	Tricine
8.20	Glycine amide, hydrochloride
8.30	Tris
8.35	Bicine
8.40	Glycylglycine
9.24	Boric acid
9.50	CHES
10.40	CAPS
12.32 (pK$_{a_3}$)	Phosphoric acid

and borate buffer is not efficient because dissolved carbon dioxide has a limited solubility. In addition, several of the currently used buffers have been shown to interact with some physiological reactions. Phosphate binds many polyvalent cations and inhibits a lot of enzymatic reactions.

Tris is known to be a potentially reactive primary amine, and borate is also known to complex several organic compounds.

The effect of temperature on dissociation constants of buffers is widely documented. Tris is notorious in this respect. For example, Tris buffer made up a 4°C to pH 7.0 drops to pH 5.95 when used at 37°C.

Therefore, the choice of an appropriate buffer is not always an easy task. Below are some useful hints which should help you.

1. It is sometime necessary to work with buffers containing no mineral cations. In these cases, tetramethylammonium hydroxide (quaternary

amine) can be used instead of potassium and sodium hydroxyde.

2. The use of systems consisting of a free acid (or base) and its corresponding salt makes a series of buffers with different pH easily obtainable by mixing the two compounds in various ratios (MOPS mixed with MOPS, sodium salt). Phosphate buffers are generally prepared by mixing precalculated amounts of monobasic and dibasic phosphates. The stock solutions (A and B) are prepared from powders and can be of any molarity compatible with the solubility of the phosphates (generally 1 M is convenient). Mixing appropriate volumes of A and B will lead to the preparation of the required buffer (see Table 12). Both stock solutions must have an identical phosphate molarity. Beware that small variations of pH may result from your precision in measuring weights and volumes. It is advisable to adjust the pH of the final buffer with stock solutions under a pH meter. Do not use other acids or bases to adjust the pH of a phosphate buffer prepared in this way. This would lead to a modification of the phosphate molarity. The final molarity of your buffer (moles of phosphate/liter of solution) will be the same as that of stock solutions.

Table 12. Preparation of Phosphate Buffers from Stock Solutions

pH	ml of solution A (monobasic sodium phosphate)	ml of solution B (dibasic sodium phosphate)
5.8	92	8
5.9	90.0	10.0
6.0	87.7	12.3
6.1	85.0	15.0
6.2	81.5	19.5
6.3	77.5	22.5
6.4	73.5	26.5
6.5	68.5	31.5
6.6	62.5	37.5
6.7	56.5	43.5
6.8	51.0	49.0
6.9	45.0	55.0
7.0	39.0	61.0
7.1	33.0	67.0
7.2	28.0	72.0
7.3	23.0	77.0
7.4	19.0	81.0
7.5	16.0	84.0
7.6	13.0	87.0
7.7	10.5	89.5
7.8	8.5	91.5

3. Select the buffer that has a pK_a value as close as possible to the required pH value of the working solution (see Table 11).

4. Adjust the pH of the buffer at the working temperature. This is achieved by using the compensation knob on pH meters.

Proper Use of pH Meters. Wash the pH meter electrode thoroughly with distilled water before use, and blot gently with a tissue to wipe off drops. Choose a standard buffer whose pH is close to the expected final pH of the test solution. It is important that you adjust the temperature compensation knob to the temperature of the solution used as standard (see above for pH changes as a function of temperature). It is wise to check periodically the calibration of the pH meter.

Once you have set up the calibration with the standard buffer, set the pH meter to "Hold" and rinse the electrode with distilled water.

Wipe the drops and transfer the electrode to the test solution. When adjusting a pH, add the acid (or the base) slowly. It is recommended that you use the pH meter under a fume hood when working with volatile acids. Mixing is generally performed with a magnetic stirrer. Stop agitation when the correct pH is reached. Wait a few seconds and read the pH again. Adjust if necessary.

Remove the electrode from the solution, rinse it with water, and transfer it in a tube containing a KCl-saturated solution for safe storage.

AUTORADIOGRAPHY

Choosing the Correct Exposure Conditions

In order to obtain the best possible results in the shortest time, it is important to choose the conditions for exposing the sample to film carefully. The main points to consider are summarized below and are followed by a flowchart for guidance when choosing a method.

There is nothing to be gained either in speed or sensitivity by exposing at $-70°C$ unless a fluor or intensifying screen is present. Direct autoradiography is of no benefit, and, indeed, film response may well be slower. (There may, however, be an advantage when making long exposures, as backgrounds are often reduced by exposure at low temperature.)

If quantitation is required in fluorography, exposure at $-70°C$ must be combined with the use of preflashed film, or the response of the film will be nonlinear.

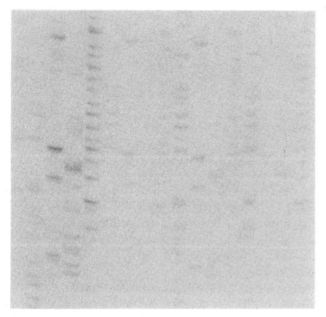 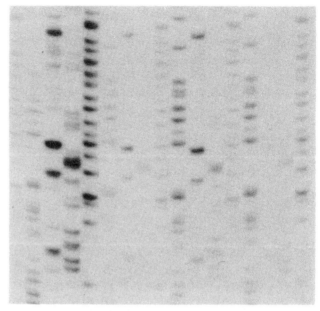

Autoradiograph, 1 hour, 20°C   **Fluorograph,** 1 hour, −70°C, p/f

Figure 9. Comparison of autoradiography and fluorography of a ^{32}P sequencing gel (courtesy of Amersham).

Some loss of resolution always occurs when using fluorography rather than direct autoradiography. The severity of this loss depends on the radioisotope and the protocol employed. The effects are often more apparent to the eye than to instruments (Figure 9).

Intensifying screens gives no benefit with ^{35}S, ^{14}C, or ^{3}H, and impregnation with fluors does not significantly improve performance with ^{32}P or ^{125}I.

Estimating Exposure Times*

The optimum exposure time will depend on many factors and is to an extent a matter of trial and error.

However, the following tables and figures are provided to give users a starting point, thus helping to cut down the number of exposures required.

The assumption has been made that samples are thin, for example, dried gels, and so do not exhibit abnormally high self-absorption of radiation. For those researchers dealing with the autoradiography of tissue sections, Pelc and Welton (1967) have reviewed self-absorption in this type of sample.

Table 13 gives the activity per 5-mm slot which must be loaded onto a gel in order to give a satisfactory image density after 24 hours exposure. The units used are disintegrations per minute, measured as acid-precipitable material, but for isotopes other than tritium this will approximate to the usual "counts per minute." For tritium, the "counts" figure is about one-third less than the "disintegrations" value.

Table 13. Optimum Radioactivity for Satisfactory Detection by Autoradiography

	Isotope			
Technique	^{32}P	^{35}S/^{14}C	^{3}H	^{125}I
Direct autoradiography at 20°C	3.5×10^2	4×10^3	$>2 \times 10^6$	1×10^3
Fluorography at −70°C (film preflashed)	3×10^1	2.5×10^2	5×10^3	6×10^1

Source: Information courtesy of Amersham.

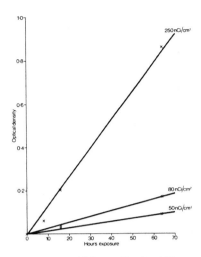

³H – Fluorography at −70°C, preflashed film

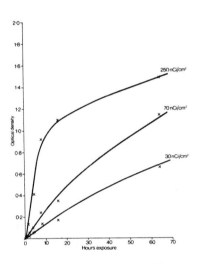

¹⁴C/³⁵S – Direct autoradiography at 20°C

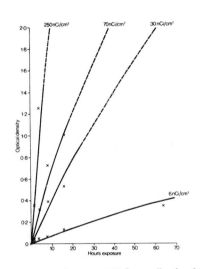

¹⁴C/³⁵S – Fluorography at −70°C, preflashed film

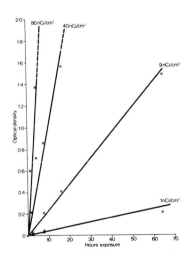

³²P – Direct autoradiography at 20°C

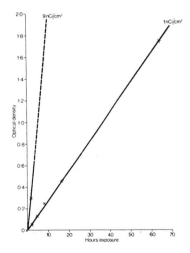

³²P – Fluorography at −70°C, preflashed film

Figure 10. Fluorography with different labeled compounds (courtesy of Amersham).

Table 14. Increased Sensitivity Obtained with Fluorography

Substrate	Isotope	Direct Autoradiography[a]	Fluorography[b] Fluor Impregnation	Intensifying Screen	Gain (By Using Fluorography)
Dried unstained, polyacrylamide gel	^{3}H	1×10^7	1×10^4	—	1000
	^{14}C/^{35}S	8×10^3	5.3×10^2	—	15
	^{32}P	7×10^2	—	7×10	10
	^{125}I	2.1×10^3	—	1.3×10^2	16
Dry TLC plate, paper chromatogram or blot	^{3}H	5×10^5	5×10^3	—	100
	^{14}C/^{35}S	8×10^3	8×10^2	—	10
	^{32}P	7×10^2	—	5×10	14
	^{125}I	2.2×10^3	—	1.3×10^2	17

Source: Information courtesy of Amersham.

[a]At ambiant temperature, circa 20°C.

[b]At -70°C, film sensitized by preflashing to 0.15 A at 540 nm, using PPO as fluor.

Note: 2220 disintegrations/minute = 1 nanoCurie (nCi). If using an end-window probe to check gels or plates, 5 counts/second of ^{125}I or ^{32}P or 20 counts/second of ^{35}S or ^{14}C will give a barely detectable image in 24 hours by direct autoradiography and a satisfactory image using an efficient fluorography technique.

Table 15. Gel Fluorography Protocols[a]

Method	PPO in DMSO	PPO in Glacial Acetic Acid	Sodium Salicylate	Amplify™
Fixing advisable	Yes	No	Yes	Yes
Recommended fixer/time	Acetic acid: 2 vol Methanol: 2 vol Water: 9 vol 1 hour	—	Acetic acid: 2 vol Methanol: 9 vol Water: 9 vol 30 minutes 1 × 30 minutes in water	7% v/v acetic acid 30 minutes[b]
Presoak required	2 × 30 minutes in DMSO	1 × 5 minutes in glacial acetic acid	1 × 30 minutes in water	None
Scintillant cocktail	4 × gel volume of 22.2% w/v PPO in DMSO	4 × gel volume of 20% w/v PPO in glacial acetic acid	10 × gel volume of 1 M sodium salicylate pH 5–7	4 × gel volume of Amplify, as supplied
Time to impregnate	3 hours	1.5 hours	0.5 hours	0.3–0.5 hours
Rinses	1 hour in 20 × gel volume of water	1 × 10 min water	None	None
Drying	Under vacuum, 1 hour	Under vacuum at 70°C, 3–5 hours	Under vacuum at 80°C, 2 hours	Under vacuum at 70–80°C, 1 hour

Source: Information courtesy of Amersham.

[a]The information given above is appropriate to polyacrylamide gels of 1 mm thickness or less. Users of thicker gels may need to increase soaking times.

[b]If, for any reason, the bromophenol blue tracker dye is not run to the bottom of the gel, this should be changed to 2 × 20 minutes, the change of fixer being advisable to rinse away soluble, low molecular weight radioactive impurities.

Fluorography Techniques*

The best known and most widely used fluor is undoubtedly 2,5-diphenyl oxazole (PPO), whether in techniques for gel impregnation such as that of Bonner and Laskey (1974), or in methods for treatment of TLC plates such as that of Randerath (1970). (Figure 10).

As a fluor, it offers great advantages of high efficiency and excellent resolution, but these are offset to some extent by the requirement that flammable, toxic, or corrosive solvents be used as carriers. Recently, Bochner and Ames have suggested an alternative method for the fluorography of TLC plates which gives efficiency comparable to that obtained with PPO, without the need for solvents.

For gel fluorography, reagents such as Amersham' "Amplify," code NAMP.100, compare favorably in efficiency and resolution with the Bonner and Laskey method, while lacking the toxicity of the sodium salicylate used by Chamberlain (1979). (Table 14).

For those researchers wishing to continue with the use of PPO in organic solvents, the method of Skinner and Griswold (1983) is of interest. It employs glacial acetic acid in place of dimethyl sulfoxide, giving some increase in convenience. If economy is of prime importance, 2-methylnaphthalene can be used in 5:1 ratio with PPO, either in DMSO or glacial acetic acid, albeit with some loss of efficiency.

If PPO is recovered from solvents for reuse, an aliquot should always be assayed by liquid scintillation counting to ensure that the fluor has not been contaminated by radioactivity leached from the gel or other substrate. Such contamination would lead to fogged films. See Table 15 for a summary of protocols.

Quantitation of Results*

A large proportion of researchers will probably wish to use film detection methods in a qualitative, or at most semiquantitative, sense. However, for those who do need to obtain reliable quantitative results there are several vital considerations.

Conditions must be absolutely standardized.

These include temperature of, and time in, developer, which can have a dramatic effect of final optical density.

Attempts should not be made to quantify very low or very high density images, as the response of the film is unlikely to be linear at its extremes.

If using fluorography, exposures must be made at $-70°C$ using preflashed film, or the response will not be linear. Low levels of activity will be underrepresented.

Intensifying X-ray Films

Despite the various improvements that have been made to increase the sensitivity of autoradiography, exposure times of several weeks are still frequently needed to obtain satisfactory results. Due to the short half-life of ^{32}P (see Chapter 16, Section I) it is sometimes useless to reexpose for longer periods of time.

Below is described a method for treating autoradiograms after exposure in such a way that a band hardly detected after several weeks exposure becomes detectable after overnight exposure.

Principle of the Method

In this method (adapted by P. W. J. Rigby from US Patent no. 4 101 780) the film is soaked in [^{35}S]thiourea, which can bind covalently with the silver grains generated during exposure of the film. The bands, therefore, become labeled with [^{35}S]thiourea and can be detected by exposure on a new X-ray film.

PROTOCOL

INTENSIFYING X-RAY FILMS

Prewashes

Seven washes are performed in large volumes of solution. Use at least 1 liter for an 18×24 cm film.

1. Wash with distilled water for 2 minutes. Pour off water.
2. Wash once more with distilled water for 2 minutes.
3. Pour off water, replace with 20% v/v methanol and wash for 2 minutes. Discard methanol.
4. Wash film in 50% v/v methanol for 2 minutes. Discard methanol.

5. Wash for 2 minutes with 20% v/v methanol. Discard methanol.
6. Wash twice in distilled water for 10 minutes each time.
7. Let the film dry.

Activation

1. Prepare a solution of [^{35}S]thiourea in 0.1 N ammonia (pH 11) to a specific radioactivity of 2 µCi/ml (500 µCi/250 ml).
2. Put the dry film (from step 7, above) in a tray. Add 0.5 ml [^{35}S]thiourea solution per square centimeter of film (215 ml for an 18 × 24 cm film). Shake for 1 hour at room temperature.
3. Pour off and save the [^{35}S]thiourea solution.

Washes

Use 1 liter washing solution/18 × 24 cm film.

1. Wash the treated film thoroughly with distilled water for 2 minutes. Discard the water in a radioactive waste bottle.
2. Repeat step 1.
3. Wash the film with 20% v/v methanol for 5 minutes.
4. Wash with 50% v/v methanol for 2 minutes.
5. Repeat step 3.
6. Wash the film with distilled water for 5 minutes.
7. Repeat step 6.
8. Let the film dry.

Exposure

Put the treated dry film in direct contact with an X-ray film, without anything between the two films. Make marks on both films to localize the bands precisely.

Notes

1. [^{35}S]thiourea can be purchased from Amersham.
2. Thiourea must be handled with care. It has been found to be a weak carcinogen in animals.
3. The thiourea activation solution can be used at least three times, with loss of activity each time.
4. If high backgrounds are obtained, dissolve the thiourea in 0.1 N ammonia, pH 8 (with HCl), or with 0.1 N NaOH, pH 8. These solutions with pH values lower than 11 should be left in contact with the film for 2 hours to obtain good activation.

5. This method is the method of choice if the original blots are no longer available, and can be used with films up to 2 years old.

Another way of intensifying autoradiograms is to use the Kodak Chromium Intensifier (Figure 11). The binding of chromium to the silver grains generated during development of the film will result in darkening of the bands after treatment with a non-staining developer such as Kodak Dektol diluted 1:3 in water. Instructions are provided with the product.

TROUBLE-SHOOTING GUIDE*

Band Resolution Worse Than Expected

Cause

1. Film and substrate not in good contact.
2. Poor fluorography technique or wrong choice of fluor.
3. Inherent characteristic of isotope (^{32}P/^{125}I).

Solution

1. Use a good X-ray cassette.
2. If using PPO, observe all washing steps. If using salicylate, change to PPO or "Amplify" or reduce soaking time.
3. Use ^{35}S/^{14}C or ^{3}H, where possible.

Film Partially or Wholly Fogged

Cause

1. Poor light tightness in darkroom/safelight too bright or too near or has wrong filter.†
2. High radiation background.
3. Old/dirty chemicals, or film badly stored.
4. Radioactive contamination of cassette.
5. Old or damaged cassette.
6. Light emission from fluor in substrate.§
7. Inadequate fixing/rinsing of gel.

Solution

1. Take suitable steps to repair/remedy (black self-adhesive sealing strip is available).

---

*Information courtesy of Amersham.
†This is particularly important when using preflashed film, which is extremely sensitive to low light levels.
§This problem is more likely to occur when using "complete" scintillants, such as 2-methylnapthalene/PPO mixtures, rather than with primary fluors alone.

2. Carry out exposure in a low background area. Segregate [32]P- and ^{125}I-containing cassettes.‡
3. Observe good housekeeping practice; change solutions frequently.
4. Wipe cassettes out after each use.
5. Replace.
6. Allow substrate 30–60 seconds "Dark Adaption" before loading into cassette.
7. See note to Table 14.

Film Shows Sharply Defined Artifacts

Cause

1. Rough handling of film before exposure/developing.
2. Chemical contamination of cassette or substrate, for example, by water, acetic acid, and so on.
3. Static electricity.

Solution

1. Handle film carefully; do not subject to pressure. Wear cotton gloves in hot weather.
2. Wipe cassettes out after each use. Ensure that gels and plates are dry. Thaw $-70°C$ exposures thoroughly before developing.
3. Remove film from box slowly. Avoid wearing nylon overalls. Remove Saranwrap from dried gels at least 1 minute before exposing to film.

---

‡It is not always realized that even soft β-emitters can fog film via the secondary Bremsstrahlung x-rays which they produce. This is particularly true of solutions at high radioactive concentration (>5 mCi/ml), and such solutions should be stored well away from film and cassettes containing film.

References

Bonner, W. M., and Laskey, R. A. (1974), *Eur. J. Biochem.,* **46**, 83.

Chamberlain, J. P. (1979), *Anal. Biochem.,* **98**, 132.

Mitchell, R. L. (1986), *Focus,* **8:3**, 10.

Pelc, S. R., and Welton, M. G. E. (1967), *Nature,* **216**, 925.

Randerath, K. (1970), *Anal. Biochem.,* **34**, 188.

Skinner, M. K., and Griswold, M. D. (1983), *Biochem. J.,* **209**, 291.

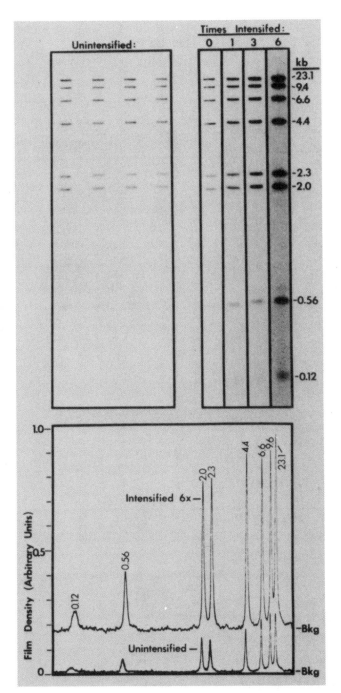

Figure 11. Autoradiogram of Hind III fragments of lambda DNA. Unintensified (a) and intensified (b), 0, 1, 3 or 6 times as described in the text. Densitometer scan (c) of autoradiogram before (unintensified) and after (intensified 6×) 6 rounds of intensification. Sizes of DNA fragments are indicated at the tops of densitometer peaks (courtesy of Bethesda Research Laboratories, Life technologies Inc.).

Physical Properties of Some Radionuclides Used in Molecular Cloning

Radionuclide	Tritium	Carbon-14
Half-life	12.43 years	5730 years
Beta energy (MeV)		
Max	0.0186	0.156
Mean	0.0057	0.049
Specific activity (maximum)	28.8 Ci/matom 1.06 TBq/matom	62.4 mCi/matom 2.31 GBq/matom
Common values for compounds	$0.1-10^2$ Ci/mmol 3.7 GBq–3.7 TBq/mmol	$1-10^2$ mCi/mol 37 MBq–3.7 GBq/mmol
Daughter nuclide (stable)	Helium-3	Nitrogen-14

Radionuclide	Sulphur-35	Phosphorous-32
Half-life	87.4 days	14.3 days
Beta energy (MeV)		
Max	0.167	1.709
Mean	0.049	0.69
Specific activity (maximum)	1.49 kCi/matom 55.3 TBq/matom	9.13 kCi/matom 338 TBq/matom
Common values for compounds	$1-10^6$ mCi/mmol 37 MBq–37 TBq/mmol	$0.01-10^3$ Ci/mmol 370 MBq–37 TBq/mmol
Daughter nuclide (stable)	Chlorine-35	Sulphur-32

Radionuclide	Iodine-131	Iodine-125
Half-life	8.04 days	60.0 days
Beta energy (MeV)		
Max	0.806	Electron capture
Mean	0.18	
Specific activity (maximum)	16.2 kCi/matom 601 TBq/matom	2.18 kCi/matom 80.5 TBq/matom
Common values for compounds	0.1–10 Ci/mmol 3.7–370 GBq/mmol	$0.1-10^3$ Ci/mmol 3.7 GBq–37 TBq/mmol
Daughter nuclide (stable)	Xenon-131	Tellurium-125

PHYSICAL CONSTANTS OF NUCLEOSIDES AND NUCLEOTIDES

Many problems encountered in molecular cloning often come from an ignorance of the physical and biochemical properties of the components used in the reactions. Among these components, the nucleosides and corresponding nucleotides play a pivotal role because they are employed in all steps involving in vitro synthesis and modification of nucleic acids. The following chart has been compiled to provide a reference guide for the adequate use and storage of these molecules (information courtesy of Boehringer

Biochemicals). Other problems may arise because old stocks of labeled nucleotides are used. Table 16 provides useful information on storage conditions and degradation of ^{32}P-labeled nucleotides.

Table 16.   Decomposition of ^{32}P Nucleotides

Compound	Formulation[a]	Specific Activity	Storage Temperature, °C	Time of Storage, days	Observed Decomposition, %
Adenosine [α-^{32}P]triphosphate	1	3	+20	10	18
	1	3	−20	7	1
	2	3	−20	7	1
	2	3000	+20	21	9
	2	3000	−20	21	3
	3	3000	−20	7	2
	3	>5000	−20	7	2
Adenosine [α-^{32}P]triphosphate	1	10	+20	7	Up to 20
	1	10	−20	7	1
	2	10	−20	7	1
	2	410	+20	7	14
	2	410	−20	7	<1
	2	3000	−20	21	<1
	3	410	−20	28	3
Guanosine [α-^{32}P]triphosphate	1	10	+20	7	30
	1	10	−20	7	2–3
	2	10	−20	7	2–3
	2	410	−20	21	<1
	3	410	−20	21	1
	2	3000	−20	14	4
Deoxycytidine [α-^{32}P]triphosphate	2	410	−20	28	4
	3	410	−20	28	4
	2	3000	−20	21	<1
	3	3000	−20	21	<1

Source: Information courtesy of Amersham.

[a]Formulation: 1 aqueous solution 1 mCi/ml; 2 aqueous ethanol (1:1) 1 mCi/ml; 3 stabilized aqueous solution [10 mCi/ml (5 *mM* 2-mercaptoethanol)].

Guide for the Adequate Use and Storage of Nucleosides and Nucleotides*

Ribonucleosides

	Purines			Pyrimidines	
	Adenosine Crystallized	Guanosine Crystallized	Inosine Crystallized	Cytidine Crystallized	Uridine Crystallized
Typical analysis (note 1)					
Purity (%)	>99%	>98%	99%	>99%	>98%
Nucleoside content					
theoretical (%)	100%	100%	100%	100%	100%
enzymatic (%)	>99%	>98%	99%	>99%	>98%
H$_2$O content (%, K. Fischer)	—	—	—	—	<1%
Formula & approx. molecular weight (note 2)	$C_{10}H_{13}N_5O_4$ 267.2	$C_{10}H_{13}N_5O_5$ 283.2	$C_{10}H_{12}N_4O_5$ 268.2	$C_9H_{13}N_3O_5$ 243.2	$C_9H_{12}N_2O_6$ 244.2
Solubility in H$_2$O (mg/ml)	Soluble (25°C)	0.75 (25°C) 30.3 (100°C)	16 (20°C)	≥50 (25°C)	≥100 (25°C)

Ribonucleosides (*continued*)

	Purines			Pyrimidines	
	Adenosine Crystallized	Guanosine Crystallized	Inosine Crystallized	Cytidine Crystallized	Uridine Crystallized
Stability in solution					
at pH 7, 4°C	1 month	1 month	1 month	Several days	Several days
at pH 7, 25°C	$\geq$ 24 hours	—	—	—	—
at pH 1	Slow hydrolysis	Slow hydrolysis	Slow hydrolysis	Stable, cold	Stable, cold
pH of compound in H_2O	Neutral	Neutral	Neutral	Neutral	Neutral
Absorptivities (ε) (note 3) $(mmol^{-1} \times$ liter $\times$ $cm^{-1})$					
ε at 260 nm (pH 7)	15.0	11.8	7.4	7.4	9.9
absorbance maximum	260 nm (pH 7)	252 nm (pH 7)	248.5 nm (pH 7)	271 nm (pH 8.2)	261 nm (pH 7.3)
ε at absorbance max.	—	13.6	12.25	9.1	10.1
Absorbance ratios (pH 7) (note 3)					
A_{250}/A_{260} (literature)	0.80	1.15	1.68	0.86	0.74
A_{250}/A_{260} (BM specif.)	0.78 $\pm$ 0.02	1.15 $\pm$ 0.03	1.68 $\pm$ 0.04	0.85 $\pm$ 0.02	0.74 $\pm$ 0.02
A_{280}/A_{260} (literature)	0.15	0.68	0.25	0.93	0.35
A_{280}/A_{260} (BM specif.)	0.15 $\pm$ 0.01	0.67 $\pm$ 0.02	0.25 $\pm$ 0.02	0.96 $\pm$ 0.02	0.36 $\pm$ 0.02
A_{290}/A_{260} (literature)	0.002	0.28	0.025	0.29	0.03
A_{290}/A_{260} (BM specif.)	0.01	0.27 $\pm$ 0.02	0.03 $\pm$ 0.01	0.31 $\pm$ 0.02	0.04
Other information	pK = 3.63, 9.80	—	—	pK = 4.22	—

Ribonucleoside-5'-Monophosphates

	Purines			Pyrimidine
	5'-AMP Disodium Salt Crystallized	5'-AMP Free Acid Crystallized	5'-IMP Disodium Salt Crystallized	5'-UMP Disodium Salt Crystallized
Typical analysis (note 1)				
Purity (%)	$\geq$ 99%	$\geq$ 98%	$\geq$ 98%	$\geq$ 99%
Nucleoside content				
theoretical (%)	70%	95%	70%	80%
enzymatic (%)	$\geq$ 69%	$\geq$ 93%	$\geq$ 68% (absorbance)	$\geq$ 76%
H_2O content (%)	23% $\pm$ 2%	Approx. 5%	25% $\pm$ 2%	Approx. 9%
Na^+ content (%)	9% $\pm$ 0.5%	—	9% $\pm$ 0.5%	10% $\pm$ 2%
Formula and approx. molecular weight (note 2)	$C_{10}H_{12}N_5O_7PNa_2 \cdot 6H_2O$ 499.2	$C_{10}H_{14}N_5O_7P \cdot H_2O$ 365.2	$C_{10}H_{11}N_4O_8PNa_2 \cdot 6H_2O$ 500.2	$C_9H_{11}N_2O_9PNa_2 \cdot 2H_2O$ 404.5
Anhydrous MW	AMP, 347.2	AMP, 347.2	IMP, 348.2	UMP, 324.2
Solubility in H_2O (mg/ml)	$\geq$ 50 (25°C)	Sparingly soluble (25°C) Very soluble (100°C) $\geq$ 50 (0.1 M NaOH, 25°C)	130 (20°C)	410 (20°C)
Stability in solution				
at pH 7, -20°C	> 1 month	—	—	—
at pH 7, 4°C	$\geq$ 1 week	$\geq$ 1 week	$\geq$ 1 week	$\geq$ 1 week
at pH 7, 25°C	$\geq$ 2 days	—	—	—
Decomposition when stored dry	Stable, $\geq$ 3 years (at RT)	Stable, $\geq$ 3 years (at RT)	Stable, $\geq$ 3 years (at RT)	Stable, $\geq$ 3 years (at RT)
Stability of phosphate ester ($t_{1/2}$, minutes, 100°C, pH 1) (note 3)	1050	1050	—	3590

Ribonucleoside-5′-Monophosphates (*continued*)

	Purines			Pyrimidine
	5′-AMP Disodium Salt Crystallized	5′-AMP Free Acid Crystallized	5′-IMP Disodium Salt Crystallized	5′-UMP Disodium Salt Crystallized
pH of compound in H_2O	—	Acid	—	—
Absorptivities (ε) (note 3) ($mmol^{-1} \times liter \times cm^{-1}$)				
ε at 260 nm (pH 7)	15.0	15.0	7.4	9.9
absorbance maximum	259 nm (pH 7)	259 nm (pH 7)	249 nm	262 nm (pH 7)
ε at absorbance maximum	15.4	15.4	12.2	10.0
Absorbance ratios (pH 7) (note 3)				
A_{250}/A_{260} (literature)	0.80	0.80	—	0.73
A_{250}/A_{260} (BM specif.)	0.78 ± 0.02	0.78 ± 0.02	1.64 ± 0.03	0.73 ± 0.02
A_{280}/A_{260} (literature)	0.15	0.15	—	0.40
A_{280}/A_{260} (BM specif.)	0.16 ± 0.01	0.16 ± 0.01	0.26 ± 0.02	0.40 ± 0.02
A_{290}/A_{260} (literature)	0.002	0.002	—	0.03
A_{290}/A_{260} (BM specif.)	0.01	0.01	0.05 ± 0.01	0.04 ± 0.01
Other information	P_i content, 0.9% Note 4	P_i content, ≤ 0.9% pK = 3.8, 6.2	pK = 2.4, 6.4 P_i content, ≤ 0.3%	pK = 6.4, 9.5

A. Ribonucleoside-5′-Diphosphates

	Purines			
	5′-ADP Potassium Salt Crystallized	5′-ADP Disodium Salt Lyophilized	5′-ADP Monosodium Salt (Amorphous)	5′-ADP Free Acid Crystallized
Typical analysis (note 1)				
Purity (%)	≥ 98%	≥ 97%	—	≥ 99%
Nucleoside content				
theoretical (%)	85%	84%	85%	98%
enzymatic (%)	≥ 84%	82%	≥ 84%	≥ 97%
H_2O content (%)	7.2 ± 1%	≤ 7%	≤ 12%	≤ 2%
Na^+/K^+ content (%)	K^+, 7.8% ± 0.5%	Na^+, 9% ± 1%	Na^+, 5%	—
Formula & approx. molecular weight (note 2)	$C_{10}H_{14}N_5O_{10}P_2K \cdot 2H_2O$ 501.3	$C_{10}H_{13}N_5O_{10}P_2Na_2 + 2H_2O$ 507.2	$C_{10}H_{13}N_5O_{10}P_2Na + 3H_2O$ 503.2	$C_{10}H_{15}N_5O_{10}P_2 + \frac{1}{2}H_2O$ 436.2
Anhydrous MW	ADP, 427.2 ADP-K, 465.3	ADP, 427.2 ADP-Na_2, 471.2	ADP, 427.2 ADP-Na, 449.2	ADP, 427.2
Solubility in H_2O (mg/ml)	≥ 50	≥ 50	≥ 50	≥ 50
Solubility in solution				
at pH 7, −20°C	Several months	Several months	Several months	Several months
at pH 7, 4°C	Several days	Several days	Several days	Several days
other	Unstable in acid	Unstable in acid	Unstable in acid	Unstable in acid
Decomposition when stored dry	Stable, ≥ 2 years (at 4°C)	Slow decay (at −20°C)	Slow decay (at −20°C)	Slow decay (at −20°C)

A. Ribonucleoside-5'-Diphosphates (*continued*)

	Purines			
	5'-ADP Potassium Salt Crystallized	5'-ADP Disodium Salt Lyophilized	5'-ADP Monosodium Salt (Amorphous)	5'-ADP Free Acid Crystallized
Stability of pyro-phosphate ester ($t_{1/2}$, minutes, 100°C, pH 1) (note 3)	< 10	< 10	< 10	< 10
pH of compound in H_2O	Acid	Acid	Acid	Acid
Absorptivities (ε) (note 3) ($mmol^{-1} \times$ liter $\times cm^{-1}$				
ε at 260 nm (pH 7)	15.0	15.0	15.0	15.0
absorbance maximum	259 nm (pH 7)	259 nm (pH 7)	259 nm (pH 7)	259 nm (pH 7)
ε at absorbance max.	15.4	15.4	15.4	15.4
Absorbance ratios (pH 7) (note 3)				
A_{250}/A_{260} (literature)	0.80	0.80	0.80	0.80
A_{250}/A_{260} (BM specif.)	0.78 ± 0.02	0.78 ± 0.02	0.78 ± 0.02	0.78 ± 0.02
A_{280}/A_{260} (literature)	0.15	0.15	0.15	0.15
A_{280}/A_{260} (BM specif.)	0.16 ± 0.01	0.16 ± 0.01	0.16 ± 0.01	0.16 ± 0.01
A_{290}/A_{260} (literature)	0.002	0.002	0.002	0.002
A_{290}/A_{260} (BM specif.)	≤ 0.01	0.01	0.01	0.01
Other information	P_i, ≤ 0.3% AMP, ≤ 1% ATP, ≤ 0.2% NH_4^+, ≤ 0.005% Free of ethanol	P_i, ≤ 0.6% AMP, ≤ 3% ATP, ≤ 1% NH_4^+, ≤ 0.01% Free of ethanol	P_i, ≤ 0.6% AMP, ≤ 3% ATP, ≤ 1% NH_4^+, ≤ 0.01% Free of ethanol	P_i, ≤ 0.6% AMP, ≤ 3% ATP, ≤ 0.3% Free of ethanol

B. Ribonucleoside-5'-Diphosphates

	Purines		Pyrimidines	
	5'-GDP Dilithium Salt (Amorphous)	5'-IDP Trisodium Salt (Amorphous)	5'-CDP Free Acid Crystallized	5'-UDP Dipotassium Salt Crystallized
Typical analysis (note 1)				
Purity (%)	≥ 91%	> 91%	98%	97%
Nucleoside content				
theoretical (%)	90%	81%	96%	76%
enzymatic (%)	≥ 80%	72%	94%	72%
H_2O content (%)	≤ 8%	8%	Approx. 4%	Approx. 10%

B. Ribonucleoside-5'-Diphosphates (*continued*)

	Purines		Pyrimidines	
	5'-GDP Dilithium Salt (Amorphous)	5'-IDP Trisodium Salt (Amorphous)	5'-CDP Free Acid Crystallized	5'-UDP Dipotassium Salt Crystallized
$Na^+/K^+/Li^+$ content (%)	Li^+, 3% $\pm$ 0.5%	Na^+, 11%	—	K^+, approx. 15%
Formula & approx. molecular weight (note 2)	$C_{10}H_{12}N_5O_{11}P_2Li_2 + 2H_2O$ 491.1	$C_{10}H_{11}N_4O_{11}P_2Na_3 + 2H_2O$ 530.1	$C_9H_{15}N_3O_{11}P_2 \cdot H_2O$ 421.2	$C_9H_{12}N_2O_{12}P_2K_2 \cdot 3H_2O$ 534.4
Anhydrous MW	GDP, 443.2 GDP-Li, 455.0	IDP, 428.2 IDP-Na$_3$, 494.1	CDP, 403.2	UDP, 404.2 UDP-K$_2$, 480.4
Solubility in H_2O (mg/ml)	$\geqslant 50$	Soluble (25°C)	—	Soluble (25°C)
Solubility in solution				
at pH 7, -20°C	Approx. 1 week	—	—	Approx. 1 week
at pH 7, 4°C	Approx. 1 day	—	Several hours	Approx. 1 day
other	Unstable in acid	Unstable in acid	Unstable in acid	Unstable in acid
Decomposition when stored dry	Not detectable in 6 months (4°C)	Stable, 6 months (at 4°C)	Stable (at 4°C)	Stable, 24 months (at 4°C)
Stability of pyrophosphate ester ($t_{1/2}$, minutes, 100°C, pH 1) (note 3)	< 10	< 10	< 15 (pH 0)	< 15 (pH 0)
pH of compound in H_2O	Acid	Acid	Acid	Acid
Absorptivities (ε) (note 3) ($mmol^{-1} \times$ liter $\times cm^{-1}$)				
ε at 260 nm (pH 7)	11.8	7.4	7.4	9.9
absorbance maximum	253 nm (pH 7)	249 nm	271 nm (pH 7)	262 nm (pH 7)
ε at absorbance max.	13.7	12.2	9.1	10.0
Absorbance ratios (pH 7) (note 3)				
A_{250}/A_{260} (literature)	1.15	—	0.84	0.73
A_{250}/A_{260} (BM specif.)	1.15 $\pm$ 0.03	1.64 $\pm$ 0.03	0.85 $\pm$ 0.02	0.73 $\pm$ 0.02
A_{280}/A_{260} (literature)	0.68	—	0.99	0.40
A_{280}/A_{260} (BM specif.)	0.66 $\pm$ 0.02	0.26 $\pm$ 0.02	0.99 $\pm$ 0.02	0.40 $\pm$ 0.02
A_{290}/A_{260} (literature)	0.28	—	0.33	0.03
A_{290}/A_{260} (BM specif.)	0.28 $\pm$ 0.01	0.05 $\pm$ 0.01	0.34 $\pm$ 0.02	0.04 $\pm$ 0.01
Other information	P_i, $\leqslant$ 1.5% GMP, $\leqslant$ 6% GTP, $\leqslant$ 0.5%	ITP, $\leqslant$ 2%	CMP, $<$ 1% CTP, not detectable	UTP, $<$ 2% pK = 6.5, 9.4

A. Ribonucleoside-5'-Triphosphates

	Purines			
	5'-ATP Disodium Salt Crystallized (Special Quality)	5'-ATP Disodium Salt Crystallized	5'-GTP Dilithium Salt (Amorphous)	5'-GTP Disodium Salt (Amorphous)
Typical analysis (note 1)				
Purity (%)	$\geqslant 99.5\%$	$\geqslant 98\%$	89%	$> 92\%$
Nucleoside content				
theoretical (%)	84%	84%	92%	84%
enzymatic (%)	84%	$> 82\%$	80%	$\geqslant 76\%$
H_2O content (%)	$9\% \pm 1\%$	$9\% \pm 1\%$	6%	$8\% \pm 2\%$
Na^+/Li^+ content (%)	Na^+, $7.5\% \pm 0.5\%$	Na^+, $7.5\% \pm 0.5\%$	Li^+, 3%	Na^+, $8\% \pm 1\%$
Formula & approx. molecular weight (note 2)	$C_{10}H_{14}N_5O_{13}P_3Na_2 \cdot 3H_2O$ 605.2	$C_{10}H_{14}N_5O_{13}P_3Na_2 \cdot 3H_2O$ 605.2	$C_{10}H_{14}N_5O_{14}P_3Li_2 + 2H_2O$ 571.1	$C_{10}H_{14}N_5O_{14}P_3Na_2 + 3H_2O$ 621.1
Anhydrous MW	ATP, 507.2 ATP-Na_2, 551.2	ATP, 507.2 ATP-Na_2, 551.2	GTP, 523.2 GTP-Li_2, 535.1	GTP, 523.2 GTP-Na_2, 567.1
Solubility in H_2O (mg/ml)	$\geqslant 50$	$\geqslant 50$	$\geqslant 50$	$\geqslant 50$
Stability in solution				
at pH 7, $-20°C$	Several months	Several months	Few days (< 1 week)	Few days (< 1 week)
at pH 7, 4°C (note 5)	1 week	1 week	1 day	1 day
other	Few hours, pH 2, 0°C	Few hours, pH 2, 0°C	Unstable, pH $\leqslant 3$	Unstable, pH $\leqslant 3$
Decomposition when stored dry	Insignificant/1 year (at 4°C)	$\leqslant 3\%$/1 year (at 4°C)	5%/6 months (at 4°C)	10%/6 months (at $-20°C$)
Stability of pyrophosphate ester ($t_{1/2}$, minutes, 100°C, pH 1) (note 3)	8	8	< 10	< 10
pH of compound in H_2O	Acid	Acid	Acid	Acid
Absorptivities (ε) (note 3) ($mmol^{-1} \times$ liter $\times$ cm^{-1})				
ε at 260 nm (pH 7)	15.0	15.0	11.8	11.8
absorbance max.	259 nm (pH 7)	259 nm (pH 7)	252 nm (pH 7)	252 nm (pH 7)
ε at absorbance max.	15.4	15.4	13.7	13.7
Absorbance ratios (pH 7) (note 3)				

A. Ribonucleoside-5′-Triphosphates (*continued*)

	Purines			
	5′-ATP Disodium Salt Crystallized (Special Quality)	5′-ATP Disodium Salt Crystallized	5′-GTP Dilithium Salt (Amorphous)	5′-GTP Disodium Salt (Amorphous)
A_{250}/A_{260} (literature)	0.80	0.80	1.15	1.15
A_{250}/A_{260} (BM specif.)	0.79 ± 0.02	0.79 ± 0.02	1.15 ± 0.03	1.15 ± 0.03
A_{280}/A_{260} (literature)	0.15	0.15	0.68	0.68
A_{280}/A_{260} (BM specif.)	0.15 ± 0.02	0.15 ± 0.02	0.66 ± 0.02	0.66 ± 0.02
A_{290}/A_{260} (literature)	0.002	0.002	0.28	0.28
A_{290}/A_{260} (BM specif.)	≤ 0.01	≤ 0.01	0.28 ± 0.01	0.28 ± 0.01
Other information	$P_i, \leq 0.05\%$ AMP + ADP, $\leq 0.05\%$ GTP, $\leq 0.01\%$ Fe, Mg, each ≤ 10 ppm Ba, V, each ≤ 1 ppm Cu, ≤ 2 ppm Al, approx. 20 ppm Notes 6, 7, 8	$P_i \leq 0.3\%$ AMP + ADP, $\leq 0.5\%$ GTP, $\leq 0.01\%$ Pb, ≤ 30 ppm Notes 6, 7	$P_i, \leq 0.9\%$ GMP + GDP, $\leq 5\%$ Notes 7, 9	$P_i, \leq 0.9\%$ GMP + GDP, $\leq 6\%$ Notes 7, 9

B. Ribonucleoside-5′-Triphosphates

	Purines	Pyrimidines	
	5′-ITP Disodium Salt (Amorphous)	5′-CTP Disodium Salt (Amorphous)	5′-UTP Trisodium Salt (Amorphous)
Typical analysis (note 1)			
Purity (%)	$\geq 94\%$	$\geq 95\%$	$\geq 95\%$
Nucleoside content			
theoretical (%)	86%	86%	80%
enzymatic (%)	78%	80%	$\geq 75\%$
H_2O content (%)	7%	6%	$8\% \pm 3\%$
Na^+ content (%)	9%	$8\% \pm 1\%$	$11\% \pm 1\%$
Formula & approx. molecular weight (note 2)	$C_{10}H_{13}N_4O_{14}P_3Na_2 + 2H_2O$ 588.1	$C_9H_{14}N_3O_{14}P_3Na_2 + 2H_2O$ 563.1	$C_9H_{12}N_2O_{15}P_3Na_3 + 3H_2O$ 604.1
Anhydrous MW	ITP, 508.2 ITP-Na$_2$, 552.1	CTP, 483.2 CTP-Na$_2$, 527.1	UTP, 484.2 UTP-Na$_3$, 550.1
Solubility in H_2O (mg/ml)	Soluble (25°C)	≥ 50	≥ 50
Stability in solution			
at pH 7, −20°C	—	Few days (< 1 week)	Few days (< 1 week)
at pH 7, 4°C (note 5)	—	Approx. 1 day	Approx. 1 day
other	Unstable in acid	Unstable in acid	Unstable in acid
Decomposition when stored dry	Approx. 3%/6 months (at 4°C)	Approx. 4%/1 year (at 4°C)	Approx. 4%/6 months (at 4°C)
Stability of pyrophosphate ester ($t_{1/2}$, minutes, 100°C, pH 1) (note 3)	< 10	< 15	< 15

B. Ribonucleoside-5'-Triphosphates (*continued*)

	Purines	Pyrimidines	
	5'-ITP Disodium Salt (Amorphous)	5'-CTP Disodium Salt (Amorphous)	5'-UTP Trisodium Salt (Amorphous)
pH of compound in H_2O	Acid	Acid	4 ± 1
Absorptivities (ε) (note 3) ($mmol^{-1} \times liter \times cm^{-1}$)			
ε at 260 nm (pH 7)	7.4	7.4	9.9
absorbance maximum	249 nm	271 nm (pH 7)	262 nm (pH 7)
ε at absorbance max.	12.2	9.1	10.0
Absorbance ratios (pH 7) (note 3)			
A_{250}/A_{260} (literatur)	—	0.84	0.73
A_{250}/A_{260} (BM specif.)	1.64 ± 0.03	0.84 ± 0.02	0.74 ± 0.02
A_{280}/A_{260} (literature)	—	0.99	0.40
A_{280}/A_{260} (BM specif.)	0.26 ± 0.02	0.98 ± 0.02	0.38 ± 0.02
A_{290}/A_{260} (literature)	—	0.33	0.03
A_{290}/A_{260} (BM specif.)	0.05 ± 0.01	0.33 ± 0.02	0.03 ± 0.01
Other information	IDP, 5% Note 7	P_i, 1.5% CDP, 2% CMP, < 0.5% Note 7	UDP, ≤ 4% UMP, ≤ 1% Note 7

A. 2'-Deoxyribonucleoside-5'-Triphosphates

	Purines			
	dATP Disodium Salt Crystallized	dGTP Dilithium Salt (Amorphous)	dGTP Disodium Salt (Amorphous)	dITP Trisodium Salt (Amorphous)
Typical analysis (note 1)				
Purity (%)	95%	≥ 92%	≥ 95%	≥ 86%
Nucleoside content				
theoretical (%)	83%	91%	86%	85%
enzymatic (%)	79%	82%	81%	72% (N content)
H_2O content (%)	$9\% \pm 2\%$	$7\% \pm 2\%$	$6\% \pm 3\%$	Approx. 2%
$Na^+/K^+/Li^+$ content (%)	Na^+, $7\% \pm 1\%$	Li^+, $3\% \pm 0.5\%$	Na^+, $8\% \pm 1\%$	Na^+, approx. 11.5%
Formula and approx. molecular weight (note 2)	$C_{10}H_{14}N_5O_{12}P_3Na \cdot 3H_2O$ 589.2	$C_{10}H_{14}N_5O_{13}P_3Li_2 + 2H_2O$ 555.1	$C_{10}H_{14}N_5O_{13}P_3Na_2 + 2H_2O$ 587.2	$C_{10}H_{12}N_4O_{13}P_3Na_3 + H_2O$ 576.2
Anhydrous MW	dATP, 491.2 dATP-Na_2, 535.2	dGTP, 507.2 dGTP-Li_2, 519.1	dGTP, 507.2 dGTP-Na_2, 551.2	dITP, 492.2 dITP-Na_3, 558.2
Solubility in H_2O (mg/ml)	≥ 10	≥ 10	≥ 10	≥ 10
Stability in solution at pH 7, $-20°C$ (note 10)	Several months	Several months	Several months	Several months
other	Unstable in acid	Unstable in acid	Unstable in acid	Unstable in acid
Decomposition when stored dry	Approx. 3%/1 year (at $-20°C$)	Approx. 10%/6 months (at $-20°C$)	Approx. 10%/6 months (at $-20°C$)	Approx. 3%/6 months (at $-20°C$)

A. 2'-Deoxyribonucleoside-5'-Triphosphates (*continued*)

	Purines			
	dATP Disodium Salt Crystallized	dGTP Dilithium Salt (Amorphous)	dGTP Disodium Salt (Amorphous)	dITP Trisodium Salt (Amorphous)
Stability of pyrophosphate ester (pH 1)	Labile (note 11)	Labile (note 11)	Labile (note 11)	Labile (note 11)
pH of compound in H_2O	Acid	Acid	Acid	Acid
Absorptivities (ε) (note 3) ($mmol^{-1} \times$ liter $\times$ cm^{-1})				
ε at 260 nm (pH 7)	15.0	11.8	11.8	7.4
absorbance maximum	259 nm (pH 7)	253 nm (pH 7)	253 nm (pH 7)	249 nm (pH 7)
ε at absorbance max.	15.4	13.7	13.7	12.2
Absorbance ratios (pH 7) (note 3)				
A_{250}/A_{260} (literature)	0.80	1.15	1.15	—
A_{250}/A_{260} (BM specif.)	0.78 ± 0.02	1.14 ± 0.03	1.14 ± 0.03	1.64 ± 0.03
A_{280}/A_{260} (literature)	0.15	0.68	0.68	—
A_{280}/A_{260} (BM specif.)	0.16 ± 0.01	0.67 ± 0.02	0.67 ± 0.02	0.26 ± 0.02
A_{290}/A_{260} (literature)	0.002	0.28	0.28	—
A_{290}/A_{260} (BM specif.)	< 0.02	0.28 ± 0.02	0.28 ± 0.02	0.05 ± 0.01
Other information	P_i, $< 0.3\%$ dAMP + dADP, $< 2\%$ Note 7	P_i, $\leq 0.6\%$ dGMP + dGDP, $\leq 2\%$ GTP, $< 3\%$ Notes 7, 9	Pzi, $\leq 0.6\%$ dGMP + dGDP, $\leq 2\%$ GTP, $< 3\%$ Notes 7, 9	dIDP, $< 8\%$ Note 7

B. 2'-Deoxyribonucleoside-5'-Triphosphates

	Pyrimidines	
	dCTP Disodium Salt (Amorphous)	dTTP Tetrasodium Salt (Amorphous)
Typical analysis (note 1)		
Purity (%)	98%	$\geq 94\%$
Nucleoside content		
theoretical (%)	88%	80%
enzymatic (%)	84% (absorbance)	76% (absorbance)
H_2O content (%)	$5\% \pm 2\%$	$6\% \pm 2\%$
$Na^+/K^+/Li^+$ content (%)	Na^+, $9\% \pm 2\%$	Na^+, $12\% \pm 1\%$
Formula and approx. molecular weight (note 2)	$C_9H_{14}N_3O_{13}P_3Na_2 + H_2O$ 529.1	$C_{10}H_{13}N_2O_{14}P_3Na_4 + 2H_2O$ 606.2
Anhydrous MW	dCTP, 467.2 dCTP-Na_2, 511.1	dTTP, 482.2 dTTP-Na_4, 570.2

B. 2'-Deoxyribonucleoside-5'-Triphosphates (*continued*)

	Pyrimidines	
	dCTP Disodium Salt (Amorphous)	dTTP Tetrasodium Salt (Amorphous)
Solubility in H_2O (mg/ml)	$\geqslant 10$	$\geqslant 10$
Stability in solution		
at pH 7, $-20°C$ (note 10)	Several months	Several months
other	Unstable in acid	Unstable in acid
Decomposition when stored dry	Approx. 5%/1 year (at $-20°C$)	Approx. 10%/1 year (at $-20°C$)
Stability of pyrophosphate ester (pH 1)	Labile (note 11)	Labile (note 11)
pH of compound in H_2O	Acid	Acid
Absorptivities (ε) (note 3)		
$(mmol^{-1} \times liter \times cm^{-1})$		
ε at 260 nm (pH 7)	7.4	8.4
absorbance maximum	271 nm (pH 7)	267 nm (pH 7)
ε at absorbance max.	9.1	9.6
Absorbance ratios (pH 7) (note 3)		
A_{250}/A_{260} (literature)	0.84	0.65
A_{250}/A_{260} (BM specif.)	0.82 ± 0.02	0.64 ± 0.02
A_{280}/A_{260} (literature)	0.99	0.73
A_{280}/A_{260} (BM specif.)	0.99 ± 0.02	0.75 ± 0.02
A_{290}/A_{260} (literature)	0.33	0.24
A_{290}/A_{260} (BM specif.)	0.33 ± 0.02	0.24 ± 0.01
Other information	$P_i, < 1.5\%$	$P_i, < 1.5\%$
	dCMP + dCDP, $< 3\%$	dTMP + dTDP, $< 5\%$
	Note 7	Note 7

A. 2',3'-Dideoxyribonucleoside-5'-Triphosphates

	Purines		
	ddATP Trilithium Salt 10 mM Solution	ddGTP Trilithium Salt 10 mM Solution	ddITP Trilithium Salt 10 mM Solution
Typical analysis (note 1)			
Purity (%)	$\geqslant 95\%$	$\geqslant 95\%$	$\geqslant 95\%$
Na^+K^+/Li^+ content (%)	Li^+, approx. 4%	Li^+, approx. 4%	Li^+, approx. 4%
Formula and approx. molecular	$C_{10}H_{13}N_5O_{11}P_3Li_3$	$C_{10}H_{13}N_5O_{12}P_3Li_3$	$C_{10}H_{12}N_4O_{12}P_3Li_3$
weight (note 2)	493.0	509.0	494.0
Anhydrous MW	ddATP, 475.2	ddGTP, 491.2	ddITP, 476.2
Solubility in H_2O (mg/ml)	$\geqslant 4.93$ (10 mM) (note 14)	$\geqslant 5.09$ (10 mM) (note 14)	$\geqslant 4.94$ (10 mM) (note 14)
Stability in solution			
at pH 7, $-20°C$ (note 10)	Several months	Several months	Several months
other	Note 12	Note 12	Note 12
Stability of pyrophosphate ester (pH 1)	Labile (note 11)	Labile (note 11)	Labile (note 11)
pH of solution (as supplied)	7	7	7
Absorptivities (ε) (note 3)			
$(mmol^{-1} \times liter \times cm^{-1})$			
ε at 260 nm (pH 7)	15.3	—	—
absorbance maximum	261 nm (pH 7)	252 nm (pH 7)	249 nm (pH 7)
ε at absorbance max.	16.9	13.7	12.3
Absorbance ratios (pH 7) (note 3)			
A_{250}/A_{260} (literature)	—	—	—
A_{250}/A_{260} (BM specif.)	0.79 ± 0.02	1.15 ± 0.03	1.64 ± 0.03
A_{280}/A_{260} (literature)	—	—	—
A_{280}/A_{260} (BM specif.)	0.15 ± 0.01	0.66 ± 0.02	0.26 ± 0.02
A_{290}/A_{260} (literature)	—	—	—

A. 2',3'-Dideoxyribonucleoside-5'-Triphosphates (*continued*)

	Purines		
	ddATP Trilithium Salt 10 mM Solution	ddGTP Trilithium Salt 10 mM Solution	ddITP Trilithium Salt 10 mM Solution
A_{290}/A_{260} (BM specif.)	< 0.01	0.28 ± 0.01	0.05 ± 0.01
Absorbance units (note 13) (for 1 μmol ddNTP/ml, pH 7.0)	15.3 A_{260} units	13.7 A $_{252}$ units	12.3 A_{249} units
Other information	ddADP, ≤ 5%	ddGDP, ≤ 5%	ddIDP, ≤ 5%

B. 2',3'-Dideoxyribonucleoside-5'-Triphosphates

	Pyrimidines	
	ddCTP Trilithium Salt 10 mM Solution	ddTTP Trilithium Salt 10 mM Solution
Typical analysis (note 1)		
Purity (%)	≥ 95%	≥ 95%
Na$^+$/K$^+$/Li$^+$ content (%)	Li$^+$, approx. 4%	Li$^+$, approx. 4%
Formula and approx. molecular weight (note 2)	$C_9H_{13}N_3O_{12}P_3Li_3$ 468.9	$C_{10}H_{14}N_2O_{13}P_3Li_3$ 484.0
Anhydrous MW	ddCTP, 451.1	ddTTP, 466.2
Solubility in H_2O (mg/ml)	≥ 4.69 (10 mM) (note 14)	≥ 4.84 (10 mM) (note 14)
Stability in solution		
at pH 7, −20°C (note 10)	Several months	Several months
other	Note 12	Note 12
Stability of pyrophosphate ester (pH 1)	Labile (note 11)	Labile (note 11)
pH of solution (as supplied)	7	7
Absorptivities (ε) (note 3) (mmol^{-1} × liter × cm^{-1})		
ε at 260 nm (pH 7)	—	—
absorbance maximum	272 nm (pH 7)	267 nm (pH 7)
ε at absorbance max.	9.6	9.5
Absorbance ratios (pH 7) (note 3)		
A_{250}/A_{260} (literature)	—	—
A_{250}/A_{260} (BM specif.)	0.84 ± 0.02	0.60 ± 0.02
A_{280}/A_{260} (literature)	—	—
A_{280}/A_{260} (BM specif.)	0.98 ± 0.02	0.90 ± 0.03
A_{290}/A_{260} (literature)	—	—
A_{290}/A_{260} (BM specif.)	0.33 ± 0.02	0.30 ± 0.02
Absorbance units (note 13) (for 1 μmol ddNTP/ml, pH 7.0)	9.6 A_{272} units	9.5 A_{267} units
Other information	ddCDP, ≤ 5%	ddTDP, ≤ 5%

Source: Information Courtesy of Boehringer Mannheim

Notes to Guide

1. Most preparations of nucleotides and nucleosides contain some water. If the preparation is supplied as a salt, part of the weight of the preparation is due to the counterion (usually Na$^+$, K$^+$ or Li$^+$). Water and counterions are not contaminants in the preparation, but integral parts of it. In the tables, % **purity** is the total percentage, by weight, of nucleotide (nucleoside) + counterion (if any) + H$_2$O (if any). On succeeding lines of the table, the proportion (by weight) of each of these three components in the preparation are listed separately. The **Na$^+$** (or other counterion) **content** is usually determined via atomic absorption spectrophotometry. The **H$_2$O content** is usually determined by iodine titra-

tion, according to Karl Fischer; for details, consult any standard text on quantitative analysis. For **nucleoside content,** both a **theoretical** value (based on the empirical formula of the compound [see note 2 below], assuming the preparation contains nothing but nucleotide, counterion and H_2O) and an assayed value (usually based on **enzymatic** analysis) are given. In those instances where the assayed value is based on non-enzymatic determinations, the alternate method (*e.g.,* absorbance) is listed in parentheses beside the assay value.

2. The table entry labelled **"Formula and approx. molecular weight"** lists the molecular formula and weight for each preparation and includes the approximate amount of water contained in the preparation at the time of manufacture. For the crystalline preparations, this water is water of crystallization, present as a part of the chemical structure of the compound (as indicated by the ".", *e.g.,* $C_{10}H_{12}N_5O_7PNa_2 \cdot 6H_2O$ for disodium adenosine-5'-monophosphate). For crystalline structures, the molecular weight given accurately reflects the water content.

    However, many of the nucleotide preparations are lyophilizates or amorphous powders. For such preparations, no precise molecular formula can be given, since the amount of water in the preparation is not precisely defined and will vary with storage of the product after the product vial is opened (since the nucleotides are hygroscopic). The formula and molecular weight given for each lyophilizate and amorphous powder are, therefore, only approximate and include, to the nearest whole number, the number of water molecules originally in the preparation (as indicated by the "+", *e.g.,* $C_{10}H_{13}N_5O_{10}P_2Na_2$ + $2H_2O$, 507.2, for lyophilized disodium adenosine-5'-diphosphate). If the exact concentration of a lyophilized or amorphous nucleotide in a preparation is vital to the success of an experiment, the concentration of the compound in solution may be determined spectrophotometrically (see note 7 below) or enzymatically.

3. The stability data (half-life at 100°C and pH 1) for phosphate and pyrophosphate bonds is according to L.F. Leloir and C.E. Cardini (1957) *Methods Enzymol.* **3,** 840.

The literature values of absorbance ratios for the ribonucleosides, ribonucleotides and deoxyribonucleotides are according to W.E. Cohn (1957) *Methods Enzymol.* **3,** 738. In general, the sugar constituent (*e.g.,* ribose, ribose-5'-triphosphate, deoxyribose-5'-triphosphate) has little effect on the spectral values of compounds.

The absorptivities at 260 nm are according to Cohn (see above). The information for absorbance maxima are drawn from several sources including Boehringer Mannheim (BM) specifications; *Molecular Cloning, A Laboratory Manual* [Maniatis, T., Fritsch, E.F. & Sambrook, J. (1982), p. 449, Cold Spring Harbor Laboratory, Cold Spring Harbor, N.Y.] and *Practical Methods in Molecular Biology* [Schleif, R.F. & Wensink, P.C. (1981), p. 112. Springer-Verlag, New York].

4. Preparations of AMP may be dried at room temperature in a vacuum dessicator over P_2O_5. AMP is stable enough to survive drying at 80°C for up to 12 h without significant decay.

5. The stability data listed for ATP and other NTPs at 4°C and −20°C are drawn from several articles in *Methods of Enzymatic Analysis* [(Bergmeyer, H.U., ed), 3rd Ed, VCH, Weinheim, W. Germany-Deerfield Beach, FL] including Vol. 2, pp. 334, 358, and 391 [Beutler, H-O. & Supp, M., authors (1984)]; Vol. 7, p. 409 [Keppler, O. & Kaiser, W. (1985)]; Vol. 7, p. 432 [Keppler, D. (1985)]; and Vol. 7, p. 439 [Keppler, D., Gawehn, K. & Decker, K. (1985)].

    ATP, even in the absence of divalent cations (see note 6) will slowly hydrolyze to AMP in solution at 4°C and pH 7. For most applications, this hydrolysis is insignificant and need not be taken into account during periods of less than one week. However, if the ATP is to be used as a reference standard [*e.g.,* in luminometric assay procedures, according to Wulff, K. & Doppen, W. (1985) in *Methods of Enzymatic Analysis* (Bergmeyer, H.U., ed), 3rd Ed, Vol. 7, p. 357, VCH, Weinheim, W. Germany-Deerfield Beach, FL], the solution should be stored no more than 8 h at 4°C before use, to insure lack of hydrolysis.

6. Ba^{2+}, Mg^{2+} and other divalent cations accelerate the hydrolysis of ATP to AMP in neutral solution. Hence, these cations should be

excluded from any stored solutions of ATP to insure maximum stability. **References:** Windholz, M., editor (1983) *The Merck Index,* 10th Ed, p. 24, Merck & Co., Inc., Rahway, NJ; Beutler, H-O. & Supp, M. (1984) in *Methods of Enzymatic Analysis* (Bergmeyer, H.U., ed), 3rd Ed, Vol. 2, p. 334, VCH, Weinheim, W. Germany-Deerfield Beach, FL.

7. To prepare and store a solution of ribonucleoside triphosphate (NTP) or deoxyribonucleoside triphosphate (dNTP) of exactly known concentration: Dissolve a weighed amount of the nucleotide in ice-cold H_2O at a concentration approx. 1.25–2.0 times as concentrated as the final solution will be. Immediately, adjust the cold solution to approx. pH 7.0 (using pH paper to determine the pH of the solution) with ice-cold 5-fold concentrated Tris base (*i.e.,* adjust the pH of 10 mM NTP solution with 50 mM Tris base). [It is better to overshoot the pH slightly on the alkaline side of pH 7 (up to pH 8.0) than to leave the solution slightly acid, since NTPs and dNTPs hydrolyze slowly in acid].

Keep the nucleotide solution on ice and dilute a small portion of it to about 50 μM with Tris–HCl buffer, pH 7.0. Place the dilute solution in a cuvette of 1.0 cm path length (b). Read the absorbance (A) of the dilute solution at 260 nm or at the absorbance maximum of the nucleotide. From the absorptivity (ε) of the nucleotide (see the tables), calculate the actual concentration (c) of the dilute solution:

$$c = A \div \varepsilon b.$$

or, for $b = 1.0$ cm,

$$c = A/\varepsilon$$

If ε is expressed in mmol$^{-1} \times$ cm^{-1}, the concentration will be expressed in mmol/l. Multiply the concentration of the dilute solution by the dilution factor to obtain the exact concentration of the stock nucleotide solution. Dilute the stock nucleotide solution to the desired concentration with ice-cold Tris–HCl, pH 7.0. Store the nucleotide solution frozen in small portions at $-20°C$ or $-70°C$. Thaw each aliquot as needed, mix well, use and discard. Do not refreeze any thawed aliquots, since this causes degradation. *Caution:* Some self-defrosting freezers can repeatedly thaw and refreeze the small samples. This is a

leading cause of nucleotide instability at "$-20°C$".

8. Vanadium (V) is an inhibitor of the Na$^+$/K$^+$-ATPase from mammalian muscle [Josephson, L. & Cantley, L.C., Jr. (1977) *Biochemistry* **16,** 4572. Aluminum (Al) is an inhibitor of hexokinase [Tornheim, K., Gilbert, T.R. & Lowenstein, J.M. (1980) *Anal. Biochem.* **103,** 87].

9. A study of the stability of GTP and dGTP in the dry state by Boehringer Mannheim laboratories revealed that these compounds hydrolyze to nucleoside monophosphate more rapidly than other nucleotides. The decay can be 1–3% per month at 4°C. It is slowed somewhat by storage at lower temperatures ($-20°C$ or $-70°C$), but not stopped.

10. The "stability" of the deoxyribonucleoside-5′-triphosphates (dNTPs) and dideoxyribonucleoside-5′-triphosphates (ddNTPs) is an operational "stability", *i.e.,* for many purposes, the preparations will remain useable for several months when stored at pH 7 and $-20°C$. In fact, the nucleoside diphosphate and nucleoside monophosphate content of dNTP and ddNTP preparation slowly, but steadily, increases when they are stored under the listed conditions. The hydrolysis is accelerated by repeated thawing and refreezing (see note 7 above). The period of time in which no detectable hydrolysis occurs in preparations of dNTP and ddNTP is probably comparable to the times listed under "Stability in solution at pH 7, $-20°C$" for the corresponding ribonucleoside triphosphate.

For each specific research application of the dNTPs and ddNTPs, the allowable length of storage time for the preparations should be determined empirically.

11. The stability of the pyrophosphate bond of dNTPs and ddNTPs in acid solution is probably comparable to that of the pyrophosphate bond of ribonucleoside-5′-triphosphate (NTPs). We do not have exact data for the dNTPs and ddNTPs.

12. The glycosidic linkage (linkage of base to sugar) in a dideoxynucleoside triphosphate (ddNTP) is much more acid labile than the glycosidic linkage in the corresponding deoxynucleoside triphosphates (dNTP). ddCTP is especially labile.

13. Quantities of dideoxynucleoside triphosphate

(ddNTP) may be conveniently expressed in **Absorbance units,** based on its absorptivity at its absorbance maximum. Thus, one A_{260} unit of ddATP is the amount of ddATP which, dissolved in 1.0 ml of buffer at pH 7.0, gives an Absorbance of 1.0 at 260 nm.

14. The dideoxynucleoside triphosphates (dd-NTPs) are also available as lyophilizates. These may be dissolved in H_2O at concentrations up to 100 mM, neutralized and stored in the same manner described above (note 7) for NTPs and ddNTPS.

Additional References Consulted

(a) Beutler, H-O. and Supp, M. (1984) in *Methods of Enzymatic Analysis,* Vol. 2, p. 333 (stability of solid ADP). See reference g.

(b) Coddington, A. (1974) in *Methods of Enzymatic Analysis,* 2nd Ed, Vol. 4, p. 1932 (stability of inosine in solution). See reference k.

(c) Coddington, A. (1985) in *Methods of Enzymatic Analysis,* Vol. 7, p. 117 (stability of guanosine in solution). See reference g.

(d) Forster, E. and Holldorf, A.W. (1974) in *Methods of Enzymatic Analysis,* 2nd Ed, Vol. 4, p. 1923 (stability of cytosine in solution). See reference k.

(e) Grassl, M. and Supp, M. (1985) in *Methods of Enzymatic Analysis,* Vol. 7, p. 449 (stability of CMP and CDP in solution). See reference g.

(f) Hampp, R. (1985) in *Methods of Enzymatic Analysis,* Vol. 7, p. 370 (stability of AMP, ADP, and ATP in solution). See reference g.

(g) Heinz, R. and Reckel, S. (1985) in *Methods of Enzymatic Analysis,* (Bergmeyer, H.U., ed), 3rd Ed, Vol. 7, p. 110, VCH, Weinheim, W. Germany-Deerfield Beach, FL (pK and stability of adenosine).

(h) Keppler, D. (1974) in *Methods of Enzymatic Analysis,* 2nd Ed, Vol. 4, p. 2088 (stability of nucleoside monophosphates in solution). See reference k.

(i) Keppler D. and Kaiser, W. in *Methods of Enzymatic Analysis,* Vol. 7, p. 409 (stability of GMP, GDP, GTP in solution). See reference g.

(j) Keppler, D., Gawehn, K. and Decker, K. (1985) in *Methods of Enzymatic Analysis,* Vol. 7, p. 439 (stability of UMP, UDP, and UTP in solution). See reference g.

(k) Moellering, H. and Bergmeyer, H.U. (1974) in *Methods of Enzymatic Analysis,* (Bergmeyer, H.U., ed), 2nd Ed, Vol. 4, p. 1919, Academic Press, New York (stability of adenosine in acid).

(l) Windholz M., editor, (1983) *The Merck Index,* 10th Ed, pp. 23, 401, 615, 722, 1411, Merck & Co., Inc., Rahway, NJ (solubility, stability and/or spectral data on guanosine, inosine, cytosine, uridine, AMP, GMP, IMP, UMP and ADP).

PHENOL EXTRACTION OF NUCLEIC ACIDS

Phenol extraction is a method of choice to purify rapidly the nucleic acids from cellular extracts. Commercial phenol generally contains impurities and therefore needs to be redistilled prior use (see Chapter 2). It also is possible to purchase redistilled phenol (nucleic acids grade).

Preparation of Phenol for Nucleic Acid Extractions

Phenol used for nucleic acid extractions is prepared in the following way (operate under a fume hood).

PROTOCOL

PHENOL PREPARATION

1. Remove a bottle of redistilled crystallized phenol from the $-20°C$ freezer and let it warm up at room temperature.
2. Place the bottle in a plastic beaker containing enough water to cover phenol and place the beaker in a 60°C water bath. Make sure that you *loosen the cap* of the bottle to avoid pressure building up.
3. When crystals are melted, add one volume of NTE buffer (about 50 ml) and a magnetic bar.
4. Place the mixture on a magnetic stirrer and mix thoroughly for 10 minutes.
5. Stop the magnetic stirrer and let the mixture separate in two phases. The lower phase is phenol.
6. Remove the upper aqueous phase by aspiration (do not mouth-pipette).
7. Add one volume of NTE buffer to the phenol solution and repeat steps 4–6.
8. Repeat one more time with fresh NTE buffer.
9. Remove most of the upper phase. Leave a layer of NTE buffer on top of the buffer-saturated phenol.
10. When a small volume of phenol is needed, it is convenient to transfer about 1 ml of equilibrated phenol solution to a polypropylene microfuge tube with a 5-ml plastic pipette and use this to take small samples with pipetman-tips. Do no use micropipettes directly through the aqueous layer on top of phenol stock solution.
11. Keep buffer-saturated phenol solution at 4°C for no more than one month. In any case, discard all solutions turning yellow or pink.

Extraction of Nucleic Acids

The following procedure can be used for the purification of both DNA or RNA species. However, when intact biologically active RNA species are needed, we recommend *not* to use plain phenol extraction, but rather follow the procedure described in Chapter 18.

PROTOCOL

EXTRACTION AND PRECIPITATION OF NUCLEIC ACIDS

1. Adjust the salt concentration of the cellular extract containing DNA so that it is in the range of 50–100 mM.

2. Prepare a 24:1 mixture of chloroform:isoamyl alcool by adding 20 ml of isoamyl alcool to 480 ml of chloroform. Homogenize on a magnetic stirrer and keep in a dark bottle.

3. To one volume of sample, add half a volume of buffer-saturated phenol and half a volume of the chloroform:isoamyl mixture.

4. Mix thoroughly for 30 seconds with a vortexer if you do not need to recover high molecular weight intact DNA (such as Lambda arms for cloning or cellular DNA for restriction analysis). Otherwise, mix by inverting slowly the tube either manually or by using a commercially available spinning wheel.

5. Centrifuge the solution at 12,000 rpm for 3 minutes at room temperature if working with microfuge tubes (volumes up to 1.5 ml) or at 5500 rpm for 15 minutes when working with larger volumes (such as 5–30 ml in clinical tubes).

6. Transfer the supernatant in a tube containing one volume of chloroform:isoamyl mixture (no phenol), and repeat steps 4 and 5.

7. Repeat step 6 once and transfer supernatant in a tube which can accept a total volume four times larger than the initial sample volume.

8. The recovery of nucleic acids (high molecular weight DNA molecules, double stranded DNA fragments, and DNA single strands or RNA) is generally achieved by precipitation with ethanol (see below) in the presence of salts. Adjust the salt concentration by adding 1/30 volume of 3 M sodium acetate (pH 5.0), or alternatively, by adding 1/2 volume of 7.5 M amonium acetate.

9. Add three volumes of cold ethanol and mix by inverting several times. Let precipitate at $-20°C$ overnight or at $-70°C$ for 20 minutes. (See below, ethanol precipitation.)

Notes

1. Precipitation of DNA with ammonium acetate has proven to be more efficient for the removal of impurities than precipitation with sodium acetate. This is especially important to consider when working with DNA solution that may contain heavy metals, detergents, or any other compounds that are

known as potent inhibitors of restriction endonucleases, as well as other enzymes, such as DNA ligase or polynucleotide kinase. Practically, when relatively large amounts of DNA are purified from eucaryotic cells, it is more convenient to use sodium acetate. However, when working with small amounts of DNA fragments (such as material recovered from gels) which may contain impurities it is recommended that ammonium acetate be used. Also, the amonium salts of nucleotides are more soluble in ethanol than the sodium salts. Therefore, if the DNA is to be used for end modifications involving precise concentrations of given nucleotides triphosphate, it is essential to use ammonium acetate to eliminate free nucleotides present in the DNA solution (beware that polynucleotide kinase activity is inhibited by ammonium ions, see Chapter 4).

2. When high molecular weight intact DNA molecules are present, they precipitate as strings when the cold absolute ethanol is added. It is then recovered by spooling out with a Pasteur pipette and transferred to the bottom of an empty tube for drying.

3. When small amounts of DNA are purified and recovered by ethanol precipitation, it is sometimes helpful to add a carrier. Many protocols call for the addition of 1 to 5 μg tRNA (from either yeast or *E. Coli*) to the samples prior to precipitation. This is fine as long as you do not intend to use your preparation for end labeling (or to prepare cDNA when RNA is precipitated). Remember also that digestion with RNAse generates many more ends which can be substrates for several enzymes such as transferase or kinase. To avoid these problems it is recommended that glycogen be used as a carrier (1 μl of a 10 mg/ml autoclaved solution, for a final volume of sample up to 500 μl).

ETHANOL PRECIPITATION OF SMALL AMOUNTS OF NUCLEIC ACIDS

Precipitation of nucleic acids with ethanol is one of the basic steps in molecular cloning. It is used to recover small or large amounts of DNA molecules whose concentration may vary from a few nanograms to a few micrograms per microliter of sample solution. It is relatively easy to obtain good yields of recovery when micrograms of DNA are precipitated. In these cases, the DNA solution is generally incubated at $-70°C$ for about 15 to 20 minutes or at $-20°C$ overnight and the precipitate collected by centrifugation at 12,000 rpm for 10 minutes in microfuge tubes or at 11,000 rpm in a SS34 type rotor for 20 minutes in Corex tubes (or equivalent). Precipitation at $-20°C$ should be preferred when polysaccharides may be present in the DNA solution because they precipitate more easily at $-70°C$.

Problems arise when small amounts of DNA are being manipulated. The effects of DNA concentration, incubation time, incubation temperature, and

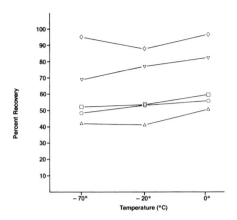

Figure 12. Effect of incubation temperature upon ethanol precipitation of DNA. All points are averages of two determinations. O = 0.6 ng, △ = 10 ng, □ = 100 ng, ▽ = 1 μg, and ◇ = 10 μg (courtesy of Bethesda Research Laboratories, Life technologies Inc.).

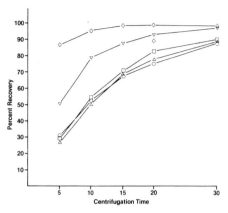

Figure 14. Effect of centrifugation time upon ethanol precipitation of DNA. Note that two values are plotted for 10-μg, 20-minute centrifugation. The lower value is believed to be due to experimental error. All other points are averages of two determinations. O = 0.6 ng, △ = 10 ng, □ = 100 ng, ▽ = 1 μg and ◇ = 10 μg (courtesy of Bethesda Research Laboratories, Life technologies Inc.).

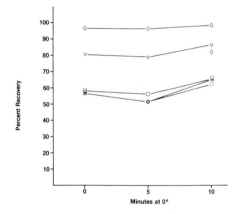

Figure 13. Effect of incubation time upon ethanol precipitation of DNA. Note that two values are plotted for 10-μg, 10-minute incubation. The lower value is believed to be due to experimental error. All other points are averages of two determinations. O = 0.6 ng, △ = 10 ng, □ = 100 ng, ▽ = 1 μg, and ◇ = 10 μg (courtesy of Bethesda Research Laboratories, Life technologies Inc.).

time of centrifugation have been studied with a 3.2 kb ^{32}P-labeled and linear double-stranded DNA (Zeugin and Hartley, 1985). The results obtained (Figures 12–14) showed that:

1. Precipitation at 4°C gave the best results at low DNA concentrations, when centrifuged 10 minutes. This may result from the fact that 75% ethanol (three volumes for one volume of sample) is viscous at −70°C and therefore re-

tards migration of the precipitated DNA. Longer centrifugation times should eliminate this problem (see below).

2. Time of incubation at 4°C had almost no effect on the DNA recovery after 10 minutes of centrifugation.

3. Long centrifugation times increase the amount of DNA recovered when the DNA solution is mixed with absolute ethanol and centrifuged immediately.

Our experience shows that fully satisfactory recoveries are obtained, even in the absence of glycogen, when cold (−20°C) ethanol is added to the samples, left at −70°C until the solution is frozen (about 15 minutes), and centrifuged for 30 minutes at 4°C in a Sigma M15 microcentrifuge. There is no need to use siliconized tubes if working with polypropylene tubes. Particular care should be taken in the choice of the microcentrifuge since the different models available on the market show considerable variations in gravitational force. (The Eppendorf and Beckman microcentrifuges also gave excellent results.) Once the precipitate is collected at the bottom of the tube, follow the protocol below.

Reference

Zeugin and Hartley (1985), Focus, **7:4**, 1.

PROTOCOL

RECOVERY OF NUCLEIC ACIDS FOLLOWING ETHANOL PRECIPITATION

1. Empty the ethanol by inverting the tube and add 1 ml (or 3–5 ml if you use clinical tubes) of cold 80% ethanol onto the pellet.
2. Vortex to resuspend the pellet and centrifuge for 10 minutes under the same conditions as above.
3. Empty the tube and invert it on a tissue to drain off most of the remaining ethanol.
4. Dry under vacuum. You may let the sample air-dry if large amounts of DNA are collected. (It has been found that in some cases, large amounts of vacuum-dried pellets are more difficult to redissolve than air-dried pellets.)
5. The dried pellet is resuspended in sterile distilled water or in TE buffer (10 mM Tris-HCl pH 7.4, 1 mM EDTA). The solution can be stored at 4°C. However, we advise that long-term storage be performed at -20°C.

Notes

1. Make sure to melt the content of the tube before centrifugation to avoid material sticking on the side of the tube.
2. Be careful not to dislodge the precipitate when pouring off the ethanol after rinsing.

SOME USEFUL HINTS AND UNITS

1 A_{260} unit of double-stranded DNA = 50 µg/ml

E_{260} for 1 mg/ml of double-stranded DNA = 20.3

E_{260} for 1 mM (in nucleotides) of double-stranded DNA = 6.7

1 A_{260} of double-stranded DNA = 0.15 mM

1 A_{260} unit of single-stranded DNA = 33 µg/ml

E_{260} for 1 mg/ml of single stranded DNA = 25.5

E_{260} for 1 mM (in nucleotides) of single-stranded DNA = 8.5

1 A_{260} of single-stranded DNA = 0.12 mM

1 A_{260} unit of single-stranded RNA = 40 µg/ml

1 µg of pBR322 = 0.36 pmol

1 pmol of pBR322 = 2.8 µg

1 pmol of linear pBR322 = 1.4 µg

1 µg of a 1000 base-pair duplex DNA = 1.52 pmol = 3.04 pmol of ends

1 pmole of a 1000 base-pair duplex DNA = 0.66 µg

1 kb of DNA = $6.5.10^5$ daltons of duplex DNA (sodium salt)

1 kb of DNA = $3.3.10^5$ daltons of single-stranded DNA (sodium salt)

1 kb of RNA = $3.4.10^5$ daltons of single-stranded RNA (sodium salt)

Average molecular weight of a deoxynucleotide base = 324.5 daltons

Average molecular weight of a deoxynucleotide base pair = 649 daltons

1 µg/ml of DNA = 3.08 µM phosphate

1 µg/ml of 1 kb of DNA = 3.08 nM 5' ends

Moles of ends (5' or 3') for circular DNA = moles of DNA × nb of cuts × 2 (ends per cut)

Moles of ends (5' or 3') for linear DNA = moles of DNA × 2 (ends per cut) + 2 (ends of linear DNA)

10.000 daltons protein = 0.27 kb of DNA

50.000 daltons protein = 1.35 kb of DNA

100.000 daltons protein = 2.7 kb of DNA

1 Becquerel (Bq) = 1 disintegration/second (dps)

1 Curie (Ci) = $3.7.10^{10}$ Bq = $2.2.10^{12}$ disintegrations/minute (dpm)

1 mCi = $3.7.10^7$ Bq = $2.2.10^9$ dpm

1 µCi = $3.7.10^4$ Bq = $2.2.10^6$ dpm

4

USEFUL MODIFYING ENZYMES FOR MOLECULAR CLONING

The enzymes listed below are available commercially. We have sometimes used preparations made in the laboratory according to published methods. However, this requires some equipment and biochemical background which may not be available everywhere, and therefore may not lead to a great money saving, as the cost of most commercial enzymes is quite reasonable for the quality provided. The use of modifying enzymes in molecular cloning is summarized in Figure 15.

T4 DNA LIGASE

This enzyme catalyzes the formation of a phosphodiester bond between adjacent 3' hydroxyl and 5' phosphate termini in duplex DNA. It can therefore ligate DNA fragments with blunt or sticky ends, such as those generated by restriction enzyme digestion (Figures 16–19 on the following pages).

T4 DNA ligase is usually purified from a sup^0 end A strain of $E.$ $coli$ 11000 lysogenic for the λ phage NM989 (λT4 lig Wam 403 E am 1100 att^+ cI857 nin 5 Sam 100) (Wilson and Murray, 1979; Murray et al., 1979) or from an $E.$ $Coli$ 600 pc1857 pPLc28lig 8 (Remaut and Fiers, unpublished). It is a single polypeptide of 68,000 daltons which requires Mg^{2+} and ATP as cofactors.

Different methods have been developed to measure the DNA ligase activity. Generally, one unit is defined as the amount of enzyme necessary to cata-

lyze the conversion, in an exchange reaction, of 1 nmol ^{32}P Pi into ^{32}P ATP in 20 minutes at 37°C (Weiss et al., 1968). However, the correct amount of DNA ligase needed in each experiment depends upon several factors such as the nature of the DNA fragments to be ligated (blunt or sticky ends, length of sticky end, stability of hydrogen bonded structure), the temperature of incubation, and the concentration of fragments (see Chapter 12). For example, it has been found that the ligation rate of pBR322 DNA restriction fragments of same length all terminated with cohesive ends, will differ with the nature of the sequence generated by the restriction endonuclease cut. Thus, ligation rates of Hind III, Pst I, Eco RI, and Sal I DNA fragments are decreasing from Hind III to Sal I, the former being 10 to 40 times faster than the latter. The same kind of conclusion is also reached with blunt end DNA fragments: ligation of pBR322 Hae III DNA fragments is four times more efficient than ligation of Hinc II DNA fragments. It was therefore convenient to define a unit of enzyme based on the ligation of restriction endonuclease fragments.

One ligation unit is defined by most suppliers as the amount of enzyme that catalyzes 50% ligation of Hind III fragments of lambda DNA in 30 minutes at 16°C under standard conditions. One such unit is equivalent to 0.015 exchange reaction unit (Weiss et al., 1968) and to 0.0025 d(A-T) circle formation unit (Modrich and Lehman, 1970). Although the activity of the T4 DNA ligase preparations is usually tested by the commercial supplier it is a good policy to per-

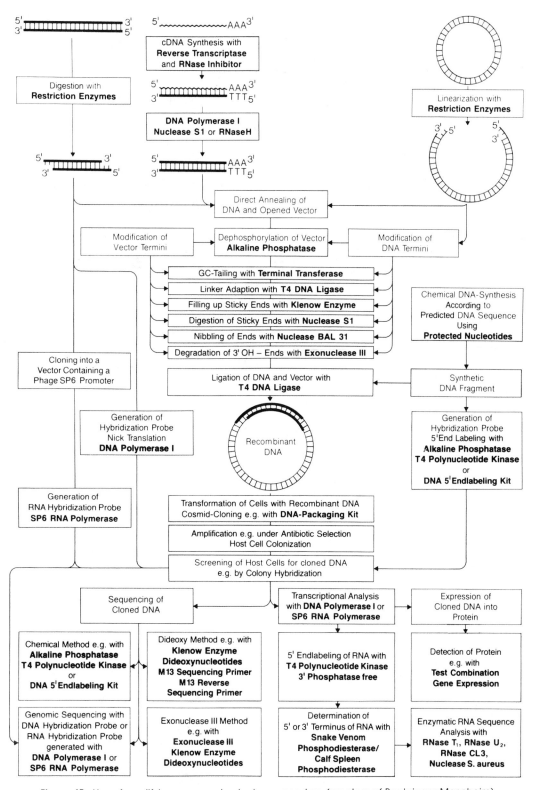

Figure 15. Use of modifying enzymes in cloning procedure (courtesy of Boehringer Mannheim).

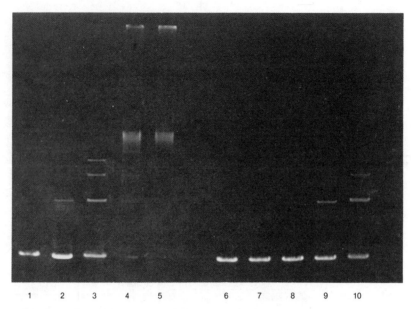

1   2   3   4   5   6   7   8   9   10

Figure 16. Ligation of linear λ*dv1* DNA obtained by cleavage with Eco RI (sticky ends, ligation at 4°C) or Sma I (blunt ends, ligation at 21°C). 1.45 μg of linear λ*dv1* DNA was incubated for 2 hours in a total volume of 10 μl according to the ligation conditions given in text. The electrophoresis was carried out on a 0.3% agarose slab gel in Tris buffer, 40 mM; sodium acetate, 20 mM; EDTA, 1mM; adjusted to pH 7.2 with acetic acid. From left to right: lane 1, Eco RI-linearized λ*dv1* DNA, without ligase; lane 2, 1.4 × 10^{-3} units ligase added; lane 3, 14 × 10^{-3} units ligase added; lane 4, 140 × 10^{-3} units ligase added; lane 5, 1400 × 10^{-3} units ligase added; lane 6, Sma I-linearized λ*dv1* DNA, without ligase; lane 7, 14 × 10^{-3} units ligase added; lane 8, 140 × 10^{-3} units ligase added; lane 9, 1400 × 10^{-3} units ligase added; lane 10, 14,000 × 10^{-3} units ligase added. The gel pattern demonstrates that with increasing ligase concentration discrete linear oligomers and circular DNA molecules are formed, the length of which depends on the amount of enzyme used. The products obtained at high enzyme concentration are so large that they cannot be resolved into single components with the gel system employed. It can also be seen clearly that blunt end ligation requires approximately 10 times more enzyme than sticky end ligation (courtesy of Boehringer Mannheim).

form a ligation test periodically with a ligase preparation being kept at −20°C for long periods of time.

A FPLC-pure cloned T4 DNA ligase that exhibits a particularly high specific activity can be purchased from Pharmacia. T4 DNA ligase from New England Biolabs was found to be most satisfactory in our opinion.

A typical test for ligation of sticky ends is performed by incubating Hind III-digested λ DNA as shown on page 83.

For blunt end ligations, the same procedure can be used with Hae III-digested DNA. In this case, it is necessary to use approximately 8–10 units of T4 DNA ligase to get satisfactory ligation.

E. COLI DNA LIGASE

This enzyme catalyzes the formation of phosphodiester bond between 5′ phosphate and 3′ hydroxyl termini in duplex DNA molecules containing cohe-

sive ends. It requires DPN as a cofactor. *E. coli* DNA ligase is purified from the *E. coli* strain 594 (*su*$^-$) lysogenic for λgt4-*lop*-11 *lig*$^+$ 57 (Panasenko et al., 1977). Cloned *E. Coli* DNA ligase is available from Pharmacia.

T4 DNA POLYMERASE

This enzyme exhibits two kinds of activity. In the absence of deoxynucleotide triphosphate a 3′ → 5′ exonuclease activity is detected, whereas in the presence of deoxynucleoside triphosphate a 5′ → 3′ chain growth activity is detected. A template and a primer are needed for nucleotide addition at the 3′ end of the DNA molecule. Unlike the *E. Coli* DNA polymerase, this enzyme does not exhibit a 5′ to 3′ exonuclease activity (Englund, 1971). Therefore, T4 DNA polymerase is often used in place of the Klenow fragment of DNA polymerase I to fill in 5′-protruding ends of DNA fragments (see Chapter 12). It is

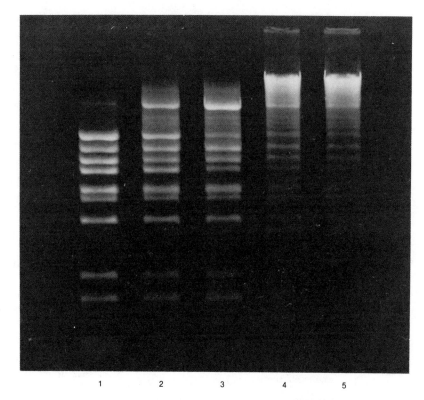

Figure 17. Ligation of λ-DNA fragments obtained by cleavage with Bst EII. 0.9 μg of λ DNA cleaved with Bst EII were incubated in a total volume of 20 μl at 22°C for 30 minutes with increasing amounts of T4 DNA ligase according to the ligation conditions given in text. The electrophoresis was carried out on a 0.6% agarose slab gel in Tris buffer, 40 mM; sodium acetate, 20 mM; EDTA, 1 mM; adjusted to pH 7.2 with acetic acid. From left to right: lane 1, λ DNA cleaved with Bst EII, without ligase; lane 2, 1.4×10^{-3} units ligase added; lane 3, 5.6×10^{-3} units ligase added; lane 4, 14×10^{-3} units ligase added; lane 5, 56×10^{-3} units ligase added. The disappearance of the original DNA fragments and the formation of higher molecular weight fragments with increasing amounts of DNA ligase can be followed clearly. This also demonstrates the difficulty in defining the point at which a certain percentage of ligation is achieved (courtesy of Boehringer Mannheim).

PROTOCOL

ASSAY FOR T4 DNA LIGASE ACTIVITY

1. Digest 5 μg of λ DNA with Hind III as described in Chapter 8.
2. Precipitate with ethanol and resuspend in 20 μl T4 DNA ligase buffer.
3. Add 0.1 unit T4 DNA ligase (one unit is defined as the amount of ligase which catalyzes 50% ligation in 30 minutes at 16°C under the conditions used here).
4. Incubate for 30 minutes at 16°C.
5. Heat at 65°C for 10 minutes to inactivate the enzyme, then load onto a gel after adding dye in glycerol.

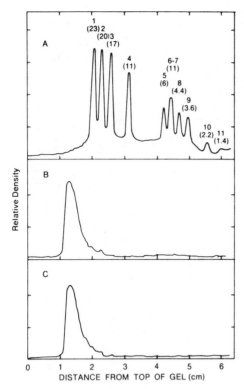

Figure 18. Scans of photographed fluorescence of agarose gels through which Hae III-digested φX174 DNA was electrophoresed after treatment with T4 DNA ligase for 0 (A), 2 (B), or 24 hours (C). Hae III fragments were produced by digesting 21 μg φX174 RFI DNA with 40 units of Hae III restriction endonuclease (P. L. Biochemicals, lot no. 866-2) in 50 μl 0.01 M Tris-HCl, 0.01 M MgCl₂ (pH 7.5), for 2 hours at 37°C. The reaction was terminated by adding 5 μl 1.0 M NaCl and 3 volumes of ethanol and mixing at room temperature. After standing at −20°C overnight the DNA was spun out, dried for 10 minutes under vacuum, and dissolved in 50 μl 0.05 M Tris-HCl, 0.01 M MgCl₂ (pH 7.8), 0.02 M dithiothreitol, 1.0 mM ATP. Samples were removed for electrophoresis before the addition of enzyme (0 time) and after the addition of 5 μl (10 Weiss units) T4 DNA ligase (lot no. 0870-7) and were stored frozen. Electrophoresis through 1.4% agarose gels containing ethidium bromide was performed and the gels photographed and scanned [Sharp, P. A., et al. (1973), *Biochemistry*, **12**, 3055]. The number in parentheses below the restriction fragment number indicates the percentage of the whole φX174 genome in that fragment (courtesy of Pharmacia).

also used as an alternative to nick translation in the replacement synthesis method for labeling DNA fragments (see Chapter 16) and to label the 3′ end of duplex DNA. The exonuclease rate on double stranded DNA is approximately 40 bases per minutes as compared to about 4000 bases on single stranded DNA, while the polymerization rate reaches 15,000 nucleotides per minutes when assayed under standard conditions.

T4 DNA polymerase is purified from *E. coli* D110

infected with T4 bacteriophage [*recA⁻*,*am* N 55⁻, *am* H39⁻ (gene 42⁻, 30⁻⁰)].

A unit of enzyme is defined as the amount which catalyzes the addition of 1 nmol of nucleotide in TCA (trichloroacetic acid)-precipitable product in 30 minutes at 37°C, using nicked DNA as template.

E. COLI DNA POLYMERASES

Three DNA polymerases, namely Pol I, Pol II, and Pol III, have been purified from *E. coli*. The enzymes Pol I and Pol II seem to be involved in DNA repair processes, whereas Pol III appears to be involved in DNA replication (see review by Kornberg, 1980).

Three kinds of enzymatic activity have been assigned to DNA polymerases. All three enzymes catalyze a 5′ → 3′ elongation of DNA strands in the presence of primer and deoxynucleoside triphosphate. A 3′ → 5′ exonuclease activity is also associated with DNA polymerases. It has been called a proofreading activity, as it is involved in the removal of an incorrectly added base before polymerization continues. Also, a 5′ → 3′ exonuclease activity of Pol I has been implicated in excision-repair in nick-translation and in removal of RNA primer from the 5′ end of DNA chains.

Proteolytic cleavage of Pol I (109,000 daltons) leads to the isolation of a large fragment (76,000 daltons), also known as a Klenow fragment or Pol1K; (Jacobsen *et al.*, 1974) containing the 5′ → 3′ polymerase and 3′ → 5′ exonuclease activities and a smaller fragment (36,000 daltons) that exhibits only the 5′ → 3′ exonuclease activity. Therefore, the large fragment of *E. coli* DNA polymerase I can be used in all cases where polymerization alone is needed (e.g., filling in cohesive ends before addition of linkers, .copying single-stranded DNA in the dideoxy method for sequencing or nick-translation).

A significant improvement in the functional quality of the Klenow fragment has been brought by cloning the sequences encoded for this fragment, therefore eliminating any contamination by residual amounts of holoenzyme in proteolytically treated preparations. A FPLC purified cloned product completely deprived of 5′ to 3′ exonuclease activity is available from Pharmacia. Cloned large fragment is also available from Bethesda Research Laboratories and International Biotechnologies, Inc. In our hands, the Klenow fragment purchased from Boehringer was found to be excellent, but lost activity upon storage at −20°C.

The native Pol I enzyme has been used success-

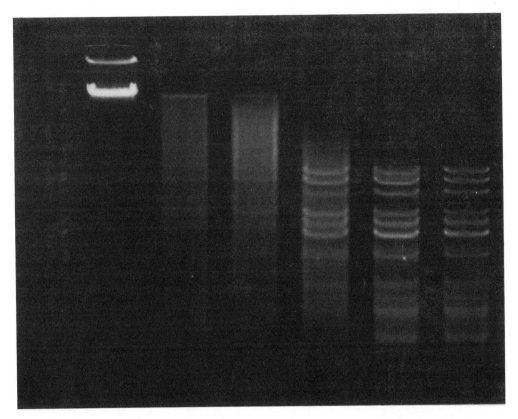

Figure 19. Efficiency of T4 ligase in joining blunt ends. The level of ligation obtained with various amounts of T4 DNA ligase. Reactions were carried out in 20-μl volume at ambient temperature for one hour. Each reaction contained 2 μg ALu I-cleaved λ DNA, 25 mM Tris–HCl (pH 7.8), 10 mM MgCl$_2$, 2 mM DTT, and 0.4 mM ATP. Lane A, no ligase; lane B, 2 U ligase; lane C, 1 U ligase; lane D, 0.5 U ligase; lane E, 0.1 U ligase; lane F, 0.05 U ligase; lane G, 0.01 U ligase (courtesy of International Biotechnologies Inc.).

fully to remove 3′-protruding DNA in the absence of deoxynucleotide triphosphate and to fill in cohesive ends in the presence of deoxynucleotide triphosphate before addition of molecular linkers (see Chapter 12).

MODIFIED BACTERIOPHAGE T7 DNA POLYMERASE

A chemically modified phage T7 DNA polymerase has been recently described by Richardson and co-workers, as an ideal tool for DNA sequencing (Tabor 1987, Tabor and Richardson, 1987; Huber et al., 1987; Tabor et al., 1987).

This modified enzyme is a one to one stoichiometric complex of two proteins: the 84 kD T7 gene 5 product and the 12 kD E. Coli thioredoxin (Modrich and Richardson, 1975; Mark and Richardson, 1976). The catalytic properties of the complex are associated to the T7 gene protein. One interesting feature of this system results from the fact that the thioredoxin protein binds the gene 5 protein to the primer template, thereby increasing the processivity of the T7 polymerase and allowing the polymerization of thousands of nucleotides without dissociating. The rate of polynucleotide synthesis is very high (> 300 nucleotides per second), more than 70 times faster than that of AMV reverse transcriptase. Because the gene 5 protein/thioredoxin complex exhibits a potent 3′ to 5′ exonuclease activity, an oxidation of the corresponding domain in the gene 5 protein was performed via the localized generation of free radicals (Tabor, 1987). The resulting modified T7 polymerase/thioredoxin complex is deprived of most exonuclease activity and can therefore be used for DNA sequencing (see Chapter 20). It is interesting to point out that the modified enzyme complex efficiently incorporates the nucleotides analogs such as 5′-(α-thio) triphosphates and dc[7] GTP or dITP used to improve the resolution of DNA sequencing gels and to avoid gel compression resulting from base pairing (see Chapter 20). The modified T7 DNA polymerase can be used advantageously for the preparation of radioactive probes and the amplification of large DNA fragments.

The modified T7 DNA polymerase is commercially available under the name of sequenase (United States Biochemical Corporation).

ALKALINE PHOSPHATASE

This enzyme, which is purified from either *E. coli* or higher organisms (e.g., calf intestine alkaline phosphatase), catalyzes the removal of 5'-phosphate groups from nucleic acids.

It does not hydrolyzes phosphodiesters. One unit of enzyme is the amount needed to hydrolyze 1 μmol of 4-nitrophenyl-phosphate in 1 minute at 37°C (in 1 M diethanolamine buffer pH 9.8, 10 mM 4-nitrophenyl-phosphate, 250 μM MgCl$_2$). Five such units of enzyme are equivalent to 1 unit of alkaline phosphatase being tested at 25°C in glycine buffer (Moessner et al., cited in Boehringer catalog).

Alkaline phosphatase is a zinc-containing metallo-protein. It is inactivated following to the chelation of zinc ions with EGTA and is strongly inhibited by low concentrations of inorganic phosphate. Since ethanol precipitation is not adequate to remove inorganic phosphate, thus in some cases it may be useful to dialyze endonuclease-cleaved DNA prior to alkaline phosphatase treatment.

Alkaline phosphatase is used to reduce the background due to recircularization of DNA vectors in transformation (see Chapter 14). Also, when 5' ^{32}P-end-labeled DNA fragments are needed (see Maxam & Gilbert sequencing procedure, Chapter 20), they are treated with alkaline phosphatase prior to incubation with polynucleotide kinase in the presence of [^{32}P]deoxynucleoside triphosphate (Figure 20).

T4 POLYNUCLEOTIDE KINASE

Polynucleotide kinase is the *pse* T gene product of bacteriophage T4 (Depew et al., 1975). It is a tetramer of four identical monomers with an apparent molecular weight of 33,000 daltons each (Panet et al., 1973; Lillehaug, 1977).

This enzyme catalyzes the transfer of a phosphate group from ATP to the 5'-hydroxyl terminus of a polynucleotide (Richardson, 1968). Polynucleotide kinase works on deoxyribonucleotide and ribonu-

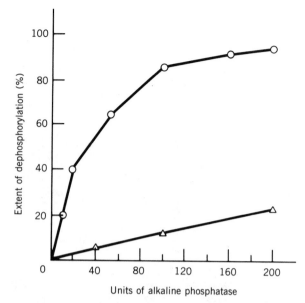

Figure 20. Dephosphorylation of 5'^{32}P-labeled DNA fragments with alkaline phosphatase. Labeled λ DNA Hae III fragments were incubated in the presence of alkaline phosphatase at 37°C (△) and at 65°C (○) for 30 minutes, in a final volume of 100 μl.

cleotide chains without any limitation on the size of the substrate molecule. It can even phosphorylate a 3'-mononucleotide. When γ[^{32}P]ATP is used, it is possible to incorporate a ^{32}P at the 5' end of the DNA molecule (Figure 21). It can also catalyze the exchange of 5'-terminal phosphate groups (Berkner and Folk, 1980; Chaconas and Van de Sande, 1980).

Maximum kinase activity is obtained at 37°C, pH 7.6, in the presence of Mg^{2+} ions and thiols reagents (DTT or 2-mercaptoethanol), with a minimum of 1 μM ATP and a 5:1 ratio of ATP over 5'-OH ends (Lillehaug et al., 1976). The addition of 6% polyethylene glycol (PEG 8000) in the reaction mixture has been found to increase the radiolabeling of recessed, protruding, and blunt ended 5'-termini of DNA (Harrison and Zimmerman, 1986). This is particularly interesting when very low concentrations of substrate are being used.

The simultaneous use of polynucleotide kinase and gamma ATP allows us to label DNA and RNA with ^{32}P residues at their 5' ends either by direct phosphorylation of 5'-hydroxyl groups generated after digestion with akaline phosphatase, or by exchange of the 5'-phosphoryl groups. Because an efficient labeling is generally obtained with dephosphorylated DNA fragments, use of polynucleotide kinase has proven to be a method of choice in labeling both DNA and RNA strands prior to base specific sequencing as described by Maxam and Gilbert (see

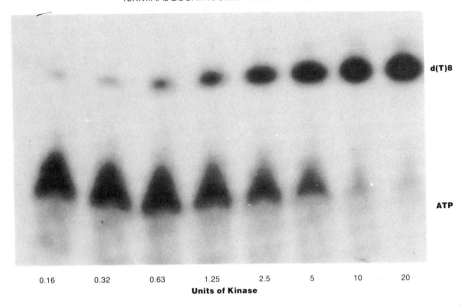

Figure 21. Transfer of the γ-phosphate group of ^{32}P-labeled ATP to the 5' hydroxyl terminus of d(T)8 (courtesy of New England Biolabs).

Chapters 20, 21). This enzyme is also being used in mapping of restriction sites by partial digestion (see Chapter 8), DNA and RNA fingerprinting (Gross et al., 1978; Reddy et al., 1981), DNA-footprinting combined to DNAse protection (Galas and Schmitz, 1978, see chapter 24), methylation protection (Johnsrud, 1978, see chapter 24), and synthesis of substrates for DNA or RNA ligase (Khorana et al., 1972; Silber et al. 1972).

In addition to the kinase activity described above, the T4 polynucleotide kinase purified from T4 phage-infected cells also exhibits a 3'-phosphatase activity (Cameron and Uhlenbeck, 1977) whose optimum pH is comprised between 5.0 and 6.0. Based on this property were developed protocols using T4 polynucleotide kinase as a specific 3'-phosphatase (Cameron et al., 1978).

A T4 polynucleotide kinase deprived of 3'-phosphatase activity has been purified from *E. Coli* infected with a mutant T4 phage producing an altered *pse* T1 gene product. It is composed of four subunits having the same apparent molecular weight (33,000 daltons) as the wild-type monomer. The resulting protein lacks the 3'-phosphatase activity and is still able to catalyse the kinase reaction under the conditions used for the wild-type T4 polynucleotide kinase. This enzyme has proven to be very useful when the 3'-exonuclease activity must be avoided (e.g., 5'-^{32}P terminal labeling of 3'-CMP in view of 3' end-labeling of RNA species prior to fingerprinting or sequencing).

TERMINAL DEOXYNUCLEOTIDYL TRANSFERASE

This enzyme is purified from thymus gland or from plants. It is a small (34,000 daltons), basic protein made up of two subunits (8,000 and 26,000 daltons) (Ratliff, 1981).

Terminal deoxynucleotidyl transferase acts as a DNA polymerase by catalyzing a 5' → 3' polymerization of 5' deoxynucleotide triphosphate. The polymerization of deoxynucleotide residues at the 3' end of DNA molecules requires a free 3' — OH group and a minimum of three residues in a single-stranded state. This enzyme is also able to catalyze a limited polymerization of ribonucleotides at the 3' end of oligodeoxynucleotides. The addition of ribonucleotides has been used successfully for primer extension, 3' end-labeling, and DNA sequencing.

Terminal deoxynucleotidyl transferase can also polymerize deoxynucleoside triphosphates analogs. Thus, incorporation of 8-(2,4-dinitrophenyl-2,6-aminohexyl) amino-adenosine-5'-triphosphate and 2'-deoxyuridine-5'-triphosphate-5-allyl-amine-biotin (biotin-11-dUTP) has been used in fluorescent-dye and avidin-conjugate labeling, respectively (Vincent et al., 1982, see chapter 16). A better polymerization rate for purine deoxyribonucleoside triphosphates is obtained in the presence of Mg^{2+} ions (usually $MgCl_2$) in cacodylate buffer, and in the presence of Co^{2+} ions ($CoCl_2$) for pyrimidines deoxyribonucleosides triphosphates (Roychoudhury et al., 1976). It should be pointed out that addition of Co^{2+} may re-

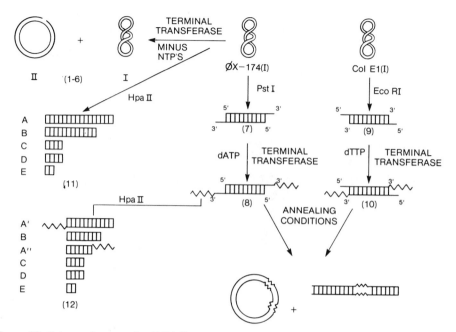

Figure 22. Scheme for assessing TdT tailing reaction. The numbers in parentheses refer to products which were electrophoresed on the agarose gels shown in Figure 23. The incubation producing no. 1 contained 0.5 μg φX174 (I) in 10 μl 0.14 M potassium cacodylate (pH 7.6), 1 mM CoCl₂, 0.1 mM dithiothreitol, and was incubated for 1 hour at 37°C; no. 2 was identical to no. 1, except that the buffer was 0.05 M Tris–HCl (pH 7.5) and 10 mM MgCl₂; no. 3 was the same as no. 2, except that the buffer also contained 5 mM 2-mercaptoethanol. Numbers 4–6 were produced under the conditions listed for 1–3, respectively, and contained 1500 units/ml of lot 730-9 TdT. The reaction conditions used to produce nos. 8 and 10 are described in the caption to Figure 24. Electrophoresis of the annealed DNA resulted in a sharp band near the top of the gel (data not shown) (courtesy of Pharmacia).

sult in a relaxation of the helical structure of the DNA, therefore allowing the tailing of internal nicks. Use of Mg^{2+} reduces this problem but results in a significant decrease of the tail length because at identical ratios of enzyme to DNA, the rate of molecules tailed in the presence of Mg^{2+} is one-fifth that tailed in the presence of Co^{2+}. Optimized incubation conditions allow addition of dG or dC tails of 15–40 nucleotides and dA or dT tails of 30–80 nucleotides, which turn to be the most suitable length for subsequent DNA annealing and cloning.

The use of [³²P]cordycepin-5′-triphosphate (3′ dATP), which is an analog of dATP, prevents further nucleotide addition. Under these conditions, the DNA molecules will have a unique ³²P-labeled 3′ end and thus can be used directly for sequencing.

This enzyme can also accept double stranded DNA as primer for the addition of ribo-or deoxyribonucleoside phosphate if the Mg^{2+} ion is replaced by Co^{2+} ion. This is a very useful method for addition of homopolymer tails at the 3′ end of any DNA molecule to be cloned (Figures 22, 23, 24), and for

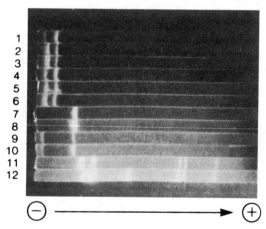

Figure 23. Agarose gel electrophoresis of the products numbered in Figure 22. The slower moving band in 1-6 is φX174 (II), and the faster moving band is φX174 (I). The rest of the bands are linear duplex DNA molecules, which migrate toward the positive electrode according to their size. The bands in no. 12 were excised, heated at 100°C with 200 μl H₂O, and counted with 10 ml Bray's solution in a liquid scintillation counter (courtesy of Pharmacia).

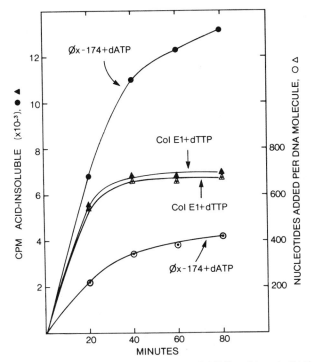

Figure 24. Time course for the addition of dAMP residues to Pst I-restricted φX174 and dTMP residues to Eco RI-restricted φX174. Reactions (0.2 ml, 37°C) contained 4–12 μg form III DNA, 8.6 nmol [^{14}C]dATP (7 × 10^4 cpm/nmol) or 5 nmol [^{14}C]dTTP (6.5 × 10^4 cpm/nmol), 0.14 M potassium cacodylate, 0.03 M Tris–HCl (pH 6.9), 1 mM CoCl$_2$, 0.1 mM dithiothreitol, 50 μg/ml bovine serum albumin, and 30 units lot 730-9 TdT (courtesy of Pharmacia).

the addition of a single nucleotide to 3' ends of DNA molecules for in vitro directed mutagenesis (Deng and Wu, 1983).

EXONUCLEASE III

Obtained from the BE 257 *E. Coli* strain which contains a thermo-inducible overproducing plasmid (pSGR3), exonuclease III is a monomeric enzyme with a molecular weight of 28,000 daltons (Weiss, 1976). This enzyme has several interesting properties. It can act as a phosphatase on a DNA chain which terminates with a 3'-phosphate group. This kind of chain is usually inert as a primer and inhibits DNA polymerase action. It can also degrade the RNA strand of DNA/RNA hybrids (RNAse H activity) (Rogers and Weiss, 1980) and exhibits an endonuclease activity, specific for apurinic and apyrimidic sites in DNA, where bases have been cleaved from the sugar phosphate backbone (Keller and Crouch, 1972).

Another significant feature of exonuclease III is its relative specificity for duplex DNA. When this enzyme acts as a nick in a DNA duplex, it will enlarge it to a gap, whereas when it acts at an end it will generate a 5'-protruding end. Exonuclease III is unique among the exonucleases in its phosphomonoesterase action on a 3'-phosphate terminus.

Optimum pH for endonucleolytic and exonucleolytic activities is between 7.6 and 8.5 while phosphatase activity is maximum at a pH comprised between 6.8 and 7.4. The presence of Mg^{2+} or Mn^{2+} ions is required for optimal activity and addition of Zn^{2+} will result in an inhibition of the enzyme activity (Richardson et al., 1964; Richardson and Kornberg, 1964; Rogers and Weiss, 1980).

Exonuclease III is often used in conjunction with the Klenow fragment of *E. Coli* DNA polymerase to generate radio-labeled DNA strands. It is also used sequentially with S1 nuclease to reduce the length of double stranded DNA. Exonuclease III has also been used to produce target deletion breakpoints in a rapid sequencing protocol (Roberts et al., 1979; Guo and Wu, 1982; Yang et al., 1983).

BAL 31 NUCLEASES

The Bal 31 nucleases are extracellular nucleases purified from the culture medium of *Alteromonas espejiana* Bal 31. They are multifunctional enzymes exhibiting a highly specific single stranded endodeoxyribonuclease activity, as well as an exonuclease activity which simultaneously degrades both 3' and 5' termini of duplex DNA without internal strand scission. Two distinct molecular species have been described as a fast (F) and a slow (S) Bal 31 nuclease. Both of them shorten DNA duplex without introducing internal nicks (Figure 25). A characterization of the DNA fragment ends generated after BAL 31 treatment has revealed that most of them have fully base paired ends which may be ligated to any blunt end DNA fragment, including molecular linkers (see Chapter 12). Another fraction of the DNA fragments is harboring 5'-protruding single strands, suggesting that BAL 31 is acting sequentially as a 3' to 5' exonuclease followed by endonucleolytic removal of the protruding strand (Talmadge et al., 1980; Shishido and Ando, 1982; Wei et al., 1983; Wei and Gray, unpublished results). It is sometimes useful to use the large fragment of *E. Coli* DNA polymerase or T4 DNA polymerase to fill the ends and therefore increase the efficiency of blunt end ligation.

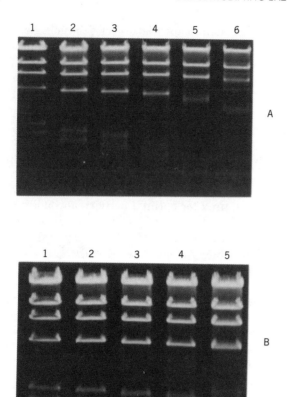

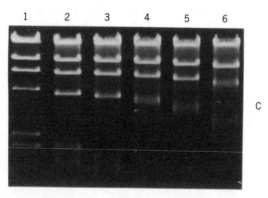

Figure 25. Digestion of λ DNA fragments by Bal 31 nucleases. 5 μg of Hind III digested λ DNA were incubated with 1 unit of Bal 31, yielding a ratio of 0.4 unit Bal 31/pmol/ends. Incubation was carried out at 30°C in 50-μl reaction containing 1x Bal 31 assay buffer. 1-μg samples of digested DNA were removed at different times and run on a 1% agarose gel for 2 hours at 100 volts. (A) Bal 31 fast. Rate of shortening : 55 bp/min/end. Lane 1, control; 2, 1 minute; 3, 2 minutes; 4, 4 minutes; 5, 8 minutes; 6, 15 minutes. (B) Bal 31 slow. Rate of shortening : 2 bp/min/end. Lane 1, 5 minutes; 2, 10 minutes; 3, 20 minutes; 4, 30 minutes; 5, 60 minutes. (C) Bal 31 mixed. Rate of shortening : 25 bp/min/end. Lane 1, control; 2, 5 minutes; 3, 10 minutes; 4, 20 minutes; 5, 30 minutes; 6, 60 minutes (courtesy of International Biotechnologies Inc.).

The purified F species of Bal 31 nuclease can be used for restriction fragment mapping, removing long stretches (up to thousands of base pairs) from duplex DNA, removing short stretches (tens to hundreds of base pairs) from duplex DNA, mapping B-Z DNA junctions, cleavage of DNA at sites of covalent lesions (such as UV-induced), and removing stretches from RNA species. Because it is a slow acting enzyme, the S species of Bal 31 can be used only for restriction mapping and removal of short stretches from DNA duplex.

Although the turnover numbers for the F and S species are very similar for single stranded DNA, they significantly differ for double stranded DNA, the turnover number for the F species being 50 times that of the S species when tested on linear duplex DNA. The fast Bal 31 nuclease can shorten duplex DNA at a rate approximately 20 times that of the slow enzyme (Talmadge et al., 1980; Wei et al., 1983). The reaction rate of digestion is dependent upon the G + C content of the substrate DNA (Kilpatrick et al., 1983). When tested under standard conditions (see Chapter 12), 1μg/ml of the F and S nuclease shortens DNA at rates of approximately 130 and 10 base pairs per terminus per minute, respectively.

EXONUCLEASE VII

This exonuclease degrades specifically single stranded DNA. Since it is able to attack either end of the DNA molecules, it is often employed to degrade long single strands protruding from duplex DNA such as those generated by restriction endonucleases, resulting in the formation of blunt end DNA. Exonuclease VII is different from exonucleases I and III in that it can degrade DNA from either its 5′ or 3′ end, it can generate oligonucleotides, and is still fully active in 8 mM EDTA (see Chapter 18).

TOBACCO ACID PYROPHOSPHATASE

This enzyme is a very useful tool in the labeling of the 5′ end of messenger RNAs, which are usually capped. In vitro ^{32}P-labeling of this 5′ terminus requires the elimination of the 7-methylguanosine and 5′-phosphate moieties of the capped end. Tobacco acid pyrophosphatase hydrolyzes the pyrophosphate bond of the cap, leading to p^{7}MeG, pp^{7}MeG, and ppN–pN–mRNA. The open cap can then be dephos-

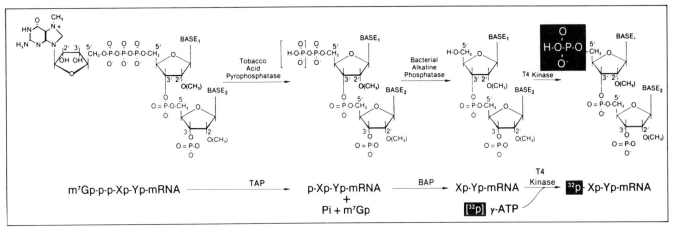

Figure 26. Labeling of RNA 5' end by tobacco acid pyrophosphatase.

phorylated by alkaline phosphatase and labeled with T4 polynucleotide kinase and $\gamma[^{32}P]ATP$ (Figure 26). The 5'-labeled RNAs may be used either as probes or directly for RNA sequencing (see chapter 21).

RIBONUCLEASE H

This enzyme specifically degrades RNA which is hybridized to DNA. It is therefore possible, by using ribonuclease H, to introduce specific cuts in RNA molecules if specific probes are used in prior hybridizations. Ribonuclease H can also be used to remove the poly-A tract at the 3' end of messenger RNAs in the presence of oligo dT.

Optimal activity of the enzyme is obtained at a pH comprised between 7.5 and 9.1 (Berkower et al., 1973) in the presence of SH-reagents. It requires Mg^{2+} ions which can be replaced partially by Mn^{2+} ions. RNAse H activity is inhibited in the presence of N-ethylmaleimide, and is not markedly affected by high ionic strength (50% activity is retained in the presence of 0.3 M NaCl).

RIBONUCLEASE Phy I

This enzyme is used for rapid sequencing of RNA (see chapter 21). It is isolated from cultures of *Physarum polycephalum*. Ribonuclease Phy I cleaves the RNA molecule at G, A, and U, but not at C residues. The products are 3' mononucleotides with 5' C termini.

RNase CL3

This enzyme is used for sequencing RNA (see chapter 21). It is isolated from chicken liver *(Gallus gallus)*.

RNase CL3 digests RNA adjacent to cytidilic acid in a ratio of 60 C residues digested for every U residue digested. The enzyme activity is inhibited by poly-A tracts if spermidine is not added.

Cereus RIBONUCLEASE

Used for RNA sequencing (Lockard et al., 1978), this endoribonuclease cleaves preferentially at U and C residues (see chapter 21).

RIBONUCLEASE PHY M

Purified from *Physarum polycephalum*, this nuclease which has also been used for sequencing of RNA (Donis-Keller, 1980) cleaves preferentially at U and A residues (see chapter 21). Occasionally cleavage at G may occur and can be reduced by using 7 M urea.

RNase U2

This enzyme, purified from *Ustilago sphaerogena*, is considered as the counterpart of RNase T1 in RNA sequencing (see chapter 21), to discriminate purines residues because it cleaves adenine residues specifically when incubated at 50°C, pH3.5 in the presence of 8 *M* urea. When incubated under standard conditions (Takahashi, 1961), RNase U2 specifically cleaves the 3′-phosphodiester bond adjacent to purines, therefore generating purines 3′-phosphates and oligonucleotides with purine 3′-phosphate terminal groups as products. It has been shown that cyclic 2′:3′ purine nucleotides are obtained as intermediates and that reversal of the final reaction step can be used to synthetize ApN and GpN. RNase U2 is thermostable (80°C for 4 minutes) in aqueous solution at pH 6.9.

RNase T1

This enzyme is purified from *Aspergillus oryzae*. It has a molecular weight of 11,000, and is specific for G residues in RNA. RNase T1 has been used widely in RNA sequencing (see chapter 21).

RNase T2

This enzyme is also purified from *Aspergillus oryzae*. It has a molecular weight of 36,000, and splits all phosphodiester bonds in RNA, with a preference for adenylic bonds. It too is used for RNA sequencing (see chapter 21).

DNAse I

Purified from bovine pancreas, deoxyribonuclease I is a nuclease which degrades both single stranded DNA and RNA in an endonucleolytic manner but double stranded DNA in an exonucleolytic manner. Maximal activity is obtained in the presence of Ca^{2+}, Mg^{2+}, Ca^{2+}, and Mn^{2+} ions (Kunitz, 1950). Depending upon the nature of the divalent cations being used in the incubation mixture, both specificity and mode of action of DNase I can be affected (Junowicz and Spencer, 1973; Campbell and Jackson, 1980).

Bovine pancreatic DNase I is a glycoprotein of molecular weight 31,000, usually purified as a mixture of four isoenzymes (A, B, C, and D).

S1 NUCLEASE

S1 nuclease is purified from *Aspergillus oryzae*. This enzyme degrades RNA or single stranded DNA into

5′ mononucleotides, but does not degrade duplex DNA or RNA–DNA hybrids in native conformation. S1 nuclease is therefore a very useful tool for measuring the extent of hybridization (DNA–DNA or DNA–RNA), probing duplex DNA regions, and removing single stranded protruding DNA of "sticky ends" generated by restriction enzymes.

This enzyme is a metalloprotein with a molecular weight of 32,000 daltons (Vogt, 1973). It requires Zn^{2+} and acts at pH 4.0–4.3. It is strongly inhibited by chelating agents such as EDTA and citrate. Co^{2+} and Hg^{2+} can replace Zn^{2+}, but are less effective as cofactors. Its activity is reduced by 50% at pH 4.9 and it is essentially inactive at any pH above 6.0. It is strongly inhibited by sodium phosphate concentrations as low as 10 m*M* and by chelating agents such as EDTA and citrate. Readdition of the metal ion cofactor restores most of the original enzyme activity. Nuclease S1 is resistant to several denaturing agents, such as urea, SDS, or formamide (Vogt, 1973; Hofstetter et al., 1976) and is thermostable (Ando, 1966). It has been shown that hydrolysis of single stranded DNA by S1 nuclease is five times more rapid than that measured for RNA and 75,000 times more efficient than with double stranded DNA.

The low level of strand breaks which can be introduced by S1 nuclease in duplex DNA (Vogt, 1973; Shenk et al., 1975; Wiegland et al., 1975; Vogt, 1980; Beard et al., 1983), is greatly reduced when incubation is performed at higher ionic strength (0.2 *M* salt). It has also been shown that this enzyme is active at S1 sensitive sites generated by negative supercoiling of the helical DNA structure (Beard et al., 1973; Godson, 1973; Mechali et al., 1973), UV irradiation (Hofstetter et al., 1976; Shishido and Ando, 1974; Heflich et al., 1979), or depurination (Shishido and Ando, 1975).

S1 nuclease has been widely used in experiments where a specific removal of single portions of duplex DNA, RNA/DNA hybrids, or RNA molecules is required. Among these applications are the mapping of spliced RNA molecules (see Chapter 18), the isolation of duplex regions in single stranded viral genomes (Shishido and Ikeda, 1970; 1971a; 1971b), probing strand breaks in duplex DNA molecules (Vogt, 1973; Germond et al., 1974; Shishido and Ando, 1975b), cleavage of regions with lesser helix stability (Shishido, 1979; Lilley, 1980; Panayotatos and Wells, 1981), localization of inverted repeated sequences (Shishido, 1979; Lilley, 1980; Panayotatos and Wells, 1981), introduction of deletion mutation at D loop sites in duplex DNA (Green and Tibbetts,

1980), and mapping of the genomic regions involved in interactions with DNA binding proteins (Meyer et al., 1980).

MUNG BEAN NUCLEASE

This enzyme is a single stranded specific DNA and RNA endonuclease purified from mung bean sprouts. It yields 5'-phosphoryl terminated mono- and oligonucleotides (Sung and Laskowski, 1962; Johnson and Laskowski, 1968; Johnson and Laskowski, 1970; Mikulski and Laskowski, 1970). When high enzyme concentrations and extended incubation times are used with duplex DNA, a complete degradation of the substrate may be observed (Johnson and Laskowski, 1970; Kroeker et al., 1976; Koreker and Kowalski, 1978) as the consequence of a two-step process: introduction of first, single stranded nicks, and then, double stranded scissions followed by an exonucleolytic digestion of the resulting fragments.

Since it preferentially degrades single stranded nucleic acids, mung bean nuclease has been used in hybridization studies, in the same way as nuclease S1, to remove protruding tails in duplex DNA, and to map transcription promoter regions.

The ability of mung bean nuclease to generate correct blunt ends has been studied (Ghangas and Wu, 1975; Hammond and D'Alessio, 1986). The data obtained revealed that proper trimming of the 5'-protruding extensions was achieved when the final blunt end contained a GC base pair at its terminus. The presence of an AT base pair at the position where the fragment would end after trimming seemed to interfere with precise removal of the protruding terminus. The nucleotide composition of the overhang did not seem to affect the efficiency or the quality of the nucleolytic digestion.

This enzyme has also proven to be very useful for excising cloned DNA fragments inserted in vectors following a dA·dT tailing (Wensink et al., 1974). Because poly(dA·dT)·poly(dT·dA) does not seem to be recognized as a typical double stranded structure (Johnson and Laskowski, 1970) and is hydrolyzed at half the rate of single stranded tails excision, there is a more efficient cleavage of these sequences as compared to the hydrolyzis of other duplex regions in the recombinant DNA.

Mung bean nuclease requires Zn^{2+} and a reducing agent such as cysteine for maximum activity and stability. It is inhibited by high salt concentrations (80 to 90% inhibition in 200–400 mM NaCl). Use

0.001% Triton X-100 when using very low concentrations of enzyme (less than 50 units/μl) because under such conditions, mung bean nuclease may adhere to surfaces and is rather unstable.

TOPOISOMERASE I

Type I topoisomerase catalyzes the relaxation of supercoiled DNA by introducing a single stranded break in the sugar–phosphate backbone and further rejoining of the free DNA ends (Champoux, 1978; Gellert, 1981; Wang, 1981). This enzyme has also been found to efficiently promote the covalent transfer of a single stranded donnor DNA to an heterologous acceptor DNA (Been and Champoux, 1981; Halligan et al., 1982). Binding of the enzyme to single stranded DNA is accompanied by the formation of a covalent DNA/enzyme complex which is catalytically active in joining the single stranded fragment to the end of an heterologous acceptor DNA containing a 5'-hydroxyl terminus (Halligan et al., 1982; Prell and Vosberg, 1980). The use of micrococcal nuclease has established that about 25 bp of DNA are covered by the bound topoisomerase I.

This enzyme runs in SDS polyacrylamide gels as a polypeptide with an apparent molecular weight of 105,000 daltons (Liu and Miller, 1981). It is an ubiquitous nuclear protein whose physiological function is unknown. It might be involved in genetic rearrangements during replication and transcription (Been and Champoux, 1981; Halligan et al., 1982).

Topoisomerase I is active at pH 7.5 in the presence of 50 to 200 mM NaCl and it is completely inactivated by 0.2% SDS. Covalent binding to DNA is stimulated by Mg^{2+} (5 to 10 mM $MgCl_2$). This enzyme has been used for the preparation in vitro of recombinant circular DNAs (Martin et al., 1983), and for studies of nucleosome assembly (Laskey et al., 1977; Germond et al., 1974) and effects of ligands on tertiary structure of DNA (Wang, 1980; Peck and Wang, 1981).

TOPOISOMERASE II (DNA GYRASE)

This enzyme, which is purified from *M. Luteus* (Klevan and Wang, 1980), catalyzes the breakage and resealing of both strands of duplex DNA, thereby changing the linking number of discrete supercoiled forms of DNA in steps of two (Brown and Cozzarelli, 1979; Liu et al., 1980). Topoisomerase II is also capable of reversibly knotting and catening intact circular DNA (Brown and Cozzarelli, 1979; Liu et al. 1980). Type II topoisomerases are multimeric and require ATP to be fully active.

GUANYLYL TRANSFERASE FROM VACCINIA VIRUS

This capping enzyme complex isolated from vaccinia virus can be used directly to label either 5' di- and triphosphate ends of RNA molecules or capped 5' ends of RNA after chemical removal of the terminal 7-methyl-guanosine (m7G) residue.

This complex has a triple enzymatic activity:

1. RNA triphosphatase:

$$pppN(pN)n \rightarrow ppN(pN)n + Pi$$

2. Guanylyl transferase:

$$\alpha[^{32}P]GTP + ppN(pN)n \leftrightarrows$$
$$G(5')^{32}PppN(pN)n + PPi$$

3. RNA (guanine-7) methyl transferase activity:

$$AdoMet + G(5')pppN(pN)n \rightarrow$$
$$m^7G(5')pppN(pN)n + AdoHcy$$

This complex does not accept monophosphate RNA as a substrate. Therefore, degraded or nicked RNA will not be labeled in any place other than the 5'-cap end. The optimal amount of enzyme to be used must be determined empirically.

In order to make sure that all the label is contained within an authentic cap structure, it is recommended to check that, after digestion of an aliquot of the treated sample with tobacco acid pyro phosphatase, all the label is released in the form of 5'-GMP. Sequential digestion with nuclease P1 and bacterial alkaline phosphatase releases all the label as a spot comigrating with GpppG.

POLYADENYLATE POLYMERASE OF *E. COLI*

Polyadenylate polymerase of *E. coli* (EC 2.7.7.19) consists of a single polypeptide chain with a molecular weight of approximately 58,000. This enzyme polymerizes adenylate residues at the 3' terminus of a large number of different polyribonucleotides. It is a rather unusual enzyme in that optimal activity requires the presence of high concentrations of monovalent cations, for example, 0.4 M NaCl. The enzyme is strongly stimulated by manganese ions and is in-

sensitive to antibiotics such as rifampicin and streptolygdin, which are transcriptional inhibitors of initiation and elongation, respectively. The enzyme is also sensitive to aurin tricarboxylic acid, an inhibitor of both RNA polymerase and the binding of mRNA to ribosomes. Poly-A polymerase polymerizes AMP residues, using ATP as a substrate. ADP and dATP are not polymerized, but both CTP and UTP are polymerized at less than 5% of the rate obtained with ATP. The enzyme does not show any activity with GTP. Phosphate and pyrophosphate ions are relatively strong inhibitors of poly-A polymerase. In contrast to polynucleotide phosphorylase, another template-independent polynucleotide synthesizing enzyme, polyadenylate polymerase does not degrade its own poly-A product by phosphorylase or pyrophosphorylase action.

The enzyme uses a wide variety of single stranded RNA species as primers. Double stranded RNA is a very poor primer, as are a number of synthetic polynucleotides, such as poly UG and poly C, and short oligonucleotides, such as di- and trinucleotides. DNA does not function as a primer. This latter property is common to virtually all other poly-A polymerases except for the enzyme isolated from plants such as maize, which shows considerable activity with natural and synthetic oligo- and polydeoxyribonucleotides.

The length of the polyadenylate sequence synthesized in vitro can be quite considerable. Depending on the conditions used for the synthesis, it can be from 20 to 2000 residues.

A number of investigators have used enzymatic polyadenylate extension of RNA as a preliminary step in the synthesis of complete cDNA copies by reverse transcriptase.

REVERSE TRANSCRIPTASE

Reverse transcriptase (RT) has been isolated from several avian and murine retroviruses. The most commonly used has been the reverse transcriptase of avian myeloblastosis virus transformed myeloblasts recovered from leukemic chickens. This enzyme has been known for several years as AMV RT. It is interesting to point out that AMV is a defective virus which needs a helper retrovirus (myeloblastosis associated virus, or MAV) to replicate efficiently in infected cells. The AMV is defective because part of its *pol* and *env* genes (encoding for reverse transcriptase and envelope proteins, respectively) are replaced by cellular-derived sequences known as *v-myb* (see review by Baluda et al., 1983). Therefore, what is called AMV reverse transcriptase is in fact the enzyme encoded by the MAV *pol* gene and should be called MAV RT (Houts et al., 1979; Hurwitz and Leis, 1972; Flugel and Wells, 1972; Canaani and Duesberg, 1972; Grandgenett et al., 1973).

It is composed of two structurally related subunits designated α and β (molecular weights of 65,000 and 95,000 daltons, respectively) assembled into an αβ holoenzyme which is thought to be generated by proteolytic cleavage of a minor less active ββ precursor (Eisenman et al., 1980; Schiff and Grandgenett, 1980).

The α subunit of the enzyme has both the RNA-directed DNA polymerase activity specific to reverse transcriptase and an RNase H activity. The RNase H activity is associated with a 24,000-dalton fragment and is generated by proteolytic cleavage of the α subunit. The polymerase activity of reverse transcriptase is dependent on the presence of a primer and a template (Verma, 1977; Leis et al., 1983).

The viral reverse transcriptase has been widely used to copy natural polyadenylated messenger RNAs previously annealed to oligo dT. It is able to use ribopolymers, particularly poly-rC·dG$_{12}$, as well as deoxyribopolymers such as poly-dC. However, poly-rU does not seem to be used efficiently.

The RNase H is a processive exoribonuclease that degrades specifically RNA strands in RNA–DNA hybrids in either $5' \rightarrow 3'$ or $3' \rightarrow 5'$ directions.

Combined use of reverse transcriptase, RNase H, and DNA polymerase activities in cDNA cloning is reported in Chapter 19.

The use of reverse transcriptase has found many applications in molecular cloning, two well-documented examples being the synthesis of complementary DNA from RNA species in the preparation of expression libraries (see Chapter 25) and in nucleotide sequencing (see Chapters 20,21). Several studies have been performed in order to establish reaction conditions optimal for the synthesis of high yields of long cDNA molecules (see for example, Retzel et al., 1980; and Berger et al., 1983).

Cloning of the *pol* gene from Moloney murine leukemia virus (M-MuLV) has allowed the purification of a reverse transcriptase deprived of DNA endonuclease activity. This enzyme was found to exhibit a lower RNAse H activity when tested under standard reaction conditions (Moelling, 1974), and has proven to be a reliable tool for cDNA synthesis (Kotewicz et al., 1985; Roth et al., 1985; Tanese et al., 1985; Gerard, 1986). The cloned murine reverse transcriptase can be purchased from Bethesda Research Laboratories and from Pharmacia (FPLC purified).

A comparative study of AMV and M-MuLV re-

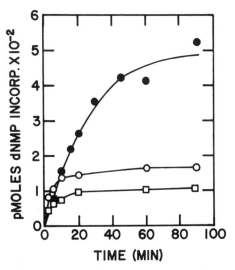

Figure 27. Time course of globin mRNA-directed DNA synthesis catalyzed by AMV and cloned M-MLV reverse transcriptase. Fifty μl aliquots were removed at the times indicated from 400-μl reaction mixtures containing either 1.5 pmol of AMV reverse transcriptase (●) or 0.9 (□) and 1.8 (○) pmol of cloned M-MLV reverse transcriptase (courtesy of Bethesda Research Laboratories, Life technologies Inc.).

verse transcriptase activity (Gerard, 1986) revealed that: (a) in 1 hour of incubation, AMV RT synthesizes four times more DNA than M-MuLV; (b) half-life of M-MuLV in the reaction mix is about 15 minutes, while AMV RT is still fully active after 1 hour of incubation under similar conditions. Therefore, comparable yields of cDNA synthesis required approximately six to eight times more M-MuLV than AMV reverse transcriptase (Figure 27). The length of cDNA species obtained with both enzymes was also examined. Addition of 4 m*M* NaPPi to the reaction mixture enhanced the synthesis of the large transcripts (6.5 kb) obtained with the AMV RT in the presence of AMV genome (7.2 kb). A minimum of 5 units (Houts et al., 1979) was required and addition of enzyme did not lead to any increase in length (Figure 28). On the contrary, NaPPi had no effect on the product size obtained with M-MuLV RT, and 100 units of enzyme were required to achieve a cDNA synthesis comparable to that obtained with AMV RT. Increasing the enzyme concentration to 200 units/μg of RNA led to the formation of 7.2 kb full-length cDNA transcripts (Figure 28).

The yields and sizes of cDNA species synthesized in the presence of reverse transcriptase from different commercial sources were also analyzed (Gerard, 1986). Striking differences were observed (Figure 29). In our hands, AMV reverse transcriptase from Life Sciences (St. Petersburg, Florida) and Boehrin-

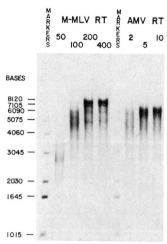

Figure 28. Autoradiogram of cDNA synthesized from AMV 35S RNA by AMV and cloned M-MLV RT. The indicated units of RT defined with $(A)_n \cdot (dT)_{12\text{-}18}$ were incubated for 1 hour at 37°C. AMV 35S RNA was substituted for globin mRNA and with AMV RT 4 mM NaPPi was substituted for actinomycin D. The ^{32}P-labeled DNA product was recovered from reaction mixtures by ethanol precipitation in the presence of carrier tRNA and was analyzed by electrophoresis on a vertical alkaline 1.4% agarose gel. The gel was autoradiographed after drying (courtesy of Bethesda Research Laboratories, Life technologies Inc.).

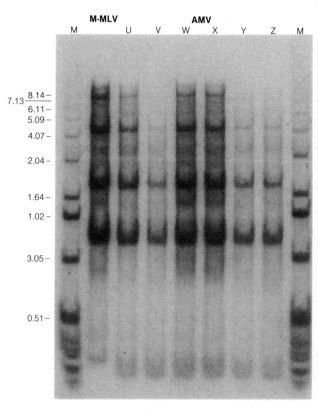

Figure 29. Autoradiogram of cDNA synthesized from BRL RNA Ladder by AMV and BRL cloned M-MLV RT. AMV RT from six sources (U–Z) and BRL M-MLV RT were used to copy the BRL RNA Ladder. Reaction mixtures (50 μl) contained 0.33 μg each of 9.49, 7.46, 4.40, 2.37, and 1.35 kb RNA, 10 μg of $(dT)_{12\text{-}18}$, 10 mM dithiothreitol, 100 μm/ml BSA, 50 μg/ml actinomycin D, 500 μM each of dATP, dGTP, dTTP and (α-^{32}P)dCTP, 50 mM Tris–HCl (pH 8.3), and either 75 mM KCl, 3 mM MgCl$_2$, and 400 units of M-MLV RT, or 100 mM KCl, 10 mM MgCl$_2$ and 10 units of AMV RT. Incubation was at 37°C for 1 hour. Incorporation was determined by TCA precipitation of an aliquot of the reaction mixture. Labeled cDNA products were recovered from reaction mixture by EtOH precipitation in presence of carrier tRNA and were analyzed by electrophoresis on a vertical alkaline 1.4% agarose gel. An autoradiogram of the dried gel was used as a template to excise portions of the gel whose ^{32}P contents were established by Cerenkov counting. The BRL 1 kb Ladder DNA was electrophoresed in lane M as a molecular weight standard (courtesy of Bethesda Research Laboratories, Life technologies Inc.).

ger (Mannheim) gave satisfactory results. Interestingly, the number of full-length cDNA transcripts obtained with either AMV or M-MuLV RT showed an inverse relationship to the logarithm of the initial RNA length. For example, reverse transcription of 1000, 500, 250, 140, and 110 fmol of 1, 2, 4, 7 and 9 kb RNA species is expected to yield about 500, 200, 50, and 10 fmol of full-length cDNA molecules, respectively (Gerard, 1986). These observations illustrate the difficulties experienced by many of us in obtaining full-length transcripts of scarce RNA species.

SP6 POLYMERASE

This enzyme is a single polypeptide (molecular weight 96,000 daltons) purified from *Salmonella typhimurium* LT2 infected with SP6 phage (Butler and Chamberlain, 1982). Commercial preparations are usually free of other RNA polymerases, DNAses, and RNAses. SP6 polymerase requires Mg^{2+} and a DNA template. It is greatly stimulated by spermidine and serum albumine (Butler and Chamberlain, 1982). Since the SP6 RNA polymerase recognizes specifically the SP6 promoter, cloning vectors have been developed which carry multiple cloning sites

directly downstream from the SP6 promoter, allowing specific transcription of the cloned sequences (see Chapters 16, 23). The high efficiency of the SP6 promoter allows the synthesis of 8 moles of RNA per mole of cloned sequence (Melton et al., 1984). The RNA produced under these conditions has been shown to be biologically active (Krieg and Melton, 1984) and can be properly spliced (Green et al., 1983). The in vitro synthesis of RNA by SP6 poly-

METHYLASES —————————————————————————————————— **97**

merase has also been used in the production of anti-sense RNA transcripts (Melton, 1985), isolation of labeled single stranded RNA for hybridization studies, and RNAse protection mapping (Zinn et al., 1983).

T7 RNA POLYMERASE

This RNA polymerase specific of T7 bacteriophage is isolated from strains of *E. Coli* containing the pAR1219 plasmid. This plasmid contains the structural gene for T7 RNA polymerase cloned downstream to the inducible lac UV5 promoter (Davanloo et al., 1984). This polymerase is a single polypeptide with a molecular weight of 98,000 daltons (Stahl and Zinn, 1981). It requires Mg^{2+} and a DNA template. Commercial preparations are usually deprived of any other RNA polymerase, DNAse, or RNAse. Cloning vectors carrying the T7 transcription promoter have been constructed, allowing the in vitro synthesis of RNA transcripts specific for cloned DNA sequences (see Chapter 25).

METHYLASES

Methylases have been described as the second component of the bacterial restriction–modification systems. They act on double stranded DNA by transferring methyl groups from S-adenosylmethionine to different residues of the DNA molecules. Most B, K, and W strains of *E. Coli* also contain two site-specific DNA methylases which are not part of the restriction modification system and are encoded by the *dam* and *dcm* genes (Pirotta, 1976). The *dam* methylase transfers the methyl group from adenosylmethionine to the N6 position of the adenine residue in the sequence GATC, while the *dcm* (also known as *mec*) methylase catalyzes the transfer of a methyl group to the internal cytosine residues in the sequences CCAGG and CCTGG (Marinus and Morris, 1973; Geier and Modrich, 1979; May and Hattman, 1975). These modifications often result in an inhibition of the restriction enzymes that recognize the corresponding sequences (see Tables 17 and 18). However, this is not the rule (McClelland, 1984). Some restriction endonucleases still cleave DNA at a recognition sequence being modified by the *dam* or *dcm* methylases (e.g., Bam HI, Sau 3AI, Bgl II, Pvu I, and BstNI). The *dam* methylation is thought to have a regulatory effect in postreplication mismatch repair and tends to occur with a high frequency in

Table 17. Inhibition of Restriction Endonucleases by the DAM Methylation

Recognition Sequence	Enzymes
GATC	Bsa PI, Bss GII, Bst EIII, Cpa I Dpn II, Fnu AII, Fnu CI, Mbo I Mno III, Mos I, Nde II
TGATCA	Atu CI, Bcl I, Cpe I
TCGA	Taq I, Tfl I, Tth HB8I
ATCGAT	Cla I
TCTAGA	Xba I
GAAGA	Mbo II
GGTGA	Hph I
TCGCGA	Nru I

Table 18. Inhibition of Restriction Endonucleases by the DCM Methylation

Recognition Sequence	Enzymes
GG(AT)CC	Ava II
CC(AT)GG	Ata BI, Atu II, Bst GII, Eca II Ecl II, Eco RII, Mph I, Scr FI, Sgr I
GGNCC	Sau 96I
AGGCCT	Stu I

the vicinity or at the ends of Okazaki fragments. All DNA isolated from *E. Coli* is not methylated to the same extent. For example, the pBR322 plasmid DNA purified from *E. Coli* after amplification with chloramphenicol (see Chapter 7) appears to be completely resistant to the digestion by Mbo I which does not cut at methylated sites, while only 50% of the lambda DNA obtained from a *dam* + strain of *E. Coli* appears to be methylated. The degree of in vivo cytosine methylation in the sequence CpG has been found to be related with the level of gene expression in eukaryotic cells, an inverse correlation being found between the amount of 5-methyl-cytosin residues and gene activity (Ehrlich and Wang, 1981; Gruenbaum et al., 1981, 1981b; Razin and Cedar, 1977; Sutter and Doerfler, 1980; Razin and Riggs, 1980).

Methylation at discrete positions within the palindromic recognition sites of restriction enzymes by specific methylases of the restriction modification system generally results as an inhibition of the endonucleolytic activity of the corresponding enzyme. For this reason, methylases have often been used in the course of molecular cloning, to protect the DNA from being digested when molecular linkers were used to clone a DNA fragment and to construct cDNA or genomic librairies (see Chapters 19). For example, a DNA fragment to be cloned at the BamH I site of a given vector is first incubated with the BamHI methylase in order to protect potential internal Bam HI sites before addition of Bam HI linkers and digestion with the Bam HI endonuclease (see Table 19 for specificity of some methylases).

DNA methylases can also be used to create new specificities for restriction endonucleases in duplex DNA (Nelson et al., 1984). For this purpose, a DNA methylase is selected whose recognition sequence overlaps only a subset of a given restriction enzyme. Two classes of overlaps can be defined. In one case, methylation sequences overlap with restriction endonuclease sequences, and alteration of cleavage specificity occurs. In the other case, cleavage of the methylated DNA by the endonuclease will be blocked at the level of the overlaps.

Table 19. Specificity of Selected Commercial Methylases

Methylases	Recognition Site (from 5' to 3')	Methylated Residue
Alu I	AGCT	C
Bam HI	GGATCC	Internal C
Cla I	ATCGAT	Internal A
Eco RI	GAATTC	Most internal A
Hae III	GGCC	Internal C
Hha I	GCGC	Internal C
Hpa II	CCGG	Internal C
Hph I	TCACC	5' C
Msp I	CCGG	External C
Pst I	CTGCAG	A
Taq I	TCGA	A
DAM methylase	GATC	A

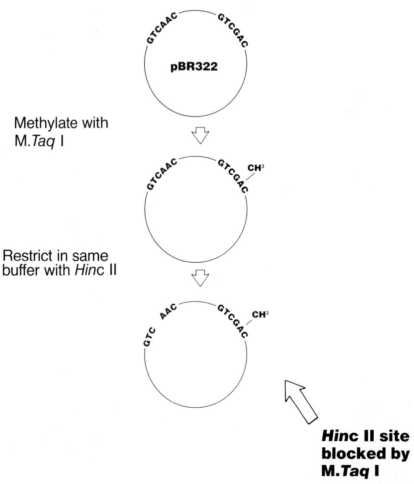

Methylate with M.*Taq* I

Restrict in same buffer with *Hinc* II

***Hinc* II site blocked by M.*Taq* I**

Figure 30. Creation of new cleavage specificity by methylases (courtesy of New England Biolabs).

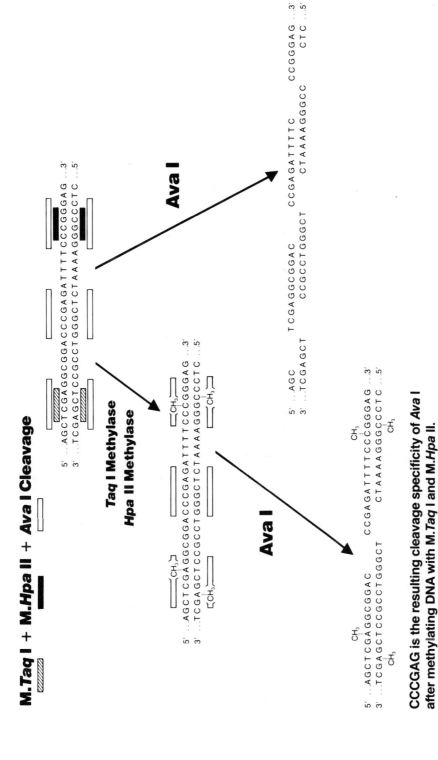

Figure 31. Modification of Ava I specificity upon incubation with methylases (courtesy of New England Biolabs).

Table 20.   DNA Cleavage Specificities Generated by Methylation at a Subset of Recognition Sequences of Degenerate Restriction Endonucleases

Restriction Endonucleases	Methylase	Cleavage Specificity
Acc I [GT(AC)(GT)AC]	M.*Taq* I [TCGmA]	GT(AC)TAC
Aha II [GPuCGPyC]	M.*Hha* I [GmCGC]	GPuCGTC
Ava I [CPyCGPuG]	M.*Taq* I [TCGmA]	CPyCGGG
Ava I [CPyCGPuG]	M.*Hpa* II [CmCGG]	CPyCGAG
Ava I [CPyCGPuG]	M.*Hpa* II [CmCGG] + M.*Taq* I [TCGmA]	CCCGAG
Ban II [GPuGCPyC]	M.*Alu* I [AGmCT]	GPuGCCC
Ban II [GPuGCPyC]	M.*Hae* III [GGmCC]	GPuGCTC
Ban II [GPuGCPyC]	M.*Hae* III [GGmCC] + M.*Alu* I [AGmCT]	GGGCTC
*Bsp*1286 [G(AGT)GC(ACT)C]	M.*Alu* I [AGmCT]	G(GT)GC(ACT)C
*Bsp*1286 [G(AGT)GC(ACT)C]	M.*Hae* III [GGmCC]	G(AGT)GC(AT)C
*Bsp*1286 [G(AGT)GC(ACT)C]	M.*Hae* III [GGmCC] + M.*Alu* I [AGmCT]	G(GT)GC(AT)C
*Hgi*A I [G(AT)GC(AT)C]	M.*Alu* I [AGmCT]	G(AT)GCAC
Hinc II [GTPyPuAC]	M.*Taq* I [TCGmA]	GTPyAAC
*Sau*96 I [GGNCC]	M.*Hae* III [GGmCC]	GGACC
*Scr*F I [CCNGG]	M.*Hpa* II [CmCGG]	CCAGG

Source: Information courtesy of New England Biolabs.

The first class of overlap is illustrated in the case of restriction endonucleases with degenerated recognition sequence and methylases being active on only one of the possible sites. For example, the Hinc II recognition sequence is GTPyPuAC and therefore corresponds to the following four combinations: GTCGAC, GTCAAC, GTTGAC, and GTTAAC. The TaqI methylase (designated M.TaqI) catalyzes the transfer of a methyl group to the adenine residue of the TCGA sequence. Thus, among the four possible combinations for the Hinc II recognition site, only those containing a TCGA sequence will be methylated and consequently become resistant to Hinc II digestion (see Figures 30 and 31 and Table 20 for examples).

The second class of overlap occurs at the boundaries of the recognition sequences for a restriction endonuclease and a methylase. For example, when the GGATCC sequence, corresponding to a Bam HI site, is followed by GG or preceded by CC, it overlaps with the CCGG M.MspI site. Since the M.MspI enzyme transfers a methyl group to the 5'cytosine residue in the CCGG sequence, it follows that the Bam HI site being methylated at its internal cytosine is now resistant to the Bam HI endonuclease (Figure 32 and Table 21).

Practically, the altered specificities are generated by a two-step in vitro procedure: (1) methylation of DNA by a site-specific methylase, followed by (2) cleavage of the treated DNA by a restriction endonuclease. DNA methylases from the type II restriction–modification systems of bacteria perform the methylation reaction under conditions similar to that used for the restriction endonucleases except that the methylases require S-adenosylmethionine (SAM) as a methyl group donnor. Therefore, it is generally acceptable to carry out the methylation reaction using standard restriction endonuclease buffers to which SAM has been added. It should be pointed out that the methylases do not require divalent cations to be active while most nucleases such as restriction endonucleases usually need a divalent cation to be effective.

The DNA fragments obtained by restriction endonuclease digestion of in vitro methylated DNA are in most respects indistinguishable from unmethylated DNA. However, beware that the methylated cytosine does not generate a band in the C channel when sequencing DNA by the Maxam and Gilbert method (see Chapter 20) and that reduced transformation efficiencies are obtained with most common strains of *E. Coli* when using DNA that contains methyl-cytosine residues. This second problem results from the existence of restrictions systems in *E. Coli* K12 responsible for a specific degradation of DNA containing methylated cytosines. These sys-

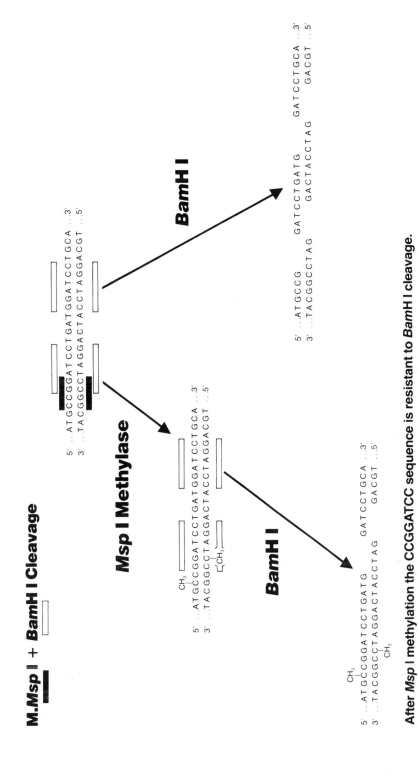

Figure 32. Modification of Bam HI site by incubation with Msp I methylase (courtesy of New England Biolabs).

Table 21. DNA Cleavage Specificities Generated as a Result of Methylation at the Boundaries of Overlapping Recognition Sequences of a Restriction Endonuclease and a Methylase

Restriction Endonuclease	Methylation(a)	Subset Blocked by Methylation(b)
Aha II [GPuCGPyC]	M.*Hpa* II [CmCGG]	CCGGCGPyC
Alu I [AGCT]	M.*Pst* I [CTGCmAG]	AGCTGCAG
Ava II [GG(AT)CC]	M.*Hpa* II [CmCGG]	CCGG(AT)CC
*Bam*H I [GGATCC]	M.*Msp* I [mCCGG]	CCGGATCC
Bgl I [GCCN5GGC]	M.*Hae* III [GGmCC]	GGCCN5GGC
*Bst*X I [CCAN6TGG]	M.*Hae* III [GGmCC]	GGCCAN5TGG
Cla I [ATCGAT]	*dam* [GmATC]	GATCGAT
Cla I [ATCGAT]	M.*Hha* II [GmANTC]	GAATCGAT
Dde I [CTNAG]	M.*Alu* I [AGmCT]	AGCTNAG
*Eco*R V [GATATC]	M.*Taq* I [TCGmA]	TCGATATC
*Fnu*D II [CGCG]	M.*Hha* I [GmCGC]	GCGCG
Hinf I [GANTC]	M.*Hph* I [TmCACC](c)	GANTCACC
Hinf I [GANTC]	M.*Mbo* II [GAAGmA](d)	GAAGANTC
Hinf I [GANTC]	M.*Taq* I [TCGmA]	TCGANTC
Hph I [GGTGA]	*dam* [GmATC]	GGTGATC
Mbo I [GATC]	M.*Cla* I [ATCGmAT]	ATCGATC
Mbo I [GATC]	M.*Taq* I [TCGmA]	TCGATC
Mbo I [GATC]	M.*Mbo* II [GAAGmA]	GAAGATC
Mbo II [GAAGA]	*dam* [GmATC]	GAAGATC
Msp I [CCGG]	M.*Bam*H I [GGATmCC]	CCGGATCC
Msp I [CCGG]	M.*Hae* III [GGmCC]	GGCCGG
Mst I [TGCGCA]	M.H1 [GCNGC + GGmCC](e)	GCTGCGCA
Nae I [GCCGGC]	M.*Hae* III [GGmCC]	GGCCGGC
Nco I [CCATGG]	M.*Hae* III [GGmCC]	GGCCATGG
Nhe I [GCTAGC]	M.*Alu* I [AGmCT]	AGCTAGC
Nru I [TCGCGA]	*dam* [GmATC]	GATCGCGA
Pst I [CTGCAG]	M.*Alu* I [AGmCT]	AGCTGCAG
Pst I [CTGCAG]	M.*H1* [GCNGC + GGmCC]	GCTGCAG
Pvu II [CAGCTG]	M.*H1* [GCNGC + GGmCC]	GCAGCTG
Sac II [CCGCGG]	M.*Hae* III [GGmCC]	GGCCGCGG
*Sau*3A I [GATC]	M.*Hph* I [TmCACC]	GATCACC
Taq I [TCGA]	M.*Cla* I [ATCGmAT]	ATCGAT
Taq I [TCGA]	*dam* [GmATC]	GATCGA
Xba I [TCTAGA]	*dam* [GmATC]	GATCTAGA
Xmn I [GAAN4TTC]	M.*Taq* I [TCGmA]	TCGAAN4TTC

Source: Information courtesy of New England Biolabs.

tems, developed at New England Biolabs (Raleigh et al., unpublished), were designated *mcrA* and *mcrB* and appear to be the same as those (*rglA* and *rglB*) previously shown to degrade DNA containing hydroxymethylcytosine. The *mcrA +* strains restrict DNA modified by the Hpa II methylase while *mcrB* restricts DNA modified by the Hae III, Alu I, Hha I, and Msp I methylases. Strains deficient in the *mcrB* system are available from New England Biolabs at no charge.

LISTING OF SOME MODIFYING ENZYMES COMMERCIALLY AVAILABLE*

Alkaline phosphatase
MAIN USE: End labeling, cloning
SUPPLIERS: AME, BOE, BRL, CAL, IBI, NEN, PHA, SIG, USB

*See key in Table 27 for abbreviations of suppliers' names.

E. coli DNA polymerase
MAIN USE: Uniform labeling, filling of 5′-protruding ends,
digestion of 3′-protruding ends
SUPPLIERS: BOE, IBI, NEB, NEN, PHA, PRB, SIG, USB

T4 DNA polymerase
MAIN USE: Uniform labeling, filling of 5′-protruding ends,
digestion of 3′-protruding ends
SUPPLIERS: AME, BEC, BOE, BRL, IBI, NEB, NEN, PHA, PRB, USB

Klenow fragment of DNA polymerase
MAIN USE: End labeling, filling of 5′-protruding ends
SUPPLIERS: AME, BEC, BOE, BRL, CAL, IBI, NEB, NEN, PHA, PRB, USB

T4 polynucleotide kinase
MAIN USE: End labeling
SUPPLIERS: AME, BEC, BOE, BRL, CAL, IBI, NEB, NEN, PHA, PRB, SIG, USB

T7 polymerase
MAIN USE: Uniform labeling, in vitro RNA synthesis
SUPPLIERS: AME, BEC, BOE, BRL, IBI, PHA, PRB, USB

SP6 polymerase
MAIN USE: Uniform labeling, in vitro RNA synthesis
SUPPLIERS: AME, BEC, BOE, BRL, IBI, NEB, PHA, PRB

T4 DNA ligase
MAIN USE: Ligation
SUPPLIERS: AME, BEC, BOE, BRL, IBI, NEB, NEN, PHA, PRB, SIG, USB

E. coli DNA ligase
MAIN USE: Ligation
SUPPLIERS: BEC, BOE, NEB, PHA

Terminal deoxyribonucleotidyl transferase
MAIN USE: End labeling, addition of homopolymers
SUPPLIERS: AME, BEC, BOE, BRL, IBI, NEN, PHA, SIG, USB

Reverse transcriptase
MAIN USE: Uniform labeling, cDNA synthesis
SUPPLIERS: AME, BEC, BOE, BRL, IBI, PHA, PRB, SIG

Exonuclease III
MAIN USE: High-resolution mapping
SUPPLIERS: AME, BEC, BOE, BRL, IBI, NEB, NEN, PHA, SIG, USB

Exonuclease VII
MAIN USE: High-resolution mapping
SUPPLIERS: BRL

BAL 31
MAIN USE: High-resolution mapping, modification of ends
SUPPLIERS: AME, BOE, BRL, CAL, IBI, NEB, NEN, PHA, PRB, USB

Mung bean nuclease
MAIN USE: High-resolution mapping
SUPPLIERS: BRL, CAL, NEB, PRB

Nuclease SI
MAIN USE: High-resolution mapping
SUPPLIERS: AME, BEC, BOE, CAL, IBI, BRL, NEN, PHA, SIG

Ribonuclease H
MAIN USE: cDNA synthesis
SUPPLIERS: AME, BRL, PHA, SIG

Ribonuclease PHY I
MAIN USE: RNA sequencing
SUPPLIERS: BRL

RNAse CL3
MAIN USE: RNA sequencing
SUPPLIERS: BOE, BRL

RNAse T1
MAIN USE: RNA sequencing
SUPPLIERS: BOE, BRL

RNAse T2
MAIN USE: RNA sequencing
SUPPLIERS: BRL

Polyadenylate polymerase
MAIN USE: End labeling, cDNA synthesis
SUPPLIERS: AME, BRL, NEN, PHA

Guanylyl transferase
MAIN USE: High-resolution mapping, end labeling of RNA
SUPPLIERS: BRL

Restriction endonucleases
MAIN USE: Digestion of DNA for mapping, cloning, sequencing
SUPPLIERS: See Table 27

Restriction endonucleases class IIS
MAIN USE: Cleavage at predetermined sites in DNA
SUPPLIERS: See Table 27

REFERENCES

Ando, T. (1966), *Biochim. Biophys. Acta,* **114**, 158.

Baluda, M. A., Perbal, B., Rushlow, K. E., and Papas T. S. (1983), *Folia Biologica (Praha),* **29**, 18.

Beard, P., Morrow, J., and Berg, P. (1973), *J. Virol.,* **12**, 1303.

Been, M. D., and Champoux, J. J. (1981), *Proc. Natl. Acad. Sci. USA,* **78**, 1883.

Berger, S. L., Wallace, D. M., Puskas, R. S., and Eschenfeldt, W. H. (1983), *Biochemistry,* **22**, 2365.

Berkner, K. L., and Folk, W. R. (1980), in *Methods in Enzymology,* L. Grossman and K. Moldave, eds., **65**, 28.

Berkower, I., Leis, J., and Hurwitz, J. (1973), *J. Biol. Chem.,* **248**, 5914.

Butler, E. T., and Chamberlain, J. (1982), *J. Biol. Chem.,* **257**, 5772.

Cameron, V., and Uhlenbeck, O. C. (1977), *Biochemistry,* **16**, 5120.

Cameron, V., Soltis, D., and Uhlenbeck, O. C. (1978), *Nucl. Acids Res.,* **5**, 825.

Campbell, V. W., and Jackson, D. A. (1980), *J. Biol. Chem.,* **255**, 3726.

Canaani, E., and Duesberg, P. (1972), *J. Virol.,* **10**, 23.

Chaconas, G., and Van de Sande, J. H. (1980), in *Methods in Enzymology,* L. Grossman and K. Moldave, eds., **65**, 680.

Champoux, J. J. (1978), *Ann. Rev. Biochem.,* **47**, 449.

Davanloo, P., Rosenberg, A. H., Dunn, J. J., and Studier, F. W. (1984), *Proc. Natl. Acad. Sci. USA,* **81**, 2035.

Deng, G. and Wu, R. (1983), in *Methods in Enzymology,* L. Grossman, K. Moldave, and R. Wu, eds., **100**, 96.

Depew, R. E., Snopek, T. J., and Cozzarelli (1975), *Virology,* **64**, 144.

Donis-Keller, H. (1980), *Nucl. Acids Res.,* **8**, 3133.

Ehrlich, M., and Wang, R. Y. H. (1981), *Science,* **212**, 1350.

Eisenman, R. N., Mason, W. S. and Linial M. (1980), *J. Virol.,* **36**, 62.

Englund, P. (1971), *J. Biol. Chem.,* **246**, 3269.

Flugel, R. M., and Wells, R. D. (1972), *Virol.,* **48**, 394.

Galas, D., and Schmitz, A. (1978), *Nucl. Acids Res.,* **5**, 3157.

Geier, G. E., and Modrich, P. (1979), *J. Biol. Chem.,* **254**, 1480.

Gellert, M. (1981), *Ann. Rev. Biochem.,* **50**, 879.

Gerard, G. F. (1986) Focus **7:1**, 1.

Germond, J., Vogt, V., and Hirt, B. (1974) *Eur. J. Biochem.,* **43**, 591.

Ghangas, G. S., and Wu, R. (1975), *J. Biol. Chem.,* **250**, 4601.

Godson, G. (1973), *Biochim. Biophys. Acta,* **308**, 59.

Grandgenett, D. P., Gerard, G. F., and Green, M. (1973), *Proc. Natl. Acad. Sci. USA,* **70**, 230.

Green, M. R., Maniatis, T., and Melton, D. A., (1983), *Cell,* **32**, 681.

Green, C., and Tibbetts, C. (1980), *Proc. Natl. Acad. Sci. USA,* **77**, 2455.

Gross, H. J., Domdey, H., Lossow, C., Jank, P., Raba, M., and Alberty, H. (1978) *Nature,* **273**, 203.

Gruenbaum, Y., Stein, R., Cedar, H. and Razin, A. (1981), *FEBS Lett.,* **124**, 67.

Guo, L., and Wu, R. (1982), *Nucl. Acids Res.,* **10**, 2065.

Halligan, B. D., Davis, J. L., Edwards, K. A., and Lice, L. F. (1982) *J. Biol. Chem.,* **257**, 3995.

Hammond, A. W., and D'Alessio, J. (1986), *Focus,* **8:4**, 4.

Harrison, B., and Zimmerman, S. B. (1986), *Analytical Biochem.,* **158**, 307.

Heflich, R., Mahoney-Leo, E., Maher, V., and McCornick, J. (1979), *Photochem. Photobiol.,* **30**, 247.

Hofstetter, H., Sambrook, A., Vandneberg, J., and Weissman, C. (1976), *Biochim. Biophys. Acta,* **454**, 587.

Houts, G. E., Miyagi, M., Ellis, C., Beard, D., and Beard, J. W. (1979), *J. Virol.,* **29**, 517.

Huber, H., Tabor, S. and Richardson, C. C. (1987), *J. Biol. Chem.,* in press.

Hurwitz, J., and Leis, J. (1972), *J. Virol.,* **9**, 116.

Jacobsen, H., Klenow, H., and Overgaard-Hansen, K. (1974), *Eur. J. Biochem.,* **45**, 623.

Johnson, P., and Laskowski, M. (1968), *J. Biol. Chem.,* **243**, 3421.

Johnson, P., and Laskowski, M. (1970), *J. Biol. Chem.,* **245**, 891.

Johnsrud, L. (1978), *Proc. Natl. Acad. Sci. USA,* **75**, 5314.

Junowicz, E., and Spencer, J. H. (1973), *Biochim. Biophys. Acta,* **312**, 85.

Keller, W., and Crouch, R. (1972), *Proc. Natl. Acad. Sci. USA,* **69**, 3360.

Khorana, H. G., Agarwall, K. L., Buchi, H., Charuthers, M. H., Gupta, N. K., Kleppe, K., Kumar, A., Ohtsuka, E., RajBhandary, U. L., Van de Sande, J. H., Sgaramella, V., Terao, T., Weber, H., and Yamada, T. (1972), *J. Mol. Biol.,* **72**, 209.

Kilpatrick, M. W., Wei, C. F., Gray, H. B., Jr., and Wells, R. D. (1983), *Nucl. Acids Res.,* **11**, 3811.

Kornberg, A. (1980), in *DNA Replication,* W. H. Freeman, San Francisco.

Kotewicz, M. L., D'Alessio, J. M., Driftmier, K. M., Blodgett, K. P., and Gerard, G. F. (1985), *Gene,* **35**, 249.

Krieg, P., and Melton, D. (1984), *Nucl. Acids Res.,* **12**, 7057.

Kroeker, W., Kowalski, D., and Laskowski, M. (1976), *Biochemistry,* **15**, 4463.

Kroeker, W., and Kowalski, D. (1978), *Biochemistry,* **17**, 3236.

Kunitz, M. (1950), *J. Gen. Physiol.,* **33**, 363.

Laskey, R. A., Mills, A. D., and Morris, N. R. (1977), *Cell,* **10**, 237.

Leis, J. P., Duyk, G., Johnson, S., Longiaru, M., and Skalka, A. (1983), *J. Virol.,* **45**, 727.

Lillehaug, J. R. (1977), *Eur. J. Biochem.,* **73**, 499.

Lillehaug, J. R., Kleppe, R. K., and Kleppe, K. (1976), *Biochemistry,* **15**, 1858.

Lilley, D. (1980), *Proc. Natl. Acad. Sci. USA,* **77**, 6468.

Liu, L. F., and Miller, K. G. (1981), *Proc. Natl. Acad. Sci. USA,* **78**, 3486.

Lockard, R. E., Alzner-Deweerd, B., Heckman, J. E., MacGee, J., Tabor, M. W., and RajBhandary, U. L. (1978), *Nucl. Acids Res.,* **5**, 37.

Marinus, M. G., and Morris, N. R. (1973), *J. Bacteriol.,* **114**, 1143.

Mark, D. F., and Richardson, C. C. (1976), *Proc. Natl. Acad. Sci. USA,* **73**, 780.

May, M. S., and Hattman, S. (1975), *J. Bacteriol.,* **123**, 768.

Martin, S. R., McCoubrey, W., Jr., McConaughy, B. L., Young, L. S., Been, M. D., Brewer, B. J., and Champoux, J. J. (1983), in *Methods in Enzymology,* L. Grossman, K. Moldave, and R. Wu, eds., **100**, 137.

McClelland, M. (1984), *Nucl. Acids Res.,* **12**, r167.

Mechali, M., Recondo, A. M., and Girard, M. (1973), *Biochem. Biophys. Res. Commun.,* **54**, 1306.

Melton, D. A., Krieg, P. A., Rebagliati, M. R., Maniatis, T., Zinn, K., and Green, M. R. (1984), *Nucl. Acids Res.,* **12**, 7035.

Melton, D. A. (1985), *Proc. Natl. Acad. Sci. USA,* **82**, 144.

Meyer, R., Grassberg, J., Scott, J., and Kronberg, A. (1980), *J. Biol. Chem.,* **255**, 2897.

Mikulski, A., and Laskowski, M. (1970), *J. Biol. Chem.,* **245**, 5026.

Modrich, P., and Lehman, I. R. (1970), *J. Biol. Chem.,* **245**, 3626.

Modrich, P., and Richardson, C. C. (1975), *J. Biol. Chem.,* **250**, 5515.

Moelling, K. (1974), *Virology,* **62**, 46.

Murray, N. E., Bruce, S. A., and Murray, K. (1979), *J. Mol. Biol.,* **132**, 493.

Nelson, M., Christ, C., and Schildkraut, I. (1984), *Nucl. Acids Res.,* **12**, 5165.

Panet, A., Van de Sande, J. H., Loewen, P. C., Khorana, H. G., Raae, H. G., Lillehaug, A. J., and Kleppe, K. (1973), *Biochemistry,* **12**, 5045.

Panasenko, S. M., Cameron, J. R., Davis, R. W., and Lehman, I. R. (1977), *Science,* **196**, 188.

Panayotatos, N., and Wells, R. (1981), *Nature,* **289**, 466.

Peck, L. J., and Wang, J. C. (1981), *Nature,* **292**, 375.

Pirotta, V. (1976), *Nucl. Acids Res.,* **3**, 1747.

Prell, B., and Vosberg, H. P. (1980), *Eur. J. Biochem.,* **108**, 389.

Ratliff, R. L. (1981), in *The Enzymes,* P. D. Boyer, ed., **14a**, 105.

Razin, A., and Cedar, H. (1977), *Proc. Natl. Acad. Sci. USA,* **74**, 2725.

Razin, A., and Riggs, A. D. (1980), *Science,* **210**, 604.

Reddy, R., Henning, D., Epstein, P., and Bush, H. (1981), *Nucl. Acids Res.,* **9**, 5645.

Retzel, E. F., Collet, M. S., and Faras, A. J. (1980), *Biochemistry,* **19**, 513.

Richardson, C. C. (1968), *Proc. Natl. Acad. Sci. USA,* **54**, 158.

Richardson, C. C., Lehman, I. R., and Kornberg, A. (1964), *J. Biol. Chem.,* **239**, 251.

Richardson, C. C., and Kornberg, A. (1964), *J. Biol. Chem.,* **239**, 242.

Roberts, T., Kacicj, R., and Ptashne, M. (1979), *Proc. Natl. Acad. Sci. USA,* **76**, 760.

Rogers, S. G., and Weiss, B. (1980), *Gene,* **11**, 187.

Roth, M. J., Tanese, N., and Goff, S. P. (1985), *J. Biol. Chem.,* **260**, 9326.

Roychoudhury, R., Jay, E., and Wu, R. (1976), *Nucl. Acids Res.,* **3**, 863.

Schiff, R. D., and Grandgenett, D. P. (1980), *J. Virol.,* **36**, 889.

Shenk, T., Rhodes, C., Rigby, P., and Berg, P. (1975), *Proc. Natl. Acad. Sci. USA,* **72**, 989.

Shishido, K. (1979), *Agric. Biol. Chem.,* **43**, 1093.

Shishido, K., and Ando, T. (1974), *Biochem. Biophys. Res. Commun.,* **59**, 1380.

Shishido, K., and Ando, T. (1975), *Agric. Biol. Chem.,* **39**, 673.

Shishido, K., and Ando, T. (1975b), *Biochim. Biophys. Acta,* **390**, 125.

Shishido, K., and Ando, T. (1982), in *Nucleases,* S. Linn and R. Roberts, eds., p. 155, Cold Spring Harbor Laboratory, New York.

Shishido, K., and Ikeda, Y. (1970), *J. Biochem.,* **67**, 759.

Shishido, K., and Ikeda, Y. (1971a), *J. Mol. Biol.,* **55**, 287.

Shishido, K., and Ikeda, Y. (1971b), *Biochim. Biophys. Res. Commun.,* **42**, 482.

Silber, R., Malathi, V. G., and Hurwitz, J. (1972), *Proc. Natl. Acad. Sci. USA,* **69**, 3009.

Stahl, S., and Zinn, K. (1981), *J. Mol. Biol.,* **148**, 481.

Sung, S. and Laskowski, M. (1962), *J. Biol. Chem.,* **237**, 506.

Sutter, D., and Doerfler, W. (1980), *Proc. Natl. Acad. Sci. USA,* **27**, 253.

Tabor, S. (1987), Dissertation (Harvard University, Boston).

Tabor, S., and Richardson, C. C. (1987), *Proc. Natl. Acad. Sci. USA,* **84**, 4767.

Tabor, S., Huber, H., and Richardson, C. C. (1987), *J. Biol. Chem.,* in press.

Takahashi, K. (1961), *J. Biochem.,* **49**, 1.

Talmadge, K., Stahl, S., and Gilbert, W. (1980), *Proc. Natl. Acad. Sci. USA,* **77**, 3369.

Tanese, N., Roth, M., and Goff, S. P. (1985), *Proc. Natl. Acad. Sci. USA,* **82**, 4944.

Verma, I. M. (1977), *Biochim. Biophys. Acta,* **473**, 1.

Vincent, C., Cohen-Solal, M. and Kourilsky, P. (1982), *Nucl. Acids Res.,* **10**, 6787.

Vogt, V. (1973), *Eur. J. Biochem.,* **33**, 192.

Vogt, V. (1980), in *Methods in Enzymology,* **65**, 248.

Wang, J. C. (1980), *Proc. Natl. Acad. Sci. USA,* **76**, 200.

Wang, J. C. (1981), in *The Enzymes,* P. D. Boyer, ed., **14**, 331.

Wei, C., Alianell, G., Bencen, G., and Gray, H. (1983), *J. Biol. Chem.,* **258**, 13506.

Weiss, B. (1976), *J. Biol. Chem.,* **251**, 1896.

Weiss, B., Jacquemin-Sablon, A., Live, T. R., Fareed, G. C., and Richardson, C. C. (1968), *J. Biol. Chem.,* **243**, 4543.

Wensink, P., Finnegan, D., Donelson, J., and Hogness, D. (1974), *Cell,* **3**, 315.

Wiegland, R., Godson, G., and Radding, C. (1975), *J. Biol. Chem.,* **250,** 8848.

Wilson, G. G., and Murray, N. E. (1979), *J. Mol. Biol.,* **132**, 471.

Yang, C., Guo, L., and Wu, R. (1983), in *Frontiers in Biochemical and Biophysical Studies of Proteins and Membranes,* T. Lie, S. Sakakibara, A. Schechter, K. Yagi, H. Yajima, and K. Tasunobu, eds., p. 5, Elsevier, New York.

Zinn, K., DiMaio, D., and Maniatis, T. (1983), *Cell,* **34**, 865.

5

RESTRICTION ENDONUCLEASES

The isolation of restriction endonucleases from *E. coli* B and K strains was first reported by Meselson and Yuan (1968) and Linn and Arber (1968). Unfortunately, these enzymes appeared to cut the DNA randomly although they recognized specific sites on the molecules. These endonucleases, which have no recognized value as molecular cloning tools, were described in several systems and have been called Class I enzymes.

The discovery of the first Class II endonuclease (those that both recognize and cleave the DNA molecule at a specific sequence) was reported by Smith and Wilcox (1970) and Kelly and Smith (1970). Since that time, the number of Class II restriction endonucleases available in a purified form has been increasing at a remarkable speed, and it is now very often possible to find one or several restriction enzymes having the specificity one desires. However, it should be pointed out that in most cases the enzyme preparations obtained (either commercially or according to the published protocols) are essentially free of contaminating nucleases but not devoid of other proteins that do not interfere with their enzymatic activity, because the final goal of purification has been the use of the enzymes as a tool in molecular cloning rather than enzyme characterization. Therefore, most of the preparations must be considered as crude extracts not suitable for enzymatic studies such as kinetic analysis or measurement of specific activity. To overcome these problems, an arbitrary unit of activity has been introduced and used to characterize the yield of a purification scheme.

One unit of enzyme is generally defined as the amount of enzyme required to digest completely 1 μg of bacteriophage λ DNA in 1 hour at 37°C. Although this definition appears to be convenient in most cases, it must be considered with care when DNAs with very different structures are digested under similar conditions, or when the number of hits will be such that the kinetic properties of the enzyme are not quite the same. Commercially available endonuclease batches are usually checked for contaminating activities. For example, the routine assay used by Pharmacia to test for the presence of contaminating exonucleases involves incubation of excess restriction endonuclease with bacterial duplex or single strand [^{3}H]DNA (sonicated) for up to 2 hours at 37°C. If less than 1.0% of the radioactivity is rendered acid soluble, the preparation is considered "functionally pure," and the exonuclease "not detectable." To test for nonspecific endonuclease, the enzyme is incubated for an extended period with substrates such as φX174 DNA, RF I. The percentage conversion to RF II (if any) is estimated by agarose gel electrophoresis. By measuring the distribution of radioactivity among the four possible bases remaining at either terminus of the restricted DNA, a precise determination of the specificity of a given restriction endonuclease can be obtained (Figure 33). Thus, any contamination by another endonuclease (including nickase) or by an exonuclease can be determined accurately. Any phosphatase contamination is revealed by the reincubation of the $5' - {}^{32}$P fragments with the restriction endonuclease. In ad-

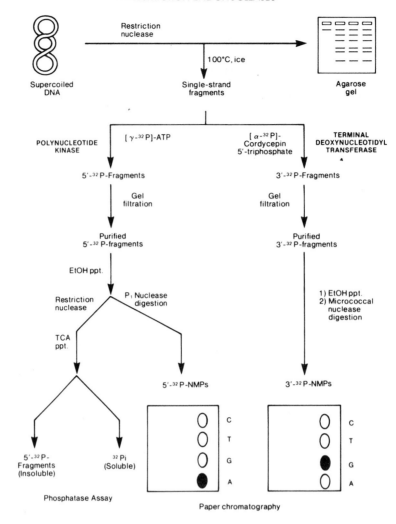

Figure 33. Scheme of a standard quality-control procedure for restriction endonuclease.

dition to the end-labeling data, densitometric scanning of the negative image of the agarose gel shown in the upper right-hand corner of the scheme provides further evidence on the functional purity of the endonuclease tested.

Tables 22 and 23 alphabetically list the restriction endonucleases with the specific nucleotide sequences they recognize and the organisms from which they are purified. Several restriction endonucleases purified from different microorganisms appear to recognize and cleave the same nucleotide sequence. These enzymes have been called isoschizomers. (Table 23 lists the restriction endonucleases with no known isoschizomers along with their recognition sequences.) In each group only one of these enzymes has been chosen as a prototype, often because it was the first to be isolated, or in some cases because its purification scheme leads to a better yield than the one obtained with the others.

At the time when the first edition of this manual was completed, about 200 different restriction endonucleases were available. More than 580 different enzymes have now been described. About 120 different specific sequences appear to be recognized, and among the new interesting specificities are the unique examples of nonpalindromic hexanucleotides recognized by the Gsu I and Bsm I enzymes. Of these, the Bsm I enzyme is the first example of a type II restriction enzyme that cleaves outside the recognition sequence on one DNA strand and within it on the complementary strand.

Also, the Aha III recognition specificity (TTT/AAA) is particularly interesting for the recovery of cloned cDNA prepared from polyadenylated eucaryotic mRNAs (see Chapter 19). A group of eleven restriction endonucleases (called class IIS or Shifter endonucleases) that cleave outside of the recognition sequence (see Table 24) appear to represent a unique

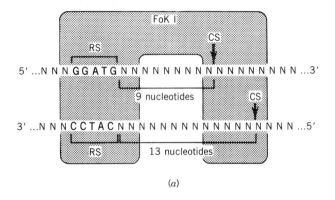

(a)

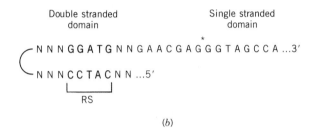

(b)

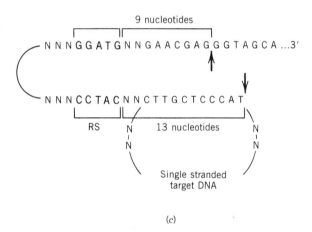

(c)

Figure 34. Class IIs adaptator. (a) Recognition and cutting sites (RS and CS) of a class IIs enzyme (Fok I). (b) Structure of an adaptator containing the specific recognition site for Fok I and the single stranded domain complementary to the sequence to be cut. (c) Hybridization of adaptator and target DNA permits generation of the cuts at predetermined positions.

tool for molecular cloning since they theoretically generate cleavage sites at predetermined locations in any given DNA (Szybalski, 1985). Among them, the enzyme Fok I has been used successfully to cut the RF DNA of M13mp7 phage at predetermined target sites (Podhajska and Szybalski, 1985). The rationale of this approach relies on the use of an adapter that carries a constant double stranded do-main (corresponding to the Fok I recognition site) and a variable single stranded domain complementary to the sequence to be cleaved (see Figure 34). There is no doubt that this method is very powerful and will give rise to many specific and general applications in molecular biology.

Table 25 lists restriction endonucleases with compatible cohesive ends, and Table 26 shows the hybrid sequences which can be constructed by the ligation of fragments created by different restriction endonucleases, as well as the enzymes which can then cut such hybrid sequences.

Different characteristics of commercial restriction endonucleases have been pooled in Table 27. This may help you to determine which enzyme is the most suitable to achieve a given project, and to find appropriate supplies. The conditions to obtain an optimal enzymatic activity during digestion of DNA are described in Chapter 8.

REFERENCES

Kelly, T. J., and Smith, H. O. (1970), *J. Mol. Biol.,* **51**, 393–410.

Linn, S., and Arber, W. (1968), *Proc. Natl. Acad. Sci. USA,* **59**, 1300–1306.

Meselson, M., and Yuan, R. (1968), *Nature,* **217**, 1110–1114.

Podhajska, A. J., and Szybalski, W. (1985), *Gene,* **40**, 175.

Smith, H. O., and Wilcox, K. W. (1970), *J. Mol. Biol.,* **51**, 379–392.

Szybalski, W. (1985), *Gene,* **40**, 169.

Table 22.   Alphabetical Listing of Restriction Endonucleases with Their Related Isoschizomers

Enzyme and Specific Recognition Sequence $(5' \rightarrow 3')^{a,b}$		Origin[c]	Isoschizomers[a,b]
AacI	(GGATCC)	*Acetobacter aceti* ssp. *liquefaciens*[180]	*Acc EB1,* Aae I, *Ali I,* Bam FI, **Bam HI,** Bam KI, Bam NI, *Bst I,* Dds I, Gdo I, Gin I, Gox I, Mle I, Nas BI, *Nsp SAIV,* Rhs I
Aae I	(GGATCC)	*Acetobacter aceti* ssp. *liquefaciens*[180]	*Acc EB1,* Aac I, *Ali I,* Bam FI, **Bam HI,** Bam KI, Bam NI, *Bst I,* Dds I, Gdo I, Gin I, Gox I, Mle I, Nas BI, *Nsp SAIV,* Rhs I
Aat I	(AGGCCT)	*Acetobacter aceti*[203]	Eco 147I, *Gdi I,* Nta SI, **Stu I**
Abr I	(CTCGAG)	*Azospirillum brasilense*[178]	Asp 703, BbI III, *Blu I,* Bss HI, Bst HI, Bsu M, Bth I, *Ccr I,* Ccr II, Dde II, Mca I, Msi I, *PaeR 7, Pan I,* PflW I, Sau 3239, Scu I, Sex I, Sga I, Sgo I, *Sla I,* Slu I, Spa PI, **Xho I,** *Xpa I*
Acc II	(CG ↓ CG)	*Acinetobacter calcoaceticus*[243]	Bce FI, Bce R, Bsu 1192 II, Bsu 1193, Bsu 6633, Bsu EII, **Fnu DII,** Fsp MI, Hin 1056 I, *Tha I*
Acc III	(T ↓ CCGGA)	*Acinetobacter calcoaceticus*[147]	**Bsp MII,** Kpn 2I
Acc EBI	(G ↓ GATCC)	*Acinetobacter calcoaceticus* EBF 65/65[97]	Aac I, Aae I, *Ali I,* Bam FI, **Bam HI,** Bam KI, Bam NI, *Bst I,* Dds I, Gdo I, Gin I, Gox I, Mle I, Nas BI, *Nsp SAIV,* Rhs I
Acy I	(GPu ↓ CGPyC)	*Anabaena cylindrica*[40]	*Aha II, Aos II, Ast WI, Asu III, Bbi II, Hgi DI, Hgi GI, Hgi HII,* Nla SII
Afl I	(G ↓ G^A_TCC)	*Anabaena flos-aquae*[230]	Asp 697, **Ava II,** Bal II, *Bam Nx,* Bme 216, Bti I, *Cau I,* Cla II, Clm II, Eco 47I, Erp I, *Fdi I,* Fsp MS1, *Hgi BI, Hgi CII, Hgi EI, Hgi HIII,* Nsp HII, Sfn I, Sin I
Aha I	(CC^C_GGG)	*Aphanothece halophytica*[229]	Bcn I, **Cau II,** Eco 51II, *Nci I,* Rsh II
Aha II	(GPu ↓ CGPyC)	*Aphanothece halophytica*[229]	**Acy I,** *Aos II, Ast WI, Asu III, Bbi II, Hgi DI, Hgi GI, Hgi HII,* Nla SII
Aha III	(TTT ↓ AAA)	*Aphanothece halophytica*[228]	*Dra I*
Ali I	(G ↓ GATCC)	*Acetobacter liquefaciens*[238]	*Acc EB1,* Aac I, Aae I, *Ali I,* Bam FI, **Bam HI,** Bam KI, Bam NI, *Bst I,* Dds I, Gdo I, Gin I, Gox I, Mle I, Nas BI, *Nsp SAIV,* Rhs I
Ali AJI	(GTGCA ↓ G)	*Acetobacter liquefaciens* AJ 2881[171]	Asp 36I, Asp 708, Bbi I, Bce 170, Bsu 1247, Bsu B, Cau III, *Cfl I,* Eae PI, Ecl 77I, Eco 36I, Eco 48I, Eco 49I, Eco 83I, Mau I, Mkr I, Nas I, Ngb I, Noc I, Pma I, Pmy I, **Pst I,** *Sal PI, Sfl I,* Ska II, Xma II, Xor I, Xph I
Alw I*	(GGATCNNNN ↓)	*Acinetobacter lwoffii*	**Bin I,** Bth II
Alu I	(AG ↓ CT)	*Arthrobacter luteus*[168]	*Mlt I,* Oxa I

Table 22.   Alphabetical Listing of Restriction Endonucleases with Their Related Isoschizomers (*continued*)

Enzyme and Specific Recognition Sequence $(5' \rightarrow 3')^{a,b}$		Origin[c]	Isoschizomers[a,b]
Ama I	(TCGCGA)	*Actinomadura madurae*[81]	**Nru I,** Sbo 13
Ani I	(GCCGGC)	*Arthrobacter nicotianae*[143]	Apr I, Eco 56I, Mis I, **Nae I,** Nba I, Nbr I, Nmu FI, Nmu I, NtaS II, Pgl I, Rlu I, Sao I, Ska I
Aoc I	(CC ↓ TNAGG) A A	*Anabaena* species[38]	*Axy I, Cvn I, Eco 81I, Mst II, Oxa NI,* **Sau I**
Aoc II	(GGGCC ↓ C) T T	*Anabaena* species[38]	*Bsp 1286, Nsp 7524 II,* **Sdu I**
Aor I	(CC ↓ $^{A}_{T}$GG)	*Acetobacter aceti sub. orleanansis*[180]	Apy I*, Atu BI, Atu II, Bin SI, Bst GII, Bst NI, Cdi 27I, Cfr 5I, Cfr 11I, Cfr 20I, Eca II, Ecl II, Ecl 66I, Eco 27I, Eco 38I, Eco 60I, Eco 61I, Eco 67I, **Eco RII,** Mph I, Mva I*, Sgr II, Taq XI*, Zan I*
Aos I	(TGC ↓ GCA)	*Anabaena oscillarioides*[41]	*Fdi II,* Fsp I, **Mst I**
Aos II	(GPu ↓ CGPyC)	*Anabaena oscillarioides*[41]	**Acy I,** *Aha II, Ast WI, Asu III, Bbi II, Hgi DI, Hgi GI, Hgi HII,* Nla SII
Apr I	(GCCGGC)	*Actinosynnema pretiosum*[231]	Ani I, Eco 56I, Mis I, **Nae I,** Nba I, Nbr I, Nmu FI, Nmu I, NtaS II, Pgl I, Rlu I, Sao I, Ska I
Apy I*	(CC$^{A}_{T}$ ↓ GG)	*Arthrobacter pyridinolis*[42]	Aor I, Atu BI, Atu II, Bin SI, Bst GII, Bst NI, Cdi 27I, Cfr 5I, Cfr 11I, Cfr 20I, Eca II, Ecl II, Ecl 66I, Eco 27I, Eco 38I, Eco 60I, Eco 61I, Eco 67I, **Eco RII,** Mph I, Mva I*, Sgr II, Taq XI*, Zan I*
Apu I	(GGNCC)	*Alteromonas putrefaciens*[97]	**Asu I,** *Bac 36I,* Cfr 4I, Cfr 8I, *Cfr 13I,* Cfr 23I, Eco 39I, Eco 47II, Mja II, Nmu EII, Nmu SI, *Nsp 7524 IV,* Psp I, *Sau 96I,* Sdy I
Aqu I	(CPyCGPuG)	*Agmenellum quadruplicatum*[113]	**Ava I,** *Avr I,* Nsp 7524 III
Asp 36I	(CTGCAG)	*Alcaligenes* species RFL 36[83]	Ali AJI, Asp 708, Bbi I, Bce 170, Bsu 1247, Bsu B, Cau III, *Cfl I,* Eae PI, Ecl 77I, Eco 36I, Eco 48I, Eco 49I, Eco 83I, Mau I, Mkr I, Nas I, Ngb I, Noc I, Pma I, Pmy I, **Pst I,** *Sal PI, Sfl I,* Ska II, Xma II, Xor I, Xph I
Asp 697	(GG$^{A}_{T}$CC)	*Achromobacter* species 697[98]	Afl I, **Ava II,** Bal II, *Bam Nx,* Bme 216, Bti I, *Cau I,* Cla II, Clm II, Eco 47I, Erp I, *Fdi I,* Fsp MS1, *Hgi BI, Hgi CII, Hgi EI, Hgi HIII,* Nsp HII, Sfn I, Sin I
Asp 700	(GAANN ↓ NNTTC)	*Achromobacter* species 700[98]	**Xmn I**
Asp 703	(CTCGAG)	*Achromobacter* species 703[98]	Abr I, Bbi III, *Blu I,* Bss HI, Bst HI, Bsu M, Bth I, *Ccr I,* Ccr II, Dde II, Mca I, Msi I, *PaeR 7, Pan I,* PflW I, Sau 3239, Scu I, Sex I, Sga I, Sgo I, *Sla I,* Slu I, Spa PI, **Xho I,** *Xpa I*
Asp 707	(ATCGAT)	*Achromobacter* species 707[98]	Ban III, *Bsc I,* **Cla I**
Asp 708	(CTGCAG)	*Achromobacter* species 708[98]	Ali AJI, Asp 36I, Bbi I, Bce 170, Bsu 1247, Bsu B, Cau III, *Cfl I,* Eae PI, Ecl 77I, Eco 36I, Eco 48I, Eco 49I, Eco 83I, Mau I, Mkr I, Nas I, Ngb I, Noc I, Pma I, Pmy I, **Pst I,** *Sal PI, Sfl I,* Ska II, Xma II, Xor I, Xph I
Asp 718*	(G ↓ GTACC)	*Achromobacter* species 718[10]	Eco 149I, **Kpn I,** Nmi I
Asp 737	(AGTACT)	*Achromobacter* species 737[97]	Asp 763
Asp 763	(AGTACT)	*Achromobacter* species 763[97]	**Asp 737**

Table 22. Alphabetical Listing of Restriction Endonucleases with Their Related Isoschizomers (*continued*)

Enzyme and Specific Recognition Sequence $(5' \rightarrow 3')^{a,b}$		Origin[c]	Isoschizomers[a,b]
Asp 742	(GGCC)	*Achromobacter* species 742[97]	Blu II, Bse I, *Bsp RI*, Bss CI, Bst CI, Bsu 1076, Bsu 1114, *Bsu RI*, Clm I, *Clt I*, Fin S1, *Fnu DI*, **Hae III,** Hhg I, Mni I, Mnn II, Ngo II, Nla I, Pai I, Pal I, Ppu I, *Sfa I*, Sul I, Ttn I, Vha I
Asp 748	(CCGG)	*Achromobacter* species 748[97]	Asp 750, Bsu 1192I, Bsu FI, *Hap II*, **Hpa II,** Mni II, *Mno I, Msp I*, Sfa GUI, Sec II
Asp 750	(CCGG)	*Achromobacter* species 750[97]	Asp 748, Bsu 1192I, Bsu FI, *Hap II*, **Hpa II,** Mni II, *Mno I, Msp I*, Sfa GUI, Sec II
Asp AI	(G ↓ GTNACC)	*Alcaligenes* species[18]	Bst 31I, **Bst EII,** *Bst PI*, Cfr7I, Cfr 19I, *Eca I*, *Nsp SAII*
Ast WI	(GPu ↓ CGPyC)	*Anabaena* strain Waterbury[39]	**Acy I,** *Aha II, Aos II, Asu III, Bbi II, Hgi DI, Hgi GI, Hgi HII*, Nla SII
Asu I	(G ↓ GNCC)	*Anabaena subcylindrica*[80]	Apu I, Bac 36I, Cfr 4I, Cfr 8I, *Cfr 13I*, Cfr 23I, Eco 39I, Eco 47II, Mja II, Nmu EII, Nmu SI, *Nsp 7524 IV*, Psp I, *Sau 96I*, Sdy I
Asu II	(TT ↓ CGAA)	*Anabaena subcylindrica*[147]	*Fsp II, Mla I*, Nsp 7524 V, Nsp BI
Asu III	(GPu ↓ CGPyC)	*Anabaena subcylindrica*[39]	*Aha II, Aos II, Ast WI, Bbi II, Hgi DI, Hgi GI, Hgi HII*, Nla SII
Atu II	(CC$_T^A$GG)	*Agrobacterium tumefaciens* 1D135[117,116]	Aor I, Apy I*, Atu BI, Bin SI, Bst GII, Bst NI, Cdi 27I, Cfr 5I, Cfr 11I, Cfr 20I, Eca II, Ecl II, Ecl 66I, Eco 27I, Eco 38I, Eco 60I, Eco 61I, Eco 67I, **Eco RII,** Mph I, Mva I*, Sgr II, Taq XI*, Zan I*
Atu BI	(CC$_T^A$GG)	*Agrobacterium tumefaciens* B$_6$[117,116]	Aor I, Apy I*, Atu II, Bin SI, Bst GII, Bst NI, Cdi 27I, Cfr 5I, Cfr 11I, Cfr 20I, Eca II, Ecl II, Ecl 66I, Eco 27I, Eco 38I, Eco 60I, Eco 61I, Eco 67I, **Eco RII,** Mph I, Mva I*, Sgr II, Taq XI*, Zan I*
Atu CI	(TGATCA)	*Agrobacterium tumefaciens* C58[179]	**Bcl I,** Bst GI, Cpe I, Fba I, Pov I, Sst IV
Ava I	(C ↓ PyCGPuG)	*Anabaena variabilis*[146]	*Avr I, Aqu I, Nsp 7524 III*
Ava II	(G ↓ G$_T^A$CC)	*Anabaena variabilis*[146]	Afl I, Asp 697, Bal II, Bam Nx, Bme 216, Bti I, Cau I, Cla II, Clm II, Eco 47I, Erp I, Fdi I, Fsp MS1, *Hgi BI, Hgi CII, Hgi EI, Hgi HIII*, Nsp HII, Sfn I
Ava III	(ATGCAT)	*Anabaena variabilis*[146]	Eco T22, Nsi I
Avr I	(C ↓ PyCGPuG)	*Anabaena variabilis*[170]	**Ava I,** Aqu I, Nsp 7524 III
Axy I	(CC ↓ TNAGG)	*Acetobacter xylinus*[241]	*Aoc I, Cvn I, Eco 81I, Mst II, Oxa NI*, **Sau I**
Bac I	(CCGCGG)	*Bacillus acidocaldarius*[132]	Cfr 37I, *Csc I*, Ecc I, Ecl 28I, Ecl 37I, Eco 55I, Eco 92I, *Gal I, Gce GLI, Gce I*, Kae 37I, Mra I, Ngo III, Nla SI, Saa I, Sab I, **Sac II,** Sbo I, Sfr I, Shy I, Sst II, Tgl I
Bac 36I	(G ↓ GNCC)	*Bacillus alcalophilus* 36[97]	Apu I, **Asu I,** Cfr 4I, Cfr 8I, *Cfr 13I*, Cfr 23I, Eco 39I, Eco 47II, Mja II, Nmu EII, Nmu SI, *Nsp 7524 IV*, Psp I, *Sau 96I*, Sdy I
Bal II	(GG$_T^A$CC)	*Brevibacterium alvidum*[235]	Afl I, Asp 697, **Ava II,** *Bam Nx*, Bme 216, Bti I, *Cau I*, Cla II, Clm II, Eco 47I, Erp I, *Fdi I*, Fsp MS1, *Hgi BI, Hgi CII, Hgi EI, Hgi HIII*, Nsp HII, Sfn I, Sin I
Bam FI	(GGATCC)	*Bacillus amyloliquefaciens* F[186]	*Acc EB1*, Aac I, Aae I, *Ali I*, **Bam HI,** Bam KI, Bam NI, *Bst I*, Dds I, Gdo I, Gin I, Gox I, Mle I, Nas BI, *Nsp SAIV*, Rhs I
Bam HI	(G ↓ GATCC)	*Bacillus amyloliquefaciens* H[232]	*Acc EB1*, Aac I, Aae I, *Ali I*, Bam FI, Bam KI, Bam NI, *Bst I*, Dds I, Gdo I, Gin I, Gox I, Mle I, Nas BI, *Nsp SAIV*, Rhs I
Bam KI	(GGATCC)	*Bacillus amyloliquefaciens* K[186]	*Acc EB1*, Aac I, Aae I, *Ali I*, Bam FI, **Bam HI,** Bam NI, *Bst I*, Dds I, Gdo I, Gin I, Gox I, Mle I, Nas BI, *Nsp SAIV*, Rhs I

Table 22. Alphabetical Listing of Restriction Endonucleases with Their Related Isoschizomers (*continued*)

Enzyme and Specific Recognition Sequence $(5' \rightarrow 3')^{a,b}$		Origin[c]	Isoschizomers[a,b]
Bam NI	(GGATCC)	*Bacillus amyloliquefaciens* N[184]	*Acc EB1*, Aac I, Aae I, *Ali I*, Bam FI, **Bam HI,** Bam KI, *Bst I*, Dds I, Gdo I, Gin I, Gox I, Mle I, Nas BI, *Nsp SAIV,* Rhs I
Bam NX	$(G \downarrow G{}^{A}_{T}CC)$	*Bacillus amyloliquefaciens* N[185]	Afl I, Asp 697, **Ava II,** Bal II, Bme 216, Bti I, *Cau I,* Cla II, Clm II, Eco 47I, Erp I, *Fdi I,* Fsp MS1, *Hgi BI, Hgi CII, Hgi EI, Hgi HIII,* Nsp HII, Sfn I, Sin I
*Ban I**	(G ↓ GPyPuCC)	*Bacillus aneurinolyticus*[203]	Eco 50 I, **Hgi CI,** Hgi HI*
Ban II	(GPuGCPy ↓ C)	*Bacillus aneurinolyticus*[203]	*Bvu I,* Eco 24I, Eco 25I, Eco 26I, Eco 35I, Eco 40I, Eco 41I, **Hgi JII**
Ban III	(ATCGAT)	*Bacillus aneurinolyticus*[203]	Asp 707, *Bsc I,* **Cla I**
Bav I	(CAG ↓ CTG)	*Bacillus alvei*[134]	**Pvu II,** Cfr 6I
Bbe I*	(GGCGC ↓ C)	*Bifidobacterium breve* YIT 4006[103]	Bbe AI, Bin SII, Eco 78I, Nam I, **Nar I,** *Nda I, Nun II,* Psp 61I, Sfo I
Bbe AI	(GGCGCC)	*Bifidobacterium breve* S 50[100]	Bbe I*, Bin SII, Eco 78I, Nam I, **Nar I,** *Nda I,* Nun II, Psp 61I, Sfo I
Bbi I	(CTGCAG)	*Bifidobacterium bifidum* YIT 4007[101]	Ali AJI, Asp 36I, Asp 708, Bce 170, Bsu 1247, Bsu B, Cau III, Cfl I, Eae PI, Ecl 77I, Eco 36I, Eco 48I, Eco 49I, Eco 83I, Mau I, Mkr I, Nas I, Ngb I, Noc I, Pma I, Pmy I, **Pst I,** *Sal PI, Sfl I,* Ska II, Xma II, Xor I, Xph I
Bbi II	(GPu ↓ CGPyC)	*Bifidobacterium bifidum* YIT 4007[101]	**Acy I,** *Aha II, Aos II, Ast WI, Asu III, Hgi DI, Hgi GI, Hgi HII,* Nla SII
Bbi III	(CTCGAG)	*Bifidobacterium bifidum* YIT 4007[101]	AbrI, Asp 703, *Blu I,* Bss HI, Bst HI, Bsu M, Bth I, *Ccr I,* Ccr II, Dde II, Mca I, Msi I, *PaeR 7, Pan I,* PflW I, Sau 3239, Scu I, Sex I, Sga I, Sgo I, *Sla I,* Slu I, Spa PI, **Xho I,** *Xpa I*
Bbr I	(AAGCTT)	*Bordetella bronchiseptica*[147]	Bpe I, Cfr 32I, Chu I, Eco VIII, Eco 65I, Hin 173, Hin JCII, Hinb III, Hinb III 1076, **Hind III,** Hinf II, Hsu I, Mki I
Bce 170	(CTGCAG)	*Bacillus cereus*[186]	Ali AJI, Asp 36I, Asp 708, Bbi I, Bsu 1247, Bsu B, Cau III, *Cfl I,* Eae PI, Ecl 77I, Eco 36I, Eco 48I, Eco 49I, Eco 83I, Mau I, Mkr I, Nas I, Ngb I, Noc I, Pma I, Pmy I, **Pst I,** *Sal PI, Sfl I,* Ska II, Xma II, Xor I, Xph I
Bce 243	(↓ GATC)	*Bacillus cereus*[37]	Bsa PI, Bss GII, Bst EIII, Bst XII, Cpa I, Dpn II, Fnu AII, *Fnu CI, Hac I,* **Mbo I,** Meu I, Mno III, Mos I, Msp 67II, Mth I, Nde II, Nfl AII, Nfl I, Nla II, Nsi AI, Nsp AI, Nsu I, *Sau 3AI,* Sin MI
Bce FI	(CGCG)	*Bacillus cereus*[157]	*Acc II,* Bce R, Bsu 1192 II, Bsu 1193, Bsu 6633, Bsu EII, **Fnu DII,** Fsp MI, Hin 1056 I, *Tha I*
Bce R	(CGCG)	*Bacillus cereus* RF *sm st*[186]	*Acc II,* Bce FI, Bsu 1192 II, Bsu 1193, Bsu 6633, Bsu EII, **Fnu DII,** Fsp MI, Hin 1056 I, *Tha I*
Bcl I	(T ↓ GATCA)	*Bacillus caldolyticus*[7]	Atu CI, Bst GI, Cpe I, Fba I, Pov I, Sst IV
Bcn I	$(CC{}^{C}_{G} \downarrow GG)$	*Bacillus centrosporus*[89]	Aha I, **Cau II,** Eco 51II, *Nci I,* Rsh II
Bgl II	(A ↓ GATCT)	*Bacillus globigii*[43]	*Nsp MAC I*
Bin I	(GGATC)	*Bifidobacterium infantis* 659[102]	Alw I*, Bth II
Bin SI	$(CC{}^{A}_{T}GG)$	*Bifidobacterium infantis* S76e[100]	Aor I, Apy I*, Atu BI, Atu II, Bst GII, Bst NI, Cdi 27I, Cfr 5I, Cfr 11I, Cfr 20I, Eca II, Ecl II, Ecl 66I, Eco 27I, Eco 38I, Eco 60I, Eco 61I, Eco 67I, **Eco RII,** Mph I, Mva I*, Sgr II, Taq XI*, Zan I*
Bin SII	(GGCGCC)	*Bifidobacterium infantis* S76e[100]	Bbe AI, Bbe I*, Eco 78I, Nam I, **Nar I,** *Nda I, Nun II,* Psp 61I, Sfo I

Table 22. Alphabetical Listing of Restriction Endonucleases with Their Related Isoschizomers (*continued*)

Enzyme and Specific Recognition Sequence $(5' \rightarrow 3')^{a,b}$		Origin[c]	Isoschizomers[a,b]
Blu I	(C ↓ TCGAG)	*Brevibacterium luteum*[59]	AbrI, Asp 703, Bbi III, Bss HI, Bst HI, Bsu M, Bth I, *Ccr I,* Ccr II, Dde II, Mca I, Msi I, *PaeR 7, Pan I,* PflW I, Sau 3239, Scu I, Sex I, Sga I, Sgo I, *Sla I,* Slu I, Spa PI, **Xho I,** *Xpa I*
Blu II	(GGCC)	*Brevibacterium luteum*[216]	Asp 742, Bse I, *Bsp RI,* Bss CI, Bst CI, Bsu 1076, Bsu 1114, *Bsu RI,* Clm I, *Clt I,* Fin SI, *Fnu DI,* **Hae III,** Hhg I, Mni I, Mnn II, Ngo II, Nla I, Pai I, Pal I, Ppu I, *Sfa I,* Spl III, Sul I, Ttn I, Vha I
Bme 216	(GG$\frac{A}{T}$CC)	*Bacillus megaterium* 216[106]	Afl I, Asp 697, **Ava II,** Bal II, *Bam Nx,* Bti I, *Cau I,* Cla II, Clm II, Eco 47I, Erp I, *Fdi I,* Fsp MS1, *Hgi BI, Hgi CII, Hgi EI, Hgi HIII,* Nsp HII, Sfn I, Sin I
Bpe I	(AAGCTT)	*Bordetella pertussis*[66]	Bbr I, Cfr 32I, Chu I, *Eco VIII,* Eco 65I, Hin 173, Hin JCII, Hinb III, Hinb III 1076, **Hind III,** Hinf II, *Hsu I,* Mki I
Bsa PI	(GATC)	*Bacillus stearothermophilus* P8[112]	*Bce 243,* Bss GII, Bst EIII, Bst XII, Cpa I, Dpn II, Fnu AII, *Fnu CI, Hac I,* **Mbo I,** Meu I, Mno III, Mos I, Msp 67II, Mth I, Nde II, Nfl AII, Nfl I, Nla II, Nsi AI, Nsp AI, Nsu I, *Sau 3AI,* Sin MI
Bsc I	(AT ↓ CGAT)	*Bacillus* species[97]	Asp 707, Ban III, **Cla I**
Bse I	(GGCC)	*Bacillus stearothermophilus*[189]	Asp 742, Blu II, *Bsp RI,* Bss CI, Bst CI, Bsu 1076, Bsu 1114, *Bsu RI,* Clm I, *Clt I,* Fin SI, *Fnu DI,* **Hae III,** Hhg I, Mni I, Mnn II, Ngo II, Nla I, Pai I, Pal I, Ppu I, *Sfa I,* Spl III, Sul I, Ttn I, Vha I
Bse II	(GTTAAC)	*Bacillus stearothermophilus*[189]	Fin II, **Hpa I**
Bse PI	(GCGCGC)	*Bacillus stearothermophilus* P6[112]	Bso PI, Bsr HI, Bss HII*
Bso PI	(GCGCGC)	*Bacillus stearothermophilus* P9[112]	**Bse PI,** Bsr HI, Bss HII*
Bsp 1286	(GGGCC ↓ C) with A A above and T T below	*Bacillus sphaericus*[186]	Aoc II, *Nsp 7524 II,* **Sdu I**
Bsp MII	(T ↓ CCGGA)	*Bacillus* species M[112]	Acc III, Kpn 2I
Bsp RI	(GG ↓ CC)	*Bacillus sphaericus*[104]	Asp 742, Blu II, Bse I, Bss CI, Bst CI, Bsu 1076, Bsu 1114, *Bsu RI,* Clm I, *Clt I,* Fin SI, *Fnu DI,* **Hae III,** Hhg I, Mni I, Mnn II, Ngo II, Nla I, Pai I, Pal I, Ppu I, *Sfa I,* Spl III, Sul I, Ttn I, Vha I
Bsr HI	(GCGCGC)	*Bacillus stearothermophilus* H4[112]	Bse PI, Bso PI, Bss HII*
Bsr PII	(GATC)	*Bacillus stearothermophilus* P5[112]	*Bce 243,* Bsa PI, Bst EIII, Bst XII, Cpa I, Dpn II, Fnu AII, *Fnu CI, Hac I,* **Mbo I,** Meu I, Mno III, Mos I, Msp 67II, Mth I, Nde II, Nfl AII, Nfl I, Nla II, Nsi AI, Nsp AI, Nsu I, *Sau 3AI,* Sin MI
Bss CI	(GGCC)	*Bacillus stearothermophilus* C11[111]	Asp 742, Blu II, Bse I, *Bsp RI,* Bst CI, Bsu 1076, Bsu 1114, *Bsu RI,* Clm I, *Clt I,* Fin SI, *Fnu DI,* **Hae III,** Hhg I, Mni I, Mnn II, Mnn II, Ngo II, Nla I, Pai I, Pal I, Ppu I, *Sfa I,* Spl III, Sul I, Ttn I, Vha I
Bss GI	(CCANNNNNTGG)	*Bacillus stearothermophilus* G6[112]	Bst TI, **Bst XI**

Table 22. Alphabetical Listing of Restriction Endonucleases with Their Related Isoschizomers (*continued*)

Enzyme and Specific Recognition Sequence $(5' \rightarrow 3')^{a,b}$		Origin[c]	Isoschizomers[a,b]
Bss GII	(GATC)	*Bacillus stearothermophilus* G6[112]	*Bce 243*, Bsa PI, Bst EIII, Bst XII, Cpa I, Dpn II, Fnu AII, *Fnu CI, Hac I*, **Mbo I**, Meu I, Mno III, Mos I, Msp 67II, Mth I, Nde II, Nfl AII, Nfl I, Nla II, Nsi AI, Nsp AI, Nsu I, *Sau 3AI*, Sin MI
Bss HI	(CTCGAG)	*Bacillus stearothermophilus* H3[112]	AbrI, Asp 703, Bbi III, *Blu I*, Bst HI, Bsu M, Bth I, *Ccr I*, Ccr II, Dde II, Mca I, Msi I, *PaeR 7, Pan I*, PflW I, Sau 3239, Scu I, Sex I, Sga I, Sgo I, *Sla I*, Slu I, Spa PI, **Xho I**, *Xpa I*
Bss HII*	(G $\downarrow$ CGCGC)	*Bacillus stearothermophilus* H3[112]	**Bse PI**, Bso PI, Bsr HI
Bst I	(G $\downarrow$ GATCC)	*Bacillus stearothermophilus* 1503-4R[24]	*Acc EB1*, Aac I, Aae I, *Ali I*, Bam FI, **Bam HI,** Bam KI, Bam NI, Dds I, Gdo I, Gin I, Gox I, Mle I, Nas BI, *Nsp SAIV*, Rhs I
Bst CI	(GGCC)	*Bacillus stearothermophilus* C1[112]	Asp 742, Blu II, Bse I, *Bsp RI*, Bss CI, Bsu 1076, Bsu 1114, *Bsu RI*, Clm I, *Ctl I, Fnu DI*, **Hae III**, Hhg I, Mni I, Mnn II, Ngo II, Nla I, Pai I, Pal I, Ppu I, *Sfa I*, Spl III, Sul I, Ttn I, Vha I
Bst EII	(G $\downarrow$ GTNACC)	*Bacillus stearothermophilus* ET[135]	Asp A1, Bst 31I, *Bst PI*, Cfr7I, Cfr 19I, *Eca I, Nsp SAII*
Bst EIII	(GATC)	*Bacillus stearothermophilus* ET[135]	*Bce 243*, Bsa PI, Bss GII, Bst XII, Cpa I, Dpn II, Fnu AII, *Fnu CI, Hac I*, **Mbo I**, Meu I, Mno III, Mos I, Msp 67II, Mth I, Nde II, Nfl AII, Nfl I, Nla II, Nsi AI, Nsp AI, Nsu I, *Sau 3AI*, Sin MI
Bst GI	(TGATCA)	*Bacillus stearothermophilus* G3[112]	Atu CI, **Bcl I**, Cpe I, Fba I, Pov I, Sst IV
Bst GII	(CC^A_TGG)	*Bacillus stearothermophilus* G3[112]	Aor I, Apy I*, Atu BI, Atu II, Bin SI, Bst NI, Cdi 27I, Cfr 5I, Cfr 11I, Cfr 20I, Eca II, Ecl II, Ecl 66I, Eco 27I, Eco 38I, Eco 60I, Eco 61I, Eco 67I, **Eco RII**, Mph I, Mva I*, Sgr II, Taq XI, Zan I*
Bst HI	(CTCGAG)	*Bacillus stearothermophilus* H1[112]	Abr I, Asp 703, Bbi III, *Blu I*, Bsu M, Bth I, *Ccr I*, Ccr II, Dde II, Mca I, Msi I, *PaeR 7, Pan I*, PflW I, Sau 3239, Scu I, Sex I, Sga I, Sgo I, *Sla I*, Slu I, Spa PI, **Xho I**, *Xpa I*
Bst NI*	(CC^A_TGG)	*Bacillus stearothermophilus*[177]	Aor I, Apy I*, Atu BI, Atu II, Bin SI, Bst GII, Cdi 271, Cfr 5I, Cfr 11I, Cfr 20I, Eca II, Ecl II, Ecl 66I, Eco 271, Eco 38I, Eco 60I, Eco 61I, Eco 67I, **Eco RII**, Mph I, Mva I*, Sgr II, Taq XI*, Zan I*
Bst PI	(G $\downarrow$ GTNACC)	*Bacillus stearothermophilus*[161]	Asp A1, Bst 31I, **Bst EII**, Cfr 71, Cfr 19I, *Eca I, Nsp SAII*
Bst TI	(CCANNNNNTGG)	*Bacillus stearothermophilus* T12[112]	Bss GI, **Bst XI**
Bst XI	(CCANNNN $\downarrow$ NTGG)	*Bacillus stearothermophilus* X1[112]	Bss GI, Bst TI
Bst XII	(GATC)	*Bacillus stearothermophilus* X1[112]	*Bce 243*, Bsa PI, Bss GII, Bst EIII, Cpa I, Dpn II, Fnu AII, *Fnu CI, Hac I*, **Mbo I**, Meu I, Mno III, Mos I, Msp 67II, Mth I, Nde II, Nfl AII, Nfl I, Nla II, Nsi AI, Nsp AI, Nsu I, *Sau 3AI*, Sin MI
Bst 31I	(GGTNACC)	*Bacillus stearothermophilus*[97]	Asp A1, **Bst EII**, *Bst PI*, Cfr7I, Cfr 19I, *Eca I, Nsp SAII*

Table 22. Alphabetical Listing of Restriction Endonucleases with Their Related Isoschizomers (*continued*)

Enzyme and Specific Recognition Sequence (5′ → 3′)[a,b]		Origin[c]	Isoschizomers[a,b]
Bsu B	(CTGCAG)	*Bacillus subtilis*[77]	Ali AJI, Asp 36I, Asp 708, Bbi I, Bce 170, Bsu 1247, Cau III, *Cfl I,* Eae PI, Ecl 77I, Eco 36I, Eco 48I, Eco 49I, Eco 83I, Mau I, Mkr I, Nas I, Ngb I, Noc I, Pma I, Pmy I, **Pst I,** *Sal PI, Sfl I,* Ska II, Xma II, Xor I, Xph I
Bsu EII	(CGCG)	*Bacillus subtilis*[164]	*Acc II,* Bce FI, Bce R, Bsu 1192 II, Bsu 1193, Bsu 6633, **Fnu DII,** Fsp MI, Hin 1056 I, *Tha I*
Bsu FI	(CCGG)	*Bacillus subtilis*[186]	Asp 748, Asp 750, Bsu 1192I, *Hap II,* **Hpa II,** Mni II, *Mno I, Msp I,* Sfa GUI, Sec II
Bsu M	(CTCGAG)	*Bacillus subtilis* Marburg 168[186]	Abr I, Asp 703, Bbi III, *Blu I,* Bss HI, Bst HI, Bth I, *Ccr I,* Ccr II, Dde II, Mca I, Msi I, *PaeR 7, Pan I,* PflW I, Sau 3239, Scu I, Sex I, Sga I, Sgo I, *Sla I,* Slu I, Spa PI, **Xho I,** *Xpa I*
Bsu RI	(GG ↓ CC)	*Bacillus subtilis* strain X5[13]	Asp 742, Blu II, Bse I, *Bsp RI,* Bss CI, Bst CI, Bsu 1076, Bsu 1114, Clm I, *Clt I,* Fin SI, *Fnu DI,* **Hae III,** Hhg I, Mni I, Mnn II, Ngo II, Nla I, Pai I, Pal I, Ppu I, *Sfa I,* Spl III, Sul I, Ttn I, Vha I
Bsu 1076	(GGCC)	*Bacillus subtilis*[186]	Asp 742, Blu II, Bse I, *Bsp RI,* Bss CI, Bst CI, Bsu 1114, *Bsu RI,* Clm I, *Clt I,* Fin SI, *Fnu DI,* **Hae III,** Hhg I, Mni I, Mnn II, Ngo II, Nla I, Pai I, Pal I, Ppu I, *Sfa I,* Spl III, Sul I, Ttn I, Vha I
Bsu 1114	(GGCC)	*Bacillus subtilis*[186]	Asp 742, Blu II, Bse I, *Bsp RI,* Bss CI, Bst CI, Bsu 1076, *Bsu RI,* Clm I, *Clt I,* Fin SI, *Fnu DI,* **Hae III,** Hhg I, Mni I, Mnn II, Ngo II, Nla I, Pai I, Pal I, Ppu I, *Sfa I,* Spl III, Sul I, Ttn I, Vha I
Bsu 1192I	(CCGG)	*Bacillus subtilis*[186]	Asp 748, Asp 750, Bsu FI, *Hap II,* **Hpa II,** Mni II, *Mno I, Msp I,* Sfa GUI, Sec II
Bsu 1192II	(CGCG)	*Bacillus subtilis*[164]	Acc II, Bce FI, Bce R, Bsu 1193, Bsu 6633, Bsu EII, **Fnu DII,** Fsp MI, Hin 1056 I, *Tha I*
Bsu 1193	(CGCG)	*Bacillus subtilis*[186]	*Acc II,* Bce FI, Bce R, Bsu 1192 II, Bsu 6633, Bsu EII, **Fnu DII,** Fsp MI, Hin 1056 I, *Tha I*
Bsu 1247	(CTGCAG)	*Bacillus subtilis*[186]	Ali AJI, Asp 36I, Asp 708, Bbi I, Bce 170, Bsu 1247, Bsu B, Cau III, *Cfl I,* Eae PI, Ecl 77I, Eco 36I, Eco 48I, Eco 49I, Eco 83I, Mau I, Mkr I, Nas I, Ngb I, Noc I, Pma I, Pmy I, **Pst I,** *Sal PI, Sfl I,* Ska II, Xma II, Xor I, Xph I
Bsu 6633	(CGCG)	*Bacillus subtilis*[186]	*Acc II,* Bce FI, Bce R, Bsu 1192 II, Bsu 1193, Bsu EII, **Fnu DII,** Fsp MI, Hin 1056 I, *Tha I*
Bth I	(CTCGAG)	*Bifidobacterium thermophilum* RU326[100]	AbrI, Asp 703, Bbi III, *Blu I,* Bss HI, Bst HI, Bsu M, *Ccr I,* Ccr II, Dde II, Mca I, Msi I, *PaeR 7, Pan I,* PflW I, Sau 3239, Scu I, Sex I, Sga I, Sgo I, *Sla I,* Slu I, Spa PI, **Xho I,** *Spa I*
Bth II	(GGATC)	*Bifidobacterium thermophilum* RU326[100]	Alw I*, **Bin I**
Bti I	(G ↓ G$_{\mathrm{T}}^{\mathrm{A}}$CC)	*Bacillus thuringiensis*[4]	Afl I, Asp 697, **Ava II,** Bal II, Bam Nx, Bme 216, *Cau I,* Cla II, Clm II, Eco 47I, Erp I, *Fdi I,* Fsp MS1, *Hgi BI, Hgi CII, Hgi EI, Hgi HIII,* Nsp HII, Sfn I, Sin I
Bvu I	(GPuGCPy ↓ C)	*Bacillus vulgatis*[5]	*Ban II,* Eco 24I, Eco 25I, Eco 26I, Eco 35I, Eco 40I, Eco 41I, **Hgi JII**
Cau I	(GG$_{\mathrm{T}}^{\mathrm{A}}$CC)	*Chloroflexis aurantiacus*[9]	Afl I, Asp 697, **Ava II,** Bal II, Bam Nx, Bme 216, Bti I, Cla II, Clm II, Eco 47I, Erp I, Fdi I, Fsp MS1, *Hgi BI, Hgi CII, Hgi EI, Hgi HIII,* Nsp HII, Sfn I, Sin I

Table 22. Alphabetical Listing of Restriction Endonucleases with Their Related Isoschizomers (*continued*)

Enzyme and Specific Recognition Sequence $(5' \rightarrow 3')^{a,b}$		Origin[c]	Isoschizomers[a,b]
Cau II	$(CC \downarrow {}^G_C GG)$	*Chloroflexis aurantiacus*[9]	Aha I, Bcn I, Eco 57II, *Nci I*, Rsh II
Cau III	(CTGCAG)	*Chloroflexus aurantiacus*[6]	Ali AJI, Asp 36I, Asp 708, Bbi I, Bce 170, Bsu 1247, Bsu B, *Cfl I*, Eae PI, Ecl 77I, Eco 36I, Eco 48I, Eco 49I, Eco 83I, Mau I, Mkr I, Nas I, Ngb I, Noc I, Pma I, Pmy I, **Pst I,** *Sal PI, Sfl I*, Ska II, Xma II, Xor I, Xph I
Ccr I	$(C \downarrow TCGAG)$	*Caulobacter crescentus* CB-13[208]	AbrI, Asp 703, Bbi III, *Blu I*, Bss HI, Bst HI, Bsu M, Bth I, Ccr II, Dde II, Mca I, Msi I, *PaeR 7, Pan I,* PflW I, Sau 3239, Scu I, Sex I, Sga I, Sgo I, *Sla I*, Slu I, Spa PI, **Xho I,** *Xpa I*
Ccr II	(CTCGAG)	*Caulobacter crescentus* CB-13[208]	AbrI, Asp 703, Bbi III, *Blu I*, Bss HI, Bst HI, Bsu M, Bth I, Ccr I, Dde II, Mca I, Msi I, *PaeR 7, Pan I,* PflW I, Sau 3239, Scu I, Sex I, Sga I, Sgo I, *Sla I*, Slu I, Spa PI, **Xho I,** *Xpa I*
Cdi 27I	$(CC^A_T GG)$	*Citrobacter diversus* RFL27[83]	Aor I, Apy I*, Atu BI, Atu II, Bin SI, Bst GII, Bst NI, Cfr 5I, Cfr 11I, Cfr 20I, Eca II, Ecl II, Ecl 66I, Eco 27I, Eco 38I, Eco 60I, Eco 61I, Eco 67I, **Eco RII**, Mph I, Mva I*, Sgr II, Taq XI*, Zan I*
Cfl I	$(CTGCA \downarrow G)$	*Cellulomonas flavigena*[74]	Ali AJI, Asp 36I, Asp 708, Bbi I, Bce 170, Bsu 1247, Bsu B, Cau III, Eae PI, Ecl 77I, Eco 36I, Eco 48I, Eco 49I, Eco 83I, Mau I, Mkr I, Nas I, Ngb I, Noc I, Pma I, Pmy I, **Pst I,** *Sal PI, Sfl I*, Ska II, Xma II, Xor I, Xph I
Cfo I	(GCGC)	*Clostridium formicoaeceticum*[32]	*Fnu DIII*, **Hha I,** Hin GUI, Hin PI*, Hin S1, Hin S2, Mnn IV, Sci NI*
Cfr I	$(Py \downarrow GGCCPu)$	*Citrobacter freundii*[93]	*Cfr 14I*, Eae I
Cfr 4I	(GGNCC)	*Citrobacter freundii* RFL 4[92]	Apu I, **Asu I,** *Bac 36I*, Cfr 8I, *Cfr 13I*, Cfr 23I, Eco 39I, Eco 47II, Mja II, Nmu EII, Nmu SI, Mnu SI, *Nsp 7524 IV,* Psp I, *Sau 96I*, Sdy I
Cfr 5I	$(CC^A_T GG)$	*Citrobacter freundii* RFL 5[92]	Aor I, Apy I*, Atu BI, Atu II, Bin SI, Bst GII, Bst NI, Cdi 27I, Cfr 11I, Cfr 20I, Eca II, Ecl II, Ecl 66I, Eco 27I, Eco 38I, Eco 60I, Eco 61I, Eco 67I, **Eco RII**, Mph I, Mva I*, Sgr II, Taq XI*, Zan I*
Cfr 6I	(CAGCTG)	*Citrobacter freundii* RFL 6[92]	**Pvu II,** *Bav I*
Cfr 7I	(GGTNACC)	*Citrobacter freundii* RFL 7[92]	Asp A1, Bst 31I, **Bst EII**, *Bst PI*, Cfr 19I, *Eca I*, *Nsp SAII*
Cfr 8I	(GGNCC)	*Citrobacter freundii* RFL8[92]	Apu I, **Asu I,** *Bac 36I*, Cfr 4I, *Cfr 13I*, Cfr 23I, Eco 39I, Eco 47II, Mja II, Nmu EII, Nmu SI, *Nsp 7524 IV,* Psp I, *Sau 96I*, Sdy I
Cfr 9I*	$(C \downarrow CCGGG)$	*Citrobacter freundii* RFL 9[92]	**Sma I,** Xcy I*, Xma I*
Cfr 11I	$(CC^A_T GG)$	*Citrobacter freundii* RFL 11[90]	Aor I, Apy I*, Atu BI, Atu II, Bin SI, Bst GII, Bst NI, Cdi 27I, Cfr 5I, Cfr 20I, Eca II, Ecl II, Ecl 66I, Eco 27I, Eco 38I, Eco 60I, Eco 61I, Eco 67I, **Eco RII**, Mph I, Mva I*, Sgr II, Taq XI*, Zan I*
Cfr 13I	$(G \downarrow GNCC)$	*Citrobacter freundii* RFL 13[90]	Apu I, **Asu I,** *Bac 36I*, Cfr 4I, Cfr 8I, Cfr 23I, Eco 39I, Eco 47II, Mja II, Nmu EII, Nmu SI, *Nsp 7524 IV,* Psp I, *Sau 96I*, Sdy I
Cfr 14I	$(Py \downarrow GGCPu)$	*Citrobacter freundii* RFL 14[90]	**Cfr I,** *Eae I*
Cfr 19I	(GGTNACC)	*Citrobacter freundii* RfL 19[97]	Asp A1, Bst 31I, **Bst EII**, *Bst PI*, Cfr 7I, *Eca I*, *Nsp SAII*

Table 22. Alphabetical Listing of Restriction Endonucleases with Their Related Isoschizomers (continued)

Enzyme and Specific Recognition Sequence $(5' \rightarrow 3')^{a,b}$		Origin[c]	Isoschizomers[a,b]
Cfr 20I	$(CC^A_T GG)$	*Citrobacter freundii* RFL 20[97]	Aor I, Apy I*, Atu BI, Atu II, Bin SI, Bst GII, Bst NI, Cdi 27I, Cfr 5I, Cfr 11I, Eca II, Ecl II, Ecl 66I, Eco 27I, Eco 38I, Eco 60I, Eco 61I, Eco 67I, **Eco RII**, Mph I, Mva I*, Sgr II, Taq XI, Zan I*
Cfr 23I	$(GGNCC)$	*Citrobacter freundii* RFL 23[97]	Apu I, **Asu I**, *Bac 36I*, Cfr 4I, Cfr 8I, *Cfr 13I*, Eco 39I, Eco 47II, Mja II, Nmu EII, Nmu SI, *Nsp 7524 IV*, Psp I, *Sau 96I*, Sdy I
Cfr 32I	$(AAGCTT)$	*Citrobacter freundii* RFL 32[97]	Bbr I, Bpe I, Chu I, *Eco VIII*, Eco 65I, Hin 173, Hin JCII, Hinb III, Hinb III 1076, **Hind III**, Hinf II, *Hsu I*, Mki I
Cfr 37I	$(CCGCGG)$	*Citrobacter freundii* RFL 37[97]	Bac I, *Csc I*, Ecc I, Ecl 28I, Ecl 37I, Eco 55I, Eco 92I, *Gal I, Gce GLI, Gce I*, Kae 37I, Mra I, Ngo III, Nla SI, Saa I, Sab I, **Sac II**, Sbo I, Sfr I, Shy I, Sst II, Tgl I
Cfr 38I	$(^T_C GGCC^A_G)$	*Citrobacter freundii* RFL 38[97]	Eco 90I
Cfr 48I	$(G^A_G GC^T_C C)$	*Citrobacter freundii* RFL 48[97]	Eco 68I
Cfu I	$(GATC)$	*Caulobacter fusiformis*[85]	**Dpn I,** Nan II, Nmu DI, Nmu EI, Nsu DI, Sau 6782
Chu I	$(AAGCTT)$	*Corynebacterium humiferum*[49]	Bbr I, Bpe I, Cfr 32I, *Eco VIII*, Eco 65I, Hin 173, Hin JCII, Hinb III, Hinb III 1076, **Hind III**, Hinf II, *Hsu I*, Mki I
Chu II	$(GTPyPuAC)$	*Corynebacterium humiferum*[49]	Hin JCI*, Hinc II, Hinc II 1160, Hinc II 1161, **Hind II,** Mnn I
Cla I	$(AT \downarrow CGAT)$	*Caryophanon latum*[146]	Asp 707, Ban III, Bsc I
Cla II	$(G \downarrow G^A_T CC)$	*Caryophanon latum*[97]	Afl I, Asp 697, **Ava II,** Bal II, *Bam Nx,* Bme 216, Bti I, *Cau I,* Clm II, Eco 47I, Erp I, *Fdi I,* Fsp MS1, *Hgi BI, Hgi CII, Hgi EI, Hgi HIII,* Nsp HII, Sfn I, Sin I
Clm I	$(GGCC)$	*Caryophanon latum*[194]	Asp 742, Blu II, Bse I, *Bsp RI,* Bss CI, Bst CI, Bsu 1076, Bsu 1114, *Bsu RI, Clt I,* Fin SI, *Fnu DI,* **Hae III,** Hhg I, Mni I, Mnn II, Ngo II, Nla I, Pai I, Pal I, Ppu I, *Sfa I,* Spl III, Sul I, Ttn I, Vha I
Clm II	$(GG^A_T CC)$	*Caryophanon latum*[194]	Afl I, Asp 697, **Ava II,** Bal II, *Bam Nx,* Bme 216, Bti I, *Cau I,* Cla II, Eco 47I, Erp I, *Fdi I,* Fsp MS1, *Hgi BI, Hgi CII, Hgi EI, Hgi HIII,* Nsp HII, Sfn I, Sin I
Clt I	$(GG \downarrow CC)$	*Caryophanon latum*[132]	Asp 742, Blu II, Bse I, *Bsp RI,* Bss CI, Bst CI, Bsu 1076, Bsu 1114, *Bsu RI,* Clm I, Fin SI, *Fnu DI,* **Hae III,** Hhg I, Mni I, Mnn II, Ngo II, Nla I, Pai I, Pal I, Ppu I, *Sfa I,* Spl III, Sul I, Ttn I, Vha I
Cpa I	$(GATC)$	*Clostridium pasteurianum*[219]	*Bce 243,* Bsa PI, Bss GII, Bst EIII, Bst XII, Dpn II, Fnu AII, *Fnu CI, Hac I,* **Mbo I,** Meu I, Mno III, Mos I, Msp 67II, Mth I, Nde II, Nfl AII, Nfl I, Nla II, Nsi AI, Nsp AI, Nsu I, *Sau 3AI,* Sin MI
Cpe I	$(TGATCA)$	*Corynebacterium petrophilum*[50]	Atu CI, **Bcl I,** Bst GI, Fba I, Pov I, Sst IV
Cpf I	$(\downarrow GATC)$	*Clostridium perfringens*[70]	*Bce 243,* Bsa PI, Bss GII, Bst EIII, Bst XII, Cpa I, Fnu AII, *Fnu CI, Hac I,* **Mbo I,** Meu I, Mno III, Mos I, Mth I, Nde II, Nfl AII, Nfl I, Nla II, Nsi AI, Nsp AI, Nsu I, *Sau 3AI,* Sin MI

Table 22. Alphabetical Listing of Restriction Endonucleases with Their Related Isoschizomers (continued)

Enzyme and Specific Recognition Sequence $(5' \rightarrow 3')^{a,b}$		Origin[c]	Isoschizomers[a,b]
Csc I	(CCGC ↓ GG)	Calothrix scopulorum[46]	Bac I, Cfr 37I, Ecc I, Ecl 28I, Ecl 37I, Eco 55I, Eco 92I, Gal I, Gce GLI, Gce I, Kae 37I, Mra I, Ngo III, Nla SI, Saa I, Sab I, **Sac II,** Sbo I, Sfr I, Shy I, Sst II, Tgl I
Cvn I	(CC ↓ TNAGG)	Chromatium vinosum[68]	Aoc I, Axy I, Eco 81I, Mst II, Oxa NI, **Sau I**
Dde II	(CTCGAG)	Desulfovibrio desulfuricans[234]	Abr I, Asp 703, Bbi III, Blu I, Bss HI, Bst HI, Bsu M, Bth I, Ccr I, Ccr II, Mca I, Msi I, PaeR 7, Pan I, PflW I, Sau 3239, Scu I, Sex I, Sga I, Sgo I, Sla I, Slu I, Spa PI, **Xho I,** Xpa I
Dds I	(GGATCC)	Desulfovibrio desulfuricans[124]	Acc EB1, Aac I, Aae I, Ali I, Bam FI, **Bam HI,** Bam KI, Bam NI, Bst I, Gdo I, Gin I, Gox I, Mle I, Nas BI, Nsp SAIV, Rhs I
Dpn I	(GA ↓ TC)	Diplococcus pneumoniae[110]	Cfu I, Nan II, Nmu DI, Nmu EI, Nsu DI, Sau 6782
Dpn II	(GATC)	Diplococcus pneumoniae[110]	Bce 243, Bsa PI, Bss GII, Bst EIII, Bst XII, Cpa I, Fnu AII, Fnu CI, Hac I, **Mbo I,** Meu I, Mno III, Mos I, Msp 67II, Mth I, Nde II, Nfl AII, Nfl I, Nla II, Nsi AI, Nsp AI, Nsu I, Sau 3AI, Sin MI
Dra I	(TTT ↓ AAA)	Deinococcus radiophilus[162]	**Aha III**
Dra II	(PuG ↓ GNCCPy)	Deinococcus radiophilus[70]	Eco 0109, Pss I*
Eae I	(Py ↓ GGCCPu)	Enterobacter aerogenes[229]	**Cfr I,** Cfr 14I
Eae PI	(CTGCAG)	Enterobacter aerogenes[157]	Ali AJI, Asp 36I, Asp 708, Bbi I, Bce 170, Bsu 1247, Bsu B, Cau III, Cfl I, Ecl 77I, Eco 36I, Eco 48I, Eco 49I, Eco 83I, Mau I, Mkr I, Nas I, Ngb I, Noc I, Pma I, Pmy I, **Pst I,** Sal PI, Sfl I, Ska II, Xma II, Xor I, Xph I
Eag I	(C ↓ GGCCG)	Enterobacter agglomerans[144]	Eco 52I, **Xma III**
Eca I	(G ↓ GTNACC)	Enterobacter cloacae[75]	Asp A1, Bst 31I, **Bst EII,** Bst PI, Cfr7I, Cfr 19I, Nsp SAII
Eca II	(CC$_T^A$GG)	Enterobacter cloacae[147]	Aor I, Apy I*, Atu BI, Atu II, Bin SI, Bst GII, Bst NI, Cdi 27I, Cfr 5I, Cfr 11I, Cfr 20I, Ecl II, Ecl 66I, Eco 27I, Eco 38I, Eco 60I, Eco 61I, Eco 67I, **Eco RII,** Mph I, Mva I*, Sgr II, Taq XI*, Zan I*
Ecc I	(CCGCGG)	Enterobacter cloacae[131]	Bac I, Cfr 37I, Csc I, Ecl 28I, Ecl 37I, Eco 55I, Eco 92I, Gal I, Gce GLI, Gce I, Kae 37I, Mra I, Ngo III, Nla SI, Saa I, Sab I, **Sac II,** Sbo I, Sfr I, Shy I, Sst II, Tgl I
Ecl II	(CC$_T^A$GG)	Enterobacter cloacae[71]	Aor I, Apy I*, Atu BI, Atu II, Bin SI, Bst GII, Bst NI, Cdi 27I, Cfr 5I, Cfr 11I, Cfr 20I, Eca II, Ecl 66I, Eco 27I, Eco 38I, Eco 60I, Eco 61I, Eco 67I, **Eco RII,** Mph I, Mva I*, Sgr II, Taq XI*, Zan I*
Ecl 28I	(CCGCGG)	Enterobacter cloacae RFL 28[83]	Bac I, Cfr 37I, Csc I, Ecc I, Ecl 37I, Eco 55I, Eco 92I, Gal I, Gce GLI, Gce I, Kae 37I, Mra I, Ngo III, Nla SI, Saa I, Sab I, **Sac II,** Sbo I, Sfr I, Shy I, Sst II, Tgl I
Ecl 37I	(CCGCGG)	Enterobacter cloacae RFL 37[97]	Bac I, Cfr 37I, Csc I, Ecc I, Ecl 28I, Eco 55I, Eco 92I, Gal I, Gce GLI, Gce I, Kae 37I, Mra I, Ngo III, Nla SI, Saa I, Sab I, **Sac II,** Sbo I, Sfr I, Shy I, Sst II, Tgl I
Ecl 66I	(CC$_T^A$GG)	Enterobacter cloacae RFL 66[97]	Aor I, Apy I*, Atu BI, Atu II, Bin SI, Bst GII, Bst NI, Cdi 27I, Cfr 5I, Cfr 11I, Cfr 20I, Eca II, Ecl II, Eco 27I, Eco 38I, Eco 60I, Eco 61I, Eco 67I, **Eco RII,** Mph I, Mva I*, Sgr II, Taq XI*, Zan I*

Table 22. Alphabetical Listing of Restriction Endonucleases with Their Related Isoschizomers (_continued_)

Enzyme and Specific Recognition Sequence $(5' \rightarrow 3')^{a,b}$		Origin[c]	Isoschizomers[a,b]
Ecl 77I	(CTGCAG)	_Enterobacter cloacae_ RFL 77[97]	Ali AJI, Asp 36I, Asp 708, Bbi I, Bce 170, Bsu 1247, Bsu B, Cau III, _Cfl I_, Eae PI, Eco 36I, Eco 48I, Eco 49I, Eco 83I, Mau I, Mkr I, Nas I, Ngb I, Noc I, Pma I, Pmy I, **Pst I**, _Sal PI, Sfl I_, Ska II, Xma II, Xor I, Xph I
Eco 24I	(GPuGCPyC)	_Escherichia coli_ RFL 24[83]	_Ban II, Bvu I_, Eco 25I, Eco 26I, Eco 35I, Eco 40I, Eco 41I, **Hgi JII**
Eco 25I	(GPuGCPyC)	_Escherichia coli_ RFL 25[83]	_Ban II, Bvu I_, Eco 24I, Eco 26I, Eco 35I, Eco 40I, Eco 41I, **Hgi JII**
Eco 26I	(GPuGCPyC)	_Escherichia coli_ RFL 26[94]	_Ban II, Bvu I_, Eco 24I, Eco 25I, Eco 35I, Eco 40I, Eco 41I, **Hgi JII**
Eco 27I	(CC$_T^A$GG)	_Escherichia coli_ RfL 27[111]	Aor I, Apy I*, Atu BI, Atu II, Bin SI, Bst GII, Bst NI, Cdi 27I, Cfr 5I, Cfr 11I, Cfr 20I, Eca II, Ecl II, Ecl 66I, Eco 38I, Eco 60I, Eco 61I, Eco 67I, **Eco RII**, Mph I, Mva I*, Sgr II, Taq XI*, Zan I*
Eco 32I	(GATATC)	_Escherichia coli_ RFL 32[94]	**EcoR V**, Nan I, Nfl AI
Eco 35I	(GPuGCPyC)	_Escherichia coli_ RFL 35[83]	_Ban II, Bvu I_, Eco 24I, Eco 25I, Eco 26I, Eco 40I, Eco 41I, _Hgi JII_
Eco 36I	(CTGCAG)	_Escherichia coli_ RFL 36[111]	Ali AJI, Asp 36I, Asp 708, Bbi I, Bce 170, Bsu 1247, Bsu B, Cau III, Cfl I, Eae PI, Ecl 77I, Eco 48I, Eco 49I, Eco 83I, Mau I, Mkr I, Nas I, Ngb I, Noc I, Pma I, Pmy I, **Pst I**, _Sal PI, Sfl I_, Ska II, Xma II, Xor I, Xph I
Eco 38I	(CC$_T^A$GG)	_Escherichia coli_ RFL 38[82]	Aor I, Apy I*, Atu BI, Atu II, Bin SI, Bst GII, Bst NI, Cdi 27I, Cfr 5I, Cfr 11I, Cfr 20I, Eca II, Ecl II, Ecl 66I, Eco 27I, Eco 60I, Eco 61I, Eco 67I, **Eco RII**, Mph I, Mva I*, Sgr II, Taq XI*, Zan I*
Eco 39I	(GGNCC)	_Escherichia coli_ RFL 39[82]	Apu I, **Asu I**, _Bac 36I_, Cfr 4I, Cfr 8I, _Cfr 13I_, Cfr 23I, Eco 47II, Mja II, Nmu EII, Nmu SI, _Nsp 7524 IV_, Psp I, _Sau 96I_, Sdy I
Eco 40I	(GPuGCPyC)	_Escherichia coli_ RFL 40[82]	_Ban II, Bvu I_, Eco 24I, Eco 25I, Eco 26I, Eco 35I, Eco 41I, **Hgi JII**
Eco 41I	(GPuGCPyC)	_Escherichia coli_ RFL 41[82]	_Ban II, Bvu I_, Eco 24I, Eco 25I, Eco 26I, Eco 35I, Eco 40I, **Hgi JII**
Eco 47I	(GG$_T^A$CC)	_Escherichia coli_ RFL 47[88]	Afl I, Asp 697, **Ava II**, Bal II, _Bam Nx_, Bme 216, Bti I, _Cau I_, Cla II, Clm II, Erp I, _Fdi I_, Fsp MS1, _Hgi BI, Hgi CII, Hgi EI, Hgi HIII_, Nsp HII, Sfn I, Sin I
Eco 47II	(GGNCC)	_Escherichia coli_ RFL 47[88]	Apu I, **Asu I**, _Bac 36I_, Cfr 4I, Cfr 8I, _Cfr 13I_, Cfr 23I, Eco 39I, Mja II, Nmu EII, Nmu SI, _Nsp 7524 IV_, Psp I, _Sau 96I_, Sdy I
Eco 48I	(CTGCAG)	_Escherichia coli_ RFL 48[87]	Ali AJI, Asp 36I, Asp 708, Bbi I, Bce 170, Bsu 1247, Bsu B, Cau III, _Cfl I_, Eae PI, Ecl 77I, Eco 36I, Eco 49I, Eco 83I, Mau I, Mkr I, Nas I, Ngb I, Noc I, Pma I, Pmy I, **Pst I**, _Sal PI, Sfl I_, Ska II, Xma II, Xor I, Xph I
Eco 49I	(CTGCAG)	_Escherichia coli_ RFL 49[87]	Ali AJI, Asp 36I, Asp 708, Bbi I, Bce 170, Bsu 1247, Bsu B, Cau III, _Cfl I_, Eae PI, Ecl 77I, Eco 36I, Eco 48I, Eco 83I, Mau I, Mkr I, Nas I, Ngb I, Noc I, Pma I, Pmy I, **Pst I**, _Sal PI, Sfl I_, Ska II, Xma II, Xor I, Xph I
Eco 50I	(GGPyPuCC)	_Escherichia coli_ RFL 50[87]	Ban I*, **Hgi CI**, Hgi HI*
Eco 51II	(CC$_G^C$GG)	_Escherichia coli_ RFL 51[87]	Aha I, Bcn I, **Cau II**, _Nci I_, Rsh II

Table 22. Alphabetical Listing of Restriction Endonucleases with Their Related Isoschizomers (*continued*)

Enzyme and Specific Recognition Sequence $(5' \rightarrow 3')^{a,b}$	Origin[c]	Isoschizomers[a,b]
Eco 52I (CGGCCG)	*Escherichia coli* RFL 52[94]	*Eag I*, **Xma III**
Eco 55I (CCGCGG)	*Escherichia coli* RFL 55[87]	Bac I, Cfr 37I, *Csc I*, Ecc I, Ecl 28I, Ecl 37I, Eco 92I, *Gal I, Gce GLI, Gce I,* Kae 37I, Mra I, Ngo III, Nla SI, Saa I, Sab I, **Sac II,** Sbo I, Sfr I, Shy I, Sst II, Tgl I
Eco 56I (GCCGGC)	*Escherichia coli* RFL 56[94]	Ani I, Apr I, Mis I, **Nae I,** Nba I, Nbr I, Nmu FI, Nmu I, NtaS II, Pgl I, Rlu I, Sao I, Ska I
Eco 60I (CC$_\text{T}^\text{A}$GG)	*Escherichia coli* RFL 60[94]	Aor I, Apy I*, Atu BI, Atu II, Bin SI, Bst GII, Bst NI, Cdi 27I, Cfr 5I, Cfr 11I, Cfr 20I, Eca II, Ecl II, Ecl 66I, Eco 27I, Eco 38I, Eco 61I, Eco 67I, **Eco RII,** Mph I, Mva I*, Sgr II, Taq XI*, Zan I*
Eco 61I (CC$_\text{T}^\text{A}$GG)	*Escherichia coli* RFL 61[94]	Aor I, Apy I*, Atu BI, Atu II, Bin SI, Bst GII, Bst NI, Cdi 27I, Cfr 51, Cfr 11I, Cfr 20I, Eca II, Ecl II, Ecl 66I, Eco 27I, Eco 38I, Eco 60I, Eco 67I, **Eco RII,** Mph I, Mva I*, Sgr II, Taq XI*, Zan I*
Eco 65I (AAGCTT)	*Escherichia coli* RfL 65[97]	Bbr I, Bpe I, Cfr 32I, Chu I, *Eco VIII,* Hin 173, Hin JCII, Hinb III, Hinb III 1076, **Hind III,** Hinf II, *Hsu I,* Mki I
Eco 67I (CC$_\text{T}^\text{A}$GG)	*Escherichia coli* RFL 67[97]	Aor I, Apy I*, Atu BI, Atu II, Bin SI, Bst GII, Bst NI, Cdi 27I, Cfr 5I, Cfr 11I, Cfr 20I, Eca II, Ecl II, Ecl 66I, Eco 27I, Eco 38I, Eco 60I, Eco 61I, **Eco RII,** Mph I, Mva I*, Sgr II, Taq XI*, Zan I*
Eco 68I (G$_\text{G}^\text{A}$GC$_\text{C}^\text{T}$C)	*Escherichia coli* RFL 68[97]	**Cfr 48I**
Eco 72I (CAC ↓ GTG)	*Pseudomonas maltophila* CB50P[97]	**Pma CI**
Eco 76I (CCTNAGG)	*Escherichia coli* RFL 76[97]	**Sec III**
Eco 78I (GGCGCC)	*Escherichia coli* RFL 78[97]	Bbe AI, Bbe I*, Bin SII, Nam I, **Nar I,** *Nda I, Nun II,* Psp 61I, Sfo I
Eco 81I (CC ↓ TNAGG)	*Escherichia coli* RFL 81[97]	*Aoc I, Axy I, Cvn I, Mst II, Oxa NI,* **Sau I**
Eco 83I (CTGCAG)	*Escherichia coli* RFL 83[97]	Ali AJI, Asp 36I, Asp 708, Bbi I, Bce 170, Bsu 1247, Bsu B, Cau III, *Cfl I,* Eae PI, Ecl 77I, Eco 36I, Eco 48I, Eco 49I, Mau I, Mkr I, Nas I, Ngb I, Noc I, Pma I, Pmy I, **Pst I,** *Sal PI, Sfl I,* Ska II, Xma II, Xor I, Xph I
Eco 90I ($_\text{C}^\text{T}$GGCC$_\text{G}^\text{A}$)	*Escherichia coli* RFL 90[97]	Cfr 381
Eco 92I (CCGCGG)	*Escherichia coli* RFL 92[97]	Bac I, Cfr 37I, *Csc I,* Ecc I, Ecl 28I, Ecl 37I, Eco 55I, Gal I, *Gce GLI, Gce I,* Kae 37I, Mra I, Ngo III, Nla SI, Saa I, Sab I, **Sac II,** Sbo I, Sfr I, Shy I, Sst II, Tgl I
Eco 136II (GAGCTC)	*Escherichia coli* RFL 136[97]	Eco ICRI, Nas SI, **Sac I,** Sco I, *Sst I*
Eco 147I (AGGCCT)	*Escherichia coli* RFL 147[97]	Aat I, *Gdi I,* Nta SI, **Stu I**
Eco 149I (GGTACC)	*Escherichia coli* RFL 149[97]	Asp 718, Eco 149I, **Kpn I,** Nmi I
Eco ICRI (GAGCTC)	*Escherichia coli* 2bt[204]	Eco 136II, Nas SI, **Sac I,** Sco I, *Sst I*
Eco 0109 (PuG ↓ GNCCPy)	*Escherichia coli* H709c[142]	**Dra II,** Pss I*
EcoR I (G ↓ AATTC)	*Escherichia coli* RY13[67]	Rsr I, *Sso I*

Table 22. Alphabetical Listing of Restriction Endonucleases with Their Related Isoschizomers (continued)

Enzyme and Specific Recognition Sequence $(5' \rightarrow 3')^{a,b}$		Origin[c]	Isoschizomers[a,b]
Eco RII	($\downarrow$ CC$_T^A$GG)	*Escherichia coli* R245[240]	Aor I, Apy I*, Atu BI, Atu II, Bin SI, Bst GII, Bst NI, Cdi 27I, Cfr 5I, Cfr 11I, Cfr 20I, Eca II, Ecl II, Ecl 66I, Eco 27I, Eco 38I, Eco 60I, Eco 61I, Eco 67I, Mph I, Mva I*, Sgr II, Taq XI*, Zan I*
EcoR V	(GAT $\downarrow$ ATC)	*Escherichia coli* J62pLG74[99]	Eco 32I, Nan I, Nfl AI
Eco T14	(CC$_{TT}^{AA}$GG)	*Escherichia coli* TB14[239]	Eco T104, **Sty I**
Eco T22	(ATGCA $\downarrow$ T)	*Escherichia coli* TB22[239]	**Ava III,** Nsi I
Eco T104	(CC$_{TT}^{AA}$GG)	*Escherichia coli* TB104[239]	Eco T14, **Sty I**
Eco VIII	(A $\downarrow$ AGCTT)	*Escherichia coli* E1585-68[140]	Bbr I, Bpe I, Cfr 32I, Chu I, Eco 65I, Hin 173, Hin JCII, Hinb III, Hinb III 1076, **Hind III,** Hinf II, *Hsu I*, Mki I
Erp I	(G $\downarrow$ G$_T^A$CC)	*Erwinia rhaponci*[97]	Afl I, Asp 697, **Ava II,** Bal II, *Bam Nx*, Bme 216, Bti I, *Cau I*, Cla II, Clm II, Eco 47I, *Fdi I*, Fsp MS1, *Hgi BI, Hgi CII, Hgi EI, Hgi HIII*, Nsp HII, Sfn I, Sin I
Fba I	(TGATCA)	*Flavobacterium balustinum*[199]	Atu CI, **Bcl I,** Bst GI, Cpe I, Pov I, Sst IV
Fbr I	(GC $\downarrow$ NGC)	*Flavobacterium breve*[199]	**Fnu 4HI**
Fdi I	(G $\downarrow$ G$_T^A$CC)	*Fremyella diplosiphon*[215]	Afl I, Asp 697, **Ava II,** Bal II, *Bam Nx*, Bme 216, Bti I, *Cau I*, Cla II, Clm II, Eco 47I, Erp I, Fsp MS1, *Hgi BI, Hgi CII, Hgi EI, Hgi HIII*, Nsp HII, Sfn I, Sin I
Fdi II	(TGC $\downarrow$ GCA)	*Fremyella diplosiphon*[215]	*Aos I*, Fsp I, **Mst I**
Fin II	(GTTAAC)	*Flavobacterium indologenes*[143]	Bse II, **Hpa I**
Fin S1	(GGCC)	*Flavobacterium indoltheticum*[199]	Asp 742, Blu II, Bse I, *Bsp RI*, Bss CI, Bst CI, Bsu 1076, Bsu 1114, *Bsu RI*, Clm I, *Clt I*, *Fnu DI*, **Hae III,** Hhg I, Mni I, Mnn II, Ngo II, Nla I, Pai I, Pal I, Ppu I, *Sfa I*, Spl III, Sul I, Ttn I, Vha I
Fnu AI	(G $\downarrow$ ANTC)	*Fusobacterium nucleatum* A[120]	Hha II, **Hinf I,** Nca I, Nov II, Nsi HI
Fnu AII	(GATC)	*Fusobacterium nucleatum* A[120]	*Bce 243*, Bsa PI, Bss GII, Bst EIII, Bst XII, Cpa I, Dpn II, *Fnu CI, Hac I*, **Mbo I,** Meu I, Mno III, Mos I, Msp 67II, Mth I, Nde II, Nfl AII, Nfl I, Nla II, Nsi AI, Nsp AI, Nsu I, *Sau 3AI*, Sin MI
Fnu CI	($\downarrow$ GATC)	*Fusobacterium nucleatum* C[120]	*Bce 243*, Bsa PI, Bss GII, Bst EIII, Bst XII, Cpa I, Dpn II, Fnu AII, *Hac I*, **Mbo I,** Meu I, Mno III, Mos I, Msp 67II, Mth I, Nde II, Nfl AII, Nfl I, Nla II, Nsi AI, Nsp AI, Nsu I, *Sau 3AI*, Sin MI
Fnu DI	(GG $\downarrow$ CC)	*Fusobacterium nucleatum* D[120]	Asp 742, Blu II, Bse I, *Bsp RI*, Bss CI, Bst CI, Bsu 1076, Bsu 1114, *Bsu RI*, Clm I, *Clt I*, Fin SI, **Hae III,** Hhg I, Mni I, Mnn II, Ngo II, Nla I, Pai I, Pal I, Ppu I, *Sfa I*, Spl III, Sul I, Ttn I, Vha I
Fnu DII	(CG $\downarrow$ CG)	*Fusobacterium nucleatum* D[120]	*Acc II*, Bce FI, Bce R, Bsu 1192 II, Bsu 1193, Bsu 6633, Bsu EII, Fsp MI, Hin 1056 I, *Tha I*
Fnu DIII	(GCG $\downarrow$ C)	*Fusobacterium nucleatum* D[120]	Cfo I, **Hha I,** Hin GUI, Hin PI*, Hin S1, Hin S2, Mnn IV, Sci NI*
Fnu 4HI	(GC $\downarrow$ NGC)	*Fusobacterium nucleatum* 4H[118]	*Fbr I*
Fok I	(GGATG)	*Flavobacterium okeanokoites*[202]	Hin GUII

Table 22. Alphabetical Listing of Restriction Endonucleases with Their Related Isoschizomers (*continued*)

Enzyme and Specific Recognition Sequence (5′ → 3′)[a,b]		Origin[c]	Isoschizomers[a,b]
Fsp I	(TGCGCA)	*Fischerella* species[209]	*Aos I, Fdi II,* **Mst I**
Fsp II	(TT ↓ CGAA)	*Fischerella* species[209]	**Asu II,** *Mla I,* Nsp 7524 V, Nsp BI
Fsp MI	(CGCG)	*Flavobacterium* species[143]	*Acc II,* Bce FI, Bce R, Bsu 1192 II, Bsu 1193, Bsu 6633, Bsu EII, **Fnu DII,** Hin 1056 I, *Tha I*
Fsp MSI	(G ↓ G^A_TCC)	*Fischerella* species[97]	Afl I, Asp 697, **Ava II,** Bal II, *Bam Nx,* Bme 216, Bti I, *Cau I,* Clm II, Eco 47I, Erp I, *Fdi I, Hgi BI, Hgi CII, Hgi EI, Hgi HIII,* Nsp HII, Sfn I, Sin I
Fsu I	(GACNNNGTC)	*Flavobacterium suaveolens* ATCC 13718[97]	Nta I, *Spl II,* Tte I, Ttr I, **Tth 111I**
Gal I	(CCGC ↓ GG)	*Gluconobacter albidus*[74]	Bac I, Cfr 37I, *Csc I,* Ecc I, Ecl 28I, Ecl 37I, Eco 55I, Eco 92I, *Gce GLI, Gce I,* Kae 37I, Mra I, Ngo III, Nla SI, Saa I, Sab I, **Sac II,** Sbo I, Sfr I, Shy I, Sst II, Tgl I
Gce I	(CCGC ↓ GG)	*Gluconobacter cerinus*[74]	Bac I, Cfr 37I, *Csc I,* Ecc I, Ecl 28I, Ecl 37I, Eco 55I, Eco 92I, *Gal I, Gce GLI,* Kae 37I, Mra I, Ngo III, Nla SI, Saa I, Sab I, **Sac II,** Sbo I, Sfr I, Shy I, Sst II, Tgl I
Gce GLI	(CCGC ↓ GG)	*Gluconobacter cerinus*[172]	Bac I, Cfr 37I, *Csc I,* Ecc I, Ecl 28I, Ecl 37I, Eco 55I, Eco 92I, *Gal I, Gce I,* Kae 37I, Mra I, Ngo III, Nla SI, Saa I, Sab I, **Sac II,** Sbo I, Sfr I, Shy I, Sst II, Tgl I
Gdi I	(AGG ↓ CCT)	*Gluconobacter dioxyacetonicus*[216]	Aat I, Eco 147I, Nta SI, **Stu I**
Gdo I	(GGATCC)	*Gluconobacter dioxyacetonicus*[167]	*Acc EB1,* Aac I, Aae I, *Ali I,* Bam FI, **Bam HI,** Bam KI, Bam NI, *Bst I,* Dds I, Gin I, Gox I, Mle I, Nas BI, *Nsp SAIV,* Rhs I
Gin I	(GGATCC)	*Gluconobacter industricus*[204]	*Acc EB1,* Aac I, Aae I, *Ali I,* Bam FI, **Bam HI,** Bam KI, Bam NI, *Bst I,* Dds I, Gdo I, Gox I, Mle I, Nas BI, *Nsp SAIV,* Rhs I
Gox I	(GGATCC)	*Gluconobacter oxydans* ssp. *melonogenes*[180]	*Acc EB1,* Aac I, Aae I, *Ali I,* Bam FI, **Bam HI,** Bam KI, Bam NI, *Bst I,* Dds I, Gdo I, Gin I, Mle I, Nas BI, *Nsp SAIV,* Rhs I
Gsb I	(CTCCAG)	*Gluconobacter suboxydans*[79]	**Gsu I**
Gsu I	(CTCCAG)	*Gluconobacter suboxydans* H-15T[84]	Gsb I
Hac I	(↓ GATC)	*Halococcus acetoinfaciens*[97]	Bce 243, Bsa PI, Bss GII, Bst EIII, Bst XII, Cpa I, Dpn II, Fnu AII, Fnu CI, **Mbo I,** Meu I, Mno III, Mos I, Msp 67II, Mth I, Nde II, Nfl AII, Nfl I, Nla II, Nsi AI, Nsp AI, Nsu I, Sau 3AI, Sin MI
Hae II	(PuGCGC ↓ Py)	*Haemophilus aegyptius*[167]	Hin HI, Ngo I
Hae III	(GG ↓ CC)	*Haemophilus aegyptius*[137]	Asp 742, Blu II, Bse I, *Bsp RI,* Bss CI, Bst CI, Bsu 1076, Bsu 1114, *Bsu RI,* Clm I, *Clt I,* Fin SI, *Fnu DI,* Hhg I, Mni I, Mnn II, Ngo II, Nla I, Pai I, Pal I, Ppu I, *Sfa I,* Spl III, Sul I, Ttn I, Vha I
Hap II	(C ↓ CGG)	*Haemophilus aphrophilus*[211]	Asp 748, Asp 750, Bsu 1192I, Bsu FI, **Hpa II,** Mni II, Mno I, Msp I, Sfa GUI, Sec II
Hgi BI	(G ↓ G^A_TCC)	*Herpetosiphon giganteus* Hpg5[76]	Afl I, Asp 697, **Ava II,** Bal II, *Bam Nx,* Bme 216, Bti I, *Cau I,* Clm II, Eco 47I, Erp I, *Fdi I,* Fsp MS1, *Hgi CII, Hgi EI, Hgi HIII,* Nsp HII, Sfn I, Sin I
Hgi CI	(↓ GGPyPUCC)	*Herpetosiphon giganteus* Hpg 9[107]	Ban I*, Eco 50I, Hgi HI*

Table 22. Alphabetical Listing of Restriction Endonucleases with Their Related Isoschizomers (*continued*)

Enzyme and Specific Recognition Sequence $(5' \to 3')^{a,b}$		Origin[c]	Isoschizomers[a,b]
Hgi CII	$(G \downarrow G^A_TCC)$	*Herpetosiphon giganteus* Hpg 9[76]	Afl I, Asp 697, **Ava II**, Bal II, *Bam Nx,* Bme 216, Bti I, *Cau I,* Clm II, Eco 47I, Erp I, *Fdi I,* Fsp MS1, *Hgi BI, Hgi EI, Hgi HIII,* Nsp HII, Sfn I, Sin I
Hgi CIII	$(G \downarrow TCGAC)$	*Herpetosiphon giganteus* Hpg 9[76]	*Hgi DII, Nop I,* Rhe I, Rhp I, Rrh I, Rro I, **Sal I,** Xam I, *Xci I*
Hgi DI	$(GPu \downarrow CGPyC)$	*Herpetosiphon giganteus* Hpa 2[76]	**Acy I,** Aha II, Aos II, Ast WI, Asu III, Bbi II, Hgi GI, Hgi HII, Nla SII
Hgi DII	$(G \downarrow TCGAC)$	*Herpetosiphon giganteus* Hpa 2[76]	Hgi CIII, *Nop I,* Rhe I, Rhp I, Rrh I, Rro I, **Sal I,** Xam I, *Xci I*
Hgi EI	$(G \downarrow G^A_TCC)$	*Herpetosiphon giganteus* Hpg 24[76]	Afl I, Asp 697, **Ava II**, Bal II, *Bam Nx,* Bme 216, Bti I, *Cau I,* Clm II, Eco 47I, Erp I, *Fdi I,* Fsp MS1, *Hgi BI, Hgi CII, Hgi HIII,* Nsp HII, Sfn I, Sin I
Hgi GI	$(GPu \downarrow CGPyC)$	*Herpetosiphon giganteus* Hpa 1[108]	**Acy I,** *Aha II, Aos II, Ast WI, Asu III, Bbi II, Hgi DI, Hgi HII,* Nla SII
Hgi HI*	$(G \downarrow GPyPuCC)$	*Herpetosiphon giganteus* HP 1049[229]	Ban I, Eco 50I, **Hgi CI**
Hgi HII	$(GPu \downarrow CGPyC)$	*Herpetosiphon giganteus* HP 1049[229]	**Acy I,** *Aha II, Aos II, Ast WI, Asu III, Bbi II, Hgi DI, Hgi GI,* Nla SII
Hgi HIII	$(G \downarrow G^A_TCC)$	*Herpetosiphon giganteus* HP 1049[229]	Afl I, Asp 697, **Ava II**, Bal II, *Bam Nx,* Bme 216, Bti I, *Cau I,* Clm II, Eco 47I, Erp I, *Fdi I,* Fsp MS1, Hgi BI, Hgi CII, Hgi EI, Nsp HII, Sfn I, Sin I
Hgi J II	$(GPuGCPy \downarrow C)^{a,b}$	*Herpetosiphon giganteus* HFS101[229]	*Ban II, Bvu I,* Eco 24I, Eco 25I, Eco 26I, Eco 35I, Eco 40I, Eco 41I
Hha I	$(GCG \downarrow C)$	*Haemophilus haemolyticus*[169]	*Cfo I, Fnu DIII,* Hin GUI, Hin PI*, Hin S1, Hin S2, Mnn IV, Sci NI*
Hha II	$(GANTC)$	*Haemophilus haemolyticus*[126]	*Fnu AI,* **Hinf I,** Nca I, Nov II, Nsi HI
Hhg I	$(GGCC)$	*Haemophilus haemoglobinophilus*[147]	Asp 742, Blu II, Bse I, *Bsp RI,* Bss CI, Bst CI, Bsu 1076, Bsu 1114, *Bsu RI,* Clm I, *Clt I,* Fin SI, *Fnu DI,* **Hae III,** Mni I, Mnn II, Ngo II, Nla I, Pai I, Pal I, Ppu I, *Sfa I,* Spl III, Sul I, Ttn I, Vha I
Hinb III	$(AAGCTT)$	*Haemophilus influenzae* Rb & serotype b, 1076[155]	Bbr I, Bpe I, Cfr 32I, Chu I, *Eco VIII,* Eco 65I, Hin 173, Hin JCII, Hinb III 1076, **Hind III,** Hinf II, *Hsu I,* Mki I
Hinb III 1076	$(AAGCTT)$	*Haemophilus influenzae* serotype b, 1076[155]	Bbr I, Bpe I, Cfr 32I, Chu I, *Eco VIII,* Eco 65I, Hin 173, Hin JCII, Hinb III, **Hind III,** Hinf II, *Hsu I,* Mki I
Hinc II	$(GTPyPuAC)$	*Haemophilus influenzae* Rc & serotype c,[155]	Chu II, Hin JCI*, Hinc II 1160, Hinc II 1161, **Hind II,** Mnn I
Hinc II 1160	$(GTPyPuAC)$	*Haemophilus influenzae* serotype c, 1160[155]	Chu II, Hin JCI*, Hinc II, Hinc II 1161, **Hind II,** Mnn I
Hinc II 1161	$(GTPyPuAC)$	*Haemophilus influenzae* serotype c, 1161[155]	Chu II, Hin JCI*, Hinc II, Hinc II 1160, **Hind II,** Mnn I
Hind II	$(GTPy \downarrow PuAC)$	*Haemophilus influenzae* Rd[193]	Chu II, Hin JCI*, Hinc II, Hinc II 1160, Hinc II 1161, Mnn I
Hind III	$(A \downarrow AGCTT)$	*Haemophilus influenzae* Rd[154]	Bbr I, Bpe I, Cfr 32I, Chu I, *Eco VIII,* Eco 65I, Hin 173, Hin JCII, Hinb III, Hinb III 1076, Hinf II, *Hsu I,* Mki I
Hinf I	$(G \downarrow ANTC)$	*Haemophilus influenzae* RF[139]	*Fnu AI,* Hha II, Nca I, Nov II, Nsi HI
Hinf II	$(AAGCTT)$	*Haemophilus influenzae* RF[127]	Bbr I, Bpe I, Cfr 32I, Chu I, *Eco VIII,* Eco 65I, Hin 173, Hin JCII, Hinb III, Hinb III 1076, **Hind III,** *Hsu I,* Mki I

Table 22. Alphabetical Listing of Restriction Endonucleases with Their Related Isoschizomers (*continued*)

Enzyme and Specific Recognition Sequence $(5' \rightarrow 3')^{a,b}$		Origin[c]	Isoschizomers[a,b]
Hin GUI	(GCGC)	*Haemophilus influenzae* GU[26]	*Cfo I, Fnu DIII,* **Hha I,** Hin PI*, Hin S1, Hin S2, Mnn IV, Sci NI*
Hin GUII	(GGATG)	*Haemophilus influenzae* GU[196]	**Fok I**
Hin HI	(PuGCGCPy)	*Haemophilus influenzae* H-1[211]	**Hae II,** Ngo I
Hin JCI*	(GTPy ↓ PuAC)	*Haemophilus influenzae* JC9[158]	Chu II, Hinc II, Hinc II 1160, Hinc II 1161, **Hind II,** Mnn I
Hin JCII	(AAGCTT)	*Haemophilus influenzae* JC9[158]	Bbr I, Bpe I, Cfr 32I, Chu I, *Eco VIII,* Eco 65I, Hin 173, Hinb III, Hinb III 1076, **Hind III,** Hinf II, *Hsu I,* Mki I
Hin PI*	(G ↓ CGC)	*Haemophilus influenzae* JC9[182]	*Cfo I, Fnu DIII,* **Hha I,** Hin GUI, Hin S1, Hin S2, Mnn IV, Sci NI*
Hin SI	(GCGC)	*Haemophilus influenzae* S1[182]	*Cfo I, Fnu DIII,* **Hha I,** Hin GUI, Hin PI*, Hin S2, Mnn IV, Sci NI*
Hin S2	(GCGC)	*Haemophilus influenzae* S2[182]	*Cfo I, Fnu DIII,* **Hha I,** Hin GUI, Hin PI*, Hin S1, Mnn IV, Sci NI*
Hin 1056I	(GCGC)	*Haemophilus influenzae* 1056[155]	*Acc II,* Bce FI, Bce R, Bsu 1192 II, Bsu 1193, Bsu 6633, Bsu EII, **Fnu DII,** Fsp MI, *Tha I*
Hin 173	(AAGCTT)	*Haemophilus influenzae* 173[196]	Bbr I, Bpe I, Cfr 32I, Chu I, *Eco VIII,* Eco 65I, Hin JCII, Hinb III, Hinb III 1076, **Hind III,** Hinf II, *Hsu I,* Mki I
Hpa I	(GTT ↓ AAC)	*Haemophilus parainfluenzae*[183]	Bse II, Fin II
Hpa II	(C ↓ CGG)	*Haemophilus parainfluenzae*[183]	Asp 748, Asp 750, Bsu 1192I, Bsu FI, Hap II, Mni II, Mno I, Msp I, Sfa GUI, Sec II
Hsu I	(A ↓ AGCTT)	*Haemophilus suis*[76]	Bbr I, Bpe I, Cfr 32I, Chu I, *Eco VIII,* Eco 65I, Hin 173, Hin JCII, Hinb III, Hinb III 1076, **Hind III,** Hinf II, Mki I
Kae 37I	(CCGCGG)	*Klebsiella aerogenes* RFL 37[82]	Bac I, Cfr 37I, *Csc I,* Ecc I, Ecl 28I, Ecl 37I, Eco 55I, Eco 92I, *Gal I, Gce GLI, Gce I,* Mra I, Ngo III, Nla SI, Saa I, Sab I, **Sac II,** Sbo I, Sfr I, Shy I, Sst II, Tgl I
Kpn I	(GGTAC ↓ C)	*Klebsiella pneumoniae* OK8[88]	Asp 718, Eco 149I, Nmi I
Kpn 2I	(TCCGGA)	*Klebsiella pneumonia* RFL2[97]	*Acc III,* **Bsp MII**
Mae I	(C ↓ TAG)	*Methanococcus aeolicus* PL-15H[182]	Mja I
Mau I	(CTGCAG)	*Micrococcus auranticus* IFO12422[204]	Ali AJI, Asp 36I, Asp 708, Bbi I, Bce 170, Bsu 1247, Bsu B, Cau III, *Cfl I,* Eae PI, Ecl 77I, Eco 36I, Eco 48I, Eco 49I, Eco 83I, Mkr I, Nas I, Ngb I, Noc I, Pma I, Pmy I, **Pst I,** *Sal PI, Sfl I,* Ska II, Xma II, Xor I, Xph I
Mbo I	(↓ GATC)	*Moraxella bovis*[56]	*Bce 243,* Bsa PI, Bss GII, Bst EIII, Bst XII, Cpa I, Dpn II, Fnu AII, *Fnu CI, Hac I,* Meu I, Mno III, Mos I, Msp 67II, Mth I, Nde II, Nfl AII, Nfl I, Nla II, Nsi AI, Nsp AI, Nsu I, *Sau 3AI,* Sin MI
Mbo II	(GAAGA)	*Moraxella bovis*[56]	Ncu I, Tce I
Mca I	(CTCGAG)	*Micromonospera carbonacea*[98]	AbrI, Asp 703, Bbi III, *Blu I,* Bss HI, Bst HI, Bsu M, Bth I, *Ccr I,* Ccr II, Dde II, Msi I, *PaeR 7,* Pan I, PflW I, Sau 3239, Scu I, Sex I, Sga I, Sgo I, *Sla I,* Slu I, Spa PI, **Xho I,** *Xpa I*
Meu I	(GATC)	*Micrococcus eurhyalis*[231]	*Bce 243,* Bsa PI, Bss GII, Bst EIII, Bst XII, Cpa I, Dpn II, Fnu AII, *Fnu CI, Hac I,* **Mbo I,** Mno III, Mos I, Msp 67II, Mth I, Nde II, Nfl AII, Nfl I, Nla II, Nsi AI, Nsp AI, Nsu I, *Sau 3AI,* Sin MI
Mfl I	(Pu ↓ GATCPy)	*Microbacterium flavum*[73]	**Xho II**

Table 22. Alphabetical Listing of Restriction Endonucleases with Their Related Isoschizomers (*continued*)

Enzyme and Specific Recognition Sequence (5′ → 3′)[a,b]		Origin[c]	Isoschizomers[a,b]
Mis I	(GCCGGC)	*Micrococcus* species[136]	Ani I, Apr I, Eco 56I, **Nae I,** Nba I, Nbr I, Nmu FI, Nmu I, NtaS II, Pgl I, Rlu I, Sao I, Ska I
Mja I	(CTAG)	*Methanococcus jannashii*[246]	**Mae I**
Mja II	(GGNCC)	*Methanococcus jannashii*[246]	Apu I, **Asu I,** *Bac 36I,* Cfr 4I, Cfr 8I, *Cfr 13I,* Cfr 23I, Eco 39I, Eco 47II, Nmu EII, Nmu SI, *Nsp 7524 IV,* Psp I, *Sau 96I,* Sdy I
Mki I	(AAGCTT)	*Moraxella kingae*[95]	Bbr I, Bpe I, Cfr 32I, Chu I, *Eco VIII,* Eco 65I, Hin 173, Hin JCII, Hinb III, Hinb III 1076, **Hind III,** Hinf II, Hsu I
Mkr I	(CTGCAG)	*Micrococcus kristinae*[231]	Ali AJI, Asp 36I, Asp 708, Bbi I, Bce 170, Bsu 1247, Bsu B, Cau III, *Cfl I,* Eae PI, Ecl 77I, Eco 36I, Eco 48I, Eco 49I, Eco 83I, Mau I, Nas I, Ngb I, Noc I, Pma I, Pmy I, **Pst I,** *Sal PI,* Sfl I, Ska II, Xma II, Xor I, Xph I
Mla I	(TT ↓ CGAA)	*Mastigocladus laminosus*[44]	**Asu II,** *Fsp II,* Nsp 7524 V, Nsp BI
Mle I	(GGATCC)	*Micrococcus luteus*[231]	*Acc EB1,* Aac I, Aae I, *Ali I,* Bam FI, **Bam HI,** Bam KI, Bam NI, *Bst I,* Dds I, Gdo I, Gin I, Gox I, Nas BI, *Nsp SAIV,* Rhs I
Mlt I	(AG ↓ CT)	*Micrococcus luteus*[231]	**Alu I,** Oxa I
Mni I	(GGCC)	*Moraxella nonliquefaciens*[95]	Asp 742, Blu II, Bse I, *Bsp RI,* Bss CI, Bst CI, Bsu 1076, Bsu 1114, *Bsu RI,* Clm I, *Clt I,* Fin SI, *Fnu DI,* **Hae III,** Hhg I, Mnn II, Ngo II, Nla I, Pai I, Pal I, Ppu I, *Saf I,* Spl III, Sul I, Ttn I, Vha I
Mni II	(CCGG)	*Moraxella nonliquefaciens*[95]	Asp 748, Asp 750, Bsu 1192I, Bsu FI, *Hap II,* **Hpa II,** *Mno I, Msp I,* Sfa GUI, Sec II
Mnn I	(GTPyPuAC)	*Moraxella nonliquefaciens*[69]	Chu II, Hin JCI*, Hinc II, Hinc II 1160, Hinc II 1161, **Hind II**
Mnn II	(GGCC)	*Moraxella nonliquefaciens*[69]	Asp 742, Blu II, Bse I, *Bsp RI,* Bss CI, Bst CI, Bsu 1076, Bsu 1114, *Bsu RI,* Clm I, *Clt I,* Fin SI, *Fnu DI,* **Hae III,** Hhg I, Mni I, Ngo II, Nla I, Pai I, Pal I, Ppu I, *Sfa I,* Spl III, Sul I, Ttn I, Vha I
Mnn IV	(GCGC)	*Moraxella nonliquefaciens*[69]	*Cfo I, Fnu DIII,* **Hha I,** Hin GUI, Hin PI*, Hin S1, Hin S2, Sci NI*
Mno I	(C ↓ CGG)	*Moraxella nonliquefaciens*[76]	Asp 748, Asp 750, Bsu 1192I, Bsu FI, *Hap II,* **Hpa II,** Mni II, *Msp I,* Sfa GUI, Sec II
Mno III	(GATC)	*Moraxella nonliquefaciens*[56]	*Bce 243,* Bsa PI, Bss GII, Bst EIII, Bst XII, Cpa I, Dpn II, Fnu AII, *Fnu CI, Hac I,* **Mbo I,** Meu I, Mos I, Msp 67II, Mth I, Nde II, Nfl AII, Nfl I, Nla II, Nsi AI, Nsp AI, Nsu I, *Sau 3AI,* Sin MI
Mos I	(GATC)	*Moraxella nonliquefaciens*[56]	*Bce 243,* Bsa PI, Bss GII, Bst EIII, Bst XII, Cpa I, Dpn II, Fnu AII, *Fnu CI, Hac I,* **Mbo I,** Meu I, Mno III, Msp 67II, Mth I, Nde II, Nfl AII, Nfl I, Nla II, Nsi AI, Nsp AI, Nsu I, *Sau 3AI,* Sin MI
Mph I	(CC$_T^A$GG)	*Moraxella phenylpyruvica*[95]	Aor I, Apy I*, Atu BI, Atu II, Bin SI, Bst GII, Bst NI, Cdi 27I, Cfr 5I, Cfr 11I, Cfr 20I, Eca II, Ecl II, Ecl 66I, Eco 27I, Eco 38I, Eco 60I, Eco 61I, Eco 67I, **Eco RII,** Mva I*, Sgr II, Taq XI*, Zan I*
Mra I	(CCGCGG)	*Micrococcus radiodurans*[225]	Bac I, Cfr 37I, *Csc I,* Ecc I, Ecl 28I, Ecl 37I, Eco 55I, Eco 92I, *Gal I, Gce GLI, Gce I,* Kae 37I, Ngo III, Nla SI, Saa I, Sab I, **Sac II,** Sbo I, Sfr I, Shy I, Sst II, Tgl I
Msi I	(CTCGAG)	*Myxococcus stipitatus* Mxs2H[139]	Abr I, Asp 703, Bbi III, *Blu I,* Bss HI, Bst HI, Bsu M, Bth I, *Ccr I,* Ccr II, Dde II, Mca I, PaeR 7, Pan I, Pflw I, Sau 3239, Scu I, Sex I, Sga I, Sgo I, *Sla I,* Slu I, Spa PI, **Xho I,** *Xpa I*

Table 22. Alphabetical Listing of Restriction Endonucleases with Their Related Isoschizomers (*continued*)

Enzyme and Specific Recognition Sequence $(5' \to 3')^{a,b}$		Origin[c]	Isoschizomers[a,b]
Msp I	(C ↓ CGG)	*Moraxella* species[217]	Bsu 1192 I, Bsu FI, Hap II, **Hpa II,** Mni II, Mno I, Sec II, Sfa GUI
Msp 67I	(CC ↓ NGG)	*Moraxella* species MS67[97]	**Scr FI**
Msp 67II	(GATC)	*Moraxella* species MS67[97]	*Bce 243,* Bsa PI, Bss GII, Bst EIII, Bst XII, Cpa I, Dpn II, Fnu AII, *Fnu CI, Hac I,* **Mbo I,** Meu I, Mno III, Mos I, Mth I, Nde II, Nfl AII, Nfl I, Nla II, Nsi AI, Nsp AI, Nsu I, *Sau 3AI,* Sin MI
Mst I	(TGC ↓ GCA)	*Microcoleus* species[36]	Aos I, *Fdi II,* Fsp I
Mst II	(CC ↓ TNAGG)	*Microcoleus* species[176]	*Aoc I, Axy I, Cvn I, Eco 81I, Oxa NI,* **Sau I**
Mth I	(GATC)	*Microbacterium thermosphactum*[112]	*Bce 243,* Bsa PI, Bss GII, Bst EIII, Bst XII, Cpa I, Dpn II, Fnu AII, *Fnu CI, Hac I,* **Mbo I,** Meu I, Mno III, Mos I, Msp 67II, Nde II, Nfl AII, Nfl I, Nla II, Nsi AI, Nsu I, *Sau 3AI,* Sin MI
Mva I*	(CC ↓ $\frac{A}{T}$GG)	*Micrococcus varians*[19]	Aor I, Apy I*, Atu BI, Atu II, Bin SI, Bst GII, Bst NI, Cdi 27I, Cfr 5I, Cfr 11I, Cfr 20I, Eca II, Ecl II, Ecl 66I, Eco 27I, Eco 38I, Eco 60I, Eco 61I, Eco 67I, **Eco RII,** Mph I, Sgr II, Taq XI*, Zan I*
Nae I	(GCC ↓ GGC)	*Nocardia aerocolonigenes*[35]	Ani I, Apr I, Eco 56I, Mis I, Nba I, Nbr I, Nmu FI, Nmu I, NtaS II, Pgl I, Rlu I, Sao I, Ska I
Nam I	(GGCGCC)	*Nocardia amarae*[119]	Bbe AI, Bbe I*, Bin SII, Eco 78I, *Nar I, Nda I, Nun II,* Psp 61I, Sfo I
Nan I	(GATATC)	*Neisseria animalis*[97]	Eco 32I, **Eco RV,** Nfl AI
Nan II	(GATC)	*Neisseria animalis*[97]	Cfu I, **Dpn I,** Nmu DI, Nmu EI, Nsu DI, Sau 6782
Nar I	(GG ↓ CGCC)	*Nocardia argentinensis*[34]	Bbe AI, Bbe I, Bin SII, Eco 78I, Nam I, Nda I, Nun II, Psp 61I, Sfo I
Nas I	(CTGCAG)	*Nocardia asteroides*[199]	Ali AJI, Asp 36I, Asp 708, Bbi I, Bce 170, Bsu 1247, Bsu B, Cau III, *Cfl I,* Eae PI, Ecl 77I, Eco 36I, Eco 48I, Eco 49I, Eco 83I, Mau I, Mkr I, Ngb I, Noc I, Pma I, Pmy I, **Pst I,** *Sal PI, Sfl I,* Ska II, Xma II, Xor I, Xph I
Nas BI	(GGATCC)	*Nocardia asteroides*[231]	Acc EB1, Aac I, Aae I, *Ali I,* Bam FI, **Bam HI,** Bam KI, Bam NI, Bst I, Dds I, Gdo I, Gin I, Gox I, Mle I, Nsp SAIV, Rhs I
Nas SI	(GAGCTC)	*Nocardia asteroides*[199]	Eco 136II, Eco ICRI, **Sac I,** Sco I, *Sst I*
Nba I	(GCCGGC)	*Nocardia brasiliensis*[164]	Ani I, Apr I, Eco 56I, Mis I, **Nae I,** Nbr I, Nmu FI, Nmu I, NtaS II, Pgl I, Rlu I, Sao I, Ska I
Nbl I	(CGAT ↓ CG)	*Nocardia blackwelli*[176]	**Pvu I,** *Rsh I,* Rsp I, Xni I, *Xor II*
Nbr I	(GCCGGC)	*Nocardia brasiliensis*[164]	Ani I, Apr I, Eco 56I, Mis I, **Nae I,** Nba I, Nmu FI, Nmu I, NtaS II, Pgl I, Rlu I, Sao I, Ska I
Nca I	(GANTC)	*Neisseria caviae*[34]	*Fnu AI,* Hba II, **Hinf I,** Nov II, Nsi HI
Nci I	(CC ↓ $\frac{C}{G}$GG)	*Neisseria cinerea*[226]	Aha I, Bcn I, **Cau II,** Eco 51II, Rsh II
Ncu I	(GAAGA)	*Neisseria cumiculi*[14]	**Mbo II,** Tce I
Nda I	(GG ↓ CGCC)	*Nocardia dassonvillei*[30]	Bbe AI, Bbe I*, Bin SII, Eco 78I, Nam I, **Nar I,** *Nun II,* Psp 61I, Sfo I
Nde II	(GATC)	*Nocardia denitrificans*[34]	*Bce 243,* Bsa PI, Bss GII, Bst EIII, Bst XII, Cpa I, Dpn II, Fnu AII, *Fnu CI, Hac I,* **Mbo I,** Meu I, Mno III, Mos I, Msp 67II, Mth I, Nfl AII, Nfl I, Nla II, Nsi AI, Nsp AI, Nsu I, *Sau 3AI,* Sin MI
Nfl I	(GATC)	*Neisseria flavescens*[34]	*Bce 243,* Bsa PI, Bss GII, Bst EIII, Bst XII, Cpa I, Dpn II, Fnu AII, *Fnu CI, Hac I,* **Mbo I,** Meu I, Mno III, Mos I, Msp 67II, Mth I, Nde II, Nfl AII, Nla II, Nsi AI, Nsp AI, Nsu I, *Sau 3AI,* Sin MI

Table 22. Alphabetical Listing of Restriction Endonucleases with Their Related Isoschizomers (continued)

Enzyme and Specific Recognition Sequence (5′ → 3′)[a,b]		Origin[c]	Isoschizomers[a,b]
Nfl AI	(GATATC)	*Neisseria flavescens*[128]	Eco 32I, **Eco RV,** Nan I
Nfl AII	(GATC)	*Neisseria flavescens*[128]	*Bce 243,* Bsa PI, Bss GII, Bst EIII, Bst XII, Cpa I, Dpn II, Fnu AII, *Fnu CI, Hac I,* **Mbo I,** Meu I, Mno III, Mos I, Msp 67II, Mth I, Nde II, Nfl I, Nla II, Nsi AI, Nsu I, *Sau 3AI,* Sin MI
Ngb I	(CTGCAG)	*Nocardia globerula*[231]	Ali AJI, Asp 36I, Asp 708, Bbi I, Bce 170, Bsu 1247, Bsu B, Cau III, *Cfl I,* Eae PI, Ecl 77I, Eco 36I, Eco 48I, Eco 49I, Eco 83I, Mau I, Mkr I, Nas I, Noc I, Pma I, Pmy I, **Pst I,** *Sal PI, Sfl I,* Ska II, Xma II, Xor I, Xph I
Ngo I	(PuGCGCPy)	*Neisseria gonorrhea*[233]	**Hae II,** Hin HI
Ngo II	(GGCC)	*Neisseria gonorrhea*[27]	Asp 742, Blu II, Bse I, *Bsp RI,* Bss CI, Bst CI, Bsu 1076, Bsu 1114, *Bsu RI,* Clm I, *Clt I,* Fin SI, *Fnu DI,* **Hae III,** Hhg I, Mni I, Mnn II, Nla I, Pai I, Pal I, Ppu I, *Sfa I,* Spl III, Sul I, Ttn I, Vha I
Ngo III	(CCGCGG)	*Neisseria gonorrhoea*[152]	Bac I, Cfr 37I, *Csc I,* Ecc I, Ecl 28I, Ecl 37I, Eco 55I, *Gal I, Gce GLI, Gce I,* Kae 37I, Mra I, Nla SI, Saa I, Sab I, **Sac II,** Sbo I, Sfr I, Shy I, Sst II, Tgl I
Nla I	(GGCC)	*Neisseria lactamica*[165]	Asp 742, Blu II, Bse I, *Bsp RI,* Bss CI, Bst CI, Bsu 1076, Bsu 1114, *Bsu RI,* Clm I, *Clt I,* Fin SI, *Fnu DI,* **Hae III,** Hhg I, Mni I, Mnn II, Ngo II, Pai I, Pal I, Ppu I, *Sfa I,* Spl III, Sul I, Ttn I, Vha I
Nla II	(GATC)	*Neisseria lactamica*[165]	*Bce 243,* Bsa PI, Bss GII, Bst EIII, Bst XII, Cpa I, Dpn II, Fnu AII, *Fnu CI, Hac I,* **Mbo I,** Meu I, Mno III, Mos I, Msp 67II, Mth I, Nde II, Nfl AII, Nfl I, Nsi AI, Nsp AI, Nsu I, *Sau 3AI,* Sin MI
Nla SI	(CCGCGG)	*Neisseria lactamica*[23]	Bac I, Cfr 37I, *Csc I,* Ecc I, Ecl 28I, Ecl 37I, Eco 55I, Eco 92I, Gal I, *Gce GLI, Gce I,* Kae 37I, Mra I, Ngo III, Saa I, Sab I, **Sac II,** Sbo I, Sfr I, Shy I, Sst II, Tgl I
Nla SII	(GPuCGPyC)	*Neisseria lactamica*[23]	**Acy I,** *Aha II, Aos II, Ast WI, Asu III, Bbi II, Hgi DI, Hgi GI, Hgi HII*
Nmi I	(GGTACC)	*Nocardia minima*[33]	Asp 718*, Eco 149I, **Kpn I**
Nmu DI	(GATC)	*Neisseria mucosa*[22]	Cfu I, **Dpn I,** Nan II, Nmu EI, Nsu DI, Sau 6782
Nmu EI	(GATC)	*Neisseria mucosa*[15]	Cfu I, **Dpn I,** Nan II, Nmu DI, Nsu DI, Sau 6782
NmuE II	(GGNCC)	*Neisseria mucosa*[15]	Apu I, **Asu I,** *Bac 36I,* Cfr 4I, Cfr 8I, *Cfr 13I,* Cfr 23I, Eco 39I, Eco 47II, Mja II, Nmu EII, Nmu SI, *Nsp 7524 IV,* Psp I, *Sau 96I,* Sdy I
Nmu FI	(GCCGGC)	*Neisseria mucosa*[22]	Ani I, Apr I, Eco 56I, Mis I, **Nae I,** Nba I, Nbr I, Nmu I, NtaS II, Pql I, Rlu I, Sao I, Ska I
Nmu I	(GCCGGC)	*Neisseria mucosa*[34]	Ani I, Apr I, Eco 56I, Mis I, **Nae I,** Nba I, Nbr I, Nmu FI, NtaS II, Pgl I, Rlu I, Sao I, Ska I
Nmu SI	(GGNCC)	*Neisseria mucosa*[32]	Apu I, **Asu I,** *Bac 36I, Cfr 4I, Cfr 8I, Cfr 13I,* Cfr 23I, Eco 39I, Eco 47II, Mja II, Nmu EII, *Nsp 7524 IV,* Psp I, *Sau 96I,* Sdy I
Noc I	(CTGCAG)	*Nocardia otitidis-caviarum*[33]	Ali AJI, Asp 36I, Asp 708, Bbi I, Bce 170, Bsu 1247, Bsu B, Cau III, *Cfl I,* Eae PI, Ecl 77I, Eco 36I, Eco 48I, Eco 49I, Eco 83I, Mau I, Mkr I, Nas I, Ngb I, Pma I, Pmy I, **Pst I,** *Sal PI, Sfl I,* Ska II, Xma II, Xor I, Xph I
Nop I	(G ↓ TCGAC)	*Nocardia opaca*[176]	*Hgi CIII, Hgi DII,* Rhe I, Rhp I, Rrh I, Rro I, **Sal I,** Xam I, *Xci I*
Nov II	(GANTC)	*Neisseria ovis*[34]	*Fnu AI,* Hha II, **Hinf I,** Nca I, Nsi HI
Nru I	(TCG ↓ CGA)	*Nocardia rubra*[33]	Ama I, Sbo 13

Table 22. Alphabetical Listing of Restriction Endonucleases with Their Related Isoschizomers (*continued*)

Enzyme and Specific Recognition Sequence $(5' \rightarrow 3')^{a,b}$		Origin[c]	Isoschizomers[a,b]
Nsi AI	(GATC)	*Neisseria sicca*[29]	*Bce 243*, Bsa PI, Bss GII, Bst EIII, Bst XII, Cpa I, Dpn II, Fnu AII, *Fnu CI*, *Hac I*, **Mbo I**, Meu I, Mno III, Mos I, Msp 67II, Mth I, Nde II, Nfl AII, Nfl I, Nla II, Nsp AI, Nsu I, *Sau 3AI*, Sin MI
Nsi HI	(GANTC)	*Neisseria sicca*[220]	*Fnu AI*, Hha II, **Hinf I**, Nca I, Nov II
Nsi I	(ATGCA ↓ T)	*Neisseria sicca*[31]	**Ava III**, Eco T 22
Nsp 7524I	(PuCATG ↓ Py) A A	*Nostoc* species[166]	*Nsp HI*
Nsp 7524II	(GGGCC ↓ C) T T	*Nostoc* species[166]	*Aoc II*, *Bsp 1286*, **Sdu I**
Nsp 7524III	(C ↓ PyCGPuG)	*Nostoc* species[166]	**Ava I**, *Avr I*, Aqu I
Nsp 7524IV	(G ↓ GNCC)	*Nostoc* species[166]	Apu I, **Asu I**, *Bac 36I*, Cfr 4I, Cfr 8I, *Cfr 13I*, Cfr 23I, Eco 39I, Eco 47II, Mja II, Nmu EII, Nmu SI, Psp I, *Sau 96I*, Sdy I
Nsp 7524V	(TTCGAA)	*Nostoc* species[166]	**Asu II**, *Fsp II*, *Mla I*, Nsp BI
Nsp AI	(GATC)	*Nocardia* species[231]	*Bce 243*, Bsa PI, Bss GII, Bst EIII, Bst XII, Cpa I, Dpn II, Fnu AII, *Fnu CI*, *Hac I*, **Mbo I**, Meu I Mno III, MosI, Msp 67II, Mth I, Nde II, Nfl AII, Nfl I, Nla II, Nsi AI, Nsu I, *Sau3AI*, Sin MI
NspBI	(TTCGAA)	*Nostoc* species[47]	**Asu II**, *Fsp II*, *Mla I*, Nsp 7524 V
NspHI	(PuCATG ↓ Py)	*Nostoc* species[47]	**Nsp 7524I**
NspHII	(GG ${}^{A}_{T}$ CC)	*Nostoc* species[47]	Afl I, Asp 697, **Ava II**, Bal II, *Bam Nx*, Bme 216, Bti I, *Cau I*, Clm II, Eco 47I, Erp I, *Fdi I*, Fsp MS1, *Hgi BI*, *Hgi CII*, *Hgi EI*, *Hgi HIII*, Sfn I, Sin I
NspMAC I	(A ↓ GATCT)	*Nostoc* species[114]	**Bgl II**
NsuDI	(GATC)	*Neisseria subflava*[22]	Cfu I, **Dpn I**, Nan II, Nmu DI, Nmu EI, Sau 6782
Nsu I	(GATC)	*Neisseria subflava*[22]	*Bce 243*, Bsa PI, Bss GII, Bst EIII, Bst XII, Cpa I, Dpn II, Fnu AII, *Fnu CI*, *Hac I*, **Mbo I**, Meu I, Mno III, Mos I, Msp 67II, Mth I, Nde II, Nfl AII, Nfl I, Nla II, Nsi AI, Nsp AI, *Sau 3AI*, Sin MI
Nsp SAII	(G ↓ GTNACC)	*Nostoc* species SA[97]	Asp A1, Bst 31I, **Bst EII**, *Bst PI*, Cfr7I, Cfr 19I, *Eca I*
Nsp SAIV	(G ↓ GATCC)	*Nostoc* species SA[97]	*Acc EB1*, Aac I, Aae I, *Ali I*, Bam FI, **Bam HI,** Bam KI, Bam NI, *Bst I*, Dds I, Gdo I, Gin I, Gox I, Mle I, Nas BI, Rhs I
Nta I	(GACNNNGTC)	*Nocordia tartaricans*[199]	Fsu I, Spl II, Tte I, Ttr I, **Tth 111I**
Nta SI	(AGGCCT)	*Nocordia tartaricans*[199]	Aat I, Eco 147I, *Gdi I*, **Stu I**
Nta SII	(GCCGGC)	*Nocordia tartaricans*[199]	Ani I, Apr I, Eco 56I, Mis I, **Nae I,** Nba I, Nbr I, Nmu FI, Nmu I, Pgl I, Rlu I, Sao I, Ska I
Nun II	(GG ↓ CGCC)	*Nocardia uniformis*[112]	Bbe AI, Bbe I*, Bin SII, Eco 78I, Nam I, **Nar I,** *Nda I,* Psp 61I, Sfo I
Oxa I	(AGCT)	*Oerskovia xanthineolytica*[200]	**Alu I,** *Mlt I*
Oxa NI	(CC ↓ TNAGG)	Okerskovia xanthineolytica[97]	*Aoc I*, *Axy I*, *Cvn I*, *Eco 81I*, *Mst II*, **Sau I**
Pae I	(GCATG ↓ C)	*Pseudomonas aeruginosa*[197]	Spa I, Spa XI, **Sph I**
Pae R7	(C ↓ TCGAG)	*Pseudomonas aeruginosa*[72]	AbrI, Asp 703, Bbi III, *Blu I*, Bss HI, Bst HI, Bsu M, Bth I, *Ccr I*, Ccr II, Dde II, Mca I, Msi I, Pan I, PflW I, Sau 3239, Scu I, Sex I, Sga I, Sgo I, *Sla I*, Slu I, Spa PI, **Xho I,** *Xpa I*
Pai I	(GGCC)	*Pseudomonas alkaligenes*[204]	Asp 742, Blu II, Bse I, *Bsp RI*, Bss CI, Bst CI, Bsu 1076, Bsu 1114, *Bsu RI*, Clm I, *Clt I*, Fin SI, *Fnu DI*, **Hae III**, Hhg I, Mni I, Mnn II, Ngo II, Nla I, Pal I, Ppu I, *Sfa I*, Spl III, Sul I, Ttn I, Vha I

Table 22. Alphabetical Listing of Restriction Endonucleases with Their Related Isoschizomers (*continued*)

Enzyme and Specific Recognition Sequence $(5' \rightarrow 3')^{a,b}$		Origin[c]	Isoschizomers[a,b]
Pal I	(GGCC)	*Providencia alcalifaciens*[58]	Asp 742, Blu II, Bse I, *Bsp RI*, Bss CI, Bst CI, Bsu 1076, Bsu 1114, *Bsu RI*, Clm I, *Ctl I*, Fin SI, *Fnu DI*, **Hae III**, Hhg I, Mni I, Mnn II, Ngo II, Nla I, Pai I, Ppu I, *Sfa I*, Spl III, Sul I, Ttn I, Vha I
Pan I	(C ↓ TCGAG)	*Pseudomonas alkanolytica*[204]	AbrI, Asp 703, Bbi III, *Blu I*, Bss HI, Bst HI, Bsu M, Bth I, *Ccr I*, Ccr II, Dde II, Mca I, Msi I, *PaeR 7*, PflW I, Sau 3239, Scu I, Sex I, Sga I, Sgo I, *Sla I*, Slu I, Spa PI, **Xho I**, *Xpa I*
Pfl WI	(CTCGAG)	*Pseudomonas fluorescens*[224]	AbrI, Asp 703, Bbi III, *Blu I*, Bss HI, Bst HI, Bsu M, Bth I, *Ccr I*, Ccr II, Dde II, Mca I, Msi I, *PaeR 7*, Pan I, Sau 3239, Scu I, Sex I, Sga I, Sgo I, *Sla I*, Slu I, Spa PI, **Xho I**, *Xpa I*
Pgl I	(GCCGGC)	*Pseudomonas glycinae*[115]	Ani I, Apr I, Eco 56I, Mis I, **Nae I**, Nba I, Nbr I, Nmu FI, Nmu I, NtaS II, Rlu I, Sao I, Ska I
Pma I	(CTGCAG)	*Pseudomonas maltophila*[176]	Ali AJI, Asp 36I, Asp 708, Bbi I, Bce 170, Bsu 1247, Bsu B, Cau III, *Cfl I*, Eae PI, Ecl 77I, Eco 36I, Eco 48I, Eco 49I, Eco 83I, Mau I, Mkr I, Nas I, Ngb I, Noc I, Pmy I, **Pst I**, *Sal PI, Sfl I*, Ska II, Xma II, Xor I, Xph I
Pma CI	(CAC ↓ GTG)	*Pseudomonas maltophila* CB50P[221]	*Eco 72I*
Pmy I	(CTGCAG)	*Proteus myxofaciens* ATCC 19692[97]	Ali AJI, Asp 36I, Asp 708, Bbi I, Bce 170, Bsu 1247, Bsu B, Cau III, *Cfl I*, Eae PI, Ecl 77I, Eco 36I, Eco 48I, Eco 49I, Eco 83I, Mau I, Mkr I, Nas I, Ngb I, Noc I, Pma I, **Pst I**, *Sal PI, Sfl I*, Ska II, Xma II, Xor I, Xph I
Pov I	(TGATCA)	*Pseudomonas ovis*[97]	Atu CI, **Bcl I**, Bst GI, Cpe I, Fba I, Sst IV
Ppu I	(GGCC)	*Pseudomonas putida C-83*[204]	Asp 742, Blu II, Bse I, *Bsp RI*, Bss CI, Bst CI, Bsu 1076, Bsu 1114, *Bsu RI*, Clm I, *Clt I*, Fin SI, *Fnu DI*, **Hae III**, Hhg I, Mni I, Mnn II, Ngo II, Nla I, Pai I, Pal I, *Sfa I*, Spl III, Sul I, Ttn I, Vha I
Psp I	(GGNCC)	*Pseudomonas* species[111]	Apu I, **Asu I**, *Bac 36I*, Cfr 4I, Cfr 8I, *Cfr 13I*, Cfr 23I, Eco 39I, Eco 47II, Mja II, Nmu EII, Nmu SI, *Nsp 7524 IV, Sau 96I*, Sdy I
Psp 61I	(GGCGCC)	*Pseudomonas* species MS61[97]	Bbe AI, Bbe I*, Bin SII, Eco 78I, Nam I, **Nar I**, *Nda I, Nun II*, Sfo I
Pss I*	(PuGGNC ↓ CPy)	*Pseudomonas species*[124]	**Dra II**, *Eco 0109*
Pst I	(CTGCA ↓ G)	*Providencia stuartii*[192]	Ali AJI, Asp 36I, Asp 708, Bbi I, Bce 170, Bsu 1247, Bsu B, Cau III, *Cfl I*, Eae PI, Ecl 77I, Eco 36I, Eco 48I, Eco 49I, Eco 83I, Mau I, Mkr I, Nas I, Ngb I, Noc I, Pma I, Pmy I, *Sal PI, Sfl I*, Ska II, Xma II, Xor I, Xph I
Pvu I	(CGAT ↓ CG)	*Proteus vulgaris*[60]	*Nbl I, Rsh I*, Rsp I, Xni I, *Xor II*
Pvu II	(CAG ↓ CTG)	*Proteus vulgaris*[60]	*Bav I*, Cfr 6I
Rhe I	(GTCGAC)	*Rhodococcus* species[81]	*Hgi CIII, Hgi DII, Nop I*, Rhp I, Rrh I, Rro I, **Sal I**, *Xam I, Xci I*
Rhp I	(GTCGAC)	*Rhodococcus* species[81]	*Hgi CIII, Hgi DII, Nop I*, Rhe I, Rrh I, Rro I, **Sal I**, *Xam I, Xci I*
Rhs I	(GGATCC)	*Rhodococcus* species[81]	*Acc EB1*, Aac I, Aae I, *Ali I*, Bam FI, **Bam HI**, Bam KI, Bam NI, *Bst I*, Dds I, Gdo I, Gin I, Gox I, Mle I, Nas BI, *Nsp SAIV*
Rlu I	(GCCGGC)	*Rhizobium lupini* 1[234]	Ani I, Apr I, Eco 56I, Mis I, **Nae I**, Nba I, Nbr I, Nmu FI, Nmu I, NtaS II, Pgl I, Sao I, Ska I
Rrh I	(GTCGAC)	*Rhodococcus rhodochrous*[111]	*Hgi CIII, Hgi DII, Nop I*, Rhe I, Rhp I, Rro I, **Sal I**, *Xam I, Xci I*
Rro I	(GTCGAC)	*Rhodococcus rhodochrous*[111]	*Hgi CIII, Hgi DIII*, Nop I, Rhe I, Rhp I, Rrh I, **Sal I**, *Xam I, Xci I*
Rsh I	(CGAT ↓ CG)	*Rhodopseudomonas sphaeroides*[122]	*Nbl I*, **Pvu I**, Rsp I, Xni I, *Xor II*

Table 22. Alphabetical Listing of Restriction Endonucleases with Their Related Isoschizomers (*continued*)

Enzyme and Specific Recognition Sequence $(5' \rightarrow 3')^{a,b}$		Origin[c]	Isoschizomers[a,b]
Rsh II	$(CC^{C}_{G}GG)$	*Rhodopseudomonas sphaeroides*[106]	Aha I, Bcn I, **Cau II,** Eco 51II, *Nci I*
Rsp I	(CGATCG)	*Rhodopseudomonas sphaeroides*[8]	*Nbl I,* **Pvu I,** *Rsh I,* Xni I, *Xor II*
Rsr I	(GAATCC)	*Rhodopseudomonas sphaeroides*[54]	**EcoR I,** *Sso I*
Saa I	(CCGCGG)	*Streptomyces alanosinicus* ATCC 15710[97]	Bac I, Cfr 37I, *Csc I,* Ecc I, Ecl 28I, Ecl 37I, Eco 55I, Eco 92I, *Gal I, Gce GLI,* Gce I, Kae 37I, Mra I, Ngo III, Nla SI, Sab I, **Sac II,** Sbo I, Sfr I, Shy I, Sst II, Tgl I
Sab I	(CCGCGG)	*Streptomyces albohelvatus* ATCC 19820[97]	Bac I, Cfr 37I, *Csc I,* Ecc I, Ecl 28I, Ecl 37I, Eco 55I, Eco 92I, *Gal I, Gce GLI,* Gce I, Kae 37I, Mra I, Ngo III, Nla SI, Saa I, **Sac II,** Sbo I, Sfr I, Shy I, Sst II, Tgl I
Sac I	(GAGCT ↓ C)	*Streptomyces achromogenes*[2]	Eco 136II, Eco ICR, Nas SI, **Sac I,** Sco I, *Sst I*
Sac II	(CCGC ↓ GG)	*Streptomyces achromogenes*[2]	Bac I, Cfr 37I, *Csc I,* Ecc I, Ecl 28I, Ecl 37I, Eco 55I, Eco 92I, *Gal I, Gce GLI,* Gce I, Kae 37I, Mra I, Ngo III, Nla SI, Saa I, Sab I, Sbo I, Sfr I, Shy I, Sst II, Tgl I
Sal I	(G ↓ TCGAC)	*Streptomyces albus* G[1]	*Hgi CIII, Hgi DII, Nop I,* Rhe I, Rhp I, Rrh I, Rro I, Xam I, *Xci I*
Sal PI	(CTGCA ↓ G)	*Streptomyces albus* G[25]	Ali AJI, Asp 36I, Asp 708, Bbi I, Bce 170, Bsu 1247, Bsu B, Cau III, *Cfl I,* Eae PI, Ecl 77I, Eco 36I, Eco 48I, Eco 49I, Eco 83I, Mau I, Mkr I, Nas I, Ngb I, Noc I, Pma I, Pmy I, **Pst I,** Sfl I, Ska II, Xma II, Xor I, Xph I
Sao I	(GCCGGC)	*Streptomyces albofaciens* ATCC 25814[97]	Ani I, Apr I, Eco 56I, Mis I, **Nae I,** Nba I, Nbr I, Nmu FI, Nmu I, NtaS II, Pgl I, Rlu I, Ska I
Sau I	(CC ↓ TNAGG)	*Streptomyces aureofaciens*[214]	Aoc I, *Axy I, Cvn I,* Eco 81I, *Mst II, Oxa NI*
Sau 3AI	(↓ GATC)	*Staphylococcus aureus* 3A[206]	Bce 243, Bsa PI, Bss GII, Bst EIII, Bst XII, Cpa I, Dpn II, Fnu AII, *Fnu CI, Hac I,* **Mbo I,** Meu I, Mno III, Mos I, Msp 67II, Mth I, Nde II, Nfl AII, Nfl I, Nla II, Nsi AI, Nsp AI, Nsu I, Sin MI
Sau 96 I	(G ↓ GNCC)	*Staphylococcus aureus* PS96[207]	Apu I, **Asu I,** *Bac 36I,* Cfr 4I, Cfr 8I, *Cfr 13I,* Cfr 23I, Eco 39I, Eco 47II, Mja II, Nmu EII, Nmu SI, *Nsp 7524 IV,* Psp I, Sdy I
Sau 3239	(C ↓ TCGAG)	*Streptomyces aureofaciens*[55]	AbrI, Asp 703, Bbi III, *Blu I,* Bss HI, Bst HI, Bsu M, Bth I, *Ccr I,* Ccr II, Dde II, Mca I, Msi I, *PaeR 7, Pan I,* PflW I, Scu I, Sex I, Sga I, Sgo I, *Sla I,* Slu I, Spa I, **Xho 1,** *Xpa I*
Sau 6782	(GATC)	*Staphylococcus aureus* 6782[3]	**Dpn I,** Cfu I, Nan II, Nmu DI, Nmu EI, Nsu DI
Sbo I	(CCGCGG)	*Streptomyces bobili*[181]	Bac I, Cfr 37I, *Csc I,* Ecc I, Ecl 28I, Ecl 37I, Eco 55I, Eco 92I, *Gal I, Gce GLI,* Gce I, Kae 37I, Mra I, Ngo III, Nla SI, Saa I, Sab I, **Sac II,** Sfr I, Shy I, Sst II, Tgl I
Sbo 13	(TCGCGA)	*Shigella boydii* 13[239]	Ama I, **Nru I**
Sci NI*	(G ↓ CGC)	*Spiroplasma citri* & *Asp 2*[198]	*Cfo I, Fnu DIII,* **Hha I,** Hin GUI, Hin PI*, Hin S1, Hin S2, Mnn IV
Sco I	(GAGCTC)	*Streptomyces coelicolor* ATCC 10174[97]	Eco 136II, Eco ICR, Nas SI, **Sac I,** *Sst I*
Scr FI	(CC ↓ NGG)	*Streptococcus cremoris* F[51]	*Msp 67I*
Scu I	(CTCGAG)	*Streptomyces cupidosporus*[187]	AbrI, Asp 703, Bbi III, *Blu I,* Bss HI, Bst HI, Bsu M, Bth I, *Ccr I,* Ccr II, Dde II, Mca I, Msi I, *PaeR 7, Pan I,* PflW I, Sau 3239, Sex I, Sga I, Sgo I, Sla I, Slu I, Spa PI, **Xho I,** Xpa I

Table 22. Alphabetical Listing of Restriction Endonucleases with Their Related Isoschizomers (*continued*)

Enzyme and Specific Recognition Sequence $(5' \rightarrow 3')^{a,b}$		Origin[c]	Isoschizomers[a,b]
Sdu I	A A (GGGCC ↓ T) T T	*Streptococcus durans*[86]	*Aoc II, Bsp 1286, Nsp 7524 II*
Sdy I	(GGNCC)	*Streptococcus dysgalactiae*[164]	Apu I, **Asu I,** *Bac 36I,* Cfr 4I, Cfr 8I, *Cfr 13I,* Cfr 23I, Eco 39I, Eco 47II, Mja II, Nmu EII, Nmu SI, *Nsp 7524 IV,* Psp I, *Sau 96I*
Sec II	(CCGG)	*Synechocystis* species 6701[21]	Asp 748, Asp 750, Bsu 1192I, Bsu FI, *Hap II,* **Hpa II,** Mni II, *Mno I, Msp I,* Sfa GUI
Sec III	(CCTNAGG)	*Synechocystis* species 6701[21]	Eco 76I
Sex I	(CTCGAG)	*Streptomyces exfoliatus*[198]	AbrI, Asp 703, Bbi III, *Blu I,* Bss HI, Bst HI, Bsu M, Bth I, *Ccr I,* Ccr II, Dde II, Mca I, Msi I, *PaeR 7, Pan I,* PflW I, Sau 3239, Scu I, Sga I, Sgo I, *Sla I,* Slu I, Spa PI, **Xho I,** *Xpa I*
Sfa I	(GG ↓ CC)	*Streptococcus faecalis* ssp. *zymogenes*[236]	Asp 742, Blu II, Bse I, *Bsp RI,* Bss CI, Bst CI, Bsu 1076, Bsu 1114, *Bsu RI,* Clm I, *Clt I,* Fin SI, *Fnu DI,* **Hae III,** Hhg I, Mni I, Mnn II, Ngo II, Nla I, Pai I, Pal I, Ppu I, Spl III, Sul I, Ttn I, Vha I
Sfa GUI	(CCGG)	*Streptococcus faecalis* GU[28]	Asp 748, Asp 750, Bsu 1192I, Bsu FI, *Hap II,* **Hpa II,** Mni II, *Mno I, Msp I,* Sec II
Sfl I	(CTGCA ↓ G)	*Streptoverticillium flavopersicum*[95]	Ali AJI, Asp 36I, Asp 708, Bbi I, Bce 170, Bsu 1247, Bsu B, Cau III, *Cfl I,* Eae PI, Ecl 77I, Eco 36I, Eco 48I, Eco 49I, Eco 83I, Mau I, Mkr I, Nas I, Ngb I, Noc I, Pma I, Pmy I, **Pst I,** *Sal PI,* Ska II, Xma II, Xor I, Xph I
Sfn I	(GG$_T^A$CC)	*Serratia fonticola*[199]	Afl I, Asp 697, **Ava II,** Bal II, Bam Nx, Bme 216, Bti I, *Cau I,* Clm II, Eco 47I, Erp I, *Fdi I,* Fsp MS1, *Hgi BI, Hgi CII, Hgi EI, Hgi HIII,* Nsp HII, Sin I
Sfo I	(GGCGCC)	*Serratia fonticola*[143]	Bbe AI, Bbe I*, Bin SII, Eco 78I, Nam I, **Nar I,** *Nda I, Nun II,* Psp 61I
Sfr I	(CCGCGG)	*Streptomyces fradiae*[187]	Bac I, Cfr 37I, *Csc I,* Ecc I, Ecl 28I, Ecl 37I, Eco 55I, Eco 92I, *Gal I, Gce GLI, Gce I,* Kae 37I, Mra I, Ngo III, Nla SI, Saa I, Sab I, **Sac II,** Sbo I, Shy I, Sst II, Tgl I
Sga I	(CTCGAG)	*Streptomyces ganmycicus*[187]	AbrI, Asp 703, Bbi III, *Blu I,* Bss HI, Bst HI, Bsu M, Bth I, *Ccr I,* Ccr II, Dde II, Mca I, Msi I, *PaeR 7, Pan I,* PflW I, Sau 3239, Scu I, Sex I, Sgo I, *Sla I,* Slu I, Spa PI, **Xho I,** *Xpa I*
Sgo I	(CTCGAG)	*Streptomyces goshikiensis*[187]	AbrI, Asp 703, Bbi III, *Blu I,* Bss HI, Bst HI, Bsu M, Bth I, *Ccr I,* Ccr II, Dde II, Mca I, Msi I, *PaeR 7, Pan I,* PflW I, Sau 3239, Scu I, Sex I, Sga I, *Sla I,* Slu I, Spa PI, **Xho I,** *Xpa I*
Sgr II	(CC$_T^A$GG)	*Streptomyces griseus*[156]	Aor I, Apy I*, Atu BI, Atu II, Bin SI, Bst GII, Bst NI, Cdi 27I, Cfr 5I, Cfr 11I, Cfr 20I, Eca II, Ecl II, Ecl 66I, Eco 27I, Eco 38I, Eco 60I, Eco 61I, Eco 67I, **Eco RII,** Mph I, Mva I*, Taq XI*, Zan I*
Shy I	(CCGCGG)	*Streptomyces hygroscopicus*[222]	Bac I, Cfr 37I, *Csc I,* Ecc I, Ecl 28I, Ecl 37I, Eco 55I, *Gal I, Gce GLI, Gce I,* Kae 37I, Mra I, Ngo III, Nla SI, Saa I, Sab I, **Sac II,** Sbo I, Sfr I, Sst II, Tgl I
Sin I	(GG$_T^A$CC)	*Staphilococcus intermedius*[121]	Afl I, Asp 697, **Ava II,** Bal II, *Bam Nx,* Bme 216, Bti I, *Cau I,* Clm II, Eco 47I, Erp I, *Fdi I,* Fsp MS1, *Hgi BI, Hgi CII, Hgi EI, Hgi HIII,* Nsp HII, Sfn I

Table 22. Alphabetical Listing of Restriction Endonucleases with Their Related Isoschizomers (*continued*)

Enzyme and Specific Recognition Sequence (5′ → 3′)[a,b]		Origin[c]	Isoschizomers[a,b]
Sin MI	(GATC)	*Staphilococcus intermedius*[22]	*Bce 243,* Bsa PI, Bss GII, Bst EIII, Bst XII, Cpa I, Dpn II, Fnu AII, *Fnu CI, Hac I,* **Mbo I,** Meu I, Mno III, Mos I, Msp 67II, Mth I, Nde II, Nfl AII, Nfl I, Nla II, Nsi AI, Nsp AI, Nsu I, *Sau 3AI*
Ska I	(GCCGGC)	*Streptomyces karnatakensis*[22]	Ani I, Apr I, Eco 56I, Mis I, **Nae I,** Nba I, Nbr I, Nmu FI, Nmu I, NtaS II, Pgl I, Rlu I, Sao I
Ska II	(CTGCAG)	*Streptomyces karnatakensis*[22]	Ali AJI, Asp 36I, Asp 708, Bbi I, Bce 170, Bsu 1247, Bsu B, Cau III, *Cfl I,* Eae PI, Ecl 77I, Eco 36I, Eco 48I, Eco 49I, Eco 83I, Mau I, Mkr I, Nas I, Ngb I, Noc I, Pma I, Pmy I, **Pst I,** *Sal PI,* Sfl I, Xma II, Xor I, Xph I
Sla I	(C ↓ TCGAG)	*Streptomyces lavendulae*[213]	AbrI, Asp 703, Bbi III, *Blu I,* Bss HI, Bst HI, Bsu M, Bth I, *Ccr I,* Ccr II, Dde II, Mca I, Msi I, *PaeR 7, Pan I,* PflW I, Sau 3239, Scu I, Sex I, Sga I, Sgo I, Slu I, Spa PI, **Xho I,** *Xpa I*
Slu I	(CTCGAG)	*Streptomyces luteoreticuli*[212]	AbrI, Asp 703, Bbi III, *Blu I,* Bss HI, Bst HI, Bsu M, Bth I, *Ccr I,* Ccr II, Dde II, Mca I, Msi I, *PaeR 7, Pan I,* PflW I, Sau 3239, Scu I, Sex I, Sga I, Sgo I, *Sla I,* Spa PI, **Xho I,** Xpa I
Sma I	(CCC ↓ GGG)	*Serratia marcescens* Sb[65]	Cfr 9I*, Xcy I*, Xma I*
Sna I	(GTATAC)	*Sphaerotilus natans* gc[160]	Xca I
Spa PI	(CTCGAG)	*Streptomyces albus* ssp. *pathocidicus*[25]	AbrI, Asp 703, Bbi III, *Blu I,* Bss HI, Bst HI, Bsu M, Bth I, *Ccr I,* Ccr II, Dde II, Mca I, Msi I, *PaeR 7, Pan I,* PflW I, Sau 3239, Scu I, Sex I, Sga I, Sgo I, *Sla I,* Slu I, **Xho I,** *Xpa I*
Spa I	(GCATGC)	*Streptomyces albus* ssp. *pathocidus*[187]	*Pae I,* Spa XI, **Sph I**
Spa XI	(GCATGC)	*Streptomyces phaeochromogenes*[204]	*Pae I,* Spa I, **Sph I**
Sph I	(GCATG ↓ C)	*Streptomyces phaeochromogenes*[53]	*Pae I,* Spa I, Spa XI
Spl II	(GACN ↓ NNGTC)	*Spirulina platensis sub. siamese*[96]	Nta I, Fsu I, Tte I, Ttr I, **Tth 111I**
Spl III	(GG ↓ CC)	*Spirulina platensis sub. siamese*[96]	Asp 742, Blu II, Bse I, *Bsp RI,* Bss CI, Bst CI, Bsu 1076, Bsu 1114, *Bsu RI,* Clm I, *Clt I,* Fin SI, *Fnu DI,* **Hae III,** Hhg I, Mni I, Mnn II, Ngo II, Nla I, Pai I, Pal I, Ppu I, *Sfa I,* Sul I, Ttn I, Vha I
Sso I	(G ↓ AATTC)	*Shigella sonnei 47ns*[97]	**Eco RI,** Rsr I
Sst I	(GAGCT ↓ C)	*Streptomyces stanford*[62]	Eco 136II, Eco ICR, Nas SI, **Sac I,** Sco I
Sst II	(CCGC ↓ GG)	*Streptomyces stanford*[62]	Bac I, Cfr 37I, Csc I, Ecc I, Ecl 28I, Ecl 37I, Eco 55I, Eco 92I, *Gal I, Gce GLI, Gce I,* Kae 37I, Mra I, Ngo III, Nla SI, Saa I, Sab I, **Sac II,** Sbo I, Sfr I, Shy I, Tgl I
Sst IV	(TGATCA)	*Streptomyces stanford*[78]	Atu CI, **Bcl I,** Bst GI, Cpe I, Fba I, Pov I
Stu I	(AGG ↓ CCT)	*Streptomyces tubercidicus*[188]	Aat I, Eco 147I, *Gdi I,* Nta SI
Sty I	(C ↓ C$^{AA}_{TT}$GG)	*Salmonella typhi 27*[141]	Eco T 104, Eco T 14
Sul I	(GGCC)	*Sulfolobus acidocaldarius*[97]	Asp 742, Blu II, Bse I, *Bsp RI,* Bss CI, Bst CI, Bsu 1076, Bsu 1114, *Bsu RI,* Clm I, *Clt I,* Fin SI, *Fnu DI,* **Hae III,** Hhg I, Mni I, Mnn II, Ngo II, Nla I, Pai I, Pal I, Ppu I, *Sfa I,* Spl III, Ttn I, Vha I
Taq I	(T ↓ CGA)	*Thermus aquaticus* YTI[174]	Tfl I, TTh HB8I

Table 22. Alphabetical Listing of Restriction Endonucleases with Their Related Isoschizomers (continued)

Enzyme and Specific Recognition Sequence $(5' \rightarrow 3')^{a,b}$		Origin[c]	Isoschizomers[a,b]
Taq XI*	(CC ↓ $\frac{A}{T}$ GG)	*Thermus aquaticus*[63]	Aor I, Apy I*, Atu BI, Atu II, Bin SI, Bst GII, Bst NI, Cdi 27I, Cfr 5I, Cfr 11I, Cfr 20I, Eca II, Ecl II, Ecl 66I, Eco 27I, Eco 38I, Eco 60I, Eco 61I, Eco 67I, **Eco RII**, Mph I, Mva I*, Sgr II, Zan I*
Tce I	(GAAGA)	*Thermococcus celer*[97]	**Mbo II**, Ncu I
Tfl I	(TCGA)	*Thermus flavus* AT62[175]	**Taq I**, Tth HB8I
Tgl I	(CCGCGG)	*Thermopolyspora glauca*[61]	Bac I, Cfr 37I, *Csc I*, Ecc I, Ecl 28I, Ecl 37I, Eco 55I, Eco 92I, *Gal I, Gce GLI, Gce I*, Kae 37I, Mra I, Ngo III, Nla SI, Saa I, Sab I, **Sac II**, Sbo I, Sfr I, Shy I, Sst II
Tha I	(CG ↓ CG)	*Thermoplasma acidophilum*[133]	*Acc II*, Bce FI, Bce R, Bsu 1192 II, Bsu 1193, Bsu 6633, Bsu EII, **Fnu DII**, Fsp MI, Hin 1056 I
Tte I	(GACNNNGTC)	*Thermus thermophilus* strain 110[190]	Nta I, Fsu I, *Spl II*, Ttr I, **Tth 111I**
Tth HB8I	(TCGA)	*Thermus thermophilus* HB8[173]	**Taq I**, Tfl I
Tth 111I	(GACN ↓ NNGTC)	*Thermus thermophilus* 111[190]	Nta I, Fsu I, *Spl II*, Tte I, Ttr I
TtnI	(GGCC)	*Tolypothrix tenuis*[201]	Asp 742, Blu II, Bse I, *Bsp RI*, Bss CI, Bst CI, Bsu 1076, Bsu 1114, *Bsu RI, Clm I, Clt I*, Fin SI, *Fnu DI*, **Hae III**, Hhg I, Mni I, Mnn II, Ngo II, Nla I, Pai I, Pal I, Ppu I, *Sfa I*, Spl III, Sul I, Vha I
Ttr I	(GACNNNGTC)	*Thermus thermophilus* 23[190]	Nta I, Fsu I, *Spl II*, Tte I, **Tth 111I**
Vha I	(GGCC)	*Vibrio harveyi*[81]	Asp 742, Blu II, Bse I, *Bsp RI*, Bss CI, Bst CI, Bsu 1076, Bsu 1114, *Bsu RI, Clm I, Clt I*, Fin SI, *Fnu DI*, **Hae III**, Hhg I, Mni I, Mnn II, Ngo II, Nla I, Pai I, Pal I, Ppu I, *Sfa I*, Spl III, Sul I, Ttn I
Xam I	(GTCGAC)	*Xanthomonas amaranthicola*[1]	*Hgi CII, Hgi DII*, Nop I, Rhe I, Rhp I, Rrh I, Rro I, **Sal I**, *Xci I*
Xca I*	(GAT ↓ TAC)	*Xanthomonas campestris*	**Sna I**
Xci I	(G ↓ TCGAC)	*Xanthomonas citrii*[227]	*Hgi CIII, Hgi DII*, Nop I, Rhe I, Rhp I, Rrh I, Rro I, **Sal I**, Xam I
Xcy I*	(C ↓ CCGGG)	*Xanthomonas cyanopsidis* 13D5[52]	Cfr 9I*, **Sma I**, Xma I*
Xho I	(C ↓ TCGAG)	*Xanthomonas holcicola*[59]	AbrI, Asp 703, Bbi III, *Blu I*, Bss HI, Bst HI, Bsu M, Bth I, *Ccr I*, Ccr II, Dde II, Mca I, Msi I, *PaeR 7, Pan I*, PflW I, Sau 3239, Scu I, Sex I, Sga I, Sgo I, *Sla I*, Slu I, Spa I, *Xpa I*
Xho II	(Pu ↓ GATCPy)	*Xanthomonas holcicola*[155]	Mfl I
Xma I*	(C ↓ CCGGG)	*Xanthomonas malvacearum*[48]	Cfr 9I*, **Sma I**, Xcy I*
Xma II	(CTGCAG)	*Xanthomonas malvacearum*[48]	Ali AJI, Asp 36I, Asp 708, Bbi I, Bce 170, Bsu 1247, Bsu B, Cau III, *Cfl I*, Eae PI, Ecl 77I, Eco 36I, Eco 48I, Eco 49I, Eco 83I, Mau I, Mkr I, Nas I, Ngb I, Noc I, Pma I, Pmy I, **Pst I**, *Sal PI*, Sfl I, Ska II, Xor I, Xph I
Xma III	(C ↓ GGCCG)	*Xanthomonas malvacearum*[109]	*Eag I*, Eco 52 I
Xmn I	(GAANN ↓ NNTTC)	*Xanthomonas manihotis* 7AS1[242]	Asp 700
Xni I	(CGATCG)	*Xanthomonas nigromaculans*[69]	*Nbl I*, **Pvu I**, *Rsh I*, Rsp I, *Xor II*

Table 22. Alphabetical Listing of Restriction Endonucleases with Their Related Isoschizomers (*continued*)

Enzyme and Specific Recognition Sequence (5′ → 3′)[a,b]		Origin[c]	Isoschizomers[a,b]
Xor I	(CTGCAG)	*Xanthomonas oryzae*[223]	Ali AJI, Asp 36I, Asp 708, Bbi I, Bce 170, Bsu 1247, Bsu B, Cau III, *Cfl I*, Eae PI, Ecl 77I, Eco 36I, Eco 48I, Eco 49I, Eco 83I, Mau I, Mkr I, Nas I, Ngb I, Noc I, Pma I, Pmy I, **Pst I**, *Sal PI*, *Sfl I*, Ska II, Xma II, Xph I
Xor II	(CGAT ↓ CG)	*Xanthomonas oryzae*[223]	*Nbl I*, **Pvu I**, *Rsh I*, Rsp I, Xni I
Xpa I	(C ↓ TCGAG)	*Xanthomonas papavericola*[59]	AbrI, Asp 703, Bbi III, *Blu I*, Bss HI, Bst HI, Bsu M, Bth I, *Ccr I*, Ccr II, Dde II, Mca I, Msi I, *PaeR 7*, *Pan I*, PflW I, Sau 3239, Scu I, Sex I, Sga I, Sgo I, *Sla I*, Slu I, Spa PI, **Xho I**
Xph I	(CTGCAG)	*Xanthomonas phaseoli*[181]	Ali AJI, Asp 36I, Asp 708, Bbi I, Bce 170, Bsu 1247, Bsu B, Cau III, *Cfl I*, Eae PI, Ecl 77I, Eco 36I, Eco 48I, Eco 49I, Eco 83I, Mau I, Mkr I, Nas I, Ngb I, Noc I, Pma I, Pmy I, **Pst I**, *Sal PI*, *Sfl I*, Ska II, Xma II, Xor I
Zan I*	(CC^A_T ↓ GG)	*Zymomonas anaerobia*[205]	Aor I, Apy I*, Atu BI, Atu II, Bin SI, Bst GII, Bst NI, Cdi 27I, Cfr 5I, Cfr 11I, Cfr 20I, Eca II, Ecl II, Ecl 66I, Eco 27I, Eco 38I, Eco 60I, Eco 61I, Eco 67I, **Eco RII**, Mph I, Mva I*, Sgr II, Taq XI*

[a]Bold: prototype enzyme. Italics: cleavage site is identical to prototype's.

[b]Asterisk indicates recognition sequence identical but cleavage site different from prototype.

[c]References for this table can be found at the end of Table 23.

Arrow indicates position of cleavage.

Table 23. Alphabetical Listing of Restriction Endonucleases with No Known Isoschizomers

Enzyme	Specific Recognition Sequence (5′ → 3′)	Origin
Aat II	(GACGT ↓ C)	*Acetobacter aceti*[72]
Acc I	(GT ↓ $^{AG}_{CT}$ AC)	*Acinetobacter calcoaceticus*[242]
Acc III	(T ↓ CCGGA)	*Acinetabacter calcoaceticus*[242]
Acc III	(T ↓ CCGGA)	*Acinetobacter calcoaceticus*[147]
Afl II	(C ↓ TTAG)	*Anabaena flos-aquae*[230]
Afl III	(A ↓ CPuPyGT)	*Anabaena flos-aquae*[230]
Apa I	(GGGCC ↓ C)	*Acetobacter pasteurianus*[181]
ApaL I	(G ↓ TGCAC)	*Acetobacter pasteurianus* sub. *pasteurianus*[205]
Avr II	(C ↓ CTAGG)	*Anabaena variabilis* uw[170]
Bal I	(TGG ↓ CCA)	*Brevibacterium albidum*[57]
Bbv I	(GCAGC)	*Bacillus brevis*[61]
Bbv II	(GAAGAC)	*Bacillus brevis* 80[128]
Bbv SI	(GC^A_TGC)	*Bacillus brevis* S[218]
Bgl I	(GCCNNNN ↓ NGGC)	*Bacillus globigii*[43]
Bsm I	(GAATGC)	*Bacillus stearothermophilus* NUB 36[147]
Bsp HI	(TCATGA)	*Bacillus* species H[97]
Bsp MI	(ACCTGC)	*Bacillus* species M[143]
Bsu FI	(CCGG)	*Bacillus subtilis*[186]
Cfr 10I	(PuCCGGPy)	*Citrobacter freundii* RFL10[91]
Dde I	(C ↓ TNAG)	*Desulfovibrio desulfuricans* Norway strain[105]
Dra III	(CACNNN ↓ GTG)	*Deinococcus radiophilus*[70]

Table 23. Alphabetical Listing of Restriction Endonucleases with No Known Isoschizomers (*continued*)

Enzyme	Specific Recognition Sequence (5′ → 3′)	Origin
Eco 47III	(AGC ↓ GCT)	*Escherichia coli* RFL47[88]
Eco D	(TTANNNNNNNGTCPy)	*Escherichia coli* E166[149]
Eco 42I	$\left(\frac{\text{GGTCTC}}{\text{CCAGAG}}\right)$	*Escherichia coli* RFL 42[97]
Eco 43I	(CCNGG)	*Escherichia coli* RFL 43[97]
Eco 64I	$\left(\text{GG}\frac{\text{TA}}{\text{CG}}\text{CC}\right)$	*Escherichia coli* RFL 64[97]
Eco 82I	(GAATTC)	*Escherichia coli* RFL 32[97]
Eco 88I	$\left(\text{C}\frac{\text{T}}{\text{C}}\text{CG}\frac{\text{A}}{\text{G}}\text{G}\right)$	*Escherichia coli* RFL 88[97]
Eco 105I	(TACGTA)	*Escherichia coli* RFL 105[97]
Eco 130I	$\left(\text{CC}\frac{\text{AT}}{\text{TA}}\text{GG}\right)$	*Escherichia coli* RFL 130[97]
Eco R124	$\left(\text{GAANNNNNN}\frac{\text{A}}{\text{G}}\text{TCG}\right)$	*Escherichia coli*[97]
Eco R 124/3	$\left(\text{GAANNNNNNN}\frac{\text{A}}{\text{G}}\text{TCG}\right)$	*Escherichia coli*[97]
Esp I	(GC ↓ TNAGC)	*Eucapsis* species[20]
Fin I	(GTCCC)	*Flavobacterium indologenes*[143]
Gdi II	(Py ↓ GGCCG)	*Gluconobacter dioxyacetonicus*[216]
Hae I	$\left(\frac{\text{A}}{\text{T}}\text{GG} ↓ \text{CC}\frac{\text{T}}{\text{T}}\right)$	*Haemophilus aegyptius*[151]
Hga I	(GACGC)	*Haemophilus gallinarum*[210]
Hgi AI	$\left(\text{G}\frac{\text{A}}{\text{T}}\text{GC}\frac{\text{A}}{\text{T}} ↓ \text{C}\right)$	*Herpetosiphon giganteus* HP 1023[17]
Hgi EII	(ACCNNNNNNGGT)	*Herpetosiphon giganteus* Hpg 24[17]
Hgi JII	(GPuGCPy ↓ C)	*Herpetosiphon giganteus* HSF101[229]
Hph I	(GGTGA)	*Haemophilus parahaemolyticus*[138]
Mae II	(A ↓ CGT)	*Methanococcus aeolicus* PL-15/H[182]
Mae III	(↓ GTNAC)	*Methanococcus aeolicus* PL-15/H[182]
Mlu I	(A ↓ CGCGT)	*Micrococcus luteus*[202]
Mnl I	(CCTC)	*Moraxella nonliquefaciens*[244]
Nco I	(C ↓ CATGG)	*Nocardia corallina*[112]
Nde I	(CA ↓ TATG)	*Neisseria denitrificans*[111]
Nhe I	(G ↓ CATGC)	*Neisseria mucosa*[33]
Nla III	(CATG ↓)	*Neisseria lactamica*[165]
Nla IV	(GGN ↓ NCC)	*Neisseria lactamica*[165]
Not I	(GC ↓ GGCCGC)	*Nocardia otitidis-caviarum*
Nsp BII	$\left(\text{C}\frac{\text{A}}{\text{C}}\text{G} ↓ \text{C}\frac{\text{G}}{\text{T}}\text{G}\right)$	*Nostoc* species[47]
Nsp SAI	$\left(\text{C} ↓ \frac{\text{T}}{\text{C}}\text{CG}\frac{\text{A}}{\text{G}}\text{G}\right)$	*Nostoc* species SA[23]
Nsp SAIII	(CCATGG)	*Nostoc* species SA[97]
Pfl MI	(CCANNNNNTGG)	*Pseudomonas fluorescens*[143]
Ple I	(GAGTC)	*Pseudomonas lemoignei*
Ppa I	(GAGACC)	*Pseudomonas paucimobilis*[143]
Ppu MI	$\left(\text{PuG} ↓ \text{G}\frac{\text{A}}{\text{T}}\text{CCPy}\right)$	*Pseudomonas putida* M[145]
Rsa I	(GT ↓ AC)	*Rhodopseudomonas sphaeroides*[123]
Rsr II	$\left(\text{CG} ↓ \text{G}\frac{\text{A}}{\text{T}}\text{CCG}\right)$	*Rhodopseudomonas sphaeroides*[153]
SB	(GAGNNNNNPuTAPyG)	*Salmonella thyphimurium*[150]
SP	(AACNNNNNNTPuc)	*Salmonella potsdam*[150]
SQ	(AACNNNNNNPuTAPyG)	*Salmonella hybrid*[148]
Sca I	(AGT ↓ ACT)	*Streptomyces caespitosus*[105]
Sec I	(C ↓ CNNGG)	*Synechocystis* species 6701[21]

Table 23. Alphabetical Listing of Restriction Endonucleases with No Known Isoschizomers (*continued*)

Enzyme	Specific Recognition Sequence (5′ → 3′)	Origin
Sfa NI	(GATC)	*Streptococcus faecalis* ND547[179]
Sfi I	(GGCCNNNN ↓ NGGCC)	*Streptomyces exfoliatus*[163]
Sna BI	(TAC ↓ GAT)	*Sphaerotilus natans*[11]
Sno I	(GTGCAC)	*Streptomyces novocastria*[97]
Spe I	(A ↓ CTAGT)	*Sphaerotilus natans*[33]
Spl I	(C ↓ GTACG)	*Spirulina platensis*[96]
Ssp I	(AAT ↓ ATT)	*Sphaerotilus natans*[64]
Taq II	(GACCGA)	*Thermus aquaticus* YT1[147]
Tth 111II	(CCAPuCA)	*Thermus thermophilus*[191]
Xba I	(T ↓ CTAGA)	*Xanthomonas badrii*[245]

Arrow indicates position of cleavage.

REFERENCES FOR TABLES 22 AND 23

[1]Arrand, J. R., Myers, P. A., and Roberts, R. J. (1978), *J. Mol. Biol.*, **118**, 127.

[2]Arrand, J. R., Myers, P. A., and Roberts, R. J., unpublished observations.*

[3]Arutyunyan, E. E., Gruber, I. M., Polyachenko, V. M., Kuachadze, L. I., Andriashuili, I. A., Chanishuili, T. V., and Nilol'skaya, I. I. (1985), *Vopr. Med. Khim. SSSR*, **31**, 127.

[4]Azizbekyan, R. R., Rebentish, B. A., Stepanova, T. V., Netyksa, E. M., and Buchkova, M. A. (1984), *Dokl. Akad. Nauk. SSSR*, **274**, 742.

[5]Beaty, J. S., McLean-Bowen, C. A., and Brown, L. R. (1982), *Gene*, **18**, 61.

[6]Bennett, S. P., and Halford, S. E., unpublished observations.

[7]Bingham, A. H. A., Atkinson, T., Sciaky, D., and Roberts, R. J. (1978), *Nucl. Acids Res.*, **5**, 3457.

[8]Bingham, A. H. A., Atkinson, A., and Darbyshire, J., unpublished observations.*

[9]Bingham, A. H. A., and Darbyshire, J., unpublished observations.*

[10]Bolton, B., Nesch, G., Comer, M., Wolf, W., and Kessler, C. (1985), *FEBS Letters*, **182**, 130.

[11]Borsetti, R., Grandoni, R., and Schildkraut, I., unpublished observations.

[12]Borsetti, R., Wise, D., and Schildkraut, I., unpublished observations.

[13]Bron, S., Murray, K., and Trautner, T. A. (1975), *Mol. Gen. Genet.*, **143**, 13.

[14]Brown, N. L., and Smith, M. (1976), *FEBS Letters*, **65**, 284.

[15]Brown, N. L., and Smith, M. (1977), *Proc. Natl. Acad. Sci. USA*, **74**, 3213.

[16]Brown, N. L., in *XIth International Congress of Biochemistry Abstracts* (1979), p. 44.

[17]Brown, N. L., McClelland, M., and Whitehead, P. R. (1980), *Gene*, **9**, 49.

[18]Brown, N. L., unpublished observations.

[19]Butkus, V., Klimasauskas, S., Kersulyte, D., Vaitkevicius, D., Lebionka, A., and Janulaitis, A. A. (1985), *Nucl. Acids Res.*, **13**, 5727.

[20]Calleja, F., Dekker, B. M. M., Coursin, T., and deWaard, A. (1984), *FEBS Letters*, **178**, 69.

[21]Calleja, F., Tandeau deMarsac, N., Coursin, T., Van Ormondt, H., and deWaard, A. (1985), *Nucl. Acids Res.*, **13**, 6745.

[22]Camp, R., and Schildkraut, I., unpublished observations.

[23]Camp, R., and Visentin, L. P., unpublished observations.

[24]Catterall, J. F., and Welker, N. E. (1977), *J. Bacteriol.*, **129**, 1110.

[25]Chater, K. F. (1977), *Nucl. Acids Res.*, **4**, 1989.

[26]Chirikjian, J. G., George A., and Smith, I. A. (1978), *Fed. Proc.*, **37**, 1415.

[27]Clanton, D. J., Woodward, J. M., and Miller, R. V. (1978), *J. Bacteriol.*, **135**, 270.

[28]Coll, E., and Chirikjian, J., unpublished observations.*

[29]Comb, D. G., unpublished observations.

[30]Comb, D. G., Hess, E. J., and Wilson, G., unpublished observations.

[31]Comb, D. G., Parker, P., Grandoni, R., and Schildkraut, I., unpublished observations.

[32]Comb, D. G., Parker, P., and Schildkraut, I., unpublished observations.

[33]Comb, D. G., and Schildkraut, I., unpublished observations.

[34]Comb, D. G., Schildkraut, I., Wilson, G., and Greenough, L., unpublished observations.

[35]Comb, D. B., and Wilson, G., unpublished observations.

[36]Comb, D. G., Schildkraut, I., and Roberts, R. J., unpublished observations.*

[37]Cruz, A., K., Kidane, G., Pires, M. Q., Rabinovitch, L., Guaycurus, T. V., and Morel, C. M. (1984), *FEBS Letters*, **173**, 99.

[38]De Waard, A., unpublished observations.

[39]De Waard, A., and Duyvesteyn, M. (1980), *Arch. Microbiol.*, **128**, 242.

[40]De Waard, A., Korsuize, J., Van Beveren, C. P., and Maat, J. (1978), *FEBS Letters*, **96**, 106.

[41]De Waard, A., Van Beveren, C. P., Duyvesteyn, M., and Van Ormondt, H. (1979), *FEBS Letters*, **101**, 71.

[42]Di Lauro, R., unpublished observations.*

[43]Duncan, C. H., Wilson, G. A., and Young, F. E. (1978), *J. Bacteriol.*, **134**, 338.

[44]Duyvesteyn, M. G. C., and De Waard, A. (1980), *FEBS Letters*, **111**, 423.

*Cited in 1987 New England Biolabs catalog.

45Duyvesteyn, M. G. C., De Waard, A., and Van Ormondt, H. (1980), *FEBS Letters,* **117,** 241. (The bacterial strain producing this enzyme has been lost.)

46Duyvesteyn, M. G. C., Korsuize, J., and De Waard, A. (1981), *Plant. Mol. Biol.,* **1,** 75.

47Duyvesteyn, M. G. C., Korsuize, J., De Waard, A., Vonshak, A., and Wolk, C. P. (1983), *Arch. Microbiol.,* **134,** 276.

48Endow, S. A., and Roberts, R. J. (1977), *J. Mol. Biol.,* **112,** 521.

49Endow, S. A., and Roberts, R. J., unpublished observations.*

50Fisherman, J., Gingeras, T. R., and Roberts, J. R., unpublished observations.*

51Fitzgerald, G. F., Daly, C., Brown, L. R., and Gingeras, T. R. (1982), *Nucl. Acids Res.,* **10,** 8171.

52Froman, B. E., Tait, R. C., Kado, C. I., and Rodriguez, R. L. (1984), *Gene,* **28,** 331.

53Fuchs, L. Y., Covarrubias, L., Escalante, L., Sanchez, S., and Bolivar, F. (1980), *Gene,* **10,** 39.

54Gardner, J. F., Cohen, L. K., Lynn, S. P., and Kaplan, S., unpublished observations.*

55Gasperik, J., Godany, A., Hostinova, E., and Zelinka, J. (1983), *Biologia (Bratislava),* **38,** 315.

56Gelinas, R. E., Myers, P. A., and Roberts, R. J. (1977), *J. Mol. Biol.,* **114,** 169.

57Gelinas, R. E., Myers, P. A., Weiss, G. A., Murray, K., and Roberts, R. J. (1977), *J. Mol. Biol.,* **114,** 433.

58Gelinas, R. E., Myers, P. A., and Roberts, R. J., unpublished observations.*

59Gingeras, T. R., Myers, P. A., Olson, J. A., Hanberg, F. A., and Roberts, R. J. (1978), *J. Mol. Biol.,* **118,** 113.

60Gingeras, T. R., Greenough, L., Schildkraut, I., and Roberts, R. J. (1981), *Nucl. Acids Res.,* **9,** 4525.

61Gingeras, T. R., and Roberts, R. J., unpublished observations.*

62Goff, S. P., and Rambach, A. (1978), *Gene,* **3,** 347.

63Grachev, S. A., Mamaev, S. V., Gurevich, A. I., Igoshin, A. V., Kolosov, M. N., and Slyusarenko, A. G. (1981), *Bioorg. Khim.,* **7,** 628.

64Grandoni, R. P., and Schildkraut, I., unpublished observations.

65Green, R., and Mulder, C., unpublished observations.*

66Greenaway, P. J. (1980), *Biochem. Biophys. Res. Commun.,* **95,** 1282.

67Greene, P. J., Betlach, M. C., Boyer, H. W., and Goodman, H. M. (1974), *Methods Mol. Biol.,* **7,** 87.

68Grosveld, G. C., unpublished observations.

69Hanberg, F., Myers, P. A., and Roberts, R. J., unpublished observations.*

70Hansen, R., unpublished observations.

71Hartman, H., and Goebel, W. (1977), *FEBS Letters,* **80,** 285.

72Hinkle, N. F., and Miller, R. V. (1979), *Plasmid,* **2,** 387.

73Hiraoka, N., Kita, K., Nakajima, H., and Obayashi, A. (1984), *J. Ferment. Technol.,* **62,** 583.

74Hiraoka, N., Kita, K., Nakajima, F., Kimizuka, F., and Obayashi, A. (1985), *J. Ferment. Technol.,* **63,** 151.

75Hobom, G., Schwaz, E., Melzer, M., and Mayer, H. (1981), *Nucl. Acids Res.,* **9,** 4823.

76Hobom, G., Mayer, H., and Schutte, H., unpublished observations.*

77Hoshino, T., Vozumi, T., Horinouchi, S., Ozaki, A., Beppu, T., and Arima, K. (1977), *Biochim. Biophys. Acta,* **479,** 367.

78Hu, A. W., Kuebbing, D., and Blakesley, R. J. (1978), *Fed. Proc.,* **38,** 780.

79Hughes, S. G., and Murray, K. (1980), *Biochem. J.,* **185,** 65.

80Hughes, S. G., Bruce, T., and Murray, K. (1980), *Biochem. J.,* **185,** 59.

81Hurlin, P., and Schildkraut, I., unpublished observations.

82Janulaitis, A. A., and Adomaviciute, L., unpublished observations.

83Janulaitis, A. A., and Bitinaite, J., unpublished observations.

84Janulaitis, A. A., Bitinaite, J., and Jaskeleviciene, B. (1983), *FEBS Letters,* **151,** 243.

85Janulaitis, A. A., Marcinkeviciene, L. U., and Petrusyte, M. P. (1982), *Dokl. Akad. Nauk. SSSR,* **577,** 241.

86Janulaitis, A. A., Marcinkeviciene, L., Petrusyte, M., and Mironov, A. (1981), *FEBS Letters,* **134,** 172.

87Janulaitis, A. A., and Petrusyte, M., unpublished observations.

88Janulaitis, A. A., Petrusyte, M., and Buktus, V. V., *FEBS Letters,* in press.

89Janulaitis, A. A., Petrusyte, M., and Buktus, V. V. (1981), *Dokl. Akad. Nauk. SSSR,* **257,** 749.

90Janulaitis, A. A., Stakenas, P. S., and Berlin, Y., unpublished observations.

91Janulaitis, A. A., Stakenas, P., and Berlin, Y. (1983), *FEBS Letters,* **161,** 210.

92Janulaitis, A. A., Stakenas, P. S., Bitinaite, J. B., and Jaskeleviciene, B. P. (1983), *Dokl. Akad. Nauk. SSSR,* **271,** 483.

93Janulaitis, A. A., Stakenas, P., Jaskeleviciene, B. P., Lebedenko, E. N., and Berlin, Y. (1980), *Bioorg. Khim.,* **6,** 1746.

94Janulaitis, A. A., Stakenas, P. S., Petrusyte, M. P., Bitinaite, J. P., Klimasauskas, S. J., and Buktus, V. V., *Molekulyarnaya Biologiya,* (Cited in New England Biolabs catalog).

95Jiang, B. D., and Myers, P. A., unpublished observations.*

96Kawamura, M., Sakakibara, M., Watanabe, T., Kita, K., Hiraoka, N., Obayashi, A., Takagi, M., and Yano, K. (1986), *Nucl. Acids Res.,* **14,** 1895.

97Kessler, C., and Höltke, H. J. (1986), *Gene,* **47,** 1.

98Kessler, C., Neumaier, P. S., and Wolf, W. (1985), *Gene,* **33,** 1.

99Kholmina, G. V., Rebentish, B. A., Skoblov, Y. S., Mironov, A. A., Yankovsky, N. K., Kozlov, Y. I., Glatman, L. I., Moroz, A. F., and Debabov, V. G. (1980), *Dokl. Akad. Nauk. SSSR,* **253,** 495.

100Khosaka, T., unpublished observations.

101Khosaka, T., and Kiwaki, M. (1984), *FEBS Letters,* **177,** 57.

102Khosaka, T., and Kiwaki, M. (1984), *Gene,* **31,** 251.

103Khosaka, T., Sakurai, T., Takahashi, H., and Saito, H. (1982), *Gene,* **17,** 117.

104Kiss, A., Sain, B., Csordas-Toth, E., and Venetianer, P. (1977), *Gene,* **1,** 323.

105Kojima, H., Takahashi, H., and Saito, H., unpublished observations.

106Kramarov, V. M., Pachkunov, D. M., and Matvienko, N. I. (1983), in A. I. Gaziev, ed., *Nek. Aspekty Fiziol. Mikroorg.,* Akad. Nauk. SSSR 22.

107Kroger, M., Hobom, G., Schutte, H., and Mayer, H. (1984), *Nucl. Acids Res.,* **12,** 3127.

[08]Kroger, M., Mayer, H., Schutte, H., and Hobom, G., unpublished observations.*

[09]Kunkel, L. M., Silberklang, M., and McCarthy, B. J. (1979), *J. Mol. Biol.,* **132,** 133.

[10]Lacks, S., and Greenberg, B. (1975), *J. Biol. Chem.,* **250,** 4060.

[11]Langdale, J. A., Myers, P. A., and Roberts, R. J. (1984), *Nucl. Acids Res.,* **12,** 167.

[12]Langdale, J. A., Myers, P. A., and Roberts, R. J., unpublished observations.

[13]Lau, R. H., and Doolittle, W. F. (1980), *FEBS Letters,* **121,** 200.

[14]Lau, R. H., Visentin, L. P., Martin, S. M., Hofman, J. D., and Doolittle, W. F. (1985), *FEBS Letters,* **179,** 129.

[15]Leary, J. V., unpublished observations.

[16]Le Bon, J. M. (1978), *Fed. Proc.,* **37,** 1413.

[17]Le Bon, J. M., Kado, C., Rosenthal, L., and Chirikjian, J. (1978), *Proc. Natl. Acad. Sci. USA,* **75,** 4097.

[18]Leung, D. W., Lui, A. C. P., Merilees, H., McBride, B. C., and Smith, M. (1979), *Nucl. Acids Res.,* **6,** 17.

[19]Lin, B. C., and Roberts, R. J., unpublished observations.

[20]Lui, A. C. P., McBride, B. C., Bovis, G. F., and Smith, M. (1979), *Nucl. Acids Res.,* **6,** 1.

[21]Lupker, H. S. C., and Dekker, B. M. M. (1981), *Biochim. Biophys. Acta,* **654,** 297.

[22]Lynn, S. P., Cohen, L. K., Gardner, J. F., and Kaplan, S. J. (1979), *J. Bacteriol.,* **138,** 505.

[23]Lynn, S. P., Cohen, L. K., Kaplan, S., and Gardner, J. F. (1980), *J. Bacteriol.,* **142,** 380.

[24]Makula, R. A., unpublished observations.*

[25]Makuka, R. A., and Meagher, R. B. (1980), *Nucl. Acids Res.,* **8,** 3125.

[26]Mann, M. B., Rao, R. N., and Smith, H. O. (1978), *Gene,* **3,** 97.

[27]Mann, M. B., and Smith, H. O., unpublished observations.*

[28]Maratea, E., and Camp, R. R., unpublished observations.

[29]Matvienko, N. I., Pachkunov, D. M., and Kramarov, V. M. (1984), *FEBS Letters,* **177,** 23.

[30]Mayer, H., Grosschedt, R., Schutte, H., and Hobom, G. (1981), *Nucl. Acids Res.,* **9,** 4833.

[31]Mayer, H., and Klaar, J., unpublished observations.*

[32]Mayer, H., and Schutte, H., unpublished observations.*

[33]McConnell, D. J., Searcy, D. G., and Sutcliffe, J. G. (1978), *Nucl. Acids Res.,* **5,** 1729.

[34]McEvoy, S., and Roberts, R. J., unpublished observations.

[35]Meagher, R. B., unpublished observations.*

[36]Meagher, R. B., unpublished observations.

[37]Middleton, J. H., Edgell, M. H., and Hutchinson, C. A. III (1972), *J. Virol.,* **10,** 42.

[38]Middleton, J. H., Stankus, P. V., Edgell, M. H., and Hutchinson, C. A. III, unpublished observations.

[39]Middleton, J. H., Stankus, P. V., Edgell, M. H., and Hutchinson, C. A. III, unpublished observations.*

[40]Mise, K., and Nakajima, K. (1984), *Gene,* **30,** 79.

[41]Mise, K., and Nkakjima, K. (1985), *Gene,* **33,** 357.

[42]Mise, K., and Nakajima, K. (1985), *Gene,* **36,** 363.

[143]Morgan, R., unpublished observations.

[144]Morgan, R., Camp, R., and Soltis, A., unpublished observations.

[145]Morgan, R., and Hempstead, S. K., unpublished observations.

[146]Murray, K., Hughes, S. G., Brown, J. S., and Bruce, S. A. (1976), *Biochem. J.,* **159,** 317.

[147]Myers, P. A., and Roberts, R. J., unpublished observations.*

[148]Nagaraja, V., Shepherd, J. C. W., and Bickle, T. A. (1985), *Nature,* **316,** 371.

[149]Nagaraja, V., Stieger, M., Nager, C., Hadi, S. M., and Bickle, T. (1985), *Nucl. Acids Res.,* **13,** 389.

[150]Nagaraja, V., Shepherd, J. C. W., Pripfl, T., and Bickle, T. A. (1985), *J. Mol. Biol.,* **182,** 579.

[151]Nardone, G., and Blakesley, R. (1981), *Fed. Proc.,* **40,** 1848.

[152]Norlander, L., Davies, J. K., Hagblom, P., and Normark, S. (1981), *J. Bacteriol.,* **145,** 788.

[153]O'Connor, C. D., Metcalf, E., Wrighton, C. J., Harris, T. J. R., and Saunders, J. R. (1984), *Nucl. Acids Res.,* **12,** 6701.

[154]Old, R., Murray, K., and Roizes, G. (1975), *J. Mol. Biol.,* **92,** 331.

[155]Olson, J. A., Myers, P. A., and Roberts, R. J., unpublished observations.*

[156]Orekhov, A. V., Rebentish, B. A., and Debavov, V. G. (1982), *Dokl. akad. Nauk. SSSR,* **263,** 217.

[157]Parker, P., and Schildkraut, I., unpublished observations.

[158]Piekarowicz, A., Stasiak, A., and Stanczak, J. (1980), *Acta Microbiol. Pol.,* **29,** 151.

[159]Piekarowicz, A., Stasiak, A., and Stanczak, J. (1980), *J. Mol. Biol.,* **157,** 373.

[160]Pope, A., Lynn, S. P., and Gardner, J. F., unpublished observations.

[161]Pugatsch, T., and Weber, H. (1979), *Nucl. Acids Res.,* **7,** 1429.

[162]Purvis, I. J., and Moseley, B. E. B. (1983), *Nucl. Acids Res.,* **3,** 5467.

[163]Qiang, B. Q., and Schildkraut, I. (1984), *Nucl. Acids Res.,* **12,** 4507.

[164]Qiang, B. Q., and Schildkraut, I., unpublished observations.

[165]Qiang, B. Q., Schildkraut, I., and Visentin, L., unpublished observations.

[166]Reaston, J., Duyvesteyn, M. G. C., and De Waard, A. (1982), *Gene,* **20,** 103.

[167]Roberts, R. J., Breitmeyer, J. B., Tabachnik, N. F., and Myers, P. A. (1975), *J. Mol. Biol.,* **91,** 121.

[168]Roberts, R. J., Myers, P. A., Morrison, A., and Murray, K. (1976), *J. Mol. Biol.,* **102,** 157.

[169]Roberts, R. J., Myers, P. A., Morrison, A., and Murray, K. (1976), *J. Mol. Biol.,* **103,** 199.

[170]Rosenvold, E. C., and Szybalski, W., unpublished observations [cited in *Gene,* **7,** 217 (1979)].

[171]Sasaki, J., and Yamada, Y. (1984), *Agric. Biol. Chem.,* **48,** 3027.

[172]Sasaki, J., Murakami, M., and Yamada, Y. (1985), *Agric. Biol. Chem.,* **49,** 3107.

[173]Sato, S., and Shinomiya, T. (1978), *J. Biochem.,* **84,** 1319.

[174]Sato, S., Hutchinson, C. A., and Harris, J. I. (1977), *Proc. Natl. Acad. Sci. USA,* **74,** 542.

[175]Sato, S., Nakazawa, K., and Shinomiya, T. (1980), *J. Biochem., 88,* 737.

[176]Schildkraut, I., unpublished observations.*

[177]Schildkraut, I., and Comb, D., unpublished observations.*

[178]Schwabe, G., Posseckert, G., and Klingmuller, W., submitted for publication to *Gene.*

[179]Sciaky, D., and Roberts, R. J., unpublished observations.*

[180]Seurink, J., and Van Montagu, M., unpublished observations.*

[181]Seurinck, J., Van de Voorde, A., and Van Montagu, M. (1983), *Nucl. Acids Res., 11,* 4409.

[182]Shen, S., Li, Q., Yan, P., Zhou, B., Ye, S., Lu, Y., and Wang, D. (1980), *Sci. Sin., 23,* 1435.

[183]Sharp, P. A., Sugden, B., and Sambrook, J. (1973), *Biochemistry, 12,* 3055.

[184]Shibata, T., and Ando, T. (1976), *Biochim. Biophys. Acta, 442,* 184.

[185]Shibata, T., and Ando, T. (1975), *Mol. Gen. Genet., 138,* 269.

[186]Shibata, T., Ikawa, S., Kim, C., and Ando, T. (1976), *J. Bacteriol., 128,* 437.

[187]Shimotsu, H., Takahashi, H., and Saito, H. (1980), *Agric. Biol. Chem., 44,* 1665.

[188]Shimotsu, H., Takahashi, H., and Saito, H. (1980), *Gene, 11,* 219.

[189]Shinomiya, T., unpublished observations.*

[190]Shinomiya, T., and Sato, S. (1980), *Nucl. Acids Res., 8,* 43.

[191]Shinomiya, T., Kobayashi, M., and Sato, S. (1980), *Nucl. Acids Res., 8,* 3275.

[192]Smith, D. I., Blattner, F. R., and Davies, J. (1976), *Nucl. Acids Res., 3,* 343.

[193]Smith, H. O., and Wilcox, K. W. (1970), *J. Mol. Biol., 51,* 379.

[194]Smith, J., and Comb, D., unpublished observations.*

[195]Smith, L., Blakesley, R., and Chirikjian, J., Cited in BRL catalog.

[196]Smith, L., Blakesley, R., and Chirikjian, J., unpublished observations.*

[197]Sokolov, N. N., Fitsner, A. B., Anikeitcheva, N. V., Choroshoutina, Yu. B., Samko, O. T., Kolosha, V. O., Fodor, I., and Votrin, I. I. (1985), *Molec. Biol. Rep., 10,* 159.

[198]Stephens, M. A., unpublished observations.*

[199]Stote, R., and Schildkraut, I., unpublished observations.

[200]Stotz, A., and Philippson, P., unpublished observations.*

[201]Streips, U., and Golemboski, B., unpublished observations.*

[202]Sugisaki, H., and Kanazawa, S. (1981), *Gene, 16,* 73.

[203]Sugisaki, H., Maekawa, Y., Kanazawa, S., and Takanami, M. (1982), *Nucl. Acids Res., 10,* 5747.

[204]Sugisaki, H., Maekawa, Y., Kanazawa, S., and Takanami, M. (1982), *Bull. Inst. Chem. Res. Kyoto Univ., 60,* 328.

[205]Sun, D. K., and Yoo, O. J., unpublished observations.

[206]Sussenbach, J. S., Monfoort, C. H., Schiphof, R., and Stoberingh, E. E. (1976), *Nucl. Acids Res., 3,* 3193.

[207]Sussenbach, J. S., Steenbergh, P. H., Post, J. A., Van Leeuwen, W. J., and Van Embden, J. D. A. (1978), *Nucl. Acids Res., 5,* 1153.

[208]Syddall, R., and Stachow, C. (1985), *Biochim. Biophys. Acta, 825,* 236.

[209]Szekeres, M., unpublished observations.

[210]Takanami, M. (1974), *Methods Mol. Biol., 7,* 113.

[211]Takanami, M. (1974), *Methods Mol. Biol., 7,* 113.

[212]Takahashi, H., unpublished observations.*

[213]Takahashi, H., Shimizu, M., Saito, H., Ikeda, Y., and Sugisaki, H. (1979), *Gene, 5,* 9.

[214]Timko, J., Horwitz, A. H., Zelinka, J., and Wilcox, G. (1981), *J. Bacteriol., 145,* 873.

[215]Van Den Hondel, C. A. M. J. J., Van Leen, R. W., Van Arkel, G. A., Duyvesteyn, M., and De Waard, A. (1983), *FEMS Microbiology Letters, 16,* 7.

[216]Van Montagu, M., unpublished observations.*

[217]Van Montagu, M., Sciaky, D., Myers, P. A., and Roberts, R. J., unpublished observations.*

[218]Vanyushin, B. F., and Dobritsa, A. P. (1975), *Biochim. Biophys. Acta, 407,* 61.

[219]Venetianer, P. unpublished observations.

[220]Visentin, L. P., unpublished observations.

[221]Walker, J. N. B., Dean, P. D. G., and Saunders, J. R. (1986), *Nucl. Acids Res., 14,* 1293.

[222]Walter, F., Hartmann, M., and Roth, M., in *Abstracts of the 12th FEBS Symposium,* Dresden (1978).

[223]Wang, R. Y. H., Shedlarski, J. G., Farber, M. B., Kuebbing, D., and Ehrlich, M. (1980), *Biochim. Biophys. Acta, 606,* 371.

[224]Wang, T. S. (1981), *Ko Hsueh Tung Pao, 26,* 815.

[225]Wani, A. A., Stephens, R. E., d'Ambrosio, S. M., and Hart, R. W. (1981), *ASM Abstracts,* Abstract 1778.

[226]Watson, R., Zueker, M., Martin, S. M., and Visentin, L. P. (1980), *FEBS Letters, 118,* 47.

[227]Whang, Y., and Too, O. J., unpublished observations.

[228]Whitehead, P. R., and Brown, N. L. (1982), *FEBS Letters, 143,* 296.

[229]Whitehead, P. R., and Brown, N. L. (1983), *FEBS Letters, 143,* 97.

[230]Whitehead, P. R., and Brown, N. L. (1985), *J. Gen. Microbiol., 131,* 951.

[231]Wickberg, L., and Schildkraut, I., unpublished observations.

[232]Wilson, G. A., and Young, F. E. (1975), *J. Mol. Biol., 97,* 123.

[233]Wilson, G. A., and Young, F. E., unpublished observations.*

[234]Winkler, K. (1979), diploma dissertation.

[235]Wirth, R., and Kessler, C., unpublished observations.

[236]Wu, R., King, C. T., and Jay, E. (1978), *Gene, 4,* 329.

[237]Yamada, Y., and Murakami, M. (1985), *Agric. Biol. Chem., 49,* 3627.

[238]Yamada, Y., Yoshioka, H., Sasaki, J., and Tahara, Y. (1983), *J. Gen. Appl. Microbiol., 29,* 157.

[239]Yoshida, Y., and Mise, K., unpublished observations.

[240]Yoshimori, R. N. (1971), Ph.D. thesis, University of California, San Francisco, CA.

[241]Yoshioka, H., Nakamura, H., Sasaki, J., Tahara, Y., and Yamada, Y. (1983), *Agric. Biol. Chem., 47,* 2871.

[242]Zabeau, M., and Roberts, R. J., unpublished observations.

[243]Zabeau, M., and Roberts, R. J., unpublished observations.*

[244]Zabeau, M., Green, R., Myers, P. A., and Roberts, R. J., unpublished observations.

[245]Zain, B. S., and Roberts, R. J. (1977), *J. Mol. Biol., 115,* 249.

[246]Zerler, B., Myers, P. A., Escalante, H., and Roberts, R. J., unpublished observations.

Table 24. Class IIs Restriction Endonucleases

Enzyme	Recognition and Cleavage Sites	Enzyme	Recognition and Cleavage
Bbv I[a]	GCAGCNNNNNNNN ↓ CGTCGNNNNNNNNNNNN ↓	Mbo II[a]	GAAGANNNNNNNN ↓ CTTCTNNNNNNN ↓
Bbv II	GAAGACNN ↓ CTTCTGNNNNNN ↓	Mnl I	CCTCNNNNNNN ↓ GGAGNNNNNNN ↓
Bin I	CGATCNNNN ↓ CCTAGNNNNN ↓	Sfa NI	GCATCNNNNN ↓ CGTAGNNNNNNNNN ↓
Fok I[a]	GGATGNNNNNNNNN ↓ CCTACNNNNNNNNNNNNN ↓	Taq II (a)	GACCGANNNNNNNNNNNN ↓ CTGGCTNNNNNNNNN ↓
Hga I	GACGCNNNNN ↓ CTGCGNNNNNNNNNN ↓	Taq II (b)	CACCCANNNNNNNNNNNN ↓ GTGGGTNNNNNNNNN ↓
Hph I[a]	GGTGANNNNNNNN ↓ CCACTNNNNNNN ↓	Tth 111II	CAAPuCANNNNNNNNNNNN ↓ GTTPyGTNNNNNNNNNN ↓

These enzymes do not exhibit endonucleolytic activity toward M13 single stranded DNA.

Table 25. Hybrid Sequences Resulting from the Ligation of Two Fragments Generated After Digestion with One or Two Different Restriction Endonucleases

Enzyme 1	Fragment Generated by Enzyme 1	Enzyme 2	Fragment Generated by Enzyme 2	Resulting Hybrid Sequence	Enzyme 3
AccI	GT ↓ CGAC	AhaII	GA ↓ CGCC	GTCGCC	AccI(*), AhaII(*)
AccI	GT ↓ CGAC	AhaII	GA ↓ CGTC	GTCGTC	AccI(*), AhaII(*)
AccI	GT ↓ CGAC	AosII	GA ↓ CGCC	GTCGCC	AccI(*), AosII(*)
AccI	GT ↓ CGAC	AosII	GA ↓ CGTC	GTCGTC	AccI(*), AosII(*)
AccI	GT ↓ CGAC	AquI	CC ↓ CGAG	GTCGAG	AccI(*), AquI(*), TaqI(2), TthHB8I(2)
AccI	GT ↓ CGAC	AquI	CC ↓ CGGG	GTCGGG	AccI(*), AquI(*)
AccI	GT ↓ CGAC	AsuII	TT ↓ CGAA	GTCGAA	AccI(*), AsuII(*), TaqI(2), TthHB8I(2)
AccI	GT ↓ CGAC	BanIII	AT ↓ CGAT	GTCGAT	AccI(*), BanIII(*), TaqI(2), TthHB8I(2)
AccI	GT ↓ CGAC	BbiII	GA ↓ CGCC	GTCGCC	AccI(*), BbiII(*)
AccI	GT ↓ CGAC	BbiII	GA ↓ CGTC	GTCGTC	AccI(*), BbiII(*)
AccI	GT ↓ CGAC	ClaI	AT ↓ CGAT	GTCGAT	AccI(*), ClaI(*), TaqI(2), TthHB8I(2)
AccI	GT ↓ CGAC	FspII	TT ↓ CGAA	GTCGAA	AccI(*), FspII(*), TaqI(2), TthHB8I(2)
AccI	GT ↓ CGAC	HapII	C ↓ CGG	GTCGG	AccI(*), Hap II(*)
AccI	GT ↓ CGAC	HinpI	G ↓ CGC	GTCGC	AccI(*), HinpI(*)
AccI	GT ↓ CGAC	HpaII	C ↓ CGG	GTCGG	AccI(*), HpaII(*)
AccI	GT ↓ CGAC	MaeII	A ↓ CGT	GTCGT	AccI(*), MaeII(*)
AccI	GT ↓ CGAC	MspI	C ↓ CGG	GTCGG	AccI(*), MspI(*)
AccI	GT ↓ CGAC	NarI	GG ↓ CGCC	GTCGCC	AccI(*), NarI(*)
AccI	GT ↓ CGAC	NunII	GG ↓ CGCC	GTCGCC	AccI(*), NunII(*)
AccI	GT ↓ CGAC	TaqI	T ↓ CGA	GTCGA	AccI(*), TaqI(2), TthHB8I(2)
AccI	GT ↓ CGAC	TthHB8I	T ↓ CGA	GTCGA	AccI(*), TthHB8I(2), TaqI(2)

Table 25. Hybrid Sequences Resulting from the Ligation of Two Fragments Generated After Digestion with One or Two Different Restriction Endonucleases (*continued*)

Enzyme 1	Fragment Generated by Enzyme 1	Enzyme 2	Fragment Generated by Enzyme 2	Resulting Hybrid Sequence	Enzyme 3
AccIII	T ↓ CCGGA	AvaI	C ↓ CCGGG	TCCGGG	AccIII(*), AvaI(*), BcnI(4), HapII(2), HpaII(2), MspI(2), NciI(3), ScrFI(3)
AccIII	T ↓ CCGGA	Cfr10I	A ↓ CCGGC	TCCGGC	AccIII(*), Cfr10I(*), HapII(2), HpaII(2), MspI(2)
AccIII	T ↓ CCGGA	Cfr10I	A ↓ CCGGT	TCCGGT	AccIII(*), Cfr10I(*), HapII(2), HpaII(2), MspI(2)
AccIII	T ↓ CCGGA	Nsp7524III	C ↓ CCGGG	TCCGGG	AccIII(*), Nsp7524III(*), BcnI(4), HapII(2), HpaII(2), MspI(2), NciI(3), ScrFI(3)
AccIII	T ↓ CCGGA	XcyI	C ↓ CCGGG	TCCGGG	AccIII(*), XcyI(*), BcnI(4), HapII(2), HpaII(2), MspI(2), NciI(3), ScrFI(3)
AccIII	T ↓ CCGGA	XmaI	C ↓ CCGGG	TCCGGG	AccIII(*), XmaI(*), BcnI(4), HapII(2), HpaII(2), MspI(2), NciI(3), ScrFI(3)
AflI	G ↓ GACC	PpuMI	AG ↓ GACCT	GGACCT	AflI(1), PpuMI(*), AsuI(1), AvaII(1), Cfr13I(1), ClaII(1), Eco47I(0), Nsp7524IV(1), Nsp7524V(1), Sau96I(1), SinI(1)
AflI	G ↓ GACC	PpuMI	AG ↓ GACCC	GGACCC	AflI(1), PpuMI(*), AsuI(1), AvaII(1), Cfr13I(1), ClaII(1), Eco47I(0), NlaIV(3), Nsp7524IV(1), Nsp7524V(1), Sau96I(1), SinI(1)
AflI	G ↓ GTCC	PpuMI	AG ↓ GTCCT	GGTCCT	AflI(1), PpuMI(*), AsuI(1), AvaII(1), Cfr13I(1), ClaII(1), Eco47I(0), Nsp7524IV(1), Nsp7524V(1), Sau96I(1), SinI(1)
AflI	G ↓ GTCC	PpuMI	AG ↓ GTCCC	GGTCCC	AflI(1), PpuMI(*), AsuI(1), AvaII(1), Cfr13I(1), ClaII(1), Eco47I(0), NlaIV(3), Nsp7524IV(1), Nsp7524V(1), Sau96I(1), SinI(1)
AflI	G ↓ GACC	RsrII	CG ↓ GACCG	GGACCG	AflI(1), RsrII(*), AsuI(1), AvaII(1), Cfr13I(1), ClaII(1), Eco47I(0), Nsp7524IV(1), Nsp7524V(1), Sau96I(1), SinI(1)
AflI	G ↓ GTCC	RsrII	CG ↓ GTCCG	GGTCCG	AflI(1), RsrII(*), AsuI(1), AvaII(1), Cfr13I(1), ClaII(1), Eco47I(0), Nsp7524IV(1), Nsp7524V(1), Sau96I(1), SinI(1)

Table 25.   Hybrid Sequences Resulting from the Ligation of Two Fragments Generated After Digestion with One or Two Different Restriction Endonucleases (*continued*)

Enzyme 1	Fragment Generated by Enzyme 1	Enzyme 2	Fragment Generated by Enzyme 2	Resulting Hybrid Sequence	Enzyme 3
AflIII	A ↓ CGCGT	BssHII	G ↓ CGCGC	ACGCGC	AflIII(*), BssHII(*), CfoI(5), FnudII(3), HhaI(5), HinpI(3), ThaI(3)
AflIII	A ↓ CATGT	EcoT14I	C ↓ CATGG	ACATGG	AflIII(*), EcoT14I(*), NlaIII(5)
AflIII	A ↓ CATGT	NcoI	C ↓ CATGG	ACATGG	AflIII(*), NcoI(*), NlaIII(5)
AflIII	A ↓ CATGT	StyI	C ↓ CATGG	ACATGG	AflIII(*), StyI(*), NlaIII(5)
AhaII	GA ↓ CGCC	AccI	GT ↓ CGAC	GACGAC	AhaII(*), AccI(*)
AhaII	GG ↓ CGCC	AccI	GT ↓ CGAC	GGCGAC	AhaII(*), AccI(*)
AhaII	GA ↓ CGCC	AhaII	GA ↓ CGTC	GACGTC	AhaII(2), AatII(5), AosII(2), BbiII(2), MaeII(2)
AhaII	GA ↓ CGCC	AhaII	GG ↓ CGCC	GACGCC	AhaII(2), AosII(2), BbiII(2), HasI(10)
AhaII	GG ↓ CGCC	AhaII	GA ↓ CGCC	GGCGCC	AhaII(2), AosII(2), BanI(1), BbeI(5), BbiII(2), CfoI(4), HaeII(5), HhaI(4), HinpI(2), NarI(2), NlaIV(3), NunII(2)
AhaII	GG ↓ CGCC	AhaII	GA ↓ CGTC	GGCGTC	AhaII(2), AosII(2), BbiII(2), HsaI(11)
AhaII	GA ↓ CGCC	AosII	GA ↓ CGTC	GACGTC	AhaII(2), AosII(2), AatII(5), BbiII(2), MaeII(2)
AhaII	GA ↓ CGCC	AosII	GG ↓ CGCC	GACGCC	AhaII(2), AosII(2), BbiII(2), HgaI(10)
AhaII	GG ↓ CGCC	AosII	GA ↓ CGCC	GGCGCC	AhaII(2), AosII(2), BanI(1), BbeI(5), BbiII(2), CfoI(4), HaeII(5), HhaI(4), HinpI(2), NarI(2), NlaIV(3), NunII(2)
AhaII	GG ↓ CGCC	AosII	GA ↓ CGTC	GGCGTC	AhaII(2), AosII(2), BbiII(2), HsaI(11)
AhaII	GA ↓ CGCC	AquI	CC ↓ CGAG	GACGAG	AhaII(*), AquI(*)
AhaII	GA ↓ CGCC	AquI	CC ↓ CGGG	GACGGG	AhaII(*), AquI(*)
AhaII	GG ↓ CGCC	AquI	CC ↓ CGAG	GGCGAG	AhaII(*), AquI(*)
AhaII	GG ↓ CGCC	AquI	CC ↓ CGGG	GGCGGG	AhaII(*), AquI(*)
AhaII	GA ↓ CGCC	AsuII	TT ↓ CGAA	GACGAA	AhaII(*), AsuII(*)
AhaII	GG ↓ CGCC	AsuII	TT ↓ CGAA	GGCGAA	AhaII(*), AsuII(*)
AhaII	GA ↓ CGCC	BanIII	AT ↓ CGAT	GACGAT	AhaII(*), BanIII(*)
AhaII	GG ↓ CGCC	BanIII	AT ↓ CGAT	GGCGAT	AhaII(*), BanIII(*)
AhaII	GA ↓ CGCC	BbiII	GG ↓ CGCC	GACGCC	AhaII(2), BbiII(2), AosII(2), HgaI(10)
AhaII	GA ↓ CGCC	BbiII	GA ↓ CGTC	GACGTC	AhaII(2), BbiII(2), AatII(5), AosII(2), MaeII(2)
AhaII	GG ↓ CGCC	BbiII	GA ↓ CGCC	GGCGCC	AhaII(2), BbiII(2), AosII(2), BanI(1), BbeI(5), CfoI(4), HaeII(5), HhaI(4), HinpI(2), NarI(2), NlaIV(3), NunII(2)
AhaII	GG ↓ CGCC	BbiII	GA ↓ CGTC	GGCGTC	AhaII(2), BbiII(2), AosII(2), HgaI(11)
AhaII	GA ↓ CGCC	ClaI	AT ↓ CGAT	GACGAT	AhaII(*), ClaI(*)
AhaII	GG ↓ CGCC	ClaI	AT ↓ CGAT	GGCGAT	AhaII(*), ClaI(*)
AhaII	GA ↓ CGCC	FspII	TT ↓ CGAA	GACGAA	AhaII(*), FspII(*)
AhaII	GG ↓ CGCC	FspII	TT ↓ CGAA	GGCGAA	AhaII(*), FspII(*)

Table 25. Hybrid Sequences Resulting from the Ligation of Two Fragments Generated After Digestion with One or Two Different Restriction Endonucleases (*continued*)

Enzyme 1	Fragment Generated by Enzyme 1	Enzyme 2	Fragment Generated by Enzyme 2	Resulting Hybrid Sequence	Enzyme 3
AhaII	GA ↓ CGCC	HapII	C ↓ CGG	GACGG	AhaII(*), HapII(*)
AhaII	GG ↓ CGCC	HapII	C ↓ CGG	GGCGG	AhaII(*), HapII(*)
AhaII	GA ↓ CGCC	HinpI	G ↓ CGC	GACGC	AhaII(*), HinpI(*), HgaI(10)
AhaII	GG ↓ CGCC	HinpI	G ↓ CGC	GGCGC	AhaII(*), HinpI(2), CfoI(4), HhaI(4)
AhaII	GA ↓ CGCC	HpaII	C ↓ CGG	GACGG	AhaII(*), HpaII(*)
AhaII	GG ↓ CGCC	HpaII	C ↓ CGG	GGCGG	AhaII(*), HpaII(*)
AhaII	GA ↓ CGCC	MaeII	A ↓ CGT	GACGT	AhaII(*), MaeII(2)
AhaII	GG ↓ CGCC	MaeII	A ↓ CGT	GGCGT	AhaII(*), MaeII(*)
AhaII	GA ↓ CGCC	MspI	C ↓ CGG	GACGG	AhaII(*), MspI(*)
AhaII	GG ↓ CGCC	MspI	C ↓ CGG	GGCGG	AhaII(*), MspI(*)
AhaII	GA ↓ CGCC	NarI	GG ↓ CGCC	GACGCC	AhaII(2), NarI(*), AosII(2), BbiII(2), HgaI(10)
AhaII	GG ↓ CGTC	NarI	GG ↓ CGCC	GGCGCC	AhaII(2), NarI(2), AosII(2), BanI(1), BbeI(5), BbiII(2), CfoI(4), HaeII(5), HhaI(4), HinpI(2), NlaIV(3), NunII(2)
AhaII	GA ↓ CGCC	NunII	GG ↓ CGCC	GACGCC	AhaII(2), NunII(*), AosII(2), BbiII(2), HgaII(10)
AhaII	GG ↓ CGTC	NunII	GG ↓ CGCC	GGCGCC	AhaII(2), NunII(2), AosII(2), BanI(1), BbeI(5), BbiII(2), CfoI(4), HaeII(5), HhaI(4), HinpI(2), NarI(2), NlaIV(3)
AhaII	GA ↓ CGCC	TaqI	T ↓ CGA	GACGA	AhaII(*), TaqI(*)
AhaII	GG ↓ CGCC	TaqI	T ↓ CGA	GGCGA	AhaII(*), TaqI(*)
AhaII	GA ↓ CGCC	TthHB8I	T ↓ CGA	GACGA	AhaII(*), TthHB8I(*)
AhaII	GG ↓ CGCC	TthHB8I	T ↓ CGA	GGCGA	AhaII(*), TthHB8I(*)
AocI	CC ↓ TNAGG	DdeI	C ↓ TNAG	CCTNAG	AocI(*), DdeI(2)
AocI	CC ↓ TNAGG	EspI	GC ↓ TNAGC	CCTNAGC	AocI(*), EspI(*), DdeI(2)
AosII	GA ↓ CGCC	AccI	GT ↓ CGAC	GACGAC	AosII(*), AccI(*)
AosII	GG ↓ CGCC	AccI	GT ↓ CGAC	GGCGAC	AosII(*), AccI(*)
AosII	GA ↓ CGCC	AhaII	GA ↓ CGTC	GACGTC	AosII(2), AhaII(2), AatII(5), BbiII(2), MaeII(2)
AosII	GG ↓ CGCC	AhaII	GA ↓ CGCC	GACGCC	AosII(2), AhaII(2), BbiII(2), HgaI(10)
AosII	GG ↓ CGCC	AhaII	GA ↓ CGCC	GGCGCC	AosII(2), AhaII(2), BanI(1), BbeI(5), BbiII(2), CfoI(4), HaeII(5), HhaI(4), HinpI(2), NarI(2), NlaIV(3), NunII(2)
AosII	GG ↓ CGCC	AhaII	GA ↓ CGTC	GGCGTC	AosII(2), AhaII(2), BbiII(2), HgaI(11)
AosII	GA ↓ CGCC	AosII	GA ↓ CGTC	GACGTC	AosII(2), AatII(5), AhaII(2), BbiII(2), MaeII(2)
AosII	GA ↓ CGCC	AosII	GG ↓ CGCC	GACGCC	AosII(2), AhaII(2), BbiII(2), HgaI(10)
AosII	GG ↓ CGCC	AosII	GA ↓ CGCC	GGCGCC	AosII(2), AhaII(2), BanI(1), BbeI(5), BbiII(2), CfoI(4), HaeII(5), HhaI(4), HinpI(2), NarI(2), NlaIV(3), NunII(2)

Table 25. Hybrid Sequences Resulting from the Ligation of Two Fragments Generated After Digestion with One or Two Different Restriction Endonucleases (continued)

Enzyme 1	Fragment Generated by Enzyme 1	Enzyme 2	Fragment Generated by Enzyme 2	Resulting Hybrid Sequence	Enzyme 3
AosII	GG ↓ CGCC	AosII	GA ↓ CGTC	GGCGTC	AosII(2), AhaII(2), BbiII(2), HgaI(11)
AosII	GA ↓ CGCC	AquI	CC ↓ CGAG	GACGAG	AosII(*), AquI(*)
AosII	GA ↓ CGCC	AquI	CC ↓ CGGG	GACGGG	AosII(*), AquI(*)
AosII	GG ↓ CGCC	AquI	CC ↓ CGAG	GGCGAG	AosII(*), AquI(*)
AosII	GG ↓ CGCC	AquI	CC ↓ CGGG	GGCGGG	AosII(*), AquI(*)
AosII	GA ↓ CGCC	AsuII	TT ↓ CGAA	GACGAA	AosII(*), AsuII(*)
AosII	GG ↓ CGCC	AsuII	TT ↓ CGAA	GGCGAA	AosII(*), AsuII(*)
AosII	GA ↓ CGCC	BanIII	AT ↓ CGAT	GACGAT	AosII(*), BanIII(*)
AosII	GG ↓ CGCC	BanIII	AT ↓ CGAT	GGCGAT	AosII(*), BanIII(*)
AosII	GA ↓ CGCC	BbiII	GG ↓ CGCC	GACGCC	AosII(2), BbiII(2), AhaII(2), HgaI(10)
AosII	GA ↓ CGCC	BbiII	GA ↓ CGTC	GACGTC	AosII(2), BbiII(2), AatII(5), AhaII(2), MaeII(2)
AosII	GG ↓ CGCC	BbiII	GA ↓ CGCC	GGCGCC	AosII(2), BbiII(2), AhaII(2), BanI(1), BbeI(5), CfoI(4), HaeII(5), HhaI(4), HinpI(2), NarI(2), NlaIV(3), NunII(2)
AosII	GG ↓ CGCC	BbiII	GA ↓ CGTC	GGCGTC	AosII(2), BbiII(2), AhaII(2), HgaI(11)
AosII	GA ↓ CGCC	ClaI	AT ↓ CGAT	GACGAT	AosII(*), ClaI(*)
AosII	GG ↓ CGCC	ClaI	AT ↓ CGAT	GGCGAT	AosII(*), ClaI(*)
AosII	GA ↓ CGCC	FspII	TT ↓ CGAA	GACGAA	AosII(*), FspII(*)
AosII	GG ↓ CGCC	FspII	TT ↓ CGAA	GGCGAA	AosII(*), FspII(*)
AosII	GA ↓ CGCC	HapII	C ↓ CGG	GACGG	AosII(*), HapII(*)
AosII	GG ↓ CGCC	HapII	C ↓ CGG	GGCGG	AosII(*), HapII(*)
AosII	GA ↓ CGCC	HinpI	G ↓ CGC	GACGC	AosII(*), HinpI(*), HgaI(10)
AosII	GG ↓ CGCC	HinpI	G ↓ CGC	GGCGC	AosII(*), HinpI(2), CfoI(4), HhaI(4)
AosII	GA ↓ CGCC	HpaII	C ↓ CGG	GACGG	AosII(*), HpaII(*)
AosII	GG ↓ CGCC	HpaII	C ↓ CGG	GGCGG	AosII(*), HpaII(*)
AosII	GA ↓ CGCC	MaeII	A ↓ CGT	GACGT	AosII(*), MaeII(2)
AosII	GG ↓ CGCC	MaeII	A ↓ CGT	GGCGT	AosII(*), MaeII(*)
AosII	GA ↓ CGCC	MspI	C ↓ CGG	GACGG	AosII(*), MspI(*)
AosII	GG ↓ CGCC	MspI	C ↓ CGG	GGCGG	AosII(*), MspI(*)
AosII	GA ↓ CGCC	NarI	GG ↓ CGCC	GACGCC	AosII(2), NarI(*), AhaII(2), BbiII(2), HgaI(10)
AosII	GG ↓ CGTC	NarI	GG ↓ CGCC	GACGCC	AosII(2), NarI(2), AhaII(2), BanI(1), BbeI(5), BbiII(2), CfoI(4), HaeII(5), HhaI(4), HinpI(2), NlaIV(3), NunII(2)
AosII	GA ↓ CGCC	NunII	GG ↓ CGCC	GACGCC	AosII(2), NunII(*), AhaII(2), BbiII(2), HgaI(10)
AosII	GG ↓ CGTC	NunII	GG ↓ CGCC	GGCGCC	AosII(2), NunII(2), AhaII(2), BanI(1), BbeI(5), BbiII(2), CfoI(4), HaeII(5), HhaI(4), HinpI(2), NarI(2), NlaIV(3)
AosII	GA ↓ CGCC	TaqI	T ↓ CGA	GACGA	AosII(*), TaqI(*)
AosII	GG ↓ CGCC	TaqI	T ↓ CGA	GGCGA	AosII(*), TaqI(*)
AosII	GA ↓ CGCC	TthHB8I	T ↓ CGA	GACGA	AosII(*), TthHB8I(*)

Table 25. Hybrid Sequences Resulting from the Ligation of Two Fragments Generated After Digestion with One or Two Different Restriction Endonucleases (*continued*)

Enzyme 1	Fragment Generated by Enzyme 1	Enzyme 2	Fragment Generated by Enzyme 2	Resulting Hybrid Sequence	Enzyme 3
AosII	GG ↓ CGCC	TthHB8I	T ↓ CGA	GGCGA	AosII(*), TthHB8I(*)
AquI	CC ↓ CGAG	AccI	GT ↓ CGAC	CCCGAC	AquI(*), AccI(*)
AquI	CT ↓ CGAG	AccI	GT ↓ CGAC	CTCGAC	AquI(*), AccI(*), TaqI(2), TthHB8I(2)
AquI	CC ↓ CGAG	AhaII	GA ↓ CGCC	CCCGCC	AquI(*), AhaII(*)
AquI	CC ↓ CGAG	AhaII	GA ↓ CGTC	CCCGTC	AquI(*), AhaII(*)
AquI	CT ↓ CGAG	AhaII	GA ↓ CGCC	CTCGCC	AquI(*), AhaII(*)
AquI	CT ↓ CGAG	AhaII	GA ↓ CGTC	CTCGTC	AquI(*), AhaII(*)
AquI	CC ↓ CGAG	AosII	GA ↓ CGCC	CCCGCC	AquI(*), AosII(*)
AquI	CC ↓ CGAG	AosII	GA ↓ CGTC	CCCGTC	AquI(*), AosII(*)
AquI	CT ↓ CGAG	AosII	GA ↓ CGCC	CTCGCC	AquI(*), AosII(*)
AquI	CT ↓ CGAG	AosII	GA ↓ CGTC	CTCGTC	AquI(*), AosII(*)
AquI	CC ↓ CGAG	AquI	CC ↓ CGGG	CCCGGG	AquI(2), AvaI(1), BcnI(4), HapII(2), HpaII(2), MspI(2), NciI(3), Nsp7524III(1), ScrFI(3), SmaI(3), XcyI(1), XmaI(1)
AquI	CC ↓ CGAG	AquI	CT ↓ CGAG	CCCGAG	AquI(2), AvaI(1), Nsp7524III(1)
AquI	CT ↓ CGAG	AquI	CC ↓ CGAG	CTCGAG	AquI(2), AvaI(1), Nsp7524III(1), PaeR7(1), SexI(1), TaqI(2), TthHB8I(2), XhoI(1)
AquI	CT ↓ CGAG	AquI	CC ↓ CGGG	CTCGGG	AquI(2), AvaI(1), Nsp7524III(1)
AquI	CC ↓ CGAG	AsuII	TT ↓ CGAA	CCCGAA	AquI(*), AsuII(*)
AquI	CT ↓ CGAG	AsuII	TT ↓ CGAA	CTCGAA	AquI(*), AsuII(*), TaqI(2), TthHB8I(2)
AquI	CC ↓ CGAG	BanIII	AT ↓ CGAT	CCCGAT	AquI(*), BanIII(*)
AquI	CT ↓ CGAG	BanIII	AT ↓ CGAT	CTCGAT	AquI(*), BanIII(*), Taq(2), TthHB8I(2)
AquI	CC ↓ CGAG	BbiII	GA ↓ CGCC	CCCGCC	AquI(*), BbiII(*)
AquI	CC ↓ CGAG	BbiII	GA ↓ CGTC	CCCGTC	AquI(*), BbiII(*)
AquI	CT ↓ CGAG	BbiII	GA ↓ CGCC	CTCGCC	AquI(*), BbiII(*)
AquI	CT ↓ CGAG	BbiII	GA ↓ CGTC	CTCGTC	AquI(*), BbiII(*)
AquI	CC ↓ CGAG	ClaI	AT ↓ CGAT	CCCGAT	AquI(*), ClaI(*)
AquI	CT ↓ CGAG	ClaI	AT ↓ CGAT	CTCGAT	AquI(*), ClaI(*), TaqI(2), TthHB8I(2)
AquI	CC ↓ CGAG	FspII	TT ↓ CGAA	CCCGAA	AquI(*), FspII(*)
AquI	CT ↓ CGAG	FspII	TT ↓ CGAA	CTCGAA	AquI(*), FspII(*), TaqI(2), TthHB8I(2)
AquI	CC ↓ CGAG	HapII	C ↓ CGG	CCCGG	AquI(*), HapII(2), BcnI(3), HpaII(2), MspI(2), NciI(2), ScrFI(2)
AquI	CT ↓ CGAG	HapII	C ↓ CGG	CTCGG	AquI(*), HapII(*)
AquI	CC ↓ CGAG	HinpI	G ↓ CGC	CCCGC	AquI(*), HinpI(*)
AquI	CT ↓ CGAG	HinpI	G ↓ CGC	CTCGC	AquI(*), HinpI(*)
AquI	CC ↓ CGAG	HpaII	C ↓ CGG	CCCGG	AquI(*), HpaII(2), BcnI(3), HapII(2), MspI(2), NciI(2), ScrFI(2)
AquI	CT ↓ CGAG	HpaII	C ↓ CGG	CTCGG	AquI(*), HpaII(*)
AquI	CC ↓ CGAG	MaeII	A ↓ CGT	CCCGT	AquI(*), MaeII(*)
AquI	CT ↓ CGAG	MaeII	A ↓ CGT	CTCGT	AquI(*), MaeII(*)

Table 25. Hybrid Sequences Resulting from the Ligation of Two Fragments Generated After Digestion with One or Two Different Restriction Endonucleases (*continued*)

Enzyme 1	Fragment Generated by Enzyme 1	Enzyme 2	Fragment Generated by Enzyme 2	Resulting Hybrid Sequence	Enzyme 3
AquI	CC ↓ CGAG	MspI	C ↓ CGG	CCCGG	AquI(*), MspI(2), BcnI(3), HapII(2), HpaII(2), NciI(2), ScrFI(2)
AquI	CT ↓ CGAG	MspI	C ↓ CGG	CTCGG	AquI(*), MspI(*)
AquI	CC ↓ CGAG	NarI	GG ↓ CGCC	CCCGCC	AquI(*), NarI(*)
AquI	CT ↓ CGAG	NarI	GG ↓ CGCC	CTCGCC	AquI(*), NarI(*)
AquI	CC ↓ CGAG	NunII	GG ↓ CGCC	CCCGCC	
AquI	CT ↓ CGAG	NunII	GG ↓ CGCC	CTCGCC	AquI(*), NunII(*)
AquI	CC ↓ CGAG	TaqI	T ↓ CGA	CCCGA	AquI(*), TaqI(*)
AquI	CT ↓ CGAG	TaqI	T ↓ CGA	CTCGA	AquI(*), TaqI(2), Tth-HB8I(2)
AquI	CC ↓ CGAG	TthHB8I	T ↓ CGA	CCCGA	AquI(*), TthHB8I(*)
AquI	CT ↓ CGAG	TthHB8I	T ↓ CGA	CTCGA	AquI(*), TthHB8I(2), TaqI(2)
Asp718	G ↓ GTACC	SpII	C ↓ GTACG	GGTACG	Asp718(*), SpII(*), RsaI(3)
AsuI	G ↓ GNCC	DraII	AG ↓ GNCCC	GGNCCC	AsuI(1), DraII(*), Cfr13I(1), NlaIV(3), Nsp7524IV(1), Nsp7524V(1), Sau96I(1)
AsuI	G ↓ GNCC	DraII	AG ↓ GNCCT	GGNCCT	AsuI(1), DraII(*), Cfr13I(1), Nsp7524IV(1), Nsp7524V(1), Sau96I(1)
AsuI	C ↓ GNCC	Eco0109	AG ↓ GNCCT	GGNCCT	AsuI(1), Eco0109(*), Cfr13I(1), Nsp7524IV(1), Nsp7524V(1), Sau96I(1)
AsuII	TT ↓ CGAA	AccI	GT ↓ CGAC	TTCGAC	AsuII(*), AccI(*), TaqI(2), TthHB8I(2)
AsuII	TT ↓ CGAA	AhaII	GA ↓ CGCC	TTCGCC	AsuII(*), AhaII(*)
AsuII	TT ↓ CGAA	AhaII	GA ↓ CGTC	TTCGTC	AsuII(*), AhaII(*)
AsuII	TT ↓ CGAA	AosII	GA ↓ CGCC	TTCGCC	AsuII(*), AosII(*)
AsuII	TT ↓ CGAA	AosII	GA ↓ CGTC	TTCGTC	AsuII(*), AosII(*)
AsuII	TT ↓ CGAA	AquI	CC ↓ CGAG	TTCGAG	AsuII(*), AquI(*), TaqI(2), TthHB8I(2)
AsuII	TT ↓ CGAA	AquI	CC ↓ CGGG	TTCGGG	AsuII(*), AquI(*)
AsuII	TT ↓ CGAA	BanIII	AT ↓ CGAT	TTCGAT	AsuII(*), BanIII(*), TaqI(2), TthHB8I(2)
AsuII	TT ↓ CGAA	BbiII	GA ↓ CGCC	TTCGCC	AsuII(*), BbiII(*)
AsuII	TT ↓ CGAA	BbiII	GA ↓ CGTC	TTCGTC	AsuII(*), BbiII(*)
AsuII	TT ↓ CGAA	ClaI	AT ↓ CGAT	TTCGAT	AsuII(*), ClaI(*), TaqI(2), TthHB8I(2)
AsuII	TT ↓ CGAA	HapII	C ↓ CGG	TTCGG	AsuII(*), HapII(*)
AsuII	TT ↓ CGAA	HinpI	G ↓ CGC	TTCGC	AsuII(*), HinpI(*)
AsuII	TT ↓ CGAA	HpaII	C ↓ CGG	TTCGG	AsuII(*), HpaII(*)
AsuII	TT ↓ CGAA	MaeII	A ↓ CGT	TTCGT	AsuII(*), MaeII(*)
AsuII	TT ↓ CGAA	MspI	C ↓ CGG	TTCGG	AsuII(*), MspI(*)
AsuII	TTCGAA	NarI	GG ↓ CGCC	TTCGCC	AsuII(*), NarI(*)
AsuII	TT ↓ CGAA	NunII	GG ↓ CGCC	TTCGCC	AsuII(*), NunII(*)
AsuII	TT ↓ CGAA	TaqI	G ↓ CGA	TTCGA	AsuII(*), TaqI(2), Tth-HB8I(2)
AsuII	TT ↓ CGAA	TthHB8I	T ↓ CGA	TTCGA	AsuII(*), TthHB8I(2), TaqI(2)
AvaI	C ↓ CCGGG	AccIII	T ↓ CCGGA	CCCGGA	AvaI(*), AccIII(*), BcnI(3), HapII(2), HpaII(2), MspI(2), NciI(2), ScrFI(2)
AvaI	C ↓ CCGGG	BspMII	T ↓ CCGGA	CCCGGA	AvaI(*), BspMII(*), BcnI(3), HapII(2), HpaII(2), MspI(2), NciI(2), ScrFI(2)

Table 25. Hybrid Sequences Resulting from the Ligation of Two Fragments Generated After Digestion with One or Two Different Restriction Endonucleases (*continued*)

Enzyme 1	Fragment Generated by Enzyme 1	Enzyme 2	Fragment Generated by Enzyme 2	Resulting Hybrid Sequence	Enzyme 3
AvaI	C ↓ CCGGG	Cfr10I	A ↓ CCGGC	CCCGGC	AvaI(*), Cfr10I(*), BcnI(3), HapII(2), HpaII(2), MspI(2), NciI(2), ScrFI(2)
AvaI	C ↓ CCGGG	Cfr10I	A ↓ CCGGT	CCCGGT	AvaI(*), Cfr10I(*), BcnI(3), HapII(2), HpaII(2), MspI(2), NciI(2), ScrFI(2)
AvaI	C ↓ TCGAG	SaII	G ↓ TCGAC	CTCGAC	AvaI(*), SaII(*), TaqI(2), TthHB8I(2)
AvaII	G ↓ GACC	PpuMI	AG ↓ GACCT	GGACCT	AvaII(1), PpuMI(*), AflI(1), AsuI(1), Cfr13I(1), ClaII(1), Eco47I(0), Nsp7524IV(1), Nsp7524V(1), Sau96I(1), SinI(1)
AvaII	G ↓ GACC	PpuMI	AG ↓ GACCC	GGACCC	AvaII(1), PpuMI(*), AflI(1), AsuI(1), Cfr13I(1), ClaII(1), Eco47I(0), NlaIV(3), Nsp7524IV(1), Nsp7524V(1), Sau96I(1), SinI(1)
AvaII	G ↓ GTCC	PpuMI	AG ↓ GTCCT	GGTCCT	AvaII(1), PpuMI(*), AflI(1), AsuI(1), Cfr13I(1), ClaII(1), Eco47I(0), Nsp7524IV(1), Nsp7524V(1), Sau96I(1), SinI(1)
AvaII	G ↓ GTCC	PpuMI	AG ↓ GTCCC	GGTCCC	AvaII(1), PpuMI(*), AflI(1), AsuI(1), Cfr13I(1), ClaII(1), Eco47I(0), NlaIV(3), Nsp7524IV(1), Nsp7524V(1), Sau96I(1), SinI(1)
AvaII	G ↓ GACC	RsrII	CG ↓ GACCG	GGACCG	AvaII(1), RsrII(*), AflI(1), AsuI(1), Cfr13I(1), ClaII(1), Eco47I(0), Nsp7524IV(1), Nsp7524V(1), Sau96I(1), SinI(1)
AvaII	G ↓ GTCC	RsrII	CG ↓ GTCCG	GGTCCG	AvaII(1), RsrII(*), AflI(1), AsuI(1), Cfr13I(1), ClaII(1), Eco47I(0), Nsp7524IV(1), Nsp7524V(1), Sau96I(1), SinI(1)
AvrII	C ↓ CTAGG	NheI	G ↓ CTAGC	CCTAGC	AvrII(*), NheI(*), MaeI(2)
AvrII	C ↓ CTAGG	SpeI	A ↓ CTAGT	CCTAGT	AvrII(*), SpeI(*), MaeI(2)
AvrII	C ↓ CTAGG	XbaI	T ↓ CTAGA	CCTAGA	AvrII(*), XbaI(*), MaeI(2)
BamHI	G ↓ GATCC	BclI	T ↓ GATCA	GGATCA	BamHI(*), BclI(*), DpnI(3), MboI(1), NdeII(1), Sau3A(1)
BamHI	G ↓ GATCC	BslII	A ↓ GATCT	GGATCT	BamHI(*), BglII(*), DpnI(3), MboI(1), MflI(1), NdeII(1), Sau3A(1), XhoII(1)
BamHI	G ↓ GATCC	MboI	↓ GATC	GGATC	BamHI(*), MboI(1), DpnI(3), NdeII(1), Sau3A(1)

Table 25. Hybrid Sequences Resulting from the Ligation of Two Fragments Generated After Digestion with One or Two Different Restriction Endonucleases (*continued*)

Enzyme 1	Fragment Generated by Enzyme 1	Enzyme 2	Fragment Generated by Enzyme 2	Resulting Hybrid Sequence	Enzyme 3
BamHI	G ↓ GATCC	MflI	A ↓ GATCC	GGATCC	BamHI(1), MflI(1), BstI(1), DpnI(3), MboI(1), NdeII(1), NlaIV(3), Sau3A(1), XhoII(1)
BamHI	G ↓ GATCC	MflI	A ↓ GATCT	GGATCT	BamHI(*), MflI(1), DpnI(3), MboI(1), NdeII(1), Sau3A(1), XhoII(1)
BamHI	G ↓ GATCC	NdeII	↓ GATC	GGATC	BamHI(*), NdeII(1), DpnI(3), MboI(1), Sau3A(1)
BamHI	G ↓ GATCC	Sau3A	↓ GATC	GGATC	BamHI(*), Sau3A(1), DpnI(3), MboI(1), NdeII(1)
BamHI	G ↓ GATCC	XhoII	A ↓ GATCC	GGATCC	BamHI(1), XhoII(1), BstI(1), DpnI(3), MboI(1), MflI(1), NdeII(1), NlaIV(3), Sau3A(1)
BamHI	G ↓ GATCC	XhoII	A ↓ GATCT	GGATCT	BamHI(*), XhoII(1), DpnI(3), MboI(1), MflI(1), NdeII(1), Sau3A(1)
BanI	G ↓ GTACC	SplI	C ↓ GTACG	GGTACG	BanI(*), SplI(*), RsaI(3)
BanIII	AT ↓ CGAT	AccI	GT ↓ CGAC	ATCGAC	BanIII(*), AccI(*), TaqI(2), TthHB8I(2)
BanIII	AT ↓ CGAT	AhaII	GA ↓ CGCC	ATCGCC	BanIII(*), AhaII(*)
BanIII	AT ↓ CGAT	AhaII	GA ↓ CGTC	ATCGTC	BanIII(*), AhaII(*)
BanIII	AT ↓ CGAT	AosII	GA ↓ CGCC	ATCGCC	BanIII(*), AosII(*)
BanIII	AT ↓ CGAT	AosII	GA ↓ CGTC	ATCGTC	BanIII(*), AosII(*)
BanIII	AT ↓ CGAT	AquI	CC ↓ CGAG	ATCGAG	BanIII(*), AquI(*), TaqI(2), TthHB8I(2)
BanIII	AT ↓ CGAT	AquI	CC ↓ CGGG	ATCGGG	BanIII(*), AquI(*)
BanIII	AT ↓ CGAT	AsuII	TT ↓ CGAA	ATCGAA	BanIII(*), AsuII(*), TaqI(2), TthHB8I(2)
BanIII	AT ↓ CGAT	BbiII	GA ↓ CGCC	ATCGCC	BanIII(*), BbiII(*)
BanIII	AT ↓ CGAT	BbiII	GA ↓ CGTC	ATCGTC	BanIII(*), BbiII(*)
BanIII	AT ↓ CGAT	FspII	TT ↓ CGAA	ATCGAA	BanIII(*), FspII(*), TaqI(2), TthHB8I(2)
BanIII	AT ↓ CGAT	HapII	C ↓ CGG	ATCGG	BanIII(*), HapII(*)
BanIII	AT ↓ CGAT	HinpI	G ↓ CGC	ATCGC	BanIII(*), HinpI(*)
BanIII	AT ↓ CGAT	HpaII	C ↓ CGG	ATCGG	BanIII(*), HpaII(*)
BanIII	AT ↓ CGAT	MaeII	A ↓ CGT	ATCGT	BanIII(*), MaeII(*)
BanIII	AT ↓ CGAT	MspI	C ↓ CGG	ATCGG	BanIII(*), MspI(*)
BanIII	AT ↓ CGAT	NarI	GG ↓ CGCC	ATCGCC	BanIII(*), NarI(*)
BanIII	AT ↓ CGAT	NunII	GG ↓ CGCC	ATCGCC	BanIII(*), NunII(*)
BanIII	AT ↓ CGAT	TaqI	T ↓ CGA	ATCGA	BanIII(*), TaqI(2), TthHB8I(2)
BanIII	AT ↓ CGAT	TthHB8I	T ↓ CGA	ATCGA	BanIII(*), TthHB8I(2), TaqI(2)
BbiII	GA ↓ CGCC	AccI	GT ↓ CGAC	GACGAC	BbiII(*), AccI(*)
BbiII	GG ↓ CGCC	AccI	GT ↓ CGAC	GGCGAC	BbiII(*), AccI(*)
BbiII	GA ↓ CGCC	AhaII	GA ↓ CGTC	GACGTC	BbiII(2), AhaII(2), AatII(5), AosII(2), MaeII(2)
BbiII	GA ↓ CGCC	AhaII	GG ↓ CGCC	GACGCC	BbiII(2), AhaII, AosII(2), HgaI(10)

Table 25. Hybrid Sequences Resulting from the Ligation of Two Fragments Generated After Digestion with One or Two Different Restriction Endonucleases (*continued*)

Enzyme 1	Fragment Generated by Enzyme 1	Enzyme 2	Fragment Generated by Enzyme 2	Resulting Hybrid Sequence	Enzyme 3
BbiII	GG ↓ CGCC	AhaII	GA ↓ CGCC	GGCGCC	BbiII(2), AhaII(2), AosII(2), BanI(1), BbeI(5), CfoI(4), HaeII(5), HhaI(4), HinpI(2), NarI(2), NlaIV(3), NunII(2)
BbiII	GG ↓ CGCC	AhaII	GA ↓ CGTC	GGCGTC	BbiII(2), AhaII(2), AosII(2), HgaI(11)
BbiII	GA ↓ CGCC	AosII	GA ↓ CGTC	GACGTC	BbiII(2), AosII(2), AatII(5), AhaII(2), MaeII(2)
BbiII	GA ↓ CGCC	AosII	GG ↓ CGCC	GACGCC	BbiII(2), AosII(2), AhaII(2), HgaI(10)
BbiII	GG ↓ CGCC	AosII	GA ↓ CGCC	GGCGCC	BbiII(2), AosII(2), AhaII(2), BanI(1), BbeI(5), CfoI(4), HaeII(5), HhaI(4), HinpI(2), NarI(2), NlaIV(3), NunII(2)
BbiII	GG ↓ CGCC	AosII	GA ↓ CGTC	GGCGTC	BbiII(2), AosII(2), AhaII(2), HgaI(11)
BbiII	GA ↓ CGCC	AquI	CC ↓ CGAG	GACGAG	BbiII(*), AquI(*)
BbiII	GA ↓ CGCC	AquI	CC ↓ CGGG	GACGGG	
BbiII	GG ↓ CGCC	AquI	CC ↓ CGAG	GGCGAG	BbiII(*), AquI(*)
BbiII	GG ↓ CGCC	AquI	CC ↓ CGGG	GGCGGG	BbiII(*), AquI(*)
BbiII	GA ↓ CGCC	AsuII	TT ↓ CGAA	GACGAA	BbiII(*), AsuII(*)
BbiII	GG ↓ CGCC	AsuII	TT ↓ CGAA	GGCGAA	BbiII(*), AsuII(*)
BbiII	GA ↓ CGCC	BanIII	AT ↓ CGAT	GACGAT	BbiII(*), BanIII(*)
BbiII	GG ↓ CGCC	BanIII	AT ↓ CGAT	GGCGAT	BbiII(*), BanIII(*)
BbiII	GA ↓ CGCC	BbiII	GG ↓ CGCC	GACGCC	BbiII(2), AhaII(2), AosII(2), HgaI(10)
BbiII	GA ↓ CGCC	BbiII	GA ↓ CGTC	GACGTC	BbiII(2), AatII(5), AhaII(2), AosII(2), MaeII(2)
BbiII	GG ↓ CGCC	BbiII	GA ↓ CGCC	GGCGCC	BbiII(2), AhaII(2), AosII(2), BanI(1), BbeI(5), CfoI(4), HaeII(5), HhaI(4), HinpI(2), NarI(2), NlaIV(3), NunII(2)
BbiII	GG ↓ CGCC	BbiII	GA ↓ CGTC	GGCGTC	BbiII(2), AhaII(2), AosII(2), HgaI(11)
BbiII	GA ↓ CGCC	ClaI	AT ↓ CGAT	GACGAT	BbiII(*), ClaI(*)
BbiII	GG ↓ CGCC	ClaI	AT ↓ CGAT	GGCGAT	BbiII(*), ClaI(*)
BbiII	GA ↓ CGCC	FspII	TT ↓ CGAA	GACGAA	BbiII(*), FspII(*)
BbiII	GG ↓ CGCC	FspII	TT ↓ CGAA	GGCGAA	BbiII(*), FspII(*)
BbiII	GA ↓ CGCC	HapII	C ↓ CGG	GACGG	BbiII(*), HapII(*)
BbiII	GG ↓ CGCC	HapII	C ↓ CGG	GGCGG	BbiII(*), HapII(*)
BbiII	GA ↓ CGCC	HinpI	G ↓ CGC	GACGC	BbiII(*), HinpI(*), HgaI(10)
BbiII	GG ↓ CGCC	HinpI	G ↓ CGC	GGCGC	BbiII(*), HinpI(2), CfoI(4), HhaI(4)
BbiII	GA ↓ CGCC	HpaII	C ↓ CGG	GACGG	BbiII(*), HpaII(*)
BbiII	GG ↓ CGCC	HpaII	C ↓ CGG	GGCGG	BbiII(*), HpaII(*)
BbiII	GA ↓ CGCC	MaeII	A ↓ CGT	GACGT	BbiII(*), MaeII(2)
BbiII	GG ↓ CGCC	MaeII	A ↓ CGT	GGCGT	BbiII(*), MaeII(*)
BbiII	GA ↓ CGCC	MspI	C ↓ CGG	GACGG	BbiII(*), MspI(*)
BbiII	GG ↓ CGCC	MspI	C ↓ CGG	GGCGG	BbiII(*), MspI(*)
BbiII	GA ↓ CGCC	NarI	GG ↓ CGCC	GACGCC	BbiII(2), NarI(*), AhaII(2), AosII(2), HgaI(10)

Table 25. Hybrid Sequences Resulting from the Ligation of Two Fragments Generated After Digestion with One or Two Different Restriction Endonucleases (*continued*)

Enzyme 1	Fragment Generated by Enzyme 1	Enzyme 2	Fragment Generated by Enzyme 2	Resulting Hybrid Sequence	Enzyme 3
BbiII	GG ↓ CGTC	NarI	GG ↓ CGCC	GGCGCC	BbiII(2), NarI(2), AhaII(2), AosII(2), BanI(1), BbeI(5), CfoI(4), HaeII(5), HhaI(4), HinpI(2), NlaIV(3), NunII(2)
BbiII	GA ↓ CGCC	NunII	GG ↓ CGCC	GACGCC	BbiII(2), NunII(*), AhaII(2), AosII(2), HgaI(10)
BbiII	GG ↓ CGTC	NunII	GG ↓ CGCC	GGCGCC	BbiII(2), NunII(2), AhaII(2), AosII(2), BanI(1), BbeI(5), CfoI(4), HaeII(5), HhaI(4), HinpI(2), NarI(2), NlaIV(3)
BbiII	GA ↓ CGCC	TaqI	T ↓ CGA	GACGA	BbiII(*), TaqI(*)
BbiII	GG ↓ CGCC	TaqI	T ↓ CGA	GGCGA	BbiII(*), TaqI(*)
BbiII	GA ↓ CGCC	TthHB8I	T ↓ CGA	GACGA	BbiII(*), TthHB8I(*)
BbiII	GG ↓ CGCC	TthHB8I	T ↓ CGA	GGCGA	BbiII(*), TthHB8I(*)
BclI	T ↓ GATCA	BamHI	G ↓ GATCC	TGATCC	BclI(*), BamHI(*), DpnI(3), MboI(1), NdeII(1), Sau3A(1)
BclI	T ↓ GATCA	BglII	A ↓ GATCT	TGATCT	BclI(*), BglII(*), DpnI(3), MboI(1), NdeII(1), Sau3A(1)
BclI	T ↓ GATCA	BstI	G ↓ GATCC	TGATCC	BclI(*), BstI(*), DpnI(3), MboI(1), NdeII(1), Sau3A(1)
BclI	T ↓ GATCA	MboI	↓ GATC	TGATC	BclI(*), MboI(1), DpnI(3), NdeII(1), Sau3A(1)
BclI	T ↓ GATCA	MflI	A ↓ GATCC	TGATCC	BclI(*), MflI(*), DpnI(3), MboI(1), NdeII(1), Sau3A(1)
BclI	T ↓ GATCA	MflI	A ↓ GATCT	TGATCT	BclI(*), MflI(*), DpnI(3), MboI(1), NdeII(1), Sau3A(1)
BclI	T ↓ GATCA	NdeII	↓ GATC	TGATC	BclII(*), NdeII(1), DpnI(3), MboI(1), Sau3A(1)
BclI	T ↓ GATCA	Sau3A	↓ GATC	TGATC	BclI9*), Sau3A(1), DpnI(3), MboI(1), NdeII(1)
BclI	T ↓ GATCA	XhoII	A ↓ GATCC	TGATCC	BclI(*), XhoII(*), DpnI(3), MboI(1), NdeII(1), Sau3A(1)
BclI	T ↓ GATCA	XhoII	A ↓ GATCT	TGATCT	BclI(*), XhoII(*), DpnI(3), MboI(1), NdeII(1), Sau3A(1)
BglII	A ↓ GATCT	BamHI	G ↓ GATCC	AGATCC	BglII(*), BamHI(*), DpnI(3), MboI(1), MflI(1), NdeII(1), Sau3A(1), XhoII(1)
BglII	A ↓ GATCT	BclI	T ↓ GATCA	AGATCA	BglII(*), BclI(*), DpnI(3), MboI(1), NdeII(1), Sau3A(1)
BglII	A ↓ GATCT	BstI	G ↓ GATCC	AGATCC	BglII(*), BstI(*), DpnI(3), MboI(1), MflI(1), NdeII(1), Sau3AI(1), XhoII(1)
BglII	A ↓ GATCT	MboI	↓ GATC	AGATC	BglII(*), MboI(1), DpnI(3), NdeII(1), Sau3A(1)

Table 25. Hybrid Sequences Resulting from the Ligation of Two Fragments Generated After Digestion with One or Two Different Restriction Endonucleases (*continued*)

Enzyme 1	Fragment Generated by Enzyme 1	Enzyme 2	Fragment Generated by Enzyme 2	Resulting Hybrid Sequence	Enzyme 3
BglII	A ↓ GATCT	MflI	A ↓ GATCC	AGATCC	BglII(*), MflI(1), DpnI(3), MboI(1), NdeII(1), Sau3A(1), XhoII(1)
BglII	A ↓ GATCT	MflI	G ↓ GATCT	AGATCT	BglII(1), MflI(1), DpnI(3), MboI(1), NdeII(1), Sau3A(1), XhoII(1)
BglII	A ↓ GATCT	NdeII	↓ GATC	AGATC	BglII(*), NdeII(1), DpnI(3), MboI(1), Sau3A(1)
BglII	A ↓ GATCT	Sau3A	↓ GATC	AGATC	BglII(*), Sau3A(1), DpnI(3), MboI(1), NdeII(1)
BglII	A ↓ GATCT	XhoII	A ↓ GATCC	AGATCC	BglII(*), XhoII(1), DpnI(3), MboI(1), MflI(1), NdeII(1), Sau3A(1)
BglII	A ↓ GATCT	XhoII	G ↓ GATCT	AGATCT	BglII(1), XhoII(1), DpnI(3), MboI(1), MflI(1), NdeII(1), Sau3A(1)
BspHI	T ↓ CATGA	Afl III	A ↓ CATGT	TCATGT	BspHI(*), AflIII(*), NlaIII(5)
BspHI	T ↓ CATGA	EcoT14I	C ↓ CATGG	TCATGG	BspHI(*), EcoT14I(*), NlaIII(5)
BspHI	T ↓ CATGA	NcoI	C ↓ CATGG	TCATGG	BspHI(*), NcoI(*), NlaIII(5)
BspHI	T ↓ CATGA	StyI	C ↓ CATGG	TCATGG	BspHI(*), StyI(*), NlaIII(5)
BspMII	T ↓ CCGGA	AvaI	C ↓ CCGGG	TCCGGG	BspMII(*), AvaI(*), BcnI(4), HapII(2), HpaII(2), MspI(2), NciI(3), ScrFI(3)
BspMII	T ↓ CCGGA	Cfr10I	A ↓ CCGGC	TCCGGC	BspMII(*), Cfr10I(*), HapII(2), HpaII(2), MspI(2)
BspMII	T ↓ CCGGA	Cfr10I	A ↓ CCGGT	TCCGGT	BspMII(*), Cfr10I(*), HapII(2), HpaII(2), MspI(2)
BspMII	T ↓ CCGGA	Nsp7524III	C ↓ CCGGG	TCCGGG	BspMII(*), Nsp7524III(*), BcnI(4), HapII(2), HpaII(2), MspI(2), NciI(3), ScrFI(3)
BspMII	T ↓ CCGGA	XcyI	C ↓ CCGGG	TCCGGG	BspMII(*), XcyI(*), BcnI(4), HapII(2), HpaII(2), MspI(2), NciI(3), ScrFI(3)
BspMII	T ↓ CCGGA	XmaI	C ↓ CCGGG	TCCGGG	BspMII(*), XmaI(*), BcnI(4), HapII(2), HpaII(2), MspI(2), NciI(3), ScrFI(3)
BssHII	G ↓ CGCGC	AflIII	A ↓ CGCGT	GCGCGT	BssHII(*), AflIII(*), CfoI(3), FnudII(3), HhaI(3), HinpI(1), ThaI(3)
BssHII	G ↓ CGCGC	MluI	A ↓ CGCGT	GCGCGT	BssHII(*), MluI(*), CfoI(3), FnudII(3), HhaI(3), HinpI(1), ThaI(3)
BstI	G ↓ GATCC	BclI	T ↓ GATCA	GGATCA	BstI(*), BclI(*), DpnI(3), MboI(1), NdeII(1), Sau3A(1)
BstI	G ↓ GATCC	BslII	A ↓ GATCT	GGATCT	BstI(*), BslII(*), DpnI(3), MboI(1), MflI(1), NdeII(1), Sau3A(1), XhoII(1)
BstI	G ↓ GATCC	MboI	↓ GATC	GGATC	BstI(*), MboI(1), DpnI(3), NdeII(1), Sau3A(1)

Table 25. Hybrid Sequences Resulting from the Ligation of Two Fragments Generated After Digestion with One or Two Different Restriction Endonucleases (*continued*)

Enzyme 1	Fragment Generated by Enzyme 1	Enzyme 2	Fragment Generated by Enzyme 2	Resulting Hybrid Sequence	Enzyme 3
BstI	G↓GATCC	MflI	A↓GATCC	GGATCC	BstI(1), MflI(1), BamHI(1), DpnI(3), MboI(1), NdeII(1), NlaIV(3), Sau3A(1), XhoII(1)
BstI	G↓GATCC	MflI	A↓GATCT	GGATCT	BstI(*), MflI(1), DpnI(3), MboI(1), NdeII(1), Sau3A(1), XhoII(1)
BstI	G↓GATCC	NdeII	↓GATC	GGATC	BstI(*), NdeII(1), DpnI(3), MboI(1), Sau3A(1)
BstI	G↓GATCC	Sau3A	↓GATC	GGATC	BstI(*), Sau3A(1), DpnI(3), MboI(1), NdeII(1)
BstI	G↓GATCC	XhoII	A↓GATCC	GGATCC	BstI(1), XhoII(1), BamHI(1), DpnI(3), MboI(1), MflI(1), NdeII(1), NlaIV(3), Sau3A(1)
BstI	G↓GATCC	XhoII	A↓GATCT	GGATCT	BstI(*), XhoII(1), DpnI(3), MboI(1), MflI(1), NdeII(1), Sau3A(1)
BstEII	G↓GTNACC	MaeIII	↓GTNAC	GGTNAC	BstEII(*), MaeIII(1)
Cfr10I	A↓CCGGC	AccIII	T↓CCGGA	ACCGGA	Cfr10I(*), AccIII(*), HapII(2), HpaII(2), MspI(2)
Cfr10I	G↓CCGGC	AccIII	T↓CCGGA	GCCGGA	Cfr10I(*), AccIII(*), HapII(2), HpaII(2), MspI(2)
Cfr10I	A↓CCGGC	AvaI	C↓CCGGG	ACCGGG	Cfr10I(*), AvaI(*), BcnI(4), HapII(2), HpaII(2), MspI(2), NciI(3), ScrFI(3)
Cfr10I	G↓CCGGC	AvaI	C↓CCGGG	GCCGGG	Cfr10I(*), AvaI(*), BcnI(4), HapII(2), HpaII(2), MspI(2), NciI(3), ScrFI(3)
Cfr10I	A↓CCGGC	BspMII	T↓CCGGA	ACCGGA	Cfr10I(*), BspMII(*), HapII(2), HpaII(2), MspI(2)
Cfr10I	G↓CCGGC	BspMII	T↓CCGGA	GCCGGA	Cfr10I(*), BspMII(*), HapII(2), HpaII(2), MspI(2)
Cfr10I	A↓CCGGC	Cfr10I	G↓CCGGC	ACCGGC	Cfr10I(1), HapII(2), HpaII(2), MspI(2)
Cfr10I	A↓CCGGC	Cfr10I	A↓CCGGT	ACCGGT	Cfr10I(1), HapII(2), HpaII(2), MspI(2)
Cfr10I	G↓CCGGC	Cfr10I	A↓CCGGC	GCCGGC	Cfr10I(1), HapII(2), HpaII(2), MspI(2), NaeI(3)
Cfr10I	G↓CCGGC	Cfr10I	A↓CCGGT	GCCGGT	Cfr10I(1), HapII(2), HpaII(2), MspI(2)
Cfr10I	A↓CCGGC	Nsp7524III	C↓CCGGG	ACCGGG	Cfr10I(*), Nsp7524III(*), BcnI(4), HapII(2), HpaII(2), MspI(2), NciI(3), ScrFI(3)
Cfr10I	G↓CCGGC	Nsp7524III	C↓CCGGG	GCCGGG	Cfr10I(*), Nsp7524III(*), BcnI(4), HapII(2), HpaII(2), MspI(2), NciI(3), ScrFI(3)
Cfr10I	A↓CCGGC	XcyI	C↓CCGGG	ACCGGG	Cfr10I(*), XcyI(*), BcnI(4), HapII(2), HpaII(2), MspI(2), NciI(3), ScrFI(3)

Table 25. Hybrid Sequences Resulting from the Ligation of Two Fragments Generated After Digestion with One or Two Different Restriction Endonucleases (*continued*)

Enzyme 1	Fragment Generated by Enzyme 1	Enzyme 2	Fragment Generated by Enzyme 2	Resulting Hybrid Sequence	Enzyme 3
Cfr10I	G↓CCGGC	XcyI	C↓CCGGG	GCCGGG	Cfr10I(*), XcyI(*), BcnI(4), HapII(2), HpaII(2), MspI(2), NciI(3), ScrFI(3)
Cfr10I	A↓CCGGC	XmaI	C↓CCGGG	ACCGGG	Cfr10I(*), XmaI(*), BcnI(4), HapII(2), HpaII(2), MspI(2), NciI(3), ScrFI(3)
Cfr10I	G↓CCGGC	XmaI	C↓CCGGG	GCCGGG	Cfr10I(*), XmaI(*), BcnI(4), HapII(2), HpaII(2), MspI(2), NciI(3), ScrFI(3)
CfrI	C↓GGCCA	CfrI	C↓GGCCG	CGGCCG	CfrI(1), BspRI(3), EaeI(1), EasI(1), Eco52I(1), HaeIII(3), PalI(3), XmaIII(1)
CfrI	C↓GGCCA	CfrI	T↓GGCCA	CGGCCA	CfrI(1), BspRI(3), EagI(1), HaeIII(3), PalI(3)
CfrI	T↓GGCCA	CfrI	C↓GGCCA	TGGCCA	CfrI(1), BalI(3), BspRI(3), EaeI(1), HaeIII(3), PalI(3)
CfrI	T↓GGCCA	CfrI	C↓GGCCG	TGGCCG	CfrI(1), BspRI(3), EaeI(1), HaeIII(3), PalI(3)
CfrI	C↓GGCCA	EaeI	C↓GGCCG	CGGCCG	CfrI(1), EaeI(1), BspRI(3), EagI(1), Eco52I(1), HaeIII(3), PalI(3), XmaIII(1)
CfrI	C↓GGCCA	EaeI	T↓GGCCA	CGGCCA	CfrI(1), EaeI(1), BspRI(3), HaeIII(3), PalI(3)
CfrI	T↓GGCCA	EaeI	C↓GGCCA	TGGCCA	CfrI(1), EaeI(1), BalI(3), BspRI(3), HaeIII(3), PalI(3)
CfrI	T↓GGCCA	EaeI	C↓GGCCG	TGGCCG	CfrI(1), EaeI(1), BspRI(3), HaeIII(3), PalI(3)
CfrI	C↓GGCCA	EagI	C↓GGCCG	CGGCCG	CfrI(1), EagI(1), BspRI(3), EaeI(1), Eco52I(1), HaeIII(3), PalI(3), XmaIII(1)
CfrI	T↓GGCCA	EagI	C↓GGCCG	TGGCCG	CfrI(1), EagI(*), BspRI(3), EaeI(1), HaeIII(3), PalI(3)
CfrI	C↓GGCCA	Eco52I	C↓GGCCG	CGGCCG	CfrI(1), Eco52I(1), BspRI(3), EaeI(1), EagI(1), HaeIII(3), PalI(3), XmaIII(1)
CfrI	T↓GGCCA	Eco52I	C↓GGCCG	TGGCCG	CfrI(1), Eco52I(*), BspRI(3), EaeI(1), HaeIII(3), PalI(3)
CfrI	C↓GGCCA	NotI	GC↓GGCCGC	CGGCCGC	CfrI(1), NotI(*), BspRI(3), EaeI(1), EagI(1), Eco52I(1), Fnu4HI(4), HaeIII(3), PalI(3), XmaIII(1)
CfrI	T↓GGCCA	NotI	GC↓GGCCGC	TGGCCGC	CfrI(1), NotI(*), BspRI(3), EaeI(1), Fnu4HI(4), HaeIII(3), PalI(3)
CfrI	C↓GGCCA	XmaIII	C↓GGCCG	CGGCCG	CfrI(1), XmaIII(1), BspRI(3), EaeI(1), EagI(1), Eco52I(1), HaeIII(3), PalI(3)

Table 25. Hybrid Sequences Resulting from the Ligation of Two Fragments Generated After Digestion with One or Two Different Restriction Endonucleases (*continued*)

Enzyme 1	Fragment Generated by Enzyme 1	Enzyme 2	Fragment Generated by Enzyme 2	Resulting Hybrid Sequence	Enzyme 3
CfrI	T ↓ GGCCA	XmaIII	C ↓ GGCCG	TGGCCG	CfrI(1), XmaIII(*), BspRI(3), EaeI(1), HaeIII(3), PalI(3)
Cfr13I	G ↓ GNCC	DraII	AG ↓ GNCCC	GGNCCC	Cfr13I(1), DraII(*), AsuI(1), NlaIV(3), Nsp7524IV(1), Nsp7524V(1), Sau96I(1)
Cfr13I	G ↓ GNCC	DraII	AG/ ↓ GNCCT	GGNCCT	Cfr13I(1), DraII(*), AsuI(1), Nsp7524IV(1), Nsp7524V(1), Sau96I(1)
Cfr13I	G ↓ GNCC	Eco0109	AG ↓ GNCCT	GGNCCT	Cfr13I(1), Eco0109(*), AsuI(1), Nsp7524IV(1), Nsp7524V(1), Sau96I(1)
ClaI	AT ↓ CGAT	AccI	GT ↓ CGAC	ATCGAC	ClaI(*), AccI(*), TaqI(2), TthHB8I(2)
ClaI	AT ↓ CGAT	AhaII	GA ↓ CGCC	ATCGCC	ClaI(*), AhaII(*)
ClaI	AT ↓ CGAT	AhaII	GA ↓ CGTC	ATCGTC	ClaI(*), AhaII(*)
ClaI	AT ↓ CGAT	AosII	GA ↓ CGCC	ATCGCC	ClaI(*), AosII(*)
ClaI	AT ↓ CGAT	AosII	GA ↓ CGTC	ATCGTC	ClaI(*), AosII(*)
ClaI	AT ↓ CGAT	AquI	CC ↓ CGAG	ATCGAG	ClaI(*), AquI(*), TaqI(2), TthHB8I(2)
ClaI	AT ↓ CGAT	AquI	CC ↓ CGGG	ATCGGG	ClaI(*), AquI(*)
ClaI	AT ↓ CGAT	AsuII	TT ↓ CGAA	ATCGAA	ClaI(*), AsuII(*), TaqI(2), TthHB8I(2)
ClaI	AT ↓ CGAT	BbiII	GA ↓ CGCC	ATCGCC	ClaI(*), BbiII(*)
ClaI	AT ↓ CGAT	BbiII	GA ↓ CGTC	ATCGTC	ClaI(*), BbiII(*)
ClaI	AT ↓ CGAT	FspII	TT ↓ CGAA	ATCGAA	ClaI(*), FspII(*), TaqI(2), TthHB8I(2)
ClaI	AT ↓ CGAT	HapII	C ↓ CGG	ATCGG	ClaI(*), HapII(*)
ClaI	AT ↓ CGAT	HinpI	G ↓ CGC	ATCGC	ClaI(*), HinpI(*)
ClaI	AT ↓ CGAT	HpaII	C ↓ CGG	ATCGG	ClaI(*), HpaII(*)
ClaI	AT ↓ CGAT	MaeII	A ↓ CGT	ATCGT	ClaI(*), MaeII(*)
ClaI	AT ↓ CGAT	MspI	C ↓ CGG	ATCGG	ClaI(*), MspI(*)
ClaI	AT ↓ CGAT	NarI	GG ↓ CGCC	ATCGCC	ClaI(*), NarI(*)
ClaI	AT ↓ CGAT	NunII	GG ↓ CGCC	ATCGCC	ClaI(*), NunII(*)
ClaI	AT ↓ CGAT	TaqI	T ↓ CGA	ATCGA	ClaI(*), TaqI(2), TthHB8I(2)
ClaI	AT ↓ CGAT	TthHB8I	T ↓ CGA	ATCGA	ClaI(*), TthHB8I(2), TaqI(2)
ClaII	G ↓ GACC	PpuMI	AG ↓ GACCT	GGACCT	ClaII(1), PpuMI(*), AflI(1), AsuI(1), AvaII(1), Cfr13I(1), Eco47I(0), Nsp7524IV(1), Nsp7524V(1), Sau96I(1), SinI(1)
ClaII	G ↓ GACC	PpuMI	AG ↓ GACCC	GGACCC	ClaII(1), PpuMI(*), AflI(1), AsuI(1), AvaII(1), Cfr13I(1), Eco47I(0), NlaIV(3), Nsp7524IV(1), Nsp7524V(1), Sau96I(1), SinI(1)
ClaII	G ↓ GTCC	PpuMI	AG ↓ GTCCT	GGTCCT	ClaII(1), PpuMI(*), AflI(1), AsuI(1), AvaII(1), Cfr13I(1), Eco47I(0), Nsp7524IV(1), Nsp7524V(1), Sau96I(1), SinI(1)

Table 25. Hybrid Sequences Resulting from the Ligation of Two Fragments Generated After Digestion with One or Two Different Restriction Endonucleases (*continued*)

Enzyme 1	Fragment Generated by Enzyme 1	Enzyme 2	Fragment Generated by Enzyme 2	Resulting Hybrid Sequence	Enzyme 3
ClaII	G ↓ GTCC	PpuMI	AG ↓ GTCCC	GGTCCC	ClaII(1), PpuMI(*), AflI(1), AsuI(1), AvaII(1), Cfr13I(1), Eco47I(0), NlaIV(3), Nsp7524IV(1), Nsp7524V(1), Sau96I(1), SinI(1)
ClaII	G ↓ GACC	RsrII	CG ↓ GTCCG	GGACCG	ClaII(1), RsrII(*), AflI(1), AsuI(1), AvaII(1), Cfr13I(1), Eco47I(0), Nsp7524IV(1), Nsp7524V(1), Sau96I(1), SinI(1)
ClaII	G ↓ GTCC	RsrII	CG ↓ GTCCG	GGTCCG	ClaII(1), RsrII(*), AflI(1), AsuI(1), AvaII(1), Cfr13I(1), Eco47I(0), Nsp7524IV(1), Nsp7524V(1), Sau96I(1), SinI(1)
CvnI	CC ↓ TNAGG	DdeI	C ↓ TNAG	CCTNAG	CvnI(*), DdeI(2)
CvnI	CC ↓ TNAGG	EspI	GC ↓ TNAGC	CCTNAGC	CvnI(*), EspI(*), DdeI(2)
DdeI	C ↓ TNAG	AocI	CC ↓ TNAGG	CTNAGG	DdeI(1), AocI(*)
DdeI	C ↓ TNAG	CvnI	CC ↓ TNAGG	CTNAGG	DdeI(1), CvnI(*)
DdeI	C ↓ TNAG	Eco81I	CC ↓ TNAGG	CTNAGG	DdeI(1), Eco81I(*)
DdeI	C ↓ TNAG	EspI	GC ↓ TNAGC	CTNAGC	DdeI(1), EspI(*)
DdeI	C ↓ TNAG	MstII	CC ↓ TNAGG	CTNAGG	DdeI(1), MstII(*)
DdeI	C ↓ TNAG	OxaNI	CC ↓ TNAGG	CTNAGG	DdeI(1), OxaNI(*)
DdeI	C ↓ TNAG	SauI	CC ↓ TNAGG	CTNAGG	DdeI(1), SauI(*)
DraII	AG ↓ GNCCC	AsuI	G ↓ GNCC	AGGNCC	DraII(*), AsuI(2), Cfr13I(2), Nsp7524IV(2), Nsp7524V(2), Sau96I(2)
DraII	GG ↓ GNCCC	AsuI	G ↓ GNCC	GGGNCC	DraII(*), AsuI(2), Cfr13I(2), NlaIV(3), Nsp7524IV(2), Nsp7524V(2), Sau96I(2)
DraII	AG ↓ GNCCC	Cfr13I	G ↓ GNCC	AGGNCC	DraII(*), Cfr13I(2), AsuI(2), Nsp7524IV(2), Nsp7524V(2), Sau96I(2)
DraII	GG ↓ GNCCC	Cfr13I	G ↓ GNCC	GGGNCC	DraII(*), Cfr13I(2), AsuI(2), NlaIV(3), Nsp7524IV(2), Nsp7524V(2), Sau96I(2)
DraII	AG ↓ GNCCC	DraII	AG ↓ GNCCT	AGGNCCT	DraII(*), AsuI(2), Cfr13I(2), Eco0109(2), Nsp7524IV(2), Nsp7524V(2), Sau96I(2)
DraII	AG ↓ GNCCC	DraII	GG ↓ GNCCC	AGGNCCC	DraII(2), AsuI(2), Cfr13I(2), NlaIV(4), Nsp7524IV(2), Nsp7524V(2), PssI(5), Sau96I(2)
DraII	GG ↓ GNCCC	DraII	AG ↓ GNCCC	GGGNCCC	DraII(*), AsuI(2), Cfr13I(2), NlaIV(4), Nsp7524IV(2), Nsp7524V(2), Sau96I(2)
DraII	GG ↓ GNCCC	DraII	AG ↓ GNCCT	GGGNCCT	DraII(*), AsuI(2), Cfr13I(2), NlaIV(3), Nsp7524IV(2), Nsp7524V(2), Sau96I(2)

Table 25. Hybrid Sequences Resulting from the Ligation of Two Fragments Generated After Digestion with One or Two Different Restriction Endonucleases (*continued*)

Enzyme 1	Fragment Generated by Enzyme 1	Enzyme 2	Fragment Generated by Enzyme 2	Resulting Hybrid Sequence	Enzyme 3
DraII	AG ↓ GNCCC	Eco0109	AG ↓ GNCCT	AGGNCCT	DraII(*), Eco0109(2), AsuI(2), Cfr13I(2), Nsp7524IV(2), Nsp7524V(2), Sau96I(2)
DraII	GG ↓ GNCCC	Eco0109	AG ↓ GNCCT	GGGNCCT	DraII(*), Eco0109(*), AsuI(2), Cfr13I(2), NlaIV(3), Nsp7524IV(2), Nsp7524V(2), Sau96I(2)
DraII	AG ↓ GNCCC	Nsp7524IV	G ↓ GNCC	AGGNCC	DraII(*), Nsp7524IV(2), AsuI(2), Cfr13I(2), Nsp7524V(2), Sau96I(2)
DraII	GG ↓ GNCCC	Nsp7524IV	G ↓ GNCC	GGGNCC	DraII(*), Nsp7524IV(2), AsuI(2), Cfr13I(2), NlaIV(3), Nsp7524V(2), Sau96I(2)
DraII	AG ↓ GNCCC	Nsp7524V	G ↓ GNCC	AGGNCC	DraII(*), Nsp7524V(2), AsuI(2), Cfr13I(2), Nsp7524IV(2), Sau96I(2)
DraII	GG ↓ GNCCC	Nsp7524V	G ↓ GNCC	GGGNCC	DraII(*), Nsp7524V(2), AsuI(2), Cfr13I(2), NlaIV(3), Nsp7524IV(2), Sau96I(2)
DraII	AG ↓ GNCCC	Sau96I	G ↓ GNCC	AGGNCC	DraII(*), Sau96I(2), AsuI(2), Cfr13I(2), Nsp7524IV(2), Nsp7524V(2)
DraII	GG ↓ GNCCC	Sau96I	G ↓ GNCC	GGGNCC	DraII(*), Sau96I(2), AsuI(2), Cfr13I(2), NlaIV(3), Nsp7524IV(2), Nsp7524V(2)
EaeI	C ↓ GGCCA	CfrI	C ↓ GGCCG	CGGCCG	EaeI(1), CfrI(1), BspRI(3), EagI(1), Eco52I(1), HaeIII(3), PalI(3), XmaIII(1)
EaeI	C ↓ GGCCA	CfrI	T ↓ GGCCA	CGGCCA	EaeI(1), CfrI(1), BspRI(3), HaeIII(3), PalI(3)
EaeI	T ↓ GGCCA	CfrI	C ↓ GGCCA	TGGCCA	EaeI(1), CfrI(1), BalI(3), BspRI(3), HaeIII(3), PalI(3)
EaeI	T ↓ GGCCA	CfrI	C ↓ GGCCG	TGGCCG	EaeI(1), CfrI(1), BspRI(3), HaeIII(3), PalI(3)
EaeI	C ↓ GGCCA	EaeI	C ↓ GGCCG	CGGCCG	EaeI(1), BspRI(3), CfrI(1), EagI(1), Eco52I(1), HaeIII(3), PalI(3), XmaIII(1)
EaeI	C ↓ GGCCA	EaeI	T ↓ GGCCA	CGGCCA	EaeI(1), BspRI(3), CfrI(1), HaeIII(3), PalI(3)
EaeI	T ↓ GGCCA	EaeI	C ↓ GGCCA	TGGCCA	EaeI(1), BalI(3), BspRI(3), CfrI(1), HaeIII(3), PalI(3)
EaeI	T ↓ GGCCA	EaeI	C ↓ GGCCG	TGGCCG	EaeI(1), BspRI(3), CfrI(1), HaeIII(3), PalI(3)
EaeI	C ↓ GGCCA	EagI	C ↓ GGCCG	CGGCCG	EaeI(1), EagI(1), BspRI(3), CfrI(1), Eco52I(1), HaeIII(3), PalI(3), XmaIII(1)

Table 25. Hybrid Sequences Resulting from the Ligation of Two Fragments Generated After Digestion with One or Two Different Restriction Endonucleases (*continued*)

Enzyme 1	Fragment Generated by Enzyme 1	Enzyme 2	Fragment Generated by Enzyme 2	Resulting Hybrid Sequence	Enzyme 3
EaeI	T ↓ GGCCA	EagI	C ↓ GGCCG	TGGCCG	EaeI(1), EagI(*), BspRI(3), CfrI(1), HaeIII(3), PalI(3)
EaeI	C ↓ GGCCA	Eco52I	C ↓ GGCCG	CGGCCG	EaeI(1), Eco52I(1), BspRI(3), CfrI(1), EagI(1), HaeIII(3), PalI(3), XmaIII(1)
EaeI	T ↓ GGCCA	Eco52I	C ↓ GGCCG	TGGCCG	EaeI(1), Eco52I(*), BspRI(3), CfrI(1), HaeIII(3), PalI(3)
EaeI	C ↓ GGCCA	NotI	GC ↓ GGCCGC	CGGCCGC	EaeI(1), NotI(*), BspRI(3), CfrI(1), EagI(1), Eco52I(1), Fnu4HI(4), HaeIII(3), PalI(3), XmaIII(1)
EaeI	T ↓ GGCCA	NotI	GC ↓ GGCCGC	TGGCCGC	EaeI(1), NotI(*), BspRI(3), CfrI(1), Fnu4HI(4), HaeIII(3), PalI(3)
EaeI	C ↓ GGCCA	XmaIII	C ↓ GGCCG	CGGCCG	EaeI(1), XmaIII(1), BspRI(3), CfrI(1), EagI(1), Eco52I(1), HaeIII(3), PalI(3)
EaeI	T ↓ GGCCA	XmaIII	C ↓ GGCCG	TGGCCG	EaeI(1), XmaIII(*), BspRI(3), CfrI(1), HaeIII(3), PalI(3)
EagI	C ↓ GGCCG	CfrI	C ↓ GGCCA	CGGCCA	EagI(*), CfrI(1), BspRI(3), EaeI(1), HaeIII(3), PalI(3)
EagI	C ↓ GGCCG	CfrI	T ↓ GGCCG	CGGCCG	EagI(1), CfrI(1), BspRI(3), EaeI(1), Eco52I(1), HaeIII(3), PalI(3), XmaIII(1)
EagI	C ↓ GGCCG	EaeI	C ↓ GGCCA	CGGCCA	EagI(*), EaeI(1), BspRI(3), CfrI(1), HaeIII(3), PalI(3)
EagI	C ↓ GGCCG	EaeI	T ↓ GGCCG	CGGCCG	EagI(1), EaeI(1), BspRI(3), CfrI(1), Eco52I(1), HaeIII(3), PalI(3), XmaIII(1)
EagI	C ↓ GGCCG	NotI	GC ↓ GGCCGC	CGGCCGC	EagI(1), NotI(*), BspRI(3), CfrI(1), EaeI(1), Eco52I(1), Fnu4HI(4), HaeIII(3), PalI(3), XmaIII(1)
EcoT14I	C ↓ CATGG	AflIII	A ↓ CATGT	CCATGT	EcoT14I(*), AflIII(*), NlaIII(5)
EcoT14I	C ↓ CTAGG	NheI	G ↓ CTAGC	CCTAGC	EcoT14I(*), NheI(*), MaeI(2)
EcoT14I	C ↓ CTAGG	SpeI	A ↓ CTAGT	CCTAGT	EcoT14I(*), SpeI(*), MaeI(2)
EcoT14I	C ↓ CTAGG	XbaI	T ↓ CTAGA	CCTAGA	EcoT14I(*), XbaI(*), MaeI(2)
Eco0109	AG ↓ GNCCT	AsuI	G ↓ GNCC	AGGNCC	Eco0109(*), AsuI(2), Cfr13I(2), Nsp7524IV(2), Nsp7524V(2), Sau96I(2)
Eco0109	AG ↓ GNCCT	Cfr13I	G ↓ GNCC	AGGNCC	Eco0109(*), Cfr13I(2), AsuI(2), Nsp7524IV(2), Nsp7524V(2), Sau96I(2)

Table 25. Hybrid Sequences Resulting from the Ligation of Two Fragments Generated After Digestion with One or Two Different Restriction Endonucleases (*continued*)

Enzyme 1	Fragment Generated by Enzyme 1	Enzyme 2	Fragment Generated by Enzyme 2	Resulting Hybrid Sequence	Enzyme 3
Eco0109	AG ↓ GNCCT	DraII	AG ↓ GNCCC	AGGNCCC	Eco0109(*), DraII(2), AsuI(2), Cfr13I(2), NlaIV(4), Nsp7524IV(2), Nsp7524V(2), PssI(5), Sau96I
Eco0109	AG ↓ GNCCT	DraII	GG ↓ GNCCT	AGGNCCT	Eco0109(2), DraII(*), AsuI(2), Cfr13I(2), Nsp7524IV(2), Nsp7524V(2), Sau96I(2)
Eco0109	AG ↓ GNCCT	Nsp7524IV	G ↓ GNCC	AGGNCC	Eco0109(*), Nsp7524IV(2), AsuI(2), Cfr13I(2), Nsp7524V(2), Sau96I(2)
Eco0109	AG ↓ GNCCT	Nsp7524V	G ↓ GNCC	AGGNCC	Eco0109(*), Nsp7524V(2), AsuI(2), Cfr13I(2), Nsp7524IV(2), Sau96I(2)
Eco0109	AG ↓ GNCCT	Sau96I	G ↓ GNCC	AGGNCC	Eco0109(*), Sau96I(2), AsuI(2), Cfr13I(2), Nsp7524IV(2), Nsp7524V(2)
Eco52I	C ↓ GGCCG	CfrI	C ↓ GGCCA	CGGCCA	Eco52I(*), CfrI(1), BspRI(3), EaeI(1), HaeIII(3), PalI(3)
Eco52I	C ↓ GGCCG	CfrI	T ↓ GGCCG	CGGCCG	Eco52I(1), CfrI(1), BspRI(3), EaeI(1), EagI(1), HaeIII(3), PalI(3), XmaIII(1)
Eco52I	C ↓ GGCCG	EaeI	C ↓ GGCCA	CGGCCA	Eco52I(*), EaeI(1), BspRI(3), CfrI(1), HaeIII(3), PalI(3)
Eco52I	C ↓ GGCCG	EaeI	T ↓ GGCCG	CGGCCG	Eco52I(1), EaeI(1), BspRI(3), CfrI(1), EagI(1), HaeIII(3), PalI(3), XmaIII(1)
Eco52I	C ↓ GGCCG	NotI	GC ↓ GGCCGC	CGGCCGC	Eco52I(1), NotI(*), BspRI(3), CfrI(1), EaeI(1), EagI(1), Fnu4HI(4), HaeIII(3), PalI(3), XmaIII(1)
Eco81I	CC ↓ TNAGG	DdeI	C ↓ TNAG	CCTNAG	Eco81I(*), DdeI(2)
Eco81I	CC ↓ TNAGG	EspI	GC ↓ TNAGC	CCTNAGC	Eco81I(*), EspI(*), DdeI(2)
EspI	GC ↓ TNAGC	AocI	CC ↓ TNAGG	GCTNAGG	EspI(*), AocI(*), DdeI(2)
EspI	GC ↓ TNAGC	CvnI	CC ↓ TNAGG	GCTNAGG	EspI(*), CvnI(*), DdeI(2)
EspI	GC ↓ TNAGC	DdeI	C ↓ TNAG	GCTNAG	EspI(*), DdeI(2)
EspI	GC ↓ TNAGC	Eco81I	CC ↓ TNAGG	GCTNAGG	EspI(*), Eco81I(*), DdeI(2)
EspI	GC ↓ TNAGC	MstII	CC ↓ TNAGG	GCTNAGG	EspI(*), MstII(*), DdeI(2)
EspI	GC ↓ TNAGC	OxaNI	CC ↓ TNAGG	GCTNAGG	EspI(*), OxaNI(*), DdeI(2)
EspI	GC ↓ TNAGC	SauI	CC ↓ TNAGG	GCTNAGG	EspI(*), SauI(*), DdeI(2)
Fnu4HI	GC ↓ NGC	ScrFI	CC ↓ NGG	GCNGG	Fnu4HI(*), ScrFI(*)
Fnu4HI	GC ↓ NGC	Tth111I	GACN ↓ NNGTC	GCNNGTC	Fnu4HI(*), Tth111I(*)
FspII	TT ↓ CGAA	AccI	GT ↓ CGAC	TTCGAC	FspII(*), AccI(*), TaqI(2), TthHB8I(2)
FspII	TT ↓ CGAA	AhaII	GA ↓ CGCC	TTCGCC	FspII(*), AhaII(*)
FspII	TT ↓ CGAA	AhaII	GA ↓ CGTC	TTCGTC	FspII(*), AhaII(*)
FspII	TT ↓ CGAA	AosII	GA ↓ CGCC	TTCGCC	FspII(*), AosII(*)
FspII	TT ↓ CGAA	AosII	GA ↓ CGTC	TTCGTC	FspII(*), AosII(*)
FspII	TT ↓ CGAA	AquI	CC ↓ CGAG	TTCGAG	FspII(*), AquI(*), TaqI(2), TthHB8I(2)

Table 25. Hybrid Sequences Resulting from the Ligation of Two Fragments Generated After Digestion with One or Two Different Restriction Endonucleases (*continued*)

Enzyme 1	Fragment Generated by Enzyme 1	Enzyme 2	Fragment Generated by Enzyme 2	Resulting Hybrid Sequence	Enzyme 3
FspII	TT ↓ CGAA	AquI	CC ↓ CGGG	TTCGGG	FspII(*), AquI(*)
FspII	TT ↓ CGAA	BanIII	AT ↓ CGAT	TTCGAT	FspII(*), BanIII(*), TaqI(2), TthHB8I(2)
FspII	TT ↓ CGAA	BbiII	GA ↓ CGCC	TTCGCC	FspII(*), BbiII(*)
FspII	TT ↓ CGAA	BbiII	GA ↓ CGTC	TTCGTC	FspII(*), BbiII(*)
FspII	TT ↓ CGAA	ClaI	AT ↓ CGAT	TTCGAT	FspII(*), ClaI(*), TaqI(2), TthHB8I(2)
FspII	TT ↓ CGAA	HapII	C ↓ CGG	TTCGG	FspII(*), HapII(*)
FspII	TT ↓ CGAA	HinpI	G ↓ CGC	TTCGC	FspII(*), HinpI(*)
FspII	TT ↓ CGAA	HpaII	C ↓ CGG	TTCGG	FspII(*), HpaII(*)
FspII	TT ↓ CGAA	MaeII	A ↓ CGT	TTCGT	FspII(*), MaeII(*)
FspII	TT ↓ CGAA	MspI	C ↓ CGG	TTCGG	FspII(*), MspI(*)
FspII	TT ↓ CGAA	NarI	GG ↓ CGCC	TTCGCC	FspII(*), NarI(*)
FspII	TT ↓ CGAA	NunII	GG ↓ CGCC	TTCGCC	FspII(*), NunII(*)
FspII	TT ↓ CGAA	TaqI	T ↓ CGA	TTCGA	FspII(*), TaqI(2), TthHB8I(2)
FspII	TT ↓ CGAA	TthHB8I	T ↓ CGA	TTCGA	FspII(*), TthHB8I(2), TaqI(2)
HapII	C ↓ CGG	AccI	GT ↓ CGAC	CCGAC	HapII(*), AccI(*)
HapII	C ↓ CGG	AhaII	GA ↓ CGCC	CCGCC	HapII(*), AhaII(*)
HapII	C ↓ CGG	AhaII	GA ↓ CGTC	CCGTC	HapII(*), AhaII(*)
HapII	C ↓ CGG	AosII	GA ↓ CGCC	CCGCC	HapII(*), AosII(*)
HapII	C ↓ CGG	AosII	GA ↓ CGTC	CCGTC	HapII(*), AosII(*)
HapII	C ↓ CGG	AquI	CC ↓ CGAG	CCGAG	HapII(*), AquI(*)
HapII	C ↓ CGG	AquI	CC ↓ CGGG	CCGGG	HapII(1), AquI(*), BcnI(3), HpaII(1), MspI(1), NciI(2), ScrFI(2)
HapII	C ↓ CGG	AsuII	TT ↓ CGAA	CCGAA	HapII(*), AsuII(*)
HapII	C ↓ CGG	BanIII	AT ↓ CGAT	CCGAT	HapII(*), BanIII(*)
HapII	C ↓ CGG	BbiII	GA ↓ CGCC	CCGCC	HapII(*), BbiII(*)
HapII	C ↓ CGG	BbiII	GA ↓ CGTC	CCGTC	HapII(*), BbiII(*)
HapII	C ↓ CGG	ClaI	AT ↓ CGAT	CCGAT	HapII(*), ClaI(*)
HapII	C ↓ CGG	FspII	TT ↓ CGAA	CCGAA	HapII(*), FspII(*)
HapII	C ↓ CGG	HinpI	G ↓ CGC	CCGC	HapII(*), HinpI(*)
HapII	C ↓ CGG	MaeII	A ↓ CGT	CCGT	HapII(*), MaeII(*)
HapII	C ↓ CGG	NarI	GG ↓ CGCC	CCGCC	HapII(*), NarI(*)
HapII	C ↓ CGG	NunII	GG ↓ CGCC	CCGCC	HapII(*), NunII(*)
HapII	C ↓ CGG	TaqI	T ↓ CGA	CCGA	HapII(*), TaqI(*)
HapII	C ↓ CGG	TthHB8I	T ↓ CGA	CCGA	HapII(*), TthHB8I(*)
HinpI	G ↓ CGC	AccI	GT ↓ CGAC	GCGAC	HinpI(*), AccI(*)
HinpI	G ↓ CGC	AhaII	GA ↓ CGCC	GCGCC	HinpI(1), AhaII(*), CfoI(3), HhaI(3)
HinpI	G ↓ CGC	AhaII	GA ↓ CGTC	GCGTC	HinpI(*), AhaII(*), HgaI(10)
HinpI	G ↓ CGC	AosII	GA ↓ CGCC	GCGCC	HinpI(1), AosII(*), CfoI(3), HhaI(3)
HinpI	G ↓ CGC	AosII	GA ↓ CGTC	GCGTC	HinpI(*), AosII(*), HgaI(10)
HinpI	G ↓ CGC	AquI	CC ↓ CGAG	GCGAG	HinpI(*), AquI(*)
HinpI	G ↓ CGC	AquI	CC ↓ CGGG	GCGGG	HinpI(*), AquI(*)
HinpI	G ↓ CGC	AsuII	TT ↓ CGAA	GCGAA	HinpI(*), AsuII(*)
HinpI	G ↓ CGC	BanIII	AT ↓ CGAT	GCGAT	HinpI(*), AquI(*)
HinpI	G ↓ CGC	BbiII	GA ↓ CGCC	GCGCC	HinpI(1), BbiII(*), CfoI(3), HhaI(3)
HinpI	G ↓ CGC	BbiII	GA ↓ CGTC	GCGTC	HinpI(*), BbiII(*), HgaI(10)
HinpI	G ↓ CGC	ClaI	AT ↓ CGAT	GCGAT	HinpI(*), ClaI(*)

Table 25. Hybrid Sequences Resulting from the Ligation of Two Fragments Generated After Digestion with One or Two Different Restriction Endonucleases (*continued*)

Enzyme 1	Fragment Generated by Enzyme 1	Enzyme 2	Fragment Generated by Enzyme 2	Resulting Hybrid Sequence	Enzyme 3
HinpI	G ↓ CGC	FspII	TT ↓ CGAA	GCGAA	HinpI(*), FspII(*)
HinpI	G ↓ CGC	HapII	C ↓ CGG	GCGG	HinpI(*), HapII(*)
HinpI	G ↓ CGC	HpaII	C ↓ CGG	GCGG	HinpI(*), HpaII(*)
HinpI	G ↓ CGC	MaeII	A ↓ CGT	GCGT	HinpI(*), MaeII(*)
HinpI	G ↓ CGC	MspI	C ↓ CGG	GCGG	HinpI(*), MspI(*)
HinpI	G ↓ CGC	NarI	GG ↓ CGCC	GCGCC	HinpI(1), NarI(*), CfoI(3), HhaI(3)
HinpI	G ↓ CGC	NunII	GG ↓ CGCC	GCGCC	HinpI(1), NunII(*), CfoI(3), HhaI(3)
HinpI	G ↓ CGC	TaqI	T ↓ CGA	GCGA	HinpI(*), TaqI(*)
HinpI	G ↓ CGC	TthHB8I	T ↓ CGA	GCGA	HinpI(*), TthHB8I(*)
HpaII	C ↓ CGG	AccI	GT ↓ CGAC	CCGAC	HpaII(*), AccI(*)
HapII	C ↓ CGG	AhaII	GA ↓ CGCC	CCGCC	HpaII(*), AhaII(*)
HpaII	C ↓ CGG	AhaII	GA ↓ CGTC	CCGTC	HpaII(*), AhaII(*)
HpaII	C ↓ CGG	AosII	GA ↓ CGCC	CCGCC	HpaII(*), AosII(*)
HpaII	C ↓ CGG	AosII	GA ↓ CGTC	CCGTC	HpaII(*), AosII(*)
HpaII	C ↓ CGG	AquI	CC ↓ CGAG	CCGAG	HpaII(*), AquI(*)
HpaII	C ↓ CGG	AquI	CC ↓ CGGG	CCGGG	HpaII(1), AquI(*), BcnI(3), HapII(1), MspI(1), NciI(2), ScrFI(2)
HpaII	C ↓ CGG	AsuII	TT ↓ CGAA	CCGAA	HpaII(*), AsuII(*)
HpaII	C ↓ CGG	BanIII	AT ↓ CGAT	CCGAT	HpaII(*), BanII(*)
HpaII	C ↓ CGG	BbiII	GA ↓ CGCC	CCGCC	HpaII(*), BbiII(*)
HpaII	C ↓ CGG	BbiII	GA ↓ CGTC	CCGTC	HpaII(*), BbiII(*)
HpaII	C ↓ CGG	ClaI	AT ↓ CGAT	CCGAT	HpaII(*), ClaI(*)
HpaII	C ↓ CGG	FspII	TT ↓ CGAA	CCGAA	HpaII(*), FspII(*)
HpaII	C ↓ CGG	HinpI	G ↓ CGC	CCGC	HpaII(*), HinpI(*)
HpaII	C ↓ CGG	MaeII	A ↓ CGT	CCGT	HpaII(*), MaeII(*)
HpaII	C ↓ CGG	NarI	GG ↓ CGCC	CCGCC	HpaII(*), NarI(*)
HpaII	C ↓ CGG	NunII	GG ↓ CGCC	CCGCC	HpaII(*), NunII(*)
HpaII	C ↓ CGG	TaqI	T ↓ CGA	CCGA	HpaII(*), TaqI(*)
HpaII	C ↓ CGG	TthHB8I	T ↓ CGA	CCGA	HpaII(*), TthHB8I(*)
MaeI	C ↓ TAG	NdeI	CA ↓ TATG	CTATG	MaeI(*), NdeI(*)
MaeII	A ↓ CGT	AccI	GT ↓ CGAC	ACGAC	MaeII(*), AccI(*)
MaeII	A ↓ CGT	AhaII	GA ↓ CGCC	ACGCC	MaeII(*), AhaII(*)
MaeII	A ↓ CGT	AhaII	GA ↓ CGTC	ACGTC	MaeII(1), AhaII(*)
MaeII	A ↓ CGT	AosII	GA ↓ CGCC	ACGCC	MaeII(*), AosII(*)
MaeII	A ↓ CGT	AosII	GA ↓ CGTC	ACGTC	MaeII(1), AosII(*)
MaeII	A ↓ CGT	AquI	CC ↓ CGAG	ACGAG	MaeII(*), AquI(*)
MaeII	A ↓ CGT	AquI	CC ↓ CGGG	ACGGG	MaeII(*), AquI(*)
MaeII	A ↓ CGT	AsuII	TT ↓ CGAA	ACGAA	MaeII(*), AsuII(*)
MaeII	A ↓ CGT	BanIII	AT ↓ CGAT	ACGAT	MaeII(*), BanIII(*)
MaeII	A ↓ CGT	BbiII	GA ↓ CGCC	ACGCC	MaeII(*), BbiII(*)
MaeII	A ↓ CGT	BbiII	GA ↓ CGTC	ACGTC	MaeII(1), BbiII(*)
MaeII	A ↓ CGT	ClaI	AT ↓ CGAT	ACGAT	MaeII(*), ClaI(*)
MaeII	A ↓ CGT	FspII	TT ↓ CGAA	ACGAA	MaeII(*), FspII(*)
MaeII	A ↓ CGT	HapII	C ↓ CGG	ACGG	MaeII(*), HapII(*)
MaeII	A ↓ CGT	HinpI	G ↓ CGC	ACGC	MaeII(*), HinpI(*)
MaeII	A ↓ CGT	HpaII	C ↓ CGG	ACGG	MaeII(*), HpaII(*)
MaeII	A ↓ CGT	MspI	C ↓ CGG	ACGG	MaeII(*), MspI(*)
MaeII	A ↓ CGT	NarI	GG ↓ CGCC	ACGCC	MaeII(*), NarI(*)
MaeII	A ↓ CGT	NunII	GG ↓ CGCC	ACGCC	MaeII(*), NunII(*)
MaeII	A ↓ CGT	TaqI	T ↓ CGA	ACGA	MaeII(*), TaqI(*)
MaeII	A ↓ CGT	TthHB8I	T ↓ CGA	ACGA	MaeII(*), TthHB8I(*)

Table 25. Hybrid Sequences Resulting from the Ligation of Two Fragments Generated After Digestion with One or Two Different Restriction Endonucleases (*continued*)

Enzyme 1	Fragment Generated by Enzyme 1	Enzyme 2	Fragment Generated by Enzyme 2	Resulting Hybrid Sequence	Enzyme 3
MaeIII	↓ GTNAC	BstEII	G ↓ GTNACC	GTNACC	MaeIII(0), BstEII(*)
MboI	↓ GATC	BamHI	G ↓ GATCC	GATCC	MboI(0), BamHI(*), DpnI(2), NdeII(0), Sau3A(0)
MboI	↓ GATC	BclI	T ↓ GATCA	GATCA	MboI(0), BclI(*), DpnI(2), NdeII(0), Sau3A(0)
MboI	↓ GATC	BglII	A ↓ GATCT	GATCT	MboI(0), BglII(*), DpnI(2), NdeII(0), Sau3A(0)
MboI	↓ GATC	BstI	G ↓ GATCC	GATCC	MboI(0), BstI(*), DpnI(2), NdeII(0), Sau3A(0)
MboI	↓ GATC	MflI	A ↓ GATCC	GATCC	MboI(0), MflI(*), DpnI(2), NdeII(0), Sau3A(0)
MboI	↓ GATC	MflI	A ↓ GATCT	GATCT	MboI(0), MflI(*), DpnI(2), NdeII(0), Sau3A(0)
MboI	↓ GATC	XhoII	A ↓ GATCC	GATCC	MboI(0), XhoII(*), DpnI(2), NdeII(0), Sau3A(0)
MboI	↓ GATC	XhoII	A ↓ GATCT	GATCT	MboI(0), XhoII(*), DpnI(2), NdeII(0), Sau3A(0)
MflI	A ↓ GATCC	BamHI	G ↓ GATCC	AGATCC	MflI(1), BamHI(*), DpnI(3), MboI(1), NdeII(1), Sau3A(1), XhoII(1)
MflI	G ↓ GATCT	BamHI	G ↓ GATCC	GGATCC	MflI(1), BamHI(1), BstI(1), DpnI(3), MboI(1), NdeII(1), NlaIV(3), Sau3A(1), XhoII(1)
MflI	A ↓ GATCC	BclI	T ↓ GATCA	AGATCA	MflI(*), BclI(*), DpnI(3), MboI(1), NdeII(1), Sau3A(1)
MflI	G ↓ GATCC	BclI	T ↓ GATCA	GGATCA	MflI(*), BclI(*), DpnI(3), MboI(1), NdeII(1), Sau3A(1)
MflI	A ↓ GATCC	BglII	A ↓ GATCT	AGATCT	MflI(1), BglII(1), DpnI(3), MboI(1), NdeII(1), Sau3A(1), XhoII(1)
MflI	G ↓ GATCC	BglII	A ↓ GATCT	GGATCT	MflI(1), BglII(*), DpnI(3), MboI(1), NdeII(1), Sau3A(1), XhoII(1)
MflI	A ↓ GATCC	BstI	G ↓ GATCC	AGATCC	MflI(1), BstI(*), DpnI(3), MboI(1), NdeII(1), Sau3A(1), XhoII(1)
MflI	G ↓ GATCT	BstI	G ↓ GATCC	GGATCC	MflI(1), BstI(1), BamHI(1), DpnI(3), MboI(1), NdeII(1), NlaIV(3), Sau3A(1), XhoII(1)
MflI	A ↓ GATCC	MboI	↓ GATC	AGATC	MflI(*), MboI(1), DpnI(3), NdeII(1), Sau3A(1)
MflI	G ↓ GATCC	MboI	↓ GATC	GGATC	MflI(*), MboI(1), DpnI(3), NdeII(1), Sau3A(1)
MflI	A ↓ GATCC	MflI	A ↓ GATCT	AGATCT	MflI(1), BglII(1), DpnI(3), MboI(1), NdeII(1), Sau3A(1), XhoII(1)
MflI	A ↓ GATCC	MflI	G ↓ GATCC	AGATCC	MflI(1), DpnI(3), MboI(1), NdeII(1), Sau3A(1), XhoII(1)

Table 25. Hybrid Sequences Resulting from the Ligation of Two Fragments Generated After Digestion with One or Two Different Restriction Endonucleases (*continued*)

Enzyme 1	Fragment Generated by Enzyme 1	Enzyme 2	Fragment Generated by Enzyme 2	Resulting Hybrid Sequence	Enzyme 3
MflI	G ↓ GATCC	MflI	A ↓ GATCC	GGATCC	MflI(1), BamHI(1), BstI(1), DpnI(3), MboI(1), NdeII(1), NlaIV(3), Sau3A(1), XhoII(1)
MflI	G ↓ GATCC	MflI	A ↓ GATCT	GGATCT	MflI(1), DpnI(3), MboI(1), NdeII(1), Sau3A(1), XhoII(1)
MflI	A ↓ GATCC	NdeII	↓ GATC	AGATC	MflI(*), NdeII(1), DpnI(3), MboI(1), Sau3A(1)
MflI	G ↓ GATCC	NdeII	↓ GATC	GGATC	MflI(*), NdeII(1), DpnI(3), MboI(1), Sau3A(1)
MflI	A ↓ GATCC	Sau3A	↓ GATC	AGATC	MflI(*), Sau3A(1), DpnI(3), MboI(1), NdeII(1)
MflI	G ↓ GATCC	Sau3A	↓ GATC	GGATC	MflI(*), Sau3A(1), DpnI(3), MboI(1), NdeII(1)
MflI	A ↓ GATCC	XhoII	A ↓ GATCT	AGATCT	MflI(1), XhoII(1), BglII(1), DpnI(3), MboI(1), NdeII(1), Sau3A(1)
MflI	A ↓ GATCC	XhoII	G ↓ GATCC	AGATCC	MflI(1), XhoII(1), DpnI(3), MboI(1), NdeII(1), Sau3A(1)
MflI	G ↓ GATCC	XhoII	A ↓ GATCC	GGATCC	MflI(1), XhoII(1), BamHI(1), BstI(1), DpnI(3), MboI(1), NdeII(1), NlaIV(3), Sau3A(1)
MflI	G ↓ GATCC	XhoII	A ↓ GATCT	GGATCT	MflI(1), XhoII(1), DpnI(3), MboI(1), NdeII(1), Sau3A(1)
MluI	A ↓ CGCGT	BssHII	G ↓ CGCGC	ACGCGC	MluI(*), BssHII(*), CfoI(5), FnudII(3), HhaI(5), HinpI(3), ThaI(3)
MspI	C ↓ CGG	AccI	GT ↓ CGAC	CCGAC	MspI(*), AccI(*)
MspI	C ↓ CGG	AhaII	GA ↓ CGCC	CCGCC	MspI(*), AhaII(*)
MspI	C ↓ CGG	AhaII	GA ↓ CGTC	CCGTC	MspI(*), AhaII(*)
MspI	C ↓ CGG	AosII	GA ↓ CGCC	CCGCC	MspI(*), AosII(*)
MspI	C ↓ CGG	AosII	GA ↓ CGTC	CCGTC	MspI(*), AosII(*)
MspI	C ↓ CGG	AquI	CC ↓ CGAG	CCGAG	MspI(*), AquI(*)
MspI	C ↓ CGG	AquI	CC ↓ CGGG	CCGGG	MspI(1), AquI(*), BcnI(3), HapII(1), HpaII(1), NciI(2), ScrFI(2)
MspI	C ↓ CGG	AsuII	TT ↓ CGAA	CCGAA	MspI(*), AsuII(*)
MspI	C ↓ CGG	BanIII	AT ↓ CGAT	CCGAT	MspI(*), BanIII(*)
MspI	C ↓ CGG	BbiII	GA ↓ CGCC	CCGCC	MspI(*), BbiII(*)
MspI	C ↓ CGG	BbiII	GA ↓ CGTC	CCGTC	MspI(*), BbiII(*)
MspI	C ↓ CGG	ClaI	AT ↓ CGAT	CCGAT	MspI(*), ClaI(*)
MspI	C ↓ CGG	FspII	TT ↓ CGAA	CCGAA	MspI(*), FspII(*)
MspI	C ↓ CGG	HinpI	G ↓ CGC	CCGC	MspI(*), HinpI(*)
MspI	C ↓ CGG	MaeII	A ↓ CGT	CCGT	MspI(*), MaeII(*)
MspI	C ↓ CGG	NarI	GG ↓ CGCC	CCGCC	MspI(*), NarI(*)
MspI	C ↓ CGG	NunII	GG ↓ CGCC	CCGCC	MspI(*), NunII(*)
MspI	C ↓ CGG	TaqI	T ↓ CGA	CCGA	MspI(*), TaqI(*)
MspI	C ↓ CGG	TthHB8I	T ↓ CGA	CCGA	MspI(*), TthHB8I(*)
MstII	CC ↓ TNAGG	DdeI	C ↓ TNAG	CCTNAG	MstII(*), DdeI(2)

Table 25. Hybrid Sequences Resulting from the Ligation of Two Fragments Generated After Digestion with One or Two Different Restriction Endonucleases (*continued*)

Enzyme 1	Fragment Generated by Enzyme 1	Enzyme 2	Fragment Generated by Enzyme 2	Resulting Hybrid Sequence	Enzyme 3
MstII	CC ↓ TNAGG	EspI	GC ↓ TNAGC	CCTNAGC	MstII(*), EspI(*), DdeI(2)
NarI	GG ↓ CGCC	AccI	GT ↓ CGAC	GGCGAC	NarI(*), AccI(*)
NarI	GG ↓ CGCC	AhaII	GA ↓ CGCC	GGCGCC	NarI(2), AhaII(2), AosII(2), BanI(1), BbeI(5), BbiII(2), CfoI(4), HaeII(5), HhaI(4), HinpI(2), NlaIV(3), NunII(2)
NarI	GG ↓ CGCC	AhaII	GA ↓ CGTC	GGCGTC	NarI(*), AhaII(2), AosII(2), BbiII(2), HgaI(11)
NarI	GG ↓ CGCC	AosII	GA ↓ CGCC	GGCGCC	NarI(2), AosII(2), AhaII(2), BanI(1), BbeI(5), BbiII(2), CfoI(4), HaeII(5), HhaI(4), HinpI(2), NlaIV(3), NunII(2)
NarI	GG ↓ CGCC	AosII	GA ↓ CGTC	GGCGTC	NarI(*), AosII(2), AhaII(2), BbiII(2), HgaI(11)
NarI	GG ↓ CGCC	AquI	CC ↓ CGAG	GGCGAG	NarI(*), AquI(*)
NarI	GG ↓ CGCC	AquI	CC ↓ CGGG	GGCGGG	NarI(*), AquI(*)
NarI	GG ↓ CGCC	AsuII	TT ↓ CGAA	GGCGAA	NarI(*), AsuII(*)
NarI	GG ↓ CGCC	BanIII	AT ↓ CGAT	GGCGAT	NarI(*), BanIII(*)
NarI	GG ↓ CGCC	BbiII	GA ↓ CGCC	GGCGCC	NarI(2), BbiII(2), AhaII(2), AosII(2), BanI(1), BbeI(5), CfoI(4), HaeII(5), HhaI(4), HinpI(2), NlaIV(3), NunII(2)
NarI	GG ↓ GGCC	BbiII	GA ↓ CGTC	GGCGTC	NarI(*), BbiII(2), AhaII(2), AosII(2), HgaI(11)
NarI	GG ↓ CGCC	ClaI	AT ↓ CGAT	GGCGAT	NarI(*), ClaI(*)
NarI	GG ↓ CGCC	FspII	TT ↓ CGAA	GGCGAA	NarI(*), FspII(*)
NarI	GG ↓ CGCC	HapII	C ↓ CGG	GGCGG	NarI(*), HapII(*)
NarI	GG ↓ CGCC	HinpI	G ↓ CGC	GGCGC	NarI(*), HinpI(2), CfoI(4), HhaI(4)
NarI	GG ↓ CGCC	HpaII	C ↓ CGG	GGCGG	NarI(*), HpaII(*)
NarI	GG ↓ CGCC	MaeII	A ↓ CGT	GGCGT	NarI(*), MaeII(*)
NarI	GG ↓ CGCC	MspI	C ↓ CGG	GGCGG	NarI(*), MspI(*)
NarI	GG ↓ CGCC	TaqI	T ↓ CGA	GGCGA	NarI(*), TaqI(*)
NarI	GG ↓ CGCC	TthHB8I	T ↓ CGA	GGCGA	NarI(*), TthHB8I(*)
NcoI	C ↓ CATGG	AflIII	A ↓ CATGT	CCATGT	NcoI(*), AflIII(*), NlaIII(5)
NdeI	CA ↓ TATG	MaeI	C ↓ TAG	CATAG	NdeI(*), MaeI(*)
NdeII	↓ GATC	BamHI	G ↓ GATCC	GATCC	NdeII(0), BamHI(*), DpnI(2), MboI(0), Sau3A(0)
NdeII	↓ GATC	BclI	T ↓ GATCA	GATCA	NdeII(0), BclI(*), DpnI(2), MboI(0), Sau3A(0)
NdeII	↓ GATC	BglII	A ↓ GATCT	GATCT	NdeII(0), BglII(*), DpnI(2), MboI(0), Sau3A(0)
NdeII	↓ GATC	BstI	G ↓ GATCC	GATCC	NdeII(0), BstI(*), DpnI(2), MboI(0), Sau3A(0)
NdeII	↓ GATC	MflI	A ↓ GATCC	GATCC	NdeII(0), MflI(*), DpnI(2), MboI(0), Sau3A(0)
NdeII	↓ GATC	MflI	A ↓ GATCT	GATCT	NdeII(0), MflI(*), DpnI(2), MboI(0), Sau3A(0)
NdeII	↓ GATC	XhoII	A ↓ GATCC	GATCC	NdeII(0), XhoII(*), DpnI(2), MboI(0), Sau3A(0)

Table 25. Hybrid Sequences Resulting from the Ligation of Two Fragments Generated After Digestion with One or Two Different Restriction Endonucleases (*continued*)

Enzyme 1	Fragment Generated by Enzyme 1	Enzyme 2	Fragment Generated by Enzyme 2	Resulting Hybrid Sequence	Enzyme 3
NdeII	↓ GATC	XhoII	A ↓ GATCT	GATCT	NdeII(0), XhoII(*), DpnI(2), MboI(0), Sau3A(0)
NheI	G ↓ CTAGC	AvrII	C ↓ CTAGG	GCTAGG	NheI(*), AvrII(*), MaeI(2)
NheI	G ↓ CTAGC	EcoT14I	C ↓ CTAGG	GCTAGG	NheI(*), EcoT14I(*), MaeI(2)
NheI	G ↓ CTAGC	SpeI	A ↓ CTAGT	GCTAGT	NheI(*), SpeI(*), MaeI(2)
NheI	G ↓ CTAGC	StyI	C ↓ CTAGG	GCTAGG	NheI(*), StyI(*), MaeI(2)
NheI	G ↓ CTAGC	XbaI	T ↓ CTAGA	GCTAGA	NheI(*), XbaI(*), MaeI(2)
NotI	GC ↓ GGCCGC	CfrI	C ↓ GGCCA	GCGGCCA	NotI(*), CfrI(2), BspRI(4), EaeI(2), Fnu4HI(2), HaeIII(4), PalI(4)
NotI	GC ↓ GGCCGC	CfrI	C ↓ GGCCG	GCGGCCG	NotI(*), CfrI(2), BspRI(4), EaeI(2), EagI(2), Eco52I(2), Fnu4HI(2), HaeIII(4), PalI(4), XmaIII(2)
NotI	GC ↓ GGCCGC	EaeI	C ↓ GGCCA	GCGGCCA	NotI(*), EaeI(2), BspRI(4), CfrI(2), Fnu4HI(2), HaeIII(4), PalI(4)
NotI	GC ↓ GGCCGC	EaeI	C ↓ GGCCG	GCGGCCG	NotI(*), EaeI(2), BspRI(4), CfrI(2), EagI(2), Eco52I(2), Fnu4HI(2), HaeIII(4), PalI(4), XmaIII(2)
NotI	GC ↓ GGCCGC	EagI	C ↓ GGCCG	GCGGCCG	NotI(*), EagI(2), BspRI(4), EaeI(2), Eco52I(2), Fnu4HI(2), HaeIII(4), PalI(4), XmaIII(2)
NotI	GC ↓ GGCCGC	Eco52I	C ↓ GGCCG	GCGGCCG	NotI(*), Eco52I(2), BspRI(4), CfrI(2), EaeI(2), EagI(2), Fnu4HI(2), HaeIII(4), PalI(4), XmaIII(2)
NotI	GC ↓ GGCCGC	XmaIII	C ↓ GGCCG	GCGGCCG	NotI(*), XmaIII(2), BspRI(4), CfrI(2), EaeI(2), EagI(2), Eco52I(2), Fnu4HI(2), HaeIII(4), PalI(4)
Nsp7524III	C ↓ CCGGG	AccIII	T ↓ CCGGA	CCCGGA	Nsp7524III(*), AccIII(*), BcnI(3), HapII(2), HpaII(2), MspI(2), NciI(2), ScrFI(2)
Nsp7524III	C ↓ CCGGG	BspMII	T ↓ CCGGA	CCCGGA	Nsp7524III(*), BspMII(*), BcnI(3), NapII(2), HpaII(2), MspI(2), NciI(2), ScrFI(2)
Nsp7524III	C ↓ CCGGG	Cfr10I	A ↓ CCGGC	CCCGGC	Nsp7524III(*), Cfr10I(*), BcnI(3), HapII(2), HpaII(2), MspI(2), NciI(2), ScrFI(2)
Nsp7524III	C ↓ CCGGG	Cfr10I	A ↓ CCGGT	CCCGGT	Nsp7524III(*), Cfr10I(*), BcnI(3), HapII(2), HpaII(2), MspI(2), NciI(2)
Nsp7524III	C ↓ TCGAG	SalI	G ↓ TCGAC	CTCGAC	Nsp7524III(*), SalI(*), TaqI(2), TthHB8I(2)

Table 25. Hybrid Sequences Resulting from the Ligation of Two Fragments Generated After Digestion with One or Two Different Restriction Endonucleases (*continued*)

Enzyme 1	Fragment Generated by Enzyme 1	Enzyme 2	Fragment Generated by Enzyme 2	Resulting Hybrid Sequence	Enzyme 3
Nsp7524IV	G ↓ GNCC	DraII	AG ↓ GNCCC	GGNCCC	Nsp7524IV(1), DraII(*), AsuI(1), Cfr13I(1), NlaIV(3), Nsp7524V(1), Sau96I(1)
Nsp7524IV	G ↓ GNCC	DraII	AG ↓ GNCCT	GGNCCT	Nsp7524IV(1), DraII(*), AsuI(1), Cfr13I(1), Nsp7524V(1), Sau96I(1)
Nsp7524IV	G ↓ GNCC	Eco0109	AG ↓ GNCCT	GGNCCT	Nsp7524IV(1), Eco0109(*), AsuI(1), Cfr13I(1), Nsp7524V(1), Sau96I(1)
Nsp7524V	G ↓ GNCC	DraII	AG ↓ GNCCC	GGNCCC	Nsp7524V(1), DraII(*), AsuI(1), Cfr13I(1), NlaIV(3), Nsp7524IV(1), Sau96I(1)
Nsp7524V	G ↓ GNCC	DraII	AG ↓ GNCCT	GGNCCT	Nsp7524V(1), DraII(*), AsuI(1), Cfr13I(1), Nsp7524IV(1), Sau96I(1)
Nsp7524V	G ↓ GNCC	Eco0109	AG ↓ GNCCT	GGNCCT	Nsp7524V(1), Eco0109(*), AsuI(1), Cfr13I(1), Nsp7524IV(1), Sau96I(1)
NunII	GG ↓ CGCC	AccI	GT ↓ CGAC	GGCGAC	NunII(*), AccI(*)
NunII	GG ↓ CGCC	AhaII	GA ↓ CGCC	GGCGCC	NunII(2), AhaII(2), AosII(2), BanI(1), BbeI(5), BbiII(2), CfoI(4), HaeII(5), HhaI(4), HinpI(2), NarI(2), NlaIV(3)
NunII	GG ↓ CGCC	AhaII	GA ↓ CGTC	GGCGTC	NunII(*), AhaII(2), AosII(2), BbiII(2), HgaI(11)
NunII	GG ↓ CGCC	AosII	GA ↓ CGCC	GGCGCC	NunII(2), AosII(2), AhaII(2), BanI(1), BbeI(5), BbiII(2), CfoI(4), HaeII(5), HhaI(4), HinpI(2), NarI(2), NlaIV(3)
NunII	GG ↓ CGCC	AosII	GA ↓ CGTC	GGCGTC	NunII(*), AosII(2), AhaII(2), BbiII(2), HgaI(11)
NunII	GG ↓ CGCC	AquI	CC ↓ CGAG	GGCGAG	NunII(*), AquI(*)
NunII	GG ↓ CGCC	AquI	CC ↓ CGGG	GGCGGG	NunII(*), AquI(*)
NunII	GG ↓ CGCC	AsuII	TT ↓ CGAA	GGCGAA	NunII(*), AsuII(*)
NunII	GG ↓ CGCC	BanIII	AT ↓ CGAT	GGCGAT	NunII(*), BanIII(*)
NunII	GG ↓ CGCC	BbiII	GA ↓ CGCC	GGCGCC	NunII(2), BbiII(2), AhaII(2), AosII(2), BanI(1), BbeI(5), CfoI(4), HaeII(5), HhaI(4), HinpI(2), NarI(2), NlaIV(3)
NunII	GG ↓ CGCC	BbiII	GA ↓ CGTC	GGCGTC	NunII(*), BbiII(2), AhaII(2), AosII(2), HgaI(11)
NunII	GG ↓ CGCC	ClaI	AT ↓ CGAT	GGCGAT	NunII(*), ClaI(*)
NunII	GG ↓ CGCC	FspII	TT ↓ CGAA	GGCGAA	NunII(*), FspII(*)
NunII	GG ↓ CGCC	HapII	C ↓ CGG	GGCGG	NunII(*), HapII(*)
NunII	GG ↓ CGCC	HinpI	G ↓ CGC	GGCGC	NunII(*), HinpI(2), CfoI(4), HhaI(4)
NunII	GG ↓ CGCC	HpaII	C ↓ CGG	GGCGG	NunII(*), HpaII(*)
NunII	GG ↓ CGCC	MaeII	A ↓ CGT	GGCGT	NunII(*), MaeII(*)
NunII	GG ↓ CGCC	MspI	C ↓ CGG	GGCGG	NunII(*), MspI(*)

Table 25. Hybrid Sequences Resulting from the Ligation of Two Fragments Generated After Digestion with One or Two Different Restriction Endonucleases (*continued*)

Enzyme 1	Fragment Generated by Enzyme 1	Enzyme 2	Fragment Generated by Enzyme 2	Resulting Hybrid Sequence	Enzyme 3
NunII	GG ↓ CGCC	TaqI	T ↓ CGA	GGCGA	NunII(*), TaqI(*)
NunII	GG ↓ CGCC	TthHB8I	T ↓ CGA	GGCGA	NunII(*), TthHB8I(*)
OxaNI	CC ↓ TNAGG	DdeI	C ↓ TNAG	CCTNAG	OxaNI(*), DdeI(2)
OxaNI	CC ↓ TNAGG	EspI	GC ↓ TNAGC	CCTNAGC	OxaNI(*), EspI(*), DdeI(2)
PaeR7	C ↓ TCGAG	SalI	G ↓ TCGAC	CTCGAC	PaeR7(*), SalI(*), TaqI(2), TthHB8I(2)
PpuMI	AG ↓ GACCT	AflI	G ↓ GACC	AGGACC	PpuMI(*), AflI(2), AsuI(2), AvaII(2), Cfr13I(2), ClaII(2), Eco47I(1), Nsp7524IV(2), Nsp7524V(2), Sau96I(2), SinI(2)
PpuMI	AG ↓ GTCCT	AflI	G ↓ GTCC	AGGTCC	PpuMI(*), AflI(2), AsuI(2), AvaII(2), Cfr13I(2), ClaII(2), Eco47I(1), Nsp7524IV(2), Nsp7524V(2), Sau96I(2), SinI(2)
PpuMI	GG ↓ GACCT	AflI	G ↓ GACC	GGGACC	PpuMI(*), AflI(2), AsuI(2), AvaII(2), Cfr13I(2), ClaII(2), Eco47I(1), NlaIV(3), Nsp7524IV(2), Nsp7524V(2), Sau96I(2), SinI(2)
PpuMI	GG ↓ GTCCT	AflI	G ↓ GTCC	GGGTCC	PpuMI(*), AflI(2), AsuI(2), AvaII(2), Cfr13I(2), ClaII(2), Eco47I(1), NlaIV(3), Nsp7524IV(2), Nsp7524V(2), Sau96I(2), SinI(2)
PpuMI	AG ↓ GACCT	AvaII	G ↓ GACC	AGGACC	PpuMI(*), AvaII(2), AflI(2), AsuI(2), Cfr13I(2), ClaII(2), Eco47I(1), Nsp7524IV(2), Nsp7524V(2), Sau96I(2), SinI(2)
PpuMI	AG ↓ GTCCT	AvaII	G ↓ GTCC	AGGTCC	PpuMI(*), AvaII(2), AflI(2), AsuI(2), Cfr13I(2), ClaII(2), Eco47I(1), Nsp7524IV(2), Nsp7524V(2), Sau96I(2), SinI(2)
PpuMI	GG ↓ GACCT	AvaII	G ↓ GACC	GGGACC	PpuMI(*), AvaII(2), AflI(2), AsuI(2), Cfr13I(2), ClaII(2), Eco47I(1), NlaIV(3), Nsp7524IV(2), Nsp7524V(2), Sau96I(2), SinI(2)
PpuMI	GG ↓ GTCCT	AvaII	G ↓ GTCC	GGGTCC	PpuMI(*), AvaII(2), AflI(2), AsuI(2), Cfr13I(2), ClaII(2), Eco47I(1), NlaIV(3), Nsp7524IV(2), Nsp7524V(2), Sau96I(2), SinI(2)

Table 25. Hybrid Sequences Resulting from the Ligation of Two Fragments Generated After Digestion with One or Two Different Restriction Endonucleases (*continued*)

Enzyme 1	Fragment Generated by Enzyme 1	Enzyme 2	Fragment Generated by Enzyme 2	Resulting Hybrid Sequence	Enzyme 3
PpuMI	AG ↓ GACCT	ClaII	G ↓ GACC	GGGACC	PpuMI(*), ClaII(2), AflI(2), AsuI(2), AvaII(2), Cfr13I(2), Eco47I(1), Nsp7524IV(2), Nsp7524V(2), Sau96I(2), SinI(2)
PpuMI	AG ↓ GTCCT	ClaII	G ↓ GTCC	AGGTCC	PpuMI(*), ClaII(2), AflI(2), AsuI(2), AvaII(2), Cfr13I(2), Eco47I(1), Nsp7524IV(2), Nsp7524V(2), Sau96I(2), SinI(2)
PpuMI	GG ↓ GACCT	ClaII	G ↓ GACC	GGGACC	PpuMI(*), ClaII(2), AflI(2), AsuI(2), AvaII(2), Cfr13I(2), Eco47I(1), NlaIV(3), Nsp7524IV(2), Nsp7524V(2), Sau96I(2), SinI(2)
PpuMI	GG ↓ GTCCT	ClaII	G ↓ GTCC	GGGTCC	PpuMI(*), ClaII(2), AflI(2), AsuI(2), AvaII(2), Cfr13I(2), Eco47I(1), NlaIV(3), Nsp7524IV(2), Nsp7524V(2), Sau96I(2), SinI(2)
PpuMI	AG ↓ GACCT	PpuMI	AG ↓ GACCC	AGGACCC	PpuMI(2), AflI(2), AsuI(2), AvaII(2), Cfr13I(2), ClaII(2), DraII(2), Eco47I(1), NlaIV(4), Nsp7524IV(2), Nsp7524V(2), PssI(5), Sau96I(2), SinI(2)
PpuMI	AG ↓ GACCT	PpuMI	GG ↓ GACCT	AGGACCT	PpuMI(2), AflI(2), AsuI(2), AvaII(2), Cfr13I(2), ClaII(2), Eco0109(2), Eco47I(1), Nsp7524IV(2), Nsp7524V(2), Sau96I(2), SinI(2)
PpuMI	AG ↓ GTCCT	PpuMI	AG ↓ GTCCC	AGGTCCC	PpuMI(2), AflI(2), AsuI(2), AvaII(2), Cfr13I(2), ClaII(2), DraII(2), Eco47I(1), NlaIV(4), Nsp7524IV(2), Nsp7524V(2), PssI(5), Sau96I(2), SinI(2)
PpuMI	AG ↓ GTCCT	PpuMI	GG ↓ GTCCT	AGGTCCT	PpuMI(2), AflI(2), AsuI(2), AvaII(2), Cfr13I(2), ClaII(2), Eco0109(2), Eco47I(1), Nsp7524IV(2), Nsp7524V(2), Sau96I(2), SinI(2)

Table 25. Hybrid Sequences Resulting from the Ligation of Two Fragments Generated After Digestion with One or Two Different Restriction Endonucleases (*continued*)

Enzyme 1	Fragment Generated by Enzyme 1	Enzyme 2	Fragment Generated by Enzyme 2	Resulting Hybrid Sequence	Enzyme 3
PpuMI	GG↓GACCT	PpuMI	AG↓GACCT	GGGACCT	PpuMI(2), AflI(2), AsuI(2), AvaII(2), Cfr13I(2), ClaII(2), Eco47I(1), NlaIV(3), Nsp7524IV(2), Nsp7524V(2), Sau96I(2), SinI(2)
PpuMI	GG↓GACCT	PpuMI	AG↓GACCC	GGGACCC	PpuMI(2), AflI(2), AsuI(2), AvaII(2), Cfr13I(2), ClaII(2), Eco47I(1), NlaIV(4), Nsp7524IV(2), Nsp7524V(2), Sau96I(2), SinI(2)
PpuMI	GG↓GTCCT	PpuMI	AG↓GTCCT	GGGTCCT	PpuMI(2), AflI(2), AsuI(2), AvaII(2), Cfr13I(2), ClaII(2), Eco47I(1), NlaIV(3), Nsp7524IV(2), Nsp7524V(2), Sau96I(2), SinI(2)
PpuMI	GG↓GTCCT	PpuMI	AG↓GTCCC	GGGTCCC	PpuMI(2), AflI(2), AsuI(2), AvaII(2), Cfr13I(2), ClaII(2), Eco47I(1), NlaIV(4), Nsp7524IV(2), Nsp7524V(2), Sau96I(2), SinI(2)
PpuMI	AG↓GACCT	RsrII	CG↓GACCG	AGGACCG	PpuMI(*), RsrII(*), AflI(2), AsuI(2), AvaII(2), Cfr13I(2), ClaII(2), Eco47I(1), Nsp7524IV(2), Nsp7524V(2), Sau96I(2), SinI(2)
PpuMI	AG↓GTCCT	RsrII	CG↓CTCCG	AGGTCCG	PpuMI(*), RsrII(*), AflI(2), AsuI(2), AvaII(2), Cfr13I(2), ClaII(2), Eco47I(1), Nsp7524IV(2), Nsp7524V(2), Sau96I(2), SinI(2)
PpuMI	GG↓GACCT	RsrII	CG↓GACCG	GGGACCG	PpuMI(*), RsrII(*), AflI(2), AsuI(2), AvaII(2), Cfr13I(2), ClaII(2), Eco47I(1), NlaIV(3), Nsp7524IV(2), Nsp7524V(2), Sau96I(2), SinI(2)
PpuMI	GG↓GTCCT	RsrII	CG↓GTCCG	GGGTCCG	PpuMI(*), RsrII(*), AflI(2), AsuI(2), AvaII(2), Cfr13I(2), ClaII(2), Eco47I(1), NlaIV(3), Nsp7524IV(2), Nsp7524V(2), Sau96I(2), SinI(2)

Table 25. Hybrid Sequences Resulting from the Ligation of Two Fragments Generated After Digestion with One or Two Different Restriction Endonucleases (*continued*)

Enzyme 1	Fragment Generated by Enzyme 1	Enzyme 2	Fragment Generated by Enzyme 2	Resulting Hybrid Sequence	Enzyme 3
PpuMI	AG ↓ GACCT	SinI	G ↓ GACC	AGGACC	PpuMI(*), SinI(2), AflI(2), AsuI(2), AvaII(2), Cfr13I(2), ClaII(2), Eco47I(1), Nsp7524IV(2), Nsp7524V(2), Sau96I(2)
PpuMI	AG ↓ GTCCT	SinI	G ↓ GTCC	AGGTCC	PpuMI(*), SinI(2), AflI(2), AsuI(2), AvaII(2), Cfr13I(2), ClaII(2), Eco47I(1), Nsp7524IV(2), Nsp7524V(2), Sau96I(2)
PpuMI	GG ↓ GACCT	SinI	G ↓ GACC	GGGACC	PpuMI(*), SinI(2), AflI(2), AsuI(2), AvaII(2), Cfr13I(2), ClaII(2), Eco47I(1), NlaIV(3), Nsp7524IV(2), Nsp7524V(2), Sau96I(2)
PpuMI	GG ↓ GTCCT	SinI	G ↓ GTCC	GGGTCC	PpuMI(*), SinI(2), AflI(2), AsuI(2), AvaII(2), Cfr13I(2), ClaII(2), Eco47I(1), NlaIV(3), Nsp7524IV(2), Nsp7524V(2), Sau96I(2)
RsrII	CG ↓ GACCG	AflI	G ↓ GACC	CGGACC	RsrII(*), AflI(2), AsuI(2), AvaII(2), Cfr13I(2), ClaII(2), Eco47I(1), Nsp7524IV(2), Nsp7524V(2), Sau96I(2), SinI(2)
RsrII	CG ↓ GTCCG	AflI	G ↓ GTCC	CGGTCC	RsrII(*), AflI(2), AsuI(2), AvaII(2), Cfr13I(2), ClaII(2), Eco47I(1), Nsp7524IV(2), Nsp7524V(2), Sau96I(2), SinI(2)
RsrII	CG ↓ GACCG	AvaII	G ↓ GACC	CGGACC	RsrII(*), AvaII(2), AflI(2), AsuI(2), Cfr13I(2), ClaII(2), Eco47I(1), Nsp7524IV(2), Nsp7524V(2), Sau96I(2), SinI(2)
RsrII	CG ↓ CTCCG	AvaII	G ↓ GTCC	CGGTCC	RsrII(*), AvaII(2), AflI(2), AsuI(2), Cfr13I(2), ClaII(2), Eco47I(1), Nsp7524IV(2), Nsp7524V(2), Sau96I(2), SinI(2)
RsrII	CG ↓ GACCG	ClaII	G ↓ GACC	CGGACC	RsrII(*), ClaII(2), AflI(2), AsuI(2), AvaII(2), Cfr13I(2), Eco47I(1), Nsp7524IV(2), Nsp7524V(2), Sau96I(2), SinI(2)

Table 25. Hybrid Sequences Resulting from the Ligation of Two Fragments Generated After Digestion with One or Two Different Restriction Endonucleases (*continued*)

Enzyme 1	Fragment Generated by Enzyme 1	Enzyme 2	Fragment Generated by Enzyme 2	Resulting Hybrid Sequence	Enzyme 3
RsrII	CG ↓ CTCCG	ClaII	G ↓ GTCC	CGGTCC	RsrII(*), ClaII(2), AflI(2), AsuI(2), AvaII(2), Cfr13I(2), Eco47I(1), Nsp7524IV(2), Nsp7524V(2), Sau96I(2), SinI(2)
RsrII	CG ↓ GACCG	PpuMI	AG ↓ GACCT	CGGACCT	RsrII(*), PpuMI(*), AflI(2), AsuI(2), AvaII(2), Cfr13I(2), ClaII(2), Eco47I(1), Nsp7524IV(2), Nsp7524V(2), Sau96I(2), SinI(2)
RsrII	CG ↓ GACCG	PpuMI	AG ↓ GACCC	CGGACCC	RsrII(*), PpuMI(*), AflI(2), AsuI(2), AvaII(2), Cfr13I(2), ClaII(2), Eco47I(1), NlaIV(4), Nsp7524IV(2), Nsp7524V(2), Sau96I(2), SinI(2)
RsrII	CG ↓ GTCCG	PpuMI	AG ↓ GTCCT	CGGTCCT	RsrII(*), PpuMI(*), AflI(2), AsuI(2), AvaII(2), Cfr13I(2), ClaII(2), Eco47I(1), Nsp7524IV(2), Nsp7524V(2), Sau96I(2), SinI(2)
RsrII	CG ↓ GTCCG	PpuMI	AG ↓ GTCCC	CGGTCCC	RsrII(*), PpuMI(*), AflI(2), AsuI(2), AvaII(2), Cfr13I(2), ClaII(2), Eco47I(1), NlaIV(4), Nsp7524IV(2), Nsp7524V(2), Sau96I(2), SinI(2)
RsrII	CG ↓ GACCG	SinI	G ↓ GACC	CGGACC	RsrII(*), SinI(2), AflI(2), AsuI(2), AvaII(2), Cfr13I(2), ClaII(2), Eco47I(1), Nsp7524IV(2), Nsp7524V(2), Sau96I(2)
RsrII	CG ↓ GTCCG	SinI	G ↓ GTCC	CGGTCC	RsrII(*), SinI(2), AflI(2), AsuI(2), AvaII(2), Cfr13I(2), ClaII(2), Eco47I(1), Nsp7524IV(2), Nsp7524V(2), Sau96I(2)
SalI	G ↓ TCGAC	AvaI	C ↓ TCGAG	GTCGAG	SalI(*), AvaI(*), TaqI(2), TthHB8I(2)
SalI	G ↓ TCGAC	Nsp7524III	C ↓ TCGAG	GTCGAG	SalI(*), Nsp7524III(*), TaqI(2), TthHB8I(2)
SalI	G ↓ TCGAC	PaeR7	C ↓ TCGAG	GTCGAG	SalI(*), PaeR7(*), TaqI(2), TthHB8I(2)
SalI	G ↓ TCGAC	SexI	C ↓ TCGAG	GTCGAG	SalI(*), SexI(*), TaqI(2), TthHB8I(2)
SalI	G ↓ TCGAC	XhoI	C ↓ TCGAG	GTCGAG	SalI(*), XhoI(*), TaqI(2), TthHB8I(2)

Table 25. Hybrid Sequences Resulting from the Ligation of Two Fragments Generated After Digestion with One or Two Different Restriction Endonucleases (*continued*)

Enzyme 1	Fragment Generated by Enzyme 1	Enzyme 2	Fragment Generated by Enzyme 2	Resulting Hybrid Sequence	Enzyme 3
SauI	CC ↓ TNAGG	DdeI	C ↓ TNAG	CCTNAG	SauI(*), DdeI(2)
SauI	CC ↓ TNAGG	EspI	GC ↓ TNAGC	CCTNAGC	SauI(*), EspI(*), DdeI(2)
Sau3A	↓ GATC	BamHI	G ↓ GATCC	GATCC	Sau3A(0), BamHI(*), DpnI(2), MboI(0), NdeII(0)
Sau3A	↓ GATC	BclI	T ↓ GATCA	GATCA	Sau3A(0), BclI(*), DpnI(2), MboI(0), NdeII(0)
Sau3A	↓ GATC	BglII	A ↓ GATCT	GATCT	Sau3A(0), BglII(*), DpnI(2), MboI(0), NdeII(0)
Sau3A	↓ GATC	BstI	G ↓ GATCC	GATCC	Sau3A(0), BstI(*), DpnI(2), MboI(0), NdeII(0)
Sau3A	↓ GATC	MflI	A ↓ GATCC	GATCC	Sau3A(0), MflI(*), DpnI(2), MboI(0), NdeII(0)
Sau3A	↓ GATC	MflI	A ↓ GATCT	GATCT	Sau3A(0), MflI(*), DpnI(2), MboI(0), NdeII(0)
Sau3A	↓ GATC	XhoII	A ↓ GATCC	GATCC	Sau3A(0), XhoII(*), DpnI(2), MboI(0), NdeII(0)
Sau3A	↓ GATC	XhoII	A ↓ GATCT	GATCT	Sau3A(0), XhoII(*), DpnI(2), MboI(0), NdeII(0)
Sau96I	G ↓ GNCC	DraII	AG ↓ GNCCC	GGNCCC	Sau96I(1), DraII(*), AsuI(1), Cfr13I(1), NlaIV(3), Nsp7524IV(1), Nsp7524V(1)
Sau96I	G ↓ GNCC	DraII	AG ↓ GNCCT	GGNCCT	Sau96I(1), DraII(*), AsuI(1), Cfr13I(1), Nsp7524IV(1), Nsp7524V(1)
Sau96I	G ↓ GNCC	Eco0109	AG ↓ GNCCT	GGNCCT	Sau96I(1), Eco0109(*), AsuI(1), Cfr13I(1), Nsp7524IV(1), Nsp7524V(1)
ScrFI	CC ↓ NGG	Fnu4HI	GC ↓ NGC	CCNGC	ScrFI(*), Fnu4HI(*)
ScrFI	CC ↓ NGG	Tth111I	GACN ↓ NNGTC	CCNNGTC	ScrFI, Tth111I(*)
SexI	C ↓ TCGAG	SalI	G ↓ TCGAC	CTCGAC	SexI(*), SalI(*), TaqI(2), TthHB8I(2)
SinI	G ↓ GACC	PpuMI	AG ↓ GACCT	GGACCT	SinI(1), PpuMI(*), AflI(1), AsuI(1), AvaII(1), Cfr13I(1), ClaII(1), Eco47I(0), Nsp7524IV(1), Nsp7524V(1), Sau96I(1)
SinI	G ↓ GACC	PpuMI	AG ↓ GACCC	GGACCC	SinI(1), PpuMI(*), AflI(1), AsuI(1), AvaII(1), Cfr13I(1), ClaII(1), Eco47I(0), NlaIV(3), Nsp7524IV(1), Nsp7524V(1), Sau96I(1)
SinI	G ↓ GTCC	PpuMI	AG ↓ GTCCT	GGTCCT	SinI(1), PpuMI(*), AflI(1), AsuI(1), AvaII(1), Cfr13I(1), ClaII(1), Eco47I(0), Nsp7524IV(1), Nsp7524V(1), Sau96I(1)
SinI	G ↓ GTCC	PpuMI	AG ↓ GTCCC	GGTCCC	SinI(1), PpuMI(*), AflI(1), AsuI(1), AvaII(1), Cfr13I(1), ClaII(1), Eco47I(0), NlaIV(3), Nsp7524IV(1), Nsp7524V(1), Sau96I(1)

Table 25. Hybrid Sequences Resulting from the Ligation of Two Fragments Generated After Digestion with One or Two Different Restriction Endonucleases (*continued*)

Enzyme 1	Fragment Generated by Enzyme 1	Enzyme 2	Fragment Generated by Enzyme 2	Resulting Hybrid Sequence	Enzyme 3
SinI	G ↓ GACC	RsrII	CG ↓ GACCG	GGACCG	SinI(1), RsrII(*), AflI(1), AsuI(1), AvaII(1), Cfr13I(1), ClaII(1), Eco47I(0), Nsp7524IV(1), Nsp7524V(1), Sau96I(1)
SinI	G ↓ GTCC	RsrII	CG ↓ GTCCG	GGTCCG	SinI(1), RsrII(*), AflI(1), AsuI(1), AvaII(1), Cfr13I(1), ClaII(1), Eco47I(0), Nsp7524IV(1), Nsp7524V(1), Sau96I(1)
SpeI	A ↓ CTAGT	AvrII	C ↓ CTAGG	ACTAGG	SpeI(*), AvrII(*), MaeI(2)
SpeI	A ↓ CTAGT	EcoT14I	C ↓ CTAGG	ACTAGG	SpeI(*), EcoT14I(*), MaeI(2)
SpeI	A ↓ CTAGT	NheI	G ↓ CTAGC	ACTAGC	SpeI(*), NheI(*), MaeI(2)
SpeI	A ↓ CTAGT	StyI	C ↓ CTAGG	ACTAGG	SpeI(*), StyI(*), MaeI(2)
SpeI	A ↓ CTAGT	XbaI	T ↓ CTAGA	ACTAGA	SpeI(*), XbaI(*), MaeI(2)
SplI	C ↓ GTACG	Asp718	G ↓ GTACC	CGTACC	SplI(*), Asp718(*), RsaI(3)
SplI	C ↓ GTACG	BanI	G ↓ GTACC	CGTACC	SplI(*), BanI(*), RsaI(3)
StyI	C ↓ CATGG	AflIII	A ↓ CATGT	CCATGT	StyI(*), AflIII(*), NlaIII(5)
StyI	C ↓ CTAGG	NheI	G ↓ CTAGC	CCTAGC	StyI(*), NheI(*), MaeI(2)
StyI	C ↓ CTAGG	SpeI	A ↓ CTAGT	CCTAGT	StyI(*), SpeI(*), MaeI(2)
StyI	C ↓ CTAGG	XbaI	T ↓ CTAGA	CCTAGA	StyI(*), XbaI(*), MaeI(2)
TaqI	T ↓ CGA	AccI	GT ↓ CGAC	TCGAC	TaqI(1), AccI(*), TthHB8I(1)
TaqI	T ↓ CGA	AhaII	GA ↓ CGCC	TCGCC	TaqI(*), AhaII(*)
TaqI	T ↓ CGA	AhaII	GA ↓ CGTC	TCGTC	TaqI(*), AhaII(*)
TaqI	T ↓ CGA	AosII	GA ↓ CGCC	TCGCC	TaqI(*), AosII(*)
TaqI	T ↓ CGA	AosII	GA ↓ CGTC	TCGTC	TaqI(*), AosII(*)
TaqI	T ↓ CGA	AquI	CC ↓ CGAG	TCGAG	TaqI(1), AquI(*), TthHB8I(1)
TaqI	T ↓ CGA	AquI	CC ↓ CGGG	TCGGG	TaqI(*), AquI(*)
TaqI	T ↓ CGA	AsuII	TT ↓ CGAA	TCGAA	TaqI(1), AsuII(*), TthHB8I(1)
TaqI	T ↓ CGA	BanIII	AT ↓ CGAT	TCGAT	TaqI(1), BanIII(*), TthHB8I(1)
TaqI	T ↓ CGA	BbiII	GA ↓ CGCC	TCGCC	TaqI(*), BbiII(*)
TaqI	T ↓ CGA	BbiII	GA ↓ CGTC	TCGTC	TaqI(*), BbiII(*)
TaqI	T ↓ CGA	ClaI	AT ↓ CGAT	TCGAT	TaqI(1), ClaI(*), TthHB8I(1)
TaqI	T ↓ CGA	FspII	TT ↓ CGAA	TCGAA	TaqI(1), FspII(*), TthHB8I(1)
TaqI	T ↓ CGA	HapII	C ↓ CGG	TCGG	TaqI(*), HapII(*)
TaqI	T ↓ CGA	HinpI	G ↓ CGC	TCGC	TaqI(*), HinpI(*)
TaqI	T ↓ CGA	HpaII	C ↓ CGG	TCGG	TaqI(*), HpaII(*)
TaqI	T ↓ CGA	MaeII	A ↓ CGT	TCGT	TaqI(*), MaeII(*)
TaqI	T ↓ CGA	MspI	C ↓ CGG	TCGG	TaqI(*), MspI(*)
TaqI	T ↓ CGA	NarI	GG ↓ CGCC	TCGCC	TaqI(*), NarI(*)
TaqI	T ↓ CGA	NunII	GG ↓ CGCC	TCGCC	TaqI(*), NunII(*)
Tth111I	GACN ↓ NNGTC	Fnu4HI	GC ↓ NGC	GACNNGC	Tth111I(*), Fnu4HI(*)
Tth111I	GACN ↓ NNGTC	ScrFI	CC ↓ NGG	GACNNGG	Tth111I(*), ScrFI(*)
TthHB8I	T ↓ CGA	AccI	GT ↓ CGAC	TCGAC	TthHB8I(1), AccI(*), TaqI(1)
TthHB8I	T ↓ CGA	AhaII	GA ↓ CGCC	TCGCC	TthHB8I(*), AhaII(*)
TthHB8I	T ↓ CGA	AhaII	GA ↓ CGTC	TCGTC	TthHB8I(*), AhaII(*)
TthHB8I	T ↓ CGA	AosII	GA ↓ CGCC	TCGCC	TthHB8I(*), AosII(*)
TthHB8I	T ↓ CGA	AosII	GA ↓ CGTC	TCGTC	TthHB8I(*), AosII(*)
TthHB8I	T ↓ CGA	AquI	CC ↓ CGAG	TCGAG	TthHB8I(*), AquI(*), TaqI(1)
TthHB8I	T ↓ CGA	AquI	CC ↓ CGGG	TCGGG	TthHB8I(*), AquI(*)
TthHB8I	T ↓ CGA	AsuII	TT ↓ CGAA	TCGAA	TthHB8I(1), AsuII(*), TaqI(1)

Table 25. Hybrid Sequences Resulting from the Ligation of Two Fragments Generated After Digestion with One or Two Different Restriction Endonucleases (*continued*)

Enzyme 1	Fragment Generated by Enzyme 1	Enzyme 2	Fragment Generated by Enzyme 2	Resulting Hybrid Sequence	Enzyme 3
TthHB8I	T ↓ CGA	BanIII	AT ↓ CGAT	TCGAT	TthHB8I(1), BanIII(*), TaqI(1)
TthHB8I	T ↓ CGA	BbiII	GA ↓ CGCC	TCGCC	TthHB8I(*), BbiII(*)
TthHB8I	T ↓ CGA	BbiII	GA ↓ CGTC	TCGTC	TthHB8I(*), BbiII(*)
TthHB8I	T ↓ CGA	ClaI	AT ↓ CGAT	TCGAT	TthHB8I(1), ClaI(*), TaqI(1)
TthHB8I	T ↓ CGA	FspII	TT ↓ CGAA	TCGAA	TthHB8I(1), FspII(*), TaqI(1)
TthHB8I	T ↓ CGA	HapII	C ↓ CGG	TCGG	TthHB8I(*), HapII(*)
TthHB8I	T ↓ CGA	HinpI	G ↓ CGC	TCGC	TthHB8I(*), HinpI(*)
TthHB8I	T ↓ CGA	HpaII	C ↓ CGG	TCGG	TthHB8I(*), HpaII(*)
TthHB8I	T ↓ CGA	MaeII	A ↓ CGT	TCGT	TthHB8I(*), MaeII(*)
TthHB8I	T ↓ CGA	MspI	C ↓ CGG	TCGG	TthHB8I(*), MspI(*)
TthHB8I	T ↓ CGA	NarI	GG ↓ CGCC	TCGCC	TthHB8I(*), NarI(*)
TthHB8I	T ↓ CGA	NunII	GG ↓ CGCC	TCGCC	TthHB8I(*), NunII(*)
XbaI	T ↓ CTAGA	AvrII	C ↓ CTAGG	TCTAGG	XbaI(*), AvrII(*), MaeI(2)
XbaI	T ↓ CTAGA	EcoT14I	C ↓ CTAGG	TCTAGG	XbaI(*), EcoT14I(*), MaeI(2)
XbaI	T ↓ CTAGA	NheI	G ↓ CTAGC	TCTAGC	XbaI(*), NheI(*), MaeI(2)
XbaI	T ↓ CTAGA	SpeI	A ↓ CTAGT	TCTAGT	XbaI(*), SpeI(*), MaeI(2)
XbaI	T ↓ CTAGA	StyI	C ↓ CTAGG	TCTAGG	XbaI(*), StyI(*), MaeI(2)
XcyI	C ↓ CCGGG	AccIII	T ↓ CCGGA	CCCGGA	XcyI(*), AccIII(*), BcnI(3), HapII(2), HpaII(2), MspI(2), NciI(2), ScrFI(2)
XcyI	C ↓ CCGGG	BspMII	T ↓ CCGGA	CCCGGA	XcyI(*), BspMII(*), BcnI(3), HapII(2), HpaII(2), MspI(2), NciI(2), ScrFI(2)
XcyI	C ↓ CCGGG	Cfr10I	A ↓ CCGGC	CCCGGC	XcyI(*), Cfr10I(*), BcnI(3), HapII(2), HpaII(2), MspI(2), NciI(2), ScrFI(2)
XcyI	C ↓ CCGGG	Cfr10I	A ↓ CCGGT	CCCGGT	XcyI(*), Cfr10I(*), BcnI(3), HapII(2), HpaII(2), MspI(2), NciI(2), ScrFI(2)
XhoI	C ↓ TCGAG	SalI	G ↓ TCGAC	CTCGAC	XhoI(*), SalI(*), TaqI(2), TthHB8I(2)
XhoII	A ↓ GATCC	BamHI	G ↓ GATCC	AGATCC	XhoII(1), BamHI(*), DpnI(3), MboI(1), MflI(1), NdeII(1), Sau3A(1)
XhoII	G ↓ GATCT	BamHI	G ↓ GATCC	GGATCC	XhoII(1), BamHI(1), BstI(1), DpnI(3), MboI(1), MflI(1), NdeII(1), NlaIV(3), Sau3A(1)
XhoII	A ↓ GATCC	BclI	T ↓ GATCA	AGATCA	XhoII(*), BclI(*), DpnI(3), MboI(1), NdeII(1), Sau3A(1)
XhoII	G ↓ GATCC	BclI	T ↓ GATCA	GGATCA	XhoII(*), BclI(*), DpnI(3), MboI(1), NdeII(1), Sau3A(1)
XhoII	A ↓ GATCC	BglII	A ↓ GATCT	AGATCT	XhoII(1), BglII(1), DpnI(3), MboI(1), MflI(1), NdeII(1), Sau3A(1)
XhoII	G ↓ GATCC	BglII	A ↓ GATCT	GGATCT	XhoII(1), BglII(*), DpnI(3), MboI(1), MflI(1), NdeII(1), Sau3A(1)
XhoII	A ↓ GATCC	BstI	G ↓ GATCC	AGATCC	XhoII(1), BstI(*), DpnI(3), MboI(1), MflI(1), NdeII(1), Sau3A(1)

Table 25.   Hybrid Sequences Resulting from the Ligation of Two Fragments Generated After Digestion with One or Two Different Restriction Endonucleases (*continued*)

Enzyme 1	Fragment Generated by Enzyme 1	Enzyme 2	Fragment Generated by Enzyme 2	Resulting Hybrid Sequence	Enzyme 3
XhoII	G ↓ GATCT	BstI	G ↓ GATCC	GGATCC	XhoII(1), BstI(1), BamHI(1), DpnI(3), MboI(1), MflI(1), NdeII(1), NlaIV(3), Sau3A(1)
XhoII	A ↓ GATCC	MboI	↓ GATC	AGATC	XhoII(*), MboI(1), DpnI(3), NdeII(1), Sau3A(1)
XhoII	G ↓ GATCC	MboI	↓ GATC	GGATC	XhoII(*), MboI(1), DpnI(3), NdeII(1), Sau3A(1)
XhoII	A ↓ GATCC	MflI	A ↓ GATCT	AGATCT	XhoII(1), MflI(1), BglII(1), DpnI(3), MboI(1), NdeII(1), Sau3A(1)
XhoII	A ↓ GATCC	MflI	G ↓ GATCC	AGATCC	XhoII(1), MflI(1), DpnI(3), MboI(1), NdeII(1), Sau3A(1)
XhoII	G ↓ GATCC	MflI	A ↓ GATCC	GGATCC	XhoII(1), MflI(1), BamHI(1), BstI(1), DpnI(3), MboI(1), NdeII(1), NlaIV(3), Sau3A(1)
XhoII	G ↓ GATCC	MflI	A ↓ GATCT	GGATCT	XhoII(1), MflI(1), DpnI(3), MboI(1), NdeII(1), Sau3A(1)
XhoII	A ↓ GATCC	NdeII	↓ GATC	AGATC	XhoII(*), NdeII(1), DpnI(3), MboI(1), Sau3A(1)
XhoII	G ↓ GATCC	NdeII	↓ GATC	GGATC	XhoII(*), NdeII(1), DpnI(3), MboI(1), Sau3A(1)
XhoII	A ↓ GATCC	Sau3A	↓ GATC	AGATC	XhoII(*), Sau3A(1), DpnI(3), MboI(1), NdeII(1)
XhoII	G ↓ GATCC	Sau3A	↓ GATC	GGATC	XhoII(*), Sau3A(1), DpnI(3), MboI(1), NdeII(1)
XhoII	A ↓ GATCC	XhoII	A ↓ GATCT	AGATCT	XhoII(1), BglII(1), DpnI(3), MboI(1), MflI(1), NdeII(1), Sau3A(1)
XhoII	A ↓ GATCC	XhoII	G ↓ GATCC	AGATCC	XhoII(1), DpnI(3), MboI(1), MflI(1), NdeII(1), Sau3A(1)
XhoII	G ↓ GATCC	XhoII	A ↓ GATCC	GGATCC	XhoII(1), BamHI(1), BstI(1), DpnI(3), MboI(1), MflI(1), NdeII(1), NlaIV(3), Sau3A(1)
XhoII	G ↓ GATCC	XhoII	A ↓ GATCT	GGATCT	XhoII(1), DpnI(3), MboI(1), MflI(1), NdeII(1), Sau3A(1)
XmaI	C ↓ CCGGG	AccIII	T ↓ CCGGA	CCCGGA	XmaI(*), AccIII(*), BcnI(3), HapII(2), HpaII(2), MspI(2), NciI(2), ScrFI(2)
XmaI	C ↓ CCGGG	BspMII	T ↓ CCGGA	CCCGGA	XmaI(*), BspMII(*), BcnI(3), HapII(2), HpaII(2), MspI(2), NciI(2), ScrFI(2)
XmaI	C ↓ CCGGG	Cfr10I	A ↓ CCGGC	CCCGGC	XmaI(*), Cfr10I(*), BcnI(3), HapII(2), HpaII(2), MspI(2), NciI(2), ScrFI(2)
XmaI	C ↓ CCGGG	Cfr10I	A ↓ CCGGT	CCCGGT	XmaI(*), Cfr10I(*), BcnI(3), HapII(2), HpaII(2), MspI(2), NciI(2), ScrFI(2)

Table 25. Hybrid Sequences Resulting from the Ligation of Two Fragments Generated After Digestion with One or Two Different Restriction Endonucleases (*continued*)

Enzyme 1	Fragment Generated by Enzyme 1	Enzyme 2	Fragment Generated by Enzyme 2	Resulting Hybrid Sequence	Enzyme 3
XmaIII	C ↓ GGCCG	CfrI	C ↓ GGCCA	CGGCCA	XmaIII(*), CfrI(1), BspRI(3), EaeI(1), HaeIII(3), PalI(3)
XmaIII	C ↓ GGCCG	CfrI	T ↓ GGCCG	CGGCCG	XmaIII(1), CfrI(1), BspRI(3), EaeI(1), EagI(1), Eco52I(1), HaeIII(3), PalI(3)
XmaIII	C ↓ GGCCG	EaeI	C ↓ GGCCA	CGGCCA	XmaIII(*), EaeI(1), BspRI(3), CfrI(1), HaeIII(3), PalI(3)
XmaIII	C ↓ GGCCG	EaeI	T ↓ GGCCG	CGGCCG	XmaIII(1), EaeI(1), BspRI(3), CfrI(1), EagI(1), Eco52I(1), HaeIII(3), PalI(3)
XmaIII	C ↓ GGCCG	NotI	GC ↓ GGCCGC	CGGCCGC	XmaIII(1), NotI(*), BspRI(3), CfrI(1), EaeI(1), EagI(1), Eco52I(1), Fnu4HI(4), HaeIII(3), PalI(3)

Notes: (*) The combination results in loss of the recognition site for this enzyme. ( ) Numbers in parentheses indicate cleavage position in the hybrid sequence. (0) means that cleavage will occur in front of the first base; (1) means that cleavage will occur to the right of the first nucleotide, etc. Only commercial enzymes have been listed in this table.

Table 26. Examples of DNA Protruding Ends Which Can Be Converted into New Recognition Sequences

Nature of Cohesive Ends	Restriction Endonuclease Generating Cohesive Ends	Recognition Sequence Generated by Ligation of Pol1k-Treated Ends[a]	Selection of Restriction Endonuclease(s) Recognizing Ligated Pol1k-Products[b,d]	Recognition Sequence of Second Nuclease(s)	Recognition Sequence Generated by Ligation of S 1-Treated Ends	Selection of Restriction Endonuclease(s) Recognizing Ligated S 1-Products[c,d]	Recognition Sequence of Third Nuclease(s)[c]
5'-CG	MaeI	ACGCGT	FnuDII ThaI	CG ↓ CG	—		—
			MluI AflIII	A ↓ CGCGT	—	—	—
	HinPII SciNI	GCGCGC	FnuDII ThaI	CG ↓ CG	—	—	—
			BssHII	G ↓ CGCGC	—	—	—
	NarI NdaI NunII	GGCGCGCC	FnuDII ThaI	CG ↓ CG	GGCC[c]	BsuRI CltI FnuDI HaeIII SfaI	GG ↓ CC
			BssHII	G ↓ CGCGC	GGCC	BsuRI CltI FnuDI HaeIII SfaI	
	AcyI AhaII AosII	G$^{(A)}_{(G)}$CGCG$^{(T)}_{(C)}$C	FnuDII ThaI	CG ↓ CG		—	—

Table 26. **Examples of DNA Protruding Ends Which Can Be Converted into New Recognition Sequences (*continued*)**

Nature of Cohesive Ends	Restriction Endonuclease Generating Cohesive Ends	Recognition Sequence Generated by Ligation of Pol1k-Treated Ends[a]	Selection of Restriction Endonuclease(s) Recognizing Ligated Pol1k-Products[b,d]	Recognition Sequence of Second Nuclease(s)	Recognition Sequence Generated by Ligation of S 1-Treated Ends	Selection of Restriction Endonuclease(s) Recognizing Ligated S 1-Products[c,d]	Recognition Sequence of Third Nuclease(s)[e]
	AstWI AsuIII BbiII HgiDI HgiGI HgiHII		MluI	A↓CGCGT	GATC	Bce243 BspAI Cpf1 FnuCI FnuEI HacI MboI NdeII NlaII Sau3AI	↓GATC
			BssHII	G↓CGCGC	GGCC	BsuRI CltI FnuDI HaeIII SfaI	GG↓CC
	HapII HpaII MnoI MspI	CCGCGG	FnuDII ThaI	CG↓CG	—	—	
			CscI GalI GceI GceGLI SacII SstII	CCGC↓GG	CCGG	HapII HpaII MnoI MspI	C↓CGG
	TaqI TthHB8I	TCGCGA	FnuDII ThaI	CG↓CG	—	—	—
			NruI	TCG↓CGA	—	—	—
	ClaI AsuII FspII MlaI	ATCGCGAT TTCGCGAA					
5'-TA	NdeI	CATATATG	—	CATG[c]	NlaIII	CTAG↓	
	MaeI	CTATAG	—	—	—	—	
			—				
			—				
5-AATT	EcoRI RsrI	GAATTAATTC	—	—	—	—	—
5'-GATC	Bce243 BspAI CpfI	GATCGATC	TaqI TthHB8I	T↓CGA	GATATC	EcoRV	GAT↓ATC
	FnuCI FnuEI HacI MboI NdeII NlaII Sau3AI		ClaI	AT↓CGAT	GATATC		
	BglII NspMACI	AGATCGATCT					
	AliI BamHI BstI	GGATCGATCC					
	MflI XhoII	(A)(G)GATCGATC(T)(C)					
	BclI	TGATCGATCA					

Table 26. Examples of DNA Protruding Ends Which Can Be Converted into New Recognition Sequences (*continued*)

Nature of Cohesive Ends	Restriction Endonuclease Generating Cohesive Ends	Recognition Sequence Generated by Ligation of Pol1k-Treated Ends[a]	Selection of Restriction Endonuclease(s) Recognizing Ligated Pol1k-Products[b,d]	Recognition Sequence of Second Nuclease(s)	Recognition Sequence Generated by Ligation of S1-Treated Ends	Selection of Restriction Endonuclease(s) Recognizing Ligated S1-Products[c,d]	Recognition Sequence of Third Nuclease(s)[e]
5'-CATG	NcoI	CCATGCATGG	EcoT22 NsiI	ATGCA↓T	CCATGG	NcoI	C↓CATGG
						StyI	C↓C (A)(T)/(T)(A) GG
5'-AGCT	EcoVIII HinDIII HsuI	AAGCTAGCTT	MaeI	C↓TAG	AAGCGCTT	HinPII SciNI	G↓CGC
						CfoI HhaI	GCG↓C
			NheI	G↓CTAGC	AAGCTT	Eco47III AluI MltI	AGC↓GCT AG↓CT
						EcoVIII HindIII HsuI	A↓AGCTT
5'-GGCC	EagI Eco52I XmaIII	CGGCCGGCCG	HapII HpaII MnoI MspI	C↓CGG	CGGCGCCG	HinPII SciNI	G↓CGC
						CfoI HhaI	GCG↓C
			NaeI	GCC↓GGC	—	—	—
	NotI	GCGGCCGGCCGC					
	CfrI Cfr14I EaeI	(T)/(C) GGCCGGCC (A)/(G)					
	GdiII	(T)/(C) GGCCGGCCG (A)/(G) CCGGCCGGC					
5'-CGCG	MluI	ACGCGCGCGT	FnuDII ThaI	CG↓CG	—	—	—
			BssHII	G↓CGCGC	ACGCGT	MluI	A↓CGCGT
	BssHII	GCGCGCGCGC					
5'-TGCA	ApaLI	GTGCATGCAC	NlaIII	CATG↓	GTGCAC	ApaLI	G↓TGCAC
						HgiAI	G (A)/(T) GC (T)/(A) C
						AocII BspI286 Nsp7524II	(A)(T) G(G)GC(C)(T) ↓ C (T)(A)
			PaeI SphI	GCATG↓C	GTGCAC		
			Nsp7524I NspHI	(A)/(G) CATG (T)/(G)	GTGCAC		
5'-CCGG	Cfr10I	(A)/(G) CCGGCCGG (T)/(C)	BsuRI CltI FnuDI HaeIII SfaI	GG↓CC	—	—	—
			EagI EcoI XmaIII	C↓GGCCG	(A)/(G) CCGG (T)/(C)	NaeI	GCC↓GGC
						Cfr10I	(A)/(G) ↓ CCGG (T)/(C)

Table 26. Examples of DNA Protruding Ends Which Can Be Converted into New Recognition Sequences (continued)

Nature of Cohesive Ends	Restriction Endonuclease Generating `Cohesive Ends	Recognition Sequence Generated by Ligation of Pol1k-Treated Ends[a]	Selection of Restriction Endonuclease(s) Recognizing Ligated Pol1k-Products[b,d]	Recognition Sequence of Second Nuclease(s)	Recognition Sequence Generated by Ligation of S 1-Treated Ends	Selection of Restriction Endonuclease(s) Recognizing Ligated S 1-Products[c,d]	Recognition Sequence of Third Nuclease(s)[c]
5′-TCGA	Cfr9I XcyI XmaI AccIII BspMII HgiCII HgiDII NopI SalI	CCCGGCCGGG TCCGGCCGGA GTCGATCGAC	Bce243 BspAI CpfI FnuCI FnuEI HacI MboI NdeII NlaII Sau3AI	↓ GATC	GTCGAC	TaqI TthHB8I HgiCII HgiDII NopI SalI	T ↓ CGA G ↓ TCGAC
			NblI PvuI RshI XorII	CGAT ↓ CG	GTCGCGAC	FnuDII ThaI NruI	CG ↓ CG TCG ↓ CGA
	AbrI BluI CcrI PaeR7 PanI Sau3239 SexI SlaI XhoI XpaI	CTCGATCGAG					
5′-GTAC	Asp718	GGTACGTACC	MaeII SnaBI	A ↓ CGT TAC ↓ GTA	GGTATACC	SnaI	GTATAC
5′-CTAG	SpeI	ACTAGCTAGT	AluI MltI	AG ↓ CT	—	—	—
	NheI AvrII XbaI	GCTAGCTAGC CCTAGCTAGG TCTAGCTAGA					
5′-TTAA	AflII	CTTAATTAAG	—	—	—	—	—

[a]5′-Cohesive termini are converted to blunt ends by filling in of the protruding ends with complementary deoxynucleoside triphosphates and Klenow enzyme. After ligation of the blunt ended fragments with T4 DNA ligase, the original restriction sites are destroyed, and new symmetric sequences are created which include new four- and six-nucleotide recognition sequences.

[b]After ligation with T4 DNA ligase of the blunt ended fragments generated by treatment with nuclease S1, the restriction sites are destroyed, and new restriction sites are created.

[c]The new recognition sequence is generated by direct nuclease S1 treatment of the original 5′-producing cohesive ends ligation with T4 DNA ligate omitting the Klenow reaction.

[d]See also tables 22, 23 and 27.

Table 27. Alphabetical Listing of Commercially Available Restriction Endonuclease (Winter 1987)

Enzymes	Recognition Sequence and Cleavage Site	Nature of Ends Generated by Cut	Number of Known Isoschizomers	Other Interesting Features	Suppliers
Aat I	↓ 5′-AGGCCT-3′ 3′-TCCGGA-5′ ↓	Blunt ends	4	Level of activity depends upon nature of DNA substrate	GEN, IBI, USB
Aat II	↓ 5′-GACGTC-3′ 3′-CTGCAG-5′ ↑	Cohesive ends, 3′ extension	0	Unstable if stored in a frost-free freezer	BOE, GEN, IBI, NEB, PHA, SIG, USB
Acc I	↓ 5′-GT$\left(\begin{smallmatrix}C\\A\end{smallmatrix}\right)\left(\begin{smallmatrix}G\\T\end{smallmatrix}\right)$AC-3′ 3′-CA$\left(\begin{smallmatrix}G\\T\end{smallmatrix}\right)\left(\begin{smallmatrix}C\\A\end{smallmatrix}\right)$TG-5′ ↑	Cohesive ends, 5′ extension	0	Activity is increased by 5 at 55°C. Sensitive to methylation	ANB, AME, APL, BEC, BOE, BRL, GEN, IBI, NEB, NEN, PHA, PRB, TAS, USB
Acc II	↓ 5′-CGCG-3′ 3′-GCGC-5′ ↑	Blunt ends	10		ANB, AME, BEC, TAS
Acc III	↓ 5′-TCCGGA-3′ 3′-AGGCCT-5′ ↑	Cohesive ends, 5′ extension	2		AME, TAS
Afl I	↓ 5′-GG$\left(\begin{smallmatrix}A\\T\end{smallmatrix}\right)$CC-3′ 3′-CC$\left(\begin{smallmatrix}T\\A\end{smallmatrix}\right)$GG-5′ ↑	Cohesive ends, 5 ′ extension	18		ANB, BEC
Afl II	↓ 5′-CTTAAG-3′ 3′-GAATTC-5′ ↑	Cohesive ends, 5′ extension	0		AME, TAS
Afl III	↓ 5′-ACPuPyGT-3′ 3′-TGPyPuCA-5′ ↑	Cohesive ends, 5′ extension	0		ANB, BEC
Aha II	↓ 5′-GPuCGPyC-3′ 3′-CPyGCPuG-5′ ↑	Cohesive ends, 5′ extension	9		NEB
Aha III	↓ 5′-TTTAAA-3′ 3′-AAATTT-5′ ↑	Blunt ends	1		ANB, BEC, PHA

Table 27. Alphabetical Listing of Commercially Available Restriction Endonuclease (Winter 1987) (*continued*)

Enzymes	Recognition Sequence and Cleavage Site	Nature of Ends Generated by Cut	Number of Known Isoschizomers	Other Interesting Features	Suppliers
Alu I	↓ 5'-AGCT-3' 3'-TCGA-5' ↑	Blunt ends	2	Level of activity depends upon nature of DNA substrate. Methylation sensitive	ANB, AME, APL, BEC, BOE, BRL, GEN, IBI, MIL, NEB, NEN, PHA, PRB, SIG, TAS, USB
Alw I	↓ 5'-GGATC(N)4-3' 3'-CCTAG(N)5-5' ↑	Cohesive ends, 5' extension	2		NEB
Alw NI	↓ 5'-CAGNNNCTG-3' - - - - 3'-GTCNNNGAC-5' ↑	Cohesive ends, 3' extension	0		NEB
Aoc I	↓ 5'-CCTNAGG-3' 3'-GGANTCC-5' ↑	Cohesive ends, 5' extension	6		ANB, BEC
Aos I	↓ 5'-TGCGCA-3' 3'-ACGCGT-5' ↑	Blunt ends	3		ANB, BEC
Aos II	↓ 5'-CPuCGPyC-3' 3'-CPyGCPuG-5' ↑	Cohesive ends, 5' extension	9		ANB
Apa I	↓ 5'-GGGCCC-3' 3'-CCCGGG-5' ↑	Cohesive ends, 3' extension	0	Sensitive to salt concentrations > 25 m*M*	BOE, BRL, GEN, IBI, NEB, PHA, PRB
Apa LI	↓ 5'-GTGCAC-3' 3'-CACGTG-5' ↑	Cohesive ends, 5' extension	0		NEB
Apy I	↓ 5'-CC(A/T)GG-3' 3'-GG(T/A)CC-5' ↑	Cohesive ends, 5' extension	24	Star activity	BOE
Aqu I	↓ 5'-CPyCGPuG-3' - - - - - - 3'-GPuGCPyC-5' ↑	Cohesive ends, 5' extension	3		ANB, BEC

Table 27. Alphabetical Listing of Commercially Available Restriction Endonuclease (Winter 1987) (continued)

Enzymes	Recognition Sequence and Cleavage Site	Nature of Ends Generated by Cut	Number of Known Isoschizomers	Other Interesting Features	Suppliers
Asp 700	5'-GAANNNNTTC-3' 3'-CTTNNNNAAG-5'	Blunt ends	1		BOE
Asp 718	5'-GGTACC-3' 3'-CCATGG-5'	Cohesive ends, 5' extension	3		BOE
Asu I	5'-GGNCC-3' 3'-CCNGG-5'	Cohesive ends, 5' extension	15		PRB
Asu II	5'-TTCGAA-3' 3'-AAGCTT-5'	Cohesive ends, 5' extension	4		ANB, BEC, GEN, PRB
Ava I	5'-CPyCGPuG-3' 3'-GPuGCPyC-5'	Cohesive ends, 5' extension	3	Fairly stable. Needs very clean DNA for good activity	ANB, AME, APL, BEC, BOE, BRL, GEN, IBI, NEB, PHA, PRB, SIG, TAS, USB
Ava II	5'-GG(A/T)CC-3' 3'-CC(T/A)GG-5'	Cohesive ends, 5' extension	18	Needs very clean DNA for good activity. Methylation sensitive	ANB, AME, APL, BEC, BOE, BRL, GEN, IBI, NEB, PHA, SIG, TAS
Ava III	5'-ATGCAT-3' 3'-TACGTA-5'	Blunt ends	2		ANB, BEC
Avr I	5'-CPyCGPuG-3' 3'-GPuGCPyC-5'	Cohesive ends, 5' extension	3		ANB
Avr II	5'-CCTAGG-3' 3'-GGATCC-5'	Cohesive ends, 5' extension	0		NEB
Bal I	5'-TGGCCA-3' 3'-ACCGGT-5'	Blunt ends	0	Needs very clean DNA for good activity. Methylation sensitive	AME, BRL, GEN, IBI, NEB, SIG, TAS

Table 27. Alphabetical Listing of Commercially Available Restriction Endonuclease (Winter 1987) (*continued*)

Enzymes	Recognition Sequence and Cleavage Site	Nature of Ends Generated by Cut	Number of Known Isoschizomers	Other Interesting Features	Suppliers
Bam HI	↓ 5′-GGATCC-3′ 3′-CCTAGG-5′ ↑	Cohesive ends, 5′ extension	16	Star activity if glycerol concentration > 5%. Enzyme aggregates upon aging. This results in loss of specificity. Aggregation may also prevent ligation of ends. Loss of specificity if no-salt. Avoid storing for long periods (6 months). Needs very clean DNA fragments for good activity. Methylation sensitive	ANB, AME, APL, BEC, BOE, BRL, GEN, IBI, MIL, NEB, NEN, PHA, PRB, SIG, TAS, USB
Ban I	↓ 5′-GGPyPuCC-3′ 3′-CCPuPyGG-5′ ↑	Cohesive ends, 5′ extension	3	Fairly stable enzyme	BOE, GEN, IBI, NEB, PHA, USB
Ban II	↓ 5′-GPuGCPyC-3′ ------ 3′-CPyCGPuG-5′ ↑	Cohesive ends, 3′ extension	8	Fairly stable enzyme	ANB, APL, BEC, BOE, GEN, IBI, NEB, PHA, SIG, USB
Ban III	↓ 5′-ATCGAT-3′ 3′-TAGCTA-5′ ↑	Cohesive ends, 5′ extension	3		GEN, IBI, USB
Bbe I	↓ 5′-GGCGCC-3′ 3′-CCGCGG-5′ ↑	Cohesive ends, 3′ extension	9		AME, TAS
Bbi II	↓ 5′-GPuCGPyC-3′ 3′-CPyGCPuG-5′ ↑	Cohesive ends, 5′ extension	9		AME, TAS
Bbv I	↓ 5′-GCAGC(N)8-3′ 3′-CGTCG(N)12-5′ ↑	Cohesive ends, 5′ extension	0	Fill-in reactions for end labeling not recommended because any of the 4 bases may be missing from the overhang. Ligation suboptimal due to end structure variability. Activity varies with nature of the DNA substrate	NEB, GEN, IBI

Table 27. Alphabetical Listing of Commercially Available Restriction Endonuclease (Winter 1987) (continued)

Enzymes	Recognition Sequence and Cleavage Site	Nature of Ends Generated by Cut	Number of Known Isoschizomers	Other Interesting Features	Suppliers
Bcl I	5'-TGATCA-3' 3'-ACTAGT-5'	Cohesive ends, 5' extension	6	Needs very clean DNA for good activity. Methylation sensitive	ANB, BEC, BOE, BRL, GEN, IBI, NEB, NEN, PHA, SIG, USB
Bcn I	5'-CC(C/G)GG-3' 3'-GG(G/C)CC-5'	Cohesive ends, 3' extension	5		AME, TAS
Bgl I	5'-GCCNNNNNGGC-3' 3'-CGGNNNNNCCG-5'	Cohesive ends, 3' extension	0	Useful to exchange wild type and mutant fragments because of variability of ends structure. However, this feature renders fragments difficult to clone. Purity of DNA substrate is essential. Activity stimulated 5 times at pH 9.5 (instead of 7.4). Very stable enzyme	ANB, BEC, BOE, BRL, GEN, IBI, NEB, NEN, PHA, PRB, SIG
Bgl II	5'-AGATCT-3' 3'-TCTAGA-5'	Cohesive ends, 5' extension	1	Activity stimulated 5 times at pH 9.5 (instead of 7.4). Needs very clean DNA for good activity. Methylation sensitive	ANB, AME, APL, BEC, BOE, BRL, GEN, IBI, MIL, NEB, NEN, PHA, PRB, SIG, TAS, USB
Bsm I	5'-GAATGCN-3' 3'-CTTACGN-5'	Cohesive ends, 3' extension	0		ANB, BEC, NEB
Bsp 1286	5'-G(G/A/T)GC(C/A/T)C-3' 3'-C(C/T/A)CG(G/T/A)G-5'	Cohesive ends, 3' extension	3		NEB
Bsp HI	5'-TCATGA-3' 3'-AGTACT-5'	Cohesive ends, 5' extension	0		NEB

Table 27. Alphabetical Listing of Commercially Available Restriction Endonuclease (Winter 1987) (*continued*)

Enzymes	Recognition Sequence and Cleavage Site	Nature of Ends Generated by Cut	Number of Known Isoschizomers	Other Interesting Features	Suppliers
Bsp MI	↓ 5′-ACCTGC(N)4-3′ 3′-TGGACG(N)8-5′ ↑	Cohesive ends, 5′ extension	0	Single site in pBR322 is resistant to cleavage	NEB
Bsp MII	↓ 5′-TCCGGA-3′ 3′-AGGCCT-5′ ↑	Cohesive ends, 5′ extension	2	Poor cleavage of pBR322 at 37°C	NEB
Bsp NI	↓ 5′-CC(A/T)GG-3′ 3′-GG(T/A)CC-5′ ↑	Cohesive ends, 5′ extension	0		ANB, BEC
Bsp RI	↓ 5′-GGCC-3′ 3′-CCGG-5′ ↑	Blunt ends	26		GEN
Bss HII	↓ 5′-GCGCGC-3′ 3′-CGCGCG-5′ ↑	Cohesive ends, 5′ extension	3		NEB
Bst I	↓ 5′-GGATCC-3′ 3′-CCTAGG-5′ ↑	Cohesive ends, 5′ extension	16	Star activity	ANB, APL, BEC, PHA
Bst EII	↓ 5′-GGTNACC-3′ 3′-CCANTGG-5′ ↑	Cohesive ends, 5′ extension	7	One of the three enzymes giving more than 4 bases extension	ANB, APL, BEC, BOE, BRL, NEB, PHA, SIG, USB
Bst NI	↓ 5′-CC(A/T)GG-3′ 3′-TT(T/A)CC-5′ ↑	Cohesive ends, 5′ extension	24	Generates fragments with single base 5′ extension poorly ligated by T4 DNA ligase. Possibly blocked by N4-methylcytosine	ANB, BEC, NEB
Bst XI	↓ 5′-CCANNNNNNTGG-3′ 3′-GGTNNNNNNACC-5′ ↑	Cohesive ends, 3′ extension	2		ANB, BEC, NEB
Cfo I	↓ 5′-GCGC-3′ 3′-CGCG-5′ ↑	Cohesive ends, 3′ extension	8	Very stable. Activity varies with nature of DNA substrate	ANB, BEC, BOE, BRL, GEN, IBI, PRB

Table 27. Alphabetical Listing of Commercially Available Restriction Endonuclease (Winter 1987) (*continued*)

Enzymes	Recognition Sequence and Cleavage Site	Nature of Ends Generated by Cut	Number of Known Isoschizomers	Other Interesting Features	Suppliers
Cfr I	↓ 5′-PyGGCCPu-3′ 3′-PuCCGGPy-5′ ↑	Cohesive ends, 5′ extension	2		AME, TAS
Cfr 10I	↓ 5′-PuCCGGPy-3′ 3′-PyGGCCPu-5′ ↑		0		AME, TAS
Cfr 13I	↓ 5′-GGNCC-3′ 3′-CCNGG-5′ ↑	Cohesive ends, 5′ extension	15		AME, TAS
Cla I	↓ 5′-ATCGAT-3′ 3′-TAGCTA-5′ ↑	Cohesive ends, 5′ extension	3	Single site in pBR322 is in Tet gene promoter	AME, ANB, APL, BEC, BOE, BRL, NEB, PHA, SIG, TAS
Cvn I	↓ 5′-CCTNAGG-3′ 3′-GGANTCC-5′ ↑	Cohesive ends, 5′ extension	6	Fragments poorly ligated with T4 DNA ligase	BRL
Dde I	↓ 5′-CTNAG-3′ 3′-GANTC-5′ ↑	Cohesive ends, 5′ extension	0	Activity varies with nature of DNA substrate. Star activity	ANB, BEC, BOE, BRL, GEN, IBI, NEB, PHA, PRB, USB
Dpn I	↓ 5′-GATC-3′ 3′-CTAG-5′ ↑	Blunt ends	6	Cleaves only if DNA is methylated at the adenine residue useful for methylation studies in conjunction with Mbo I and Sau 3A which cleave the same site when methylation is not present (Mbo I) or regardless of methylation (Sau 3A). Dpn I fragments cloned at blunt ends sites are difficult to recover by restriction endonuclease digestion.	ANB, AME, BEC, BOE, BRL, GEN, IBI, PRB, NEB, TAS

Table 27. Alphabetical Listing of Commercially Available Restriction Endonuclease (Winter 1987) (continued)

Enzymes	Recognition Sequence and Cleavage Site	Nature of Ends Generated by Cut	Number of Known Isoschizomers	Other Interesting Features	Suppliers
Dra I	↓ 5'-TTTAAA-3' 3'-AAATTT-5' ↑	Blunt ends	1	One of rare enzymes recognizing pure hexameric AT sites. DraI is preferred over Aha III for cloning or sequencing. Looses as much as 50% activity upon dilution	AME, ANB, APL, BEC, BOE, BRL, GEN, IBI, NEB, PHA, PRB, TAS, USB
Dra II	↓ 5'-PuGGNCCPy-3' 3'-PyCCNGGPu-5' ↑	Cohesive ends, 5' extension	2		ANB, BEC, BOE
Dra III	↓ 5'-CACNNNGTG-3' 3'-GTGNNNCAC-5' ↑	Cohesive ends, 3' extension	0		ANB, BEC, BOE, NEB
Eae I	↓ 5'-PyGGCCPu-3' 3'-PuCCGGPy-5' ↑	Cohesive ends, 5' extension	2		NEB
Eag I	↓ 5'-CGGCCG-3' 3'-GCCGGC-5' ↑	Cohesive ends, 5' extension	2		NEB
Eco 0109	↓ 5'-PuGGNCCPy-3' 3'-PyCCNGGPu-5' ↑	Cohesive ends, 5' extension	2		AME, NEB, TAS
Eco 47I	↓ 5'-GG(A/T)CC-3' 3'-CC(T/A)GG-5' ↑	Cohesive ends, 5' extension	15		USB
Eco 47III	↓ 5'-AGCGCT-3' 3'-TCGCGA-5' ↑	Blunt ends	0		AME, TAS
Eco 52I	↓ 5'-CGGCCG-3' 3'-GCCGGC-5' ↑	Cohesive ends 5' extension	2		AME, TAS

Table 27. Alphabetical Listing of Commercially Available Restriction Endonuclease (Winter 1987) (*continued*)

Enzymes	Recognition Sequence and Cleavage Site	Nature of Ends Generated by Cut	Number of Known Isoschizomers	Other Interesting Features	Suppliers
Eco 81I	5'-CCTNAGG-3' 3'-GGANTCC-5'	Cohesive ends, 5' extension	6		AME, TAS
Eco RI	5'-GAATTC-3' 3'-CTTAAG-5'	Cohesive ends, 5' extension	2	Altered specificity if low salt, glycerol > 5% or high pH (above 8.0). Replacement of Mg^{2+} by Mn^{2+} also leads to star activity. Methylation sensitive. BME inhibits star activity	ANB, AME, APL, BEC, BOE, BRL, GEN, IBI, MIL, NEB, NEN, PHA, PRB, SIG, TAS, USB
Eco RII	5'-CC$\left(\begin{smallmatrix}A\\T\end{smallmatrix}\right)$GG-3' 3'-GG$\left(\begin{smallmatrix}T\\A\end{smallmatrix}\right)$CC-5'	Cohesive ends, 5' extension	24		ANB, BEC, BRL, MIL
EcoR V	5'-GATATC-3' 3'-CTATAG-5'	Blunt ends	3	One site in pBR322 Tet gene. Very stable. Star activity	ANB, AME, APL, BEC, BOE, BRL, GEN, IBI, NEB, PHA, PRB, SIG, TAS, USB
Eco T14I	5'-CC$\left(\begin{smallmatrix}AA\\TT\end{smallmatrix}\right)$GG-3' 3'-GG$\left(\begin{smallmatrix}TT\\AA\end{smallmatrix}\right)$CC-5'	Cohesive ends, 5' extension	2		AME, TAS
Esp I	5'-GCTNAGC-3' 3'-CGANTCG-5'	Cohesive ends, 5' extension	0		ANB, BEC
Fnu DII	5'-CGCG-3' 3'-GCGC-5'	Blunt ends	10		NEB
Fnu 4HI	5'-GCNGC-3' 3'-CGNCG-5'	Cohesive ends, 5' extension	1	Fragments difficult to ligate with T4 DNA ligase	NEB
Fok I	5'-GGATG(N)9-3' 3'-CCTAC(N)13-5'	Cohesive ends, 5' extension	1	Can be used to cleave between virtually any two nucleotides by constructing a complementary oligonucleotide to the sequence to be cleaved	AME, BOE, NEB, TAS

Table 27. Alphabetical Listing of Commercially Available Restriction Endonuclease (Winter 1987) (*continued*)

Enzymes	Recognition Sequence and Cleavage Site	Nature of Ends Generated by Cut	Number of Known Isoschizomers	Other Interesting Features	Suppliers
Fsp I	↓ 5'-TGCGCA-3' 3'-ACGCGT-5' ↑	Blunt ends	3		NEB, SIG
Fsp II	↓ 5'-TTCGAA-3' 3'-AAGCTT-5' ↑	Cohesive ends, 5' extension	5		NEB
Hae II	↓ 5'-PuGCGCPy-3' 3'-PyCGCGPu-5' ↑	Cohesive ends, 3' extension	2	Should not be stored for more than 3 months. Sensitive to dessication. Do not store in frost-free freezer. DNA purity is essential for activity	ANB, AME, APL, BEC, BOE, BRL, GEN, IBI, NEB, PHA, SIG, TAS
Hae III	↓ 5'-GGCC-3' 3'-CCGG-5' ↑	Blunt ends	26	Very stable. Active up to 70°C even after urea treatment. Strongly inhibited by salt concentrations > 25 mM. Star activity	ANB, AME, APL, BEC, BOE, BRL, GEN, IBI, NEB, NEN, PHA, PRB, SIG, TAS, USB
Hap II	↓ 5'-CCGG-3' 3'-GGCC-5' ↑	Cohesive ends, 5' extension	10		AME, IBI, TAS
Hga I	↓ 5'-GACGC(N)5-3' 3'-CTGCG(N)10-5' ↑	Cohesive ends, 5' extension	0	One of the three enzymes giving extension of more than 4 bases	NEB
Hgi AI	5'-G$\binom{T}{A}$GC$\binom{T}{A}$↓C-3' 3'-C$\binom{A}{T}$CG$\binom{A}{T}$G-5' ↑	Cohesive ends, 3' extension	0		NEB
Hha I	↓ 5'-GCGC-3' 3'-CGCG-5' ↑	Cohesive ends, 3' extension	8	Star activity if glycerol > 5%	ANB, AME, APL, BEC, BRL, NEB, PHA, PRB, SIG, TAS
Hinc II	↓ 5'-GTPyPuAC-3' 3'-CAPuPyTG-5' ↑	Blunt ends	6	Stable at room temperature in storage buffer	ANB, AME, APL, BEC, BRL, GEN, IBI, NEB, NEN, PHA, PRB, SIG, TAS, USB

Table 27. **Alphabetical Listing of Commercially Available Restriction Endonuclease (Winter 1987)** (*continued*)

Enzymes	Recognition Sequence and Cleavage Site	Nature of Ends Generated by Cut	Number of Known Isoschizomers	Other Interesting Features	Suppliers
Hind II	↓ 5′-GTPyPuAC-3′ 3′-CAPuPyTG-5′ ↑	Blunt ends	6		BOE, SIG
Hind III	↓ 5′-AAGCTT-3′ 3′-TTCGAA-5′ ↑	Cohesive ends, 5′ extension	13	Star activity. Needs duplex DNA > 10 bases to cut. Sensitive to methylation. Very sensitive to DNA contaminants	ANB, AME, APL, BEC, BOE, BRL, GEN, IBI, MIL, NEB, NEN, PHA, PRB, SIG, TAS, USB
Hinf I	↓ 5′-GANTC-3′ 3′-CTNAG-5′ ↑	Cohesive ends, 5′ extension	5	Star activity. Unstable if protein < 20 µg/ml	ANB, AME, APL, BEC, BOE, BRL, GEN, IBI, NEB, NEN, PHA, PRB, SIG, TAS, USB
Hin PI	↓ 5′-GCGC-3′ 3′-CGCG-5′ ↑	Cohesive ends, 5′ extension	8	Unlike its isoschizomer Hha I, generates 5′ extensions	NEB
Hpa I	↓ 5′-GTTAAC-3′ 3′-CAATTG-5′ ↑	Blunt ends	2	Star activity if glycerol > 5%	ANB, AME, APL, BEC, BOE, BRL, GEN, IBI, NEB, NEN, PHA, PRB, SIG, TAS, USB
Hpa II	↓ 5′-CCGG-3′ 3′-GGCC-5′ ↑	Cohesive ends, 5′ extension	10	Blocked by methylation at either C	ANB, APL, BEC, BOE, BRL, GEN, NEB, NEN, PHA, PRB, SIG, USB
Hph I	↓ 5′-GGTGA(N)8-3′ 3′-CCACT(N)7-5′ ↑	Cohesive ends, 3′ extension	0		NEB
Kpn I	↓ 5′-GGTACC-3′ 3′-CCATGG-5′ ↑	Cohesive ends, 3′ extension	3	Star activity if glycerol > 5%. DNA purity is essential for satisfactory activity. Exhibits extreme substrate dependency. Gives rise to large fragments with eucaryotic DNAs	ANB, AME, APL, BEC, BOE, BRL, GEN, IBI, NEB, NEN, PHA, PRB, SIG, TAS, USB
Mae I	↓ 5′-CTAG-3′ 3′-GATC-5′ ↑	Cohesive ends, 5′ extension	1		BOE

Table 27. Alphabetical Listing of Commercially Available Restriction Endonuclease (Winter 1987) (*continued*)

Enzymes	Recognition Sequence and Cleavage Site	Nature of Ends Generated by Cut	Number of Known Isoschizomers	Other Interesting Features	Suppliers
Mae II	↓ 5′-ACGT-3′ 3′-TGCA-5′ ↑	Cohesive ends, 5′ extension	0		BOE
Mae III	↓ 5′-GTNAC-3′ 3′-CANTG-5′ ↑	Cohesive ends, 5′ extension	0		BOE
Mbo I	↓ 5′-GATC-3′ 3′-CTAG-5′ ↑	Cohesive ends, 5′ extension	24	High purity of DNA required for efficient cutting. Methylation sensitive	ANB, APL, BEC, BRL, GEN, IBI, MIL, NEB, PHA
Mbo II	↓ 5′-GAAGA(N)8-3′ 3′-CTTCT(N)7-5′ ↑	Cohesive ends, 3′ extension	2	DNA purity essential for activity. Due to the one base overhang and the unspecified bases in the recognition sequence, ligation of fragments is suboptimal. Methylation sensitive	ANB, BEC, BRL, GEN, IBI, NEB, SIG
Mfl I	↓ 5′-PuGATCPy-3′ 3′-PyCTAGPu-5′ ↑	Cohesive ends, 5′ extension	1		AME, TAS
Mlu I	↓ 5′-ACGCGT-3′ 3′-TGCGCA-5′ ↑	Cohesive ends, 5′ extension	0		AME, BOE, BRL, GEN, NEB, PHA, PRB
Mnl I	↓ 5′-CCTC(N)7-3′ 3′-GGAG(N)7-5′ ↑	Blunt ends	0	Relatively unstable. Do not store for more than six months	IBI, NEB
Msp I	↓ 5′-CCGG-3′ 3′-GGCC-5′ ↑	Cohesive ends, 5′ extension	8	Useful in conjunction with Hpa II in methylation studies	AME, APL, BOE, BRL, GEN, IBI, NEB, PHA, PRB, TAS
Mst II	↓ 5′-CCTNAGG-3′ 3′-GGANTCC-5′ ↑	Cohesive ends, 5′ extension	6		NEB

Table 27. Alphabetical Listing of Commercially Available Restriction Endonuclease (Winter 1987) (*continued*)

Enzymes	Recognition Sequence and Cleavage Site	Nature of Ends Generated by Cut	Number of Known Isoschizomers	Other Interesting Features	Suppliers
Mva I	5'-CC$\binom{A}{T}$GG-3' 3'-GG$\binom{T}{A}$CC-5'	Cohesive ends, 5' extension	24		AME, TAS
Nae I	5'-GCCGGC-3' 3'-GCCCCG-5'	Blunt ends	13	Marked site preferences	BOE, NEB
Nar I	5'-GGCGCC-3' 3'-CCGCGG-5'	Cohesive ends, 5' extension	9	Marked site preferences	BRL, NEB, PHA
Nci I	5'-CC$\binom{C}{G}$GG-3' 3'-GG$\binom{C}{G}$CC-5'	Cohesive ends, 5' extension	5		BOE, BRL, NEB, USB
Nco I	5'-CCATGG-3' 3'-GGTACC-5'	Cohesive ends, 5' extension	0		AME, ANB, BEC, BOE, BRL, PHA, TAS
Nde I	5'CATATG-3' 3'-GTATAC-5'	Cohesive ends, 5' extension	0	Very sensitive to impurities in DNA preparations. DNA from minipreps is cleaved at lower rate. Very unstable (half-life at 37°C = 15 minutes)	BOE, BRL, NEB, PHA
Nde II	5'-GATC-3' 3'-CTAG-5'	Cohesive ends, 5' extension	24		BOE, BRL
Nhe I	5'-GCTAGC-3' 3'-CGATCG-5'	Cohesive ends, 5' extension	0		BOE, BRL, NEB, PHA
Nla III	5'-CATG-3' 3'-GTAC-5'	Cohesive ends, 3' extension	0		NEB

Table 27. Alphabetical Listing of Commercially Available Restriction Endonuclease (Winter 1987) (*continued*)

Enzymes	Recognition Sequence and Cleavage Site	Nature of Ends Generated by Cut	Number of Known Isoschizomers	Other Interesting Features	Suppliers
Nla IV	↓ 5'-GGNNCC-3' 3'-CCNNGG-5' ↑	Blunt ends	0		NEB
Not I	↓ 5'-GCGGCCGC-3' 3'-CGCCGGCG-5' ↑	Cohesive ends, 5' extension	0	One of the two enzymes recognizing octanucleotides	AME, ANB, BEC, BOE, NEB, PHA, PRB, TAS
Nru I	↓ 5'-TCGCGA-3' 3'-AGCGCT-5' ↑	Blunt ends	2		BOE, BRL, NEB, PHA
Nsi I	↓ 5'-ATGCAT-3' - - - - 3'-TACGTA-5' ↑	Cohesive ends, 3' extension	2		ANB, BEC, BOE, BRL, NEB
Nsp 7524I	↓ 5'-PuCATGPy-3' 3'-PyGTACPu-5' ↑	Cohesive ends, 5' extension	1		AME, TAS
Nsp 7524II	5'-G$\binom{A}{G}{T}$GC$\binom{T}{C}{A}$C-3' ↓ 3'-C$\binom{A}{C}{T}$CG$\binom{T}{G}{A}$G-5' ↑	Cohesive ends, 3' extension	3		APL, PHA
Nsp 7524III	↓ 5'-CPyCGPuG-3' 3'-GPuGCPyC-5' ↑	Cohesive ends, 5' extension	3		PHA
Nsp 7524IV	↓ 5'-GGNCC-3' 3'-CCNGG-5' ↑	Cohesive ends, 5' extension	15		PHA
Nsp 7524V	↓ 5'-TTCGAA-3' 3'-AACGTT-5' ↑	Cohesive ends, 5' extension	4		AME, PHA, TAS
Nsp BII	5'-C$\binom{A}{C}$GC$\binom{G}{T}$G-3' ↓ 3'-G$\binom{T}{G}$CG$\binom{C}{A}$C-5' ↑	Blunt ends	0		ANB, BEC

Table 27. Alphabetical Listing of Commercially Available Restriction Endonuclease (Winter 1987) (*continued*)

Enzymes	Recognition Sequence and Cleavage Site	Nature of Ends Generated by Cut	Number of Known Isoschizomers	Other Interesting Features	Suppliers
Nsp HI	↓ 5'-PuCATGPy-3' 3'-PyGTACPu-5' ↑	Cohesive ends, 3' extension	1		ANB, BEC
Nun II	↓ 5'-GGCGCC-3' 3'-CCGCGG-5' ↑	Cohesive ends, 5' extension	9		ANB
Oxa NI	↓ 5'-CCTNAGG-3' 3'-GGANTCC-5' ↑	Cohesive ends, 5' extension	6		NEB
Pae R7	↓ 5'-CTCGAG-3' 3'-GAGCTC-5' ↑	Cohesive ends, 5' extension	25	Star activity	NEB
Pal I	↓ 5'-GGCC-3' 3'-CCGG-5' ↑	Blunt ends	26		ANB, APL, BEC, PHA
Pfl MI	↓ 5'-CCANNNNNTGG-3' 3'-GGTNNNNNACC-5' ↑	Cohesive ends, 3' extension	0		NEB
Ple I	↓ 5'-GAGTC(N)4-3' 3'-CTCAG(N)5-5' ↑	Cohesive ends, 5' extension	0		NEB
Ppu MI	↓ 5'-PuGG(A/T)CCPy-3' 3'-PyCC(T/A)GGPu-5' ↑	Cohesive ends, 5' extension	0		NEB
Pss I	↓ 5'-PuGGNCCPy-3' 3'-PyCCNGGPu-5' ↑	Cohesive ends, 3' extension	2		GEN, IBI

Table 27. Alphabetical Listing of Commercially Available Restriction Endonuclease (Winter 1987) (*continued*)

Enzymes	Recognition Sequence and Cleavage Site	Nature of Ends Generated by Cut	Number of Known Isoschizomers	Other Interesting Features	Suppliers
Pst I	↓ 5'-CTGCAG-3' 3'-GACGTC-5' ↑	Cohesive ends, 3' extension	28	Stable enzyme, single site in Amp. gene of pBR322. G tailing of Pst I sites often used for cloning. Site dependent cleavage. Gives poor results with crude preparations such as mini prep. Reprecipitation of DNA prior usage is recommended. Star activity	ANB, AME, BEC, BOE, BRL, GEN, IBI, MIL, NEB, NEN, PHA, PRB, SIG, TAS, USB
Pvu I	↓ 5'-CGATCG-3' 3'-GCTAGC-5' ↑	Cohesive ends, 3' extension	5	Methylated adenosine residues in the vicinity of cleavage site inhibits ligation of fragments	ANB, AME, APL, BEC, BOE, BRL, GEN, NEB, PHA, PRB, SIG, USB
Pvu II	↓ 5'-CAGCTG-3' 3'-GTCGAC-5' ↑	Blunt ends	2	Very stable. Star activity	AME, ANB, APL, BEC, BOE, BRL, GEN, IBI, NEB, PHA, PRB, SIG, TAS, USB
Rsa I	↓ 5'-GTAC-3' 3'-CATG-5' ↑	Blunt ends	0	Fairly stable. Use of fresh buffer is recommended	ANB, BEC, BOE, BRL, GEN, IBI, NEB, PHA, USB
Rsr II	↓ 5'-CGG$\left(\begin{smallmatrix}A\\T\end{smallmatrix}\right)$CCG-3' 3'-GCC$\left(\begin{smallmatrix}T\\A\end{smallmatrix}\right)$GGC-5' ↑	Cohesive ends, 5' extension	0		ANB, BEC, NEB
Sac I	↓ 5'-GAGCTC-3' 3'-CTCGAG-5' ↑	Cohesive ends, 3' extension	6	Moderate stability. DNA purity essential for good activity	ANB, AME, APL, BEC, BOE, GEN, IBI, NEB, PHA, PRB, SIG, TAS, USB
Sac II	↓ 5'-CCGCGG-3' 3'-GGCGCC-5' ↑	Cohesive ends, 3' extension	22	Sites in lambda and φX DNA are cleaved less efficiently than in other DNAs. DNA purity is essential	ANB, BEC, GEN, IBI, NEB, PHA, PRB, USB

Table 27. Alphabetical Listing of Commercially Available Restriction Endonuclease (Winter 1987) *(continued)*

Enzymes	Recognition Sequence and Cleavage Site	Nature of Ends Generated by Cut	Number of Known Isoschizomers	Other Interesting Features	Suppliers
Sal I	↓ 5'-GTCGAC-3' 3'-CAGCTG-5' ↑	Cohesive ends, 5' extension	9	BME is not necessary. Star activity. Cleaves C-methylated DNA less efficiently than nonmethylated DNA. This results in a great variability in activity with the origin of the DNA substrate. DNA purity is essential	ANB, AME, APL, BEC, BOE, BRL, GEN, IBI, MIL, NEB, NEN, PHA, PRB, SIG, TAS, USB
Sau I	↓ 5'-CCTNAGG-3' 3'-GGANTCC-5' ↑	Cohesive ends, 5' extension	6		BOE
Sau 3AI	↓ 5'-GATC-3' - - - - - 3'-CTAG-5' ↑	Cohesive ends, 5' extension	24	6 mM BME decreases activity by as much as 50%. Star activity. Nature of DNA sequence adjacent to recognition sites may affect enzyme activity	ANB, AME, APL, BEC, BOE, BRL, GEN, IBI, MIL, NEB, NEN, PHA, PRB, SIG, TAS, USB
Sau 96I	↓ 5'-GGNCC-3' 3'-CCNGG-5' ↑	Cohesive ends, 5' extension	15	Methylation sensitive	ANB, BEC, BOE, BRL, GEN, NEB
Sca I	↓ 5'-AGTACT-3' 3'-TCATGA-5' ↑	Blunt ends	0	Unique site in the Amp. gene of pBR322. Star activity	AME, ANB, BEC, BOE, BRL, GEN, NEB, PHA, PRB, USB
Scr FI	↓ 5'-CCNGG-3' 3'-GGNCC-5' ↑	Cohesive ends, 5' extension	0	Fragments very difficult to ligate	NEB
Sdu I	5'-G(G/A/T)GC(C/A/T)C-3' ↓ - - - - - - - - 3'-C(C/T/A)CG(G/T/A)G-5' ↑	Cohesive ends, 3' extension	3		ANB, BEC
Sex I	↓ 5'-CTCGAG-3' 3'-GAGCTC-5' ↑	Cohesive ends, 5' extension	25		ANB, BEC

Table 27. Alphabetical Listing of Commercially Available Restriction Endonuclease (Winter 1987) (*continued*)

Enzymes	Recognition Sequence and Cleavage Site	Nature of Ends Generated by Cut	Number of Known Isoschizomers	Other Interesting Features	Suppliers
Sfa NI	↓ 5′-GCATC(N)5-3′ 3′-CGTAG(N)9-5′ ↑	Cohesive ends, 5′ extension	0		NEB
Sfi I	↓ 5′-GGCCNNNNNGGCC-3′ 3′-CCGGNNNNNCCGG-5′ ↑	Cohesive ends, 3′ extension	0	One of the two enzymes recognizing octanucleotides	GEN, NEB, PHA, PRB
Sin I	↓ 5′-GG$\left(\begin{smallmatrix}A\\T\end{smallmatrix}\right)$CC-3′ 3′-CC$\left(\begin{smallmatrix}T\\A\end{smallmatrix}\right)$GG-5′ ↑	Cohesive ends, 5′ extension	19		ANB, GEN, PRB, SIG
Sma I	↓ 5′-CCCGGG-3′ 3′-GGGCCC-5′ ↑	Blunt ends	3	Absolute requirement for potassium salt. DNA purity essential. No cleavage if 5 bromodeoxyuridine is in place of thymine. Half-life of 15 minutes at 37°C	ANB, AME, APL, BEC, BOE, BRL, GEN, IBI, MIL, NEB, NEN, PHA, PRB, SIG, TAS, USB
Sna BI	↓ 5′-TACGTA-3′ 3′-ATGCAT-5′ ↑	Blunt ends	0		ANB, BEC, BOE, NEB
Spe I	↓ 5′-ACTAGT-3′ 3′-TGATCA-5′ ↑	Cohesive ends, 5′ extension	0		BOE, BRL, NEB
Sph I	↓ 5′-GCATGC-3′ 3′-CGTACG-5′ ↑	Cohesive ends, 3′ extension	3	May aggregate to DNA. Phenol extraction is recommended before gel analysis. Enzyme is difficult to extract. Aggregation is enhanced by low salt concentrations. Single site in Tet. gene of pBR322	ANB, APL, BEC, BOE, BRL, GEN, IBI, NEB, PHA, PRB, SIG, USB
Spl I	↓ 5′-CGTACG-3′ 3′-GCATGC-5′ ↑	Cohesive ends, 5′ extension	0		AME, TAS

Table 27. Alphabetical Listing of Commercially Available Restriction Endonuclease (Winter 1987) (*continued*)

Enzymes	Recognition Sequence and Cleavage Site	Nature of Ends Generated by Cut	Number of Known Isoschizomers	Other Interesting Features	Suppliers
Ssp I	↓ 5'-AATATT-3' 3'-TTATAA-5' ↑	Blunt ends	0	One of the three enzymes recognizing pure AT recognition sequences	BOE, BRL, NEB
Sst I	↓ 5'-GAGCTC-3' 3'-CTCGAG-5' ↑	Cohesive ends, 3' extension	5	Star activity	BRL
Sst II	↓ 5'-CCGCGG-3' 3'-GGCGCC-5' ↑	Cohesive ends, 3' extension	22	Star activity	BRL
Stu I	↓ 5'-AGGCCT-3' 3'-TCCGGA-5' ↑	Blunt ends	4	Methylation sensitive	AME, ANB, APL, BEC, BOE, BRL, GEN, NEB, PHA, PRB, TAS
Sty I	↓ 5'-CC$\binom{AA}{TT}$GG-3' 3'-GG$\binom{TT}{AA}$CC-5' ↑	Cohesive ends, 5' extension	2		ANB, BEC, BRL, NEB
Taq I	↓ 5'-TCGA-3' 3'-AGCT-5' ↑	Cohesive ends, 5' extension	2	Incubation at 37°C results in 50% decrease of activity. If incubation longer than one hour is required, it is advisable to overlay the mix with a drop of mineral oil to avoid evaporation. Methylation sensitive	ANB, APL, BEC, BOE, BRL, GEN, IBI, MIL, NEB, NEN, PHA, PRB, SIG, USB
Tha I	↓ 5'-CGCG-3' 3'-GCGC-5' ↑	Blunt ends	10	Very stable enzyme. Incubation at 37°C results in 50% decrease in activity	ANB, BEC, BRL, GEN, IBI
Tth HB8I	↓ 5'-TCGA-3' 3'-AGCT-5' ↑	Cohesive ends 5' extension	2		AME, TAS

Table 27. Alphabetical Listing of Commercially Available Restriction Endonuclease (Winter 1987) (*continued*)

Enzymes	Recognition Sequence and Cleavage Site	Nature of Ends Generated by Cut	Number of Known Isoschizomers	Other Interesting Features	Suppliers
Tth 111I	5'-GACNNNGTC-3' 3'-CTGNNNCAG-5'	Cohesive ends, 5' extension	5	Fragments very difficult to ligate. Add a drop of mineral oil if incubation longer than one hour. Star activity	AME, APL, GEN, IBI, NEB, PHA
Xba I	5'-TCTAGA-3' 3'-AGATCT-5'	Cohesive ends, 5' extension	0	Star activity if low ionic strength, high glycerol concentration, high enzyme concentration, or presence of DMSO. Methylation sensitive	AME, ANB, BEC, BOE, BRL, GEN, IBI, NEB, PHA, PRB, SIG, USB
Xca I	5'-GTATAC-3' 3'-CATATG-5'	Blunt ends	1		NEB
Xcy I	5'-CCCGGG-3' 3'-GGGCCC-5'	Cohesive ends, 5' extension	3		PHA
Xho I	5'-CTCGAG-3' 3'-GAGCTC-5'	Cohesive ends, 5' extension	25	Fairly unstable, extremely sensitive to dessication. DNA purity is essential for good activity. When protein-free digested DNA is needed, it is recommended that the mixture be treated with proteinase K before phenol extraction because Xho I is stable to phenol. Methylation sensitive	AME, ANB, APL, BEC, BOE, BRL, GEN, IBI, NEB, PHA, PRB, SIG, TAS, USB
Xho II	5'-PuGATCPy-3' 3'-PyCTAGPu-5'	Cohesive ends, 5' extension	1	Optimal activity on methylated DNA. Poor activity if DNA is not methylated. 30 units are required to digest 1 μg of SV 40 DNA at 37°C for 16 hours	ANB, APL, BEC, BOE, NEB

Table 27. Alphabetical Listing of Commercially Available Restriction Endonuclease (Winter 1987) (*continued*)

Enzymes	Recognition Sequence and Cleavage Site	Nature of Ends Generated by Cut	Number of Known Isoschizomers	Other Interesting Features	Suppliers
Xma I	↓ 5'-CCCGGG-3' 3'-GGGCCC-5' ↑	Cohesive ends, 5' extension	3	Generates a 4 base 5' extension which is a better substrate for ligation than the blunt ends generated by its isoschizomer Sma I	ANB, BEC, GEN, IBI, NEB, PRB
Xma III	↓ 5'-CGGCCG-3' 3'-GCCGGC-5' ↑	Cohesive ends, 5' extension	2	Activity varies greatly with the nature of substrate DNA	ANB, BEC, BRL
Xmn I	↓ 5'-GAANNNNTTC-3' 3'-CTTNNNNAAG-5' ↑	Blunt ends	1	Xmn I recognition sequence is generated after filling in the cleaved Eco RI sites with DNA polymerase	NEB
Xor II	↓ 5'-CGATCG-3' 3'-GCTAGC-5' ↑	Cohesive ends, 3' extension	5		BRL

abbreviations of suppliers names:

(AME)	Amersham International plc		(MIL)	Miles Laboratories
(ANB)	Stehelin/Anglian Biotechnology		(NEB)	New England Biolabs
(APL)	Appligene		(NEN)	New England Nuclear
(BEC)	Beckman		(PHA)	Pharmacia
(BOE)	Boehringer Mannheim		(PRB)	Promega Biotec
(BRL)	Bethesda Research Laboratories		(SIG)	Sigma Chemical Co.
(CAL)	Calbiochem		(TAS)	Takara Shuzo Co. Ltd.
(GEN)	Genofit		(USB)	United States Biochemical Corp.
(IBI)	International Biotechnologies Inc.			

6

VECTORS FOR MOLECULAR CLONING

The considerable number of cloning vectors described in the literature over the past few years make it impossible to present in this manual a complete updated listing of the different cloning systems available. The reader interested in obtaining more information in this area should consult the book by Pouwels et al. (1986), which will be updated on a regular basis. We wish to present here a general overview of the various kinds of basic vectors being used in most laboratories involved with molecular cloning. The choice of a cloning system depends primarily on the goals to be reached. Cloning a DNA fragment is generally performed for at least one of the following reasons:

1. Isolation and propagation of pure DNA fragments.
2. Preparation of highly specific probes for DNA or RNA sequences.
3. Sequencing important regions of various genomes.
4. Expression and purification of large amounts of a biologically active protein.
5. Modification of the genetic background of certain living species.
6. In vitro modification of sequences of interest.

This chapter will deal mainly with the prokaryotic cloning vectors. A few cloning systems used in higher organisms also will be described briefly, with some of the interesting vectors commercially available.

Several prokaryotic and eukaryotic vectors have been developed for specialized applications, such as preparation of genomic libraries, in vitro synthesis of mRNA species, cDNA cloning, sequencing, or expression of gene products. Their use is described in more detail in the chapters devoted to these topics.

There are basically three kinds of prokaryotic cloning vectors, namely, plasmid vectors, bacteriophage-derived vectors, and cosmid vectors. The choice of a particular system is usually dictated by the availability of restriction sites and the length of the DNA fragment to be cloned.

PLASMID VECTORS

In most cases, plasmid vectors have been derived from naturally occurring plasmids that have a "relaxed" kind of replication control, meaning that they are found in multiple (1--20) copies in the cell (plasmids with "stringent" replication control are present in a single copy per bacterial genome).

Relaxed plasmids, such as Col El or pMBl, have been modified to give rise to a series of very popular cloning vectors, such as pBR 322 (Table 28, Figure 35). On addition of chloramphenicol to the culture medium, the number of plasmid copies may rise to 1000–3000, leading to an extraordinary amplification of any foreign DNA covalently linked to the plasmid DNA (see Chapter 7 for preparation of plasmid DNA).

Table 28. Cloning Vectors Derived from pBR322 Plasmid*

Plasmid	Selective Markers	Cloning Restriction Sites	Promoter	Main Features
Expression Vectors with Promoters Under Control of the lac Operator Region				
pPCφ series (5)	Ampicillin resistance	EcoRI	lac	Fusion proteins with β-galactosidase. Three reading frames.
pLG series (11)	Ampicillin resistance	HindIII	lac	Fusion proteins with β-galactosidase.
pOP series (10)	Ampicillin resistance Tetracyclin resistance	EcoRI	lac	Fusion and nonfusion proteins (with or without CAP site).
pMC series (32) pSK series (32) pFR series (32)	Ampicillin resistance Chloramphenicol resistance or kanamycin resistance	BamHI, ClaI, PstI, EcoRI, SacI, SalI, HindIII, SmaI, XbaI, or XhoI	lac	Fusion proteins with β-galactosidase C-terminus and three reading frames.
pUC8 (41) pUC9 (41)	Ampicillin resistance	HindIII, PstI, EcoRI, BamHI, SmaI, SalI	lac	Vectors, covering the three reading frames in both orientations.
pUC18 (22) pUC19 (22)	Ampicillin resistance	HindIII, PstI, EcoRI, BamHI, SmaI, SalI, SstI, SphI, XbaI, KpnI	lac	pUC8 derivatives
pIC19 series (16)	Ampicillin resistance	HindIII, PstI, EcoRI, BamHI, SmaI, SalI, BglII, XhoI, NruI, ClaI, SacI, EcoRV	lac	pUC9 derivative.
pIC20 series (16)	Ampicillin resistance	HindIII, PstI, EcoRI, BamHI, SmaI, SalI, SstI, SphI, XbaI, KpnI, BglII, XhoI, NruI, ClaI, SacI, EcoRV	lac	pUC19 derivative.
pICEM19 series (16)	Ampicillin resistance	HindIII, PstI, EcoRI, BamHI, SmaI, SalI, BglII, XhoI, XbaI, NruI, ClaI, SacI, EcoRV	lac	pEMBL8 derivative.
pHG165 (36)	Ampicillin resistance	HindIII, PstI, EcoRI, BamHI, SmaI, SalI	lac	derivative of pUC8.
pWR590 (12)	Ampicillin resistance	SacI, EcoRI, SmaI, HindIII, BamHI, XbaI	lac	a pUC18 derivative, allows β-galactosidase fusions.
pIN.II series (17, 18, 20) pIN.III series (17, 18, 20) pIC.III series (17, 18, 20)	Ampicillin resistance	EcoRI, HindIII, BamHI	lpp-lac	Options for cloning in three reading frames, to synthesize fusion, secretable proteins.
pNH7a (24, 37)	Ampicillin resistance	EcoRI, BamHI	lac tac	Requires host with inducible *Int* protein.

Table 28. Cloning Vectors Derived from pBR322 Plasmid* (*continued*)

Plasmid	Selective Markers	Cloning Restriction Sites	Promoter	Main Features
pDR540 (30)	Ampicillin resistance	BamHI	tac	Expression of galK under control of *tac* promoter.
pTac series (1)	Ampicillin resistance	PvuII, HindIII, EcoRI	tac	For cloning of genes without RBS.
pKK233-2 (2)	Ampicillin resistance	NcoI	tac	Contains *lac* RBS and an ATG for translation initiation.
pER series (4)	Ampicillin resistance	EcoRI, HindIII, ClaI	rac	Allows fusions with β-galactosidase.
pYEJ001 (29)	Ampicillin resistance	various	SCP	pBR327 derivative containing two tandem *lac* operators.

Expression Vectors with Promoters Under Control of the trp Promoter–Operator region

Plasmid	Selective Markers	Cloning Restriction Sites	Promoter	Main Features
pTrpED series (13)	Ampicillin resistance Tetracyclin resistance	HindIII, SalI, BamHI, EcoRI	trp	Fusion with *trp*D. Plasmids code for *trp*E and carry the attenuator.
pWT series (38)	Ampicillin resistance Tetracyclin resistance	HindIII, SalI, BamHI	trp	Fusion with trpE, allow cloning in three reading frames. Plasmids carry the attenuator.
pWT series (39)	Ampicillin resistance Tetracyclin resistance	HindIII, SalI, BamHI	trp	Vectors without the attenuator sequences.
pEP series (9) pHP series (9)	Ampicillin resistance Tetracyclin resistance	HindIII, BglII, SstI	trp	Vectors derived from pBR313 or pBR345. Insertional inactivation of the *trp* structural genes.
pTrpLI (8)	Ampicillin resistance	ClaI	trp	Carries *trp*L RBS.
pSTPI (43)	Ampicillin resistance Tetracyclin resistance	ClaI, SalI, TaqI, HindIII, BamHI		Carries a synthetic *trp* promoter with RBS.
pDR720 (30)	Ampicillin resistance	BamHI, SalI, SmaI	trp	*trp* promoter directs the expression *gal*K (from pKO-1).
pER103 (7)	Ampicillin resistance Tetracyclin resistance	HindIII	trp	Carries a synthetic promoter–operator and RBS.
pKYP series (21)	Ampicillin resistance Tetracyclin resistance	HindIII, SalI, BamHI, ClaI	trp	Vectors with two or three tandem *trp* promoters and RBS. Plasmids carry *lpp* transcriptional terminator.

Table 28. Cloning Vectors Derived from pBR322 Plasmid* (continued)

Plasmid	Selective Markers	Cloning Restriction Sites	Promoter	Main Features
Expression Vectors with Temperature-Inducible Transcription Promoters				
pHUB series (3)	Kanamycin resistance Ampicillin resistance	EcoRI, BamHI, SalI, or HpaI	P_L	Vectors with or without λ antiterminator gene N.
pLc series (27) pLa series (27)	Ampicillin resistance	BalI, EcoRI, PstI, BamHI, SalI, AccI or HindIII	P_L	Vectors allowing cloning of genes with their own traductional initiation signals, or fusion proteins with MS2 *pol* or *bla*.
pJL6 (23)	Ampicillin resistance	ClaI, HindIII, BamHI	P_L	Vector encoding for the N-terminus of λ cII gene. Carries an efficient traductional initiation signal.
pAS1 (33)	Ampicillin resistance	BamHI	P_L	Allows fusion of any coding sequence to the λ cII translational initiation signal.
pKC series (26)	Ampicillin resistance	BamHI, AvaI	P_L	Vectors containing λ N gene.
pFCE4 (15)	Ampicillin resistance	BamHI	P_L	Vectors containing F1 encapsidation origin, O_L P_L, *nut*L, *nut*R, t, λcII and RBS.
pEV-vrf series (6)	Ampicillin resistance	EcoRI, BamHI, ClaI	P_L	Vectors for expression in three reading frames. Contain RBS.
pEMBLex2 (34)	Ampicillin resistance	SalI, EcoRI, HindIII	P_L	pEMBL8 derivative. Contains RBS of MS2 replicase and *lac*Z.
pANH-1 (31)	Ampicillin resistance	HpaI	P_L	Allows nonfusion protein synthesis.
pANK-12 (31)	Ampicillin resistance	KpnI	P_L	Allows nonfusion protein synthesis.
pPL2 (31)	Ampicillin resistance	BamHI	P_L	Allows nonfusion protein synthesis.
pCQV2	Ampicillin resistance	BamHI	P_R	Contains an ATG codon as a translation start codon.
pEMBLex3 (34)	Ampicillin resistance	BamHI, SalI, PstI	P_R	pEMBL8 derivative. Contains RBS of λ *cro* and *lac*Z.

Table 28. Cloning Vectors Derived from pBR322 Plasmid* (*continued*)

Plasmid	Selective Markers	Cloning Restriction Sites	Promoter	Main Features
pEX series (35)	Ampicillin resistance	PstI, SalI, BamHI, SmaI, EcoRI	P_R	Allows fusion proteins with λ *Cro* and β-galactosidase. ORF DNA expression vector.
pCL series (44)	Ampicillin resistance	BamHI	P_R	Allows fusion proteins with λ *Cro* and β-galactosidase ORF DNA suppression vector.
Expression Vectors with Constitutive Promoters				
pSPA series (40)	Ampicillin resistance Tetracyclin resistance	EcoRI, SmaI, SalI, AccI, BamHI, PstI or HincII	P protA	Vectors allowing fusion with the staphylococcal protein A.
pJP1 (28)	Ampicillin resistance Tetracyclin resistance	HindIII	T5 P25	Contains synthetic early T5 promoter instead of the Tc^R promoter.
pIN.I series (18, 20)	Ampicillin resistance	EcoRI, HindIII, or BamHI	P_{LPP}	Vectors allowing cloning in any reading frames, to synthetize fusion proteins and secretable proteins.
pXJ002 (29)	Ampicillin resistance Chloramphenicol resistance	HpaII, BamHI, SalI, PstI	SCP	pBR 325 derivative conferring heparin resistance.
pY2 (29)	Ampicillin resistance Tetracyclin resistance Chloramphenicol resistance	various	SCP	pBR327 derivative. Carrying a *cat* gene cartridge.
pSP series (19, 34)	Ampicillin resistance	EcoRI, SacI, SmaI, BamHI, XbaI, SalI, PstI, HindIII	SP6	Contains the promoter of phage SP6 RNA polymerase.
pORF series (42)	Ampicillin resistance	BglII, SmaI, XmaI, AvaI, SalI, BamHI, PstI	OmpF	Allows fusion proteins with OmpF and β-galactosidase.

*After Balbás et al (1986)

REFERENCES FOR TABLE 28

[1]Amann, E., Brosius, J., and Ptashne, M. (1983), *Gene,* **25,** 167.

[2]Amann, E., and Brosius, J. (1985), *Gene,* **40,** 183.

[2b]Balbás, P., Soberón, X., Merino E., Zurita, M., Lomeli, H., Valle F., Flores, N. and Bolivar, F. (1986), *Gene* **50,** 3.

[3]Bernard, H., Remaut, E., Hershfield, M.V., Das, J.K., Helinski, D.R., Yanofsky, C., and Franklin, N. (1979), *Gene,* **5,** 59.

[4]Boros, I., Lukascovich, T., Baliko, G., and Venetianer, P. (1986), *Gene,* **42,** 97.

[5]Charnay, P., Perricaudet, M., Galibert, F., and Tiollais, P. (1978), *Nucl. Acids Res.,* **5,** 4479.

[6]Crowl, R., Seamans, C., Lomedico, P., and McAndrew, S. (1985), *Gene,* **38,** 31.

[7]Dworkin-Rastl, E., Swetly, P., and Dworkin, M.B. (1983), *Gene,* **21,** 237.

[8]Edman, J.C., Hallewell, R.A., Valenzuela, P., Goodman, H.M., and Rutter, W.J. (1981), *Nature*, **291**, 503.

[9]Enger-Valk, B.E., Heyneker, H.L., Oosterbaan, R.A., and Pouwels, P.H. (1980), *Gene*, **9**, 69.

[10]Fuller, F. (1982), *Gene*, **19**, 43.

[11]Guarente, L., Lauer, G., Roberts, T.M., and Ptashne, M. (1980), *Cell*, **20**, 543.

[12]Guo, L., Stepien, P.P., Tso, J.Y., Brousseau, R., Narang, S., Thomas, D.Y., and Wu, R. (1984), *Gene*, **29**, 27.

[13]Hallewell, R.A., and Emtage, S. (1980), *Gene*, **9**, 27.

[14]Hanna, Z., Fregeau, C., Prefontaine, G., and Brousseau, R. (1984), *Gene*, **30**, 247.

[15]Lorenzetti, R., Dani, M., Lappi, D.A., Martineau, D., Casati, M., Monaco, L., Shatzman, A., Rosenberg, M., and Soria, M. (1985), *Gene*, **39**, 85.

[16]Marsh, J.L., Erfle, M., and Wykes, E.J. (1984), *Gene*, **32**, 481.

[17]Masui, Y., Mizuno, T., and Inouye, M. (1984), *Biotechnology*, **2**, 81.

[18]Masui, Y., Coleman, J., and Inouye, M. (1983), *Experimental Manipulation of Gene Expression*, p. 15, Academic Press, New York.

[19]Melton, D.A., Krieg, P.A., Rebagliati, M.R., Maniatis, T., Zinn, K., and Green, M.R. (1984), *Nucl. Acids Res.*, **12**, 271.

[20]Nakamura, K., and Inouye, M. (1982), *EMBO J.*, **1**, 771.

[21]Nishi, T., Saito, A., Oka, T., Itoh, S., Takaoka, C., and Taniguchi, T. (1984), *Agric. Biol. Chem.*, **48**, 669.

[22]Norrander, J., Kempe, T., and Messing, J. (1983), *Gene*, **26**, 101.

[23]Oppenheim, A.B., Gottesman, S., and Gottesman, M. (1982), *J. Mol. Biol.*, **158**, 327.

[24]Podhajska, A.J., Hasan, N., and Szybalski, W. (1985), *Gene*, **40**, 163.

[25]Queen, C.A. (1983), *J. Mol. Appl. Genet.*, **2**, 1.

[26]Rao, R.N. (1984), *Gene*, **31**, 247.

[27]Remaut, E., Stanssens, P., and Fiers, W. (1981), *Gene*, **15**, 81.

[28]Rommens, J., MacKnight, D., Pomeroy-Cloney, L., and Jay, E. (1983), *Nucl. Acids Res.*, **11**, 5921.

[29]Rossi, J.J., Soberon, X., Marumoto, Y., McMahon, J., and Itakura, K. (1983), *Proc. Natl. Acad. Sci. USA*, **80**, 3203.

[30]Russell, D.R., and Bennett, G.N. (1982), *Gene*, **20**, 231.

[31]Seth, A., Lapis, P., Vande-Woude, G.F., and Papas, T. (1986), *Gene*, **42**, 49.

[32]Shapira, S.K., Chou, J., Richaud, F.V., and Casadaban, M.J. (1983), *Gene*, **25**, 71.

[33]Shatzman, A., Ho, Y., and Rosenberg, M. (1983), *Experimental Manipulation of Gene Expression*, p. 1, Academic Press, New York.

[34]Sollazzo, M., Frank, R., and Cesareni, G. (1985), *Gene*, **37**, 199.

[35]Stanley, K.K., and Luzio, P.J. (1984), *EMBO J.*, **3**, 1429.

[36]Stewart, G.S.A.B., Lubinsky-Mink, S., Jackson, C.G., Cassel, A., and Kuhn, J. (1986), *Plasmid*, **15**, 172.

[37]Szybalski, W., Brown, A.L., Hasan, N., Podhajska, A.J., and Somasekhar, G. (1987), *RNA Polymerase and the Regulation of Transcription*, p. 381, Elsevier, New York.

[38]Tacon, W., Carey, N., and Emtage, S. (1980), *Mol. Gen. Genet.*, **177**, 427.

[39]Tacon, W.C.A., Bonass, W.A., Jenkins, B., and Emtage, J.S. (1983), *Gene*, **23**, 255.

[40]Uhlen, M., Nilsson, B., Guss, B., Lindberg, M., Gatenbeck, S., and Philipson, L. (1983), *Gene*, **23**, 369.

[41]Vieira, J., and Messing, J. (1982), *Gene*, **19**, 259.

[42]Weinstock, G.M., Rhys, C., Berman, M.L., Hampar, B., Jackson, D., Silhavy, T.J., Weisemann, J., and Zweig, M. (1983), *Proc. Natl. Acad. Sci. USA*, **80**, 4432.

[43]Windass, J.D., Newton, C.R., DeMaeyer-Guignard, J., Moore, V.E., Markham, A.F., and Edge, M.D. (1982), *Nucl. Acids Res.*, **10**, 6639.

[44]Zabeau, M., and Stanley, K.K. (1982), *EMBO J.*, **1**, 1217.

Introduction of foreign DNA and selection of chimeric plasmids are made easy by the presence in the plasmid DNAs of single sites for several restriction endonucleases and of drug resistance markers. For example, if a plasmid DNA carries two drug resistance markers, bacteria containing the plasmid DNA can be selected on a culture medium supplemented with one of these drugs. If the site chosen for insertion of DNA is located within the other drug resistance locus, successful cloning will result in the loss of the resistance. Such an insertional inactivation can be used to distinguish recombinant plasmids from others.

Since cloning DNA is eased when both the recipient vector and the passenger fragment contain common restriction sites, several vectors have been engineered to include a polylinker region expanding the number of available restriction endonucleases sites. The pUC family of plasmid vectors offers a multiple cloning site (MCS) and the possibility of selecting chimaeric plasmids with a coloration test (based on a functional complentation assay for β-galactosidase). Bacteria that carry the regulatory elements and the Z-gene of the *lac* operon can be induced by isopropyl-thiogalactoside (IPTG) to express β-galactosidase. This enzyme has the ability to hydrolyze the colorless 5-Bromo-4-chloro-3-indolyl-β-D-galactoside (X-gal) and give rise to a blue insoluble indigo derivative (Davies and Jacob, 1968). This enzymatic activity is destroyed by a small deletion (M15) of the Z gene which removes amino acid residues 11 to 41 in the α peptide (Langley et al., 1975a),

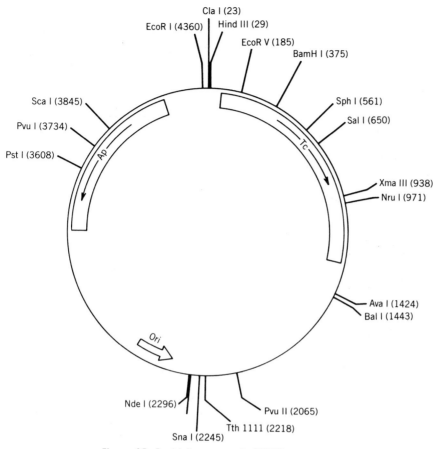

Figure 35. Restriction map of pBR322.

defined itself as the region spanning amino acid residues 3 to 92 of β-galactosidase (Langley et al., 1975b). It had been shown by Ullmann, Jacob, and Monod (1967) that α peptide intracistronic complementation may lead to a reactivation of the defective M15 β-galactosidase. The α complementation can be visualized by the appearance of blue colonies on indicator plates containing IPTG and XGal.

A whole series of plasmid vectors containing the bacterial *lac* gene with a multiple cloning site inserted at a position corresponding to the fifth amino-acid of the α peptide has been constructed by Messing and coworkers. Since these vectors contain the Pvu II-Eco RI fragment of the pBR322 plasmid, they also carry the β-lactamase gene (conferring ampicillin resistance). DNA fragments may be inserted in the unique sites of the MCS. Successful cloning is monitored by the loss of β-galactosidase activity upon transformation of appropriate host strains. When recipient bacteria harboring the M15 lac deletion are transformed by a plasmid expressing the α peptide (e.g., parental pUC), complementation will

occur and give rise to blue colonies. On the contrary, if a DNA fragment has been inserted in the multiple cloning site of pUC, transformants will appear as white colonies because there will be no α-complementation (see Figure 36). The structure and map of the latest clones in the series (pUC18 and 19) are represented in Figures 37–40. These plasmids, which contain the same polylinker (multiple cloning sites) inserted in opposite orientation, are often used to clone a DNA fragment with two different restriction ends in a forced orientation with respect to the *E. coli lac* promoter.

Among the different vectors with increased number of unique restriction sites are the vectors with restriction site banks (Davison et al., 1984) in which a large number of unique cloning sites (up to 43 in the pJRD184 plasmid) have been collected on a small DNA fragment introduced in a nonessential region of the plasmid.

Among the advantages of restriction site bank vectors are a greater flexibility in the choice of cloning sites and facilitation of subcloning. The only ap-

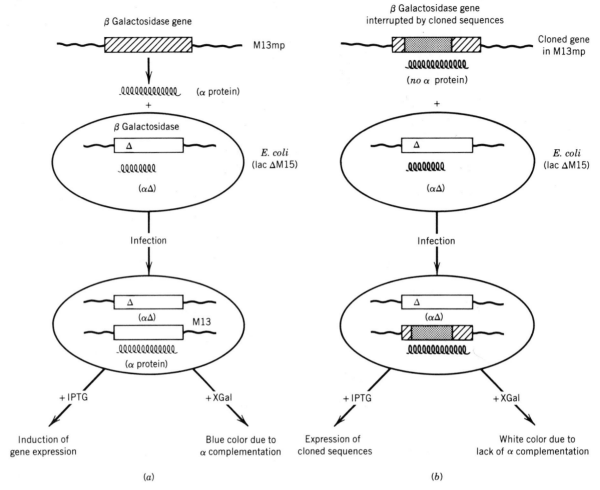

Figure 36. α-complementation. (*a*) Expression of α protein from M13mp-galactosidase gene in *E. coli* carrying a deleted β-galactosidase gene allows functional complementation (blue color). (*b*) When cloned sequences are introduced in the M13–β-galactosidase gene, no α complementation occurs (white color).

parent disadvantage, that several restriction sites are not located in regions allowing insertional inactivation, may be circumvented by cloning between sites of different nature (forced orientation cloning) or by using dephosphorylated DNA fragments (see Chapter 14).

Another interesting class of cloning plasmids is represented by vectors using the filamentous phage *fd* replication functions. Less than 250 nucleotides constitute the replication signals of the filamentous phages (see below for use of these bacteriophages as vectors). A DNA fragment of 260 nucleotides was dissected from the fd genome and linked to a gene for resistance to kanamycin (NPT I) (Meyer and Geider, 1981). Because propagation of this artificial plasmid (pfd) depends on the presence of gene 2 protein in the same cell, the gene coding for this protein

was inserted into a pBR plasmid and provided a sufficient supply of the viral replication protein. The helper plasmid could be selectively removed by cleavage with certain restriction enzymes from the pfd plasmid, which could then be used as a cloning vector.

To avoid the two-plasmid cloning system, a DNA fragment with fd gene 2 was ligated into phage λ, which was subsequently inserted into the host chromosome (Geider et al., 1985). Phage fd gene 2 was therefore propagated with the host chromosome, and the pfd plasmid was supplied with sufficient gene 2 protein to allow propagation at copy numbers of about 100 per cell. The pfd plasmids have also been constructed with various double resistances and unique restriction sites. A version with a polylinker outside a resistance gene is also available. These

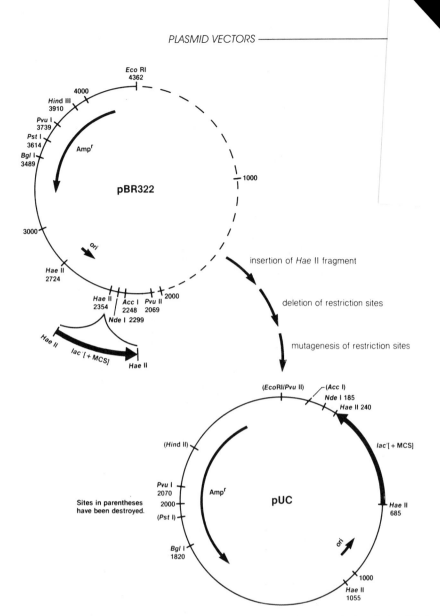

Figure 37. Construction of pUC plasmid (courtesy of Bethesda Research Laboratories, Life technologies Inc.).

plasmids are neither autonomously transferred nor can they be replicated in the absence of gene 2 protein. They are not homologous to phage λ DNA nor to many pBR plasmids and, therefore, are suitable as vectors for subcloning DNA with subsequent DNA hybridization. The heat lability of gene 2 protein allows removal of the pfd vectors from cells in the absence of selective pressure. The vectors can also be used, therefore, for transient labeling of cells with resistance markers or for transfer of a transposon followed by removal of the donor plasmid. The pfdA plasmids (Table 29) cannot be packaged upon infection with a helper phage. Like other plasmids, they can be used as vectors for DNA sequencing by chain termination methods, by applying labeled primers to denatured supercoiled DNA. Derivatives of these plasmids can be used for α-complementation; others carry fd gene 2 (pfd C1), thus creating an autonomously replicating pfd plasmid (Baldes and Geider, unpublished). Much larger inserts can be propagated in these plasmids than can be propagated with intact filamentous bacteriophage. Plasmid vectors are well adapted for cloning DNA fragments whose sizes are in the range of a few hundred base pairs up to approximately 9000 base pairs. Cloning of larger fragments should preferably be performed in cosmid vectors (see below). The choice of a particular plasmid depends essentially on the

(text continued on page 214)

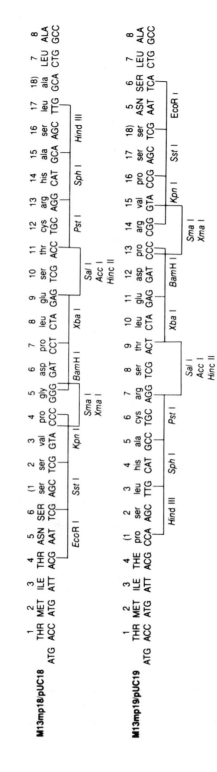

Figure 38. Sequence of the multiple cloning sites of pUC18 and pUC19 (courtesy of Bethesda Research Laboratories, Life technologies Inc.).

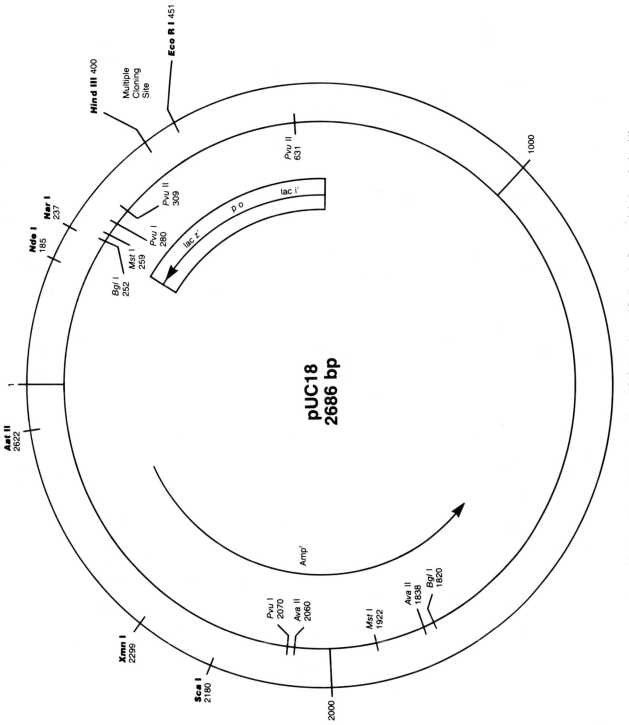

Figure 39. Simplified restriction map for pUC18 (courtesy of Bethesda Research Laboratories, Life technologies Inc.).

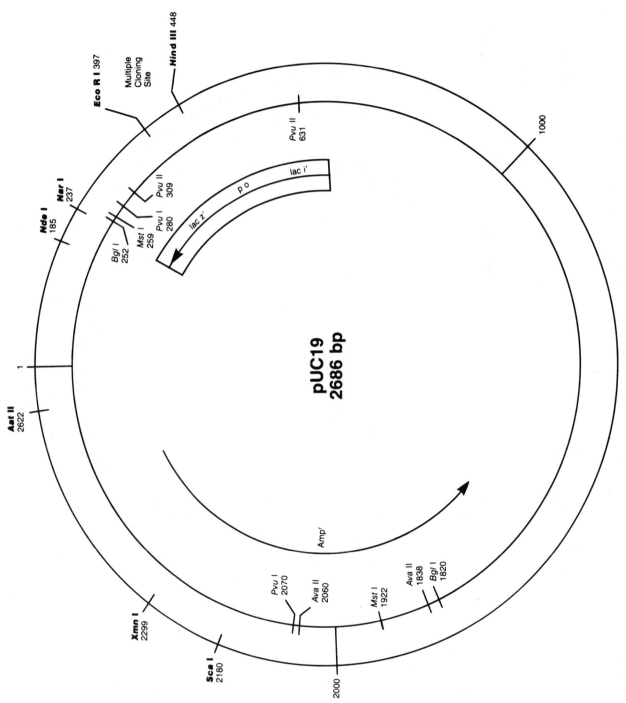

Figure 40. Simplified restriction map for pUC19 (courtesy of Bethesda Research Laboratories, Life technologies Inc.).

Table 29. A Selective List of Cloning Vectors Derived from Filamentous Bacteriophages

Vector Name	Mode of Replication	Remarks[a]
pfdA2[b], A3[b], A4[b]	fd ori, fd gene 2	Km, Km/Ap, Km/Cm
pfdA8[b]	fd ori, fd gene 2	Km, polylinker
pfdC1[b]	fd ori, fd gene 2	Km, autonomous growth
pfdB2	fd ori, fd gene 2	Km, site for cloning: EcoRI
pEMBL8, 9	ColE1 (pBR322)	Polylinker, Ap
pBS8, 9 +/−	ColE1 (pBR322)	Polylinker, Km
pEMBLY	ColE1 (pBR322)	*E. coli*-yeast shuttle vector, polylinker
pGX3804, 3805	ColE1 (pBR322)	*E. coli*-*B. subtilis* shuttle vector
pEMBLex2	ColE1 (pBR322)	P_R, RBS (MS2 *pol*)
pEMBLex3	ColE1 (pBR322)	P_R, c1857, RBS (MS2 *pol*)
pSP65ss	ColE1 (pBR322)	P_{SP6}
pSP64/65.fl + −	ColE1 (pBR322)	P_{SP6}
pFEC4	ColE1 (pBR322)	With P_L promoter
pKUN9	ColE1 (pBR322)	(+) and (−) strand packaging
pSS24/25	ColE1 (pBR322)	cDNA cloning

Source: Courtesy of K. Geider, with Permission of Society for General Microbiology.

[a]Abbreviations: Km, Ap, C, resistance to kanamycin, ampicillin, or chloramphenicol; RBS, ribosome binding site; pol, polymerase.

[b]Vectors without packaging signal. The other vectors can be packaged after infection of carrier cells by a helper phage. The yield is increased by using a natural (Dotto and Horiuchi, 1981) or an artificially created (Viera and Messing, 1986) interference resistance helper phage.

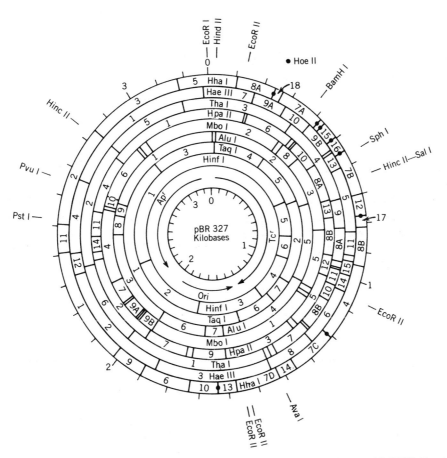

Figure 41. Restriction map of pBR327. [With permission of F. Bolivar. Copyright (1980) Elsevier Biomedical Press.]

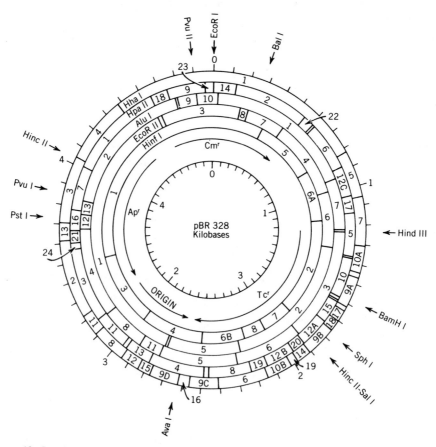

Figure 42. Restriction map of pBR328. [With permission of F. Bolivar. Copyright (1980) Elsevier Biomedical Press.]

availability of restriction sites compatible with those present in the fragment to be cloned and of drug resistance markers for selection of recombinants.

Figures 41–45 show the restriction maps of a number of common plasmid vectors. Tables 30 and 31 list the location of restriction sequences in the plasmid pBR322, and pUC19.

REFERENCES

Davies, J., and Jacob, F. (1968), *J. Mol. Biol.,* **36**, 413.

Davison, J., Heusterpreute, M., Merchez, M., and Brunel, F. (1984), *Gene,* **28**, 311.

Geider, K., Hohmeyer, C., Haas, R., and Meyer, T. F. (1985), *Gene,* **33**, 341.

Heusterpreute, M., and Davison, J. (1984), *DNA,* **3**, 259.

Heusterpreute, M., Oberto, J., Ha Thi, V. and Davison, J. (1985), *Gene,* **34**, 363.

Heusterpreute, M., Ha Thi, V., Emery, S., Tournis-Gamble, S., Kennedy, N., and Davison, J. (1985), *Gene,* **39**, 299.

Langley, K. E., Villajero, M. R., Fowler, A. V., Zamenhof, P. J., and Zabin, I. (1975a), *Proc. Natl. Acad. Sci. USA* **72**, 1254.

Langley, K. E., Fowler, A. V., and Zabin, I. (1975b), *J. Biol. Chem.,* **250** 2587.

Meyer, T. F., and Geider, K. (1981), *Proc. Natl. Acad. Sci. USA,* **78**, 5416.

Norrander, J., Kempe, T., and Messing, J. (1983), *Gene,* **26**, 101.

Pouwels, P. H., Enger Walk, B. E., and Bramar, W. J. (1986), *Cloning Vectors. A Laboratory Manual,* Elsevier, Amsterdam.

Ullman, A., Jacob F., and Monod, J. (1967), *J. Mol. Biol.,* **24**, 339.

Viera, J., and Messing, J. (1982), *Gene,* **19**, 259.

Yanisch-Perron, C., Viera, J., and Messing, J. (1985), *Gene,* **33**, 103.

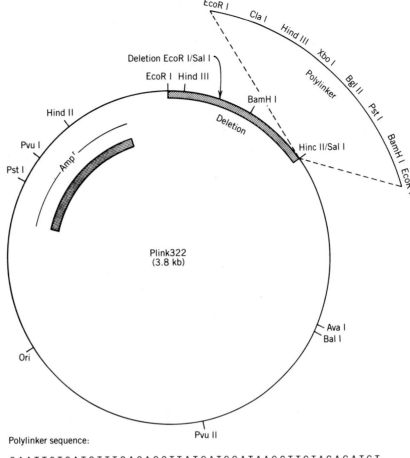

Polylinker sequence:

G A A T T C T C A T G T T T G A C A G C T T A T C A T C G A T A A G C T T C T A G A G A T C T
⌐EcoR I⌐   ⌐Cla I⌐ ⌐Hind III⌐ ⌐Xba I⌐ ⌐Bgl II⌐

T C C A T A C C T A C C A G T T C T C C G C C T G C A G C A A T G G C A A C A A C G T T G C C
                                      ⌐Pst I⌐

C G G A T C C G G T C G C G C G A A T T C
⌐BamH I⌐   ⌐EcoR I⌐

Figure 43. Restriction map of plink322.

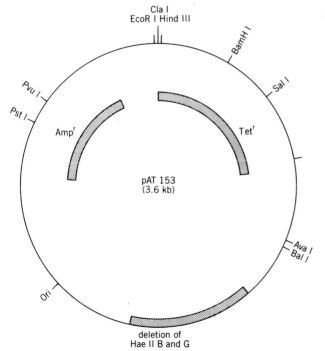

Figure 44. Restriction map of pAT153.

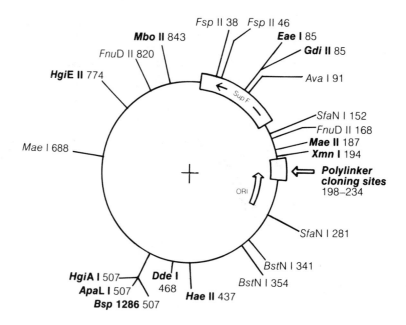

Scale:
60 base pairs/cm.

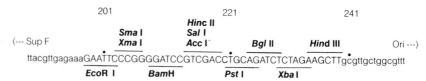

Figure 45. Restriction map of PiAN7. PiAN7 is a small *E. coli* plasmid cloning vehicle. The molecule is a double stranded DNA circle 885 base pairs in length. It was constructed in vitro by H. Huang and B. Seed from a synthetic Sup F tRNA gene, a synthetic polylinker, and the pBR322 origin of replication. The plasmid has a high copy number, and it is chloramphenicol-amplifiable. It was developed to replace PiVX. PiAN7 is a "recombination-probe" vector; it is used to select individuals from within phage libraries that carry fragments homologous to that inserted in the plasmid. Nucleotide numbering begins close to the junction of the pBR322 and Sup F: the first T in the sequence ... CTTTCGGA ... is designated as number 1. Numbering proceeds clockwise around the molecule in the direction Sup F to Polylinker. The map shows the restriction sites of those enzymes that cut the molecule once or twice; the unique sites are shown in bold type. The polylinker is shown below the map (courtesy of New England Biolabs).

Table 30. Number and Location of Cutting Sites for Commercial Restriction Endonucleases in pBR322 DNA

0 Site

AatI	BglII	FspII	PaeR7	SpeI
AflII	BspMI	HpaI	RsrII	SplI
AocI	BssHII	KpnI	SacI	SstI
ApaI	BstEII	MluI	SacII	SstII
Asp718	BstXI	MstII	SauI	StuI
AsuII	CvnI	NcoI	SexI	XbaI
AvaIII	DraIII	NotI	SfiI	XcyI
AvrII	Eco81I	NsiI	SmaI	XhoI
BclI	EspI	OxaNI	SnaBI	XmaI

	Location	Fragment Ends	Fragment Length
		1 Site	
AatII	4290	4291–4290	4363
AccIII	1664	1665–1664	4363
AflIII	2475	2476–2475	4363
AlwNI	2891	2892–2891	4363
AquI	1426	1427–1426	4363
AvaI	1425	1426–1425	4363
BalI	1446	1447–1446	4363
BamHI	375	376–375	4363
BanIII	24	25–24	4363
BsmI	1359	1360–1359	4363
BspMII	1664	1665–1664	4363
BstI	375	376–375	4363
ClaI	24	25–24	4363
EagI	939	940–939	4363
EcoRI	4361	4362–4361	4363
EcoRV	187	188–187	4363
EcoT141	1369	1370–1369	4363
Eco0109	4344	4345–4344	4363
Eco52I	939	940–939	4363
HindIII	29	30–29	4363
NdeI	2298	2299–2298	4363
NheI	229	230–229	4363
NruI	974	975–974	4363
Nsp7524III	1425	1426–1425	4363
PstI	3613	3614–3613	4363
PvuI	3738	3739–3738	4363
PvuII	2068	2069–2068	4363
SalI	651	652–651	4363
ScaI	3848	3849–3848	4363
SphI	566	567–566	4363
SspI	4172	4173–4172	4363
StyI	1369	1370–1369	4363
Tth111I	2222	2223–2222	4363
XcaI	2248	2249–2248	4363

	Location	Fragment Ends	Fragment Length
Xma III	939	940–939	4363
XorII	3738	3739–3738	4363
		2 Sites	
AccI	2247	2248–652	2768
	652	653–2247	1595
Asp700	3967	3968–2035	2431
	2035	2036–3967	1932
BanII	489	490–475	4349
	475	476–489	14
HincII	653	654–3909	3256
	3909	3910–653	1107
HindII	653	654–3909	3256
	3909	3910–653	1107
PflMI	1370	1371–1321	4314
	1321	1322–1370	49
PleI	635	636–2849	2214
	2849	2850–635	2149
PpuMI	1481	1482–1439	4321
	1439	1440–1481	42
XmnI	3967	3968–2035	2431
	2035	2036–3967	1932
		3 Sites	
AhaIII	3945	3946–3234	3652
	3253	3254–3945	692
	3234	3235–3253	19
ApaLI	4035	4036–2291	2619
	2789	2790–4035	1246
	2291	2292–2789	498
BglI	1169	1170–3488	2319
	3488	3489–935	1810
	935	936–1169	234
DraI	3945	3946–3234	3652
	3253	3254–3945	692
	3234	3235–3253	19
RsaI	165	166–2283	2118
	2283	2284–3848	1565
	3848	3849–165	680
		4 Sites	
AosI	1456	1457–3590	2134
	262	263–1358	1096
	3590	3591–262	1035
	1358	1359–1456	98
BbeI	1209	1210–417	3571
	552	553–1209	657
	438	439–552	114
	417	418–438	21

Table 30. Number and Location of Cutting Sites for Commercial Restriction Endonucleases in pBR322 DNA (*continued*)

	Location	Fragment Ends	Fragment Length		Location	Fragment Ends	Fragment Length
BspHI	489	490–3195	2706	Tth111II	23	24–1938	1915
	3195	3196–4203	1008		3098	3099–23	1288
	4308	4309–489	544		1938	1939–3065	1127
	4203	4204–4308	105		3079	3080–3098	19
DraII	1481	1482–4344	2863		3065	3066–3079	14
	524	525–1439	915				
	4344	4345–524	543				
	1439	1440–1481	42			*6 Sites*	
Eco47III	1729	1730–234	2868	AhaII	1206	1207–3905	2699
	777	778–1729	952		549	550–1206	657
	496	497–777	281		4287	4288–414	490
	234	235–496	262		3905	3906–4287	382
FspI	1456	1457–3590	2134		435	436–549	114
	262	263–1358	1096		414	415–435	21
	3590	3591–262	1035	AlwI	1670	1671–3119	1449
	1358	1359–1456	98		3216	3217–3996	780
MstI	1456	1457–3590	2134		3996	3997–378	745
	262	263–1358	1096		378	379–1100	722
	3590	3591–262	1035		1100	1101–1670	570
	1358	1359–1456	98		3119	3120–3216	97
NaeI	1285	1286–403	3481	AosII	1206	1207–3905	2699
	403	404–771	368		549	550–1206	657
	931	932–1285	354		4287	4288–414	490
	771	772–931	160		3905	3906–4287	382
NarI	1206	1207–414	3571		435	436–549	114
	549	550–1206	657		414	415–435	21
	435	436–549	114	ApyI	2637	2638–131	1857
	414	415–435	21		1443	1444–2503	1060
Nsp7524I	2479	2480–566	2450		131	132–1060	929
	566	567–1820	1254		1060	1061–1443	383
	2114	2115–2479	365		2503	2504–2624	121
	1820	1821–2114	294		2624	2625–2637	13
NspHI	2479	2480–566	2450	BbiII	1206	1207–3905	2699
	566	567–1820	1254		549	550–1206	657
	2114	2115–2479	365		4287	4288–414	490
	1820	1821–2114	294		3905	3906–4287	382
NunII	1206	1207–414	3571		435	436–549	114
	549	550–1206	657		414	415–435	21
	435	436–549	114	BspNI	2637	2638–131	1857
	414	415–435	21		1443	1444–2503	1060
PssI	1484	1485–4347	2863		131	132–1060	929
	527	528–1442	915		1060	1061–1443	383
	4347	4348–527	543		2503	2504–2624	121
	1442	1443–1484	42		2624	2625–2637	13
				BstNI	2637	2638–131	1857
		5 Sites			1443	1444–2503	1060
					131	132–1060	929
MaeI	1489	1490–2970	1481		1060	1061–1443	383
	230	231–1489	1259		2503	2504–2624	121
	3558	3559–230	1035		2624	2625–2637	13
	3223	3224–3558	335	CfrI	1444	1445–3756	2312
	2970	2971–3223	253		3756	3757–295	902

Table 30. Number and Location of Cutting Sites for Commercial Restriction Endonucleases in pBR322 DNA (*continued*)

	Location	Fragment Ends	Fragment Length		Location	Fragment Ends	Fragment Length
	939	940–1444	505		1268	1269–2575	1307
	531	532–939	408		652	653–1127	475
	399	400–531	132		4019	4020–24	368
	295	296–399	104		24	25–339	315
EaeI	1444	1445–3756	2312		339	340–652	313
	3756	3757–295	902		1127	1128–1268	141
	939	940–1444	505				
	531	532–939	408			*8 Sites*	
	399	400–531	132	AflI	1760	1761–3506	1746
	295	296–399	104		3728	3729–799	1434
EcoRII	2635	2636–129	1857		1136	1137–1439	303
	1441	1442–2501	1060		1481	1482–1760	279
	129	130–1058	929		887	888–1136	249
	1058	1059–1441	383		3506	3507–3728	222
	2501	2502–2622	121		799	800–887	88
	2622	2623–2635	13		1439	1440–1481	42
MvaI	2637	2638–131	1857	AvaII	1760	1761–3506	1746
	1443	1444–2503	1060		3728	3729–799	1434
	131	132–1060	929		1136	1137–1439	303
	1060	1061–1443	383		1481	1482–1760	279
	2503	2504–2624	121		887	888–1136	249
	2624	2625–2637	13		3506	3507–3728	222
NspBII	4003	4004–1141	1501		799	800–887	88
	3062	3063–4003	941		1439	1440–1481	42
	1141	1142–2068	927	ClaII	1760	1761–3506	1746
	2187	2188–2817	630		3728	3729–799	1434
	2817	2818–3062	245		1136	1137–1439	303
	2068	2069–2187	119		1481	1482–1760	279
SdnI	2793	2794–3954	1161		887	888–1136	249
	4039	4040–591	915		3506	3507–3728	222
	591	592–1469	878		799	800–887	88
	1469	1470–2295	826		1439	1440–1481	42
	2295	2296–2793	498	DdeI	4291	4292–1581	1653
	3954	3955–4039	85		1743	1744–2285	542
					3325	3326–3865	540
		7 Sites			2285	2286–2750	465
Cfr10I	1283	1284–3448	2165		3865	3866–4291	426
	3448	3449–160	1075		2750	2751–3159	409
	410	411–769	359		3159	3160–3325	166
	929	930–1283	354		1581	1582–1743	162
	160	161–401	241	Eco47I	1759	1760–3505	1746
	769	770–929	160		3727	3728–798	1434
	401	402–410	9		1135	1136–1438	303
TaqI	2575	2576–4019	1444		1480	1481–1759	279
	1268	1269–2575	1307		886	887–1135	249
	652	653–1127	475		3505	3506–3727	222
	4019	4020–24	368		798	799–886	88
	24	25–339	315		1438	1439–1480	42
	339	340–652	313	HgiAI	2793	2794–3954	1161
	1127	1128–1268	141		1469	1470–2295	826
TthHB8I	2575	2576–4019	1444		4039	4040–280	604

Table 30. Number and Location of Cutting Sites for Commercial Restriction Endonucleases in pBR322 DNA (*continued*)

	Location	Fragment Ends	Fragment Length
	591	592–1178	587
	2295	2296–2793	498
	280	281–591	311
	1178	1179–1469	291
	3954	3955–4039	85
MflI	1667	1668–3116	1449
	375	376–1667	1292
	3225	3226–3993	768
	4010	4011–375	728
	3127	3128–3213	86
	3993	3994–4010	17
	3213	3214–3225	12
	3116	3117–3127	11
SinI	1760	1761–3506	1746
	3728	3729–799	1434
	1136	1137–1439	303
	1481	1482–1760	279
	887	888–1136	249
	3506	3507–3728	222
	799	800–887	88
	1439	1440–1481	42
XhoII	1667	1668–3116	1449
	375	376–1667	1292
	3225	3226–3993	768
	4010	4011–375	728
	3127	3128–3213	86
	3993	3994–4010	17
	3213	3214–3225	12
	3116	3117–3127	11

9 Sites

	Location	Fragment Ends	Fragment Length
BanI	1289	1290–3316	2027
	3316	3317–76	1123
	766	767–1205	439
	119	120–413	294
	548	549–766	218
	434	435–548	114
	1205	1206–1289	84
	76	77–119	43
	413	414–434	21

10 Sites

	Location	Fragment Ends	Fragment Length
AocII	2793	2794–3954	1161
	1469	1470–2295	826
	4039	4040–280	604
	591	592–1178	587
	2295	2296–2793	498
	1178	1179–1469	291
	280	281–475	195
	489	490–591	102
	3954	3955–4039	85
	475	476–489	14

	Location	Fragment Ends	Fragment Length
BcnI	536	537–1260	724
	2157	2158–2856	699
	2856	2857–3552	696
	3903	3904– 172	632
	172	173–536	364
	3552	3553–3903	351
	1486	1487–1814	328
	1814	1815–2122	308
	1260	1261–1486	226
	2122	2123–2157	35
Bsp1286	2793	2794–3954	1161
	1469	1470–2295	826
	4039	4040–280	604
	591	592–1178	587
	2295	2296–2793	498
	1178	1179–1469	291
	280	281–475	195
	489	490–591	102
	3954	3955–4039	85
	475	476–489	14
HinfI	3363	3364–632	1632
	2846	2847–3363	517
	1525	1526–2031	506
	2450	2451–2846	396
	2031	2032–2375	344
	1006	1007–1304	298
	1304	1305–1525	221
	632	633–852	220
	852	853–1006	154
	2375	2376–2450	75
MaeII	4287	4288–901	977
	2228	2229–3178	950
	957	958–1546	589
	1800	1801–2228	428
	3178	3179–3594	416
	3594	3595–3967	373
	3967	3968–4287	320
	1570	1571–1800	230
	901	902–957	56
	1546	1547–1570	24
NciI	535	536–1259	724
	2156	2157–2855	699
	2855	2856–3551	696
	3902	3903–171	632
	171	172–535	364
	3551	3552–3902	351
	1485	1486–1813	328
	1813	1814–2121	308
	1259	1260–1485	226
	2121	2122–2156	35
Nsp7524II	2793	2794–3954	1161
	1469	1470–2295	826
	4039	4040–280	604

Table 30. Number and Location of Cutting Sites for Commercial Restriction Endonucleases in pBR322 DNA (*continued*)

	Location	Fragment Ends	Fragment Length		Location	Fragment Ends	Fragment Length
	591	592–1178	587		1757	1758–1835	78
	2295	2296–2793	498		1694	1695–1757	63
	1178	1179–1469	291		99	100–146	47
	280	281–475	195		1000	1001–1045	45
	489	490–591	102	HphI	2106	2107–3212	1106
	3954	3955–4039	85		446	447–1300	854
	475	476–489	14		1521	1522–2097	576
					3439	3440–3854	415
		11 Sites			4095	4096–119	387
					119	120–401	282
HaeII	2723	2724–236	1876		3212	3213–3439	227
	1731	1732–2353	622		1300	1301–1521	221
	1209	1210–1648	439		3854	3855–4061	207
	779	780–1209	430		401	402–446	45
	2353	2354–2723	370		4061	4062–4095	34
	552	553–779	227		2097	2098–2106	9
	236	237–417	181				
	1648	1649–1731	83			*15 Sites*	
	438	439–498	60	AsuI	1949	1950–3410	1461
	498	499–552	54		3728	3729–4344	616
	417	418–438	21		172	173–524	352
HgaI	3895	3896–399	867		1481	1482–1760	279
	3164	3165–3895	731		524	525–799	275
	1380	1381–2013	633		887	888–1136	249
	2586	2587–3164	578		3506	3507–3728	222
	2171	2172–2586	415		4344	4345–172	191
	639	640–953	314		1760	1761–1949	189
	985	986–1230	245		1260	1261–1439	179
	399	400–639	240		1136	1137–1260	124
	2013	2014–2171	158		799	800–887	88
	1230	1231–1380	150		3410	3411–3489	79
	953	954–985	32		1439	1440–1481	42
MboII	2347	2348–3137	790		3489	3490–3506	17
	3209	3210–3964	755	Cfr13I	1949	1950–3410	1461
	1594	1595–2347	753		3728	3729–4344	616
	1002	1003–1594	592		172	173–524	352
	4347	4348–476	492		1481	1482–1760	279
	731	732–1002	271		524	525–799	275
	476	477–731	255		887	888–1136	249
	4151	4152–4347	196		3506	3507–3728	222
	4042	4043–4151	109		4344	4345–172	191
	3964	3965–4042	78		1760	1761–1949	189
	3137	3138–3209	72		1260	1261–1439	179
					1136	1137–1260	124
		12 Sites			799	800–887	88
FokI	2163	2164–3335	1172		3410	3411–3489	79
	146	147–1000	854		1439	1440–1481	42
	3803	3804–99	659		3489	3490–3506	17
	1045	1046–1694	649	Nsp7524IV	1949	1950–3410	1461
	3516	3517–3803	287		3728	3729–4344	616
	1835	1836–2022	187		172	173–524	352
	3335	3336–3516	181		1481	1482–1760	279
	2022	2023–2163	141		524	525–799	275

Table 30. Number and Location of Cutting Sites for Commercial Restriction Endonucleases in pBR322 DNA (*continued*)

Enzyme	Location	Fragment Ends	Fragment Length
	887	888–1136	249
	3506	3507–3728	222
	4344	4345–172	191
	1760	1761–1949	189
	1260	1261–1439	179
	1136	1137–1260	124
	799	800–887	88
	3410	3411–3489	79
	1439	1440–1481	42
	3489	3490–3506	17
Nsp7524V	1949	1950–3410	1461
	3728	3729–4344	616
	172	173–524	352
	1481	1482–1760	279
	524	525–799	275
	887	888–1136	249
	3506	3507–3728	222
	4344	4345–172	191
	1760	1761–1949	189
	1260	1261–1439	179
	1136	1137–1260	124
	799	800–887	88
	3410	3411–3489	79
	1439	1440–1481	42
	3489	3490–3506	17
Sau96I	1949	1950–3410	1461
	3728	3729–4344	616
	172	173–524	352
	1481	1482–1760	279
	524	525–799	275
	887	888–1136	249
	3506	3507–3728	222
	4344	4345–172	191
	1760	1761–1949	189
	1260	1261–1439	179
	1136	1137–1260	124
	799	800–887	88
	3410	3411–3489	79
	1439	1440–1481	42
	3489	3490–3506	17

16 Sites

Enzyme	Location	Fragment Ends	Fragment Length
ScrFI	2855	2856–3551	696
	3902	3903–131	592
	535	536–1060	525
	171	172–535	364
	3551	3552–3902	351
	2156	2157–2503	347
	1485	1486–1813	328
	1813	1814–2121	308
	2637	2638–2855	218
	1060	1061–1259	199
	1259	1260–1443	184
	2503	2504–2624	121
	1443	1444–1485	42
	131	132–171	40
	2121	2122–2156	35
	2624	2625–2637	13

17 Sites

Enzyme	Location	Fragment Ends	Fragment Length
AluI	1090	1091–2000	910
	3720	3721–16	659
	31	32–687	656
	3036	3037–3557	521
	687	688–1090	403
	2136	2137–2417	281
	2779	2780–3036	257
	2417	2418–2643	226
	3557	3558–3657	100
	2643	2644–2733	90
	3657	3658–3720	63
	2000	2001–2057	57
	2068	2069–2117	49
	2733	2734–2779	46
	2117	2118–2136	19
	16	17–31	15
	2057	2058–2068	11
MaeIII	212	213–880	668
	1147	1148–1807	660
	2224	2225–2831	607
	4023	4024–124	464
	3293	3294–3624	331
	3010	3011–3293	283
	880	881–1147	267
	1916	1917–2129	213
	3835	3836–4023	188
	3682	3683–3835	153
	2894	2895–3010	116
	2129	2130–2224	95
	124	125–212	88
	1830	1831–1916	86
	2831	2832–2894	63
	3624	3625–3682	58
	1807	1808–1830	23

21 Sites

Enzyme	Location	Fragment Ends	Fragment Length
BbvI	3797	3798–214	780
	785	786–1418	633
	2386	2387–2805	419
	214	215–603	389
	1673	1674–2056	383
	3103	3104–3407	304
	2897	2898–3103	206
	3596	3597–3797	201
	3407	3408–3596	189
	603	604–785	182

Table 30. Number and Location of Cutting Sites for Commercial Restriction Endonucleases in pBR322 DNA (continued)

	Location	Fragment Ends	Fragment Length		Location	Fragment Ends	Fragment Length
	2223	2224–2368	145		1130	1131–1145	15
	1550	1551–1673	123		3215	3216–3227	12
	1442	1443–1547	105		3118	3119–3129	11
	2126	2127–2223	97		3129	3130–3137	8
	2805	2806–2894	89	HaeIII	3758	3759–4345	587
	2077	2078–2126	49		1950	1951–2490	540
	1418	1419–1442	24		1446	1447–1950	504
	2056	2057–2077	21		2953	2954–3411	458
	2368	2369–2386	18		2519	2520–2953	434
	2894	2895–2897	3		3491	3492–3758	267
	1547	1548–1550	3		597	598–831	234
					1049	1050–1262	213
	22 Sites				4345	4346–174	192
					1262	1263–1446	184
BspRI	3758	3759–4345	587		401	402–525	124
	1950	1951–2490	540		174	175–297	123
	1446	1447–1950	504		297	298–401	104
	2953	2954–3411	458		831	832–920	89
	2519	2520–2953	434		3411	3412–3491	80
	3491	3492–3758	267		533	534–597	64
	597	598–831	234		992	993–1049	57
	1049	1050–1262	213		941	942–992	51
	4345	4346–174	192		920	921–941	21
	1262	1263–1446	184		2501	2502–2519	18
	401	402–525	124		2490	2491–2501	11
	174	175–297	123		525	526–533	8
	297	298–401	104	MboI	1667	1668–3041	1374
	831	832–920	89		4046	4047–348	665
	3411	3412–3491	80		466	467–825	359
	533	534–597	64		3330	3331–3671	341
	992	993–1049	57		1143	1144–1460	317
	941	942–992	51		825	826–1097	272
	920	921–941	21		3735	3736–3993	258
	2501	2502–2519	18		1460	1461–1667	207
	2490	2491–2501	11		3225	3226–3330	105
	525	526–533	8		375	376–466	91
DpnI	1669	1670–3043	1374		3135	3136–3213	78
	4048	4049–350	665		3041	3042–3116	75
	468	469–827	359		3689	3690–3735	46
	3332	3333–3673	341		4010	4011–4046	36
	1145	1146–1462	317		1097	1098–1128	31
	827	828–1099	272		348	349–375	27
	3737	3738–3995	258		3671	3672–3689	18
	1462	1463–1669	207		3993	3994–4010	17
	3227	3228–3332	105		1128	1129–1143	15
	377	378–468	91		3213	3214–3225	12
	3137	3138–3215	78		3116	3117–3127	11
	3043	3044–3118	75		3127	3128–3135	8
	3691	3692–3737	46	NdeII	1667	1668–3041	1374
	4012	4013–4048	36		4046	4047–348	665
	1099	1100–1130	31		466	467–825	359
	350	351–377	27		3330	3331–3671	341
	3673	3674–3691	18		1143	1144–1460	317
	3995	3996–4012	17				

Table 30. Number and Location of Cutting Sites for Commercial Restriction Endonucleases in pBR322 DNA (*continued*)

	Location	Fragment Ends	Fragment Length		Location	Fragment Ends	Fragment Length
	825	826–1097	272		348	349–375	27
	3735	3736–3993	258		3671	3672–3689	18
	1460	1461–1667	207		3993	3994–4010	17
	3225	3226–3330	105		1128	1129–1143	15
	375	376–466	91		3213	3214–3225	12
	3135	3136–3213	78		3116	3117–3127	11
	3041	3042–3116	75		3127	3128–3135	8
	3689	3690–3735	46	SfaNI	2572	2573–3624	1052
	4010	4011–4046	36		4064	4065–125	424
	1097	1098–1128	31		1035	1036–1430	395
	348	349–375	27		649	650–1024	375
	3671	3672–3689	18		3816	3817–4064	248
	3993	3994–4010	17		1430	1431–1673	243
	1128	1129–1143	15		414	415–649	235
	3213	3214–3225	12		1919	1920–2142	223
	3116	3117–3127	11		2352	2353–2572	220
	3127	3128–3135	8		3624	3625–3816	192
PalI	3758	3759–4345	587		238	239–402	164
	1950	1951–2490	540		2142	2143–2276	134
	1446	1447–1950	504		1682	1683–1778	96
	2953	2954–3411	458		125	126–213	88
	2519	2520–2953	434		1778	1779–1856	78
	3491	3492–3758	267		1856	1857–1919	63
	597	598–831	234		2313	2314–2352	39
	1049	1050–1262	213		2276	2277–2313	37
	4345	4346–174	192		213	214–238	25
	1262	1263–1446	184		402	403–414	12
	401	402–525	124		1024	1025–1035	11
	174	175–297	123		1673	1674–1682	9
	297	298–401	104				
	831	832–920	89			*23 Sites*	
	3411	3412–3491	80	FnudII	2522	2523–3103	581
	533	534–597	64		3433	3434–3926	493
	992	993–1049	57		4258	4259–347	452
	941	942–992	51		1635	1636–2007	372
	920	921–941	21		347	348–703	356
	2501	2502–2519	18		2181	2183–2522	341
	2490	2491–2501	11		3926	3927–4258	332
	525	526–533	8		3103	3104–3433	330
Sau3A	1667	1668–3041	1374		1245	1246–1390	145
	4046	4047–348	665		1106	1107–1235	129
	466	467–825	359		818	819–947	129
	3330	3331–3671	341		1416	1417–1538	122
	1143	1144–1460	317		703	704–818	115
	825	826–1097	272		2078	2079–2181	103
	3735	3736–3993	258		1538	1539–1635	97
	1460	1461–1667	207		2007	2008–2076	69
	3225	3226–3330	105		1040	1041–1106	66
	375	376–466	91		979	980–1040	61
	3135	3136–3213	78		947	948–974	27
	3041	3042–3116	75		1390	1391–1416	26
	3689	3690–3735	46		1235	1236–1245	10
	4010	4011–4046	36		974	975–979	5
	1097	1098–1128	31		2076	2077–2078	2

Table 30. Number and Location of Cutting Sites for Commercial Restriction Endonucleases in pBR322 DNA (*continued*)

	Location	Fragment Ends	Fragment Length		Location	Fragment Ends	Fragment Length
ThaI	2522	2523–3103	581		3045	3046–3449	404
	3433	3434–3926	493		1812	1813–2121	309
	4258	4259–347	452		3660	3661–3902	242
	1635	1636–2007	372		1020	1021–1258	238
	347	348–703	356		170	171–387	217
	2181	2183–2522	341		1284	1285–1485	201
	3926	3927–4258	332		2855	2856–3045	190
	3103	3104–3433	330		1485	1486–1665	180
	1245	1246–1390	145		770	771–930	160
	1106	1107–1235	129		534	535–694	160
	818	819–947	129		2682	2683–2829	147
	1416	1417–1538	122		1665	1666–1812	147
	703	704–818	115		411	412–534	123
	2078	2079–2181	103		3550	3551–3660	110
	1538	1539–1635	97		930	931–1020	90
	2007	2008–2076	69		694	695–770	76
	1040	1041–1106	66		3483	3484–3550	67
	979	980–1040	61		3449	3450–3483	34
	947	948–974	27		2121	2122–2155	34
	1390	1391–1416	26		2829	2830–2855	26
	1235	1236–1245	10		1258	1259–1284	26
	974	975–979	5		387	388–402	15
	2076	2077–2078	2		402	403–411	9
					161	162–170	9
		24 Sites		HpaII	3902	3903–161	622
					2155	2156–2682	527
NlaIV	2546	2547–3318	772		3045	3046–3449	404
	1762	1763–2507	745		1812	1813–2121	309
	3664	3665–4254	590		3660	3661–3902	242
	889	890–1207	318		1020	1021–1258	238
	1483	1484–1762	279		170	171–387	217
	550	551–768	218		1284	1285–1485	201
	3453	3454–3664	211		2855	2856–3045	190
	121	122–331	210		1485	1486–1665	180
	4254	4255–78	187		770	771–930	160
	768	769–889	121		534	535–694	160
	1326	1327–1440	114		2682	2683–2829	147
	3318	3319–3412	94		1665	1666–1812	147
	436	437–526	90		411	412–534	123
	1207	1208–1256	49		3550	3551–3660	110
	331	332–377	46		930	931–1020	90
	1440	1441–1483	43		694	695–770	76
	78	79–121	43		3483	3484–3550	67
	3412	3413–3453	41		3449	3450–3483	34
	2507	2508–2546	39		2121	2122–2155	34
	377	378–415	38		2829	2830–2855	26
	1291	1292–1326	35		1258	1259–1284	26
	1256	1257–1291	35		387	388–402	15
	526	527–550	24		402	403–411	9
	415	416–436	21		161	162–170	9
		26 Sites		MnlI	3741	3742–4335	594
					2907	2908–3307	400
HapII	3902	3903–161	622		1472	1473–1803	331
	2155	2156–2682	527		2112	2113–2374	262

Table 30. Number and Location of Cutting Sites for Commercial Restriction Endonucleases in pBR322 DNA (continued)

Enzyme	Location	Fragment Ends	Fragment Length
	2657	2658–2907	250
	389	390–608	219
	2374	2375–2583	209
	3535	3536–3741	206
	185	186–389	204
	974	975–1177	203
	1900	1901–2082	182
	608	609–790	182
	1303	1304–1472	169
	4335	4336–125	153
	3388	3389–3535	147
	875	876–974	99
	790	791–875	85
	3307	3308–3388	81
	2583	2584–2657	74
	1177	1178–1238	61
	125	126–185	60
	1803	1804–1861	58
	1861	1862–1900	39
	1238	1239–1276	38
	2082	2083–2112	30
	1276	1277–1303	27
MspI	3902	3903–161	622
	2155	2156–2682	527
	3045	3046–3449	404
	1812	1813–2121	309
	3660	3661–3902	242
	1020	1021–1258	238
	170	171–387	217
	1284	1285–1485	201
	2855	2856–3045	190
	1485	1486–1665	180
	770	771–930	160
	534	535–694	160
	2682	2683–2829	147
	1665	1666–1812	147
	411	412–534	123
	3550	3551–3660	110
	930	931–1020	90
	694	695–770	76
	3483	3484–3550	67
	3449	3450–3483	34
	2121	2122–2155	34
	2829	2830–2855	26
	1258	1259–1284	26
	387	388–402	15
	402	403–411	9
	161	162–170	9
NlaIII	2479	2480–3199	720
	3199	3200–3690	491
	3814	3815–4207	393
	8	9–355	347
	2219	2220–2479	260
	1595	1596–1820	225

Enzyme	Location	Fragment Ends	Fragment Length
	1254	1255–1461	207
	751	752–937	186
	1949	1950–2114	165
	566	567–712	146
	355	356–493	138
	1461	1462–1595	134
	1054	1055–1182	128
	937	938–1054	117
	4207	4208–4312	105
	2114	2115–2219	105
	3700	3701–3778	78
	493	494–566	73
	1884	1885–1949	65
	1820	1821–1884	64
	4312	4313–8	59
	1197	1198–1254	57
	712	713–751	39
	3778	3779–3814	36
	1182	1183–1197	15
	3690	3691–3700	10

31 Sites

Enzyme	Location	Fragment Ends	Fragment Length
CfoI	3105	3106–3498	393
	1730	1731–2078	348
	3591	3592–3928	337
	3928	3929–4260	332
	2385	2386–2655	270
	949	950–1208	259
	4260	4261–103	206
	1457	1458–1647	190
	2822	2823–2996	174
	263	264–416	153
	551	552–703	152
	1208	1209–1359	151
	2211	2212–2352	141
	103	104–235	132
	818	819–949	131
	2996	2997–3105	109
	2078	2079–2181	103
	2722	2723–2822	100
	3498	3499–3591	93
	1647	1648–1730	83
	703	704–778	75
	2655	2656–2722	67
	1359	1360–1421	62
	437	438–497	60
	497	498–551	54
	778	779–818	40
	1421	1422–1457	36
	2352	2353–2385	33
	2181	2182–2211	30
	235	236–263	28
	416	417–437	21
HhaI	3105	3106–3498	393

Table 30. Number and Location of Cutting Sites for Commercial Restriction Endonucleases in pBR322 DNA (continued)

	Location	Fragment Ends	Fragment Length		Location	Fragment Ends	Fragment Length
	1730	1731–2078	348		435	436–495	60
	3591	3592–3928	337		495	496–549	54
	3928	3929–4260	332		776	777–816	40
	2385	2386–2655	270		1419	1420–1455	36
	949	950–1208	259		2350	2351–2383	33
	4260	4261–103	206		2350	2351–2383	33
	1457	1458–1647	190		2179	2180–2209	30
	2822	2823–2996	174		233	234–261	28
	263	264–416	153		414	415–435	21
	551	552–703	152				
	1208	1209–1359	151			*42 Sites*	
	2211	2212–2352	141	Fnu4HI	4110	4111–227	480
	103	104–235	132		3092	3093–3420	328
	818	819–949	131		1767	1768–2066	299
	2996	2997–3105	109		301	302–579	278
	2078	2079–2181	103		3881	3882–4110	229
	2722	2723–2822	100		2886	2887–3092	206
	3498	3499–3591	93		3420	3421–3609	189
	1647	1648–1730	83		774	775–939	165
	703	704–778	75		2520	2521–2675	155
	2655	2656–2722	67		3609	3610–3759	150
	1359	1360–1421	62		2675	2676–2818	143
	437	438–497	60		1431	1432–1560	129
	497	498–551	54		1563	1564–1686	123
	778	779–818	40		1288	1289–1407	119
	1421	1422–1457	36		2402	2403–2520	118
	2352	2353–2385	33		2265	2266–2381	116
	2181	2182–2211	30		616	617–723	107
	235	236–263	28		2115	2116–2212	97
	416	417–437	21		3786	3787–3881	95
HinpI	3103	3104–3496	393		939	940–1024	85
	1728	1729–2076	348		1024	1025–1107	83
	3589	3590–3926	337		1686	1687–1767	81
	3926	3927–4258	332		1209	1210–1288	79
	2383	2384–2653	270		227	228–298	71
	947	948–1206	259		2818	2819–2883	65
	4258	4259–101	206		1107	1108–1164	57
	1455	1456–1645	190		2212	2213–2265	53
	2820	2821–2994	174		723	724–774	51
	261	262–414	153		2069	2070–2115	46
	549	550–701	152		1164	1165–1209	45
	1206	1207–1357	151		582	583–616	34
	2209	2210–2350	141		3759	3760–3786	27
	101	102–233	132		2381	2382–2399	18
	816	817–947	131		1417	1418–1431	14
	2994	2995–3103	109		1410	1411–1417	7
	2076	2077–2179	103		2883	2884–2886	3
	2720	2721–2820	100		2399	2400–2402	3
	3496	3497–3589	93		2066	2067–2069	3
	1645	1646–1728	83		1560	1561–1563	3
	701	702–776	75		1407	1408–1410	3
	2653	2654–2720	67		579	580–582	3
	1357	1358–1419	62		298	299–301	3

Table 31. Number and Location of Cutting Sites for Commercial Restriction Endonucleases in pUC19 DNA

0 Site

AatI	BglII	EcoT14I	NheI	SexI
AccIII	BsmI	Eco47III	NotI	SfiI
AflII	BspMII	Eco52I	NruI	SnaBI
AocI	BssHII	Eco81I	NsiI	SpeI
ApaI	BstEII	EspI	OxaNI	SplI
AsuII	BstXI	FspII	PaeR7	SstII
AvaIII	ClaI	HpaI	PflMI	StuI
AvrII	CvnI	MluI	PpuMI	StyI
BalI	DraIII	MstII	RsrII	Tth111I
BanIII	EagI	NaeI	SacII	XcaI
BclI	EcoRV	NcoI	SauI	XhoI
				XmaIII

	Location	Fragment Ends	Fragment Length
		1 Site	
AatII	2621	2622–2621	2686
AccI	430	431–430	2686
AflIII	806	807–806	2686
AlwNI	1222	1223–1222	2686
AquI	413	414–413	2686
Asp700	2298	2299–2298	2686
Asp718	408	409–408	2686
AvaI	412	413–412	2686
BamHI	417	418–417	2686
BanII	406	407–406	2686
BbeI	239	240–239	2686
BspMI	444	445–444	2686
BstI	417	418–417	2686
Cfr10I	1779	1780–1779	2686
DraII	2675	2676–2675	2686
EcoRI	396	397–396	2686
Eco0109	2675	2676–2675	2686
HincII	431	432–431	2686
HindII	431	432–431	2686
HindIII	447	448–447	2686
KpnI	412	413–412	2686
NarI	236	237–236	2686
NdeI	184	185–184	2686
Nsp7524III	412	413–412	2686
NunII	236	237–236	2686
PssI	2678	2679–2678	2686
PstI	439	440–439	2686
SacI	406	407–406	2686
SalI	429	430–429	2686
ScaI	2179	2180–2179	2686
SmaI	414	415–414	2686
SphI	445	446–445	2686

	Location	Fragment Ends	Fragment Length
SspI	2503	2504–2503	2686
SstI	406	407–406	2686
XbaI	423	424–423	2686
XcyI	412	413–412	2686
XmaI	412	413–412	2686
XmnI	2298	2299–2298	2686
		2 Sites	
AflI	2059	2060–1837	2464
	1837	1838–2059	222
AosI	258	259–1921	1663
	1921	1922–258	1023
AvaII	2059	2060–1837	2464
	1837	1838–2059	222
BglI	251	252–1819	1568
	1819	1820–251	1118
ClaII	2059	2060–1837	2464
	1837	1838–2059	222
Eco47I	2058	2059–1836	2464
	1836	1837–2058	222
FspI	258	259–1921	1663
	1921	1922–258	1023
MstI	258	259–1921	1663
	1921	1922–258	1023
PleI	1180	1181–430	1936
	430	431–1180	750
PvuI	279	280–2069	1790
	2069	2070–279	896
PvuII	630	631–308	2364
	308	309–630	322
SinI	2059	2060–1837	2464
	1837	1838–2059	222
XorII	279	280–2069	1790
	2069	2070–279	896
		3 Sites	
AhaII	236	237–2236	2000
	2236	2237–2618	382
	2618	2619–236	304
AhaIII	2276	2277–1565	1975
	1584	1585–2276	692
	1565	1566–1584	19
AosII	236	237–2236	2000
	2236	2237–2618	382
	2618	2619–236	304
ApaLI	1120	1121–2366	1246
	177	178–1120	943
	2366	2367–177	497

Table 31. Number and Location of Cutting Sites for Commercial Restriction Endonucleases in pUC19 DNA (*continued*)

	Location	Fragment Ends	Fragment Length		Location	Fragment Ends	Fragment Length
BbiII	236	237–2236	2000		430	431–906	476
	2236	2237–2618	382		400	401–430	30
	2618	2619–236	304	TthHB8I	906	907–2350	1444
BspHI	2639	2640–1526	1573		2350	2351–400	736
	1526	1527–2534	1008		430	431–906	476
	2534	2535–2639	105		400	401–430	30
CfrI	645	646–2087	1442				
	2087	2088–388	987			*5 Sites*	
	388	389–645	257	AocII	1124	1125–2285	1161
DraI	2276	2277–1565	1975		406	407–1124	718
	1584	1585–2276	692		2370	2371–181	497
	1565	1566–1584	19		181	182–406	225
EaeI	645	646–2087	1442		2285	2286–2370	85
	2087	2088–388	987	ApyI	968	969–355	2073
	388	389–645	257		546	547–834	288
HaeII	1054	1055–239	1871		355	356–546	191
	239	240–684	445		834	835–955	121
	684	685–1054	370		955	956–968	13
Nsp7524I	810	811–41	1917	Bsp1286	1124	1125–2285	1161
	41	42–445	404		406	407–1124	718
	445	446–810	365		2370	2371–181	497
NspHI	810	811–41	1917		181	182–406	225
	41	42–445	404		2285	2286–2370	85
	445	446–810	365	BspNI	968	969–355	2073
RsaI	410	411–2179	1769		546	547–834	288
	2179	2180–169	676		355	356–546	191
	169	170–410	241		834	835–955	121
Tth111II	1429	1430–1396	2653		955	956–968	13
	1410	1411–1429	19	BstNI	968	969–355	2073
	1396	1397–1410	14		546	547–834	288
					355	356–546	191
		4 Sites			834	835–955	121
AlwI	420	421–1450	1030		955	956–968	13
	1547	1548–2327	780	EcoRII	966	967–353	2073
	2327	2328–420	779		544	545–832	288
	1450	1451–1547	97		353	354–544	191
BanI	1647	1648–235	1274		832	833–953	121
	550	551–1647	1097		953	954–966	13
	235	236–408	173	FokI	334	335–1666	1332
	408	409–550	142		2134	2135–90	642
HgaI	98	99–917	819		1847	1848–2134	287
	1495	1496–2226	731		90	91–334	244
	917	918–1495	578		1666	1667–1847	181
	2226	2227–98	558	HgiAI	1124	1125–2285	1161
MaeI	1889	1890–424	1221		406	407–1124	718
	424	425–1301	877		2370	2371–181	497
	1554	1555–1889	335		181	182–406	225
	1301	1302–1554	253		2285	2286–2370	85
TaqI	906	907–2350	1444	MaeII	374	375–1509	1135
	2350	2351–400	736		2618	2619–374	442
					1509	1510–1925	416

Table 31. Number and Location of Cutting Sites for Commercial Restriction Endonucleases in pUC19 DNA (*continued*)

	Location	Fragment Ends	Fragment Length		Location	Fragment Ends	Fragment Length
	1925	1926–2298	373		1837	1838–2059	222
	2298	2299–2618	320		1741	1742–1820	79
MvaI	968	969–355	2073		1820	1821–1837	17
	546	547–834	288	Nsp7524V	286	287–1741	1455
	355	356–546	191		2059	2060–2675	616
	834	835–955	121		2675	2676–286	297
	955	956–968	13		1837	1838–2059	222
Nsp7524II	1124	1125–2285	1161		1741	1742–1820	79
	406	407–1124	718		1820	1821–1837	17
	2370	2371–181	497	Sau96I	286	287–1741	1455
	181	182–406	225		2059	2060–2675	616
	2285	2286–2370	85		2675	2676–286	297
SdnI	1124	1125–2285	1161		1837	1838–2059	222
	406	407–1124	718		1741	1742–1820	79
	2370	2371–181	497		1820	1821–1837	17
	181	182–406	225				
	2285	2286–2370	85				
						7 Sites	
		6 Sites		BcnI	415	416–1187	772
AsuI	286	287–1741	1455		1187	1188–1883	696
	2059	2060–2675	616		2234	2235–49	501
	2675	2676–286	297		1883	1884–2234	351
	1837	1838–2059	222		84	85–414	330
	1741	1742–1820	79		49	50–84	35
	1820	1821–1837	17		414	415–415	1
Cfr13I	286	287–1741	1455	HphI	33	34–1543	1510
	2059	2060–2675	616		1770	1771–2185	415
	2675	2676–286	297		2426	2427–24	284
	1837	1838–2059	222		1543	1544–1770	227
	1741	1742–1820	79		2185	2186–2392	207
	1820	1821–1837	17		2392	2393–2426	34
DdeI	171	172–1081	910		24	25–33	9
	1656	1657–2196	540	MboII	678	679–1468	790
	2196	2197–2622	426		1540	1541–2295	755
	1081	1082–1490	409		2482	2483–284	488
	2622	2623–171	235		284	285–678	394
	1490	1491–1656	166		2373	2374–2482	109
HinfI	1694	1695–427	1419		2295	2296–2373	78
	1177	1178–1694	517		1468	1469–1540	72
	781	782–1177	396	MflI	417	418–1447	1030
	427	428–641	214		1556	1557–2324	768
	706	707–781	75		2341	2342–417	762
	641	642–706	65		1458	1459–1544	86
NspBII	1393	1394–2334	941		2324	2325–2341	17
	630	631–1148	518		1544	1545–1556	12
	2334	2335–114	466		1447	1448–1458	11
	308	309–630	322	NciI	414	415–1186	772
	1148	1149–1393	245		1186	1187–1182	696
	114	115–308	194		2233	2234–48	501
Nsp7524IV	286	287–1741	1455		1882	1883–2233	351
	2059	2060–2675	616		83	84–413	330
	2675	2676–286	297		48	49–83	35
					413	414–414	1

Table 31. Number and Location of Cutting Sites for Commercial Restriction Endonucleases in pUC19 DNA (*continued*)

	Location	Fragment Ends	Fragment Length		Location	Fragment Ends	Fragment Length
XhoII	417	418–1447	1030		850	851–1284	434
	1556	1557–2324	768		2676	2677–288	298
	2341	2342–417	762		1822	1823–2089	267
	1458	1459–1544	86		390	391–647	257
	2324	2325–2341	17		647	648–821	174
	1544	1545–1556	12		288	289–390	102
	1447	1448–1458	11		1742	1743–1822	80
					832	833–850	18
		8 Sites			821	822–832	11
SfaNI	903	904–1955	1052	MaeIII	367	368–1162	795
	238	239–903	665		2354	2355–56	388
	2395	2396–69	360		1624	1625–1955	331
	2147	2148–2395	248		56	57–347	291
	1955	1956–2147	192		1341	1342–1624	283
	69	70–162	93		2166	2167–2354	188
	199	200–238	39		2013	2014–2166	153
	162	163–199	37		1225	1226–1341	116
FnudII	853	854–1434	581		1162	1163–1225	63
	108	109–653	545		1955	1956–2013	58
	1764	1765–2257	493		347	348–367	20
	2257	2258–2589	332	NlaIII	810	811–1530	720
	1434	1435–1764	330		1530	1531–2021	491
	655	656–853	198		41	42–445	404
	5	6–108	103		2145	2146–2538	393
	2589	2590–3	100		464	465–810	346
	653	654–655	2		2538	2539–2643	105
	3	4–5	2		2643	2644–41	84
					2031	2032–2109	78
ThaI	853	854–1434	581		2109	2110–2145	36
	108	109–653	545		445	446–464	19
	1764	1765–2257	493		2021	2022–2031	10
	2257	2258–2589	332	NlaIV	877	878–1649	772
	1434	1435–1764	330		1995	1996–2585	590
	655	656–853	198		2585	2586–237	338
	5	6–108	103		552	553–838	286
	2589	2590–3	100		1784	1785–1995	211
	653	654–655	2		237	238–410	173
	3	4–5	2		419	420–552	133
					1649	1650–1743	94
		11 Sites			1743	1744–1784	41
BspRI	2089	2090–2676	587		838	839–877	39
	1284	1285–1742	458		410	411–419	9
	850	851–1284	434	PalI	2089	2090–2676	587
	2676	2677–288	298		1284	1285–1742	458
	1822	1823–2089	267		850	851–1284	434
	390	391–647	257		2676	2677–288	298
	647	648–821	174		1822	1823–2089	267
	288	289–390	102		390	391–647	257
	1742	1743–1822	80		647	648–821	174
	832	833–850	18		288	289–390	102
	821	822–832	11		1742	1743–1822	80
HaeIII	2089	2090–2676	587		832	833–850	18
	1284	1285–1742	458		821	822–832	11

Table 31. Number and Location of Cutting Sites for Commercial Restriction Endonucleases in pUC19 DNA (*continued*)

	Location	Fragment Ends	Fragment Length		Location	Fragment Ends	Fragment Length
		12 Sites			48	49–82	34
					1160	1161–1186	26
BbvI	2128	2129–53	611	MnlI	2072	2073–2666	594
	717	718–1136	419		1238	1239–1638	400
	1738	1739–2128	390		39	40–299	260
	1434	1435–1738	304		988	989–1238	250
	315	316–618	303		431	432–655	224
	1228	1229–1434	206		705	706–914	209
	53	54–242	189		1866	1867–2072	206
	1136	1137–1225	89		1719	1720–1866	147
	618	619–699	81		299	300–431	132
	242	243–315	73		1638	1639–1719	81
	699	700–717	18		914	915–988	74
	1225	1226–1228	3		2666	2667–39	59
ScrFI	1186	1187–1882	696		655	656–705	50
	2233	2234–48	501	MspI	2233	2234–48	501
	1882	1883–2233	351		524	525–1013	489
	546	547–834	288		1376	1377–1780	404
	83	84–355	272		82	83–413	331
	968	969–1186	218		1991	1992–2233	242
	414	415–546	132		1186	1187–1376	190
	834	835–955	121		1013	1014–1160	147
	355	356–413	58		413	414–524	111
	48	49–83	35		1881	1882–1991	110
	955	956–968	13		1814	1815–1881	67
	413	414–414	1		1780	1781–1814	34
					48	49–82	34
					1160	1161–1186	26
		13 Sites					
HapII	2233	2234–48	501			*15 Sites*	
	524	525–1013	489	AluI	2051	2052–44	679
	1376	1377–1780	404		1367	1368–1888	521
	82	83–413	331		1110	1111–1367	257
	1991	1992–2233	242		63	64–308	245
	1186	1187–1376	190		748	749–974	226
	1013	1014–1160	147		974	975–1110	136
	413	414–524	111		630	631–748	118
	1881	1882–1991	110		1888	1889–1988	100
	1814	1815–1881	67		308	309–404	96
	1780	1781–1814	34		471	472–566	95
	48	49–82	34		566	567–630	64
	1160	1161–1186	26		1988	1989–2051	63
HpaII	2233	2234–48	501		404	405–449	45
	524	525–1013	489		449	450–471	22
	1376	1377–1780	404		44	45–63	19
	82	83–413	331	DpnI	419	420–1374	955
	1991	1992–2233	242		2379	2380–278	585
	1186	1187–1376	190		1663	1664–2004	341
	1013	1014–1160	147		2068	2069–2326	258
	413	414–524	111		278	279–419	141
	1881	1882–1991	110		1558	1559–1663	105
	1814	1815–1881	67		1468	1469–1546	78
	1780	1781–1814	34				

Table 31. Number and Location of Cutting Sites for Commercial Restriction Endonucleases in pUC19 DNA (*continued*)

	Location	Fragment Ends	Fragment Length		Location	Fragment Ends	Fragment Length
	1374	1375–1449	75			*17 Sites*	
	2022	2023–2068	46	CfoI	1436	1437–1829	393
	2343	2344–2379	36		1922	1923–2259	337
	2004	2005–2022	18		2259	2260–2591	332
	2326	2327–2343	17		259	260–590	331
	1546	1547–1558	12		716	717–986	270
	1449	1450–1460	11		1153	1154–1327	174
	1460	1461–1468	8		108	109–238	130
MboI	417	418–1372	955		1327	1328–1436	109
	2377	2378–276	585		5	6–108	103
	1661	1662–2002	341		2591	2592–5	100
	2066	2067–2324	258		1053	1054–1153	100
	276	277–417	141		1829	1830–1922	93
	1556	1557–1661	105		986	987–1053	67
	1466	1467–1544	78		590	591–655	65
	1372	1373–1447	75		683	684–716	33
	2020	2021–2066	46		655	656–683	28
	2341	2342–2377	36		238	239–259	21
	2002	2003–2020	18				
	2324	2325–2341	17	HhaI	1436	1437–1829	393
	1544	1545–1556	12		1922	1923–2259	337
	1447	1448–1458	11		2259	2260–2591	332
	1458	1459–1466	8		259	260–590	331
NdeII	417	418–1372	955		716	717–986	270
	2377	2378–276	585		1153	1154–1327	174
	1661	1662–2002	341		108	109–238	130
	2066	2067–2324	258		1327	1328–1436	109
	276	277–417	141		5	6–108	103
	1556	1557–1661	105		2591	2592–5	100
	1466	1467–1544	78		1053	1054–1153	100
	1372	1373–1447	75		1829	1830–1922	93
	2020	2021–2066	46		986	987–1053	67
	2341	2342–2377	36		590	591–655	65
	2002	2003–2020	18		683	684–716	33
	2324	2325–2341	17		655	656–683	28
	1544	1545–1556	12		238	239–259	21
	1447	1448–1458	11	HinpI	1434	1435–1827	393
	1458	1459–1466	8		1920	1921–2257	337
Sau3A	417	418–1372	955		2257	2258–2589	332
	2377	2378–276	585		257	258–588	331
	1661	1662–2002	341		714	715–984	270
	2066	2067–2324	258		1151	1152–1325	174
	276	277–417	141		106	107–236	130
	1556	1557–1661	105		1325	1326–1434	109
	1466	1467–1544	78		3	4–106	103
	1372	1373–1447	75		2589	2590–3	100
	2020	2021–2066	46		1051	1052–1151	100
	2341	2342–2377	36		1827	1828–1920	93
	2002	2003–2020	18		984	985–1051	67
	2324	2325–2341	17		588	589–653	65
	1544	1545–1556	12		681	682–714	33
	1447	1448–1458	11		653	654–681	28
	1458	1459–1466	8		236	237–257	21

Table 31.   Number and Location of Cutting Sites for Commercial Restriction Endonucleases in pUC19 DNA (continued)

	Location	Fragment Ends	Fragment Length		Location	Fragment Ends	Fragment Length
	19 Sites				42	43–151	109
Fnu4HI	1751	1572–2090	339		151	152–255	104
	1423	1424–1751	328		2117	2118–2212	95
	328	329–631	303		631	632–712	81
	2441	2442–42	287		255	256–328	73
	2212	2213–2441	229		1149	1150–1214	65
	1217	1218–1423	206		2090	2091–2117	27
	851	852–1006	155		712	713–730	18
	1006	1007–1149	143		1214	1215–1217	3
	733	734–851	118		730	731–733	3

BACTERIOPHAGE-DERIVED VECTORS

E. coli bacteriophage λ and filamentous phages f1, fd, and M13 have proven to be extremely useful for molecular cloning.

Bacteriophage λ

Bacteriophage λ (Figures 46 and 47) has been one of the most studied viruses, and very quickly became a basic vector for molecular cloning because of a very interesting feature of its genetic organization. Readers interested in the molecular biology of bacteriophage λ should read the excellent books cited in the references; we will mention only that bacteriophage λ cannot only multiply like any other infectious bacteriophage in *E. coli* by using *lytic functions* but also can exist as an integrated "silent form" in the bacterial genome. This phenomenon has been termed *lysogenization*. The lysogen *E. coli* replicates the integrated λ DNA as a part of its own genetic material so long as the expression of viral lytic functions is repressed.

Many of the λ genes involved in recombination and lysogenization are not essential for phage multiplication, and can therefore be deleted and replaced by foreign DNA without impairment of the lytic phage functions. This provides the basis for construction of λ-derived cloning vectors (Figures 48–51).

Replacement-Type Vectors. It has been known for a long time that λ capsids cannot accommodate DNA molecules smaller than 38 kb or greater than 53 kb.

Elimination of nonessential functions would therefore lead to DNA molecules which could not be propagated. To avoid this problem, λ vectors have been constructed in which a removable piece of "stuffer DNA" has been inserted in place of λ nonessential functions, thus allowing efficient packaging and multiplication. Cloning of foreign DNA into these vectors involves three steps:

1. Physical elimination of the stuffer DNA after digestion with restriction endonucleases.
2. Ligation of the foreign DNA fragment to the λ arms prepared above.
3. Packaging and multiplication of the recombinant DNA molecule to give rise to infectious λ recombinant phages.

Among the most popular vectors of this kind are the Charon derivatives of bacteriophage λ (Blattner and coworkers). These vectors have been successfully used in the past few years for the preparation of genomic libraries (see Chapter 17) and still represent a very useful and versatile system for cloning large DNA fragments (up to 22 kb) generated by several different restriction endonucleases (see Tables 32 and 33).

By careful choice of stuffer DNA, it has been possible to develop selection methods for recombinant DNA molecules based on the appearance of a particular phenotype.

For example, λ bacteriophages carrying functional *Red* and *gam* genes cannot grow on bacterial strains lysogenic for phage P2. This sensitivity to P2 (*spi*[+] phenotype) can be eliminated after insertion

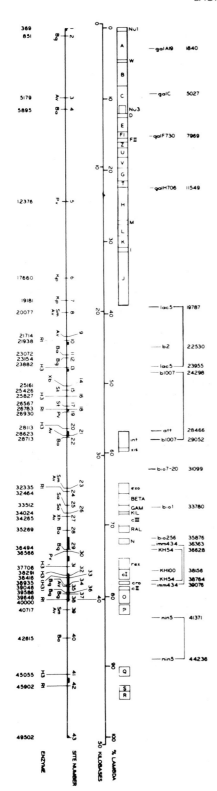

Figure 46. Genetic map of lambda DNA. (With permission of F. Blattner.)

of foreign DNA or replacement of stuffer DNA carrying these genes. The recombinant phages, being *Red⁻ gam⁻*, are able to grow in P2 lysogen strains, whereas parental phages cannot multiply under these conditions.

The λ 1059 (Karn et al., 1980) has been the first example of bacteriophage-derived vector designed to allow selection of recombinant molecules by plating on *E. coli* lysogenic for P2. This vector was originally engineered to accept DNA fragments produced by random digestion with Sau 3A (recognition sequence 5′-GATC-3′) to facilitate the construction of genomic libraries. However, the presence of sequences from the mini Col EI plasmid pacI 29 in the middle of λ 1059 genome did not allow the direct screening of libraries with DNA probes cloned in plasmids.

A series of replacement vectors have been derived from the λ 1059 bacteriophage to combine the advantages of (a) large cloning capacity (up to 23 kb), (b) genetic selection on the basis of *spi* phenotype, and (c) use of polylinker with multiple cloning sites.

Among them, the EMBL3 and EMBL4 bacteriophages (Frischauf et al., 1983) have proven to be particularly useful in the construction of genomic libraries. These two vectors, which do not contain plasmid sequences, carry an asymmetrical polylinker sequence which allows cloning of DNA fragments generated by Eco RI, Bam HI, Bgl II, Bcl I, Xho II, Sal I, Xho I, and Ava I restriction endonucleases (see Chapter 17). The SalI-Bam HI-Eco RI polylinker sequence of the EMBL3 vector has been replaced by the Sal I-Eco RI polylinker fragment of pUC12 (see above) to generate a new vector (EMBL 12) containing restriction sites for Sal I, Xba I, Bam HI, Sst I and Eco RI, and conserving the same cloning capacity (Natt and Scherer, 1986). This vector is particularly useful for cloning DNA fragments generated by the Avr II, Nhe I, and Spe I enzymes, since they can be inserted at the Xba I site of the vector (see enzyme compatibilities in Table 25) and can be recovered by cutting at the flanking Sal I sites.

Another derivative of λ 1059 was also constructed (Karn et al., 1984). This bacteriophage (λ 2001) can accommodate 10 to 23 kb DNA inserts. It contains a polylinker with cleavage sites for Xba I, Sst I, Xho I, Eco RI, Hind III, and Bam HI restriction endonucleases. The insertion of foreign DNA fragments into λ 2001 can be monitored by the *spi*-phenotype. Cloning at the Bam HI or Xho I sites of DNA fragments with compatible ends may result in the loss of both recognition sites (see Table 25). In these cases, the cloned fragments can be recovered by excision at flanking sites in the polylinker.

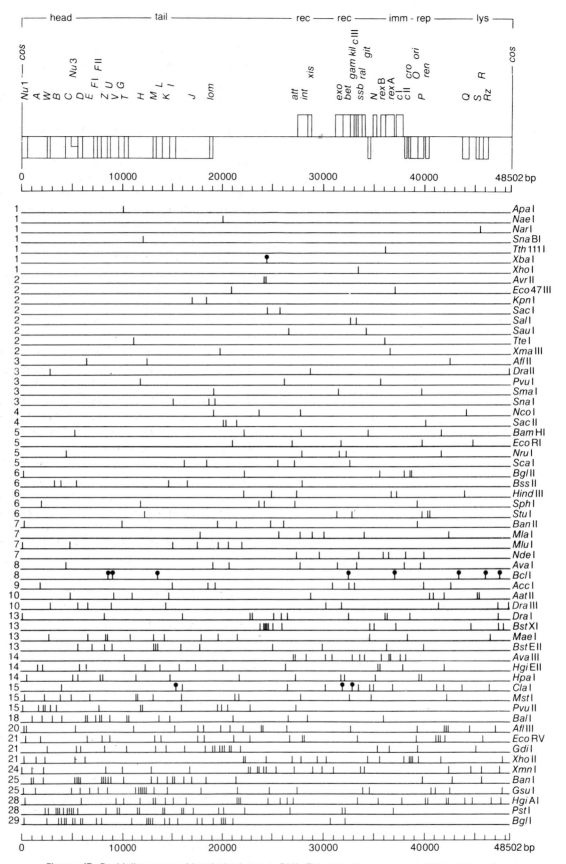

Figure 47. Restriction map of bacteriophage λ DNA. This map has been established from the published sequence of λ CI *indl ts* 857 Sam 7 (courtesy of Boehringer Mannheim).

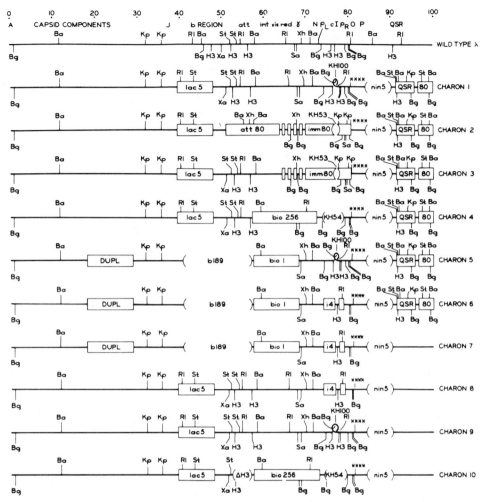

Figure 48. Restriction maps of λ Charon phages 1 to 40. Compiled from DeWet et al. (1980), Rimm et al. (1980), Loenen and Blattner (1983), and Dunn and Blattner (1987).

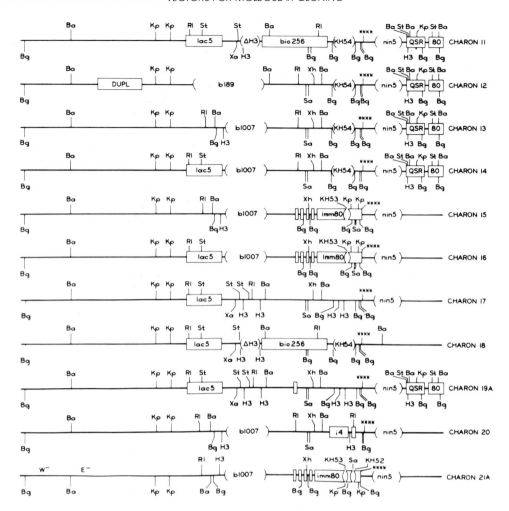

Figure 48. (continued)

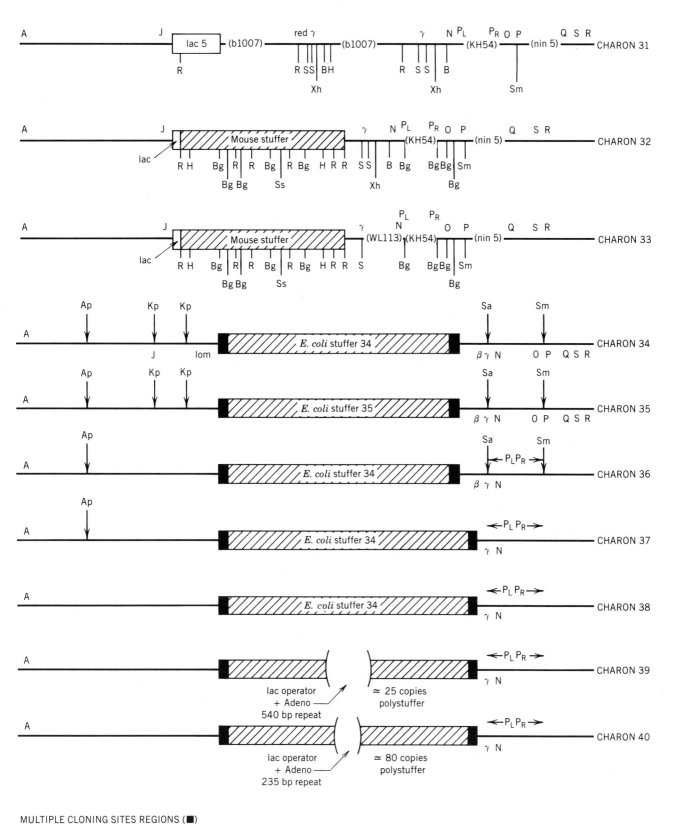

MULTIPLE CLONING SITES REGIONS (■)
■ CH-34/35 = EcoRI, SstI, XbaI, SalI, HindIII, BamHI
■ CH-36 = EcoRI, SstI, KpnI, XbaI, SalI, HindIII, BamHI
■ CH-37 = EcoRI, SstI, KpnI, SmaI, XbaI, SalI, HindIII, BamHI
■ CH-38 = EcoRI, SstI, KpnI, SmaI, XbaI, SalI, HindIII, BamHI
■ CH-39/40 = EcoRI, SstI, KpnI, SmaI, XbaI, SalI, HindIII, XmaIII, NotI, AvrII, SpeI, XhoI, ApaI, BamHI, SfiI, NaeI

Figure 48. (continued)

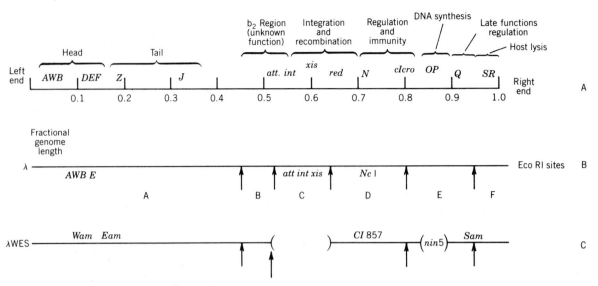

Figure 49. Schematic physical map of a λ WES. The position of the different λ genes (A) and Eco RI sites (B) are given for wild-type λ DNA. Parentheses in λ WES DNA represent deletions. Arrows indicate location of Eco RI sites (A, B, C, D, E, F are DNA fragments generated after Eco RI digestion).

REFERENCES

Brammar, W. J. (1982), in *Genetic Engineering,* R. Williamson, ed., p. 53, Academic Press, New York.

De Wet, J. R., Daniels, D. L., Schroeder, J. L., Williams, B. G., Denniston-Thompson, K., Moore, D. D. and Blattner, F. R. (1980), *J. Virol.,* **33,** 401.

Dunn, I., and Blattner, F. R. (1987), *Nucl. Acids Res.,* **15,** 2677.

Frischauf, A. M., Lehrach, H., Poustka, A., and Murray, N. (1983), *J. Mol. Biol.,* **170,** 827.

Karn, J., Brenner, S., Barnett, L., and Cesareni, G. (1980), *Proc. Natl. Acad. Sci. USA,* **77,** 5172.

Karn, J. (1983), in *Techniques in Life Sciences, Nucleic Acid Biochemistry,* R. A. Flavell, ed., p. 1, Elsevier, Shannon.

Karn, J., Matthes, H. W. D., Gait, M. J. and Brenner, S., (1984), *Gene,* **32,** 217.

Loenen, W. A. M., and Blattner, F. R. (1983), *Gene,* **26,** 171.

Murray, N. E. (1983), in *Lambda II,* R. W. Hendrix, J. W. Roberts, F. W. Sthal, and R. A. Weisberg, eds., p. 395, Cold Spring Harbor Laboratory, Cold Spring Harbor.

Natt, E., and Scherer, G. (1986), *Nucl. Acids Res.,* **14,** 7128.

Rimm, D. L., Horness, D., Kucera, J., and Blattner, F. R. (1980), *Gene,* **12,** 301.

Williams, B. G., and Blattner, F. R. (1980), in *Genetic Engineering,* J. K. Setlow and A. Hollaender, eds., Vol. 2, p. 201, Plenum, New York.

Insertion-Type Vectors. Cloning of foreign DNA in this kind of vector exploits the insertional inactivation of a biological function.

Inactivation of an Immunity Function. The insertion of foreign DNA at a single target within the immunity region destroys the capacity of the phage to produce an active repressor and leads to the isolation of λ recombinants, which give clear plaques; the parental phages give turbid plaques.

Several immunity vectors, such as NM 607, 641, 1149, and 1150 have been used to clone rather small DNA fragments (Murray et al., 1977; Scherer et al., 1981). However, the recovery of recombinant molecules was generally low because in vitro packaging turned out to be more efficient with wild-type λ DNA than with shorter molecules generated by restriction endonucleases for the construction of genomic libraries. The recombinant phage λ gt10 (Huynh et al., 1985) was constructed to overcome the problems related to the size selectivity of in vitro packaging extracts. Because λ gt10 genome is directly packageable, it is necessary to select against the nonrecombinant parental phage when amplifying a library. This is usually performed by plating the mixture of phages on an *E. coli* strains carrying the *high-frequency lysogeny* mutation *hfl*150 (Hoyt et al., 1982). Only the bacteriophages having integrated DNA

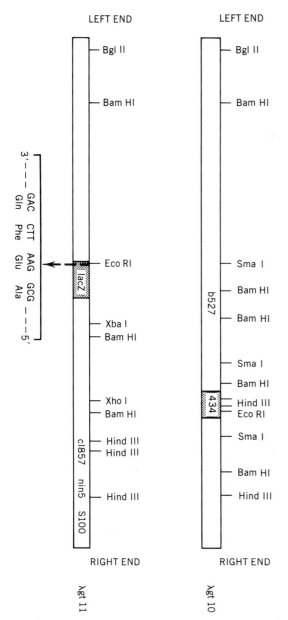

Figure 50. Physical maps of λ gt10 and λ gt11 cloning vectors.

toside) to the medium allows formation of blue plaques on a *lac⁻* indicator bacteria. Insertion of foreign DNA in *lac* 5 will interrupt the coding sequence for β-galactosidase and lead to the formation of colorless plaques. (When a *lac⁺ E. coli* indicator is used, the colors obtained are dark blue for intact *lac* 5 and pale blue for interrupted *lac* 5 function.)

In the expression vector λ gt11 (Young and Davis, 1983), the restriction site used for insertion of foreign DNA is a unique Eco RI site which is located 53 base pairs upstream to the β-galactosidase termination codon in the Lac Z gene (see figure 50). Insertion of DNA at this site results in an inactivation of the β-galactosidase activity. This vector can accommodate DNA fragments with a size of up to 8.3 kb. Recombinant and parental phages are distinguished on the basis of the colored reaction with X-gal. Since λ gt11 contains a temperature sensitive repressor (cI857) and an amber mutation (S100) that renders its lysis defective (Neubauer and Calef, 1970; Goldberg and Howe, 1969), bacteria lysogen for λ gt11 derivatives can be induced by a temperature shift to accumulate large amounts of phage products, in the absence of lysis. Particularly efficient lysogeny is obtained by using strains that carry the *high-frequency lysogeny* mutation (hfl-A150). Upon induction, the cloned products can be expressed as fusion polypeptides containing a β-galactosidase moiety which can bind tightly to diethylaminoethyl cellulose (DEAE) and therefore allow purification of the hybrid molecules. The use of recipient hosts carrying a *lon* mutation, and therefore altered in their protein degradation pathways (Mount, 1980), results in increased amounts of hybrid products accumulated in the induced cells. This system has been widely used to analyze cDNA library expression products, with immunological assays performed directly on nitrocellulose filters (see Chapter 25).

fragments in the repressor gene (cI⁻) can grow on such a host, therefore allowing a positive selection of the recombinants. Cloning in λ gt10 and 11 has been used mainly for the construction of cDNA libraries (see Chapter 19).

Inactivation of E. coli *β-galactosidase.* Several vectors, including λ vectors, contain within their genome the *lac* 5 substitution, which carries the *E. coli* gene for β-galactosidase. Addition of the chromogenic substrate X-Gal (5-Bromo-4-chloro-3-indoly-β-D-galac-

REFERENCES

Goldberg, A. R., and Howe, M. (1969), *Virology,* **38**, 200.

Hoyt, M. A., Knight, D. M., Das, A., Miller, H. I., and Echols, H. (1982), *Cell,* **31**, 565.

Huynh, T. V., Young, R. A., and Davis, R. W. (1985), in *DNA Cloning Techniques: A Practical Approach* D. Glover, ed., p. 49, IRL Press, Oxford.

Mount, D. W. (1980), *Ann. Rev. Genet.,* **14**, 279.

Murray, N. E., Brammar, W. J., and Murray, K. (1977), *Mol. Gen. Genet.,* **150**, 53.

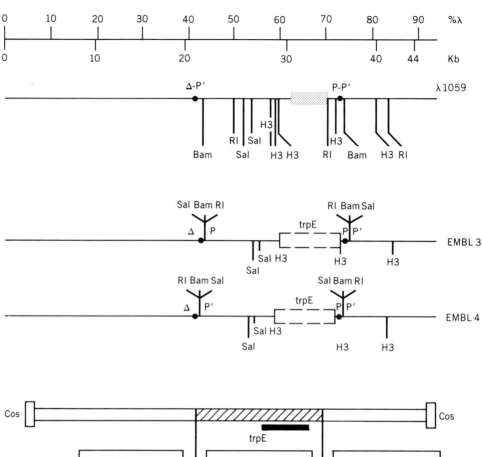

Figure 51. Physical maps of λ 1059, EMBL3 and EMBL4 cloning vectors.

Neubauer, Z., and Calef, E. (1970), *J. Mol. Biol.*, **51**, 1.

Scherer, G., Telford, J., Baldari, L., and Pirotta, V. (1981), *Dev. Biol.*, **86**, 438.

Young, R. A., and Davis, R. W. (1983), *Proc. Natl. Acad. Sci. USA*, **80**, 1194.

Limitations of the Different Vector Systems. The cloning capacity of the different λ vectors varies considerably with respect to the nature of the restriction sites available and the size of the clonable fragments. The properties of several λ vectors are summarized in Tables 32 and 33 and in Figures 48–51.

The minimum size limit for foreign DNA in a vector can be a very powerful indicator of whether cloning will be successful, as the ability to form a plaque is strongly related to the replacement of the dispensable portion of DNA previously removed. Thus, replacement-type vectors are very well adapted for cloning relatively large DNA fragments.

On the other hand, cloning of small fragments requires the use of vectors having no minimum size limit, and is generally achieved by using insertion within a nonessential gene, the function of which can be monitored (see above).

The choice of vector also depends on the availability of restriction sites that allow easy insertion of foreign DNA. A considerable effort has been made by several groups over the past few years to construct λ vectors with many different usable cloning sites. More than 100 different λ vectors have now

Table 32.   Positions of Restriction Sites in λ Charon Phages 1 to 40

Enzyme	Location (bp)	Enzyme	Location (bp)	Enzyme	Location (bp)	Enzyme	Location (bp)
Ch1		*Ch2*		*Ch3*		*Ch4*	
L-end	0	L-end	0	L-end	0	L-end	0
Bgl II	482	Bgl II	482	Bgl II	482	Bgl II	482
Bam HI	5,526	Bam HI	5,526	Bam HI	5,526	Bam HI	5,526
Kpn I	17,290	Kpn I	17,290	Kpn I	17,290	Kpn I	17,290
Kpn I	18,812	Kpn I	18,812	Kpn I	18,812	Kpn I	18,812
Eco RI	19,801	Eco RI	19,801	Eco RI	19,801	Eco RI	19,801
Sst I	20,905	Sst I	20,905	Sst I	20,905	Sst I	20,905
Xba I	25,072	Bam HI	26,923	Xba I	25,072	Xba I	25,072
Sst I	25,336	Xho I	27,228	Sst I	25,336	Sst I	25,336
Hind III	25,737	Bam HI	29,998	Hind III	25,737	Hind III	25,737
Sst I	26,477	Bgl II	31,993	Sst I	26,477	Sst I	26,477
Eco RI	26,693	Xho I	32,749	Eco RI	26,693	Eco RI	26,693
Hind III	28,023	Bgl II	32,875	Hind III	28,023	Hind III	28,023
Bam HI	28,623	Kpn I	35,620	Bam HI	28,623	Bam HI	29,026
Eco RI	32,374	Bgl II	36,351	Bgl II	33,920	Bgl II	32,613
Sal I	33,422	Sal I	36,642	Xho I	34,676	Bgl II	32,747
Sal I	33,934	Sal I	36,795	Bgl II	34,802	Eco RI	34,524
Xho I	34,175	Kpn I	37,430	Kpn I	37,547	Bgl II	36,071
Bam HI	35,179	Bgl II	38,379	Bgl II	38,278	Bgl II	36,373
Bgl II	36,403	****	38,394	Sal I	38,569	Bgl II	37,024
Hind III	37,616	Bam HI	40,441	Sal I	38,722	Bgl II	37,084
Bgl II	38,277	Sst I	40,591	Kpn I	39,357	****	37,438
Eco RI	39,132	Hind III	40,663	Bgl II	40,306	Bam HI	39,485
Hind III	39,377	Bam HI	40,947	****	40,321	Sst I	39,635
Hind III	39,502	Kpn I	42,833	Bam HI	42,368	Hind III	39,677
Bgl II	40,021	Bgl II	42,897	Sst I	42,518	Bam HI	39,991
Bgl II	40,672	Sst I	44,453	Hind III	42,560	Kpn I	41,877
Bgl II	40,732	Bam HI	44,875	Bam HI	42,874	Bgl II	41,941
****	41,086	Bgl II	45,052	Kpn I	44,760	Sst I	43,497
Bam HI	43,133			Bgl II	44,824	Bam HI	43,919
Sst I	43,283	R-end	46,366	Sst I	46,380	Bgl II	44,096
Hind III	43,325			Bam HI	46,802		
Bam HI	43,639			Bgl II	46,979	R-end	45,410
Kpn I	45,525						
Bgl II	45,589			R-end	48,293		
Sst I	47,145						
Bam HI	47,567						
Bgl II	47,744						
R-end	49,058						

been described. Some of them have been commercially available for a few years and have turned out to be quite reliable, making them of great utility in molecular cloning.

Filamentous Bacteriophages as Vectors*. From the numerous isolates of filamentous bacteriophages, three have been well characterized in the last 20 years: fd, f1, and M13. The genomes of these phages have been

sequenced and only minor deviations have been detected (see Van Wezenbeek et al., 1980; Beck and Zink, 1981). They are very similar in their gene functions and molecular mode of propagation (reviewed by Zinder and Horiuchi, 1981; Baas, 1985).

The filamentous bacteriophages infect cells via F-

---

*Courtesy of K. Geider, with permission of Society for General Microbiology.

Table 32 *(continued)*

Enzyme	Location (bp)	Enzyme	Location (bp)	Enzyme	Location (bp)	Enzyme	Location (bp)
Ch5		*Ch6*		*Ch7*		*Ch8*	
L-end	0	L-end	0	L-end	0	L-end	0
Bgl II	482	Bgl II	482	Bgl II	482	Bgl II	482
Bam HI	5,526	Bam HI	5,526	Bam HI	5,526	Bam HI	5,526
Kpn I	23,815	Kpn I	23,815	Kpn I	23,815	Kpn I	17,290
Kpn I	25,337	Kpn I	25,337	Kpn I	25,337	Kpn I	18,812
Bam HI	27,338	Bam HI	27,338	Bam HI	27,338	Eco RI	19,801
Sal I	30,105	Sal I	30,105	Sal I	30,105	Sst I	20,905
Xho I	30,346	Xho I	30,346	Xho I	30,346	Xba I	25,072
Bam HI	31,350	Bam HI	31,350	Bam HI	31,350	Sst I	25,336
Bgl II	32,574	Hind III	33,360	Hind III	33,360	Hind III	25,737
Hind III	33,787	Eco RI	33,596	Eco RI	33,596	Sst I	26,477
Bgl II	34,448	Bgl II	34,499	Bgl II	34,499	Eco RI	26,693
Eco RI	35,303	Bgl II	34,559	Bgl II	34,559	Hind III	28,023
Hind III	35,548	****	34,913	****	34,913	Bam HI	28,623
Bgl II	36,192	Bam HI	36,960			Eco RI	32,374
Bgl II	36,843	Sst I	37,110	R-end	41,550	Sal I	33,422
Bgl II	36,903	Hind III	37,152			Sal I	33,934
****	37,257	Bam HI	37,466			Xho I	34,175
Bam HI	39,304	Kpn I	39,352			Bam HI	35,179
Sst I	39,454	Bgl II	39,416			Hind III	37,189
Hind III	39,496	Sst I	40,972			Eco RI	37,425
Bam HI	39,810	Bam HI	41,394			Bgl II	38,328
Kpn I	41,696	Bgl II	41,571			Bgl II	38,388
Bgl II	41,760					****	38,742
Sst I	43,316	R-end	42,885				
Bam HI	43,738					R-end	45,379
Bgl II	43,915						
R-end	45,229						

pili, although high multiplicities of the phage can also cause rate infections of F⁻ cells (Geider, unpublished). The first mature phages appear within 15 minutes. Phages can also be expected at a similar time after cell transformation with double stranded DNA. Filamentous bacteriophages are not lytic; they coexist with the infected cells for many generations. If the products of cloned genes cause toxic effects in cells they can be inserted into filamentous bacteriophage and thus brought into cells by phage infection at any growth stage of the cells.

The genome of filamentous bacteriophages has a size of 6408 (fd) or 6407 bp (f1, M111), representing nine genes and an intergenic region (IG) Figure 52). A tenth gene is translated from the region of gene 2 coding for the C-terminal end of the protein. Mutation analysis and gene complementation demonstrated the necessity of gene 10 for viral DNA synthesis (Fulford and Model, 1984). Genes 2, 10, 5, 7,

9, and 8 are strongly expressed, increasingly in that order, and gene 8 is followed by a strong terminator that allows some readthrough for the expression of gene 3. Genes 3, 6, 1, and 4 are infrequently expressed. This also occurs in a cascade system, which is efficiently terminated by the intergenic region (Smits et al., 1984) located between gene 4 and gene 2. It has a size of 507 nucleotides and neither codes for known peptides nor for RNA or readthrough transcripts. For phage fd the longest possible reading frame solely located within this region comprises fewer than 200 nucleotides. The intergenic region contains four hairpin-like structures, which are thought to be involved in replication and packaging of the viral genome (Figure 53).

The first major hairpin carries the cleavage site for gene 2 protein (Meyer et al., 1979; Meyer and Geider, 1979), which initiates the synthesis of the viral strand. The second and third hairpins are in-

Table 32 (continued)

Enzyme	Location (bp)	Enzyme	Location (bp)	Enzyme	Location (bp)	Enzyme	Location (bp)
Ch9		*Ch10*		*Ch11*		*Ch12*	
L-end	0	L-end	0	L-end	0	L-end	0
Bgl II	482	Bgl II	482	Bgl II	482	Bgl II	482
Bam HI	5,526	Bam HI	5,526	Bam HI	5,526	Bam HI	5,526
Kpn I	17,290	Kpn I	17,290	Kpn I	17,290	Kpn I	23,815
Kpn I	18,812	Kpn I	18,812	Kpn I	18,812	Kpn I	25,337
Eco RI	19,801	Eco RI	19,801	Eco RI	19,801	Bam HI	26,819
Sst I	20,905	Sst I	20,905	Sst I	20,905	Eco RI	30,570
Xba I	25,072	Sst I	25,336	Xba I	25,072	Sal I	31,618
Sst I	25,336	Hind III	25,739	Sst I	25,336	Sal I	32,130
Hind III	25,737	Bam HI	26,742	Hind III	25,739	Xho I	32,371
Sst I	26,477	Bgl II	30,329	Bam HI	26,742	Bam HI	33,375
Eco RI	26,693	Bgl II	30,463	Bgl II	30,329	Bgl II	34,599
Hind III	28,023	Eco RI	32,240	Bgl II	30,463	Bgl II	34,901
Bam HI	28,623	Bgl II	33,787	Eco RI	32,240	Bgl II	35,552
Eco RI	32,374	Bgl II	34,089	Bgl II	33,787	Bgl II	35,612
Sal I	33,422	Bgl II	34,740	Bgl II	34,089	****	35,966
Sal I	33,934	Bgl II	34,800	Bgl II	34,740 .	Bam HI	38,013
Xho I	34,175	****	35,154	Bgl II	34,800	Sst I	38,163
Bam HI	35,179			****	35,154	Hind III	38,205
Bgl II	36,403	R-end	41,791	Bam HI	37,201	Bam HI	38,519
Hind III	37,616			Sst I	37,351	Kpn I	40,405
Bgl II	38,277			Hind III	37,393	Bgl II	40,469
Eco RI	39,132			Bam HI	37,707	Sst I	42,025
Hind III	39,377			Kpn I	39,593	Bam HI	42,447
Hind III	39,502			Bgl II	39,657	Bgl II	42,624
Bgl II	40,021			Sst I	41,213		
Bgl II	40,672			Bam HI	41,635	R-end	43,938
Bgl II	40,732			Bgl II	41,812		
****	41,086						
				R-end	43,126		
R-end	47,723						

volved in the start of complementary strand synthesis (Geider et al., 1976). The primer for initiation of the complementary strand is synthesized at the beginning of the second hairpin, but binding of the initiating RNA polymerase (Geider and Kornberg, 1974) also protects the third hairpin (Schaller et al., 1976). Mutations in this region cause a drastic drop in replication efficiency (Cleary and Ray, 1981; Kim et al., 1981; Geider et al., 1985). Initiation of complementary strand synthesis may then occur at other sites on the plasmid. Ray and coworkers (1982) constructed an M13 vector with deletions of sequences from the complementary strand origin. This phage forms small plaques and can be used to select for inserts with an initiation signal for complementary strand synthesis (Nomura et al., 1982; Strathearn et al., 1984). The fourth hairpin has been implicated in packaging (Schaller, 1979; Dotto et al., 1981; Dotto and Zinder, 1983). To obtain phage formation, the orientation of the morphogenetic signal, not its proximity to the viral strand origin, is important.

The infecting viral strand is prepared for DNA replication by an RNA primer of 26 to 30 nucleotides synthesized by the host RNA polymerase (Geider et al., 1976). It is then converted into double stranded DNA by *E. coli* DNA polymerase III holoenzyme. The primer RNA is removed by RNase H and the 5'-3' exonuclease activity of DNA polymerase I, the gap is filled by incorporation by the same polymerase, the strand is sealed by *E. coli* DNA ligase, and the DNA is then converted into the supercoiled form, RFI, by DNA gyrase (*E. coli* topoisomerase II). This form is used for activation of viral strand replication

Table 32 *(continued)*

Enzyme	Location (bp)	Enzyme	Location (bp)	Enzyme	Location (bp)	Enzyme	Location (bp)
Ch13		*Ch14*		*Ch15*		*Ch16*	
L-end	0	L-end	0	L-end	0	L-end	0
Bgl II	482	Bgl II	482	Bgl II	482	Bgl II	482
Bam HI	5,526	Bam HI	5,526	Bam HI	5,526	Bam HI	5,526
Kpn I	17,290	Kpn I	17,290	Kpn I	17,290	Kpn I	17,290
Kpn I	18,812	Kpn I	18,812	Kpn I	18,812	Kpn I	18,812
Eco RI	21,569	Eco RI	19,801	Eco RI	21,569	Eco RI	19,801
Bam HI	22,703	Sst I	20,905	Bam HI	22,703	Sst I	20,905
Bgl II	22,784	Eco RI	27,642	Bgl II	22,784	Bgl II	29,188
Hind III	23,513	Sal I	28,690	Hind III	23,513	Xho I	29,944
Eco RI	27,354	Sal I	29,202	Bgl II	28,898	Bgl II	30,070
Sal I	28,402	Xho I	29,443	Xho I	29,654	Kpn I	32,815
Sal I	28,914	Bam HI	30,447	Bgl II	29,780	Bgl II	33,546
Xho I	29,155	Bgl II	31,671	Kpn I	32,525	Sal I	33,837
Bam HI	30,159	Bgl II	31,973	Bgl II	33,256	Sal I	33,990
Bgl II	31,383	Bgl II	32,624	Sal I	33,547	Kpn I	34,624
Bgl II	31,685	Bgl II	32,684	Sal I	33,700	Bgl II	35,574
Bgl II	32,336	****	33,038	Kpn I	34,335	****	35,589
Bgl II	32,396	Bam HI	35,085	Bgl II	35,284		
****	32,750	Sst I	35,235	****	35,299	R-end	42,226
Bam HI	34,797	Hind III	35,277				
Sst I	34,947	Bam HI	35,591	R-end	41,939		
Hind III	34,989	Kpn I	37,477				
Bam HI	35,303	Bgl II	37,541				
Kpn I	37,189	Sst I	39,097				
Bgl II	37,253	Bam HI	39,519				
Sst I	38,809	Bgl II	39,696				
Bam HI	39,231						
Bgl II	39,408	R-end	41,010				
R-end	40,722						

by the phase gene 2 protein. Viral strands are replicated in the rolling circle mode and circularized via cleavage of the single stranded tails and strand sealing by gene 2 protein (Meyer and Geider, 1982). Gene 2 of filamentous bacteriophage has been cloned into pBR plasmids (Meyer and Geider, 1981), thus increasing the level of the protein far beyond that normally found in phage-infected cells. A further increase of gene 2 protein in cells was obtained by expressing the gene under the control of the strong, inducible *areB* promoter of *Salmonella typhimurium* (Johnston et al., 1985). Plasmids having the replication origin of phage fd seem to stimulate the expression of cloned gene 2 (Meyer and Geider, 1981). Gene 5 protein prevents initiation of complementary strand synthesis by binding to the single stranded tails of the rolling circles. At the bacterial membrane these complexes are converted into phage particles which contain the proteins encoded by genes 3, 6, 7, 8, and 9 (Grant and Webster, 1984). Viral gene 1 protein and cellular thioredoxin may interact in the assembly process (Russel and Model, 1985). The hairpin nearest to gene 4 in the intergenic region is positioned opposite the attachment protein (gene 3 protein) and binds to the minor coat proteins encoded by gene 7 and gene 9 (Armstrong et al., 1983). The mature phages are extruded through the cell envelope without cell lysis; membrane adhesion sites of the infected cells seem to be preferred for the extrusion process (Bayer and Bayer, 1986).

If propagated at high multiplicities, filamentous bacteriophages produce high levels of smaller phage particles, called miniphages (Griffith and Kornberg, 1974; Enea and Zinder, 1975). Their genomes contain the whole intergenic region with parts of the

Table 32 *(continued)*

Enzyme	Location (bp)	Enzyme	Location (bp)	Enzyme	Location (bp)	Enzyme	Location (bp)
Ch17		*Ch18*		*Ch19*		*Ch20*	
L-end	0	L-end	0	L-end	0	L-end	0
Bgl II	482	Bgl II	482	Bgl II	482	Bgl II	482
Bam HI	5,526	Bam HI	5,526	Bam HI	5,526	Bam HI	5,526
Kpn I	17,290	Kpn I	17,290	Kpn I	17,290	Kpn I	17,290
Kpn I	18,812	Kpn I	18,812	Kpn I	18,812	Kpn I	18,812
Eco RI	19,801	Eco RI	19,801	Eco RI	19,801	Eco RI	21,569
Sst I	20,905	Sst I	20,905	Sst I	20,905	Bam HI	22,703
Xba I	25,072	Xba I	25,072	Xba I	25,072	Bgl II	22,784
Sst I	25,336	Sst I	25,336	Sst I	25,336	Hind III	23,513
Hind III	25,737	Hind III	25,739	Hind III	25,737	Eco RI	27,354
Sst I	26,477	Bam HI	26,742	Sst I	26,477	Sal I	28,402
Eco RI	26,693	Bgl II	30,329	Eco RI	26,693	Sal I	28,914
Hind III	28,023	Bgl II	30,463	Hind III	28,023	Xho I	29,155
Bam HI	28,623	Eco RI	32,240	Bam HI	28,623	Bam HI	30,159
Sal I	33,422	Bgl II	33,787	Sal I	33,422	Hind III	32,169
Sal I	33,934	Bgl II	34,089	Sal I	33,934	Eco RI	32,405
Xho I	34,175	Bgl II	34,740	Xho I	34,175	Bgl II	33,308
Bam HI	35,179	Bgl II	34,800	Bam HI	35,179	Bgl II	33,368
Bgl II	36,403	****	35,154	Bgl II	36,403	****	33,722
Hind III	37,616	BamHI	37,769	Hind III	37,616		
Hind III	38,201			Hind III	38,201	R-end	40,359
Bgl II	38,845	R-end	44,656	Bgl II	38,845		
Bgl II	39,496			Bgl II	39,496		
Bgl II	39,556			Bgl II	39,556		
****	39,910			****	39,910		
				Bam HI	41,957		
R-end	46,547			Sst I	42,107		
				Hind III	42,149		
				Bam HI	42,463		
				Kpn I	44,349		
				Bgl II	44,413		
				Sst I	45,969		
				Bam HI	46,391		
				Bgl II	46,568		
				R-end	47,882		

surrounding genes 2 and 4 and can also be used as cloning vectors. Sequences from the intergenic region can occur amplified in these miniphages (Enea et al., 1977) and also in intact filamentous bacteriophage (Schaller, 1979). Replication of miniphages depends on the presence of a helper phage in the same cell, because they lack complete genes. Construction of cloning vectors derived from filamentous bacteriophage demands insertion in nonessential regions without inactivation of the genes required for genome propagation. The intergenic region is quite suitable for artificial DNA inserts, al-

though they may interfere with packaging functions or DNA replication. The favored sites for inserts are therefore between the packaging signal and the complementary strand replication origin (Figure 53). Other inserts have been located in the first hairpin of the complementary strand origin (Boeke et al., 1979; Barnes and Bevan, 1983) just after the termination sequence of the RNA primer (Geider et al., 1976). Inserts in the second complementary strain hairpin (Geider et al., 1985) or deletion of this part (Cleary and Ray, 1981) reduce the copy number of the genome as mentioned above. Insertions into the

(text continued on page 258)

Table 32 *(continued)*

Enzyme	Location (bp)	Enzyme	Location (bp)	Enzyme	Location (bp)	Enzyme	Location (bp)
Ch21		*Ch27*		*Ch28*		*Ch30*	
L-end	0	L-end	0	L-end	0	L-end	0
Bgl II	482	Bgl II	482	Bgl II	472	Bgl II	472
Bam HI	5,526	Kpn I	17,290	Hpa I	747	Hpa I	747
Kpn I	17,290	Kpn I	18,812	Ava I	4,826	Ava I	4,826
Kpn I	18,812	L-vl	19,700	Hpa I	5,326	Hpa I	5,326
Eco RI	21,569	Eco RI	21,569	Hpa I	5,771	Hpa I	5,771
Bam HI	22,703	Bam HI	22,703	Hpa I	6,015	Hpa I	6,015
Bgl II	22,784	Bgl II	22,784	Hpa I	8,262	Hpa I	8,262
Hind III	23,513	Hind III	23,513	Bcl I	8,894	Bcl I	8,894
Bgl II	28,898	Bgl II	28,898	Bcl I	9,397	Bcl I	9,397
Xho I	29,654	Xho I	29,654	Hpa I	11,713	Hpa I	11,713
Bgl II	29,780	Bgl II	29,780	Pvu I	12,055	Pvu I	12,055
Kpn I	32,525	Kpn I	32,525	Bcl I	13,998	Bcl I	13,998
Bgl II	33,256	Bgl II	33,256	Kpn I	17,288	Kpn I	17,288
Sal I	33,547	Sal I	33,547	Kpn I	18,790	Kpn I	18,790
Kpn I	34,182	R-vl	33,552	L-vl	19,600	L-vl	19,600
Bgl II	35,131	Kpn I	34,182	Sma I	19,660	Sma I	19,660
****	35,146	Bgl II	35,131	Sst II	20,597	Sst II	20,597
		****	35,146	Sst II	20,812	Sst II	20,812
R-end	41,783			Ava I	21,293	Ava I	21,293
		R-end	41,783	Sst II	22,817	Eco RI	21,524
				Hpa I	23,101	Sst II	21,917
				Bam HI	23,567	Hpa I	22,201
				Bgl II	23,656	Bam HI	22,667
				Hind III	24,362	Bgl II	22,756
				Sma I	28,083	Hind III	23,462
				Eco RI	28,212	Sma I	27,183
				Hpa I	28,279	Eco RI	27,312
				Hpa I	28,680	Hpa I	27,379
				Bcl I	29,198	Hpa I	27,780
				Sal I	29,235	Bcl I	28,298
				Sal I	29,739	Sal I	28,335
				Xho I	29,999	Sal I	28,839
				Bam HI	30,996	Xho I	29,099
				Hpa I	31,763	Bam HI	30,096
				Bgl II	32,215	Bgl II	30,184
				Pvu I	32,281	Hind III	30,890
				R-vl	32,296	Sma I	34,611
				Bgl II	32,511	Eco RI	34,740
				Ava I	32,622	Hpa I	34,807
				Bgl II	33,162	Hpa I	35,208
				Bgl II	33,222	Bcl I	35,726
				****	33,576	Sal I	35,763
				Hpa I	34,016	Sal I	36,267
				Hpa I	34,243	Xho I	36,527
				Sma I	34,297	Bam HI	37,524
				Sst II	34,829	Hpa I	38,291
				Bcl I	35,308	Bgl II	38,743
				Bcl I	38,077	Pvu I	38,809
				Bcl I	39,646	R-vl	38,852
				R-end	40,229	Bgl II	39,039
						Ava I	39,150
						Bgl II	39,690

Table 32 *(continued)*

Enzyme	Location (bp)	Enzyme	Location (bp)	Enzyme	Location (bp)	Enzyme	Location (bp)
Ch30 (continued)		*Ch32*		*Ch32 (continued)*		*Ch32 (continued)*	
Bgl II	39,750	L-end	1	Ava I	4,390	Xmn I	11,909
****	40,104			Bst XI	4,475	Aha III	11,964
Hpa I	40,544	Eco RI	1	Ava III	4,482	Hind II	12,071
Hpa I	40,771	Hpa I	63	Nde I	4,534	Pvu II	12,083
Sma I	40,825	Hind II	63	Hind II	4,724	Bgl I	12,282
Sst II	41,357	Cla I	245	Xho II	4,840	Bgl II	12,287
Bcl I	41,836	Pst I	263	Bgl II	4,930	Xho II	12,287
Bcl I	44,605	Hpa I	473	Xho II	4,930	Afl III	12,329
Bcl I	46,174	HindII	473	Bgl II	4,990	Mst I	12,337
		Pst I	510	Xho II	4,990	Afl III	12,499
R-end	46,757	Bgl I	583	Aha III	5,011	Eco RV	12,526
		Nru I	663	Eco RV	5,532	Hpa I	12,606
		Aha III	959	Afl III	5,570	Hind II	12,606
		Bcl I	983	Sph I	5,598	Xmn I	13,027
		Sal I	999	Xho II	5,752	Bal I	13,200
		Acc I	1,000	Hpa I	5,784	Xho II	13,478
		Hind II	1,001	Hind II	5,784	Pvu II	13,791
		Afl III	1,017	Hpa I	6,012	Eco RV	13,960
		Sca I	1,055	Hind II	6,012	Acc I	14,063
		Cla I	1,218	Xma I	6,064	Bal I	14,080
		Ava III	1,220	Ava I	6,064	Sph I	14,088
		Stu I	1,253	Stu I	6,170	Xmn I	14,195
		Sal I	1,488	Bst EII	6,225	Pvu II	14,259
		Acc I	1,499	Nde I	6,308	Mst I	14,377
		Hind II	1,500	Acc I	6,378	Pvu II	14,400
		Hgi A	1,725	Hgi A	6,396	Xho II	14,403
		Xho I	1,752	Sst II	6,565	Pst I	14,432
		Ava I	1,752	Hgi A	6,669	Bgl I	14,538
		Cla I	1,839	Bcl I	7,052	Pst I	14,696
		Eco RV	1,845	Afl III	7,131	Bst XI	14,734
		Nde I	1,934	Cla I	7,195	Pvu II	14,932
		Ava III	1,935	Hgi A	7,551	Bal I	15,134
		Xmn I	2,069	Nco I	7,618	Pst I	15,501
		Xmn I	2,443	Afl III	7,870	Pvu II	15,511
		Ava III	2,461	Xmn I	8,101	Pst I	15,516
		Mst II	2,573	Hgi A	8,220	Bgl I	15,676
		Asu II	2,586	Aat II	8,937	Pst I	15,732
		Bam I	2,753	Aat II	8,966	Bal I	16,067
		Xho II	2,753	Nar I	9,050	Cla I	16,071
		Bst XI	2,857	Xmn I	9,115	Mst I	16,144
		Cla I	2,951	Eco RV	9,200	Bgl I	16,238
		Mst I	3,077	Bcl I	9,736	Pst I	16,246
		Cla I	3,305	Cla I	9,809	Bgl I	16,329
		Hpa I	3,515	Bst XI	9,811	Bgl I	16,455
		Hind II	3,515	Hgi A	10,072	Nru I	16,464
		Hgi A	3,841	Afl III	10,351	Pst I	16,585
		Hind II	3,869	Aha III	10,801	Ava I	16,592
		Bgl II	3,965	Xmn I	10,938	Pst I	16,785
		Xho II	3,965	Hgi A	11,034	Aat II	16,981
		Pvu I	4,044	Xho II	11,143	Pst I	16,996
		Hgi A	4,113	Hind II	11,308	Mst I	17,029
		Hind II	4,165	Bcl I	11,312	Pst I	17,090
		Bgl II	4,279	Hind II	11,668	Bgl I	17,124
		Xho II	4,279	*Cos*	11,872		

Table 32 *(continued)*

Enzyme	Location (bp)	Enzyme	Location (bp)	Enzyme	Location (bp)	Enzyme	Location (bp)
Ch32 *(continued)*		Ch32 *(continued)*		Ch32 *(continued)*		Ch33/34/35	
Apa I	17,141	Bst XI	22,794	Pst I	28,107	L-end	1
Hind II	17,141	Bgl I	22,936	Aha III	28,168		
Bgl I	17,310	Tth I	23,077	Sca I	28,292	Bam HI	1
Afl III	17,419	Aat II	23,119	Rru I	28,295	Xho II	1
Hgi A	17,495	Afl III	23,152	Hgi A	28,392	Hind III	7
Pst I	17,558	Mst I	23,437	Xmn I	28,785	Pst I	20
Bst EII	17,559	Hpa I	23,457	Kpn I	28,929	Sal I	22
Hpa I	17,582	Hind II	23,457	Hind II	28,948	Acc I	23
Hind II	17,582	Mst I	23,564	Pst I	29,266	Xba I	28
Hgi A	17,878	Pst I	23,639	Bgl I	29,516	Cla I	37
Bgl I	17,931	Pst I	23,711	Eco RV	29,643	Xma I	43
Bgl I	17,982	Pvu I	23,808	Afl III	29,662	Ava I	43
Xho II	18,294	Hgi AI	23,826	Mlu I	29,663	Sst I	54
Bal I	18,370	Sph I	23,878	Bst EII	29,813	Hgi A	54
Afl II	18,412	Pvu II	23,973	Bst XI	29,908	Eco RI	56
Eco RV	18,557	Pvu II	24,036	Asu II	29,921	Hpa I	118
Bst XI	18,585	Stu I	24,308	Bgl I	29,963	Cla I	300
Bal I	18,751	Afl II	24,490	Afl III	30,155	Pst I	318
Mst I	18,853	Bgl I	24,586	Eco RV	30,261	Hpa I	528
Bst EII	18,930	Bgl I	24,595	Kpn I	30,432	Pst I	565
Bgl I	19,428	Bgl I	24,710	Sca I	30,555	Bgl I	638
Bal I	19,458	Xmn I	24,978	Rru I	30,558	Nru I	718
Pvu II	19,705	Bgl I	25,076	Hind II	30,628	Mst I	994
Hpa I	19,822	Bst XI	25,142	Acc I	30,707	Aha III	1,014
Hind II	19,822	Hgi A	25,165	Nco I	31,201	Bcl I	1,038
Bal I	19,852	Bst EII	25,220	Bgl I	31,212	Sal I	1,054
Bgl I	19,927	Mst I	25,229	Pvu II	31,339	Acc I	1,055
Bal I	19,930	Eco RI	25,311			Afl III	1,072
Eco RV	19,960	Bst EII	25,444	Mouse stuffer		Sca I	1,110
Hpa I	20,073	Bst EII	25,561	Eco RI	31,343	Rru I	1,113
Hind II	20,073	Hind II	25,657	Hind III	31,849	Cla I	1,273
Bst EII	20,194	Bcl I	25,692	Bgl II	34,347	Ava III	1,275
Bst XI	20,292	Bal I	25,808	Bgl II	34,997	Stu I	1,308
Aha III	20,334	Eco RV	25,899	Eco RI	35,528	Xho II	1,557
Xmn I	20,366	Pst I	26,170	Bgl II	35,651	Bst X	1,661
Pst I	20,396	Bst XI	26,217	Eco RI	36,681	Cla I	1,755
Eco RV	20,698	Pst I	26,257	Bgl II	37,928	Mst I	1,881
Bcl I	20,716	Bgl I	26,279	Sst I	38,862	Cla I	2,109
Bst XI	20,729	Hgi A	26,350	Eco RI	39,303	Hpa I	2,319
Bal I	20,733	Bgl I	26,768	Bgl II	40,523	Hgi A	2,645
Bst EII	20,896	Bal I	26,777	Hind III	42,391	Bgl II	2,769
Hind II	20,928	Aat II	26,850	Eco RI	42,870	Xho II	2,769
Bcl I	21,233	Hpa I	26,865	Eco RI	43,802 =	Pvu I	2,848
Aat II	21,270	Hind II	26,865	Eco RI	1	Hgi A	2,917
Hgi A	21,361	Bgl I	27,035			Bgl II	3,083
Pst I	21,489	Hgi A	27,087			Xho II	3,083
Hind II	21,498	Acc I	27,133			Ava I	3,194
Pst I	21,653	Afl III	27,243			Bst X	3,279
Apa I	21,962	Cla I	27,456			Ava III	3,286
Xmn I	21,987	Bst EII	27,884			Nde I	3,338
Hgi A	22,171	Mst I	27,920			Xho II	3,644
Ava III	22,196	Pvu II	27,952			Bgl II	3,734
Bal I	22,483	Pst I	27,957			Xho II	3,734
Bal I	22,651	Cla I	27,993			Bgl II	3,794

Table 32 *(continued)*

Enzyme	Location (bp)	Enzyme	Location (bp)	Enzyme	Location (bp)	Enzyme	Location (bp)
Ch33/34/35 (continued)		*Ch33/34/35 (continued)*		*Ch33/34/35 (continued)*		*Ch33/34/35 (continued)*	
Xho II	3,794	Hpa I	11,410	Afl II	17,216	Stu I	23,112
Aha III	3,815	Xmn I	11,831	Eco RV	17,361	Hge II	23,188
Eco RV	4,336	Bal I	12,004	Bst XI	17,389	Afl II	23,294
Afl III	4,374	Xho II	12,282	Bal I	17,555	Bgl I	23,390
Sph I	4,402	Pvu II	12,595	Mst I	17,657	Bgl I	23,399
Bvu I	4,437	Eco RV	12,764	Bst EII	17,734	Bgl I	23,514
Xho II	4,556	Acc I	12,867	Bgl I	18,232	Xmn I	23,782
Hpa I	4,588	Bal I	12,884	Bal I	18,262	Bgl I	23,880
Hpa I	4,816	Sph I	12,892	Pvu II	18,509	Bst XI	23,946
Xma I	4,868	Xmn I	12,999	Hpa I	18,626	Hgi AI	23,969
Ava I	4,868	Pvu II	13,063	Bal I	18,656	Bst EII	24,024
Stu I	4,974	Mst I	13,181	Bgl I	18,731	Mst I	24,033
Bst EII	5,029	Pvu II	13,204	Bal I	18,734	Eco RV	24,115
Nde I	5,112	Xho II	13,207	Eco RV	18,764	Gdi II	24,161
Acc I	5,182	Pst I	13,236	Hpa I	18,877	Hgi AI	24,172
Hgi A	5,200	Bgl I	13,342	Bst EII	18,998	Bst EII	24,248
Sst II	5,369	Pst I	13,500	Bst XI	19,096	Bst EII	24,365
Hgi A	5,473	Bst XI	13,538	Hgi CI	19,117	Bcl I	24,496
Bcl I	5,856	Pvu II	13,736	Aha III	19,138	Bal I	24,612
Afl III	5,935	Bal I	13,938	Xmn I	19,170	Eco RV	24,703
Cla I	5,999	Pst I	14,305	Pst I	19,200	Pst I	24,974
Hgi A	6,355	Pvu II	14,315	Hgi CI	19,440	Bst XI	25,021
Nco I	6,422	Pst I	14,320	Eco RV	19,502	Pst I	25,061
Afl III	6,674	Bgl I	14,480	Bcl I	19,520	Bgl I	25,083
Xmn I	6,905	Pst I	14,536	Bst XI	19,533	Hgi AI	25,154
Hgi A	7,024	Bal I	14,871	Bal I	19,537	Bgl I	25,572
Aat II	7,741	Cla I	14,875	Bst EII	19,700	Bal I	25,581
Aat II	7,770	Mst I	14,948	Bcl I	20,037	Aat II	25,654
Nar I	7,854	Bgl I	15,042	Aat II	20,074	Hpa I	25,669
Xmn I	7,919	Pst I	15,050	Hgi AI	20,165	Bgl I	25,839
Eco RV	8,004	Bgl I	15,133	Pst I	20,293	Hgi AI	25,891
Bcl I	8,540	Bgl I	15,259	Pst I	20,457	Sna I	25,935
Cla I	8,613	Nru I	15,268	Bvu I	20,766	Acc I	25,937
Bst XI	8,615	Pst I	15,389	Apa I	20,766	Afl III	26,047
Hgi A	8,876	Ava I	15,396	Xmn I	20,791	Mlu I	26,048
Afl III	9,155	Pst I	15,589	Hgi AI	20,975	Cla I	26,260
Aha III	9,605	Aat II	15,785	Ava III	21,000	Bst EII	26,688
Xmn I	9,742	Pst I	15,800	Bal I	21,287	Mst I	26,724
Hgi A	9,838	Mst I	15,833	Bal I	21,455	Pvu II	26,756
Xho II	9,947	Pst I	15,894	Bst XI	21,598	Pst I	26,761
Bcl I	10,116	Bgl I	15,928	Bgl I	21,740	Cla I	26,797
Cos	10,676	Hpa I	15,945	Tth I	21,881	Pst I	26,911
Xmn I	10,713	Bgl I	16,114	Aat II	21,923	Aha III	26,972
Aha III	10,768	Afl III	16,223	Afl III	21,956	Sca I	27,096
Pvu II	10,887	Mlu I	16,224	Mst I	22,241	Rru I	27,099
Bgl I	11,086	Hgi A	16,299	Hpa I	22,261	Hgi AI	27,196
Bgl II	11,091	Pst I	16,362	Mst I	22,368	Xmn I	27,589
Xho II	11,091	Bst EII	16,363	Pst I	22,443	Hgi CI	27,729
Afl III	11,133	Hpa I	16,386	Pst I	22,515	Kpn I	27,733
Mlu I	11,134	Hgi A	16,682	Pvu I	22,612	Pst I	28,070
Mst I	11,141	Bgl I	16,735	Hgi AI	22,630	Bgl I	28,320
Bvu I	11,261	Bgl I	16,786	Sph I	22,682	Eco RV	28,447
Afl III	11,303	Xho II	17,098	Pvu II	22,777	Afl III	28,466
Eco RV	11,330	Bal I	17,174	Pvu II	22,840	Mlu I	28,467

Table 32 *(continued)*

Enzyme	Location (bp)	Enzyme	Location (bp)	Enzyme	Location (bp)	Enzyme	Location (bp)
Ch33/34/35 (continued)		*Ch33/34/35 (continued)*		*Ch40*		*Ch40 (continued)*	
Bst EII	28,617	Sal I	39,117	Left arm		Dra I	8,462
Bst XI	28,712	Bgl II	43,082			Ssp I	8,471
Asu II	28,725	Hind III	44,358	L-end	1	Xmn I	8,494
Bgl I	28,767	Sal I	44,384	Xmn I	37	Eco RV	8,824
Afl III	28,959	R-end	46,602	Dra I	92	Bcl I	8,844
Eco RV	29,065	Stuffer 35		Pvu II	211	Bst XI	8,857
Kpn I	29,236	(*E. coli*)		Bgl II	415	Bal I	8,861
Sca I	29,359	Bam HI	30,202	Mlu I	458	Dra I	9,004
Rru I	29,362	Eco RI	30,295	Eco RV	652	Bst EII	9,024
Sna I	29,509	Bgl II	30,928	Hpa I	734	Bcl I	9,361
Acc I	29,511	Bgl II	31,399	Xmn I	1,155	Aat II	9,398
Nco I	30,005	Xma I	33,022	Bal I	1,328	Xmn I	10,115
Bgl I	30,016	Eco RI	33,264	Bsp II	1,826	Esp I	10,298
Pvu II	30,143	Bam HI	33,762	Pvu II	1,919	Nsi I	10,329
Eco RI	30,147	Eco RI	34,673	Eco RV	2,086	Bal I	10,611
Sst I	30,157	Bgl II	34,868	Bal I	2,208	Esp I	10,683
Hgi AI	30,157	Sst I	35,653	Sph I	2,216	Bal I	10,779
Bvu I	30,157	Sst I	36,408	Xmn I	2,323	Bst XI	10,922
Xma I	30,160	Sal I	37,120	Pvu II	2,387	Tth I	11,205
Ava I	30,160	Xma I	37,932	Pvu II	2,528	Aat II	11,247
Cla I	30,168	Bam HI	39,469	Bst XI	2,862	Hpa I	11,585
Xba I	30,175	Bgl II	40,430	Dra III	2,959	Esp I	11,662
Sal I	30,181	Sal I	40,493	Pvu II	3,060	Pvu I	11,936
Acc I	30,182	Sal I	43,777	Bal I	3,262	Sph I	12,006
Pst I	30,191	Sal I	44,750	Pvu II	3,639	Pvu II	12,101
Hind III	30,196	Sal I	45,272	Rsr II	3,801	Pvu II	12,164
Bam HI	30,202	Sal I	45,512	Bal I	4,195	Sna B	12,190
Xho II	30,202	R-end	45,769	Cla I	4,199	Stu I	12,436
Mouse stuffer				Nru I	4,592	Xmn I	13,106
Eco RI	30,147			Aat II	5,109	Bst XI	13,270
Hind III	30,653			Hpa I	5,269	Bst EII	13,348
Bgl II	33,151			Mlu I	5,548	Eco RV	13,437
Bgl II	33,801			Dra III	5,618	Bst EII	13,572
Eco RI	34,332			Bst EII	5,687	Bst EII	13,689
Bgl II	34,455			Hpa I	5,710	Bcl I	13,820
Eco RI	35,485			Rsr II	6,042	Bal I	13,936
Bgl II	36,732			Bal I	6,498	Rsr II	13,984
Sst I	37,666			Dra III	6,640	Eco RV	14,025
Eco RI	38,107			Eco RV	6,683	Bst XI	14,345
Bgl II	39,327			Bst XI	6,713	Dra III	14,482
Hind III	41,195			Bal I	6,879	Bal I	14,905
Eco RI	41,674			Bst EII	7,058	Aat II	14,978
Eco RI	42,606			Bal I	7,586	Hpa I	14,993
Eco RI	56,000			Pvu II	7,833	Mlu I	15,372
Stuffer 34				Hpa I	7,950	Cla I	15,584
(*E. coli*)				Bal I	7,980	Bst EII	16,012
Bam HI	30,202			Eco RV	8,086	Bsp II	16,040
Sal I	32,304			Hpa I	8,201	Pvu II	16,080
Xma I	33,220			Bst EII	8,322	Cla I	16,121
Xho I	34,527			Bst XI	8,420	Dra I	16,296
Xba I	38,423					Sca I	16,423

Table 32 *(continued)*

Enzyme	Location (bp)	Enzyme	Location (bp)	Enzyme	Location (bp)
Ch40 (continued)		*Ch40 (continued)*		*Ch40 (continued)*	
Esp I	16,519	Pvu II	19,759	Hpa I	23,502
Xmn I	16,913	Nco I	19,766	Hpa I	23,730
Eco RV	17,781	Polystuffer		Stu I	23,888
Mlu I	17,803			Bst EII	23,943
Bst EII	17,953	Polylinker		Nde I	24,026
Bst XI	18,048	Nae I	19,820	Sst II	24,283
Asu II	18,061	Stu I	19,826	Bcl I	24,770
Eco RV	18,399	Sfi I	19,832	Cla I	24,913
Sca I	18,698	Bam HI	19,843	Nco I	25,336
		Apa I	19,853	Xmn I	25,819
Left arm Polylinker		Nar I	19,856	Aat II	26,655
Eco RI	19,180	Bbe I	19,859	Aat II	26,684
Sst I	19,190	Nco I	19,861	Nar I	26,768
Asp I	19,192	Nhe I	19,867	Bbe I	26,771
Xma I	19,196	Xho I	19,873	Xmn I	26,833
Kpn I	19,196	Spe I	19,879	Eco RV	26,916
Sma I	19,198	Avr II	19,885	Bcl I	27,454
Cla I	19,204	Xma III	19,892	Cla I	27,527
Xba I	19,211	Not I	19,892	Bst XI	27,529
Sal I	19,217	Hind III	19,899	Dra III	28,405
Sph I	19,233	Sph I	19,909	Dra I	28,519
Hind III	19,235	Sal I	19,917	Xmn I	28,656
Xma III	19,242	Xba I	19,923	Bcl I	29,030
Not I	19,242	Cla I	19,932	Dra III	29,527
Avr II	19,249	Xma I	19,938	Ppu I	29,562
Spe I	19,255	Sma I	19,940	R-end	29,590
Xho I	19,261	Asp I	19,942		
Nhe I	19,267	Kpn I	19,946		
Nco I	19,273	Sst I	19,952		
Nar I	19,280	Eco RI	19,954		
Bbe I	19,283	Polylinker			
Apa I	19,289				
Bam HI	19,291	Right arm			
Sfi I	19,309	Sca I	20,027		
Stu I	19,312	Cla I	20,187		
Nae I	19,318	Nsi I	20,194		
		Stu I	20,222		
Polylinker		Bst XI	20,575		
		Nhe I	20,651		
Polystuffer		Cla I	20,669		
Nco I	19,333	Cla I	21,023		
Tth I	19,380	Hpa I	21,233		
Sst II	19,406	Bcl II	21,683		
Xma III	19,444	Pvu I	21,762		
Xma III	19,473	Bcl II	21,997		
Nae I	19,477	Bst XI	22,193		
Pvu II	19,520	Nsi I	22,205		
Nco I	19,527	Nde I	22,252		
Nco I	19,572	Bcl II	22,648		
Tth I	19,619	Bcl II	22,708		
Sst II	19,645	Dra I	22,729		
Xma III	19,683	Eco RV	23,248		
Xma III	19,712	Sph I	23,316		
Nae I	19,716	Esp I	23,345		

Table 33.   Cloning Capacities of the λ Charon Phages 1 to 40

		Cloning Capacity (kb)	
Phage	Restriction Enzyme(s)	Maximum Size	Minimum Size
Ch1	Eco RI	20.3	8.3
	Xba I	0.9	0
	Sal I	1.5	0
	Xho I	0.9	0
	Xba I–Sal I	9.8	0
	Xba I–Xho I	10.0	0
	Sal I–Xho I	1.7	0
Ch2	Eco RI	3.6	0
	Xho I	3.6	0
	Sal I	3.8	0
	Eco RI–Xho I	16.6	4.6
Ch3	Eco RI	8.6	0
	Xba I	1.7	0
	Xho I	1.7	0
	Sal I	1.9	0
	Eco RI–Xho I	16.6	4.6
	Eco RI–Sal I	20.6	8.6
	Xba I–Xho I	11.3	0
	Xba I–Sal I	15.4	3.4
	Xba I–Sal I	5.8	0
Ch4	Eco RI	19.3	7.3
	Xba I	4.6	0
Ch5	Sal I	4.8	0
	Xho I	4.8	0
	Eco RI	4.8	0
	Sal I–Xho I	5.0	0
	Sal I–Eco RI	10.0	0
	Xho I–Eco RI	9.7	0
Ch6	Sal I	7.1	0
	Xho I	7.1	0
	Eco RI	7.1	0
	Sal I–Xho I	7.4	0
	Sal I–Eco RI	10.6	0
	Xho I–Eco RI	10.4	0
Ch7	Sal I	8.5	0
	Xho I	8.5	0
	Eco RI	8.5	0
	Sal I–Xho I	8.7	0
	Sal I–Eco RI	12.0	0
	Xho I–Eco RI	11.7	0
	Hind III	8.5	0
	Sal I–Hind III	11.7	0
	Xho I–Hind III	11.5	0
	Hind III–Eco RI	8.7	0
Ch8	Eco RI	22.2	10.2
	Sst I	10.2	0
	Xba I	4.6	0
	Hind III	16.1	4.1
	Sal I	5.1	0
	Xho I	4.6	0
	Sst I–Hind III	20.9	8.9
	Sst I–Sal I	17.6	5.6
	Sst I–Xho I	17.9	5.9

Table 33. **Cloning Capacities of the λ Charon Phages 1 to 40** (*continued*)

Phage	Restriction Enzyme(s)	Cloning Capacity (kb)	
		Maximum Size	Minimum Size
	Xba I–Hind III	16.7	4.7
	Xba I–Sal I	13.5	1.5
	Xba I–Xho I	13.7	1.7
	Sal I–Xho I	5.4	0
Ch9	Eco RI	21.6	9.6
	Sst I	7.8	0
	Xba I	2.3	0
	Hind III	15.9	3.9
	Sst I	2.8	0
	Xho I	2.3	0
	Eco RI–Hind III	21.9	9.9
	Sst I–Sal I	15.3	3.3
	Sst I–Xho I	15.5	3.5
	Sst I–Hind III	20.7	8.7
	Xba I–Sal I	11.1	0
	Xba I–Xho I	11.4	0
	Xba I–Hind III	16.6	4.6
	Sal I–Xho I	3.0	0
Ch10	Eco RI	20.6	8.6
	Sst I	12.6	0.6
	Xba I	8.2	0
	Hind III	8.2	0
	Sst I–Hind III	13.0	1.0
	Xba I–Hind III	8.9	0
Ch11	Eco RI	19.3	7.3
	Xba I	6.9	0
Ch12	Eco RI	6.1	0
	Sal I	6.6	0
	Xho I	6.1	0
	Eco RI–Sal I	7.6	0
	Eco RI–Xho I	7.9	0
	Sal I–Xho I	6.8	0
Ch13	Eco RI	15.1	3.1
	Sal I	9.8	0
	Xho I	9.3	0
	Eco RI–Sal I	16.6	4.6
	Eco RI–Xho I	12.9	4.9
	Sal I–Xho I	10.0	0
Ch14	Eco RI	16.8	4.8
	Sal I	9.5	0
	Xho I	9.0	0
	Eco RI–Sal I	18.4	6.5
	Eco RI–Xho I	18.6	6.6
	Sal I–Xho I	9.7	0
Ch15	Eco RI	8.1	0
	Hind III	8.1	0
	Xho I	8.1	0
	Sal I	8.2	0
	Eco RI–Hind III	10.0	0
	Eco RI–Xho I	16.1	4.1
	Eco RI–Sal I	20.2	8.2
	Hind III–Xho I	14.2	2.2
	Hind III–Sal I	18.3	6.3

Table 33.   Cloning Capacities of the λ Charon Phages 1 to 40 (*continued*)

Phage	Restriction Enzyme(s)	Cloning Capacity (kb)	
		Maximum Size	Minimum Size
	Xho I–Sal I	12.1	0.1
Ch16	Eco RI	7.8	0
	Sst I	7.8	0
	Xho I	7.8	0
	Sal I	7.9	0
	Eco RI–Sst I	8.9	0
	Eco RI–Xho I	17.9	5.9
	Eco RI–Sal I	22.0	10.0
	Sst I–Xho I	16.8	4.8
	Sst I–Sal I	20.9	8.9
	Xho I–Sal I	11.8	0
Ch17	Eco RI	10.3	0
	Sst I	9.0	0
	Xba I	3.5	0
	Hind III	15.9	3.9
	Sal I	4.0	0
	Xho I	3.5	0
	Eco RI–Sal I	17.6	5.6
	Eco RI–Xho I	17.8	5.8
	Sst I–Hind III	20.7	8.7
	Sst I–Sal I	16.5	4.5
	Sst I–Xho I	16.7	4.7
	Xba I–Hind III	16.6	4.6
	Xba I–Sal I	12.3	0.3
	Xba I–Xho I	12.6	0.6
	Sal I–Xho I	4.2	0
	Eco RI–Hind III	21.9	9.9
Ch18	Eco RI	17.8	5.8
	Sst I	9.8	0
	Xba I	5.3	0
	Hind III	5.3	0
	Sst I–Hind III	10.2	0
	Xba I–Hind III	6.0	0
Ch19	Eco RI	9.0	0
	Xba I	2.1	0
	Sal I	2.6	0
	Xho I	2.1	0
	Eco RI–Sal I	16.3	4.3
	Eco RI–Xho I	16.5	4.5
	Xba I–Sal I	11.0	0
	Xba I–Xho I	11.2	0
	Sal I–Xho I	2.9	0
Ch20	Eco RI	20.5	8.5
	Hind III	18.3	6.3
	Sal I	10.2	0
	Xho I	9.6	0
	Sal I–Xho I	10.4	0
Ch21	Eco RI	8.2	0
	Hind III	8.2	0
	Xho I	8.2	0
	Sal I	8.2	0
	Eco RI–Hind III	10.2	0
	Eco RI–Xho I	16.3	4.3

Table 33. Cloning Capacities of the λ Charon Phages 1 to 40 (*continued*)

Phage	Restriction Enzyme(s)	Cloning Capacity (kb)	
		Maximum Size	Minimum Size
	Eco RI–Sal I	20.2	8.2
	Hind III–Xho I	14.4	2.4
	Hind III–Sal I	18.3	6.3
	Xho I–Sal I	12.1	0.1
Ch27	Bam HI	9.217	0
	Hind III	9.217	0
	Eco RI	9.217	0
	Sal I	9.217	0
	Xho I	9.217	0
	Bam HI–Hind III	10.027	0
	Eco RI–Bam HI	10.351	0
	Bam HI–Sal I	20.061	7.061
	Bam HI–Xho I	16.168	3.168
	Eco RI–Hind III	11.161	0
	Hind III–Sal I	19.251	6.251
	Hind III–Xho I	15.358	2.358
	Eco RI–Sal I	21.195	8.195
	Eco RI–Xho I	17.302	4.302
	Xho I–Sal I	13.110	0.110
Ch28	Bam HI	18.200	5.200
	Hind III	10.771	0
	Eco RI	10.771	0
	Sal I	11.275	0
	Xho I	10.771	0
	Hind III–Eco RI	14.621	1.621
	Hind III–Sal I	16.148	3.148
	Hind III–Xho I	16.408	3.408
	Eco RI–Sal I	12.298	0
	Eco RI–Xho I	12.558	0
	Sal I–Xho I	11.535	0
Ch30	Bam HI	19.100	6.100
	Hind III	11.671	0
	Eco RI	17.459	4.459
	Sal I	12.175	0
	Xho I	11.671	0
	Eco RI–Bam HI	20.243	7.243
	Hind III–Sal I	17.048	4.048
	Hind III–Xho I	17.308	4.308
	Eco RI–Sal I	18.986	5.986
	Eco RI–Xho I	19.246	6.246
	Sal I–Xho I	12.435	0
	Eco RI	19	9
	Eco RI–Sal I		
Ch32	Sal I–Xho I		
	Bam HI–Eco RI		
Ch33	Same as Ch32	20	9
Ch34/35	Eco RI	22.915	7.915
	Bam HI		
	Sst I		
	Hind III		
	Xba I		
	Sal I		
	Eco RI–Sal I		

Table 33. Cloning Capacities of the λ Charon Phages 1 to 40 (*continued*)

Phage	Restriction Enzyme(s)	Cloning Capacity (kb)	
		Maximum Size	Minimum Size
	Sst I–Sal I		
	Xba I–Sal I		
Ch36	Xba I	22.915	7.915
	Eco RI		
	Bam HI		
	Sst I		
	Hind III		
	Kpn I		
	Sal I		
Ch37	Eco RI	23.896	8.896
	Sst I		
	Kpn I		
	Sma I		
	Xba I		
	Sal I		
	Hind III		
	Bam HI		
Ch38	Eco RI	24.207	9.207
	Sst I		
	Kpn I		
	Sma I		
	Xba I		
	Sal I		
	Hind III		
	Bam HI		
Ch39/40	Eco RI	24.207	9.207
	Sst I		
	Kpn I		
	Sma I		
	Xba I		
	Sal I		
	Hind III		
	Xma III		
	Not I		
	Avr II		
	Spe I		
	Xho I		
	Apa I		
	Bam HI		
	Sfi I		
	Nae I		

replication origin of the viral strand or deletion in this region have been employed for detailed analysis of this region. The viral strand origin can be divided into (a) a core sequence of about 40 nucleotides with the gene 2 protein recognition sequence, which seems to coincide with the termination signal, and (b) a larger area for viral strand initiation (Dotto et al., 1981, 1984). Deletion mutants have defined the core region as being from 12 nucleotides in the 5′ direction from the gene 2 protein cleavage site (Dotto et al., 1982) to 29 nucleotides in the 3′ direction (Dotto and Zinder, 1984a; Kim and Ray, 1985). An area of about 100 nucleotides adjacent to the core region enhances viral strand initiation. The effects of disruptions in this area can be suppressed by structural changes in viral gene 2 protein (Dotto and

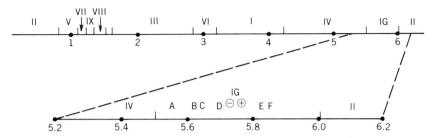

Figure 52. Map of the filamentous phage genome, location of inverted DNA in various filamentous vectors. The upper line represents the genetic map of phage f1: genes are identified by roman numerals and the intergenic (IG) region by capital letters. The second line shows a fivefold expansion of the IG region: DNA has been inserted at positions A–F to make various filamentous cloning vectors (Table 35). Circled − and + denote the positions of the origins of complementary and viral strand DNA synthesis, respectively. Vertical bars denote gene boundaries. Arabic numerals specify kb. [With permission of N. Zinder. Copyright (1982) Elsevier Biomedical Press.]

Zinder, 1984a; Kim and Ray, 1985) or by higher intracellular levels of the protein (Dotto and Zinder, 1984b); the latter can be caused by mutation in viral gene 5 protein or by changes in the gene 2 mRNA leader. Gene 5 protein from filamentous bacteriophage is not only a repressor of complementary strand synthesis, but also a regulator for the translation of gene 2 (Model et al., 1982). Owing to their single stranded genome, filamentous bacteriophages cannot correct errors in the base sequence, which may occur in the virus particle. On the other hand, the various mechanisms of suppression of gene 2 mutations indicate an ability to overcome distortions in virus propagation.

Insertions of DNA into the genome of a filamentous bacteriophage result in large phage particles, which can eventually become fragile and less effective in DNA replication. Early attempts to create restriction sites for insertion into filamentous bacteriophage DNA were carried out with phage M13 (Messing et al., 1977). The introduction of a DNA fragment with an indicator of insertion has rendered M13 a more convenient cloning vector. This was achieved by insertion of a short stretch of the lac operon (Gronenborn and Messing, 1978). The insert, located on an HaeII fragment, comprises an early segment of the lacZ gene which produces a protein fragment, the α-protein, which is able to complement a defect at the N terminus of β-galactosidase. To obtain complementation with such an M13 vector, the *E. coli* strain must have an appropriate mutation in the lacZ gene, for instance the ΔM15 deletion. Complementation is visualized as blue plaques on plates containing 5-bromo-4-chloro-3-indolyl-β-D-galactopyranoside (X-gal) for color production and isopropyl-β-D-thiogalactopyranoside (IPTG) for

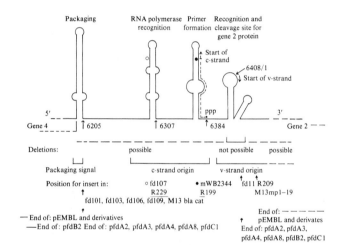

Figure 53. Use of the intergenic region of filamentous bacteriophages for the construction of cloning vectors. The approximate sites of insertions are given for phage vectors. For pFd plasmids and pBR-ori vectors, the borders are indicated (with permission of G. Geider, courtesy of Society for General Microbiology).

gene induction (Messing, 1983). Insertion of foreign DNA into the lac Z′ gene, which carries a polylinker region, disrupts the coding sequence for the α-protein. Consequently, these phages give rise to white plaques. Phages without insertions or some "in frame" additions retain the blue plaque phenotype. The blue plaque phenotype for the latter insertions can be obtained if the DNA fragment provides a ribosome binding site and an initiation codon distal to the lac promoter of the lacZ′ segment (Close et al., 1983). Development of the M13 cloning system has created phages with numerous unique cloning sites (Norrander et al., 1983; Yanisch-digestion. "In frame" inserts have been expressed as inducible fu-

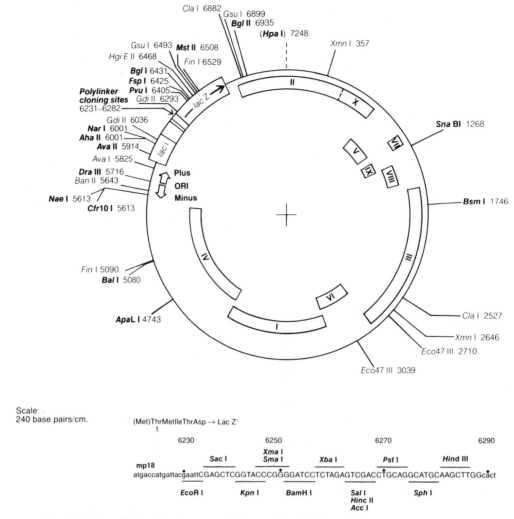

Figure 54. Simplified restriction map of M13mp18 RF DNA (courtesy of New England Biolabs).

sion proteins with the N terminus of the α-protein and a modified human interferon α gene (Fuke et al., 1984) or parts of the α-protein with an adapter sequence joined to the synthetic presequence for human proinsulin (Georges et al., 1984) or with α-protein and interferon α2 (Slocombe et al., 1982). Further, the vector M13mp7 has been changed in such a way that in three versions the HindIII site can be used to translate an insert in phase with the N terminus of β-galactosidase (Lathe et al., 1984) (Figure 54, Table 35).

Visualization of *lac* complementation can only be achieved when the host cell has a suitable genetic background. The first host strains and M13 vectors were designed on the basis of biological safety requirements; indeed, suppressor functions of the bacterium can occasionally be desirable for restriction

of the propagation of the viral genome to those cells. But since survival of artificially created potential pathogens and their growth potential outside laboratory conditions seem to be very low, this restrictive genetic background is no longer required, although one should be aware of the existence of mutant phages depending on host suppressors. On the other hand, M13mp8-*his* phages defective in gene 2 were used to introduce promoter mutations into a deficient *his* operon of an F⁺ *S. typhimurium* strain that did not suppress the viral amber mutation (Artz et al., 1983). A further improvement of *E. coli* hosts for M13 phages was the transfer of the gene with the lacZ′ deletion to an F episome together with the *proAB* genes and with the *lacI* gene which constitutively expresses the lac repressor. To prevent recombination with the *lac* region of the chromosome,

Table 34. Examples of Host Strains Used for Propagation of Filamentous Phage Vectors

Strain	Genotype	Reference[a]
JM83	*ara*, Δ*(lac-proAB)*, *strA*, φ80d*lacZ*ΔM15	1
JM101	*supE*, Δ*(lac-proAB)*, *thi*, [F', *traD36*, *proAB*, *lacI^qZ*ΔM15]	1
JM107	*hsdR17*, *supE*, Δ*(lac-proAB)*, *endA1*, *gyrA96*, *relAl*, *thi*, [F', *traD36*, *proAB*, *lacI^qZ*ΔM15]	1
JM109	*recA1*, *hsdR17*, *supE*, Δ*(lac-proAB)*, *endA1*, *gyrA96*, *relA1*, *thi*, [F', *traD36*, *proAB*, *lacI^qZ*ΔM15]	1
JM110	*dam*, *dcm*, *rpsL*, *thr*, *leu*, *supE44*, Δ*(lac-proAB)*, *lacY*, *galK*, *relA1*, *galT*, *ara*, *tonA*, *tsx*, *thi*, [F', *traD36*, *proAB*, *lacI^qZ*ΔM15]	1
BHB2600	*supE*, *supF* lysogenic for λ CH616[b]	2
1101	F⁺ *supE*	3
KB35	F⁺ *supE*	4

Strains 1101 and KB35 (F donors) can be used directly for the transfer of the F episome by conjugation with F⁻ strains (recipient); the JM strains require complementation of the *traD* gene for F transfer. To verify the F⁻ character, cells from single colonies are plated to a lawn and droplets of male-specific DNA phages (fd, f1, M13) or RNA phage (fr, f2, Qβ, R17, MS2) are applied to the surface, or they are infected in culture with a filamentous phage carrying an antibiotic resistance gene for 30 min and then plated on selective agar. The JM strains can also be screened on minimal plates.

Source: Courtesy of K. Geider, with permission of Society for General Microbiology.

[a]1, Yanisch-Peron *et al.* (1985); 2. Geider *et al.* (1985); 3. Collection of H. Hoffmann-Berling, Heidelberg; 4. Collection of N. Zinder, New York.

[b]λ NM616 with an insert of fd gene 2.

this chromosomal area (*lac-pro*AB) was deleted. The presence of the F episome can be detected by screening for proline prototrophy on minimal agar plates. Furthermore, this specific F' episome cannot be spontaneously transferred to other strains because of its *traD* mutation. Strain JM101 has the *pro-lac* deletion and a suppressor for amber mutations (Table 34). An additional feature of strain JM107 is the hspR17 gene, which produces a deficient host restriction endonuclease ($r_{K12}-$) without affecting the corresonding methylase ($m_{K12}+$). This genetic background is important for cloning DNA that is not derived from *E. coli* K12 strains, because the host-specific restriction system would otherwise degrade "foreign" inserts. To avoid spontaneous deletions of DNA, particularly of inserts with repetitive sequences, the recA mutation was introduced (JM109). Since cleavage by some restriction enzymes is abolished by the methylation of certain bases, genes encoding methylases (dam, dcm) were mutated in another strand (JM110). There are also strains such as JM83 carrying the lacZ M15 deletion in the chromosome in a lysogenic defective ø 80 phage (Table 35).

Other filamentous bacteriophages like fd or fl can also be used as cloning vectors (Table 36). Indicator functions for phage fd vectors were mainly based on insertion of genes from pACYC- or pBR-based plasmids such as those for kanamycin, ampicillin, or chloramphenicol resistance. The genes were mostly inserted into the double HaeII site of the intergenic region at positions 6187/6195, or a HaeIII site was converted into an EcoRI site without the introduction of a resistance gene (Herrmann et al., 1980). Double resistance allows inactivation of one gene and screening for the intact one. There are also fl phages with artificial EcoRI sites (Boeke et al., 1979; Boeke, 1981) or with ampicillin–tetracycline double resistance (Boeke et al., 1979). After transformation of competent cells, some time should be allowed (15 to 30 minutes) for the phage to establish in the cell before the cells are plated on selective agar. Longer incubation will result in the onset of phage production. Expansion of the genome by in-

Table 35. Filamentous Phage DNA Cloning Vectors

Strain	Parent(s)	Useful Restriction Sites	Additional Phenotypes[a]	Propagation[b]
R199	fl	Eco RI		P
R209	fl	Eco RI		P
R229	fl	Eco RI		P
CGF1	fl R229	Xho I		P
CGF2	fl R229	Sal I		P
CGF3	fl R229	Pst I		P
CGF4	fl R229	Hind III		P
CGF8	fl R229	Sst I (Sac I)		P
R208	fl R199	Hind III, Pst I, Sal I	Ap^r, TC^r	C
mp2	M13, pMG1106	Eco RI	Lac	P/C
mp5	M13, pMG1106	Hind III	Lac	P/C
mp7	M13, pMG1106	Eco RI, Hind III, Pst I	Lac	P/C
mp8/9	M13, pMG1106	Eco RI, Bam HI, Pst I, Hind III, Sal I	Lac	P/C
mp18/19	M13, pMG1106	Eco RI, SacI, Kpn I, Xma I, Sma I Bam HI, Xba I, Sal I, HincII Acc I, Pst I, Sph I, Hind III	Lac	P/C
M13Hol76	M13, λh80dhis	Eco RI, Sal I, Pst I		C
M13Goril	M13, G4	Eco RI, Sst I, Xho I, Kpn I		P
M13bla	M13, Tn3	Pst I	Ap^r	C
M13bla/cat	M13, Tn3, pACYC184	Eco RI, Pst I	Ap^r, Cm^r	C
fdtet	fd, Tn10	Eco RI, Hind III	Tc^r	C
fKN16	fd-tet	Eco RI, Hind III	Tc^r	C
fd11	fd, pKB252	Eco RI		P
fd101	fd, pACYC177	Pst I, Hind III, Sma I	Ap^r, Km^r	C
fd103	fd, pACYC184/177	Pst I, Eco RI	Ap^r, Cm^r	C
fd104	fd, pACYC177	Xho I, Hind III, Sma I	Km^r	C
fd106	fd, pACYC184/177	Eco RI, Xho I, Hind III, Sma I	Cm^r, Km^r	C
fd107	fd, pBR322	Pst I, Hind III, Sal I, Eco RI	Ap^r	C
fd109	fd, pBR322	Pst I, Hind III, Eco RI	Ap^r	C

[a]Ap^r, ampicillin resistance; Km^r, kanamycin resistance; Lac, *lac* promoter–operator; Tc^r, tetracycline resistance.

[b]P, phage are propagated by plaquing; C, phage may be propagated as colonies (infected cells).

clusion of DNA fragments conferring antibiotic resistance is disadvantageous for further DNA insertion, because very large genomes have a tendency to undergo spontaneous deletion, although some vector constructs appear to propagate large DNA inserts stably in certain strains (Barnes and Bevan, 1983). The rolling circle mode of replication may be unfavorable for transposon insertion, as Tn5 has been found to be rather labile in the fd genome (Hermann et al., 1978).

The packaging signal of bacteriophage fd is located in a hairpin structure in the intergenic region near gene 4 (Schaller, 1979; Dotto et al., 1981; Dotto and Zinder, 1983). Insertion of sequences with this signal into a pfdA plasmid allows ssDNA of the vector to be packaged in the presence of a helper phage (Geider et al., 1985). Spontaneously derived miniphages with antibiotic resistances can also be used as cloning and packaging vectors in cells lysogenic

for a λ phage carrying fd gene 2 (Beck, Herrmann, and Geider, unpublished). Successful infection of cells by a helper phage can be selected via drug resistance (Geider et al., 1985). Miniphages do not rely only on the presence of the helper phage, but also interfere with the propagation of the helper. In the barely visible plaques interference-resistant phages arise and produce a normal plaque morphology (Dotto and Horiuchi, 1981). Studies with a complete and with a partially deleted viral replication origin (Zagursky and Berman, 1984) suggest competition for gene 2 protein in the amplification of each genome.

Mixed plasmids with the ColEl replication origin and part of the intergenic region from filamentous bacteriophage including the packaging signal have been described (Dente et al., 1983; Levinson et al., 1984; Zagursky and Berman, 1984; Peeters et al., 1986; see Table 35). They can stably propagate long

Table 36. Restriction Fragment Compatibility for M13mp Cloning

Cloning Site in M13mp Vector		Compatible Restriction Fragments	
Enzyme	Cut Site	Enzyme	Cut Site
Acc I	GT ↓ CGAC	HinP I	G ↓ CGC
		Hpa II/Msp I	C ↓ CGG
		Mae II	A ↓ CGT
		Taq I	T ↓ CGA
		Aha II	GPu ↓ CGPyC
		Asu II	TT ↓ CGAA
		Cla I	AT ↓ CGAT
		Nar I	GG ↓ CGCC
BamH I	G ↓ GATCC	Mbo I/Sau3A I	↓ GATC
		BamH I	G ↓ GATCC
		BclI	T ↓ GATCA
		BglII	A ↓ GATCT
		Xho II	Pu ↓ GATCPy
EcoR I	G ↓ AATTC	EcoR I	G ↓ AATTC
		EcoR I star	↓ AATT
Hinc II	GTC ↓ GAC	Any blunt ended fragment	
Hind III	A ↓ AGCTT	Hind III	A ↓ AGCTT
Kpn I	GGTAC ↓ C	Kpn I	GGTAC ↓ C
Pst I	CTGCA ↓ G	Nsi I	ATGCA ↓ T
		Pst I	CTGCA ↓ G
Sac I	GAGCT ↓ C	Sac I	GAGCT ↓ C
Sal I	G ↓ TCGAC	Sal I	G ↓ TCGAC
		Xho I	C ↓ TCGAG
Sma I	CCC ↓ GGG	Any blunt ended fragment	
Sph I	GCATG ↓ C	Nla III	CATG ↓
		Nsp 7524 I	PuCATG ↓ Py
		Sph I	GCATG ↓ C
Xba I	T ↓ CTAGA	Avr II	C ↓ CTAGG
		Nhe I	G ↓ CTAGC
		Spe I	A ↓ CTAGT
		Xba I	T ↓ CTAGA
Xma I	C ↓ CCGGG	Xma I	C ↓ CCGGG

Source: Courtesy of New England Biolabs.

inserts of DNA due to their ColEl replication mode, and they are efficiently packaged upon infection with an interference-resistant helper phage. This type of fl ori-pBR plasmid has been further developed, and shuttle vectors (pEMBLY) for cloning and packaging in *E. coli* and subsequent transformation of yeast cells have been constructed (Baldari and Cesareni, 1985). Other vectors express cloned DNA via phage SP6 promoters (Mead et al., 1985; Sollazo et al., 1985) or phage λ promoters (Lorenzetti et al., 1985; Sollazo et al., 1985), or they can be used for cDNA cloning by the Okayama–Berg procedure (Kowalski et al., 1985). Packaging of both strands of an insert can be done with vectors carrying parts of the intergenic region of phage M13 and of phage Ike (Peeters et al., 1986). The latter is a filamentous bacteriophage which infects *E. coli* cells carrying plasmids of the I and N incompatibility group and whose gene 2 protein and packaging proteins recognize only its own intergenic region and not that of phage M13. Joint cloning of the intergenic region fragments of each phage in opposite orientations on a pBR plasmid and superinfection with one phage activate only the corresponding packaging signal.

REFERENCES

Armstrong, J., Hewitt, J. A., and Perham, R. N. (1983), *EMBO J.,* **2**, 1641.

Artz, S., Holzschu, D., Blum, P., and Shand, R. (1983), *Gene,* **26**, 147.

Baas, P. D. (1985), *Biochim. Biophys. Acta,* **825**, 111.

Baldari, C., and Cesareni, G. (1985), *Gene,* **35**, 27.

Barnes, W. M., and Bevan, M. (1983), *Nucl. Acids Res.,* **11**, 349.

Bayer, M. E., and Bayer, M. H. (1986), *J. Virol.,* **57**, 258.

Beck, E. and Zink, B. (1981), *Gene* **16**, 35.

Boeke, J. D. (1981), *Mol. Gen. Genet.,* **181**, 288.

Boeke, J. D., Vovis, G. F. and Zinder, N. D. (1979), *Proc. Natl. Acad. Sci. USA,* **76**, 2699.

Cleary, J. M. and Ray, D. S. (1981), *J. Virol.,* **40**, 197.

Close, T. J., Christman, J. L. and Rodriguez, R. L. (1983), *Gene,* **23**, 131.

Dente, L., Cesarini, G., and Cortese, R. (1983), *Nucl. Acids Res.,* **11**, 1645.

Dotto, G. P. and Horiuchi, K. (1981), *J. Mol. Biol.,* **153**, 169.

Dotto, G. P., and Zinder, N. D. (1983), *J. Virol.,* **130**, 252.

Dotto, G. P. and Zinder, N. D. (1984a), *Nature,* **311**, 279.

Dotto, G. P., and Zinder, N. D. (1984b), *Proc. Natl. Acad. Sci. USA,* **81**, 1336.

Dotto, G. P., Enea, V. and Zinder, N. D. (1981), *Virology,* **114**, 463.

Dotto, G. P., Horiuchi, K., Jakes, K., and Zinder, N. D. (1982), *J. Mol. Biol.,* **162**, 335.

Dotto, G. P., Horiuchi, K., and Zinder, N. D. (1984), *J. Mol. Biol.,* **172**, 507.

Enea, V., and Zinder, N. D. (1975), *Virology,* **68**, 105.

Enea, V., Horiuchi, K., Turgeon, B. G., and Zinder, N. D. (1977), *J. Mol. Biol.,* **111**, 395.

Fuke, M., Hendrix, L. C., and Bollon, A. P. (1984), *Gene,* **32**, 135.

Fulford, W., and Model, P. (1984), *J. Mol. Biol.,* **178**, 137.

Geider, K., and Kornberg, A. (1974), *J. Biol. Chem.,* **249**, 3999.

Geider, K., Beck, E., and Schaller, H. (1976), *Proc. Natl. Acad. Sci. USA,* **75**, 645.

Geider, K., Hohmeyer, C., Haas, R., and Meyer, T. F. (1985), *Gene,* **33**, 341.

Georges, F., Brousseau, R., Michniewicz, J., Prefontaine, G., Stawinski, J., Sung, W., Wu, R., and Narang, S. A. (1984), *Gene,* **27**, 201.

Grant, R., and Webster, R. E. (1984), *Virology,* **133**, 315.

Griffith, J., and Kornberg, A. (1974), *Virology,* **59**, 139.

Gronenborn, B. and Messing, J. (1978), *Nature,* **272**, 375.

Herrmann, R., Neugebauer, K., Zentgraf, H. and Schaller, H. (1978), *Mol. Gen. Genet.,* **159**, 171.

Herrmann, R., Neugebauer, K., Pirkl, E., Zentgraf, H., and Schaller, H. (1980), *Mol. Gen. Genet.,* **177**, 231.

Johnston, S., Lee, J. H., and Ray, D. S. (1985), *Gene,* **34**, 137.

Kim, M. H . and Ray, D. S. (1985), *J. Virol.,* **53**, 871.

Kim, M. H., Hines, J. C., and Ray, D. S. (1981), *Proc. Natl. Acad. Sci. USA,* **78**, 6784.

Kowalski, J., Smith, J. H., Ng, N., and Denhardt, D. T. (1985), *Gene,* **35**, 45.

Lathe, R. F., Kieny, M. P., Schmitt, D., Curtis, P., and Lecocq, J. P. (1984), *J. Mol. Appl. Genet.,* **2**, 331.

Levinson, A., Silver, D., and Seed, B. (1984), *J. Mol. Appl. Genet.,* **2**, 507.

Mead, D. A., Skorupa, E. S., and Kemper, B. (1985), *Nucl. Acids Res.,* **13**, 1103.

Messing, J. (1983) *Methods Enzymol.,* **101**, 20.

Messing, J., Gronenborn, B., Müller-Hill, B., and Hofschneider, P. H. (1977), *Proc. Natl. Acad. Sci. USA,* **74**, 3642.

Meyer, T. F. and Geider, K. (1979), *J. Biol. Chem.,* **252**, 12642.

Meyer, T. F., and Geider, K. (1981), *Proc. Natl. Acad. Sci. USA,* **78**, 5416.

Meyer, T. F., and Geider, K. (1982), *Nature,* **296**, 828.

Meyer, T. F., Geider, K., Kurz, C., and Schaller, H. (1979), *Nature,* **278**, 365.

Model, P., McGill, C., Mazur, B., and Fulford, W. D. (1982), *Cell,* **29**, 329.

Nomura, N., Low, R. L., and Ray, D. S. (1982), *Gene,* **18**, 239.

Norrander, J., Kempe, T., and Messing, J. (1983), *Gene,* **26**, 101.

Peeters, B. P. H., Schoenmakers, J. G. G., and Konings, R. N. H. (1986), *Gene,* **41**, 39.

Ray, D. S., Hines, J. C., Kim, M. H., Imber, R., and Nomura, N. (1982), *Gene,* **18**, 231.

Russel, M., and Model, P. (1985), *Proc. Natl. Acad. Sci. USA,* **82**, 29.

Schaller, H. (1979), *Cold Spring Harbor Symp. Quant. Biol.,* **43**, 401.

Schaller, H., Uhlmann, A., and Geider, K. (1976), *Proc. Natl. Acad. Sci. USA,* **73**, 49.

Slocombe, P., Easton, A., Boseley, P., and Burke, D. C. (1982), *Proc. Natl. Acad. Sci. USA,* **79**, 5455.

Smits, M. A., Jansen, J., Konings, R. N. H. and Schoenmakers, J. G. G. (1984), *Nucl. Acids Res.,* **12**, 4071.

Sollazzo, M., Frank, R., and Cesareni, G. (1985), *Gene,* **37**, 199.

Strathearn, M. D., Low, R. L., and Ray, D. S. (1984), *J. Virol.,* **49**, 178.

Van Wezenbeek, P. M. G. F., Hulsebos, T. J. M ., and Schoenmakers, J. G. G. (1980), *Gene,* **11**, 129.

Yanisch-Perron, C., Vieira, J., and Messing, J. (1985), *Gene,* **33**, 103.

Viera, J., and Messing, J. (1987), *Methods in enzymology* in press

Zagursky, R. J., and Berman, M. L. (1984), *Gene,* **27**, 183.

Zinder, N. D., and Horiuchi, K. (1985), *Microbiol. Rev.,* **49**, 101.

COSMID VECTORS

The Use of Cosmids as Cloning Vehicles*

The basic scheme for cosmid cloning that has originally been developed by Collins and Hohn (1978) and Collins and Brüning (1978) is shown in Figure 55. Cosmid cloning is carried out by restriction of the vector DNA at a single site and ligation to large fragments of foreign DNA to form long concatemeric structures which mimic the natural packaging substrate. By packaging the ligated concatemeric mixture and subsequent transduction, the molecules containing two cohesive end sites in the same orientation separated by 37–50 kb of DNA are selected. A strong selection for hybrid molecules containing large inserts can therefore be exerted when small cosmid vectors are used. A number of small vectors has been especially constructed so as to use most of the packaging capacity for the transduction of foreign DNA (see Table 37). In order to obtain complete and faithful representation of the genome to be cloned, several parameters have to be considered.

1. Since large fragments are preferentially cloned, the foreign DNA to be used should be large (> 150 kbp) to ensure that the fragments to be cloned have specific "sticky" tails on both ends after partial digestion with restriction endonuclease.

2. To produce a random population of overlapping fragments, partial digests with enzymes that cut frequently, such as Sau3A and MboI for BamHI vectors or TaqI, MspI, and HpaII for ClaI vectors, are convenient (Hohn and Collins, 1980; Seed et al., 1982). In addition, fragments produced by mechanical shearing have been used in connection with tailing or synthetic linker methods (Meyerowitz et al., 1980). When synthetic linkers are used, internal cleavage sites must be fully protected by a DNA-methylase before cleaving the attached linkers.

3. Ligations have to yield cosmid-insert concatemers and should therefore be performed at high DNA concentration to make intermolecular linkage more likely than intramolecular ligation.

The ideal cosmid library should contain no vector without an insert, no hybrid cosmids (i.e., with an insert) smaller than 37 kb, and no rearranged insert. However, deviations from the ideal in all these three aspects have been reported (Grosveld et al., 1981; Meyerowitz et al., 1980). Most of the problems seem to be due to the stochastic process of formation

*With permission of J Collins; Courtesy of Marcel Dekker Inc.

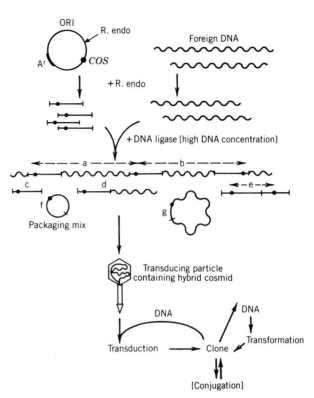

Figure 55. A scheme illustrating the general method of gene cloning with a cosmid vector containing a selectable marker (e.g., antibiotic resistance, A^r), the cohered λ cohesive ends (cos), a replication origin (Ori), and a site (↓) for a restriction endonuclease (R.endo). See text for details. [With permission of J. Collins. Copyright (1978) Elsevier Biomedical Press.]

of the packaging substrate that yields a complex mixture of packageable structures.

Formation of polycosmids should be avoided for two reasons, first, it unnecessarily reduces the insert size, and second, perhaps more importantly, it renders hybrids less stable (Collins and Brüning, 1978; De Saint Vincent et al., 1981) both by reduction of copy number (Scangos and Ruddle, 1981) and by budding-off (in recA⁺ hosts) of nonhybrid molecules. Formation of clones that contain fragments unlinked in the starting material must be excluded since this would lead to faulty linkage maps. In the early days of cosmid cloning these difficulties were overcome by dephosphorylation of the vector DNA and size selection (Lin et al., 1980; Meyerowitz et al., 1980) of the insert fragments.

Quite a number of cosmid gene banks representing the DNA of complex genomes [e.g., *Drosophila* (Lin et al., 1980; Meyerowitz et al., 1980), mouse (Cattaneo et al., 1981), and human (Grosveld et al., 1981; Lindenmaier et al., 1985a; Groffen et al.,

Table 37. Cosmid Cloning Vectors (Prokaryotes and Lower Eukaryotes)

Name	Origin of Replication	Resistance Marker(s)[a]	Size (kb)	Cloning Sites[b]	Special Properties
pJC 720	Col E$_1$	rif, Col E	25	H	Dimers not produced, simple cloning procedure.
pJC 75–58	Col E$_1$ts	amp	11.4	E, B, Bg	biological containment: low copy number, temperature sensitive, nonmobilizable.
pHC 79	pMB 1	amp, tet	6.4	B, C, E, E5, H, S, P	First widely used cosmid.
pV 34	pMB 1	amp, tet	5.7	B, Bg, C, E, E5, H, S, P	Pvu II deletion of pHC 79, higher copy number.
pJB 8	pMB1	amp	5.4	B, H, S	Two Eco RI sites border Bam HI site, facilitate excision of insert.
pcos2 EMBL	RGK	neo, tet	6.1	B, Xh, S	Double *cos* vector, simple preparation of vector arms, recombination screening with pUC8 hybrid probes.
pAN 513	R1 ts	neo	7.6	Xb, C, B	Temperature inducible recombination screening with pBR322 hybrid probes.
MUA-10	pMB 1	tet	4.8	E, B$_a$, R1, R2	0.4 kb Pst I *cos* fragment.
c2 XB	pMB 1	amp, neo	6.8	B, C, E, H	Double *cos,* simple vector arm preparation.
loric	λ	neo	6.3	B, C, H, Sa, Xb	Copy number independent of hybrid size.
lorist β	λ	neo	5.5	B, H	As above, SP6 and T7 promoter bordering cloning sites, easy preparation of probes for chromosome walking.
pHSG 262	Col E$_1$	neo	2.8	B, H2, E	High cloning capacity, simple vector arm preparation.
pHSG 422	pSC 101 ts	neo, amp, cm	8.8	E, H, P, Xh, Xm	Biological containment: low copy number, temperature sensitive, nonmobilizable.
pKT 247	RSF 1010	amp, su, strep	11.5	E, Ss	Broad host range cosmid (Gram-negatives), mobilizable.
pMMB 33	RSF 1010	neo	13.8	B, E, Ss	As above.
pLAFR 1	RK 2	tet	21.6	E	As above.
pKB 42	pMB 1	amp, cm	9.9	B	Shuttle vector for *Aspergillus nidulans:* carries *trp*C$^+$ gene for transfer and integration in *Aspergillus.*
pYC 1	Col E$_1$, 2	amp	10.0	B, E, Xh, S	Shuttle vector for *E. coli* and yeast, *his* 3 selection in yeast.

Source: Reprinted from *Principles of Recombinant DNA* (1986), courtesy of Marcel Dekker, Inc., with permission of J. Collins.

[a]Abbreviations for antibiotics: ampicillin, amp; chloramphenicol, cm; colicin E1, ColE; neomycin/kanamycin, neo; rifampicin, rif; sulfonamide, su; streptomycin, strep; tetracyclin, tet.

[b]Abbreviations for restriction enzymes: *Bam*HI, B; *Bal*I, Ba; *Bgl*II, Bg; *Cla*I, C; *Eco*RI, E; *Eco*RV, E5; *Hind*III, H; *Hinc*II, H2; *Kpn*I, K; *Pst*I, P; *Pvu*I, PV1; *Pvu*II, PV2; *Sac*II, Sa; *Sal*I, S; *Sts*I, Ss; *Xba*I, Xb; *Xho*I, Xh; *Xma*I, Xm.

1982)] have been established by a combination of the procedures described above.

An elegant and straightforward cosmid cloning procedure that eliminates the need to size the insert DNA and prevents formation of clones containing short or multiple inserts has been reported by Ish-Horowitz and Burke (1981). In this procedure (Figure 56), the cosmid vector is digested such that left-hand and right-hand cos-fragments are produced, which cannot be ligated in tandem but will accept

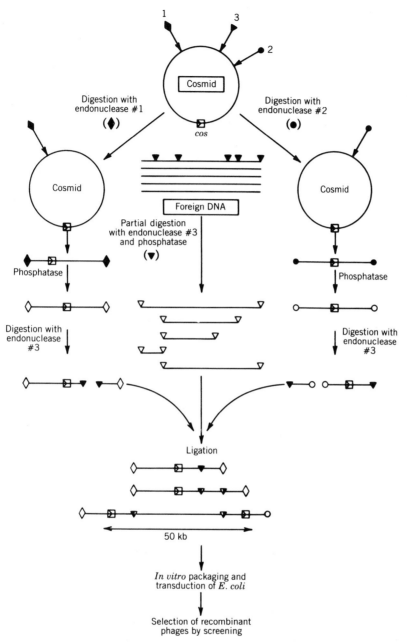

Figure 56. Cosmid cloning using left- and right-hand fragments prepared by digestion with two different restriction endonucleases (see text for details).

dephosphorylated target DNA. In the example described by Ish-Horowitz and Burke, the left-hand cos-fragment is prepared by cutting pJB8 with HindIII, dephosphorylating the HindIII ends, and digesting with BamHI to create a phosphorylated cloning site for Sau3A digested dephosphorylated target DNA. Similarly, the right-hand cos-fragment is formed by first digesting with SalI and removing terminal phosphates, followed by a second digestion with BamHI. By using this method polycosmid for-

mation is avoided because the semiphosphorylated vector fragments cannot be ligated in tandem. Ligation of unlinked target DNA is prevented by dephosphorylation. Size selection is imposed by the in vitro packaging reaction. This method can be used with any cosmid or pair of cosmids from which a left- and right-hand fragment can be produced (Groffen et al., 1982; Brady et al., 1984). The method has been further simplified by construction of special cosmid vectors which allow easy physical isolation

Table 38.   Cosmid Vectors for Use in Higher Eukaryotic Cells

Name	Markers	Size [Kb]	Cloning Sites	Marker in Eukaryotic Cell[a]	Special Properties
pHC 79/2 cos	tk amp	11.8	C, H	tk[a]	Herpes simplex thymidine kinase, HAT selection in TK⁻ cells, double *cos* for packaging in vivo.
pRT	amp	6.9	B, C	tk	HAT selection in TK⁻ cells.
pOPF	amp	8.0	C, B, K	tk	As above, SV 40 origin.
pTM	amp	7.6	B, C, H	G 418	TK promoter fused to neo gene.
pCV 103	amp		B	gpt	SV 40 original promoter fused to *E. coli* HGPRT gene.
pCV 107	amp		B	dhfr	SV 40 *ori* and promoter fused to mouse *dhfr*.
pCV 108	amp		B	G 418	SV 40 *ori* and promoter fused to *E. coli* neo gene.
pHSG 274	neo	3.4	B, S	G 418	TK promoter fused to *E. coli* neo gene.
cEUK	amp, neo	12.5	B	tk	HAT selection in TK⁻ cells, LTR enhancer of RAV.
cos P	amp	7.5	B	Terminal repeats of P-element	Transposition of cloned sequences in *Drosophila* embryos, triple *cos,* simple preparation of vector arms.
cos P neo	amp, neo	9.2	B	Terminal repeats of P-element, G 418	As above, in addition neo gene for G 418 selection in *Drosophila* cells.
pC 22	amp, spec	17.5	Kana	B, Xb	Ori V of pRiHR 1 for stable maintenance in *Agrobacterium,* hybrid–kanamycin resistance gene, *cos* and polylinker between T-DNA borders. OriV, *bom* for replication in *E. coli* and mobilization into *Agrobacterium.*
pPCV 003	amp, cm, tet	12.9	Kana	E, Bg, H, B, S	OriV, oriT of RK2 for maintenance in *Agrobacterium.* pBR 322 ori, *bom* for replication and mobilization in *E. coli*, hybrid kanamycin resistance gene, polylinker and Pg 5-promoter between T-DNA borders.

Source: Reprinted from *Principles of Recombinant DNA* (1986), courtesy of Marcel Dekker, Inc., with permission of J. Collins. Abbreviations for antibiotic resistance in *E. coli* and for restriction enzymes are as in Table 37.

[a]tk: allows selection of transformed TK⁻ cells via growth in HAT-medium (hypoxanthine, aminopterin, thymidine). G418: transformants are resistant to the toxic neomycin analogue G418 due to expression of the amino glycoside phosphotransferase gene of transposon Tn5. gpt: expression of the *E. coli* xanthine/guanine phosphoribosyltransferase allows selection of transformed HGPRT-cells in HAT medium, normal cells can be selected against resistance to mycophenolic acid in a special xanthine containing medium because the *E. coli* enzyme can use xanthine to overcome an inhibition of the *de novo* synthesis of GMP. dhfr: this dihydrofolate reductase gene confers dominant methotrexate resistance on recipient cells. Kana: kanamycin/neomycin resistance gene of Tn5 under control of the nopaline synthase promoter allows selection in plants.

of the vector arms (Brady et al., 1984) or even better provide the two arm fragments by double digestion of a single double cos vector (Poustka et al., 1984; Bates and Swift, 1983).

Cosmid Vectors

Some of the cosmid vectors currently used are listed in Tables 37 and 38. It is impossible to give a complete list of available cosmid vectors, since any plasmid vector can be converted to a cosmid by inserting the cohesive end site. This is very conveniently done by inserting the 1.65-kb BglII-fragment of pHC79 or the 0.4 kb PstI-fragment of MUA-10 into the plasmid (Hohn and Collins, 1980; Meyerowitz et al., 1980). Many of the recently developed cosmids are small plasmids constructed to allow cloning of very large (40–50 kb) insert fragments. Some of them were designed specifically to facilitate cloning, according to Ish-Horowitz and Burke (1981), by inserting two cos sites in tandem so that digestion with one enzyme between the cos site, dephosphorylation of the ends, and digestion with the second enzyme deliver the cosmid arms (Poustka et al., 1984; Bates and Swift, 1983). Use of different origins of replication and selection markers allow construction of vector pairs without sequence homology for recombination screening and accelerated chromosome walking (see below) (Poustka et al., 1984; Lindenmaier et al., 1985b). The vector lorist B is designed so that marked probes (RNA) can be prepared for hybridization screening during genome walking experiments. This is achieved by having T7 and SP6 promotors in the vectors bordering the cloning site. RNA-probes labeled to a very high specificity can then be prepared in vitro from either end of the inserted region (personal communication, Peter Little). Although the application of this method is still in its infancy it would seem to have much potential. Another line of development is to construct cosmids that allow cloning in organisms other than *E. coli*. These vectors are "shuttle vectors," which allow cosmid cloning in *E. coli* and selection and maintenance of hybrid cosmids in other organisms. Such vectors have been constructed for *pseudomonas* (Gautier and Bonewald, 1980), *bacillus* (Fouet et al., 1982), yeast (Hohn and Hinnen, 1980), *Aspergillus* (Yelton et al., 1985), plants (Simoens et al., 1986; Koncz and Schell, 1986), *Drosophila* (Haenlin et al., 1985; Steller and Pirrotta, 1985), and even mammalian cells (Lindenmaier et al., 1982; Groffen et al., 1982; Brady et al., 1984) (see below).

This versatility of the cosmid system is also doc-umented by the large number of cloning sites available for fragments created by digestion with different restriction endonucleases. This is especially apparent if enzymes are included which recognize different sites but create identical sticky ends. BamHI- or BglII-cloning sites can be used directly to insert fragments generated by digestion with BamHI, BglII, Sau3A, MboI, and XhoII. ClaI-cloning sites can be used in connection with ClaI, MspI, HpaII, and TaqI (Hohn and Collins, 1980). The screening of cosmid clones by colony hybridization methods presents no additional problems and can be compared in efficiency with screening of plasmid clones or bacteriophage plaque hybridization (Cami and Kourilsky, 1978; Grunstein and Hogness, 1975; Hanahan and Messelson, 1980).

Host Strains and Cosmid Stability

The host strains used for transduction of cosmids must be sensitive to infection by bacteriophage λ and should be recombination deficient to increase the stability of the hybrid cosmid. In most cases recA⁻ strains like HB101 (Boyer and Roulland-Dus-soix, 1969), ED8767 (Lund et al., 1982), or *E. coli* 1400 (Cami and Kourilsky, 1978) have been used. Before transduction the recA⁻ phenotype should be tested by assaying UV sensitivity (De Saint Vincent et al., 1981; Ish-Horowitz and Burke, 1981). To increase the transduction efficiency the bacteria should be grown in medium containing maltose to induce the production of phage receptors (Boyer and Roulland-Dussoix, 1969). In addition Mg starvation can be used (Grosveld et al., 1981; Ish-Horowitz and Burke, 1981). After transduction nonselective medium is added and the cells are incubated to allow expression of antibiotic resistance. The time required for full expression of resistance is dependent on the selection used (Collins, 1979; Lund et al., 1984).

Even a low frequency of recombination can lead to loss of larger plasmids via the replicative advantage of smaller hybrids which in general have a higher copy number. It has been noted that palindromic and inverted repeat DNA is unstable in *E. coli* (Collins and Hohn, 1978; Perricaudet et al., 1977; Collins, 1981; Collins et al., 1982; Hagen and Warren, 1982, 1983) in both lambda and plasmid vectors. Such deletion events are *recA* independent (Perricaudet et al., 1977; Collins et al., 1982) but greatly reduced in *recBC sbcB* hosts for both lambda (Leach and Stahl, 1983; Wyman et al., 1985) and plasmid hybrids (Collins et al., 1982; Nader et al.,

1985; Boissy and Astell, 1985). Wyman et al. (1985) estimate that some 9% of human genomic clones are probably lost due to deletion of palindromes and the small burst sizes or inviability of the subsequent miniaturized hybrids.

Although the effect of the *recBC sbcB* host on stabilizing cosmids has not been thoroughly investigated it is likely to be positive. Again it should be emphasized that by repacking the original bank as soon as possible after establishing the bank, keeping the bank as a packaged cosmid particle suspension, and subsequently screening with the recombination screening method, even relatively unstable clones can be isolated.

In Vivo Packaging of Cosmids

Cosmids can be packaged in vivo into phage heads to yield transducing particles if packaging functions are provided by superinfection or induction of a lysogenic phage (Lindenmaier et al., 1982, 1985; Poustka et al., 1984; Umene et al., 1978; Vollenweider et al., 1980a,b; Feiss et al., 1982). Efficiency of the in vivo packaging process depends on size and structure of the cosmids as well as on the host-phage system, induction protocol, and culture condition used. Cosmids too small to make up a headful are packaged as oligomers (Gitschier et al., 1984; Feiss et al., 1982). Since circular λ-molecules containing a single cos-site are not packaged in vivo (Feiss and Margulies, 1973) and in vitro (Feiss et al., 1977), cosmids with two cos-sites in tandem have been constructed to improve the yield of transducing particles (Lindenmaier et al., 1982). However, in contrast to the reports on packaging of monomeric λ circles, λ-sized cosmids with a single cos-site are packaged quite efficiently in vivo even in the absence of known recombination functions. In the experiments concerning packaging, replication did not occur. This is in contrast to the situation with cosmid in vivo packaging (Lindenmaier et al., 1982; Feiss and Margulies, 1973). For smaller cosmids that are packaged in oligomeric form, a dependence of packaging efficiency on the *gam*-function has been reported (Feiss et al., 1982). The *gam* gene of λ encodes an inhibitor of the recBC nuclease of *E. coli* (Sakaki et al., 1973) which otherwise prevents the transition from monomer circle to rolling circle replication (Enquist and Skalka, 1973). This suggested that a rolling circle replication mechanism is involved in the creation of a packageable (concatemeric?) substrate.

Cos-site cutting by λ-terminase in vitro is not a very effective process (Feiss and Widner, 1982; Becker and Gold, 1978), but in vivo packaging allows the use of the very specific ter-function which functions like a specific endonuclease. Since the sticky end sequence is not palindromic, asymmetric labeling of the ends is possible (Nichols and Donelson, 1978; Rackwitz et al., 1984; Zehetner and Lehrach, 1986). This allows rapid restriction enzyme mapping of cosmids (Little and Cross, Lehrach, personal communication) according to the method of Smith and Birnstiel (1976).

It has been shown that when λ-sized cosmids are packaged in vivo the relative abundance of specific genes within the gene bank was not altered on transducing the bank into a new host (Lindenmaier et al., 1985). Thus, in vivo packaging allows reproducible preparation of cosmid gene banks in the form of a cosmid lysate (Poustka et al., 1984; Lindenmaier et al., 1985). With highly efficient packaging strains, for example, BHB 3175 (Poustka et al., 1984), high titer lysates containing $5–10 \times 10^9$ transducing particles/ml may be produced. An *E. coli*–DNA gene bank and a human genomic gene bank have been maintained as cosmid lysates for more than a year without significant loss of infectivity. The hybrid cosmids can be transduced easily into *E. coli* strains of different genetic backgrounds at a preselected multiplicity of infection to facilitate colony screening of specific clones. The possibility of preparing and maintaining cosmid gene banks as cosmid lysates also provides the basis for more sophisticated techniques (e.g., recombination screening and cosmid shuttle) described below.

Recombination Screening

Homologous recombination in *E. coli* is a very specific process. This specificity can be used to screen for the presence of specific sequences if the probe sequences are linked to a selectable marker. Cointegrates formed by recombination carry markers of both components in *cis*. These recombined molecules must be selected from the original components. The selectivity of this process must be extremely high. With recombination frequencies of 10^{-2} to 10^{-4} for homologous sequences, screening for a complex animal DNA library will yield recombinant clones with frequencies of 10^{-7} to 10^{-9}. The λ packaging system is selective and efficient enough to isolate specific recombinants from complex libraries as has first been shown by Seed (1985) by the isolation of clones from λ libraries. A protocol based on similar principles has been used by the groups of Lehrach (Poustka et al., 1984) and Collins (Lindenmaier et al., 1985b) for the isolation of genes from cosmid libraries (Figure 57). The development of this simple genetic screening technique was made possible once the in vivo packaging of cosmid libraries had been established (Lindenmaier et al., 1982). Cosmid–plasmid vector pairs without homologous sequences are used. A recombination proficient host strain with heat inducible λ-prophage is transformed by the recombination screening plasmid carrying the sequence of interest. After infection with the cosmid gene bank, and recombination, recombinant clones are selected by in vivo packaging and transduction to a new host. Only those transductants that formed cointegrates before packaging will carry the antibiotic resistance markers of both vectors and as such can be easily selected. To restore the original cosmid the process can be reversed (Poustka et al., 1984). Chromosomal genes for the mouse t-complex (Poustka et al., 1984), human IL 2 (Lindenmaier et al., 1985) parathyroid hormone (Mayer, unpublished results), and a protease inhibitor (unpublished) have been isolated using this method.

Recombination can be used for gene modificaton (e.g., deletion of introns and construction of dimeric genes between members of a gene family) as well as selection of homologous sequences.

REFERENCES

Bates, P. F., and Swift, R. A. (1983), *Gene* **26**, 137.

Becker, A., and Gold, M. (1978), *Proc. Natl. Acad. Sci. USA,* **75**, 4199.

Boissy, R., and Astell, C. R. (1985), *Gene,* **35**, 179.

Boyer, H. W., and Roulland-Dussoix, D. (1969), *J. Mol. Biol.,* **41**, 459.

Brady, G., Jantzen, H. M., Bernard, H. U., Brown, R., Schütz, G., and Hashimoto-Gotoh, T. (1984), *Gene,* **27**, 223.

Cami, B., and Kourilsky, P. (1978), *Nucl. Acids Res.,* **5**, 2381.

Cattaneo, R., Gorski, J., and Mach, B. (1981), *Nucl. Acids Res.,* **9**, 2777.

Collins, J., and Hohn, B. (1978), *Proc. Natl. Acad. Sci. USA,* **75**, 4242.

Collins, J., and Brüning, H. J. (1978), *Gene,* **4**, 85.

Collins, J. (1979), in *Methods in Enzymology,* P. Wu, ed., Vol. 68, p. 309, Academic Press, New York.

Collins, J. (1981), *Cold Spring Harbor Symp. Quant. Biol.,* **45**, 409.

Collins, J., Volckaert, G., and Nevers, P. (1982), *Gene,* **19**, 139.

De Saint-Vincent, B. R., Delbrück, S., Eckhart, W., Meinkoth, J., Vitto, L., and Wahl, G. (1981), *Cell,* **27**, 267.

Enquist, L. W., and Skalka, A. (1973), *J. Mol. Biol.,* **75**, 185.

Feiss, M., and Margulies, T. (1973), *Mol. Gen. Genet.,* **127**, 285.

Feiss, M., Fisher, R. A., Crayton, M. A., and Egner, C. (1977), *Virology,* **77**, 281.

Feiss, M., Siegele, D. A., Rudolph, C. F., and Frackman, S. (1982), *Gene,* **17**, 123.

Feiss, M., and Widner, W. (1982), *Proc. Natl. Acad. Sci. USA,* **79**, 3498.

Fouet, A., Klier, A., and Rapoport, G. (1982), *Mol. Gen. Genet.,* **186**, 399.

Gautier, F., and Bonewald, R. (1980), *Mol. Gen. Genet.,* **178**, 375.

Gitschier, J., Wood, W. I., Goralka, T. M., Wion, K. L., Chen, E. Y., Eaton, D. H., Vehar, G. A., Capon, D. I., and Lawn, R. M. (1984), *Nature (London),* **312**, 326.

Groffen, J., Heisterkamp, N., Grosveld, F., van den Wen, W., and Stephenson, J. R. (1982), *Science,* **216**, 1136.

Grosveld, F. G., Dahl, H. H. M., de Boer, E., and Flavell, R. A. (1981), *Gene,* **13**, 227.

Grunstein, M., and Hogness, D. (1975), *Proc. Natl. Acad. Sci. USA,* **72**, 3961.

Haenlin, M., Steller, H., Pirotta, V., and Mohier, E. (1985), *Cell,* **40**, 827.

Hagen, C. E., and Warren, G. J. (1982), *Gene,* **19**, 147.

Hagen, C. E., and Warren, G. J. (1983), *Gene,* **24**, 317.

Hanahan, D., and Meselson, M. (1980), *Gene,* **10**, 63.

Hohn, B., and Collins, J. (1980), *Gene,* **11**, 291.

Hohn, B., and Hinnen, A. (1980), in *Genetic Engineering: Principles and Methods,* J. K. Setlow and A. Hollaender, eds., Vol. 2, p. 169, Plenum Press, New York.

Ish-Horowitz, D., and Burke, J. F. (1981), *Nucl. Acids Res.,* **9**, 2989.

Koncz, C., and Schell, J. (1986), *Mol. Gen. Genet.,* **204**, 383.

Leach, D. R. F., and Stahl, F. W. (1983), *Nature,* **305**, 448.

Lin, C. P., Tucker, P. W., Mushinski, J. F., and Blattner, F. R. (1980), *Science,* **209**, 1348.

Lindenmaier, W., Hauser, H., Gresier de Wilke, I., and Schütz, G. (1982), *Nucl. Acids Res.,* **10**, 1243.

Lindenmaier, W., Hauser, H., and Collins, J. (1985a), in *The Impact of Gene Transfer Techniques in Eukaryotic Cell Biology,* J. S. Schell and P. Starlinger, eds., p. 17, Springer Verlag, Berlin, Heidelberg.

Lindenmaier, W., Dittmar, K. E. J., Hauser, H., Necker, A., and Sebald, W. (1985b), *Gene,* **39**, 33.

Little, P. F. R., and Cross, S. U. (1985), *Proc. Natl. Acad. Sci. USA,* **82**, 3159.

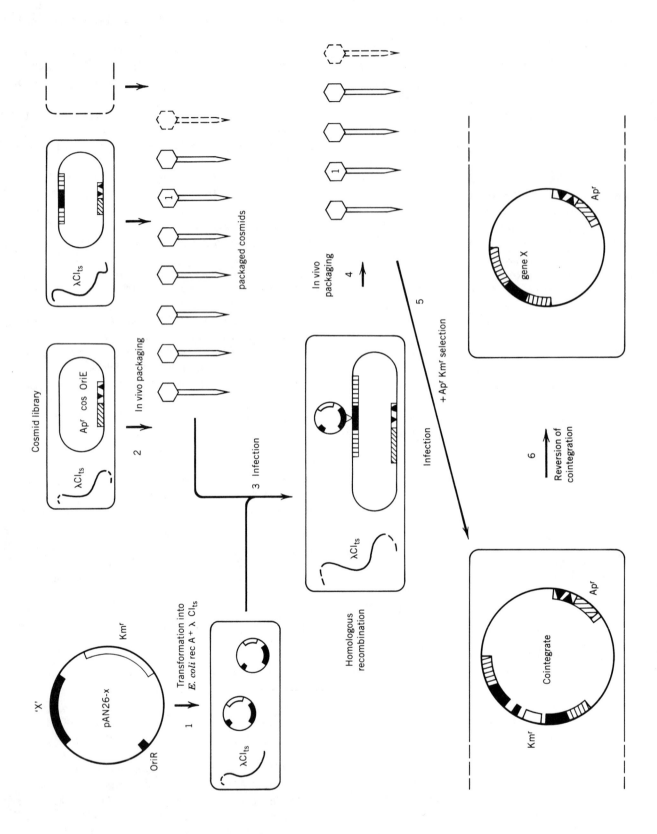

Figure 57. Schematic representation of the "recombination screening" method.

1. The oligonucleotide or gene fragment probe ("x") is cloned into the vector pAN26. This vector has no homology to the vector used to create the cosmid library which is to be screened. pAN26 containing the probe DNA x is transformed into a recombination proficient lambda lysogenic strain of *E. coli* (see below).

2. A cosmid library constructed in the ampicillin resistance cosmid (e.g., pHC79 indicated as a box containing the ampicillin resistance gene, cos and Ori E) is packaged in vivo by a temperature shift which induces the synthesis of packaging functions from the resident (lysogenous) lambda prophage (= 10⁹ particles).

3. The packaged cosmids are transduced into the pAN26x carrying strain. During the following hour or so homologous recombination can take place between pAN26x and any cosmid having extensive homology (i.e., those cosmids containing the homologous gene X).

4. Step 2 is repeated, whereby the cosmid library is again packaged, but now includes some particles containing a cointegrate of pAN26x and the homologous cosmid.

5. The cointegrates are simply selected by transducing the packaged particles from 4 to a fresh *E. coli* strain and simultaneously selecting for marker genes specific to the two vectors, that is, kanamycin and ampicillin resistance. Note that the integration of pAN26x leads to a duplication of probe region x.

6. As a final step the cosmid containing the uninterrupted gene x may be obtained by reversal of the cointegration process (again via homologous recombination in the duplicated region x), by screening for kanamycin sensitivity. Abbreviations: Ori R, Ori E: origin of replication for the plasmids R1 and ColE1 respectively. Km^r: kanamycin resistance. Ap^r: ampicillin resistance. λ cl$_{ts}$: lambda prophage integrated into the *E. coli* genome which can be induced by a temperature shift (inactivation of the temperature sensitive -repressor cl). This prophage is in addition defective in its ability to provide a packageable genome (e.g. b2 nin 5 deletions, red-, xis- or cos-). cos: cohesive end-site of λ; the site recognized and cleaved by the activated lambda prehead, as prerequisite to packaging into the lambda head. rec A: denotes the gene required, together with the rec BC-nuclease (exonuclease V) for homologous recombination in *E. coli*. [Reprinted from Principles of Recombinant DNA (1986), courtesy of Marcel Dekker Inc., with permission of J. Collins].

◄——————————

Lund, T., Grosveld, F. G., and Flavell, R. A. (1982), *Proc. Natl. Acad. Sci. USA,* **79**, 520.

Meyerowitz, E. M., Guild, G. M., Prestidge, L. S., and Hogness, D. S. (1980), *Gene,* **11**, 271.

Nader, W. F., Edling, T. D., Huetterman, A., and Sauer, H. W. (1985), *Proc. Natl. Acad. Sci. USA,* **82**, 2698.

Nichols, B. P., and Donelson, J. E. (1978), *J. Virol.,* **26**, 429.

Perricaudet, M., Fritsch, A., Petterson, U., Philipson, L., and Tiollais, P. (1977), *Science,* **196**, 208.

Poustka, A., Rackwitz, H. R., Frischauf, A. M., Hohn, B., and Lehrach, H. (1984), *Proc. Natl. Acad. Sci. USA,* **81**, 4129.

Rackwitz, H. R., Zehetner, G., Frischauf, A. M., and Lehrach, H. (1984), *Gene,* **30**, 195.

Sakaki, Y., Karn, A. E., Lim, S., and Echols, H. (1973), *Proc. Natl. Acad. Sci. USA,* **70**, 2215.

Scangos, G., and Ruddle, F. H. (1981), *Gene,* **14**, 1.

Seed, B., Parker, C. R., and Davidson, N. (1982), *Gene,* **19**, 201.

Seed, B. (1985), *Nucl. Acids Res.,* **11**, 2427.

Simoens, C., Alliotte, T. H., Mendel, R., Müller, A., Schiemann, J., Van Lijsebettens, M., Schell, J., Van Montagu, M., and Inzé, D. (1986), *Nucl. Acids Res.,* **14**, 8073.

Smith, H. O., and Birnstiel, M. L. (1976), *Nucl. Acids Res.,* **3**, 2387.

Steinmetz, M., Winoto, A., Minrad, K., and Hood, L. (1982), *Cell,* **28**, 489.

Steller, H., and Pirrotta, V. (1985), *EMBO J.,* **4**, 167.

Umene, K., Shimada, K., and Takagi, Y. (1978), *Mol. Gen. Genet.,* **159**, 39.

Vollenweider, J. J., Fiandt, M., and Szybalski, W. (1980a), *Gene,* **9**, 157.

Vollenweider, J. J., Fiandt, M., Rosenvold, E. C., and Szybalski, W. (1980b), *Gene,* **9**, 171.

Wyman, A. R., Wolfe, L. B., and Botstein, D. (1985), *Proc. Natl. Acad. Sci. USA,* **82**, 2880.

Yelton, M. M., Timberlake, W. E., and van den Hondel, C. A. M. J. J. (1985), *Proc. Natl. Acad. Sci. USA,* **82**, 834.

Zehetner, G., and Lehrach, H. (1986), *Nucl. Acids Res.,* **14**, 335.

OTHER CLONING VECTORS

A tremendous effort has been made in the past few years to apply the cloning technology developed with *E. coli* to the dissection of more complex genomes. Thus, aside from the three basic kinds of prokaryotic cloning vectors that have been described above, a considerable number of vectors have been developed to meet experimental needs specific to these systems. Without going into a detailed analysis of the various vectors and their utilization, we would like to outline here, through a few typical examples, the considerable potential of molecular cloning when applied to the manipulation of genomes from higher organisms. Expression vectors for mammalian cells will be described in more detail in Chapter 25.

Prokaryotic Vectors

Because of many different economical or fundamental reasons, an increasing number of cloning systems are now being used to introduce foreign genes in prokaryotic hosts other than *E. coli*. There have been several reports concerning the expression in

Bacillus subtilis of eukaryotic gene products such as hepatitis B core antigen, foot and mouth disease virus major antigen (Hardy et al., 1981), mouse dihydrofolate reductase (Williams et al., 1981, Schoner et al., 1983), human interferon alpha (Palva et al., 1983, Band and Henner, 1984, Yansura and Henner, 1984), rat proinsulin (Mosbach et al., 1983), human urogastrone (Flock et al., 1984), human interleukin 2, and human gamma interferon (Nakahama et al., 1985). It is particularly interesting to note that heterologous DNA sequences can be inserted directly in the genome of *Bacillus subtilis,* provided they are ligated to DNA sequences homologous to the recipient chromosome (Harris-Warrick and Lederberg, 1978). This observation has been the basis for the development of specialized phage vectors in *B. subtilis.*

Pseudomonas is a particularly interesting microorganism because it is able to metabolize various organic chemicals including toluene, terpenes, aromatic compounds, and long-chain hydrocarbons. It can also perform detoxification reactions involving organic and inorganic heavy metals (Clark and Richmond, 1975). Versatile cloning vectors for *Pseudomonas* species have been derived from the nonconjugative IncP-4 plasmids RSF 1010 and R300B (Bagdasarian et al., 1979, 1981, 1983; Barth, 1979; Sharpe, 1984). Low copy number vectors have also been prepared from the conjugative IncP-1 plasmids RP4 and RK2 (Barth, 1979; Ditta et al., 1980) and used for the construction of cosmids (Friedman et al., 1982; Knauf and Nester, 1982). Some of these cosmids have been utilized in *Pseudomonas* species (Darzins and Chakrabarty, 1984; Moores et al., 1984). The 30 kb *Pseudomonas* plasmid pVS1 is a nonconjugative mobilizable plasmid which can also replicate in *Rhizobium leguminosarum,* and *Agrobacterium tumefaciens,* but not in *E. coli.* Several cloning vectors were derived from pVS1 (Itoh and Haas, 1985). Two nonmobilizable plasmids, pME260 (6.3 kb) and pME290 (6.8 kb), carry the Tn801 *bla* gene encoding the carbenicillin resistance (Cb), and the Tn903 *aph* gene specifying the kanamycin resistance (Km) in *E. coli* (Nomura et al., 1978) and *Pseudomonas* (Bagdasarian et al., 1981). The pME290 plasmid could be introduced by transformation and maintained in *P. aeruginosa, P. fluorescens, P. putida, P. acidovorans, P. stutzeri, P. mendocina, P. cepacia,* and *P. syringae.* Since *Pseudomonas* is generally resistant to many antibiotics and antimetabolites, usable selective markers are scarce. To alleviate this lack of suitable markers, Itoh and Haas (1985) constructed the *Mob* + vector pME285 (10.6

kb) which carries, in addition to the *aph* gene, the Tn501-derived *merRTCA* genes conferring mercuric ion resistance.

The construction of shuttle vectors able to replicate in both *E. coli* and *Pseudomonas* (Werneke et al., 1985) has also been undertaken to provide a tool in the characterization of the *Pseudomonas* genes governing the biotransformation reactions unique to this organism. Thus, several vectors obtained after fusion of pUC13 (Norrande et al., 1983) and pUC4K (Viera and Messing, 1982) have been constructed and used for cloning, expression, and selection of *Pseudomonas* genes by complementation (Werneke et al., 1985).

Numerous attempts have been made in the past to improve industrial bacterial strains used for the production of biological compounds such as amino acids. Among these strains are the coryneform glutamic acid-producing bacteria *Brevibacterium lactofermentum* and *Corynebacterium glutamicum.* The development of molecular cloning has brought tools for both basic research and improvement of industrial productivity. The different cloning vectors used with these bacteria are either cryptic plasmids (Kaneko and Sakaguchi, 1979; Sandoval et al., 1984; Miwa et al., 1984) or plasmids carrying selective markers (Katsumata et al., 1984; Santamaria et al., 1984; Miwa et al., 1985). A cosmid vector that can be packaged and transduced through phage infection has also been described (Miwa et al., 1985).

Several useful cloning vectors have been constructed from the *streptomyces* plasmid SCP2* (Lydiate et al., 1985). This plasmid can give rise to recombination frequencies as high as 10^{-3} (Bibb and Hopwood, 1981), is present at a ratio of one to two copies per chromosome (Bibb et al., 1977: Schrempf and Goebel, 1977; Ikeda et al., 1984), and is stably inherited (Bibb et al., 1980). Chimeric plasmids able to replicate in both *E. coli* and *Streptomyces* (shuttle vectors) have also been described (Schrempf and Goebel, 1975; Larson and Hershberger, 1984; Lydiate, 1984; Lydiate et al., 1985).

Yeast Cloning Vectors

Yeast Integrating Vectors. These vectors are composed of a bacterial cloning vector (plasmid, bacteriophage λ, or cosmid) carrying yeast genes that can be used in the selection of yeast transformants. The bacterial cloning vector sequences allow amplification of the ligated vector in *E. coli,* and therefore compensate for the low efficiencies of yeast transformation usually obtained (1–10 yeast transformants/ μg DNA). Selectable yeast markers used include *his3*, *leu2*, and *ura3*.

ars Vectors

These vectors contain a yeast replicator sequence (*ars*) combined with a yeast selectable marker and a bacterial vector. The efficiency of transformation obtained with such vectors is 10^2–10^4/μg DNA.

It should be noted that such shuttle vectors (for example, YRp7) are not stably maintained in budding yeast cultures, unless they contain functional yeast centromere DNA (Clarke and Carbon, 1980; Fitzgerald-Hayes et al., 1982). Similarly, *ars* elements have been isolated from *Kluyveromyces lactis* chromosomes (Das and Hollenberg, 1982; Fabiani and Frontali, 1986). Hybrid vectors also containing pBR322 sequences were found to transform both *Saccharomyces cerevisiae* and *Kluyveromyces lactis* with a relatively high frequency (up to 10^3 transformants per μg of DNA). these *ars* elements were found to differ from those previously described. They were also found to be unstably maintained in yeast cultures performed in nonselective media.

2-Micron Plasmid Hybrid Vectors

The 2-micron plasmid (*Scp*) is found as small circular DNA of 2-μm contour length in almost all strains of *S. cerevisiae*. There may be up to 100 copies of this plasmid per cell.

It contains 6318 base pairs with two inverted repeats of about 600 base pairs (see Figure 58). Two isomeric forms are generated following recombination within these repeats.

Successful transformation of *S. cerevisiae* (Broach, 1982) and *Schizosaccharomyces pombe* (Beach and Nurse, 1981) has been obtained with 2μ-derived plasmids.

For a long time the 2μ plasmid has been the only one described in yeast. Circular DNA molecules with comparable sizes have been characterized in osmophilic yeasts (Toh-e et al., 1982; Toh-e et al., 1984; Araki et al., 1985). Although they do not exhibit sequence homology with the 2μ circle of *S. cerevisiae,* the plasmids pSR1 from *Zygosaccharomyces rouxii,* pSB1 and pSB2 from *Zygosaccharomyces bailii,* and pSB3 and pSB4 from *Zygosaccharomyces bisporus* share the same dimorphic organization.

A screening of 70 yeast strains belonging to various yeast genera has led to the isolation (Falcone et al., 1986; Chen et al., 1986) of a new plasmid (pKD1) which has further been used to construct by far the most efficient and practical vector system for *K. lactis* (Bianchi et al., 1987). The pKD1 plasmid is a circular molecule with a structure very comparable to that of the 2μ circle. Although they share little sequence homology, they exhibit different host specificity. Initially isolated from *K. drosophilarium,* the pKD1 plasmid has been transferred successfully to *K. lactis* where it can replicate stably.

Two types of transforming vectors have been constructed from pKD1. The first type contained the total pKD1 sequence (4757 base pairs) into which the pBR322 sequences as well as the yeast selection markers were introduced through the unique Eco RI site. This vector is highly stable in *K. lactis* hosts. The other vector, which contains only a portion of the pKD1 sequence, including the replication origin linked to a yeast marker gene and to the bacterial plasmid, was often found to be unstable. It can be stabilized by the presence of resident pKD1 (cir + hosts).

The pKD1 vectors add new possibilities in manipulating the yeasts that are important in the milk industry, while 2μ vectors have been most useful in the yeasts used in the bakery and brewery industries.

The pKD1-derived vectors are unstable in *S. cerevisiae*. Shuttle vectors containing both the 2μ and the pKD1 origins of replication were stable in cir + *S. cerevisiae* and in cir + *K. lactis*. Derivatives of pKD1 with the G418 resistance marker are also available. A chromosomal gene required for the production of the linear-DNA-coded killer toxin of *K. lactis* has been successfully cloned (Wesolowski-Louvel et al., personal communication) by complementation in the KEp6 vector derived from pKD1 (see Figure 59).

Linear DNA Plasmids

The linear DNA of killer plasmids isolated from *K. lactis* has also been used as a transformation vector but turned out to be less satisfactory than the circular systems described above.

Plant Vectors

Within the last few years there has been an increasing interest in the development of vectors allowing gene transfer in high plants, not only because of their extraordinary economical potential, but also because molecular studies performed with these systems will undoubtedly lead to the discovery of new mechanisms governing gene replication and expression.

The tumor-inducing (Ti) plasmids are responsible for the pathogenic effect of *Agrobacterium tumefaciens* in most dicotyledonous plants (De Cleene and

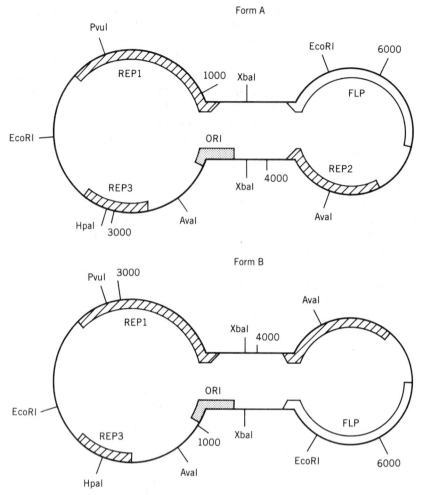

Figure 58. Two forms of 2 μ-plasmid.

De Ley, 1976). Derivatives of these plasmids have been used as cloning vectors for plant engineering. With the root-inducing (Ri) plasmids of *Agrobacterium rhizogenes* (Chilton et al., 1982), the Ti plasmids have been for a long time the only agents known to insert foreign DNA into the genome of higher plants (Caplan et al., 1983), and more recently in some monocotyledonous species (Hooykaas-Van Slogteren et al., 1984; Hernalsteens et al., 1984). Two regions (T-DNA and *vir* region) of these very large plasmids (160–240 kb) have been shown to be essential for tumor induction (see the review by Gheysen et al., 1985). In tumor cells, the T-region is integrated in host nuclear DNA. Two direct repeats of 25 base pairs (Simpson et al., 1982; Yadav et al., 1982; Zambryski et al., 1982) flanking the T-region are thought to play a crucial role in gene transfer and integration into the host plant genome (Shaw et al., 1984; Wang et al., 1984). The expression of T-DNA genes which results in hormone auxotrophy of crown gall cells and in the synthesis of new metabolites called opines (Schroder et al., 1981; Murrai and Kemp, 1982; Petit et al., 1970; Bomhoff et al., 1976; Tempe and Schell, 1977; Leemans et al., 1981) in transformed cells is not required for T-DNA transfer and integration. The direct introduction of DNA fragments in T-DNA restriction sites of Ti plasmids has been hampered mainly because of the enormous size of the plasmids, and the use of Ti plasmids as cloning vectors has required subcloning of the T-DNA sequences and fusion with selective marker genes. For example, several vectors containing transcriptional control elements from T-DNA fused to the coding region of the neomycin phosphotransferase II (NPTII) gene from Tn5 have been constructed and used for gene transfer in different plant species (Velten and Schell, 1985). These vectors conferred to the transformed cells a high level of resistance to the aminoglycoside antibiotics kananmycin and G418. Aside from the *A. tumefaciens* Ti plasmids, the develop-

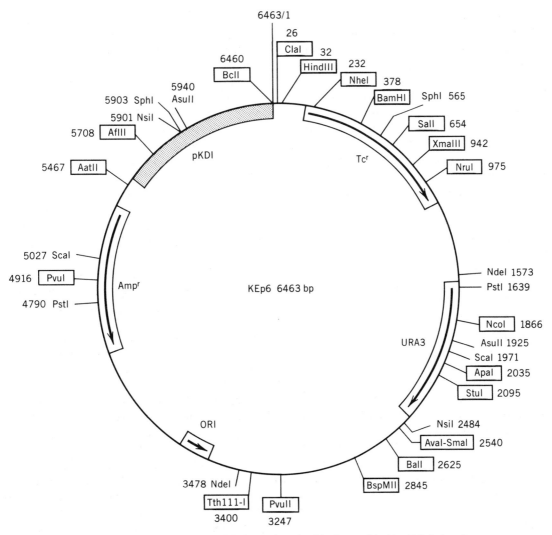

Figure 59. Restriction map of KEp6 yeast vector (kindly provided by H. Fukuhara).

ment of protoplast transformation systems using bacterial plasmid DNAs with selectable plant marker genes has offered new ways of introducing foreign genes into plant cells (Paszkowski et al., 1984; Shillito et al., 1985). High levels of expression are generally achieved by placing the foreign genes under the control of plant transcription signals. Promoter sequences from cauliflower mosaic virus have been used, in conjunction with appropriate plant selective markers, for plant cell transformation (Koziel et al., 1984; Odell et al., 1985; Balazs et al., 1985). More recently, a very convenient vector (pDH51) containing the strong cauliflower mosaic virus 35S RNA promoter and a transcription terminator separated by a polylinker sequence has been successfully used for the expression of several genes in plants (Pietrzak et al., 1986). The introduction of

the neomycin phosphotransferase type III gene from streptococcal plasmid pJH1 into the polylinker of pDH51 has also been found to confer kanamycin resistance to plant cells.

Viral Vectors

Aside from a whole variety of eukaryotic host–vector systems which have been engineered to allow an elevated expression of cloned genes in eucaryotic cells (see Chapter 25), several viral vectors have been developed for efficient introduction of foreign sequences in the genome of mammalian species. These vectors can be classified in two categories, depending on whether or not they require a helper virus to ensure their propagation. For the sake of simplicity in experimental design, helper-independent viral vectors are often preferred, especially when the wild-type helper virus is tumorigenic in vivo. Recently, for example, a helper-independent human adenovirus vector has been used to transfer the herpes simplex virus thymidine Kinase (HSV-tk) gene into mammalian cells (Haj-Ahmad and Graham, 1986). The human adenoviruses are medium-sized DNA tumor viruses whose genome is a double stranded DNA molecule of about 36,000 base pairs. The expression of adenovirus in infected cells proceeds through two major steps known as early and late phases. The bulk of early genes (E1a, E1b, E2a, E2b, E3, and E4) lying in noncontiguous regions of the genome are expressed sequentially, before viral DNA replication (see review by Sharp, 1984). The E1a gene encodes the first transcripts expressed after infection of permissive cells and is required to activate most, if not all, of the remaining viral genes (Berk et al., 1979; Jones and Shenk, 1979; Nevins, 1981) while the E3 gene is not essential for viral growth (Anderson et al., 1976; Berkner and Sharp, 1983; Kelly and Lewis, 1973; Saito et al., 1985). Since only a small portion of the viral genome is required *in cis* for adenovirus multiplication, cell lines have been constructed to provide most of the essential viral functions *in trans,* therefore allowing multiplication of partial viral genomes in which foreign genes have been introduced. Studies performed to evaluate the packaging constraints of adenoviruses have suggested that the wild-type adenovirus type 5 may accommodate about 2000 base pairs of foreign DNA without notably interfering with packaging. Deletion mutants such as dlE3 and dl E1, 3, lacking most E3 or E1 and E3 genes, are expected to accommodate up to 7500 base pairs of foreign DNA (Haj-Ahmad and Graham, 1986). The dl E1,3 mutant contains a unique Xba I cloning site and is efficiently propagated in 293 cells which constitutively express the E1 gene products (Aiello et al., 1979; Berk et al., 1979; Graham et al., 1977).

Similar systems have also been developed for the use of retroviral vectors. Thus, the ψ 2 cell line is a NIH-3T3-derived cell line which contains an endogenous murine leukemia virus (MuLV) lacking the sequences required for viral RNA packaging (Mann et al., 1983). These cells provide *in trans* the retroviral proteins involved in retrotranscription and packaging of the defective retroviral genomes that are used as cloning vectors, and therefore permit acquisition of helper-free stocks of replication defective recombinant retroviruses carrying foreign genes.

However, the fact that multiplication of the recombinant genomes is only possible in the competent cells is a major limitation of such systems.

Infectious retroviruses have been used for some years as efficient tools to obtain a stable integration of foreign sequences in the mammalian DNA. Although the prototype virus for these constructs has been the highly leukemogenic Moloney strain of Murine leukemia virus (M.MuLV), several other retroviral vectors have been used to introduce cloned genes in cultured cells (Shimitohno and Temin, 1981; Wei et al., 1981; Tabin et al., 1982; Joyner and Berstein, 1983, Joyner et al., 1983; Miller et al., 1983; Perkins et al., 1983; Cepko et al., 1983; Williams et al., 1984; Hellerman et al., 1984). These systems have proven to transduce foreign genes at very high frequencies in cells of different origins. Insertion of MuLV in the mouse germ line has been reported for some time (Jaenisch et al., 1975; Jaenisch, 1976; Jaenisch, 1980; Jahner and Jaenisch, 1980; Jaenisch et al., 1981; Stuhlman et al., 1984). More recently, strategies have been developed to insert genes of interest into the mouse germ line via replication competent or defective retroviral vectors (Huszar et al., 1985; Van der Putten et al., 1985; Jensen et al., 1986), while a helper-dependent retroviral cloning vector has allowed the transfer and expression of the human gene encoding for hypoxanthine phosphoribosyl-transferase (HPRT) into human bone marrow cells (Gruber et al., 1985).

Although the retroviral vectors offer several advantages over other cloning systems, it should be kept in mind when working with such systems that the strong promoters and enhancers that are located in the long terminal repeats (LTR) of the proviral genomes may activate proto-oncogenes after integration in the cellular genome, and may therefore induce tumors in susceptible hosts.

CLONING VECTORS FROM COMMERCIAL SOURCES

This section is intended to provide the reader with a brief review of some useful vectors that are commercially available. The material that we have used from the firms listed below turned out to be of very good quality in all instances and perfectly adapted to most of our cloning requirements. The properties of these vectors are described in Table 39 (see also Figures 60–65). Expression vectors are described in more detail in Chapter 25.

The most striking progress in this area has been the introduction of the "cartridge," or "cassette" concept. In other words, instead of purchasing cloning vectors offering somewhat limited potential applications, it is now possible to buy carrier plasmids from which DNA fragments containing genes useful

(text continued on page 288)

Table 39. A Selection of Commercial Cloning Vectors

Name	Size (kbp)	Selective and Indicator Markers	Features and Applications	Suppliers
		Plasmids		
pBR322 (4)	4.36	Tetracycline resistance Ampicillin resistance	Most widely used cloning vector. Unique sites for SalI, BamHI, SpaI, EcoRV, and XmaIII in the tetracycline resistance gene. Unique site for PvuI, ScaI, and PstI in ampicilline resistance gene.	AME, BOE, BRL, IBI, PRB, PHA
pBR325 (5)	5.99	Tetracyclin resistance Ampicillin resistance Chloramphenicol resistance	Derivative of pBR322. Cloning in the single EcoRI site inactivates chloramphenicol resistance gene.	AME, BRL
pBR328 (28)	4.90	Tetracyclin resistance Ampicillin resistance Chloramphenicol resistance	Derivative of pBR322. Contains the chloramphenicol resistance gene. Cloning at EcoRI leads to inactivation of chloramphenicol resistance gene.	BOE
pBR329	4.50	Tetracyclin resistance Ampicillin resistance Chloramphenicol resistance	Deletion derivative of pBR325. Contains single sites for BalI, EcoRI, PvuII in the chloramphenicol resistance gene.	IBI
pBR322 SmaI		Same as pBR322	Derivative of pBR322 containing a unique SmaI site into the EcoRI site. Insertion of prokaryotic promoters at the EcoRI and SmaI sites restores tetracycline resistance.	PHA
Col E1 (14)	6.36	Colicin resistance	Unique sites for EcoRI and SmaI. Expression of DNA cloned at the EcoRI site. May be induced at high levels by mytomicine.	AME
pAT153 (31)	3.6	Ampicillin resistance Tetracyclin resistance	Derivative of pBR322 deleted between nucleotides 1647 to 2352. Lacks the *nic* site and cannot be mobilized. Higher copy number than pBR322.	AME
pUC8 (32)	2.73	Lack of β-galactosidase activity Ampicillin resistance	Derivative of pBR322 contains polylinker identical to that of M13mp8. Allows forced cloning.	AME, IBI
pUC9 (32)	2.73	Lack of β-galactosidase activity Ampicillin resistance	Same as pUC8 except that polylinker is inserted in reverse orientation (antiparallel to M13mp8) relative to the *lac* gene.	AME, IBI
pUC12 (32)	2.68	Lack of β-galactosidase activity Ampicillin resistance	Derivative of pBR322 containing polylinker identical to that of M13mp12. Allows forced cloning.	BOE, PHA
pUC13 (32)	2.68	Lack of β-galactosidase activity Ampicillin resistance	Same as pUC12 except that polylinker is inserted in reverse orientation relative to the *lac* gene.	BOE, PHA
pUC18 (33)	2.68	Lack of β-galactosidase activity Ampicillin resistance	Derivative of pBR322 containing the *lac* gene and polylinker from M13mp18. Provides single sites for SstI, KpnI, XbaI, and SphI in addition to those present in pUC8. Allows forced cloning.	BOE, BRL, IBI, PHA

Table 39. A Selection of Commercial Cloning Vectors (*continued*)

Name	Size (kbp)	Selective and Indicator Markers	Features and Applications	Suppliers
pUC19 (33)	2.68	Lack of β-galactosidase activity Ampicillin resistance	Same as pUC18 except that polylinker is inserted in reverse orientation relative to the *lac* gene.	BOE, BRL, IBI, PHA
pFBI 1 (2) pFBI 2 (2) pFBI 13 (2) pFBI 14 (2) pFBI 15 (2)	3.2	Ampicilline resistance Tetracyclin resistance	Derivatives of pFB69 obtained by TAB linker-mutagenesis. Unique cloning sites for EcoRI (pFBI 1), SalI (pFBI 15), located in the Tet gene. Insertion of foreign DNA at these sites leads to tetracyclin sensitivity.	PHA
pNO1523 (2, 8)	5.2	Ampicillin resistance Streptomycin resistance	Unique sites for SmaI, HpaI and SstII in streptomycin gene.	PHA
pDR540 (24)	4.0	Ampicillin resistance	Prokaryotic expression vector containing the *tac* (*trp - lac*) promoters and *lac* ribosome binding site. Insertion is usually made at the unique BamHI site.	PHA
pDR720 (24)	4.0	Ampicillin resistance	Prokaryotic expression vector containing the strong *trp* promoter operator. Insertion is usually made at the unique SalI, BamHI or SmaI sites.	PHA
pNEO	5.5	Ampicillin resistance Neomycin resistance	Neomycin gene can be used as selective marker in mammalian and plant cells. Contains a convenient single BglII cloning site in the neomycin resistance gene.	PHA
pJDB207 (3)	6.9	Ampicillin resistance Tetracyclin resistance Fonctional *leu2* yeast gene	Yeast *E. coli* shuttle vector. Derivative of pAT153. Contains the 2μ DNA from yeast. Efficient complementation of *leu⁻* yeast strains.	AME
BLUR8 (15)	4.7	Ampicillin resistance	Derivative of pBR322 containing a 300 base pair repeat sequence from human DNA. Suitable as probe for detection of human *alu* sequences.	AME
BPV-1/pML2d (26)			*E. coli* mammalian cells shuttle vector. Derivative of bovine papilloma virus and pML2. Contains unique HindIII and SalI sites. High copy number of stable non-integrated episomal form in *E. coli* and mouse cells.	IBI
pHSV-106 (19)	7.8	Ampicillin resistance Thymidine kinase activity	Allows expression of herpes simplex virus thymidine kinase in prokaryotes and eukaryotes. May be used as marker in transfection of mammalian cells. Unique HindIII, KpnI, SalI and SphI cloning sites.	BRL
pKH47 (11)	4.4	Ampicillin resistance Tetracyclin resistance	Derivative of pBR322 containing a 100 base pairs poly (dA-dT) insertion. Allows separation of single strands of complementary DNA by chromatography an oligo (dA) or oligo (dT) columns.	BRL

Table 39. A Selection of Commercial Cloning Vectors (*continued*)

Name	Size (kbp)	Selective and Indicator Markers	Features and Applications	Suppliers
pUR222 (25)	2.7	Ampicillin resistance Lack of β-galactosidase activity	Especially suitable for chemical DNA sequencing without isolation of labeled fragments.	BOE
pTZ18R	2.9	Ampicillin resistance Lack of β-galactosidase activity	Contains polylinker of M13mp18. Contains f1 and pBR322 origins of replication for generation of single and double stranded DNA. Allows rapid dideoxy-sequencing. Also contains T7 promoter for efficient transcription of cloned genes.	PHA
pTZ19R	2.9	Ampicillin resistance Lack of β-galactosidase activity	Same as pTZ18R except that contains polylinker is reverse orientation relative to the *lac* gene.	PHA
pIBI20 (9) pIBI21 (9) pIBI24 pIBI25	4.2 2.9	Ampicillin resistance Lack of β-galactosidase activity Ampicillin resistance Lack of β-galactosidase activity	Derived from pEMBL plasmids. Contains both plasmid and f1 origins of replication for generation of single and double stranded DNA. Also contain polylinker and T7 promoter for efficient in vitro transcription of cloned genes.	IBI
pIBI76 (9)	4.2	Ampicillin resistance Lack of β-galactosidase activity	Same as pIBI20, 21 except that it contains both SP6 and T7 RNA polymerase promoters for efficient in vitro transcription of both strands.	IBI
pIBI30 pIBI31	2.8	Ampicillin resistance Lack of β-galactosidase activity	Derived from pEMBL plasmids. Contain both plasmid and f1 origins of replication for generation of single and double stranded DNA. Contain the same cloning sites as M13mp18 plus XhoI, ApaI and MluI sites, as well as T7 and T3 RNA polymerase promoters on either side of the polylinker for efficient in vitro transcription of both strands.	IBI
pSP64 (20, 16)			Derived from pUC12. Contain SP6 RNA polymerase promoter allowing efficient in vitro transcription of genes cloned in the M13 derived polylinker.	BOE, PRB
pSP65 (20, 16)			Same as pSP64 except that orientation of polylinker is inserted with respect to the bacteriophage promoter.	BOE, PRB
pSP18 (20, 16)/ pSP68 (20, 16)	3.0	Ampicillin resistance	Contains SP6 RNA polymerase promoter upstream to the M13mp18 multiple cloning site region, for efficient in vitro transcription.	BRL PHA
pSP19 (20, 16)/ pSP69 (20, 16)	3.0	Ampicillin resistance	Same as pSP18 except that polylinker is in reverse orientation with respect to the promoter sequence.	BRL PHA
pT712 (7)	2.8	Ampicillin resistance	Contains T7 RNA polymerase promoter upstream to the polylinker region from pUC12. Allows efficient in vitro transcription of cloned genes.	BRL

Table 39. A Selection of Commercial Cloning Vectors (*continued*)

Name	Size (kbp)	Selective and Indicator Markers	Features and Applications	Suppliers
pT713 (7)	2.8	Ampicillin resistance	Same as pT712 except that it contains the polylinker region of pUC13.	BRL
pSPT18/ pGEM3	3.1	Ampicillin resistance	Contains both SP6 and T7 transcription promoters on either side of the multiple cloning sites region from pUC18.	BOE, PHA/ PRB
pSPT19/ pGEM4	3.1	Ampicillin resistance	Same as pSPT18 except that polylinker is in reverse orientation.	BOE, PHA/ PRB
pT7/T318	2.8	Ampicillin resistance Lack of β-galactosidase activity	Multipurpose vector containing T3 and T7 RNA polymerase promoters on either site of a multiple cloning sites region from pUC18. Also contains the β galactosidase peptide sequences and the intergenic region from phage f1. Useful for cloning, in vitro transcription, mutagenesis, and sequencing.	BRL
pT7/T319, pT3T7	2.8	Ampicillin resistance Lack of β-galactosidase activity	Same as pT7/T318 except that polylinker is inserted in reverse orientation with respect to the promoters.	BRL, BOE
Bluescript	2.95	Ampicillin resistance Lack of β-galactosidase activity	Contain a polylinker with 21 unique restriction sites flanked by T3 and T7 RNA polymerase promoters for efficient in vitro transcription. Useful vector for site specific mutagenesis and construction of nested deletions in DNA sequencing. Also contains the β-galactosidase sequences for easy selection and f1 IG sequences for production of single stranded DNA on F′ strains. Use of helper phages VCS-M13 and R408 is recommended for efficient rescue of single stranded progeny.	STR
pUCF1	3.2	Ampicillin resistance	Contains the f1 IG as a gene cassette. Introduction of this cassette in a vector allows recovery of single strands for DNA sequencing or mutagenesis.	PHA
pKSV-10	7.2	Ampicillin resistance	*E. coli* mammalian cells shuttle vector derived from pKB111. Contains pBR322 origin of replication, an SV40 transcriptional unit. Analogous to pSV2. Allows insertion of genes downstream from SV40 early promoter, upstream to large T antigen intervening sequence and early region polyadenylation site. Unique cloning sites: BglII, KpnI, BamHI, SalI, EcoRI. Replicates in both *E. coli* and *COS* cells useful for gene transfer and expression in mammalian cells.	PHA
pSVL	4.8	Ampicillin resistance	Very efficient expression vector for eukaryotic cells. Derivative of	PHA

Table 39. **A Selection of Commercial Cloning Vectors (continued)**

Name	Size (kbp)	Selective and Indicator Markers	Features and Applications	Suppliers
			pJC119. Multiple cloning sites downstream to SV40 promoter and upstream to polyadenylation signals. DNA cloned in the polylinker is translated from the first ATG that it contains.	
pCH110 (13, 18)		Ampicillin resistance	Insertion of potential promoter sequences at the unique HindIII site adjacent to the lacZ gene allows testing of their efficiency for transcription. Contains pBR322 "poison sequences."	PHA
pPL-Lambda (10)	5.2	Ampicillin resistance	Contains thermosensitive CI_{857} repressor sequences. Allows inducible expression of sequences inserted at the unique HpaI site located in the N gene, 321 base pairs downstream to the start of P_L transcription.	PHA
pYEJ001	4.1	Ampicillin resistance	Derivative of pBR327 containing a synthetic consensus *E. coli* RNA polymerase promoter and tandem synthetic lactose operators upstream to a unique EcoRI cloning site. Cloning at the HindIII site located close to the promoter site allows monitoring of expression by CAT gene activity.	PHA
pKK223–3 (6)	4.5	Ampicillin resistance	Expression vector. Derivative of pBR322. Contains the *tac* (*trp-lac*) promoter upstream to the multiple cloning sites region from pUC18. Also contains transcription terminator sequences.	PHA
pKK233–2 (1)	4.6	Ampicillin resistance	Expression vector. Contains the *trc* promoter upstream to the *lac* Z ribosome binding site followed by an ATG initiation codon contained in a unique NcoI site. Other unique sites include PstI and HindIII. Useful vector for cloning cDNA with homopolymer tails, or carrying linkers.	PHA
PMSG	7.6	Ampicillin resistance Hypoxanthine, aminopterin, and mycophenolic acid resistance	Very useful vector for stable integration in mammalian cells. Powerful inducible-expression vector. Contains the dexamethasone-inducible transcription promoter from the mouse mammary tumor virus (MMTV) 5′LTR, upstream to a polylinker region with unique cloning sites for NheI, XcyI, SmaI, SalI, and XhoI, The SV40 splicing and polyadenylation sequences located downstream to the multiple cloning sites region complete the functional transcription unit.	PHA

Table 39. A Selection of Commercial Cloning Vectors (continued)

Name	Size (kbp)	Selective and Indicator Markers	Features and Applications	Suppliers
pRIT2T	4.2	Ampicillin resistance	Protein A-gene fusion vector for purification of proteins by IgG-sepharose affinity column (see Chapter 3). Contains λ P$_R$ promoter upstream to the fraction of protein A gene coding for the IgG binding domain and to a polylinker region with unique cloning sites for EcoRI, SmaI, BamHI, SalI, and PstI. Also contains the protein A gene transcription termination sequence inserted downstream to the multiple cloning sites region.	PHA
pRIT5	6.9	Ampicillin resistance Chloramphenicol resistance	*E. coli–S. aureus* shuttle protein A fusion vector. Contains protein A promoter and sequences coding for the IgG binding region, a multiple cloning site as in pRIT2T, origin of replication from the staphylococcal plasmid pC194, and the *E. coli* origin of replication.	PHA
pMC1871	7.4	Tetracyclin resistance	β-Galactosidase-fusion vector. Derivative of pBR322. Contains a promoterless *lac* Z gene deleted at its 5′ terminus. Unique SmaI site to allow fusion to the N terminus of β-galactosidase.	PHA
Okayama and Berg cDNA cloning vectors (22, 23)			See Chapter 19.	PHA
Heidecker and Messing cDNA cloning vectors (12)			See Chapter 19.	PHA

Cosmids

Name	Size (kbp)	Selective and Indicator Markers	Features and Applications	Suppliers
pHC79 (35)	6.4	Ampicillin resistance Tetracyclin resistance	Derivative of pBR322. Contains cohesive ends of λ (*cos*) necessary for phage packaging. Can accept DNA fragments up to 40 kb long. Unique cloning sites: EcoRI, ClaI, HindIII, BamHI, SalI, and PstI.	BOE, BRL
pBTI-1 (37)	11.1	Ampicillin resistance Tetracyclin resistance	*E. coli*–yeast shuttle vector. Contains *cos* region from replication origin from λ, and pBR322, with the 2μ and *leu2* gene from yeast. Allows cloning of DNA fragments up to 40 kb long. The yeast sequences allow complementation screening and autonomous replication. Contains a unique BamHI cloning site.	BOE
pBTI-10 (37)	7.7	Ampicillin resistance Tetracyclin resistance	*E. coli*–yeast shuttle vector. Contains *cos* region from λ and replication origin from pBR322, with the *ars* and	BOE

Table 39. A Selection of Commercial Cloning Vectors (*continued*)

Name	Size (kbp)	Selective and Indicator Markers	Features and Applications	Suppliers
			trp1 gene from yeast. Allows cloning of 40 kb long DNA fragments. Contains both BamHI and SalI unique cloning sites.	
Homer III (34)	5.3	Ampicillin resistance	*E. coli*–mammalian cells shuttle vector. Derivative of pBR322. Contains the *cos* region from λ charon 4A, the SV40 origin of replication and the 72 base pairs repeats, which are essential for in vitro transcription of viral early genes. Very useful vector for reintroduction of cloned fragments (up to 40 kb) into mammalian cells.	AME
pJB8 (36)	5.4	Ampicillin resistance	Derivative of pBR322. Contains *cos* sequences from λ charon 4A. Allows cloning of large fragments (40 kb). Unique BamHI site for cloning of DNA fragments with compatible ends (i.e., sau3AI). Purified arms available.	AME

Bacteriophages

Name	Selection of Recombinants		Features and Applications	Suppliers
λ Charon 4A (38)	Use of purified arms	*lac-* and *bio-* phenotypes	Replacement vector. Very useful for the construction of genomic libraries. Purified EcoRI arms available. Cloning capacity 8–22 kb needs supF host strains (see text).	AME
λ Charon 28A (39)	Use of purified arms	*lac-* and *bio-* phenotypes	Replacement vector. Useful for the construction of genomic libraries. Cloning sites: EcoRI, BamHI, HindIII, SalI, and XhoI (see text). No purified arms available. Needs supF host strains.	PHA
λ gtWES.λB (44)	Use of purified arms		Replacement vector. Derived from λgt, containing amber mutations in the W, E, and S genes. Cloning sites: Eco RI, XhoI, SalI, SstI. Needs supF host strains.	AME, BRL
λ L47 (45)	*spi*	phenotype	Replacement vector. Allows cloning of DNA fragments (generated by EcoRI (8–2 kb), HindIII (7–20 kb), and BamHI (4–7 kb).	AME
λ 1059 (42)	*spi*	phenotype	Replacement vector. Contains ColE1 sequences. Avoid screening of plaques with plasmid-derived probes (see text).	BRL
λ EMBL 3 (40)	*spi*	phenotype	Replacement vector. Derivatives of λ 1059. Cloning capacity 9–23 kb. Contains a multiple cloning sites region allowing insertion of fragments generated by BamHI, EcoRI, and SalI. Does not contain ColE1 sequences. Purified BamHI arms available.	AME, PRB

Table 39. A Selection of Commercial Cloning Vectors (*continued*)

Name	Size (kbp)	Selective and Indicator Markers	Features and Applications	Suppliers
λ EMBL 4 (40)		*spi⁻* phenotype	Same as λ EMBL 3 except that polylinker is inverted. Purified BamHI arms available.	AME, PRB
λ gt10 (41)		Clear plaques	Insertion vector. Useful for construction of cDNA libraries. Purified EcoRI phosphatased arms provided.	PRB
λ gt11 (50)		Lack of β-galactosidase activity	Insertion vector. Useful for construction and rapid screening of cDNA and genomic libraries. Allows expression of cloned fragments. Can accommodate up to 8.0 kb of insert DNA. Purified EcoRI phosphatased arms provided by PRB.	IBI, PRB
M13mp8 (46)		Lack of β-galactosidase activity	Cloning and sequencing system for Sanger's dideoxy nucleotide sequencing method. Contains a multiple cloning sites region. Provided as replicative form double stranded DNA (RF).	AME, BOE
M13mp9 (46)		Lack of β-galactosidase activity	Same as M13mp8 with multiple cloning site region in reverse orientation with respect to the β-galactosidase gene.	AME, BOE
M13mp10 (47)		Lack of β-galactosidase activity	Derivative of M13mp8. Polylinker contains additional sites for SstI and XbaI restriction endonucleases.	AME
M13mp11 (47)		Lack of β-galactosidase activity	Same as M13mp10 with polylinker in reverse orientation.	AME
M13mp18 (48, 49)		Lack of β-galactosidase activity	Derivative of M13mp8 with polylinker containing sites for: AccI, BamHI, EcoRI, HincII, HindIII, KpnI, PstI, SalI, SmaI, SphI, SstI, XbaI, XmaI.	BRL, IBI
M13mp19 (48, 49)		Lack of β-galactosidase activity	Same as M13mp18 with polylinker in reverse orientation.	BRL, IBI
M13tg130 (43)		Lack of β-galactosidase activity	Derivative of M13mp7 containing sites for: AccI, BamHI, EcoRI, EcoRV, HincII, HindIII, KpnI, PstI, SalI, SmaI, SphI, SstI, XbaI.	AME
M13tg 131 (43)		Lack of β-galactosidase activity	Same as M13tg130 except that polylinker is in reverse orientation and it contains an additional BglII restriction site adjacent to the cloning region. Digestion with BglII alleviates the need of direct sequencing to recognize a recombinant clone from its inverted counterpart.	AME
M13um20		Lack of β-galactosidase activity	In addition to new cloning sites for ApaI, MluI, and XhoI restriction fragments, this vector contains T7 and T3 RNA polymerase promoters on either site of the multiple cloning sites region.	IBI

Table 39.   A Selection of Commercial Cloning Vectors (*continued*)

Name	Size (kbp)	Selective and Indicator Markers	Features and Applications	Suppliers
M13um21		Lack of β-galactosidase activity	Same as M13um20 with polylinker in reverse orientation.	IBI

Abbreviations of suppliers names: (AME) Amersham International; (ANB) Stehelin/Anglian Biotechnology; (APL) Appligene; (BEC) Beckman; (BOE) Boehringer Mannheim; (BRL) Bethesda Research Laboratories; (CAL) Calbiochem; (GEN) Genofit; (IBI) International Biotechnologies, Inc.; (MIL) Miles Laboratories; (NEB) New England Biolabs; (NEN) New England Nuclear; (PHA) Pharmacia; (PRB) Promega Biotec; (SIG) Sigma Chemical Co.; (STR) Stratagene; (TAS) Takara Shuzo Co. Ltd.; (USB) United States Biochemical Corp. See Appendix for addresses.

REFERENCES FOR TABLE 39: COMMERCIAL VECTORS

References for Plasmids

[1]Amann, E., and Brosius, J. (1985), *Gene,* **40:** 183.

[2]Barany, F. (1985), *Proc. Natl. Acad. Sci. USA,* **82,** 4202.

[3]Beggs, J.D. (1981), in *Molecular Genetic in Yeast,* Alfred Benzon Symposium, **16,** 383.

[4]Bolivar, F., Rodriguez, R.L., Betlach, M.C., and Boyer, H.N. (1977), *Gene,* **2,** 95.

[5]Bolivar, F. (1978), *Gene,* **4:** 121.

[6]Brosius, J. (1984), *Gene,* **27,** 151.

[7]Davanloo, P., Rosenberg, A.H., Dunn, J.J., and Studier, F.W. (1984), *Proc. Natl. Acad. Sci. USA,* **81,** 2035.

[8]Dean, D. (1981), *Gene,* **15,** 99.

[9]Dente, L., Cesareni, G., and Cortese, R. (1983), *Nucl. Acids Res.,* **11,** 1645.

[10]Drahos, D., and Szybalski, W. (1981), *Gene,* **16,** 261.

[11]Hayashi, K. (1980), *Gene,* **11,** 109.

[12]Heidecker, G., and Messing, J. (1983), *Nucl. Acids Res.* **11,** 4891.

[13]Herbomel, P., Boucharot, B., and Yaniv, M. (1984), *Cell,* **39,** 653.

[14]Hershfield, V., Boyer, H.W., Yanofsky, C., Lovett, M.A., and Helinski, D.R. (1974), *Proc. Natl. Acad. Sci. USA,* **71,** 3255.

[15]Jelinek, W.R., Toomey, T.P., Leinwand, L., Duncan, C.H., Biro, P.A., Choudary, P.V., Weissman, S.M., Rubin, C.M., Houck, C.M., Deininger, P.L., and Schmid, C.W. (1980), *Proc. Natl. Acad. Sci. USA,* **77,** 1398.

[16]Kassavetis, G.A., Butler, E.T., Roulland, D., and Chamberlin, M.J. (1982), *J. Biol. Chem.,* **257,** 5779.

[17]Lee, F., Mulligan, R., Berg, P., and Ringold, G. (1981), *Nature,* **294,** 228.

[18]Lee, F., Hall, C.V., Ringold, G.M., Dobson, D.E., Luh, J. and Jacob, P.E. (1984), *Nucl. Acids Res.* **12,** 4191.

[19]McKnight, S.L., and Grace, E.R. (1980), *Nucl. Acids Res.,* **8,** 5931.

[20]Melton, D.A., Kreig, P.A., Rebagliatai, M.R., Maniatis, T., Zinn, K., and Green, M.R. (1984), *Nucl. Acids Res.,* **12,** 7035.

[21]Nilsson, B., Abrahmsen, L. and Uhlén, M. (1985), *EMBO J.,* **4,** 1075.

[22]Okayama, H., and Berg, P. (1982), *Mol. Cell Biol.* **2,** 161.

[23]Okayama, H., and Berg, P. (1983), *Mol. Cell Biol.,* **3,** 280.

[24]Russel, D.R., and Bennett, G.N. (1982), *Gene,* **20,** 231.

[25]Rüther, U., Koenen, M., Otto, K., and Müller-Hill, B. (1981), *Nucl. Acids Res.* **9,** 4087.

[26]Sarver, N., Byrne, J.C., and Howley, P.M. (1982), *Proc. Natl. Acad. Sci.* **79,** 7147.

[27]Shapira, S.K., Chou, J., Richaud, F.V., and Casadaban, M.J. (1983), *Gene,* **25,** 71.

[28]Soberon, X., Covarrubias, L., and Bolivar, F. (1980), *Gene,* **9,** 287.

[29]Sprague, J., Condra, J.H., Arnheiter, H., and Lazzarini, R.A. (1983), *J. Virol.,* **45,** 773.

[30]Templeton, D., and Eckhart, W. (1984), *Mol. Cell. Biol.,* **4,** 817.

[31]Twigg, A.T., and Sherratt, D. (1980), *Nature,* **283,** 216.

[32]Vieira, J., and Messing, J. (1982), *Gene,* **19,** 259.

[33]Yanisch-Perron, C., Vieira, J., and Messing, J. (1975), *Gene,* **33,** 103.

References for Cosmids

[34]Chia, A., Scott, M.R.D., and Rigby, P.N.J. (1982), *Nuc. Acids Res.,* **10,** 2503.

[35]Hohn, B., and Collins, J. (1980), *Gene,* **11,** 291.

[36]Ish-Horowicz, D., and Burke, J.F. (1981), *Nuc. Acids Res.,* **9,** 2989.

[37]Morris, D.W., Noti, J.D., Osborne, F.A., and Szalay, A.A. (1981), *DNA,* **1,** 27.

References for Bacteriophages

[38]Blattner, F.R., Williams, B.G., Blechl, A.E., Denniston-Thompson, K., Faber, H.E., Furlong, L.A., Grunwald, D.J., Kiefer, D.O., Moore, D.D., Schumm, J.W., Sheldon, E.L., and Smithies, O. (1977), *Science,* **196,** 161.

[39]De Wet, J.R., Daniels, D.L., Schroeder, J.L., Williams, B.G., Denniston-Thompson, K., Moore, D.D., and Blattner, F.R. (1980), *J. Virol.,* **33,** 401.

[40]Frischauf, A.M., Lehrach, H., Poustka, A., and Murray, N. (1983), *J. Mol. Biol.,* **170,** 827.

[41]Huyhn, T.V., Young, R.A., and Davis, R.N. (1985), in *DNA Cloning Techniques: A Practical Approach,* D. Glover, ed., p. 49, IRL Press, Oxford.

[42]Karn, J., Brenner, S., Barnett, L., and Cesareni, G. (1980). *Proc. Natl. Acad. Sci. USA,* **77,** 5172.

[43]Kieny, M.P., Lathe, R., and Lecocq, J.P. (1983), *Gene,* **26,** 91.

[44]Leder, P., Tiemeier, D., and Enquist, L. (1977), *Science,* **196,** 175.

[45]Loenen, W.A.M., and Brammar, W.J. (1980), *Gene,* **20,** 249.

[46]Messing, J., and Viera, J. (1982), *Gene,* **19,** 269.

[47]Messing, J. (1983), in *Methods in Enzymology,* R. Wu, L.

Grossman, and K. Moldave, eds., Vol. 101, Part C, p. 20, Academic Press, New York.

[48]Norrander, J., Kempe, T., and Messing, J. (1983), *Gene,* **26,** 101.

[49]Yanisch-Perron, C., Viera, J., and Messing, J. (1985), *Gene,* **33,** 103.

[50]Young, R.A., and Davis, R.W. (1983), *Proc. Natl. Acad. Sci. USA,* **80,** 1194.

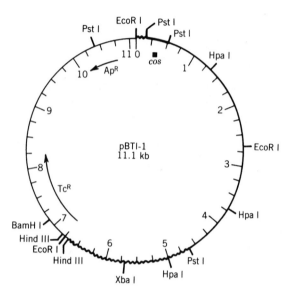

Figure 60. Structure of the pBTI-1 *E. coli*-yeast shuttle vector. This vector, produced and sold by Boehringer Mannheim, carries the yeast origin of replication and 2-μm circular DNA and allows the use of the *leu* phenotype for selection in yeast (courtesy of Boehringer Mannheim).

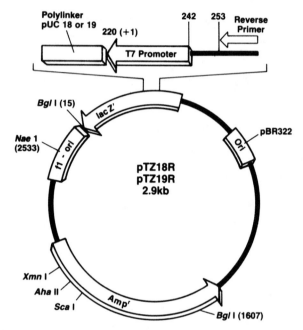

Figure 62. Simplified restriction map of pTZ vectors (courtesy of Pharmacia).

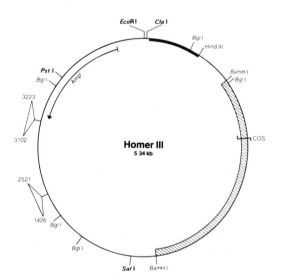

Figure 61. Simplified restriction map of Homer III cosmid (courtesy of Amersham).

for cloning can be excised and used as genetic building blocks to construct unique vectors tailored to the specific needs of a given experimental strategy. For example, a series of cartridges coding for kanamycin resistance are available from Pharmacia S.A. They can be generated from pUC-derived plasmids and inserted in any other cloning vector at various restriction sites, including Apa I, Bam HI, Eco RI, Hind III, Kpn I, Pst I, Sac I, Sal I, Sma I, Sph I, Xma I, and Xho I. Similarly, there are also cartridges containing the ampicillin-resistance gene (Amp) of pBR322, the tetracycline-resistance gene (Tet) from pUC9, the chloramphenicol acetyl transferase gene (CAT) from transposon Tn9, the multiple cloning sites from pUC 13 and pUC 18, and a modified lacZ gene allowing fusions to the N-terminal part of the β-galactosidase (see Figure 66). The commercial cassettes providing regulatory sequences for gene

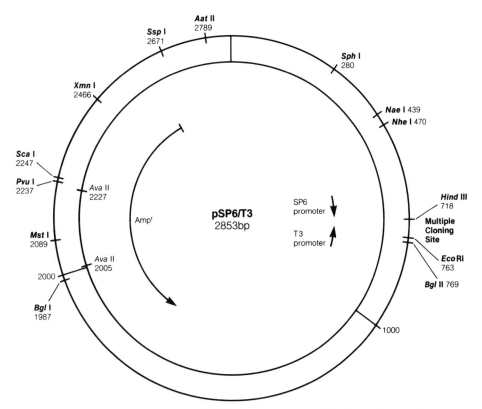

Figure 63. Simplified restriction map of pSP6/T3 vector (courtesy of Bethesda Research Laboratories Life Technologies Inc.).

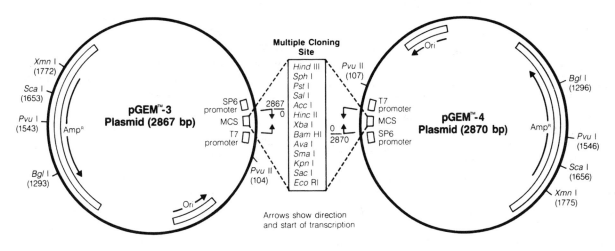

Figure 64. Simplified restriction map of pGEM vector (courtesy of Promega Biotec).

expression, such as transcription promoters (tac, trp, λpl, T7), ribosomal binding site, and universal translation terminator, are extremely convenient complements to these elements (see Figure 217). Undoubtedly, such cartridges are of considerable interest in constructing the ideal cloning or expression vector from a barely simple plasmid such as pBR322; however, the relatively high cost of these products might hinder the generalization of their use.

Considerably more progress has been realized through the development of several "multipurpose"

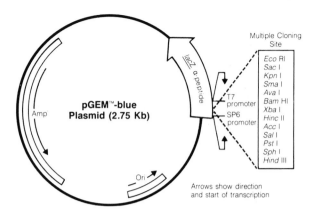

Figure 65. Simplified restriction map of pGEM-Blue vector (courtesy of Promega Biotec).

vectors which can be used for (a) easy cloning of DNA fragments (in a polylinker–β-galactosidase region allowing rapid screening of recombinants), (b) in vitro transcription of the two cloned DNA strands by T7, T3, or SP6 RNA polymerases, (c) direct dideoxy sequencing of the double stranded cloned DNA insert, and (d) production of single stranded DNA template for sequencing or mutagenesis. The use of these "all-in-one" versatile vectors should certainly expand in the near future.

REFERENCES FOR BACILLUS SUBTILIS VECTORS

Band, L., and Henner, D. J. (1984), *DNA*, **3**, 17.

Claverys, J. P., Louarn, J. M., and Sicard, A. M. (1981), *Gene,* **13**, 65.

Flock J.-I., Flotheringham, I., Light, J., Bell, L., and Derbyshire, R. (1984), *Mol. Gen. Genet.* **195**, 246.

Furasato, T., Takano, J. I., Jigami, Y., Tanaka, H., Yamane, K. (1986), *J. Biochem.,* **155**, 577.

Gallizzi, A., Scofoone, F., Milanesi, G., and Albertini, A. M. (1981), *Mol. Gen. Genet.,* **182**, 99.

Hardy, D., Stahl, S., and Küpper, H. (1981), *Nature,* **293**, 481.

Harris-Warrick, R. M., and Lederberg, J. (1978), *J. Bacteriol.,* **133**, 1246.

Mosbach, K., Birnbaum, S., Hardy, K., Davies, J., and Bülow, L. (1983), *Nature,* **302**, 543.

Nakahama, K., Miyazaki, T., and Kikuchi, M. (1985), *Gene,* **36**, 179.

Palva, I., Lehtovaara, P., Kääriäinen, L., Sibakov, M., Cantell, K., Schein, C. H., Kashiwagi, K., and Weissmann, C. (1983), *Gene,* **22**, 229.

Prozorov, A. A., Savchenko, G. V., Lakomova, N. M., and Poluektova, E. U. (1983), *Gene,* **22**, 41.

Prozorov, A. A., Bashkirov, V. I., Khasanov, F. K., Glumova, E. F., and Irich, V. (1985), *Gene,* **34**, 39.

Schoner, R. G., Williams, D. M., and Lovett, P. S. (1983), *Gene,* **22**, 47.

Shimotsu, H., and Henner, D. J. (1986), *Gene,* **43**, 85.

Shiroza, T., Nakazawa, K., Tashiro, N., Yamane, K., Yanagi, K., Yamasaki, M., Tamura, G., Saito, H., Kawade, Y., and Taniguchi, T. (1985), *Gene,* **34**, 1.

Sibakov, M. (1986), *Eur. J. Biochem.,* **155**, 577.

Stuy, J. H., and Walter, R. B. (1981), *J. Bacteriol.,* **148**, 565.

Williams, D. M., Schoner, R. G., Duvall, E. J., Preis, L. H., and Lowett, P. S. (1981), *Gene,* **16**, 199.

Yansura, D. G., and Henner, D. J. (1984), *Proc. Natl. Acad. Sci. USA,* **81**, 439.

REFERENCES FOR PSEUDOMONAS VECTORS

Bagdasarian, M., Bagdasarian, M. M., Coleman, S., and Timmis, K. N. (1979), in Timmis, K. N., and Pühler, A. (eds.), *Plasmids of Medical, Environmental and Commercial Importance,* Elsevier, Amsterdam, p 411.

Bagdasarian, M., Lurz, R., Ruckert, B., Franklin, F. C. H., Bagdasarian, M. M., Frey, T., and Timmis, K. N. (1981), *Gene,* **16**, 237.

Bagdasarian, M., and Timmis, K. N. (1982), *Current Topics in Microbiology and Immunology,* Springer Verlag, Berlin, p. 47.

Bagdasarian, M. M., Amann, E., Lurz, R., Ruckert, B., and Bagdasarian, M. (1983), *Gene,* **26**, 273.

Barth, P. T. (1979), in *Plasmids of Medical, Environmental and Commercial Importance,* Elsevier, Amsterdam, pp. 399.

Clark, P. H., and Richmond, M. H. (1975), *Genetics and Biochemistry of Pseudomonas,* Wiley, New York.

Darzin, A., and Chakrabarty, A. M. (1984), *J. Bacteriol.,* **159**, 9.

Ditta, G., Stanfield, S., Corbin, D., and Helinski, D. R. (1980), *Proc. Natl. Acad. Sci. USA,* **77**, 7347.

Friedman, A. M., Long, S. R., Brown, S. E., Buikema, W. J., and Ausubel, F. M. (1982), *Gene,* **18**, 289.

Itoh, Y., and Haas, D. (1985), *Gene,* **36**, 27.

Knauf, V. C., and Nester, E. W. (1982), *Plasmid,* **8**, 45.

Moores, J. C., Magazin, M., Ditta, G. S., and Leong, J. (1984), *J. Bacteriol.,* **157**, 53.

Morales, V. M., and Sequeira, L. (1985), *J. Bacteriol.,* **163**, 1263.

Nomura, N., Yamagishi, H. and Oka, A. (1978), *Gene,* **3**, 39.

Norrander, J., Kempe, T., and Messing, J. (1983), *Gene,* **26**, 101.

Sharpe, G. S. (1984), *Gene,* **29**, 93.

Vierira, J., and Messing, J. (1982), *Gene,* **19**, 259.

Werneke, J. M., Sligar, S. G., and Schuler, M. A. (1985), *Gene,* **38**, 73.

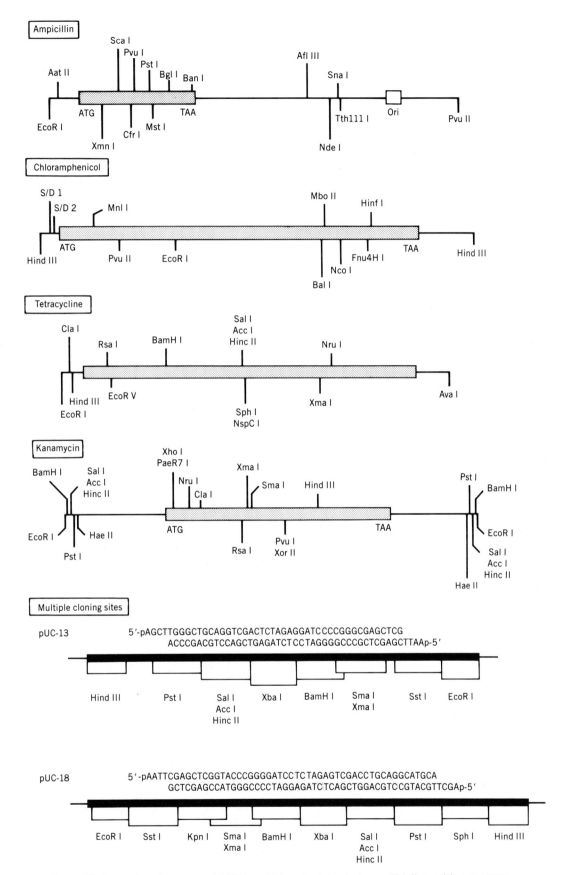

Figure 66. Examples of commercial DNA cartridges for introducing antibiotics resistance genes and multiple cloning sites in cloning vectors.

REFERENCES FOR BREVIBACTERIUM LACTOFERMENTUM AND CORYNEBACTERIUM GLUTAMICUM

Kaneko, H., and Sakaguchi, K. (1979), *Agric. Biol. Chem.,* **48**, 867.

Katsumata, R., Ozaki, A., Oka, T., and Furuya, A. (1984), *J. Bacteriol.* **159**, 306.

Miwa, K., Matsui, H., Terabe, M., Nakamori, S., Sano, K. and Momose, H. (1984), *Agric. Biol. Chem.,* **48**, 2901.

Miwa, K., Matsui, K., Terabe, M., Ito, K., Ishida, M., Takagi, H., Nakamori, S., and Sano, K. (1985), *Gene,* **39**, 281.

Sandoval, H., Aguilar, A., Paniagua, C., and Martin, J. F. (1984), *Appl. Microbiol. Biotechnol.,* **19**, 409.

Santamaria, R., Gil, J. A., Mesas, J. M., and Martin, J. F. (1984), *J. Gen. Microbiol.,* **130**, 2237.

Yoshihama, M., Higashiro, K., Rao, E. A., Akedo, M., Shanabruch, W. G., Follettie, M. F., Walker, G. C., and Sinskey, A. J. (1985), *J. Bacteriol.,* **162**, 591.

REFERENCES FOR STREPTOMYCES VECTORS

Bibb, M. J., Freeman, R. F., and Hopwood, D. A. (1977), *Mol. Gen. Genet.,* **154**, 155.

Bibb, M. J., Schottel, J. L., and Cohen, S. N. (1980), *Nature,* **284**, 526.

Bibb, M. J., Ward, J. M., and Hopwood, D. A. (1980), *Dev. Ind. Microbiol.,* **21**, 55.

Bibb, M. J., and Hopwood, D. A. (1981), *J. Gen. Microbiol.,* **126**, 427.

Heusterspreute, M., Oberto, J., Ha-Thi, V., and Davidson, J. (1985), *Gene,* **34**, 363.

Ikeda, H., Seno, E. T., Bruton, C. J., and Chater, K. F. (1984), *Mol. Gen. Genet.,* **19**, 501.

Larson, J. L., and Hersberger, C. L. (1984), *J. Bacteriol.,* **157**, 314.

Lydiate, D. J. (1984), Ph. D. Thesis, University of East Anglia, Norwich.

Lydiate, D. J., Malpartida, F., and Hopwood, D. A. (1985), *Gene,* **35**, 223.

Schrempf, H. and Goebel, W. (1975), *J. Bacteriol.,* **121**, 416.

REFERENCES FOR YEAST VECTORS

Araki, H., Jearnpipatkul, A., Tatsumi, H., Sakurai, T., Ushio, K., Muta, T., and Oshima, Y. (1985), *J. Mol. Biol.,* **182**, 191.

Beach, D., and Nurse, P. (1981), *Nature,* **290**, 123.

Bianchi, M. M., Falcone, C., Chen, X. J., Wesolonski-Louvel, M., Frontali, L., and Fukuhara, H. (1987), *Curr. Genet.,* **12**, 185.

Broach, J. R. (1982), in *The Molecular Biology of the Yeast Saccharomyces. Life Cycle and Inheritance,* J. N. Strathern, E. W. Jones, and J. R. Broach, eds., p. 445, Cold Spring Harbor Laboratory, Cold Spring Harbor, N.Y.

Chen, X. J., Saliola, M., Falcone, C., Bianchi, M. M., and Fukuhara, H. (1986), *Nucl. Acids Res.,* **14**, 4471.

Clarke, L., and Carbon, J. (1980), *Proc. Natl. Acad. Sci. USA,* **77**, 2173.

Das, S., and Hollenberg, C. P. (1982), *Curr. Genet.,* **6**, 123.

Fabiani, L., and Frontali, L. (1986), *Yeast,* **2**, S104.

Fagan, M. C., and Scott, J. F. (1985), *Gene,* **40**, 217.

Falcone, C., Saliola, M., Chen, X. J., Frontali, L., and Fukuhara, H. (1986), *Plasmid,* **15**, 248.

Fitzgerald-Hayes, M., Buhler, J. M., Cooper, T. G., and Carbon, J. (1982), *Mol. Cell. Biol.,* **2**, 82.

Toh-e, A., Tada, S., and Oshima, Y. J. (1982), *J. Bacteriol.,* **151**, 1380.

Toh-e, A., Araki, H., Utatsu, I., and Oshima, Y. J. (1984), *J. Gen. Microbiol.,* **130**, 2527.

REFERENCES FOR PLANT VECTORS

An, G., Watson, B. D., Stachel, S., Gordon, M. P., and Nester, E. W. (1985), *EMBO J.,* **4**, 277.

Balazs, E., Guilley, H., Jonard, G., Paszkowski, J., and Richards, K. (1985), *Gene,* **40**, 273.

Balazs, E., Bouzoubaa, S., Guilley, H., Jonard, G., Paszkowski, J., and Richards, K. (1985), *Gene,* **40**, 343.

Barton, K. A., and Chilton, M.-D. (1983), in *Methods in Enzymology,* R. Wu, L. Grossman, and K. Moldave, eds., Vol. 101 part C, p. 527, Academic Press, New York.

Bevan, M. W., Flavell, R. B., and Chilton, M. D. (1983), *Nature,* **304**, 184.

Bevan, M. (1984), *Nucl. Acids Res.,* **12**, 8711.

Bomhoff, G., Klapwijk, P. M., Kester, H. C. M., Schilperoort, R. A., Hernalsteens, J. P., and Schell, J. (1976), *Mol. Gen. Genet.,* **145**, 177.

Caplan, A., Herrera-Estrella, L., Inzé, D., Van Haute, E., Van Montagu, M., Schell, J., and Zambryski, P. (1983), *Science,* **222**, 816.

Chilton, M.-D., Tepfer, D. A., Petit, A., David, C., Casse-Delbart, F., and Tempe, J. (1982), *Nature,* **295**, 432.

Deblaere, R., Bytebier, B., De Greve, H., Deboeck, F., Schell, J., Van Montagu, M., and Leemans, J. (1985), *Nucl. Acids Res.,* **13**, 4777.

De Cleene, M., and DeLey, J. (1976), *Botan. Rev.,* **42**, 389.

Fraley, R. T., Rogers, S. G., Horsch, R. B., Sanders, P. R., Flick, J. S., Adams, S. P., Bittner, M. L., Brand, L. A., Fink, C. L., Fry, J. S., Galluppi, G. R., Goldberg, S. B.,

Hoffmann, N. L., and Woo, S. C. (1983), *Proc. Natl. Acad. Sci. USA,* **80,** 4803.

Gheysen, G., Dhaese, P., Van Montagu, M., and Schell, J. (1985), in *Genetic Flux in Plants (Advances in Plant Gene Research, vol. 2),* B. Hohn, and E. S. Dennis, eds., p. 11, Springer Verlag, Wien.

Herrera-Estrella, L., De Block, M., Messens, E., Hernalsteens, J.-P., Van Montagu, M., and Schell, J. (1983), *EMBO J.,* **2,** 987.

Herrera-Estrella, L., Depicker, A., Van Montagu, M., and Schell, J. (1983), *Nature,* **303,** 209.

Hernalsteens, J.-P., Thia-Toong, L., Schell, J., and Van Montagu, M. (1984), *EMBO J.,* **3,** 3039.

Hille, J., Wullems, G., and Schilperoort, R. (1983), *Plant Mol. Biol.,* **2,** 155.

Hoekema, A., Hirsch, P. R., Hooykaas, P. J. J., and Schilperrort, R. A. (1983), *Nature,* **303,** 179.

Hooykaas-Van Slogteen, G. M. S., Hooykas, P. J. J., and Schilperoort, R. A. (1984), *Nature,* **311,** 763.

Joos, H., Inzé, D., Caplan, A., Sormann, M., Van Montagu, M., and Schell, J. (1983), *Cell,* **32,** 1057.

Klee, H. J., Yanofsky, M. F., and Nester, E. W. (1985), *Biotechnology,* **3,** 637.

Koncz, C., and Schell, J. (1986), *Mol. Gen. Gen.,* **204,** 383.

Koziel, M. G., Adams, T. L., Hazlet, M. A., Damm, D., Miller, J., Dahlbeck, D., Jayne, S., and Staskawicz, B. J. (1984), *J. Mol. Appl. Genet.,* **2,** 549.

Leemans, J., Deblaere, R., Willmitzer, L., De Greve, H., Hernalsteens, J. P., Van Montagu, M., and Schell, J. (1982), *EMBO J.,* **1,** 147.

Leemans, J., Shaw, C., Deblaere, R., Degreve, H., Hernalsteens, J. P., Maes, M., Van Montagu, M., and Schell, J. (1981), *J. Mol. Appl. Genet.,* **1,** 149.

Murrai, N., and Kemp, J. D. (1982), *Proc. Natl. Acad. Sci. USA,* **79,** 86.

Nester, E. W., Gordon, M. P., Amasino, R. M., and Yanofsky, M. F. (1984), *Annu. Rev. Plant Physiol.,* **35,** 387.

Odell, J. T., Nagy, F., and Chua, N.-M. (1985), *Nature,* **313,** 810.

Paszkowski, J., Shillito, R. D., Saul, M., Mandak, V., Hohn, T., Hoh, B., and Potrykus, I. (1984), *EMBO J.,* **3,** 2717.

Petit, A., Delhaye, S., Tempe, J., and Morel, G. (1970), *Physiol. Veg.,* **8,** 205.

Pietrzak, M., Shillito, R. D., Hohn, T., and Potrykus, I. (1986), *Nucl. Acids Res.,* **14,** 5857.

Schoffl, F., and Baumann, G. (1985), *EMBO J.,* **4,** 1119.

Schroder, J., Schroder, G., Huisman, H., Schilperoort, R. A., and Schell, J. (1981), *FEBS Lett.,* **129,** 166.

Shaw, C. H., Watson, M. D., Carter, G. H., and Shaw, C. H. (1984), *Nucl. Acids Res.,* **12,** 6031.

Shillito, R. D., Saul, M. W., Paszkowski, J., Müller, M., and Potrykus, I. (1985), *Biotechnology,* **3,** 1099.

Simoens, C., Alliotte, Th., Mendel, R., Müller, A., Schiemann, J., Van Lijsebettens, M., Schjell, J., Van Montagu, M., and Inzé, D. (1986), *Nucl. Acids Res.,* **14,** 8073.

Simpson, R. B., O'Hara, P. J., Kwok, W., Montoya, A. L., Lichtenstein, C., Gordon, M. P., and Nester, E. W. (1982), *Cell,* **29,** 1005.

Tempe, J., and Schell, J. (1977), in *Translation of Natural and Synthetic Polynucleotides* A. B. Legocki, ed., p. 416, University of Agriculture, Poznan.

Van Den Elzen, P., Lee, K. Y., Townsend, J., and Bedbrook, J. (1985), *Plant Mol. Biol.,* **5,** 149.

Velten, J., Velten, L., Hain, R., and Schell, J. (1984), *EMBO J.,* **12,** 2723.

Velten, J., and Schell, J. (1985), *Nucl. Acids Res.,* **13,** 6981.

Wang, K., Herrera-Estrella, L., Van Montagu, M., and Zambryski, P. (1984), *Cell,* **38,** 455.

Yadav, N. S., Vanderleyden, J., Bennett, D. R., Barnes, W. M., and Chilton, M. D. (1982), *Proc. Natl. Acad. Sci. USA,* **79,** 6322.

Zambryski, P., Depicker, A., Kruger, K., and Goodman, H. (1982), *J. Mol. Appl. Genet.,* **1,** 361.

Zambryski, P., Joos, H., Genetello, C., Leemans, J., Van Montagu, M., and Schell, J. (1983), *EMBO J.,* **2,** 2143.

REFERENCES FOR VIRAL VECTORS

Adenovirus Vectors

Aiello, L., Guilfoyle, R., Huebner, K., and Weinmann, R. (1979), *Virology,* **94,** 460.

Anderson, C. W., Lewis, J. B., Baum, P. R., and Gesteland, R. F. (1976), *J. Virol.,* **18,** 685.

Berk, A. J., Lee, F., Harrison, T., Williams, J., and Sharp, P. A. (1979), *Cell,* **17,** 935.

Berkner, K. L., and Sharp, P. A. (1983), *Nucl. Acids Res.,* **11,** 6003.

Graham, F. L., Smiley, J., Russel, W. C., and Nairn, R. (1977), *J. Gen. Virol.,* **36,** 56.

Haj-Ahmad, Y., and Graham, F. L. (1986), *J. Virol.,* **57,** 267.

Hayward, W. S., Neel, B. G., and Astrin, S. M. (1981), *Nature,* **290,** 475.

Jones, N., and Shenk, T. (1979), *Proc. Natl. Acad. Sci. USA,* **76,** 3665.

Kelly T. J., and Lewis, A. M. (1973), *J. Virol.,* **12,** 643.

Mansour, S. L., Grodzicker, T., and Tjian, R. (1985), *Proc. Natl. Acad. Sci. USA,* **82,** 1359.

Neel, B., Hayward, W., Robinson, H. L., Fang, L., and Astrin, S. (1981), *Cell,* **23,** 323.

Nevins, J. R. (1981), *Cell,* **26,** 213.

Payne, G., Courtneidge, S., Crittendon, L., Fadly, A., and Bishop, J. (1981), *Cell*, **23**, 311.

Ruether, J. E., Maderious, A., Lavery, D., Logan, J., Fu, S. M., and Chen-Kiang, S. (1986), *Mol. Cell. Biol.*, **6**, 123.

Saito, I., Oya, Y., Yamamoto, T., Yuasa, T., and Shimojo, H. (1985), *J. Virol.*, **54**, 711.

Saito, I., Oya, Y., and Shimojo, H. (1986), *J. Virol.*, **58**, 554.

Sharp, P. A. (1984), in *The Andenoviruses*, H. S. Ginsberg, ed., Plenum, New York.

Yamada, M., Lewis, J. A. and Grodzicker, T. (1985), *Proc. Natl. Acad. Sci. USA*, **82**, 3567.

Retroviral vectors

Brown, A. M. C., Wildin, R. S., Prendergast, T. J., and Varmus, H. E. (1986), *Cell*, **46**, 1001.

Cepko, C. L., Robert, B. E., and Mulligan, R. C. (1983), *Cell*, **37**, 1053.

Friedman, R. L. (1985), *Proc. Natl. Acad. Sci. USA*, **82**, 703.

Gazit, A., Pierce, J. H., Kraus, M. H., Di Fiore, P. P., Pennington, C. Y. and Aaronson, S. A. (1986), *J. Virol.*, **60**, 19.

Gruber, H. E., Finley, K. D., Hershberg, R. M., Katzman, S. S., Laikind, P. K., Seegmiller, J. E., Friedman, T., Yee, J. K., and Jolly, D. J. (1985), *Science*, **230**, 1057.

Hayward, W. S., Neel, B. G., and Astrin, S. M. (1981), *Nature*, **290**, 475.

Hellerman, J. G., Cone, R. C., Potts, J. T., Rich, A., Mulligan, R. C., and Kronenberg, H. M. (1984), *Proc. Natl. Acad. Sci. USA*, **81**, 5340.

Huszar, D., Balling, R., Kothary, R., Magli, M. C., Hozumi, N., Rossant, J., and Bernstein, A. (1985), *Proc. Natl. Acad. Sci. USA*, **82**, 8587.

Jaenisch, R., Fan, H., and Croker, B. (1975), *Proc. Natl. Acad. Sci. USA*, **72**, 4008.

Jaenisch, R. (1976), *Proc. Natl. Acad. Sci. USA*, **73**, 1260.

Jaenisch, R. (1980), *Cell*, **19**, 181.

Jaenisch, R., Jähner, D., Nobis, P., Simon, I., Löhler, J., Harbers, K., and Grotkopp, D. (1981), *Cell*, **24**, 519.

Jahner, D., and Jaenisch, R. (1980), *Nature*, **287**, 456.

Jensen, N. A., Jorgensen, P., Kjeldgaard, N. O., and Pedersen, F. S. (1986), *Gene*, **41**, 59.

Jhappan, C., Vande Woude, G. R., and Robins, T. S. (1986), *J. Virol.*, **60**, 750.

Jolly, D. J., Willis, R. C., and Friedman, T. (1986), *Mol. Cell. Biol.*, **6**, 1141.

Joyner, A. L., and Bernstein, A. (1983), *Mol. Cell. Biol.*, **3**, 2180.

Joyner, A., Keller, G., Phillips, R. A., and Bernstein, A. (1983), *Nature*, **305**, 556.

Kornbluth, S., Cross, F. R., Harbison, M. and Hanafusa, H. (1986), *Mol. Cell. Biol.*, **6**, 1545.

Mann, R., Mulligan, R. C., and Baltimore, D. (1983), *Cell*, **33**, 153.

Miller, A. D., Jolly, D. J., Friedmann, T., and Verma, I. M. (1983), *Proc. Natl. Acad. Sci. USA*, **80**, 4709.

Neel, B., Hayward, W., Robinson, H. L., Fang, L., and Astrin, S. (1981), *Cell*, **23**, 323.

Payne, G., Courtneidge, S., Crittenden, L., Fadly, A., Bishop, J., and Varmus, H. (1981), *Cell*, **23**, 311.

Perkins, A. S., Kirschmeler, P. T., Gattoni-Celli, S., and Weinstein, J. B. (1983), *Mol. Cell. Biol.*, **3**, 1123.

Robertson, E., Bradley, A., Kuehn, M. and Evans, M. (1986), *Nature*, **323,** 445.

Shimitohno, K., and Temin, H. M. (1981), *Cell*, **26**, 67.

Sorge, J., and Hughes, S. H. (1982), *J. Mol. Appl. Genet.*, **1**, 547.

Stewart, C. L., Vanek, M., and Wagner, E. F. (1985), *EMBO J.*, **4**, 3701.

Stuhlmann, H., Cone, R., Mulligan, R. and Jaenisch, R. (1984), *Proc. Natl. Acad. Sci. USA*, **81**, 7151.

Tabin, C. J., Hoffmann, J. W., Goff, S. P., and Weinberg, R. A. (1982), *Mol. Cell. Biol.*, **2**, 426.

Uchida, N., Cone, R. D., Freeman, G. J., Mulligan, R. C. and Cantor, H. (1986), *J. Immunol.*, **136**, 1876.

Van Der Putten, H., Botteri, F. M., Miller, A. D., Rosenfeld, M. G., Fan, H., Evans, R. M., and Verma, I. M. (1985), *Proc. Natl. Acad. Sci. USA*, **82**, 6148.

Wei, C. M., Gibson, M., Spear, P. G., and Scolnick, E. M. (1981), *J. Virol.*, **39**, 935.

Williams, D. A., Lemischka, I. R., Nathan, D. G., and Mulligan, R. C. (1984), *Nature,* **310**, 476.

RECENT REFERENCES FOR VECTORS OF INTEREST IN A FEW OTHER SYSTEMS

Aspergillus nidulans

Ballance, D. J., and Turner, G. (1985), *Gene*, **36**, 321.

Myxococcus xanthus

Furuichi, T., Inouye, M., and Inouye, S. (1985), *J. Bacteriol.*, **164**, 270.

Streptococcus (pneumoniae and foecalis)

Prats, H., Martin, B., Pognonec, P., Burger, A. C., and Claverys, J. P. (1985), *Gene*, **39**, 41.

Wirth, R., An, F. Y., and Clewell, D. B. (1986), *J. Bacteriol.*, **165**, 831.

Staphylococcus aureus

Zyprian, E., and Matzura, H. (1986), *DNA,* **3,** 219.

Cyanobacterium

Buzby, J. S., Porter, R. D., and Stevens, S. E. (1985), *Science,* **230,** 805.

Chauvat, F., De Vries, L., Van Der Ende, A., Van Arkel, G. A. (1986), *Mol. Gen. Genet.,* **204,** 185.

REFERENCES FOR VECTORS USED IN THE DETECTION OF REGULATORY SIGNALS

Brosius, J. (1984), *Gene,* **27,** 151.

Brosius, J. (1984), *Gene,* **27,** 161.

Burke, J. F., and Mogg, A. E. (1985), *Nucl. Acids Res.,* **13,** 1317.

Casadaban, M. J., Chou, J., and Cohen, S. N. (1980), *J. Bacteriol.,* **143,** 971.

Chak, K. F., and James, R. (1985), *Nucl. Acids Res.,* **13,** 2519.

de Boer, H. A. (1984), *Gene,* **30,** 251.

de Crombrugghe, B., Mudryj, M., DiLauro, R., and Gottesman, M. (1979), *Cell,* **18,** 1145.

Enger-Valk, B. E., Van Rotterdam, J., Kos, A., and Pouwels, P. H. (1981a), *Gene,* **15,** 297.

Enger-Valk, B. E., Van Rotterdam, J., and Pouwels, P. H. (1981b), *Nucl. Acids Res.,* **9,** 1973.

Honigman, A., Mahajna, J., Altuvia, S., Koby, S., Teff, D., Locker-Giladi, H., Hyman, H., Kronman, C., and Oppenheim, A. B. (1985), *Gene,* **36,** 131.

Kaufman, R. J. (1985), *Proc. Natl. Acad. Sci. USA,* **82,** 689.

Linn, T., and Ralling, G. (1985), *Plasmid,* **14,** 134.

Masui, Y., Coleman, J., and Inouye, M. (1983), in *Experimental Manipulation of Gene Expression,* M. Inouye, ed., p. 15, Academic Press, New York.

Mc Kenney, K., Shimatake, H., Court, D., Schmeissner, U., Brady, C., and Rosenberg, M. (1981), *Gene Amplification and Analysis, Vol. 2,* p. 383, *Structural Analysis of Nucleic Acids,* Elsevier, Amsterdam.

Mashko, S. V., Lebedeva, M. I., Podkovyrov, S. M., Kashlev, M. V., Trukhan, M. E., Rebentish, B. A., Kozlov, Y. I., and Debabov, V. G. (1985), *Mol. Biol.,* **19,** 973.

Minton, N. P. (1984), *Gene,* **31,** 269.

Prost, E., and Moore, D. D. (1986), *Gene,* **45,** 107.

Rechinskii, V. O., Savochkina, L. P., and Bibilashvili, R. (1981), *Mol. Biol.,* **15,** 737.

Ruether, J. E., Maderrious, A., Lavery, D., Logan, J., Fu, S. M., and Chen-Kiang, S. (1986), *Mol. Cell. Biol.,* **6,** 277.

Schneider, K., and Beck, C. F. (1986), *Gene,* **42,** 37.

Soberon, X., Rossi, J. J., Larson, G. P., and Itakura, K. (1982), *Structure and Function,* Praeger, New York, p. 407.

Thomas, D. Y., Dubuc, G., and Narang, S. (1982), *Gene,* **19,** 211.

West, R. W., Jr., Neve, R. L., and Rogriguez, R. L. (1979), *Gene,* **7,** 271.

West, R. W., and Rodriguez, R. L. (1982), *Gene,* **20,** 291.

Wyckoff, E., Sampson, L., Hayden, M., Parr, R., Mun Huang, W., and Casjen, S. (1986), *Gene,* **43,** 281.

REFERENCES FOR USE OF SHUTTLE VECTORS

Glazer, P. M., Sarkar, S. N., and Summers, W. C. (1986), *Proc. Natl. Acad. Sci USA,* **83,** 1041.

Heusterspreute, M., Oberto, J., Ha-Thi, V., and Davison, J. (1985), *Gene,* **34,** 363.

Jhappan, C., Vande Woude, G. F., and Robins, T. S. (1986), *J. Virol.,* **60,** 750.

Lazo, P. A. (1985), *Gene,* **39,** 41.

Lebkowski, J. S., Miller, J. H., and Calos, M. P. (1986), *Mol. Cell. Biol.,* **6,** 1838.

Piwnica-Worms, H., Kaplan, D. R., Whitman, M., and Roberts, T. M. (1986), *Mol. Cell. Biol.,* **6,** 2033.

Seidman, M. M., Dixon, K., Razzaque, A., Zagursky, R. J., and Berman, M. L. (1985), *Gene,* **38,** 233.

Wirth, R., An, F. Y., and Clewell, D. B. (1986), *J. Bacteriol.,* **165,** 831.

RECENT REFERENCES FOR A FEW OTHER VIRAL VECTORS

Bovine Papilloma Vectors

Braam-Markson, J., Jaudon, C., and Krug, R. M. (1985), *Proc. Natl. Acad. Sci. USA,* **82,** 4891.

Vaccina Virus Vectors

Buller, R. M. L., Smith, G. L., Cremer, K., Notkins, A. L., and Moss, B. (1985), *Nature,* **317,** 813.

Stott, E. J., Ball, L. A., Young, K. K., Furze, J., and Wertz, G. W. (1986), *J. Virol.,* **60,** 607.

T4 Bacteriophage Vectors

Henrich, B., and Plapp, R. (1986), *Gene,* **42,** 345.

Kern, F. G., and Basilico, C. (1986), *Gene,* **43,** 237.

Shub, D. A., and Casna, N. J. (1985), *Gene,* **37,** 31.

Baculovirus Vectors

Kuroda, K., Hausa, C., Rott, R., Klenk, H. D., and Doerfler, W. (1986), *EMBO J.,* **5**, 1359.

Maeda, S., Kawai, T., Obinata, M., Fujiwara, H., Horiuchi, T., Saeki, Y., Sato, Y., and Furusawa, M. (1985), *Nature,* **315**, 592.

Smith, G. E., Ju, G., Ericson, B. L., Moschera, J., Lahm, H. W., Chizzonite, R., and Summers, M. D. (1985), *Proc. Natl. Acad. Sci. USA,* **82**, 8404.

7

PURIFICATION AND CHARACTERIZATION OF VECTOR AND PASSENGER DNA

Several different procedures have been developed over the past few years for the isolation of plasmid DNA free of contaminating RNA species. These methods involve cesium choride equilibrium density centrifugation (Clewel and Helinski, 1969; see below), lithium chloride precipitation followed by a chromatography on glass powder column (Marko et al., 1982), chromatography through Sepharose columns CL-4B (Cornelis et al., 1981; see also Chapter 3), and Sephacryl S-300 or S-100 (Norgard, 1981; see also Chapter 3) or Ultrogel A2 (Micard et al., 1985). Other methods involve chromatography on hydroxyapatite columns (Colman et al., 1978) or precipitation with calcium chloride (Mukhopadhyay and Mandal, 1983). In all cases, the complete removal of RNA species is achieved by treatment with RNAse A. Because RNAse treatment must be considered with caution when the DNA is to be employed for the subsequent purification of biologically active RNA species, we describe techniques in which no RNAse treatment is performed. When DNA is to be used for cloning purposes only, incubation with RNAse followed by phenol extraction and reprecipitation allows quantification of DNA yields. If DNA preparations free of RNA are required (for example, when labeling 5′ ends of DNA fragments with polynucleotide kinase, or for sequencing), we recommend performing two successive cycles of cesium choride centrifugation. It is also possible to remove RNA by lithium chloride precipitation followed by sedimentation through high salt solution (Lev, 1987, see Protocol below).

ISOLATION OF PLASMID DNA USING CESIUM CHLORIDE GRADIENT CENTRIFUGATION

Figure 67 schematically illustrates the steps of a procedure for the isolation of plasmid DNA following mild lysis of spheroplasts by osmotic shock.

PROTOCOL

ISOLATION OF PLASMID DNA FOLLOWING MILD LYSIS OF SPHEROPLASTS

Inoculation and Amplification

1. In a sterile tube put 10 ml 2X TYE medium, supplemented with 0.1% bacto dextrose and antibiotics (depending on plasmid used) as desired.

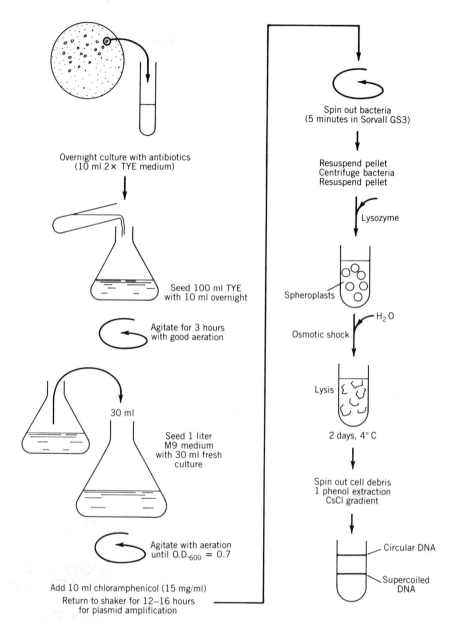

Figure 67. Amplification and purification of plasmid DNA. See text for details.

2. Seed with a single colony from a freshly streaked plate and let grow overnight on a shaker at 37°C.
3. The next morning, inoculate 100 ml prewarmed TYE medium with the 10-ml overnight culture.
4. Grow at 37°C for 3 hours to get organisms out of the stationary phase.
5. Add 30 ml of the 100-ml culture (now containing exponentially growing bacteria) to 1 liter M9 minimal medium.
6. Read the absorbance at 600 nm; let grow until the A_{600} reaches 0.7 when read on a Beckman spectrophotometer.

7. Add 10 ml 15 mg/ml chloramphenicol solution (in 100% ethanol).

8. Put back on shaker for 12–16 hours at 37°C, to allow amplification of plasmid DNA.

Purification of Plasmid DNA

1. Harvest cells by centrifugation for 30 minutes at 4200 rpm in a Beckman JS4.2 rotor or 10 minutes in a Sorvall GS3 rotor.

2. Resuspend the cells from each liter of culture in 150 ml resuspension buffer. Transfer to a 250-ml centrifuge bottle. Rinse the 1-liter bottle with 100 ml resuspension buffer, then add to the 250-ml bottle.

3. Spin out the cells by centrifugation at 4200 rpm in a Beckman JS4.2 rotor for 30 minutes.

4. Resuspend the cells in 15 ml of resuspension buffer for each liter of original culture. Put in a Beckman SW28 polyallomer tube, or a 50-ml polypropylene tube.

5. Prepare a fresh solution (5–10 ml) of lysozyme (3.2 mg/ml) in resuspension buffer.

6. Add 1 ml of the lysozyme solution to the cell suspension. Invert gently to mix, and incubate at 20°C for 20 minutes.

7. Put the spheroplast mix on ice.

8. Add 16 ml ice-cold sterile distilled water. Let stand on ice for 5 minutes.

9. Incubate the mixture at 68°C for 5 minutes.

10. Put on ice until cold.

11. Leave at 4°C for two days.

12. Centrifuge the mixture for 2 hours at 25,000 rpm in a Beckman SW28 rotor at 4°C. (Equilibrate content of the tubes with resuspension buffer.)

13. Carefully pour off the supernatant into a 50-ml polypropylene tube. Avoid taking any of the viscous material which may be present just above the cell debris pellet.

14. Add an equal volume of a 1:1 mixture of phenol:chloroform/isoamyl alcohol (24:1) to the cell extract. Mix semivigorously for 5 minutes. Spin at 6000 rpm for 10 minutes, in low speed centrifuge (i.e. Heraeus).

15. Pipette out the upper phase without taking any of the thick white interface. Transfer to a 250-ml centrifuge bottle.

16. Add 1/10 volume of sodium acetate (2.5 M) and three volumes of cold ethanol. Let stand at -20°C for 2 hours.

17. Spin out the nucleic acid precipitate at 4200 rpm for 20 min-

utes at 4°C in a Beckman JS4.2 rotor, or at 10500 rpm in the Sorvall GSA rotor for 20 minutes.

18. Pour off the supernatant and dry the pellet under vacuum for 15 minutes.

19. Resuspend the pellet in 10 ml gradient buffer. Wash the bottle with 5 ml of the same buffer and add to the DNA solution.

20. Take *exactly* 15 ml of the DNA solution and add 16.17 g of CsCl (1.078 g CsCl/ml resuspended DNA). It is convenient at this step to use SW28 polyallomer tubes.

21. Add 1.5 ml of 10 mg/ml ethidium bromide solution.

22. Adjust the refractive index of the solution to 1.3875 if necessary. The average density of the gradient should be 1.571 g/cm³ (see Table 40).

Table 40. Relationship Between Density, Refractive Index, and Concentration of Cesium Chloride at 25°C[a]

Percentage by Weight	Density (g/cm³)	Refractive Index	Concentration (mg/ml)	Molarity
1	1.0047	1.3333	10.0	0.056
2	1.0125	1.3340	20.2	0.119
3	1.0204	1.3348	30.6	0.182
4	1.0284	1.3356	41.1	0.244
5	1.0365	1.3364	51.8	0.308
6	1.0447	1.3372	62.8	0.373
7	1.0531	1.3380	73.7	0.438
8	1.0615	1.3388	84.9	0.504
9	1.0700	1.3397	96.3	0.572
10	1.0788	1.3405	107.9	0.641
11	1.0877	1.3414	119.6	0.710
12	1.0967	1.3423	131.6	0.782
13	1.1059	1.3432	143.8	0.854
14	1.1151	1.3441	156.1	0.927
15	1.1245	1.3450	168.7	1.002
16	1.1340	1.3459	181.4	1.077
17	1.1437	1.3468	194.4	1.155
18	1.1536	1.3478	207.6	1.233
19	1.1637	1.3488	221.1	1.313
20	1.1739	1.3498	234.8	1.395
21	1.1843	1.3508	248.7	1.477
22	1.1948	1.3518	262.9	1.561
23	1.2055	1.3529	277.3	1.647
24	1.2164	1.3539	291.9	1.734
25	1.2275	1.3550	306.9	1.823
26	1.2387	1.3561	322.1	1.913
27	1.2502	1.3572	337.6	2.005
28	1.2619	1.3584	353.3	2.098
29	1.2738	1.3596	369.4	2.194
30	1.2858	1.3607	385.7	2.291
31	1.2980	1.3619	402.4	2.390
32	1.3110	1.3631	419.5	2.492
33	1.3240	1.3644	436.9	2.595

Table 40. Relationship Between Density, Refractive Index, and Concentration of Cesium Chloride at 25°C[a] (continued)

Percentage by Weight	Density (g/cm³)	Refractive Index	Concentration (mg/ml)	Molarity
34	1.3360	1.3657	454.2	2.698
35	1.3496	1.3670	472.4	2.806
36	1.3630	1.3683	490.7	2.914
37	1.3770	1.3696	509.5	3.026
38	1.3910	1.3709	528.6	3.140
39	1.4060	1.3722	548.3	3.257
40	1.4196	1.3735	567.8	3.372
41	1.4350	1.3750	588.4	3.495
42	1.4500	1.3764	609.0	3.617
43	1.4650	1.3778	630.0	3.742
44	1.4810	1.3792	651.6	3.870
45	1.4969	1.3807	673.6	4.001
46	1.5130	1.3822	696.0	4.134
47	1.5290	1.3837	718.6	4.268
48	1.5460	1.3852	742.1	4.408
49	1.5640	1.3868	766.4	4.552
50	1.5825	1.3885	791.3	4.700
51	1.6010	1.3903	816.5	4.849
52	1.6190	1.3920	841.9	5.000
53	1.6380	1.3937	868.1	5.156
54	1.6580	1.3955	895.3	5.317
55	1.6778	1.3973	922.8	5.481
56	1.6990	1.3992	951.4	5.651
57	1.7200	1.4012	980.4	5.823
58	1.7410	1.4032	1009.8	5.998
59	1.7630	1.4052	1040.2	6.178
60	1.7846	1.4072	1070.8	6.360
61	1.8080	1.4093	1102.9	6.550
62	1.8310	1.4115	1135.8	6.746
63	1.8560	1.4137	1167.3	6.945
64	1.8800	1.4160	1203.2	7.146
65	1.9052	1.4183	1238.4	7.355

Source: Bruner, R., and Vinograd, J. (1965), *Biochim. Biophys. Acta,* **108,** 18–29.
[a]Data from International Critical Tables.

23. Set up tubes for ultracentrifugation.

24. Spin for at least 40 hours at 40,000 rpm at 20°C in a Beckman Ti 70 rotor. Fixed angle rotors provide better band separation (see Chapter 3 and Figure 68).

25. Visualize DNA bands after the run with a long-wave ultraviolet lamp.

26. Remove the bottom DNA band by side puncture of the tube with an 18 gauge needle. Transfer this band to a 50-ml polypropylene tube. Be sure to loosen set screw at top of the tube before pulling off the band.

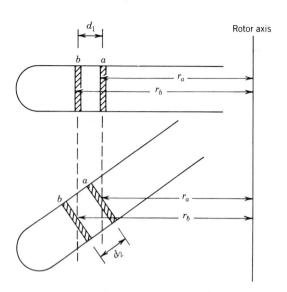

Figure 68. Relative positions of DNA bands in fixed angle and swinging bucket rotors. Two components *a* and *b* that band at the same radial distances in fixed angle and swinging bucket rotors are separated with much better resolution in fixed angle rotors. In swinging buckets, the distance between the centers of bands *a* and *b* after centrifugation (d_1) is smaller than the distance obtained in fixed angle rotor after centrifugation (d_2). During the run, bands *a* and *b* are vertically oriented and separated by distance d_1 in both cases, since the particle separation during centrifugation is identical (see Chapter 3).

27. Add an equal volume of *n*-butanol saturated with 10 m*M* tris–HCl (pH 7.5), 1 m*M* EDTA. Shake vigorously for 5 minutes to extract ethidium bromide.

28. Pipette out the upper phase, which should be pink. Discard in a special container.

29. Add fresh *n*-butanol to the DNA solution and repeat extraction at least five times.

30. After the last extraction, dialyze the DNA solution three times against 2 liters of dialysis buffer.

31. The dialyzed sample, which usually contains 0.1–0.2 mg/ml of plasmid DNA, can be used directly without further purification, but remember that the preparation is not free of RNA.

Notes

1. It is very critical that the amplification step be 12–16 hours long to obtain efficient amplification and to avoid lysis of the cells.

2. We have found that when amplification is performed under the same conditions in TYE medium, the final yield of pure plasmid DNA does not exceed 150–200 μg/liter, compared with about 750 μg/liter when M9 is used.

3. Instead of lysing the cells by osmotic shock, it is possible to add 1/10 volume of 5% Triton X100, 0.2 M Na$_2$EDTA, 50 mM Tris–HCl, pH 8.0. Mix very gently, then leave on ice 10–30 minutes until the lysis is complete.

4. Purification of the plasmid DNA can be achieved by chromatography on agarose A50 (Bio-Rad). Prepare a column (at least 2 × 30 cm) of agarose A50, equilibrate, and wash with several volumes of A50 buffer. The plasmid DNA is usually loaded as a 2-ml sample. Then add 1 ml A50 buffer and connect the column to a buffer reservoir. Collect 2.5-ml fractions, monitoring the presence of DNA either by reading the O.D. at 260 nm or by running samples on agarose gels. When a 2.5 × 30 cm column is used, the plasmid DNA elutes between the 25th and the 35th fractions with an O.D.$_{260}$ peak of approximately 0.1–0.4. Fractions can be pooled and the DNA precipitated.

5. A critical step in plasmid purification is the lysis–centrifugation. Any large chromosomal DNA sheared during the preparation and contained in the supernatant of the lysis–centrifugation will copurify with the plasmid DNA on an agarose A50 column. This DNA can be eliminted by extraction of the deproteinized DNA in 50 mM sodium acetate (pH 5.0), 75 mM NaCl. Nicked plasmid DNA will also be removed under these conditions.

6. The phenol used for extraction of DNA from the lysis supernatant must be saturated with NTE buffer (0.1 M NaCl).

7. Precipitation at $-20°C$ is preferable because more polysaccharides precipitate at $-70°C$.

8. The use of vertical rotors is not recommended when 1-liter cultures are processed because of the large amounts of supercoiled DNA obtained with this procedure. The content of a 1-liter culture should be split at least into two tubes when the vertical rotor 65 VTi is employed.

Buffers and Solutions

Resuspension buffer	*1 liter*
20 mM Tris–HCl (pH 8.0)	20 ml 1 M Tris–HCl (pH 8.0)
20 mM EDTA	40 ml 0.5 M EDTA
0.14 M NaCl	28 ml 5 M NaCl
	912 ml H$_2$O

NTE buffer	*1 liter*
20 mM Tris–HCl (pH 8.5)	20 ml 1 M Tris–HCl (pH 8.5)
100 mM NaCl	20 ml 5 M NaCl
1 mM EDTA	2 ml 0.5 M EDTA
	958 ml H$_2$O

Gradient buffer
10 mM Tris–HCl (pH 7.5)
1 mM EDTA

100 ml
1 ml 1 M Tris–HCl (pH 7.5)
0.2 ml 0.5 M EDTA

TYE culture medium
For 1 liter:
10 g Bacto tryptone
5 g Yeast extract
5 g NaCl

M9 culture medium (10X salts)
60 g Na_2HPO_4
30 g KH_2PO_4
5 g NaCl
10 g NH_4Cl
1 liter H_2O

M9 culture medium (1X)
For 1 liter:
100 ml 10 × M9 (salts)
20 ml 20% Glucose
5 g Casamino acids
860 ml H_2O
Autoclave at 120°C for 30
minutes and add
10 ml Sterile 0.1 M $MgSO_4$
10 ml Sterile 0.01 M $CaCl_2$
0.2 ml 1% B_2 vitamin (Thiamine)

A50 buffer
10 mM Tris–HCl (pH 8.0)
1 mM EDTA
300 mM NaCl

1 liter
10 ml 1 M Tris–HCl (pH 8.0)
2 ml 0.5 M EDTA
60 ml 5 M NaCl
928 ml H_2O

RAPID PROCEDURE FOR ISOLATION OF PLASMID DNA

The protocol described below permits the rapid purification of plasmid DNA from overnight bacterial cultures obtained in the presence of an appropriate selective antibiotic (e.g., ampicillin). It gives relatively high amounts of supercoiled DNA when 100 ml cultures are manipulated, and therefore represents a convenient way of obtaining large quantities of clean DNA needed for some cloning steps. However, when larger amounts of clean supercoiled DNA are required (for example, in transfection experiments described in Chapter 25), we do recommend that previous protocol be followed.

PROTOCOL

ISOLATION OF PLASMID DNA FOLLOWING ALKALINE-SDS TREATMENT

1. Grow the transformed cells in 100 ml of TYE medium in the presence of ampicillin (100 μg/ml if using HB 101 or equivalent) for 18 hours.
2. Split the culture in two 50-ml polypropylene tubes.
3. Centrifuge at 4000 rpm for 10 minutes at 4°C.
4. Discard the supernatant. However, to allow a better resuspension of the cells, leave about 0.5 ml of medium on top of the bacterial pellet.
5. Leave the tubes on ice for 5 minutes.
6. In each tube, add 5 ml of lysozyme solution (5 mg/ml).
7. Resuspend the pellet thoroughly with a pipette-aid and leave the tubes on ice for 15 minutes.
8. Transfer the contents of each tube to a 30-ml Corex tube which has been carefully checked (these tubes break upon centrifugation if they are damaged).
9. Add 10 ml of alkaline–SDS solution to each tube and invert gently until the mixture is homogenous.
10. Incubate the tubes at 68°C for 10 minutes and cool in wet ice.
11. Add 7.5 ml of the high salt solution per tube. Mix by inverting gently a few times and set on ice for about 1 hour until the white precipitate that formed is floating on the surface.
12. Centrifuge for 20 minutes at 4°C in the Sorvall SS34 rotor (or equivalent) at 10,500 rpm.
13. Save the supernatant (about 25 ml) in a 50-ml polypropylene tube.
14. Add half a volume of NTE-saturated phenol and half a volume of chloroform:isoamyl alcohol (24:1) mixture.
15. Mix vigorously and centrifuge at room temperature for 15 minutes at 5500 rpm.
16. Repeat steps 13 to 15.
17. Transfer the upper phase to a clean tube and add 1 volume of chloroform:isoamyl alcohol mixture.
18. Centrifuge at room temperature for 15 minutes at 5500 rpm.
19. Repeat steps 17 and 18.
20. Transfer the supernatant in two 50-ml polypropylene tubes, add three volumes of cold absolute ethanol, and let the DNA precipitate for 1 hour at −70°C.
21. Centrifuge at 5500 rpm for 20 minutes at 4°C.

22. Discard the supernatant and leave the tube dry (upside down) on Kimwipes tissues.

23. Resuspend the DNA pellet from each tube in 100 μl of sterile distilled water. The pellets should dissolve immediately.

28. Combine the contents of the tubes in a microfuge tube.

29. Rinse the polypropylene tubes with 100 μl of water. First add the water to one tube and swirl or pipette up and down several times. Use this solution to rinse the other tubes in the same way.

30. Add the 100 μl of wash to the 400 μl of sample contained in the microfuge tube. Tilt the tube a few times to homogenize the solution.

31. Put 20 ml of TE buffer in a Petri dish, and carefully layer four millipore filters on the liquid surface.

32. With a pipetman, carefully place 100 to 150 μl of the plasmid solution onto the surface of each filter. Drops should not spread on the filters.

33. Set the filters in a safe place for 2 hours to allow efficient dialysis to proceed.

34. Collect the drops with tips and transfer the desalted DNA solution in a microfuge tube.

35. Add 50 μl of RNAse solution to the DNA solution and incubate for 1 hour at 37°C.

36. Extract DNA as described in Chapter 3.

37. Transfer the DNA solution in two microfuge tubes (about 300 μl in each tube).

38. Add 150 μl of 7.5 M ammonium acetate and 900 μl of cold absolute ethanol to each tube. Mix and let the DNA precipitate for 1 hour at $-70°C$.

39. Centrifuge the tubes for 30 minutes at 12,000 rpm (4°C).

40. Discard the supernatant and add 1 ml of 80% cold ethanol to the tubes. Vortex to resuspend the DNA pellets.

41. Centrifuge the tubes for 10 minutes at 12,000 rpm (4°C).

42. Empty the tubes and dry the pellets under vacuum.

43. Resuspend each pellet in 300 μl of sterile distilled water, and combine the two samples in one microfuge tube.

Notes

1. The DNA prepared in this way can be digested by restriction endonucleases without any trouble if the final volume of reaction is 10 times greater than the volume of DNA to be tested. With lower ratios, partial digestion may occur with some en-

zymes that are sensitive to high salt concentrations (see chapter 8).

2. The DNA solution obtained with this procedure is suitable for cesium chloride centrifugation in vertical rotors. The dialysis step is critical if equilibrium centrifugation in cesium chloride is performed. High salt concentrations will ruin the separation. If this happens with one of your preparations, treat your sample in the same way as DNA fractions collected from large-volume gradients (steps 27 to 30 of previous protocol).

3. Cesium chloride centrifugation is strongly recommended when the plasmid DNA is to be used as vector in subsequent experiments. This step will considerably lower the transformation background.

4. Buffers and Solutions can be found under Miniscreen for Recombinant Plasmids in Chapter 15.

Removal of RNA Species from DNA Preparations

The following protocol, originally described by Lev (1987), can be successfully used to remove high and low molecular weight RNA species from plasmid and lambda-phage DNA preparations. It can be performed conveniently within 24 hours with several samples of different origins.

PROTOCOL

REMOVAL OF RNA SPECIES FROM DNA PREPARATIONS

1. Pellet 200-ml samples of an overnight bacterial culture at 6000 g for 10 minutes at 4°C in 500-ml polypropylene centrifuge bottles.

2. Discard the supernatant and add 40 ml of cold STE buffer per bottle.
   The remainder of the protocol applies to each pelleted sample.

3. Resuspend thoroughly and centrifuge at 4°C for 10 minutes at 6000 g.

4. Resuspend washed cells in 10 ml of STE buffer and transfer to 50-ml pyrex flask.

5. Add 1 ml of lysozyme solution (10 mg/ml) and let sit on the bench for 10 minutes.

6. Heat the lysed cell suspension over an open flame with constant swirling until it boils.

7. Transfer to a boiling water bath for 40 seconds.

8. Cool the viscous solution by swirling in a wet iced bath and leave on ice for 10 minutes.

9. Transfer the solution to a 50-ml polypropylene centrifuge tube and spin at 17,000 g for 20 minutes at 4°C.

10. Transfer supernatant to a clean 50-ml tube and add 10 ml of isopropanol. Incubate at -20°C for 1 hour.

11. Centrifuge the tube at 17,000 g for 20 minutes.

12. Discard supernatant and invert the tube to let the precipitate dry off. Dissolve the dry pellet in 0.7 ml of TE buffer.

13. Transfer the solution to a microfuge tube and add 0.35 ml of a 15 M lithium chloride solution.

14. Incubate for 10 minutes at room temperature and split the sample in two microfuge tubes.

15. Fill both tubes with ethanol and let precipitate at -20°C for 30 minutes.

16. Centrifuge at 12,000 rpm for 15 minutes at 4°C. During this time, prepare two SW 50.1 Beckman tubes containing 4.3 ml of high salt buffer.

17. Resuspend the pellets in 0.3 ml of TE buffer and combine the two aliquots. Spin at 12,000 rpm for 30 seconds to pellet unsoluble material.

18. Carefully pipette 0.5 ml of the supernatant (without resuspending the pellet unsoluble particles) and layer onto the surface of the high salt solution (from step 17).

19. Centrifuge at 35,000 rpm (20°C) for 14 to 16 hours.

20. Carefully remove the supernatant with a Pasteur pipette (from top to bottom), and resuspend the plasmid DNA pellet in 0.4 ml of TE buffer.

21. Transfer to a microfuge tube and add successively 10 μl of 4M NaCl and 0.1 ml of cold ethanol. Let the DNA precipitate for 1 hour at -20°C.

22. Centrifuge at 12,000 rpm for 20 minutes (4°C), discard the supernatant, and wash the precipitate with 1 ml of 80% cold ethanol.

23. Centrifuge at 12,000 rpm for 10 minutes (4°C), discard the supernatant, and dry the pellet under vacuum (e.g., Speedvac).

24. Resuspend the pellet in 0.2 ml of TE buffer.

Notes

1. High molecular weight bacterial DNA is usually eliminated after step 8 (Holmes and Quigley, 1981).

2. Precipitation with lithium chloride eliminates most of the high molecular weight RNA species (steps 13–15).

3. Low molecular weight RNA species are eliminated by sedimentation through the high salt solution (step 19).

Buffers and Solutions

STE buffer	*1 liter*
0.15 *M* NaCl	30 ml 5 *M* NaCl
10 m*M* EDTA	20 ml 0.5 *M* EDTA
25 m*M* Tris–HCl (pH 8.0)	25 ml 1 *M* Tris–HCl (pH 8.0)
	925 ml H$_2$O

Lysozyme solution
10 mg of lysozyme/ml of STE buffer. Prepare fresh each time.

TE buffer	*1 liter*
10 m*M* Tris–HCl (pH 8.0)	10 ml 1 *M* Tris–HCl (pH 8.0)
0.1 m*M* EDTA	200 μl 0.5 *M* EDTA
	990 ml H$_2$O

High salt buffer	*100 ml*
3M NaCl	60 ml 5 *M* NaCl
0.1 m*M* EDTA	20 μl 0.5 *M* EDTA
20 m*M* Tris–HCl (pH 8.0)	2 ml 1 *M* Tris–HCl (pH 8.0)
	38 ml H$_2$O

REFERENCES

Clewell, D. B., and Helinski, D. R. (1969), *Proc. Natl. Acad. Sci. USA*, **62**, 1159.

Colman, A., Byers, M. J., Primrose, S. B., and Lyons, A. (1978), *Eur. J. Biochem.*, **91**, 303.

Cornelis, P., Digneffe, C., Willemot, K., and Colson, C. (1981), *Plasmid*, **5**, 221.

Holmes, D. S., and Quigley, M. (1981), *Anal. Biochem.*, **114**, 193.

Lev, Z. (1987), *Anal. Biochem.*, **160**, 332.

Marko, M. A., Chipperfield, R., and Birboim, H. C. (1982), *Anal. Biochem.*, **121**, 382.

Micard, D., Sobrier, M. L., Couderc, J. L., and Dastugue, B. (1985), *Anal. Biochem.*, **148**, 121.

Mukhopadhyay, M., and Mandal, N. C. (1983), *Anal. Biochem.*, **133**, 265.

Norgard, M. V. (1981), *Anal. Biochem.*, **113**, 34.

CsTFA FOR ISOLATION AND SEPARATION OF NUCLEIC ACIDS BY ISOPYCNIC CENTRIFUGATION*

CsTFA is a standardized solution of cesium trifluoroacetate intended for the isolation and separation of nucleic acids by isopycnic centrifugation. CsTFA is used in a fashion similar to cesium chloride (CsCl) and cesium sulfate (Cs$_2$SO$_4$). The trifluoroacetate anion, however, imparts properties to CsTFA that result in higher quality nucleic acid preparations than do traditional cesium density gradient media.

CsTFA is unique in that it: (a) isopycnically bands all types of RNA; (b) solubilizes and denatures proteins, resulting in their removal from nucleic acids;

*Information courtesy of Pharmacia.

Table 41. Physical Properties of CsTFA vs. CsCl

Property	CsTFA	CsCl
Maximum density (g/ml)	~2.6	1.9
Maximum solubility (g/ml)	2.5	1.23
Maximum solubility (g/g)	89%	60%
Maximum molarity (M)	~10	7.36
Formula weight (g)	245.93	168.37

and (c) inhibits nuclease activity, resulting in undegraded DNA and RNA.

Simultaneous banding in CsTFA gradients of the DNA, RNA, and proteins from rat embryos has been reported by Mirkes (1985).

Product Description

CsTFA is an aqueous solution of cesium trifluoroacetate with a density of 2.0 ± 0.05 g/ml. The exact density of each lot is indicated on the manufacturer's label. CsTFA contains approximately 134 g of cesium trifluoroacetate per 100 ml of solution.

CsTFA is supplied at approximately neutral pH, free of preservatives and buffers. Table 41 compares the physical properties of CsTFA and CsCl.

Specifications for CsTFA

Density: 2.0 ± 0.05 g/ml

Purity of cation: > 99.99% Cs

Purity of anion: > 99% TFA

Total heavy metal ion content: < 12 ppm

pH: 4.9 (unbuffered)

Ultraviolet absorption (2.0 g/ml solution):
$A_{260} < 0.3$
$A_{280} < 0.075$

Instructions for Use

Preparation of Sample. CsTFA solubilizes and denatures proteins, assisting in their removal from nucleic acids during isopycnic centrifugation. This property reduces and may even eliminate the need for prior deproteinization. Samples relatively low in protein content, such as viruses and ribosomes, usually do not require prior deproteinization. Samples containing high amounts of protein should be partially deproteinized prior to centrifugation, using conventional purification techniques such as phenol extraction or protease treatment.

Samples particularly adversely susceptible to phenol extraction are good candidates for direct application onto CsTFA gradients.

CsTFA inhibits nuclease activity during centrifugation. Nuclease inhibitors should be present, however, during sample preparation to minimize nuclease activity prior to CsTFA centrifugation.

Sodium dodecyl sulfate (SDS) forms insoluble salts in the presence of cesium and should be removed prior to sample application. An alternative detergent compatible with CsTFA is sodium-*N*-lauroylsarcosinate.

Estimation of Nucleic Acid Density. CsTFA hydrates nucleic acids, causing them to band at lower densities than in CsCl (Figure 69a). It is therefore important to estimate the density of the nucleic acid in CsTFA for correct gradient preparation.

Most double stranded DNAs band at densities between 1.60 and 1.63 g/ml in CsTFA. Single stranded DNA bands in CsTFA at a density approximately 0.08 g/ml greater than the corresponding native DNA.

Banding densities of RNAs are influenced by the degree of secondary structure. RNAs with a high degree of secondary structure, such as ribosomal RNA and transfer RNA, normally band near 1.65 g/ml in CsTFA. RNAs with little secondary structure, such as messenger RNA, will typically band at densities around 1.90 g/ml (Figure 69b).

Proteins are found at 1.20–1.50 g/ml in CsTFA.

Preparation of Gradient

1. The chemical resistance of centrifuge tubes varies according to the material used. Tube materials such as polyallomer, polyethylene, polypropylene, polycarbonate, and glass are generally satisfactory. **DO NOT USE CELLULOSE NITRATE TUBES.** Temperature, the use of denaturants or organic solvents, and extremes of pH may greatly affect the resistance of tube material (see Table 2). The chemical resistance should be confirmed under actual running conditions.

2. CsTFA should be diluted with buffers normally used for storing and maintaining nucleic acids. Buffers containing EDTA, pH 7–8, are recommended. A typical buffer to use is 0.1 M Tris–HCl, pH 8.0, 0.1 M KCl, 1 mM EDTA.

3. The following formula is used to calculate the volume required to obtain a solution of desired density:

$$V_x = V_0 \frac{(p_0 - p)}{(p - p_x)}$$

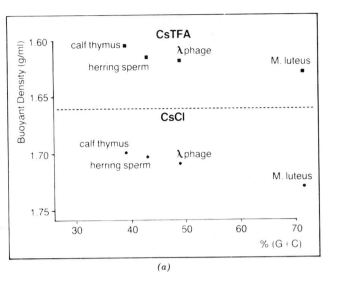

(a)

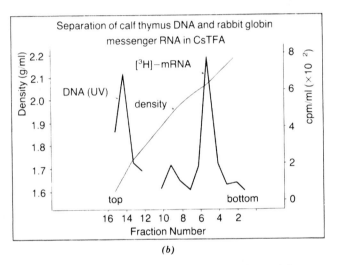

(b)

Figure 69. (a) Buoyant density of DNA from various origins centrifuged in CsTFA and CsCl. Since the hydration of CsTFA is 0.410 g H₂O/g DNA versus 0.267 for CsCl, the DNA molecules will band at correspondingly lower densities in CsTFA than they do in CsCl (courtesy of Pharmacia). (b) Simultaneous banding of DNA and RNA in CsTFA gradient (courtesy of Pharmacia).

the density of the diluted solution should be checked and readjusted before use.

4. CsTFA gradients are prepared using the same methods as CsCl gradients. By diluting CsTFA to the estimated nucleic acid density and subjecting the solution to high *g* forces, CsTFA will self-generate a gradient capable of isopycnically banding the nucleic acid of interest.

Application of Sample. Sample application techniques are the same as for CsCl. Samples are most conveniently combined with the entire CsTFA solution. Diluting the sample in this manner enhances the effectiveness of CsTFA in solubilizing proteins and inhibiting nuclease activity.

Ultracentrifugation Conditions. Recommended running conditions for CsTFA are very similar to those for CsCl. CsTFA forms steeper gradients than CsCl when run under identical conditions. These gradients permit isopycnic banding of nucleic acids which differ greatly in density (e.g., DNA and mRNA).

CsTFA gradients may be generated in vertical, fixed angle, and swinging bucket rotors. If desired, the gradient steepness may be altered by changing to a rotor of different geometry. Fixed angle and vertical rotors produce shallower gradients than swinging bucket rotors (see Figure 68, page 302), and are recommended for increasing the resolving power of CsTFA gradients during isopycnic centrifugation.

Typical centrifugation conditions are 100,000 × *g* for 36–72 hours. Running temperatures of 4–25°C are recommended. Deproteinization is more effective at higher running temperatures. The buoyant density of nucleic acids varies only slightly with temperature unless temperatures approach the melting point of the nucleic acid.

It is the responsibility of the user to ensure that the centrifuge rotor is operated within the design limits of the machine. Most manufacturers recommend that a derating factor be used whenever the average gradient density exceeds 1.2 g/ml.

Fractionation of Gradients. CsTFA gradients can be fractionated by any of the common methods used for cesium salts. For gradients containing large amounts of protein, it is recommended that fractionation originate from the bottom of the tube. This avoids the possibility of contaminating the nucleic acid band with protein.

where V_x is the volume of the diluting medium in ml, V_0 is the volume of the original CsTFA solution in ml, p_0 is the density of the CsTFA in g/ml (see bottle label), p_x is the density of the diluting medium (water = 0.998 g/ml at 25°C) in g/ml, and p is the density of the desired solution in g/ml.

For example, to obtain a density of 1.600 g/ml, add 0.654 ml water to 1.000 ml CsTFA (p_0 = 2.00 g/ml).

CsTFA supplied at a predetermined density is diluted to the desired density. For very precise work,

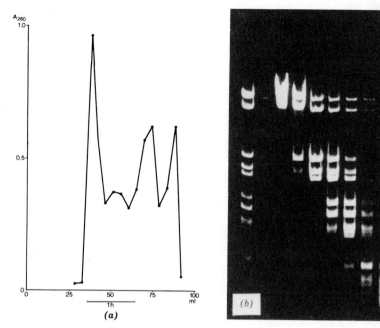

Figure 70. Separation of restriction fragments of λ DNA on Sephacryl S-1000 Superfine. (*a*) λ DNA was digested with Bgl I restriction endonuclease. The sample was applied to a 1.3 × 76 cm column packed with Sephacryl S-1000 Superfine. Eluant, 20 m*M* Tris–HCl, pH 7.5, containing 1 m*M* EDTA; flow rate, 20 ml/h (20 cm/h). (*b*) Electrophoresis was performed in 1% agarose gels (courtesy of Pharmacia).

Determination of Density. Densities of CsTFA solutions after fractionation can be determined using any method applicable to cesium salts. These methods include use of calibrated micropipettes, densitometry, conductivity determination, and refractive index measurements. The most convenient method is to measure the refractive index. After fractionation, the refractive indices of CsTFA solutions should be measured as soon as possible, as they are susceptible to changes in density on standing. The density of CsTFA can be determined by using the equation:

$$p^{25} = 163.559 - 262.271\,(n_D^{25}) + 105.281\,(n_D^{25})$$

where p^{25} is the density of CsTFA at 25°C and n_D^{25} is the refractive index of CsTFA at 25°C.

Determination and Recovery of Nucleic Acids. CsTFA possesses a low degree of absorbance in the ultraviolet region, thus facilitating detection of nucleic acids. Fractions from CsTFA gradients may be monitored spectrophotometrically at 260 nm.

Nucleic acids may be separated from CsTFA by traditional techniques such as gel permeation chromatography, ethanol precipitation, and dialysis. Unlike CsCl, CsTFA is freely soluble in ethanol, so coprecipitation with nucleic acids cannot occur.

CsTFA does not interfere with the intercalation of dyes such as ethidium bromide in DNA. However, since CsTFA is soluble in *n*-butanol and isopropanol, these commonly used solvents should not be employed for the removal of ethidium bromide from DNA subsequent to centrifugation. As an alternative, one can use ethanol precipitation, which removes most of the dye. If further purification is required, an ion exchange column such as DEAE–Sephacel can be used.

Sterilization. CsTFA exhibits bacteriostatic and bacteriocidal properties, so sterilization is not generally necessary. If desired, CsTFA can be sterilized by sterile filtration or autoclaving. Do not use cellulose nitrate filters. Filters made of mixed esters of cullulose are acceptable. Since autoclaving may alter the density of CsTFA solutions, the density should be reconfirmed.

Practical Notes. CsTFA is for in vitro laboratory use only. Spills should be cleaned using copious amounts

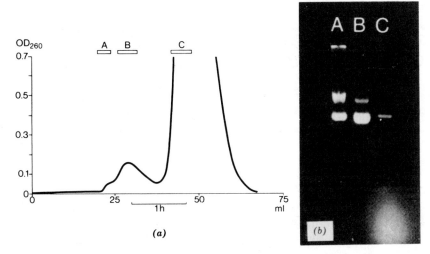

Figure 71. Purification of plasmids on Sephacryl S-1000 Superfine. (*a*) pBR322 plasmids were obtained from *E. coli* 259. After lysis and removal of cell debris the bulk of the chromosomal DNA was removed by precipitation at alkaline pH. The supernatant was treated with ribonuclease A followed by phenol extraction. The sample (250 μl), containing approximately 100 μg plasmid material, was applied to a column of Sephacryl S-1000 Superfine. Column, k 16/40; bed height, 30 cm; eluant: Tris–phosphate, pH 8, containing 1 *M* NaCl; flow rate: 16 ml/h. Recovery of plasmid DNA was greater than 90%. (*b*) Electrophoresis was performed in 0.8% agarose. Lane A shows residual amounts of chromosomal DNA, open plasmids, and closed plasmids. Lane B shows a distinct band corresponding to closed plasmids and a fainter band corresponding to open plasmids. Lane C shows mainly RNA and nucleotides (courtesy of Pharmacia).

of water. Metal surfaces or instrument parts exposed to CsTFA should be rinsed thoroughly with water or alcohol immediately after use. Skin exposed to CsTFA should be washed with soap and water. There is no evidence that CsTFA is significantly more toxic than other cesium salts. Good hygiene should be practiced at all times.

Storage Conditions. It is recommended that CsTFA be stored in its original container, tightly sealed, at 4°C. Density changes wil be minimized if CsTFA solutions are withdrawn by a syringe needle through the septum closure. Since solutions stored for long periods in open containers or transferred to other containers may change slightly in density, CsTFA solutions should be routinely checked to confirm density.

REFERENCE

Mirkes, P. E. (1985), *Analytical Biochem.*, **148**, 376.

PURIFICATION OF DNA RESTRICTION FRAGMENTS, PLASMID DNA, AND CHROMOSOMAL DNA BY SEPHACRYL GEL CHROMATOGRAPHY

Sephacryl S-1000 Superfine (Pharmacia) is a highly porous and mechanically stable matrix allowing rapid enrichment of restriction fragments (Figure 70) and good resolution of DNA (Figure 71) without the use of ethidium bromide. Because of the excellent bed stability and flow properties of this matrix, scaling up for larger sample volumes is straightforward.

PREPARATION OF λ PHAGE STOCKS AND DNA

Titration of Initial Seed Stock, and Preparation of Large Stocks.

The preparation of phage stocks requires the use of a particular multiplicity of infection (m.o.i.), that is, a known ratio of viral infectious particles to bacteria. Therefore, the first step is the titration of the initial seed stock on the chosen host.

PROTOCOL

CORRELATION BETWEEN ABSORBANCE AT 600 NM AND THE NUMBER OF BACTERIA PER MILLILITER OF CULTURE

1. Seed an overnight culture (5 ml) of the host in TYE medium.
2. The next day pipette 0.1 ml of the overnight culture and seed 10 ml of fresh, prewarmed rich medium.
3. Follow the variation of absorbance at 600 nm. At regular intervals, pipette 0.1 ml from the culture and perform serial dilutions in tubes containing 1–10 ml TYE medium (prepare 10^{-2}, 10^{-4}, 10^{-6}, and 10^{-7} dilutions). Plate 0.1 ml of the 10^{-4}, 10^{-6}, and 10^{-7} dilutions onto Petri dishes containing TYE agar medium. Prepare duplicate plates.
4. Incubate the plates at 37°C overnight. Count the colonies. Plot the number of bacteria versus absorbance at 600 nm on semi-log paper.

PROTOCOL

TITRATION OF SEED STOCK

1. Grow the host in 10 ml TYE for approximately 10 hours.
2. Spin the bacteria at 3000 rpm for 5 minutes.
3. Discard the supernatant and resuspend the pellet in 10 ml of 10 mM magnesium sulfate. Hosts such as DP50 and LE392 can be kept for several days at 4°C in 10 mM MgSO$_4$ without problem.
4. Prepare tubes containing 0.5–1.0 ml (depending on dilution) of TMG buffer.
5. Make serial logarithmic dilutions of the initial phage stock. For example, take 5 μl of phage and add to 500 μl TMG (10^{-2} dilution). Mix well. With a new tip, pipette 5 μl of the 10^{-2} dilution and add to 500 μl TMG to obtain a 10^{-4} dilution. It is advised, when the stock is of high titer, to go down to a 10^{-7} dilution.
6. Pipette 0.1 ml of each dilution into a sterile glass tube with metal cap.

7. Incubate at 37°C for 15 minutes to remove any residual chloroform (phage stocks are kept in chloroform).
8. Add to each tube 0.1 ml of bacteria resuspended in 10 mM $MgSO_4$.
9. Incubate at room temperature for 15 minutes.
10. Prepare a series of glass tubes containing 3 ml 0.8% bacto agar in TYE medium. Keep at 45°C.
11. When adsorption of phage is realized (step 9), add the content of one agar tube (step 10) to each sample and pour immediately on 1.2% agar plates (TYE medium). Let the overlay solidify and incubate upside-down at 37°C overnight.
12. Count the plaques which appeared during overnight incubation. Calculate the titer. For example, if you count 150 plaques for a 10^{-7} dilution, the original titer was

$$\underset{\text{(counted)}}{150} \times \underset{\substack{\text{(0.1 ml of} \\ \text{1 ml used)}}}{10} \times \underset{\text{(dilution)}}{10^7} = 1.5 \times 10^{10} \text{ phages/ml}$$

PROTOCOL

PREPARATION OF VIRAL MINISTOCKS

1. With a Pasteur pipette pick a single isolated plaque (you can also aspirate the underlying agar).
2. Transfer the plaque (and agar) into a tube containing 100 μl TMG.
3. Add 1 drop of chloroform and shake gently.
4. Let sit at room temperature for 15 minutes.
5. Incubate at 37°C for 15 minutes with occasional shaking to remove chloroform.
6. For one plaque, prepare one host plate in the following way: In a sterile glass tube, mix 0.1 ml of bacteria resuspended in 10 mM $MgSO_4$ (cf. above) and 3 ml of 0.8% bacto agar in TYE (45°C). Pour immediately on a 1.2% TYE–agar plate. Let solidify.
7. Pour the 100 μl of TMG containing the eluted phage onto the

solidified agar overlay containing bacteria. Be careful to obtain a single area of liquid. Do not spread all over the plate.

8. Incubate the plate at 37°C overnight without putting it upside-down.

9. Scrape with a scalpel or a glass slide the lysis zone which appeared where the phage was spotted.

10. Transfer scraping to a glass tube (5 ml) and add 1 ml TMG.

11. Add three drops of chloroform.

12. Vortex gently. Spin at 4000 rpm for 5 minutes at 4°C.

13. Carefully pipette the supernatant, trying not to pipette any agar from the bottom. Transfer to an microfuge tube (1.5 ml).

14. Add a few drops of chloroform and store at 4°C.

PROTOCOL

PREPARATION OF LARGE PLATE STOCKS

1. Prepare 1 ml of a 10^{-2} dilution of the corresponding ministock.

2. Incubate the diluted phage suspension at 37°C for 15 minutes to remove residual chloroform.

3. Put 0.1 ml of the 10^{-2} phage dilution into each of 10 sterile glass tubes (5 ml).

4. Add to each tube 0.1 ml of bacteria in 10 mM MgSO$_4$.

5. Let sit at room temperature for 15 minutes to allow phage to adsorb onto host.

6. To each tube add 3 ml 0.8% TYE–agar and overlay immediately on 1.2% TYE–agar plates, freshly prepared. It is advised to use plates prepared the same day and not dried. Run a control without phage.

7. Incubate plates (without putting them upside-down) at 37°C overnight.

8. On the next day, scrape the surface of each plate and collect the overlays into one 50-ml tube.

9. Add 5 ml TMG and vortex gently.

10. Add 5 drops of chloroform and vortex gently.

11. Spin at 4000 rpm for 10 minutes at 4°C.

12. Carefully collect the supernatant.

13. Add 5 drops of chloroform. Invert several times. Keep at 4°C.
14. Titrate the stock as described above. Titers of stocks prepared under these conditions are usually $2–8 \times 10^{10}$ phages/ml of stock.

PROTOCOL

PREPARATION OF LARGE LIQUID STOCKS

The following protocol is designed for a 1-liter culture. If larger volumes are to be used, it is possible to scale the procedure up.

1. Seed an overnight culture of the appropriate host bacterium in 10 ml TYE medium.
2. In a 50-ml plastic tube mix 0.1 ml Ca–Mg solution and 0.9 ml of the bacterial culture. Add the phage at a multiplicity of infection of 0.1 (1 phage/10 bacteria).
3. Incubate for 15 minutes at 37°C in a water bath (without shaking).
4. Add 18 ml of prewarmed TYE medium to the 50-ml tube containing the bacteria–phage mixture. Split in two 10-ml aliquots to seed two 500-ml cultures (prewarmed TYE medium in 2-liter flasks containing 10 mM MgCl$_2$).
5. Incubate at 37°C for 10 hours with vigorous shaking (200–250 rpm).
6. At this time, pipette an aliquot of the culture and check that cell debris are visible.
7. Add 2.5 ml chloroform to each flask and let shake for 15 additional minutes.
8. Spin the culture at $5000 \times g$ for 30 minutes (6200 rpm, in GSA Sorvall rotor) at 4°C to pellet out cellular debris.
9. Save the supernatant in a 2-liter flask.
10. Add 35 g NaCl; swirl to dissolve.
11. Add 80 g polyethylene glycol (PEG 6000) and transfer the flask to 4°C.
12. Shake the mixture gently by hand, and then mix with a magnetic stirrer until PEG is completely dissolved (do not mix too forcefully—PEG will dissolve in 30–45 minutes).

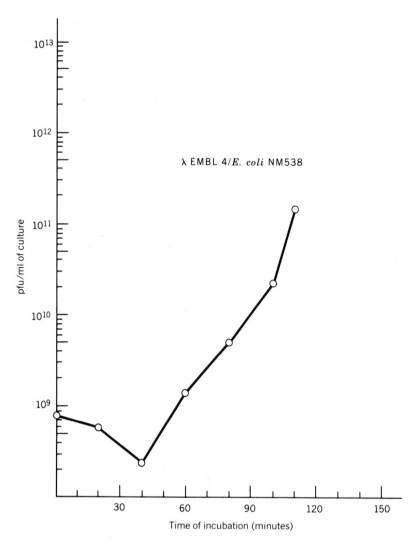

Figure 72. Yield of phages obtained with stocks of different origins. 20 ml culture of the appropriate host cell strains were grown until the optical density reached a value comprised between 0.6 and 0.7. After addition of $MgCl_2$ and $CaCl_2$ (10 mM final) the cultures were infected with different stocks of λ, so as to obtain a multiplicity of infection of 10 pfu/cell. At regular intervals of time, aliquots were taken and used for phage titration (courtesy of V. Maloisel).

13. Let the mixture sit at 4°C overnight, unstirred, to allow complete phage precipitation.

14. Pellet the PEG precipitate for 40 minutes at 5000 × g at 4°C (if 1-liter bottles are used).

15. Discard supernatant. A white pellet should be visible. Draw off all remaining liquid without loosening pellet.

16. Put the bottle on ice and add 10 ml PEG resuspension buffer.

17. Let sit for 10–20 minutes on ice before resuspending the pellet carefully, without generating foam.

18. When the pellet is completely resuspended, transfer to a 50-ml tube. Rinse the bottle with 2 ml PEG resuspension buffer. Add to the 10-ml stock.

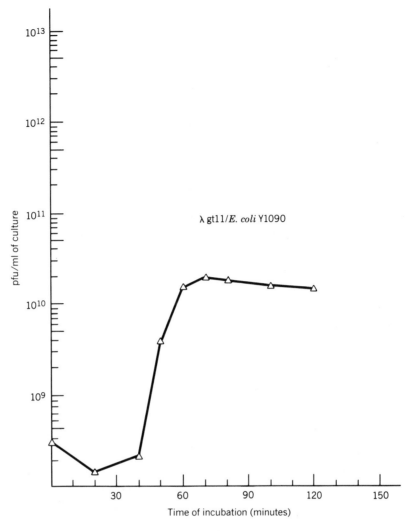

λ gt11/*E. coli* Y1090

Figure 72. *(continued)*

19. Let the resuspended PEG pellet sit at 4°C for about 12 hours before purifying the phage particles (if necessary).

The stock obtained at this point should have a titer of approximately $8-9 \times 10^{11}$ phages/ml.

Notes

1. It is very important that a correct evaluation of the number of bacteria per milliliter of culture is available to adjust the m.o.i. to 0.1.

2. The time of incubation (10 hours) is crucial. This time is given for the DP50 host and λ charon phages. If you use other phages and hosts, you may have to adjust the time of incubation. This is usually done by comparing pilot cultures (100 ml or less). The yield varies with the phage strain (Figure 72).

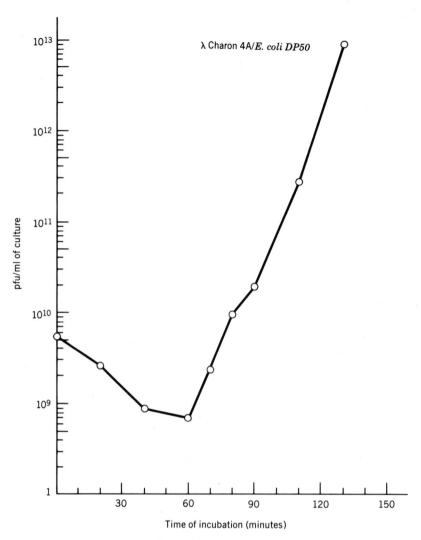

λ Charon 4A/*E. coli DP50*

Figure 72. *(continued)*

3. If you spin the culture in 250-ml fractions, you may centrifuge for only 20 minutes at 5000 × *g*. In such a case, resuspend one pellet and use this mixture to resuspend the second pellet, and so on, so as to use a minimum volume for resuspension.

4. When you resuspend the PEG pellet, it is easier to use an automatic pipette-aid. We have found that resuspension of the pellet is easier if the resuspension buffer covers the pellet for about 30 minutes before resuspension.

5. PEG stocks are much more stable than TYE plate stocks when kept at 4°C.

Buffers for λ Stocks

TMG buffer   *100 ml*
10 m*M* Tris–HCl (pH 7.4)   1 ml 1 *M* Tris–HCl (pH 7.4)
10 m*M* MgCl$_2$ 1 ml 1 *M* MgCl$_2$

100 μg/ml gelatin	10 ml gelatin (1 mg/ml)
	88 ml H_2O
	Sterilize at 120°C for 20 minutes

Ca–Mg solution — *100 ml*
10 mM $CaCl_2$ — 1 ml 1 M $CaCl_2$
10 mM $MgCl_2$ — 1 ml 1 M $MgCl_2$
— 98 ml H_2O

PEG resuspension buffer — *100 ml*
10 mM Tris–HCl (pH 7.4) — 1 ml 1 M Tris–HCl (pH 7.4)
10 mM $MgCl_2$ — 1 ml 1 M $MgCl_2$
100 mM NaCl — 2 ml 5 M NaCl
— 96 ml H_2O

PROTOCOL

PURIFICATION OF λ VIRUS FROM LIQUID STOCKS

As liquid stocks give high yields of λ phage, they are an excellent source for phage purification.

1. Start with resuspended PEG–phage precipitate.
   Add DNase I to the resuspended phage to a concentration of 100 μg/ml. Use a 50-ml polypropylene tube.
2. Incubate in a water bath for 30 minutes at 37°C; swirl occasionally.
3. Add an equal volume of chloroform to the tube and mix semi-vigorously for 10 minutes.
4. Spin the mixture for 10 minutes at 2000 × g at 20°C.
5. Save the upper layer. Record its exact volume (at this step it should be between 6 and 7 ml).
6. Add 0.77 g cesium chloride per milliliter of upper layer.
7. Fill an SW41 ultraclear tube with the sample in cesium chloride (one tube). Prepare a balance tube with cesium chloride in water.
8. Centrifuge in a Beckman SW41 rotor at 35,000 rpm for 40 hours at 4°C.

9. Collect the phage band (the thin "blue" band in the middle of the gradient) by puncturing the tubes with a 25-gauge needle. (You may wish to put a little piece of adhesive tape on the tube where you insert the needle in order to avoid leaking.)

10. Transfer the band to a dialysis bag and dialyze twice against 2 liters of dialysis buffer (4°C).

The dialyzed phage is used in the purification of λ DNA.

PROTOCOL

PURIFICATION OF λ DNA

1. To the dialyzed phage (1–2 ml) add 1/10 volume of 10X buffer C, 20% SDS (to obtain a final concentration of 0.2%), 0.5 M EDTA (to obtain a final concentration of 10 mM), and 100–200 μg/ml of proteinase K.

2. Incubate 1 hour at 37°C in a water bath.

3. Add one volume of a 1:1 phenol:chloroform/isoamyl alcohol (24/1) mixture and mix gently for 5 minutes. (Any vigorous shaking at this point will generate broken λ DNA, unsuitable for most subsequent experiments.)

4. Centrifuge 5 minutes at 1000 $\times$ g at 20°C.

5. Collect the upper phase. Add an equal volume of a chloroform:isoamyl alcohol mixture. Mix gently for 5 minutes (cf. step 3).

6. Centrifuge 5 minutes at 1000 $\times$ g at 20°C.

7. Save the upper phase. Transfer to a 15-ml conical tube. Adjust to 0.1 M sodium acetate. Add three volumes of cold (-20°C) ethanol slowly, without shaking.

8. Invert the tube slowly. The DNA precipitates in long strands. Continue to mix slowly, without breaking the DNA, until it is completely precipitated.

9. With a Pasteur pipette, spool out the DNA from the ethanol. Transfer to another tube.

10. Dry off the remaining ethanol by placing DNA under vacuum for a few minutes.

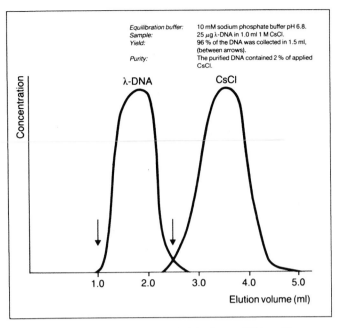

Equilibration buffer:	10 mM sodium phosphate buffer pH 6.8.
Sample:	25 µg λ-DNA in 1.0 ml 1 M CsCl.
Yield:	96 % of the DNA was collected in 1.5 ml, (between arrows).
Purity:	The purified DNA contained 2 % of applied CsCl.

Figure 73. Removal of cesium chloride from λ DNA on Pharmacia NAP-10 column (courtesy of Pharmacia).

11. Add a small volume of water (about 200 µl for 1 liter of initial culture) to the DNA.

12. Let the DNA dissolve slowly at 4°C overnight. The next day, mix the solution with a cutoff micropipette tip.

13. Dilute 1/400 to read the optical density at 260 nm against water.

Notes

1. Preparations of λ recombinants grown on DP50 under the conditions described here yield approximately 400 µg DNA/liter of initial culture.

2. Isolation of λ phage DNA can also be performed by hydroxylapatite chromatography (Ivanov and Gigova, 1985).

3. Removal of CsCl from λ-DNA can be achieved by passage through a Sephadex column (Figure 73).

Buffers for λ DNA Purification

λ *Dialysis buffer*
10 m*M* Tris–HCl (pH 7.4)
10 m*M* MgSO$_4$

2 liters
20 ml 1 *M* Tris–HCl (pH 7.4)
20 ml 1 *M* MgSO$_4$
1960 ml H$_2$O

10 × buffer C
500 mM Tris–HCl (pH 8.0)
100 mM NaCl
100 mM EDTA

100 ml
50 ml 1 M Tris–HCl (pH 8.0)
2 ml 5 M NaCl
20 ml 0.5 M EDTA (pH 7.5)
28 ml H$_2$O

REFERENCE

Ivanov, I., and Gigova, L., (1985), *Analytical Biochem.*, **146**, 389.

PURIFICATION OF HIGH MOLECULAR WEIGHT DNA FROM EUCARYOTIC CELLS

In many instances, intact chromosomal DNA is needed when analyzing specific sequences contained in eucaryotic genomes. Although the protocol described below is dealing with cells in cultures it can be applied without problems to the preparation of high molecular DNA from whole organs. In this case, it is necessary to mince the tissues in liquid nitrogen so as to obtain an homogenous powder which can then be resuspended directly in buffer for DNA purification.

PROTOCOL

PURIFICATION OF HIGH MOLECULAR WEIGHT CELLULAR DNA

1. Wash the cells twice with PBSA.
2. Prepare a solution of proteinase K (1 mg/ml) in NTE, pH 7.4. Let it incubate for 15 minutes at 37°C.
3. Scrape the cells and resuspend them in NTE, pH 7.4, to approximately 10^7 cells/ml.
4. Add the pretreated proteinase K to the cell suspension to a final concentration of 100 µg/ml (1/10 dilution).

5. Add 20% SDS to a final concentration of 0.5% (1/40 dilution).

6. Incubate in a water bath at 37°C overnight.

7. Add 1 volume of phenol previously saturated with NTE, pH 8.5.

8. Rotate for 20 minutes at 1 rotation/second.

9. Centrifuge at room temperature for 10 minutes at 5000 rpm in a JS4-2 Beckman rotor (or equivalent).

10. Pipette out and save the upper phase, without disturbing the interface.

11. Mix the upper phase with one volume of a chloroform:isoamyl alcohol (24:1) mixture.

12. Centrifuge for 10 minutes at 5000 rpm.

13. Pipette out and save the upper phase.

14. Repeat steps 11–13 until there is no material visible at the interface between aqueous and organic phases.

15. Pipette out the upper phase. Add to it 1/20 volume of 3 M sodium acetate and two volumes of cold 100% ethanol.

16. Spool out the DNA precipitate with a glass rod.

17. Collect the precipitate in a tube. Dry off the ethanol under vacuum and add a small volume of sterile distilled water.

18. Let the DNA dissolve in water at 4°C without resuspending if intact high-molecular-weight DNA is needed.

Note

A rapid and simple method for isolating high molecular weight cellular and chromosome specific DNA without the use of organic solvents has been described by Longmire et al. (1987).

NTE (pH 7.4)
For 1 liter:
20 ml 5 M NaCl
20 ml 1 M Tris–HCl (pH 7.4)
2 ml 0.5 M EDTA (pH 7.5)
958 ml H_2O

PBSA
For 1 liter:
8 g NaCl
0.2 g KCl
1.44 g $Na_2HPO_4 \cdot 2 H_2O$
0.2 g KH_2PO_4
H_2O to 1 liter

REFERENCE

Longmire, J. L., Albright, K. L., Lewis, A. K., Meincke, L. J., and Hildebrand, C. E. (1987), *Nucl. Acids Res.*, **15**, 859.

DIGESTION OF DNA WITH RESTRICTION ENDONUCLEASES

ENZYMATIC UNITS

It is the general convention that one unit of restriction endonuclease corresponds to the amount of enzyme required to completely digest 1 μg of λ DNA in 1 hour of incubation under optimal assay conditions. Most suppliers (see Table 27, for list of suppliers) use this definition.

CONDITIONS FOR OPTIMAL ENZYMATIC ACTIVITY

1. The nature of the DNA to be restricted is also important, since supercoiled or complex DNA may not be comparable to the λ DNA used for unit definition. It is therefore advisable to titrate each enzyme preparation with your own DNA preparation and recalculate the unit number before proceeding with large-scale experiments. The enzyme quantities needed for complete digestion of 1 μg of various DNAs are given in Table 42. Activity on single stranded DNA is reported in Table 43. Under certain particular circumstances, such as high endonuclease concentrations, substitution of manganese for magnesium, low ionic strength, high pH, or presence of organic solvents, several restriction endonucleases have been found to cleave nucleotide sequences which are similar but not identical to their specific recognition sequence. This altered specificity has been called "star" activity and has been observed to occur under a variety of conditions for different enzymes (see Table 44). A well-documented example is the star activity of Eco RI. There have been contra-

dictory reports concerning the recognition site for Eco RI star activity, probably because of different incubation conditions being used. Polisky et al., (1975) claimed that a N↓AATTN sequence could be cleaved as a result of star activity (the specific recognition site for Eco RI is G↓AATTC), while Gardner et al. (1982) reported that the cleavage sites could be any hexanucleotide sequence that contains no more than one base substitution from the normal recognition site, providing that the substitution does not result in an A to T or T to A change in the central portion of the palindrome. The Eco RI star fragments could be successfully cloned at the Eco RI site of M13mp2 phage (see Chapter 6) in spite of the apparent mismatched bases in the overlapping sequences. In another report by Woodbury et al. (1980), it was suggested that the Eco RI star activity was taking place at GGATTT, AAATTT, GAATTT, and GAATTA sites.

It is also important to remember that what is often called "pure" restriction enzyme refers only to the absence of contaminant endo- and exonucleases (see below). When you change incubation conditions, you may "activate" or detect nucleases which were not detected under the conditions used by the supplier. The most critical factors are the concentration of glycerol present in the incubation mixture and the nature of the DNA used.

Glycerol is used as an antifreeze agent in many enzyme purification protocols. Restriction endonucleases are usually stored in the presence of 50% glycerol. It has been shown by Polisky et al. (1975),

Table 42.   Enzymatic Units Needed for Complete Digestion of Various DNAs

Enzyme	pBR322 (Supercoiled)	pBR322 (Linear)	SV40	ADENO	Lambda
Acc I	4		1	1	1
Aat II	0.2		—	—	1
Alu I	1		4.0	1	1
Ava I	1		—	2	1
Ava II	2		1	2	1
Bal I	>10		—	>10	1
Bam HI	2		4.0	2	1
Bcl I	—		1	1	>10
Bgl I	8	0.5	1	1	1
Bgl II	—	—	—	2	1
Bsp MII	1		—	—	1
Bst EII	—		—	1	1
Cfo I	4		0.5	2	1
Cla I	1		—	—	1
Cla II	1		3	1	1
Dde I	1		4	10	1
Dra I	4		—	—	1
Eco RI	0.5		3	—	1
Eco RII	>10		2	1	>10
Eco RV	2		—	—	1
Hae II	2		0.3	1.5	1
Hae III	2		3	2	1
Hha I	4		2	10	1
Hinc II	4	4	2	1	1
Hind III	0.5		10	3	1
Hinf I	4		1	1	1
Hpa I	—		1	1	1
Hpa II	1		10	—	1
Mbo I	>10		1	1	>10
Mbo II	4	2	1	3	1
Msp I	0.5		1	1	1
Nar I	2		—	—	1
Nci I	4		—	1.5	1
Nde I	4		—	—	1
Nde II	>10	>10	—	—	1
Nhe I	1		—	—	1
Nru I	2		—	—	1
Pst I	4	0.5	—	2	1
Pvu I	1–2		—	—	1
Pvu II	1		4	4	1
Sac I	—		—	—	1
Sal I	8	0.5	—	—	1
Sau 3AI	2		1.5	—	1
Sau 96I	1	2	1	1	1
Sca I	4		—	—	1
Sma I	—		—	1	1
Sph I	2–3		1	1	1
Ssp I	1		—	—	1
Sst I	—		—	1	1
Sst II	—		—	1	1
Sty I	8				
Taq I	2		0.5	—	1
Tha I	10		—	4	1
Xba I	—		—	1	1
Xho I	—		—	1	1
Xma III	>8	>8	—	1	1
Xor II	>8	>8	—	—	1

Table 43. Cleavage of Single Stranded DNA by a Selection of Restriction Endonucleases

Enzymes	Activity on Single Stranded DNA Compared to Double Stranded DNA
Alu I	None
Bbv I	None
BstN I	1%
Dde I	1%
Dpn I	None
FnuD II	None
Fok I	None
Hae II	10%
Hga I	1%
Hha I	50%
Hinf I	1%
HinP I	50%
Hpa II	None
Hph I	None
Mbo I	None
Mbo II	None
Mnl I	50%
Msp I	None
Sau 3AI	None
Sfa NI	None
Taq I	1%

Table 44. Conditions Altering Activity of Restriction Endonucleases

Enzymes	Conditions Altering Enzymatic Specificity
Apy I	Glycerol, enzyme
Ava I	Glycerol, enzyme
Bam HI	Glycerol, enzyme, low salt, Mn^{2+} substituted for Mg^{2+}
Bst I	Glycerol, enzyme
Dde I	Glycerol, enzyme
Eco RI	Glycerol, enzyme, low salt, Mn^{2+} substituted for Mg^{2+}
Eco RV	Glycerol
Hae III	Glycerol, enzyme
Hha I	Glycerol, enzyme
Hind III	Low salt, Mn^{2+} substituted for Mg^{2+}
Hinf I	Glycerol, Mn^{2+} substituted for Mg^{2+}
Hpa I	Glycerol, enzyme
Kpn I	Glycerol, enzyme
Pae R7	Glycerol, enzyme, low salt
Pst I	Glycerol, enzyme
Pvu II	Glycerol, enzyme
Sal I	Glycerol, enzyme
Sau 3AI	Glycerol, enzyme
Sca I	Glycerol, enzyme, low salt, Mn^{2+} substituted for Mg^{2+}
Sst I	Glycerol, enzyme
Sst II	Glycerol, enzyme
Tth 111I	Glycerol
Xba I	Glycerol, enzyme

Malyguine et al. (1980) and George et al. (1980) that the dielectric constant of the incubation medium can alter the specificity of cleavage for several restriction endonucleases (Table 44). Therefore a final glycerol concentration greater than 5% should not be used in incubation mixtures.

Several studies have shown that the mean composition of nucleotides adjacent to the recognition site may affect the rate of enzyme cutting.

2. The use of frost-free freezers is not recommended for storage of enzymes because of the heat generated during the defrost cycle. We have found that vertical turbo freezers (such as ARB-9470 from Phillips) very conveniently avoid these problems. We also store our enzyme stocks in polystyrene racks in order to provide short-term protection against power failures and to reduce temperature variations upon opening and closing of the freezer door.

3. Use only autoclaved siliconized glassware or, better yet, presterilized plastic tubes and pipettes. Never allow the pipette tips to come in contact with anything else than the inside of the enzyme and reaction tubes. Avoid wiping the outside of the tips with tissues to blot residual enzyme. Instead, try to pipette from the surface of the enzyme solution or remove the tip up the inside of the tube in a circular motion. This serves to remove the enzyme solution

droplets which may cling to the outside of the tip. Always wear protective gloves when manipulating the enzyme tubes. Because of the small size of the microtubes being used by most suppliers, fingers can come in contact with the vial cap. This may occur without being noticed and can have unexpected consequences. The "Decap it" provided by New England Biolabs has proved to be helpful in this matter as well as others suggested by the supplier . . . (see Figure 74 for description).

4. Addition of bovine serum albumin (final concentration, 100 µg/ml) is often suggested to reduce nonspecific adsorption of macromolecules onto the glassware.

REFERENCES

George, J., Blakesley, R. W., and Chirikjian, J. G. (1980), *J. Biol. Chem.*, **255**, 6521.

Gardner, R. C., Howarth, A. J., Messing, J., and Shepherd, R. (1982), *DNA*, **1**, 109.

Malyguine, E., Vannier, P., and Yot, P. (1980), *Gene*, **8**, 163.

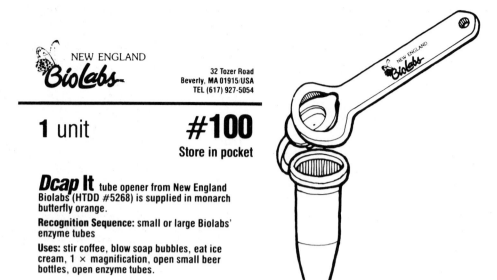

NEW ENGLAND
Biolabs

32 Tozer Road
Beverly, MA 01915/USA
TEL (617) 927-5054

1 unit **#100**

Store in pocket

Dcap It tube opener from New England Biolabs (HTDD #5268) is supplied in monarch butterfly orange.

Recognition Sequence: small or large Biolabs' enzyme tubes

Uses: stir coffee, blow soap bubbles, eat ice cream, 1 × magnification, open small beer bottles, open enzyme tubes.

Figure 74. A useful device from New England Biolabs.

Polisky, B., Greene, P., Garfin, D. E., McCarthy, B. J., Goodman, H. M., and Boyer, H. W. (1975), *Proc. Natl. Acad. Sci. USA*, **72**, 3310.

Woodbury, C. P., Hagenbuchle, O., and von Hippel, P. H., (1980), *J. Biol. Chem.*, **225**, 11534.

PROTOCOL

SETTING UP A DIGESTION WITH RESTRICTION ENDONUCLEASES

1. Reactions are usually performed in microfuge polypropylene tubes (1.5 ml). In most cases it is not necessary to siliconize these tubes. However, experience shows that some batches significantly retain the DNA. In some other cases DNA pellets will not stick to the bottom of the tube. Beckman tubes have tighter caps but lead to some nonspecific sticking of DNA, whereas Eppendorf tubes have often been found to have leaking caps, especially when used with phenol. Conclusion: Be careful and test your tubes.

2. To a microfuge tube on ice add successively:
   a. Sterile distilled water
   b. 5X or 10X reaction buffer (so as to give a 1X final concentration)
   c. DNA in sterile distilled water
   d. Restriction endonuclease.

The total volume is kept to a minimum, usually between 10 and 25 μl.

3. Flip the tube gently and spin at 12,000 rpm in a microfuge for 1–2 seconds to collect all liquid at the bottom of the tube.

4. Incubate at the desired temperature. For most enzymes the incubation temperature is 37°C. Some of them (e.g., Bst EII, Taq I) require higher temperatures (60–65°C). When long periods of incubation are required, spin the tubes occasionally for 1–2 seconds at 12,000 rpm to collect evaporated water that condenses on the tube cap. For elevated temperatures, you may perform the reactions under paraffin oil. (See section on restriction endonucleases stability below).

5. Add half a volume of 2X endo R stop solution [50 mM EDTA, 0.2% SDS, 50% glycerol, 0.05 bromophenol blue (w/v)]. Mix well and spin for 2 minutes at 12,000 rpm to remove bubbles before loading on gel.

MULTIPLE DIGESTIONS OF DNA WITH RESTRICTION ENDONUCLEASES

Very often, multiple digestions of DNA are required, either to establish a restriction map (see Chapter 10), to perform forced orientation cloning of DNA fragments in cloning vectors (see Chapter 6), or to delete specific portions of DNA molecules. Care should be taken not to mix several restriction endonucleases, as this may lead to inhibition of the reactions because of competitive binding of different enzymes at the same or overlapping recognition sites. Also, because of steric considerations (remember that enzymes are fairly large proteins), it is sometimes difficult to obtain a complete digestion of restriction sites in close proximity. Another critical aspect to take into account when performing double (or multiple) digestion of DNA is the compatibility of the reaction mixtures recommended for the optimal activity of the different enzymes being used. Although slight variations in the composition of buffers may be made to meet conditions acceptable for both enzymes, this often results in loss of activity or modification of enzyme specificity (see above). It is common for some researchers to use an excessive amount of enzyme or extended incubation times to overcome the apparent decrease in activity. Be aware that such procedures may yield unexpected enzymatic activity (see above). The ionic strength of the incubation mixture is certainly one of the most important factors to be checked prior to a double digestion, since some enzymes will be completely inactivated by high-salt or low-salt concentrations (see Table 45).

Although several attempts have been made in the past few years to introduce "universal" buffers for restriction endonucleases, we highly recommend that restriction enzymes be used under the standard conditions provided by the suppliers and that you proceed according to the following protocol when double digestions are needed.

Table 45. Effect of Salt Concentration on Restriction Endonucleases Activity[a]

Enzyme	0	50	100	150
AatI	+ + +	+ + +	+ + +	+ +
AatII	+	+ +	+ +	+
AccI	+ + +	+ + +	+	+
AhaII	+	+ +	+ + +	+ + +
AhaIII		+ + +	+ + +	+ +
AluI	+	+ + +	+ + +	+ +
ApaI	+ + +	+ + +	+ +	+
ApaLI	+ + +	+ +	+	+
AsuII	+ + +	+ + +	+ +	
AvaI	+ + +	+ + +	+ + +	+ + +
AvaII	+ + +	+ + +	+ +	+

Table 45. Effect of Salt Concentration on Restriction Endonucleases Activity[a] (continued)

Enzyme	0	50	100	150
AvrII	+ + +	+ + +	+ + +	+ +
BalI	+ + +	+ +	+ +	+
BamHI	+	+ +	+ + +	+ + +
BanI	+ + +	+ + +	+ +	+ +
BanII	+ + +	+ + +	+ + +	+ + +
BanIII	+ +	+ +	+ + +	+ + +
BbvI	+ + +	+ + +	+ + +	+ + +
BclI	+	+ + +	+ + +	+
BglI	+	+ + +	+ + +	+ + +
BglII	+ +	+ + +	+ + +	+ + +
BsmI	+ + +	+ + +	+ + +	+ + +
Bsp1286	+ + +	+ + +	+ +	+
BspMI	+ +	+ +	+ + +	+ + +
BspMII	+ +	+ +	+ + +	+ + +
BssHI	+ + +	+ + +	+ + +	+ + +
BssHII	+ + +	+ + +	+ + +	+ + +
BstEII	+	+ +	+ + +	+ + +
BstNI	+ +	+ +	+ + +	+ + +
BstXI	+ +	+ + +	+ + +	+ + +
CfoI	+ + +	+ + +		
ClaI	+ + +	+ + +	+ + +	+ +
DdeI	+ +	+ + +	+ + +	+ + +
DpnI	+	+ +	+ + +	+ + +
DraI	+ +	+ + +	+	+
DraIII	+ +	+ + +	+ + +	+ +
EaeI	+ +	+ + +	+ +	+
EagI	+ +	+ + +	+ + +	+ + +
Eco0109	+ + +	+ + +	+ + +	+ + +
EcoRI		+ + +	+ + +	+ + +
EcoRII		+ + +	+ + +	+ + +
EcoRV	+	+	+	+ + +
FnuDII	+ + +	+ +	+ +	+
Fnu4HI	+ + +	+ + +	+ +	+
FokI	+ + +	+ + +	+ + +	+ + +
FspI	+	+ + +	+ +	+ +
HaeII	+ + +	+ + +	+ + +	+ +
HaeIII	+ + +	+ + +	+ + +	+ + +
HgaI	+ + +	+ + +	+	+
HgiAI	+	+	+ +	+ + +
HhaI	+	+	+	+ +
HincII	+ +	+ + +	+ + +	+ + +
HindIII	+ +	+ + +	+ + +	+ +
HinfI	+ +	+ + +	+ + +	+ + +
HinPI	+ + +	+ + +	+ + +	+ + +
HpaI	+	+ + +	+ + +	+
HpaII	+ + +	+ + +	+ +	+
HphI	+ + +	+ + +	+ + +	+
KpnI	+ + +	+	+	+
MboI	+ +	+ + +	+ + +	+ + +
MboII	+ + +	+ + +	+ + +	+ + +
MluI	+ +	+ + +	+ + +	+ +
MnlI	+ + +	+ + +	+ + +	+ +
MspI	+ + +	+ + +	+ + +	+ + +
MstI		+ + +	+ + +	+ + +

Table 45. Effect of Salt Concentration on Restriction Endonucleases Activity[a] continued)

Enzyme	0	50	100	150
MstII	+	+ +	+ + +	+ + +
NaeI	+ + +	+ + +	+ + +	+
NarI	+ + +	+ + +	+	+
NciI	+ + +	+ + +	+ +	+
NcoI	+	+ +	+ + +	+ + +
NdeI	+	+	+ +	+ + +
NheI	+ + +	+ + +	+ + +	+ +
NlaIII	+	+	+	+
NlaIV	+	+	+	+
NotI	+	+ + +	+ + +	+ + +
NruI	+	+ + +	+ + +	+ + +
NsiI	+ +	+ +	+ +	+ +
PaeR7I	+ + +	+ + +	+ + +	+
PflMI	+ +	+ + +	+ + +	+ +
PpuMI	+ + +	+ + +	+ +	+
PstI	+ + +	+ + +	+ + +	+ + +
PvuI	+	+ +	+ + +	+ + +
PvuII	+ + +	+ + +	+ + +	+ + +
RsaI	+ + +	+ + +	+ + +	+ + +
RsrII	+ + +	+ +	+	+
SacI	+ + +	+ + +	+ +	+
SacII	+ + +	+ + +	+ +	+
SalI	+	+	+ +	+ + +
Sau3AI	+ + +	+ + +	+ + +	+ + +
Sau96I	+ + +	+ + +	+ + +	+ + +
ScaI	+	+ + +	+ + +	+ +
ScrFI	+ +	+ + +	+ + +	+ + +
SfaNI	+	+	+ + +	+ + +
SinI	+ + +	+ +		
SmaI	+	+	+	+
SnaBI	+ + +	+ + +	+ +	+
SpeI	+ +	+ + +	+ + +	+ +
SphI	+	+	+ + +	+ + +
SspI	+ +	+ + +	+ + +	+ +
SstI	+ + +	+ + +	+ +	
StuI	+ + +	+ + +	+ + +	+ + +
StyI	+	+ +	+ + +	+ + +
TaqI	+ + +	+ + +	+ + +	+ +
ThaI	+ + +	+ + +	+ +	
Tth111I	+ + +	+ + +	+ + +	+
XbaI	+	+ + +	+ + +	+ + +
XhoI	+ +	+ + +	+ + +	+ + +
XhoII	+ + +	+ + +	+ +	+
XmaI	+ + +	+ + +	+ +	+
XmaIII	+ + +	+ + +	+ + +	+ +
XmnI	+ + +	+ + +	+	+
XorI	+ + +			

[a]The activity of the listed enzymes has been assayed in the presence of various NaCl concentrations (0 mM to 150 mM) under the buffer and temperature conditions recommended by suppliers. The scale for relative activity over that obtained in the recommended conditions is (+ + +) > 50%, (+ +) 10 to 30%, (+) < 10%.

PROTOCOL

DIGESTION OF DNA WITH TWO DIFFERENT RESTRICTION ENDONUCLEASES

1. Check reaction mixtures for salt concentration and temperature of incubation. Also check for stability of enzymes if extended times of incubation are going to be used. Check for heat denaturation (Table 46) as well.

2. Set up the first digestion as described above. The choice of the first enzyme to be used is primarily dictated by salt requirements. Most of the time we start the digestion with the enzyme for which the lowest salt concentration is needed.

Table 46. Heat Inactivation of Restriction Endonucleases

Enzyme	Heat Inactivation[a]	Enzyme	Heat Inactivation[a]	Enzyme	Heat Inactivation[a]
Aat I	PR	Fok I	S	Rsa I	S
AatII	S	Hae II	S	Sac I	S
Acc I	R	Hae III	R	Sac II	PR
Acc II	R	Hap II	R	Sal I	R
Alu I	S	Hha I	S	Sau 3AI	S
Apa I	S	Hinc II	S	Sau 96 I	R
Asu II	S	Hind III	R	Sca I	S
Ava I	PR	Hinf I	R	Sin I	S
Ava II	S	Hpa I	R	Sma I	S
Bal I	S	Hpa II	R	Sph I	S
Bam HI	S	Kpn I	S	Sst I	S
Ban I	S	Mbo I	S	Sst II	R
Ban II	S	Mbo II	PR	Stu I	S
Ban III	S	Mlu I	R	Taq I	R
Bcl I	R	Msp I	S	Tha I	R
Bgl I	S	Nar I	S	Tth HB8I	R
Bgl II	R	Nci I	PR	Tth 111I	R
Bst I	R	Nco I	S	Xcy I	S
Bst EII	R	Nde I	R	Xba I	R
Cfo I	PR	Nde II	S	Xho I	R
Cla I	PR	Nru I	PR	Xma III	PR
Cvn I	S	Nsp 7524II	PR	Xor II	R
Dde I	PR	Nsp 7524III	PR		
Dpn I	S	Nsp 7524IV	PR		
Dra I	S	Pal I	PR		
EcoR I	PR	Pst I	PR		
Eco RII	S	Pvu I	R		
EcoR V	S	Pvu II	PR		

[a]Activity of restriction endonucleases was tested after 10 minutes incubation at 68°C. (R) resistant, (PR) partially resistant, (S) sensitive, to heat.

3. Perform incubation until the first digestion is expected to be complete. Without removing the tube from the waterbath, take out a small sample (usually 1 or 2 μl) of the mixture and add it to a microfuge tube containing 2 μl of sterile distilled water and 2 μl of endo R stop solution.

4. Load the sample on an agarose minigel containing ethidium bromide (see Chapter 9) and run at 80 mA for about 10 minutes.

5. Check under the UV lamp that complete digestion occurred.

6. If digestion is satisfactory, you may proceed to step 7. Otherwise, incubate further until complete digestion is obtained.

7. Depending upon the heat stability of the first enzyme (refer to Table 46), you may inactivate it by a 10-minute incubation at 68°C. If the enzyme is resistant to thermal denaturation, perform a phenol extraction under the following conditions (steps 8–18).

8. Add NTE buffer (1X) to the digested DNA solution to obtain a final volume of 200 μl.

9. Add successively 100 μl of equilibrated phenol and 100 μl of chloroform/isoamyl alcohol (24:1). Vortex for 15 seconds and spin at 12,000 rpm for 3 minutes at room temperature.

10. Carefully pipette the supernatant and transfer it to a tube containing 200 μl of chloroform/isoamyl alcohol (24:1). Vortex for 15 seconds and spin at 12,000 rpm for 3 minutes at room temperature.

11. Carefully pipette the supernatant and repeat step 10.

12. Add the final supernatant to a tube containing 100 μl of 7.5 M ammonium acetate (pH 7.5).

13. Add 600 μl of absolute ethanol and let the DNA precipitate at -70°C for 20 minutes.

14. Spin the DNA pellet at 12,000 rpm for 20 minutes at 4°C.

15. Carefully discard the ethanol to avoid losing any material and add 500 μl of 80% ethanol to the tube.

16. Spin at 12,000 rpm for 10 minutes at 4°C.

17. Carefully discard the ethanol and dry the precipitate under vacuum.

18. Resuspend the dry precipitate in water and buffer to carry out the second restriction endonuclease digestion.

If more than two digestions are being performed, repeat the extraction procedure when the buffers for different restriction endonucleases are not compatible or when enzymes are resistant to heat inactivation.

Notes

1. Many of us have experienced difficulties in obtaining complete digestions of restriction sites which are in close proximity, such as the multiple cloning site (MCS) regions of several vectors (see Chapters 6 and 20). In this case we recommend performing a phenol extraction of the first enzyme before running the second reaction.

2. Precipitation of DNA with ammonium acetate has proved to be more efficient for the removal of impurities than precipitation with sodium acetate. This is especially important to consider when working with DNA solutions that may contain heavy metals, detergents, or any other compounds that are known as potent inhibitors of restriction endonucleases, as well as other enzymes, such as DNA ligase or polynucleotide kinase. Be aware that ammonium ions inhibit polynucleotide kinase activity.

3. If the DNA concentration in the samples is lower than 50 μg/ml, an overnight ethanol precipitation at −20°C is recommended in order to obtain good recoveries of DNA. Above this concentration, a 20-minute precipitation at −70°C is sufficient.

4. When short DNA fragments are generated during the first digestion, it is advisable to perform the precipitation in the presence of glycogen (see Chapter 3).

STABILITY OF RESTRICTION ENDONUCLEASE ACTIVITY DURING EXTENDED DIGESTIONS

It is frequently observed that complete digestion of high molecular weight cellular (or viral) DNA requires more than the theoretical amount of restriction endonucleases needed according to the definition of standard units. In such cases and when large amounts of DNA are being digested, we often increase incubation time to avoid using considerable amounts of enzymes. It is therefore important to have some information about the stability of restriction endonucleases when they are incubated over extended periods of time in the reaction buffer. Most of the 60 restriction endonucleases tested (Crouse and

Amorese, 1986) could not digest a standard DNA to completion after 1 hour of incubation in the absence of DNA (listed as 1 hour complete activity in Table 47). Among them, BamH I, Kpn I, Pst I, and Sma I were among the least stable. On the contrary, 18 enzymes were found to be fully active after 5 hours incubation in the reaction buffer containing no DNA. As expected, some enzymes may be more stable in the presence of substrate DNA than in the buffer alone. This is the case for Bal I which shows increased activity after 5 hours incubation in the presence of DNA.

In practice, when we perform a digestion of large amounts of high molecular weight DNA we use twice as many units as theoretically required and incubate

Table 47. Restriction Endonuclease Stability[a]

Restriction Endonucleases	Hours[a] Complete Activity	Hours[b] Partial Activity
AccI	5	N.D.
AluI	5	N.D.
ApaI	5	N.D.
AvaI	5	N.D.
AvaII	3	5
BalI	1	5
BamHI	1	2
BclI	5	N.D.
BglI	1	5
BglII	1	5
BstEII	1	5
CfoI	1	1
ClaI	5	N.D.
CvnI	1	2
DdeI	1	5
DpnI	1	5
DraI	1	3
EcoRI	5	N.D.
EcoRII	1	2
EcoRV	1	4
HaeII	1	5
HaeIII	2	5
HhaI	1	4
HincII	1	5
HindIII	2	5
HinfI	1	3
HpaI	1	5
HpaII	5	N.D.
KpnI	1	2
MboI	1	5
MboII	3	5
MluI	5	N.D.
MspI	1	2
NarI	5	N.D.
NciI	5	N.D.
NcoI	3	5
NdeI	1	5
NdeII	5	N.D.
NruI	1	5
NsiI	1	2
PstI	1	2
PvuI	3	5
PvuII	1	4
RsaI	1	2
SalI	2	5
Sau3AI	1	2
Sau96I	5	N.D.
ScaI	1	1
SmaI	1	4
SphI	1	1
SstI	5	N.D.
SstII	5	N.D.

Table 47. Restriction Endonuclease Stability[a] (*continued*)

Restriction Endonucleases	Hours[a] Complete Activity	Hours[b] Partial Activity
StuI	5	N.D.
StyI	1	1
TaqI	3	5
ThaI	1	2
XbaI	2	5
XhoI	5	N.D.
XmaIII	5	N.D.
XorII	1	5

N.D. Not determined. Restriction endonucleases that showed 5 hour stability were not tested in the presence of sheared salmon sperm DNA.

[a]The values refer to the total period of time the enzyme was incubated at the reaction temperature.

[b]The values were determined in the presence of sheared salmon sperm DNA; see text.

Source: Courtesy of Bethesda Research Laboratories, Life technologies Inc.

the reaction mixtures for 5 hours (this represents a ten-fold excess in terms of conventional units). In addition, we mix the incubation mixture several times during digestion (by inverting the tube) in order to facilitate the diffusion of the enzyme in the viscous solution. The results obtained show that under these conditions no nonspecific degradation of DNA is observed.

REFERENCE

Crouse, J., and Amorese, D. (1986), *Focus*, **8:3**, 1.

PARTIAL DIGESTION OF DNA WITH RESTRICTION ENDONUCLEASES

Partial digestion of DNA is achieved, either by performing the incubation in the presence of limited amounts of enzyme for a shorter period of time than that required to obtain complete digestion (see protocol below), or by performing the incubation in the presence of ethidium bromide.

PROTOCOL

INCUBATION IN THE PRESENCE OF A LIMITING AMOUNT OF ENZYME FOR A SHORT PERIOD OF TIME

Adjust the enzyme/DNA ratio to approximately 1 unit/μg DNA.

1. In a microfuge tube mix successively:
   a. 82 μl sterile distilled water
   b. 22 μl 5X restriction endonuclease buffer
   c. 5.5 μl DNA (1 μg/μl)
   d. 0.5 μl restriction endonuclease (10 units/μl)
   Keep at 4°C.

2. Spin 2 seconds at 12,000 rpm. Take 10 μl and immediately add to 5 μl endo R stop mix. This will be time 0 of incubation.

3. Incubate the mixture at 37°C (unless another specific temperature is required).

4. At 2, 5, 10, 15, 20, 30, 40, 50, 60, and 90 minutes, pipette out 10 μl of the incubation mixture without taking the tube out of the water bath. Mix with 5 μl endo R stop mix. Keep all samples on ice until the last one is done.

5. Determine extent of digestion as a function of time by running the samples on an agarose gel (Figure 75).

6. When you have determined the time of incubation corresponding to the amount of digestion required, you may scale up the

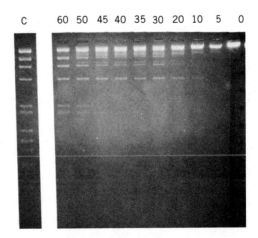

Figure 75. Partial digestion of λ DNA. 5 μg of λ DNA were incubated in presence of 5 units of Hind III restriction endonuclease for increasing periods of time. Aliquots were analyzed on a 0.8% agarose gel. Fragments were stained with ethidium bromide. C: Complete digestion.

quantities of the different components if you need large amounts of partially digested DNA. Remember that the population of molecules having a single nick is heterogeneous, all of the different sites being theoretically equally susceptible to the enzyme.

Incubation in the Presence of Ethidium Bromide

Ethidium bromide binds to DNA and therefore interferes with the enzymatic nicking activity of restriction endonucleases (Parker et al., 1977). It is necessary to adjust the quantity of ethidium bromide present during digestion for each experiment, since the number of sites for different restriction enzymes will be different.

We usually incubate 1 μg DNA with 5 units of enzyme in the presence of ethidium bromide quantities varying from 0.5 to 10 μg in a total volume of 30 μl for 2 hours at 37°C. The digested DNA is then extracted with a 1:1 mixture of phenol:chloroform/isoamyl alcohol (24/1) and analyzed by gel electrophoresis.

When the correct ratio of DNA to ethidium bromide has been determined, the concentrations can be scaled up for preparative experiments.

NOTES

1. If no cleavage occurs after incubation with a given restriction endonuclease, first make sure that you did not forget one of the components essential to the reaction (enzyme or buffer). In most cases absence of digestion is due to the presence of impurities or traces of organic solvent in the DNA preparations. A new round of extraction and precipitation will lead to clean preparations suitable for endonuclease digestion. If in spite of these precautions no digestion occurs, check that:

    a. The enzyme stock that you used is still active on a reference DNA substrate (i.e., λ or pBR322 DNA).

    b. There is no inhibition of enzyme activity because of methylation (or any other modification) of the DNA substrate (see Chapter 4 and Tables 17–18). You may certainly find an isoschizomer that is not sensitive to methylation (see Tables 17, 18, 27) to test this possibility.

2. If a partial cleavage of DNA is observed, it may result from:

    a. A loss of restriction enzyme activity (too old preparation, enzyme kept in frost-free freezers, etc.).

    b. The presence of inhibitory compounds in the reaction mix. Prepare new buffers and use new pipette tips.

    c. The topology of the DNA substrate. You may try to relax the supercoiled DNA with topoisomerase I prior to restriction enzyme digestion. When working with high molecular weight cellular DNA (see Chapter 7), use excess enzyme and increase incubation time.

d. Partial methylation of the DNA substrate.

3. If more than expected DNA fragments are obtained, check for possible star activity (see Table 44).

4. If a smearing of the DNA fragments is observed (fuzzy bands or broad staining), it can be due to the binding of enzyme to the DNA fragments or to the presence of nucleases. Reextract the DNA preparation with phenol to get rid of proteins. The presence of nucleases can be detected by incubation of the DNA preparation for 1 hour at 37°C in the absence of enzyme, followed by gel electrophoresis and staining.

5. Because DNA is a charged biopolymer it is often associated with ionic compounds that may reduce endonuclease activity. Avoid using DNA concentrations greater than 250 μg/ml when performing a restriction endonuclease digestion.

6. The presence of residual salt in DNA preparations may result in a severe inhibition of enzyme activity (see Table 45). To avoid these problems, the DNA pellets must be carefully washed with 80% ethanol, or dialysed onto filters as described in step 31, protocol for isolation of plasmid DNA (page 306).

7. Some enzymes appear to be very sensitive to small variations of temperature. Make sure your thermometers are accurate.

REFERENCE

Parker, R. C., Watson, R. M., and Vinograd, J. (1977), *Proc. Natl. Acad. Sci. USA,* **74**, 851.

9

SEPARATION OF DNA FRAGMENTS BY ELECTROPHORESIS

Two kinds of gels (agarose and polyacrylamide) are widely used to separate electrophoretically the DNA fragments generated after endonuclease digestion or to purify DNA fragments before and after enzymatic modification, ligation with other fragments, or sequencing.

AGAROSE GELS

Gel Concentration

The gel concentration used depends on the size of the DNA fragments to be analyzed and the degree of separation needed. Most of the DNA fragments currently used for further manipulations are of such a size that they are usually resolved in agarose gels with concentrations between 0.8 and 1.0%. Separation of fragments smaller than 0.5 kb may require gel concentrations of 1.2–1.5%, whereas separation of fragments longer than 10 kb should be performed in 0.5–0.3% gels (Table 48).

Preparation of the Agarose Solution

Two kinds of agarose are currently used. Both are agar derivatives that form gels of high polymeric strength and low electroendosmosis.

Regular agarose powder is melted in electrophoresis buffer by boiling (either on a Bunsen burner or in a microwave oven for a few minutes). Make sure that you mix the suspension properly before it boils

to prevent the agar from sticking to the bottom of the flask. Transfer the melted agar to a water bath at 45°C, where it can be kept without solidifying. Before pouring the gel, you may add the nucleic acid stain directly to the slurry.

Low melting point agarose is an agar derivative with a melting point of 62–65°C. This agarose, once melted, will remain liquid for several hours at a temperature of 37°C and for about 10 minutes at 25°C. The advantages of low melting agarose are the following: (a) It can be used to recover DNA fragments without electroelution or smashing the gel. Heating to 65°C will allow a direct DNA phenol extraction on the liquefied agar; after centrifugation, the upper phase contains DNA and is deprived of agarose. (b) Once the agarose is melted and brought to 37°C it is possible to perform certain enzymatic reactions. For instance, one can perform a second restriction digest on DNA fragments already separated by electrophoresis. The mixture is then loaded on another gel, and when the agarose has solidified in the well, the second electrophoresis is run.

Gel Beds for Horizontal Electrophoresis

A considerable variety of gel beds have been described in the literature. Several systems are available commercially (Figure 76). Homemade gel beds are often better adapted to the requirements of the user (size of the gel bed, size of the comb, volume of buffer tanks, etc.) and cheaper.

The simplest way to prepare a gel is to tape a

Table 48. Choice of Agarose Concentration for DNA Gels[a]

Agarose Concentration (%)	Separation Range (Size in bp)
0.3	5,000 to 60,000
0.6	1,000 to 20,000
0.8	800 to 10,000
1.0	400 to 8,000
1.2	300 to 7,000
1.5	200 to 4,000
2.0	100 to 3,000

[a]The table shows the range of separation for linear double stranded DNA molecules in TAE agarose gels with regular power sources. Note that these values may be affected if another running buffer is used and if voltage is over 5 v/cm. Use of reverse field electrophoresis increases separation range (Figures 84 and 85).

glass or plastic plate all around with tape (such as Magic Scotch) and fold half of the tape underneath the plate to form a chamber into which the gel can be poured. Plastic combs (see below) are used to form the sample wells in the gels. The comb is usually placed a few centimeters away from one end of the gel bed (corresponding to the minus pole) before pouring the melted slurry. When the agar is solidified, the tape is carefully removed to avoid damaging the gel. The plate is then put at the bottom of a container equipped with two electrodes and the gel is covered with electrophoresis buffer. The samples (in a 10% glycerol or Ficoll loading buffer) are loaded directly into the buffer. Bromphenol blue is the usual tracking dye.

Another very simple way to prepare a gel is to use a glass or plastic plate with raised edges in place of

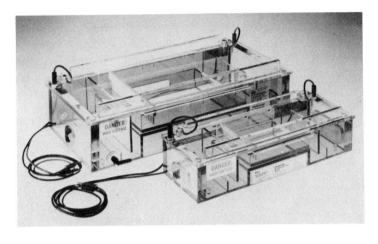

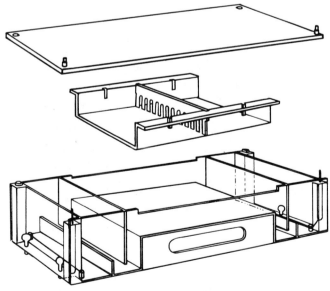

Figure 76. Horizontal gel electrophoresis system manufactured by BRL.

Table 49.   Relationship Between Voltage and Current in Agarose Gels Run with TAE and TBE Buffers[a]

Total Volume of Buffer in Tanks: 900 ml			Agarose 0.8%		Agarose 1.2%	
Volume of Gel (ml)	Volume of Buffer Covering the Gel (ml)	Voltage (volts)	Current with TAE Buffer (mA)	Current with TBE Buffer (mA)	Current with TAE Buffer (mA)	Current with TBE Buffer (mA)
150	96	20	15	5	16	8
		50	40	12	41	18
		100	82	25	85	36
		150	126	39	129	54
250	85	20	22	6	18	6
		50	57	16	51	16
		100	117	32	105	34
		150	178	49	160	52
400	76	20	23	8	25	9
		50	63	21	68	23
		100	130	43	141	47
		150	199	65	210	72

[a]Experimental values obtained with a U-shaped gel bed.

the tape. The melted agarose slurry is poured directly onto the plate and allowed to solidify. In this method the gel is connected with the buffer tanks containing the electrodes by two pieces of Whatman 3MM paper. Pieces are cut in such a way that one dimension matches with the width of the gel and the other allows the paper to dip in the buffer tanks. The two pieces of Whatman paper are wetted with electrophoresis buffer, applied firmly onto the gel at both ends, and dipped in the buffer by placing the plate directly on the buffer tanks. In this case, the samples are loaded without flooding the gel. Once the DNA samples have entered the gel (this can be checked with a portable UV lamp if the gel contains a nucleic acid stain), the wells are filled with melted agarose and the gel is covered with Saran wrap (or equivalent) to prevent dessication.

A third simple way to prepare agarose gels is to use a U-shaped bed with two "legs" dipping into the buffer tanks. Before the gel is poured, a piece of tape is applied at the base of each leg. The slurry is poured and the comb inserted in such a way that the top of the solidified gel does not reach the top of the gel bed edges. When the gel has solidified, the tape is removed and the legs dipped directly into the buffer tanks. The samples can be loaded with or without buffer on top of the gel. In either case the gel should be covered with plastic wrap after loading to prevent desiccation.

The migration rate of DNA fragments is a function of the electric field in the gel. It will vary with the potential value at each point. In practice, the resistance of the gel and the intensity of current deliv-

ered are to be considered. Most commercial power supplies allow delivery of constant voltage (V) or constant current (I). Some of them may work at a constant power. In agarose gels, when a constant voltage (V) is applied, the current tends to increase during the run because the overall resistance is decreasing [remember that $V_{(i)} = R_{(\Omega)} \times I_{(A)}$]. On the contrary, setting up a constant current (I) will result in a decreasing migration rate, as the resistance value will drop. The resistance of the system is slightly affected by the quantity and concentration of agarose used. It will depend essentially on the ionic strength of the buffer being used for electrophoresis. The values reported in Table 49 show that samples run in identical gels but in the presence of TAE or TBE running buffer will not migrate in the same way. At a given voltage, about twice the current is obtained with TAE. Therefore, if electrophoresis is performed at a constant current, migration will be faster in TBE buffer (Table 49, Figure 77).

Be aware that the anode buffer becomes rapidly acidic when the gels are run in TAE at a relatively high amperage. If acidification reaches the bottom of the gel, DNA will be degraded and no band will be visible in this area after staining. This is often observed when gels are run overnight at a constant voltage (Figure 78).

To avoid this kind of problem, you may recirculate the buffer between the two tanks with a peristaltic pump, and work at constant amperage rather than constant voltage. Also keep in mind that if you run submarine gels, the current will increase with the depth of buffer above the gel.

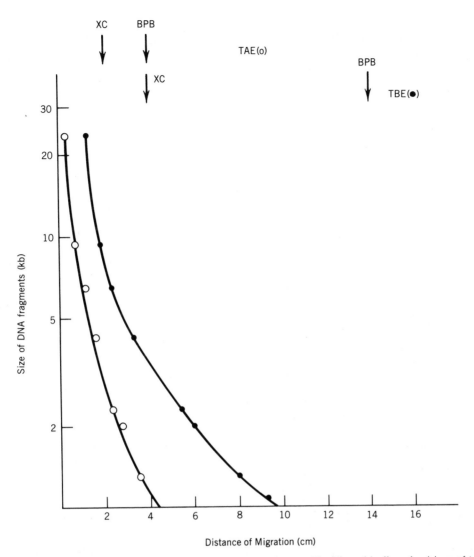

Figure 77. Migration of DNA fragments in agarose gels run with different buffers. A mixture of λ Hind III DNA fragments was electrophoresed on 1% agarose gels prepared and run in the presence of TAE (o) or TBE (●) buffer. The position of bromophenol blue (BPB) and xylene cyanol (XC) dyes is indicated for each gel system. Both gels were run at 16 mA for 15 hours.

The gel systems described above are well adapted for the simultaneous analysis of many samples or for preparative electrophoresis of specific DNA fragments. However, in most cases, there is a need not for analysis of many samples or for large amounts of DNA but rather for quick analytical answers (e.g., to check electroelution, to check DNA preparation, or to follow a restriction digest). In such cases minigels should be used.

To prepare a minigel, melt the agarose as for a regular gel, and, when the gel has cooled to 45°C, add ethidium bromide. Pour the melted agarose directly onto a glass microscope slide of the desired size. We currently use 4 × 6 cm slides and pour 7–

8-ml gels with a 10-ml pipette. Be careful not to hit the bench or the plate to prevent the agarose from running off the slide. It is essential to position the comb before pouring the gel; this will also help to maintain the agarose on the slide. When the agarose has solidified (this requires a few minutes), remove the comb and place the gel in a small electrophoresis apparatus equipped with two small buffer chambers (Figure 79). An example of commercial setting for minigels is shown in Figure 80.

We usually cover the gel with buffer and load 8 μl of sample (in 10% glycerol) per well. The gel is run at 75 mA for 15 minutes. Higher amperage can generate so much heat that the gel melts. These small gels can be used for DNA blotting as well as regular gels. Transfer requires no more than 2 hours. Alternatively, the DNA fragments run in minigels can be successfully hybridized with specific probes after denaturation within the gels. After washing to eliminate unreacted probe, the gel can be dried and set for autoradiography.

Combs

A great variety of combs are being used. Some examples are shown in Figure 81. When forming sam-

ple wells, do not put the comb in contact with the bottom of the gel bed; leave some space (0.5–1 mm) so that the gel bed will be covered with a film of agarose that will prevent the samples from leaking from one well to another. Let the gel solidify at room temperature for about 2 hours (or at least 45 minutes) before removing the comb. Addition of electrophoresis buffer to the surface of the gel eases removal of the comb. Pull the comb straight up to avoid wrecking the wells.

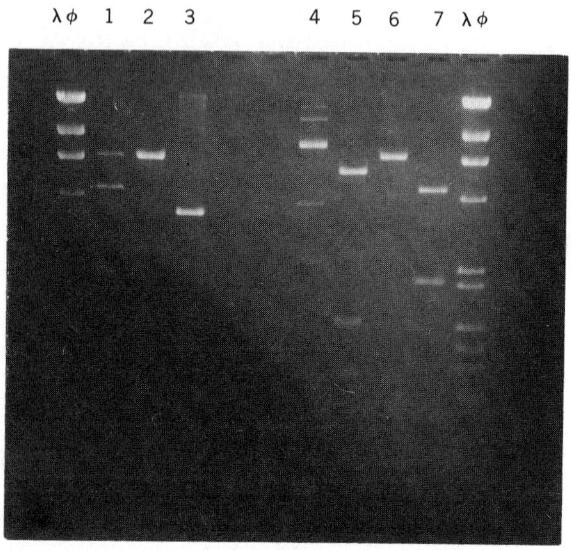

Vertical Agarose Gels

Although restriction fragments are generally separated on horizontal agarose gels, the vertical agarose gels are especially useful for isolating large plasmids which are sometimes difficult to isolate with other techniques (see Chapter 11) (Figure 82).

In vertical units, the gel should be cooled on both sides to eliminate "smiling" of the bands. Because glass plates are opaque to UV light, ethidium bromide is not added to the agarose solution and the gels are stained after the run. The main disadvantage of the technique is that no monitoring of the run is possible.

Separation of DNA Fragments by Pulsed and Reversed Field Electrophoresis

Conventional gel electrophoresis performed as described above allows the separation of DNA fragments whose size in the range of 15 to 20 kb. The use of low percentage gels makes it possible to improve slightly the resolution of the system and to resolve DNA fragments up to 50 kb.

The first successful attempt to separate DNA fragments larger than 50 kb was described by

Figure 78. Example of gel acidification during electrophoresis. Note on the left side that two λ DNA fragments (2.2 and 2.0 kb) and all φX174 RF DNA fragments were completely hydrolyzed. Also, a DNA fragment could not be detected in lane 3 because of acidification.

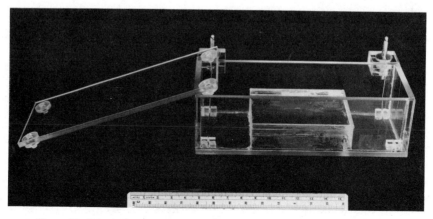

Figure 79. "Homemade" electrophoresis apparatus for agarose minigels.

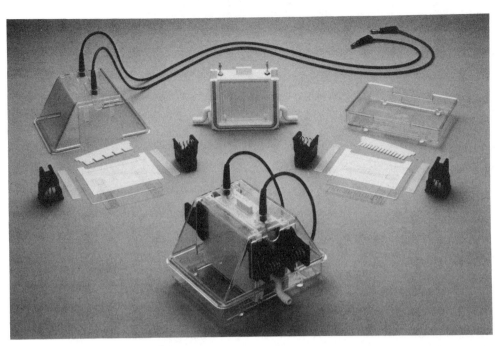

Figure 80. Mighty small vertical slab gel unit (courtesy of Hoefer Scientific).

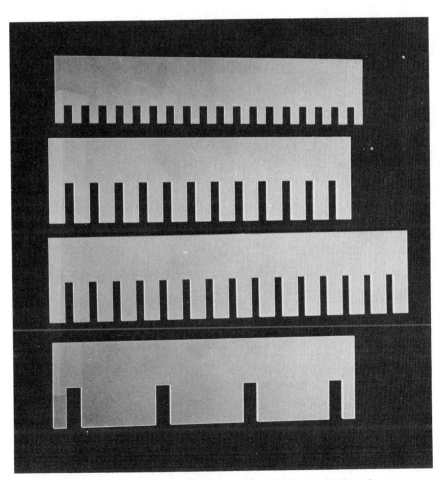

Figure 81. Various "homemade" combs used for electrophoresis.

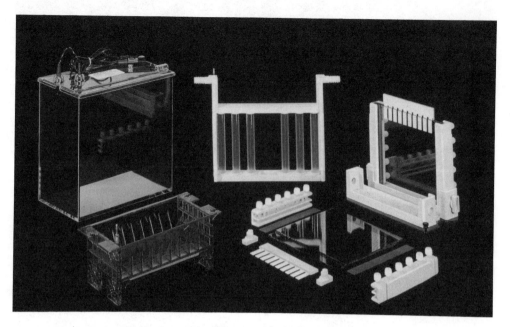

Figure 82. Exploded view of SE600 vertical slab gel unit (courtesy of Hoefer Scientific).

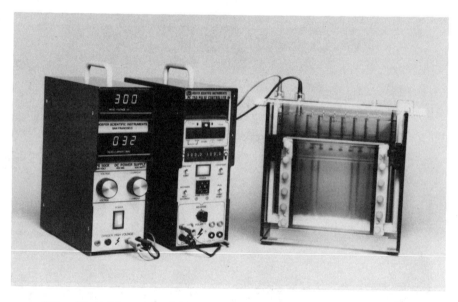

Figure 83. Unit for reverse field electrophoresis (courtesy of Hoefer Scientific).

Schwartz and Cantor (1984), who suggested forcing large DNA molecules to turn corners periodically in the gel matrix, thereby introducing a size dependence effect which does not occur in the one dimension migration. Pulsed field gel electrophoresis was applied to separate intact chromosomal DNA molecules from yeast and several other protozoans (Carle and Olson, 1984; Carle and Olson, 1985; Van der Ploeg et al., 1985). Several commercial control units have been commercialized (see for example, Figure 83).

More recently, Carle et al. (1986) have discovered that the migration of large molecules of DNA can be made strongly size dependent by performing a periodic inversion of the electric field in one dimension. In their original paper, Carle et al. used a 1% agarose gel and obtained a good separation of DNA molecules ranging in size between 2 and 400 kb. Since

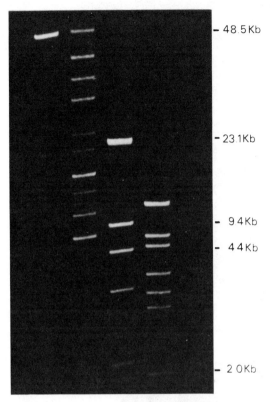

Figure 84. Separation of high molecular weight DNA fragments by reverse field electrophoresis (courtesy of Hoefer Scientific).

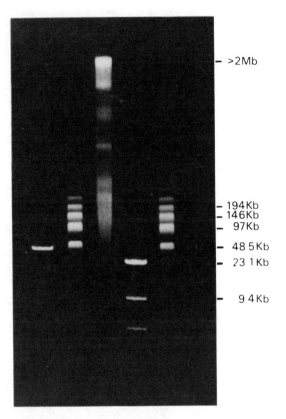

Figure 85. Separation of very large size DNA fragments by reverse field electrophoresis (courtesy of Hoefer Scientific).

given pulse times affect DNAs within a limited size range, it is necessary to change pulse times during a run to resolve fully a broader range of DNA molecules. The results obtained with a Hoefer PC 750 Pulse Controller Unit (Figures 84, 85) show that separation of DNAs up to 1000 kb is made possible by this technique. It is interesting to note that pulsed field electrophoresis also improves separation of smaller DNA fragments (Hoefer, personal communication).

This improvement in gel electrophoresis is extremely promising and opens new ways of analyzing large DNA molecules which could not be analyzed and manipulated by conventional techniques.

REFERENCES

Carle, G. F., and Olson, M. V. (1984), *Nucl. Acids. Res.,* **12**, 5647.

Carle, G. F., and Olson, M. V.. (1985), *Proc. Natl. Acad. Sci. USA,* **82**, 3756.

Carle, G. F., Frank, M., and Olson, M. V. (1986), *Science,* **232**, 65.

Schwartz, D. C., and Cantor, C. R. (1984), *Cell,* **37**, 67.

Van der Ploeg, L. H. T., Schwartz, D. C., Cantor, C. R., and Borst, P. (1984), *Cell,* **37**, 77.

Staining of DNA Fragments

Ethidium bromide is the most convenient dye for detecting DNA fragments in agarose gels, as it intercalates between stacked bases and emits fluorescent radiation on UV illumination. Although DNA fragment migration is reduced by the presence of ethidium bromide (by 10–15%), it is very convenient to include the dye directly in the gel. This allows one to check loading and migration of the fragments in the gel (Figure 86).

When preparing an ethidium bromide-containing gel, use a stock solution of ethidium bromide (10 mg/ml in water). The final concentration of dye in the gel should be between 0.5 and 1 µg/ml. If you run submarine gels, add ethidium bromide to the buffer.

Because ethidium bromide is a very powerful mutagen and a potential carcinogen, it should be handled with great care. Always wear gloves when manipulating gels and solutions containing ethidium

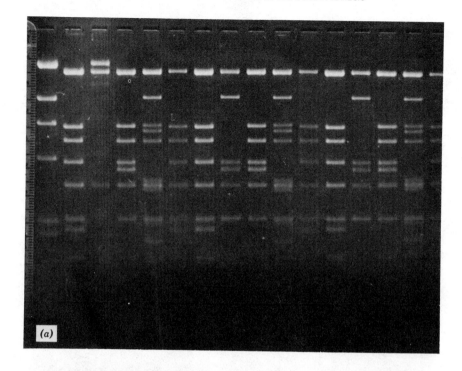

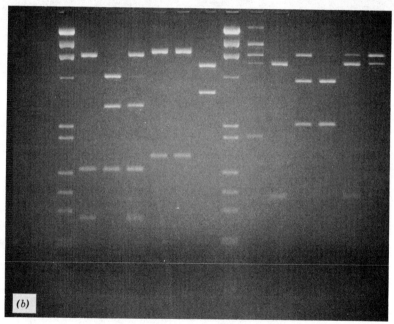

Figure 86. Ethidium bromide staining of DNA fragments. 0.5 μg DNA from various λ recombinants (a) or plasmid DNAs (b) were digested with restriction endonuclease and run in 0.8% agarose gels containing ethidium bromide (1 μg/ml or gel). Polaroid 3000 ASA instant print film was used. A mixture of Hind III-digested λ DNA and Hae III-digested φX174 RF DNA was run in tracks 1 and 8 (b) to provide marker DNA fragments of known molecular weight.

bromide. Do not dispose of solutions containing ethidium bromide in the sink. Keep ethidium bromide solutions away from the light by wrapping bottles in aluminum foil or by using dark brown containers.

Photography of DNA Fragments

The most convenient system for photography of gels is the Polaroid MP4 system, which produces excellent pictures with both regular and minigels. We recommend the use of type 57 film (3000 ASA), with exposures of a few seconds at ƒ4.5 using a long-wave transilluminator. When middle-range UV and short-wave UV are used, very small amounts (10 ng) of DNA can easily be detected. Combined use of a minigel and type 57 film gives excellent sensitivity with long-wave UV illumination.

Buffers For Agarose Gels

10X TAE buffer	*2 liters*
0.4 *M* Tris–base	96.88 g Tris–base
0.05 *M* Sodium acetate	13.60 g Sodium acetate · $3H_2O$
0.01 *M* EDTA	7.44 g EDTA
pH 7.6	Dissolve in 1500 ml H_2O.
	Bring pH to 7.6 with 12 *N* HCl.
	Adjust to 2000 ml.

10X TBE buffer	*2 liters*
0.89 *M* Tris–borate	216 g Tris–base
0.025 *M* EDTA	110 g Boric acid
	18.6 g EDTA

POLYACRYLAMIDE GELS

Polyacrylamide gels are used to obtain good separation of DNA fragments less than 1 kb in length and to analyze sequencing reactions (Chapter 20).

Since polymerization of acrylamide is inhibited by oxygen, gels are usually poured between two glass plates. One of the plates is usually cut in such a way that two "rabbit ears" are present at both sides on one end of the gel. Although relatively fragile, such plates are still commonly used. The thickness of the gel is determined by the thickness of the spacers placed between the plates.

Gel Concentration

In most cases 5% gels are sufficient for analysis of DNA fragments generated by restriction endonucleases (Table 50). Sequencing gels use 8 or 12% acrylamide concentrations.

Table 50. Choice of Polyacrylamide Concentration for DNA Gels[a]

Acrylamide Concentration (%)	Separation Range (Size in bp)
3.5	100 to 1,000
5.0	80 to 500
8.0	60 to 400
12.0	40 to 200
20.0	10 to 100

[a]The indicated figures are referring to gels run in TBE buffer. Voltages over 8 v/cm may affect these values.

Preparation of Gels

Since a 5% acrylamide gel is rather sticky, it is convenient to siliconize one of the glass plates used to prepare the gels so that is can be removed easily, with the gel sticking only to the other plate.

Reagents should be mixed in a side arm flask kept at 4°C to reduce the speed of polymerization, then deaerated with a vacuum. Shake occasionally to help deaeration. When no more bubbles appear remove the flask from vacuum and pour the gel.

The gel is poured between the two plates directly from the flask. Do not mouth-pipette the mixture. Bubbles which appear while pouring the gel may be removed by inclining the plates from one side to the other.

Insert the comb after the mixture has been poured. Be careful not to trap air bubbles underneath the comb teeth. This can be avoided by inserting the comb on one side first, and then sliding it carefully until it is straight.

Let polymerization continue until a Schlieren pattern is visible just beneath the comb teeth. Remove the comb carefully and immediately add running buffer to fill the wells. Samples (5–10 µl) are usually loaded through the running buffer with a capillary pipette or a Hamilton syringe. Bromphenol blue is used as a tracking dye and glycerol (10%) is usually added to the sample to help loading. Flush the wells with buffer before loading.

The gels are usually run at 1–10 V/cm. It is possible to increase the voltage to 15 V/cm without much distortion of the patterns obtained. Under such conditions, running a 5% acrylamide gel takes a few hours.

Acrylamide in solution is a powerful neurotoxic which is rapidly absorbed through the skin. Never mouth-pipette acrylamide solutions. Wear gloves when manipulating acrylamide powder or solutions.

The chart below indicates quantities (in ml) to mix together to obtain gels with acrylamide concentrations ranging from 3.5 to 20%.

Final acryla-mide concen-tration (%)	3.5	5.0	8.0	12.0	20.0
Acrylamide so-lution (30%)	11.6	16.6	26.6	40.0	66.6
10 × TBE buffer	10.0	10.0	10.0	10.0	10.0
H_2O	77.8	72.8	62.8	49.4	22.8
Amonium per-sulfate (10%)	0.6	0.6	0.6	0.6	0.6
TEMED	0.005	0.005	0.005	0.005	0.005

Once ammonium persulfate has been added, the polymerization reaction starts. Do not wait too long after adding ammonium persulfate and TEMED catalyst to pour the gel.

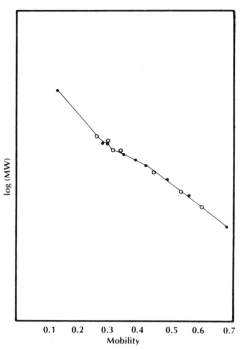

Figure 87. Plot of mobility versus log (MW) for pBR322 Hinf I fragments (filled circles) and pSP64 Hae III fragments (open circles) electrophoresed on a 10% polyacrylamide gel (courtesy of Promega Biotec).

Anomalous Mobilities of DNA Restriction Fragments on Polyacrylamide Gels*

High percentage polyacrylamide gels are frequently used to size small DNA fragments (0.1–1.5 kb). However, DNA fragments often do not migrate in these gels as would be expected based on their molecular weights. In particular, it is not infrequently observed that a DNA fragment migrating more quickly than another is in fact larger in size.

In Figure 87, mobility versus log (MW) for pSP64 Hae III fragments and pBR322 HinfI fragments electrophoresed on a 10% polyacrylamide gel have been plotted. While an approximately straight line can be fitted through the collection of points, it is notable that the slope of the line generated by connecting any two adjacent points varies widely. Thus, while mobility is usually a linear function of log (MW), in individual cases the change in mobility for a given change in log (MW) is not constant and can be seen here to vary up to a factor of 6. A consequence of this variation is that certain fragments can assume an inverted order of migration. This can

*Lewis, K. Courtesy of Promega.

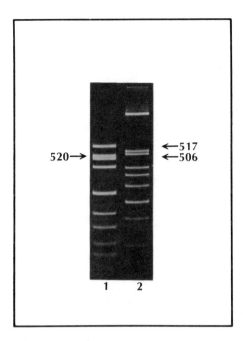

Figure 88. pSP64 Hae III fragments (lane 1) and pBR322 Hinf I fragments (lane 2) electrophoresed on a 10% polyacrylamide gel. 1.6 μg of each digestion, run for 4 hours, 150 volts, then stained with 1.5 μg/ml ethidium bromide. Numbers on the sides refer to fragment sizes in base pairs (courtesy of Promega Biotec).

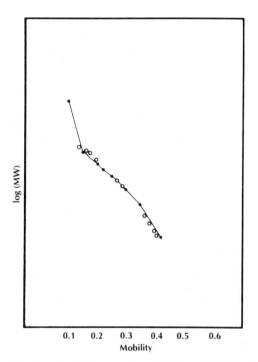

Figure 89. Plot of mobility versus log (MW) for pBR322 Hinf I fragments (filled circles) and pBR322 Hae III fragments (open circles) electrophoresed on a 10% polyacrylamide gel (courtesy of Promega Biotec).

be seen in the corresponding gel shown in Figure 88. Here it is seen that both the 517 base pair (bp) and 506 pb HinfI fragments of pBR322 (lane 2, fragments 2 and 3) travel more slowly than the 520 bp HaeIII fragment of pSP64 (lane 1, fragment 2).

It might be thought that these anomalous mobilities are due to the presence of three base single stranded extensions on the HinfI fragments (HaeIII gives blunt ends) which could conceivably retard the mobility of these fragments in the gel. To address this issue, mobility versus log (MW) for HaeIII fragments and for HinfI fragments of pBR322 electrophoresed on a 10% gel have been plotted (Figure 89). It is noteable that the HaeIII points lie on both sides of the curve formed by joining the HinfI points. Thus, it is not that there is a systematic retardation in mobility of the HinfI fragments due to the single stranded extensions unaccounted for in these fragment sizes. If any such effect did exist, it would presumably be greater for the smaller fragments and, as seen by the plot (in the mobility range of 0.35-0.43), the blunt ended HaeIII fragments actually travel slower (underneath the HinfI line) than would be expected based on the mobilities of HinfI fragments of similar sizes. The presence of three base single stranded extensions on both ends of the

HinfI fragments appears to have no systematic effect on mobility, and it appears that the sizes of these fragments are well represented by their double stranded length alone.

Additional cases of inverted fragment mobilities can be found. Figure 90 shows that on a 10% PAG the pBR322 HinfI 344 bp fragment migrates more slowly than the HinfI 354 bp fragment of pSP64. Also, the 220 and 221 bp HinfI fragments of pBR322 can be seen to migrate more quickly than the 218 bp pSP64 HinfI fragment (Figure 90a). Even more surprising is the huge separation observed between the 659 and 655 bp AluI fragments of pBR322 upon electrophoresis in a 10% PAG (Figure 90b).

The same digests seen in Figures 88 and 90 (a and b) have been electrophoresed on 4% NuSieve agarose (NuSieve is a registered trademark of FMC Corp.). The anomalous mobilities seen upon electrophoresis in 10% PAG are not seen when the digests are electrophoresed in 4% NuSieve agarose (Figure 91). It appears that accurate sizing of small DNA restriction fragments, which includes small inserts into pSP64, cannot be reliably accomplished by electrophoresis in high percentage polyacrylamide gels. The use of NuSieve agarose appears to be a more reliable alternative.

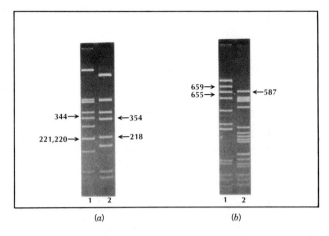

(a)   (b)

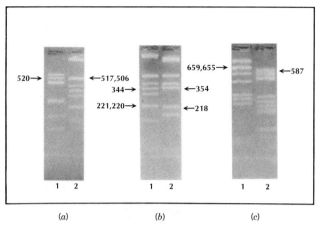

(a)   (b)   (c)

Figure 90. (a) pBR322 Hinf fragments (lane 1) and pSP64 Hinf I fragments (lane 2) electrophoresed on a 10% polyacrylamide gel. 1.6 µg of each digestion, run for 4 hours, 150 volts, then stained with 1.5 µl/ml ethidium bromide. Numbers on the sides refer to fragment sizes in base pairs. (b) pBR322 Alu I fragments (lane 1) and pBR322 Hae III fragments (lane 2) electrophoresed on a 10% polyacrylamide gel. 1.6 µg of each digestion, run for 4 hours, 150 volts, then stained with 1.5 µg/ml ethidium bromide (courtesy of Promega Biotec).

Figure 91. (a) pSP64 Hae III fragments (lane 1) and pBR322 Hinf I fragments (lane 2) electrophoresed on 4% NuSieve agarose. 0.8 µg of each digestion, run for 2 hours, 75 volts. Numbers on the sides refer to fragment sizes in base pairs. (b) pBR322 Hinf I fragments (lane 1) and pSP64 Hinf I fragments (lane 2) electrophoresed in 4% NuSieve agarose. 0.8 µg of each digestion, run for 2 hours, 75 volts. Numbers on the sides refer to fragment sizes in base pairs. (c) pBR322 Alu I fragments (lane 1) and pBR322 Hae III fragments (lane 2) electrophoresed in 4% NuSieve agarose. 0.8 µg of each digestion, run for 2 hours, 75 volts. Numbers on the sides refer to fragment sizes in base pairs (courtesy of Promega Biotec).

PROTOCOL

SEPARATION OF DNA SINGLE STRANDS BY ELECTROPHORESIS ON POLYACRYLAMIDE GELS

1. Prepare a nondenaturing polyacrylamide gel (in ½X TBE buffer) as described above. The polyacrylamide percentage in the gel is adjusted according to Table 51. We currently use a homemade gel apparatus. The gel (1.5 mm width) is poured between two glass plates (16.5 cm × 27 cm). You may siliconize one of the two plates to facilitate disassembling of the gel sandwich.

2. Let the gel polymerize for at least 1 hour before using it. Remove the comb. Fill the buffer tanks with running buffer (½X TBE).

3. Transfer the gel in the cold room (or use a cooling unit). To avoid renaturation of the DNA fragments, do not allow the gel to warm up during electrophoresis.

4. Perform a preelectrophoresis for 30 minutes at 300 volts.

5. Resuspend the dry DNA sample in 70 µl of denaturation buffer (30% v/v DMSO, 1 mM EDTA).

Table 51. Choice of Polyacrylamide Gel Concentration for the Separation of Single Stranded DNA Fragments[a]

Size of the Fragments (bp)	Gel Concentration (% acrylamide)					
	3	3.5	4	5	6	8
50–100	—	—	—	—	—	1,500
100–150	—	—	—	—	2,000	3,000
150–200	—	—	—	—	3,000	4,000
200–250	—	—	—	2,500	3,500	5,000
250–300	—	—	—	3,000	4,000	5,500
300–350	—	—	—	3,500	4,500	6,500
350–400	—	—	3,000	4,000	5,000	7,500
400–450	—	—	3,500	4,500	5,500	
450–500	—	—	4,000	5,000	7,000	
500–600	—	—	4,500	6,000	8,500	
600–700	—	—	5,000	7,000	10,000	
700–800	—	—	5,500	8,000	12,000	
800–900	4,000	5,000	6,500	9,000	—	
900–1,000	5,000	6,000	8,000	10,000	—	

[a]To calculate the appropriate voltage, proceed as follows: 1. Localize fragment size on the left column in the table. 2. Localize in the top line the percentage of acrylamide used in the gel. 3. Divide the cross value by the time of the run (in hours); the result is the voltage to apply (in volts). For example, to separate strands of a 600-bp fragments in a 5% gel, run for 15 hours, apply 6000/15 = 400 volts.

Source: Data provided by F. Sor

6. Heat for 2 minutes in a water bath at 90°C and rapidly cool the denatured sample in iced water.

7. Flush the wells with running buffer to clean the surface of the gel use a 5-ml syringe).

8. Load the gel immediately. Run for about 15 minutes at 100 volts above the voltage value that will be applied for electrophoresis. The required voltage for a 15-cm migration under our experimental conditions can be calculated from Table 51. The running voltage should not be greater than 400 volts to avoid heating up.

9. When dyes have moved about 1 cm inside the gel, turn down the voltage to the chosen electrophoresis value. Let the migration proceed for 16 to 20 hours.

10. After electrophoresis, remove the siliconized plate and carefully immerse the other plate with gel in 1 liter of TBE buffer containing ethidium bromide (1 µg/ml). Leave at room temperature for 1 hour, with occasional gentle shaking of the buffer.

11. Empty the staining solution by aspiration. Do not mouth-pipette the ethidium bromide solution.

12. Add 1 liter of fresh TBE buffer and leave at room temperature for about 1 hour with occasional gentle shaking.

13. Carefully aspirate TBE buffer. Do not let the gel slip off the plate. Cover the plate with Saran wrap.

14. Flip the plate on the transilluminator so that the Saran wrap is in contact with the UV screen. Remove the glass plate carefully. Localize the DNA fragments with UV.

15. Cut out the gel slices containing fragments of interest and proceed for elution (see Chapter 11).

Solutions and Buffers

Solution A
Acrylamide 40%   40 g in 100 ml of distilled
                                                   water

Solution B
Bis-acrylamide 2%   2 g in 100 ml of distilled water

To prepare a 5% polyacrylamide gel:

10% Base gel (50 ml)
12.5 ml of solution A
2.5 ml of solution B
2.5 ml of TBE 10X
32.5 ml of distilled water
250 µl of ammonium persulfate 10%
50 µl of TEMED

Resolving gel (200 ml)
25 ml of solution A
10 ml of solution B
10 ml of TBE 10X
154 ml of distilled water
1 ml of ammonium persulfate 10%
200 µl of TEMED

Running buffer 0.5X TBE

10X TBE	*2 liters*
0.89 M Tris–borate	216 g Tris–base
0.02 M EDTA	110 g Boric acid
	18.6 g EDTA

Staining of DNA Fragments

After running the gel, carefully remove the siliconized glass plate. The gel remains stuck to the other plate. Immerse the plate and gel in a solution of ethidium bromide (0.5 µg/ml in running buffer) for about 30 minutes. Remove the plate and cover the gel with Saran wrap. Invert the plate on the transilluminator so that the Saran wrap is beneath the gel. Remove the glass plate and take the picture.

It should be noted that the detection of DNA fragments in polyacrylamide gels by ethidium bromide staining is somewhat less sensitive than in agarose gels.

Silver staining is a highly sensitive method for detecting nucleic acids in polyacrylamide slab gels. It is not recommended for agarose gels. Both double stranded DNA and single stranded DNA and RNA can be detected by silver staining. A typical protocol for nucleic acid silver staining involves incubation in fixating solutions (30 minutes in 50% methanol, 10% acetic acid; 30 minutes in 50% methanol, 7% acetic acid; 30 minutes in 10% glutaraldehyde) followed by washing with water and incubation in 0.1% silver nitrate solution for 30 minutes. The gel is quickly rinsed and soaked in developer (50 µl 37% formaldehyde in 100 ml 3% sodium carbonate) until the desired level of staining is obtained. Staining is stopped by adding 5 ml 2.3 *M* citric acid directly to the developer and agitating for 10 minutes. The gel is then washed in distilled water for at least 30 minutes. For storage, it is best to soak the gel for 10 minutes in 0.03% sodium carbonate (to prevent bleaching) and then wrap the gel in cellophane or keep it in a sealed bag.

Bio-Rad has developed a commercial kit for nucleic acid and protein silver-staining. Results obtained with this kit are shown in Figures 92 and 93. The use of deionized water of less than 1 µmho conductivity is recommended. Contaminants such as chloride ions will precipitate silver ions and reduce the sensitivity of the method. Water suspected of containing interfering ions can be thoroughly deionized by passage over a column of Bio-Rad AG 501 × 8 (D) mixed-bed ion exchange resin.

REFERENCES

Laemmli, U. K. (1970), *Nature,* **213**, 1133.

Merril, C. R., Goldman, D., Sedman, S. A., and Ebert, M. H. (1981), *Science,* **211**, 1437.

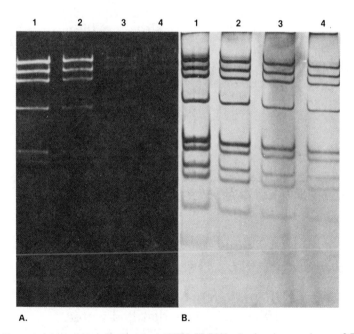

Figure 92. Silver staining of Hae III-digested φX174 RF DNA electrophoresed on a 0.75-mm thick 6% acrylamide slab gel containing Tris–borate–EDTA (TBE), pH 8.3. Amount of digest per lane: (1) 120 ng, (2) 50 ng, (3) 25 ng, (4) 10 ng. Gels were stained with ethidium bromide (A), or with the Bio-Rad silver stain (B) (courtesy of Bio-Rad).

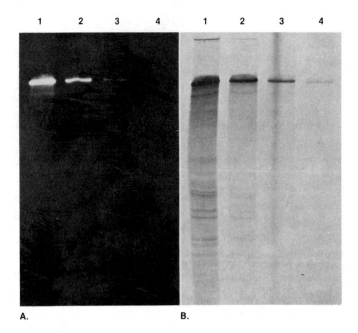

Figure 93. Silver staining of ribosomal RNA. Sucrose-gradient-purified *E. coli* 16S ribosomal RNA was electrophoresed on a 0.75-mm thick 4% acrylamide slab gel containing TBE buffer with 8 *M* urea. Amount of rRNA per lane: (1) 500 ng, (2) 100 ng, (3) 25 ng, (4) 5 ng. RNA sample kindly provided by Doug Black and Dr. Harry Noller, University of California, Santa Cruz. (A) Ethidium bromide staining; (B) silver staining (courtesy of Bio-Rad).

SIZE MARKERS FOR DNA FRAGMENTS RUN IN AGAROSE AND POLYACRYLAMIDE GELS

Agarose Gels

We essentially use a mixture of Hind III-digested λ DNA and Hae III-digested φX174 RF DNA. The DNA fragments contained in this mixture cover a satisfactory range of sizes (Figures 86, 94 and Table 52) for most experiments performed in 0.8 to 1.2% agarose gels. When a more accurate evaluation of DNA fragments size is needed in the range of 1 to 8 kb, we use the mixture of fragments generated by a Bst EII digestion of λ DNA (Figure 94 and Table 52). The preparation of these mixtures is performed as described in the protocols below.

PROTOCOL

PREPARATION OF λ-HIND III / φX174RF-Hae III MARKER

In this example, the concentrations of λ DNA and φX174 RF DNA are 0.68 and 0.56 µg/µl respectively.

1. In a microfuge tube, mix successively:
   193 µl of sterile distilled water
   40 µl of 5X Hind III buffer
   147 µl of λ DNA (100 µg)
   20 µl of Hind III (100 units)
2. Incubate at 37°C for 3 hours.

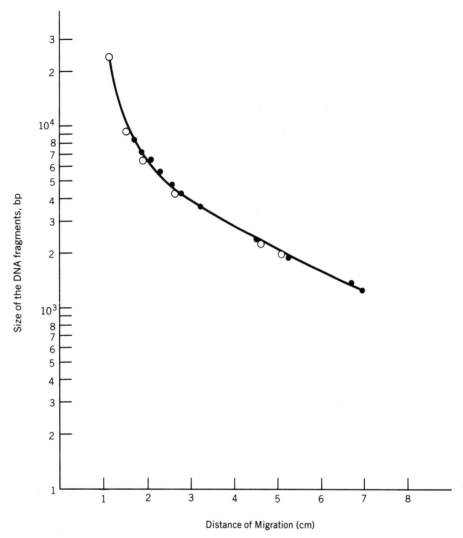

Figure 94. Migration of Bst EII and Hind III digested λ DNA in a 1% agarose gel. (o) Hind III, (●) Bst EII.

3. In another microfuge tube, mix successively:
   45 μl of sterile distilled water
   10 μl of 5X Hae III buffer
   35 μl of φX174 RF DNA (20 μg)
   10 μl of Hae III (50 units)

4. Incubate at 37°C for 3 hours.

5. Take a 1-μl sample of each incubation mixture, and add to microfuge tubes containing 3 μl of distilled water and 4 μl of EndoR stop buffer. Heat the λ digested DNA sample at 68°C for 10 minutes.

6. Run a minigel in TAE buffer for 5 to 10 minutes at 75 mA to check if digestion is complete.

Table 52.   Restriction DNA Fragments as Size Markers for Gel Electrophoresis

Size of the Fragments (bp)	Fraction of Total	Mass of Fragment (ng) in 1 μl of Mixture
λ Hind III	*100 μg/750 μl*	*0.133 μg/μl*
23,130	0.476	63.5
9,416	0.194	25.8
6,557	0.135	18.0
4,361	0.090	11.9
2,322	0.048	6.3
2,027	0.041	5.6
564	0.012	1.5
125	0.002	0.3
Total: 48,502		
φX174 Hae III	*20 μg/750 μl*	*0.026 μg/μl*
1,353	0.251	6.7
1,078	0.200	5.3
872	0.161	4.3
603	0.112	2.9
310	0.057	1.5
281	0.052	1.38
271	0.050	1.33
234	0.043	1.15
194	0.036	0.96
118	0.022	0.58
72	0.013	0.35
Total: 5,386		
Lambda Bst EII	*100 μg/750 μl*	*0.133 μg/μl*
8,454	0.174	23.2
7,242	0.149	19.9
6,369	0.131	17.5
5,686	0.117	15.6
4,822	0.099	13.2
4,324	0.089	11.9
3,675	0.075	10.0
2,323	0.047	6.4
1,929	0.039	5.3
1,371	0.028	3.7
1,264	0.026	3.5
702	0.014	1.9
224	0.005	0.62
117	0.002	0.32
Total: 48,502		

7. If digestion is not completed leave the sample for a longer period of time at 37°C, and run another minigel.

8. When digestion is complete, add 200 μl of Endo R stop solution to the digested λ DNA and 50 μl of Endo R stop solution to the digested φX174 RF DNA.

9. Transfer the content of one tube in the other and mix by in-

verting several times. Heat the mixture at 68°C for 10 minutes and store 200 μl aliquots at −20°C.

We use 6 μl of this mixture on our regular agarose gels (300 ml) and 2 μl for minigels. One microliter of this mixture contains of 133 ng of λ DNA fragments and 27 ng of φX174 RF DNA. The relative amount of each fragment can be used for quantification of DNA in other samples (see below).

PROTOCOL

PREPARATION OF λ-Bst EII MARKER

In this example, the same source as above of λ DNA was used (0.68 μg/ml).

1. Set up a Bst EII digestion as described above, except that incubation is performed at 60°C.
2. Check that digestion is complete (as above, steps 5–7).
3. Add 100 μl of sterile distilled water and 250 μl of Endo R stop solution to the digested DNA and heat to 68°C for 10 minutes before aliquoting and freezing.

One micro liter of this mixture contains also 133 ng of λ DNA fragments.

Since we currently handle these enzymes and DNAs, we find it more convenient to prepare the mixtures in the laboratory. However, one must be aware that such mixtures are commercially available (see, for example, Bethesda Research Laboratories, New England Biolabs, Pharmacia, International Biotechnologies Inc.).

Markers for High Molecular Weight DNA Fragments and for Supercoiled DNA Molecules. The BRL high molecular weight DNA fragments mixture, which has been used to study (Uher, 1986) the factors affecting mobility and separation of large DNA fragments in agarose gels (Figure 95 and Table 53), is very convenient for evaluating more precisely the size of large fragments in the range of 8 to 50 kb (Figure 96).

It has long been established that supercoiled

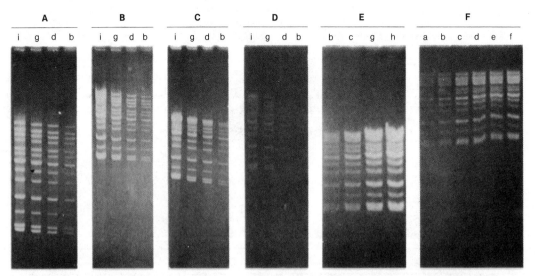

Figure 95. Gel electrophoresis conditions. Various parameters (see text) of gel electrophoresis conditions were examined. The lanes contained the following amounts of high molecular weight DNA markers: a = 25 ng; b = 50 ng; c = 100 ng; d = 150 ng; e = 200 ng; f = 250 ng; g = 300 ng; h = 500 ng and i = 600 ng (courtesy of Bethesda Research Laboratories, Life technologies Inc.).

Table 53. Conditions Used for Electrophoresis of High-Molecular DNA Fragments[a]

Panel	Time	Voltage	Temperature	Wavelength	Apparatus
A	19 hours	20	Room temperature	254 nm	Model H3 (11 × 14 cm gel bed)
B	19 hours	20	4°C	254 nm	Model H3
C	6 hours	50	4°C	254 nm	Model H3
D	19 hours	20	4°C	366 nm	Model H3
E	3 hours	25	Room temperature	254 nm	Model H6 ("Baby Gel") (50 × 75 mm gel bed)
F	16 hours	5	Room temperature	254 nm	Model H6

[a]*Source:* Courtesy of Bethesda Research Laboratories, Life technologies Inc.

DNA molecules move faster in agarose gels than the corresponding linear form and that relaxed close circles (molecules with nicks) migrate more slowly than the linear form (Figure 97). Therefore, it is sometimes difficult to estimate the real molecular weight of supercoiled DNA molecules in gels. A set of 11 supercoiled DNA plasmids (available from BRL) with molecular weights ranging from 2 to 16 kb can be used as size markers to alleviate these problems.

Polyacrylamide Gels

Several restriction enzymes generate small fragments when incubated with vector DNAs such as pBR 322 (see Table 30). Mixtures of this type can be conveniently used as size markers. For example, the Msp I digest of pBR 322 DNA (also commercialized by New England Biolabs) gives rise to 26 fragments whose sizes range from 9 base pairs to 622 base pairs (Table 54). The molecular weight ladders from

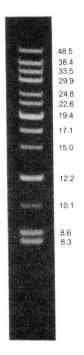

Figure 96. Electrophoresis of high molecular weight DNA markers. A 0.4% agarose gel (3 mm thick, 11 × 14 cm gel bed) was electrophoresed for 18 hours at 25 volts at room temperature with 200 ng of DNA (courtesy of Bethesda Research Laboratories, Life technologies Inc.).

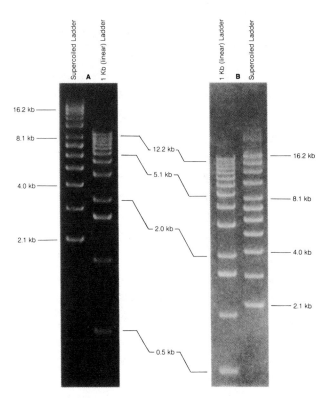

Figure 97. Results of an experiment in which the supercoiled ladder and the 1-kb ladder were electrophoresed at 8 V/cm on agarose gels with Tris–borate (A) or Tris–acetate (B) electrophoresis buffer (courtesy of Bethesda Research Laboratories, Life technologies Inc.).

BRL are also very useful for precise size determination in polyacrylamide gels. The 123 base pair ladder is prepared from a plasmid containing 34 repeats of a 123 base pair fragment (Hartley and Gregori, 1981). It therefore contains 34 DNA fragments whose sizes range from 123 to 4182 base pairs. The oligo (dT) 4-22 ladder is more appropriate for sequencing gels or characterization of oligonucleotides synthesized in vitro. It is made of 19 single stranded oligonucleotides with sizes ranging from 4 to 22 nucleotides (Thompson et al., 1983).

QUANTIFICATION OF DNA FRAGMENTS ON AGAROSE GELS

The use of DNA markers allows a rough quantification of the amount of a given DNA fragment in a gel. For this purpose, you need to run in parallel with your sample of unknown concentration, a set of marker DNA, which contains at least one fragment whose size is comparable to that of the sample. Record the exact amount of marker DNA that has been run on the gel and the relative amount of each fragment in your marker preparation (Table 52). Since the amount of ethidium bromide bound to a DNA fragment is a function of its size, it is possible to compare the coloration obtained with a known mass of a particular DNA fragment to the coloration obtained with another fragment whose mass is unknown. In practice, the intensity of coloration is determined by the picture of the gel for each DNA band. If they give a comparable signal, the same amount of the two fragments has been run on the gel. This method is useful when a rough estimation of a DNA concentration is required. For example, we often use this method to check the yield of recovery after electroelution or the concentrations of DNA fragments and vectors in ligations.

Table 54. pBR322 DNA Fragments Generated by Incubation with Msp I Restriction Endonuclease[a]

Fragment #	Number of Base Pairs	Daltons
1	622	4.04×10^5
2	527	3.42×10^5
3	404	2.62×10^5
4	309	2.00×10^5
5	242	1.57×10^5
6	238	1.55×10^5
7	217	1.41×10^5
8	201	1.31×10^5
9	190	1.24×10^5
10	180	1.17×10^5
11	160	1.04×10^5
12	160	1.04×10^5
13	147	0.955×10^5
14	147	0.955×10^5
15	123	0.799×10^5
16	110	0.715×10^5
17	90	0.585×10^5
18	76	0.494×10^5
19	67	0.436×10^5
20	34	0.221×10^5
21	34	0.221×10^5
22	26	0.169×10^5
23	26	0.169×10^5
24	15	0.097×10^5
25	9	0.059×10^5
26	9	0.059×10^5

[a]*Source:* Courtesy of New England Biolabs

REFERENCES

Hartley, J. L., and Gregori, T. J. (1981), *Gene,* **13**, 347.

Thompson, J. A., Blakesley, R. W.., Doran, K., Hough, C. J., and Wells, R. D. (1983), in *Methods in Enzymology,* R. Wu, L. Grossman, and K. Moldave, eds., Vol. 100, p. 368, Academic Press, New York.

Uher, L. (1986), *Focus,* **8:1**, 10.

10

ORDERING THE RESTRICTION DNA FRAGMENTS IN PHYSICAL MAPS

Molecular cloning of a particular DNA fragment often requires some basic knowledge of the physical organization of the genomic region surrounding the fragment. When little or no information is available, restriction maps must be established. Several methods exist for doing this. They are often complementary to each other. This is useful, as a single method may lead to unresolved ambiguities.

DOUBLE DIGESTION OF DNA FRAGMENTS

Double digestion is the most commonly used method of mapping the specific sites for restriction endonucleases. The DNA is usually first digested with one endonuclease at a time and then with pairs of enzymes. The rationale behind this method may be schematized as follows:

1. Digestion with enzyme E_1 leads to different fragments. We will consider fragment AB (size 2 kb):

$$\text{A} \quad \overset{2 \text{ kb}}{\underset{(E_1)}{\rule{4cm}{0.4pt}}} \quad \text{B} \qquad (a)$$

2. Digestion of the same DNA region with enzyme E_2 generates different fragments, one of them, CD (5 kb), being delineated by a site C within AB, and a site D lying outside AB. The problem is to map this site with respect to A and B. There are two possible positions for the E_2 site within AB:

$$\text{A} \ \overset{0.5 \text{ kb}}{\underset{(E_1)}{\rule{0pt}{0pt}}} \ \text{C} \ \overset{1.5 \text{ kb}}{\underset{(E_2)}{\rule{0pt}{0pt}}} \ \text{B} \ \overset{3.5 \text{ kb}}{\underset{(E_1)}{\rule{0pt}{0pt}}} \ \text{D} \quad (b)$$

$$\text{A} \ \overset{1.5 \text{ kb}}{\underset{(E_1)}{\rule{0pt}{0pt}}} \ \text{C} \ \overset{0.5 \text{ kb}}{\underset{(E_2)}{\rule{0pt}{0pt}}} \ \text{B} \ \overset{4.5 \text{ kb}}{\underset{(E_1)}{\rule{0pt}{0pt}}} \ \text{D} \quad (c)$$

In either case double digestion with E_1 and E_2 will lead to three fragments: AC, CB, and BD. None of the fragments will be the same size as the fragments AB and CD generated by single digestion with E_1 and E_2, respectively. The sizes of the fragments obtained allow us to distinguish between possibilities (b) and (c). In the latter, a fragment of 4.5 kb will be generated from CD, whereas in the former case, a fragment of 3.5 kb will be obtained.

Use of enzyme combinations of this kind generally allow the mapping of several restriction sites on a piece of DNA. This kind of analysis is much easier when the location of one restriction site is already established, as with cloned fragments in vectors such as pBR322 that have single restriction sites for several endonucleases. When this is not the case, one can purify the fragments obtained with the first enzyme and analyze redigestion products.

Mapping by Further Digestion of Purified DNA Fragments

Let us consider two overlapping DNA fragments: AB (6 kb), generated by Hind III, and CD (5 kb), generated by Eco RI:

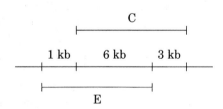

The fragment AC is common to both and will therefore be detected after Hind III digestion of CD and Eco RI digestion of AB. This allows us to position the fragments with respect to each other.

Digestion of a given DNA with Hind III and Eco RI generates the following fragments:

Hind III	EcoRI
A 6 kb	D 5 kb
B 7 kb	E 7 kb
C 9 kb	F 9 kb
X 1 kb	Y 2 kb

Double digestions give the following fragments:

Initial Fragment	Second Enzyme Digest Products
A	2 kb + 4 kb
B	2 kb + 5 kb
C	3 kb + 6 kb
D	2 kb + 3 kb
E	1 kb + 6 kb
F	4 kb + 5 kb
X	1 kb
Y	2 kb

A search for fragments common to both sets of digestions reveals that only E and C give rise to a 6-kb fragment, and therefore they must overlap.

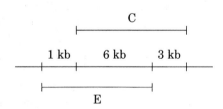

Similarly, C and D share a 3-kb fragment and we can therefore order E, C, and D as follows:

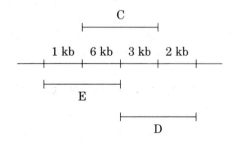

Since there is a 2.0-kb fragment common to both

A and B, we cannot yet map these fragments. However, F and A can be arranged as follows:

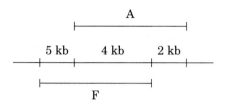

or

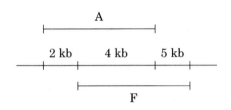

F and B share a 5-kb fragment. Therefore, the two possible arrangements become:

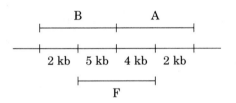

and

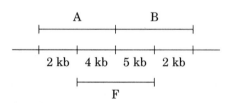

This situation does not allow us to establish an unambiguous map, as two possibilities remain:

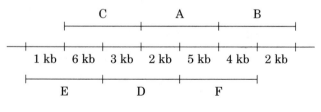

or

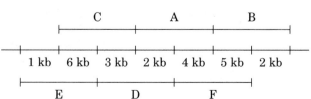

This kind of problem can be resolved by identifying the fragments corresponding to the DNA termini (see below).

MAPPING BY PARTIAL DIGESTION OF END-LABELED DNA

This mapping procedure involves three steps: labeling of the DNA ends, asymmetric cleavage of the labeled DNA, and partial digestion of the purified cleaved fragments. Figure 98 schematically illustrates this procedure.

Labeling of the DNA Ends

The 5′ end of the DNA molecule is enzymatically labeled after dephosphorylation with alkaline phosphatase (see Chapter 16). The DNA is then incubated in the presence of T4 polynucleotide kinase and γ[^{32}P]ATP.

Asymmetric Digestion of Labeled DNA

Asymmetric digestion is necessary to generate fragments labeled at only one end. An enzyme should be chosen that will generate fragments of such a size that they are easily separated by electrophoresis on agarose gels. If a single site is not available, one can use enzymes that give several fragments, so long as the end fragments can be identified.

Partial Digestion of Purified Labeled Fragments

The labeled fragments purified after electrophoresis on an agarose gel (see Chapter 11.) are digested with restriction enzymes under conditions that on average yield only one cleavage per DNA molecule (partial digestion). This can be achieved by using limiting amounts of enzyme, reducing incubation times, or performing digestion in the presence of ethidium bromide (see Chapter 8.) The resulting DNA fragment population forms a "ladder" of discrete labeled molecules whose sizes can be determined accurately after electrophoresis on acrylamide gels. The difference in size between two neighboring fragments is a measure of the distance between two adjacent sites for the enzyme under consideration.

This procedure, first described by Smith and Birnstiel (1976), has been used widely to establish restriction site maps.

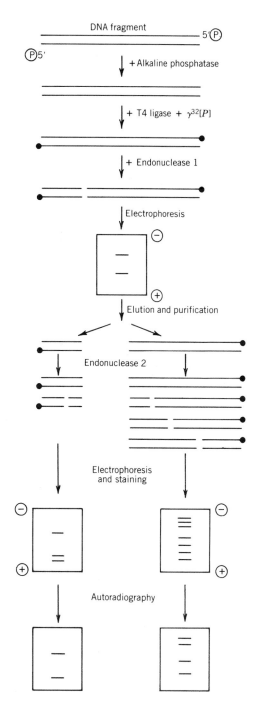

Figure 98. Mapping restriction sites by partial digestion of end-labeled DNA (see text for details).

MAPPING RESTRICTION SITES WITH Bal 31

The enzyme Bal 31 progressively digests double stranded DNA (see Chapter 4.) Under controlled incubation conditions, it is possible to use this enzyme for mapping restriction endonuclease sites.

First, prepare the DNA under conditions such

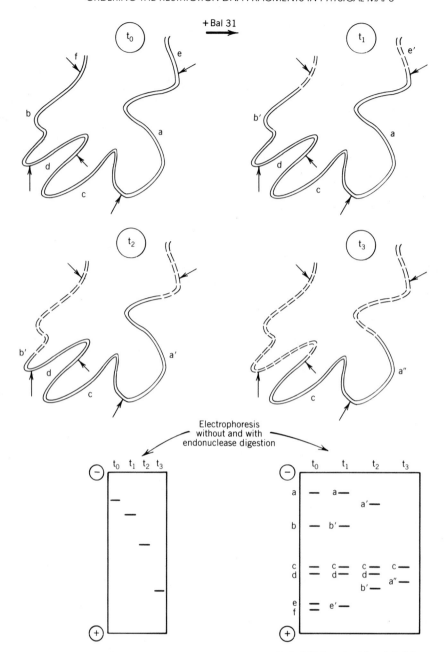

Figure 99. Mapping restriction sites by digestion with Bal 31 (see text for details).

that the initial point is known (e.g., one unique restriction site). Incubate the DNA in the presence of Bal 31 for increasing periods of time and run an aliquot of each sample on a gel to determine the extent of digest. The remainder of each sample is digested with one or more restriction enzymes. When the restriction sites have been eliminated by Bal 31, corresponding fragments disappear from the digestion pattern, allowing one to determine the location of these sites on the DNA (see Figure 99).

MAPPING BY TWO-DIMENSIONAL ANALYSIS OF END-LABELED DNA

In this technique, originally described by Kovacic and Wang (1979), the DNA is linearized, labeled at both ends, and partially digested with the enzyme whose sites are being mapped. The partially digested DNA is electrophoresed on an agarose gel. The position of the labeled fragments is detected by autoradiography, and a complete digestion of the

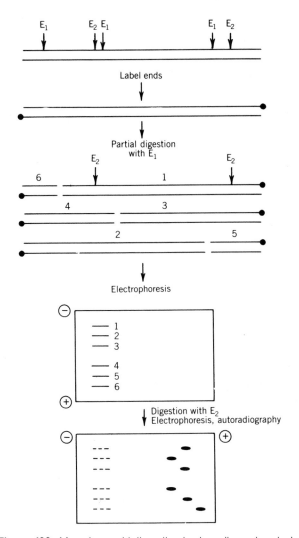

Figure 100. Mapping restriction sites by two-dimensional electrophoresis (see text for details).

fragments obtained is performed directly in the gel matrix with a second restriction enzyme. After the second digestion, the gel is rotated 90° and electrophoresis is again performed.

Labeled spots corresponding to the right and left ends of the DNA molecule are detected by autoradiography. The sizes of the DNA fragments obtained in the partial digestion are deduced from the migration in the first dimension, and correspond to the distances of restriction sites from one of the labeled ends. The mobilities of the spots in the second dimension allow one to establish from which end of the DNA the first-dimension bands are derived, and therefore to map the corresponding restriction sites.

Figure 100 schematically illustrates this procedure.

REFERENCES

Kovacic, R. T., and Wang, J. C. (1979), *Plasmid*, **2**, 394.

Smith, H. O., and Birnstiel, M. L. (1976), *Nucleic Acids Res.*, **3**, 2387.

11

PURIFICATION OF DNA FRAGMENTS

PURIFICATION OF DNA FRAGMENTS FROM REGULAR AGAROSE GELS

The protocol described below can be conveniently used to recover DNA fragments of various sizes from standard agarose gels. It allows handling of several different samples at the same time and is easy to set up.

PROTOCOL

ELECTROELUTION OF DNA FROM AGAROSE GELS

1. Stain the DNA with ethidium bromide by soaking the gel in 2 volumes of gel buffer containing 1 μg/ml of ethidium bromide for 15 minutes. Wash with water twice.

2. Locate the DNA bands with a transilluminator. Cut out the agar containing the DNA to be recovered with a sterile razor blade.

3. Put the piece of agar in a piece of dialysis tubing prepared as described in note no. 7. Add 1X electrophoresis buffer in such a way that the piece of agar is completely surrounded by buffer without bubbles when the dialysis tubing is tight. Try to keep the volume of buffer used to a minimum. For a piece of agar whose size is about 2 × 0.5 × 0.5 cm, we currently use 2–3 ml of buffer.

4. Put the dialysis bag containing the piece of agar at the bottom of a gel apparatus filled with buffer. Place the dialysis bag so that the piece of agar is in the same position with respect to the electrodes as it was in the gel.

5. Apply 150 mA for 90 minutes.

6. Without moving the dialysis bag, reverse the current for 20 seconds. This will mobilize any DNA which may be stuck to the dialysis membrane.

7. You can check at this point that the DNA is properly electroeluted by illuminating the dialysis bag with a UV transilluminator.

8. Pipette out the buffer containing the electroeluted DNA. Rinse the dialysis bag and the piece of gel with 1 ml of electrophoresis buffer. Combine the two solutions in a 15-ml polypropylene snap-cap tube.

9. Add half a volume of equilibrated phenol and half a volume of chloroform:isoamyl alcohol (24:1) to the DNA solution. Mix thoroughly. Spin at 4000 rpm for 5 minutes.

10. Pipette out the upper (aqueous) phase and add to it one volume of chloroform: isoamyl alcohol (24:1). Mix and spin at 4000 rpm for 5 minutes.

11. Pipette out the upper phase, being very careful not to take any of the material located at the interface. Repeat step 10.

12. Transfer the aqueous phase to a Corex tube. Add sodium acetate (0.15 M final concentration) and 2 volumes of ethanol. Leave at $-20°C$ overnight.

13. Spin the DNA pellet in a Sorvall SS34 rotor (or Beckman JA20) for 20 minutes at 10,000 rpm.

14. Carefully pour off the ethanol. Let the tube dry after inverting on a paper tissue.

15. Resuspend the DNA in 200 μl sterile distilled water. At this stage the DNA precipitate may be very faintly visible or not at all. Locate the precipitate from the position of the tube in the rotor.

16. Transfer the 200 μl to a microfuge tube.

17. Rinse the Corex tube once more with 200 μl sterile distilled water. Combine with the 200 μl from step 16.

18. Add 15 μl 3 M sodium acetate and 1 ml cold ethanol to the 400 μl of electroeluted DNA. Leave at $-70°C$ for 1 hour or $-20°C$ overnight.

19. Precipitate the DNA by spinning at 12,000 rpm for 10 minutes.

20. Pour off the ethanol and, without resuspending the precipitate (which should be visible at this stage), add 1 ml cold ethanol.

21. Spin for 5 minutes at 12,000 rpm. Pour off the ethanol. Invert the tube over a tissue, then dry the DNA under a vacuum before resuspending in sterile distilled water.

Notes for Electroelution

1. We have found that the quality of the agarose used may be critical for the recovery of usable electroeluted DNA. Some batches of agarose apparently contain impurities that combine irreversibly with DNA either in the gel or during electroelution, leading to fragments that cannot be ligated to other DNAs, restricted, or nick-translated efficiently.

2. It also seems very important to extract the electroeluted DNA fragment immediately after electroelution, in a relatively large volume. When we have precipitated the DNA first and extracted subsequently, we have very often recovered a DNA fragment unusable for further cloning manipulations. It is probable that some component present in the agarose can bind to the DNA fragment irreversibly during ethanol precipitation. Prior phenol extraction will prevent this kind of problem.

3. Do not forget to reverse the current at the end of the electroelution. If you do forget, you may be surprised by the amounts of DNA finally recovered!

4. When you empty the dialysis bag use siliconized Pasteur pipettes or plastic pipettes.

5. After chloroform:isoamyl alcohol extractions, significant amounts of agar or other solid residues will be present at the interface. Be careful not to aspirate these; it is better to use larger volumes and leave some aqueous phase behind.

6. The DNA recovered by this method can be quantitated by electrophoresis in agarose minigels very quickly. We usually get very good yields and DNA fragments in excellent shape.

7. Dialysis bags should be sterilized for 30 minutes at 120°C in a solution of 5% w/v sodium carbonate, 1 mM EDTA, and kept at 4°C in sterile distilled water.

PURIFICATION OF DNA FRAGMENTS FROM LOW-MELTING AGAROSE GELS

When small quantities of DNA are loaded on gels, it is sometimes more convenient to use low-melting agarose for efficient recovery. Since the method relies on direct phenol extraction of the DNA from melted gel slices, it is very important that the gel be completely melted and warm phenol used, at least for the first extraction.

PROTOCOL

PURIFICATION OF DNA FRAGMENTS FROM LOW-MELTING AGAROSE GELS

1. After taking a picture of the stained gel, cut out with a razor blade the band(s) corresponding to the DNA fragment(s) that you wish to extract.

2. Transfer the gel piece to a 5-ml polypropylene tube and add 5 volumes of buffer C.

3. Heat the sample at 65°C in a water bath until the agar is completely melted. This requires between 15 and 45 minutes, depending on the size of the agar piece.

4. Add 1 volume of prewarmed (37°C) phenol previously equilibrated with NTE buffer (pH 8.5). Shake vigorously for 5 minutes.

5. Spin the tube at 5000 rpm in a Beckman JS4.2 rotor (or equivalent) for 5 minutes at room temperature.

6. Take the upper phase, being very careful not to pipette any of the interphase. Discard the lower phase.

7. Repeat steps 4–6.

8. Mix the upper phase with one volume of chloroform:isoamyl alcohol (24:1). Shake vigorously for 5 minutes. Centrifuge as above and pipette upper phase.

9. Repeat step 8.

10. Mix upper phase with one volume of ethyl ether. Shake vigorously for 2 minutes. Spin as above and *discard upper phase*. Keep lower (aqueous) phase.

11. Add one volume of ethyl ether and treat as in step 10.

12. Blow compressed air onto the surface of the sample with a plugged Pasteur pipette for a few minutes to remove the remaining ether from the aqueous phase.

13. Transfer to a 15-ml Corex tube. Add sodium acetate to 0.2 *M* and two volumes of cold ethanol. Leave at −20°C overnight.

14. Collect the precipitate by centrifugation (15 minutes, 11,000 rpm).

15. Pour off the ethanol and dry the precipitate under vacuum for 10 minutes at room temperature.

16. Resuspend the precipitate in 400 μl sterile distilled water. Transfer to a microfuge tube.

17. Add 15 μl 3 *M* sodium acetate and 1 ml ethanol. Let precipitate at −20°C overnight or for a few hours at −70°C.

18. Spin at 12,000 rpm for 10 minutes at 4°C.

19. Pour off the ethanol. Add 1 ml ethanol, and spin for 5 additional minutes.

20. Pour off the ethanol, dry the precipitate under vacuum, and resuspend in the desired volume of sterile buffer or water.

PURIFICATION OF DNA FROM AGAROSE OR POLYACRYLAMIDE GELS USING DEAE PAPER

This method, introduced by Dretzen et al. (1981), allows good recovery of large DNA fragments from gels. It is fast and versatile. It is a method of choice for purification of intact long DNA molecules, such as λ arms (see Chapter 7.)

PROTOCOL

PURIFICATION OF DNA FROM AGAROSE OR POLYACRYLAMIDE GELS USING DEAE PAPER

1. Run the gel as described previously (see Chapter 9) in the presence of ethidium bromide.

2. Monitor the migration of the DNA fragments with a long-wave UV lamp.

3. During migration of the DNA, soak small pieces of Whatman DE81 (DEAE cellulose) paper for about 2 hours at 4°C in 10 ml 2.5 *M* NaCl. The paper should be cut wider than the bands and a little bit taller than the gel.

4. Wash the paper pieces three times in sterile distilled water and once in 5 ml of electrophoresis running buffer for 5 minutes.

5. locate the DNA bands to be eluted. Make an incision in the gel on the side of each band and insert the DE81 paper until it touches the bottom of the gel plate.

6. There are two ways of inserting the pieces of paper. The incision may be parallel or perpendicular to the DNA band. In the latter case, the gel plate is rotated by 90° before DNA transfer onto paper by electrophoresis. Which method is used depends on the separation of the different bands on the gel and the amount of material electrophoresed. Perpendicular migration on paper is performed when more concentrated solutions are needed because small amounts of DNA were electrophoresed.

7. Monitor the migration of the DNA band with a UV lamp.

8. When all the DNA has been transferred to the paper, carefully remove the paper with sterile tweezers.

9. Wash each piece of paper three times in 0.5 ml sterile distilled water, taking care that the paper does not come apart.

10. Eliminate excess water by blotting the DE81 paper on Whatman 3MM paper (gently apply the edge of the DE81 to the 3MM paper).

11. With a 25 gauge sterile needle, punch a small hole at the bottom of a 0.75-ml microfuge tube, and tightly pack the bottom of this tube with nylon wool (or siliconized glass wool).

12. Place this small tube in a 1.5-ml microfuge tube and fill the small tube with 200 μl elution buffer.

13. Spin at 12,000 rpm for 5 minutes and discard the 1.5-ml tube in which the buffer was collected.

14. Place the small tube in a new 1.5-ml microfuge tube. Transfer the DE81 paper (with bound DNA) to the small tube and add 200 μl elution buffer.

15. Incubate at 37°C for 1 hour. Spin for 5 minutes at 12,000 rpm and save the buffer.

16. Add 200 μl elution buffer to the small tube and spin for 5 minutes at 12,000 rpm. Combine the eluate with that of step 15.

17. Add 100 μl elution buffer to the small tube and spin for 5 minutes at 12,000 rpm. Add the buffer to the previous buffer samples. This makes a total volume of 500 μl.

18. Add 1 ml n-butanol equilibrated with H_2O. Mix thoroughly. Spin for 5 seconds at 12,000 rpm. Remove n-butanol (upper phase).

19. Add 1 ml of unequilibrated n-butanol. Mix and spin for 5 seconds at 12,000 rpm. Remove n-butanol and repeat step 18.

20. Add 1 ml cold ethanol to precipitate the DNA (at $-70°C$ for 1 hour or $-20°C$ overnight).

Elution buffer	*10 ml*
$20\ mM$ Tris–HCl (pH 8.0)	200 μl 1 M Tris–HCl (pH 8.0)
2 mM EDTA	40 μl 0.5 M EDTA
1.5 M NaCl	3 ml 5 M NaCl
	6.76 ml H_2O

ELUTION OF DNA SPECIES FROM POLYACRYLAMIDE GELS

The separation of DNA fragments by electrophoresis in polyacrylamide gels is described in Chapters 3 and 9. The following protocol allows a satisfactory recovery of individual species following electrophoresis.

PROTOCOL

RECOVERY OF DNA FRAGMENTS AFTER ELUTION FROM POLYACRYLAMIDE GELS

1. Transfer the gel slice to a microfuge tube, and crush it with a siliconized glass rod.

2. Add 600 μl of elution buffer (see page 373) containing heterologous tRNA (10 μg/ml). If labeling is to be performed on the eluted DNA, do not add tRNA in elution buffer.

3. Close the tube and incubate for 10 hours at 37°C in a water bath.

4. Aspirate carefully the eluate with a pipettman and split in two microfuge tubes.

5. In each tube, add 150 μl of 7.5 M ammonium acetate.

6. Add 900 μl of cold absolute ethanol and let the DNA precipitate for 20 minutes at $-70°C$.

7. Centrifuge for 20 minutes at 12,000 rpm (4°C) and discard supernatant.

8. Resuspend each pellet in 250 μl of 0.3 M sodium acetate, and combine the content of the two tubes.

9. Add 1.5 ml of absolute cold ethanol and precipitate DNA at $-70°C$ for 20 minutes.

10. Centrifuge for 20 minutes at 12,000 rpm (4°C) and discard supernatant.

11. Resuspend the DNA pellet in 25 μl of sterile distilled water, and add 1 ml of cold absolute ethanol. Incubate for 20 minutes at $-70°C$.

12. Centrifuge for 20 minutes at 12,000 rpm (4°C) and discard supernatant.

13. Dry the DNA pellet under vacuum and resuspend in a small volume (20 to 30 μl) of sterile distilled water.

REFERENCES

Danner, D. B. (1982), *Anal. Biochem.*, **125**, 139.

Dretzen, G., Bellard, M., Sassone Corsi, P., and Chambon, P. (1981), *Anal. Biochem.*, **112**, 295.

12

MODIFICATION OF DNA FRAGMENTS WITH COHESIVE TERMINI

MODIFICATION OF 3′ AND 5′ DNA TERMINI WITH DNA POLYMERASE

DNA fragments with protruding 5′ termini can be treated with DNA polymerase to generate directly blunt-ended fragments, which can be ligated in the presence of T4 DNA ligase. In some cases, a recognition site for endonuclease cleavage can be regenerated (Figure 101).

Modification of Cohesive Ends with 3′ Extension

In this case, the protruding extension is first removed by the exonuclease activity of DNA polymerase I in the absence of deoxynucleotide triphosphates (see Chapter 4). Because removal of more nucleotides may occur (therefore generating a 5′ protruding end), repair is performed after addition

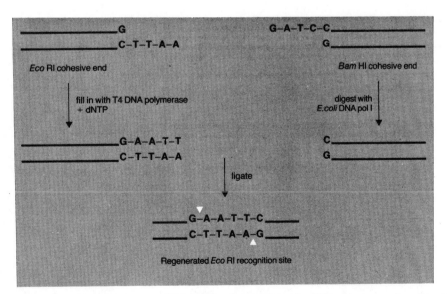

Figure 101. Ligation of blunt-ended DNA fragments generated by DNA polymerase (courtesy of Amersham).

PROTOCOL

MODIFICATION OF COHESIVE ENDS WITH 3' EXTENSION

1. In a microfuge tube, prepare a 1 mM mixture of the four deoxynucleotide triphosphates (dATP, dCTP, dGTP, TTP):
   5 μl 10 mM dATP
   5 μl 10 mM dCTP
   5 μl 10 mM dGTP
   5 μl 10 mM TTP
   30 μl H$_2$O

2. In another microfuge tube mix:
   8.5 μl H$_2$O
   5 μl DNA solution (containing 1–5 μg DNA)
   1.5 μl 1 M Tris–HCl (pH 7.5)
   2.5 μl 0.1 M MgCl$_2$
   2.5 μl 0.1 M 2-mercaptoethanol
   0.5 μl DNA polymerase I (2.5 units)
   Incubate at 12°C for 10 minutes.

3. Add 5 μl of the 1 mM deoxynucleotide triphosphates mixture to the DNA.

4. Incubate for 1 hour at 12°C.

5. Heat at 68°C for 10 minutes to inactivate the DNA polymerase.

PROTOCOL

MODIFICATION OF COHESIVE ENDS WITH 5' EXTENSION

1. Mix in a microfuge tube:
   8.5 μl H$_2$O
   5 μl DNA (1–5 μg)
   1.5 μl 1 M Tris–HCl (pH 7.5)
   2.5 μl 0.1 M MgCl$_2$
   2.5 μl 0.1 M 2-mercaptoethanol
   5 μl 1 mM deoxynucleotide triphosphates mixture (see protocol above, step 1)
   0.5 μl DNA polymerase I (2.5 units)
   Incubate at 12°C for 1 hour

2. Heat at 68°C for 10 minutes to inactivate the DNA polymerase.

of deoxynucleotide triphosphates, in order to generate a blunt end.

Modification of Cohesive Ends with 5′ Extension

The DNA mixture is incubated directly in the presence of the four deoxynucleotide triphosphates.

DIGESTION OF DNA WITH Bal 31 NUCLEASE

The enzyme Bal 31 nuclease has both 3′ → 5′ and 5′ → 3′ exonuclease activities, and usually tends to generate blunt ends. This property of Bal 31 has made the enzyme useful in cloning DNA fragments with linkers, preparing deletion mutants at specific sites in vitro, and mapping restriction sites in DNA fragments (see Chapter 10).

Before using Bal 31 to prepare deletion mutants, it is recommended to run a time course of DNA degradation by the enzyme with your own DNA preparation, since some preparations have been found to be more sensitive than others to Bal 31. Also, keep in mind that asymmetric digestion is often obtained, one end of a linear duplex DNA being degraded more rapidly than the other end.

PROTOCOL

TIME COURSE DIGESTION OF DNA WITH BAL 31

1. Prepare your DNA sample on cesium chloride gradients, if working with a plasmid.
2. Digest about 5–10 μg DNA with a restriction endonuclease giving a single linear fragment (i.e., with a unique restriction site).
3. Eliminate enzymes and other proteins by phenol extraction and precipitation of DNA.
4. Resuspend the digested DNA in 125 μl sterile distilled water.
5. Add 125 μl 2X running buffer.
6. In each of six microfuge tubes put 20 μl TE 50 buffer and 50 μl phenol (equilibrated in NTE).
7. Dilute the enzyme stock in storage buffer to a concentration of about 0.1–0.2 units/μl (one unit is usually defined as the amount of enzyme which releases 1 μg of acid-soluble material in 1 minute at 30°C).
8. Warm the DNA solution (250 μl) at 30°C.
9. Take 35 μl of this DNA solution and mix with one phenol tube (step 6). This will give the "time 0" point.
10. Add 1 μl diluted Bal 31 to the remaining DNA and incubate at 30°C.

11. At 2, 5, 10, 30, and 60 minutes, pipette 35 µl of the incubation mixture and mix immediately with phenol–TE 50 (tubes from step 6). You can leave these tubes at room temperature until the last point (60 minutes) has been taken.

12. Spin tubes and pipette upper phase.

13. Add the upper phase of each tube to fresh tubes containing 100 µl chloroform:isoamyl alcohol (24:1). Mix, spin, and save upper phase.

14. Repeat step 13.

15. Add 8 µl 3 *M* sodium acetate to each tube and precipitate DNA after addition of three volumes of cold ethanol.

16. Resuspend DNA in a small volume of distilled water (10 µl).

17. Digest 1 µl of each Bal 31-treated DNA sample with another restriction enzyme whose recognition site is located about 2 kb from the first one used. This will generate fragments with a common end (enzyme 2) and different lengths (due to different incubation times with Bal 31).

18. Run second-digest samples on agarose or acrylamide gels to estimate sizes of the fragments generated.

19. When ligating molecular linkers to Bal 31-treated fragments (i.e., for cloning each size), add a ten-fold molar excess of phosphorylated linkers to the treated DNAs (step 16), ligase buffer, and ligase (see below and Chapter 13).

The same kind of protocol can be used to determine the location of a restriction site with respect to another one (enzyme 2 in the example described above).

Note. For concentrations of DNA ends of $5–8 \times 10^{-9}$ *M* and enzyme concentrations of 0.25 units/ml, a rate of degradation of about 1200 bp/hour/end has been reported (Mock, personal communication).

Buffers for Bal 31 Digestion

2X Running buffer	*10 ml*
24 m*M* CaCl$_2$	240 µl 1 *M* CaCl$_2$
24 m*M* MgCl$_2$	240 µl 1 *M* MgCl$_2$
400 m*M* NaCl	800 µl 5 *M* NaCl
40 m*M* Tris–HCl (pH 8.0)	400 µl 1 *M* Tris–HCl (pH 8.0)
2 m*M* EDTA	40 µl 0.5 *M* EDTA
	8.28 ml H$_2$O

Bal 31 storage buffer　　　　　　　*10 ml*
100 mM NaCl　　　　　　　　200 μl 5 M NaCl
5 mM MgCl$_2$　　　　　　　　50 μl 1 M MgCl$_2$
5 mM CaCl$_2$　　　　　　　　50 μl 1 M CaCl$_2$
20 mM Tris–HCl (pH 6.8)　　200 μl 1 M Tris–HCl (pH 6.8)
　　　　　　　　　　　　　　9.5 ml H$_2$O

TE 50 buffer　　　　　　　　*10 ml*
50 mM Tris–HCl (pH 8.0)　　500 μl 1 M Tris–HCl (pH 8.0)
50 mM EDTA　　　　　　　　1 ml 0.5 M EDTA
　　　　　　　　　　　　　　8.5 ml H$_2$O

ADDITION OF MOLECULAR LINKERS TO DNA FRAGMENTS

The use of synthetic oligonucleotide sequences which contain the recognition and cleavage sites for specific restriction endonucleases is widespread in molecular cloning.

Procedures for addition of linkers to passenger DNA are illustrated in Figures 102 and 103. Formation of a new Eco RI linker from preexisting linkers is illustrated in Figure 104.

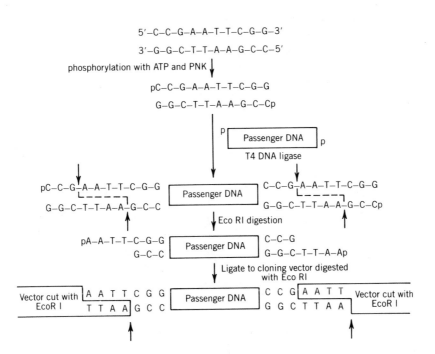

Figure 102. Cloning passenger DNA with Eco RI molecular linkers. PNK, polynucleotide kinase (courtesy of Amersham).

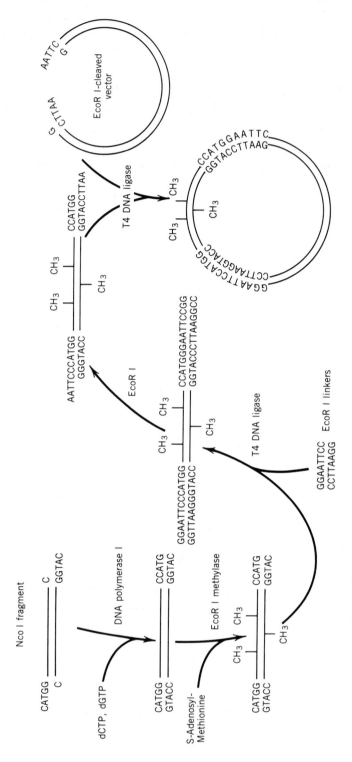

Figure 103. Insertion of Nco I fragments at the Eco RI site of a cloning vector. The staggered ends of fragments produced by Nco I digestion of a target DNA are filled in with DNA polymerase. Prior to ligation to the Eco RI linker molecules, the target DNA is made resistant to Eco RI digestion by treatment with Eco RI methylase. The "adapted" fragments with Eco RI linkers are digested with Eco RI and ligated to an Eco RI cloning vehicle.

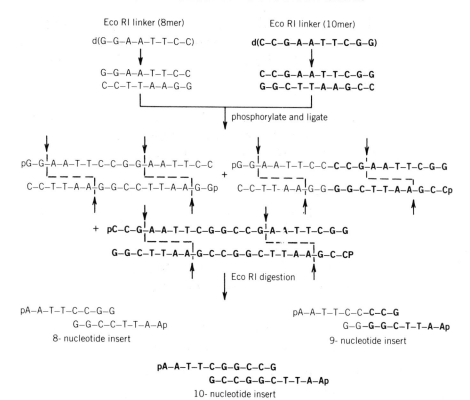

Figure 104. Formation of a new Eco RI linker after ligation of two preexisting Eco RI linkers. The linkers obtained after ligation and digestion with Eco RI will allow the insertion of 8, 9, or 10 nucleotides between passenger DNA and vector DNA ends (courtesy of Amersham).

PROTOCOL

PHOSPHORYLATION OF MOLECULAR LINKERS

1. Dissolve 1 unit of linkers in 250 µl 1 mM Tris–HCl (pH 7.4). Keep frozen at −20°C.

2. In a microfuge tube mix:
   6 µl H_2O
   5 µl linker solution
   2 µl 1 M Tris–HCl (pH 7.4)
   3 µl 0.1 M $MgCl_2$
   3 µl 0.1 M 2-mercaptoethanol
   10 µl 0.2 mM ATP
   5–10 units polynucleotide kinase (sequencing grade, 10 units/µl)

3. Incubate 1–2 hours at 37°C.

4. Heat at 70°C for 10 minutes to denature the polynucleotide kinase.

5. Store phosphorylated linkers at −20°C.

It is recommended that the efficiency of the phosphorylation reaction be checked before the linkers are used. An easy way is to follow the rate of 5′ phosphorylation using P^{32}-labeled ATP. The protocol is then modified as follows.

1. In a microfuge tube, put 10–50 μCi of γ[^{32}P]-labeled ATP. Evaporate the liquid by blowing air very carefully on the surface of the radioactive solution.

2. To the dried ATP add:
   1 μl H$_2$O
   5 μl linker solution
   2 μl 1 M Tris–HCl (pH 7.4)
   3 μl 0.1 M MgCl$_2$
   3 μl 0.1 M 2-mercaptoethanol
   5 units polynucleotide kinase

3. Incubate for 20 minutes at 37°C.

4. Add 2 μl 1 mM ATP.

5. Incubate for 40 minutes at 37°C.

6. Heat at 70°C for 10 minutes to denature the polynucleotide kinase.

PROTOCOL

LIGATION OF THE PHOSPHORYLATED LINKERS TO THE DNA FRAGMENTS WITH BLUNT ENDS

1. Mix blunt-ended DNA fragments with 30 μl of the phosphorylated linkers prepared as described above. Bring to 50 μl.

2. Add 5.5 μl 10 mM ATP to the 50-μl incubation mixture.

3. Add 1 μl T4 DNA ligase (50 units).

4. Incubate for 16 hours at 12°C.

PROTOCOL

PURIFICATION OF THE DNA FRAGMENTS WITH MOLECULAR LINKERS AT THEIR ENDS

1. Add 350 μl 1X buffer C to the incubation mixture from step 4 of the previous protocol.

2. Add 200 μl phenol (equilibrated with 1X buffer C) and 200 μl of chloroform:isoamyl alcohol (24:1). Vortex for 20 seconds. Spin for 5 minutes in a microfuge.

3. Pipette out the upper phase and using it, repeat step 2.

4. Put the upper phase in a fresh microfuge tube containing 40 μl 2.5 M sodium acetate. Add 1 ml cold absolute ethanol. Let stand at $-70°C$ for 20 minutes (or $-20°C$ for 2 hours).

5. Centrifuge for 10 minutes at 4°C in a microfuge to recover the DNA precipitate.

6. Pour off the ethanol, add 1 ml cold absolute ethanol, and centrifuge at 4°C for 15 minutes.

7. Pour off the ethanol and dry the pellet under vacuum.

8. Resuspend the DNA pellet in 20 μl H_2O and set up for digestion with the restriction enzymes corresponding to the molecular linkers used.

9. When the DNA has been digested, run it on 0.8% agarose gel to separate the DNA from the unreacted linkers.

10. Elute the DNA having the desired molecular weight (Chapter 11) and use for ligation in a vector.

Buffers

Buffer C (1X)	*100 ml*
50 mM Tris–HCl (pH 8.0)	5 ml 1 M Tris–HCl (pH 8.0)
10 mM EDTA (pH 7.5)	2 ml 0.5 M EDTA (pH 7.5)
10 mM NaCl	0.2 ml 5 M NaCl
	92.8 ml H_2O

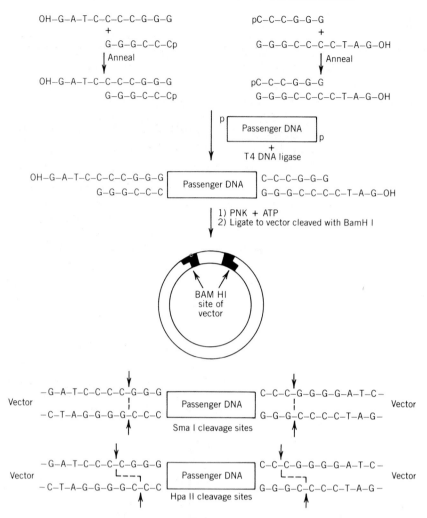

Figure 105. Formation and use of a Bam HI adapter. Two single-stranded oligonucleotides, a Bam HI adapter and a Hpa II linker, are annealed to obtain a preformed adapter. The 5' end of the linker was previously phosphorylated so that the preformed adapter could be spliced to the passenger DNA. The Bam HI adapter is not phosphorylated (5'-hydroxyl form) in order to prevent self-polymerization. The preformed adapters are ligated to the passenger DNA and the 5' ends are then phosphorylated. The passenger DNA now has two Bam HI cohesive ends and may be ligated to a vector previously digested with Bam HI endonuclease. The passenger DNA may now be recovered intact by digestion with Sma I or Hpa II. PNK, polynucleotide kinase (courtesy of Amersham).

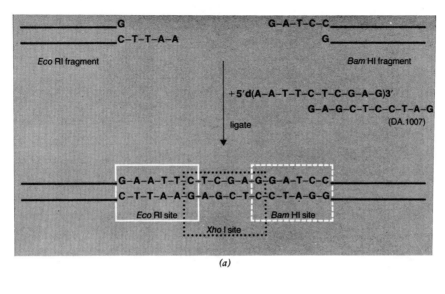

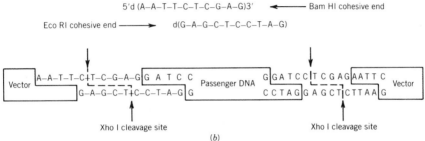

Figure 106. Use of conversion adapters. An Eco RI-Bam HI conversion adapter is used to ligate fragments with noncomplementary cohesive ends. (*a*) Formation of an Xho I restriction site after annealing. (*b*) Introduction of a passenger DNA digested with Bam HI into a vector DNA digested with Eco RI. The passenger DNA can be recovered by Xho I digestion (courtesy of Amersham).

USE OF ADAPTERS FOR CLONING

Cloning of DNA fragments sometimes requires the use of nonidentical cohesive ends or the use of cloning sites corresponding to sequences represented within the passenger DNA. Adapters can be used in both cases to overcome the problems encountered.

For example, cloning of a DNA fragment containing one or several Bam HI sites into the Bam HI site of a vector can be achieved by using a Bam HI adapter to introduce a new restriction site not represented in the passenger DNA, thereby allowing retrieval of the intact cloned fragment from the vector

(Figure 105.) In another case, cloning of a passenger DNA with Bam HI cohesive ends into the Eco RI site of a vector is made possible by the use of an Eco RI–Bam HI conversion adapter that contains an Xho I site (Figure 106). Two general methods for generating conversion adapters are illustrated in Figure 107.

Cloning of DNA fragments with 3′ protruding cohesive ends can also be achieved by using a single stranded conversion adapter, as shown in Figure 108.

A listing of some commercially available linkers and adapters is presented in Table 55.

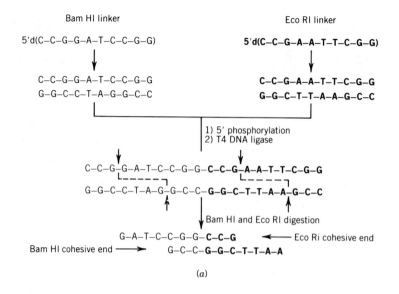

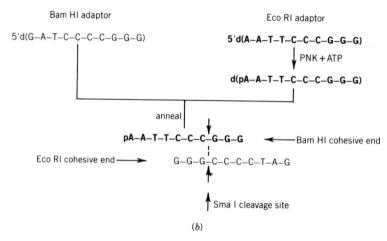

Figure 107. Preparation of conversion adapters. (*a*) Ligation of two self-annealed linkers an oligonucleotide that can be digested by both enzymes to generate the required cohesive ends. (*b*) Annealing of two single-stranded oligonucleotides to prepare a conversion adapter containing a Sma I cleavage site. PNK, polynucleotide kinase (courtesy of Amersham).

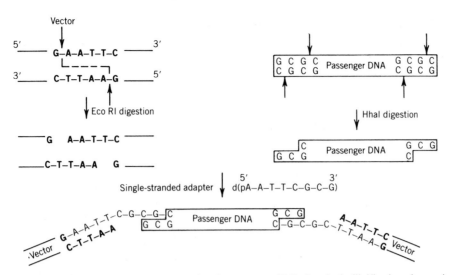

Figure 108. Use of a single-stranded adapter. A passenger DNA digested with Hha I and a vector DNA digested with Eco RI are ligated to a single-stranded adapter. The 2-nucleotide gap is filled by repair synthesis in vitro or in vivo, leading to the regeneration of both original restriction sites. The cloned passenger DNA can therefore be readily retrieved (courtesy of Amersham).

Table 55.   Listing of Commercial Linkers and Adaptors

Linker	Sequence	Supplier
	Nonphosphorylated Linkers	
AatII	5'-GGACGTCC-3'	New England Biolabs
AatII	5'-CGACGTCG-3'	Pharmacia
AluI	5'-AGCT-3'	Pharmacia
ApaI	5'-GGGCCC-3'	Pharmacia
ApaI	5'-GGGGCCCC-3'	New England Biolabs
		Boehringer, Mannheim
ApaLI	5'-GGTGCACC-3'	New England Biolabs
AsuII	5'-CTTCGAAG-3'	Pharmacia
AsuII	5'-GAATTCGAATTC-3'	International Biotechnologies, Inc.
AvrII	5'-GCCTAGGC-3'	New England Biolabs
BalI	5'-GTGCCCAC-3'	Bethesda Research Laboratories
BalI	5'-GTGGCCAC-3'	Boehringer, Mannheim
BamHI	5'-CCGGATCCGG-3'	Bethesda Research Laboratories
		Pharmacia
		Amersham
BamHI	5'-CGGATCCG-3'	New England Biolabs
		Pharmacia
		Amersham
BamHI	5'-CGGGATCCCG-3'	New England Biolabs
		Boehringer, Mannheim
BamHI	5'-CGCGGATCCGCG-3'	New England Biolabs
BamHI	5'-GGGATCCC-3'	Boehringer, Mannheim
BamHI	5'-CCCGGATCCGGG-3'	Pharmacia
		Amersham
BamHI,Sau3A	5'-GGATCC-3'	International Biotechnologies, Inc.
BamHI	5'-GGATCCGGATCC-3'	International Biotechnologies, Inc.
BclI	5'-CTGATCAG-3'	New England Biolabs
		Pharmacia
		Amersham

Table 55. Listing of Commercial Linkers and Adaptors (*continued*)

Linker	Sequence	Supplier
BglII	5'-CAGATCTG-3'	New England Biolabs
		Boehringer, Mannheim
		Pharmacia
		Amersham
BglII (MboII-MboII)	5'-GAAGATCTTC-3'	New England Biolabs
BglII	5'-GGAAGATCTTCC-3'	New England Biolabs
BglII,Sau3A	5'-AGATCT-3'	International Biotechnologies, Inc.
BglII	5'-AGATCTAGATCT-3'	International Biotechnologies, Inc.
BglII	5'-TCTAGATCTAGA-3'	International Biotechnologies, Inc.
BspMII	5'-CTCCGGAG-3'	New England Biolabs
BsshII	5'-GCGCGC-3'	Pharmacia
BsshII	5'-CGCGCGCG-3'	New England Biolabs
ClaI,TaqI	5'-CATCGATG-3'	New England Biolabs
		Bethesda Research Laboratories
		Boehringer, Mannheim
		Pharmacia
		Amersham
ClaI (substrate for dam methylase)	5'-GATCGATC-3'	New England Biolabs
ClaI	5'-CCATCGATGG-3'	New England Biolabs
		Pharmacia
ClaI	5'-CCCATCGATGGG-3'	New England Biolabs
		Pharmacia
EcoRI	5'-GAATTC-3'	International Biotechnologies, Inc.
EcoRI	5'-GGAATTCC-3'	New England Biolabs
		Boehringer, Mannheim
		Pharmacia
		Amersham
EcoRI	5'-CGGAATTCCG-3'	New England Biolabs
		Boehringer, Mannheim
EcoRI	5'-CCGGAATTCCGG-3'	New England Biolabs
		Boehringer, Mannheim
EcoRI	5'-CCGAATTCGG-3'	Bethesda Research Laboratories
		Pharmacia
		Amersham
EcoRI	5'-CCCGAATTCGGG-3'	Pharmacia
		Amersham
EcoRI	5'-GAATTCGAATTC-3'	International Biotechnologies, Inc.
EcoRV	5'-GGATATCC-3'	Boehringer, Mannheim
HaeIII	5'-GGCC-3'	Pharmacia
HhaI	5'-GCGC-3'	Pharmacia
HhaI	5'-GCGGC-3'	Amersham
HindIII	5'-CAAGCTTG-3'	New England Biolabs
		Amersham
		Pharmacia
HindIII	5'-CCAAGCTTGG-3'	New England Biolabs
		Bethesda Research Laboratories
		Amersham
		Pharmacia
HindIII	5'-CCCAAGCTTGGG-3'	New England Biolabs
		Amersham
		Pharmacia
HindIII	5'-GAAGCTTC-3'	Boehringer, Mannheim
HindIII	5'-CGAAGCTTCG-3'	Boehringer, Mannheim

Table 55. Listing of Commercial Linkers and Adaptors (*continued*)

Linker	Sequence	Supplier
HindIII	5'-CGCAAGCTTGCG-3'	Boehringer, Mannheim
HpaI	5'-GTTAAC-3'	Pharmacia
		Amersham
HpaI	5'-CCGTTAACGG-3'	Amersham
HpaII,MspI	5'-CCGG-3'	Pharmacia
HpaII (blunt end),SmaI	5'-CCCGGG-3'	Pharmacia
		International Biotechnologies, Inc.
		Amersham
KpnI	5'-GGGTACCC-3'	New England Biolabs
KpnI,RsaI	5'-GGTACC-3'	International Biotechnologies, Inc.
KpnI	5'-CGGTACCG-3'	Amersham
		Boehringer, Mannheim
		Pharmacia
MluI	5'-GACGCGTC-3'	New England Biolabs
MspI	5'-CCCGGG-3'	Amersham
MaeI	5'-CACTAGTG-3'	Boehringer, Mannheim
NcoI	5'-GCCATGGC-3'	Boehringer, Mannheim
NcoI	5'-CCCATGGG-3'	New England Biolabs
NcoI* (methionine codon)	5'-CATGCCATGGCATG-3'	New England Biolabs
NcoI	5'-CCATGG-3'	Pharmacia
NdeI	5'-CCATATGG-3'	New England Biolabs
NheI	5'-GGCTAGCC-3'	New England Biolabs
NheI	5'-CGGCTAGCCG-3'	New England Biolabs
NheI* (nonsense codon)	5'-CTAGCTAGCTAG-3'	New England Biolabs
NotI (XmaIII)	5'-GCGGCCGC-3'	New England Biolabs
NotI (XmaIII)	5'-AGCGGCCGCT-3'	New England Biolabs
NotI (XmaIII)	5'-TTGCGGCCGCAA-3'	New England Biolabs
NsiI* (cysteine codon)	5'-TGCATGCATGCA-3'	New England Biolabs
PstI	5'-GCTGCAGC-3'	New England Biolabs
		Amersham
		Pharmacia
PstI	5'-CCTGCAGG-3'	Bethesda Research Laboratories
		Boehringer, Mannheim
PstI* (cysteine codon)	5'-TGCACTGCAGTGCA-3'	New England Biolabs
PvuI	5'-CCGATCGG-3'	New England Biolabs
PvuI (ClaI-ClaI)	5'-ATCGATCGAT-3'	New England Biolabs
PvuI (NruI-NruI)	5'-TCGCGATCGCGA-3'	New England Biolabs
PvuI	5'-GCGATCGC-3'	Bethesda Research Laboratories
PvuI	5'-GCGATCGC-3'	Boehringer, Mannheim
PvuII	5'-CCAGCTGG-3'	New England Biolabs
PvuII	5'-GCAGCTGC-3'	Bethesda Research Laboratories
PvuII	5'-GCAGCTGC-3'	Boehringer, Mannheim
SacI,AluI	5'-GAGCTC-3'	International Biotechnologies, Inc.
SacI (SstI)	5'-CGAGCTCG-3'	New England Biolabs
		Boehringer, Mannheim
SacI	5'-GAGCTCGAGCTC-3'	International Biotechnologies, Inc.
SacI	5'-CTCGAGCTCGAG-3'	International Biotechnologies, Inc.
SacII	5'-GCCGCGGC-3'	New England Biolabs
SacII* (β-turn proline codon)	5'-TCCCCGCGGGGA-3'	New England Biolabs
SalI,TaqI	5'-GGTCGACC-3'	New England Biolabs
		Amersham
		Pharmacia
SalI	5'-CGGTCGACCG-3'	New England Biolabs
SalI	5'-CCGGTCGACCGG-3'	New England Biolabs

Table 55. Listing of Commercial Linkers and Adaptors (continued)

Linker	Sequence	Supplier
SalI	5′-CGTCGACG-3′	Bethesda Research Laboratories Boehringer, Mannheim
ScaI* (cationic lysine codon)	5′-AAAAGTACTTTT-3′	New England Biolabs
SmaI	5′-CCCGGG-3′	International Biotechnologies, Inc.
SmaI	5′-CCCCGGGG-3′	New England Biolabs
SmaI	5′-CCCCCGGGGG-3′	New England Biolabs
SmaI	5′-GCCCGGGC-3′	Bethesda Research Laboratories Boehringer, Mannheim
SmaI* (β-turn proline codon)	5′-TCCCCCGGGGGA-3′	New England Biolabs
SpeI	5′-GACTAGTC-3′	New England Biolabs
SpeI	5′-GGACTAGTCC-3′	New England Biolabs
SpeI	5′-CGGACTAGTCCG-3′	New England Biolabs
SpeI* (nonsense codon)	5′-CTAGACTAGTCTAG-3′	New England Biolabs
SphI	5′-GGCATGCC-3′	New England Biolabs
SphI	5′-GGCATGCC-3′	Bethesda Research Laboratories Boehringer, Mannheim
SphI* (methionine codon)	5′-CATGCATGCATG-3′	New England Biolabs
SphI* (met. or cys. codon)	5′-ACATGCATGCATGT-3′	New England Biolabs
SphI	5′-CGCATGCG-3′	Pharmacia
SstI,SacI,AluI	5′-CGAGCTCG-3′	Pharmacia Amersham
ThaI	5′-CGCG-3′	Pharmacia Amersham
XbaI	5′-CTCTAGAG-3′	New England Biolabs Amersham Pharmacia
XbaI	5′-AGATCTAGATCT-3′	International Biotechnologies, Inc.
XbaI	5′-GCTCTAGAGC-3′	New England Biolabs
XbaI	5′-TGCTCTAGAGCA-3′	New England Biolabs
XbaI	5′-TCTAGA-3′	Pharmacia International Biotechnologies, Inc. Amersham
XbaI	5′-TCTAGATCTAGA-3′	International Biotechnologies, Inc.
XbaI* (nonsense codon)	5′-CTAGTCTAGACTAG-3′	New England Biolabs
XhoI,TaqI,AvaI	5′-CCTCGAGG-3′	New England Biolabs Amersham Pharmacia
XhoI	5′-CCCTCGAGGG-3′	New England Biolabs
XhoI	5′-CCGCTCGAGCGG-3′	New England Biolabs
XhoI,TaqI	5′-CTCGAG-3′	International Biotechnologies, Inc.
XhoI	5′-GAGCTCGAGCTC-3′	International Biotechnologies, Inc.
XhoI	5′-CTCGAGCTCGAG-3′	International Biotechnologies, Inc.
XmaI	5′-CCCGGG-3′	International Biotechnologies, Inc.
XmaIII	5′-CCGGCCGG-3′	New England Biolabs
(ClaI-MboII) requires complement	5′-ATCGATCTTC-3′	New England Biolabs
(ClaI-MboII) requires complement	5′-GAAGATCGAT-3′	New England Biolabs

double stranded nonphosphorylated linker with multiple sites.

SacII,NotI,SfiI,SpeI	5′-GATCCGCGGCCGCATAGGCCACTAGTG-3′ 3′-GCGCCGGCGTATCCGGTGATCACCTAG-5′	Boehringer, Mannheim

Table 55. Listing of Commercial Linkers and Adaptors (*continued*)

Linker	Sequence	Supplier
	Phosphorylated Linkers	
AatII	5′-CGACGTCG-3′	Pharmacia
AatII	5′-GGACGTCC-3′	New England Biolabs
AluI	5′-AGCT-3′	Pharmacia
ApaI	5′-GGGCCC-3′	Pharmacia
ApaI	5′-GGGGCCCC-3′	New England Biolabs
AsuII	5′-CTTCGAAG-3′	Pharmacia
BamHI	5′-CGGATCCG-3′	New England Biolabs
		Pharmacia
BamHI	5′-CGGGATCCCG-3′	New England Biolabs
BamHI	5′-CGCGGATCCGCG-3′	New England Biolabs
BamHI	5′-CCGGATCCGG-3′	Pharmacia
BamHI	5′-CCCGGATCCGGG-3′	Pharmacia
BclI	5′-CTGATCAG-3′	New England Biolabs
		Pharmacia
BglII	5′-CAGATCTG-3′	New England Biolabs
		Pharmacia
BglII (MboII-MboII)	5′-GAAGATCTTC-3′	New England Biolabs
BglII	5′-GGAAGATCTTCC-3′	New England Biolabs
BssHII	5′-GCGCGC-3′	Pharmacia
ClaI,TaqI	5′-CATCGATG-3′	New England Biolabs
		Pharmacia
ClaI	5′-CCATCGATGG-3′	New England Biolabs
		Pharmacia
ClaI	5′-CCCATCGATGGG-3′	New England Biolabs
		Pharmacia
EcoRI	5′-GGAATTCC-3′	New England Biolabs
		Pharmacia
EcoRI	5′-CGGAATTCCG-3′	New England Biolabs
EcoRI	5′-CCGGAATTCCGG-3′	New England Biolabs
EcoRI	5′-CCGAATTCGG-3′	Pharmacia
EcoRI	5′-CCCGAATTCGGG-3′	Pharmacia
HaeIII	5′-GGCC-3′	Pharmacia
HhaI	5′-GCGC-3′	Pharmacia
HindIII	5′-CAAGCTTG-3′	New England Biolabs
		Pharmacia
HindIII	5′-CCAAGCTTGG-3′	New England Biolabs
		Pharmacia
HindIII	5′-CCCAAGCTTGGG-3′	New England Biolabs
		Pharmacia
HpaI	5′-GTTAAC-3′	Pharmacia
HpaII,MspI	5′-CCGG-3′	Pharmacia
HpaII (blunt end)	5′-CCCGGG-3′	Pharmacia
KpnI	5′-GGGTACCC-3′	New England Biolabs
KpnI	5′-CGGTACCG-3′	Pharmacia
MluI	5′-GACGCGTC-3′	New England Biolabs
MluI	5′-CGACGCGTCG-3′	New England Biolabs
NcoI	5′-CCCATGGG-3′	New England Biolabs
NcoI	5′-CCATGG-3′	Pharmacia
NdeI	5′-CCATATGG-3′	New England Biolabs
NheI	5′-GGCTAGCC-3′	New England Biolabs
NotI (XmaIII)	5′-GCGGCCGC-3′	New England Biolabs
NotI (XmaIII)	5′-AGCGGCCGCT-3′	New England Biolabs
NotI (XmaIII)	5′-TTGCGGCCGCAA-3′	New England Biolabs

Table 55. Listing of Commercial Linkers and Adaptors (*continued*)

Linker	Sequence	Supplier
PstI	5'-GCTGCAGC-3'	New England Biolabs
		Pharmacia
PvuI	5'-CCGATCGG-3'	New England Biolabs
PvuII	5'-CCAGCTGG-3'	New England Biolabs
SacI (SstI)	5'-CGAGCTCG-3'	New England Biolabs
SalI,TaqI	5'-GGTCGACC-3'	New England Biolabs
		Pharmacia
SmaI	5'-CCCCGGGG-3'	New England Biolabs
SpeI	5'-GACTAGTC-3'	New England Biolabs
SpeI	5'-GGACTAGTCC-3'	New England Biolabs
SphI	5'-GGCATGCC-3'	New England Biolabs
SphI	5'-CGCATGCG-3'	Pharmacia
SstI,SacI,AluI	5'-CGAGCTCG-3'	Pharmacia
ThaI	5'-CGCG-3'	Pharmacia
XbaI	5'-CTCTAGAG-3'	New England Biolabs
		Pharmacia
XbaI	5'-GCTCTAGAGC-3'	New England Biolabs
XbaI	5'-TGCTCTAGAGCA-3'	New England Biolabs
XbaI	5'-TCTAGA-3'	Pharmacia
XhoI,TaqI,AvaI	5'-CCTCGAGG-3'	New England Biolabs
		Pharmacia
XhoI	5'-CCCTCGAGGG-3'	New England Biolabs
XhoI	5'-CCGCTCGAGCGG-3'	New England Biolabs
XmaIII	5'-CCGGCCGG-3'	New England Biolabs

Nonphosphorylated Adaptors

Linker	Sequence	Supplier
BamHI to PstI	5'-GATCCCTGCAGG-3'	Pharmacia
BamHI to SmaI	5'-GATCCCCGGG-3'	Pharmacia
		Amersham
BamHI to SmaI	5'-GATGCCCGGG-3'	New England Biolabs
BamHI to XmnI	5'-GATCCGAAGGGGTTCG-3'	New England Biolabs
BamHI to XmnI	5'-GATCCGAACCCCTTCG-3'	New England Biolabs
EcoRI to BamHI	5'-AATTCTCGAG-3'	Amersham
EcoRI to HhaI	5'-AATTCGCG-3'	Amersham
EcoRI to PstI	5'-AATTCCTGCAGG-3'	Pharmacia
EcoRI to SmaI	5'-AATTCCCGGG-3'	Pharmacia
		Amersham
		New England Biolabs
EcoRI to ThaI	5'-AATTCGCG-3'	Pharmacia
EcoRI to XmnI	5'-AATTCGAACCCCTTCG-3'	New England Biolabs
HindIII to PstI	5'-AGCTCCTGCAGG-3'	Pharmacia
HindIII to SmaI	5'-AGCTCCCGGG-3'	Pharmacia
		Amersham
		New England Biolabs
HindIII to XmnI	5'-AGCTCGAAGGGGTTCG-3'	New England Biolabs
HpaII/TaqI to SmaI	5'-CGCCCGGG-3'	Pharmacia
SalI to SmaI	5'-TCGACCCGGG-3'	New England Biolabs
XbaI to SmaI	5'-CTAGCCCGGG-3'	Pharmacia
XcyI to SmaI	5'-CCGGCCGGG-3'	Pharmacia
XmaIII to SmaI	5'-GGCCCCCGGG-3'	Pharmacia

Table 55. Listing of Commercial Linkers and Adaptors (*continued*)

Linker	Sequence	Supplier
	Phosphorylated Adaptors	
BamHI to PstI	5′-GATCCCTGCAGG-3′	Pharmacia
BamHI to SmaI	5′-GATCCCCGGG-3′	Pharmacia
EcoRI to PstI	5′-AATTCCTGCAGG-3′	Pharmacia
EcoRI to SmaI	5′-AATTCCCGGG-3′	Pharmacia
EcoRI to ThaI	5′-AATTCGCG-3′	Pharmacia
HindIII to PstI	5′-AGCTCCTGCAGG-3′	Pharmacia
HindIII to SmaI	5′-AGCTCCCGGG-3′	Pharmacia
HpaII/TaqI to SmaI	5′-CGCCCGGG-3′	Pharmacia
XbaI to SmaI	5′-CTAGCCCGGG-3′	Pharmacia
XcyI to SmaI	5′-CCGGCCCGGG-3′	Pharmacia
XmaIII to SmaI	5′-GGCCCCCGGG-3′	Pharmacia

13

LIGATION

The in vitro construction of chimeric DNA molecules has been and remains a pivotal aspect of recombinant DNA technology. The covalent joining, or *ligation,* of vector and passenger duplex DNA molecules is generally performed by DNA ligases of either cellular *(E. coli)* or viral (T4 bacteriophage) origin (see Chapter 4). Since this step is one of the most critical in molecular cloning, several studies have been conducted to determine the optimum conditions for efficient ligation.

Many different factors have been found to affect both the rate of ligation and the nature of the end products. Among them are the temperature of incubation, the enzyme concentration, and the molar ratio of vector and passenger DNAs.

One can intuitively understand that the concentration of the DNAs present in the reaction mixture will affect the nature of the ligation products. The higher the concentration, the more likely two DNA ends may be in the vicinity of each other and be ligated. Also, at low concentrations, the probability for joining two ends of the same fragment will be of the same order of magnitude as the probability for joining two ends from different fragments present in the ligation mixture. It is also obvious that the probability for joining two ends of the same fragment will be dependent on the length of this fragment. Therefore, linear and circular molecules can be formed as the result of two competing reactions: end to end joining of distinct molecules and joining of opposite ends from the same molecule.

Because circular DNA (such as plasmid, or SV40

hybrids) and linear DNA (such as bacteriophage, cosmids, or retroviral genomes) are currently used in molecular cloning (see Chapters 6 and 25), it is important to know whether the ligation conditions employed favor the formation of closed circular molecules or the formation of linear products.

THEORY

In 1975, Dugaczyk et al. performed an elegant study of the parameters governing the formation of linear and circular molecules in the ligation reaction. They used, as a model, different EcoR I fragments obtained after digestion of SV40 and φX174 RF DNA and the circular plasmid pSC101. The different DNA fragments were ligated under various conditions of temperature, concentration, and time. The ligation products were separated by electrophoresis in agarose gels and quantitated by densitometry.

They proposed that two factors should be considered in order to predict whether the ligation products will be linear or circular. These two factors are (a) the total concentration of DNA ends (designated i) and (b) the effective concentration of one terminus of a DNA molecule in the immediate vicinity of the other terminus of the same molecule (designated j) (Jacobson and Stockmayer, 1950). In considering the parameter (i) Dugaczyk et al. distinguish between self-complementary ends (identical and complementary) and cohesive ends which are not identical.

The total concentration of ends per milliliter for non-identical cohesive termini is

$$i = N_0 M \times 10^{-3} \qquad (1)$$

while it is

$$i = 2N_0 M \times 10^{-3} \qquad (2)$$

for self-complementary (as well as blunt end) termini. In each formula, N_0 is the Avogadro number (6.022×10^{23}) and M is the molar concentration of the DNA molecules.

The effective concentration of one end in the neighborhood (or volume) of the other end of the same DNA molecule is given by the relation

$$j = \left(\frac{3}{2\pi lb}\right)^{3/2} \text{(ends/ml)} \qquad (3)$$

where l represents the contour length and b the random coil segment length of a DNA molecule. Considering the rigidity of duplex DNA, b is the minimal length of DNA which can bend to form a closed circle.

For λ phage DNA, l is 13.2 μm and the value for b (taken from the interpretation of sedimentation coefficients) is 7.17 μm. The value calculated for $j\lambda$ from equation (3) is 3.6×10^{11} ends/ml. It appears to be very close to the experimental value of 3.4×10^{11} reported by Wang and Davidson (1966).

It is possible to write equation (3) so that the j value can be calculated for any DNA molecule from the j value of phage λ and the molecular weight of the DNA molecule considered. The new equation is

$$j = j\lambda \left(\frac{MW_\lambda}{MW}\right)^{3/2} \text{(ends/ml)} \qquad (4)$$

where $j\lambda = 3.6 \times 10^{11}$ ends/ml and MW $\lambda = 30.8 \times 10^6$ (Davidson and Szybalski, 1971; Blattner et al., 1977).

Note that the j value is constant for a linear DNA molecule of a given length, and is independent of the DNA concentration. However, it will vary with the ionic strength of the solution because b is dependent upon this parameter.

For pBR322 whose molecular weight (2.9×10^6) can be calculated from the nucleotide sequence (Sutcliffe, 1978), one calculates

$$j \, pBR = 1.25 \, 10^{13} \text{ (ends/ml)}$$

It then appears that the shorter the DNA molecule, the higher the effective concentration of one terminus of a DNA molecule in the volume occupied by the other terminus of the same molecule.

When the j and i values are identical ($j/i = 1$) the probability that the terminus of a specific DNA molecule comes into association with its opposite end is the same as the probability that this terminus interacts with one end of another DNA molecule. In other words, when $j = i$ the linear and circular forms are equally likely to occur upon ligation.

By combining the equations (2) and (4) it is possible to calculate the molar concentration of DNA at which $j = i$

$$M = \frac{j^\lambda}{2N_0 \times 10^{-3}} \left(\frac{MW_\lambda}{MW}\right)^{3/2} \qquad (5)$$

and express it as the DNA concentration (in g/l) for which $j = i$

$$[DNA] = \frac{51.1}{(MW)^{0.5}} \qquad (6)$$

Above this DNA concentration (i.e., when $i > j$ and $j/i < 1$), linear n mers are expected to be favored, because the ligation event is more likely to occur between the ends of two different molecules rather than between opposite ends of the same molecule. Below this DNA concentration (i.e., when $i < j$ and $j/i > 1$ the formation of circular ligation products is expected to be favored.

It is possible to rearrange equations (1) and (3) to calculate the j/i ratio for a DNA whose molecular weight and concentrations are known.

$$j/i = \frac{51.1}{[DNA](MW)^{0.5}} \qquad (7)$$

This equation can be rearranged into

$$MW = \left(\frac{51.1}{j/i \, [DNA]}\right)^2 \qquad (8)$$

In a graph of equation (8) j/i ratios can be used to predict the nature of the ligation products expected for the ligation of DNAs of different molecular weights as a function of various DNA concentrations. Figure 109 shows a typical graph obtained with five j/i ratios for DNAs of molecular weight $< 26 \times 10^6$ daltons, at concentrations ranging between 10 to 100 μg/ml of solution.

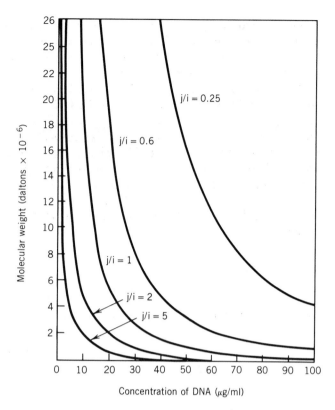

Figure 109. Graph illustrating the relation between molecular weight and DNA concentration at five different values of j/i (5, 2, 1, 0.6 and 0.25).

Dugaiczyk et al. (1975) have shown that these predictions can be verified when analyzing the ligation products obtained with EcoR I fragments under different experimental conditions (Table 56). When $j/i < 1$ linear oligomers are formed, whereas circular ligation products are readily obtained when $j/i > 2$. For values of j/i comprised between 1 and 2, they could check that according to equation (1) the shorter the contour length of the molecule (l), the more readily it circularizes upon ligation. Furthermore, they did not find any circular monomers among the products resulting from ligation of a small $\phi80$ DNA fragment whose l value (contour length) is less than the random coil segment b, as predicted by equation (1).

In cross ligation reactions involving SV 40 or pSC101 DNA and a $\phi80$ DNA fragment, all the expected combinations were observed. Due to the vectorial character of the pSC101 DNA, linear dimers were represented by three types of molecules. Covalently closed circular structures were obtained among ligation products when j/i was 2 or higher, while linear oligomers were obtained when j/i was < 1. In these cases, the j value for a given species is calculated as above, but the total concentration of identical ends of the DNA species is obtained by summation of the i values for each species in the reaction mixture.

Table 56. Products Formed During Ligation

Origin of Fragments	M Size (Kb)	DNA Concentration (µg/ml)	j/i	l/b	Monomer[a]	Dimer[a]	Trimer[a]	>Trimer[a]
PSC 101	9.1	30	0.71	35	40 L	32 L	20 L	8 L
	5.2	167	0.164	19	3 C	7 L	3 L	85 L
	5.2	1	28	19	39 L	6 L	1 L	
SV40					50 C	4 C		
	5.2	16	1.68	19	9 L	18 L	15 L	28 L
					19 C	3 C		
	5.2	26	1.08	19	30 L	25 L	11 L	8 L
	0.72	48.4	1.56	2.75	35 C	20 C	5 L	30 L
						5 L	5 C	
$\phi80$	0.72	6.05	12.4	2.75	64 C	12 C	4 C	8 L
						8 L	4 L	
	0.21	47.8	2.8	0.84	<1 L	12 C	21 L	54 L
						12 L		

Source: Information from Dugaiczyk et al. (1975) and De Vries et al. (1976).
[a]L (linear) and C (circular) n-mers, given as percentage of total.

PRACTICE

In practice, two basic types of cloning vectors are used: either circular molecules such as bacterial plasmids or linear DNA derived from bacteriophage λ and animal viruses (see Chapter 6). Therefore, it is important to use DNA concentrations that favor the formation of hybrid molecules having the correct conformation.

In order to determine whether the predictions described by Dugaiczyk et al. are also verified when using the λ-derived vectors, the optimal conditions which lead to the successful construction of representative genomic libraries in charon phages have been analyzed. The experimental parameters considered below were taken from work published by Maniatis et al. (1978) and Lawn et al. (1978). For the construction of their rabbit library, Maniatis et al. used 20 μg of rabbit DNA with a mean size of 20 kb and 55 μg of the cohered λ arms (31 kb) from λ charon 4A (Blattner et al. 1977) in a total volume of 300 μl. The ligation products were then packaged in vitro into viable phage particles (Sternberg et al., 1977; Hohn and Murray, 1977).

Taking these different parameters into account, it is possible to determine what type of hybrids the theoretical considerations described above would lead to predict.

One can calculate the j value for the rabbit DNA fragments of 20 kb (1.25×10^7 daltons)

$$j_{20kb} = J_\lambda \, (MW_\lambda/MW_{20kb})^{3/2} \text{ (ends/ml)}$$

$$= 3.6 \times 10^{11} \, [(30.8 \times 10^6)/(12.5 \times 10^6)]^{3/2}$$

$$\text{(ends/ml)}$$

$$= 1.4 \times 10^{12} \text{ (ends/ml)}$$

The same value would have been obtained if the ratio of sizes in kb had been considered (49.4/20) instead of the molecular weight ratio.

The concentration of these fragments in the ligation mixture is 20 μg in 300 μl (i.e., 0.067 g/l). This is equivalent to $5.33 \times 10^{-9} \, M$. It is now possible to calculate the i value for the 20-kb fragments

$$i_{20kb} = 2N_0 M \times 10^{-3} \text{ (ends/ml)}$$

$$= 6.42 \times 10^{12} \text{ (ends/ml)}$$

To calculate the parameters for the cloning vector, one must consider the fact that in this experiment, the λ charon 4A vector is composed of two EcoR I

arms of different length with cohesive termini. The right arm is 12 kb long (7.48×10^6 daltons), while the left arm is 19 kb long (1.18×10^7 daltons). Therefore the ratio j/i must be calculated for each arm. The relative proportions of each fragment (61.3% of the left arm and 38.7% of the right arm) allow us to calculate that 33.7 μg of the 19-kb fragment and 21.3 μg of the 12-kb fragment are present in the 300-μl reaction mixture. The molar concentration of each fragment can be calculated.

Since there is 0.112 g/l of 19-kb fragment (33.7 μg in 300 μl), the molar concentration is

$$0.112 \text{ g/l}:(1.18 \times 10^7 \text{ g/mol}) = 9.49 \times 10^{-9}$$

Similarly, the molar concentration of the 12 kb fragment is

$$0.071 \text{ g/l}:(7.48 \times 10^6 \text{ g/mol}) = 9.49 \times 10^{-9}$$

One can then calculate the i values for each arm. It is necessary to take into account that the termini of the vector are cohesive but not identical (see theory above) and apply the formula

$$i = N_0 M \times 10^{-3} \text{ (ends/ml)}$$

$$i_{19kb} = (6.022 \times 10^{23}) (9.49 \times 10^{-9}) (10^{-3})$$

$$= 5.71 \times 10^{12} \text{ (ends/ml)}$$

and

$$i_{12kb} = (6.022 \times 10^{23}) (9.49 \times 10^{-9}) (10^{-3})$$

$$= 5.71 \times 10^{12} \text{ (ends/ml)}$$

It is important to realize that each of the Eco RI arms of λ charon 4A represent a particular situation since they have dissimilar cohesive termini (*cos* at one end and Eco RI at the other end) and cannot therefore recircularize even by high-efficiency ligation. There is a need for one left arm and one right arm to form a DNA fragment able to circularize. The size of this fragment is 31 kb and its j value can be calculated as follows:

$$j_{31kb} = 7.24 \times 10^{11} \text{ (ends/ml)}$$

and

$$J/i_{31kb} = 7.24 \times 10^{11}/5.71 \times 10^{12} = 0.126$$

As stated above (see theory) the concentration of the ligatable ends of the DNA species present in the ligation mixture is given by the sum of i values for the three different DNA fragments (left and right λ arms and rabbit DNA fragments).

$$i = i_{20kb} + i_{19kb} + i_{12kb}$$

$$i = (6.42 \times 10^{12}) + (5.71 \times 10^{12}) + (5.71 \times 10^{12})$$

$$i = 1.78 \times 10^{13} \text{ (ends/ml)}$$

The analysis of these parameters shows that the concentration of ends (i) for the three different DNA species is close to 1 and that the ratio of vector ends to passenger DNA ends is close to 2:1. These values are in agreement with the fact that two ligation events are required to generate a correct hybrid molecule, while reformation of vector only needs one ligation event. Also, the formation of linear oligomers needed for efficient packaging (Feiss et al., 1977) is favored under the conditions used, because the concentration of each fragment is such that the j/i values for all DNA molecules are $<$ 1 (see theory above).

Thus, cloning of DNA fragments in λ vectors such as charon 4A is optimal when the concentration of each DNA species (a) is high enough so that the j/i ratio for each DNA is less than one, and (b) is such that the ratio of ligatable ends (i) for the three different DNA species is close to 1:1:1.

For cloning DNA fragments in plasmid vectors, the situation is slightly different in that the optimal ligation conditions must initially favor the joining of the DNA fragment to the vector, and secondly, lead to the circularization of the chimeric molecule (efficient transformation of E. coli is obtained only with circular molecules).

If we want to establish the conditions leading to a high number of hybrid-transforming molecules, we must determine the concentrations of plasmid and DNA fragment for which the j/i value at t_0 is minimal (in order to favor the oligomer formation) and is high for the expected hybrid monomer as the reaction proceeds (to favor circularization).

In their time course studies, Dugaiczyk et al. found that mixtures in which j/i is less than 0.8 lead essentially to the formation of linear products upon ligation. For j/i values contained in the range of 1 to 1.7 linear oligomers formed after 15 minutes subsequently circularize. From the relative amounts of various linear oligomers that can be predicted (Table 56), it becomes possible to calculate the concentrations of DNA plasmid and insert leading to high values of j (hybrid monomer)/i at an early time (t_{15}) in the ligation mixture. The data reported in Table 57 show a few sample calculations of j(hybrid monomer)/i at t_{15} for chimeric monomer molecules which form early during ligation between the 4.3-kb plasmid pBR322 and a 4.9-kb DNA fragment.

In Table 58 we have summarized the data obtained with DNA fragments whose size ranged between 1.6 and 21 kb under conditions resulting in high values of j(hybrid monomer)/i at t_{15}.

According to these studies, circularization of the hybrid molecules is optimal when the values of j(hybrid monomer)/i at t_{15} are greater than 0.9, and should therefore lead to maximal efficiency in transformation.

In practice, the situation is often complicated by

Table 57. Composition of the Ligation Reaction for Two Values of $j/i(t_0)$

Type of Molecule	Molecular Weight (kb)	Percentage of Total	Concentration (μg/μl) for $j/i(t_0)$ =		j ($\times 10^{-12}$)	i ($\times 10^{-11}$) for $j/i(t_0)$ =		$j/i(t_{15})$ for $j/i(t_0)$ =	
			1.0	1.5		1.0	1.5	1.0	1.5
Precursors									
pBR322	4.8	17.2	5.15	3.44	12.5	21.3	14.2	1.89	2.84
Insert	5.0	17.8	5.3	3.56	11.8	21.3	14.2	1.79	2.69
Hybrid									
Monomer	9.8	25.0	7.48	5.0	4.29	15.2	10.0	0.65	0.95
Dimer	19.6	18.5	5.39	3.6	1.52	5.48	3.66	0.23	0.35
Trimer	29.4	13.0	3.89	2.6	0.54	1.98	1.32	0.081	0.12
Tetramer	39.2	6.0	1.8	1.2	0.38	0.73	0.49	0.058	0.09

Table 58. Ligation Reaction Parameters Resulting in High Values of j(hybrid monomer)/i at t_{15} for DNA Fragments of Different Sizes

Fragment Size (kb)	DNA (μg/ml)		$j/i(t_0)$		ipBR322/ $i$frag.	j(hybrid monomer)/$i(t_{15})$
	pBR322	Fragment	pBR322	Fragment		
1.66	10.42	10.08	0.5	2.46	0.25	1.15
4.99	9.84	10.20	1.5	1.5	1.0	0.97
11.66	4.01	9.68	3.77	1.0	1.0	1.08
	2.68	6.48	5.65	1.5	1.0	1.63
16.66	2.35	8.11	12.85	1.0	1.0	1.19
	3.14	10.83	9.62	0.75	1.0	0.93
	3.54	4.07	8.54	1.0	4.0	1.73
21.66	3.55	2.38	9.54	1.0	4.0	1.97
	1.56	7.11	19.0	1.0	1.0	1.35

the fact that several other factors, such as ATP concentration, ligase concentration, temperature and time of incubation, or molar ratio of insert to vector may significantly affect the efficiency of ligation. In an attempt to optimize DNA ligations for transformation, the number of transformants obtained per microgram of input DNA was measured as a function of these different factors (King and Blakesley, 1986). Their standard ligation mixture contained, in a final volume of 20 μl:

50 mM Tris–HCl pH 7.6

10 mM MgCl$_2$

1 mM ATP, 170 ng (60 fmol) of dephosphorylated Eco RV or Pvu II cleaved pBR322 DNA (blunt ends)

13 ng (20 fmol) of a 1 kb Tha I DNA fragment (blunt ends)

1 unit of T4 DNA ligase

Incubation was performed at 4°C or at room temperature (23–26°C) for 4 hours, and 1 μl of 0.5 M sodium EDTA was added to stop the reaction. Transformation assays were performed on 2-μl samples of the ligase reaction, diluted to 10 μl with 8 μl of TE buffer pH 7.2. This dilution is important to avoid inhibition of transformation which occurs at high DNA concentrations (Jesse, 1984).

The temperature of ligation was found to influence dramatically the efficiency of transformation. When the ligation was performed at room temperature for 4 hours, the number of transformants generated was about 25 greater than that obtained after 4 hours at 4°C and represented about 90% of the number of transformants obtained after 23 hours of incubation (Figure 110). On the other hand, ligation

of blunt end fragments performed at 14°C overnight results in at least four times more transformants than ligations performed at room temperature for 4 hours.

The amount of T4 DNA ligase needed in blunt end ligation was found to be 10 times greater than the amount needed to achieve sticky end ligation. (It had been empirically determined previously with other systems that blunt end ligation required about 80 times the amount of enzyme required for sticky end ligation.)

The ATP concentration could be varied from 10 μM to 1 mM, without affecting the efficiency of blunt end ligation, while the recircularization of dephosphorylated vector without insert was maximum at 0.1 mM ATP concentration.

Increasing the ratio of vector to insert did not influence greatly the efficiency of transformation. Less than a three-fold increase was obtained when the ratio varied from 0.3 to 0.03.

The stimulating effects of polyethylene glycol (PEG) 8000 (Pheiffer and Zimmerman, 1983; Zimmerman and Harrison, 1985) were determined by using the standard conditions described above. As shown in Figure 111, the addition of PEG up to 5% increased the efficiency of transformation by a factor of 3 to 6. Above this value, PEG showed an inhibitory effect. Finally, be aware that phosphate concentrations greater than 25 mM, and more generally, salt concentrations greater than 50mM will considerably inhibit ligation of blunt end fragments.

As a general conclusion, it seems that the theoretical considerations described by Dugaiczyk et al. (1975) are useful in determining the optimal conditions that lead to the formation of linear or circular hybrid molecules, therefore increasing the probability of success in cloning with λ- or plasmid-derived

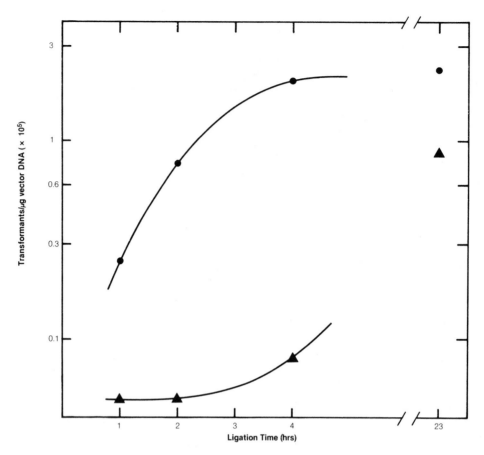

Figure 110. The effect of time and temperature of ligation on transformation. Dephosphorylated EcoR V-cleaved pBR322 DNA and a 1-kb Tha I fragment were incubated in the standard ligation reaction either at 26°C (●) or 4°C (▲). Aliquots (2 µl) were removed at 1, 2, 4, and 23 hours of incubation. The ligation products were analyzed by the small-scale transformation assay (courtesy of Bethesda Research Laboratories, Life technologies Inc.).

vectors. In addition to these parameters, it is important to consider the other conditions that have been shown to affect notably the successfulness of bacterial transformation, which in most cases is the final goal to be reached when cloning.

REFERENCES

Blattner, F. R., Williams, B. G., Blechl, A. E., Denniston-Thompson, K., Faber, H. E., Furlong, L. A., Grunwals, D. J., Keifer, D. O., Moore, D. D., Schumm, J. W., Sheldon, E. L., and Smithies, O. (1977), *Science,* **196**, 161.

Davidson, N., and Szybalski, W., in *Cold Spring Harbor laboratory symposia,* Cold Spring Harbor, New York (1971), p. 45.

De Vries, F. A. J., Collins, C. J., and Jackson, D. A. (1976), *Biochim. Biophys. Acta,* **435**, 213.

Dugaiczyk, A., Boyer, H. W., and Goodman, H. M. (1975), *J. Mol. Biol.,* **96**, 174.

Feiss, M., Fisher, R. A., Clayton, M. A., and Enger, C. (1977), *Virology,* **77**, 281.

Hohn, B., and Murray, K. (1977), *Proc. Natl. Acad. Sci. USA,* **74**, 3263.

Jacobson, H., and Stockmayer, W. H. (1950), *J. Chem. Phys.,* **18**, 1600.

Jesse, J. (1984), *Focus,* **6:4**, 4.

King, P. V., and Blakesley, R. W. (1986), *Focus,* **8:1**, 1.

Lawn, R. M., Fritsch, E. F., Parker, R. C., Blake, G., and Maniatis, T. (1978), *Cell,* **15**, 1157.

Maniatis, T., Hardison, R. C., Lacy, E., Lauer, J., O'Connell, C., and Quon, D. (1978), *Cell,* **15**, 687.

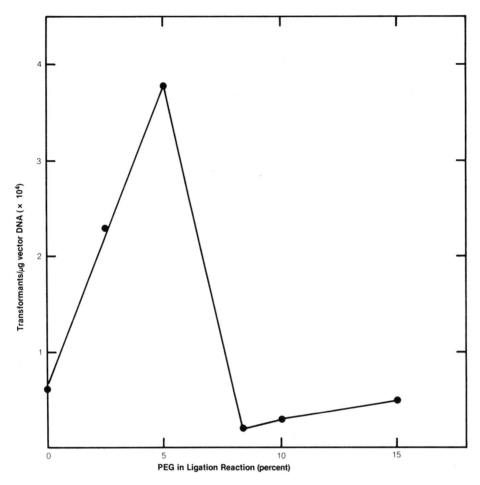

Figure 111. The effect of PEG concentration in the ligation reaction on transformation. Dephosphorylated Pvu II-cleaved pBR322 DNA and a 1-kb Tha I fragment in the standard ligation reaction were incubated at 23.5°C for only 1 hour. The ligation products were analyzed by the small-scale transformation assay (courtesy of Bethesda Research Laboratories, Life technologies Inc.).

Pheiffer, B. H., and Zimmerman, S. B. (1983), *Nucl. Acids Res.,* **11**, 7853.

Sternberg, N., Tiemeier, D., and Enquist, L. (1977), *Gene,* **1**, 255.

Sutcliffe, J. G. (1978), *Nucleic Acids Res.,* **5**, 2721.

Timmis, K. M., Cohen, S. M., and Cabello, F. C., in *Progress in Molecular and Subcellular Biology,* Vol. 6, F. Hahn, Ed., Springer-Verlag, New York (1978), p. 1.

Zimmerman, S. B., and Harrison, B. (1985), *Nucl. Acids Res.,* **13**, 2241.

14

PROPAGATION OF RECOMBINANT DNA MOLECULES

REDUCTION OF TRANSFORMATION BACKGROUND DUE TO VECTOR RECIRCULARIZATION—USE OF ALKALINE PHOSPHATASE

One of the most critical steps in the propagation of recombinant DNA molecules after ligation to plasmid vectors is selection of transformants. When linear vector and DNA molecules to be cloned are ligated and used to transform adequate hosts, it is very frequently the case that the great majority of transformants appear to contain only vector DNA, without any insert. This is due to the ability of the linear plasmid DNA molecules generated after digestion with restriction endonucleases to recircularize when incubated in the presence of DNA ligase, leading to transformants containing the parental circular plasmid. The resulting background can sometimes be a quite limiting step in the isolation of transformants carrying recombinant DNA molecules.

Because dephosphorylated 5′ ends of DNA molecules are not substrates for DNA ligase, a drastic reduction of the background due to recircularization can be achieved by incubating the restriction-endonuclease-digested vector DNA with alkaline phosphatase. This enzyme removes the 5′-terminal phosphate of linear DNA molecules, preventing recircularization. On the other hand, the 3′-hydroxyl end of the treated vector can be ligated efficiently to the 5′-phosphorylated end of the DNA molecules to be cloned. This will result in the formation of nicked recombinant circles that are fully transforming.

PROTOCOL

DIGESTION OF VECTOR DNA

1. In a microfuge tube mix:
29 μl sterile distilled H_2O
10 μl 5X reaction buffer

> 10 µl DNA (0.5 µg/µl)
> 1 µl restriction endonuclease (10 units)
> 50 µl total volume

2. Incubate for 2 hours at 37°C.
3. Heat at 70°C for 10 minutes to denature enzyme.
4. Add 5 µl 2.5 *M* sodium acetate.
5. Add 150 µl 100% cold ethanol.
6. Leave at -70°C for 20 minutes (or -20°C for 2 hours).
7. Spin in a microfuge for 10 minutes at 4°C.
8. Pour off the ethanol carefully.
9. Rinse the pellet once with absolute ethanol.
10. Spin in a microfuge for 3 minutes at 4°C.
11. Pour off the ethanol carefully.
12. Dry the pellet under vacuum for 10 minutes.

PROTOCOL

TREATMENT WITH ALKALINE PHOSPHATASE

1. Resuspend the digested-DNA pellet (see above) in 44 µl of 10 m*M* Tris–HCl (pH 8.0).
2. Add 5 µl of 10X CIP buffer.
3. Add 1 µl of calf intestine alkaline phosphatase.
4. Incubate at 37°C for 30 minutes.
5. Add 1 µl of alkaline phosphatase, and incubate for 30 additional minutes.
6. Add 300 µl stop buffer.
7. Add 150 µl phenol and 150 µl chloroform: isoamyl alcohol (24:1). Vortex. Spin in a microfuge for 3 minutes.
8. Transfer the upper phase to a new tube.
9. Add 300 µl chloroform:isoamyl alcohol (24:1). Vortex and spin for 3 minutes in a microfuge. Transfer upper phase to a new tube.
10. Repeat step 9.
11. Transfer the supernatant to a tube containing 30 µl 2.5 *M* sodium acetate. Add 1 ml cold absolute ethanol. Leave at -70°C for 20 minutes (or -20°C for 2 hours).

12. Spin 10 minutes at 4°C in a microfuge.
13. Pour off the ethanol, then add 1 ml ethanol.
14. Spin 10 minutes at 4°C in a microfuge.
15. Pour off the ethanol. Dry the pellet under vacuum for 10 minutes.
16. Resuspend in 50 μl sterile distilled H$_2$O. Store at −20°C.

PROTOCOL

CHECK FOR EFFICIENCY OF DEPHOSPHORYLATION

1. In a microfuge tube, mix:
   19 μl sterile distilled H$_2$O
   1 μl treated vector (100 ng)
   5 μl 5X ligation buffer
   1 unit T4 DNA ligase
2. Incubate for 16–18 hours at 14°C.
3. Heat-denature T4 DNA ligase at 70°C for 10 minutes.
4. Use this mixture to transform competent cells as described in this chapter (pages 411–417)

A control experiment run under the same conditions with untreated DNA should be included. The ratio of the number of colonies obtained with the treated vector to the number of colonies obtained with the untreated vector should be approximately 10^{-4} to 10^{-5}.

Notes

1. If bacterial alkaline phosphatase is used to dephosphorylate the DNA, the reaction mix (5–10 units/μg DNA) should be incubated at 65°C.
2. Better results have been obtained in our hands with calf intestine alkaline phosphatase than with bacterial alkaline phosphatase.
3. Alkaline phosphatase activity has been reported to be inhibited by inorganic phosphatase. We therefore recommend dialyzing DNA before treating with alkaline phosphatase.
4. Particular care should be taken not to use considerable excess of enzyme. We have experienced loss of restriction sites upon dephosphorylation of some endonuclease-digested vectors.

5. 0.01 units of enzyme are sufficient to dephosphorylate 1 pmol of 5′ ends.

6. When 5′-recessed ends are used as substrate, it is recommended to perform the incubation twice, at 37°C for 15 minutes and then at 56°C for 15 minutes. The higher temperature ensures accessibility of the recessed end.

Buffers

10X CIP buffer
0.5 M Tris–HCl (pH 9.0)
10 mM MgCl$_2$
1 mM ZnCl$_2$
10 mM Spermidine

1 ml
500 µl 1 M Tris–HCl (pH 9.0)
10 µl 1 M MgCl$_2$
1 µl 1 M ZnCl$_2$
100 µl 0.1 M Spermidine
389 µl H$_2$O

Stop buffer
10 mM Tris–HCl (pH 7.5)
1 mM EDTA (pH 7.5)
200 mM NaCl
0.5% Sodium dodecyl sulfate (SDS)

100 ml
1 ml 1 M Tris–HCl (pH 7.5)
200 µl 0.5 M EDTA (pH 7.5)
4 ml 5 M NaCl
2.5 ml 20% SDS
92.3 ml H$_2$O
Store at room temperature.

5X Ligation buffer
250 mM Tris–HCl (pH 7.5)
50 mM MgCl$_2$
50 mM 2-mercaptoethanol
5 mM ATP

500 µl
125 µl of 1 M Tris–HCl (pH 7.5)
50 µl 0.5 M MgCl$_2$
50 µl 0.5 M 2-mercaptoethanol
125 µl 20 mM ATP
150 µl H$_2$O
Store at −20°C.

CLONING IN TAILED VECTOR

PROTOCOL

CLONING DNA (OR cDNA) IN dG-TAILED VECTOR

This procedure is schematically illustrated in Figure 112.

dG Tailing of Vector DNA (pBR322)

1. Mix successively in a microfuge tube:
   117 µl sterile distilled water

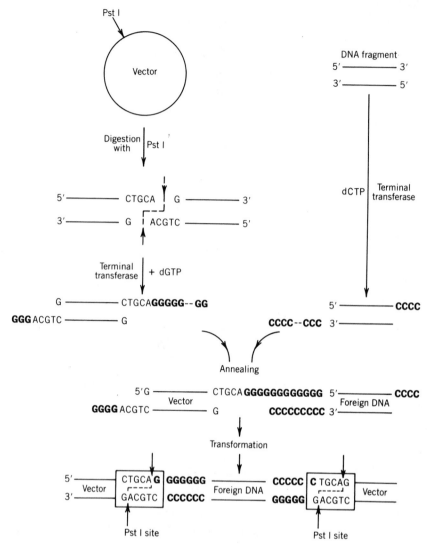

Figure 112. Cloning of a DNA fragment with dC tails into a vector whose cloning site has been elongated with dG tails. The method shown allows the recovery of the cloned foreign DNA because the Pst I recognition sites are regenerated after ligation.

32 μl purified pBR322 DNA digested twice with Pst I endonuclease (~10 μg pBR322)

10 μl 5X tailing buffer

1.5 μl 23 m*M* dGTP

2.0 μl 0.1 *M* CoCl$_2$

8 μl terminal deoxynucleotidyl transferase (8 units/μl)

2. Incubate for 7 minutes at 37°C.

3. Add 40 μl 5X buffer C and 2 μl 20% SDS.

4. Add 100 μl phenol equilibrated in NTE (pH 8.5) and 100 μl chloroform:isoamyl alcohol (24:1).

5. Mix vigorously and centrifuge at 12,000 rpm for 2 minutes.

6. Pipette out the upper phase and run it over a 10-ml column of Sephadex G75 equilibrated in NTE buffer. Collect 10 to 12 0.5-ml fractions.

7. Run 10-μl aliquots of each fraction on an 0.8% agarose gel to locate the dG-tailed vector.

dC Tailing of DNA (or cDNA)

1. Mix successively in a microfuge tube:
   102 μl sterile distilled water
   40 μl 5X tailing buffer
   3 μl 10 mM dCTP
   50 μl DNA (100 ng)
   2 μl 10 mM dithiothreitol
   10 μCi α[^{32}P]dCTP (if you wish to check incorporation)

2. Leave at 37°C for 5 minutes to warm up.

3. Add 2 μl 0.1 M CoCl$_2$ and 1 μl terminal deoxynucleotidyl transferase (30 units/μl).

4. Incubate for 10 minutes at 37°C.

5. Add 40 μl 5X buffer C.

6. Add 2 μl 20% SDS.

7. Add 200 μl phenol equilibrated in NTE (pH 8.5) and 100 μl chloroform:isoamyl alcohol (24:1).

8. Mix thoroughly and spin at 12,000 rpm for 2 minutes.

9. Pipette out the upper phase and mix it with 200 μl chloroform:isoamyl alcohol (24:1). Centrifuge at 12,000 rpm. Carefully collect the upper phase.

10. Add 120 μl (~600 ng) of tailed vector to obtain a 5:1 excess of vector over tailed DNA.

11. Add 10 μl 3 M sodium acetate and precipitate DNAs by addition of 2 ml cold ethanol. Leave for 2 hours at −20°C or 20 minutes at −70°C.

12. Precipitate DNA by spinning at 12,000 rpm for 10 minutes.

13. Pour off the ethanol without disturbing the DNA pellet. Add 1 ml ethanol without resuspending the DNA pellet.

14. Spin at 12,000 rpm for 5 minutes. Pour off the ethanol. Dry the precipitate. Resuspend in 100 μl annealing buffer.

15. Precipitate the remaining tailed pBR322 from above with ethanol and resuspend in 50 μl annealing buffer.

Annealing of Tailed DNA and Vector

1. Incubate the resuspended DNA mixture (100 μl from step 14 above) at 68°C for 5 minutes in a water bath.
2. Switch off the water bath and leave the sample to cool slowly overnight with the bath lid on.

Buffers for Tailing

Tailing Buffer (5X)
0.7 *M* Cacodylic acid
0.15 *M* Tris base
0.55 *M* KOH
Bring pH to 7.6

Buffer C (5X)	*100 ml*
250 m*M* Tris–HCl (pH 8.0)	25 ml 1 *M* Tris–HCl (pH 8.0)
50 m*M* EDTA	10 ml 0.5 *M* EDTA
50 m*M* NaCl	1 ml 5 *M* NaCl
	64 ml H$_2$O

NTE (pH 8.5)	*100 ml*
100 m*M* NaCl	2.0 ml 5 *M* NaCl
10 m*M* Tris–HCl (pH 8.5)	2.0 ml 1 *M* Tris–HCl (pH 8.5)
1 m*M* EDTA	0.2 ml 0.5 *M* EDTA
	95.8 ml H$_2$O

Annealing buffer	*100 ml*
10 m*M* Tris–HCl (pH 7.4)	1 ml 1 *M* Tris–HCl (pH 7.4)
100 m*M* NaCl	2 ml 5 *M* NaCl
0.25 m*M* EDTA	0.05 ml 0.5 *M* EDTA

TAILING CONDITIONS AND CALCULATIONS*

The following protocol outlines a method that determines the appropriate conditions for incorporating a specified number of dG residues (tails) on a linearized DNA vector with terminal transferase (TdT). The example presented here was performed with Pst I-cleaved pBR322 DNA as the substrate.

*Courtesy of Bethesda Research Laboratories, Life Technologies, Inc.

PROTOCOL

CALIBRATION OF dG-TAILING CONDITIONS

Prepare a series of assays containing 5 to 50 μM dGTP as shown in Table 59. The component volume may be altered for varying concentrations of TdT and DNA. THe final reaction volume is brought to 50 μl with water. It is recommended that each assay be performed in duplicate.

1. Combine all components in 1.51 ml microcentrifuge tubes. Add the TdT last.
2. Vortex tubes gently to mix.
3. Centrifuge tubes briefly at 4°C in a microcentrifuge.
4. Incubate tubes for 30 minutes at 37°C.
5. Place tubes on ice.
6. Spot 10 μl of each assay on numbered GF/C filters (a pencil or India pink pen may be used to number the filters).
7. Precipitate the DNA with cold (4°C) 10% TCA + 1% NaPPi (10 ml/filter). This may be performed batchwise in a beaker.
8. Wash the filters four times with 5% TCA (10 ml/filter).
9. Wash the filters two times with absolute alcohol (5 ml/filter).
10. Spot 5 μl of assays 1 and 2 on separate filters. These will be used to calculate total counts per minute (cpm) of ^{3}H-dGTP per assay and will be referred to as 1T and 2T.
11. Place all filters under a heat lamp to dry.
12. Place filters in vials with scintillation fluid and count on ^{3}H-channel.

Table 59.   TdT Tailing Assay

	1	2	3	4	5	6	7	8	9	10	11	12
Component	(μl)	(μl)	(μl)	(μl)	(μl)	(μl)	(μl)	(μl)	(μl)	(μl)	(μl)	(μl)
5X reaction buffer[a]	10	10	10	10	10	10	10	10	10	10	10	10
Pst I-cleaved pBR322 DNA (1 μg/μl)	—	—	3	3	3	3	3	3	3	3	3	3
^{3}H-dGTP (1 mCi/ml)	2	2	2	2	2	2	2	2	2	2	2	2
dGTP (100 μM)	25	25	2.5	2.5	5	5	10	10	15	15	25	25
Sterile distilled H$_2$O	13	13	31	31	28.5	28.5	25.5	25.5	18.5	18.5	8.5	8.5
TdT (10 U/μl)	—	—	1.5	1.5	1.5	1.5	1.5	1.5	1.5	1.5	1.5	1.5

Information Courtesy of Bethesda Research Laboratories, Life Technologies.

[a]5X reaction buffer: 100 mM cacodylate phosphate, pH 7.2, 10 mM CoCl$_2$, 1.0 mM DTT.

Calculations

NUMBER OF 3'-ENDS

1. General formula:

$$\text{pmol 3'-ends/assay} = \frac{\mu g \text{ DNA}}{\text{MW}} \times 2 \times 10^6$$

$$\text{MW} = (\text{number of base pairs}) \times 666 \text{ g/mol})$$

$$(\text{assume } 50\% \text{ A } + \text{ T content})$$

2. Sample calculation. All sample calculations are from actual data for tailing Pst I-cut pBR322.

$$\text{pmol 3'-ends/assay} = \frac{3 \mu g}{2.9 \times 10^6 \mu g/\mu mol} \times 2 \times 10^6$$

$$= 2 \text{ pmol of 3' ends/assay}$$

SPECIFIC ACTIVITY (SA) OF dGTP IN ASSAY

1. General formula:

$$\text{SA (cpm/pmol)} = \frac{\text{cpm } - \text{ cpm background}}{\text{pmol dGTP}}$$

   a. cpm = average cpm of filters 1T and 2T
   b. cpm background = average cpm of filters 1 and 2
   c. pmol dGTP = (μM dGTP) $\times$ (volume spotted)
   d. μM dGTP =
   $$\frac{(\mu M \text{ conc. of dGTP stock}) \times (\text{volume dGTP added})}{\text{total volume of assay}}$$

2. Sample calculations. The cpm obtained from each filter were as follows:

Filter No.	1	2	1T	2T	3	4
cpm	38	38	114,313	129,783	32,660	35,436
average cpm	38		122,048		34,061	

$$\mu M \text{ dGTP} = \frac{(100 \ \mu M) \times (2.5 \ \mu l)}{50 \ \mu l} = 5 \ \mu M$$

$$\text{pmol dGTP} = (5 \ \mu M) \times (5 \ \mu l) = 25 \text{ pmol}$$

$$\text{SA} = \frac{(122,048 \text{ cpm } - \text{ 38 cpm})}{25 \text{ pmol}} = 4,880 \text{ cpm/pmol}$$

LENGTH OF dG TAILS

1. General formula:

$$\text{length of dG tail} = \frac{\text{pmol dGTP}}{(\text{pmol 3}'\text{-ends of DNA})}$$

$$= \frac{\text{cpm} - (\text{cpm background})}{\text{SA}}$$

$$\times \frac{(\text{total assay volume})}{(\text{volume spotted})}$$

$$\times \frac{(1 \text{ pmol 3}' \text{ ends})}{(\text{pmol 3}'\text{-ends/assay})}$$

2. Sample calculation:

$$\frac{(34061 - 38 \text{ cpm})}{4880 \text{ cpm/pmol}} \times \frac{50 \text{ }\mu\text{l}}{10 \text{ }\mu\text{l}} \times \frac{1}{2} = 17$$

TRANSFORMATION OF *E. COLI* STRAINS WITH PLASMID DNA

See the following protocols.

PROTOCOL

TRANSFORMATION OF E COLI DH1 AND MC1061 STRAINS

1. Disperse four isolated colonies from a freshly streaked plate in 5 ml of LB medium.
2. Shake the culture at 37°C for about 2 hours until the absorbance at 550 nm reaches 0.3.
3. Inoculate a prerinsed flask (see note 1) with 2 ml of this preculture in 40 ml of prewarmed LB medium.
4. Shake the culture at 37°C until the absorbance at 550 nm reaches 0.45. This usually requires 1½ to 2 hours.

For the preparation of competent cells steps 5–12 should be performed in a cold room with cold pipettes and tubes.

5. Collect the cells into 50-ml polypropylene tubes (e.g., Falcon 2070) and place on ice for 10 to 15 minutes.
6. Centrifuge the culture at 2500 rpm for 15 minutes at 4°C.

7. Resuspend the cell pellet in 2 ml of TFB1 by gently tilting the tube, then bring the volume to 16 ml with TFB1.

8. Leave the suspension on ice for 15 minutes.

9. Spin the cells again at 2500 rpm for 15 minutes at 4°C.

10. Resuspend the cell pellet in 1.6 ml of TFB2.

11. Leave the suspension on ice for 15 minutes.

12. Place samples (200 µl) into chilled screw-cap polypropylene tubes (e.g., Nunc) and freeze at −70°C slowly (in a cardboard box).

13. For the transformation assay, thaw the competent cells in ice.

14. Transfer gently 200 µl of cells in 5-ml Falcon tubes (2054).

15. Add DNA (in a volume less than 10 µl), swirl the mixture, and incubate on ice for 30–45 minutes.

16. Transfer to 42°C for 90 seconds and place on ice for 5 minutes.

17. Add 5 volumes of LB medium (1 ml) and shake slowly at 37°C for 1 hour.

18. Pipet a fraction of the culture (100–200 µl) and spread gently with a (L-shape) Pasteur pipette on an LB plate with the appropriate antibiotic.

19. Incubate plates at 37°C overnight.

Notes

1. Preparation of flask for culture:
   Rinse three times with the most purified water available (water purified by reverse osmosis Millipore-Milla).
   Sterilize at 120°C for 20 minutes.
   Shake overnight.
   Rinse again three times with sterile purified water.
   Fill with 40 ml of LB medium.
   Prewarm at 37°C.

2. If 50-ml polypropylene tubes (Falcon 2070) are used in the transformation assay, the optimal time for the heat pulse is 210 seconds.

3. This protocol is derived from that originally described by Hanahan (1985). It is currently used in our laboratory with *E. coli* strains MC1061 and DH1 (rec⁺ and rec⁻, respectively) and with JM strains. Transformation efficiencies are in the range of 10^8 transformants/µg of plasmid DNA such as pBR322. It has also been claimed by BRL that up to 5×10^8 transformants were obtained per microgram of pUC19 plasmid DNA with their DH5 competent cells prepared in this way. Beware that this procedure may not work with all cell strains

and that the presence of Mg^{2+} ions is inhibitory to MC1061 transformation.

Solutions and Buffers

LB culture medium
For 1 liter:
10 g Bactotryptone
5 g Yeast extract
5 g NaCl
The pH should be naturally 6.8 – 7.0. Do not adjust to pH 7.5 because it is inhibitory in the transformation assay.
Sterilize at 120°C for 30 minutes.

MOPS 1 M [3-(*N*-morpholino) propane sulfonic acid]
11.51 g for 50 ml
Filter through a prerinsed 0.22-μm filter unit. Store at 4°C.

MES 1 M [Morpholino ethane sulfonic acid]
19.5 g for 100 ml
Adjust to pH 6.2 using KOH, sterilize filter, and store at −20°C.

TFB1	*500 ml*
MES 10 mM	5 ml MES 1 M (pH 6.2)
RbCl$_2$ 100 mM	6.045 g
CaCl$_2$, 2H$_2$O 10 mM	0.735 g
MnCl$_2$, 4H$_2$O 50 mM	4.94 g
	H$_2$O to 500 ml

All salts are added as solids. Adjust the pH to 5.8 with acetic acid, filter the solution through a prerinsed 0.22-μm filter unit, and store at 4°C. TFB1 is stable at 4°C for more than 1 year.

TFB2	*100 ml*
MOPS 10 mM	1 ml MOPS 1 M
CaCl$_2$, 2H$_2$O 75 mM	1.102 g
RbCl$_2$ 10 mM	0.120 g
Glycerol 15%	15 ml

Bring the pH to 6.5. Sterilize filter and store at 4°C.

Reference

Hanahan, D. (1983), *J. Mol. Biol.* **166,** 557.

PROTOCOL

TRANSFORMATION OF E. COLI STRAIN HB101

This protocol is more general, but less efficient (2×10^6 transformants/µg of DNA, at the most). The following quantities are for five transformation assays.

1. Seed 5 ml TYE medium with one isolated colony of the host to be used for transformation. Shake overnight at 37°C.
2. The next morning, seed 6 ml TYE medium with 0.1 ml of the overnight culture.
3. Shake at 37°C until the optical density at 600 nm of the culture reaches a value between 0.5 and 0.6. This usually requires 2–3 hours of incubation at 37°C.
4. Centrifuge the culture at $2000 \times g$ for 10 minutes at 4°C.
5. Resuspend the cell pellet in 2.5 ml 50 mM CaCl$_2$.
6. Let the suspension stand at 4°C for 15 minutes.
7. Spin the suspension at $2000 \times g$ for 10 minutes at 4°C.
8. Resuspend the cells in 500 µl 40 mM CaCl$_2$.
9. In a sterile tube mix 50 µl TCM buffer with 20–30 µl of the DNA preparation (0.1–100 ng) to be used for transformation.
10. Add to the DNA–TCM mixture 100 µl of the cell suspension obtained in step 8.
11. Incubate for 15 minutes on ice.
12. Transfer to 42°C for 2 minutes.
13. Let stand for 10 minutes at room temperature.
14. Add 1 ml prewarmed TYE medium to the mixture and incubate without shaking for 30 minutes at 37°C.
15. Spread 0.1 ml of the mix directly onto agar plates containing the medium for selection of transformed cells.
16. Mix the remaining 0.9 ml of cells with 3 ml 1% agar and pour the mixture onto agar plates containing the selective medium.
17. Incubate plates at 37°C overnight.
18. Pick colonies and test them for the presence of recombinant DNA.

Buffers and Culture Media

TCM buffer   *100 ml*
10 mM Tris–HCl (pH 7.4)   1 ml 1 M Tris–HCl (pH 7.4)

10 mM CaCl$_2$ 1 ml 1 M CaCl$_2$
10 mM MgCl$_2$ 1 ml 1 M MgCl$_2$
                                   97 ml H$_2$O

TYE culture medium
For 1 liter:
10 g Bacto tryptone
5 g Yeast extract
5 g NaCl

Ampicillin Plates

1.2% agar
30–100 µg/ml ampicillin (depending on host used)

1. To 1 liter TYE add 12 g bacto agar.
2. Sterilize at 120°C for 20 minutes.
3. Cool to 50°C in a water bath.
4. Add 1 ml of a 1000X concentrated solution of ampicillin (sodium salt, 30–100 mg/ml). The antibiotic solution can be kept frozen at −20°C for several months. Mix.
5. Pour the medium into sterile plates (~30 ml/plate). Let solidify at room temperature. Dry overnight at 37°C before use.

Tetracycline Plates

Prepare as for ampicillin plates, except to 1 liter of autoclaved TYE–agar medium cooled to 50°C, add 2 ml of a 500X concentrated solution of tetracycline hydrochlorate (7.5 mg/ml). The final tetracycline concentration in the plates will be 15 µg/ml. The stock solution of tetracycline can be kept at −20°C for several months. Note that Mg^{2+} is an antagonist of tetracycline.

Chloramphenicol Plates

Prepare as for ampicillin plates, except to 1 liter of autoclaved TYE–agar medium cooled to 50°C, add 0.5–5 ml of a 20 mg/ml chloramphenicol solution (in 100% ethanol). The final chloramphenicol concentration in the plates will be 10–100 µg/ml. The stock solution of chloramphenicol can be kept at −20°C for several months.

Kanamycin Plates

Prepare as for ampicillin plates, except to 1 liter of autoclaved TYE–agar medium cooled to 50°C, add 2 ml of a 500X concentrated solution of kanamycin sulfate (25 mg/ml). The final kanamycin concentration will be 50 μg/ml. The stock solution of kanamycin can be kept at −20°C for several months.

Although these antibiotic-containing plates can be stored for 1–2 weeks at 4°C, we recommend only using freshly made plates dried for one night at 37°C.

TRANSFORMATION OF COMPETENT CELLS FOLLOWING DIRECT LIGATION IN LOW MELTING POINT AGAROSE

This protocol is particularly useful when rapid cloning is needed with only small amounts of material available. It alleviates the need to elute the DNA fragments from the gel. A typical ligation reaction is performed as follows:

PROTOCOL

TRANSFORMATION FOLLOWING LIGATION IN MELTED AGAROSE

1. Once the separation of the digested DNA fragments is performed in low melting point agarose, cut out the portion of gel containing the DNA fragment (20–50 ng) to be ligated.

2. Melt the agarose (10–20 μl) by heating at 68°C and bring the temperature to 37°C by immersing the tube in a water bath. At this temperature, the low melting point agarose remains molten.

3. Transfer 10 μl of the molten agarose to a tube containing:
   2 μl of sterile distilled water
   1 μl (20 ng) of vector
   4 μl of 5X ligation buffer
   1 μl of 10 mM ATP
   1 μl of 0.5 mg/ml bovin serum albumin

4. Add 1 μl of T4 DNA ligase (New England Biolabs).

5. Incubate overnight at 16°C.

6. Heat the mixture at 68°C for 10 minutes.

7. Transfer 10 μl of the mixture to 50 μl of TCM and proceed for transformation as described above in this chapter.

Notes

1. We generally use HB101 when screening is performed on the basis of ampicillin, tetracyclin, or chloramphenicol resistance.

The JM 107 strain is generally used for transformation with the pUC plasmids allowing a selection of the recombinants on the basis of β-galactosidase alpha complementation (see Chapters 6 and 20). In this case, the bacteria are plated in the presence of IPTG and X-Gal (40 and 100 μl of 2% stock solutions).

2. When using the Hanahan method, the pH of the S.O.C. medium added to the cells following transformation should be between 6.9 and 7.0 A ten to twenty-fold decrease in transformation efficiency may occur with S.O.C. medium at pH 7.5.

3. It has been claimed that extraction of ligated DNA is necessary following direct ligation in agarose in order to avoid inhibition of transformation by some impurities present in the agarose. We have never encountered such problems when using the low melting point agarose from Sigma under the conditions described above.

4. Genotype of *E. coli* strains used in this study:

DHI: F⁻, *endA1*, *hsdR17* (r_k^-, m_k^+), *supE44*, thi*1*, recA*1*, gyrA96, relA*1*

DH5: F⁻, *endA1*, *hsdR17* (r_k^-, m_k^+), *supE44*, thi*1*, gyrA96, relA*1*, 080*d*lacZ M15

HB 101: F⁻, *pro, leu, lac4, ara14, galK2, xyl5, mtl1, endA1, hsdR, supE44,* thi*1,* recA*1,* rpsL20

MC 1061: F⁻, (*lac* IPOZYA)X74, *galK, galU, hsdR, araD139, strA* (use 25 μg/ml streptomycin), (*ara,leu*)7697

JM 107: see page 588

Solutions and Buffers.

5X Ligation buffer	*500* μ*l*
250 m*M* Tris–HCl (pH 7.5)	125 μl of 1 *M* Tris–HCl (pH 7.5)
50 m*M* MgCl₂	50 μl 0.5 *M* MgCl₂
5 m*M* ATP	125 μl 20 m*M* ATP
5 m*M* DTT	5 μl 0.1 *M* DTT
250 μg/ml BSA	25 μl of 5mg/ml BSA
	170 μl of H₂O
	Store at −20°C.

15

CHARACTERIZATION OF RECOMBINANT CLONES

Once bacterial transformants have been obtained, it is necessary to analyze their DNA content to make sure that they harbor the desired recombinant DNA molecules. This is usually achieved either by rapid purification of plasmid DNA followed by digestion with appropriate endonucleases (mini-screen) or by direct hybridization of bacterial DNA to specific probes (colony hybridization).

Alternatively, a more complete analysis can be performed by the Southern transfer technique, which involves transferring the electrophoretically separated DNA fragments onto nitrocellulose (or activated paper) before hybridizing to specific probes.

RAPID PURIFICATION OF RECOMBINANT PLASMID DNA

The following protocol is based on the method of Birnboim and Doly (1979).

PROTOCOL

MINISCREEN FOR RECOMBINANT PLASMIDS

1. Grow the bacterial clones carrying the recombinant plasmid to be tested in 5 ml of TYE medium containing the appropriate antibiotic at the same concentration as that used for the selection of transformants. The clones are usually grown at 37°C for 18 hours with agitation.

2. The next day, transfer 1.5 ml of the overnight cultures to microfuge tubes and pellet the cells by centrifugation for 30 seconds. Freeze the remaining cell culture after adding glycerol (to a 30% final concentration).

3. After centrifugation, discard the supernatant and resuspend the cell pellet in 100 μl lysozyme solution. To allow thorough resuspension without generating foam, vortex the cell pellet

for 5 seconds before adding the lysozyme solution. Incubate the resuspended cells for 15 minutes at 0°C.

4. Add 200 µl alkaline–SDS solution to the treated cells and vortex gently to mix. Maintain the tubes at 0°C for 5 minutes.

5. Add 150 µl high salt solution and invert several times to allow complete mixing. Keep the tubes at 0°C for at least 60 minutes.

6. Spin down the precipitate formed after addition of the high salt solution by centrifugation for 10 minutes at 4°C.

7. Carefully pipette the supernatant and transfer to fresh tubes.

8. Add 200 µl phenol equilibrated with NTE buffer and 200 µl of a 24:1 mixture of chloroform:isoamyl alcohol. Vortex for 15 seconds and spin in a microfuge for 3 minutes at room temperature.

9. Carefully aspirate the supernatant of each preparation and transfer to a new tube.

10. Add 400 µl chloroform:isoamyl alcohol (24:1) to each tube. Vortex for 10 seconds. Centrifuge for 3 minutes at room temperature.

11. Transfer the supernatant to new tubes and add 400 µl chloroform:isoamyl alcohol (24:1) to each tube. After vortexing for 10 seconds and spinning for 3 minutes, transfer the final supernatant to a fresh set of tubes.

12. Precipitate the nucleic acids by adding 1 ml cold absolute ethanol and incubating at −20°C for 2 hours or −70°C for 30 minutes.

13. Pellet the nucleic acid precipitate by centrifugation for 10 minutes at 4°C. Remove the supernatant of each preparation.

14. Dry the pellets under vacuum for 10 minutes.

15. Resuspend the nucleic acids in 60 µl sterile distilled water.

16. Use 5–10 µl of this suspension to run endonuclease digests in the presence of 1 µg RNase A.

Solutions and Buffers

Lysozyme solution
Prepared fresh each time.
5 mg/ml lysozyme in
lysozyme buffer

Lysozyme buffer	*100 ml*
25 mM Tris (pH 8.0)	2.5 ml 1 M Tris–HCl (pH 8.0)
50 mM Glucose	5 ml 1 M Glucose
10 mM EDTA	2 ml 0.5 M EDTA (pH 7.5)
	90.5 ml H$_2$O

Alkaline–SDS solution   *100 ml*
0.2 N NaOH   93 ml H_2O
1% SDS   2 ml 10 N NaOH
                                  5 ml 20% SDS

High salt solution
3 M Sodium acetate (pH 4.8)

Sodium acetate should be dissolved in the minimal amount of water, as much concentrated acetic acid is needed to bring the pH down to 4.8.

Note for Miniscreen. Two other methods have been used for rapid extraction and analysis of plasmid DNA. One involves cell lysis with phenol, centrifugation, phenol extraction, ethanol precipitation, and RNase digestion (Klein et al., 1980). In the other, plasmid DNA is recovered by isopropanol precipitation after the bacteria have been boiled for 15–40 seconds and the resulting clot centrifuged (Holmes and Quigley, 1981).

REFERENCES

Birnboim, H. C., and Doly, J. (1979), *Nucl. Acids Res.*, **7**, 1513.

Klein, R. D., Selsing, E., and Wells, R. (1980), *Plasmid*, **3**, 88.

Holmes, D. S., and Quigley, M. (1981), *Anal. Biochem.*, **114**, 193.

DIRECT ANALYSIS OF RECOMBINANT PLASMIDS BY VERTICAL ELECTROPHORESIS ON AGAROSE GELS

This protocol is particularly useful for isolating large plasmid DNAs (Trevors, 1985). It has also proved to be helpful for screening a large number of recombinant clones within a few hours. The SE 600 vertical slab gel unit commercialized by Hoeffer Scientific Instruments is very convenient because it allows several gels to be run at the same time at uniform low temperatures.

PROTOCOL

CHARACTERIZATION OF RECOMBINANT DNA BY VERTICAL ELECTROPHORESIS IN AGAROSE GELS

Preparation of the Agarose Gel

1. Assembly of the SE 600 unit is performed by using one frosted and one plain glass plate, 3.0 mm spacers, and a polyethylene transverse spacer.
2. Adjust a 3.0-mm comb, and warm the whole system at 50°C in an oven to facilitate gel pouring.
3. Prepare 100 ml of a 0.8% agarose gel in TBE buffer. Melt in a microwave oven and let equilibrate in a 50°C water bath.
4. Pipette the warm agarose solution between the glass plates of

the prewarmed SE 600 unit until the level reaches the top of the glass plates.

5. When the agarose is solidified, loosen the two top screws of each clamp and carefully remove the comb. Retighten the screws from the bottom ones, going up.

Sample Preparation and Electrophoresis

1. Centrifuge 1.0-ml samples of overnight bacterial cultures for 30 seconds at 12,000 rpm in a microfuge tube.
2. Aspirate the supernatant and vortex the wet pellet for a few seconds in order to resuspend the cells.
3. Add 50 μl of lysozyme solution per tube and mix thoroughly.
4. Incubate the cell suspension at 37°C for 45 minutes.
5. In each well of the agarose gel, add 30 μl of the SDS solution and let stand for 10 minutes before use.
6. With a micropipette, take 20 μl of the mixture from step 4 and layer on the surface of the gel, underneath the 30 μl of SDS previously placed in each well.
7. Carefully overlay with TBE buffer the SDS solution in the wells to fill them up.
8. Complete the assembly of the electrophoresis unit by installing the upper buffer reservoir.
9. Fill the upper tank with TBE buffer. Be careful not to disturb the content of the wells.
10. Set up the power and cooling units.
11. Run the gel for 90 minutes at 14 volts to allow gentle lysis of the spheroplasts.
12. Increase voltage to 150 volts and run for an additional 90-minute period of time.
13. When electrophoresis is performed, staining of the DNA fragments is achieved by incubation in TBE buffer containing 1 μg/ml of ethidium bromide.
14. Rinse the gel in TBE for 1 hour to take a picture.

Notes   Rock the comb back and forth when you remove it. Be careful not to move the comb side to side.

REFERENCE

Trevors, J. T. (1985), *Microbiol. Meths.*, **3**, 259.

SCREENING OF RECOMBINANT BACTERIA ON NITROCELLULOSE FILTERS

In the direct screening of bacterial colony content, the cells are grown and lyzed directly on the nitrocellulose filters. The DNA of each individual colony is then bound to the membrane following denaturation and the filter can be used for hybridization with labeled probes (see chapter 16). In the protocol described below, nitrocellulose filters are used. Other types of membranes (see below) have also been successfully employed.

PROTOCOL

BACTERIAL COLONY HYBRIDIZATION

1. *Preparation of nitrocellulose filters.* Place nitrocellulose filters between pieces of Whatman 3MM paper. Hold together with clamps. Sterilize at 120°C for 20 minutes. Let the filters dry for 10 minutes.

2. Place a filter on the surface of an agar plate with forceps. The filter should become wet uniformly.

3. Pick isolated colonies from original plate containing colonies to be tested, and inoculate both a new plate and the filter. This should be done in such a way that for each colony on the filter there is a corresponding colony on the new plate. The use of sterile toothpicks is recommended when a large number of colonies is to be picked.

4. Do not seed more than 50 colonies on a 10-cm diameter filter to avoid mixing in subsequent steps.

5. Incubate the plates overnight, upside-down, at 37°C.

6. *Lysis and denaturation.* Place the filters (on which colonies should be visible) with forceps (colonies uppermost) on a piece of Whatman 3MM paper previously soaked with 0.5 N NaOH. Let the filters sit for 8 minutes. The colonies should become moist and syrupy and have a shiny aspect. If this is not the case, let the filters stand in the denaturation solution longer.

7. *Neutralization.* Set the filters for 3 minutes on a piece of Whatman 3MM paper soaked with 1 M Tris–HCl (pH 7.5). Repeat once on a new piece of soaked paper for the same length of time.

8. Set the filters for 5 minutes on a piece of Whatman 3MM paper soaked with 0.5 M Tris–HCl (pH 7.5), 1.5 M NaCl.

9. Blot the filters on Whatman 3MM paper until dry.

10. Dip the filters five times in absolute ethanol (see note 4).

11. Let the filter dry on Whatman 3MM paper. The colonies should be on the upper side.

12. Dip the filters five times in chloroform.

13. Let the filters dry on Whatman 3MM paper.

14. Dip the filters in 0.3 M NaCl for 3 minutes.

15. Let the filters dry at room temperature.

16. Bake the filters at 80°C for 2 hours.

The filters are now ready to be prehybridized and hybridized.

Notes

1. If the filters are to be seeded on an antibiotic-containing plate, it is not necessary to sterilize the filters as long as you handle them carefully with forceps.

2. You should take care to avoid trapping air bubbles beneath the filters when setting them on the plates.

3. Bacteria can also be spread on the filter for direct selection. In this case, be sure to spread the 0.1–0.2 ml bacteria sample immediately.

4. Millipore HA 0.45 μm, pore filters and Schleicher & Schuell BA85 filters have been used interchangeably. However, when Schleicher & Schuell filters are used, care should be taken not to dip them in absolute alcohol, as these filters are not resistant to ethanol. The lipid extraction step with chloroform can be performed directly after the use of aqueous solution, without dehydration. Therefore, when using BA85 filters, skip from step 8 to step 12.

Buffers

Denaturation buffer	*100 ml*
0.5 N NaOH	5 ml 10 N NaOH
	95 ml H_2O
Neutralization buffer	*100 ml*
0.5 M Tris–HCl (pH 7.5)	50 ml 1 M Tris–HCl (pH 7.5)
1.5 M NaCl	30 ml 5 M NaCl
	20 ml H_2O

REFERENCES

Grunstein, M., and Hogness, D. S. (1975), *Proc. Natl. Acad. Sci. USA*, **72**, 3691.

CAPILLARY TRANSFER OF DNA FRAGMENTS TO NITROCELLULOSE MEMBRANES

Since single-stranded nucleic acids bind to nitrocellulose, denatured DNA fragments can be transferred to filters for subsequent hybridization. In the protocol described below, the buffer and sample are drawn out of the agarose gel by placing it in contact with blotting paper. A nitrocellulose membrane layered between the gel and the blotting paper binds the DNA (or RNA) strands as they flow out of the gel.

PROTOCOL

SOUTHERN DNA TRANSFER FROM AGAROSE GELS ONTO NITROCELLULOSE

Figure 113 illustrates the setup described below. Figure 114 shows a commercial apparatus for Southern transfer.

Materials Needed

Plastic tray (of such a size that the gel pieces will fit on the bottom)

Plastic spacers (at least as long as the gel and 5 mm in width and height)

Whatman 3MM paper

Nitrocellulose filters

Forceps

Oven (To bake the filters at 80°C. A vacuum oven is recommended but not absolutely necessary.)

1. After the gel is run, cut out the lane containing the molecular weight markers and stain for 30 minutes in 100 ml TAE buffer to which 10 μl ethidium bromide (10 mg/ml) has previously been added. Wash the gel lane for 30 minutes in distilled water. Visualize the DNA fragments by UV light and photograph the gel. It is wise to include a ruler by the side of the gel. The picture provides a means for direct comparison between migrations and the locations of hybridizing fragments.

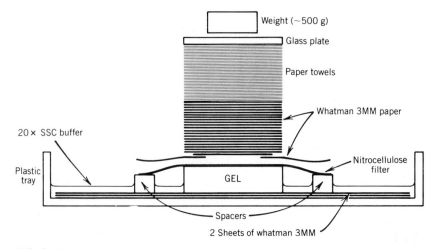

Figure 113. Southern transfer of DNA fragments. This drawing schematizes one of the several ways to set up a DNA fragment transfer onto nitrocellulose. Many other variations of the original description by Southern (1975, 1979) have been published.

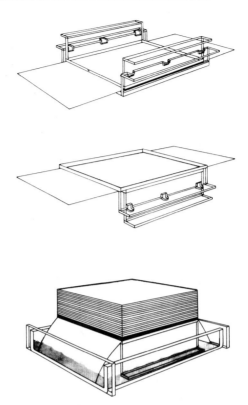

Figure 114. Commercial apparatus for DNA/RNA blot transfer. BRL has designed an apparatus system to facilitate the Southern filter blot transfer method for DNA fragments and the modifications developed by other laboratories for the transfer of RNA and proteins. This system is specifically designed to take advantage of the removable UV-transparent tray exclusive to all models of the BRL horizontal gel system. After electrophoresis of DNA fragments, for example, the gel is lifted out of the horizontal gel unit in the removable UV-transparent tray, which is then placed in the buffer tray for staining and subsequent in situ denaturation and neutralization of the DNA fragments. To assemble the blotting stack, the neutralized gel (still in the UV-transparent tray) is removed from the buffer tray. The precut wicking sheet is laid over the gel and the tray insert is placed over the wicking sheet. The entire assembly is inverted and the UV-transparent tray is lifted off to expose the gel, which is now lying on the wicking sheet, supported by the tray insert. A precut nitrocellulose filter is placed over the gel and a stack of sized blotting pads is placed over the nitrocellulose filter. The entire assembly is then placed in the buffer tray to conduct the blot transfer (courtesy of Bethesda Research Laboratories).

2. Cut out the lanes containing the digested DNA samples to be transferred (after staining with ethidium bromide and photography). We find it easier, when a large number of samples is run on the same gel, to cut out blocks of three to six lanes, depending on the relationship of the samples and on the probes to be used. We try to avoid manipulation of large pieces of gel. Blocks of three to six lanes of 0.8% agarose gel can be manipulated easily.

3. Place the gel pieces in a plastic tray containing 300 ml

buffer I and shake for 40 minutes at room temperature. Be careful, as too violent a shaking may break the gel.

4. Carefully pour off the buffer I and rinse the gel blocks twice with distilled water.

5. Add 300 ml buffer II to the gel blocks and shake at room temperature for 1 hour. This time can be increased to 2 hours, when necessary, without loss of material.

6. Prepare the nitrocellulose sheets during the buffer II wash. Never touch nitrocellulose with your bare hands, as fingerprints will lead to background problems. Nitrocellulose strips are cut out from large rolls. The strips should be the same length as and 3 cm wider than the gel pieces. Wet the nitrocellulose strips by placing them carefully *on the surface* of a 2X SSC solution contained in a plastic tray cover. When the nitrocellulose is wet, the sheet can then be dipped into the 2X SSC solution. Immersing the dry nitrocellulose directly could lead to an uneven wetting of the filter, which could in turn affect the DNA transfer in some areas of the nitrocellulose strip.

7. Cut strips of Whatman 3MM chromatography paper of the same length as the gel and 3 cm wider. Soak these strips with 2X SSC. Two strips are needed per gel.

8. When the gel blocks have been treated with buffer II for the indicated time, the transfer may be performed. Place a double layer of Whatman 3MM paper on the bottom of the tray. Soak with 20X SSC and put the gel blocks on the Whatman paper. Place the spacers 1 cm away from the sides of the gel (see Figure 113).

9. Remove the nitrocellulose sheets from the 2X SSC solution and carefully layer them on the gel blocks in such a way that the edges of the filters touch the spacers and that **no air bubbles** form between the gels and the nitrocellulose sheets.

10. Place the Whatman 3MM strips on the nitrocellulose filter in such a way that they cover the edges of the gels and the edges of the spacers.

11. Cover with five to ten dry sheets of Whatman 3MM paper and with paper towels cut to the desired dimensions. Cover with a glass plate and a weight (~500 g).

12. Add 20X SSC to the bottom of the tray in such a way that a meniscus is formed between the gel and the spacers. Be careful not to fill the space between the gel and spacers to avoid formation of "buffer bridges." The buffer should diffuse through the gel to obtain transport of the DNA fragments.

13. When the level of 20X SSC becomes low, add some more. Keep adding buffer for 5–6 hours. Change the paper towels when they become wet.

14. Leave overnight without adding buffer. The gel will dry and stick to the nitrocellulose membrane.

15. The next day, cut out the nitrocellulose filters to the dimensions of the gels and carefully remove the gels with forceps. The gels can be stained as described in step 1 to check the efficiency of transfer if the DNA fragments were not stained prior to transfer.

16. Label the top, left, and right of the nitrocellulose strips.

17. Place the nitrocellulose filters between two pieces of Whatman 3MM paper and incubate at 80°C in an oven for 2 hours. Baking the filter prevents the denatured DNA from being displaced from the nitrocellulose when prehybridization and hybridization are performed.

Buffers

TAE buffer (10X concentrated)
For 2 liters:
96.88 g Tris–base
13.60 g Sodium acetate ($3H_2O$)
7.44 g EDTA
Dissolve in 1500 ml H_2O
Adjust pH to 7.6 with concentrated
   HCl (about 40 ml).
Adjust volume to 2 liters with H_2O.

Buffer I — *1 liter*
1.5 M NaCl — 300 ml 5 M NaCl
0.5 M NaOH — 50 ml 10 N NaOH
   — 650 ml H_2O

Buffer II — *1 liter*
0.5 M Tris–HCl (pH 7.4) — 500 ml 1 M Tris–HCl (pH 7.4)
3.0 M NaCl — 174 g NaCl
   — Bring to 1 liter with H_2O.

20X SSC

		1 liter	*5 liters*
3 M NaCl	5 M NaCl	600 ml	3000 ml
0.3 M Sodium citrate (pH 7.0)	1 M Sodium citrate	300 ml	1500 ml
	H_2O	100 ml	500 ml

2X SSC
Dilute 20X SSC 10-fold with H_2O.

Notes

1. The method presented here is a variation of the original technique described by Southern (1975, 1979). It gave the best results in our hands and allowed the use of rather thick gels.

2. Fragmentation of the DNA to be transferred will help to obtain comparable transfer efficiencies for the different species considered. Treatment of the gel with 0.25 M HCl will result in depurination and base cleavage in the DNA molecules. The gel is usually soaked for 10–30 minutes in the HCl solution prior to treatment for blotting (denaturation and renaturation as described in steps 3 and 5 above). It has been reported that 10 minutes of moderate shaking in 0.25 M HCl allowed a subsequent complete transfer of DNA fragments (0.5–50 kb in size) within 2 hours from a 9-mm-thick 0.3–0.7% gel. Another way of generating partial cleavage of DNA consists of irradiating the ethidium bromide-stained DNA with short-wave UV light. An energy of 100 μW/cm^2 for 10 minutes is recommended.

3. Different types of nitrocellulose have been reported to have different DNA binding properties. For instance, 0.22-μm porosity was found to be better than 0.45-μm porosity for binding of small fragments (Laub et al., 1979). Also, different transfer patterns have been obtained with nitrocellulose batches from Millipore and from Schleicher & Schuell (Hartmann, personal communication). We have never encountered such a problem in our experiments.

4. The salt concentration used to transfer DNA seems to be critical, as different retentions of fragments were obtained at concentrations lower than 10X SSC (Nagamine et al., 1980).

5. A phosphate-containing buffer (SSPE) can also be used instead of SSC if necessary. 1X SSPE is 0.18 M NaCl, 10 mM sodium phosphate (pH 7.7), 1 mM EDTA.

6. The smallest number of base pairs of DNA that can be transferred to nitrocellulose from a gel by the Southern procedure was reported to be about 200. It has been shown since, that fragments of 72 bp could be transferred to Schleicher & Schuell BA85 filter.

REFERENCES

Laub, O., Bratosin, S., Horowitz, M., and Aloni, Y. (1979), *Virology*, **92**, 310.

Nagamine, Y. A., Sentenac, A., and Fromageot, P. (1980), *Nucl. Acids Res.*, **8**, 2453.

Southern, E. (1975), *J. Mol. Biol.*, **98**, 503.

Southern, E. (1979), in *Methods in Enzymology*, R. Wu, ed., Vol. 68, p. 152, Academic Press, New York.

PROTOCOL

BIDIRECTIONAL TRANSFER OF DNA FRAGMENTS ON NITROCELLULOSE

1. After running and photographing the gel, cut out the lanes to be transferred.

2. Place the gel block in a plastic box containing two volumes of 0.25 M HCl. Shake moderately for 15 minutes to induce chemical cleavage of the DNA.

3. Pour off the solution, add two volumes of fresh 0.25 M HCl and shake for an additional 10 minutes at room temperature.

4. Rinse the gel twice with distilled water.

5. Denature the DNA fragments by incubation in two volumes of buffer I for 15 minutes with moderate shaking at room temperature.

6. Pour off the denaturing buffer. Add two volumes of fresh buffer I and repeat step 5.

7. Pour off the denaturing buffer and add two volumes of neutralization buffer (buffer II). Shake for 30 minutes at room temperature.

8. Pour off the buffer. Add two volumes of fresh buffer II and shake for 30 minutes at room temperature.

9. For a bidirectional transfer, two nitrocellulose sheets are needed and should be soaked with the neutralization buffer.

10. The setup is drawn schematically in Figure 115. From the bottom to the top are placed: a flat smooth surface such as a plastic tray, about a 5-cm thickness of dry paper towels, four sheets of Whatman 3MM paper saturated with the neutralization buffer and cut to the dimensions of the gel, a nitrocellulose filter soaked with buffer, the piece of gel treated as described in steps 1–8, a nitrocellulose filter soaked with buffer, four sheets of Whatman 3MM paper saturated with the neutralization buffer and cut to the dimensions of the gel, about a 5-cm thickness of dry paper towels, and a weight, which need not be more than 400 g for a thick gel. Too much weight will smash the gel.

11. Let the transfer take place for 1–2 hours at room temperature.

12. Wash the filter two times in 4X SSC.

13. Let the filter dry and bake it at 80°C for 2 hours as described in Protocol on Bacterial Colony Hybridization.

Solutions and Buffers

Acid solution	*1 liter*
0.25 M HCl	20 ml concentrated HCl (12.2 N)
	980 ml H_2O
Buffer I	*1 liter*
1.5 M NaCl	300 ml 5 M NaCl
0.5 M NaOH	50 ml 10 N NaOH
	650 ml H_2O

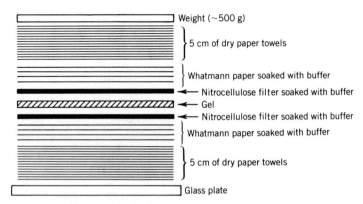

Figure 115. Setup for bidirectional transfer of DNA fragments to nitrocellulose.

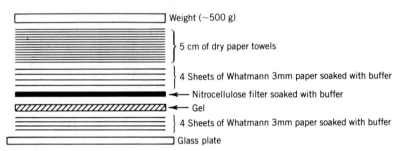

Figure 116. Setup for unidirectional transfer of DNA fragments to nitrocellulose.

Buffer II	*1 liter*
1 M Ammonium acetate	77.1 g Ammonium acetate
0.02 M NaOH	2 ml 10 N NaOH
	H_2O to 1 liter

Notes

1. The ammonium acetate buffer (II) should be made fresh each time.

2. Avoid putting too much weight on for the transfer. We usually use a glass plate on which we put a glass bottle containing about 300 ml water.

3. Chemical breakage of the large DNA molecules with the HCl solution is necessary because of the short time of transfer used.

4. This method can be used for unidirectional as well as bidirectional transfer. When unidirectional transfer is performed, the setup in Figure 116 is used. Note that no dry paper towels are put at the bottom.

REFERENCE

Smith, G. E., and Summers, M. D. (1980), *Anal. Biochem.*, **109**, 123.

ELECTROPHORETIC DNA TRANSFER TO NITROCELLULOSE MEMBRANES

The blotting methods described above are very inefficient for transferring DNA fragments from acry-

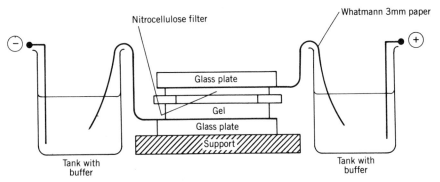

Figure 117. Setup for electrophoretic DNA transfer to nitrocellulose filters.

lamide and acrylamide–agarose composite gels. Although electrophoretic transfer performed at low salt concentration may affect the retention of some fragments, it seems to be the most efficient method of transferring DNA fragments from acrylamide gels.

Figure 117 illustrates a setup for electrophoretic transfer of DNA fragments to nitrocellulose. (See also commercial devices in Chapter 1.)

REFERENCES

Bittner, M., Kupferer, P., and Morris, C. F. (1980), *Anal. Biochem.*, **102**, 459.

Kutateladze, T. V., Axelrod, V. D., Gorbulev, V. G., Belzhelarskaya, S. N., and Vartikyan, R. M. (1979), *Anal. Biochem.*, **100**, 129.

Vaessen, R. T. M. J., Kreike, J., and Groot, G. S. P. (1981), *FEBS Letters,* **124,** 193.

PREPARATION AND ACTIVATION OF CHEMICALLY MODIFIED PAPERS FOR NUCLEIC ACID TRANSFER

Figure 118 schematizes the chemistry of the following protocol.

PROTOCOL

PREPARATION OF DIAZOBENZYLOXYMETHYL (DBM) PAPER

1. Cut out a piece of Whatman 540 paper (10 × 10 cm) and place it at the bottom of a stainless steel, enamel, or glass pan.

2. Prepare the NBPC (1-[(*m*-nitrobenzyloxy)methyl] pyridinium chloride) solution. For 100 cm² of paper prepare a solution of 230 mg NBPC and 70 mg sodium acetate trihydrate in 2.85 ml water.

3. Pour the freshly prepared solution over the piece of paper as evenly as you can, and using rubber gloves, push out all air bubbles. Rub the paper with your gloved hand to squeeze out excess liquid.

4. Dry the paper at 60°C in an oven for 10 minutes.

5. Set the oven to 135°C.

6. Incubate the treated paper for 30–40 minutes at 135°C.

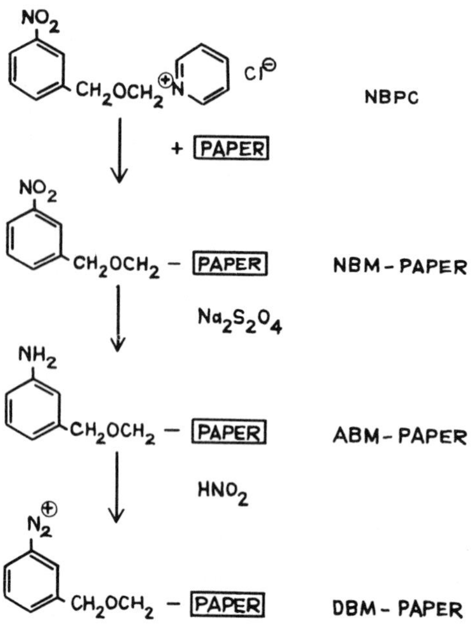

Figure 118. Steps in the preparation of DBM paper. Reaction of nitrobenzyloxymethylpyridinium (NBPC) with the paper leads to the formation of nitrobenzyloxymethyl (NBM) paper, which in turn is reduced in the presence of sodium bisulfite to aminobenzyloxymethyl (ABM) paper. Reaction with nitrous acid converts the amine into a diazo group, forming DBM paper.

7. Wash the filter several times with H_2O over 20 minutes. Dry at 60°C.

8. Wash the filter with acetone twice, for 10 minutes each time. Let dry at room temperature. The paper obtained at this step is nitrobenzyloxymethyl (NBM) paper. You can keep it for many months at 4°C.

9. Incubate the NBM paper for 30 minutes at 60°C in the presence of 0.4 ml/cm² of 20% w/v sodium dithionite in H_2O, with occasional shaking. This step leads to the formation of aminobenzyloxymethyl (ABM) paper, which can be activated to react with nucleic acids.

10. Wash the ABM paper several times with large amounts of H_2O until no odor of H_2S remains.

11. Wash the paper once with 50 ml 1.2 *M* HCl (for a 10 × 10 cm piece).

12. Prepare 300 ml ice-cold 1.2 *N* HCl.

13. Prepare a fresh solution of 1% (10 mg/ml) sodium nitrite ($NaNO_2$) in water.

14. Mix 8.1 ml of the nitrite solution with 300 ml cold HCl (1.2 *N*).

15. Pour over the washed ABM paper, which is then kept on ice.
16. Leave the paper to react for 30 minutes. Swirl occasionally.
17. Rinse rapidly twice with cold sterile distilled water and once with cold transfer buffer.
18. Use at once for transfer.

Notes

1. Whatman 540 paper was chosen by Alwine et al. (1977) because of its excellent mechanical strength and resistance to chemicals. They also stated that Schleicher & Schuell 589 WH paper can be used with good results.
2. The ABM paper can be stored easily at 4°C under vacuum for a year.
3. It is possible to check that activation of ABM to DBM paper (step 16) is taking place by putting a drop of the solution on starch–iodide paper. A dark spot due to the presence of nitrous acid in the solution should be obtained.
4. As diazonium groups are not stable, it is important to use the activated paper right away (see protocol on page 435 for timing of transfer).
5. NPBC is available commercially (from British Drug Houses or Pierce Chemical Company) or can be synthesized according to published methods.
6. NBM and ABM papers are commercially available (Schleicher & Schuell Transa-Bind).

PROTOCOL

PREPARATION OF DIAZOPHENYLTHIOETHER (DPT) PAPER

1. Cut pieces of Whatman no. 50 paper to desired sizes.
2. Weight out 20 g of these pieces and place them in a sealable freezer bag. Be sure to use a good quality bag, with no holes or cracks, as the chemicals used in this procedure are harmful. Double-seal your bags. Carry out all subsequent steps under a fume hood. Wear gloves.
3. Add to the bag 70 ml 0.5 N NaOH containing 2 mg/ml sodium borohydride.
4. Add 30 ml 1,4-butanediol diglycidyl ether to the NaOH so-

lution in the bag and seal, with enough room left at the top of the bag to allow further opening and resealing.

5. Agitate the bag for 8–16 hours at room temperature. It is very important that the bag be rotated end to end. Horizontal agitation is not adequate.

6. Open the bag and carefully pour the diglycidyl ether solution into a flask or beaker containing 1 M NH_4OH. Leave to react for 24 hours in a fume hood before washing down a drain. Add 10 ml 2-aminothiophenol in 40 ml acetone to the paper.

7. Reseal the bag and agitate as previously for 10 hours.

8. Open the bag and collect the 2-aminothiophenol for organic waste. Remove the paper with forceps.

9. Wash twice with 500 ml acetone. Stir the filters with forceps during each wash.

10. Wash two times with 500 ml 0.1 N HCl.

11. Wash four times with 500 ml sterile distilled water.

12. Repeat steps 10 and 11 and air-dry the filters.

The paper obtained at this point is aminophenylthioether (ATP) paper, which is stable for several months at room temperature. APT paper has been reported to be more stable in storage than ABM paper (see above).

13. The diazotation of ATP paper to DPT paper is conducted as for ABM paper (see steps 14–17 of the DBM paper protocol).

Notes

1. 1,4-Butanediol diglycidyl ether and 2-aminothiophenol are commercially available (Aldrich 12,419-2 and 12,313-7).

2. 1,4-Butanediol diglycidyl ether is an irritant and may also be carcinogenic. Handle it with care.

3. The APT paper is stable for several months at room temperature. It should be kept in a lightproof container. We used to keep ATP paper in a sealed bag at $-20°C$ under nitrogen.

4. Upon diazotation, APT paper will turn bright yellow.

5. The DPT paper is reasonably stable in the diazo form, as long as the pH is kept low.

6. APT paper is commercially available (Schleicher & Schuell Transa-Bind).

PROTOCOL

TRANSFER OF DNA TO DBM OR DPT PAPER FROM AGAROSE GELS

Figure 119 illustrates the chemistry of this protocol.

1. Incubate the gel pieces containing DNA to be transferred for 30 minutes in denaturing buffer (buffer I) as in the Southern method and after fragmentation.

2. If the DNA species are to be transferred to DBM paper, soak the gel in 300 ml 1 M sodium acetate buffer (pH 4.0) for 30 minutes.

3. Pour off the buffer and replace with 300 ml fresh buffer. Incubate the gel for 30 additional minutes with moderate shaking.

4. If the transfer is performed on DPT paper, soak the gel once in 1 M sodium acetate buffer (pH 4.0) for 30 minutes and once in 20 mM sodium acetate buffer (pH 4.0) for 30 minutes.

5. Transfer is done at 4°C.

6. Place across a plastic tray containing the transfer buffer a glass plate larger than the gel.

7. Cut three pieces of Whatman 3MM paper large enough that when they are placed on the glass plate the two ends will dip into the transfer buffer. Place the paper on the plate.

8. Soak the Whatman paper with transfer buffer and carefully slide the gel onto the surface of the wet paper.

9. Place two plastic spacers on the sides of the gel, parallel to the length of the Whatman paper.

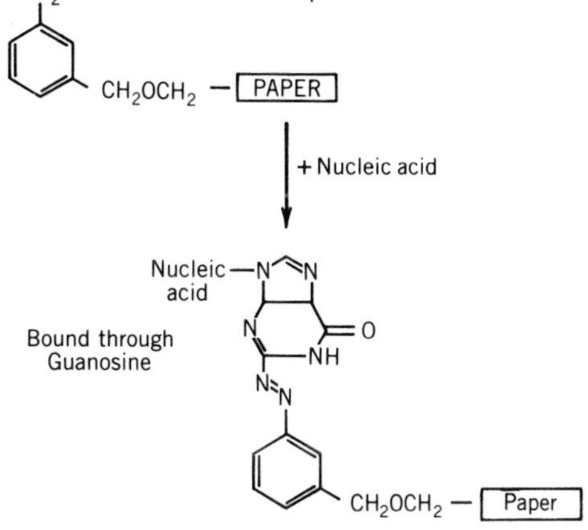

Figure 119. Covalent coupling of nucleic acids to DBM paper.

10. Place activated paper (DBM or DPT) saturated with the blotting buffer on top of the gel.

11. Cover with a few layers of Whatman 3MM paper and with paper towels (about 10 cm thick). Put a weight on top.

12. Leave overnight (14 hours) at 4°C.

REFERENCES

Alwine, J. C., Kemp, D. J., Parker, B. A., Reiser, J., Renart, J., Stark, G. R., and Wahl, G. M. (1979), in *Methods in Enzymology,* R. Wu, ed., Vol. 68, p. 220, Academic Press, New York.

Alwine, J. C., Kemp, D. J., and Stark, G. R. (1977), *Proc. Natl. Acad. Sci. USA,* **74,** 5350.

Goldberg, M. L., Lifton, R. P., Stark, G. R., and Williams, J. G. (1979), in *Methods in Enzymology,* R. Wu, ed., Vol. 68, p. 206, Academic Press, New York.

Wahl, G. M., Stern, M., and Stark, G. R. (1979), *Proc. Natl. Acad. Sci. USA,* **76,** 3683.

DNA TRANSFER ON MODIFIED NYLON MEMBRANES

The use of modified nylon-66 membranes is particularly recommended when several rounds of hybridization–dehybridization are being performed. These membranes are mechanically resistant, and therefore provide a greater ease of handling. They can be used to transfer DNA fragments with sizes ranging from 75 to 23,000 base pairs from 1-mm thick agarose gels, and allow hybridization with radioactive and biotinylated probes. The Biodyne (Pall Process Filtration Limited, Portsmouth, England), Nytran (Schleicher and Schuëll GmbH, Dassel, Germany), GeneScreen *Plus* (New England Nuclear, Boston, USA), and Hybond (Amersham International, Buckingamshire, England) membranes have been found to give satisfactory results.

Although this kind of membrane need not be moistened before use (they are inherently hydrophilic) we do recommend performing this step to facilitate their manipulation. This is particularly true for the GeneScreen *Plus* membrane which is naturally curled when dry.

Be aware that both sides of the membrane are not equivalent; therefore, it is important to differentiate them. When looking at the dry curled membrane, the convex side is usually referred to as side A, while the concave side is referred to as side B.

PROTOCOL

DNA TRANSFER ON MODIFIED NYLON MEMBRANES

Capillary Transfer

1. Proceed as described in protocol on Southern transfer (steps 1–8) for the preparation of gel blocks prior to transfer.

2. Wet the membrane in distilled water.

3. Lay the wet membrane onto a 10X SSC solution and let it soak for approximately 15 minutes.

4. Carefully place the wet membrane on the gel, so that side B is in contact with the gel. Make sure that no air bubbles are trapped between the membrane and the gel surface.

5. Cover with five to ten dry sheets of Whatman 3MM paper and with paper towels cut to the desired dimensions. Cover with a glass plate and a weight (500 g).

6. Proceed as in Southern transfer (steps 12–16).

7. Let the membrane dry at room temperature. There is no need

to fix DNA by baking when using the nylon-derived membranes.

Electrotransfer

1. Place the gel piece in a tray containing 300 ml of 0.4 *N* NaOH and shake for 30 minutes at room temperature.
2. Carefully pour off the solution and quickly rinse the gel with distilled water.
3. Add 300 ml of E-buffer to the gel block and shake at room temperature for 30 minutes to neutralize the gel.
4. Cut the membrane to the dimensions of the gel piece (mark side B of GeneScreen *Plus*) and soak it in E-buffer for 15 minutes.
5. Cut two Scotch-Brite (3M Company) pads and four pieces of 3MM Whatman paper to the dimensions of the gel and wet them with E-buffer.
6. Place a wet Scotch-Brite pad on a glass plate.
7. Lay two pieces of wet Whatman paper on top of the Scotch-Brite pad.
8. Place the gel on the Whatman paper.
9. Carefully lay the wet membrane on the surface of the gel (side B of GeneScreen *Plus* against the gel).
10. Place the two other sheets of wet Whatman paper on top of the membrane.
11. Put the other pad of Scotch-Brite on the top, and insert the complete "gel sandwich" in the electroblot unit.
12. Transfer is performed at 4°C in E-buffer generally for 1 hour at 1 V/cm, followed by 2 hours at 4 V/cm. Small fragments transfer more effectively at lower voltage.
13. When transfer is completed, turn off the power, remove the gel sandwich, and let the membrane dry at room temperature.

Notes

1. Do not use sandblasted plates, which will cause the agarose sticking to the membrane, leading to background problems.
2. Although suppliers claim that salt is not required for binding of DNA to their membranes, it appears that efficient transfer of DNA fragments greater than 7 kb is only obtained in the presence of 10X SSC. For transfer of small DNA fragments (< 7 kb) one can use 25 m*M* sodium phosphate buffer, pH 6.5 (Dandoy, personal communication).
3. The NEN protocol calls for a treatment of the membrane with 0.4 *N* NaOH for 30–60 seconds in order to ensure complete

denaturation of immobilized DNA, followed by an immersion in 0.2 *M* Tris–HCl pH 7.5, 2X SSC. These steps can be skipped and replaced by one rinse with 2X SSC (Dandoy, personal communication).

Solutions and Buffers

E-buffer
12 m*M* Tris
6 m*M* Sodium acetate
0.3 m*M* EDTA pH 7.5

DOT BLOTTING

In the dot blotting procedure, the transfer of nucleic acids (denatured DNA or RNA) is performed by spotting small volumes of solutions directly onto nitrocellulose or activated paper. This technique enables rapid screening for specific sequences. However, you must keep in mind that a positive test may sometimes be the result of the nonspecific reaction that often occur with degraded DNA or RNA preparations. Therefore, it is a good policy to always check the intactness of the preparations by gel and regular transfer.

Different apparatus have been commercialized to allow dot blotting of large amounts (up to 100 μl) of diluted RNA (or single-stranded DNA) solutions (0.1 μg/μl). Minifold apparatus from Schleicher and Schuell appears to give the best results (Nahon, personal communication). Samples in 0.2 *M* sodium acetate (pH 4.0) or 10 m*M* sodium phosphate (pH

6.5) are denatured for 5 minutes at 100°C and rapidly cooled on ice before spotting. After overnight incubation at 4°C filters are hybridized with specific labeled probes (Nahon, personal communication).

When only a few samples are being tested, it is possible to perform dot blotting without using a special device. In this case, it is recommended to use tips or glass micropipettes delivering 5–25 μl volumes. Spot small aliquots at a time and wait until dry between each loading.

It is also possible to spot double-stranded DNA solution directly on nitrocellulose. In this case, once the samples are applied, layer the dry nitrocellulose membrane on a Whatman paper soaked with Southern buffer I for 2–3 minutes. Transfer the wet membrane to another Whatman paper soaked with Southern buffer II for 2–3 minutes. Dry and bake before using it for prehybridization and hybridization.

PROTOCOL

DOT BLOTTING ON NITROCELLULOSE

1. Wet the nitrocellulose sheets or filters with H_2O and 20X SSC.
2. Dry the nitrocellulose by blotting on Whatman 3MM paper.
3. Spot small volumes of denatured nucleic acid (1–5 μl) on the dry nitrocellulose.
4. Dry under a heat lamp and bake for 2 hours at 80°C (under vacuum) before prehybridizing.

PROTOCOL

DOT BLOTTING ON DBM OR DPT PAPER

1. Cut small pieces of ABM or APT paper (~1 cm²).
2. Activate them as described for DNA transfer on DBM or DPT paper (see above protocols.) Dry them by blotting on Whatman paper.
3. Place the activated pieces of paper on a sheet of Parafilm.
4. Spot the sample in a maximum volume of 25 µl. Usually poly-A⁺ RNA is used at a concentration of 1 µg/µl in either 10 mM sodium phosphate buffer (pH 4.0) or 10 mM acetate buffer (pH 6.5).
5. Let sit for 10 minutes at room temperature.
6. Wash the filters in batches. For each filter use 5 ml of 0.2 M sodium acetate (pH 4.0) or 0.2 M sodium phosphate (pH 6.5) buffer. Wash for 5 minutes at room temperature.
7. Repeat the washing for 5 minutes at room temperature.
8. Prehybridize and hybridize under standard conditions.

PROTOCOL

DOT BLOTTING ON NYLON MEMBRANES

1. Spot the samples onto the membrane in a volume of approximately 1 to 5 µl. Multiple pipetings can be performed to apply the required amount of material.
2. Let the samples dry at room temperature.
3. Bake the membrane at 80°C for 1 hour before prehybridizing.

SLOT BLOTTING

Because the dot blot procedure leads to round spots which cannot be precisely quantified by densitometry, filtration manifold with slot-shaped wells has been designed. In these wells, the samples are concentrated in a small surface area and give rise to uniform deposition on the membrane. Scanning densitometry of the resulting bands allows accurate quantitation of the labeled or stained samples (Figure 120). Therefore, the slot blot technique has been found to be five times more sensitive than the dot blot technique. The BRL Hybri-Slot™ and the Bio-

rad Bio-Dot are 24- and 48-well filtration units, while the Schleicher & Schuell Minifold II offers a slot configuration of three rows of 24 slots, therefore providing 72 slots in all. Both of them have given satisfactory results.

ANALYSIS OF RECOMBINANT DNA BY HYBRIDIZATION WITH SPECIFIC PROBES

The detection of specific sequences in DNA fragments separated by electrophoresis and immobilized on membranes is made possible by hybridization under the conditions described below. This protocol can also be used for hybridization of the filters obtained in screening of recombinant bacteria (see above) and for RNA characterization (see chapter 18)

PROTOCOL

HYBRIDIZATION OF NITROCELLULOSE FILTERS

Figures 121 and 122 show typical autoradiography results from Southern blots.

1. Place filters in plastic bags (such as Seal-a-Meal) and add 5–40 ml of the prehybridization solution (depending on the number of filters) before sealing. Be careful not to leave any air bubbles in the bag.

2. Incubate at 42°C in a water bath for at least 3 hours.

3. Clip off a corner and empty the bag. Add the hybridization

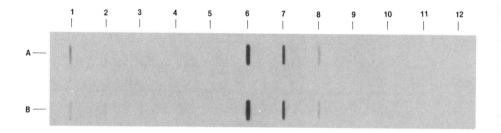

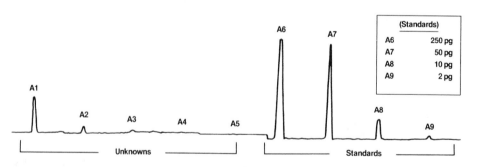

Figure 120. (a) Quantitation of viral DNA by hybridization. Serial fourfold dilutions of a virus stock were denatured with NaOH, neutralized, and filtered in duplicate in slots 1–5. Slots 6–9 contain 250 pg, 50 pg, 10 pg, and 2 pg respectively, of cloned, purified viral DNA. Slots 10–12 contain negative controls. After filtration onto nitrocellulose, the filter was baked at 80°C, then blocked with 10× Denhardt's solution. The hybridization probe was a cloned viral DNA labeled with $\alpha[^{35}S]dATP$ at a specific activity of 1.3×10^8 dpm/μg. Hybridization was carried out overnight at 42°C. After washing, the filter was dried, immersed in 20% PPO in toluene, dried, and fluorographed at −70°C overnight. (b) Densitometer measurement of autoradiogram.

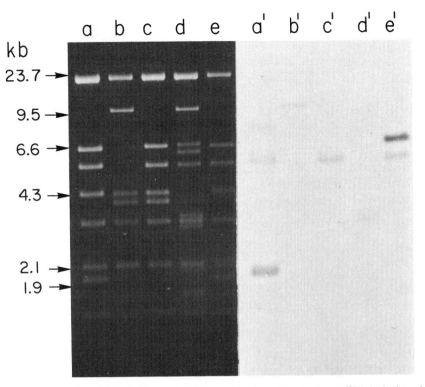

Figure 121. Autoradiography of a λ DNA Southern blot hybridized with a ^{32}P-labeled probe. 0.5 μg of various λ recombinant DNAs were digested with an endonuclease and the stained gel was photographed (*left*). After transfer and hybridization with a ^{32}P-nick-translated probe, the blot was exposed for autoradiography. The picture shown (*right*) is from a 2-hour exposure at room temperature. Markers are λ DNA Hind III fragments.

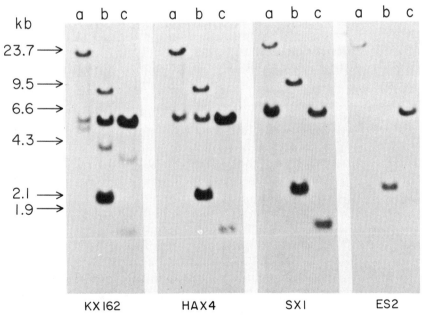

KX162   HAX4   SXI   ES2

Figure 122. Autoradiography of a total-DNA Southern blot hybridized with ^{32}P-labeled probe. 20 μg chicken DNA were digested with Bam HI (a), Eco RI (b), or Hind III (c) and electrophoresed on a 0.8% agarose gel before Southern blotting on Millipore nitrocellulose. The different digests were run on the same gel and blotted in four separate sets. Different nick-translated probes (KX162, HAX4, SX1, and ES2) were used for hybridization. Exposure was for 2 days at −70°C with an intensifying screen. λ DNA digested with Hind III was run in parallel and used for size markers.

solution (5–10 ml) containing the radioactive probe. Seal the bag after eliminating any air bubbles.

4. Incubate at 42°C in a water bath for 15 hours.

5. Clip off a corner to empty the radioactive solution into a special waste container. Cut the plastic bag and remove the filters with forceps.

6. Put the filters in 1 liter washing buffer. Agitate for 10 minutes at room temperature.

7. Pour off the washing buffer; replace with 1 liter fresh solution. Agitate for 10 minutes at room temperature.

8. Repeat step 7.

9. Take the filters out with forceps and place them in 1 liter heated (68°C) washing buffer. Agitate occasionally for 20 minutes at 68°C.

10. Change the washing buffer. Repeat wash at 68°C for 20 minutes with 1 liter preheated buffer.

11. Take the filters out with forceps; put them in 1 liter 2X SSC. Mix. Empty out the solution and rinse with 1 liter fresh 2X SSC.

12. Take the filters out. Blot them on Whatman 3MM paper until dry.

13. Tape the filters to a piece of Whatman 3MM paper and set up for autoradiography with (or without) intensifying screen.

Solutions and Buffers

Prehybridization solution
50% Formamide (deionized)
5X SSC
5X Denhardt's solution
50 mM Sodium phosphate (pH 6.5)
250 µg/ml Single stranded DNA

10 ml
5.0 ml 100% Formamide
2.5 ml 20X SSC
0.5 ml 100X Denhardt's solution
0.5 ml 1 M sodium phosphate buffer (pH 6.5)
0.8 ml DNA (3.3 mg/ml)
0.7 ml H$_2$O

Hybridization solution
50% Formamide (deionized)
5X SSC
1X Denhardt's solution
20 mM Sodium phosphate (pH 6.5)
10% Dextran sulfate
100 µg/ml Single stranded DNA

10.5 ml
5.0 ml 100% Formamide
2.5 ml 20X SSC
0.1 ml 100X Denhardt's solution
0.2 ml 1 M Sodium phosphate buffer (pH 6.5)
2.0 ml 50% Dextran sulfate
0.3 ml DNA (3.3 mg/ml)
Labeled probe in minimum volume

Washing buffer
For 2 liters:
200 ml 20X SSC
20 ml 20% SDS
1780 ml H_2O

Mix first 20X SSC with H_2O and complete with SDS. If you mix 20X SSC and SDS, a heavy white precipitate will form.

20X SSC	*1 liter*
3 M NaCl	600 ml 5 M NaCl
0.3 M Sodium citrate (pH 7.0)	300 ml 1 M Sodium citrate
	100 ml H_2O

100X Denhardt's solution	*100 ml*
2% Bovine serum albumin (Pentex)	2 g BSA
2% Ficoll	2 g Ficoll
2% Polyvinylpyrrolidone	2 g Polyvinylpyrrolidone
	H_2O to 100 ml

This solution should be kept at 4°C and mixed before use.

50% *Dextran Sulfate*

This is a 50% (w/v) solution. Fifty g dry Dextran sulfate occupies a volume very close to 100 cm³. Therefore, water should be added carefully when preparing the solution. To make, proceed as follows: Put 100 ml distilled water in a 125-ml glass bottle containing a magnetic bar and mark the level. Empty the bottle. With a funnel, pour 50 g Dextran sulfate (Pharmacia) into the empty bottle (still containing the magnetic bar). Add a small amount of distilled water to the powder, which will slowly dissolve. Be careful that the final volume does not exceed the 100-ml mark. Mix the suspension with a magnetic stirrer until it becomes a clear solution. Then adjust the level to 100 ml. Store the resulting solution at 4°C. It is recommended that the solution be taken out of the refrigerator and mixed for some time before use.

Single Stranded DNA

The nature of the heterologous DNA to be used depends on the nature of the blotted DNA. We currently use salmon sperm DNA.

Extract the heterologous DNA once with a 1:1 mixture of phenol:chloroform/isoamyl alcohol (24:1) and with chloroform/isoamyl alcohol until no interface is visible (see DNA purification protocol). It is important to perform the extractions in the presence of SDS to help remove the proteins present in most commercial DNA preparations.

Sonicate the purified DNA to reduce its average size. This is usually achieved by using the microtip of a Bronson sonifier cell disrupter W350 (setting 7, continuous mode) for several 30-second cycles until the DNA solution is no longer viscous. Alternatively, the long DNA strands can be sheared randomly by forcing with a syringe through a 25 gauge ⅝ needle. The size of the DNA fragments is usually checked by electrophoresis on agarose gels.

Preparation of Deionized Formamide

Melt the recrystallized formamide and add 10–20% w/v mixed bed resin AG 501-X8 (D), 20–50 mesh, fully regenerated (Bio-Rad). Mix for 1 hour at room temperature. Filter through a Whatman 3MM paper on a Büchner funnel and store in 50-ml fractions at $-20°C$.

Notes

1. Phosphate buffer can be replaced by Tris–HCl buffer. For instance, Dawid's hybridization buffer is 0.6 M NaCl, 20 mM EDTA, 0.02 M Tris–HCl (pH 7.5).

2. The use of formamide has been reported to help reduce background. Some loss of DNA from the nitrocellulose filters may occur in the presence of formamide. Retardation of hybridization rate has also been reported.

3. Use of Denhardt's solution also helps to reduce background because it prevents sticking of single-stranded DNA to nitrocellulose. However, at concentrations greater than 5X, hybridization may be inhibited.

4. The rate of hybridization is increased in the presence of 10% Dextran sulfate when 50% formamide is used.

5. It is necessary to heat-denature DNA probes to obtain single stranded labeled DNA. Usually, the probes are denatured by incubation in boiling water for 10 minutes, followed by rapid cooling in ice-cold water.

REFERENCE

Wahl, G. M., Stern, M., and Stark, G. R. (1979), *Proc. Natl. Acad. Sci. USA*, **76**, 3683.

REHYBRIDATION OF FILTERS

It is often necessary to use the same blot with different probes, especially when the material analyzed is scarce (for example, DNA samples from human tumors, or embryonic source). In these cases we advise the use of nylon membranes, or activated papers, rather than nitrocellulose filters. We have had good results after rehybridization of nitrocellulose blots,

but they tend to tear off upon repeated cycles of drying and washing. In all cases:

1. Dehybridize the blot before reusing it.
2. Use purified inserts as probes (containing no vector sequences).
3. Use 1% SDS in the hybridization mix when using nylon membranes in order to avoid high backgrounds.
4. *Do not let the nylon membranes dry* at any step between the first and the subsequent hybridizations since irreversible binding of the probe may occur (you can place the wet membrane in a sealed plastic bag to perform autoradiography).

PROTOCOL

DEHYBRIDIZATION OF NYLON MEMBRANES, PAPER FILTERS AND NITROCELLULOSE MEMBRANES

1. Incubate the membranes at 42°C for 30 minutes with gentle agitation in 200 ml of 0.4 N NaOH.
2. Discard the solution; rinse the membrane with DH buffer.
3. Incubate the membrane in 200 ml of DH buffer for 30 minutes at 42°C with gentle agitation.

When nitrocellulose filters are being used, dehybridization is achieved by incubation in boiling sterile water for 10 minutes.

Solutions and Buffers

0.4 N NaOH	16 g NaOH
	H$_2$O to 1 liter

DH buffer	*2 liters*
0.1X SSC	20 ml 20X SSC
0.1% SDS	20 ml SDS 20%*
0.2 M Tris–HCl (pH 7.5)	200 ml 1 M Tris–HCl (pH 7.5)
	1760 ml H$_2$O

*Add about 1 liter of H$_2$O to 20X SSC before adding SDS 20% to avoid precipitation.

Notes

1. It is recommended that the efficiency of dehybridization be checked by performing an autoradiography of the treated blot.
2. To reduce background problems when rehybridizing nylon-derived membranes, do not use more than 10 ng of probe/ml of hybridization solution. Also, do not add dextran sulfate if using less than 10 ml of solution per filter (Dandoy, personal communication).

DIRECT HYBRIDIZATION OF LABELED DNA TO DNA IN AGAROSE GELS

This fast and simple protocol permits hybridization of radioactive probes directly to DNA fragments separated by electrophoresis on agarose gels.

PROTOCOL

"NO BLOT" HYBRIDIZATION

1. Digest the DNA to be analyzed as described in Chapter 6, and run the fragments on an agarose gel (see Chapter 9).
2. If the gel does not contain ethidium bromide, stain the DNA fragments by placing the gel in electrophoresis buffer containing 20 μl of 10 mg/ml ethidium bromide per 200 ml buffer for 20 minutes. Wash with buffer for 15 minutes.
3. Photograph the stained fragments under UV light.
4. Place the gel on a porous polyethylene sheet (35-μm opening, BOLAB, Inc., Arizona) or on Whatman 17 filter paper.
5. Cover the gel with Saran wrap and dry the gel in a gel-slab dryer (e.g., Bio Rad dryer) under vacuum at 25°C for 8–16 hours or at 60°C for 2 hours.
6. If you use Whatman paper, the dried gel will stick to the filter. Remove it carefully.
7. Soak the dried gel in buffer I for 30 minutes at room temperature.
8. Quickly rinse the gel in sterile distilled water.

9. Place the gel in buffer II (neutralizing buffer) for 30 minutes at room temperature, with occasional shaking.

10. Wash the gel with 2X SSCP buffer and dry under vacuum for 3–4 hours.

11. Place the dried gel in a sealing bag and add prehybridization buffer in such a way that the gel is surrounded by buffer and is not in contact with the bag. Let incubate at 37°C for at least 4 hours.

12. Discard prehybridization buffer and replace by hybridization buffer containing the labeled probe. Add approximately 1 ml for a 4 × 5 cm minigel (i.e., 50 μl/cm² gel).

13. Let hybridize for 48–72 hours at 37°C.

14. Discard hybridization mix, and wash the gel six to eight times with washing buffer I for 15 minutes at 65°C and four times in washing buffer II for 15 minutes at 65°C.

15. Dry the gel, cover with Saran wrap, and expose for autoradiography at −70°C with an intensifying screen.

Buffers

Buffer I (denaturing buffer) | *1 liter*
0.5 *M* NaOH | 50 ml 10 *N* NaOH
1 *M* NaCl | 200 ml 5 *M* NaCl
| 750 ml H$_2$O

Buffer II (neutralizing buffer) | *1 liter*
0.5 *M* Tris–HCl (pH 7.4) | 500 ml 1 *M* Tris–HCl (pH 7.4)
0.3 *M* NaCl | 60 ml 5 *M* NaCl
| 440 ml H$_2$O

2X SSCP buffer | *1 liter*
0.3 *M* NaCl | 60 ml 5 *M* NaCl
0.03 *M* Sodium citrate | 30 ml 1 *M* Sodium citrate
0.05 *M* Phosphate (pH 7.0) | 50 ml 1 *M* Phosphate buffer pH 7.0)
| 860 ml H$_2$O

Washing buffer I | *1 liter*
2X SSCP | 60 ml 5 *M* NaCl
0.1% SDS | 30 ml 1 *M* Sodium citrate
| 50 ml 1 *M* Phosphate buffer (pH 7.0)
| 5 ml 20% SDS
| 855 ml H$_2$O

Washing buffer II	*1 liter*
0.1X SSCP	3 ml 5 M NaCl
0.1% SDS	1.5 ml 1 M Sodium citrate
	2.5 ml 1 M Phosphate buffer (pH 7.0)
	5 ml 20% SDS
	988 ml H_2O

REFERENCE

Purello, M., and Balazs, I. (1983), *Anal. Biochem.*, **128**, 393.

HYBRIDIZATION HISTOCHEMISTRY*

The critical elements for optimizing the hybridization reaction have been carefully studied using DNA and RNA bound to nitrocellulose membranes or activated paper supports as the experimental model system (Southern, 1975; Alwine et al., 1977). Some of the variables that have to be considered are the length and concentration of probe, the degree of sequence homology with the target nucleotide, duration and temperature of hybridization, as well as salt concentrations, and so on, and a number of options such as the presence of denaturants (e.g., formamide) or dextran sulfate (Whal et al., 1979).

When applying these conditions to the detection of DNA or RNA in fixed sections of tissues, it became obvious that the denatured cellular debris contributed significantly to the rate of hybridization. The nucleic acid is physically entrapped in the cell debris and this reduces probe accessibility. "Background" hybridization results from the probe binding to various cell components and further reduces the signal-to-noise ratio. Each laboratory has optimized slightly different hybridization conditions and this probably reflects differences in the tissue structure under study as well as the chemical method of fixation.

Preparation of Tissue

Study of frozen tissue sectioned on a cryostat allows detection of the unstable mRNA in eukaryotic cells. Fresh tissues are taken as soon as possible after an-

*Courtesy of Academic Press, with permission of J.P. Coghlan

imal death or surgical excision and immersed in OCT compound (Lab-Tek Products, Naperville, Ill.) at 0°C in aluminium foil embedding moulds and then frozen immediately by slow immersion in hexane/dry ice (-70°C). Tissues can be frozen within 30 seconds of animal death by this technique. Sometimes it is necessary to select an alternative procedure more appropriate for the experiment being undertaken. Specimens such as a whole mouse or other large samples should first be perfusion fixed (Gee et al., 1983), or frozen in a large volume of freezing mixture (Rall et al., 1985), such as hexane/dry ice prior to freeze embedding and cryostat sectioning. Immersion fixation in an appropriate fixative (Godard and Jones, 1980; Angerer and Angerer, 1981) for a short period, and subsequent paraffin embedding, is possible for very small samples (Angerer and Angerer, 1981) but prolonged exposure to fixatives adversely affects hybridization (Brigati et al., 1983).

Preparation of Sections

Sections are cut 5 to 10 μm thick at -10 to -15°C using a cryostat and then thawed onto slides coated with 1% gelatine hardened with 0.25% formaldehyde. Placing slides on a dry-ice block and leaving for at least 30 minutes reduces ribonuclease activity and helps in adherence of the sections. Slides are then dipped in an appropriate fixative, such as 4% glutaraldehyde in 0.1 M phosphate, pH 7.3, with 20% ethylene glycol at 0°C, for 5 minutes or in Carnoy's fluid (60% ethanol, 30% chloroform, 10% acetic acid) at 0°C, for 10 seconds. The final choice of fixative depends on the tissue under study and the particular features that need to be retained for accurate staining at the last stage in the procedure.

Sections are then prehybridized for at least 1 hour at 40°C in Denhardt's solution containing 50% formamide, 0.6 M sodium chloride, and herring sperm DNA. To remove the prehybridization solution slides are rinsed in ethanol and allowed to dry; at this point they may be stored at -20°C in ethanol vapor

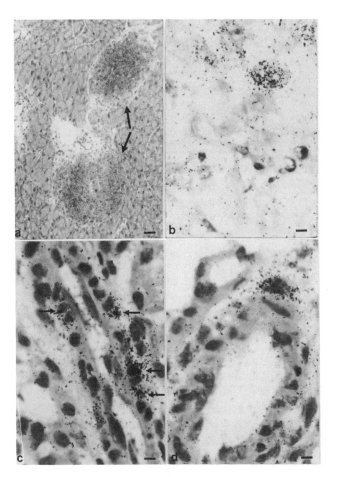

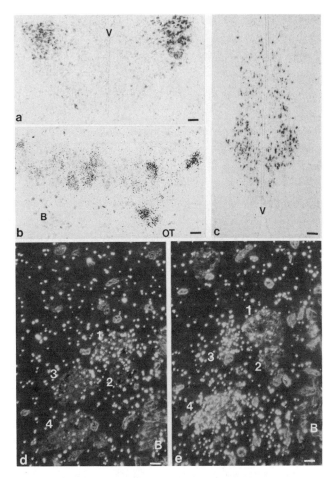

Figure 123. Liquid emulsion autoradiographs after 1 week exposure of 5-μm frozen sections stained with hematoxylin and eosin after hybridization with ^{32}P-labeled cDNA probes. (a) Rat insulin cDNA (Ulrich et al., 1977) localized to the islets of Langherhans (arrowed) of mouse pancreas. Bar = 25 μm. (b) Labeled decidual cells of human placenta after hybridization with human prolactin cDNA (Cooke et al., 1981). Bar = 5 μm. (c and d) mRNA for renin localized by sheep renin cDNA (Aldred et al., 1984; Rougeon et al., 1981; Darby et al., 1985) to cells (arrowed) along an afferent arteriole (c) in longitudinal section and an interlobular artery (d) in transverse section of kidney cortex from a 5-day-old lamb. Bar = 5 μm.

Figure 124. Liquid emulsion autoradiographs after 1 week exposure of frozen 10-μm coronal sections localizing mRNA in the hypothalamus of female rat (a,b) and sheep (c–e) brain using ^{32}P-labeled synthetic 30-mers oligodeoxyribonucleotides probes. (a and b) Paraventricular (a) and suproptic (b) nuclei from a Sprague-Dawley rat after hybridization with a probe corresponding to amino acids 123 to 132 inclusive of rat prepro-AVP-neurophysin II (Schmale et al., 1983; Land et al., 1983). Third ventricle (V) bisects field. Neutral red. Bar = 320 μm. (d and e) Darkfield photomicrographs of adjacent sections of supraoptic nucleus from lactating merino ewe after hybridization with discriminating synthetic oligodeoxyribonucleotide probes corresponding to amino acids 125 to 134 inclusive of bovine AVP-Np II (d) or oxytocin-Np I (e) (Land et al., 1983). With blood vessel (B) mainly out of field as landmark 4 cells (Newmark, 1983; Dugaiczyk et al., 1978; Nygaard and Hall, 1964) appear consecutively in both sections. 1 contains AVP mRNA, 3 and 4, oxytocin mRNA, and 2, neither. Neutral red. Bar = 5 μm.

for several weeks without noticeable loss of either mRNA content or tissue morphology.

Additional treatments have been suggested (Harrison et al., 1973; Brahic and Haase, 1978; Gee et al., 1983; Godard and Jones, 1980; Angerer and Angerer, 1981; Burrell et al., 1982; Gee and Roberts, 1983) which seem to be particularly relevant for tissues fixed by perfusion or immersion and then sectioned in a paraffin block. Frozen sections, although adequate for most single cell localizations (Figs. 123b, 124c, and 124d), are generally unsuitable for the retention of intracellular granules which is

sometimes necessary for morphological integrity or antigen localization. The procedure developed can be used for tissue specimens up to 5mm[3] and is compatible with hybridization histochemistry and the peroxidase–antiperoxidase method of antigen localization, as well as some special histological stains. Fresh tissue specimens are frozen in propane cooled

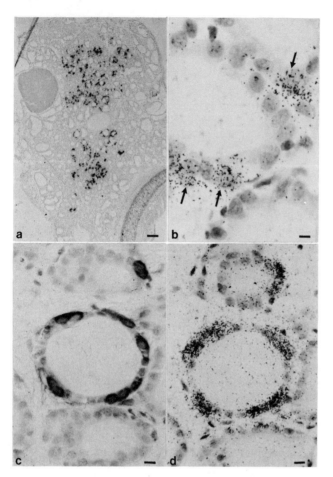

Figure 125. Haematoxylin-stained sections showing localization of calcitonin mRNA by hybridization histochemistry with a 945-bp ^{32}P-labeled cDNA for rat calcitonin (Jacobs et al., 1983; Jacobs et al., 1981) and of calcitonin by immunoperoxidase (d) to C cells of rat thyroid. (a) A 7-μm frozen section of a whole rat thyroid lobe with attached trachea (T). Probe localized to parafollicular areas in central portion of lobe. AR-10 stripping film; 1 week exposure. Bar = 175 μm. (b) A 3-μm paraffin section showing single labeled C cells (arrowed). G5 liquid emulsion; 1 week exposure. Bar = 5 μm. (c and d) Adjacent 3-μm paraffin sections colocalizing in C cells calcitonin mRNA by hybridization histochemistry (c) and immunoreactive calcitonin with rabbit antiserum to human calcitonin and the peroxidase–antiperoxidase procedure (d). Bar = 10 μm.

by liquid nitrogen, freeze dried, fixed with formaldehyde vapor at 37°C, and vacuum embedded in paraplast. Sections can be cut as thin as 2 to 3 μm on a conventional rotary microtome, permitting accurate localization of ^{32}P-labeled probes (Figs. 125b, 125c) and antibody localization on an adjacent section if required (Fig. 125d). Handling of paraffin sections is conventional, except that prolonged exposure to heat during drying is avoided and storage is at −20°C, if necessary, for a limited period. It must be stressed

that the simplicity and convenience of frozen sections with their adequate morphology (Figs. 123, 124, 125a) should not be overlooked, but alternative procedures may be necessary to obtain the desired result.

cDNA Probe Labeling

A number of methods are currently available for the preparation of radiolabeled recombinant cDNA probes to a high specific activity (up to 10^9 cpm/μg for ^{32}P-nucleotide incorporation). Greatest success is obtained starting with an isolated double stranded fragment of recombinant cDNA which is then labeled with ^{32}P by either nick translation (Rigby et al., 1977) or primed synthesis after denaturation with random primers or calf thymus DNA (Taylor et al., 1976; Hudson et al., 1983). The resultant probe is first purified to remove unincorporated nucleotide and then precipitated with ethanol at −80°C and finally taken up in a small volume of the appropriate hybridization buffer. The double stranded product, in which one strand has radiolabeled nucleotide incorporated, has to be denatured (usually by heating) before use as a single stranded hybridization probe.

The use of probes prepared from DNA still attached to the plasmid vector (pBR322) gave high background due to entrapment of nonhomologous bacterial DNA sequences in the tissue. Probes made from M13 vectors are currently under evaluation. The presence of tails [e.g., 20–30 bp poly (dG·dC) from the cloning procedure; or longer poly (dA·dT) tails from the mRNA] definitely increases background labeling in some tissues and should be avoided if possible.

Synthesis of DNA Probes

Oligodeoxyribonucleotide probes can be conveniently prepared by the solid-phase phosphoramidite procedure (Caruthers et al., 1982). Syntheses are generally commenced with 33 mg of the appropriate nucleotide coupled by the 3′-hydroxyl to controlled pore glass support at a loading of 30 μmol/g (total 1 μmol nucleotide). The standard phosphite chemistry program is used with ten-fold excess to protected diisopropyl phosphoramidites. Following chain assembly and removal of the phosphite methyl ester protecting groups with thiophenol, the oligodeoxyribonucleotide is cleaved from the silica support in aqueous ammonia solution.

Additional concentrated ammonia solution is added and the oligodeoxyribonucleotide solution treated for a further 16 hours at 55°C to remove the base-protecting groups. After removal of the ammonia solution by rotary evaporation the residue is made up to a final volume of 3 ml in freshly purified water. Aliquots (200 µl of this aqueous solution) are purified by ion-exchange HPLC on a Whatman Partisil 10-SAX column using a phosphate formamide gradient system or by gel electrophoresis on 18% acrylamide gels. The sequence of the purified probes is confirmed by the Maxam and Gilbert procedure (Maxam and Gilbert, 1980).

The synthetic oligodeoxyribonucleotide probes are 5'-end-labeled with $[\delta - {}^{32}]$ATP using T4 polynucleotide kinase to a specific activity of 6×10^6 cpm/pmol.

Washing Conditions

These are standardized on the use of SSC (standard saline citrate: 1X SSC = 0.15 M sodium chloride, 0.015 M sodium citrate), but the other components of the hybridization buffer could be used in addition. Washing is normally performed by reducing the SSC concentration to 1X SSC at 40°C and incubating for at least 30 minutes to remove most of the background labeling. Increasing the washing temperature or reducing the SSC concentration depends on the length of probe and degree of sequence homology between the probe and mRNA species (Wallace et al., 1981).

Autoradiography

Using ^{32}P-labeled probes and Kodak X-Omat AR film was found to be adequate for obtaining a fast result with reasonable definition. If the specimens are adequately labeled they are exposed then to Cronex MRF 32 (Dupont), a single coated fine-grain film which gives good resolution but requires a longer exposure (approximately four times). See Table 60.

Higher resolution can be achieved using stripping film (Kodak AR-10, Leaflet Nos. P1 1157 and Sc-10, the only type available, or liquid emulsions (e.g., Ilford G5) which come in a variety of grain sizes and are much easier to handle. Precoating sections with subbing solution (Pardue and Gall, 1975) assists in obtaining an even emulsion layer, thereby minimizing background irregularities with high-energy ^{32}P isotopes. To stabilize the hybrid, up to 1X SSC can be used either in the liquid emulsion or in the water bath for the floating of the stripping film. Exposure times for stripping film and liquid emulsion autoradiography must be increased by a factor of about 15 over those for X-Omat AR when using isotopes with the decay rate of ^{32}P.

Table 60. Hybridization Histochemistry Procedure

Day 1

Immediately freeze fresh tissue in hexane/dry ice.
Cut frozen sections -10 to -15°C.
Fix section at 4°C.
Rinse–prehybridize at 40°C in formamide.
Rinse in ethanol and dry.
Apply labeled probe, then coverslip.
Incubate at 40°C in humidified chamber (for probes >30 nucleotides).

Day 2

Rinse; wash in salt solution at 40°C.
Rinse in ethanol and dry.
Apply fast X-ray film to slides in cassettes overnight.

Day 3

Develop X-ray film.
Evaluate result, estimate time for emulsion.
Apply liquid emulsion for 1 to 14 days.

Source: Courtesy of Academic Press, with permission of J.P. Coghlan.

Staining

Most of the photographs shown here are taken from slides that have been stained with hemotoxylin and/or eosin, which resolves most of the features we wish to visualize. Other stains are possible but the effect of the formamide-containing hybridization buffer can prevent some common staining reactions and special stains are rarely successful.

Preparation of Whole Mouse Sections

Adult Swiss male mice are killed by an overdose of pentobarbitone by intraperitoneal injection. The mice are frozen immediately by gradual immersion in hexane/dry ice and subsequently freeze embedded in a 2% carboxymethylcellulose gel. Near-midline sagittal sections (30–40 µm thick) are cut at -20°C using a PMV cryomicrotome (LKB, Sweden). Transparent tape (3M 688) is stuck to the block prior to cutting the sections. This prevents shattering and the cut section comes off adherent to the tape. These sections are immediately fixed at room temperature by immersion in 2.5% glutaraldehyde in 0.1 M phosphate, pH 7.3, and rinsed at 40°C in hybridization buffer (Hudson et al., 1981). The sections are then prehybridized at 40°C for 2 hours, rinsed in ethanol, and air-dried. Hybridization with ^{32}P-labeled probe is carried out under glass slides for 3 days at 40°C within a sealed humidified chamber (Coghlan et al., 1984). Following hybridization the sections are washed initially in 2X SSC, followed by a final wash for 45 minutes in 1X SSC at 40°C. Prior to autoradiography, sections are rinsed in ethanol and air-dried. Autoradiography is carried out by direct exposure to X-ray film (Cronex MRF32 medical recording film, DuPont).

REFERENCES

Aldred, G. P., Fernley, R. T., Hudson, P. J., Niall, H. D., and Coghlan, J. P. (1984), *Proc. Endocr. Soc. Aust.,* **27,** 9.

Alwine, J. C., Kemp, D. J., and Stark, G. R. (1977), *Proc. Natl. Acad. Sci. USA,* **74,** 5350.

Angerer, L. M., and Angerer, R. C. (1981), *Nucl. Acids Res.,* **9,** 2819.

Brahic, M., and Haase, A. T. (1978), *Proc. Natl. Acad. Sci. USA,* **75,** 6125.

Brigati, D. J., Myerson, D., Leary, J. J., Spalholz, B., Travis, S. Z., Fong, C. K. Y., Hsiung, G. D., and Ward, D. C. (1983), *Virology,* **126,** 32.

Burrell, C. J., Gowans, E. J., Jilbert, A. R., Lake, J. R., and Marmion, B. P. (1982), *Hepatology,* **2,** 85.

Caruthers, M. H., Beaucage, S. L., Efcavitch, J. W., Fisher, E. F., Goldman, R. A., De Haseth, P. L., Mandecki, W., Matteucci, M. D., Rosendahl, M. S., and Stabinsky, Y. (1982), *Cold Spring Harbor Symp. Quant. Biol.,* **47,** 411.

Coghlan, J. P., Penschow, J. D., Hudson, P. J., and Niall, H. D. (1984), *Clin. Exp. Hypertens.,* **A6,** 63.

Coghlan, J. P., Penschow, J. D., Tregear, G. W., and Niall, H. D. (1984), in *Receptors, Membranes and Transport Mechanisms in Medicine,* A. Doyle and F. Mendelsohn, eds., Excerpta Medica, Amsterdam, pp. 1.

Cooke, N. E., Cost, D., Shine, J., Baxter, J. D., and Matial, A. (1981), *J. Biol. Chem.,* **256,** 4007.

Darby, I. A., Aldred, G. P., Coghlan, J. P., Fernley, R. T., Penschow, J. D., and Ryan, G. B. (1985), *J. Hypertension,* **3,** 9.

Dugaiczyk, A., Woo, S. L. C., Lai, E. C., Mace, M. L., Reynolds, L., and O'Malley, B. W. (1978), *Nature (London),* **274,** 328.

Gee, C. E., and Roberts, J. L. (1983), *DNA,* **2,** 155.

Gee, C. E., Chen, C. L. C., Roberts, J. L., Thompson, R., and Whatson, S. J. (1983), *Nature (London),* **306,** 374.

Godard, C. M., and Jones, K. W. (1980), *Histochemistry,* **65,** 291.

Harrison, P. R., Conkie, D., Paul, J., and Jones, K. (1973), *FEBS Lett.,* **32,** 109.

Hudson, P. J., Penschow, J. D., Shine, J., Ryan, G., Niall, H., and Coghlan, J. (1981), *Endocrinology,* **108,** 353.

Hudson, P., Haley, J., John, M., Cronk, M., Crawford, R., Haralambidis, J., Tregearn, G., Shine, J., and Niall, H. (1983), *Nature (London),* **301,** 628.

Jacobs, J. W., Chen, W. W., Dee, P. C., Habener, O. F., Goodman, R. H., Bell, N. H., and Potts, J. T., Jr. (1981), *Science,* **213,** 457.

Jacobs, J., Simpson, E., Penschow, J. D., Hudson, P. J., Coghlan, J., and Niall, H. D. (1983), *Endocrinology,* **113,** 1616.

Land, H., Grez, M., Ruppert, S., Schmale, H., Rehbein, M., Richter, D., and Schultz, G. (1983), *Nature (London),* **302,** 342.

Maxam, A. M., and Gilbert, W. (1980), in *Methods in Enzymology,* L. Grossman and K. Moldave, eds., Vol. 65, Part I, pp. 499, Academic Press, New York.

Newmark, P. (1983), *Nature (London),* **303,** 655.

Nygaard, A. P., and Hall, B. D. (1964), *J. Mol. Biol.,* **9,** 125.

Pardue, M. L., and Gall, J. G. (1975), in *Methods in Cellular Biology,* D. M. Prescott, ed., Vol 10, pp. 1, Academic Press, New York.

Rall, L. B., Scott, J., Bell, G. I., Crawford, R. J., Penschow, J. D., Niall, H. D., and Coghlan, J. P. (1985), *Nature (London),* **313,** 228.

Richards, R. I., Catanzaro, D. F., Mason, A. J., Morris, B. J., Baxter, J. D., and Shine, J. (1982), *J. Biol. Chem.,* **257,** 2758.

Rigby, P. W. J., Dieckmann, M., Rhodes, C., and Berg, P. (1977), *J. Mol. Biol.,* **113,** 237.

Rougeon, F., Chambraud, B., Foote, S., Pantheir, J. J., Nageotte, R., and Corvol, P. (1981), *Proc. Natl. Acad. Sci. USA,* **78,** 6367.

Schmale, H., Heinsohn, S., and Richter, D. (1983), *EMBO,* **2,** 763.

Southern, E. M. (1975), *J. Mol. Biol.,* **98,** 503.

Taylor, J. M., Illmensee, R., and Summers, J. (1976), *Biochim. Biophys. Acta,* **442,** 324.

Ulrich, A., Shine, J., Chirgwin, J., Picktet, R., Tischer, E., Rutter, W. J., and Goodman, H. M. (1977), *Science,* **196,** 1313.

Wallace, R. B., Johnson, M. J., Hirose, T., Miyake, T., Kawashima, E. H., and Itakura, K. (1981), *Nucl. Acids Res.,* **9,** 879.

Whal, G. M., Stern, M., and Stark, G. R. (1979), *Proc. Natl. Acad. Sci. USA,* **76,** 3683.

16

LABELING OF NUCLEIC ACIDS

Many different and reliable methods for labeling DNA fragments have been developed over the past few years to generate probes for the detection of DNA and RNA sequences by hybridization. The choice of the method to be used depends primarily on the levels of sensitivity and resolution which are required for a particular experiment. It is obvious, for example, that searching for single copies of genes in complex genomes by the blot-hybridization procedure is only feasible with very specific and sensitive probes while the detection of RNA or DNA sequences by in situ hybridization requires both a high sensitivity and a high resolution. In most of these cases, uniformly labeled probes prepared by nick translation are satisfactory. In some cases, when higher specific activities and increased sensitivities are needed, the use of the in vitro synthesis procedures turn out to be the best choice. In other cases where intact probes are required, for example, in S1 mapping experiments, the end-labeling procedures should be used.

Although the use of radionuclides (especially ^{32}P-labeled compounds) remains the method of choice in most laboratories for the generation of highly sensitive probes, a considerable effort has been made in the past years to develop methods combining satisfactory sensitivity and resolution and based on the use of colored reactions. We present in this section a brief review of the methods available for nonradioactive labeling of probes because they may be useful in places where the use of radiolabeled components is not allowed or not recommended.

PREPARATION OF RADIOACTIVE PROBES

Nick Translation

This procedure remains the most useful for producing labeled probes with DNA fragments which are

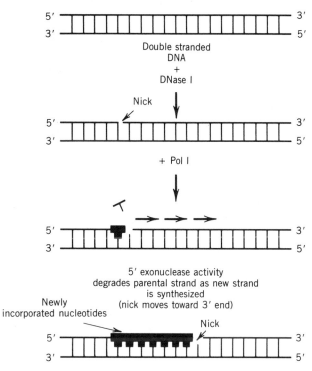

Figure 126. Schematic representation of probe labeling by nick-translation.

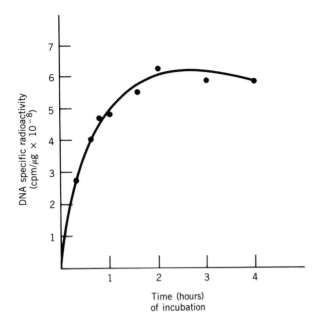

Figure 127. Incorporation of labeled nucleotides during nick-translation.

not available in a cloned form or with fragments eluted from gels after cloning. This method is particularly recommended when high probe yields are required (with radioactive or nonradioactive percursors) and when large quantities of probes are needed.

DNA polymerase I can bind at nicks in DNA and extend the primer terminus in the presence of deoxynucleotide triphosphates. Polymerization at a nick is coupled with the $5' \to 3'$ exonuclease activity of DNA polymerase I. The use of limiting amounts of DNase I generates nicks in DNA, and use of radioactive deoxynucleotide triphosphate allows the labeling of the newly synthesized strand. (Figures 126 and 127).

The following protocol is based on that of Rigby et al. (1977)

REFERENCE

Rigby, P. W. J., Dieckmann, M., Rhodes, C., and Berg, P. (1977), *J. Mol. Biol.*, **113**, 237

PROTOCOL

NICK TRANSLATION

1. In a microfuge tube, mix:
   1 μl 10 mM dGTP
   1 μl 10 mM dATP
   1 μl 10 mM dTTP
   27 μl sterile distilled water
   This constitutes the diluted dNTP stock.

2. Into a microfuge tube pipette 1 μl DNase I (5 mg/ml). Add 500 μl water. Mix and put on ice. Pipette 5 μl of this solution, put it in a tube containing 500 μl water, mix, and use immediately.

3. Mix in a microfuge tube (1.5-ml) at 4°C:
   64.5 μl sterile distilled water
   10 μl 10X nick-translation buffer
   7.5 μl diluted dNTP (step 1)
   4.0 μl DNA (0.5 μg/4 μl)
   3.0 μl freshly diluted DNase (step 2)
   10 μl α[^{32}P]dCTP (10–100 μCi)

4. Incubate for 10 minutes at 12°C.

5. Add 1 μl DNA polymerase I (Boehringer Mannheim).

6. Incubate for 75 minutes at 12°C.

7. Add 100 μl stop buffer, 100 μl phenol equilibrated in NTE, and 100 μl chloroform:isoamyl alcohol (24:1).

8. Vortex for 15 seconds. Spin for 30 seconds at 12,000 rpm.

9. Pipette out the upper phase and extract it with 100 μl phenol and 100 μl chloroform:isoamyl alcohol.

10. Load the upper phase on a G75 Sephadex column equilibrated with 0.1X SSC. It is convenient before loading the column to mix the sample with some buffer containing bromphenol blue, since this dye will be included in the gel and elutes just after the unreacted nucleotide triphosphates.

11. After the sample has been applied to the column, carefully add 0.1X SSC on the top of the gel to prevent the Sephadex from drying out.

12. Connect the column to a small buffer reservoir and collect fractions of about 0.5 ml each. The labeled DNA will not be included in the gel and will therefore come out with the void volume of the column. Follow the migration of labeled DNA through the column with a Geiger counter.

13. Collect fractions until the bromphenol blue has been eluted. Count an aliquot (1 μl) of each fraction to localize the peak of labeled DNA and the peak of nonincorporated nucleotide triphosphates.

14. Pool the fractions containing labeled DNA (generally 1–1.5 ml in total). Denature the labeled DNA by incubation for 10 minutes in boiling water before using it as a probe for hybridization. Freeze the DNA quickly and keep at −20°C in a lead container.

Buffers for Nick Translation

10X Nick translation buffer	*10 ml*
0.5 M Tris–HCl (pH 7.8)	5 ml 1 M Tris–HCl (pH 7.8)
0.05 M 2-Mercaptoethanol	1 ml 0.5 M 2-Mercaptoethanol
0.05 M MgCl$_2$	1 ml 0.5 M MgCl$_2$
0.5 mg/ml Bovine serum albumin	100 μl Bovine serum albumin (50 mg/ml)
	2.9 ml H$_2$O

Store in 0.5-ml aliquots at −20°C.

20X SSC
175.32 g NaCl
88.23 g Sodium citrate
H$_2$O to 1 liter
Dilute 200-fold for 0.1X SSC.

Stop buffer   *10 ml*
10 mM Tris–HCl (pH 7.8)   0.1 ml 1 M Tris–HCl (pH 7.8)
1 mM EDTA   0.02 ml 0.5 M EDTA
0.2 M NaCl   0.4 ml 5 M NaCl
0.5% SDS   0.25 ml 20% SDS
                                      9.23 ml H$_2$O

Notes for Nick Translation

1. G75 Sephadex must be swollen in 0.1X SSC before pouring the column.

2. The column can conveniently be made of a 10-ml plastic disposable sterile pipette (e.g., Corning) plugged with siliconized glass wool.

3. When collecting the fractions, use of siliconized glass tubes (disposable) or plastic tubes is advised to prevent sticking and loss of labeled DNA.

4. Incorporation of the labeled nucleotide triphosphate usually follows a pattern of rapid linear increase, plateau, and then decline (see Figure 127). The reaction is stopped before the decline phase, generally at the beginning of the plateau.

5. α[^{32}P]-deoxynucleotide triphosphates are usually prepared by the supplier from the deoxynucleotide 5′-monophosphate using ATP and polynucleotide kinase.

6. Because of the short ^{32}P half-life (see Table 61), [^{35}S]nucleotides have sometimes been used, without modification of existing protocols. New England Nuclear has introduced such nucleotides, with phosphate replaced by [^{35}S]thiophosphate in the α or γ position (ATPγS, GTPγS, dATPαS, ATPαS) (Figure 128). Since ^{35}S has six times the half-life of ^{32}P (87 days versus 14 days), nick-translated probes prepared with dATPαS are usa-

Table 61. Decay of ^{32}P Radioactivity as a Function of Time[a]

Days Elapsed	Amount of Activity Remaining per 6 Hours at $t =$					
	0	6	12	18	24	30 Hours
0	1.0000	.9879	.9759	.9641	.9524	.9409
1.5	.9295	.9182	.9071	.8961	.8852	.8746
3.0	.8639	.8534	.8431	.8329	.8228	.8129
4.5	.8030	.7933	.7837	.7742	.7648	.7556
6.0	.7464	.7374	.7284	.7196	.7109	.7023
7.5	.6938	.6854	.6771	.6689	.6608	.6528
9.0	.6448	.6370	.6293	.6217	.6142	.6067
10.5	.5994	.5921	.5849	.5779	.5709	.5640
12.0	.5571	.5504	.5437	.5371	.5306	.5242
13.5	.5178	.5116	.5054	.4992	.4932	.4872

[a]Amounts are expressed as the fraction of radioactivity remaining per 6-hour period. $T_{1/2} =$ 14.22 days. (Courtesy of Amersham.)

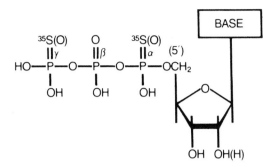

Figure 128. Structure of [^{35}S]-nucleotide triphosphate (courtesy of New England Nuclear).

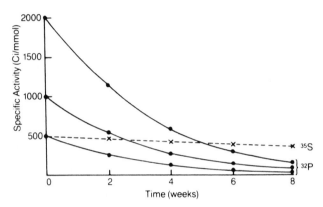

Figure 129. Decay of specific activity of [^{35}S]-thiophosphate-labeled molecules as compared with that of [^{32}P] phosphate-labeled compounds (courtesy of New England Nuclear).

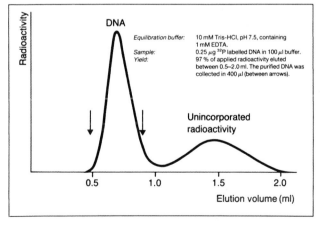

Figure 130. Separation of nick-translated fragments of pBR322 DNA from unincorporated α[^{32}P]dATP on Pharmacia NICK-column (courtesy of Pharmacia).

ble for longer periods of time (Figure 129). This can be very useful when the isolation of specific DNA requires several preparative steps.

7. Instead of using a regular column to separate labeled DNA and free nucleotides, it is possible to centrifuge directly the nick-translation mixture through the filtration gel. Take a 1-ml plastic syringe (or a 1-ml tip) and add a plug of siliconized wool at the bottom (it is also possible to use small disks of porous polyethylene with the syringes). Add Sephadex G50 previously swollen in TE buffer to obtain 1 ml packed gel. Centrifuge in a clinical centrifuge at top speed for 2 minutes. The level of Sephadex should lower to about 0.7 ml. Place the syringe (or the tip) in another tube and add 50–100 μl of the nick-translation mix per gel. Centrifuge for 2 minutes and collect the fraction at bottom of the tube, which contains the labeled DNA.

8. Prepacked, ready to use, sephadex G50 columns (Nick columns from pharmacia) are also very convenient (Figure 130).

Labeling DNA by the Replacement Synthesis Method

This protocol takes advantage of the exonuclease and polymerase activities of T4 DNA polymerase (see Figure 131). It can be used with DNA fragments that are not cloned in pBR-derived vectors (see below). Its main advantage over nick translation is that it allows labeling of a DNA fragment around a specific restriction site, or preferential 3′ end labeling (O'Farrell et al., 1980). It can also conveniently be used to prepare labeled size standards for agarose or polyacrylamide gel analysis. We will describe below two applications of this method.

REFERENCE

O'Farrell, P. H., Kutter, E., and Nakanishi, M. (1980), *Molec. Gen. Genet.*, **179**, 421

PROTOCOL

PREPARATION OF LABELED SUBFRAGMENTS AS PROBES FOR HYBRIDIZATION

In this example, a λ recombinant bacteriophage has been isolated from a library (see Chapter 17). It contains an insert with four internal Eco RI sites, cloned in the Eco RI arms of charon 4A. The goal of the experiment is to obtain probes for each of the five resolvable Eco RI fragments containing the insert sequences.

1. In a microfuge tube, add successively:
   14 μl of sterile distilled water
   2 μl of DNA (1 mg/ml)
   2 μl of 10X TA buffer
   2 μl of Eco RI (10 units/μl)

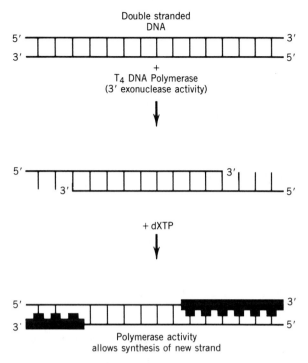

Figure 131. Exonuclease and polymerase activities of T4 DNA polymerase in the absence and in the presence, respectively, of deoxynucleotide triphosphates.

2. Mix and incubate for 30 minutes at 37°C.
3. Add 2.5 units of T4 DNA polymerase
4. Incubate for 25 minutes (see note).
5. Add successively:
   1 μl of 2 mM dATP
   1 μl of 2 mM dGTP
   1 μl of TTP
   2.5×10^{-10} mol of α[^{32}P]dCTP
6. Incubate at 37°C for 12 minutes.
7. Add 1 μl of 2 mM dCTP and incubate for 40 minutes at 37°C. This will ensure a complete regeneration of the duplex.
8. Collect each band by preparative electrophoresis (see Chapter 11).

Note. If we assume that the smallest Eco RI fragment is likely to be larger than 1 kb, one can excise 500 nucleotides from each end without losing any fragment.

Calculation of the number of moles excised:

$$\text{total DNA (as moles of nucleotides)} = 6 \times 10^9 \text{ mol}$$

$$\text{number of ends} = 12 \text{ (corresponding to 6 Eco RI sites)} + 2 \text{ (arms)}$$

fraction of the DNA being degraded =
$$\frac{500 \text{ nucleotides per end} \times 14 \text{ ends}}{50,000 \text{ base pair } (\lambda \text{ DNA})}$$

moles of nucleotides being excised
$$= 6 \times 10^{-9} \times 0.07 = 4.2 \times 10^{-10}$$

This corresponds to 1×10^{-10} mol of each nucleotide.

PROTOCOL

PREPARATION OF LABELED SIZE STANDARDS

Labeling Eco RI λ Fragments

1. In a microfuge tube containing 20 μg of Eco RI digested λ DNA (digestion performed in TA buffer and heated at 68°C for 10 minutes) add 10 μl of T4 DNA polymerase (30 units/μl) and 1X TA buffer to 1 ml.
2. Incubate for 5 minutes at 37°C.
3. Add successively:
   10 μl of 10 m*M* dATP
   10 μl of 10 m*M* dGTP
   10 μl of 10 m*M* TTP
   and store as 50-μl aliquots.
4. For labeling, thaw a 50-μl sample.
5. Add 1 μl (10 μCi, 0.37 MBq) of α[^{32}P]dCTP (800 Ci/mmol).
6. Incubate 5 minutes at 37°C.
7. Add 1 μl of 10 m*M* dCTP and incubate for 10 minutes at 37°C.
8. Stop the reaction by adding EDTA to a final concentration of 10 m*M* and heating at 68°C for 5 minutes.

Labeling the 1-kb Ladder (BRL)

1. Put in a microfuge tube about 15 μg of the 1-kb ladder with 38 units of T4 DNA polymerase.
2. Add 100 μl of 10X TA buffer.
3. Add sterile distilled water up to 1 ml.
4. Incubate for 5 minutes at 37°C.
5. Add successively:
   10 μl of 10 m*M* dCTP

   10 μl of 10 m*M* dGTP

   10 μl of 10 m*M* TTP

   and 110 μl of bidistilled sterile glycerol.

6. Mix by inverting several times and store as 50-μl aliquots at −20°C.

7. For labeling, thaw a 50-μl aliquot on ice.

8. Add 1 μl (10 μCi, 0.37 MBq) of α[^{32}P]dATP (800 Ci/mmol).

9. Incubate 5 minutes at 37°C.

10. Add 1 μl of 10 m*M* dATP and incubate for 10 minutes at 37°C.

11. Stop the reaction by adding EDTA to a final concentration of 10 m*M* and heating at 68°C for 5 minutes.

Solutions and Buffers

10X TA Buffer

330 m*M* Tris–acetate pH 7.9

660 m*M* Potassium acetate

100 m*M* Magnesium acetate

5 m*M* DTT

1 mg/ml Nuclease-free BSA

(or human serum albumin from Worthington)

This buffer should not be kept at 4°C. Keep frozen at −20°C. Stock solution of dithiothreitol should be kept frozen in small aliquots since it degrades upon repeated cycles of freezing/defreezing.

End Labeling of DNA Fragments

End labeling of DNA fragment is required in many different instances (for example, in the mapping of RNA species on DNA templates, or in nucleotide sequencing).

5′ End Labeling with Polynucleotide Kinase and γ *[*^{32}*P] ATP.* This labeling can be performed by using either the forward phosphorylation of 5′-hydroxyl ends or by exchange phosphorylation of the 5′-phosphate. See protocols below.

PROTOCOL

DEPHOSPHORYLATION OF 5′ ENDS

1. In a microfuge tube containing 25 pmol of DNA (equivalent to 3 μg of a 500-bp fragment) as a dry pellet, add successively:

100 μl of 20 m*M* Tris–HCl pH 8.0

1 μl of calf intestine alkaline phosphatase (Boehringer Mannheim)

2. Incubate at 37°C for 1 hour if 5′ end is protruding. (When 5′ end is recessed or blunt, incubate at 60°C for 30 to 45 minutes.)

3. Add successively 100 μl of Stop buffer, 100 μl of buffer-saturated phenol, and 100 μl of chloroform:isoamyl alcohol mixture (24:1).

4. Vortex for 30 seconds, and centrifuge at 12,000 rpm (room temperature) for 3 minutes.

5. Discard supernatant, add 150 μl of buffer-saturated phenol, and 150 μl of chloroform:isoamyl alcohol mixture.

6. Vortex and centrifuge for 3 minutes at 12,000 rpm.

7. Transfer the supernatant in a tube containing 100 μl of 7.5 *M* ammonium acetate.

8. Mix and add 600 μl of cold absolute ethanol.

9. Incubate at −70°C until the solution is frozen.

10. Centrifuge for 30 minutes at 12,000 rpm (4°C).

11. Discard supernatant and add 1 ml of 80% cold ethanol.

12. Vortex to resuspend the DNA pellet and centrifuge for 10 minutes at 12,000 rpm (4°C).

13. Discard the supernatant and dry the pellet in a Speedvac vacuum centrifuge (Savant).

Notes

1. It is important that the phosphatase be eliminated by phenol extraction under the conditions described above. Residual phosphatase activity interferes with subsequent labeling.

2. Precipitation is performed in the presence of ammonium acetate to efficiently eliminate potent inhibitors of polynucleotide kinase.

3. Some of us prefer to add an equal amount of phosphatase after the first 30 minutes of incubation.

PROTOCOL

LABELING OF PROTRUDING 5′-HYDROXYL ENDS

1. In a microfuge tube, transfer at least 50 pmol of γ[^{32}P]ATP (>5000 Ci/mmol, >180 TBq/mmol).

2. Place the tube in a vacuum centrifuge (Speedvac, Savant) and let the solution dry down. This is achieved in a few minutes.

3. In the tube containing the dried $\gamma[^{32}P]$ATP, add successively:
   10 μl of dephosphorylated DNA (25 pmol) dissolved in sterile distilled water
   1 μl of 10X FW. Kinase buffer (pH 7.6)
   10 units of T4 polynucleotide kinase (Boehringer, 5000 to 20,000 units/ml)

4. Mix well and incubate at 37°C for 30 minutes.

5. Stop the reaction by adding 200 μl of 2.5 M ammonium acetate.

6. Add 750 μl of cold absolute ethanol. Store at -70°C until frozen.

7. Close the cap and wrap it with parafilm (autoclaved microfuge tubes often leak).

8. Invert the tube a few times and centrifuge for 10 minutes at 15,000 g (4°C).

9. Remove the supernatant carefully with a Pasteur pipette and discard the radioactive solution in an adequate container.

10. Add 250 μl of 0.3 M of sodium acetate and dissolve the precipitate by vortexing.

11. Add 750 μl of cold absolute ethanol, close, wrap in parafilm, invert, and centrifuge for 5 minutes at 15,000 g (4°C).

12. Discard the supernatant, and add 1 ml of cold absolute ethanol.

13. Centrifuge at 15,000 g for 10 seconds.

14. Remove supernatant and dry in a Speedvac vacuum centrifuge (Savant).

PROTOCOL

LABELING OF RECESSED AND BLUNT 5'-HYDROXYL ENDS

1. Dry down at least 100 pmol of $\gamma[^{32}P]$ATP ($>$5000 Ci/mmol) as described in steps 1 and 2 above.

2. Add to this tube 10 μl of 2X FW. Kinase buffer (pH 9.5); keep on ice.

3. Resuspend the ATP by vortexing, centrifuge for a few seconds at 12,000 rpm to collect drops, and put on ice.

4. In a microfuge tube add 25 pmol of DNA resuspended in 10

μl of resuspension buffer (pH 9.5), and incubate for 2 minutes at 90°C.

5. Add immediately the 10 μl of radioactive mixture prepared in step 3, and 10 units of T4 polynucleotide kinase (Boehringer 5000 to 20,000 units/ml).

6. Mix carefully and incubate at 37°C for 30 minutes.

7. Stop the reaction by adding 200 μl of 2.5 *M* ammonium acetate.

8. Add 750 μl of cold absolute ethanol. Store at −70°C until frozen.

9. Close the cap and wrap it with parafilm (autoclaved microfuge tubes often leak).

10. Invert the tube a few times and centrifuge for 10 minutes at 15,000 *g* (4°C).

11. Remove the supernatant carefully with a Pasteur pipette and discard the radioactive solution in an adequate container.

12. Add 250 μl of 0.3 *M* of sodium acetate and dissolve the precipitate by vortexing.

13. Add 750 μl of cold absolute ethanol, close, wrap in parafilm, invert, and centrifuge for 5 minutes at 15,000 *g* (4°C).

14. Discard the supernatant, and add 1 ml of cold absolute ethanol.

15. Centrifuge at 15,000 *g* for 10 seconds.

16. Remove supernatant, and dry in a Speedvac vacuum centrifuge (Savant).

PROTOCOL

POLYNUCLEOTIDE KINASE ECHANGE REACTION FOR LABELING

When the DNA fragments have 5′ phosphate protruding ends, the exchange kinase activity can also be used to label the termini in the presence of ADP and high concentrations of γ[^{32}P]ATP (optimal concentration 10 μ*M*, minimum 2 μ*M*). This method alleviates the need for dephosphorylation prior labeling.

1. In a microfuge tube, transfer 100 pmol of γ[^{32}P]ATP (>5000 Ci/mmol).

2. Place the tube in a vacuum centrifuge (Speedvac, Savant) and let the solution dry down. This is achieved in a few minutes.

3. Add successively in the tube containing the dried $\gamma[^{32}P]ATP$:
   1.5 µl of 10X EX. Kinase buffer.
   10 µl of DNA fragment (25 pmol of 5′ ends; e.g., 3 µg of a 500 base pair DNA fragment)
   2 µl of 2.5 mM ADP

4. Mix carefully by vortexing.

5. Add 10 units of polynucleotide kinase (Boehringer 5000 to 20,000 units/ml).

6. Incubate for 30 minutes at 37°C.

7. Stop the reaction by adding 200 µl of 2.5 M ammonium acetate.

8. Add 750 µl of cold absolute ethanol. Store at −70°C until frozen.

9. Close the cap and wrap it with parafilm.

10. Invert the tube a few times and centrifuge for 10 minutes at 15,000 g (4°C).

11. Remove the supernatant carefully with a Pasteur pipette and discard the radioactive solution in an adequate container.

12. Add 250 µl of 0.3 M of sodium acetate and dissolve the precipitate by vortexing.

13. Add 750 µl of cold absolute ethanol, close, wrap in parafilm, invert, and centrifuge for 5 minutes at 15,000 g (4°C).

14. Discard the supernatant, and add 1 ml of cold absolute ethanol.

15. Centrifuge at 15,000 g for 10 seconds.

16. Remove supernatant, and dry in a Speedvac vacuum centrifuge (Savant).

Notes. ADN–spermidine complexes tend to precipitate when cold.

Solutions and Buffers

10X EX. Kinase buffer
0.5 M Immidazole–HCl
0.1 M MgCl$_2$
50 mM DTT
1 mM Spermidine
1 mM EDTA

10X FW. Kinase buffer (pH 7.6)
0.5 M Tris–HCl (pH 7.6)
0.1 M MgCl$_2$
50 mM DTT
1 mM Spermidine
1 mM EDTA

Resuspension buffer (pH 9.5)
10 mM Tris–HCl (pH 9.5)
1 mM Spermidine
0.1 mM EDTA

2X FW. Kinase buffer (pH 9.5)
0.1 M Tris–HCl (pH 9.5)
20 mM MgCl$_2$
10 mM DTT
10% Glycerol

Fill-in End Labeling. This method described in the following protocol is fast and very convenient for preparing labeled markers for agarose and polyacrylamide gels. The procedure relies on the $5'\rightarrow3'$ elongation of 3'-recessed DNA ends by DNA polymerase I, in the presence of α-labeled deoxynucleotide triphosphate(s). The nature of the labeled dXTP used will depend of the base composition of the fragment's end.

3' End Labeling with Terminal Transferase and Cordycepin Triphosphate. Terminal transferase can be used to extend 3'-ends of duplex DNA molecules with single stranded homopolymers (see Chapter 4). The protocol described on page 468 relies upon the addition of a nucleotide analog to label DNA ends.

PROTOCOL

FILL-IN LABELING OF DNA

1. In a microfuge tube containing 21 µl of distilled sterile water add 3 µl of each cold dXTP according to the following table.

$\alpha[^{32}P]dXTP$	*3 µl dATP*	*3 µl dCTP*	*3 µl dGTP*	*3 µl TTP*
dATP	–	+	+	+
dCTP	+	–	+	+
dGTP	+	+	–	+
TTP	+	+	+	–

2. In another microfuge tube containing about 10 pmol of DNA (equivalent to about 1.5 μg of a 500-bp fragment) as a dried pellet, add successively:
   5 μl of the dXTP mix prepared in step 1 (500 μM final)
   1 μl of 10X Pol. buffer
   1 μl of the α[^{32}P]dXTP chosen for the labeling
   $(3 - x)$ μl of sterile distilled water depending on the concentration of the enzyme used
   x μl of *E. coli* DNA polymerase I Klenow fragment (so as to obtain 0.5 units)

3. Incubate for 20 minutes at 20°C.

4. Stop the reaction by phenol extraction (rather than heating at 68°C for 10 minutes which may stimulate the exonuclease activity of Klenow fragment; see notes).

5. Separate the labeled DNA from excess of dXTP by passage through a G75 Sephadex column.

Notes

1. Labeling may decrease if longer times of incubation or higher concentrations of enzyme are used, due to the 3′→5′ exonuclease activity of the Klenow fragment.

2. The reaction may also be performed with *E. coli* DNA polymerase I. In this case, the salt concentration should be raised to 70 mM, instead of 50 mM, and the dXTP concentration should be 25 to 100 μM final. Up to 10 μg of DNA can be labeled in this way with 5 to 10 units of DNA polymerase. Incubation should be performed at 4°C on ice for 5–10 minutes. The 5′→3′ exonuclease activity of DNA polymerase is greatly inhibited by salt.

3. It is possible to use a mixture of the four cold deoxynucleotide triphosphates when working with an α[^{32}P]dXTP with higher specific radioactivity (e.g., 3000 Ci/mmol, 110 TBq/mmol).

Solutions and Buffers

10X Pol. buffer
0.5 M NaCl
66 mM Tris–HCl pH 7.4
66 mM MgCl$_2$
10 mM DTT

10 mM dXTP
See Chapter 3.

PROTOCOL

3' END LABELING WITH TERMINAL TRANSFERASE

1. In a microfuge tube containing 5 µl of DNA solution (containing 2 pmol of 3' ends), add 10 µl of sterile distilled water.
2. Heat at 90°C for 1 minute and cool quickly in iced water.
3. Centrifuge a few seconds to collect all material.
4. Add successively:
   5 µl of transferase buffer
   5 µCi (1.85 MBq) of α[^{32}P]cordycepin triphosphate
   1 µl (12 units) of terminal transferase (Pharmacia or International Biotechnologies Inc.)
5. Add sterile distilled water up to 50 µl.
6. Mix and incubate at 37°C for 30 minutes.
7. Add 25 µl of ammonium acetate 7.5 M, mix.
8. Add 150 µl of cold absolute ethanol and let the DNA precipitate at −70°C for 15 minutes.
9. Centrifuge at 12,000 rpm for 30 minutes, discard the supernatant, and dissolve the pellet in 50 µl of sterile distilled water.
10. Repeat steps 7–9 and dry the precipitate in a Speedvac vacuum centrifuge (Savant).

Notes

1. The use of cordycepin 5' triphosphate (3' deoxyadenosine triphosphate) instead of dATP allows addition of a single residue to the 3' end of DNA fragments, with either a blunt or a 3' protruding terminus.
2. After precipitation, the specific radioactivity should reach 2 × 10^6 cpm/pmol of ends.
3. Free labeled compounds represent 2% after ethanol precipitation with ammonium acetate.
4. Activity of terminal transferase is strongly inhibited by the presence of sodium ions. Check origin and purity of the DNA fragments.
5. High quality cacodylic buffer should be used. Be aware that cacodylate is harmful.
6. About 40% of blunt ends and 90% of 3' extension are labeled by using these conditions.

Solutions and Buffers

Transferase buffer
250 mM Potassium cacodylate (pH 7.2)
20 mM CoCl$_2$
10 mM DTT

When preparing the cacodylate buffer, it is essential that cacodylic acid and cobalt chloride do not come in direct contact prior to hydrating in order to avoid precipitation.

Procedures for Nonradioactive Labeling of Probes

The nonradioactive labeling of probes is especially useful under conditions where the use of radioactive precursors is not possible or is to be avoided. The development of such methods therefore permits hybridization technology to be extended to routine diagnosis performed daily in many laboratories, without the need of any special containment required for the manipulation of the radioactive probes. Following the first report by Langer et al. (1981) that biotinylated dUTP analogs (bio-4-dUTP) could be enzymatically incorporated into DNA molecules and subsequent colorimetric detection of DNA–DNA hybrids using avidin and biotinylated horseradish peroxidase and alkaline phosphatase (Leary et al., 1983; Ruth, 1984), several improvements have been brought to the original labeling protocols which result in an increased specificity and sensitivity of such probes in the detection of nucleic acids immobilized on filters or after in situ hybridization.

Biotinylation of the nucleic acids can be performed by using nucleotide analogs instead of radioactive precursors in the labeling reactions described above. For example, biotin-11-dUTP or biotin-7-dATP are currently used to prepare biotinylated DNA by the nick translation procedure. Slight modifications to the labeling procedures are described in the manuals provided by the suppliers and will not be discussed here. Make sure that the deoxynucleotide concentration is 20 μM in the reaction and be aware that phenol extraction of biotinylated probes should be avoided since partitioning of the biotinylated DNA in the phenol phase results in severe loss of material (BRL instruction manual). Once the probe is synthesized, it is usually purified by gel filtration and heat denatured at 95°C for 10 minutes before use. It is not advisable to perform an additional alkaline denaturation of the biotinylated probes because the amide bond in the 11 atom spacer arm is sensitive to alkali, and loss of the biotin moiety may result. Hybridization is usually performed at probe concentrations in the range of 100 to 200 ng/ml. A concentration of 20 ng/ml is recommended for M13 DNA probes.

In order to avoid high backgrounds due to nonspecific fixation of the probes to the filters, it is usually recommended that the filter be pretreated by incubation with bovin serum albumin (BSA). It is important that the filter be baked under vacuum to ensure efficient drying following blocking with BSA. Nitrocellulose filters usually give lower backgrounds than nylon-derived filters. Efficient blocking of the nylon membranes such as Genescreen may require as much as 60 minutes of incubation at 65°C in the presence of BSA to ensure low backgrounds, while 20 minutes of incubation at 42°C has been found to be sufficient for nitrocellulose filters.

The biotin derivatives have also been successfully used in the replacement synthesis method, in the synthesis of labeled cDNA with reverse transcriptase, and in the in vitro synthesis of labeled RNA probes.

The colorimetric detection of biotinylated nucleic acids is usually based on the enzymatic activity of alkaline phosphatase or peroxidase coupled to a carrier able to recognize specifically the modified probe. Avidin was first chosen because of its high affinity for biotin ($kd = 10^{-15} M$) but was soon replaced by streptavidin, which exhibits a similar affinity toward biotin and does not contain any glycosylated residue responsible for the high backgrounds obtained in the detection of DNA molecules bound on nylon or nitrocellulose filters with avidin derivatives.

For the revelation of the hybridized molecules, the BSA solution is removed from the bags and the filters are incubated for 10 minutes in the presence of streptavidin and then with a polymer of alcaline phosphatase. After washing, the treated filters are incubated in the presence of a substrate dye solution containing nitro blue tetrazolium (NBT) and 5-bromo-4-chloro-3-indolyl phosphate (BCIP). This solution is usually made fresh before experimentation (it can also be stored in frozen aliquots at $-20°C$ for 1 to 2 weeks). The color development should preferably be performed in the dark or under subdued light in order to minimize the nonspecific background. The coloration of the target DNA may require from 5 minutes to several hours, depending on the amount of biotinylated probe hybridized. Single copy sequences are usually detected in about 4 hours of incubation. The coloration is stable and cannot be washed out. Biotinylated probes have been used to detect both DNA restriction fragments and mRNA species immobilized on filters (Figures 132–135).

The use of a of streptavidin–alkaline phosphatase conjugate (BRL BluGENE nonradioactive detection system) is supposed to permit the detection of as little as 0.5 pg of DNA. However, it would seem that in practice such a level of sensitivity is not easily reached. This system has been used to study the limits of RNA detection on nylon and nitrocellulose membranes and different denaturation methods. The results obtained (Figures 136–138) revealed that this system allowed the detection of viral mRNAs representing 0.2% abundance in 500 ng of polyladenylated RNA (Amorese, 1986).

The direct coupling of biotin to DNA and RNA molecules can also be achieved by photoactivation of photobiotin acetate (Forster et al., 1985). One biotin molecule can be coupled per 100–400 nucleotides fol-

Figure 132. Detection of β-globin plasmid by BluGENE™. A plasmid containing the 1.1-kb Mst II fragment of the human β-globin gene in the Eco RI site of pBR322 (from K.W. Kan, UCSF) was nick-translated in the presence of biotin-11-dUTP and hybridized to Southern blots in 45% formamide and 5% dextran sulfate at 42°C for 18 hours. Lanes 1 through 5 contain Eco RI-digested plasmid DNA (0.3125, 0.625, 1.25, 2.5, and 5.0 pg) and sheared herring sperm DNA (1 µg). The upper vector bands correspond to 0.25, 0.5, 1.0, 2.0, and 4.0 pg in lanes 1 through 5, respectively. The lower insert bands correspond to 0.0625, 0.125, 0.25, 0.5, and 1.0 pg in lanes 1 through 5, respectively (courtesy of Bethesda Research Laboratories, Life technologies Inc.).

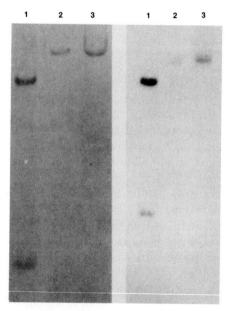

Figure 133. Detection of single copy human genomic β-globin gene by nonradioactive (A) and radioactive (B) DNA detection procedures. A plasmid containing the 1.1-kb Mst II fragment of the human β-globin gene in the Eco RI site of pBR322 (from Y.W. Kan, UCSF) was nick-translated in the presence of biotin-11-dUTP or α[^{32}P]dATP and hybridized to Southern blots in 45% formamide and 5% dextran sulfate at 42°C for 18 hours. Lane 1 contains Eco RI-digested plasmid DNA (5 pg) and sheared herring sperm DNA (1 µg). The upper vector and lower insert bands correspond to approximately 4 and 1 pg, respectively. Lanes 2 and 3 contain Eco RI-digested human genomic DNA 2 µg and 10 µg, respectively. (A) The biotinylated probe concentration in the hybridization mix was 100 µg/ml. The filter of the ^{32}P-labeled DNA probe was 1.5×10^8 dpm/µg, and the probe concentration was 2 ng/ml. The filter was autoradiographed with an intensifying screen for 3 hours. (courtesy of Bethesda Research Laboratories, Life technologies Inc.)

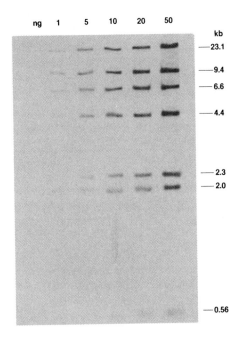

Figure 134. Biotinylated λ DNA/Hind III fragments. The indicated amounts were electrophoresed on a 1% agarose gel, transferred to nitrocellulose, and developed with the BRL Nonradioactive DNA Detection System (courtesy of Bethesda Research Laboratories, Life technologies Inc.).

lowing exposition to a strong visible light (250 to 500 Watts) for 10 to 30 minutes. A photobiotin labeling and detection kit is commercialized by Biotechnology Research Enterprises S.A. (Australia). An improvement in the nonradioactive labeling of nucleic acids by direct sulfonation of the cytidine residues has been described by Lebacq et al. (1987). This method allows the detection of a few picograms of DNA (Figures 139 and 140) Commercial Kits (Chemiprobe) are available from Orgenics Ltd. (Israel) and PBS Orgenics (France). Several methods are also available for labeling directly the DNA and RNA molecules with peroxydase, without the need of nucleotide analogs. For example, oligonucleotide probes labeled with peroxidase (Jablonski et al., 1986) are available from Molecular Biosystems, Inc. and Gen-Probe (California). The use of such probes permits performance of the hybridization and washing steps in a short period of time (15 to 30 minutes at 70°C) and permits detection of picograms of DNA bound to filters by the conventional blotting methods. The considerable advantages of these probes come from the fact that they are single stranded molecules (no reassociation possible) and that their small size (15 to 35 nucleotides) leads to a quasi immediate hybridization with complementary sequences.

REFERENCES

Amorese, D. (1986), *Focus,* **8:4**, 7.

Forster, A. C., McInnes, J. L., Skingle, D. C., and Symons, R. H. (1985), *Nucl. Acids Res.,* **13**, 745.

Langer, P. R., Waldrop, A. A., and Ward, D. C. (1981), *Proc. Natl. Acad. Sci. USA,* **78**, 6633.

Lebacq, P., Squalli, D., Duchenne, M., Pouletty, P., and Joannes, M. (1987), *Biophysica and Biochemica Methods,* (in press).

Leary, J., Brigati, D., and Ward, D. C. (1983), *Proc. Natl. Acad. Sci. USA,* **80**, 4045.

Ruth, J. (1984), *DNA,* **3**, 123.

Seyfert, H. M. (1985), *Focus,* **7:4**, 3.

Soochan, P. (1986), *Focus,* **8:2**, 10.

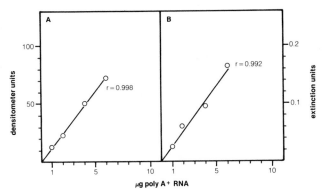

Figure 135. Quantitation of α-tubulin mRNA in poly(A⁺)RNA with a biotinylated gene probe. Increasing amounts (1, 2, 4, and 6 µg) of poly (A⁺)RNA were dotted onto nitrocellulose and hybridized with a biotinylated gene probe. Method I (panel A) shows increased hybridization of the gene probe by increasing intensity of the spots (insert). Densitometry of the filter demonstrates that the amount of biotinylated probe that is bound is highly correlated with the amount of RNA applied (r = coefficient of correlation). Method II (panel B) measures increased hybridization of increasing absorbance of the substrate product in solution. No permanent record of the spots is produced. However, after the procedure, the filters may be removed and exposed to the substrates of method I, and the staining intensity will be proportional to the amount of alkaline phosphatase bound (not shown) (courtesy of Bethesda Research Laboratories, Life technologies Inc.).

IN VITRO SYNTHESIS OF LABELED RNA PROBES

Several vectors have been constructed for efficient in vitro synthesis of RNA (see Chapter 6). These systems offer the possibility of obtaining from a single recombinant clone highly specific probes for each strand of the inserted DNA fragment.

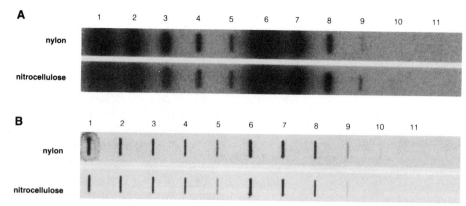

Figure 136. Comparison of radioactive and nonradioactive detection of RNA. RNA samples were denatured in formaldehyde and hybridizations were performed as described in text. Samples in slots 6 to 11 contained 10 m*M* VRC. The amount of the 1.35-kb RNA in each slot is 100 ng (1 and 6), 10 ng (2 and 7), 1 ng (3 and 8), 100 pg (4 and 9), 10 pg (5 and 10), and 1 pg (11). Slot 12 has 100 pg of total HeLa RNA. Panel A: Autoradiograph from a 24-hour exposure without an intensifying screen. Panel B: Nonradioactive detection with BluGENE™ using a 30-minute color development. Both detections were performed on nylon and nitrocellulose as indicated (courtesy of Bethesda Research Laboratories, Life technologies Inc.).

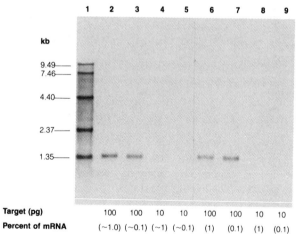

Target (pg)   100   100   10   10   100   100   10   10

Percent of mRNA   (~1.0) (~0.1) (~1) (~0.1) (1) (0.1) (1) (0.1)

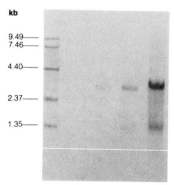

Figure 137. Nonradioactive detection of RNA on Northern blots. RNA samples were denatured and electrophoresed on a 1% agarose gel containing formaldehyde. Hybridization and washing were performed as described in the BRL Blu-GENE™ Manual. The probe (200 ng/ml) was hybridized overnight. Color was developed for 1 hour. Lanes: (1) 1 ng BRL RNA Ladder; (2) 250-ng HeLa RNA containing 100 pg 1.35 kb RNA (target); (3) 2.5 μg HeLa RNA containing 100 pg target RNA; (4) 25 ng HeLa RNA containing 10 pg target RNA; (5) 250 ng HeLa RNA containing 10 pg target RNA; (6) 10 ng HeLa poly(A)⁺ RNA containing 100 pg target RNA; (7) 100 ng HeLa poly(A)⁺ RNA containing 100 pg target RNA; (8) 1 ng HeLa poly(A)⁺ RNA containing 10 pg target RNA; and (9) 10 ng HeLa poly(A)⁺ RNA containing 10 pg target RNA (courtesy of Bethesda Research Laboratories, Life technologies Inc.).

Figure 138. Nonradioactive detection of HPV-18 mRNA on Northern blots. RNA samples were denatured and electrophoresed on a 1% agarose gel containing formaldehyde. Hybridization and washing conditions were as described in the BRL BluGENE™ Manual. Hybridization was performed with 100 ng/ml biotinylated BRL 1-kb ladder and 100 ng/ml biotinylated HPV-18 DNA clone. Color was developed for 2 hours. Lanes: (1) 500 pg BRL RNA ladder; (2) 5 ng HeLa poly(A)⁺ RNA; (3) 50 ng HeLa poly(A)⁺ RNA; (4) 500 ng HeLa poly(A)⁺ RNA; and (5) 5 μg HeLa poly(A)⁺ RNA (courtesy of Bethesda Research Laboratories, Life technologies Inc.).

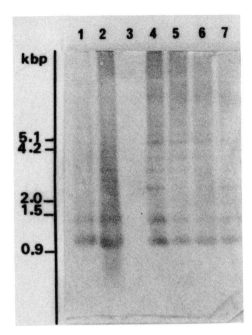

Figure 139. Southern blot hybridization of native mitochondrial DNA from different lines of sugar beet with cloned minicircle of one sugar beet line as sulfonated probe. Lanes 1, 4, 6: 200 ng of native mitochondrial DNA from different male-sterile sugar beet lines. Lanes 2, 5, 7: 200 ng of native mitochondrial DNA from different male-fertile sugar beet lines. Lane 3: λ phage DNA digested with Hind III and Eco RI (courtesy of P. LEBACQ).

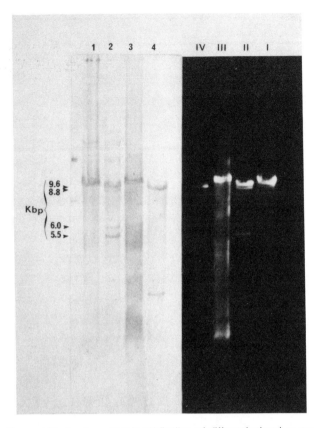

Figure 140. Southern blot hybridization of different wheat or rye nuclear DNAs with wheat spacer rDNA sulfonated or ^{32}P-radiolabeled probe. In each lane, 2 μg of DNA have been digested by Eco RI. Lanes 1, 2, 3, 4 correspond respectively to DNA of *Triticum aestivum, Triticum turgidum, Aegilops umbellulata, Secale cereale* hybridized with sulfonated probe. Hybridization was performed at a probe concentration of 0.5 μg/ml for 16 hours at 42°C in 50% formamide. Roman numerals correspond to the same DNA species and concentration but hybridization has been carried out with ^{32}P-radiolabeled DNA probe. Hybridization was performed at a probe concentration of 10 ng/ml (specific radioactivity: 10^8 cpm/μg of DNA). Autoradiography was carried out for 16 hours without intensifying screen (courtesy of P. LEBACQ).

Because of the high thermal stability of RNA/DNA duplexes, it is possible to perform the posthybridization washes at higher temperatures, therefore leading to a notably decreased background with high-specific activity probes. The maximum sensitivity of the hybridization method is obtained with probes whose length are in the range of 50–200 bases. Reduction of the RNA length is easily achieved by limited alkaline hydrolysis.

Radiolabeled RNA Probe Synthesis with SP6, T3 and T7 RNA Polymerases

In the example given below, SP6 RNA polymerase is used.

PROTOCOL

SYNTHESIS OF RADIOLABELED RIBOPROBES WITH SP6 RNA POLYMERASE

1. Prepare a 2.5 m*M* rXTP mix according to the following chart:

Labeled Nucleotide Triphosphate	10 m*M* rATP (μl)	10 m*M* rCTP (μl)	10 m*M* rGTP (μl)	10 m*M* UTP (μl)	Sterile H$_2$O (μl)
[^{32}P]-ATP	—	10	10	10	10
[^{32}P]-CTP	10	—	10	10	10
[^{32}P]-GTP	10	10	—	10	10
[^{32}P]-UTP	10	10	10	—	10

2. In a microfuge tube, mix successively:
   4.0 μl of 5X transcription buffer
   2.0 μl of 100 m*M* DDT
   0.8 μl of RNasin (25 units/μl)
   4.0 μl of the 2.5 m*M* rXTP mix (step 1)
   2.4 μl of 100 μ*M* XTP used for labeling (i.e., CTP when using ^{32}P CTP; 12 μ*M* final)
   1.0 μl of linearized DNA (2 mg/ml)
   5.0 μl of [^{32}P] XTP (50 μCi; >400 Ci/mmol)
3. Add 0.8 μl of SP6 polymerase (8 units).
4. Incubate for 1 hour at 37°C.
5. Add 2 μl of DNase I (1 unit/μl).
6. Incubate 15 minutes at 37°C.
7. Add 50 μl of a 1:1 phenol/chloroform:isoamyl alcohol (24:1) mixture.
8. Extract and recover RNA as described in Chapter 3.

Notes

1. The reaction can be performed in the absence of the unlabeled nucleotide corresponding to the one used as tracer. In this case use 100 μCi of a nucleotide triphosphate with a specific radioactivity of 400 Ci/mmol (or equivalent) to obtain a concentration of 12 μ*M*.
2. Lowering the XTP concentration to 10–20 μ*M* (necessary to obtain satisfactory incorporation) is accompanied by a reduction of full-length transcripts length, due to premature transcription termination. Under such conditions, the yield of RNA

Table 62. Yield and Size of RNA Transcripts Produced in SP6 RNA Polymerase Reactions Containing Varying Concentrations of UTP[a]

		RNA Yield				RNA Product Size Distribution					
Reaction	[UTP] (μM)	Size of Template (kb)	Yield of RNA (μg)	Specific Radio-activity of RNA (dpm/μg)	% ^{32}P-UTP Incor-porated	7.5–10 kb (%)	6–7.5 kb (%)	4–6 kb (%)	2–4 kb (%)	1–2 kb (%)	<1 kb (%)
1	10	10	0.10	1×10^8	79	4	7	15	34	30	10
2	13	10	0.12	1×10^9	76	0.2	0.6	4	27	43	25
3	25	10	0.23	3×10^7	93	11	10	22	32	19	7
4	400	10	1.8	1.5×10^6	42	23	52	8	7	6	4
5	10	1.3	0.10	4×10^8	82	—	—	—	—	71	29

Source: Courtesy of Bethesda Research Laboratories, Life Technologies.

[a]All reactions were performed in a 10-μl volume, terminated after 1 hour incubation at 37°C, extracted and glyoxylated as described in chapter 18. Samples were electrophoresed on a 1.2% agarose gel, the gel was dried, exposed to Kodak XAR-5 film, cut into size regions, and counted (Cerenkov) to assess the relative size distribution. One μl of each reaction was removed at the end of the incubation and diluted 100-fold. Aliquots were spotted on GF/C filters for determination of specific radioactivity, acid precipitable counts, and % of ^{32}P-UTP incorporated.

synthesis is proportional to the XTP concentration (Table 62). At 10 μM and 25 μM UTP concentration, only 11 and 21%, respectively, of the transcripts were greater than 6 kb, with a 10-kb template (Table 62 and Figure 141). About 40% of the transcripts synthesized in the presence of 10 μM UTP were less than 2 kb long.

3. The use of short templates (less than 2 kb) allows the synthesis of heavily labeled RNA probes with the correct length. If representative probes are needed for DNA regions longer than 2 kb, subcloning of short DNA pieces is advised, to avoid the synthesis of probes enriched with sequences representative for sequences proximal to the SP6 promoter (see note 2).

4. When specific radioactivity is increased to 10^9 cpm/μg, radiochemical damage of the RNA transcripts may be observed (Table 62 and Figure 141).

5. While [^{35}S]uridine 5′ (α-thio) triphosphate can advantageously replace α[^{32}P]uridine triphosphate for in vitro synthesis of labeled probes, it has been found that [^{35}S]adenosine 5′ (α-thio) triphosphate is not an efficient substrate for SP6 RNA polymerase.

6. Specific radioactivity of RNA probes is dependent upon the labeled to unlabeled nucleotide concentration ratio (Table 62).

7. T3 and T7 RNA polymerases may be used as well as SP6 RNA polymerase to synthesize labeled RNA probes (Figures 142, and 143).

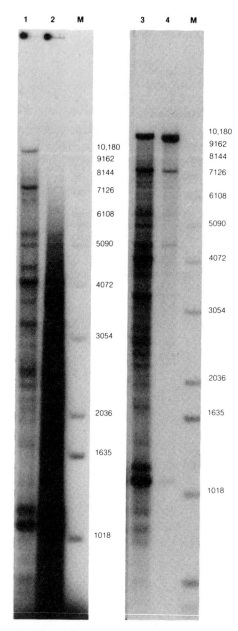

Figure 141. Size distribution of SP6 RNA polymerase transcripts prepared at varying UTP concentrations. Ten μl SP6 RNA polymerase reactions were performed as described in text: (lane 1) 10 μM UTP, 1 × 10⁸ dpm/μg RNA; (lane 2) 13 μM UTP, 1 × 10⁹ dpm/μg RNA; (lane 3) 25 μM UTP, 3 × 10⁷ dpm/μg RNA; (lane 4) 400 μMUTP, 1 × 10⁶ dpm/μg RNA. M indicates DNA molecular weight markers. Reaction products were extracted, glyoxylated, and electrophoresed on a 1.2% agarose gel. The dried gel was exposed to Kodak XAR-5 film. Because the film was not preflashed prior to exposure, the autoradiogram does not give a quantitative representation of size distribution of RNA transcripts. Size regions of the gel were cut and counted (Cerenkov) to accurately assess size distribution of products (Table 62) (courtesy of Bethesda Research Laboratories, Life technologies Inc.).

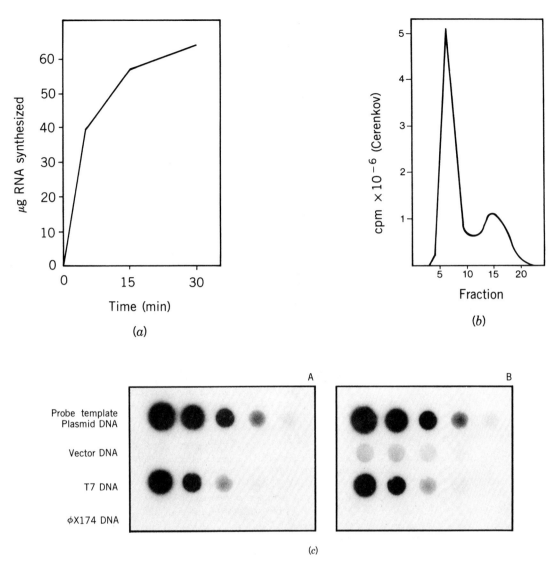

Figure 142. (*a*) Synthesis of RNA using T7 RNA polymerase and GeneScribe™ vector pT7-1. Reactions (0.1 ml) contained 40 m*M* Tris–HCl (pH 8.0); 15 m*M* MgCl$_2$; 5 m*M* DTT; 0.5 mg/ml BSA; 1 m*M* each ATP, CTP, GTP, and (³H)UTP (5 cpm/pmol); and 2 μg super-coiled plasmid pT7-1 DNA. Reactions were started by the addition of T7 RNA polymerase (25 units) and incubated at 37°C for the times indicated after which acid-insoluble radioactivity was determined. After 30 minutes, 50% of the ribonucleoside–triphosphates had been incorporated into RNA. (*b*) Gel filtration chromatography of ³²P-labeled RNA probe. Radiolabeled RNA was synthesized in a reaction (25 μl) containing 1 m*M* each of ATP, CTP, and GTP, 0.065 m*M* α[³²P]UTP (25 μCi), 0.5 μg linear probe template DNA, and 50 units T7 RNA polymerase. After incubation at 37°C for 20 minutes, the reaction was stopped by the addition of excess EDTA. The entire reaction volume was then applied to a Sephadex G-75 column (1.2 ml bed vol.) equilibrated with 0.15 *M* NaCl, 30 m*M* Tris (pH 8.0), 1 m*M* EDTA. Fractions (0.13 ml) were collected and radioactivity determined by Cerenkov counting. Over 70% of the radioactivity was incorporated into probe RNA. (*c*) Filter hybridization of DNA using radiolabeled RNA probe. Denatured DNA was immobilized on nitrocellulose filters (Kafatos et al., 1979) using serial, fivefold dilutions for adjacent dots. Filters were prehybridized in a standard solution containing 50% formamide at 45°C, 5 hours, hybridized in the same solvent (5 ml) with probe added for 16 hours, then washed 4 times without probe. Panel A, probe was pooled fractions 5–8 of Figure 142(*b*); Panel B, probe was 10 μl of reaction mix similar to that of Figure 142(*b*), using 0.5 μg of supercoiled plasmid as template. Probe template DNA contained sequences of bacteriophage genes 1.1 and 1.2 inserted into plasmid pBR322. Exposure to X-ray film was 90 minutes at room temperature (courtesy of United States Biochemical Corporation).

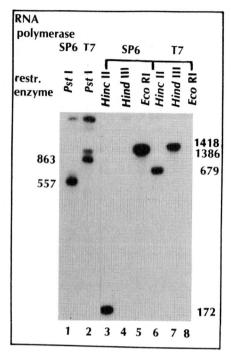

Figure 143. Transcription of Riboprobe Gemini control plasmid DNA linearized with different restriction enzymes by SP6 and T7 RNA polymerases. Template DNAs (0.5 μg) were transcribed using 7 units of SP6 or T7 RNA polymerase as indicated. Reactions contained, in a total volume of 10 μl, 40 mM Tris–HCl, pH 7.9, 10 mM NaCl, 10 mM DTT, 6 mM MgCl$_2$, 2 mM spermidine, 0.5 mM each of ATP, CTP, GTP, and UTP, 2 μCi α[^{32}P]CTP, template DNA, and SP6 RNA polymerase. The mixtures were incubated at 37°C for 60 minutes followed by the addition of 10 μl formamide containing 10% sucrose and tracking dyes. The samples were heated at 60°C for 5 minutes, allowed to cool, and applied to a 4% polyacrylamide gel containing 7 M urea. Following electrophoresis the undried gel was exposed to X-ray film with intensifying screens for 10 minutes (courtesy of Promega Biotec).

Buffer

5X Transcription buffer (for SP6)	1 ml
200 mM Tris–HCl (pH 7.5 at 37°C)	200 μl 1 M Tris–HCl (pH 7.5)
30 mM MgCl$_2$	30 μl 1 M MgCl$_2$
10 mM Spermidine	100 μl 0.1 M Spermidine
50 mM NaCl	10 μl 5 M NaCl

Store as 100 μl aliquots at -20°C.

In Vitro Synthesis of Nonradioactive Labeled RNA Probes with SP6, T3, and T7 RNA Polymerases

SP6, T3, and T7 RNA polymerases can be used to prepare biotinylated RNA probes for nonradioactive detection of DNA (see Chapter 15). This method may find interesting applications in the field of in situ hybridization because of the increased detection sensitivity and reduction of noise levels associated with the use of DNA probes.

The reaction mixture is prepared as described for the in vitro synthesis of radiolabeled probes except that it contains 1 mM of the modified nucleotide biotin 11 UTP (analogue of UTP with a biotin moiety attached through a 11-atom linker at the 5′ position of the pyrimidine rind instead of UTP. The T3 RNA polymerase was found to incorporate biotin 11-UTP three times as much as SP6 RNA polymerase and twice as much as T7 RNA polymerase (Table 63).

The allylamine UTP derivative has also been found to be a good substrate for SP6, T3, and T7 RNA polymerases. This compound was used efficiently by all three enzymes. Interestingly, the allylamine derivative was incorporated more effectively in RNA than UTP (138% of the control reaction) with SP6 RNA polymerase (Table 64).

Note. As stated before for biotynilated DNA, phenol extraction should be avoided. After DNase treatment, separate RNA from nucleotides on small columns.

Table 63. Incorporation of Biotin-11-UTP by Phage RNA Polymerases

Enzyme	Units[a]	RNA Yield[b] (μg)	
		UTP	Biotin-11-UTP
SP6	15	4.8 (4)	0.6 (2)
T7	20	4.1 (4)	0.9 (2)
T3	20	6.0 (4)	1.9 (2)
T3	100	9.5 (2)	4.5 (1)

Source: Courtesy of Bethesda Research Laboratories, Life Technologies.

[a]One unit incorporates 1 nmol of labeled nucleotide into acid precipitable material in 1 hour at 37°C.

[b]The number of determinations is shown in parentheses.

REFERENCES

Church, G. M., and Gilbert, N. (1984), *Proc. Natl. Acad. Sci. USA,* **81**, 1991.

Green, M. R., Maniatis, T., and Melton, D. A. (1983), *Cell,* **32**, 681.

Krainer, A. R., Maniatis, T., Ruskin, B., and Green, M. R. (1984), *Cell,* **36**, 993.

Krieg, P. A., and Melton, D. A. (1984), *Nature,* **308**, 203.

Krieg, P. A., and Melton, D. A. (1984), *Nucl. Acids Res.,* **12**, 7057.

Melton, D. A., Krieg, P., Rebagliati, M., Maniatis, T., Zinn, K., and Green, M. (1984), *Nucl. Acids Res.,* **12**, 7035.

Melton, D. A. (1985), *Proc. Natl. Acad. Sci. USA,* **82**, 144.

Zin, K., DiMaio, D., and Maniatis, T. (1984), *Cell,* **34**, 865.

Table 64. Preparation of RNA Transcripts Using Biotin-11-UTP and Allylamine-UTP Nucleotide Analogues[a]

Reaction	Nucleotide Analogue	Units SP6	μg RNA Produced	% Yield
1	UTP (400 μM)	30	24	100
2	Allylamine UTP (400 μM)	30	33	138
3	UTP (400 μM)	30	24	100
4	Biotin-11-UTP (1 mM)	30	3.3	13
5	Biotin-11-UTP (1 mM)	90	6.6	27
6	Biotin-11-UTP (1 mM)	200	11	45

Courtesy of Bethesda Research Laboratories, Life Technologies.

[a]Reactions were performed in 200-μl volumes containing 400 μM UTP, 400 μM AA-UTP, or 1 mM biotin-11-UTP using [^{3}H] GTP as a trace label. Reactions were incubated at 37°C for 90 minutes except for tubes 1 and 2 which were incubated for 60 minutes. Aliquots were removed at intervals during the incubation to follow the time course of incorporation into acid precipitable RNA.

17

PREPARATION OF GENOMIC LIBRARIES

Because conventional biochemical methods did not allow the isolation of single copy genes from complex genomes, methods have been developed for preparing libraries from partially digested DNA (Maniatis et al., (1978)). Random fragments are cloned and if the number of clones is large enough, it is hoped that any gene can be isolated by screening the library. Construction and screening of libraries have been made possible by the use of λ-derived cloning vectors (see Chapter 6), in vitro packaging of recombinant DNA (Hohn and Murray, 1977; Sternberg et al., 1977) and in situ hybridization for screening recombinant plaques (Benton and Davis, 1977).

References

Benton, W. D., and Davis, R. W. (1977), *Science,* **196,** 180.

Hohn, B., and Murray, K. (1977), *Proc. Natl. Acad. Sci. USA,* **74,** 3529.

Maniatis, T., Hardison, R. C., Lacy, E., Lauer, J., O'Connell, C., and Quon, D. (1978), *Cell,* **15,** 687.

Sternberg, N., Tiemeier, D., and Enquist, L. (1977), *Gene,* **1,** 255.

THEORETICAL STUDY OF THE FRACTION OF A LONG-CHAIN DNA THAT CAN BE INCORPORATED IN A RECOMBINANT DNA PARTIAL-DIGEST LIBRARY*

Theory

The Basic Assumption and Some Simple Results. We assume that the probability of occurrence of a restriction enzyme cleavage site is the same for all internucleotide bonds. This requires that cleavage sites occasionally fall on adjacent bonds, even though the minimum distance between sites should be four to six base pairs in most cases (because enzyme recognition sequences cannot overlap). In the calculations presented below, the contribution of terms due to sites at adjacent bonds has been neglected, on the assumption that the average distance between sites will be large. This accords well with the observations of Botchan et al. (1973) and Hamer and Thomas (1975).

Following the complete digestion of a complex DNA with a restriction enzyme, the fraction of DNA that can be incorporated in the gene library (the clonable fraction) is the weight fraction of DNA carried by restriction fragments smaller than the max-

---

*Reproduced with permission from Seed, B. (1982), *Biopolymers,* **21,** 1793.

imum and larger than the minimum length chosen for attachment to the viral carrier DNA. This fraction can be written

$$(pL + 1)e^{-pL} - [p(L + r) + 1]e^{-p(L + r)} \qquad (1)$$

where $L + 1$ and $L + r$ are the minimum and maximum lengths in base pairs, and p is the frequency of occurrence of the restriction enzyme cleavage site. When $L = 0$, the fraction clonable is

$$1 - (pr + 1)e^{-pr} \qquad (2)$$

We note here for future comparison that equation (2) is an approximation: the exact result is

$$1 - (pr + 1)(1 - p)^r \qquad (2')$$

which is negligibly different from equation (2) when p is small.

A far more difficult problem is the fraction of sequences that can be found among restriction enzyme partial-digest products having lengths between $L + 1$ and $L + r$ base pairs. Obviously, partial digestion allows some of the complete-digest fragments shorter than $L + 1$ base pairs to be incorporated in the library. Fragments larger than $L + r$ can never be cloned, and so

$$1 - [p(L + r) + 1](1 - p)^{(L + r)} \qquad (3)$$

is an overestimate of the fraction clonable when $L = 0$. We will return to this formula later in this section.

Background, Nomenclature, and Conventions. We have derived two expressions for the fraction of sequences that can be found among restriction enzyme partial-digest products having lengths between $L + 1$ and $L + r$ base pairs. The first is a product of recursively defined conditional probabilities, representing the fraction of sequences that cannot be cloned. The second is a probability sum representing the union of all clonable sequences. The clonable fraction is not explicitly computable from either expression, but rigorous under- and overestimates can be formed and computed from both formulas.

For convenience we will briefly review here the properties of conditional probabilities. If the probability of A given B is $P(A|B)$ and the probability of the intersection of A and B is $P(AB)$, then $P(AB) = P(A|B)P(B)$. We shall frequently use this formula in

the form $P(A|B) = P(AB)/P(B)$. When A and B are independent events $P(A|B) = P(A)$, and $P(AB) = P(A)P(B)$. The probability of the union of two events, $P(A \cup B)$ is the sum of the probabilities of the two individual events minus the probabillity of their intersection: $P(A \cup B) = P(A) + P(B) - P(AB)$. The conditional probability analog of this equation is $P(A \cup B|C) = P(A|C) + P(B|C) - P(AB|C)$.

As above, let $L + 1$ and $L + r$ be the minimum and maximum lengths in base pairs of the inserted DNA, and let p be the mean frequency of occurrence of a restriction site. We choose a particular nucleotide at random and derive the probability that the nucleotide cannot be cloned. The probability that the nucleotide cannot be cloned is also the fraction of all nucleotides that cannot be cloned. To derive this fraction, we first examine the bond immediately adjacent to the chosen nucleotide (e.g., on the left—see Figure 144) and calculate the probability that a clonable partial-digest fragment cannot terminate at that bond. We then calculate the probability that a clonable partial-digest fragment cannot terminate at the second bond, given failure at the first, and so on. As a matter of terminology, we call the successively enumerated bonds "adjacent bonds." To each adjacent bond there correspond r "window" bonds, which represent the acceptable end points for clonable partial-digest fragments terminating at the adjacent bond. The product of the conditional probabilities of the successive failure of the first $L + r$ trials is the probability that the nucleotide cannot be cloned at all.

Figure 144 is a schematic diagram of the indexing system we have adopted. The first adjacent bond falls immediately to the left of the index nucleotide. The first window consists of the r bonds which fall immediately to the right of the $(L + 1)$th nucleotide.

The Fraction of Sequences That Are Not Clonable. Our first approach will lead to a probability product of the form

$$\prod_{i=0}^{L+r} K(i,r,L)$$

representing an intersection of failures at every trial. $K(i,r,L)$ is the probability that a clonable partial-digest fragment cannot arise at the ith adjacent bond, given $i - 1$ previous unsuccessful attempts. $K(0,r,L) = 1$ by definition.

When $L + 1 > r$, there are three forms of recur-

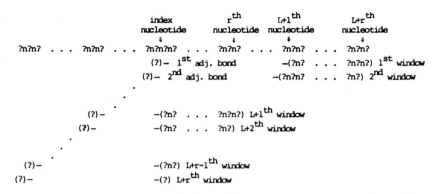

Figure 144. Schematic diagram illustrating the labeling conventions of the text. Nucleotides (the DNA monomers) are represented by the symbol _n_, internucleotide bonds by the symbol "?" (signifying no information about the presence or absence of a cleavage site). The first "window" contains _r_ bonds _L_ + 1 nucleotides to the left of the first "adjacent" bond. Successive trials initiate at progressively more leftward adjacent bonds. (With permission of B. Seed.)

sion relation for $K(i,r,L)$. The first is applicable when $1 < i < r$, the second when $r + 1 < i < L + 1$, and the third when $L + 2 < i < L + r$. Similarly, for $L + 1 < r$ there are three forms applicable: when $1 < i < L + 1$, when $L + 2 < i < r$, and when $r + 1 < i < L + r$. For our first case we take $L + 1 > r$.

The Fraction Not Clonable When L + 1 > r. In general $K(i,r,L)$ is the union of the conditional probabilities that either (a) no site will be found at the _i_th adjacent bond, given $i - 1$ previous failures, or (b) no site will be found in the _i_th window, given $i - 1$ previous failures. (a) is just $(1 - p)$ and (b) we will call $w(i)$. The union is the sum of both conditional probabilities minus their intersection, or

$$K(i,r,L) = 1 - p + w(i) - (1 - p)w(i) = 1 - p + pw(i) \quad (4)$$

When $i = 1$ there have been no previous failures, so $w(1)$ is the unconditional probability that no sites fall in the first window, or $(1 - p)^r$. Hence,

$$K(1,r,L) = 1 - p + p(1 - pi)^r$$
$$= 1 - p[1 - (1 - p^r)] \quad (5)$$

On the second trial, the window is displaced one bond to the left (Figure 144). The probability that a site falls on the leftmost bond of the second window shows no dependence on the distribution of sites in the first window, and is just $(1 - p)$. Hence, we can write $w(2) = (1 - p)d(2)$, where $d(2)$ is the probability that no sites will be found among the $r - 1$

rightmost bonds of the second window, given failure on the first trial. In general, for $i < L + 2$,

$$w(i) = (1 - p)d(i) \quad (6)$$

where $d(i)$ is the conditional probability that there will be no sites in the $r - 1$ rightmost bonds of the _i_th window, given $i - 1$ previous failures. Since the $r - 1$ rightmost bonds of the _i_th window are identical to the $r - 1$ leftmost bonds of the $(i - 1)$th window, $d(i)$ can be written as

$$d(i) = \frac{s(i - 1)}{K(i - 1,r,L)}$$

where $(i - 1)$ is the probability that no sites fall among the $r - 1$ leftmost bonds of the $(i - 1)$th window, given $i - 2$ previous failures, and $K(i - 1,r,L)$ is the probability that the $(i - 1)$th trial fails, given $i - 2$ previous failures.

There are two ways to form a window that has no sites in its $r - 1$ leftmost bonds. One form has no sites at all in the window and the other a single site at the rightmost bond. Since the two possible arrangements are mutually exclusive, we can write $s(i - 1)$ as a sum of $w(i - 1)$ and $t(i - 1)$. $t(i - 1)$ is the probability that both (a) the only site in the $(i - 1)$th window falls on the rightmost bond and (b) the $(i - 1)$th trial fails, given $i - 2$ failures. $t(1)$ thus equals $p(1 - p)^r$; $p(1 - p)^{r-1}$ is the probability that the sole site in the first window falls on the rightmost bond; and $(1 - p)$ is the probability that the first adjacent bond has no site. $t(1)$ is an unconditional probability.

For $i > 2$, however, $t(i - 1)$ depends on the existence of $i - 1$ previous failures. In particular, because $t(i - 1)$ describes the probability of a site existing at the rightmost bond of the $(i - 1)$th window, it is necessary to stipulate that none of the preceding adjacent bonds whose windows overlap that site can themselves bear restriction sites. For example, $t(2) = p(1 - p)^{r + 1}/K(1,r,L)$. The extra factor of $(1 - p)$ represents the probability that no site falls on the first adjacent bond, and the divisor of $K(1,r,L)$ converts an unconditional probability into a conditional one, given failure on the first trial. In general, in the region defined by $2 < i < r$,

$$t(i - 1) = \frac{p(1 - p)^{r - 1}(1 - p)^{i - 1}}{\prod_{j=0}^{i-2} K(j,r,L)} \qquad (7)$$

The factor of $(1 - p)^{i - 1}$ ensures that the $i - 1$ preceding adjacent bonds will not give rise to clonable fragments terminating at the last bond of the $(i - 1)$th window. The product of $K(i,r,L)$ in the denominator converts the numerator into a conditional probability, given $i - 2$ previous failures. Collecting results, we find

$$K(i,r,L) = \frac{1 - p + p(1 - p)[w(i - 1) + t(i - 1)]}{K(i - 1,r,L)} \qquad (8)$$

and since

$$w(i - 1) = \frac{[K(i - 1,r,L) - (1 - p)]}{p} \qquad (9)$$

we have

$$K(i,r,L) = (1 - p)$$
$$\times \frac{2 - (1 - p)/K(i - 1,r,L) + p^2(1 - p)^{r + i - 2}}{\prod_{j=0}^{i-1} K(j,r,L)} \qquad (10)$$

in the region defined by $2 < i < r$.
When $r + 1 \leq i \leq L + 1$,

$$t(i - 1) = \frac{p(1 - p)^{2r - 1}}{\prod_{j=i-r}^{i-2} K(j,r,L)} \qquad (11)$$

and

$$K(i,r,L) = (1 - p)$$
$$\times \frac{2 - (1 - p)/K(i - 1,r,L) + p^2(1 - p)^{2r - 1}}{\prod_{j=i-r}^{i-1} K(j,r,L)} \qquad (12)$$

in a fairly simple extension of equation (10).

In the region defined by $L + 2 < i < L + r$, the window contracts on successive trials without exposing any new bonds (Figure 144). Thus, $w(i) = d(i)$, and the factor of $1 - p$ in equation (6) describing the absence of a site at the leftmost bond of the window is lost. To derive $(i - 1)$ here, we note first that if $i = L + k$, there are $r - (k - 1)$ bonds in the ith window, and the rightmost bond of the $(i - 1)$th window is $r - k + 2$ bonds distant from the index nucleotide. $p(1 - p)^{r - k + 1} = p(1 - p)^{L + r + 1 - i}$ is the probability that the sole site in the $(i - 1)$th window falls on the rightmost bond. Since the rightmost bond also falls in the $r - 1$ preceding windows, we must include a factor of $(1 - p)^r$ for the probability that none of these adjacent bonds contain restriction sites. Hence,

$$t(i - 1) = \frac{p(1 - p)^{L + 2r + 1 - i}}{\prod_{j=i-r}^{i-2} K(j,r,L)} \qquad (13)$$

and

$$K(i,r,L) =$$
$$\frac{2 - p - \dfrac{(1 - p)}{K(i - 1,r,L)} + p^2(1 - p)^{L + 2r + 1 - i}}{\prod_{j=i-r}^{i-1} K(j,r,L)},$$
$$L + 2 \leq i \leq L + r \qquad (14)$$

The Fraction Not Clonable When $L + 1 < r$. For $L + 1 < r$, a different treatment is necessary. In the region defined by $L + 2 < i < r + 1$, successive windows contract, making $p(1 - p)^{L + r + 1 - i}$ the a priori probability that the sole site of the $(i - 1)$th window falls on the rightmost bond; however, there are only $i - 1$ adjacent bonds that can be said to have no sites [with probability $(1 - p)^{i - 1}$], and so

$$t(i-1) = \frac{p(1-p)^{r+L}}{\displaystyle\prod_{j=0}^{i=2} K(j,r,L)} \qquad (15)$$

Now, $w(i) = s(i-1)/K(i-1)$ as in the previous analysis of a contracting window; thus,

$$K(i,r,L) =$$
$$2 - p - \frac{(1-p)}{K(i-1,r,L)} + \frac{p^2(1-p)^{r+L}}{\displaystyle\prod_{j=0}^{i-1} K(j,r,L)},$$
$$L + 2 \leqslant i \leqslant r + 1 \qquad (16)$$

For $i < L + 1$, equation (10) holds, and for $i > r + 1$, equation (12) holds.

The preceding analysis can be applied to the case $L = 0$; it yields the correct form for the fraction clonable, equation (2) (Seed, 1981). An exact result can also be obtained when $r = 1$, and the formulas above yield the correct equation in this case as well.

The Fraction of Clonable Sequences. The preceding formulas have described the fraction of sequences that cannot be incorporated in a partial digest library. In the following, we describe the complement of this fraction, the union of all clonable sequences. We will find it convenient to choose a set of mutually exclusive events to describe the desired union. The probability of the union will then be the sum of the probabilities of the individual events. To implement this, at the ith adjacent bond we derive the probability that the first $i - 1$ trials fall, and the ith trial then succeeds. No two such combinations of different trial outcomes can ever occur simultaneously. The sum over all i of the probability that the ith trial yields the first success is the probability that at least one trial succeeds.

Formally, we seek the sum

$$\prod_{i=1}^{L+r} pb(i) \qquad (17)$$

where $pb(i)$ is the probability that a clonable fragment containing the index nucleotide will terminate at the ith adjacent bond and that $i - 1$ failures have already occurred. p is the probability that a site falls on the ith adjacent bond, and $b(i)$ is the probability that at least one site falls in the ith window and that $i - 1$ failures have already occurred.

The Fraction Clonable When $L + 1 > r$. For $i < L + 2$ we follow a reasoning similar to that employed in the preceding section and write

$$b(i) = p \prod_{j=0}^{i-1} K(j,r,L) + (1 - p)c(i) \qquad (18)$$

where $p \; \Pi \; K(j,r,L)$ is the probability that the leftmost bond of the ith window contains a site and that $i - 1$ failures have occurred. $c(i)$ is the probability that the $r - 1$ rightmost bonds of the ith window contain at least one site and that $i - 1$ failures have occurred. Because the definition of $c(i)$ incorporates the failure of all prior trials, we are required, for $i > 1$, to ensure that no prior successes arise from sites falling in the ith window. Thus, whenever a restriction site falls in the ith window, there must be no sites at any of the adjacent bonds whose windows overlap that site. For example, when $i = 2$, if a restriction site falls anywhere in the rightmost $r - 1$ bonds of the second window, the first adjacent bond must not have borne a site. For $i > 2$, the situation is more complicated. For simplicity we will derive $c(i)$ in the region defined by $r + 1 < i < L + 1$ first. Figure 145 is a schematic diagram illustrating our reasoning.

The probability that the rightmost site in the ith window falls on the nth bond from the right end is $p(1 - p)^{n-1}$. When this occurs, the previous $r - n$ adjacent bonds must not have contained any restriction sites, with probability $(1 - p)^{r-n}$. The probability that the rightmost site falls on the nth bond and the preceding $r - n$ trials have failed is then $p(1 - p)^{n-1}(1 - p)^{r-n} = p(1 - p)^{r-1}$.

Now consider the $(r - n + 1)$th preceding trial (Figure 145). The window for this trial does not contain the site at the nth bond from the right end, but it does contain $n - 1$ bonds at which we know there are no restriction sites. In fact, from the $(i - r)$th trial to the $(i - r + n - 1)$th, the number of possible bonds at which sites may be found decreases by one for each trial. The effective window for each trial thus decreases with each successive trial. We have treated this situation before in deriving equation (14), which described the contracting window region $L + 2 < i < L + r$. Replacing $K(i,r,L)$ in formula (14) by $K(j,r,i - r)$, we can say that the conditional probability that the jth trial fails, given $j - 1$ failures, is, for $i - r + 1 < j < i - r + n - 1$,

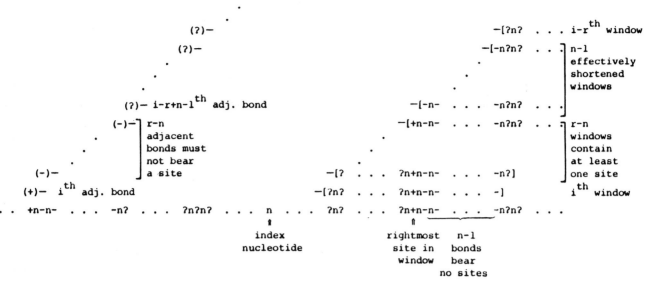

Figure 145. Schematic diagram illustrating the derivation of $c(i)$, $r + 1 < i < L + 1$. Nucleotides are represented by the symbol n as before. Internucleotide bonds are represented by the symbol + when the bond must bear an enzyme cleavage site, by the symbol − when the bond must not bear a site, and by the symbol "?" when there is no information. When the rightmost site in the *i*th window falls on the *n*th bond from the right end, the absence of sites in the $n − 1$ bonds to the right of the rightmost site affects the preceding $n − 1$ windows. The presence of the rightmost site in the *i*th window also affects the $r − n$ preceding adjacent bonds. (With permission of B. Seed.)

$$K(j,r,i - r) =$$

$$2 - p - \frac{(1 - p)}{K(j - 1,r,i - r)} + \frac{p^2(l - p)^{r+i-j}}{\prod_{m=j-r}^{j-1} K(m,r,i - r)} \tag{19}$$

The complete probability that the rightmost site in the *i*th window falls on the *n*th bond from the right end and that $i − 1$ failures have already occurred is

$$p(1 - p)^{n - 1}(1 - p)^{r - n} \prod_{j=0}^{i - r + n - 1} K(j,r,i - r) \tag{20}$$

where for the first $i − r$ trials, $K(j,r,i − r) = K(j,r,L)$. Summing over n, $1 < n < r − 1$, we get the probability that at least one site falls in the $r − 1$ rightmost bonds of the *i*th window and that $i − 1$ failures have occurred:

$$p(1 - p)^{r - 1} \sum_{n=1}^{r-1} \prod_{j=0}^{i - r + n - 1} K(j,r,i - r) \tag{21}$$

Using equations (17) and (18), the probability that the *i*th trial succeeds and all previous trials have failed is then

$$pb(i) = p^2 \prod_{j=0}^{i-1} K(j,r,L)$$

$$+ p^2(1 - p)^r \sum_{n=1}^{r-1} \prod_{j=0}^{i-r+n-1} K(j,r,i - r) \tag{22}$$

In the region defined by $1 < i < r$, formula (18) still applies. The probability that the rightmost site in the *i*th window falls on the *n*th bond from the right end is $p(1 - p)^{n-1}$. However, it is not necessarily true that there must be $r − n$ previous adjacent bonds that cannot have borne sites, since there have only been $i − 1$ prior trials (Figure 146a). If $r − n > i − 1$, the probability that the rightmost bond falls on the *n*th bond from the right end and that the previous $i − 1$ trials have failed is $p(1 - p)^{n-1} (1 - p)^{i-1}$. If $r − n < i − 1$, the probability that the rightmost site falls on the *n*th bond from the right end and that the previous $r − n$ trials have failed is $p(1 - p)^{n-1} (1 - p)^{r-n}$, as in the case first treated (Fig-

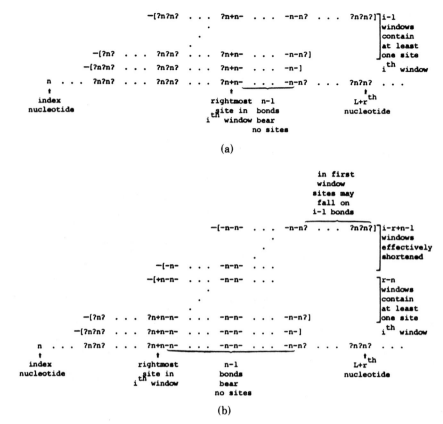

(a)

(b)

Figure 146. Schematic diagram illustrating the derivation of c(i) $1 < i < r, r - n > i - 1$. Symbols are described in Figure 145, and only the first *i* windows are shown. For $r - n > i - 1$, all prior windows contain the rightmost site of the *i*th window, and hence all prior adjacent bonds cannot bear sites. (b) Schematic diagram illustrating the derivation of c(i), $1 > i < r, r - n < i - 1$. Symbols are described in Figure 145, and only the windows are shown. The first window contains only $i - 1$ bonds that may contain restriction sites. The $(i - r + n)$th window is the first to contain the rightmost site of the *i*th window. (With permission of B. Seed.)

ure 146b). The trials preceding the $i - (r - n)$th must be described by a new formula, however. The effective window for the first trial contains only $i - 1$ bonds; for the second, $i - 2$; and so on. There are $i - 1 - (r - n)$ windows that do not contain the site at the *n*th bond of the *i*th window. The probability of the successive failure of these $i - r + n - 1$ trials is the same as the product of the first $i - r + n - 1$ terms of the fraction not clonable when $L = 0$ and $r = i - 1$, so that

$$K(j, i - 1, 0) =$$
$$2 - p - \frac{(1 - p)}{K(j - 1, i - 1, 0)} + \frac{p^2(1 - p)^{i-1}}{\prod\limits_{m=0}^{j-1} K(m, i - 1, 0)}$$

$$(23)$$

is the probability that the *j*th trial fails ($1 < j < i - r + n - 1$), given $j - 1$ previous failures. Collecting results, we find, for $r - n < i - 1$, the probability that the rightmost bond falls on the *n*th bond from the right end of the *i*th window and that $i - 1$ prior failures take place is, using expression (19),

$$p(1 - p)^{r-1} \prod_{j=0}^{i-r+n-1} K(j, i - 1, 0) = p(1 - p)^{r-1}$$

$$\times [(1 - p)^{i-r+n-1} + p(i - r + n - 1)(1 - p)^{i-1}]$$

$$(24)$$

Taking into account the trials for which $r - n > i - 1$, we find

$$c(i) = \sum_{n=1}^{r-i+1} p(1-p)^{n-1}(1-p)^{i-1} + \sum_{n=r-i+2}^{r-1}$$

$$p(1-p)^{r-1} \times \prod_{j=0}^{i-r+n-1} K(j,i-1,0) \quad (25)$$

and, using (18),

$$pb(i) = p^2 \prod_{j=0}^{i-1} K(j,r,L) + p\,(1-p)^i$$

$$\left[1 - (1-p)^{i-2} + \frac{[1-(1-p)^{r-i+1}] + p(1-p)^{r+1}}{2} \frac{p^2(i-2)(i-1)(1-p)^{i-2}}{2}\right]$$

$$(26)$$

Here, $pb(1) = p[1 - (1-p)^r]$, as expected.

Within the region defined by $L + 2 < i < L + r$, the index n can only rise to $L + r + 1 - i$, the width of the window. Here we find

$$c(i) = p(1-p)^{r-1} \sum_{n=1}^{L+r+1-i-r+n-1} \prod_{j=0}^{} K(j,r,i-r)$$

$$(27)$$

and

$$pb(i) = p^2(1-p)^{r-1} \sum_{n=1}^{L+r+1-i} \prod_{j=0}^{i-r+n-1} K(j,r,i-r)$$

$$(28)$$

There is no p^2 term because no new window bonds are unmasked.

The Fraction Clonable When L + 1 < r. In the event that $L + 1 < r$, a new treatment is needed for the region defined by $L + 1 < i < r$. As in the preceding cases, the index n can only rise to $L + r + 1 - i$. At the same time, the effective window for the rightmost adjacent bond is only $i - 1$ bonds wide. For $r - n > i - 1$, all adjacent bonds are counted in the product $p(1-p)^{n-1}(1-p)^{i-1}$, as before. The sum over n gives

$$c(i) = \sum_{n=1}^{r-i+1} p(1-p)^{n+i-1} + p(1-p)^{r-1}$$

$$\times \sum_{n=r-i+2}^{L+r+1-i-r+n-1} \prod_{j=0}^{} K(j,i-1,0) \quad (29)$$

and

$$pb(i) = p(1-p)^i[1 - (1-p)^{r-i+1}]$$

$$+ p^2(1-p)^{r-1} \times \sum_{n=r-i+2}^{L+r+1-i-r+n-1} \prod_{j=0}^{} K(j,i-1,0)$$

$$(30)$$

For $i < L + 1$, equation (26) holds, and for $i > L + 1$, formula (28) holds.

The analysis above can easily be turned to libraries for which $L = 0$, and the exact result, equation (2'), emerges (Seed, 1981).

Derivation of Computable Under- and Overestimates

An Underestimate of the Fraction Not Clonable, L + 1 > r. On examining equations (10), (12), and (14), it should be clear that a recursion formula that eliminates or underestimates the terms derived from $t(i - 1)$ must also underestimate $K(i,r,L)$. By replacing

$$\prod_{j=1}^{i-2} K(j,r,L)$$

in equation (10) by $(1 - p)^{i-r-1}$, $2 < i < r$, we form an underestimate for $K(i,r,L)$, since $1 < (1-p)^{i-r-1}$ for all i in this region, and

$$\prod_{j=1}^{i-2} K(j,r,L) < 1$$

Labeling the underestimate $K'(i,r,L)$, we find

$$K'(i,r,L) = (1-p)\{2 - [1 - p - p^2 (1-p)^{2r-1}]/[K'(i-1,r,L)]\} \quad (31)$$

a tractable one-term recursion. Substituting $K'(1,r,L) = 1 - p[1 - (1-p)^r] = K(1,r,L)$, we find $K'(2,r,L) = 1 - p[1 - (1-p)^r]$. Hence, for all $1 < i < r$, $K'(i,r,L) = K(1,r,L)$. We similarly modify (12), so that

$$\prod_{j=i-r}^{i-2} K(j,r,L)$$

is replaced by unity. Then,

$$\prod_{j=0}^{L} K'(j,r,L) = \{1 - p[1 - (1-p)^r]\}^L \quad (32)$$

In the region defined by $L + 2 \leq i \leq L + r$, we form an underestimate by replacing (14) with

$$K_u(i,r,L) = 2 - p - \frac{(1 - p)}{K_u(i - 1,r,L)} \qquad (33)$$

It is easy to show that

$$\prod_{i=0}^{j} K_u(n - i,r,L)$$
$$= \frac{[1 - (1 - p)^{j+1}]K_u(n - j,r,L)}{p - \dfrac{(1 - p)[1 - (1 - p)^j]}{p}} \qquad (34)$$

by induction (Seed, 1981). To construct the complete underestimate, we want

$$\prod_{i=1}^{r} K_u(L + r - i) \prod_{j=0}^{L} K'(j,r,L) \qquad (35)$$

Since $K'(j,r,L) = 1 - p[1 - (1 - p)^r]$, $1 \leq j \leq L + 1$, we find, after some algebra,

$$\prod_{i=0}^{r-1} K_u(L + r + 1 - i,r,L) \prod_{j=0}^{L} K'(j,r,L)$$
$$= [2(1 - p)^r - (1 - p)^{2r}]$$
$$\{1 - p[1 - (1 - p)^r]\}^L$$
$$\approx (2e^{-pr} - e^{-2pr})e^{-Lp[1 - \exp(-pr)]} \qquad (36)$$

An Underestimate of the Clonable Fraction, L + 1 < r. Using the results just derived, it is a simple matter to form an underestimate of the fraction clonable by substituting the underestimates (31) and (33) for their exact counterparts in the formulas for the fraction clonable ($L + 1 < r$). Only two substitutions need be made. The first term of formula (26) becomes $p^2 K_1^{i-1}$, where $K_1 = K(1,r,L)$; and (28) becomes

$$p^2(1 - p)^{r-1} \sum_{n=1}^{L+r+1-i} K_1^{i-r-1} \prod_{j=0}^{n-1} K_u$$
$$(i - r + n - 1 - j,r,i - r) \qquad (37)$$

is an application of (31) and (33) that parallels the construction of (35). Using equation (34), and summing (37) over i, $i = r + 1$ to $L + 1$, we get

$$p(r - L - 1)(1 - p)^r[1 - (1 - p^L]$$
$$+ \frac{p^2(1 - p)^{L+r}(L + 1)(L)[1 - (1 - p)^{r-L-1}]}{2}$$

Combining this with the remaining formulas for the clonable fraction ($L + 1 < r$), neglecting terms of order p, factors of order $(1 - p)^2$, and using the exponential approximation, we arrive at the following underestimate for the fraction clonable:

$$\sum_{i=1}^{L+r} pb'(i) \approx 1 - (pr + 1)e^{-p(L+r)}$$
$$- \frac{(pL)^2 e^{-2pr}}{2} + e^{-pL} - e^{pL[1 - \exp(-pr)]} \qquad (38)$$
$$+ e^{-2pr}\{e^{-pL[1 - \exp(;pr)]} + \frac{Lp(1 - e^{-pr}) - 1\}}{(1 - e^{-pr})^2}$$

when $L + 1 \leq r$.

A similar result has been derived from the case $L + 1 > r$, but this estimate is not as good as that described below (Seed, 1981).

An Underestimate of the Clonable Fraction, L +1 > r. A different approach to an underestimate of the clonable fraction exploits the fact that $K(i,r,L)$ must be a monotone increasing function in i for $r > 1$. This allows us to conclude that

$$K_a(i,r,L) = (1 - p)$$

$$\left[2 - \frac{(1 - p)}{K_a(i,r,L)} + \frac{p^2(1 - p)^{2r-1}}{\displaystyle\prod_{j=i-r}^{i-1} K_a(j,r,L)} \right] \qquad (39)$$

is an overestimate for $K(i,r,L)$ in the region defined by $r + 1 < i < L + 1$, since $(1 - p)/K(i - 1,r,L)$ in (12) has been replaced by the smaller $(1 - p)/K(i,r,L)$. Since $K(j,r,L) > K_1^r$,

$$K_0(i,r,L) = (1 - p)$$

$$\left[\frac{2 - (1 - p)}{K_0(i,r,L) + p^2(1 - p)^{2r-1} K_1^r} \right] \qquad (40)$$

is also an overestimate. Equation (40) can be solved for $K_0(i,r,L)$ taking the larger root of the quadratic equation (the smaller root yields $K_0 < 1 - p$). Thus,

$$K_0(i,r,L) = 1 - p + \frac{p^2 (1 - p)^{2r-1}}{2K_1^r} +$$

$$p(1 - p)^r \left[\frac{4K_1^r + p^2(1 - p)^{2r-2}}{4K_1^{2r}} \right]^{1/2} \qquad (41)$$

and

$$K_0(i,r,L) = K_0(r,L) \qquad (42)$$
$$= 1 - p + p(1 - p)^r K_1^{-r/2} + O(p^2)$$

This treatment can be extended to $i > L + 1$. We first note that for $i < L + 1$,

$$w(i) = (1 - p) \frac{[w(i - 1) + t(i - 1)]}{K(i - 1, r, L)} \qquad (43)$$

which may be written

$$w(i) = \frac{[w(i - 1) + t(i - 1)]}{(1 + x)}$$

Since $w(i) \geqslant w(i - 1)$, $t(i - 1) \geqslant xw(i - 1)$ or $t(i - 1) \geqslant [w(i - 1)]^2 p/(1 - p)$. Then, since

$$t(i) = \frac{p(1 - p)^{2r-1}}{\displaystyle\prod_{j=i-r+1}^{i=1} K(j,r,L)} \qquad (44)$$

and because it can be shown that

$$w(i) = \frac{(1 - p)^r \displaystyle\prod_{j=0}^{i-1} K(j,r,i - r)}{\displaystyle\prod_{j=0}^{i-1} K(j,r,L)} \qquad (45)$$

it follows that

$$\prod_{j=i-r+1}^{i-1} K(j,r,L) \geqslant \left[\prod_{j=i-r+1}^{i-1} K(j,r,i - r) \right]^2 \qquad (46)$$

and so

$$[K_0(r,L)^{L+r/2} > \prod_{i=0}^{L+r} K(i,r,L) \qquad$$

For computational convenience we will refer to a simpler form for $K_0(r,L)$ in which the term of order p^2 is ignored, and the exponential approximation is applied:

$$[K_0(r,L)^{L+r/2} \approx \exp\left\{ - p\left(\frac{L + r}{2}\right) 1 \right.$$
$$\left. - \exp\left(\frac{-pr(1 + e^{-pr})}{2}\right) \right\} \qquad (47)$$

For completeness we mention one additional approximation. This formula is similar to equation (36) and is used in the derivation of the fraction of sequences incorporated in a finite partial-digest library (see below). Under the conditions to be described there,

$$\exp\left[- p\left(L + \frac{r}{2}\right)(1 - e^{-pr}) \right] \qquad (48)$$

is an underestimate of the fraction not clonable.

Comparison of the Estimates

In Figure 147 we have plotted three different approximations for the fraction of sequences that cannot be cloned in two different types of DNA cloning experiments. The decimal logarithm of the fraction not clonable is plotted as a function of the size range of acceptable inserts, r. For the upper three curves, the minimum length was chosen to be 14 kilobase pairs (kbp), and for the lower three, 40 kbp (typical values for two different types of recombinant DNA

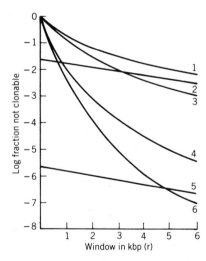

Figure 147. Logarithm of the fraction not clonable calculated from formulas (3), (36) or (47) of the text is plotted against the size range, r, of the partial cleavage fragments to be cloned. The cleavage site frequency is once per 2.5 kilobase pairs (kbp), which is the observed frequency of the site for the commonly used Eco RI enzyme in both fruit fly (Hamer and Thomas, 1975) and mouse (Botchan et al, 1973) DNAs. The upper three curves represent a library with minimum fragment length $L = 14$ kbp. The lower three curves represent a library with minimum length $L = 40$ kbp. Curves 1 and 4, 2 and 5, and 3 and 6 are derived from formulas (47), (3), and (36) respectively. (With permission of B. Seed.)

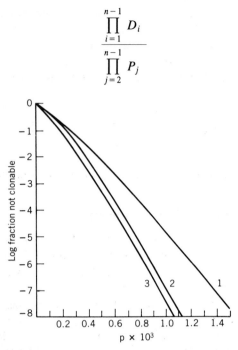

$$\frac{\prod_{i=1}^{n-1} D_i}{\prod_{j=2}^{n-1} P_j}$$

Figure 148. Effect of cleavage-site frequency on the fraction of DNA that can be cloned. Curve 1 is a plot of the logarithm of the fraction not clonable in a virus accepting DNA fragments ranging from 1 to 14 kbp (Tiemeier et al, 1976). Over- and underestimates cannot be resolved in this curve. Curves 2 and 3 are the over- and underestimates of the logarithm of the fraction not clonable in a virus accepting fragments ranging from 15 to 20 kbp. The abscissa is the frequency of occurrence of the cleavage site, *p*. (With permission of B. Seed.)

experiments). Figure 147 demonstrates that the best approximation varies with the length of the inserted DNA. Curves 1 and 4 are overestimates derived from equation (47). Curves 2 and 5 are underestimates derived from equation (3), and curves 3 and 6 are underestimates from equation (36). In all plots shown here, the influence of the exponential approximation is undetectable. In Figure 148 we have plotted the best approximations describing the fraction of a DNA that cannot be cloned as a function of the frequency of occurrence of the restriction site to be cleaved in the DNA. Curve 1 describes the fraction not clonable when $L = 1$ and $r = 14$ kbp (values appropriate to the viral cloning vehicle λgt WES) (Tiemeier et al., 1976), and the curve shown is the overestimate of the fraction not clonable [equation (38)]. The sharpest underestimate in this case is given by equation (3) and is indistinguishable from the overestimate at the resolution of this graph. Curves 2 and 3 describe over- and underestimates of the fraction not clonable when the inserted DNA

ranges from 15 to 20 kbp in length. The overestimate is given by equation (47), and the underestimate by equation (3).

Figures 147 and 148 together satisfy the simple intuitive expectation that the clonable fraction should increase with increasing L, r, or p. Of practical interest are the bounds computed for the clonable fraction in currently popular cloning vehicles. In Figure 147 we show that partial-digest fragments of mammalian DNAs created by Eco RI restriction enzyme digestion theoretically will include between 99.3 and 99.7% of all sequences when inserted in standard-capacity viral cloning vehicles. Similar larger fragments can accommodate all but 4×10^{-6} to 2×10^{-7} of a mammalian DNA. The equations derived here have been used to analyze the fraction of DNA sequences expected in libraries of finite size (see below).

In principle, the probability of any restriction enzyme recognition sequence can be calculated from the base compositions of the target DNA and the recognized sequence: for example, the frequency of occurrence of the Eco RI recognition sequence GAATTC should be $p_A^4 p_G^2$, where p_A and p_G are the frequencies of occurrence of A and G nucleotides in the target DNA ($p_A = p_T$; $p_G = p_C$). In practice, this calculation may not accurately reflect the true prevalence of sites. A mild example of this inaccuracy is the discrepancy between the measured frequency of Eco RI cleavage sites in mammalian DNAs, 1/2.5 kbp (Botchan et al., 1973), and the predicted frequency, 1/3.2 kbp. The predicted frequency can be brought in closer accord with experiment by taking dinucleotide pair probabilities into account. With this correction, the probability of an n nucleotide sequence can be written

$$\frac{\prod_{i=1}^{n-1} D_i}{\prod_{j=2}^{n-1} P_j}$$

where D_i is the probability of the ith dinucleotide, and P_j is the probability of the jth nucleotide. For example, the Eco RI frequency will be

$$\frac{D_{GA} D_{AA} D_{AT} D_{TT} D_{TC}}{P_A P_A P_T P_T}$$

and the Bam HI (GGATCC) frequency

$$\frac{D_{GG} D_{GA} D_{AT} D_{TC} D_{CC}}{P_G P_A P_T P_T}$$

Using these formulas and the dinucleotide frequency data of Swartz et al. (1962), the Eco RI frequency should be 1/2.7 kbp, which is fairly close to the observed value. Thus, with the appropriate precautions it should be possible in many cases to use the equations presented here to obtain a working estimate of the genome fraction clonable with different enzymes or DNAs.

References

Botchan, M., McKenna, G., and Sharp, P. (1973), *Cold Spring Harbor Symp. Quant. Biol.,* **38,** 383.

Hamer, D. H., and Thomas, G. A. (1975), *Chromosoma,* **49,** 243.

Kuhn, W. (1930), *Ber. Dtsch. Chem. Ges.,* **63,** 1503.

Maniatis, T., Hardison, R. C., Lacy, E., Lauer, J., O'Connell, C., Quon, D., Sim, G. K., and Efstratiadis, A. (1978), *Cell* **15,** 687.

Seed, B. (1981), Thesis, California Institute of Technology, Pasadena, California.

Seed, B., Parker, R. C., and Davidson, N. D. (1982), *Gene,* **19,** 201.

Swartz, M. N., Trautner, T. A., and Kornberg, A. (1962), *J. Biol. Chem.,* **237,** 1961.

Tiemeier, D., Enquist, L. E., and Leder, P. (1976), *Nature,* **263,** 526.

REPRESENTATION OF DNA SEQUENCES IN RECOMBINANT DNA LIBRARIES PREPARED BY RESTRICTION-ENZYME PARTIAL DIGESTION*

As previously, we will assume that every internucleotide bond has the same prospective probability, *p,* of being the location of a restriction enzyme cleavage site. This allows cleavage sites to occasionally fall on adjacent bonds, even though the minimum distance between sites should usually be 4 to 6 bp. We have neglected the exclusion of sites from adjacent bonds since the average distance between sites will be large. In an earlier study, Hamer and Thomas (1975) compared the mass distributions of restriction-enzyme-cleaved DNAs with the theoretical distributions predicted by the random dispersion of restriction sites throughout a genome. They found the random dispersion hypothesis valid within experimental error for both *E. coli* DNA and a nonrefractory component of *D. melanogaster* DNA. A

highly reiterated low complexity fraction of *D. melanogaster* DNA did not display a random cleavage pattern, however. Although the presence of randomly dispersed repeated sequences should not significantly compromise the validity of these calculations, the results presented below will not in general apply to tandemly repeated sequences (see below).

In the present section we also assume that the rates of restriction-enzyme cleavage are invariant from site to site in the DNA to be cloned. Although there is evidence for heterogeneity in the rate of Eco RI endonuclease cleavage at different sites in bacteriophage λ DNA (Thomas and Davis, 1975), no estimate yet exists for the distribution of cleavage rates in genomic DNAs from higher organisms. If significant heterogeneity is found in the genomic cleavage rates, the effective frequency of occurrence of the restriction site used for cloning will be smaller than that measured by complete digestion. We will discuss some of the consequences of rate inhomogeneity later in this section.

In the previous section, we derived over- and underestimates for the fraction of a genomic DNA which cannot be included in an infinitely large partial-digest library. We also showed how to use dinucleotide frequency data to calculate the frequency of occurrence of a restriction site with improved accuracy. In the appendix of this section we have extended an overestimate [formula (47)] or an underestimate [formula (48)] to obtain an approximate expression for the fraction of a genome expected in a finite sample of clones. Each of the two existing estimates is taken to be the null term of a Poisson distribution. The basic notion is that a distribution of fragments having lengths between L and $L + r$ bp may be replaced by a single equivalent fragment having a length $L + r/2$ bp. On the surface we expect that the validity of the approximation will decrease as the size range of permissible insert lengths, r, increases. Certainly the infinite-library under- and overestimates diverge with increasing r (Seed, 1982). However, for the rather small libraries studied here, we have found that because the total fraction clonable increases as r increases, the approximations can often converge toward the simple Poisson (Clarke and Carbon, 1976) limit.

In the appendix we have also exploited a computational simplification. We assume that all possible multiple clones containing a particular common nucleotide can be replaced by an identical number of clones each bearing the same number of restriction sites. This averaging procedure greatly simplifies the computation, but also introduces an error of uncertain magnitude in the under- and overestimates.

We do not believe this error will have a large effect on the final result, since the genome fraction represented in a library is computed from a sum over all possible clonable site configurations. On average, for each clonable fragment misrepresented by a fictitious average fragment, there should be another clonable fragment (containing wholly different sequences) which is accurately described by the fictitious fragment but is itself misrepresented by a fictitious fragment having the same characteristics as the first real fragment. The probability that the chosen nucleotide falls on one or the other type of fragment depends only on the probability of the particular combination of parameters describing that fragment. The net probability that a given nucleotide will not be found in a sample of T clones of average length $L + r/2$ bp is [Appendix, equation (A14)]

$$\sum_{m=0}^{\infty} \sum_{k=0}^{\infty} P_i(m, k) \exp\left(- \frac{F_i(u, m)G_i(u, k)mT}{Np} \right) \quad (1)$$

where N is the length in bp of the target genome, p is the frequency of occurrence of the restriction site in the genomic DNA, u is the fraction of restriction sites cleaved, and P_i, F_i, and G_i are probability functions described in the appendix. The index i takes on the value 1 or 2 depending on whether the approximation is based on the underestimate or the overestimate.

In Figure 149 we have used formula (1) to plot the decimal logarithm of the fraction of a genome absent from a library as a function of the number of library clones. The unit of the abscissa, the genome equivalent, is the number of clones required for the incorporation of one haploid complement of DNA in the library; it is defined as the number of bp in a haploid complement of DNA divided by the average number of bp per clone of the inserted DNA. The various curves simulate the representation in λ and cosmid vector libraries constructed under optimal conditions with either Eco RI or tetranucleotide-recognition-site restriction enzymes.

Figure 149 shows that it is possible to obtain good representation of genomic sequences in partial-digest libraries which have a high intrinsic capacity for those sequences. Equations presented elsewhere (Seed, 1982) show that an infinitely large cosmid library of the sort modeled by curve 3 could accommodate all but 1.6×10^{-7} to 4.3×10^{-8} of the target DNA. Similarly, the phage library modeled by curve 4 could contain all but 10^{-26} to 10^{-33} of the target DNA at saturation. Representation is appreciably poorer in low-capacity libraries, such as the phage

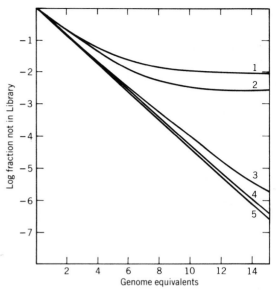

Figure 149. The fraction of genomic sequences absent from a recombinant DNA partial-digest library as a function of the size of the library. Curve 1 is the overestimate form of equation (1) applied to a library in which $L = 15$ kb, $r = 5$ kb, and $p = 1(2.5$ kb) (a model Eco RI partial phage library). Curve 2 is the underestimate for the same values. For curve 3, $L = 40$, $r = 10$, and $p = 1/(2.5$ kb) (Eco RI cosmid library); for curve 4, $L = 15$, $r = 5$ and $p = 1/(2.5$ kb) (Mbo I phage library). Only the underestimate is shown for curves 3 and 4 since the estimates differ little (see Figures 150 and 151). Curve 5 is the shear-library formula. [With permission of B. Seed. Copyright (1982) Elsevier Biomedical Press.]

library modeled by curves 1 and 2. Here saturation allows only 99.1 to 99.7% of the target sequences to be incorporated.

In Figures 150 and 151 we use formula (1) to plot the logarithm of the fraction of a genome absent from a library as a function of the degree of restriction enzyme partial digestion. The number of clones in the library is held constant for each curve. The abscissa is the number average length of the DNA prepared for cloning, expressed as a multiple of the average insert length, $L + r/2$. Optimal incorporation of sequences can be found in a library prepared by partial digestion of the sample DNA to a number average length equal to the desired average length of the cloned insert. Overdigestion beyond this optimum degree profoundly reduces the fraction of sequences incorporated, while underdigestion has little effect.

We will discuss three practical considerations related to the result shown in Figures 150 and 151: (a) underdigestion will result in less DNA available in the appropriate size range to be cloned; (b) heterogeneity in cleavage rates may have the effect of decreasing the degree of digestion required for optimal

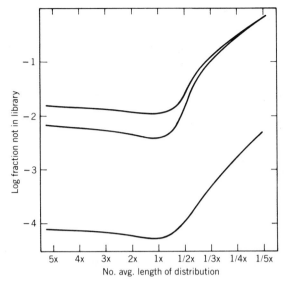

Figure 150. Effect of digestion on representation. The upper two curves represent a model Eco RI phage library (L = 15 kb, r = 5, p = 1/2.5); the lower curve represents a model Mbo I phage library (L = 15, r = 5, p = 1/0.25). Under- and overestimates cannot be resolved for the lower curve. The library size is 10 genome equivalents in each case. The abscissa is the number average length of the distribution of partially digested DNA, expressed as a multiple of the final average insert length (17.5 kb). The scale of the abscissa has been changed at 1 × to allow the effects of overdigestion to be more visible. [With permission of B. Seed. Copyright (1982) Elsevier Biomedical Press.]

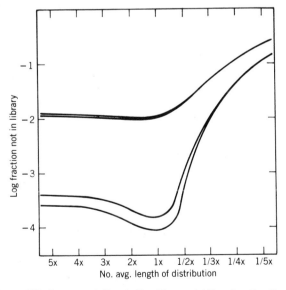

Figure 151. Representation in Eco RI cosmid libraries. The library size is five genome equivalents in the upper two curves, and 10 equivalents in the lower two. L = 40, r = 10, p = 1/(2.5 kb). Plotting coordinates are as described in Figure 150. [With permission of B. Seed. Copyright (1982) Elsevier Biomedical Press.]

representation; and (c) the degree of digestion is likely to be underestimated if the digestion is monitored by gel electrophoresis.

The fraction of a DNA sample that falls within predetermined size limits L and $L + r$ for cloning is (Kuhn, 1930)

$$(sL + 1)e^{-sL} - [s(L + r) + 1]e^{-s(L + r)}$$

where s is the reciprocal of the number average length of the digested target DNA. s is the probability that a bond chosen at random will be found cleaved in the digested DNA. $1/s$, the number average length, is the number of bp in the genome divided by the average number of cleaved bonds per genome. We have found that optimal representation results from a DNA fragment distribution having a number average length approximately equal to the desired average length of the insert, that is: $s = 1/(L + r/2)$. The distribution which optimizes the mass of DNA falling between lengths L and $L + r$, however, yields a number average length approximately one-half the desired insert length, or $s = 2/(L + r/2)$. For example, 10.4% of the mass of a digested DNA falls between 15 and 20 kb when the number average length of the DNA is 17.5 kb and 15.5% falls between the same bounds when the number average length is 8.75 kb. The same figures used in this example hold when all dimensions are increased by a factor of two, a situation likely to be encountered in cosmid cloning.

Two considerations suggest that underdigestion is likely to improve representation when restriction site cleavage rates are heterogeneous: (a) our studies show underdigestion does not strongly influence representation (Figures 150 and 151); (b) underdigestion allows fragments formed at different rates to be represented more equally in the DNA to be cloned. Consider the hypothetical DNA containing two types of restriction sites, a susceptible type designated "A," and a resistant type designated "B." We associate rates of cleavage k_A and k_B to the two types of sites, so that the fraction of uncut sites remaining as a function of time is either $e^{-k_A t}$ or $e^{-k_B t}$ at constant enzyme concentration. Now we consider two types of partial-digest fragments. One has the site sequence *ABB . . . BA*, in which the A sites on each end are cleaved, and the internal B sites are uncleaved; the second has the analogous sequence *BAA . . . AB*. Both fragments have the same number, n, of internal sites which must not be cleaved, and in each case all other possible cleavages do not yield a clonable fragment. Comparing the rates of formation of the two types of fragments allows us to discern the effects of increasing digestion on representation in the presence of heterogeneous cleavage rates.

The probability of formation of the first fragment is

$$F_1(t) = (1 - e^{-k_A t})^2 e^{-nk_B t}$$

whereas the probability of formation of the second is

$$F_2(t) = (1 - e^{-k_B t})^2 e^{-nk_A t}$$

The ratio of the two probabilities, $F_1(t)/F_2(t)$, describes the relative abundance of the two fragments, and should be as close to unity as possible for good representation. Since A is cleaved more rapidly than B, $F_1(t)/F_2(t) > 1$ for all $t > 0$. Furthermore, $F_1(t)/F_2(t)$ increases monotonically with time for any $n > 1$, which means that increasing digestion gives decreasing parity between the two types of fragments. The monotonicity of the ratio can be simply seen by expanding $F_1(t)/F_2(t)$ in the power series form

$$\left(\frac{e^{-(n-2)k_B t}}{e^{-(n-2)k_A t}}\right) \left(\frac{e^{-k_B t} + e^{-2k_B t} + \cdots}{e^{-k_A t} + e^{-2k_A t} + \cdots}\right)^2$$

When $n = 0$, $F_1(t)/F_2(t)$ approaches unity as t goes to infinity, which is an appropriate description for the relative abundance of complete digest products. This simple model demonstrates that underdigestion promotes more equal representation of partial-digest fragments when cleavage rates are heterogeneous. However, underdigestion also reduces the fraction of target DNA that falls within the desired insert range. For the 15 to 20 kb range discussed above, twofold underdigestion [$s = 1/2(L + r/2)$] leaves only 4.3% of the target in the 15 to 20 kb range, and fourfold underdigestion [$s = 1/4(L + r/2)$] leaves only 1.4% in the same range.

Confronted with the task of preparing partially digested DNA for cloning, many find it convenient to assay a number of different small-scale digestion conditions by gel electrophoresis, scaling up whichever condition yields a local ethidium fluorescence maximum at the desired length. The average length of the sample DNA is invariably overestimated by this procedure, and the size distributions that result tend to be too small for optimal representation, or even optimal utilization of the target DNA.

After electrophoresis, the high-M_r fractions of a DNA sample are compressed into a smaller length of gel than the low-M_r fractions. This compression lends undue emphasis to the fraction of nucleic acid at high M_r values. However, existing theory allows one to describe a simple relationship between the number average length of a randomly degraded nucleic acid sample and the fragment length corresponding to the maximum value of the ethidium bromide fluorescence of the fractionated sample.

If a sample of randomly cleaved DNA is fractionated so that the M_r value is linearly dependent on the fraction number, the concentration of DNA in each fraction will follow the weight fraction distribution (Kuhn, 1930)

$$w(x)dx = xs^2 e^{-sx} dx \qquad (2)$$

where x is the molecular weight and s is the frequency of cleavage (the reciprocal of the number average length). Now let y be the fraction number or distance migrated in an agarose gel or other nonlinear fractionating device. For agarose gels, for example, low-M_r DNA obeys the well-known law

$$y = b[d - \log(x)] \qquad (3)$$

where b and d are constants; isokinetic sucrose gradients yield

$$y = cS = kx^a \qquad (4)$$

where S is the sedimentation coefficient, k and c are constants, and a is generally taken to be about 0.4 for duplex DNA (Bloomfield et al., 1974). Each of these relationships may be inverted to yield $x = f(y)$. The concentration as a function of distance migrated then becomes

$$w(x)dx = f(y)s^2 - e^{-sf(y)} df(y) = W(y)dy \qquad (5)$$

Since we are interested in the physical position of the maximum of the concentration distribution, we wish to find y such that $dW(y)/dy = 0$. From equations (3) and (4) we find the maximum intensity corresponds to $x = 2/s$ and $(2 - a)/s$, respectively. Thus when the maximum value of the fluorescence distribution falls within the logarithmic migration range of an agarose gel, the length corresponding to the maximum fluorescence intensity is twice the number average length of the distribution. For very high M_r DNA, we have noticed that duplex DNA migrates on dilute agarose gels according to a power law given roughly by

$$y = k_g x^{-2} \qquad (6)$$

(k_g is a constant), which is a form of equation (4). In this case the brightest region of the gel falls at a molecular length about four times the number average length of the distribution.

The discussion above shows that a serious overestimate of the number average length of a sample can be generated by uncritically examining the distribution of fluorescence in a dilute agarose gel. It is likely that the resulting overdigestion has been resonsible for the underrepresentation of sequences in some genomic libraries, for example, the human/charon 4A library of Lawn et al. (1978) (Fritsch et al., 1980). Our calculations indicate that this library contains 90% or less of the genomic target, rather than the 99% projected on the basis of a shear library calculation (Lawn et al., 1978).

While the relationship between the number average length and the location of the maximum fluorescence can be applied to evaluate partial digestion conditions for cloning in phage libraries, the fragment lengths encountered in cosmid library construction are usually too large for this approach to be effective. However, a simple alternative allows the use of gel data to estimate the appropriate degree of digestion for cloning in cosmid (and phage) vectors. One prepares an enzymatic concentration series or time course of digestion to measure the susceptibility of the target DNA to the restriction enzyme chosen for cloning. Equal aliquots of each of the samples are loaded on a dilute agarose gel and electrophoresed under the appropriate conditions [low-voltage gradient, long time (Fangman, 1978), necessary to resolve high-M_r fragments of DNA. Blocking off all regions of the gel not containing the desired insert range, the various tracks are compared to estimate the degree of digestion that leads to the greatest amount of DNA at the desired size. As discussed above, these conditions correspond to twice the extent of digestion necessary for optimal representation.

Our final consideration is the fraction of DNA fragments bearing restriction sites at both ends after digestion. This fraction can be small if the initial DNA sample is degraded, and thus can markedly reduce the efficiency of cloning. If the number average length of a randomly sheared initial DNA is L_1, and that of the enzyme-cleaved DNA is L_2, $(1 - L_2/L_1)^2$ is the fraction of fragments which bear restriction sites at both ends. This fraction is independent of the length of molecules chosen, so that smaller fragments have the same fraction of restriction site ends as do larger fragments. To improve on the fraction of sites at each end of a distribution of fragments, it is necessary to increase the extent of digestion: choos-

ing smaller fragments from a population or partially cleaved DNAs will not improve the fraction of restriction site ends.

APPENDIX

The frequency with which a particular nucleotide appears in a finite sample of clones depends both on the number of distinguishable partial-digest fragments which may contain the nucleotide and the extent of partial digestion. Although we have not obtained an exact formula for the genome fraction present in a finite sample of clones, we have derived approximate formulas based on the over- and underestimates [formulas (47) and (48)]. While the underestimate (48) is not as good as equation (3) in this section, it shares with (47) a common structure which is useful in describing the representation of finite libraries.

In the paper by Seed (1982), the fraction clonable was calculated from a sequential analysis of the probabilities of being able or unable to form a clonable partial-digest fragment from each successive bond flanking a given index nucleotide. The resulting formulas yielded those probabilities through a series of recursion relations. In contrast, the approximations (47) and (48) can be viewed as ascribing to each bond the same probability of failure, independent of all previous trials. For approximation (48) the probability of failure at each trial is $1 - p(1 - e^{-pr})$; for (47), $1 - p(1 - e^{-pr(1 + \exp(-pr))/2})$. The number of bonds described corresponds to the approximate average length of an insert, $L + r/2$, and hence the probability of failure at every bond is either

$$[1 - p(1 - e^{-pr})]^{L + r/2}$$
$$\approx \exp\left[-p\left(L + \frac{r}{2}\right)(1 - e^{-pr})\right] \quad [A1(48)]$$

or

$$[1 - p(1 - e^{-pr[1 + \exp(-pr)]/2})]^{L + r/2} \approx \exp$$
$$\times \left[-p\left(L + \frac{r}{2}\right)(1 - e^{-pr[1 + \exp(-pr)]/2})\right] \quad [A2(47)]$$

We can now extend this treatment to describe the probability distribution of the number of restriction sites which can or cannot act as termini for clonable partial digest fragments. This distribution is the

Poisson limit of a multinomial distribution, and is given either by the formula

$$P_1(m, k) = \left[p\left(L + \frac{r}{2}\right)(1 - e^{-pr})\right]^m$$
$$+ \left[p\left(L + \frac{r}{2}\right)e^{-pr}\right]^k e^{-p(L + r/2)}/m!k! \quad \text{(A3)}$$

derived from equation (A1), or by the formula

$$P_2(m, k) = \left[p\left(L + \frac{r}{2}\right)(1 - e^{-pr[1 + \exp(-pr)]/2})\right]^m$$
$$\times \left[p\left(L + \frac{r}{2}\right)e^{-pr[1 + \exp(-pr)/2}\right]^k \times e^{-p(L + r/2)}m!k!$$
$$\text{(A4)}$$

derived from equation (A2).

Equations (A3) and (A4) represent the approximate probabilities that a contiguous sequence of $L + r/2$ bonds will contain m restriction sites than can act as a termini for clonable partial digest fragments, and k sites that cannot. Summing over k and setting $m = 0$, we recover the probabilities that the nucleotide cannot be cloned,

$$\sum_{k=0}^{\infty} P_1(0,k) = e^{-p(L + r/2)[1 - \exp(-pr)]} = e^{-X_1} \quad \text{(A5)}$$

and

$$\sum_{k=0}^{\infty} P_2(0,k) = \exp\left[-p\left(L + \frac{r}{2}\right)\right.$$
$$\left.(1 - e^{-pr[1 + \exp(-pr)]/2})\right] = e^{-X_2} \quad \text{(A6)}$$

as expected. The average number of clonable fragments that terminate at a given restriction site is either pr or $pr(1 + e^{-pr})/2$ in this model. If there are m sites giving rise to at least one clonable fragment in the first $L + r/2$ bonds flanking the chosen nucleotide, on average there will be $m - 1$ sites on each clonable partial-digest fragment which can also act as termini for clonable partial-digest fragments. This relationship merely reflects the symmetry of sites on either side of the chosen nucleotide.

The a priori probability that a given clonable nucleotide can be found on random sampling of an equally represented pool of clonable fragments is the number of distinguishable clonable fragments con-

taining the nucleotide divided by the total number of distinguishable clonable fragments in the genome. Our two approximations yield the same result for this value, m/Np, where N is the number of nucleotides in the genome. As expected, the a priori probability of finding a chosen nucleotide increases as m increases. However, as m (or, more precisely, $m + k$) increases, the probability of forming a partial-digest fragment of exactly the right length decreases. This is because the probability of forming a fragment of the desired length is proportional to $(1 - u)^{m-1}$, where u is the fraction of restriction sites cleaved. We will incorporate this effect by multiplying m/Np by a factor F, normalized so that the fraction of nucleotides which cannot be cloned is invariant to any shift in u, $0 < u < 1$. Thus we find

$$\sum_{m=1}^{\infty} P_i(m,k)F_i(u,m) = 1 - e^{-X_i} \quad \text{(A7)}$$

for the fraction clonable. From this we deduce

$$F_i(u,m) = (e^{uX_i} - e^{(u-1)X_i})\frac{(1 - u)^m}{1 - e^{(u-1)X_i}} \quad \text{(A8)}$$

where the X_i are described by either (A5) or (A6). Similar considerations for the k sites not usable for cloning cause us to introduce a second multiplicative factor G, such that

$$G_i(u,k) = (1 - u)^{ke^{uY_i}} \quad \text{(A9)}$$

where

$$Y_1 = p\left(L + \frac{r}{2}\right)e^{-pr} \quad \text{(A10)}$$

and

$$Y_2 = p\left(L + \frac{r}{2}\right)e^{-pr[1 + \exp(-pr)]/2} \quad \text{(A11)}$$

Taking all of the above factors into account, the probability of sampling a nucleotide which is known to have m acceptable sites in the $L + r/2$ leftward bonds adjacent to the chosen nucleotide is approximately

$$\sum_{j=1}^{m} \frac{F_i(u,m)G_i(u,k_i)}{Np} \quad \text{(A12)}$$

where the k_j are the average numbers of nonproductive internal sites on each set of fragments originating from a common adjacent site. The k_j are not independent with respect to j. However, assuming that all the k_j for a given nucleotide are equal (see the text for a discussion of this), the probability of not encountering the nucleotide at least once in a pool of T clones is

$$\left[\frac{1 - mF_i(u,m)G_i(u,k)}{Np} \right]^T \qquad \text{(A13)}$$

The probability that the nucleotide will be described by the variables m and k is either (A1) or (A2), depending on the approximation chosen. The net probability that a given nucleotide will not be found in a sample of T clones is

$$\sum_{m=0}^{\infty} \sum_{k=0}^{\infty} P_i(m,k) \left[\frac{1 - mF_i(u,m)G_i(u,k)}{Np} \right]^T \approx \sum_{m=0}^{\infty}$$

$$\times \sum_{k=0}^{\infty} P_i(m,k) e^{-mTF(u,m)G(u,k)/Np}$$

$$\text{(A14)}$$

References

Bloomfield, V. A., Crothers, D. M., and Tinoco, I., Jr. (1974), *Physical Chemistry of Nucleic Acids*, Harper & Row, New York.

Clarke, L., and Carbon, J. (1976), *Cell*, **9**, 91.

Fritsch, E. F., Lawn, R. M., and Maniatis, T. (1980), *Cell*, **19**, 379.

Fangman, W. L. (1978), *Nucl. Acids Res.*, **5**, 653.

Hamer, D. H., and Thomas, C. A. (1975), *Chromosoma*, **49**, 243.

Kuhn, W. (1930), *Ber. Dtsch. Chem. Ges.*, **63**, 1503.

Lawn, R. M., Fritsch, E. F., Parker, R. C., Blake, G., and Maniatis, T. (1978), *Cell*, **15**, 1157.

Seed, B. (1982), *Biopolymers*, **21**, 1793.

Thomas, M., and Davis, R. W. (1975), *J. Mol. Biol.*, **91**, 315.

PREPARATION OF LIBRARIES WITH λ CHARON PHAGES

Depending on the restriction sites used to generate the DNA fragment to be cloned, a whole series of λ vectors can be used (see Chapter 6). The procedure described below is derived from that used for Eco RI cloning with the λ charon phages.

Outline of the Procedure

1. Isolation of the DNA to be cloned.
2. Preparation of λ charon arms.
3. Ligation of the DNA fragments with the purified λ charon arms.
4. In vitro packaging.
5. Plating and amplification.

Isolation of the DNA to Be Cloned

The quantity of total DNA to be digested is largely dependent on the size of the fragments which will be cloned and the method used to purify the DNA fragments. Gel elution chromatography and centrifugation yield different efficiencies of recovery.

In our experience, centrifugation of digested DNA has led to recovered DNA of better quality. In some cases DNA fragments obtained from agarose gels were not ligated to λ arms as well as we expected.

Figure 152 outlines the procedure for the isolation of λ DNA.

Setup of the Digestion Mixture. DNA is digested as described in Chapter 8, according to the endonuclease chosen. A total of 400 μg eucaryotic DNA is generally digested with a ten-fold excess of restriction endonuclease if complete digestion is necessary. However, in most cases, as stated above, incomplete digestion must be performed to increase the probability of cloning DNA fragments representing intact copies of eucaryotic genes. When partial digestions are necessary, they are run in pilot samples as described in Figure 153. The time of incubation is determined after analysis of the digestion products by electrophoresis in agarose gels and staining. Conditions should be such that maximal staining is obtained in the region of the gel corresponding to the desired DNA fragment size.

After the appropriate conditions have been determined by pilot runs, they are scaled up for the digestion of the 400-μg DNA sample (scale up by using the same DNA/enzyme ratio as was determined in the pilot run, not by increasing the incubation time).

Inactivation of Restriction Endonucleases. Most endonuclease digestions can be stopped by the three-step procedure described previously: chilling at 0°C, addition of EDTA (20 mM final concentration), and heating at 65°C for 20 minutes.

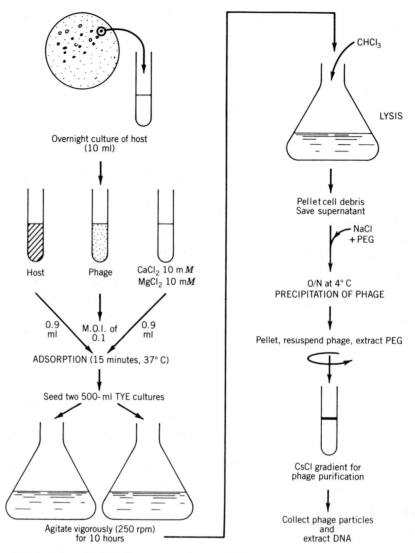

Figure 152. Isolation of λ DNA. The virus stock and λ DNA are prepared as described in Chapter 5, page 317.

It is also possible to inactivate the endonucleases by addition of 1/10 volume of a freshly prepared 1% diethylpyrocarbonate solution in ethanol. After 10 minutes incubation at 37°C, it is necessary to eliminate the residual CO_2 and ethanol under vacuum (5–10 minutes).

Isolation of DNA Fragments. Usually, 10–40% sucrose gradients can conveniently be used for preparation of DNA fragments and λ charon arms. The lengths and speeds of the runs may have to be adapted to particular cases. In most cases, satisfactory separation of DNA fragments is obtained in 17–20 hour runs at 25,000 rpm in a Beckman SW41 rotor.

PROTOCOL

PURIFICATION OF DNA FRAGMENTS BY CENTRIFUGATION IN SUCROSE GRADIENTS

Preparation of Sucrose Gradients

Prepare solutions in volumetric flasks. Prepare two 200-ml stock solutions containing 10% and 40% of highly purified sucrose as follows:

	10%	40%
Sucrose	20 g	80 g
5 M NaCl	40 ml	40 ml
0.5 M EDTA	4 ml	4 ml
1 M Tris-HCl (pH 8.0)	1 ml	1 ml

Many gradient-forming devices are commercially available, or they can easily be homemade. Some suppliers (such as Beckman) sell a gradient former delivering up to three gradients at the same time.

Remember that the more concentrated solution must be at the bottom of the tube. If you pour the gradient down the side of the

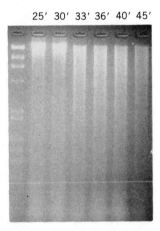

Figure 153. Partial digestion of cellular DNA. High molecular weight DNA (> 30 kb) was digested in the presence of limiting amounts of restriction endonuclease for increasing periods of time. Aliquots were analyzed by electrophoresis on a 0.8% agarose gel containing 1 μg/ml ethidium bromide. A mixture of Hind III-digested λ DNA and Hae III-digested φX174 RF DNA was used as size markers.

tube, the more concentrated sucrose solution should be poured first and should be in the well located near the outlet of the gradient former.

On the other hand, if you use a capillary tube to pour the gradient from the bottom of the tube, the less concentrated solution should be poured first and should be the one near the outlet of the gradient former.

We use an SW41 rotor with 14-ml capacity tubes in our experiments. Gradients are made with 6 ml 10% sucrose and 6 ml 40% sucrose. Polyallomer tubes are preferred to ultra clear tubes if the gradients are poured with a capillary pipette. Ultra clear tubes are used if the gradient is poured down the side of the tube.

Centrifugation of DNA Sample

Load approximately 50 μg DNA fragments in a total volume of 200–300 μl onto each gradient. Adjust the DNA buffer to 10 mM Tris–HCl (pH 8.0), 10 mM EDTA, and heat the sample for 10 minutes at 68°C before layering onto the gradient.

Run the centrifugation at 25°C to avoid reassociation of sticky ends generated by endonuclease digestion, and stop the run with

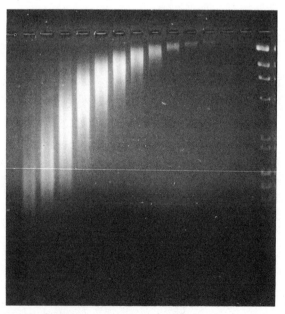

Figure 154. Purification of DNA fragments by ultracentrifugation. Partially digested DNA was run on a 10–40% sucrose gradient as described in text. An aliquot from each fraction was run on a 1% agarose gel. Staining with ethidium bromide allows recognition of the fraction(s) containing DNA fragments with appropriate sizes. A mixture of Hind III-digested λ DNA and Hae III-digested φX174 RF DNA was used as size markers.

the brake. A 25-hour run at 25,000 rpm in the SW41 rotor usually allows isolation of DNA fragments with sizes ranging between 4.5 and 7.5 kb.

Collection and Testing of the Fractions

Puncture the bottom of the tube and collect fractions of about 0.5 ml.

Add 200 µl of distilled water to each fraction.

Add 4 ml cold isopropanol to each tube and let the DNA precipitate at −70°C overnight.

Spin the fractions at 4°C to pellet the DNA precipitate. Centrifugation at 4200 rpm in a Beckman JS2 rotor for 20 minutes has been found to be sufficient. Resuspend the DNA pellet (which is often not visible) in 250 µl distilled water and analyze 5 µl directly on an agarose gel. Run the gel as described before (Chapter 7).

Stain the gel with ethidium bromide. You should obtain good separation of the DNA fragments in approximately 10 of the fractions (Figure 154).

At this step, you may wish to blot the DNA fractions from your gel and hybridize them with a specific probe to determine exactly which fraction to keep for further cloning.

Preparation of λ Charon Arms

The λ arms are prepared by circularization of purified λ DNA in the presence of ligase and elimination of the "stuffer" piece of DNA.

PROTOCOL

ELIMINATION OF STUFFER DNA AND ISOLATION OF λ ARMS FOR CLONING

Ligation of the Cohesive Ends of λ DNA

Purified λ DNA (250 µg) is diluted to a concentration of 0.5–0.75 µg/µl in 1X ligation buffer and incubated for 2 hours at 16°C in the presence of T4 DNA ligase.

Ligation Buffer

For 500 μl:
125 μl 1 *M* Tris–HCl (pH 7.5)
50 μl 0.5 *M* MgCl$_2$
50 μl 0.5 *M* 2-Mercaptoethanol
125 μl 0.02 *M* ATP
150 μl H$_2$O

The viscosity of the DNA solution should increase greatly as ligation proceeds.

Digestion of Ligated λ DNA with Endonuclease

Because the ligated λ DNA is very viscous, it is necessary to mix reagents (by inverting the tube frequently) during endonuclease digestion. We usually choose the enzyme concentration such that a 2-hour incubation time is sufficient for complete digestion. After the first hour of incubation, an equal amount of fresh enzyme is added and the incubation is run for another hour.

Centrifugation of Digested λ Arms

After digestion with restriction endonuclease (e.g., Eco RI), the λ DNA is treated once with phenol and chloroform:isoamyl alcohol (24:1) to remove proteins. The DNA is precipitated with ethanol, resuspended in 1 ml 10 m*M* Tris–HCl (pH 8.0), 10 m*M* EDTA, and heated at 68°C for 10 minutes before being layered onto the gradients. Four tubes are used, each loaded with 250 μl digested λ DNA. After 20 hours of centrifugation at 25,000 rpm, the tubes are punctured on the bottom and 0.5-ml fractions are collected.

Treat each fraction as for total digested DNA (see above), and resuspend in 250 μl after precipitation with isopropanol. Run a sample (5 μl) of each fraction on a 0.8% agarose gel. Stain with ethidium bromide to detect the fractions containing the λ insert and the fractions containing λ arms without the insert.

Ligation of Digested DNA to λ Charon Arms

Theoretical Considerations. In the present case, ligation of digested DNA to λ arms will lead to the formation of long concatamers. If we consider the two parameters j and i introduced by Dugaiczyk et al. (see Chapter 13), we find that we must choose relative DNA concentrations such that $j << i$ for both the λ arms and the DNA to be inserted. Let us recall that j represents the effective concentration of the two ends of the same molecule and that

$$j = \frac{6.1 \times 10^{22}}{MW^{3/2}} \text{ (ends/ml)}$$

We can assume that the λ arms account for 31 kb; this allows us to calculate j. The j value for the insert will depend on the size of the fragment to be cloned.

The parameter i represents the total concentration of complementary DNA in the reaction. We want i to equal $10 \times j$, which should favor formation of concatamers, and we know that

$$i = 2N_0 M \times 10^{-3} \text{ (ends/ml)}$$

where N_0 is Avogadro's number (6.02×10^{23}) and M is the molar concentration of DNA. Since

$$i = i_{arms} + i_{insert}$$

and we have the constraint of a 2:1 molar ratio of the λ arms to the insert DNA, we can deduce that

$$i = (2N_0 M_{insert} + 2N_0 M_{arms}) \times 10^{-3} \text{ (ends/ml)}$$

$$i = (2N_0 M_{insert} + 2N_0 2M_{insert}) \times 10^{-3} \text{ (ends/ml)}$$

$$i = 6N_0 M_{insert} \times 10^{-3} \text{ (ends/ml)}$$

The value of i is given by the equation

$$i = 10 \times j_{insert}$$

This allows us to determine M_{insert} and M_{arms}

These, however, are theoretical values.

Practice. In practice, the λ arms and the DNA to be inserted may behave somewhat differently from theoretical predictions, probably because the DNAs are not intact after purification.

To test the efficiency of the reaction, run several test ligations using molar ratios of 2:1, 1:1, and 0.5:1 (λ arms/insert DNA). In each case resuspend 2.5 μg DNA in 10 μl 1X ligation buffer. Incubate overnight at 14°C after adding T4 DNA ligase (0.5 μl from New England Biolabs). Include a control test in which no insert DNA is added; this will give you an estimate of the background due to the presence of contaminating λ insert in the purified arms.

Lambda In Vitro Packaging System

The in vitro packaging of recombinant λ DNA molecules into mature phage particles results in 2–5 logs greater plating efficiencies [10^7–10^8 plaque forming units (pfu)/μg of intact wild type λ DNA] than do calcium chloride ($CaCl_2$) transfection procedures (Mandel and Higa, 1978; Hohn and Murray, 1977; Enquist and Sternberg, 1979; Blattner et al., 1978; Maniatis et al., 1978). In addition, the packaged phage mixture can be stored at 4°C as a stable phage preparation (Hohn and Murray, 1979; Enquist and Sternberg, 1979).

Phage packaged by in vitro methods may be plated at a 100-fold greater density (pfu/plate) than a similar transfection mixture (Hohn and Murray 1977; Enquist and Sternberg, 1979; Blattner et al., 1978; Maniatis et al., 1978), can be easily amplified a millionfold by cleared plate lysis (Maniatis et al., 1978), and contains a high percentage of recombinant molecules (Maniatis et al., 1978). [For additional information, see Williams and Blattner (1980) and Enquist and Sternberg (1979)].

Successful packaging of recombinant molecules is dependent on the physical nature of the substrate DNA. There is a strict requirement for the presence of the λ cohesive termini (*cos* sites). For efficient packaging, these *cos* sites must be separated by approximately 35–52 kb of DNA (75–105% of the length of wild type λ DNA). Although linear, multimeric linear (concatamers), and multimeric circular DNAs are all packaged, monomeric circular λ molecules are not packaged either in vivo or in vitro. Thus, ligations for packaging should be performed under conditions which limit monomeric circle formation.

The protocol presented here for the preparation of in vitro packaging extracts, is that of Faber, Kiefer, and Blattner which has been successfully used in several systems. Although this procedure relies on the difficult purification of the protein A (involved in site-specific recognition of λ DNA), it gives excellent efficiency.

The procedure calls for the preparation of three components known as Freeze Thaw Lysate (FTL), Sonic Extract (SA), and λ A protein (pA). In each case, thermoinduction of lysogenes is a basic and very critical step.

PROTOCOL

PREPARATION OF IN VITRO PACKAGING EXTRACTS ·

Preparation of Freeze Thaw Lysate

1. Grow 1000 ml NS428 *E. coli* strain in two 2-liter flasks (500 ml/flask) at 30°C until the optical density at 600 nm reaches 0.4 (on a Beckman spectrophotometer).

2. Quickly transfer the two flasks to a 45°C water bath with agitation for 15 minutes. This step ensures thermodenaturation of repressor cI_{857}.

3. Transfer cultures to a 39°C water bath for 2 more hours with vigorous agitation. At this step you can check to see that induction has proceeded by pipetting a few milliliters of the culture into a tube containing two or three drops of chloroform. If the culture remains turbid when you shake with chloroform, the cells were not induced properly. The cells should lyse completely.

4. Centrifuge cultures in a large volume rotor (such as Sorvall GSA or Beckman JS 4.2) for 10 minutes at 9000 rpm (GSA) or 25 minutes at 4200 rpm (JS 4.2). Perform centrifugation at 4°C.

5. Drain all culture media carefully.

6. Resuspend each cell pellet in 0.8 ml cold FTL buffer.

7. Add 50 µl lysozyme solution to each tube.

8. Mix thoroughly and immediately freeze in liquid nitrogen.

9. Keep in closed cap tubes at −180°C or at −70°C.

Sonic Extract

1. Grow and induce 1000 ml NS428 cells as described above.

2. Resuspend pellets from 200-ml aliquots in 0.5 ml buffer A.

3. To each resuspended pellet transferred to a plastic test tube, add about 1.3 ml buffer A to a final volume of 2 ml.

4. Place the tube in a salt–ice bath and sonicate at maximum power (microtip) with three 5-second blasts. The solution should become clear.

5. Spin out all debris at 6000 rpm in a Sorvall SS34 rotor for 6 minutes. If cells are sonicated properly, a small pellet should be obtained.

6. Transfer the supernatant to small polypropylene tubes and store in liquid nitrogen.

Isolation of Protein A

1. Grow and induce 6 liters of λ*dg*805 cells as described above, but perform induction at an O.D.$_{600}$ of 0.8 (on a Spec 20).
2. Harvest cells by centrifugation in a large rotor.
3. Resuspend cells in 50 ml buffer A and sonicate in a salt–ice bath until cells are completely broken.
4. Spin out cell debris in an SS34 rotor for 20 minutes at 10,000 rpm.
5. Collect supernatant and gradually add 10% polyethyleneimine in buffer A with stirring until no more precipitate forms. This requires about 5 ml of a 10% solution.
6. Leave on ice for 15 minutes with stirring.
7. Collect precipitate by centrifugation at 10,000 rpm for 20 minutes.
8. Discard supernatant.
9. Resuspend pellet in 250 ml 50 mM ammonium succinate (pH 6) containing 10^{-2} M 2-mercaptoethanol and 10% glycerol. We advise using an electric tissue homogenizer to help resuspension.
10. Centrifuge at 10,000 rpm for 20 minutes. Discard supernatant.
11. Resuspend pellet in 250 ml 100 mM ammonium succinate (pH 6) containing 10^{-2} M 2-mercaptoethanol and 10% glycerol.
12. Centrifuge and discard supernatant.
13. Resuspend pellet in 250 ml 200 mM ammonium succinate (pH 6) containing 10^{-2} M 2-mercaptoethanol and 10% glycerol.
14. Centrifuge and save supernatant.
15. Add to the supernatant 100 ml ammonium sulfate buffer and stir at 4°C for 20 minutes.
16. Spin out precipitate at 10,000 rpm for 30 minutes.
17. To the supernatant add 150 ml ammonium sulfate buffer and 7.5 g solid ammonium sulfate. Stir for 30 minutes at 4°C.
18. Collect precipitate by centrifugation at 10,000 rpm for 30 minutes.
19. Dissolve pellet in 2–3 ml buffer A deprived of MgCl$_2$. Dialyze twice against the same buffer.
20. Store in liquid nitrogen as small aliquots.

PROTOCOL

IN VITRO PACKAGING

1. Thaw a sample of Freeze Thaw Lysate and place it on ice for 45 minutes. Do not allow sample to warm to more than 4°C.
2. Add 50 μl buffer M and mix well.
3. Centrifuge in a precooled Beckman Ti 50 rotor at 35,000 rpm for 25 minutes (at 4°C). Save the supernatant, which can be used for several days if stored frozen at $-20°C$.
4. During the spin, thaw an aliquot of sonic extract. Do not warm up to more than 4°C.
5. Mix in a tube:
   30 μl buffer A
   4 μl ligated DNA (<1 μg)
   4 μl buffer M
   20 μl sonic extract
   1–2 μl protein A
6. Incubate for 15 minutes at room temperature.
7. Add 75–150 μl Freeze Thaw Lysate.
8. Incubate for 60 minutes at room temperature.
9. Dilute and place as for regular phage. You can expect a titer between 10^5 and 10^8 infectious particles/ml.

An estimated efficiency of packaging is given by the number of plaques obtained for a given number of phage genomes. If we assume that 1 μg DNA is equivalent to 2×10^{10} phages and 20 O.D. units (at 260 nm) equals 1 mg DNA/ml, we can estimate efficiencies as being comprised of between 5×10^{-3} and 5×10^{-4}.

AMPLIFICATION OF LAMBDA LIBRARIES

PROTOCOL

AMPLIFICATION OF A λ LIBRARY GROWN ON AGAR PLATES

1. Prepare 150-mm plates of TYE-agar.
2. Mix 10^4 phage particles from in vitro packaging with 0.2 ml of indicator bacteria (for titration and preparation of λ stocks, see Chapter 7.
3. Pour with an overlay onto plates.

4. Incubate for no more than 8–10 hours at 37°C.

5. Scrape and treat as described in Chapter 7 for preparation of regular λ stocks. The library obtained under these conditions, if kept in the presence of chloroform at 4°C, will be stable for several years.

Buffers for In Vitro Packaging

FTL buffer
10% Sucrose
50 mM Tris–HCl (pH 7.4)

100 ml
10 g Sucrose
5 ml 1 M Tris–HCl (pH 7.4)
H_2O to 100 ml

Lysozyme solution
2 mg/ml Lysozyme in 0.25 M
Tris–HCl (pH 7.4)
Make fresh each time.

Buffer A
20 mM Tris–HCl (pH 8.0)
3 mM MgCl$_2$
5 mM 2-Mercaptoethanol
1 mM EDTA
10% Glycerol

100 ml
2 ml 1 M Tris–HCl (pH 8.0)
0.3 ml 1 M MgCl$_2$
0.2 ml 0.5 M EDTA
10 ml Glycerol
85 ml H_2O
Add 14 M 2-mercaptoethanol
(35 μl/100 ml) just before
use.

Ammonium sulfate buffer
500 ml Saturated ammonium
   sulfate
10 ml 1 M Tris–HCl (pH 6.5)
0.5 ml 2-Mercaptoethanol

Buffer M
Mix in the order given:
110 μl sterile H_2O
6 μl 0.5 M Tris–HCl (pH 7.4)
300 μl 0.05 M Spermidine; 0.1 M
   putrescine (adjusted to pH 7
   with Tris–base)
9 μl 1.0 M MgCl$_2$
75 μl 0.1 M ATP (adjusted to pH
   7.0 with NH$_4$OH)
1 μl 2-Mercaptoethanol

References

Blattner, F. R., Blechl, A. E., Denniston-Thompson, K., Faber, H. E., Richards, J. E., Slightom, J. L., Tucker, P. W., and Smithies, O. (1978), *Science*, **202**, 1279.

Enquist, L., and Sternberg, N. (1979), in *Methods in Enzymology*, Vol. 68, R. Wu, ed., Academic Press, New York, pp. 281.

Hohn, B., and Murray, K. (1977), *Proc. Natl. Acad. Sci. USA*, **74**, 3259.

Mandel, W., and Higa, A. (1970), *J. Mol. Biol.*, **53**, 159.

Maniatis, T., Hardison, R. C., Lacy, E., Lauer, J., O'Connell, C., Quon, D., Sim, D. K., and Efstratiadis, A. (1978), *Cell*, **15**, 687.

Williams, B. G., and Blattner, F. R. (1980), in *Genetic Engineering*, J. K. Setlow and A. Hollaender, eds., Plenum Press, New York, pp. 201.

EFFICIENCY OF COMMERCIAL PACKAGING EXTRACTS

Bacterial extracts for in vitro packaging of λ DNA are now available from several commercial sources. Because it is not always easy to choose the products sold by the different suppliers, and because these extracts are not cheap, we have decided to perform a comparative study with V. Maloisel in our laboratory under the following conditions:

1. Packaging extracts were purchased from the local representatives of **Amersham** (Les Ulis, France), **Bethesda Research Laboratories** (Herblay, France), **Boehringer** (Meylan, France), **Promega Biotec** (c/o Coger, Paris, France), and **Stratagene** (c/o Genofit, Grand-Lancy, Switzerland).

2. The control λ DNA provided with each batch was used to test the packaging efficiency of the different extracts under the exact experimental conditions recommended by the suppliers. The extracts were purchased at the same time and used within a week following receipt. All extracts were stored for a few days at $-70°C$, until the recommended recipient strains were grown out. Two suppliers (Promega and Genofit) provided the bacterial strains with the extracts.

3. High molecular weight cellular DNA extracted from tumor cells was partially digested with Bam HI and centrifuged in a sucrose gradient as described on page 500. DNA fragments with sizes in the range of 10 to 20 kb were purified and ligated to λ EMBL4 arms. Aliquots of the ligation mixture were then used to test the efficiency of in vitro packaging obtained with each commercial extract.

The results obtained are reported in Table 65.

Parameters Affecting Efficiency of the Packaging Reaction

Several different parameters are known to affect the efficiency of in vitro packaging.

Table 65. Efficiency of Commercial in Vitro Packaging Extracts

Suppliers		Efficiency of Packaging According to Suppliers (pfu/μg) of DNA		Efficiency of Packaging in Our Hands (pfu/μg of DNA)		
		λ DNA	no DNA	λ DNA from supplier[a]	DNA from EMBL4[b] Library	No DNA[c]
Stratagene	Gigapack Plus	1.05×10^9	—	2.53×10^8	3.9×10^4	nt[d]
	Gigapack Gold	2.1×10^9	—	6.22×10^8	2.0×10^4	nt
Amersham		7.58×10^8	<25	8×10^8	$9.0 \quad 10^3$	0
Boehringer		$>1 \times 10^8$	0	6.1×10^8	1.3×10^4	0
Promega biotec		3.5×10^8	<100	1.98×10^8	$5.5 \quad 10^3$	1
Bethesda Research Laboratories		$>10^7$	<100	5.54×10^6	$7.5 \quad 10^2$	0

Source: V. Maloisel (unpublished data).

[a]Based upon DNA concentrations given by suppliers, 0.5 μg of λSam7 DNA was used in each assay except for BRL, which provided less than 0.5 μg of λSam7 DNA as control (in this case 0.05 μg of λSam7 DNA was used).

[b]Aliquots of a genomic library prepared with EMBL4 DNA were used to test the different extracts. In each case the vector: insert DNA ratio was (1:0.4). Results are expressed as p.f.u./extract.

[c]Extracts tested without any DNA. Results are expressed as p.f.u./extract.

[d]nt: non tested.

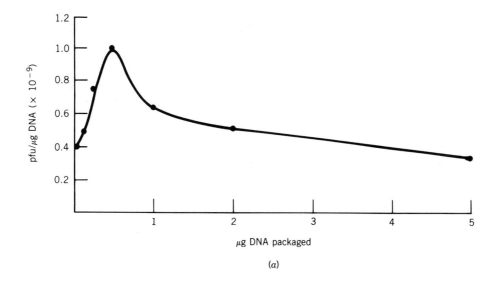

(a)

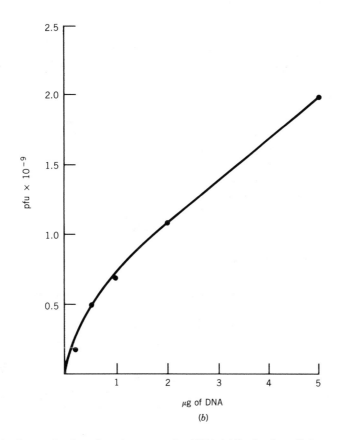

(b)

Figure 155. In vitro packaging of varying amounts of DNA. (*a*) Packaging efficiency versus amount of DNA. (*b*) Plaque forming units per reaction versus amount of DNA. Highest efficiencies are therefore obtained at approximately 0.5 μg of wild type DNA (*a*). This is the level at which the highest proportion of added DNA is incorporated into infectious phage particles. Above the optimal amount of DNA the number of plaques formed per reaction continues to increase with the addition of up to 5 μg of wild-type DNA (*b*) (courtesy of Amersham).

Table 66. Packaging of Varying Amounts of DNA

Amount of DNA Packaged (μg)	Plaque Forming Units Obtained per Reaction (pfu)	Packaging Efficiency (pfu/μg)
0.05	2.0×10^7	4.0×10^8
0.1	5.0×10^7	5.0×10^8
0.25	1.9×10^8	7.5×10^8
0.5	5.0×10^8	1.0×10^9
1.0	6.5×10^8	6.5×10^8
2.0	1.1×10^9	5.5×10^8
5.0	2.0×10^9	4.0×10^8

Source: Information courtesy of Amersham.

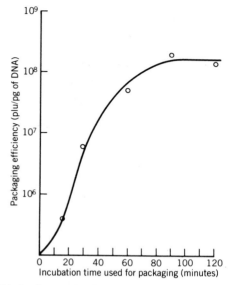

Figure 156. In vitro packaging efficiency as a function of incubation time, using wild-type λ DNA as the vector (courtesy of Amersham).

1. Amount of DNA to be packaged.
   The results reported in Figure 155 and Table 66 show that best efficiencies are obtained with approximately 0.5 μg of wild-type DNA in the assay.

2. Time course of the packaging reaction.
   As shown in Figure 156 the packaging reaction should be performed for at least 2 hours to obtain optimum efficiencies.

3. Size of the DNA to be packaged.
   The in vitro packaging efficiency has been shown to decrease with the size of the DNA used in the reaction (Sternberg et al., 1977 and Figure 157). Therefore, a strong selection for high molecular weight DNA fragments may help in constructing DNA libraries with large inserts.

Note.   In addition to the factors listed above, you must keep in mind that both the purity of the DNA, the quality of the cell preparations used for plating, and the way you handle the mixtures (shake, mix, etc.) are critical in obtaining satisfactory packaging.

References

Enquist, L., and Sternberg, N. (1979), in *Methods in Enzymology,* R. Wu, ed., Vol. 68, Academic Press, New York, p. 281.

Karn, J., Brenner, S., Barnett, L., Cesareni, G. (1980), *Proc. Natl. Acad. Sci. USA,* **77,** 5172.

Sternberg, N., Tiemeier, D., and Enquist, L. (1977), *Gene,* **1,** 255.

SCREENING OF λ LIBRARIES BY HYBRIDIZATION WITH SPECIFIC PROBES

Screening of the λ libraries for recombinants carrying the desired sequences is usually performed by the plaque hybridization assay (Figure 158) originally introduced by Benton and Davis (1977). The procedure is as follows.

PROTOCOL

PLAQUE HYBRIDIZATION ASSAY

1. A few days before screening, prepare large plates (15-cm diameter) containing 1.5% TYE–agar medium, complemented as needed depending on the host bacterial strain used to

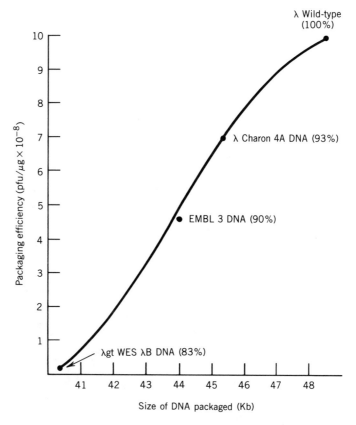

Figure 157. Relationship between DNA size and in vitro packaging efficiency: non-isogenic vector DNA. Size selectivity is particularly useful when a small vector phage is used, as the packaging system then provides a strong selection for phage with large inserted DNA fragments (courtesy of Amersham).

grow the library. Dry the plates at 37°C and keep them at room temperature until used.

2. Seed an overnight culture (10 ml) of the host bacteria. The next day, spin the cells at 4000 rpm for 10 minutes and re-suspend them in 10 ml 10 mM MgSO$_4$. Bacteria can be kept for several days at 4°C in 10 mM MgSO$_4$.

3. In 10-ml glass tubes, mix 0.3 ml of bacteria and 0.3 ml of diluted phage (3 × 10^5 phages). Let stand on the bench for 15 minutes.

4. To each tube, add 8 ml 1.0% melted agar (in TYE) kept at 45°C, and immediately pour the mixture on large Petri dishes, being careful to cover the surface evenly with top agar. We have found it helpful to prewarm the plates before pouring the overlay.

5. Incubate at 37°C until the lysis is visible (usually 14–16 hours). Under the conditions used, a semiconfluent or confluent lysis should be obtained.

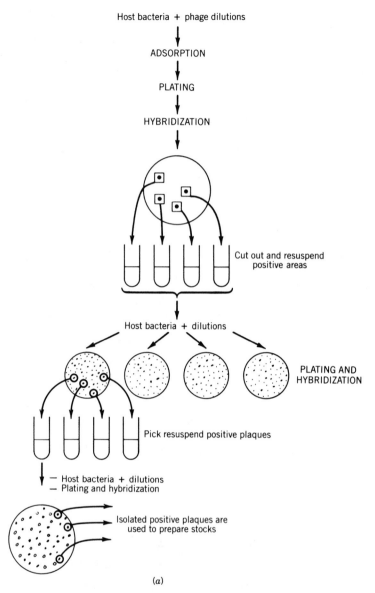

Figure 158. Plaque hybridization assay for screening λ libraries. (a) Outline of the procedure.

6. Put the agar plates in a refrigerator for about 2 hours to harden the agar overlay.

7. In the meantime, prepare solutions for DNA transfer and label nitrocellulose filters with a pencil. Up to 6 filters/plate can be prepared if necessary. We usually put three pencil marks on the filter and three matching marks on the agar plates to orient the filters precisely when going back to the plate after hybridization (see below).

8. With tweezers, carefully applied a nitrocellulose filter to the surface of the top agar. Make sure you do not trap any air

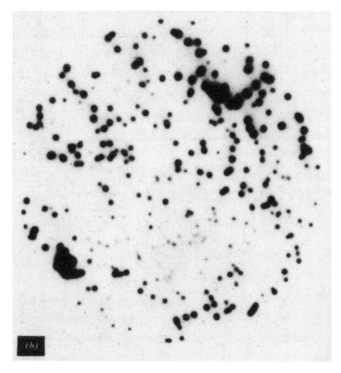

Figure 158 (*continued*). (*b*) autoradiography obtained in screening a λ charon 4A library for the presence of a specific DNA fragment.

bubbles between filter and plate. Once the filter is wet, let it sit for 1 minute to allow transfer of phage DNA onto the nitrocellulose. Remove the filter carefully without peeling off the top agar overlay. You should not encounter any problem if you use dry plates and 1% top agar and cool the plates before transfer.

9. Put the filter in denaturing buffer for 30 seconds (see note).

10. Transfer to neutralizing buffer for 30 seconds.

11. Transfer to 10X SSC and wash off any pieces of agar stuck to the filter.

12. Repeat the same protocol (steps 8–11) with other filters. It is recommended that you increase the time of contact with the overlay by 30 seconds for each additional filter, that is, leave filter 2 on the plate for 90 seconds, filter 3 for 2 minutes, and so on.

13. When all the filters have been rinsed with 10X SSC, blot them on Whatman 3MM paper until dry.

14. Bake the filters and hybridize with the specific probes, as described in Chapter 15.

15. On the autoradiograms, positive hybridization is indicated by black dots on a clear background (see figure 158). It is pos-

sible to assign each of these dots to areas corresponding to a mixture of many different plaques on the plate.

16. Areas are picked as for isolated plaques (Chapter 7) and re-suspended in 500 μl λ SM. Add a few drops of chloroform and vortex for 10 seconds. Spin at 3500 rpm for 10 minutes. Take out the supernatant, transfer it to a microfuge tube, and add two drops of chloroform. Keep at 4°C.

17. Dilute and plate the phage stock to obtain about 200–400 isolated plaques per plate. At this stage, regular Petri dishes (100-mm diameter) or square plates (which can accommodate a larger number of plaques) are used.

18. Prepare nitrocellulose filters (Chapter 15) and treat them for hybridization as described above. This step should allow you to obtain isolated positive plaques. If no positive plaques are obtained, either you did not pick up the correct area on the original plate, or you picked too large an area and the proportion of positive recombinants is too small to be detected among 200–300 plaques. You can check this possibility by plating lower dilutions of the stock obtained at step 16.

19. Pick up a few isolated plaques, replate them twice, and prepare stocks of the λ recombinants as described before (Chapter 7) for further analysis.

Buffers for Screening Libraries

Denaturing buffer (1 liter)
300 ml 5 *M* NaCl
50 ml 10 *N* NaOH
650 ml H_2O

Neutralizing buffer (1 liter)
500 ml 1 *M* Tris–HCl (pH 7.4)
174 g NaCl
H_2O to 1 liter

10X SSC (1 liter)
87.7 g NaCl
44.2 g Sodium citrate
H_2O to liter

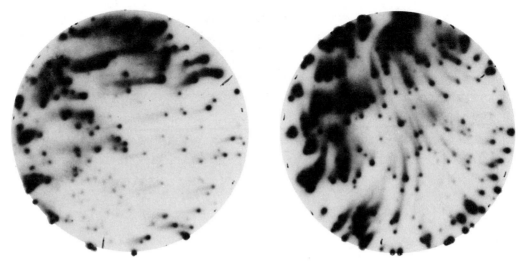

Figure 159. Diffusion of λ DNA from plaques, as a result of inappropriate treatment of the filters.

SM (1 liter)
5.8 g NaCl
1 ml 1 *M* MgSO₄ (add after sterilization)
100 mg Gelatin
20 ml 1 *M* Tris–HCl (pH 7.5)
H₂O to 1 liter

Note. To avoid spreading of the DNA from the plaques (Figure 159) the filters should not be immersed in the denaturing solution. Instead, allow the solution to slowly diffuse through the filter by placing it carefully on the surface of the liquid.

Reference

Benton, W. D., and Davis, R. W. (1977), *Science*, **196,** 180.

18

PURIFICATION AND CHARACTERIZATION OF RNA SPECIES

PURIFICATION OF INTACT RNA SPECIES BY ULTRACENTRIFUGATION

The protocol described below should enable you to obtain intact eucaryotic or procaryotic RNA species when needed for accurate sizing and further char-acterization or for preparing cDNA libraries (see chapter 19). When working with bacteria, make sure that the cell wall be digested in a small volume of buffer with lyzozyme (see chapter 15) before adding the guanidine thiocyanate solution.

PROTOCOL

WHOLE CELL RNA ISOLATION WITH GUANIDINE THIOCYANATE

1. Wash cells on Petri dishes with phosphate-buffered saline.
2. Resuspend the cells in guanidine thiocyanate buffer to a final ratio of about 18 ml buffer/gram of packed cells.
3. Put 3 ml 7.5 M cesium chloride, 25 mM sodium acetate in the bottom of a Beckman SW41 tube.
4. Layer 9 ml of the guanidine thiocyanate-treated cells onto the 3-ml cesium chloride cushion.
5. Centrifuge for at least 15 hours at 30,000 rpm, and 4°C.
6. After centrifugation, the RNA is pelleted at the bottom of the tube. Just above the RNA pellet is a viscous layer of DNA and above that is the cesium chloride interface.
7. Draw off supernatant, cesium chloride, and DNA with a Pasteur pipette. Leave some solution on the pellet.
8. Pour off the remaining solution and cut off the bottom of the tube with a razor blade.

9. Resuspend each pellet in 400 µl sterile distilled water.
10. Heat for 5 minutes at 68°C.
11. Add 2.6 ml guanidine–HCl solution; mix.
12. Add 75 µl 1 *N* acetic acid and 1.5 ml ethanol.
13. Let sit at − 20°C for at least 3 hours.
14. Centrifuge at 15,000 rpm in an SS34 rotor (Sorvall) for 30 minutes.
15. Resuspend pellet in 1 ml sterile distilled water.
16. Centrifuge as in step 14. Save supernatant.
17. Resuspend pellet in 0.5 ml H$_2$O.
18. Repeat step 16.
19. Repeat steps 17 and 18.
20. Combine the three supernatants.
21. Add 285 µl 3 *M* sodium acetate (pH 5.0) and 7 ml cold ethanol and let sit at −20°C overnight to precipitate RNA.
22. Centrifuge at 10,500 rpm for 30 minutes in an SS34 rotor.
23. Resuspend the pellet in 400 µl sterile distilled water per original 9-ml fraction (one SW41 tube).
24. Dilute 1:100 to read the optical density and calculate yield.

Buffers and Solutions

GUANIDINE THIOCYANATE SOLUTION

For 100 ml.

1. Dissolve 50 g guanidine thiocyanate and 0.5 g *n*-lauryl sarcosine in 30 ml of sterile distilled water. Heat and stir to help dissolution.
2. Add 2.5 ml 1 *M* sodium citrate buffer (pH 7.0).
3. Add 0.7 ml 2-mercaptoethanol.
4. Bring the pH of the mixture to 7.0 with a few drops of 0.1 *N* NaOH.
5. Bring up to 100 ml with sterile distilled water.
6. Filter the solution on a millipore filter.

Do not keep this solution more than two weeks.

GUANIDINE–HCL SOLUTION

For 100 ml:

1. Dissolve 72 g guanidine–HCl in 36 ml sterile distilled water; heat and stir to help dissolution. The pH of the solution is approximately 4.5.
2. Add 2.5 ml 1 *M* sodium citrate buffer (pH 7.0).

3. Add 5 ml 0.1 *M* dithiothreitol solution.

4. Bring up the pH of the solution to 7.0 with 0.1 *N* NaOH.

5. Add water to 100 ml.

6. Filter the solution on a 0.45-μm pore size membrane.

Make fresh every month.

CESIUM CHLORIDE SOLUTION

Dissolve 63 g cesium chloride in 49 ml water and add 1.25 ml 1 *M* sodium acetate (pH 5.0).

Notes

1. *Do not* use polycarbonate tubes when working with guanidine thiocyanate.

2. Use only 2-mercaptoethanol as reducing agent. Dithiothreitol chemically reacts with thiocyanate.

3. You may expect to obtain about 1 to 2 mg of RNA from 10^8 cells.

REFERENCE

Chirgwin, J. M., Przygyla, A. E., MacDonald, R. J., and Rutter, W. J. (1979), *Biochemistry,* **18,** 5294.

PURIFICATION OF RNA SPECIES FOR QUICK BLOT

This protocol can be used to screen several different cell samples for their content, in particular RNA species by dot blot hybridization.

PREPARATION OF CYTOPLASMIC RNA FROM EUCARYOTIC CELLS

Because the vast majority of eucaryotic mRNAs are generated by splicing of large nuclear precursors, it is often necessary to isolate cytoplasmic RNA species. In the protocol which is described on pages 520, 521, phenol extraction is used to prepare the RNAs. When needed, it is also possible to add directly the proper amount of guanidine thiocyanate to the cytoplasmic preparation and then proceed as described above for RNA purification.

PROTOCOL

PURIFICATION OF RNA SPECIES FOR QUICK BLOT

1. Collect 1.0×10^6 cells samples by low-speed centrifugation (600 *g*) for 5 minutes.

2. Resuspend the cells from each sample in 1 ml of PBS and transfer in a microfuge tube.

3. Centrifuge at 12,000 rpm (15,000 g) for 15 seconds.

4. Discard the supernatant and resuspend the cell pellet in 45 μl of TE buffer.

5. Add 5 μl of 5% NP40 onto the cell pellet to lyse the cells.

6. Mix gently and leave on ice for 5 minutes.

7. Add 5 μl of 5% NP40 and leave on ice for 5 more minutes.

8. Centrifuge the cell extracts for 2.5 minutes at 15,000 g. The supernatant contains cytoplasmic RNA. The pellet contains nuclei.

9. Without disturbing the nuclei pellet, carefully pipette 50 μl of the supernatant in a tube containing:
   30 μl of 20X SSC
   20 μl of 37% formaldehyde

10. Incubate for 15 minutes at 60°C.

11. You may keep the treated sample at − 70°C.

12. Prepare serial dilutions of the sample to perform dot or slot blotting (see Chapter 15). For this purpose, mix 5 to 20 μl of sample with 15X SSC so as to obtain a final volume of 150 μl to transfer.

Notes

1. This technique is well adapted for rapid screening. However, the RNA species prepared in this way may not be suitable for accurate sizing on gels. Keep in mind that a positive signal in dot (or slot) blotting only reflects the presence of complementary sequences. Preparations of native and degraded RNA species may give comparable signals.

2. The technique relies on the elimination of nuclei. In many cases, the cell nuclei are very fragile and break upon centrifugation. In such cases, the preparation may be heavily contaminated by cellular DNA.

PROTOCOL

LYSIS OF GROWING EUCARYOTIC CELLS

Monolayers Grown on Plastic or Glass

1. Wash cell monolayers twice with ice-cold phosphate-buffered saline (PBS). Cells grown in 100-mm Petri dishes are washed

twice with 20 ml PBS, whereas cells grown in roller bottles are washed twice with 250 ml PBS.

2. Use rubber policemen to scrape the cells off the surface. Resuspend cells in ice-cold PBS ($\sim$100 ml/10^8 cells) and collect by low-speed centrifugation ($\sim$1000 $\times$ *g* for 5 minutes at 4°C).

3. Resuspend cells in high pH isotonic buffer (Buffer I) to approximately 1 $\times$ 10^8 cells/4 ml buffer.

Cells Grown in Suspension

Some eucaryotic normal or transformed cells have the ability to grow in suspension in a liquid medium. Such cells should be washed two times with ice-cold PBS after collection by centrifugation of the culture medium. After the second wash, resuspend the cells in Buffer I (1 $\times$ 10^8 cells/4 ml buffer).

PROTOCOL
REMOVAL OF CELL NUCLEI

1. Add 80 μl 5% of NP40 for each 4 ml of cell suspension.
2. Vortex twice for 15 seconds.
3. Pellet the nuclei by centrifugation (2000 $\times$ g for 8 minutes at 4°C).
4. Carefully aspirate the supernatant containing the cytoplasmic RNA. (The pellet of nuclei can be used to prepare nuclear RNAs.)
5. Mix the supernatant with a one-fifth volume of buffer II (containing SDS).

PROTOCOL
EXTRACTION OF RNAS

1. Add one volume (4.8 ml) of a 1:1 mixture of phenol:chloroform/isoamyl alcohol (24:1) to the supernatant containing SDS.

Shake for 5 minutes at room temperature.

2. Spin at 2000 $\times$ g for 5 minutes at room temperature.

3. Aspirate the upper (aqueous) phase and mix it with 5 ml chloroform:isoamyl alcohol (24:1). Shake for 5 minutes at room temperature.

4. Spin at 2000 $\times$ g for 5 minutes at room temperature.

5. Aspirate the upper phase and repeat steps 3–5 until a clear interface is obtained. Usually, two chloroform:isoamyl alcohol extractions are sufficient.

6. Mix the supernatant ($\sim$4.5 ml) with 0.18 ml sterile 5 M NaCl. Add three volumes of absolute ethanol (12 ml) to the RNA preparation and leave the mixture at $-20°C$ overnight.

7. Recover the precipitated RNAs by centrifugation at 20,000 g for 15 minutes. Pour off the ethanol and wash the pellet once with absolute ethanol.

8. Dissolve the RNA pellet in sterile distilled water and store at $-70°C$ (or $-20°C$).

Notes

1. Some of us include 0.02 M vanadyl ribonucleoside complexes (potent inhibitors of RNase; available from BRL) in all buffers used in the preparation of cytoplasmic RNAs. However, this inhibitor should be used with caution since it cannot be used to inhibit nucleases when performing in vitro translation. Another ribonuclease inhibitor (RNAguard, commercialized by Pharmacia and RNAsin commercialized by Promega Biotec) isolated from human placenta (Blackburn, 1979) has also proved to be very efficient for maintaining the RNA integrity needed in experiments such as cDNA synthesis, or in vitro transcription and translation.

2. Before pelleting nuclei, it is recommended to check by microscopic examination that the majority of the cells are broken. Additional disruption may be generated with a Dounce homogenizer (B pestle).

3. Some cells have larger than average nuclei. In this case, vortexing or homogenization with a Dounce must be very gentle to prevent disruption of the nuclei and liberation of DNA, which could very much increase the viscosity of the preparation.

Buffers

Buffer I	*100 ml*
10 mM Tris–HCl (pH 8.5)	1 ml 1 M Tris–HCl (pH 8.5)
1.5 mM MgCl$_2$	150 µl 1 M MgCl$_2$

140 mM NaCl

2.8 ml 5 M NaCl
96 ml H$_2$O

Buffer II
50 mM Tris–HCl (pH 8.0)
10 mM EDTA (pH 7.5)
10 mM NaCl
0.5% SDS

100 ml
5 ml 1 M Tris–HCl (pH 8.0)
2 ml 0.5 M EDTA (pH 7.5)
200 μl 5 M NaCl
2.5 ml 20% SDS
90.3 ml H$_2$O

5% NP40
5 ml Nonidet NP40
95 ml H$_2$O

PBS
For 1 liter:
20 ml 5 M NaCl
20 ml 1 M Tris–HCl (pH 7.4)
2 ml 0.5 M EDTA (pH 7.5)
958 ml H$_2$O

REFERENCE FOR USE OF VANADYL RIBONUCLEOSIDE COMPLEX

Berger, S. L., and Birkenmeir, C. S. (1979), *Biochemistry*, **18**, 5143.

REFERENCE FOR USE OF PLACENTA INHIBITOR

Blackburn, P. (1979), *J. Biol. Chem.*, **254**, 12484.

ISOLATION OF CYTOPLASMIC POLYADENYLATED RNA SPECIES

Most eucaryotic messenger RNAs possess a stretch of polyadenylic acid at their 3′ end. The proportion of cytoplasmic mRNA that can be isolated by its possession of the poly(A) track is substantial and sometimes approach 90%.

PROTOCOL

POLY-A RNA SELECTION ON OLIGO-dT CELLULOSE

1. Set up a column with approximately 1-ml bed volume in buffer A and wash it several times with the same buffer. Sterilize the column as described in step 17.
2. Denature the RNA in water (0.5–1.0 mg RNA/0.4 ml H$_2$O) at 62–65°C for 1–2 minutes. Cool rapidly in ice.
3. Adjust to buffer A (i.e., add 5 μl 1 M Tris–HCl, pH 7–7.2, 25 μl 10% SDS, 20 μl H$_2$O, 50 μl 4 M NaCl).

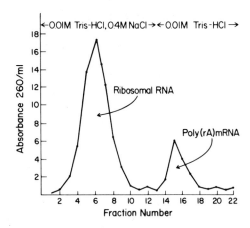

Figure 160. Purification of poly-A mRNA on oligo-dT cellulose.

4. Apply the 0.5-ml sample to the column.

5. Wash the tube with 0.5 ml buffer A and apply to the column.

6. Collect the 1 ml and pass it through the column two to three more times.

7. Wash the column with 6–7 ml buffer A, collecting 6–7 fractions of 1 ml each (poly A⁻).

8. Wash the column with 3–4 ml buffer B, collecting 6–8 fractions of 0.5 ml each (poly A⁺).

9. Read the optical density at 260 and 280 nm of each fraction in acid-washed cuvettes (it is advised that you dilute the first two poly A⁻ fractions 1:100) (Figure 160).

10. Collect the fractions of poly A⁺ and poly A⁻ of each group which have optical densities ≥0.044 at 260 nm.

11. Precipitate in 0.15 *M* NaOAc (pH 5–5.5) with 2.5 volumes of absolute ethanol in microfuge tubes (at least for the poly A⁺ fractions) overnight.

12. Reequilibrate the column in buffer A and store it at 4°C.

13. Spin the poly A⁺ in a microfuge at 4°C for 30 minutes.

14. Wash the precipitate twice with absolute ethanol.

15. Lyophilize to eliminate residual ethanol and water.

16. Resuspend in sterile distilled water about 1 mg/ml. Bring to 10 m*M* Tris–HCl (pH 7–7.2) and to 1 μg proteinase K/ml. Freeze.

17. Before reusing the column:

   a. Wash it with a few milliliters of 0.1 *M* NaOH.

   b. Fill it up with 0.1 *M* NaOH.

   c. Let it stand at room temperature for 20–30 minutes.

   d. Wash it several times with sterile distilled water.

   e. Wash it several times with buffer A. Check the pH.

Notes:

1. Use siliconized (or plastic) columns.
2. Warming the elution buffer to 30°C prior to elution may sometimes improve the recovery of poly A$^+$ RNA.
3. The yield of poly A$^+$ RNA is about 1 to 5% of the total RNA applied to the column.

Buffers

Buffer A	*100 ml*
10 mM Tris–HCl (pH 7.2)	1 ml 1 M Tris–HCl (pH 7.2)
0.4 M NaCl	8 ml 5 M NaCl
0.3% SDS	1.5 ml 20% SDS
	89.5 ml H$_2$O
Buffer B	*100 ml*
10 mM Tris–HCl (pH 7.2)	1 ml 1 M Tris–HCl (pH 7.2)
0.3% SDS	1.5 ml 20% SDS
	97.5 ml H$_2$O

Keep these buffers at room temperature: SDS precipitates at 4°C.

REFERENCE

Aviv, H., and Leder, P. (1972), *Proc. Natl. Acad. Sci. USA*, **69**, 1408.

ISOLATION OF SPECIFIC RNA SPECIES

In many cases the proportion of mRNA species encoding for a specific gene product only represents a few percent of the whole poly(A)-RNA population. Therefore, it is frequent that an enrichment in particular species of interest is being performed. The protocol described below allows a quite satisfactory recovery of translatable mRNA species from total RNA preparations obtained as described above.

PROTOCOL

*PURIFICATION OF SPECIFIC RNA SPECIES BY HYBRIDIZATION
WITH DNA BOUND ON NITROCELLULOSE*

1. Prepare a 2 µg/ml DNA solution in 0.1X SSC.
2. Add 3 M NaOH to a 0.3 M final concentration.

3. Heat 10 minutes in a water bath at 100°C.
4. Adjust to pH 7.5 with concentrated HCl.
5. Dilute to 1 μg/ml with 6X SSC.
6. Apply denatured DNA solution onto nitrocellulose filters (1 μg/15 mm² nitrocellulose) under slight vacuum.
7. Wash with 5 ml 6X SSC.
8. Bake filters 2 hours at 80°C.
9. Wash filters in hybridization buffer for 1 hour at 50°C.
10. Incubate filters with 500 μg/ml poly A⁺ RNA (1 ml for 10 μg DNA) in hybridization buffer.
11. Wash filters 10 times by vortexing in hot washing buffer (50°C).
12. Wash 3 times with hot 2 mM EDTA pH 8.0 (60°C).
13. Transfer filters in 1.5 microfuge tubes and add 400 μl of H_2O.
14. Incubate at 100°C for 1 minute.
15. Cool quickly in a dry-ice-ethanol bath and remove filters (or aspirate liquid).
16. Add 5 μl of 1 mg/ml carrier tRNA, 15 μl of 3 M potassium acetate pH 6.5, 1 ml of cold absolute ethanol and let precipitate RNAs at −20°C.

Solutions and Buffers

3 M NaOH 100 ml
Weight 12 g of NaOH
Dissolve with mixing in 50 ml
   H_2O
Add H_2O to 100 ml

6X SSC 1 liter
52.60 g NaCl
26.47 g Sodium citrate
H_2O to 1 liter

Hybridization Buffer   *10 ml*
65% Formamide   6.5 ml Deionized formamide
10 mM PIPES   0.1 ml, 1 M PIPES (pipera-
1 M NaCl   zine-N-N′-bis-2-ethanesul-
                                   fonic acid)
                             2 ml 5 M NaCl
                             RNA solution and H_2O to 10
                                   ml

Washing buffer   *100 ml*
0.15 *M* NaCl   3 ml 5 *M* NaCl
0.015 *M* Sodium citrate   1.5 ml 1 *M* Sodium citrate
0.5% SDS   2.5 ml 20% SDS
                                          93 ml H$_2$O

Notes

1. It is also possible to use DNA covalently bound on chemically modified paper (see Chapter 15).
2. RNA purified in this way can be used for in vitro translation (see Chapter 25).

REFERENCE

Riccardi, R. P., Miller, J. S., and Roberts, B. E. (1979), *Proc. Natl. Acad. Sci. USA,* **76**, 4927.

FORMALDEHYDE GELS FOR RNA

Formaldehyde is a dangerous reagent. Manipulate it only under a recently checked chemical hood.

PROTOCOL

ELECTROPHORESIS OF RNA SPECIES IN FORMALDEHYDE AGAROSE GELS

Preparation of the gel

For a 1% gel:

1. Weigh 2 g agarose and melt in 147 ml sterile distilled water.
2. Cool the melted agarose solution to about 60°C.
3. Add successively 20 ml 10X buffer F and 33 ml 37% formaldehyde. Mix well.
4. Pour the agarose onto a gel bed. Run *nonsubmarine* gels.
5. Use a peristaltic pump to recirculate buffer between the two tanks.
6. After the samples are loaded, cover the gel with Saran wrap and run at 50 volts overnight (18 hours).

Preparation of the RNA Samples

1. Carefully wash RNA pellets from ethanol precipitation with absolute ethanol to remove salts.
2. Resuspend the dried pellets in sterile distilled water to a final concentration of 10µg RNA in 9.3 µl.

3. Add 20 μl deionized formamide; mix.
4. Add 6.7 μl 37% formaldehyde (a final concentration of 2.2 *M*). Mix.
5. Add 4 μl 10X gel buffer; mix.
6. Heat for 5 minutes at 60°C.
7. Cool on ice.
8. Add 10 μl loading solution; mix.
9. Load the gel with the 50 μl sample solution and run immediately.

Staining of RNA Markers

1. Cut out the lane containing the RNA markers.
2. Incubate for 1 hour in 200 ml water containing 140 μl 2-mercaptoethanol and 20 μl 10 mg/ml ethidium bromide.

Buffers for Formaldehyde Gels

Buffer F (10X) — *1 liter*
200 m*M* MOPS (sodium salt) — 46.26 g MOPS [3-(*N*-morpholino) propanesulfonic acid]
50 m*M* Sodium acetate (pH 7.0) — 50 ml 1 *M* Sodium acetate (pH 7.0)
10 m*M* EDTA — 20 ml 0.5 *M* EDTA
H₂O to 1 liter

Loading solution — *10 ml*
20% Ficoll — 2 g Ficoll
20 m*M* EDTA — 400 μl 0.5 *M* EDTA
0.15% Bromphenol blue — 15 mg Bromphenol blue
H₂O to 10 ml

Formaldehyde. This compound oxidizes when in contact with the air. Deionize as described for formamide (see Chapter 15) until a neutral pH is obtained. Recrystallize at 0°C and store aliquots at −20°C in tightly capped tubes.

REFERENCES

Rave, N., Crkvenjakov, R., and Boedtker, H. (1979), *Nucleic Acids Res.*, **6**, 3559.
Thomas, P. (1980), *Proc. Natl. Acad. Sci. USA*, **77**, 5201.

METHYL MERCURY GELS FOR RNA

In our hands methyl mercury gels always led to sharp RNA bands and often provided a better resolution than glyoxal or formaldehyde gels. Unfortunately, the potential biological hazards associated with the manipulation of methyl mercury has hampered the wide use of this system. In response to the widespread anxiety about this compound, Dr. R. P. Junghans has reviewed the toxicity of methyl mercury compounds (see Chapter 2) and has reached the conclusions that (1) working with gels and diluted solutions does not require hood facilities and (2) disposal of methyl mercury materials does not require exceptional precautions (see Chapter 2).

Our general laboratory policy is to recommend the use of a fume hood when working with hazardous chemicals. Methyl mercury powder is one among them, but it is important to note that mouth pipetting a polyacrylamide solution or organic solvants such as phenol and chloroform is certainly more hazardous than working with diluted methyl mercury.

PROTOCOL

PREPARING AND RUNNING METHYLMERCURY GELS

1. Melt a 1% agarose solution (e.g., 150 ml) in 1X EM buffer and cool it to 60–62°C until used.

2. Add 750 μl 1 M methyl mercury solution to the 150 ml of gel to obtain a 5 mM final concentration. It is recommended but not absolutely necessary (see Chapter 2) that this step and all steps up to 14 be performed under a fume hood.

3. Pour a horizontal gel and insert the comb.

4. Allow the agarose to solidify; remove the comb. Fill the wells with 1X EM buffer. *Run nonsubmarine gels* (see Chapter 9, see also note 9).

5. Prepare a 1-μg/μl RNA solution.

6. For each sample, mix in a microfuge tube:
   5 μl RNA solution from step 5
   5 μl 2X sample buffer

7. Prepare a ¹⁄₁₀ dilution of the 1 M stock solution of methyl mercury in sterile distilled water.

8. Add 1 μl 0.1 M methyl mercury solution to each sample.

9. Close the caps tightly, vortex, and spin for 1 second at 12,000 rpm.

10. Load samples quickly onto the gel to avoid diffusion. Have the gel connected to the power supply and completely set up before you begin to load the gel.

11. Run gel at a constant voltage between 60 and 100 V.

12. Use a peristaltic pump to recirculate the buffer between the two tanks. If you don't need to blot the gel, proceed to step 13. If you need to blot the gel, see pages 533–535.

13. Place the gel in 200 ml of 0.05 M ammonium sulfate and add 20 μl of ethidium bromide solution (10 mg/ml in water).

Shake the gel for 20 minutes. Dispose the solution.

14. Repeat step 13. Photograph the gel under UV light.

Notes for Methyl Mercury Gels

1. As discussed above and in Chapter 2, working with diluted solutions of methyl mercury is no more hazardous than working with several other chemicals currently used in the laboratory. We do recommend, however, wearing gloves when manipulating methyl mercury gels and wrapping waste before discarding.

2. Ammonium phosphate, ammonium sulfate or sodium thiosulfate will inactivate methyl mercury in case of a spill.

3. All pieces of gel in which methyl mercury has not been complexed should be soaked in a solution of 0.5 M ammonium phosphate for 30 minutes before being dumped in a waste basket or wrapped and disposed of as organic waste.

4. All glassware, gel bed, and comb can be rinsed with water after use, before being processed as regular glassware for washing.

5. Since EDTA forms complexes with methyl mercury, it is essential to maintain a sufficient excess of methyl mercury to be denaturing for nucleic acids. EDTA concentration should be between 0.1 and 1 mM.

6. Chloride ions should not be used in buffers because they also form complexes with methyl mercury.

7. There is no need to put methyl mercury in the buffer tanks for electrophoresis.

8. 0.1% Xylene cyanol dye can be run as a marker for 18S RNA. This dye runs just behind 18S RNA in 1% gels. Since the mobility of the xylene cyanol dye is not affected by the size of the agarose matrix, the relative migrations of xylene cyanol and 18S RNA will vary with the agarose concentration in the gels.

9. If you wish to run a submarine gel, you should use 10 mM methyl mercury as diffusion of methyl mercury out of the gel during a long run may reduce the concentration of the denaturing agent below some critical value. A methyl mercury concentration ranging between 5 and 10 mM is satisfactory for vertical gels.

10. Better resolution of small RNA species can be achieved by running polyacrylamide gels in the presence of methyl mercury (Chandler et al., 1979).

Buffers for Methyl Mercury Gels

All buffers are made with sterile distilled water in baked glassware.

EM buffer (10X)   *1 liter*
0.05 M Sodium borate   19.0 g Sodium borate $\cdot$ 10H$_2$O
0.1 M Sodium sulfate   14.2 g Sodium sulfate
0.01 M EDTA   3.7 g EDTA
0.15 M Boric acid   30.9 g Boric acid
                                    H$_2$O to 1 liter

2X Sample buffer   *10 ml*
20% Glycerol   2 ml Glycerol
0.05% Bromphenol blue   5 mg Bromophenol blue
2X EM buffer   2 ml 10X EM buffer
                                    H$_2$O to 10 ml

0.5 M Ammonium acetate
38.55 g Ammonium acetate
H$_2$O to 1 liter
(Store solid powder at 4°C.)

NaOH–mercaptoethanol   *400 ml*
                                    200 ml H$_2$O
50 mM NaOH   0.8 g NaOH
5.7 mM 2-Mercaptoethanol   160 µl 14.3 M 2-Mercaptoethanol
                                    H$_2$O to 400 ml

Citrate phosphate buffer
(CPB, 0.14 M)
13.5 g Citric acid
21.5 g Sodium phosphate
(disodic) $\cdot$ 7H$_2$O
H$_2$O to 1 liter

Citrate phosphate buffer
(CPB, 20 mM)
98 ml 140 mM CPB
H$_2$O to 1400 ml

REFERENCES

Bailey, J. M., and Davidson, N. (1976), *Anal. Biochem.,* **70**, 75.

Chandler, P. M., Rimkus, D., and Davidson, N. (1979), *Anal. Biochem.,* **99**, 200.

Junghans, R. P. (1983), *Environ. Res.,* **31**, 1.

GLYOXAL GELS FOR RNA

Since glyoxal is much safer than the chemicals used in the preceding protocols, it's use as a denaturing agent for RNA electrophoresis alleviates the need for running gels in fume hoods. However, note that to obtain satisfactory results, it is essential that glyoxal be properly deionized prior use.

PROTOCOL

GLYOXAL AGAROSE GELS

1. Prepare the RNA sample as described on pages 516–526.

2. Bring RNA to 10 mM phosphate buffer (pH 7.0) by addition of concentrated stock buffer.

3. For each sample to be run on the gel, mix in a microfuge tube:
   8 µl sterile distilled water
   10 µl RNA (0.5 µg/µl) in phosphate buffer (10 mM)
   3 µl 7 M Glyoxal (prepared as described below)
   The pH of the mix is now 6.5.

4. Heat at 50°C for 60 minutes in a water bath. Spin the samples periodically (1 second at 12,000 rpm) to collect the droplets of evaporated water that form during incubation.

5. Place samples on ice and add a half volume (10 µl) of dye buffer.

6. Load samples onto the gel and run. (One can use voltages up to 120 V with good resolution.)

Buffer for Glyoxal Gels

Dye buffer	*10 ml*
0.05% Bromphenol blue	5 mg Bromphenol blue
20% Glycerol	2 ml Glycerol (double-distilled and sterile)
20 mM EDTA	400 µl 0.5 M EDTA
	H$_2$O to 10 ml

Preparation of Deionized Glyoxal

1. The stock solution of 7 M glyoxal should be stored refrigerated in the presence of mixed bed resin (Mallinckrodt Amberlite MB-3).

2. Mix a small volume of glyoxal (10 ml) with a ¹⁄₁₀ volume of resin and swirl on ice for 5 minutes.

3. Run through a 15-ml syringe equipped with a Whatman GFC filter.

4. Repeat the filtration six times.

5. The pH of the solution should rise from about 3 to 5.5, and the glyoxal solution should no longer be as yellow as the untreated solution.

6. Store the deionized glyoxal solution in aliquots of 100–150 μl at −70°C.

Note. On thawing, a white precipitate may appear. If this occurs, do not use the sample.

REFERENCE

McMaster, G. K., and Carmichael, G. G. (1977), *Proc. Natl. Acad. Sci. USA*, **74**, 4835.

SIZE MARKERS FOR RNA SPECIES RUN IN DENATURING AGAROSE GELS

The accurate sizing of RNA species run under denaturing conditions in agarose gels has been ham-

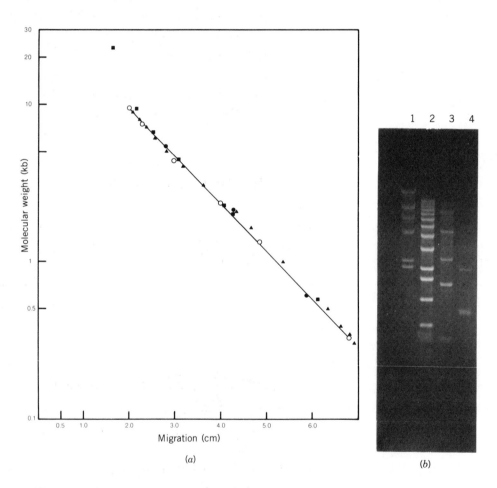

Figure 161. (*a*) Electrophoretic mobility of RNA molecular weight standards. The RNA ladder (○) was used to plot a standard curve for molecular weight. The mobility of lambda DNA/Hind III fragments (■), the 1 kb ladder (▲), and 9S, 18S and 28S RNA species (●) from panel (*b*) are shown. (*b*) Comparison of gel electrophoretic mobilities. All samples were treated with 1 *M* glyoxal and 50% DMSO in 10 m*M* sodium phosphate (pH 7.0). Electrophoresis was performed in a 1.2% agarose gel in 10 m*M* sodium phosphate (pH 7.0) for 2 hours at 160 volts. Bands were detected by staining with ethidium bromide. Lane 1 = 2 μg of lambda DNA/Hind III fragments; lane 2 = 3 μg of 1 kb ladder; lane 3 = 3 μg of RNA ladder; and lane 4 = 2.5 μg total of globin 9S RNA and 18S and 28S rRNA(courtesy of Bethesda Research Laboratories, Life technologies Inc.).

pered by the fact that until recently most laboratories used to express the size of their standards in Svedberg (S). There is no problem converting a sedimentation value to a molecular weight unit by using the empirically established formula

$$n = 4.697 \times S_{20W}^{2.1}$$

where n is the number of nucleotides, and S_{20W} the sedimentation coefficient of the RNA expressed in Svedbergs.

The most currently used markers are ribosomal RNA species from *E. coli* (16S and 23S) and from eucaryotic cells (18S and 28S), but be aware that all eucaryotic ribosomal RNAs do not have the exact same sedimentation coefficient.

To circumvent these problems, it is possible to use restriction fragments of known size. It is essential that the DNA fragments are denatured prior to electrophoresis. Use at least five µg of DNA because single stranded DNA does not bind stain as well as double stranded DNA. Proceed exactly as described for RNA samples to denature the restriction endonuclease-digested DNA (you may use any of the denaturing conditions described in this section).

In some cases, a smear is obtained in place of the expected bands of single stranded DNA. The most frequent cause for this problem is due to the fact that the double stranded DNA fragments that are used content several nicks at random. Under non-denaturing conditions (i.e., when you check the digestion), there is enough strong attraction between the two DNA strands to stabilize the double helix structure. Under denaturing conditions, each single strand falls in pieces and a smear is observed. Therefore, when you plan on using DNA fragments as markers for denaturing gels, make sure that they are in good shape.

All these problems can be eliminated by the use of RNA species of known size. The preparation of such markers can be performed by in vitro synthesis with the bacteriophage-derived transcription systems (see Chapters 25 and 6). These transcripts can be synthesized in the presence of radioactive precursor (see Chapter 16) and therefore provide a simple way to determine precisely the size of RNAs in gels. The RNA ladder commercialized by BRL is prepared under such conditions and provides a reliable set of markers (Figure 161).

TRANSFER OF RNA TO NITROCELLULOSE

RNA transfer to nitrocellulose or to chemically modified paper (see below) is generally known as Northern blotting, by analogy to Southern blotting for DNA.

Methyl Mercury Gels

PROTOCOL

TRANSFER OF RNA SPECIES FROM METHYL MERCURY GELS TO NITROCELLULOSE

After RNA has been run on methyl mercury gels, blot the gel without treating it. Do all manipulations under a fume hood.

1. Cut off the lane of the gel in which size markers have been run and stain it by incubation in H_2O containing 2-mercaptoethanol and ethidium bromide (page 527). Do not stain the whole gel, as staining decreases the efficiency of the transfer.

2. Cut a piece of nitrocellulose filter, four sheets of Whatman 3MM paper, and paper towels to the same size as the gel.

3. Soak the nitrocellulose in water and then in 20X SSC.

4. Soak the sheets of Whatman paper in 20X SSC.

5. Set up the transfer as described for DNA transfer (see Chapter 15, Figure 113). From the bottom to the top, place two sheets of Whatman paper, the gel, the nitrocellulose sheet, the 2 remaining sheets of Whatman paper and the paper towels (about a 15-cm thickness).

6. Cover with Saran wrap. Put a Plexiglass or glass plate on top, and on that a weight of about 1 kg.

7. Let the transfer proceed overnight at room temperature under the hood.

8. The following day, remove the paper towels and indicate slot positions with a pencil, through the gel, onto the nitrocellulose filter. Cut a corner to allow orientation of the filter after hybridization.

9. You can now remove and stain the gel. In most cases, some RNA, especially high molecular weight species, will be left, giving you some information about how the gel ran.

10. Bake the nitrocellulose filter at 80°c for 2 hours before prehybridization and hybridization (Chapter 15).

Formaldehyde Gels

PROTOCOL

TRANSFER OF RNA SPECIES FROM FORMALDEHYDE GELS TO NITROCELLULOSE

After the gels have finished running, simply soak the gels in 250 ml 20X SSC for 30 minutes and transfer to nitrocellulose sheets as described for methyl mercury gels above. Do not treat gels with NaOH, as treatment with alkali and subsequent neutralization reduces transfer efficiency by about 50%.

TRANSFER OF RNA TO DBM OR DPT PAPER

Methyl Mercury Gels

PROTOCOL

TRANSFER OF RNA SPECIES FROM METHYL MERCURY GELS TO DBM OR DPT PAPER

Preparation of the Gel

1. Remove the gel from the bed and place it in 400 ml 5 mM 2-mercaptoethanol, 50 mM NaOH (Wash 1). Shake for 45 minutes at room temperature under the hood.

2. Pour off wash 1 and replace with 300 ml 0.14 M CPB, 1 μg/ml ethidium bromide. Shake for 15 minutes at room temperature.

3. Take a picture of the gel. Cut away the size marker lane and measure the gel (for next step).

4. Shake the gel at room temperature in 300 ml 0.14 M CPB, 7 mM iodoacetate for 30 minutes.

5. Pour off buffer and repeat step 4.

6. Pour off and shake the gel for 30 minutes in 300 ml ice-cold 20 mM CPB.

7. Pour off and let the gel sit for 30 minutes in 300 ml ice-cold 20 mM CPB. Activate the paper at this point.

Activation of ABM or APT Paper

1. Cut a piece of paper to the dimensions of the gel.

2. Prepare 10 ml 1% sodium nitrite (NaNO$_2$) in H$_2$O.

3. Prepare 200 ml 1.2 N HCl. Cool in ice.

4. Add 5.4 ml 1% NaNO$_2$ (fresh solution) to the 200 ml of 1.2 N HCl.

5. Pour over the paper. The paper will turn bright yellow. It is possible to check the generation of nitrous acid with starch iodide paper (which should turn black).

6. Let sit on ice for 30 minutes (or longer).

7. Just prior to the transfer of paper to the surface of the gel, rinse the activated paper twice with sterile cold water and once with cold 20 mM CPB.

Transfer of RNA

1. Set up the transfer at 4°C. From the bottom of the tray to the top, place:
   a. The gel.
   b. The activated paper (use RNase-free tweezers to move paper). Transfer will proceed immediately, so DO NOT MOVE THE PAPER ONCE IT IS IN FULL CONTACT WITH THE GEL.
   c. Dry sheets of Whatman 3MM paper cut to the size of the gel.
   d. A stack of paper towels.
   e. A weight (500 g–1 kg).
2. Let sit at 4°C for 12–18 hours.

Glyoxal or Formaldehyde Gels

PROTOCOL

TRANSFER OF RNA SPECIES FROM GLYOXAL OR FORMALDEHYDE GELS TO DBM OR DPT PAPER

The denaturing agent must be removed from the gel prior to transfer.

Glyoxal Gels

1. Wash the gel in 200 ml 50 mM NaOH containing 1 μg/ml ethidium bromide for 60 minutes at room temperature.
2. Neutralize by washing twice with 200 ml 200 mM sodium acetate (pH 4.0) for 30 minutes, each time at room temperature.
3. Wash once with 200 ml 20 mM sodium acetate (pH 4.0) for 30 minutes.

Formaldehyde Gels

1. Stain the RNA with acridine orange (33 μg/ml in 10 mM sodium phosphate, pH 7.4, 500 mM formaldehyde) for 10 minutes at room temperature with shaking.

2. Destain in 200 ml 10 mM sodium phosphate (pH 7.4), 500 mM) formaldehyde.

3. Wash the stained gel with 200 ml 50 mM NaOH for 60 minutes at room temperature.

4. Wash the gel twice with 200 ml 200 mM sodium acetate (pH 4.0) for 15 minutes at room temperature to neutralize residual NaOH.

TRANSFER OF RNA TO NYLON MEMBRANES

PROTOCOL

TRANSFER OF RNA TO NYLON MEMBRANES

1. The transfer of RNA from denaturing gels is carried out essentially as described for transfer to nitrocellulose filters.

2. *Do not rinse* the membrane after transfer since washing may result in significant losses of RNA.

3. Baking the membrane will increase the retention of the RNA species by the membrane.

Notes

1. The Biodyne A membranes have been successfully derivatized with diazobenzyloxymethyl groups for covalent binding of macromolecules, as described on page 432.

2. For prehybridization, use 4 ml of the hybridization solution per 100 cm^2 of membrane.

3. For hybridization, use 2 ml of the hybridization solution per 100 cm^2 of membrane.

4. To avoid background problems do not use more than 10^6 cpm of probe per milliliter of hybridization mixture.

5. Washing is performed as described on page 442.

HYBRIDIZATION TO mRNA IMMOBILIZED ON HYBOND mAP*

Hybond-mAP (messenger affinity paper) has become established as a medium for the rapid and convenient isolation of polyA⁺ mRNA from a total RNA population. It allows elution of biologically active mRNA by simply heating in a salt-free solution (Wreschner and Hersberg, 1984; Werner et al., 1984; Bludau et al., 1986). Hybond-mAP can also be used as a convenient medium for the detection of specific mRNA species by nucleic acid hybridization (Wreschner and Herzberg, 1984). For example, by applying total RNA extracts to Hybond-mAP, it is possible to screen a number of samples for the presence of a particular message by hybridization with an appropriate labeled probe. A quantitative estimate of relative levels of expression can also be obtained by autoradiography followed by a densitometric scan, or by elution of bound RNA plus probe followed by scintillation counting. The use of Hybond-mAP in hybridization provides a convenient means of immobi-

*Courtesy of Amersham International plc.

lization and subsequent elution if required. It also avoids problems that might be caused by the presence of an excess of ribosomal RNA, or by less than quantitative binding to nitrocellulose or nylon membranes.

To screen a large number of samples, it is convenient to use a commercial "dot blot" or "slot blot" apparatus for immobilizing the mRNA. The protocol below has been found to work well at Amersham, although a number of variations on this basic procedure might be envisaged. If fewer samples are to be screened, then it is possible to use the standard isolation protocol, omitting the elution step. The protocol is supplied with the product.

Several hybridization and washing procedures have been used successfully with Hybond-mAP using either nick translated or cDNA probes. Two specific protocols that have been shown to work well are detailed below, but it is possible that, for any given system, further optimization of one or more parameters may be desirable.

PROTOCOL

HYBRIDIZATION TO mRNA BOUND ON NYLON-DERIVED MEMBRANES

Transfer of RNA to Hybond Membranes

1. Isolate total RNA by standard procedure and ethanol precipitate.

2. Resuspend in sterile distilled water at a concentration greater or equal to 1 mg/ml (generally, as concentrated a solution as possible is desirable).

3. Heat at 60–65°C for 5 minutes and cool rapidly on ice.

4. Add sterile NaCl to a final concentration of 0.5 M.

5. Prewet the Hybond-mAP in 0.5 M NaCl. Allow to air dry (optional). Assemble slot/dot blot apparatus as described by the manufacturer. Usually a filter paper backing is used which should not be prewet.

6. Pipette the total RNA into the wells, loading less than 50 µl if possible. The solution should filter through Hybond-mAP without vacuum but if more liquid is to be applied than the

wells can easily hold, apply a *very* gentle vacuum. Slow filtration is essential for proper binding.

7. Dismantle the apparatus and agitate the Hybond-mAP gently for 5 minutes in 0.5 M NaCl. Air dry thoroughly.

Hybridization in the Presence of Formamide at 42°C

1. Prehybridize Hybond-mAP in 0.2 ml/cm^2 prehybridization solution at 42°C for at least 4 hours.
   Prehybridization solution:
   5X SSPE
   50% deionized formamide
   5X Denhardt's reagent
   10% dextran sulfate
   100 μg/ml polyadenylic acid

2. Discard prehybridization solution and add 0.2 ml/cm^2 fresh prehybridization solution plus denatured probe. Hybridize at 42°C overnight.

3. Wash filter at 35°C as follows:
   4X SSC, 0.1% SDS, 15 minutes
   1X SSC, 0.1% SDS, 30 minutes (repeat two further times)
   0.1X SSC, 0.1% SDS, 30 minutes

4. Dry filter and expose to film.

Results obtained using this protocol are illustrated in Figure 162.

Hybridization with No Formamide at 65°C

1. Prehybridize Hybond-mAP in 0.2 ml/cm^2 prehybridization solution at 65°C for at least 4 hours.
   Prehybridization solution:
   6X SSPE
   0.2% w/v Ficoll
   0.2% w/v polyvinylpyrollidone
   0.1% SDS
   25 μg/ml salmon sperm DNA (denatured)

2. Discard prehybridization solution and add 0.2 ml/cm^2 fresh prehybridization solution plus denatured probe. Hybridize at 65°C overnight.

3. Wash filter as follows:
   4X SSC, 0.1% SDS, 15 minutes
   1X SSC, 0.1% SDS, 30 minutes. Repeat two further times.
   0.1X SSC, 0.1% SDS, 15 minutes.

4. Dry filter and expose to film.

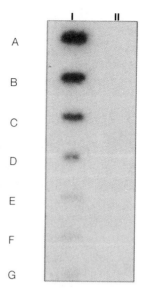

Figure 162. Hybridization to mRNA immobilized on Hybond™ map. Column I. Total RNA from a *Saccharomyces cerevisiae* strain carrying a plasmid containing the Leu-2 gene. The yeast was grown in leucine-minus minimal medium. Total RNA was extracted by grinding the yeast in a hypotonic solution, followed by proteinase-K/SDS digestion, phenol chloroform extraction, and ethanol precipitation. The hybridization solution contained 5–6 × 10⁶ dpm/ml of a Leu-2 DNA probe labeled with ^{32}P by nick-translation. The film was exposed for 22 hours at −70°C. A to G represent serial dilutions of total yeast RNA. The amount of mRNA in each slot has been calculated from yields of mRNA to be (a) 300 ng; (b) 150 ng; (c) 75 ng; (d) 37 ng; (e) 18 ng; (f) 9 ng; and (g) 5 ng. Column II. 3 μg of tobacco mosaic virus mRNA. Although this RNA does not have a poly A tail it binds efficiently to Hybond-mAP via an internal poly A tract (courtesy of Amersham).

Solutions

20X SSPE
3.6 *M* NaCl
200 m*M* sodium phosphate (pH 7.7)
20 m*M* Na₂EDTA

20X SSC
3.0 *M* NaCl
300 m*M* trisodium citrate (pH 7.0)

100X Denhardt's
2% w/v Ficoll
2% w/v polyvinyl pyrollidone
2% w/v bovine serum albumin

Notes

When carrying out hybridizations on Hybond-mAP, the following points should be kept in mind.

1. It is not possible to wash blots at high levels of stringency (for example, 0.1X SSC, 65°C), as this will elute the target mRNA from the filter.

2. When hybridizing to dot or slot blots, a negative control DNA dot should be included, to ensure hybridization is specific. If a quantitative comparison of samples is to be carried out, it is desirable to include a series of dilutions of control (or sample) DNAs to ensure response is linear over the required range.

3. If a heavy background across the entire paper is obtained, there are several possible causes:

   a. Probe concentration is too high. This is particularly applicable if dextran sulfate is used. The situation can be remedied by a lower probe concentration, or a shorter hybridization time, or by the omission of dextran sulfate.

   b. Check quality of probe, for example, probe size and presence of unincorporated label.

   c. Check quality of hybridization reagents, particularly the heterologous blocking DNA or RNA. Improved results might be obtained by using more than one blocking nucleic acid (for example, calf thymus DNA and long-chain poly A).

4. If nonspecific hybridization occurs, again several causes are possible:

   a. As DNA–RNA hybrids are more stable than DNA–DNA hybrids, it is possible that some probes may show a high level of hybridization to a variety of related RNA species. It may be possible to overcome this by carrying out the hybridization and washing steps in the presence of formamide at a higher temperature (for example, 50°C or above). However, beware of eluting target mRNA during the most stringent washes.

   b. Again, check quality of probe and hybridization reagents.

5. If a nonlinear response is seen over a range of dilutions, in addition to the considerations in 3 and 4 above, it may be necessary to load a lower amount of target RNA if the Hybond-mAP is saturated. Alternatively, use of a larger area of mAP may help, or a slower application of RNA to the paper. Sample that have been filtered through Hybond-mAP may also be reapplied if necessary.

6. If a lower intensity signal is obtained, again a number of factors may be responsible:

   a. Check that the RNA has been correctly applied to the paper (further guidelines are supplied with the product).

   b. Check quality of probe and hybridization reagents, particularly for nuclease contamination. For example, some sources of BSA may be significantly contaminated. The presence of SDS or formamide should help to minimize this problem, although incompletely deionized formamide may also lead to reduced signal.

c. Check that the washes have not been too stringent, for example, by autoradiography after the penultimate wash or by repeating with lower washing temperatures (if initially carried out above room temperature).

7. Hybond-mAP blots cannot be reprobed, as any procedure used to remove bound probe is also likely to elute target RNA from the paper.

REFERENCES

Bludau, H., Kopun, M., and Werner, D. (1986), *Exp. Cell Res.*, **165**, 269.

Werner, D., Chemla, Y., and Herzberg, M. (1984), *Anal. Biochem.*, **141**, 329.

Wreschner, D. H., and Herzberg, M. (1984), *Nucl. Acids Res.*, **12**, 1349.

MAPPING RNAs BY DIGESTION OF RNA–DNA HYBRIDS

Among the different single strand specific nucleases that are now available, nuclease S1, exonuclease VII, and mung bean nuclease have proved to be useful in recombinant DNA technology. Several studies have shown the high specificity of these nucleases. While nuclease S1 and mung bean nuclease are both endodeoxyribonuclease and endoribonuclease, exonuclease VII is specific for single stranded DNA and is able to digest from either 3′ and 5′ termini in both 3′ and 5′ directions.

Based on the specific digestion of single stranded portions of RNA–DNA hybrids by nuclease S1, Berk and Sharp (1977, 1978) described a method for mapping RNA molecules colinear with DNA templates. The rationale of their experiments is described in Figure 163. An RNA transcript homologous to n nucleotides of a DNA sequence is hybridized to the coding strand of a restriction fragment containing this sequence (fragment A). Upon incubation of the hybrid molecule with nuclease S1, the single stranded portions of the hybrid are digested and the S1-resistant hybrid DNA can be recovered by ethanol precip-

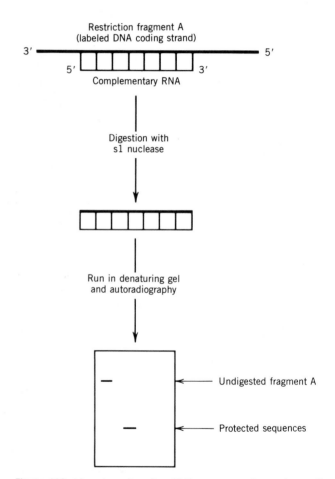

Figure 163. Mapping of coding DNA sequences by nuclease S1 digestion of DNA–RNA hybrids.

itation. Sizing of the complementary portion is then made possible by running the denatured DNA hybrid in an alkaline agarose gel or in a polyacrylamide gel containing urea (see Chapters 3 and 9). If the DNA is labeled with ^{32}P, it can be detected by autoradiography. By choosing different overlapping restriction fragments, it becomes possible to map the RNA rather precisely on the DNA.

S1 nuclease does not remove the poly-A track or the capped 5′-end of eucaryotic messenger RNAs when they are base paired (Haegeman et al., 1979), but the presence of the 5′ cap structure has been thought to sterically hinder S1 digestion at the phosphodiester bond adjacent to the first base-paired nucleotide (Weaver and Weissman, 1979). When mapping is performed with an end-labeled probe one can determine the polarity as well as the map position of the RNA on the corresponding DNA sequences (Weaver and Weissman, 1979) (Figure 164). S1 nuclease has also been used in the mapping of eucaryotic-spliced RNA molecules. The procedure is outlined below and is illustrated in Figure 165. The first step consists in a mapping of the exons under the conditions described above. The second step is based on the observation that digestion of single stranded RNA by nuclease S1 is sevenfold less effective than digestion of single stranded DNA under similar conditions (Vogt, 1973). Therefore, digestion of the RNA–DNA hybrid molecule will lead in most cases to a duplex molecule containing an intact RNA strand and migrating in nondenaturing gels as a double stranded DNA with the size of the exons represented in the RNA. However, under these conditions, some hybrid molecules are being cut at the level of the gap corresponding to the splice, where the two exons are joined on the DNA strand. When run on neutral gels, these fragments migrate with the same mobility as the DNA fragments containing the exons; therefore their length can be determined. The fraction of molecules cut at this level vary with the nature of the hybrids considered (Berk and Sharp, 1977; 1978).

Exonuclease VII can trim the single stranded DNA in a RNA–DNA heteroduplex but is not able to digest loops of single stranded DNA. Digestion with exonuclease VII must be performed at 45°C in the presence of low salt concentrations (e.g., 30 mM KCl) in order to favor the disruption of the single stranded DNA secondary structure (Goff and Berg, 1978). Exonuclease VII has proved to be very useful in the mapping of short exons in which no restriction sites had been mapped (Berk and Sharp, 1978). The combined use of exonuclease VII and nuclease S1

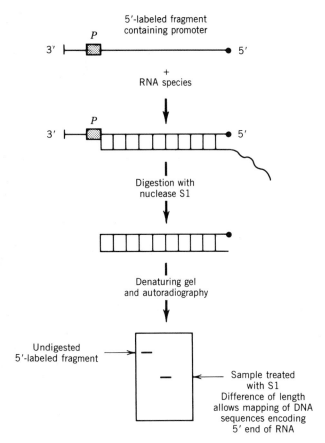

Figure 164. Nuclease S1-mapping of DNA regions encoding the 5′ end of RNA species.

has permitted precise maps of exons represented in complex spliced mRNA species to be obtained (see, for example, Berk and Sharp, 1977, 1978).

In molecular cloning, the single strand specific nucleases are mostly used (1) to localize the DNA sequences homologous to the 5′ terminus of well-defined RNA species, and (2) to create or eliminate restriction sites in DNA after trimming of the protruding ends, and (3) to cleave hairpin loops in cloning cDNA (see Chapter 19).

Nuclease S1 Mapping of DNA Sequences Homologous to RNA Species

Mapping of DNA sequences complementary to RNA species requires that the experiment be performed in such a way that the RNA molecules can first hybridize with the coding strand of the DNA region involved in the expression of the considered RNA species. Isolation of single stranded DNA can be fa-

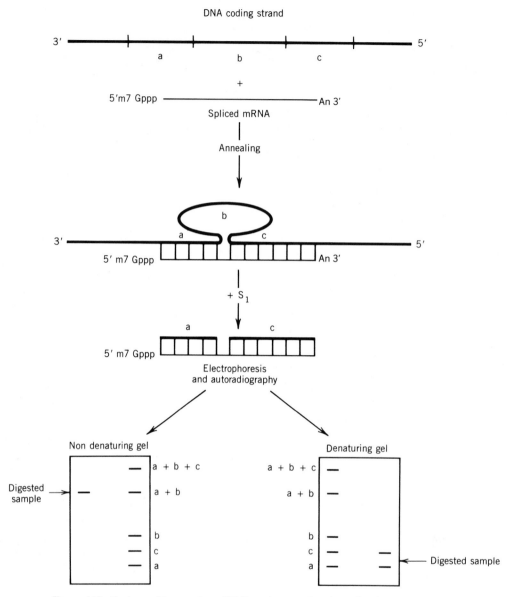

Figure 165. Nuclease S1-mapping of DNA regions coding for spliced mRNA species.

cilitated by using filamentous phages or plasmid vectors harboring the IG region of such phages (see Chapter 6). It is also possible to purify single strands of denatured DNA on neutral agarose gels (Herrman et al, 1980) or on polyacrylamide gels (see Chapter 9), and it has been reported that hybridization of poly UG to the denatured DNA strands may increase markedly the difference in mobility of the DNA strands in neutral agarose gels (Hayward, 1972; Goldback et al., 1978).

In aqueous buffers, the kinetics of hybridization of RNA molecules to single strands of DNA is similar to the kinetics of DNA single strand renaturation

(Wetmur and Davidson, 1968; Britten and Kohne, 1968). Therefore, the rate of hybridization is maximal in the presence of a high salt concentration (1 M NaCl) and at a temperature corresonding to the melting temperature (T_m) of 25–30°C. In practice, we often perform the hybridization in the presence of formamide (generally 50%) in order to decrease the value of the T_m, and consequently perform the hybridization at a lower temperature (Bonner et al., 1967; McConaughy et al., 1969). The effect of formamide on the melting temperature of duplex DNA molecules can be calculated by the following equation established by Casey and Davidson (1977):

$$T_m = 81.5 + 0.5(\%G + C)$$
$$+ 16.6 \log(\mathrm{Na}^+) - 0.6\ (\% \text{ formamide})$$

At high concentrations of formamide (70–80%), the T_m of a DNA–DNA duplex is lower (5–10°C) than the T_m of the corresponding RNA–DNA hybrid (Casey and Davidson, 1977). Under such conditions, one can use a hydridization temperature comprised between the two T_m to hybridize RNA directly to the coding strand of a duplex DNA. This is particularly useful when the separation of the two DNA strands is not possible for some technical reason but it should not systematically replace strand separation which often leads to better results. Since most RNA–DNA hybridizations are performed at a temperature ranging between 40 and 60°C, Casey and Davidson suggested using a buffer that provides constant pH and ionic strength at these temperatures. They found that a 0.04 M, pH 6.4 PIPES buffer (see Chapter 3) containing 0.4 M NaCl and 1 mM EDTA was particularly suitable. To establish the optimal conditions required for hybridization, it is necessary to first determine the T_m of the DNA duplex utilized. Then, hybridization is performed at a temperature 2 to 4°C above that of T_m. It is generally admitted that best results will be obtained with excess of DNA. Under these conditions, RNA–DNA hybridization is nearly complete in 3 hours, with Adenovirus 2 DNA (35 kb) at a concentration of 10 μg/ml corresponding to a DNA excess (Sharp et al., 1980). This experimental observation can be used to predict that under similar conditions nearly complete hybridization for a DNA of n kb will be obtained at a DNA concentration of n/35 in 3 hours.

In practice, the hybridization is performed with labeled DNA fragment. Depending on the nature of the experiment being performed, the DNA may be labeled uniformly or specifically at the 5′ or the 3′ end as described Chapter 16. The 5′ end labeling of DNA is usually performed with polynucleotide kinase (see Chapters 16 and 4) in the presence of γ ^{32}P ATP while the 3′ end labeling of DNA is achieved either by incubation with the Klenow fragment of $E.\ coli$ polymerase I (see Chapters 16 and 4) in the presence of α[^{32}P]dNTP or with the terminal transferase (see Chapters 16 and 4) in the presence of α[^{32}P]ATP.

PROTOCOL

NUCLEASE S1 MAPPING FOR THE IDENTIFICATION OF DNA SEQUENCES ENCODING RNA SPECIES

Hybridization of RNA to the Coding DNA Strand

1. In a microfuge tube add successively:
   the labeled DNA (at least 10,000 to 50,000 cpm)
   the RNA (5 to 25 μg)
   distilled water up to 100 μl
2. Add 150 μl of buffer-saturated phenol.
3. Add 150 μl of chloroform:isoamyl alcohol mixture (24:1).
4. Vortex and centrifuge at 12,000 rpm for 10 minutes.
5. Transfer the aqueous phase in a clean tube, and add 50 μl of distilled water to the organic phase.
6. Vortex, centrifuge at 12,000 rpm for 10 minutes, and mix the supernatant to the aqueous phase from previous step.
7. Add 15 μl of NaCl 4M to the aqueous phase and invert several times to mix.
8. Add 375 μl of cold absolute ethanol, mix, and transfer to −70°C until frozen.

9. Centrifuge at 12,000 rpm for 10 minutes, discard the supernatant, and add 1 ml of cold 80% ethanol.

10. Centrifuge at 12,000 rpm for 5 minutes, discard the supernatant, and dry the pellet in a speedvac.

WHEN USING DOUBLE STRANDED LABELED DNA

11ds. Resuspend the pellet from step 10 in 25 μl of D.S. hybridization mix. Let the tube stand on the bench at room temperature for 30 minutes with occasional vortexing.

12ds. Centrifuge for a few seconds to recover liquid that is on the side of the tube.

13ds. Aspirate the mixture in a 50-μl class capillary pipette (Corning) and seal it with a bunsen burner.

14ds. Transfer the tube in a glass tube with cap and immerse in a 65°C water bath for 10 minutes.

15ds. For hybridization, transfer to a water bath whose temperature is 10°C above the hybridization temperature (the hybridization temperature that is determined by using the graph on Figure 166 depends on the $G + C$ content of the labeled DNA), and let the water in the bath slowly reach the hybridization temperature. (This should take several hours.)

WHEN USING SINGLE STRANDED LABELED DNA

11ss. Resuspend the pellet from step 10 in 17.5 μl of distilled water and add 8 μl of S.S. hybridization mix.

12ss. Centrifuge for a few seconds to recover liquid that is on the side of the tube.

13ss. Aspirate the mixture in a 50-μl glass capillary pipette (Corning) and seal it with a bunsen burner.

14ss. Transfer the tube in a glass tube with cap and immerse in a 65°C water bath for the required hybridization time (see above). We generally perform an 8-hour incubation.

Digestion With Nuclease S1

1. In an ice bucket, place a tube containing cold (4°C) ethanol.

2. Quickly immerse the capillary tubes containing the hybridized samples in the cold ethanol, and let stand on ice for about 5 minutes.

3. Remove the tubes and carefully transfer the samples in tubes containing 275 μl of cold S1 mix. Aspirate the samples several times up and down to allow a thorough mixing.

4. Centrifuge a few seconds at 12,000 rpm, and incubate at 37°C for 30 to 45 minutes.

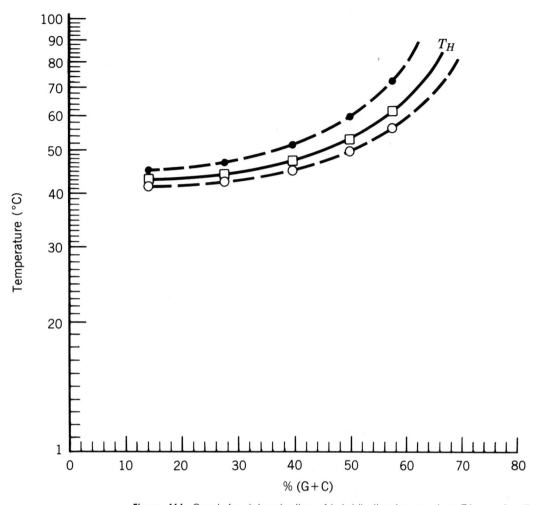

Figure 166. Graph for determination of hybridization temperature (T_H) as a function of G + C content. (● - - - ●) hybridization of DNA-RNA hybrids. (○ - - - ○) hybridization of DNA–DNA hybrids (kindly provided by M. Simon).

5. Cool the tubes on ice, and add 5 μl of *E. coli* or yeast tRNA (1 mg/ml) per sample.

6. Add 700 μl of cold absolute ethanol in each tube and let the RNA–DNA hybrid precipitate at −70°C for 15 minutes.

7. Centrifuge at 12,000 rpm for 15 minutes and discard the supernatant.

8. Dry the pellet in a Speedvac and resuspend in 12 μl of denaturing buffer.

9. Centrifuge a few seconds at 12,000 rpm and heat at 65–70°C for 10 minutes.

10. Add 12 μl of dye solution and mix.

11. Load 4 to 6 µl samples on a sequencing gel to determine the size of the labeled DNA sequences protected by the hybridized RNA.

Notes

1. Recrystallized formamide should be deionized before use.

2. It is advisable to perform a negative control hybridization in which tRNA is used instead of the homologous RNA species.

3. When running the treated sample in the denaturing gel, run in parallel a labeled size marker (see Chapters 9 and 16) and an aliquot of the untreated sample. This will allow a clearer visualization of the effect of S1 nuclease digestion. Usually, one-tenth of the initial labeled mixture is sufficient.

4. If small DNA fragments (size < 500 base pairs), are used, hybridization temperature is given by the formula

$$T_{\text{hyb}} = T_{\text{graph}} - \frac{500}{\text{length of fragment (in base pairs)}}$$

Buffers and Solutions

NaCl 4M
23, 37g of NaCl
Dissolve in 100 ml H$_2$O

NaOH 5 M
10 g of NaOH
Dissolve in 50 ml of H$_2$O.

D.S. Hybridization solution
50 m*M* PIPES (pH 6.4)
1 m*M* EDTA pH 7.5
NaCl 0.4 *M*
80% Formamide

1 ml
50 µl of 1 *M* PIPES (pH 6.4)
2 µl of 0.5 *M* EDTA pH 7.5
100 µl of 4 *M* NaCl
800 µl formamide
48 µl of H$_2$O

S.S. Hybridization solution
0.3 *M* PIPES (pH 6.4)
30 m*M* EDTA pH 7.5
2.5 *M* NaCl

250 µl of 1 *M* PIPES (pH 6.4)
50 µl of 0.5 *M* EDTA (pH 7.5)
500 µl of 4 *M* NaCl

S1 Mix
90 µl of 1 *M* Zn SO$_4$
60 µl of carrier DNA (salmon sperm or calf thymus)

300 µl of 10X S1 Buffer
150 µl of NaCl 4 M
2.4 ml of H_2O

10X S1 Buffer	*40 ml*
0.3 M Sodium acetate	4.0 ml of 3 M Sodium acetate
10 mM Zn SO$_4$	0.4 ml of 1 M Zn SO$_4$
0.5 M NaCl	3.2 ml of NaCl 4 M
	32.4 ml of H_2O

Denaturing buffer	*5 ml*
0.1 M NaOH	0.10 ml NaOH 5 M
5 mM EDTA	0.25 ml EDTA 0.1 M (pH 7.0)
	4.65 ml of H_2O

REFERENCES

Berk, A. J., and Sharp, P. A. (1977), *Cell,* **12**, 721.

Berk, A. J., and Sharp, P. A. (1978), *Proc. Natl. Acad. Sci. USA,* **75**, 1274.

Bonner, J., Kugn, G., and Beklov, I. (1967), *Biochemistry,* **6**, 3650.

Britten, R. J., and Kohne, D. E. (1968), *Science,* **161**, 529.

Casey, J., and Davidson, N. (1977), *Nucl. Acids. Res.,* **5**, 1539.

Goff, S. P., and Berg, P. (1978), *Proc. Natl. Acad. Sci. USA,* **75**, 1763.

Goldback, R. W., Evens, R. F., and Borst, P. (1978), *Nucl. Acids. Res.,* **5**, 2743.

Hamaguchi, K., and Geiduschek, P. (1962), *J. Am. Chem. Soc.,* **84**, 1329.

Hayward, G. S. (1972), *Virology,* **49**, 342.

McConaughy, B. L., Laird, C. D., and McCarthy, B. I. (1969), *Biochemistry,* **8**, 3289.

Sharp, P. A., Berk, A. J., and Berget, S. M., (1980), in *Methods in Enzymology* **65**, L. Grossman and K. Moldave, eds., Academic Press, New York, p. 750.

Weaver, R. F., and Weissman, C. (1979), *Nucl. Acids Res.,* **7**, 1175.

Wetmur, J. G., and Davidson, N. (1968), *J. Mol. Biol.,* **31**, 349.

19

CLONING OF cDNA SPECIES

SYNTHESIS OF cDNA SPECIES WITH REVERSE TRANSCRIPTASE

The synthesis of cDNA species has proved to be an essential step in the characterization of the regulatory processes involved in RNA synthesis. Also, the precise identification and sequencing of many important genes in different species has been made possible after isolation of cDNAs. The end products in cDNA synthesis are double stranded DNA molecules whose length is a direct function of the reverse transcription efficiency. Different sources of reverse transcriptase are commercially available (see chapter 4). By using the following protocol, cDNA molecules with a mean size of 1kb are currently obtained.

PROTOCOL

SYNTHESIS OF cDNA

Figure 167 schematically illustrates the steps in cDNA synthesis.

Synthesis of the First Strand

1. Mix successively in an Eppendorf tube:
   4.6 µl H_2O
   0.5 µl 12–18 oligo-dT
   0.4 µl 1 M KCl
   1.0 µl 10 mM dXTP
   1.0 µl 10X reverse transcriptase buffer
   1.0 µl 250 µg/ml RNase inhibitor
2. Add 0.5 µl reverse transcriptase (4 units/ml).

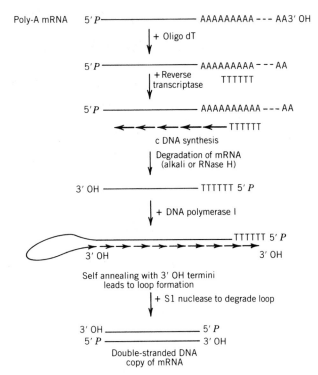

Figure 167. Synthesis of cDNA from poly mRNA.

3. Incubate for a few minutes at room temperature to allow RNase inhibitor to react with RNase.

4. Add 1.0 μl mRNA (0.5 mg/ml in H_2O).

5. Incubate for 1 hour at 43°C.

6. Incubate for 3 minutes at 100°C to melt off and partially degrade the RNA.

7. Cool on ice and spin for 2 seconds at 12,000 rpm.

Synthesis of the Second Strand

8. Add:
   5.2 μl sterile distilled water
   2.0 μl 1 *M* HEPES–KOH (pH 6.9)
   1.6 μl 1 *M* KCl
   0.6 μl 0.1 *M* $MgCl_2$
   0.4 μl 0.1 *M* DTT
   0.2 μl 10 m*M* dXTP

9. Add 4 units DNA polymerase I.

10. Incubate at 15°C for 30 minutes.

11. Incubate for 5 minutes at 65°C.

12. Cool on ice and spin for 2 seconds at 12,000 rpm.

S1 Nuclease Treatment

Intact cDNA is S1-resistant.

13. Add to the incubation mixture 100 μl S1 buffer.

14. Add 5 μl S1 nuclease (1 unit/μl).

15. Incubate at 37°C for 30 minutes.

16. Add 5 μl 0.5 *M* EDTA (pH 7.5).

17. Add 0.6 μl 20% SDS.

18. Separate cDNA and nucleotides on a Sephadex G75 column equilibrated with TES buffer. cDNA constitutes the first peak.

Notes for cDNA Synthesis

1. Oligo dT is usually extracted once with phenol and twice with ether before being used.

2. The concentration of dXTP (deoxyadenosine, -cytidine, -thymidine, and -guanosine triphosphates) is checked by UV absorption.

3. The RNase inhibitor can be RNase Sepharose, RNasin, or vanadyl–ribonucleoside complex.

4. RNA should be kept at $-20°C$ and adjusted to the desired concentration just before use. Before adding the RNA to the incubation mixture, heat it at 65°C for about 30 seconds.

5. The G75 column is prerun with approximately 20 μg of sonicated salmon sperm DNA before use.

6. This method has been reported to give about 30–70% of the input RNA as S1-resistant material of up to 8 kb.

7. This cDNA can be used for tailing or for linker addition and cloning.

Buffers for cDNA Synthesis

Reverse transcriptase buffer
(10X)   *5 ml*

40 m*M* $MgCl_2$ 200 μl 1 *M* $MgCl_2$

500 m*M* Tris–HCl (pH 8.2)   2.5 ml 1 *M* Tris–HCl (pH 8.2)

1 mg/ml Gelatin   0.5 ml Gelatin (10 mg/ml)

20 m*M* DTT   0.5 ml 200 m*M* DTT

                                         1.3 ml H_2O

S1 nuclease buffer   *5 ml*

30 m*M* Sodium acetate (pH 4.5)   150 μl *M* Sodium acetate (pH 4.5)

250 m*M* NaCl   250 μl 5 *M* NaCl

1 mM Zinc sulfate	50 µl 0.1 M ZnSO$_4$
5% Glycerol	250 µl Glycerol
	4.3 ml H$_2$O

TES buffer *100 ml*
10 mM Tris–HCl (pH 7.5) 1 ml 1 M Tris–HCl (pH 7.5)
0.5 mM EDTA (pH 7.5) 100 µl 0.5 M EDTA (pH 7.5)
0.1% SDS 0.5 ml 20% SDS
98.4 ml H$_2$O

CLONING OF cDNA SEQUENCES IN PLASMID VECTORS

When cDNA is available (see above) it is possible to treat it as DNA to introduce it into plasmid vectors. Molecular linkers or adapters can be ligated to cDNA, which can then be introduced at unique cloning sites in a plasmid vector such as pBR322. Alternatively, the cDNA can be tailed with dC residues and introduced at the dG-tailed Pst I site of pBR322 (see Chapter 14).

A limitation of this approach resides in the fact that cDNA often represents incomplete copies of the RNA sequences because of reverse transcriptase quality or the reaction conditions used. Furthermore, the second strand synthesis leads to the formation of a hairpin double stranded DNA with the 5′ end of the mRNA sequence forming a single stranded loop of variable size and location. Digestion with S1 nuclease (which is performed before tailing of the ends) invariably removes portions of the cDNA corresponding to the 5′ proximal sequences of the original mRNA.

The method described by Okayama and Berg (1982) has proved to be very effective for the preparation of cDNA libraries from polyadenylated eucaryotic mRNAs. The protocol presented below has been adapted from that described by G. Freeman (courtesy of Pharmacia). This protocol was used for the construction of several cDNA libraries from different eucaryotic tumor cells and permitted the expression of specific proteins such as interleukin and interferon. Both oligo-dT-tailed plasmid primer and oligo-dG-tailed linker DNA are now available from PHARMACIA. In addition, we also describe preparation of pBR322-derived vector and linker. The cloning protocol is outlined in Figure 168.

PROTOCOL

DIRECT CLONING OF cDNA SPECIES IN PLASMID VECTORS

Preparation of dT-Tailed pBR322-Derived Vector

1. Digest pBR322 (200 µg) twice with Pst I. Extract with phenol, precipitate with ethanol, and resuspend DNA in 100 µl sterile distilled water.

2. In a microfuge tube, mix successively:
   28 µl sterile distilled water
   40 µl 5X tailing buffer
   3 µl 0.01 M dTTP
   2 µl of 0.01 M DTT
   100 µl of Pst I-digested vector (step 1)

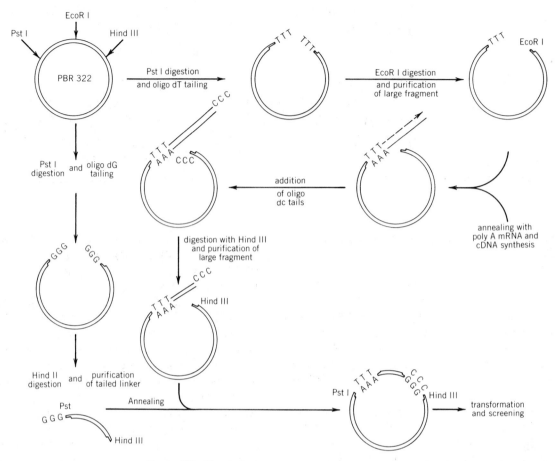

Figure 168. Direct cloning of poly-A RNA species in pBR322 (after Okayama and Berg, 1982).

3. Incubate for 5 minutes at 37°C to warm the mixture and add:
 2 µl 0.1 _M_ CoCl$_2$
 25 µl terminal deoxynucleotidyl transferase (200 units)

4. Incubate for 30 minutes at 37°C.

5. Add 20 µl 0.25 _M_ EDTA (pH 8.0) and 5 µl 20% SDS.

6. Add 100 µl phenol equilibrated with NTE buffer and 100 µl chloroform:isoamyl alcohol (24:1); mix and spin at 12,000 rpm for 2 minutes.

7. Pipette out upper phase and mix with 200 µl chloroform:isoamyl alcohol; spin at 12,000 rpm for 2 minutes.

8. Repeat step 7.

9. Add 7 µl 3 _M_ sodium acetate and 500 µl cold ethanol to upper phase. Leave at −70°C for 2 hours, or −20°C overnight.

10. Spin the DNA down at 12,000 rpm for 10 minutes at 4°C. Pour off the ethanol. Add 1 ml cold ethanol to the tube without resuspending the pellet.

11. Spin for 3 minutes at 12,000 rpm at 4°C. Pour off the ethanol and dry the pellet under vacuum.

12. Resuspend the pellet in 100 μl sterile distilled water.

13. Adjust the salt concentration for Eco RI digestion. Perform the Eco RI digest as described in Chapter 8 and run the digested DNA on an agarose gel.

14. Electroelute the large fragment as described in Chapter 11. This fragment contains sequences between the Eco RI and Pst I (dT) sites of pBR322 (see Figure 168).

15. Resuspend the electroeluted DNA in 1 ml A buffer.

16. Prepare a small column (0.6 × 2.5 cm) of oligo-dA cellulose. Wash it several times at 4°C with A buffer (see protocol for oligo-dT cellulose columns, Chapter 18).

17. Apply the DNA sample to the column and wash the column with two volumes of A buffer.

18. Elute the oligo-dT-containing DNA fragment with sterile distilled water at room temperature.

19. Collect fractions containing oligo-dT DNA and precipitate the DNA with two volumes of ethanol.

20. Wash the DNA pellet obtained after centrifugation with cold ethanol and resuspend in 50 μl TE buffer.

Preparation of dG-Tailed Linker DNA

1. Digest 100 μg pBR322 twice with Pst I. Extract with phenol, precipitate with ethanol, and resuspend DNA in 20 μl sterile distilled water.

2. Mix in a microfuge tube:
   11.75 μl sterile distilled water
   10 μl 5X tailing buffer
   0.75 μl 0.01 M dGTP
   0.5 μl 10 mM DTT
   20 μl Pst I digested DNA

3. Incubate for 5 minutes at 37°C and add:
   7.5 μl terminal deoxynucleotidyl transferase (60 units)
   0.5 μl 0.1 M CoCl$_2$

4. Incubate for 20 minutes at 37°C.

5. Add 5 μl 0.25 M EDTA (pH 8.0) and 1.5 μl 20% SDS.

6. Extract with phenol and chloroform:isoamyl alcohol (24:1) (steps 6–11 above).

7. Resuspend the DNA in 150 μl Hind III buffer. Add 150 units Hind III and incubate for 2 hours at 37°C.

8. Run the digested DNA on an agarose gel and electroelute the small fragment as described before (Chapter 11).

9. Resuspend in 200 μl sterile distilled water.

Assay for Intactness of Poly A⁺ mRNA

The intactness of the poly A⁺ RNA and its ability to serve as a template for full-length cDNA transcription are measured in a small-scale reaction using oligo $p(dT)_{12-18}$ as primer.

1. Add into a siliconized 0.5-ml microfuge tube:
   1 μl of poly A⁺ RNA (1 mg/ml)
   2 μl of 15 mM Tris–HCl (pH 7.5)

2. Heat at 65°C for 3 minutes.

3. Chill quickly on wet ice.

4. Add 0.4 μl of 20 U/μl RNAguard (or RNasin) inhibitor.

5. Mix gently.

6. Add:
   0.5 μl of oligo $p(dT)_{12-18}$ (20 A_{260} units/ml)
   1 μl of a 20 mM dXTP mix 20 mM dATP, dCTP, dGTP, TTP)
   0.5 μl of α[^{32}P] dCTP (800 Ci/mmol)
   1 μl of 10X RT buffer
   0.5 μl of 10 mM dithiothreitol
   2.3 μl of sterile distilled H$_2$O

7. Mix gently.

8. Add 1 μl of reverse transcriptase (10 U/μl).

9. Mix moderately, spin for 1 second in a microcentrifuge, and incubate at 42°C for 60 minutes.

10. Dilute 1 μl of the reaction into 100 μl of 10 mM Tris-HCl (pH 7.5), 1 mM EDTA. Spot 2 μl onto a GFC filter and count. TCA precipitate the remaining 99 μl by adding 1 ml of cold 10% TCA. After 10 minutes at 4°C, filter the sample through a prewetted GFC filter and count. This determines the TCA precipitable (i.e., incorporated) cpm.

11. Add 1 μl 0.25 M EDTA (pH 8.0) and 0.5 μl 10% SDS to the undiluted reaction mix. Vortex. Add 10 μl of chloroform:isoamyl alcohol:phenol (48:2:50). Vortex. Spin for 1 minute in a microcentrifuge.

12. Remove the aqueous layer with a micropipeting device. Add 10 μl of 4 M ammonium acetate and 40 μl of absolute ethanol to the aqueous phase.

13. Chill for 1 hour on dry ice. Warm to room temperature. Gently vortex. Spin for 15 minutes at 4°C in a microcentrifuge.

14. Remove the supernatant with a micropipeting device and immediately add 200 μl of 75% ethanol. Do not agitate the tube. Spin for 1 minute in a microcentrifuge and remove the supernatant. Be careful not to disturb the pellet. Briefly dry in a speedvac.

15. Dissolve the pellet in 10 μl of 10 mM Tris–HCl (pH 7.5), 1 mM EDTA.

16. Run 5–10,000 TCA-precipitable cpm (as determined above) on a 1.4% alkaline agarose gel. Dry gel onto DE-81 paper and autoradiograph. A good RNA will yield cDNA transcripts from 350 to 6000 nucleotides in length with an average of 1300 to 1600 nucleotides.

cDNA Synthesis Using Oligo (dT) Tailed Plasmid Primer

1. In a siliconized microfuge tube, add successively:
   10 μl of poly A$^+$ RNA (6–10 μg)
   8 μl of 15 mM Tris–HCl (pH 7.5)

2. Heat at 65°C for 3 minutes.

3. Quickly chill on wet ice.

4. Bring the tube to room temperature.

5. Add 1.5 μl of RNAguard (or RNasin) inhibitor (20 U/μl).

6. Mix gently.

7. Add:
   2 μl of oligo dT tailed plasmid primer (1 μg/μl) (as prepared on page 553 or from any other source)
   1 μl of 20 mM dATP
   1 μl of 20 mM dCTP
   1 μl of 20 mM dGTP
   1 μl of 20 mM TTP
   2 μl of α[^{32}P] dCTP (800 Ci/mmol)
   4 μl of 10X RT buffer
   2 μl of 20 mM dithiothreitol
   2.5 μl of sterile distilled H$_2$O

8. Mix gently.

9. Add 4 μl of reverse transcriptase (10 U/μl; 1000 U/ml final).

10. Mix moderately, spin for 1 second in a microcentrifuge, and incubate at 42°C for 60 minutes.

11. Add 4 μl of 0.25 M EDTA (pH 8.0) and 2 μl of 10% SDS. Vortex.

12. Take a 1 μl sample and dilute with 100 μl of 10 mM Tris (pH 7.5), 1 mM EDTA.

13. Spot 2 μl of this diluted sample onto a GFC filter for the

determination of the total cpm. TCA precipitate the remaining 99 μl from step 12 to determine the cpm incorporated.

14. Add 45 μl of chloroform:isoamyl alcohol:phenol (48:2:50) to the reaction. Vortex. Spin for 1 minute in a microcentrifuge.

15. Transfer the aqueous phase to a new, siliconized microcentrifuge tube. Add 45 μl of 4 *M* ammonium acetate and 180 μl of ethanol.

16. Chill on dry ice for 1 hour. Warm to room temperature, and gently vortex. Spin for 15 minutes at 4°C in a microcentrifuge.

17. Remove supernatant with a micropipeting device or finely drawn out Pasteur pipet.

18. Immediately add 1 ml of 75% ethanol, 25% 0.2 *M* NaCl to the pellet. Do not agitate the tube. Spin for 1 minute in the microcentrifuge. Remove the supernatant while being careful not to disturb the pellet.

19. Dissolve the pellet in 40 μl of 10 m*M* Tris–HCl (pH 7.5), 1 m*M* EDTA.

20. Remove two 5000 cpm TCA precipitable samples (as determined from step 13). Save one sample for the analysis described in step 23 and one for step 10 on page 562.

21. To the remaining solution, add an equal volume (39 μl) of 4 *M* ammonium acetate and four volumes (156 μl) of ethanol. Chill for 1 hour on dry ice. Store at −20°C pending the results of the gel electrophoresis.

22. From the total and TCA precipitable cpm determined in step 13, determine the total nanograms of cDNA synthesized and the percent conversion of mRNA into cDNA as follows.

Moles of dXTP per reaction
= (concentration of dXTP) × (reaction volume)
= $(2 \times 10^{-3}$ mol/liter) × $(4 \times 10^{-5}$ liters)
= 8×10^{-8} mol or 80 nmol each dNTP

Moles of dXTP incorporated
$= \dfrac{\text{(TCA precipitable cpm)}}{\text{total cpm}} \times$ mol of dXTP per reaction

Amount of cDNA synthesized (g)
= (mol of dXTP incorporated) × (4 dXTPs) × (MW dXTP)
= (mol of dXTP incorporated) × (4) × (325 g/mol)

%conversion
= (μg cDNA synthesized)/(μg poly A$^+$ RNA)

Starting with 1.03 pmol of pcDV-1 plasmid primer, the addition of a 1500 nucleotide cDNA to every plasmid primer would correspond to 0.502 μg of cDNA or a 6.37% conversion of 8 μg poly A$^+$ RNA into cDNA. Expected conversion is 3.4–5.3% using the oligo(dT) plasmid primer.

23. Run 5000 TCA precipitable cpm in a 1% alkaline agarose gel with ^{32}P-labeled Hind III fragments as markers (see Chapter 9). Dry onto DE-81 paper and autoradiograph. A good preparation will have a heavy streak from the position of the plasmid primer to approximately 6000 nucleotides.

Addition of Oligo-dC to DNA

1. Warm the ethanol precipitate from step 21, above to room temperature. Vortex gently. Spin for 15 minutes at 4°C in a microcentrifuge. Two cycles of ethanol precipitation in the presence of 2 *M* ammonium acetate should remove greater than 99% of the unreacted dXTPs (see Chapter 3). It is critical that all dXTPs be removed so they cannot be incorporated into the oligo-dC-tails in the next step. This can be checked by determining the total and TCA precipitable cpm in equal small aliquots.

2. Remove the supernatant with a micropipet.

3. Immediately add 0.5 ml of cold 75% ethanol. Do not agitate the tube. Spin for 1 minute in the microfuge. Remove the supernatant and dry the pellet.

4. To the siliconized 0.5-ml microfuge tube on ice, add 4.3 μl sterile distilled H$_2$O to dissolve cDNA plasmid pellet.

5. Add in order:
   1 μl of poly(rA) (200 μg/ml)
   2 μl of 10 m*M* CoCl$_2$
   2.8 μl of 0.5 m*M* dCTP
   0.5 μl of α[^{32}P] dCTP (800 Ci/mmol)
   2.3 μl of 1.2 *M* sodium cacodylate (pH 6.8)
   0.6 μl of 1 *M* Tris–HCl (pH 6.8)
   1 μl of 2 m*M* dithiothreitol

6. Mix gently.

7. Pipette 0.5 μl, TCA precipitate, filter, and count.

8. Add 6 μl of terminal transferase (5 U/μl; 100 U/ml final).

9. Mix and spin for 1 second in a microcentrifuge. Incubate at 37°C for 4–5 minutes.

10. Put the tube on wet ice until the number of dC residues added per 3′-end is determined in steps 11–13 below.

11. Remove a 0.5-µl sample and add to 99.5 µl of 10 mM Tris–HCl (pH 7.5), 1 mM EDTA.

12. Spot 2 µl on a GFC filter for the determination of total cpm. TCA precipitate the remainder, filter, and count.

13. Determine the number of dC residues added per 3'-end as follows:

cpm of dCMP added
$$= \frac{(\text{TCA precipitable cpm after tailing ; step 12})}{(\text{TCA precipitable cpm before tailing ; step 7})}$$

mol de dCMP added
$$= \frac{(\text{cpm dC added}) \times (\text{mol of dCTP reaction})}{(\text{total cpm of dCTP in reaction})}$$

mol of 3'-ends
$$= (2 \text{ ends/plasmid primer})$$
$$\times \frac{(2 \times 10^{-6} \text{ g plasmid primer})}{(\text{MW g/mol plasmid primer})}$$

MW of pcDV-1 $= 2.01 \times 10^6$ g/mol
MW of pSV7186 $= 1.88 \times 10^6$ g/mol

dC residues per 3'-end
$$= \frac{(\text{mol of dCMP added})}{(\text{mol of 3'-ends})}$$

14. If 8–12 dC residues have been added per 3'-end, the reaction is terminated. However, if fewer dCs have been added, the reaction is incubated further at 37°C. The time of this incubation is determined from the equation

$$\frac{\text{time}}{(\text{minutes})} = \frac{(10 - \text{no. of dC added}) \times (\text{incubation time})}{10}$$

Oligo-dC-tails longer than 15 are to be avoided as the subsequent long G:C stretch may lead to decreased levels of expression of the cDNA in the pcDV-1 vector.

15. Terminate the reaction by adding 2 µl of 0.25 M EDTA (pH 8.0) and 1 µl of 10% SDS. Vortex. Add 23 µl of chloroform:isoamyl alcohol:phenol (48:2:50). Vortex. Spin for 1 minute in a microcentrifuge.

16. Transfer the aqueous phase to a new, siliconized microcentrifuge tube. Add 23 µl of 4 M ammonium acetate and 92 µl of ethanol.

17. Chill for 1 hour on dry ice. Warm to room temperature and gently vortex. Spin for 15 minutes at 4°C in a microcentrifuge.

18. Remove supernatant with a micropipeting device. Immediately add 500 μl of 75% ethanol, 25% 0.2 M NaCl. Do not agitate the tube. Spin for 1 minute in a microcentrifuge. Remove the supernatant while being careful not to disturb the pellet.

19. Dissolve the pellet in 40 μl of 10 mM Tris–HCl (pH 7.5), 1 mM EDTA. Add 40 μl of 4 M ammonium acetate and 160 μl of ethanol. Chill for 1 hour on dry ice. Store at −20°C.

Digestion of Tailed cDNA with HindIII Endonuclease

This is a surprisingly difficult step in the procedure. Impurities (poly A$^+$ RNA?) from previous steps can inhibit this restriction enzyme.

1. Warm the ethanol precipitate from step 19 above to room temperature. Vortex gently. Spin for 15 minutes at 4°C in a microcentrifuge.

2. Remove the supernatant with a micropipette.

3. Immediately add 0.5 ml of cold 75% ethanol. Do not agitate the tube. Spin for 1 minute in the microcentrifuge. Remove the supernatant. Dry the pellet briefly under vacuum.

4. Dissolve the pellet in 37.5 μl of H$_2$O. Remove an aliquot of about 4000 cpm for gel electrophoresis (step 10).

5. Four microliters of 10X Hind III buffer [200 mM Tris–HCl (pH 7.4), 70 mM MgCl$_2$, 600 mM NaCl, and 1 mg/ml BSA] are added.

6. Add 20 units of Hind III. Mix moderately. Microcentrifuge for 1 second. Incubate at 37°C for 2 hours. Avoid overdigestion since Hind III will cleave DNA/RNA hybrids at a low rate, thereby producing truncated inserts.

7. Remove an aliquot of about 6000 cpm for gel electrophoresis at step 10. Terminate the reaction by adding 4 μl of 0.25 M EDTA (pH 8.0) and 2 μl of 10% SDS. Vortex.

8. Add 46 μl of chloroform: isoamyl alcohol:phenol (48:2:50) to the reaction. Vortex. Spin for 1 minute in a microcentrifuge.

9. Transfer the aqueous phase to a new, siliconized microfuge tube. Add 46 μl of 4 M ammonium acetate and 184 μl of ethanol. Chill for 1 hour on dry ice. Store at −20°C pending the results of gel electrophoresis (step 10).

10. Check the completeness of the Hind III digest by running the

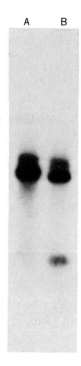

Figure 169. Hind III digestion of olido-dC-extended plasmid primer/cDNA hybrid. (*a*) 1700 cpm of undigested sample. (*b*) 1000 cpm of Hind III digested sample (kindly provided by C. Martinerie).

samples taken at steps 20 (page 558), and 4, 7 (page 561) on a 1% alkaline agarose gel (Figure 169) with ^{32}P-labeled λ-Hind III and pBR322-HinfI digests as size markers. After electrophoresis, neutralize the gel by soaking in 0.1 *M* ammonium acetate for 15 minutes. Stain the gel with ethidium bromide in 0.1 *M* ammonium acetate until the bands are visible under UV light (15–30 minutes). The amounts of DNA are very small but you should be able to see if the digestion worked. Photograph the gel and dry it onto DE-81 paper (so the small fragments are not lost). Autoradiograph the bands on the DE-81 paper. A complete digest is indicated by the appearance of a new band and a corresponding decrease in the size of the leading edge of the plasmid-cDNA smear; the size of the new band should correspond to the Hind III–EcoRI fragment of pcDV-1 (530 bp) or the Hind III–HpaI fragment of pSV7186 (527 bp). The intensities of the new band to the plasmid-cDNA smear should have the same ratio as

(cpm dC added/2): [cpm cDNA + (cpm dC added/2)]
Some lots of Hind III are contaminated with a single strand specific nuclease that digests away the oligo-dC-tails. If the digestion has worked, as judged by staining, but the small fragment is not radiolabeled, then nuclease activity may be present.

Ligation of cDNA with Linker DNA and Recircularization (¹⁄₁₀ Scale Reaction)

In this step we only use ¹⁄₁₀ of the cDNA:mRNA-plasmid to check that it can be circularized with the linker fragment, repaired, and used to transform *E. coli.*

Prepare three 0.5-ml siliconized microfuge tubes. Tube A is a background control for the linker fragment. Tube B undergoes the complete set of reactions while tube C is a control for the repair reactions.

	Tubes		
	A	B	C
cDNA:mRNA-plasmid	+	+	+
Circularization	+	+	+
Ligation	+	+	+
Linker	−	+	+
Repair	−	+	+
Transformation	+	+	+

1. Warm the ethanol precipitate from step 9 above to room temperature. Spin for 15 minutes at 4°C in a microcentrifuge.

2. Remove the supernatant with a micropipeting device.

3. Immediately add 0.5 ml of cold 75% ethanol. Do not agitate the tube. Spin for 1 minute in a microcentrifuge. Remove the supernatant. Dry the pellet briefly under vacuum.

4. Dissolve the cDNA:mRNA-plasmid in 10 μl of 10 m*M* Tris–HCl (pH 7.5), 1 m*M* EDTA.

5. In the circularization reaction, 2 mol of linker are used per mole of primer. For pcDV-1/pL1, use 0.5 μg of the pL1 linker for each 1.6 μg of pcDV-1 primer plasmid (from step 4). For pSV7186/pSV1932, use 0.25 μg of the pSV1932 linker per 1.4 μg of pSV7186 primer plasmid. Freeze the remainder of the cDNA:mRNA-plasmid from step 4 and store at −20°C for use in step 1, on page 565.

6. To the siliconized 0.5-ml microfuge tube A, add:
1 μl	cDNA:mRNA-plasmid (from step 4)
0.5 μl	2 *M* NaCl
7.8 μl	10 m*M* Tris–HCl (pH 7.5), 1 m*M* EDTA

7. Transfer 4.65 μl from tube A into tube B.

8. Add 0.35 μl of 10 m*M* Tris–HCl (pH 7.5), 1 m*M* EDTA to tube A. Add 0.35 μl of 0.1 μg/μl oligo-dG-tailed linker

DNA (pL 1, 524 bp or pSV1932, 272 bp) to tube B. Tube B now contains approximately 0.1 pmol of linker and 0.05 pmol cDNA:mRNA-plasmid.

9. Incubate tubes A and B at 65°C for 5 minutes, followed by 42°C for 60 minutes. Transfer the tubes to a 50-ml beaker of water at 42°C and allow to cool to room temperature (about 15 minutes). Transfer the tubes to a bucket of ice and cool to 0°C.

10. Prepare the ligation mix. Keep at 4°C.

11. Add 45 µl of ligation mix to tube A and B.

12. Add 0.3 µg of *E. coli* DNA ligase to tube A and B. Mix gently by pipeting up and down. Do not vortex.

13. Incubate overnight at 12°C.

Replacement of RNA Strand by a DNA Strand (¹/₁₀ Scale Reaction)

1. After the overnight incubation, tube A is stored at −20°C for use in step 2. Ten microliters are taken from tube B and put into tube C. Freeze tube C in a dry ice bath and store at −20°C.

2. Prepare a 2.5 mM dXTP solution by mixing:
   10 µl of 10 mM dATP
   10 µl of 10 mM dGTP
   10 µl of 10 mM dCTP
   10 µl of 10 mM TTP
   Put on ice. Store at −70°C.

3. To the remaining 40 µl in tube B, on ice, add:
   0.7 µl of the dXTP mix (from step 2)
   4.3 µl of 1.4 mM NAD
   1.0 µl of *E. coli* DNA ligase (200 µg/ml)
   1.0 µl of RNase H (450 U/ml; 9.4 U/ml final)
   10 µl of DNA polymerase I (800 U/ml; 16.7 U/ml final)

4. Mix gently by pipeting up and down several times.

5. Incubate tubes B and C successively at 12°C, 15°C, room temperature, and 25°C for one-half hour each to promote optimal repair synthesis and nick-translation by DNA polymerase I.

Transformation of E. Coli (¹/₁₀ Scale Reaction)

1. Thaw a 0.5-ml tube of transformation competent MC1061 on ice (see chapter 14). The cells should yield 10×10^7 transformants/µg pBR322.

2. Transfer 0.1 ml of competent cells to each of five sterile tubes on ice. Add the following DNAs to the competent cells.

Tube no.	DNA
1	10 μl of tube A
2	6 μl of tube B
3	5 μl of tube C
4	2 ng of pBR322
5	No DNA

3. Allow the tubes to stand on ice for 15 minutes.
4. Transfer to a 37°C water bath for 5 minutes.
5. Add 1 ml of L broth to each tube and return them to the 37°C bath for 60 minutes.
6. Prepare 1:10 and 1:100 serial dilutions of each transformation tube.
7. Plate 0.1 ml of the undiluted, 1:10 dilution, and 1:100 dilution, in duplicate, on LB plates containing 50 μg/ml ampicillin.
8. The transformation results of a typical experiment are given below:

Tube no.	*Typical transformants/ml*
1	10
2	1,600
3	200
4	60,000
5	<10

These results indicate that the mRNA:cDNA-plasmid gives very few transformants unless the linker fragment is present (compare tubes 1 and 3). The repair reaction produces a DNA:cDNA-plasmid with an eight-fold higher transformation efficiency than the unrepaired mRNA:cDNA-plasmid (compare tubes 2 and 3). Following the repair reaction, seven- to seventeen-fold simulations of transformation efficiency have been obtained. From this data, the expected number of independent transformants in the library would be

$$(1600)\left(\frac{48}{6.0}\right)\left(\frac{50}{40}\right)\left(\frac{9.3}{4.65}\right)\left(\frac{9}{1}\right) = 288,000$$

Ligation of cDNA with Linker DNA and Recircularization (Large-Scale Reaction)

1. To a siliconized 1.5-ml microfuge tube, add:
   9 μl (1.6 μg) of mRNA:cDNA-plasmid from step 5 on page 563

    4.5 µl of 2 *M* NaCl

    70.7 µl of TE

    5.8 µl of 0.1 µg/ml oligo-dG-tailed linker DNA (0.5 µg of pL1 ; 0.25 µg of pSV1932 linker)

2. Incubate in a 65°C water bath for 5 minutes, followed by 42°C for 60 minutes. Transfer the tube to a 50-ml beaker of water at 42°C and allow to cool to room temperature (about 15 minutes). Finally, transfer the tube to a bucket of ice.

3. Prepare the ligation mix and cool it to 4°C.

4. Add 810 µl of ligation mix to the 90 µl of circularized plasmid (step 2). Mix gently.

5. Add 6 µg of *E. coli* DNA ligase. Mix gently by pipeting up and down. Do not vortex.

6. Incubate overnight at 12°C.

Replacement of RNA Strand by a DNA Strand (Large-Scale Reaction)

1. Remove 5 µl from the tube (step 6, above) and save for the transformation assay (step 3, below).

2. To the remainder of the ligation from step 6 above (895 µl), on ice, add:

    14.4 µl of the 2.5 m*M* dXTP mix (step 2, page 564)

    96.4 µl of 1.4 m*M* NAD

    4 µl of *E. coli* DNA ligase (1 mg/ml; 3.9 µg/ml final)

    7.5 µl of RNase H (1200 U/ml; 8.8 U/ml final)

    2.7 µl of DNA polymerase I (6000 U/ml; 16 U/ml final)

3. The mixture (1.020 ml) is incubated successively at 12°C, 15°C, room temperature, and 25°C for one-half hour each to promote optimal repair synthesis and nick translation by DNA polymerase I.

4. Remove 5.0 µl of the complete reaction and save for the transformation assay (step 3, below).

Transformation of E. Coli (Large-Scale Reaction)

1. The large scale reaction should have approximately 1800 ng of plasmid in 1.015 ml. Up to 250 ng DNA or 0.1 ml (whichever is less) can be used to transform 1 ml of competent MC1061.

2. Put 22 0.5-ml aliquots of transformation competent MC1061 on ice (1.1 ml of plasmid requires 10.5 ml of MC1061).

3. The 5 µl aliquots saved at step 1 and step 4, above, should be used in the 0.1-ml transformation assays as described on page 564.

4. Add 50 μl of repaired plasmid to each of the 21 0.5-ml aliquots of transformation competent MC1061.

5. Mix and let stand on ice for 15 minutes.

6. Transfer the tubes to a 37°C bath for 5 minutes.

7. Transfer the contents of all 21 tubes to 500 ml of L broth in a 2-liter flask.

8. Take a 0.2-ml sample from the flask and plate.

9. Incubate the flask in a 37°C water bath for 1 hour with occasional swirling.

10. After 1 hour, take a 0.1-ml sample and 0.1 ml of a 1/10 dilution and plate on ampicillin plates in order to determine the number of transformed cells.

11. Add ampicillin to the flask to a final concentration of 50 μg/ml. Incubate at 37°C with moderate swirling.

12. After 10 hours, transfer 20 ml of the culture to a flask containing 500 ml of L broth + 50 μg/ml ampicillin. Reserve the remainder for step 15.

13. Incubate at 37° with swirling for 1–2 hours.

14. Store the cDNA library by adding 7% (final concentration) dimethyl sulfoxide (spectro quality, non oxidized) to a 1.5-ml aliquot of the culture. Prepare a good number of 1.5-ml aliquots and quick-freeze in a dry ice bath and store at −70°C.

15. Prepare plasmid DNA from the culture in step 12 and purify it on a CsCl gradient. When preparing plasmid DNA from a saturated culture, double the volume of solutions used for preparing plasmids from chloramphenicol-amplified culture. This DNA can then be used in the construction of size-selected libraries.

Notes

1. Polyadenylated mRNA should be prepared by two cycles of adsorption and elution from oligo-dT cellulose column.

2. The molar ratio of polyadenylated mRNA to vector DNA (carrying oligo-dT tails) should be greater than 1.0 and preferably 3.0. This will minimize the dC tailing of unreacted poly-dT tails in subsequent steps.

3. Homopolymer tails of about 60 dT residues should be added to the vector DNA.

4. The optimal amount of reverse transcriptase to use should be determined by titration for each mRNA preparation. Pilot reactions should be run with varying amounts of material. Generally, between 0.5 and 5 μl of reverse transcriptase at 10 units/μl is sufficient for 1 μg of RNA.

5. Addition of dC residues to ligated cDNA should be performed so as to obtain approximately 10–15 residues of dCMP/end of DNA.

6. When annealing the linker DNA to the cDNA–vector hybrid, use a twofold molar excess of linker DNA over the quantity of vector DNA.

7. If the RNA strand is not replaced by a DNA strand in the recombinant molecule, the efficiency of *E. coli* transformation is reduced approximately fivefold.

8. *E. coli* DNA ligase is used instead of T4 DNA ligase because it cannot join adjacent RNA and DNA segments which arise during the second strand synthesis.

9. Do not add carrier RNA at any step. To maximize recoveries, do ethanol precipitations in siliconized microcentrifuge tubes, remove the supernatants carefully, and save them until you verify, with a geiger counter, that the pellet is still there.

10. Successful library construction follows from the careful verification of each step of the procedure. TCA precipitation or gel electrophoresis should be used to analyze the efficiency of each reaction.

11. When growing large-scale cultures from bacteria frozen in dimethyl sulfoxide, centrifuge the bacteria and resuspend in media. Culturing in the presence of DMSO can lead to a loss of the plasmid. Bacteria in DMSO can be diluted and plated on agar or nitrocellulose without deleterious effects.

Buffers

10X E. coli DNA ligase buffer — *10 ml*
200 mM Tris–HCl (pH 7.5) — 2 ml 1 M Tris–HCl (pH 7.5)
40 mM MgCl$_2$ — 400 μl 1 M MgCl$_2$
100 mM Ammonium sulfate — 132 mg Ammonium sulfate
1 M KCl — 745 mg KCl
H$_2$O to 10 ml
Store at −20°C as 100 μl aliquots.

Ligation mix	*135 μl*	*1.35 ml*
10X *E. coli* DNA ligase buffer	15 μl	150 μl
1.4 mM β-Nicotinamide adenine dinucleotide (NAD)	10.7 μl	107 μl
1 mg/ml BSA (nuclease free)	7.5 μl	75 μl
Sterile distilled water	101.8 μl	1.018 ml

REFERENCES

Atushi Miyajima, A., Nakayama, N., Miyajima, I., Arai, N., Okayama, H., and Arai, K. I. (1984), *Nucl. Acids Res.*, **12**, 6397.

Chirgwin, J. M., Przybla, A. E., MacDonald, R. J., and Rutter, W. J. (1979), *Biochemistry*, **18**, 5294.

Freeman, G. J., Clayberger, C., De Kruyff, R., Rosenblum, D. S., and Cantor, H. (1983), *Proc. Natl. Acad. Sci. USA*, **80**, 4094.

Okayama, H., and Berg, P. (1982), *Mol. Cell. Biol.*, **2**, 161.

Okayama, H., and Berg, P. (1983), *Mol. Cell. Biol.*, **3**, 280.

PREPARATION OF A cDNA LIBRARY IN LAMBDA GT 11 VECTOR

Lambda derived vectors also provide a good mean for preparing cDNA libraries enriched in molecules representing the 3′ end of the corresponding RNA species. The main advantages of lambda derived vector come from the very large number of recombinants that can be screened at the same time, and the possibility to select recombinants on the basis of a colored reaction when using λgt11 (see chapter 6).

PROTOCOL

CLONING A cDNA LIBRARY IN λgt11

First Strand Synthesis of cDNA Species

Two reactions are run in parallel. In one of them only a ^{32}P-labeled nucleotide triphosphate is added to follow the successfulness of the synthesis. In each case, the starting material is composed of polyadenylated RNA purified by two passages on an oligo-dT cellulose column (see Chapter 18).

1. Mix in a microfuge tube:
   2.5 μl of polyadenylated RNA (1 μg/μl)
   2 μl of oligo-dT (1 μg/μl)
   2 μl of dATP, dCTP, dGTP, and TTP (10 mM each)
   2 μl of 10X RT buffer
   1 μl of RNAsin (30 units/μl diluted in H_2O) (Promega)
   1 μl of 0.2 M DTT
   1.5 μl of reverse transcriptase (Life Sciences)
   2 μl (10 μCi) of α[^{32}P]dCTP if labeled reaction **or**
   2 μl of H_2O if cold reaction

2. Incubate at 42°C for 90 minutes.

3. Add 80 μl of sterile distilled water.

4. Take 5 μl of the reaction mixture. TCA precipitate 2 μl of the radioactive sample, and load 3 μl on gel.

5. Add 50 μl of buffer-saturated phenol and 50 μl of chloroform:isoamyl alcohol mixture (24:1).

6. Vortex, centrifuge for 5 minutes at 12,000 rpm.

7. Transfer the supernatant in another tube, and add 90 μl of H_2O to the organic phase.

8. Repeat step 6.
9. Combine the two aqueous phases (200 μl).
10. Add 200 μl of chloroform:isoamyl alcohol mixture.
11. Vortex and centrifuge for 5 minutes at 12,000 rpm.
12. Transfer the supernatant to a tube and add 200 μl of ethyl ether.
13. Vortex and centrifuge for 5 minutes at 12,000 rpm.
14. Discard *upper phase* in a container *in a fume hood*.
15. Repeat steps 13 and 14.
16. Add 200 μl of 4 *M* ammonium acetate to the aqueous phase and 400 μl of absolute ethanol.
17. Put 15 minutes in a dry ice–ethanol bath.
18. Centrifuge for 15 minutes at 12,000 rpm (4°C).
19. Discard the supernatant and add 1 ml of 70% ethanol (*at room temperature*).
20. Vortex to wash the pellet and centrifuge for 10 minutes at 12,000 rpm.
21. Discard the supernatant and dry the pellet.

Second Strand Synthesis of cDNA Species

1. Add to the dry pellet
   10 μl of 10X Pol I buffer
   0.5 μl of dATP, dCTP, dGTP, TTP (10 m*M* each)
   5 μl of bovine serum albumin (1 μg/μl)
   79 μl of sterile distilled water
   1 μl of RNase H (1000 units/ml) (BRL)
   4.5 μl of Pol I (endonuclease free) (Boehringer)
2. Incubate for 1 hour at 14°C.
3. Incubate for 1 hour at 20°C.
4. At this stage, it is recommended to take 2 μl for TCA precipitation and 3 μl for gel analysis.
5. Precipitate the DNA as described in steps 16–21 above.
6. Resuspend the DNA pellet in 36.5 μl of H₂O.

Polishing of cDNA Ends with T4 DNA Polymerase

1. In the tube containing 36.5 μl of DNA in H₂O, add:
   5 μl of T4Pol 10X buffer
   2.5 μl of bovine serum albumin (1 μg/μl)
   2 μl of dATP, dCTP, dGTP, TTP (10 m*M* each)
   1 μl of T4 DNA polymerase

2. Incubate for 1 hour at 37°C.
3. Add 100 μl of sterile distilled water, 150 μl of 4 *M* ammonium acetate, and 1.2 ml of cold ethanol.
4. Centrifuge and wash the DNA precipitate with 70% ethanol.

Methylation of Potential Internal Eco RI Sites in cDNA Species

1. Resuspend the DNA pellet from step 4 above, in 9 μl of sterile distilled water and add:
   2 μl of 1 *M* Tris–HCl pH 8.0
   2 μl of 0.1 *M* EDTA pH 7.5
   1 μl of bovine serum albumin (1 μg/μl)
   1 μl of 0.1 *M* DTT
   2 μl of 1 m*M* SAM (*S. adenosylmethionine*)
   1 μl of Eco RI methylase (20 units/μl) (New England Biolabs)
2. Incubate at 37°C for 1 hour.
3. Add 1 μl of Eco RI methylase (20 units/μl) and incubate 1 hour further.
4. Add 80 μl of sterile distilled water.
5. Add 100 μl of phenol/chloroform:isoamyl alcohol and extract DNA as described above.

Addition of EcoRI Linkers to cDNA Species and Digestion with Eco RI

1. Resuspend the DNA pellet in 9 μl of sterile distilled water and add:
   2 μl of 10X ligation buffer
   7.5 μl of phosphorylated linkers (see chapter 12 for linkers preparation)
   1.5 μl of T4 DNA ligase (400 units/μl) (New England Biolabs)
2. Incubate at 14°C for 16 hours.
3. Heat inactivate for 10 minutes at 68°C. At this stage, it is possible to check efficiency of ligation by gel analysis.
4. Add to the ligation mix:
   15 μl of Eco RI 10X buffer
   1 μl of NaCl 5*M*
   100 μl of sterile distilled water
   7.5 μl of Eco RI (20 units/μl)
5. Incubate for 1 hour at 37°C.

6. Add 7.5 µl of Eco RI, and incubate for 1 hour further.
7. Add 7.5 µl of 3 *M* sodium acetate (pH 5.0) and 300 µl of cold absolute ethanol.
8. Precipitate at −70°C for 15 minutes and centrifuge at 12,000 rpm for 20 minutes.
9. Discard the supernatant and wash the pellet in 1 ml of 70% ethanol.
10. Centrifuge at 12,000 rpm for 10 minutes, discard the supernatant, and dry the pellet under vacuum.
11. Resuspend the pellet in 100 µl of elution buffer.

Purification of cDNA and Elimination of Linkers

The separation of cDNA from linkers is achieved by molecular sieve chromatography on a column of Sepharose CL-4B (see Chapter 3).

PREPARATION OF THE COLUMN

In a 5-ml sterile plastic-pipette, pluged with siliconized glass-wool, pour about 2 ml of preswollen Sepharose CL-4B (Pharmacia) and wash with 20 ml of elution buffer.

SEPARATION OF DNA FROM LINKERS

1. Mix the radioactive and the cold sample.
2. Load the column, and apply the elution buffer.
3. Collect 2-drop fractions (about 100 µl each) in microfuge tubes.
4. Count a few microliters of each fraction, to determine in which tubes the DNA is located.
5. Run 5 µl of each radioactive fraction on a 1% agarose gel.
6. Dry the gel and expose for autoradiography at −70°C with an intensifying screen.
7. Estimate the size of the products and pool fractions with satisfactory cDNA species (generally > 300 bp).
8. Add 1/30th volume of 3 *M* sodium acetate pH 5.0 and 2 volumes of cold absolute ethanol.
9. Let precipitate at −70°C for 20 minutes.
10. Centrifuge at 12,000 rpm for 20 minutes, and discard the supernatant.
11. Add 1 ml of 70% ethanol. Vortex and centrifuge for 10 minutes at 12,000 rpm.
12. Discard supernatant, dry the pellet under vacuum, and resuspend in 2–10 µl of sterile distilled water (depending on the amount of cDNA recovered, see below: ligation).

Ligation of cDNA Species to λGT-11 Arms and In Vitro Packaging

The ligation is performed in a 5-μl volume. The molar ratio of λ arms to insert is 2:1. One μl of λ DNA is used.

1. Mix in a microfuge tube:
   1 μl of dephosphorylated Eco RI-digested λ arms (1 μg)
   0.5 μl of 10X ligation buffer
   0.5 μl of T4 DNA ligase (400 units/μl) (New England Biolabs)
   2 μl of insert
   1 μl of sterile distilled water
2. Incubate at 14°C overnight.
3. Heat to inactivate DNA ligase at 68°C for 10 minutes.
4. The in vitro packaging is performed as described in Chapter 17.

Buffers and Solutions

10X RT Buffer   *100 μl*
0.7 M Tris–HCl pH 8.3 (at 42°C)   70 μl 1 M Tris–HCl (pH 8.3)
0.1 M MgCl$_2$ 10 μl 1 M MgCl$_2$
0.5 M KCl   10 μl 5 M NaCl
                                         10 μl H$_2$O

10X Pol I buffer   *100 μl*
0.67 M Tris–HCl (pH 8.8)   67 μl 1 M Tris–HCl (pH 8.8)
0.067 M MgCl$_2$ 6.7 μl 1 M MgCl$_2$
0.1 M 2-Mercaptoethanol   1.4 μl 7 M 2-Mercaptoethanol
0.067 mM EDTA (pH 8.5)   1.3 μl 5 mM EDTA (pH 8.5)
0.166 M Ammonium sulfate   4.15 μl 4 M Ammonium
1 mg/ml Bovine serum albumine   sulfate
                                         10 μl bovine serum albumin
                                           (10 mg/ml)
                                         9.45 μl H$_2$O

Elution buffer   *100 ml*
10 mM Tris–HCl (pH 8.0)   1 ml 1 M Tris–HCl (pH 8.0)
1 mM EDTA   0.2 ml 500 mM EDTA (pH 8.0)
                                         98.8 ml H$_2$O

Eco RI 10X Modified buffer   *100 μl*
0.06 M Tris–HCl pH 7.5   6 μl 1 M Tris–HCl pH 7.5
0.06 M MgCl$_2$ 6 μl 1 M MgCl$_2$
0.06 M 2-Mercaptoethanol   6 μl 1 M 2-Mercaptoethanol
0.6 M NaCl   12 μl 5 M NaCl
                                         70 μl H$_2$O

10X Ligation buffer	100 μl
0.5 M Tris–HCl pH 7.8	50 μl 1 M Tris–HCl pH 7.8
0.1 M MgCl$_2$	10 μl 1 M MgCl$_2$
0.2 M DTT	20 μl 1 M DTT
0.01 M ATP	10 μl 0.1 M ATP
500 μg/ml Bovine serum albumin	10 μl Bovine serum albumin (5 mg/ml)

CHARACTERIZATION OF THE RECOMBINANT CLONES

Recombinant clones are detected either on the basis of a colored reaction (when the cloning site is in the β-galactosidase gene of λ-derived vectors) or by hybridization with specific probes. The screening of λ libraries can be performed as described in chapter 17.

The method described below, which is a modification of the colony hybridization technique introduced by Grunstein and Hogness (1975), permits easy screening of a very large number of colonies (up to 15,000/100-mm plate) by hybridization with specific probes. The high sensitivity of the method makes it useful for the screening of cDNA libraries prepared in plasmid vectors as described in the preceding section and allows shotgun cloning directly into plasmids.

PROTOCOL

SCREENING OF PLASMID LIBRARIES BY HYBRIDIZATION WITH SPECIFIC PROBES

1. Prepare sterile nitrocellulose filters and mark with a pencil for future orientation and identification.
2. With sterile forceps, lay a nitrocellulose filter on the surface of a 1-day-old TYE–agar plate.
3. Once the filter is wet, peel it off, turn it over, and put it back on the plate.
4. With a bent glass rod, plate the transformation mixture directly onto the nitrocellulose filter. For a 100-mm Petri dish, we usually plate 0.1–0.2 ml of bacteria. It is possible to plate up to 0.8–1 ml if necessary.
5. Incubate at 37°C until very small colonies (about 0.1 mm in diameter) have appeared.
6. Peel off the nitrocellulose filter and lay it on a few Whatman 3MM paper flat sheets.

7. Apply a sterile, wet nitrocellulose filter to the template (original) filter. The pencil marks should match.

8. Press the two filters against each other gently and evenly with a replica-plating tool covered with velvet.

9. Peel the two filters apart and place them, colonies up, on fresh plates.

10. Let the filters incubate until small (0.5 mm) colonies have appeared.

11. Select one plate as a master and keep it at 4°C until hybridizations are complete and the results obtained.

12. Transfer the other filter onto a chloramphenicol-containing plate if amplification is needed. Incubate overnight at 37°C.

13. Treat the nitrocellulose filters for lysis of the colonies and hybridization as described in Chapter 15.

14. Positive hybridization allows one to locate on the master filter small areas corresponding to a few colonies.

15. Scrape positive colonies with a sterile loop and resuspend them in 0.5 ml TYE medium.

16. Plate dilutions of the suspension to obtain 100–200 colonies on nitrocellulose filters. Treat the filters as described above. This should lead to the isolation of single colonies giving positive hybridization with the specific probes used.

Notes for Screening of Plasmid Libraries

1. It is recommended that the filters be prewet before sterilization. First float the filters on H_2O and immerse when wet. Prepare sandwiches of nitrocellulose filters between Whatman filters and wrap them in aluminum foil. Sterilize for 20 minutes at 120°C on liquid cycle in an autoclave. Immediately after sterilization, place the filters in a sealed bag to maintain them at the proper level of humidity. This eliminates the need to wet the filters before placing them on the template filter and therefore eliminates the risks of using a filter that is too wet.

2. The plating efficiency on nitrocellulose filters has been found to be very similar to that obtained when plating directly on agar, when Millipore filters were used. However, variable results have been obtained with some batches of Schleicher & Schuell filters, which needed to be boiled in large volumes of 100 mM EDTA and then washed repeatedly. Washing the filters in boiling water eliminates detergents that interfere with some highly sensitive bacterial strains.

3. It is possible to make two replicas from a master filter without any problems. If additional replicas are needed, the master

plate should be returned to 37°C for further incubation after the first two replicas have been made.

4. The degree of moisture of the replica nitrocellulose filters is very critical. If you use filters that are too wet, the colonies will smear. If the filters are too dry, the transfer will not be even.

5. After template and replica plates have been returned to 37°C for incubation, the colonies on the template will grow more rapidly because of the high quantity of bacteria retained on the template. Replica plates may need longer incubation times for the colonies to reach the desired size.

6. The radioactive probes used for screening should not have any homology with plasmid sequences. Usually inserts are purified from vector sequences by endonuclease digestion and preparative electrophoresis. Be aware that such preparations are cross-contaminated and that labeling of DNA purified in this way may lead to a high background if carrier plasmid DNA is not used. In addition to the salmon sperm carrier DNA used in prehybridization and hybridization mixtures (see Chapter 15, page 442), we use nicked plasmid DNA (25 μg/hybridization) to eliminate any possible interfering background.

REFERENCE

Grunstein, M., and Hogness, D. S. (1975), *Proc. Natl. Acad. Sci. USA*, **72**, 3691.

FREEZING BACTERIAL COLONIES DIRECTLY ON NITROCELLULOSE FILTERS

This method allows colonies grown on nitrocellulose filters to be kept indefinitely by quick freezing and storage at −70°C.

PROTOCOL

FREEZING BACTERIAL COLONIES

1. Prepare 1.5% agar plates in TYE–freezing medium.

2. Lay nitrocellulose filters on plates and grow bacteria either as patches (Chapter 15) or by direct spreading.

3. Proceed as for replication under conditions described in steps 5 to 7 above. Do not separate the filters.

4. Place the nitrocellulose filters between Whatman 3MM filters to form a sandwich. On the top of the sandwich put a damp

filter to maintain humidity, and place the sandwich in a sealing bag.

5. Seal the bag and place it in a dry-ice–ethanol bath to freeze it. Store at $-70°C$.

6. When the frozen colonies are needed, thaw the sandwich, peel the filters apart, and grow each on a fresh plate until colonies appear.

Note. When preexisting filters are used for freezing, it is advised to incubate them on freezing medium–agar plates for several hours before replication.

Solutions. TYE–freezing medium is 1X TYE, 1X freezing medium.

2X TYE (1 liter)
20 g Bacto tryptone
10 g Yeast extract
10 g NaCl
H_2O to 1 liter
Sterilize at 120°C for 20 minutes.

2X Freezing medium (1 liter)
7.2 mM K_2HPO_4
2.6 mM KH_2PO_4
4 mM Sodium citrate
1 mM $MgSO_4 \cdot 7H_2O$
90 ml Glycerol
H_2O to 1 liter
Sterilize at 120°C for 20 minutes.

20

SEQUENCING OF DNA

SEQUENCING OF DNA BY BASE-SPECIFIC CLEAVAGE

This method, originally described by Maxam and Gilbert (1977, 1980), is based on chemical reactions that lead to DNA strand cleavage at specific positions. It does not involve any in vitro copying of the DNA.

The general outline of the procedure is as follows (Figure 170).

---

1. Large-scale production of unique DNA fragments.
2. Labeling of DNA with ^{32}P.
3. Separation of labeled ends.
4. Chemical cleavage reactions.
5. Running a sequencing gel.
6. Autoradiography and interpretation.

Large-Scale Production of Unique DNA Fragments

Large quantities of purified DNA can be obtained by the plasmid preparation procedures described in Chapter 7. Amounts of 5–10 µg of plasmid DNA containing the insert to be sequenced have been used successfully for sequencing. When you quantitate your starting material, be careful to eliminate RNA to prevent overestimation of the amount of DNA contained in the sample.

Digestion of DNA and purification of the insert are performed as described in Chapters 8 and 9. The purity of the DNA can be checked by agarose or acrylamide electrophoresis.

End Labeling of the DNA Fragments

Depending on the nature of the ends generated by endonuclease digestion (see Chapters 8 and 5), different systems can be used for DNA labeling (see Chapter 16).

Recent improvements to the methods have consisted in labeling selectively one of the four ends created by some restriction enzymes in special plasmids (Volckaert et al., 1984), and more recently, in the use of a synthetic oligonucleotide to label specifically the 3' protruding end of Pst I–DNA fragments with ^{32}P–dCTP instead of radioactive cordicepin 5' phosphate (Chiral and Wakil, 1986).

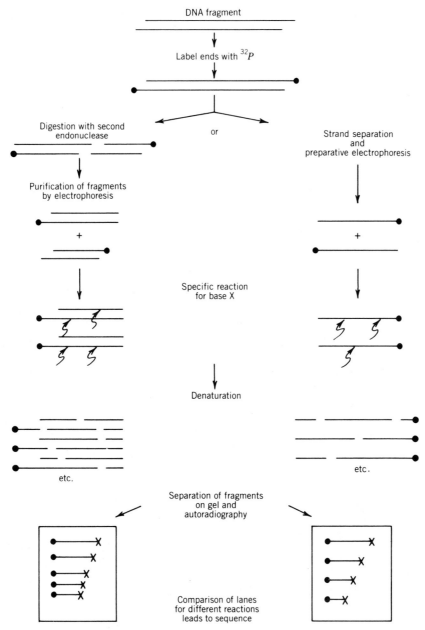

Figure 170. Scheme for sequencing by-base specific cleavage.

Separation of Labeled Ends

Depending on the restriction enzymes used to generate the fragment to be sequenced, the ends are either of identical or different cohesive (or blunt) type.

If the ends are of different types, and if only one type of reaction (5′ and 3′ labeling) has been used, the labeled DNA can usually be sequenced directly.

If the ends are of identical type or if both kinds of labeling (5′ *and* 3′) have been performed, it is nec-

essary to separate the two ends physically, because sequencing gels do not allow reading of the sequence from two ends.

Two methods can be used to separate labeled ends. The first method is digestion of the labeled purified DNA fragment with a second restriction enzyme. Care should be taken at this stage because some endonuclease preparations may contain contaminant phosphatases. It is therefore recommended that the second cleavage be performed in

Table 67. Summary of the Five Basic Reactions for Base-Specific Cleavage of DNA

	G	G + A	A > C	C + T	C
End-labeled DNA	5 μl	10 μl	5 μl	10 μl	5 μl
Buffer	pH 8.0	−	−	−	−
H₂O	−	+	−	+	+
High salt	−	−	−	−	+
NaOH	−	−	+	−	−
Dimethyl sulfate (DMS)	+	−	−	−	−
Formic acid (FA)	−	+	−	−	−
Hydrazine (HA)	−	−	−	+	+
T (incubation) (°C)	20	20	90	20	20
	Eliminate DMS Precipitate DNA	Eliminate FA by Lyophilization	Neutralize Precipitate DNA	Eliminate HA Precipitate DNA	Eliminate HA Precipitate DNA
Piperidine	+	+	+	+	+

For all reactions: Heat at 90°C; lyophilize; resuspend in 80% deionized formamide; load on sequencing gel.

the presence of 1 m*M* orthophosphate. After this second digestion is performed, DNA fragments are separated by electrophoresis on agarose or acrylamide gels (see Chapter 11) and then recovered.

The second method consists of denaturing the DNA to generate single strands, which are then separated by electrophoresis. Denaturation is usually performed in 30% dimethyl sulfoxide (DMSO) at 90°C for 2 minutes followed by immediate chilling in ice water. Strand separation is achieved by electrophoresis on either agarose or acrylamide gels (see Chapter 11).

Chemical Cleavage Reactions

Specific cleavage of DNA molecules relies on the use of such chemicals as acids, alkalis, dimethyl sulfate, and hydrazine, which are known to be able to substitute and open base rings in DNA. After treatment with one of these agents, the DNA molecule becomes more susceptible to the action of piperidine, which can thus be used to induce cleavage of the molecule

at the modified base. For further details on the nature of these chemical reactions, consult the review by Maxam and Gilbert (1980).

Limited cleavage of the DNA molecule is central to the methodology of this sequencing procedure. As shown in Figure 170 limited cleavage will generate a family of DNA fragments having a common labeled end and varying in size. These fragments are separated ultimately in polyacrylamide gels (see below). The time of incubation in the presence of alkali, acid, dimethyl sulfate, or hydrazine is therefore very important. The times indicated below are appropriate for sequencing approximately 250 nucleotides if a 0.4-mm-thick gel is used (see sequencing gels, page 619).

Eight different chemical procedures have been described for partially cleaving DNA at one or two of the four bases. Five of them (summarized in Table 67) are currently used and allow accurate determination of nucleotide sequences.

Detailed Procedures

PROTOCOL

REACTION 1 (CLEAVAGE AT **G***)*

1. In a microfuge tube mix:
   200 μl cacodylate (pH 8.0) buffer
   1 μl sonicated carrier DNA (1 mg/ml in water)
   5 μl ³²P-end-labeled DNA

2. Put tube on ice.

3. When cold, add 1 μl dimethyl sulfate (DMS).

4. Incubate with cap closed at 20°C for 10 minutes.

5. Add 50 μl 1.5 M sodium acetate (pH 7.0), 1.0 M 2-mercapto-ethanol, 100 μg/ml tRNA.

6. Add 750 μl cold absolute ethanol.

7. Precipitate DNA at -70°C for 10 minutes in a dry-ice–ethanol bath.

8. Spin for 10 minutes at 4°C.

9. Remove the supernatant carefully with a Pasteur pipette and transfer it to a flask containing 5 M sodium hydroxide (waste container).

10. Resuspend DNA in 250 μl 0.3 M sodium acetate.

11. Precipitate DNA by adding 750 μl cold ethanol.

12. Let the tube sit for 5 minutes at -70°C.

13. Spin for 10 minutes at 4°C and remove supernatant.

14. Add 1.5 ml cold ethanol without resuspending precipitate.

15. Spin for 5 minutes at 4°C.

16. Remove ethanol and dry under vacuum for 10 minutes.

17. Add 100 μl freshly prepared 1.0 M piperidine and dissolve precipitate.

18. Close the tube tightly and incubate at 90°C for 30 minutes. It is necessary to put a weight on top of the tube to prevent it from popping open.

19. Spin for a few seconds in a microfuge.

20. Punch holes in the tube cap. Freeze the sample and lyophilize.

21. Redissolve the DNA in 10 μl water, freeze, and lyophilize.

22. Repeat step 21.

23. Add 10 μl of formamide mix (loading buffer).

24. Heat at 90°C for 1 minute. Quickly chill in ice water and load immediately onto a sequencing gel.

PROTOCOL

REACTION 2 (CHEMICAL CLEAVAGE AT **G** + **A**)

1. In a microfuge tube mix:
10 μl sterile distilled water

1 μl sonicated carrier DNA (1 mg/ml in water)
10 μl ^{32}P-end-labeled DNA
2. Chill on ice and add 2 μl 1.0 *M* piperidine formate (pH 2.0).
3. Incubate at 20°C for 60 minutes.
4. Punch holes in the cap of the tube, freeze, and lyophilize.
5. Dissolve DNA in 20 μl water and repeat step 4.
6. Add 100μl freshly prepared 1.0 *M* piperidine and dissolve precipitate.
7. Follow steps 18–24 from Reaction 1.

PROTOCOL

REACTION 3 (CLEAVAGE OF **C** + **T**)

1. In a microfuge tube mix:
   10 μl sterile distilled water
   1 μl sonicated carrier DNA (1 mg/ml in water)
   10 μl ^{32}P-end-labeled DNA
2. Chill on ice.
3. Add 30 μl 95% hydrazine.
4. Incubate capped tube at 20°C for 10 minutes.
5. Add 200 μl 0.3 *M* sodium acetate, 0.1 m*M* EDTA, 25 μg/ml tRNA.
6. Add 750 μl cold absolute ethanol.
7. Precipitate DNA at −70°C for 10 minutes in a dry-ice–ethanol bath.
8. Spin at 4°C for 10 minutes.
9. Carefully remove the supernatant and discard it in a waste container containing 2 *M* ferric chloride.
10. Follow steps 10–24 of Reaction 1.

PROTOCOL

REACTION 4 (CLEAVAGE AT **C**)

1. In a microfuge tube mix:
   15 μl 5 *M* sodium chloride

1 μl sonicated carrier DNA (1 mg/ml in water)
5 μl ³²P-end-labeled DNA

2. Follow steps 2–10 of Reaction 3.

PROTOCOL

REACTION 5 (CLEAVAGE **A** > **C**)

1. In a microfuge tube mix:
   100 μl 1.2 *N* sodium hydroxide, 1 m*M* EDTA
   2 μl sonicated carrier DNA (1 mg/ml in water)
   5 μl ³²P-end-labeled DNA

2. Close the tube tightly and incubate at 90°C for 10 minutes under a weight.

3. Add:
   150 μl 1 *N* acetic acid
   5 μl tRNA (1 mg/ml in water)
   750 μl cold ethanol

4. Chill at −70°C for 10 minutes in a dry-ice–ethanol bath to precipitate DNA.

5. Follow steps 13–24 of Reaction 1.

Solutions and Buffers for Base Cleavage

Cacodylate Buffer	*2 ml*
50 m*M* Sodium cacodylate (pH 8.0)	100 μl 1 *M* Sodium cacodylate (pH 8.0)
1 m*M* EDTA	4 μl 0.5 *M* EDTA
10 m*M* MgCl₂	20 μl 1 *M* MgCl₂
	1876 μl H₂O

DMS stop (Reaction 1)
1.5 *M* Sodium acetate (pH 7.0)
1.0 *M* 2-Mercaptoethanol
100 μg/ml Yeast tRNA

Formamide mix (loading buffer)
80% v/v Formamide
10 m*M* NaOH
1 m*M* EDTA
0.1% w/v Xylene cyanol
0.1% w/v Bromphenol blue

1.0 M piperidine formate (pH 2.0)
(Reaction 2)
4% v/v Formic acid
Adjust to pH 2.0 with 1.0 *M*
    piperidine.

Hydrazine stop (Reaction 3)
0.3 *M* Sodium acetate
0.1 m*M* EDTA
25 μg/ml Yeast tRNA

Modified Maxam–Gilbert Sequencing Procedure for Use with Oligodeoxynucleotides

Refer to Table 68 for a summary of the procedure. The experimental detailed protocol is available from Pharmacia.

Gel Electrophoresis and Autoradiography

These steps are described on pages 619–624.

Interpretation of the Autoradiograms. Each lane of the autoradiograms contains several dark, horizontal bands corresponding to DNA fragments separated on the basis of relative size (Figure 171). In the lane corresponding to the specific reaction for G, all fragments have a terminal deoxyguanosine, indicating that a guanosine lies at these positions in the sequenced DNA. Similar conclusions can be drawn from other lanes. Since the DNA fragments are separated according to size, it is possible, by comparing the positions of the bands in the four lanes, to deduce the sequence of a particular DNA fragment. The easiest way to proceed is as follows: First, go to the lane which contains the smallest oligonucleotide. Identify the lane and the nature of its end. Go to the next smallest oligonucleotide, identify the corresponding lane and therefore a second nucleotide. The nucleotide order obtained represents the sequence. Accurate sequence reading is limited by the gel resolution (generally 200–400 nucleotides).

Notes for Sequencing by Base-Specific Cleavage

End Labeling of DNA

1. If double stranded DNA has blunt or recessed 5′ ends, it is recommended to first heat-denature the DNA for 2 minutes at 90°C and then immediately to chill the sample before adding polynucleotide kinase.

2. Spermidine has been shown to stimulate label incorporation and to inhibit a nuclease present in some polynucleotide kinase preparations. It is therefore added to the kinase buffer.

3. An exchange reaction has been described for 5′ protruding ends in which phosphatase treatment is avoided. This method is reported to be faster but less efficient.

4. Addition of ammonium acetate to stop the kinase reaction also prevents protein precipitation after addition of ethanol to the treated samples.

5. Refer to chapter 16 for detailed experimental procedures.

Base-Specific Modification and Cleavage

1. Reagents for chemical modification of bases are poisonous and volatile. Hydrazine is flammable. Handle them only under a chemical hood. Use latex or plastic gloves.

2. Dimethyl sulfate is inactivated by 5 *M* sodium hydroxide and hydrazine is inactivated by 3 *M* ferric chloride.

3. The chemicals used in these reactions are labile and should be stored in aliquots away from the atmosphere. The hydrazine solution must be made fresh every day.

4. Carrier DNA is prepared as described for hybridizations (see Chapter 15).

5. Temperatures and times of incubation can be varied. If a large amount of undigested DNA is observed at the level of the gel well after electrophoresis, the time of incubation should be

Table 68.   Modified Maxam and Gilbert Procedure for Oligonucleotide Sequencing

T	A + C	C	C + T	G	A + G
³²P-DNA	³²P-DNA	³²P-DNA	³²P-DNA	³²P-DNA	³²P-DNA
5 µl H₂O	5 µl H₂O	5 µl H₂O	20 µl H₂O	5 µl H₂O	20 µl H₂O
10 µl 5% OsO₄		15 µl 5 *M* NaCl		160 µl DMS buffer	
1 µl 98% Pip	80 µl 1.2 *N* NaOH	30 µl Hydrazine	30 µl Hydrazine	1 µl DMS	3 µl Pip-HOF
25°C/75 minutes	90°C/60 minutes	37°C/35 minutes	37°C/35 minutes	37°C/2 minutes	37°C/25 minutes
100 µl NaOAc	120 µl 1 *N* HOAc	50 µl H₂O	50 µl NaOAc	30 µl 1.0 *M* NaOAc	Evaporate
5 µl C.T. DNA	5 µl C.T. DNA	5 µl C.T. DNA	5 µl C.T. DNA	5 µl C.T. DNA	Wash evaporate

4 volumes EtOH	300 µl KPi
Chill − 70°C, spin 12,000 *g*	C-18 column
30 µl H₂O	100 µl H₂O column wash (4 times)
120 µl EtOH *or*	25° AcCN or 40% MeOH
Chill − 70°C, spin 12,000 *g*	Dry
100 µl EtOH	Wash
Spin 12,000 *g*	Dry
Dry	

50 µl 1 *M* Piperidine
90°C/30 minutes
Dry
Wash
Dry
|
Dissolve in Formamide/dyes
|
Sequence gel

Source: Courtesy of Pharmacia.
DMS = dimethyl sulfate
Pip = piperidine
Pip-HOF = piperidinium formate
NaOAc = 0.3 *M* sodium acetate, pH 4.5
C.T. DNA = 1 mg/ml calf thymus DNA
EtOH = 95–100% ethanol
KPi = 50 m*M* KPO₄ pH 5.9
AcCN = acetonitrile
MeOH = methanol
HOAc = acetic acid

increased. Do not change the concentrations of hydrazine (17–18 *M*) and alkali (1.2 *N*), which have been established to maximize base specificity and minimize side reactions.

6. Piperidine is somewhat hard to handle. Rinse the pipettes used carefully and mix well when diluting piperidine in water.

References

Chirala, S. S., and Wakil, S. J., (1986), *Gene*, **47**, 297.

Maxam, A. M., and Gilbert, W. (1977), *Proc. Natl. Acad. Sci. USA*, **74**, 560.

Maxam, A. M., and Gilbert, W., (1980) in *Methods in Enzymology,* Vol. 65, L. Grossman and K. Moldave, eds., Academic Press, New York, p. 499.

Volckaert, G., De Vleeschouwer, E., Blocker, H., and Frank, R. (1984), *Gene Anal. Techn.*, **1**, 52.

DNA CLONING IN M13 PHAGE FOR SEQUENCING

Figure 172 schematically illustrates this procedure.

Preparation of Vector DNA

Host Strains. The JM 101, 105, 107, 109 series of *E. coli* K12 hosts strains have been constructed to be

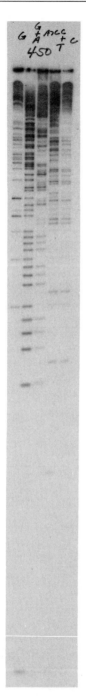

Figure 171. Sequencing gel run after base-specific chemical cleavage. A 20% acrylamide gel (19% acrylamide, 1% bis-acrylamide) was run at 2500 V until the bromophenol blue dye was 15 cm from the origin. (Courtesy of K. Rushlow.)

fore leading to a cell population with an increasing number of cells resistant to phage infection. To avoid these problems, the F′ JM strains should be maintained in minimal medium before use in TYE-rich medium. The genotype of a few JM strains is reported in Table 69. Note that JM 101 does not carry the K12 restriction-modification system (see Chapter 4). Therefore, DNA should be obtained in a K12 m⁺ strain before transformation in JM 101. The JM 103 strains originally described as r⁻ (Messing et al., 1981) has been reported to be r⁺m⁺ (Yanisch-Perron et al., 1985) and the JM 105, 107, and 109 strains are r⁻m⁺ (Yanisch-Perron et al., 1985).

Since the conditions used for growing cells and measuring absorbancy of the cultures may vary from one laboratory to another it is a good idea to determine for each bacterial strain the relationship existing between the number of cells and the absorbancy of the culture under the conditions that you are going to use. This is performed by plating (in duplicate or triplicate) dilutions of the cultures at different intervals of time following inoculation, and counting the number of colonies obtained after an overnight incubation at 37°C.

Sequencing Vectors. The M13-derived vectors developed by Messing and his collaborators have been widely used over the past few years for dideoxy-nucleotide chain termination method of sequencing. Although they have the disadvantage that certain inserts are unstable at the cloning site used (Barnes, 1979), they offer the considerable advantage that the blue plaque assay can be used to select recombinants (Gronenborn and Messing, 1978). Several cloning sites have been introduced within the sequence coding for β-galactosidase (see Chapter 6) to form the M13mp2 to M13mp19 derivatives (Figure 173). Because each of these cloning sites occur only once within the gene coding for β-galactosidase, it is possible to insert DNA fragments with different termini in a given orientation ("forced orientation cloning").

Since the cloning sites in the M13mp 8/9, 10/11, and 18/19 pairs of vectors are in opposite orientation, the cloning of a DNA fragment in both orientations allows cloning and sequencing to be performed in either direction. The M13 mp18 and mp19 vectors do not contain amber mutations present in other vectors and can therefore be propagated either in cells carrying a suppressor mutation (JM 101, JM 107, JM 109) or in cells deprived of suppressor mutation (JM 105). A whole variety of synthetic primers homologous to different regions surrounding the polylinker cloning region have been synthe-

used in conjunction with the M13-derived vectors as indicators for the β-galactosidase α-complementation (see Chapter 6). Since the M13 phage is specific for male bacteria, these strains carry the F factor on an episome which also harbors the proline region deleted in the bacterial chromosome (they are called F′). Continuous growth in rich medium containing proline may result in the loss of the episome, there-

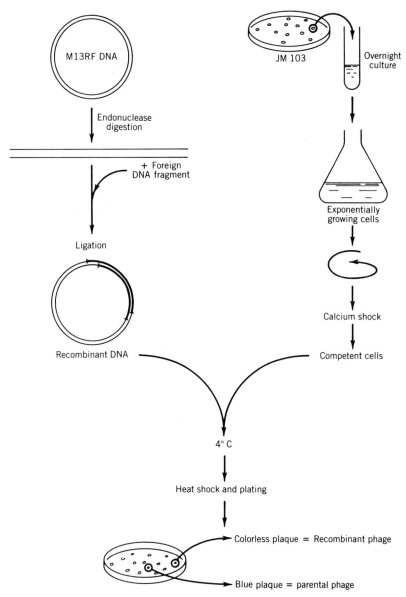

Figure 172. Outline of the procedure used for M13 cloning.

sized and are commercially available. Since all DNA fragments are cloned in the same region of M13mp vectors, a universal primer has been constructed to be used for sequencing of cloned DNA inserted in any of the M13mp derivatives (Heidecker et al., 1980). Also, reverse sequencing primers, which anneal only to the (−) strand of M13 mp vectors, can be used to sequence the opposite end of a cloned DNA fragment.

Several other sequencing vectors have been derived from plasmids by inserting the intergenic region of filamentous phage fl so that they replicate in a single stranded mode in cells infected with a

helper phage. These vectors have proved to be useful because single stranded DNA can be prepared easily in sufficient amounts for sequencing from 1.5 ml of culture in less than 1 hour. They also alleviate the need for recloning genes from plasmid to single stranded phages and back to plasmid. Moreover, it appears that long DNA inserts are more stable in plasmid derivatives such as pEMBL than in M13 vectors (Dente et al., 1983). Practical difficulties are related to the relatively low yields of single stranded DNA generated and to the fact that ambiguous results may be obtained because of contaminant helper phage single stranded DNA.

Table 69. Genotypes of JM Strains

E. coli Strain	Genotype
JM83	*ara*, Δ(*lac-proAB*), *rpsL*(= *strA*), φ80, *lacZ*ΔM15
JM101	*supE*, *thi*, Δ(*lac-proAB*), [F′, *traD*36, *proAB*, *lacI*ᵍZΔM15]
JM105	*thi*, *rpsL*, *endA*, *sbcB*15, *hspR*4, Δ(*lac-proAB*), [F′, *traD*36, *proAB*, *lacI*ᵍZΔM15]
JM106	*endA*1, *gyrA*96, *thi*, *hsdR*17, *supE*44, *relA*1, λ⁻, Δ(*lac-proAB*)
JM107	*endA*1, *gyrA*96, *thi*, *hsdR*17, *supE*44, *relA*1, λ⁻, Δ(*lac-proAB*), [F′, *traD*36, *proAB*, *lacI*ᵍZΔM15]
JM108	*recA*1, *endA*1, *gyrA*96, *thi*, *hsdR*17, *supE*44, *relA*1, Δ(*lac-proAB*)
JM109	*recA*1, *endA*1, *gyrA*96, *thi*, *hsdR*17, *supE*44, *relA*1, λ⁻, Δ(*lac-proAB*), [F′, *traD*36, *proAB*, *lacI*ᵍZΔM15]
JM110	*rpsL*, *thr*, *leu*, *thi*, *lacY*, *galK*, *galT*, *ara*, *tonA*, *tsx*, *dam*, *dcm*, *supE*44, Δ(*lac-proAB*), [F′, *traD*36, *proAB*, *lacI*ᵍZΔM15]

For example, it has been observed with the pEMBL system that almost no packaging occurs when transformed colonies are left for several hours at 4°C prior to superinfection with the helper phage, while abundant production of pEMBL DNA is obtained when the cells are kept at 37°C. Derivatives of pBR322 containing the intergenic region of f1 have also been constructed and used for the production of single stranded DNA (Zagursky and Berman, 1984).

The most frequently used helper phages are mutants of M13 called *rvl* (Levinson et al. 1984), M13K07 (Viera and Messing, unpublished; available from Pharmacia), and VCS-M13 (Stratagene) and a mutant of f1 designate d IR1 (Ena and Zinder, 1982). More recently, an improved filamentous helper phage has been derived from f1 bacteriophage (Russel et al., 1986). The genome of this mutant (R408) contains a deletion of the morphogenetic signal (PS) region which is involved in the packaging of single stranded DNA (see chapter 6). Therefore, this phage provides the proteins required for packaging but is not itself efficiently packaged. The yields of single stranded DNA obtained with these phages are in the range of 1 to 10 µg DNA/ml culture and may vary quite a lot with the efficiency of helper superinfection.

Commercially available systems include Bluscribe from Stratagene, and pTZ18R and 19R from Pharmacia. More recently, "all in one" plasmids (pT7/T3 -18 and 19 from BRL) have been constructed to combine several of the recent improvements introduced in different independent systems (see Chapter 6). These plasmids contain the intergenic region of f1 phage and allow both cloning and sequencing of either strand of DNA fragments in addition to in vitro synthesis of corresponding RNA species.

Preparation of Single Stranded Plasmid DNA. The following protocol was found to yield reasonable amounts of single stranded recombinant DNA with the pTZ (Pharmacia) and pEMBL vectors in combination with the M13K07, IR1, or R408 helper phages.

PROTOCOL

PREPARATION OF SINGLE STRANDED PLASMID DNA

1. Set up a transformation assay with the plasmid of your choice, the DNA to be cloned, and the recipient competent host cells. Keep in mind that host cells should be F⁺ to allow further superinfection with f1 derived helper phages. You may choose, for example, the JM 103 strain or the HB101(F⁺::Tn5) strain constructed by D. Ish-Horowicz (cited in Russel et al., 1986).

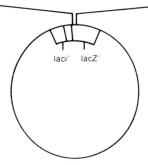

Figure 173. Sequences of the cloning regions in the M13mp series vectors (courtesy of New England Biolabs).

2. The next morning, pick cells from individual colonies on the plates which have been left at 37°C.

3. Immediately seed 1 ml of 2X prewarmed (37°C) TYE medium containing 0.1% glucose and the appropriate antibiotic (usually 100 µg/ml ampicillin).

4. Let the culture grow with shaking until a visible turbidity is obtained. This requires about 60 to 90 minutes. At this stage the absorbancy at 600 nm is close to 0.2 and the number of cells in the range of 10^8.

5. Add 10^{10} helper phages in a small volume (up to 50 µl) of TYE medium, so as to obtain a multiplicity of infection in the range of 50 to 100.

6. Incubate the culture with vigorous agitation for 4 to 5 hours at 37°C.

7. Transfer the infected cultures in microfuge tubes and centrifuge at 12,000 rpm for 30 seconds to pellet the cells (room temperature).

8. Transfer each supernatant to microfuge tubes containing 0.2 ml of 2.5 M NaCl and 20% PEG 6000.

9. Mix well, and let stand at room temperature for 15 minutes.

10. Centrifuge at 12,000 rpm for 5 minutes at room temperature to pellet the phage particles and remove the supernatant by aspiration.

11. Centrifuge a few seconds to collect droplets and remove all traces of liquid.

12. Resuspend the pellets in 100 µl of 10 mM Tris–HCl pH 8.0, 1 mM EDTA.

13. Add to the DNA 50 µl of phenol previously equilibrated with 10 mM Tris–HCl pH 8.0, 1 mM EDTA, vortex, centrifuge at 12,000 rpm at room temperature.

14. Transfer the upper phase to a microfuge tube, and add 50 µl of a 1:1 mixture of phenol:chloroform/isoamyl alcohol (24/1). Vortex and centrifuge at 12,000 rpm.

15. Transfer the aqueous phase to a microfuge tube containing 4 µl of 5 M NaCl, and add 80 µl of cold absolute ethanol.

16. Mix and incubate at -70°C for 30 minutes (or at -20°C overnight).

17. Centrifuge at 12,000 rpm for 30 minutes at 4°C and pour off the ethanol.

18. Add 1 ml of 80% cold ethanol, vortex, and centrifuge for 10 minutes at 12,000 rpm (4°C).

19. Pour off the ethanol, allow the tube to drain on a tissue, and dry the pellet in a vacuum centrifuge.

Notes

1. Filamentous phage infection and production are sensitive to temperature. Below 33–34°C the formation of bacterial F pili is inhibited (Novotny and Fives-Taylor, 1974) and infection does not proceed, while the yield of phage obtained above 42°C is about tenfold (Russel and Model, 1983) lower due to a drastic reduction of the single stranded DNA level at this temperature (Linn and Pratt, 1972; Horiuchi et al., 1978).

2. Vigorous agitation is necessary for efficient aeration. Milder agitation may be partially compensated by a longer incubation time (up to 6 hours).

3. Remaining traces of PEG after phage precipitation may inhibit reactions catalyzed by PolIk.

References

Barnes, W. M. (1979), *Gene,* **5,** 127.

Dente, L., Cesareni, G., and Cortese, R. (1983), *Nucl. Acids Res.,* **11,** 1645.

Ena, V., and Zinder, N. D. (1982), *Virology,* **122,** 222.

Gronenborn, B., and Messing, J. (1978), *Nature* **272,** 375.

Heidecker, G., Messing, J., and Gronenborn, B. (1980). *Gene* **10,** 69.

Horiuchi, K., Vovis, G. F., and Model, P. (1978), in *The Single Stranded DNA Phages,* D. T. Denhardt, D. Dressler, and D. S. Ray, eds. p. 113, Cold Spring Harbor Laboratory, Cold Spring Harbor, NY.

Levinson, A., Silver, D., and Seed, B. (1984), *J. Molec. Applied Genet.,* **2,** 507.

Lin, N. S. -C., and Pratt, D. (1972), *J. Mol. Biol.,* **72,** 37.

Messing, J., Crea, R., and Seeburg, P. H. (1981), *Nucl. Acids Res.* **9,** 309.

Novotny, C. P., and Fives-Taylor, P. (1974), *J. Bacteriol.,* **117,** 1306.

Russel, M., Kidd, S., and Kelley, M. R. (1986), *Gene* **45,** 333.

Russel, M., and Model, P. (1983), *J. Bacteriol.,* **154,** 1064.

Yanisch, Perron, C., Viera, J., and Messing, J. (1985), *Gene* **33,** 103.

Zagursky, R., and Berman, M. (1984), *Gene,* **27,** 183.

Isolation of Large Quantities of M13 RF DNA. Following infection of male hosts, the M13 (+) strand of DNA is used as a template for the synthesis of a complementary (−) strand. About 200 copies of the double stranded replicative form (RF) of the M13 genome accumulate in the infected cells. Cloning of DNA fragments in M13 is performed with the RF DNA purified from infected cells.

PROTOCOL

PREPARATION OF A M13 VIRUS STOCK

1. Seed 10 ml of TYE medium with a single colony of the appropriate strain previously streaked on a minimal medium agar plate.

2. Grow at 37°C overnight with agitation.

3. Seed 90 ml of prewarmed TYE medium with 10 ml of the over-

night culture and incubate with agitation at 37°C until the absorbancy at 600 nm reaches 0.1.

4. Add the M13 virus so as to obtain a multiplicity of infection of 0.1 (refer to the standard curve for number of bacteria over absorbancy).

5. Incubate for 12 to 14 hours with agitation at 37°C.

6. Centrifuge the culture at 5000 g in two 50-ml polypropylene tubes at room temperature for 10 minutes.

7. Aliquot the supernatant (viral stock) which contains about 10^{12} phages/ml in several tubes that will be stored at 4°C *without adding chloroform*.

PROTOCOL

PURIFICATION OF M13 RF DNA

1. Inoculate 20 ml of TYE medium with a single colony of the appropriate bacterial strain, taken from a fresh streak on a minimal medium agar plate.

2. Grow overnight at 37°C with agitation.

3. Transfer 10 ml of the overnight culture in a 2-liter flask containing 1 liter of TYE medium and incubate at 250 rpm until the absorbancy at 600 nm is 0.6 (this should correspond to about 5×10^8 cells/ml; check with your standard curve).

4. Infect cells by adding M13 at a multiplicity of 10 phages per cell (about 10^{12} phages/liter).

5. Return to 37°C and incubate with agitation (250 rpm) for 4 hours.

6. Transfer the cell culture in two 500-ml polypropylene centrifuge bottles and harvest cells by centrifugation for 30 minutes at 4200 rpm in a Beckman JS4.2 rotor or 10 minutes in a Sorvall GS3 rotor.

7. Purify M13 RF DNA from infected cells as described for plasmid DNA (see Chapter 7). About 200 to 500 μg of double stranded DNA replicative form should be obtained from 1 liter of culture.

Digestion of M13 RF DNA. Digestion of M13 RF DNA is carried out under the standard conditions already described in this manual (see Chapter 8.) The choice of the enzyme used for digestion is usually dictated by the nature of the fragment to be sequenced. The multiple cloning sites available in most M13 vectors offer several possibilities for easy cloning (see Chapter 6). Blunt end DNA fragments are usually inserted at the Sma I or the Hinc II sites of the polylinkers, while forced orientation cloning can be performed with DNA fragments harboring two different ends. In this case, refer to Chapter 8 for double digestion of DNA at restrictions sites that are close together.

PROTOCOL

DIGESTION OF M13 RF DNA WITH RESTRICTION ENDONUCLEASES

1. In a microfuge tube, mix:
   38 μl of sterile distilled water
   5 μl of 10X restriction endonuclease buffer
   5 μl of purified M13 RF DNA (0.1 μg/μl)
   2 μl of restriction endonuclease (10 units/ml)

2. Incubate at 37°C (or other appropriate temperature) for 1 hour and heat-denature the restriction endonuclease for 10 minutes at 65°C.

3. Although it is not absolutely necessary (see note 7 on page 599), you may wish at this step to dephosphorylate the vector as described for plasmid DNA (see Chapter 14) in order to reduce the self-ligation of DNA (which would lead to a high percentage of blue plaques lacking inserts).

4. Store dephosphorylated digested DNA at −20°C in sterile distilled water (a convenient DNA concentration for further use is 10 μg/ml).

Preparation of DNA Fragments to Be Inserted

Once the length of DNA to be sequenced has been determined and the restriction sites have been chosen, two possibilities must be considered: (a) Only one DNA fragment will be inserted and cloned. (b) A collection of subfragments will be cloned for sequencing. In either case, the ends of the molecules to be inserted have to harbor a restriction site complementary to the restriction sites generated in the vector DNA.

Several possibilities must be considered, depending on the origin of the DNA fragment to be inserted.

1. The DNA fragment has, at both ends, a restriction site analogous to the vector restriction site. The procedure is then ligation of insert and vector without prior modification.

2. The DNA fragment has one or no suitable restriction sites for insertion into vector DNA. Cloning will then be possible after addition of molecular linkers at one or both ends (see Chapter 12) or after trimming the cohesive ends to generate blunt ends (depending on the restriction sites used).

The conditions used for modification of the DNA fragment ends are identical to those described earlier (Chapter 12). It is important to remember that no separation of DNA fragments is needed when subdivided DNA is used, since the various fragments will automatically be segregated during cloning.

Due to the resolution of the sequencing gels, the average size of DNA fragments cloned for sequencing should not exceed 500 base pairs, although it is possible to insert up to 3 kb of foreign sequences without problems in the M13 vectors. Also, keep in mind that digestion must be performed only by restriction endonucleases having one site in the polylinker region, or producing compatible ends with those generated by enzymes cutting once in this region (see Chapters 5 and 8).

Ligation of DNA Fragments to Vector DNA

The ligation of DNA fragments to vector DNA previously digested with restriction endonuclease is performed as described previously (Chapters 13 and 14). The concentration of ends is a critical parameter to consider when performing ligations (see Chapter 13). Schematically, the concentration of DNA insert ends should be equal or greater than that of the vector ends. When the vector DNA is not dephosphorylated, the best results are obtained with concentrations of 0.8–3.2 fmol of insert ends, and 0.8–1.6 fmol of vector ends (2–4 ng) per microliter, corresponding to 1:1 and 2:1 ratios. When large fragments are used, it is recommended that DNA fragments with length larger than 2 kb be in 3 × molar excess over the vector DNA in the reaction mix.

In most cases, 10 ng of vector per ligation reaction has been found to give satisfactory results.

A control experiment involving all reagents except insert DNA should be included in each set of ligations. Also, a small aliquot of each sample (ligation, control, and linearized DNA) should be run on an 0.8% agarose slab gel as described in Chapter 9.

PROTOCOL

LIGATION OF DNA FRAGMENTS TO M13-DIGESTED DNA

1. In a microfuge tube mix:
   2 μl sterile distilled water
   1 μl insert DNA (0.002–0.010 pmol)
   0.5 μl 10X ligation buffer
   1 μl M13 RF DNA (digested with endonuclease and diluted to 10 ng/μl; see page 593)
   0.5 μl T4 DNA ligase (250 units)
2. Incubate at 16°C for 2 to 16 hours depending on the nature of the DNA ends (see Chapter 13).
3. Add 2 μl of 250 mM EDTA (pH 7.5) to stop the reaction and 18 μl of sterile distilled water.

The ligated DNA mixture can be kept at 4°C (or on ice) until transformation is performed, or stored at −20°C for long periods of time.

Preparation of Competent Cells Transformation

If F⁻ cell strains (such as HB 101, DH1, DH5, or MC1061) are used in the transformation assay, ma-ture infectious phage particles will be secreted from transfected cells but will not infect the other sur-rounding cells. We currently use the JM 107 strain for M13 transformation.

PROTOCOL

PREPARATION OF COMPETENT CELLS (FOR 20 TRANSFORMATION ASSAYS)

1. Streak the JM 107 cells on a minimum medium agar plate and incubate at 37°C for at least 24 hours.
2. Inoculate 5 ml of TYE medium with a single isolated colony, and incubate overnight with shaking at 37°C.
3. Seed 100 ml of prewarmed (37°C) TYE medium (in a 500-ml flask) with 0.5 ml of the overnight culture and incubate at 37°C with shaking (250 rpm) until the absorbancy at 600 nm is in the range of 0.3 to 0.4. This usually requires 2–3 hours.
4. Chill the cells by swirling the flask in an iced water bath.
5. Centrifuge the cells at 2000 g for 10 minutes at 4°C.
6. Discard the supernatant and resuspend the cells gently in 20 ml of ice-cold 0.1 M MgCl$_2$ by tilting the tube.
7. Let the resuspended cells stand on ice for 15 minutes.
8. Repeat step 5.
9. Resuspend the cells gently in 2 ml of ice-cold 0.1 M CaCl$_2$.
10. Let the cells stand in ice for at least 1 hour before use for transformation. Use the competent cells within the next 8 hours for best results.

PROTOCOL

TRANSFORMATION OF COMPETENT CELLS

1. Prepare a stock of lawn host cells (plating cells) by seeding 10 ml of TYE medium with 100 μl of an overnight culture, and growing cells at 37°C with agitation for 2–3 hours (A_{600} about 1.0).
2. This culture can be stored at 4°C for a week without any problem. Resuspend cells by swirling when they have not been used for a few days.

3. In an ice-cold microfuge tube mix 100 µl of competent cells with 15 µl of ligation mixture (containing 0.4 to 1.5 ng vector/µl).

4. Let stand on ice for 30 minutes.

5. Meanwhile, prepare six sterile glass tubes (13 × 100 mm) containing 0.2 ml of lawn host cells culture, 100 µl of 2% × Gal and 20 µl of 2% IPTG per transformation assay. Label from 1 to 6, and keep at room temperature.

6. Incubate transformation mixture from step 4 at 42°C for 2 minutes.

7. Immediately tilt the tube in order to resuspend the cells and transfer 1 µl of the transformation mixture to tube 1 containing the lawn cells, 10 µl to tube 2, and 100 µl to tube 3.

8. Add 6 ng of undigested vector DNA in tube 4 (control for transformation efficiency).

9. Add 6 ng of digested dephosphorylated vector DNA in tube 5 (control for background if digestion is not complete).

10. Add 6 ng of vector DNA digested, and religated without insert DNA, in tube 6 (control for religation efficiency).

11. In all six tubes, add quickly 3 ml of molten soft agar (kept at 45°C).

12. Immediately pour on TYE plates, and let harden at room temperature before incubating upside down at 37°C until the plaques are visible (6–12 hours).

Notes for M13 Cloning

1. When you digest M13 RF DNA to prepare the vector DNA, you may use less than 2 µg of DNA as long as you keep a final DNA concentration of approximately 10 µg/ml after digestion.

2. The transformation procedure calls for the addition of fresh, exponentially growing cells to the transfected competent cells to reveal the phage production (plaques). An easy way to obtain exponentially growing cells at the appropriate time is to dilute the remainder of the primary culture used to prepare the competent cells by approximately 1:100 with fresh medium.

3. Uncut RF DNA used as control must give blue plaques only. Also it is crucial to make sure that no other DNA is present in the M13 RF DNA preparation, to avoid the generation of false positive following digestion, ligation, and transformation.

4. Religated control phage DNA (relaxed circular form) should give about half the number of blue plaques obtained with the same amount of uncut DNA.

5. Remember that any "contaminant" DNA harboring one available endonuclease site will compete with the insert for ligation to the vector and may therefore decrease significantly the number of plaques obtained with the ligation mixture. Also, an inefficient ligation reaction will lead to a small number of plaques. The correct efficiency of ligation is checked in the control experiment using circular relaxed DNA.

6. Ligation of blunt-ended fragments is inefficient. Therefore, when Hind II-digested M13mp DNA is used as a vector for blunt-end ligation, it is recommended that you use about 25 ng of cleaved vector and no more than a 40-fold molar excess of fragment in a total volume of 10 μl. After ligation, plate directly half of the incubation mixture (~12 ng vector) at a time.

7. We have observed that dephosphorylation of the vector M13 DNA often results in an increased number of false positive (white plaques) corresponding to vector DNA with no insert.

8. It is possible to store competent cells at 4°C for a few days. However, a significant decrease in transformation efficiency may be observed as time increases.

9. Long-term storage of competent cells is sometimes performed in the presence of glycerol (30–40% final) at −70°C. We do not recommend using frozen competent cells when high efficiencies of transformation are required.

10. Some of us prefer to add IPTG and XGal to the top agar and then add the transfected cells before pouring on the plate in order to avoid a possible decrease in transformation efficiency. We have not encountered any problem when addition of the reagents is performed quickly, and the mixture poured on the plate immediately.

11. It has been shown that variation in size of white plaques is often due to the different sizes of the cloned inserts. Large inserts usually give rise to small plaques.

Isolated plaques will be selected on the basis of their colorless aspect to grow the recombinant phages and perform sequencing reactions.

The procedure described above deals with a situation in which well-defined restriction fragments are cloned for sequencing. However, this is not always the case, and when working with large por-tions of DNA, one often needs to perform shotgun cloning and sequencing prior to the detailed analysis of a particular region.

The strategy for shotgun cloning uses randomly fragmented DNA generated by controlled digestion with DNase I (Anderson, 1981) or sheared DNA obtained by sonication (Deininger, 1983). In either

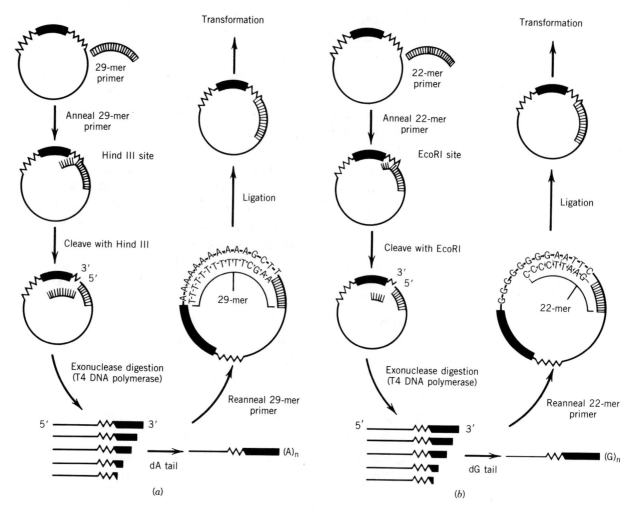

Figure 174. Outline of the single stranded cloning procedure for (*a*) mp8, 10, or 18 and (*b*) mp9, 11, or 19. The solid bar represents the insert and the sawtooth line the polylinker region of M13 (courtesy of International Biotechnologies Incorporated).

case, a precise calibration of the experimental conditions to be used is necessary. Once the DNA is fragmented, the fragments having an appropriate size (300 to 800 base pairs) are fractionated and purified as described in Chapters 7 and 9. The ends of the DNA fragments can be modified by the methods described in Chapter 12, prior to cloning at the correct site in the vector, and the white plaques are picked randomly for further analysis. This method can be applied even when no restriction mapping has been performed on the DNA of interest.

Alternative approaches for sequencing large DNA regions include the use of Bal 31 exonuclease (see Chapter 12) and exonuclease III (see Chapter 12). The first step in the Bal 31 method (Poncz et al.,

1982) is the digestion of the vector at a site located near one end of the cloned DNA fragment to sequence. Digestion with Bal 31 is then performed as described in Chapter 12 and aliquots are taken at regular intervals of time. After inactivation of Bal 31, each aliquot is digested by a second restriction endonuclease in order to separate the vector DNA from the series of shortened inserts resulting from exonuclease treatment. These fragments are subsequently introduced at the corresponding restriction sites in an M13 vector and sequenced. When using exonuclease III (Henikoff, 1984) the recombinant vector is digested at two restriction sites located between the end of the insert and the sequencing primer annealing site. It is necessary that the enzyme whose site is located close to the primer annealing region generate a 3 or 4 nucleotide 3' protruding end. Also, the second restriction site has to be as close as possible to the insertion site. Exonuclease III digestion is then performed and aliquots are taken at regular intervals of time. Following treatment with S1 nuclease (see Chapter 18), the shortened vectors are recircularized by ligase and sequencing is performed on the resulting clones.

More recently, a rapid and very elegant method for generating a nested series of templates for sequencing from a single cloned DNA fragment has been developed by Dale et al. (1985).

Rapid Deletion Subcloning System for Sequencing*

A rapid and convenient method for producing a sequential series of overlapping deleted clones has been developed for use in DNA sequencing (Dale et al., 1985).

This method is based on the use of M13-derived single stranded DNA and complementary oligonucleotides to form specific cleavage and ligation sites. It does not require any information concerning the restrictions sites present in the cloned DNA fragment and is extremely rapid and easy (2–4 kb of DNA can be subcloned and sequenced in about one week).

The first step involves hybridization of an oligomer to single stranded recombinant M13 or pIBI DNA at the 3' end of the cloned insert (see Figure 174) (for example, RD 29 to mp8, 10, 18 or um20; RD 22 to mp9, 11, 19 or um21). A different oligomer is required for inserts cloned into the even-numbered series (M13 mp8, 10, 18, and um20) and for inserts cloned into the odd-numbered series (M13 mp9, 11, 19, and um21).

The second step employs the 3' to 5' specific exonuclease activity of T4 polymerase to digest the linearized DNA from the 3' end. The polymerase will digest the DNA at a rate of approximately 40–60 bases/minute. The enzyme rapidly heat inactivates and if aliquots of the reaction mixture are removed at intervals, a series of overlapping deletions may be generated. The deletion products are tailed with either dA's (mp8, 10, 18, or um20) or dG's (mp9, 11, 19, or um21) using terminal deoxyribonucleotidyl transferase (TdT), thus producing a short homopolymer tail at the 3' end. Fresh oligomer is annealed to the deletion products joining the two ends of the molecule. The remaining nick is sealed with T4 DNA ligase. The product is then used to transform competent *E. coli* cells; the deleted phage DNA is harvested and used as a substrate for DNA sequencing (see Figure 174).

Estimated Time Frame. In one day, the procedure can be easily carried through to the point of plating transformed cells. Sizing gels can be run on day 2, templates prepared on day 3, and sequencing carried out on day 4. The rate-limiting step in a sequencing project is often the number of templates that can be sequenced per day. Since this procedure makes it easier to generate the necessary templates and simultaneously reduces the number of templates required to obtain a complete sequence, it shortens the time of sequencing projects considerably.

---

*Courtesy of International Biotechnologies, Inc.

PROTOCOL

PREPARATION OF PHAGE TEMPLATE FOR RAPID DELETION SUBCLONING

1. In a 125-ml Erlenmeyer flask, combine:
   12 ml 2X YT medium
   50 μl of an overnight culture of *E. coli* JM 101
2. Grow for 1 hour at 37°C shaking at 125 rpm.
3. Add 50 μl of phage stock and grow for 8–12 hours.

4. Pellet cells and cell debris by centrifugation.
5. To the supernatant, add:
   0.67 ml 40% polyethylene glycol
   0.67 ml 5 M NaOAc (pH 7.0)
6. Vortex solution and incubate on ice for 45 minutes.
7. Pellet phage at 10,000 g for 15 minutes.
8. Pour off supernatant and allow to drain for 30–60 minutes.
9. Carefully remove any remaining liquid with cotton swabs.
10. Resuspend pellets in 600 µl 20 mM Tris–HCl (pH 7.2).
11. Perform the following:
    one extraction: with phenol saturated with an equal volume of Tris–HCl (pH 8.0)
    two extraction: with phenol:chloroform (1:1/v:v)
    one extraction: with chloroform (see Chapter 3)
    Note: At each extraction, it is essential that none of the precipitate at the interface be transferred with the aqueous phase as this material will interfere with subsequent use of the template.
12. Combine the aqueous phase with the following:
    0.05 volume 2.5 M NaOAc (pH 5.2)
    2 volumes ethanol
13. Mix well and incubate at $-70°C$ for 30 minutes or at $-20°C$ overnight.
14. Pellet DNA: wash once with 70% ethanol and once with 95% ethanol.
15. Resuspend in 60 µl sterile water.
16. Determination of yield: Dilute 10 µl into 1000 µl and measure A_{260}. Multiply reading by 40 µg (i.e., 100 $\times$ 40 µg per 1.0 A_{260}/µl). Results: Anticipated yield = 60 µg of high-quality template. Yields are lower with JM 103, 105, 107, and 109.

PROTOCOL

LINEARIZATION OF CIRCULAR SINGLE STRANDED DNA

Most problems encountered in the procedure can be traced back to incomplete linearization at step I which can almost always be traced back to problems with the DNA. Before proceeding, it is advisable to confirm that the DNA of interest can be cut.

Test Cutting

In a reaction tube, combine:

1 µg DNA
2.5 µl 10X reaction buffer
1 µl RD22-mer or 2 µl RD29-mer (20 µg/ml)
1 µl DTT (100 m*M*)
1 µl Eco RI or Hind III (10–20 units/µl)
Add water to 20 µl final volume.

Incubate Eco RI reactions for 1 hour at 42°–50°C and Hind III digests for 2 hours at 50°C. Add a stop loading dye and electrophorese the digested sample as well as an undigested control on a 1.0% TBE agarose minigel for 2–3 hours at 100 volts. The digestion should be at least 90% complete. If cutting is essentially complete, proceed with the protocol. If not, you will need to reprepare your template.

Cleavage at the Eco RI Site of mp9, 11, 19 or um21

1. In a reaction tube, combine:
   4 µg DNA
   2.5 µl 10X reaction buffer
   4 µl RD22-mer (20 µg/ml)
   1 µl DTT (100 m*M*)
   2 µl of Eco RI (10–20 units/µl)
   Add water to 20 µl final volume.
2. Incubate the reaction at 42°C to 50° for 1 hour.
3. Inactivate the Eco RI by heating the reaction mixture at 65°C for 10 minutes.
4. Allow the reaction to cool to room temperature.

Cleavage at the Hind III Site of mp8, 10, 18 and um20

1. In a reaction tube, combine:
   4 µg DNA
   2.5 µl 10X reaction buffer
   8 µl RD29-mer (20 µg/ml)
   1 µl DTT (100 m*M*)
   2 µl Hind III (10–20 units/µl)
   Add water to 20 µl final volume.
2. Incubate at 50°C for 2 to 12 hours.
3. Inactivate the Hind III by heating the reaction mixture at 85°C for 5 minutes.
4. Allow the reaction to cool to room temperature.

It is recommended that verification of complete digestion be performed by running a sample on a 1% agarose minigel at 100 volts for 2 to 3 hours in TBE buffer. The digestion should be at least 90% complete.

PROTOCOL

EXONUCLEASE DIGESTION OF LINEARIZED DNA

The rate of digestion is proportional to the amount of T4 DNA polymerase added (Figure 175).

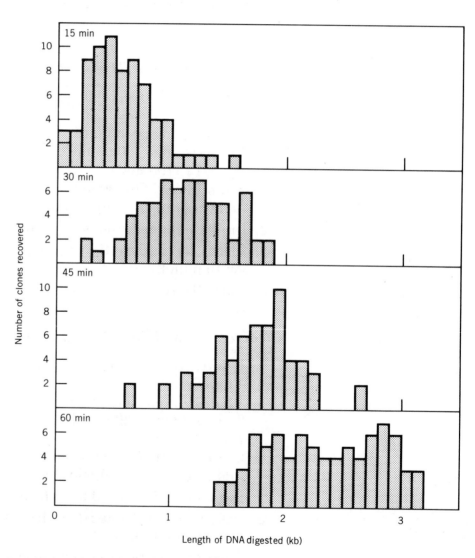

Figure 175. Histogram showing the number of clones recovered with different size deletions (courtesy of International Biotechnologies Incorporated).

Units of T4 DNA Polymerase per Microgram of DNA	Approximate Rate of Digestion
1.5	45 bases/minute
2	60 bases/minute
2.5	75 bases/minute
3	90 bases/minute

1. To the above reaction mixture, add the following:
   2.5 μl DTT (100 mM)
   1 μl Bovin serum albumin (10 mg/ml)
   6–12 units of T4 DNA polymerase (1–2 μl)
2. Incubate at 37°C.
3. Remove aliquots at appropriate intervals. Digestion occurs at a rate of approximately 50 bp/minute.
4. Terminate the reaction at each time point by heating the reaction mixture at 65°C for 10 minutes. If multiple time points are pulled, they may be stored on ice after heat denaturation. After the final time point all the digests are combined before going to the next step.

PROTOCOL

TAILING REACTION EXONUCLEASE DIGESTED DNA

1. To the reaction mixture obtained at step 4 in protocol above, add the following:
   3.0 μl 50 μM dGTP for inserts in mp9, 11, 19, or um21 or
   3.0 μl 50 μM dATP for inserts in mp8, 10, 18, or um20
   1.0 μl of water
   1.0 μl of TdT (at 15 units/μl)
2. Incubate at 37°C for 10 minutes.
3. Heat inactivate for 10 minutes at 65°C.
4. Remove 10 μl for use as an unligated control in the transformation step.

PROTOCOL

LIGATION OF OLIGOMER TO TAILED-DNA

1. Add 1 μl of the appropriate oligomer (RD22 for inserts in mp9, 11, 19, or um21 and RD29 for inserts in mp8, 18, or um20), to the samples prepared in protocol above.
2. Anneal by placing the tube(s) in a heat block at 65°C for 10 minutes, removing the heat block and allowing the reaction to cool to 40°C (30 minutes).
3. Add the following:
   3.0 μl 10 mM ATP
   4.0 μl of water
   1.0 μl of T4 DNA ligase (1–3 units)
4. Incubate at room temperature for 1 hour or longer. A control reaction without ligase should be prepared using the 10 μl saved earlier (tailing reaction, step 4).
5. Transform competent *E. coli* JM101 with 2 to 4 μl of the ligation mix. Let the balance of the ligation reaction sit at room temperature overnight. Results: Transformation efficiency in the range of 10^4 to 10^5 transformants per microgram of DNA are routinely observed.

PROTOCOL

SIZING SUBCLONED FRAGMENTS

1. Pick isolated colonies or plaques with sterile toothpicks and grow in 2 ml of 2X YT media in 18 × 150 mm test tubes shaking in a 37°C incubator at a 45° angle at 150 rpm for 8–12 hours.
2. Transfer cultures to 1.5-ml centrifuge tube.
3. Pellet cells in a microfuge at 12,000 g for 2 minutes.
4. Transfer supernatants to new tubes and store at 4°C for phage preparation. The cell pellet may be saved if you wish to examine the double stranded RF form of the phage DNA.
5. For phage, combine 25 μl of the supernatant with 5 μl of the SDS loading dye. Electrophoresis is carried out in 0.7% agarose gels run in 1X Tris–borate–EDTA gel buffer. Large 20-cm gels may be double combed and run for 12–14 hours at 100

volts. The outside lanes on either side should contain markers consisting of the initial nondeleted ssDNA as well as some ss vector DNA without an insert.

6. Measure the migration distance of the two markers and plot distance migrated versus insert size on regular graph paper. Plotting the distance migrated of the DNAs containing various deletions will give a good estimate of the size of fragment remaining in each clone.

7. Select a group of clones showing an ordered decrease in size. Prepare the templates and sequence.

Note: The RD primers can be purchased from International Biotechnologies Inc. and New England Biolabs.

References

Anderson, S. (1981), *Nucl. Acids Res.,* **9,** 3015.

Dale, R. M., McClure, B. A., and Houchins, J. P. (1985), *Plasmid,* **12,** 31.

Deininger, P. L. (1983), *Anal. Biochem.,* **129,** 216.

Henikoff, S. (1984), *Gene,* **28,** 351.

Messing, J., Crea, R., and Seeburg, P. H. (1981), *Nucl. Acids Res.,* **9,** 309.

Poncz, M., Solowiejczyk, D., Ballantine, M., Schwartz, E., and Surrey, S. (1982), *Proc. Natl. Acad. Sci. USA,* **79,** 4298.

Yanisch-Perron, C., Viera, J., and Messing, J. (1985), *Gene,* **33,** 103.

Characterization of M13 DNA Recombinants

The inactivation of β-galactosidase activity resulting from the insertion of foreign DNA into the M13mp7 to 19 genomes does not give any information about the sizes or the orientations of the fragments cloned. Thus, biochemical methods need to be used to further characterize the DNA recombinants.

Three techniques can be used to obtain rapidly information about the different clones (Figure 176). Such information is important to keep for subsequent sequencing of recombinants carrying different inserts. Analysis of the DNA content of the recombinant phage may be carried out by *direct agarose electrophoresis,* Southern transfer, and hybridization. Alternatively, a miniscreen procedure can be used to obtain enough RF DNA to perform an *endonuclease analysis.* A third method is based on *partial sequence screening* of the recombinants. Only one dideoxynucleotide triphosphate is used, under conditions similar to those used in complete sequencing. The identity of patterns suggests that different clones harbor the same insert, either in a similar or in an opposite orientation.

Direct Agarose Electrophoresis. This test relies on the comparison of the size of different recombinant DNAs obtained after infection of host cells with single plaques.

PROTOCOL

DIRECT AGAROSE ELECTROPHORESIS OF M13 DNA RECOMBINANTS

1. Grow host cells in complete medium as described before and put 2 ml of the bacteria suspension in sterile tubes (i.e., Corn-

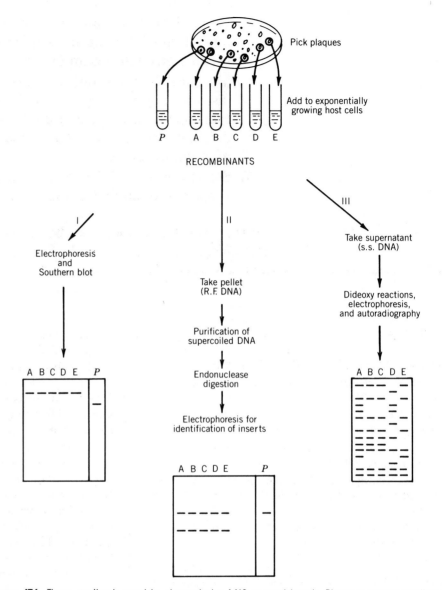

Figure 176. Three methods used to characterize M13 recombinants. Plaques corresponding to M13 recombinants are grown and tested either by direct agarose electrophoresis (I), by restriction endonuclease analysis (II), or by partial sequence screening (III). P is for parental DNA from phage alone with no insert. See text for experimental details.

ing 15-ml plastic tubes). Add individual colorless plaques to the cultures and incubate at 37°C for 6–7 hours with vigorous shaking.

2. Spin cells down in a microfuge tube.

3. Pipette out 20 μl of the culture supernatant containing the single stranded DNA of the recombinant phages. Add 5 μl endonuclease stop buffer.

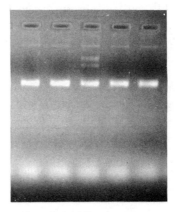

Figure 177. Formation of eight-like structures following annealing of M13 recombinant DNA harboring inserts in the same orientation. Middle track shows superstructures migrating more slowly in agarose gel (kindly provided by J. Soret).

4. Run the sample on a 0.7% agarose gel for 5 hours at 100 volts.
5. Take a picture of the gel (see Chapter 9) and compare migrations of the different DNA molecules.

Notes

1. Direct agarose electrophoresis is performed on a small aliquot (20 μl) of the supernatant obtained after infection of host cells. The remainder of the supernatant can be used for sequencing after ligation to single stranded templates. This supernatant can be stored at 4°C and used as M13 stock for the corresponding recombinant. It is advised that you spin this stock before using to sediment any cells that may have grown during storage.

2. Direct agarose electrophoresis is usable only for DNA recombinants containing inserts of at least 200 bp. Otherwise, differences in migration will not be significant.

3. Clones harboring identical inserts can be detected by cross-hybridization of single stranded DNA contained in culture supernatants. For this procedure, mix 20 μl of each supernatant, incubate at 65°C for 1 hour, and run on an agarose gel. Hybridization of common sequences will generate complex structures (figure-eight-like) that will migrate more slowly in the gel. This method is very useful for recognizing clones that harbor inserts in opposite orientations (Figure 177).

This method does not tell you if different clones contain the same inserts. If specific probes for subfragments are available, you may wish to perform a Southern blotting of your analytical gel and hybridize the corresponding blot to your probes (see Chapter 15). This may enable you to differentiate clones. In many cases, however, the preparation and use of probes specific for small fragments is a laborious and tedious procedure, and it is easier to perform an endonuclease analysis of the DNA recombinants (see below).

Restriction Endonuclease Analysis. This method uses a quick procedure originally developed for plasmid purification. The high yield of RF DNA, and the low background of host DNA, make the method adequate for recombinant phages.

PROTOCOL

RESTRICTION ENDONUCLEASE ANALYSIS OF M13 DNA RECOMBINANTS

1. Spin down 2 ml of infected cells grown at 37°C for 7 hours, as described above, in a microfuge tube.
2. Resuspend the cells in 60 μl of resuspension buffer.
3. Put on ice.
4. Add 15 μl lysozyme solution.
5. Let sit on ice for 5 minutes.
6. Add 30 μl resuspension buffer.
7. Add 1.3 μl RNase A solution.
8. After 5 minutes of incubation add 120 μl Triton solution.
9. Incubate for 10 minutes.
10. Spin down the cell debris.
11. Pipette out the supernatant. Extract once with phenol and twice with a mixture of phenol and chloroform as described previously (Chapter 3).
12. Precipitate and resuspend DNA as described before (see Chapter 3).
13. Use 10 μl of each sample for endonuclease digestions.
14. Electrophoresis is performed as described before (Chapter 9).

Buffers and Solutions for Restriction Endonuclease Analysis

Resuspension buffer
20 mM Tris–HCl (pH 7.5)
10 mM NaCl
0.1 mM EDTA

100 ml
2 ml 1 M Tris–HCl (pH 7.5)
0.2 ml 5 M NaCl
20 μl 0.5 M EDTA
98.8 ml H$_2$O

Lysozyme solution
5 mg/ml lysozyme in resuspension
    buffer

RNase A solution
10 mg RNase A/ml 10 mM sodium
    acetate buffer (pH 5.0)
Boil in water bath for 10 minutes.
Keep at $-20°C$.

3 M Sodium acetate buffer (pH 5.0)
408 g Sodium acetate
Dissolve in ~700 ml H_2O.
Bring to pH 5.0 with glacial acetic
    acid.
Adjust to 1 liter with H_2O.

A more complete restriction endonuclease analysis is needed to differentiate between clones carrying similar inserts.

Partial Sequence Screening. This method allows quick analysis of several recombinant DNAs for their A, T, G, or C content. A similar pattern for two different clones indicates that they carry the same insert (though not necessarily in the same orientation).

The supernatant of infected cells prepared as for direct agarose electrophoresis analysis (above) is a good source of viral DNA.

PROTOCOL

PARTIAL SEQUENCE SCREENING OF M13 DNA RECOMBINANTS

1. In a microfuge tube mix:
   1.2 μl H_2O
   1 μl supernatant (containing ~200–500 ng viral DNA)
   0.5 μl of M13 sequencing primer (1.25 ng)
   0.3 μl 10X polymerase buffer
2. Place the microfuge tubes in test tubes filled with water at 85–90°C in a water bath. Incubate for 2 minutes.
3. Transfer and incubate the tube in a 58°C water bath for 2 hours.
4. Prepare an incubation mixture as follows:
   10 μl dXTP mix (see page 616)
   10 μl ddX (see page 617)

1 μl α[^{32}P]dXTP (10 μCi)

2 μl DNA polymerase I, large fragment (1 unit/μl)

5. Add 2 μl of this incubation mixture to the 3-μl samples prepared in steps 1–3. Incubate at 20°C for 15 minutes.

6. Add to each reaction sample 1 μl 0.5 m*M* cold dXTP corresponding to that used for labeling (if α[^{32}P]dATP is used to label, add 1 μl 0.5 m*M* dATP to each reaction). This will prevent any interference that could result from the limiting amount of dXTP contained in α[^{32}P]dXTP (to keep high specific radioactivity).

7. Incubate further for 15 minutes at 20°C.

8. Stop the reaction by adding 1 μl 0.5 *M* EDTA. Store at -20°C if necessary.

9. Add 5 μl dye mix to each sample before loading 2–3 μl on a sequencing gel (see page 619).

10. Run electrophoresis as described in Chapter 9.

11. Autoradiography is performed as described on page 624.

When identical patterns are obtained, one can assume that identical M13 recombinant clones were used.

Sequencing of Recombinant DNA Obtained with M13 Cloning

Purification of DNA Strands

PROTOCOL

PURIFICATION OF SINGLE STRANDED DNA TEMPLATES FROM M13 RECOMBINANTS

1. In a 15-ml tube, seed 2 ml of TYE medium with 10 μl of a fresh exponential culture of JM 107 cells, and add a toothpick that has just been inserted in the center of a single isolated recombinant plaque.

2. Let grow with agitation at 37°C for 5–8 hours. (Do not incubate more than 8 hours.)

3. Spin 1 ml of the culture in a microfuge tube for 2 minutes at 12,000 rpm (4°C) to pellet cells.

4. Transfer supernatant in a 1.5-ml microfuge tube and centrifuge for 2 more minutes to remove all remaining cells.

5. Transfer the supernatant in another 1.5-ml microfuge tube and process it as follows. (The pellet can be saved as stock for the corresponding clone.)

6. Add 200 μl of 2.5 *M* NaCl containing 20% PEG 6000, mix, and incubate for 30 minutes on ice.

7. Spin down the precipitate for 30 minutes at 12,000 rpm and 4°C.

8. Pour off the supernatant, and spin for 2 additional minutes. Remove the supernatant with a drawn out Pasteur pipette.

9. Resuspend the precipitate by vortexing in 100 μl of 10 m*M* Tris–HCl (pH 7.5), 0.1 m*M* EDTA.

10. Add 50 μl of phenol saturated with 10 m*M* Tris–HCl pH 8.0, 1 m*M* EDTA. Vortex to mix.

11. Centrifuge at 12,000 rpm for 10 minutes at room temperature.

12. Transfer the top aqueous phase to a microfuge tube, and add 10 μl of 3.0 *M* sodium acetate (pH 6.0). Mix.

13. Add 200 μl of cold absolute ethanol, mix by inverting several times, and incubate at −70°C for 30 minutes.

14. Centrifuge at 12,000 rpm for 30 minutes at 4°C; carefully pour off the supernatant.

15. Add 1 ml of 80% ethanol, mix, and centrifuge at 12,000 rpm for 10 minutes at 4°C.

16. Carefully pour off the supernatant without dislodging the pellet, drain the tube, and dry the DNA pellet in a vacuum centrifuge (Speed Vac, Savant).

17. Resuspend the dry precipitate in 50 μl of 10 m*M* Tris–HCl (pH 7.5), 0.1 m*M* EDTA. The template concentration should be about 0.1 μg/μl.

Primer Annealing Reaction. As in the partial sequence screening test, the next step consists of annealing a DNA primer to the purified templates. This primer allows the DNA polymerase large fragment to copy the single stranded DNA.

Several different synthetic primers are now commercially available and provide a great flexibility to sequencing strategy (see, for example, the position and structure of New England Biolabs primers on M13 vectors in Figure 173).

PROTOCOL

PRIMER ANNEALING REACTION

1. In a microfuge tube mix:
   12.5 µl distilled sterile water
   5 µl purified single stranded DNA (~0.5 µg)
   2 µl primer DNA fragment (4–6 ng)
   1 µl 10X polymerase buffer
2. Transfer and incubate the tube at 58°C for 2 hours.

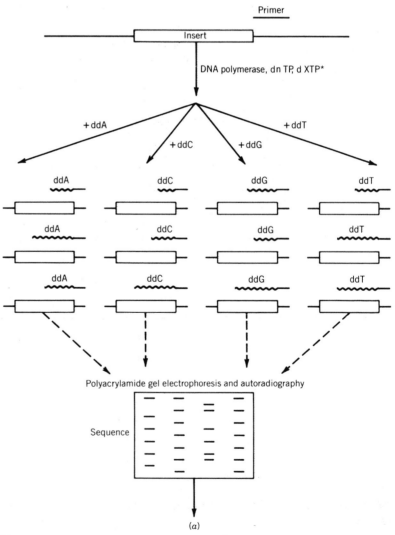

Figure 178. Dideoxy reaction for sequencing. (*a*) Copying of the insert DNA by DNA polymerase is inhibited at specific sites in the presence of dideoxynucleotides.

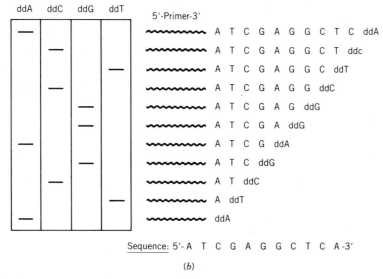

Sequence: 5'- A T C G A G G C T C A -3'

(b)

Figure 178 *(continued).* *(b)* the different fragments obtained are separated according to size by electrophoresis. Sequence can then be determined. (See text for details.)

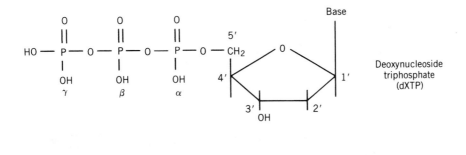

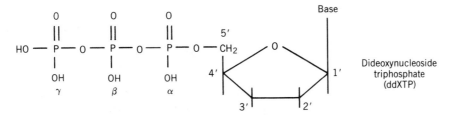

Figure 179. Structure of the deoxy- and dideoxynucleoside triphosphates. The base can be A, C, G, or T. The dideoxynucleoside lacks the 3' − OH group present in the ribose moeity of the deoxynucleoside. This hydroxyl group is involved in the formation of 5' − 3' phosphodiester bond of the DNA molecule.

Dideoxy Reaction. Figure 178 illustrates the rationale of this sequencing method. The dideoxy chain termination sequencing method is based on the random enzymatic incorporation of deoxynucleotides analogues (dideoxynucleosides triphosphate) into newly synthesized strands, in the presence of a radioactive tracer. The dideoxynucleosides triphosphate lack the 3' hydroxyl group involved in the 5'–3' phosphodiester bonds of DNA strands, as well as the 2' hydroxyl group on the ribose moiety (Figure 179). The incorporation of a dideoxynucleotide in a growing DNA strand results in termination of synthesis. The extent of DNA synthesis will depend on the ratio of deoxy/dideoxy nucleotides (dXTP/

ddXTP) present in the reaction mixture. An elevated concentration of dideoxynucleotides will give rise to short fragments while longer DNA chains will be synthesized if the concentration of analogue is lowered.

To determine the nucleotide sequence of a cloned DNA fragment, four separate reactions are performed. In each of them polymerization is performed in the presence of one of the four ddXTP. A random incorporation of ddXTP is achieved by using a well-defined concentration of each ddXTP over the concentration of the corresponding dXTP in the reactions. The length of the DNA chains obtained in these conditions will be a function of the nucleotide sequence of a particular fragment. For example, all DNA strands whose synthesis has been stopped in the presence of ddCTP will contain a C at their end, while all strands whose elongation has been stopped in the presence of ddGTP will terminate at a G. Thus, by analyzing the length of the DNA strands synthesized in each reaction, it will be possible to

determine the relative position of each nucleotide in the sequence (see Figure 178).

Once the primer is annealed, enzymatic copy of the single stranded recombinant M13 DNA is performed by incubation in the presence of either the large fragment of DNA polymerase (Pol1K) sequenase, or reverse transcriptase, and DNA synthesis is monitored by addition of ^{32}P- or ^{35}S-deoxynucleotides. Starting from the primer, the enzyme performs a 5′ to 3′ elongation through the polylinker region and extends into the cloned DNA fragment. After incubation, each reaction mixture contains a nested set of DNA fragments whose 5′ end is common and corresponds to the primer 5′ end.

To obtain reliable results, it is highly recommended that all volumes are dispersed accurately and that the different reagents are mixed thoroughly and collected by quick centrifugation to the bottom of the tubes.

Take advantage of the 2-hour annealing period to prepare the dideoxy mixtures.

PROTOCOL

SEQUENCING WITH DNA POLYMERASE

Labeling the Newly Synthesized Strands with ^{32}P Deoxynucleotides

1. Label four microfuge tubes as A, G, C, and T.
2. Into each of these tubes add 1 μl of the corresponding dideoxynucleoside triphosphate (ddA, ddG, ddC, ddT).
3. Depending on the α[^{32}P]dXTP to be used for labeling, prepare four dXTP mixes, each corresponding to a different dideoxynucleoside triphosphate (refer to the chart in **Buffers,** below). To each of the A-, G-, C-, and T-labeled tubes, add 1 μl of the corresponding dXTP mix.
4. When the 2-hour annealing period (protocol above) is over, add to the DNA–primer mix:
   1 μl α[^{32}P]dXTP (10 μCi)
   2 μl 0.1 M DTT
   1 μl DNA polymerase I, large fragment (1 unit)
   Mix gently by pipetting solution into and out of the tip.
5. Add 3 μl of the mixture prepared in step 4 to each of the tubes labeled A, G, C, and T (which contain 1 μl dXTP mix and 1 μl dideoxynucleoside triphosphate).

6. After mixing, incubate at 25°C (or room temperature) for 15 minutes.

7. To each tube, add 1 μl of the dXTP chase mixture.

8. Incubate at 25°C for 15 minutes.

9. Stop the reaction by freezing in a dry-ice ethanol bath and store at −70°C.

10. Just before use, thaw the content of the tube, add 10 μl of dye mix, and denature the samples by heating at 90°C for 3–5 minutes.

Labeling the Newly Synthesized Strands with ^{35}S deoxynucleotides

The use of ^{35}S-labeled thionucleotides in sequencing reactions (Biggin et al., 1983) offers considerable advantages over the use of ^{32}P-labeled nucleotides (see also Chapter 16). First of all, ^{35}S has a longer half-life than ^{32}P (see Figure 129). Therefore, labeled compounds are more stable and sequencing reaction samples or nucleotides stocks can be kept for longer periods of time at −70°C. Also, because the decay of labeled thionucleotides does not involve a breakage of the sugar–phosphate bonds in the DNA molecules, sequencing samples can be kept for at least a week at −20°C without significant degradation. Since ^{35}S is a "soft" β-particle emitter, the radiation dose to the researcher is considerably reduced and there is no real need for protective screens when performing sequencing reactions. This is an important factor to take into consideration when long-term intensive sequencing programs are being developed. Because the thio-deoxynucleotides are not incorporated as efficiently as the normal deoxynucleotide, it is necessary to increase the incubation time slightly.

1. In a microfuge tube containing 0.5 to 1.0 μg of single stranded DNA annealed to 2.5 ng of sequencing primer (in a final volume of 12 μl), add 2 μl (40 pmol) of α[^{35}S]dATP (New England Nuclear, 500 Ci/mmol) (see note in interpretation of autoradiograms).

2. Mix and centrifuge briefly at 12,000 rpm (room temperature).

3. Add 1 μl (1 unit) of Pol1k polymerase (Boehringer) and mix by pipetting up and down several times.

4. Add 3 μl of the mixture from step 3 (above) to each of the tubes labeled A, G, C, and T (which as above, contain 1 μl of dXTP mix and 1 μl of dideoxynucleoside triphosphate).

5. Incubate at 25°C for 20 minutes.

6. Add to each tube 1 μl of dXTP chase mixture.

7. Incubate at 25°C for 20 minutes.

8. Stop reaction and store the samples as described in step 9 above.

Buffers and Solutions for Sequencing

Polymerase buffer (10X)	*10 ml*
70 m*M* Tris–HCl (pH 7.5)	0.7 ml 1 *M* Tris–HCl (pH 7.5)
70 m*M* MgCl$_2$	0.7 ml 1 *M* MgCl$_2$
500 m*M* NaCl	1 ml 5 *M* NaCl
	7.6 ml H$_2$O

dXTP mixes. These mixes contain low concentrations of the nucleotide triphosphate whose analog (dideoxy) is used to stop the polymerization, but do not contain the nucleotide triphosphate used for labeling. For instance, when α[^{32}P]dATP and ddTTP are used, the incubation mix will contain only 1 µl 0.5 m*M* TTP and no dATP. Refer to the following chart to prepare the relevant mixtures.

In addition to the compounds indicated in this chart, each sample should contain 20 µl 10X polymerase buffer.

	ddXTP solution	
[^{32}P]dXTP	ddATP	ddCTP
[^{32}P]dATP	20 µl 0.5 m*M* dCTP	1 µl 0.5 m*M* dCTP
	20 µl 0.5 m*M* dGTP	20 µl 0.5 m*M* dGTP
	20 µl 0.5 m*M* TTP	20 µl 0.5 m*M* TTP
[^{32}P]dCTP	1 µl 0.5 m*M* dATP	20 µl 0.5 m*M* dATP
	20 µl 0.5 m*M* dGTP	20 µl 0.5 m*M* dGTP
	20 µl 0.5 m*M* TTP	20 µl 0.5 m*M* TTP
[^{32}P]dGTP	1 µl 0.5 m*M* dATP	1 µl 0.5 m*M* dCTP
	20 µl 0.5 m*M* dCTP	20 µl 0.5 m*M* dATP
	20 µl 0.5 m*M* TTP	20 µl 0.5 m*M* TTP
[^{32}P]TTP	1 µl 0.5 m*M* dATP	1 µl 0.5 m*M* dCTP
	20 µl 0.5 m*M* dCTP	20 µl 0.5 m*M* dATP
	20 µl 0.5 m*M* dGTP	20 µl 0.5 m*M* dGTP
	ddGTP	ddTTP
[^{32}P]dATP	1 µl 0.5 m*M* dGTP	1 µl 0.5 m*M* TTP
	20 µl 0.5 m*M* dCTP	20 µl 0.5 m*M* dCTP
	20 µl 0.5 m*M* TTP	20 µl 0.5 m*M* dGTP
[^{32}P]dCTP	1 µl 0.5 m*M* dGTP	1 µl 0.5 m*M* TTP
	20 µl 0.5 m*M* dATP	20 µl 0.5 m*M* dATP
	20 µl 0.5 m*M* TTP	20 µl 0.5 m*M* dGTP

[^{32}P]dGTP	20 μl 0.5 mM dATP	1 μl 0.5 mM TTP
	20 μl 0.5 mM dCTP	20 μl 0.5 mM dATP
	20 μl 0.5 mM TTP	20 μl 0.5 mM dCTP
[^{32}P]TTP	1 μl 0.5 mM dGTP	20 μl 0.5 mM dATP
	20 μl 0.5 mM dATP	20 μl 0.5 mM dCTP
	20 μl 0.5 mM dCTP	20 μl 0.5 mM dGTP

ddXTP solutions
ddATP: 1.0 mM
ddCTP: 0.35 mM
ddGTP: 0.7 mM
ddTTP: 2 mM

It is convenient to prepare 10 mM stock solutions in distilled water.
These can be stored at $-20°C$ for a few weeks.

PEG solution
20 g Polyethylene glycol 6000
10 g NaCl
H_2O to 100 ml

Dithiothreitol stock solution   *Working solution (10 m*M*)*
   (0.1 M)
154 mg dithiothreitol (DDT)   Dilute stock solution 1:10.
H_2O to 10 ml

Triton X-100 solution (2%)
2 ml Triton X-100
98 ml Resuspension buffer

IPTG (isopropyl-β-D-thiogalacto-
   pyranoside) solution (1 M,
   100 ml)
2.38 g IPTG
H_2O to 100 ml

2% X-gal (5-bromo-4-chloro-3-in-
   dolyl-β-D-galactoside) solution
   (5 ml)
100 mg X-gal
Dissolve in dimethylformamide
   to a final volume of 5 ml.

2% Bluogal (5-bromo-3-indolyl-β-
   D-galactoside) solution (5 ml)
Proceed as for X-gal solution

*Sequencing gel stock solution
(for 8% gel)*
76 g Acrylamide
4 g Bis-acrylamide
420 g Urea
100 ml 10X Tris–borate
H₂O to 1 liter

To prepare 8% gel:

Take 100 ml stock solution.
Add 0.6 ml 10% ammonium
    persulfate (prepared fresh).
Add 0.12 ml TEMED.
Mix and pour immediately into
    glass "sandwich" gel mold.

Tris–borate (10X)
1 *M* Tris–base
1 *M* Boric acid
20 m*M* EDTA
Adjust pH to 8.3.

REFERENCES

Biggin, M. D., Gibson, T. J., and Hong, G. F. (1983), Proc. Natl. Acad. Sci USA, **80,** 3963.

Sanger, F., and Coulson, A. R. (1975), *J. Mol. Biol.,* **94,** 441.

Sanger, F., Nicklen, S., and Coulson, A. R. (1977), *Proc. Natl. Acad. Sci. USA,* **74,** 5463.

Procedure for Performing Dideoxy Sequencing Reactions in Microtiter Plates

The procedure outlined below is a modification of the method of Dr. Alan Bankier, from the MRC Laboratory of Molecular Biology, in Cambridge, England and has been kindly provided to us by International Biotechnologies Inc.

PROTOCOL

PERFORMING DIDEOXY SEQUENCING REACTIONS IN MICROTITER PLATES

1. Use a 96 well, U-bottom microtiter plate for this procedure. Assign a nucleotide base (A, C, G, T) to each vertical column of 12 wells. Assign a horizontal row of bases for each template to be sequenced.

2. Add the following components to a 1.5-ml microfuge tube:
   2 μl of sequencing primer (2.4 ng)
   1 μl of 10X sequencing buffer
   5 μl of sterile water

This provides sufficient material for one template. To scale up, multiply by the number of templates to be sequenced.

3. Add 2 μl of the sequencing primer prepared in step 2 to each well of the microtiter plate.

4. Add 2 μl of each template DNA (0.12–0.20 μg/ml diluted in

1X sequencing buffer) to the appropriate horizontal row of bases (A, C, G, T).

5. Cover the wells with a layer of plastic wrap. Centrifuge the plate briefly to mix and then place the plate in a 55°C oven for at least 45 minutes. For centrifugation, use a microtiter plate carrier in a bench top centrifuge.

6. Centrifuge the plate briefly to concentrate any condensation and then remove the plastic wrap.

7. Dispense 2 μl of the appropriate XTP mix close to the rim of each sample well in the appropriate vertical column (e.g., add reaction mix A to each vertical "A column").

8. For each template to be sequenced, mix the following components in a small siliconized glass tube:
   1.0 μl Klenow fragment (2.5 units)
   1.5 μl ^{35}S–dATP (600–800 Ci/mmol, 10 μCi/μl)
   4.5 μl 1X sequencing buffer

To scale up, multiply by the number of templates to be sequenced.

9. *Immediately,* add 2 μl of the mixture prepared in step 8 to each well of the microtiter plate, being careful to avoid the nucleotide drop. Centrifuge as briefly as possible to mix and then incubate at 37°C for 20 minutes.

10. Add 2 μl of case solution to each sample well. Centrifuge briefly to mix and then incubate at 37°C for 20 minutes.

11. Add 2 μl of stop solution to each sample well and centrifuge to mix.

12. Just before loading the sequencing gel, heat-denature the reaction by placing the microtiter plate in an 80°C oven for 15 minutes (or until the volume decreases to 3–4 μl).

13. Load 2–3 μl into each sample well of the sequencing gel.

Alternate Method. If you prefer *not* to load the entire sample onto the sequencing gel, you may increase the volume of the stop solution added in step 11 to 8 μl. Place the microtiter plate in an 80°C oven just prior to loading your sequencing gel for 10 minutes to denature the reactions.

POLYACRYLAMIDE GEL ELECTROPHORESIS

The general principles for gel electrophoresis are described in Chapter 9.

1. For sequencing 8%, 10%, or 20% polyacrylamide gels are routinely used. These gels contain 8 *M* urea as a denaturing agent and are usually very thin (typical dimensions: 34 × 40 × 0.04 cm.)

2. Regular sequencing gels are poured between two glass plates, as described in Chapter 9. The size of the glass plates can vary a lot depending on the particular needs of the user. Generally, 20 × 40 cm and 40 × 40 cm gel units are satisfactory. In some

cases, however, the use of very long gels (one meter or more) has been described. The plates must be absolutely clean before use. Wash them with detergent, water, and then ethanol. Rinse with distilled water and dry with tissues. Usually, the notched plate is siliconized with a 2% dimethyl-dichlorosilane in 1,1,1-, trichloroethane in a fume hood (this product is harmful; wear gloves and handle with care). It is also possible to treat the regular plate with an antisilane to allow firm adhesion of the gel to the plate and direct drying (Garoff and Ansorge, 1981).

In this method, the plate is covered with a freshly prepared mixture composed of 2.5 ml 0.5% antisilane (in ethanol) and 75 μl 10% acetic acid. The plate is washed thoroughly with ethanol after the antisilane is dry. The plates are taped together with a waterproof vinyl tape, and the gel solution is poured slowly without forming bubbles. If bubbles are trapped between the plates, they can be removed by knocking the plates gently with a piece of wood, or by tilting the plates. Fill to the top, insert the comb, and clamp the plates with fold back spring clips. Polymerization requires about 1 hour at room tempearture.

3. Very thin gels (0.2 mm) can be easily prepared with the Macrophor unit commercialized by Pharmacia-LKB. Although it gave excellent results in our hands, we found that setting up the Macrophor is somewhat tricky and time consuming.

4. Two recent approaches have been reported for altering the migration of DNA fragments in gels, in order to increase by as much as 20% the number of bands resolved by electrophoresis. In the wedge thickness gel method (Olsson et al., 1984), the thickness of the spacers varies linearly from the top to the bottom of the gel (for example, from 0.4 mm to 1.2 mm). No modification of the gel apparatus or gel recipe is needed when using wedge gels. The other method to improve sequencing gel resolution was based on the use of buffer gradient gels (Biggin et al., 1983).

Buffer gradient gels* offer several advantages over conventional gels. They have a higher buffer concentration at the bottom of the gel than at the top. This retards the more quickly moving smaller fragments of DNA. The mobility of a DNA fragment in an electrical field on a normal gel has a logarithmic relationship to molecular weight. Thus, while the bands at the bottom of the gel are very widely spaced, those at the top of the gel are also so tightly packed as to be unreadable. In the case of a buffer gradient gel, the band pattern is no longer logarithmic, but instead bears an almost arithmetic relationship to molecular weight.

As a result, the bands at the top of the gel are now more evenly spaced and the sequence of a clone can be read further into this region. With conventional gels it is necessary to run each sample twice, for different lengths of time, in order to determine the maximum number of bases possible. With buffer gradient gels each sequencing reaction need be run only once. In addition, these gels require an electrophoresis time of only 2 hours compared to the 4½ hours necessary for normal gels. Although a single buffer gradient gel will not give quite as much information as a double run on a conventional gel (250 compared to 300 nucleotides), it does allow more samples to be analyzed in a shorter time.

This technique is particularly powerful when used with ^{35}S-labeled nucleotides (page 615). In this case, the high-definition, evenly spaced band pattern is much easier and more convenient to read and generates a great deal of sequencing information from each gel. The quality of the final autoradiograph makes this a most valuable technique.

*Courtesy of Amersham Plc.

PROTOCOL

PREPARATION OF A BUFFER GRADIENT GEL

The procedure is described in Figure 180.

For a 40 × 20 × 0.04 cm gel prepare the following acrylamide solution:

Pouring the gel

1) To 30ml of top gel (in a 50ml beaker) add 60µl of 25% ammonium persulphate and 60µl of TEMED.

2) To 7ml of bottom gel (in a 50ml beaker) add 14µl of 25% ammonium persulphate and 14µl of TEMED.

3) Take up 22ml of top gel in a 50ml syringe or pipette. Place to one side.

4) Take up 6ml of top gel into a 20ml syringe or pipette.

5) Next, carefully take up 6ml of the bottom gel into the same 20ml syringe.

6) Draw up one or two small air bubbles into the syringe. This will cause slight mixing to give a more diffuse gradient.

7) Pour this gradient down the side of the gel plates to fill slowly from the bottom without forming bubbles. Incline the plates at an angle of 45° to aid pouring the gel, but keep the plates level, so that an even gradient of blue coloured acrylamide is formed.

8) Follow this by adding the top gel from the 50ml syringe. Decrease the angle of the plates to prevent the lower layer from being unduly disturbed. Continue pouring this half of the gel, at first from the side of the gel plates, and then slowly moving to the centre, in order to prevent swirling of the lower layers which may disturb the gradient.

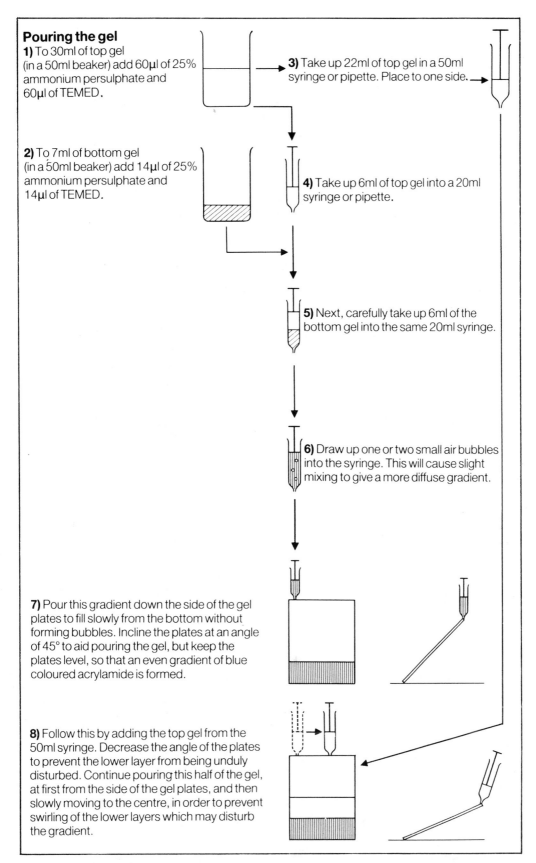

Figure 180. Preparation of a buffer-gradient polyacrylamide gel (courtesy of Amersham).

1. Top gel (for 4 gels)
   76.8 g urea
   24 ml acrylamide stock (38% acrylamide, 2% bisacrylamide)
   8 ml 10 × TBE
   Add H_2O to 160 ml.
   Store in the dark at 4°C.
2. Bottom gel (for 4 gels)
   14.4 g urea
   3.0 g sucrose
   7.5 ml 10 × TBE
   4.5 ml acrylamide stock
   0.3 ml bromophenol blue solution (0.01 g/ml)
   Add H_2O to 30 ml.
   Store in the dark at 4°C.

5. We recommend the use of sharktooth combs (see Figure 181). We found that the BRL combs gave excellent results.

6. In the absence of thermostatation of the plates, temperature differences across the gel during electrophoresis result in a relative acceleration of the sample migration in the center of the gel. This effect is often referred as "smiling." To avoid these problems, many manufacturers have included a thermostating plate in their gel units.

7. The DNA samples are usually heat-denatured for 3 minutes in boiling water and then immediately loaded onto the gel. The wells should be rinsed with fresh 1X running buffer before samples are applied.

A typical sequencing reaction provides enough material to perform several independent analyses. It is possible to load up to 2 µl of sample when using sharktooth combs or up to 5 µl of sample when using the regular combs. Remember that the smaller the volume of the sample, the sharper the bands on the autoradiogram. The remaining volume of the sample is kept at -20°C. We find it convenient to use Hamilton syringes for loading samples. Be sure to rinse the syringe in a large amount of water between each loading.

8. Electrophoresis is carried out at a very high voltage: 1600 volts for 8% gels, 1800–2000 volts for 10% gels, and 2500 volts for 20% gels. Prior to sample loading, the gels should be preelectrophoresed for 30 minutes at the chosen voltage. Use constant power sources to avoid excessive heating during the run. We usually perform our electrophoresis at 70 watts for 6% and 8% regular gels, and 50 watts for gradient gels (when using the BRL SO apparatus).

Under such conditions, the gel temperature is about 50–55°C and bromophenol blue reaches the bottom of an 8% gel in 1.5–2 hours. This dye runs with 20-bp fragments for DNA in a denatured state. Mixtures of size markers for sequence analysis are commercially available (e.g., from BRL). One of the BRL DNA gel marker kits is a 123-bp-fragment ladder generated by multiple ligation. The molecular weights covered by the 34 fragments are multiples of 123 (123, 246, 369, etc.) up to 4182. Band 1 migrates just behind the bromophenol blue marker on 4% polyacrylamide gels with Tris–borate (pH 8.0) as the running buffer. On a 1.5% agarose gel with Tris–acetate (pH 7.6) as the running buffer, bromophenol blue migrates just ahead of the 492-bp fragment. On high-resolution (8% polyacrylamide) gels, fragments 1 and 2 migrate slightly slower than blunt-ended 123- and 246-bp fragments, respectively. Another gel marker kit is a 1 kbp ladder which covers fragment size from 1.018 kb to 12,216 kb (1.018, 2.036, 3.054, 4.072, etc.). An oligo(dT)4–22 ladder consisting of 19 fragments increasing by 1 base from 4 to 22 bases is also commercially by BRL and provides an accurate reference for determining the location and size of the smallest fragments on gels.

9. It has been reported (Williams et al., 1986) that electrophoresis at 60 watts for longer than 4 hours results in a loss of resolution at the top of the gel. This loss of resolution appears to migrate toward the bottom of the gel as the electrophoresis time increases. Long-run buffers may circumvent this kind of problem (Anderson, 1981; Barnes et al., 1983).

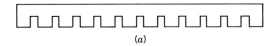

(a)

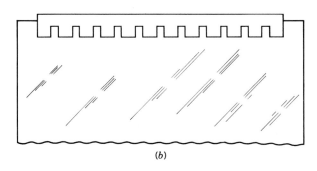

(b)

Insert comb, pour gel, let gel polymerize,
remove comb, and load samples.

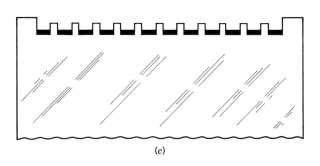

(c)

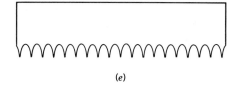

(e)

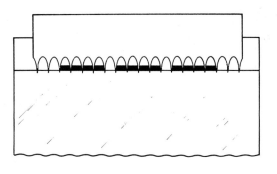

Pour gel, let gel polymerize,
insert combs and load samples.

(f)

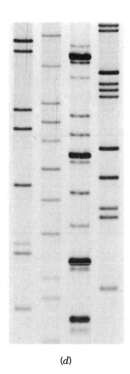

(d)

(g)

Figure 181. Use of different Sharktooth combs for sequencing.

AUTORADIOGRAPHY

Autoradiography can be performed either at room temperature or at low temperature ($-70°C$) to reduce scattering.

PROTOCOL

AUTORADIOGRAPHY OF GELS CONTAINING ^{32}P-LABELED NUCLEOTIDES (FIGURES 182, 183)

When exposure is being performed at room temperature:

1. Carefully remove the siliconized plate so that the gel remains on the other plate.
2. Transfer the gel + plate into a tank containing 2 liters of 10% acetic acid for at least 10 minutes.
3. Carefully drain the gel and gently blot with paper towels to remove excess of liquid.
4. Transfer the gel with an old X-ray film (when the film is layered on the surface of the gel, polyacrylamide sticks to gelatin).
5. Cover the other side of the gel with Saran wrap and expose in a cassette.

When exposure is being performed at low temperature, transfer the gel directly to an old film, without fixing with acetic acid.

PROTOCOL

AUTOGRADIOGRAPHY OF GELS CONTAINING ^{35}S-LABELED NUCLEOTIDES (FIGURES 184 and 185)

1. Fix the gel as described in steps 2–3 above.
2. Cut a sheet of Whatmann 3MM paper to the size of the gel and carefully layer it onto the surface of the gel.
3. Blot with paper towels in order to remove all excess of liquid which would prevent gel adhesion to the Whatmann.

(Protocol continued on page 628)

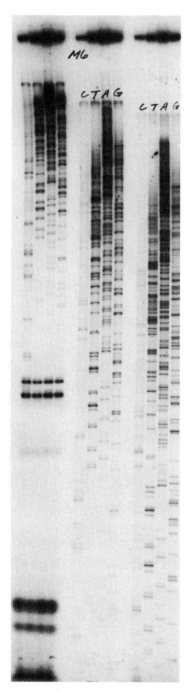

Figure 182. Sequencing gel obtained for a 350-bp fragment cloned into M13mp7. The (+) strand was isolated and the template sequenced by the standard dideoxy chain-termination method. The resulting fragments were then separated on an 8% acrylamide gel (courtesy of K. Rushlow).

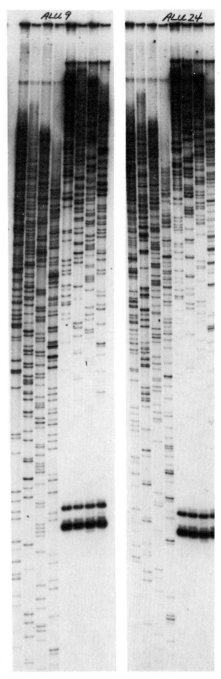

Figure 183. Sequencing of an Alu I restriction fragment cloned into the Sma I site of M13mp8. A 10% gel was used. The dark bands at the bottom are the result of the M13 primer used (29-bp biological primer, pSP14). Loading was performed twice. For the first set, the bromphenol blue dye was run to the bottom. The second set was then loaded and the dye run to 3 cm from the bottom of the gel. This staggered loading allows precise sequencing over a larger range of sizes (courtesy of K. Rushlow).

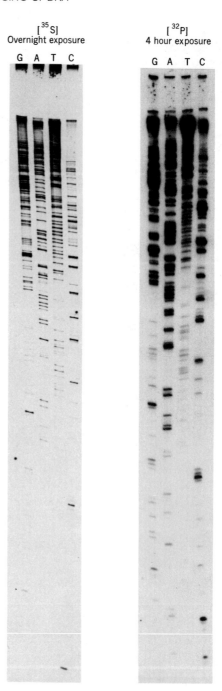

Figure 184. Comparison of use of α[³²P]-dATP and [³⁵S]-dATPαS in Sanger Sequence Analysis of bacteriophage M13mp8 DNA. Procedure and reactions were as described in the text. Nucleotide concentrations for ³⁵S labeling were as follows: for the G lane, 54 μ*M* dCTP, 5.4 μ*M* dGTP, 54 μ*M* dTTP, 300 μ*M* ddGTP; for the A lane: 3.75 μ*M* dCTP, dGTP, dTTP, 20 μ*M* ddATP; for the T lane: 54 μ*M* dCTP, 5.4 μ*M* dGTP, dTTP, 600 μ*M* ddTTP; for the C lane: 5.4 μ*M* dCTP, 54 μ*M* dGTP, dTTP; for all lanes: 1.1 μ*M* [³⁵S]-dATPαS. The initial elongation reaction (20 minutes incubation at 20°C) was followed by a chase reaction (15–20 minutes additional incubation at 20°C). Courtesy of New England Nuclear.

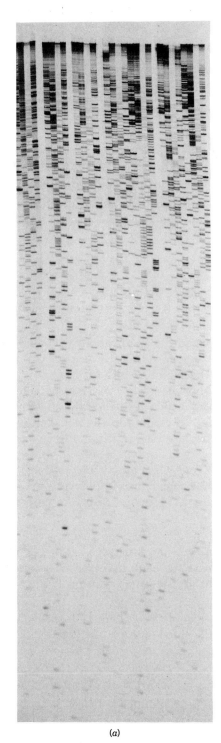

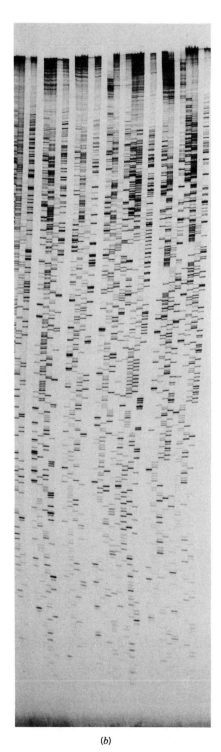

(a)   (b)

Figure 185. Comparison of results obtained with regular and buffer-gradient polyacrylamide gels. Yeast nuclear DNA was used for dideoxy-sequencing with α [^{35}S]-dATP as tracer. Samples were analyzed in parallel on a regular gel (a) and on buffer gradient gel (b). Note that buffer gradient polyacrylamide gel allows reading many more bases in the sequence (kindly provided by B. Seraphin).

4. By lifting it from the top or the bottom, carefully peel the gel off the glass plate and transfer to another dry glass plate (gel up).

5. Cover the gel with Saran wrap.

6. Transfer in a vacuum gel dryer (Hoefer) for at least 30 minutes at 80°C.

7. When the gel is dried, remove Saran wrap and expose at room temperature.

Note. When drying of the gel is to be performed in an oven, it is recommended to treat the unnotched plate with bind-silane to avoid curling and shrinking of the gel (Garoff and Ansorge, 1981).

References

Anderson, S. (1981), *Nucl. Acids Res.,* **9,** 3015.

Barnes, W. M., Bevan, M., and Son, P. H. (1983), *Methods Enzymology,* **101,** 98.

Biggin, M. D., Gibson, T. J., and Hong, G. F. (1983), *Proc. Natl. Acad. Sci. USA,* **80,** 3963.

Garoff, A., and Ansorge, W. (1981), *Anal. Biochem.* **115,** 450.

Mills, D. R., and Kramer, F. R. (1979), *Proc. Natl. Acad. Sci USA,* **76,** 2232.

Martin, R. (1987), *Focus,* **9,** 8.

Mizusawa S., Nishimura, S., and Seela F. (1986), *Nucl. Acids Res.,* **14,** 1319.

Olsson, A., Moks, T., Uhlen, M., and Gaal, A. B. (1984), *J. Biochem. Biophys. Methods,* **10,** 83.

Williams, S. A., Slatko, B. E., Moran, L. S., and De Simone, S. M. (1986), *Bio Techniques,* **4,** 138.

Zagursky, R. J., Baumeister, K., Lomax, N., and Berman, M. L. (1985), *Gene Anal. Tech.* **2,** 89.

Interpretation of the Autoradiograms

1. When reading a sequence, it is important to note spaces as well as bands in each track. Faint bands located in a region where irregular spacing is observed generally result from artifacts and should not be considered.

Dark shadow bands occur mainly with long exposure times. It is frequent that reading of sequencing data is easier on a short time exposure.

The bands corresponding to the T and A bases can leave faint shadow bands in the G and C tracks. Also, A bands can leave shadows in the T track and sometimes in the T, G, and C tracks. By examining the band spacing carefully it is generally easy to determine whether a band should be considered as a shadow or as a real band.

2. Band compression may occur when G + C rich regions are sequenced. In this case, inappropriate spacing of the bands is observed. This problem is most probably due to the formation of stable secondary structures which influence the mobility of the corresponding fragments in the gel. The use of deoxyinosine triphosphate (Mills and Kramer, 1979) or 7-deaza dGTP (Mizusawa, 1986) instead of dGTP may circumvent this migration artifacts (Figure 186). Compressions can also be overcome by using gels containing 7 M urea and 40% formamide as denaturing agents instead of 8 M urea (Martin, 1987).

3. A 1:1 primer/template ratio is the optimal value when working with [35S]–dATP. Increasing this value to 5 may result in increased shadow bands and difficulty in the interpretation of the autoradiograms.

4. When doublets of identical nucleotides are present in the same lane:

The upper C is always darker than the lower C (if a third C occurs, it is also frequently dark).

The upper G is often darker than the lower G.

The upper A is often less dark than the lower A.

5. When a GG doublet is preceded by a T the upper G is darker than the lower G (Williams et al., 1986)

Better resolution of the DNA bands on the autoradiogram can be obtained if the scattering of emitted photons is reduced. This is generally achieved by

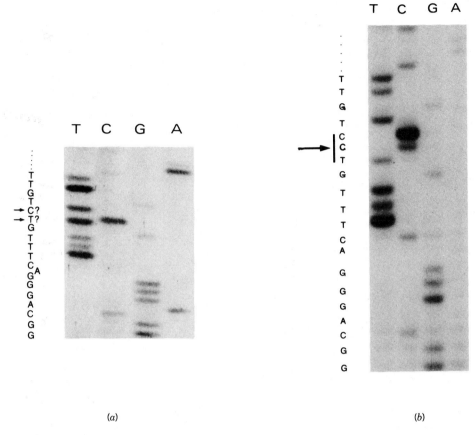

(a) (b)

Figure 186. Use of inosine instead of guanosine triphosphate to eliminate problems due to compression. (a) Sequencing gel performed in the presence of dGTP. (b) Same sample sequenced in the presence of deoxyinosine triphosphate (kindly provided by A. Hampe).

placing the X-ray film very tightly against the gel, reducing the thickness of the gels used, using single-side-coated X-ray films, and exposing the films at −70°C.

If several band patterns are superimposed, check the purity of primer and template DNAs. Contamination of primer DNA with exonuclease may result in a pattern in which each band appears as a doublet or triplet.

In some cases, the intensity of the bands varies greatly. Bands at the bottom are very dark and bands at the top are very light, or vice versa. An inappropriate ratio of ddXTP to dXTP causes such variations. A high ratio leads to more efficient termination of the polymerizing DNA chains. Therefore, long fragments will be less well represented and will hardly be detected on the autoradiogram.

6. Streches of DNA containing high percentages of G + C or T + A or palindromic regions may be difficult to sequence with the dideoxy chain termination method, due to the formation of stable secondary structures which inhibit enzyme progression on the template. A very dark band is then observed in all lanes throughout the gel at this position. This is particularly true for a 24 base pair inverted repeat located within the multiple cloning site of M13 mp vectors (including the Sal I, Bam HI and Eco RI sites) and which forms a hairpin structure responsible for stops in chain elongation, resulting in compression of the nucleotides comprised between the Pst I and Eco RI sites. Increasing the running voltage from 1300 volts to 1650 volts has been suggested to partially solve this problem (New England Biolabs) but still leads to compression between the

Bam HI and Eco RI sites. Performing the sequencing reactions at 50°C (rather than 30°C) with the Klenow fragment leads to a significant reduction of this problem (Williams, 1986; Martin, 1987) probably because of thermal destabilization of the hairpin structure. The use of reverse transcriptase may also help to alleviate this kind of problem (Zagursky et al., 1985).

SOME COMMON PROBLEMS THAT MIGHT BE ENCOUNTERED WHEN PERFORMING SEQUENCING: TROUBLE SHOOTING GUIDE

1. Because Thio-dATP is not as good substrate as dATP for elongation, the Pol1k occasionally terminates a chain before incorporating the adenosine. The resulting artifact is a shadow below each band in the A track. Reducing the ddATP concentration when using labeled thio-dATP can circumvent the problem.

2. All bands on the autoradiogram are smeared. Check age of the film. Ensure good contact between gel and film.

3. Sodium acetate left over in the DNA preparation. Can lead to complete inhibition of the deoxynucleotide incorporation.

4. Contamination of DNA preparation by PEG or RNA. Leads to high background and spurious bands.

5. Use of bad lots of Polymerase, contaminated with 5′-exonuclease activity. Produce a ladder effect by the generation of multiple images for each band.

6. Use of partially inactivated batch of Polymerase. Premature termination of strand copy due to loss of activity. Generation of dark bands at the bottom of the gel and absence of bands at the top. Uneven intensities of the bands.

7. Incubation temperature too low. Generation of dark bands corresponding to stops of the enzyme at hairpin secondary structures.

8. Inadequate dXTP/ddXTP ratios. Too much ddXTP results in dark bands at the bottom and no bands at the top.

9. Inadequate denaturation of the samples prior to loading. Very dark bands at the top of the gel, almost no label at the bottom.

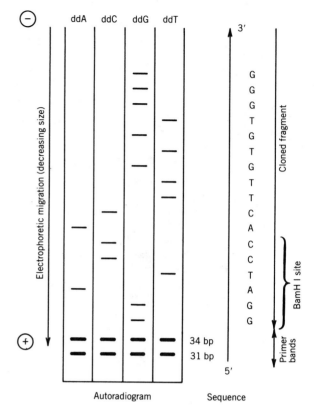

Figure 187. Reading DNA sequence from a dideoxy sequencing gel. Electrophoresis of the fragments generated by polymerization in the presence of dideoxynucleotides gives rise to ladders from which the 5′ – 3′ sequence can be established. The use of reverse sequencing primers would allow determination of the 3′ – 5′ sequence.

Notes for Dideoxy Sequencing. With some previous types of primers, two heavy bands with sizes of 31 and 34 nucleotides were usually detected on autoradiograms at the bottom of all four lanes. These two dark bands result from reannealing of the primer strands and filling in of the cohesive ends in the presence of DNA polymerase I large fragment (Figures 182, 183 and 187). These bands are not observed with single-stranded primers.

Sequencing of the (+) strand of purified M13mp8 or M13mp9 vectors can serve as a good control for the experimental procedures. The sequence found should match that represented in Figure 184. Because of sequence symmetry around the Pst I site in the cloning region of M13mp7 and derivatives, it is important that the gel be run at a high temperature

(~60°C) to prevent annealing of the complementary regions, which could lead to the formation of double stranded molecules with altered mobilities.

Since the resolution of polyacrylamide gels is generally poor for long oligonucleotides, the shotgun cloning method is appropriate for sequencing long stretches of DNA.

SEQUENCING WITH REVERSE TRANSCRIPTASE

Reverse transcriptase can be used instead of DNA polymerase for sequencing DNA under the conditions outlined in the following protocol.

PROTOCOL

SEQUENCING WITH REVERSE TRANSCRIPTASE

1. In a tube containing 20 μCi of dried α[^{35}S]dATP (500 Ci/mmol), or 40 μCi of dried α[^{32}P]dATP (>400 Ci/mmole), add the mixture of annealed template/primer DNA (precipitated with ethanol and resuspended in 13 μl of sterile distilled water).

2. Add 4 μl of 5X reverse transcriptase buffer.

3. Add 2 μl of AMV reverse transcriptase (20 units/μl) and mix.

4. Dispense 4 μl of this mixture in 0.5-ml microfuge tubes containing 1 μl of the reaction mixtures, prepared as described in **"Buffers and Solutions"** below.

5. Incubate at 42°C for 10 minutes.

6. Add 1 μl of the dXTP mixture (chase solution) and incubate for 5 additional minutes at 42°C.

7. Stop the reaction and proceed for gel analysis as described above.

Buffers and Solutions

5X Reverse transcriptase buffer   *100 μl*
300 m*M* Tris–HCl (pH 9.0)   30 μl 1 *M* Tris–HCl (pH 9.0)
375 m*M* NaCl   37.5 μl 1 *M* NaCl
37.5 m*M* MgCl$_2$ 3.75 μl 1 *M* MgCl$_2$
25 m*M* DTT   2.5 μl 1 *M* DTT
                                    26.25 μl sterile distilled water

[32]P SEQUENCING MIXTURES

	A Mixture	C Mixture	G Mixture	T Mixture
ddATP (μM)	60	—	—	—
ddCTP (μM)	—	150	—	—
ddGTP (μM)	—	—	150	—
ddTTP (μM)	—	—	—	150
dATP (μM)	1.7	1.7	1.7	1.7
dCTP (μM)	30	4.2	40	40
dGTP (μM)	30	40	4.2	40
dTTP (μM)	30	40	40	1.7
MgCl$_2$ (mM)	5	5	5	5
Tris–HCl pH 7.5 (mM)	10	10	10	10
dithiothreitol (mM)	7.5	7.5	7.5	7.5

[35]S SEQUENCING MIXTURES

	A Mixture	C Mixture	G Mixture	T Mixture
ddATP (μM)	20	—	—	—
ddCTP (μM)	—	300	—	—
ddGTP (μM)	—	—	300	—
ddTTP (μM)	—	—	—	600
dCTP (μM)	37.5	5.4	54	54
dGTP (μM)	37.5	54	5.4	54
dTTP (μM)	37.5	54	54	5.4
MgCl$_2$ (mM)	5	5	5	5
Tris–HCl pH 7.5 (mM)	10	10	10	10
dithiothreitol (mM)	7.5	7.5	7.5	7.5

REACTION MIXTURES FOR REVERSE TRANSCRIPTASE

	A	C	G	T
A mixture	12.6 μl	—	—	—
C mixture	—	8.4 μl	—	—
G mixture	—	—	8.4 μl	—
T mixture	—	—	—	8.4 μl

REACTION MIXTURES FOR REVERSE TRANSCRIPTASE

	A	C	G	T
dATP (100 μM)	24.9 µl	10.0 µl	10.0 µl	10.0 µl
dCTP (2 mM)	18.8 µl	3.0 µl	12.5 µl	12.5 µl
dGTP (2 mM)	18.8 µl	12.5 µl	3.0 µl	12.5 µl
dTTP (2 mM)	18.8 µl	12.5 µl	12.5 µl	3.0 µl
H_2O	56.1 µl	53.6 µl	53.6 µl	53.6 µl

dXTP Mixture
0.25 mM dATP
0.25 mM dCTP
0.25 mM dGTP
0.25 mM dTTP

Stop solution
0.3% Xylene cyanol FF
0.3% Bromophenol blue
0.37% EDTA (disodium) (pH 7.0)
in deionized formamide

Notes

1. The primer–template mixture can be stored on ice until sequencing reactions are performed. Alternatively, it can be stored at $-20°C$.

2. The large fragment of DNA polymerase I (Klenow) being deprived of $5' \rightarrow 3'$ exonuclease activity is particularly adapted to dideoxynucleotide sequencing. Diluted solutions of the Klenow fragment are not stable when kept for long periods of time at $-20°C$. If you need to dilute your enzyme stock for sequencing, prepare only the required amount. It is a good practice to work on aliquots instead of holding the whole stock of enzyme in ice during the experimentation. We currently use the diluted Klenow enzyme (1 unit/µl) from Boehringer which is dispensed in 100-µl aliquots, allowing us to perform 100 sequencing reactions.

3. Some adjustments of the deoxy/dideoxynucleotide ratios may be required in some instances to increase the reading range of the gels. For example, termination by ddC and ddG is expected

to occur earlier than normal for DNA containing a high G + C ratio. Also, sequencing of short fragments will require ddXTP concentrations higher than those currently used, while lower ddXTP concentrations will be used when sequencing long stretches of DNA. Concentrations may be changed by a factor of 2 to 3. The optimum concentration is determined by a process of trial and error.

4. When labeling is performed with dATP, use the low specific activity batch (600–800 Ci/mmol, 22–30 MBq/mmol).

5. Some of us include cold dATP (1.7 μM) in the reaction mixture when labeling with $\alpha[^{35}S]$ or $[^{32}P]$dATP.

DIRECT SEQUENCING OF DNA FRAGMENTS CLONED IN PLASMIDS

Since the early report by Ruther (1981) on rapid sequencing of DNA in pUR250 plasmid (Figure 188), several other attempts have been described for direct sequencing of DNA fragments without the need of purifying single stranded molecules (Wallace et al., 1981; Guo et al., 1983; Chea and Seeburg, 1985).

The dideoxy-nucleotide chain termination method of sequencing is performed following denaturation of the plasmid DNA, and annealing with standard or reverse primers to obtain sequence of both strands. This protocol can be performed on any double stranded (or linearized) vector, particularly with the multipurpose cloning vectors carrying polylinkers and transcription promoters (see Chapter 6). The standard protocol is as follows:

PROTOCOL

DIRECT SEQUENCING OF DNA FRAGMENTS CLONED IN PLASMIDS

1. In a 0.5-ml microfuge tube containing 2 μg of double stranded DNA as a dry pellet, add 20 μl of 0.2 M NaOH, containing 0.2 mM EDTA.

2. Incubate for 5 minutes at room temperature.

3. Add 2 μl of 2 M ammonium acetate (adjusted to pH 4.5 with glacial acetic acid) to neutralize the denatured DNA.

4. Add 100 μl of cold absolute ethanol and let the DNA precipitate at −70°C for 30 minutes.

5. Centrifuge at 12,000 rpm for 30 minutes at 4°C and pour off the ethanol.

6. Add 1 ml of 80% cold ethanol, vortex, and centrifuge for 10 minutes at 12,000 rpm (4°C).

7. Pour off the ethanol, allow the tube to drain on a tissue, and dry the pellet in a vacuum centrifuge.

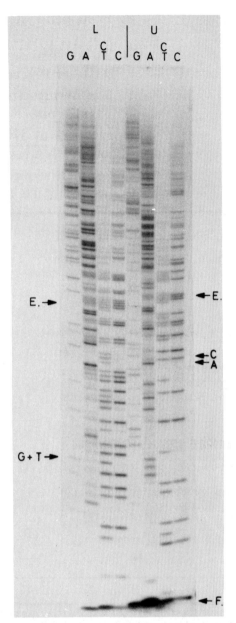

Figure 188. Rapid sequencing technique using pUR250 plasmid cloning. Direct sequence analysis of both DNA strands of an insert in pUR250 was performed. A Hae III fragment of the *lac Z* gene was cloned into the Hinc II site of the plasmid pUR250. Plasmid DNA was sequenced without isolation of the labeled fragments. The autoradiogram of the sequence on a 12% polyacrylamide gel is shown. The sequence codes for amino acid residues 881–892 of β-galactosidase. The sequence of the lower strand (L) shows a compression of two bases (G + T) at one position but the sequence of the upper strand (U) clearly shows C and A in two positions at the complementary strand. The small labeled fragments obtained by the recutting steps are presented on the autoradiogram (F). The ends of the inserted DNA fragments are indicated (E) (provided by U. Rüther).

8. Resuspend the dry DNA pellet in 7.2 µl of sterile distilled water and add:

   2 µl of sequencing primer (corresponding to 15 pmol)
   0.8 µl of 10X sequencing buffer

9. Mix by pipetting up and down; centrifuge a few seconds at 12,000 rpm.

10. Incubate at 37°C for 15 minutes for annealing of the primer to the DNA template.

11. Perform sequencing reactions with either Pol1K or reverse transcriptase, as described.

References

Chea, E. J., and Seeburg, P. H. (1985), *DNA* **4**, 165.

Guo, L. H., Yang, R. C. A., and Wu, R. (1983), *Nucl. Acids Res.* **11**, 5521.

Rüther, U. (1981), *Nucl. Acids Res.* **10**, 5765.

Wallace, R. B., Johnson, M. J., Suggs, S. Y., Myoski, K., Bhatt, R. and Itakura, K. (1981), *Gene* **16**, 21.

SEQUENCING OF SINGLE STRANDED DNA IN BOTH DIRECTIONS

The following protocol, which is a modification of the original method described by Hong (1981), has been adapted in our laboratory by J. Soret. It is based on the primer-directed synthesis of a minus-strand sense template from a single stranded DNA fragment cloned in a M13 vector and then sequencing of this new strand by the dideoxynucleotide termination method with a reverse primer. The key point of this method is to ensure that the first primer is completly removed before sequencing.

PROTOCOL

SEQUENCING OF SINGLE-STRANDED DNA IN BOTH DIRECTIONS

Preparation of template DNA

1. Grow an overnight culture of JM 107 culture in TYE medium from a single colony taken on a minimal medium agar plate.

2. Prepare a tube containing 3 ml of fresh TYE medium and 15 µl of the overnight cell culture.

3. Grow the desired recombinant phage by inserting a toothpick into the center of an isolated colorless plaque and transferring it into the 3-ml tube prepared in step 2.

4. Incubate the tube at 37°C with agitation for 5–6 hours.

5. Centrifuge 1.5 ml of each phage culture in a microfuge tube to pellet the cells at 12,000 rpm for 5 minutes at 4°C.

6. Transfer the supernatant to a microfuge tube and repeat step 5.

7. Transfer the supernatant to a fresh tube and add 200 μl of 2.5 *M* NaCl containing 20% of polyethylene glycol (PEG) 6000.

8. Mix by inverting several times and let the phage particle precipitate by keeping the tube in ice for 30 minutes.

9. Centrifuge at 12,000 rpm for 30 minutes at 4°C. A white pellet should be obtained.

10. Pour off the supernatant and spin the tube 2 more minutes under the same conditions in order to collect all droplets.

11. Aspirate the remaining supernatant with a drawn out Pasteur pipette, being careful not to dislodge the PEG-phage pellet.

12. Add 100 μl of 10 m*M* Tris–HCl pH 7.5, 0.1 m*M* EDTA buffer and vortex to dissolve the precipitate. Keep on ice for 5 minutes.

13. Centrifuge a few seconds at 12,000 rpm to collect all solution and add 50 μl of phenol (previously saturated with 10 m*M* Tris–HCl pH 8.0, 1 m*M* EDTA).

14. Vortex for 30 seconds and centrifuge at room temperature for 5 minutes (12,000 rpm).

15. Carefully pipette the upper phase, avoiding the white interface, and without disturbing the phenol phase. Transfer to a fresh tube.

16. Add 10 μl of 3 *M* sodium acetate (pH 5.0) and 200 μl of cold absolute ethanol. Mix by inverting several times.

17. Incubate at −20°C for at least 1 hour or at −70°C for 30 minutes.

18. Collect precipitated DNA by centrifugation (12,000 rpm, 4°C, for 30 minutes).

19. Pour off ethanol carefully, and add 1 ml of 80% ethanol v/v and vortex for a few seconds.

20. Centrifuge for 10 minutes at 12,000 rpm (4°C).

21. Carefully pour off the supernatant and drain the tube by inverting on a tissue.

22. Dry the pellet in a vacuum centrifuge (SpeedVac, Savant).

23. Dissolve the pellet in 50 μl of 10 m*M* Tris–HCl pH 7.5, 0.1 m*M* EDTA buffer.

Synthesis of the Complementary Strand

1. In a 0.5-ml microfuge tube mix successively:
   9.5 µl of single stranded template DNA
   1.0 µl (2.5 ng) of M13 sequencing primer
   1.5 µl of 10X sequencing buffer
2. Mix by pipetting up and down several times, and centrifuge a few seconds at 12,000 rpm to collect droplets.
3. Incubate in a boiling water bath for 3 minutes.
4. Spin briefly at 12,000 rpm and immediately incubate for 2 hours in an oven (or in a closed water bath) at 60°C.
5. Centrifuge a few seconds at 12,000 rpm and add successively:
   1 µl of dXTP mix
   1 µl of 100 mM DTT
   1 µl of Klenow DNA polymerase (Pol1K, see Chapter 4)
6. Incubate at room temperature for 15 minutes and add another 1 µl of polymerase.
7. Incubate for 15 minutes at room temperature.
8. Add 1 µl of ddXTP (100 µM each) mix and incubate for 10 minutes at room temperature.
9. Stop the reaction by incubating at 68°C for 10 minutes.

Precipitation of DNA and Removal of the First Primer

1. To the content of tube from step 9 above, add an equal volume of 1.6 M NaCl containing 13% PEG 6000 and mix.
2. Let stand on ice for 1 hour.
3. Centrifuge at 12,000 rpm for 30 minutes at 4°C, and remove the supernatant with a drawn out Pasteur pipette.
4. Wash the precipitate with 40 µl of 0.8 M NaCl containing 6.5% PEG 6000 by pipetting up and down several times.
5. Repeat step 3.
6. Resuspend the DNA precipitate in 50 µl of 10 mM Tris–HCl pH 7.5, 0.1 mM EDTA buffer.
7. Perform a phenol extraction of the DNA with 25 µl of phenol previously saturated with 10 mM Tris–HCl (pH 8.0), 0.1 mM EDTA.
8. Extract the aqueous phase once with 50 µl of chloroform/iso-amyl alcohol (24/1).
9. Transfer the aqueous phase to a microfuge tube containing 4 µl of 5 M NaCl, and add 80 µl of cold absolute ethanol.
10. Mix and incubate at −70°C for 30 minutes (or at −20°C overnight).

11. Centrifuge at 12,000 rpm for 30 minutes at 4°C and pour off the ethanol.
12. Add 1 ml of 80% cold ethanol, vortex, and centrifuge for 10 minutes at 12,000 rpm (4°C).
13. Pour off the ethanol, allow the tube to drain on a tissue, and dry the pellet in a vacuum centrifuge.

Denaturation of double strands and annealing of the reverse primer

1. Resuspend the dry pellet from step 12 above, in 8 μl of 10 mM Tris–HCl pH 7.5, 0.1 mM EDTA buffer.
2. Add 2.5 μl (12.5 ng) of reverse primer.
3. Incubate in a boiling water bath for 3 minutes.
4. Immediately freeze by immersing the bottom of the tube in a dry-ice–ethanol bath for about 1 minute.
5. Thaw the content of the tube and add 1.5 μl of 10X sequencing buffer.
6. Incubate the mixture at room temperature for 20 minutes.
7. Carry out sequencing as described on page 614.

Notes

1. Remaining traces of PEG after phage precipitation may inhibit reactions catalyzed by Pol1K.
2. Adding ddXTP 10 minutes before stopping the reaction (step 8 in synthesis of the complementary strand) has been found to be critical since it prevents further elongation of short complementary strands in the following steps.
3. To avoid long exposure when sequencing with the reverse primer, use labeled deoxynucleotides with the highest specific activity.

Solutions and Buffer

Tris/EDTA Buffer
10 mM Tris–HCl (pH 7.5)
0.1 mM EDTA

1 liter
10 ml 1 M Tris–HCl (pH 7.5)
0.2 ml 500 mM EDTA
H$_2$O to 1 liter

10X Sequencing buffer
0.1 M Tris–HCl (pH 7.5)
0.05 M MgCl$_2$
0.075M DTT

dXTP Mix (chase solution)
0.25 m*M* dATP
0.25 m*M* dCTP
0.25 m*M* dGTP
0.25 m*M* TTP

Reference

Hong, G. F., (1981), *Biosci. Rep.,* **1,** 243.

21

RNA SEQUENCING

The recent introduction of efficient in vitro systems that allow large quantities of pure RNA species to be obtained will most certainly trigger the need for improved RNA sequencing techniques. Thus, it would be interesting in many cases to obtain direct information on the primary structure of mRNA species synthesized in vitro from cloned genes.

For example, a comparison of the sequences obtained for both the cloned DNA fragment and the in vitro synthesized (and eventually spliced) mRNA species might allow us to check that the potential donor and acceptor splicing sites are correctly used, and to determine whether alternate processes may govern the regulation of gene transcription. Sequencing of RNA species might also allow us to determine the nature of domains that interact with putative protein factors playing a role in the regulation of translation and in the stability of the mRNA molecules. Such an approach could be similar to the footprint analysis described in Chapter 24. Three methods can be used, to date, for RNA sequencing. Two of them deal directly with RNA molecules and rely on the use of either (1) ribonucleases that can produce cleavages at specific sites in the RNA (see Chapter 4) or (2) chemicals that induce partial, base-

specific modifications of the ribonucleotidic chain and scission at the modified sites. The second method is very much like the Maxam–Gilbert sequencing method (see Chapter 20). The third method relies on the use of reverse transcriptase to synthesize a cDNA copy of the mRNA of interest. The cDNA is then treated as a DNA double stranded molecule and analyzed by the base-specific cleavage method of Maxam and Gilbert. We present below a brief outline of the three methods.

CHEMICAL SPECIFIC CLEAVAGE OF RNA FOR SEQUENCING

Due to the nature of products generated by the chemical strand scission of RNA molecules, only 3'-terminally labeled molecules can be sequenced in this way.

Labeling of RNA 3' Terminus

The 3' terminus of RNA can be efficiently labeled with [5'-^{32}P]pCp in the presence of T4 RNA ligase.

PROTOCOL

LABELING OF RNA 3' TERMINUS

1. In a microfuge tube, add successively:
   3 μl of 10X ligase buffer
   6 μl of dephosphorylated RNA (about 25 pmol)
   3 μl of DMSO (dimethylsulfoxide)
   3 μl of 33 μM ATP (pH 7.0)
   10 μl (100 μCi) of aqueous cytidine biphosphate [5'-^{32}P]pCp
   (3000 Ci/mmol, 110 TBq/mmol)
2. Add 5 μl of T4 RNA ligase (4 units) and mix.
3. Incubate at 4°C until the reaction reaches a plateau (4 to 6 hours). This is checked by TCA precipitation of 1-μl aliquots taken at regular intervals of time (the samples are spotted on Whatman GF-C filters and washed in ethanol to remove the TCA, prior to counting).
4. Add 25 μl of cold 4 M ammonium acetate, extract, and recover the RNA as described for DNA samples (see Chapter 3).
5. Labeled RNA is purified by electrophoresis in a denaturing urea–5% polyacrylamide sequencing gel (see Chapter 20) and recovered by elution from crushed gel slice (see Chapter 11).

BASE-SPECIFIC CLEAVAGE OF LABELED RNA

Four reactions are performed:

1. *Guanosines* are methylated with dimethylsulfate at position N-7. Subsequent ring opening occurs in the presence of sodium borohydride.
2. *Purines* rings are opened following carboxymethylation of imidazole N-7 with diethylpyrocarbonate. Since adenine reacts faster than guanine, this reaction is rather specific for *adenine*.
3. Addition of hydrazine to the C$_5$=C$_6$ double bond of *pyrimidines* results in their removal from RNA. Under aqueous solution, hydrazinolysis of *uracil* is predominant.
4. Reaction 3 performed in the presence of 3 M NaCl with anhydrous hydrazine will lead to the preferential modification on *cytosines*.

PROTOCOL

MODIFICATION OF GUANINE

1. In a microfuge tube, mix:
   300 μl of 50 mM cacodylate buffer (pH 5.5) containing 1 mM EDTA
   2 μl of carrier tRNA (5 mg/ml)
   5 μl of [^{32}P] 3'-labeled RNA (about 10^5 cpm)

2. Add 1 μl of 50% (v/v in water) dimethylsulfate fresh solution.

3. Vortex; centrifuge at 12,000 rpm a few seconds.

4. Incubate at 90°C for 1 minute.

5. Chill on ice.

6. Add 75 μl of acetate buffer (pH 7.5). Vortex.

7. Add 900 μl of cold absolute ethanol, mix, chill on dry ice for 10 minutes.

8. Centrifuge for 10 minutes at 12,000 rpm.

9. Discard the supernatant in 10 M NaOH.

10. Add 200 μl of cold 0.3 M sodium acetate to the pellet; mix by vortexing.

11. Add 900 μl of cold absolute ethanol.

12. Centrifuge for 5 minutes at 12,000 rpm.

13. Discard supernatant. Add 0.5 ml of cold ethanol.

14. Centrifuge at 12,000 rpm for 1 minute.

15. Discard the ethanol, and dry the pellet under vacuum (SpeedVac, Savant).

16. Resuspend the pellet in 10 μl of 1 M Tris–HCl (pH 8.2). Chill on ice.

17. Add 10 μl of a fresh 0.2 M solution of sodium borohydride.

18. Incubate in the dark for 30 minutes at 4°C.

19. Add 1 μg of carrier tRNA (5 mg/ml) and 200 μl of cold sodium acetate to stop the reaction. Return to ice.

20. Add 600 μl of cold absolute ethanol. Let at -70°C for 10 minutes.

21. Centrifuge at 12,000 rpm for 10 minutes (4°C). Wash the pellet with 500 μl of cold absolute ethanol.

22. Centrifuge at 12,000 rpm for 5 minutes (4°C) and discard the supernatant.

23. Dry the treated RNA under vacuum.

24. Treat with aniline just before loading the gel (see below).

PROTOCOL

MODIFICATION OF CYTOSINE

1. In a microfuge tube, mix:
   2 μl of tRNA (5 mg/ml)
   5 μl of [^{32}P] 3′-labeled RNA (about 10^5 cpm)

2. Lyophilize until completely dry.

3. Meanwhile, prepare a 3 M NaCl solution in anhydrous hydrazine. Keep on ice.

4. Add 10 μl of the NaCl/hydrazine solution from step 3 to the dry pellet prepared in step 1.

5. Mix; centrifuge quickly at 12,000 rpm for a few seconds.

6. Incubate on ice for 30 minutes.

7. Add 1 ml of cold 88% ethanol and mix to stop the reaction.

8. Incubate at $-70°C$ for 10 minutes and centrifuge at 12,000 rpm for 10 minutes (at 4°C).

9. Discard supernatant in 2 M FeCl3 and add 200 μl of cold 0.3 M sodium acetate to dissolve the RNA precipitate by vortexing.

10. Add 600 μl of cold absolute ethanol, mix, and incubate at $-70°C$ for 10 minutes.

11. Centrifuge at 12,000 rpm for 10 minutes (4°C). Discard supernatant in 2 M FeCl3.

12. Rinse the RNA pellet with 500 μl of cold absolute ethanol, without resuspending it.

13. Centrifuge at 12,000 rpm for 5 minutes (4°C). Discard the supernatant.

14. Dry the pellet in a vacuum centrifuge, and treat the sample with aniline just before loading onto the gel.

PROTOCOL

MODIFICATION OF URACIL

1. In a microfuge tube, mix:
   2 μl of tRNA
   5 μl of [^{32}P] 3'-labeled RNA (about 10^5 cpm)
   Lyophilize until completely dry.

2. Meanwhile, prepare a 50% v/v solution of hydrazine. Keep on ice.

3. Add 10 μl of the hydrazine solution to the dried RNA from step 1.

4. Incubate for 15 minutes on ice.

5. Stop the reaction by adding 200 μl of cold 0.3 M sodium acetate.

6. Add 750 µl of cold absolute ethanol, mix, and incubate at − 70°C for 10 minutes.

7. Centrifuge at 12,000 rpm for 10 minutes (4°C).

8. Discard the supernatant in 2 *M* FeCl3 and add 200 µl of cold 0.3 *M* sodium acetate to dissolve the RNA pellet by vortexing.

9. Add 600 µl of cold absolute ethanol, mix, and incubate for 10 minutes at − 70°C.

10. Centrifuge at 12,000 rpm for 10 minutes (4°C); discard the supernatant in 2 *M* FeCl3.

11. Rinse the RNA pellet with 500 µl of cold absolute ethanol without resuspending it.

12. Centrifuge at 12,000 rpm for 5 minutes (4°C) and discard the supernatant.

13. Dry the RNA pellet under vacuum (SpeedVac, Savant) and treat with aniline just before loading onto gel.

PROTOCOL

MODIFICATION OF ADENINE

1. In a siliconized microfuge tube on ice, mix:
   200 µl of 50 m*M* sodium acetate (pH 4.5) containing 1 m*M* EDTA
   2 µl of tRNA (5 mg/ml)
   5 µl of [^{32}P] 3′-labeled RNA (about 10^5 cpm)

2. Add 1 µl of diethylpyrocarbonate, mix, and centrifuge immediately for a few seconds.

3. Incubate for 10 minutes at 90°C.

4. Chill quickly on wet ice and add 50 µl of cold 1.5 *M* sodium acetate. Mix well.

5. Add 750 µl of cold absolute ethanol; mix by inverting several times.

6. Let the treated RNA precipitate at − 70°C for 10 minutes and centrifuge at 12,000 rpm for 5 minutes (4°C).

7. Discard the supernatant in a tube containing 10 *M* NaOH.

8. Add 200 µl of cold 0.3 *M* sodium acetate to dissolve the RNA pellet by vortexing.

9. Add 600 µl of cold absolute ethanol, mix, and incubate at − 70°C for 10 minutes.

10. Centrifuge at 12,000 rpm for 10 minutes (4°C) and discard supernatant in 10 M NaOH.

11. Rinse the pellet without resuspending it, by adding 500 μl of cold ethanol to the tube. Centrifuge at 12,000 rpm for 2 minutes (at 4°C).

12. Remove the ethanol and dry the pellet in a vacuum centrifuge (SpeedVac, Savant).

13. Treat with aniline just before loading onto gel (see below).

PROTOCOL

CLEAVAGE OF MODIFIED RNA BY ANILINE

The cleavage of the modified RNA by aniline is performed on the dried samples obtained as described above.

1. To each dried pellet, add 20 μl of 1 M aniline acetate buffer. Vortex and centrifuge for a few seconds at 12,000 rpm.

2. Incubate at 60°C in the dark for 20 minutes.

3. Evaporate the solutions in a vacuum centrifuge.

4. Resuspend the pellets in 20 μl of sterile distilled water.

5. Freeze the samples and lyophilize until a dry pellet is obtained.

6. Repeat steps 4 and 5.

7. Resuspend the RNA samples in 5 μl of loading buffer.

8. Melt the RNA samples for 30 seconds at 90°C just prior to loading and chill them quickly in iced water.

Buffers and Solutions

10X Ligase buffer
0.5 M Hepes–NaOH (pH 7.5)
0.1 M MgCl$_2$
33 mM DTT

Acetate buffer (pH 7.5)
1 M Tris–acetate (pH 7.5)
1 M 2-Mercaptoethanol
1.5 M Sodium acetate

Aniline–acetate buffer
0.5 ml Glacial acetic acid
7 ml H_2O
1 ml Redistilled aniline
Adjust to pH 4.5 with 1 *M* acetic
    acid.
Add H_2O to 11 ml.
Make fresh before use.
    Keep on ice.

Loading buffer
1X TBE buffer
7 *M* Urea
0.025% Xylene cyanol
0.025% Bromophenol blue

Notes

1. See notes for base-specific cleavage of DNA (pages 584, 585).
2. The sequencing gels should contain high concentrations of urea (8.3 *M*) to ensure maximal denaturation of the RNA molecules.
3. Running the gels at 50 to 60 watts constant power results in heating up to about 55 to 60°C. Electrophoresis at this temperature enhances the denaturing power of sequencing gels.

ENZYMATIC SEQUENCING OF RNA

The strategy used to determine RNA sequences by the enzymatic method is somewhat comparable to the base-specific chemical cleavage described in Chapter 20 for sequencing of DNA and above for RNA sequencing, except that base-specific ribonucleases are used instead of chemical reagents. Substrate RNA is usually labeled to high specificity with T4 polynucleotide kinase and $\gamma[^{32}P]$ATP. The T1, U2, PhyM, and *Bacillus-cereus* ribonucleases are the most appropriate for RNA sequencing.

RNases T1 and U2 cleave next to G and A, respectively, and are therefore used as guanine and adenine specific enzymes. PhyM ribonuclease hydrolyzes phosphodiester bonds adjacent to A and U residues and *Bacillus cereus* nuclease is specific for pyrimidines.

It is also possible to use RNase CL3 (from chicken liver) which has been reported to be essentially cytosine specific, although it can cleave at a uracil residue placed 5' to adenine, and RNase PhyI isolated from the culture medium of *Physarum polycephalum* which cuts anywhere, except at cytosine.

Labeling of the 5' Terminus of RNA with T4 Polynucleotide Kinase. Since 5' hydroxyl end are needed for this reaction, it is necessary to dephosphorylate the RNA molecules with alkaline phosphatase, as described in Chapter 14. In some cases, when natural RNAs are used as substrates for sequencing, it may be necessary to remove the 5'-cap structure with tobacco acid pyrophosphatase (see Chapter 4), prior to dephosphorylation. See protocol below.

PROTOCOL

LABELING OF THE 5' TERMINUS OF RNA WITH T4 POLYNUCLEOTIDE KINASE

1. In a microfuge tube, put 250 μCi of γ[^{32}P] ATP (>5000 Ci/mmol, 185 TBq/mmol) and dry it down in a vacuum centrifuge (SpeedVac, Savant).

2. Add to the dried labeled ATP:
   7 μl of sterile distilled water
   3 μl of 10X kinase buffer
   5 μl of 10 μM ATP
   10 μl of dephosphoylated RNA (about 25 pmol)
   Vortex to mix.

3. Add 5 μl of T4 polynucleotide kinase (diluted to 1 unit/μl in TE buffer containing 25% glycerol), mix by vortexing, and centrifuge a few seconds at 12,000 rpm.

4. Incubate at 37°C for 30 minutes.

5. Add 30 μl of 4 M ammonium acetate to stop the reaction.

6. Add 10 μl of carrier tRNA (1 mg/ml in H$_2$O).

7. Add 200 μl of cold absolute ethanol, mix, and incubate at −70°C for 10 minutes.

8. Centrifuge at 12,000 rpm for 10 minutes and discard the supernatant.

9. Resuspend the pellet in 100 μl of 0.5 M sodium acetate (pH 5.0) and vortex to dissolve.

10. Add 300 μl of cold absolute ethanol, mix, and incubate at −70°C for 10 minutes.

11. Centrifuge at 12,000 rpm for 5 minutes. Discard the supernatant and rinse the pellet quickly with 500 μl of cold absolute ethanol without resuspending it.

12. Remove the supernatant and dry the pellet in a vacuum centrifuge (SpeedVac, Savant).

13. Resuspend pellet to purify labeled RNA by polyacrylamide gel electrophoresis (see Chapter 11).

Notes

1. NH4$^+$ ions are potent inhibitor to T4 polynucleotide kinase.

2. Add tRNA *after* addition of ammonium acetate to avoid labeling of carrier.

Procedures for Enzymatic Sequencing

Enzymatic sequencing reactions are performed with about 2–5 10⁵ cpm of ³²P-labeled RNA (see above). The ratios of RNA to specific nucleases are determined empirically to obtain optimal digestions. Usually, serial dilutions of the enzyme solutions are prepared in the corresponding incubation buffers, just before use. Typical reactions are conducted as shown in the following protocol.

PROTOCOL

ENZYMATIC SEQUENCING OF RNA SPECIES

1. Prepare serial 1:10 dilutions of the specific nucleases being used by mixing 1 µl of the stock solutions in 9 µl of buffer, then 1 µl of the 1/10th dilution in 9 µl of buffer, and so on. Be sure to change tips for each dilution.

2. For each of the four enzyme dilutions obtained in step 1, prepare four microfuge tubes (1 to 4) according to the following chart. Also run a control experiment for each incubation condition (tubes 5 to 7).

	Buffers			Enzymes (diluted)	³²P-RNA	H₂0/tRNA	Urea/Dyes
	I	II	III				
	(2 µl)			(1 µl)	(1 µl)		
1	+	−	−	A/T1/PhyM	+	2 µl	14 µl
2	+	−	−	*B. cereus*	+	16 µl	−
3	−	+	−	U2	+	2 µl	14 µl
4	−	−	+	CL3/Phy 1	+	16 µl	−
5	+	−	−	−	+	3 µl	14 µl
6	−	+	−	−	+	3 µl	14 µl
7	−	−	+	−	+	17 µl	−

3. The 1 µl of each enzyme dilution should be added after all other components.

4. Mix by vortexing a few seconds and incubate for 15 minutes at 37°C (for RNases CL3 and Phy I), or at 55°C (for RNases A, T1, U2, PhyM, and *B. cereus*).

5. Add 5 µl of the urea/dye mix in the tubes containing RNases CL3, Phy I, and *B. cereus*.

6. Freeze all tubes at − 70°C until sequencing gel is performed.

RNA Sequencing Gels

These gels are performed under similar conditions as those described in Chapter 20. The few points to bear in mind are the following:

1. Keeping a high concentration of urea in the gel (8.3 *M*) is essential to avoid renaturation of RNA.
2. Denaturation of the RNA is also favored by running the gels at high voltages to generate heat.
3. Preelectrophorese the gels for about 30 minutes at a constant power of 20–25 watts (approximately 1000 to 1500 volts).
4. Heat the samples in a boiling water bath for 30 seconds just before loading.
5. Rinse the wells of the gels carefully to remove urea.
6. Run the gel at about 50 watts. Under such conditions, the temperature in the gel should rise up to 55–60°C.

Use of Reverse Transcriptase for RNA Sequencing

The protocol presented below has been adapted from a procedure kindly provided by I. Seif. It can be used for sequencing RNA species synthesized in vitro by the RNA polymerase from SP6, T7, and T3 bacteriophages, or RNA strands directly isolated from virions.

PROTOCOL

USE OF REVERSE TRANSCRIPTASE FOR RNA SEQUENCING

1. In a microfuge tube, mix:
   4 μl of sterile distilled water
   8 μl of 5X reverse transcriptase buffer
   2 μl of 1 m*M* dATP
   2 μl of 1 m*M* dGTP
   2 μl of 1 m*M* dCTP
   2 μl of 1 m*M* TTP
   4 μl (40 μCi) of [^{32}P] dATP (3000 Ci/mmol, 100 TBq/mmol) or [^{35}S] dATP (20 μCi)
   Keep on ice until used (at step 7).

2. In a microfuge tube containing 1 µg of RNA add:

   x ng of synthetic primer (so as to be in a 1:2 molar ratio of RNA over primer)

   sterile distilled water up to 10 µl

3. Add 1 µl of diluted methyl mercury hydroxyde (2 µl of the 1 M solution in 38 µl of water).

4. Incubate for 10 minutes at 20°C.

5. Add 2 µl of mercaptoethanol solution (10 µl of 14 M 2-mercaptoethanol in 390 µl of water).

6. Incubate for 5 minutes at 20°C.

7. Add the treated-RNA sample to the microfuge tube containing mixture prepared in step 2.

8. Add 2 µl of reverse transcriptase (Life Science, Florida).

9. Mix, and add 9 µl of this mixture in each of the four tubes labeled (G), (A), (T), (C) and containing:

   (G) 2 µl of 0.05 mM ddGTP (2 µl of 1 mM in 38 µl of H_2O)

   (A) 2 µl of 0.05 mM ddATP (2 µl of 1 mM in 38 µl of H_2O)

   (C) 2 µl of 0.05 mM ddCTP (2 µl of 1 mM in 38 µl of H_2O)

   (T) 2 µl of 0.05 mM ddTTP (2 µl of 1 mM in 38 µl of H_2O)

10. Mix and centrifuge at 12,000 rpm for a few seconds.

11. Incubate for 15 minutes at 42°C.

12. Add 1 µl of the dXTP mixture used as chase solution.

13. Add 5 µl of stop solution.

14. Heat at 90°C for 10 minutes and load on a sequencing gel.

Buffers and Solutions

5X Reverse transcriptase buffer   *100 µl*

300 mM Tris–HCl (pH 9.0)   30 µl 1 M Tris–HCl (pH 9.0)

375 mM NaCl   37.5 µl 1 M NaCl

37.5 mM MgCl$_2$ 3.75 µl 1 M MgCl$_2$

25 mM DTT   2.5 µl 1 M DTT

      26.25 µl Sterile distilled water

dXTP mixture

0.25 mM dATP

0.25 mM dCTP

0.25 mM dGTP

0.25 mM dTTP

Stop solution
0.3% Xylene cyanol FF
0.3% Bromophenol blue
0.37% EDTA (disodium) (pH 7.0)
in deionized formamide

Notes

1. Although this method works fairly well for relatively small RNA species, bear in mind that reverse transcriptase does not give rise to 100% of full-length cDNA transcripts (see Chapter 19), and that sequencing of long RNA molecules may not be possible under such conditions.

2. When using this technique for RNA sequencing, be aware that some experimental conditions may increase the error rate of reverse transcriptase during DNA synthesis.

The accuracy of in vitro DNA synthesis by reverse transcriptase (RT) has been studied with homopolymer templates such as poly$(A)_n \cdot (dT)_{12-18}$ or poly$(C)_n \cdot (dG)_{12-18}$. The results obtained with avian myeloblastosis virus RT, and Rauscher murine leukemia virus (R-MLV) RT, have shown that these enzymes incorporated one noncomplementary nucleotide for 600 bases of complementary strand synthesized (Battula and Loeb, 1974; Sirover and Loeb, 1976). Similar error rates were observed with natural single stranded RNA templates, and incorrectly incorporated nucleotides were found throughout the newly synthesized copy (Gopinathan et al., 1979; Battula and Loeb, 1975). Substitution of Mn^{2+} to Mg^{2+} increased the error rate dramatically (Mizutani and Temin, 1976; Sirover and Loeb, 1976, 1977). Studies performed on the BRL-cloned Moloney murine leukemia virus RT and AMV RT have led to similar results (Table 70) suggesting that accuracy of DNA synthesis by these enzymes is such that one in every two cDNA copies of a 2-kb mRNA of medium abundance will contain a single error (Gerard, 1986).

However, it should be noted that the error fre-

Table 70. The Level of Infidelity for AMV and Cloned M-MLV Reverse Transcriptase

Enzyme	Template–Primer	Complementary Nucleotide	Noncomplementary Nucleotide	Complementary Nucleotide Incorporated (pmol)	Noncomplementary Nucleotide Incorporated (pmol)	Error Rate
M-MLV	$(A)_n \cdot (dT)_{12-18}$	dTTP	dCTP	152.7	0.044	1/3,470
AMV	$(A)_n \cdot (dT)_{12-18}$	dTTP	dCTP	569.8	0.117	1/4,870
M-MLV	$(C)_n \cdot (dG)_{12-18}$	dGTP	dATP	442.3	0.060	1/7,372
AMV	$(C)_n \cdot (dG)_{12-18}$	dGTP	dATP	616.6	0.114	1/5,409

Courtesy of Bethesda Research Laboratories, Life Technologies Inc.

Reaction mixtures (10 μl) contained 20 mM Tris–HCl (pH 8.3), 75 mM KCl, 5 mM MgCl$_2$, 10 mM dithiothreitol, 100 μg/ml BSA, and either 100 μM (A)$_n$ 40 μM (dT)$_{12-18}$, 250 μM [^{3}H]dTTP (53 cpm/pmol), and 250 μM [α-^{32}P]dCTP (9000 cpm/pmol), or 100 μM (C)$_n$, 40 μM (dG)$_{12-18}$, 250 μM [^{3}H]dGTP (35 cpm/pmol), and 250 μM [α-^{32}P]dATP (6175 cpm/pmol). Assays contained 50 units of cloned M-MLV or 20 units of AMV reverse transcriptase as indicated and were incubated at 37°C for 30 minutes. Acid insoluble product was washed and recovered on GF/C filters. Background counts from nonincubated assay tubes (about 75 cpm for ^{3}H and 50 cpm for ^{33}P) were subtracted from the counts incorporated before incorporation was calculated.

quency of RT (1/600) is quite high as compared to the reported values (1/3000 − 1/30,000) for mammalian DNA polymerases (Seal et al., 1979; Krauss and Linn, 1980; Kunkel and Loeb, 1981).

REFERENCES

Battula, N., and Loeb, L. A. (1974), *J. Biol. Chem.,* **249,** 4086.

Battula, N., and Loeb, L. A. (1975), *J. Biol. Chem.,* **250,** 4405.

Gerard, G. F. (1986), *Focus,* **8:3,** 12.

Gopinathan, K. P., Weymouth, L. A., Kunkel, T. A., and Loeb, L. A. (1979), *Nature,* **278,** 857.

Krauss, S. W., and Linn, S. (1980), *Biochemistry,* **19,** 220.

Kunkel, T. A., and Loeb, L. A. (1981), *Science,* **231,** 765.

Mizutani, S., and Temin, H. M. (1976), *Biochemistry,* **15,** 1510.

Seal, G., Shearman, C. W., and Loeb, L. A. (1979), *J. Biol. Chem.,* **254,** 5229.

Sirover, M. A., and Loeb, L. A. (1976), *Biochem. Biophys. Res. Commun.,* **70,** 812.

Sirover, M. A., and Loeb, L. A. (1977), *J. Biol. Chem.,* **252,** 3605.

22

USE OF COMPUTERIZED PROGRAMS IN THE TREATMENT OF DNA SEQUENCE INFORMATION

Establishing the nucleotide sequence of a cloned DNA fragment is only the primary step toward a better knowledge of the basic mechanisms governing gene expression.

DNA sequencing is usually performed (1) to gain more insight about the structure and origin of given genes, (2) to search for consensus sequences acting as specific regulatory signals involved in the control of gene expression, and (3) to characterize the expression products of genes whose physiological roles are totally unknown.

For the past few years, many different programs have been developed to provide help in the search for given sets of nucleotides (such as restriction enzymes sites, regulatory signals, etc.) or in the comparison of different information to look for homology between sequences. However, the rapid development of new sequencing techniques has led to such an amount of sequence data that it is almost impossible nowadays to store and manipulate this information on the midrange microcomputers which are often being used in most laboratories. It is therefore highly recommended to be connected with one or several data banks that provide "up to date" releases of published sequences. There are nucleotide data banks and protein data banks. For example, *EMBL* Nucleotide Sequence Data Library and *GENBANK* contained, respectively, about 8800 and 10900 nucleotide sequences analyzed in their last version while the *NBRF-NIH* National Biomedical Research Foundation offers more than 4000 protein sequences.

The benefit gained in using the batch of information available in these banks is obvious, and comparison of sequence data has proven to be an invaluable tool in the search for putative gene functions. One of the most striking examples in this respect is the high degree of homology which has been found to exist between a growth factor (PDGF) and the predicted transforming (Waterfield et al., 1983, Nature **304**, 35–39 Doolittle et al., 1983, Science *221*, 275–276) protein of the simian sarcoma virus oncogene (v-sis), a discovery that has opened new roads in the fields of cancer research and cellular biology. Since it is not our aim to review in great detail all the different programs that have been used or being developed, the reader interested in this field should consult the bibliography listed at the end of this chapter. Because the amount of information stored in these banks has been constantly expanding over the past years, it is not realistic to extract the whole content of a bank to search manually for homologies between a given sequence and those contained in the banks. Several sophisticated programs have been developed to provide help in this matter. The programs that are described here represent some typical examples of the applications permitted by the use of computers in the manipulation of nucleotide sequences. They have been set up by M. M. Mugnier and Fondrat in the Centre Inter-universitaire de Traitement de l'Information (C.I.T.I. 2) in Paris, and have been made available to us by Dr. P. Le Beux who is in charge of this center.

```
          10        20        30        40        50        60
  1  GCAGCAGCGCTACTGAACTATGTCAATAAGTGTCTCGCTCAGGTCGTCTAAAGGGAGCGC
  C  GCAGCAGCGCTACTGAACTATGTCAATAAGTGTCTCGCTCAGGTCGTCTAAAGGGAGCGC

          70        80        90       100       110       120
  1  GAAGTGTGCACAATCGGTTCATGGTGGCTGCCGTGGACGAGACGTGAGTTAATATGCTAT
  2                                         GACGTGAGTTAATATGCTAT
  C  GAAGTGTGCACAATCGGTTCATGGTGGCTGCCGTGGACGAGACGTGAGTTAATATGCTAT

         130       140       150       160       170       180
  1  TGGACAGCCTACACCTTGGTCTTCGGCCAT
  2  CGGACAGCCTACACCTTGGTCTTCGGCCATTATGTTTCCATTTGCTGGCTAAGCATCCGT
  C  -GGACAGCCTACACCTTGGTCTTCGGCCATTATGTTTCCATTTGCTGGCTAAGCATCCGT

         190       200       210       220       230       240
  2  ACATTGAATAAACACGGACGATGGCACGATACACCCTCCTCTAGAGTGCTGAGGCCCGAG
         ATGGCACGATACACCCTCCTCTAGAGTGCTGAGGCCCGAG
  C  ACATTGAATAAACACGGACGATGGCACGAT--ACCCTCCTCTAGAGTGCTGAGGCCCGAG

         250       260       270       280       290       300
  2  AGCCGTTTTG
  3  AGCCGTTTTGCCAATAGAACAGAGGACCAGTTCGACTTCCCGCCTGTATTCCCCTATGGG
  C  AGCCGTTTTGCCAATAGAACAGAGGACCAGTTCGACTTCCCGCCTGTATTCCCCTATGGG
```

Figure 189. Example of computer-assisted shotgun sequencing. In this example, three sequences have been joined to each other by means of two overlapping sets of nucleotides (positions 100 to 150, and positions 200 to 250). The resulting sequence is shown on line C. A mismatch for two bases has been detected at positions 211–212.

SEQUENCING PROGRAMS FOR SHOTGUN SEQUENCING

Ordering the sequenced DNA fragments in a linear (or circular) map is not an easy task when the length of the fragment increases. As mentioned in the previous chapter, sequencing is usually performed on DNA fragments whose size is in the range of 300 to 600 base pairs, and ordering of the sequence stretches requires searching for overlapping regions. This tedious process is facilitated by the use of programs such as those described in this section.

We use basically two kinds of programs for shotgun sequencing (Staden, 1984). In the *manual sequencing programs,* the user is in charge of determining whether the sequences belong to a unique DNA region, while in the *automatic sequencing program,* relationships between sequences are automatically tested; all gels that do not fit in a given contiguity are rejected. In summary, the programs run as follows:

Manual Shotgun Sequencing

1. Initialization of a project (P) which corresponds to only one set of contiguous sequences (i.e, a given DNA fragment).
2. Read one autoradiogram and store sequence no. 1.
3. Define a consensus overlapping sequence no. 1.
4. Read new autoradiograms, store sequences.
5. Compare these sequences with consensus no. 1.
6. Enter into project P the sequences that contain consensus sequence no. 1.
7. Define a consensus sequence no. 2.
8. Read new autoradiograms.
9. Compare remaining gels with consensus sequence no. 2 and repeat process until all gels are being analyzed. Several optional functions are available for now editing, checking, and searching for particular stretches. It is also possible to scan the resulting sequence in order to detect mismatches in overlapping sequences that may result from errors introduced either at the level of the gel reading or at the level of typing, when entering the sequence in the computer. A typical example is shown in Figure 189.

Automatic Shotgun Sequencing

This program takes a batch of gels and enters them into a database for a sequencing project. It takes each gel in turn, and compares it with the current consensus for the database. It then produces an alignment for any regions of the consensus it overlaps; if this alignment is sufficiently good it then edits both the new gel reading and the gels it overlaps and adds the new gel to the database. The program then updates the consensus accordingly and carries on to the next gel reading. The user controls the editing performed by the program by specifying three parameters: (a) the maximum number of padding characters that the alignment routines may enter into the contigs for each gel reading, (b) the maximum number of padding characters the alignment routines may place in the sequence of the new gel reading, and (c) the maximum percentage of mismatch allowed between the new gel and the consensus after alignment has been achieved. All alignments are displayed and any gels which do match but which cannot be aligned sufficiently well have their names written at a file of failed gel reading names. The program works without any user intervention and can process any number of gel readings in a single run. The sequencer can periodically examine the aligned sequences to see if the alignments made by the program are correct and perform any deletions or rearrangements that are necessary.

The only operator input required is the project name, a batch of gel reading files, and a few values that define the limits in which the program must work. These limits will include such things as minimum overlap to define relatedness, maximum number of padding characters allowed during editing, and maximum amount of mismatch allowed after alignment.

Reference

Staden, R. (1984), *Nucl. Acids Res.* **12**, 499.

SEARCH FOR OCCURRENCES OF NUCLEOTIDES AND BASE SEQUENCES IN DNA

Once the primary structure of nucleic acids has been established, it becomes possible to obtain readily a whole batch of invaluable information concerning (1) the precise determination of bases and codon content of a given stretch of DNA, (2) the location of regulatory signals and restriction endonuclease recognition sites, (3) the determination of secondary structures in single stranded DNA and RNA molecules, (4) the presence of open reading frames and the nature of the corresponding polypeptides, and (5) the possible existence of homologous sequences or peptides already described in the literature.

Base Frequencies

In practice, the calculation of the base frequency in a cloned DNA fragment may be useful, for example, to determine the melting temperature of this fragment when it is used in hybridization studies (see Chapter 18), to predict the behavior of the corresponding single strands in nondenaturing polyacrylamide gels (see Chapter 9), or to adjust cesium chloride concentrations in preparative and analytical centrifugations (see Chapter 7). The codon frequencies for one to three reading frames can be expressed, either as the number, or as the percentage of triplets specifying each amino acid over the whole length of the considered DNA [see Table 71 (a) and (b)].

The programs that we currently use allow us to search for sequences of different length (up to 2000 bases long) simultaneously. They are mostly used to study DNA fragments for their content in restriction sites, and consensus sequences corresponding to regulatory elements such as transcription promoters, termination signals, enhancers, splice donnors and acceptors, ribosomal binding sites, and so on.

Restriction Endonuclease Maps

Many different programs are available to establish the position of restriction sites on DNA and to provide accurate restriction maps. The position of the different sites may be given in the form of a table or as a map (see, for examples, restriction maps provided with this manual).

These programs also appear to be very useful in order to predict the size of the DNA fragments that will be generated following either a complete or a partial digestion with different combinations of restriction endonucleases. Such studies can be performed with personal files or directly from a data bank, as described below.

In this example, search was performed in the λ-phage sequence stored in the EMBL data bank.

identification of the sequence: LAMBDA
first base: 1
last base: 48502
name of enzyme: Bam HI
    Information obtained:

Enzyme	*site*	*number of fragments*
Bam HI	GGATCC	6

Position of cuts in DNA
Bam HI 5505, 22346, 27972, 34499, 41732

Table 71. Two Ways of Expressing the Frequency of Codons Counted in a Given Sequence

Number of Codons[a]

F	TTT	6.0	S	TCT	1.0	Y	TAT	0.0	C	TGT	0.0
F	TTC	3.0	S	TCC	5.0	Y	TAC	0.0	C	TGC	5.0
L	TTA	0.0	S	TCA	0.0	*	TAA	0.0	*	TGA	0.0
L	TTG	0.0	S	TCG	4.0	*	TAG	0.0	W	TGG	0.0
L	CTT	0.0	P	CCT	0.0	H	CAT	0.0	R	CGT	0.0
L	CTC	4.0	P	CCC	0.0	H	CAC	0.0	R	CGC	0.0
L	CTA	1.0	P	CCA	2.0	Q	CAA	0.0	R	CGA	2.0
L	CTG	0.0	P	CCG	0.0	Q	CAG	3.0	R	CGG	0.0
I	ATT	4.0	T	ACT	2.0	N	AAT	0.0	S	AGT	1.0
I	ATC	2.0	T	ACC	0.0	N	AAC	2.0	S	AGC	0.0
I	ATA	0.0	T	ACA	1.0	K	AAA	0.0	R	AGA	0.0
M	ATG	1.0	T	ACG	5.0	K	AAG	4.0	R	AGG	3.0
V	GTT	0.0	A	GCT	0.0	D	GAT	0.0	G	GGT	0.0
V	GTC	1.0	A	GCC	0.0	D	GAC	0.0	G	GGC	1.0
V	GTA	0.0	A	GCA	0.0	E	GAA	0.0	G	GGA	0.0
V	GTG	0.0	A	GCG	2.0	E	GAG	0.0	G	GGG	1.0

Total codon count: 66.

Percentage of Codons[b]

F	TTT	72.0	S	TCT	2.5	Y	TAT	0.0	C	TGT	0.0
F	TTC	28.0	S	TCC	40.0	Y	TAC	0.0	C	TGC	100.0
L	TTA	20.8	S	TCA	0.0	*	TAA	0.0	*	TGA	100.0
L	TTG	0.0	S	TCG	32.5	*	TAG	0.0	W	TGG	100.0
L	CTT	0.0	P	CCT	5.0	H	CAT	0.0	R	CGT	0.0
L	CTC	41.7	P	CCC	0.0	H	CAC	100.0	R	CGC	0.0
L	CTA	33.3	P	CCA	95.0	Q	CAA	0.0	R	CGA	49.1
L	CTG	4.2	P	CCG	0.0	Q	CAG	100.0	R	CGG	1.8
I	ATT	52.6	T	ACT	18.5	N	AAT	0.0	S	AGT	25.0
I	ATC	35.1	T	ACC	3.7	N	AAC	100.0	S	AGC	0.0
I	ATA	12.3	T	ACA	29.6	K	AAA	5.0	R	AGA	5.5
M	ATG	100.0	T	ACG	48.1	K	AAG	95.0	R	AGG	43.6
V	GTT	0.0	A	GCT	0.0	D	GAT	0.0	G	GGT	0.0
V	GTC	80.0	A	GCC	30.0	D	GAC	100.0	G	GGC	92.9
V	GTA	0.0	A	GCA	0.0	E	GAA	50.0	G	GGA	0.0
V	GTG	20.0	A	GCG	70.0	E	GAG	50.0	G	GGG	7.1

Total codon count: 2000

[a]*Number:* among the four possible codons specifying glycine (6), codons GGC and GGG are found once among the total of 66 triplets counted.

[b]*Percentage:* over a total count of 2000 codons, valine-specific triplets were GTC (80%) and GTG (20%); no GTT or GTA codons were detected.

Ordering of restriction fragments by position

Restriction Fragments	Positions	Length
1. Left end–Bam HI	1–5505	5505
2. Bam HI–Bam HI	5506–22346	16841
3. Bam HI–Bam HI	22347–27972	5626
4. Bam HI–Bam HI	27973–34499	6527
5. Bam HI–Bam HI	34500–41732	7233
6. Bam HI–Right end	41733–48502	6770

Ordering of restriction fragments by size

Restriction Fragments	Positions	Length
1. Left end–Bam HI	1–5505	5505
2. Bam HI–Bam HI	22347–27972	5626
3. Bam HI–Bam HI	27973–34499	6527
4. Bam HI–right end	41733–48502	6770
5. Bam HI–Bam HI	34500–41732	7233
6. Bam HI–Bam HI	5506–23346	16841

In a second step, a Bam HI partial digestion was simulated.

name of enzymes: Bam HI, Eco RI, Hind III

Enzyme	site	number of fragments
Bam HI	GGATCC	6
Eco RI	GAATTC	6
Hind III	AAGCTT	8

Position of cuts
Bam HI 5505, 2246, 27972, 34499, 41732
Eco RI 21226, 26104, 31747, 39168, 44972
Hind III 23130, 25157, 27479, 36895, 37459, 37584, 44141

Restriction fragments generated after Bam HI partial digestion (on a circular map)

	Enzymes	Position	Length
1.	Bam HI–Eco RI	5506–21226	15721
2.	Eco RI–Hind III	21227–23130	1904*
3.	Eco RI–Bam HI	21227–22346	1120
4.	Bam HI–Hind III	22347–23130	784
5.	Hind III–Hind III	23131–25157	2027
6.	Hind III–Eco RI	25158–26104	947
7.	Eco RI–Hind III	26105–27479	1375
8.	Hind III–Eco RI	27480–31747	4268*
9.	Hind III–Bam HI	27480–27972	493
10.	Bam HI–Eco RI	27973–31747	3775
11.	Eco RI–Hind III	31748–36895	5148*
12.	Eco RI–Bam HI	31748–34499	2752
13.	Bam HI–Hind III	34500–36895	2396
14.	Hind III–Hind III	36896–37459	564
15.	Hind III–Hind III	37460–37584	125
16.	Hind III–Eco RI	37585–39168	1584
17.	Eco RI–Hind III	39169–44141	4973*
18.	Eco RI–Bam HI	39169–41732	2564
19.	Bam HI–Hind III	41733–44141	2409
20.	Hind III–Eco RI	44142–44972	831
21.	Eco RI–Bam HI	44973–5505	9035

DNA fragments resulting from the Bam HI partial digestion are marked with an asterisk (*). See also Figure 190 for the relative position of the fragments on the map.

Consensus Sequences

Each consensus sequence can be represented by a matrix in which values are given to each nucleotide as a function of its position with respect to the consensus.

Schematically, a TATA box would be defined by the following matrix in which a T at position 1 is given a 10 value, a A at position 2 is given a 10 value, while a T at position 2 is given a 0 value. The maximum score of 40 is obtained for T(1) A(2) T(3) A(4).

Position	1	2	3	4
T	10	0	10	0
C	0	0	0	0
A	0	10	0	10
G	0	0	0	0

If the AGCTATC sequence is compared to the TATA matrix, one can obtain several values whose

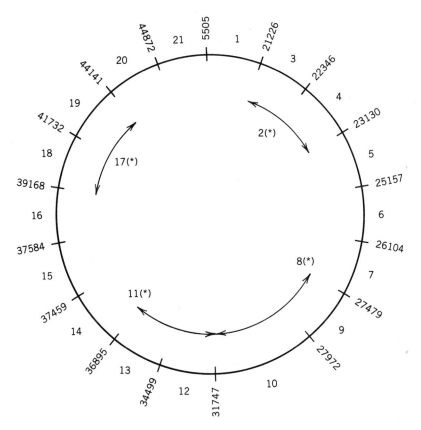

Figure 190. Predicted position of DNA fragments resulting from a computer-simulated Bam HI partial digestion.

maximum is 30 from position 4 in the considered stretch (T = 10, A = 10, T = 10, C = 0). Several matrices have been established for different consensus signals (see Table 72).

DETERMINATION OF SECONDARY STRUCTURES

A very useful program for finding all possible single stranded hairpin loop structures in a nucleic acid sequence has been described by Kanehisa and Goad (1982).

The input parameters include:

The maximum free energy for each structure to be printed.
The maximum size of a hairpin loop.
The maximum size of an internal loop.
The maximum size of a bugle loop.

Each structure is accompanied by the number of base pairs and the free energy value in kCal/mol.

An example of the results obtained with this program is described in Figure 191.

IDENTIFICATION OF PUTATIVE GENE PRODUCTS

Computer-assisted translation of nucleic acid sequences into amino acids sequences has proved to be a unique tool in molecular biology since it is sometimes the only possible way of gaining insight into the nature of gene products suspected to play key roles in cellular biology.

Once open reading frames are identified, it becomes possible, for example, (1) to predict some biochemical properties of the putative polypeptides (secondary structure, isolectric point, hydrophobicity, molecular weight etc.), (2) to predict the number

Table 72. Some Examples of Matrix for Consensus Signals

3' Ends of Introns (Acceptors)

T	58	50	57	67	75	62	62	57	57	73	75	38	40	0	0	11	48	37
C	21	28	35	27	30	38	42	35	46	46	36	28	84	0	0	23	28	42
A	17	11	11	19	8	19	14	24	15	4	13	33	5	130	0	29	22	25
G	17	24	11	13	13	7	9	11	9	6	6	31	1	0	130	67	32	26

5' Ends of Introns (Donors)

T	28	10	18	17	0	139	9	16	7	87	30	36
C	42	60	16	8	0	0	3	13	3	17	28	40
A	42	56	89	12	0	0	86	94	12	23	53	33
G	27	13	16	102	139	0	41	16	117	12	25	27

Eukaryotic Ribosome Site

T	19	24	31	12	0	18	5	0	102	0
C	20	15	32	65	5	42	52	0	0	0
A	50	27	27	19	86	36	34	102	0	0
G	6	29	12	6	11	6	11	0	0	102

E. Coli Promoter (−35 Region)

T	41	33	32	25	34	22	35	35	42	27	32	42	47	14	92	94	11	19
C	22	27	18	29	20	14	20	12	22	23	16	25	10	43	7	6	11	18
A	28	38	30	37	35	56	42	42	37	42	39	18	25	26	2	6	2	72
G	16	11	29	19	21	18	13	21	9	19	24	26	29	29	11	6	88	3

T	15	37	46	34	38	48	34
C	60	8	25	23	23	17	20
A	26	50	26	34	25	26	31
G	11	17	15	21	26	21	27

E. Coli Promoter (−10 Region)

T	35	28	28	27	39	51	34	43	26	31	89	3	49	15	19	108	31	29	21
C	34	21	24	27	12	25	20	25	20	27	10	2	16	14	22	3	13	16	30
A	20	39	33	33	39	23	29	16	23	19	2	106	29	66	57	1	35	23	31
G	23	24	27	25	22	13	29	28	43	35	11	1	18	17	14	0	33	24	30

E. Coli Promoter (+1 Region)

T	16	22	2	42	27	23	20	25	27	15	16	29
C	29	49	4	25	25	13	18	22	17	17	16	17
A	20	9	45	16	24	25	28	24	24	32	35	26
G	21	8	37	5	12	27	22	17	20	24	21	16

and size of peptides resulting from a proteolytic digestion, (3) to locate possible antigenic domains of a protein and to prepare synthetic peptides for subsequent immunological characterization of the corresponding physiological gene product, or (4) to search in data banks for proteins with related structures.

Primary Structure, Molecular Weight

The primary structure of the polypeptide can be expressed with either the one or three letter amino acid abbreviation (see Table 73), for the three possible reading frames. From the amino acid composition one can determine the corresponding molecular weight, charge, and hydrophobicity.

Proteolytic Digestion

It is possible to predict, from the amino acid sequence, the position of several protease cuts. The results can be expressed as a map (Figure 192) or in a tabular form (Table 74). Peptides can be sorted by position, by weight, or by retention.

```
LOCALLY STABLE SECONDARY STRUCTURES IN ARG-TRNA

  1(16) -18.5
                    10            20
XC     GCCC      UUAGC    UCAGUUGGAU
::     ::::      ::::      :: ::::    A
GGACGCCGGGUGCUGAAUCUUCCAG CAACGAG
   50         40         30


  2(21) -30.8
          10        20                30
CGCCCUUAGCUCAGUUGGAUA GA    GCA    ACGACCUU
:::::: ::       :::: ::     ::    :::::    C
GCGGGA  CGU     CCUAAGCUUGGACGCCGGGUGCUGAAU
  70              60        50        40


  3(11) -24.3
           50
UGG    GCC   GCAGGUUC
:::    :::   :::::    G
ACCGCGCGGGACGUCCUAA
       70        60



SEQUENCE ARG-TRNA

        10        20        30        40        50        60        70
XCGCCCUUAGCUCAGUUGGAUAGAGCAACGACCUUCUAAGUCGUGGGCCGCAGGUUCGAAUCCUGCAGGG
          !!!!!!!!!        !!!!!!!!!!         !!!!!!!!!
          1                2                  3


        80
CGCGCCA
```

Figure 191. Search for stable secondary structures in DNA sequences. The Arg tARN sequence was used in this example (kindly provided by C. Mugnier).

Localization of Potential Antigenic Epitops

Hopp and Woods (1981) described a program designed to locate possible antigenic domains of proteins based on the amino acid sequence. Assuming that antigenic domains must be on the surface of a protein in order to be accessible to antibodies, hydrophilic regions of proteins are assessed as the most likely sites of antigenic determinants. In the algorithm they described, the average hydrophilicity of each segment of a specified length is plotted versus sequence position. Peaks in the plot indicate hydrophilic regions of the protein. The hydrophilicity is calculated from thermodynamic data that were adjusted to optimize the correspondence between the predicted sites and experimentally determined antigenic domains. The algorithm is similar to an algorithm previously described by Rose and Roy (1980).

Localization of Secondary Structures

Richard Feldman suggested that a similar algorithm could be employed to predict secondary structure-forming regions. In this method, the amino acid secondary structure-forming frequencies of Chou and Fasman (1978) are substituted for the hydrophilicity data of Hopp and Woods, and the average structure-forming propensities are plotted. Indeed, the same algorithm can be used in a variety of applications simply by substituting the appropriate amino acid scoring values in the algorithm.

To take advantage of this approach, program PRPLOT was developed. In this program, user-specified amino acid scoring data are averaged over each segment of a user-specified length (window size) and the average values are plotted versus sequence position. The average scores are plotted at the position corresponding to the middle of each segment. If an

Table 73.   One and Three- Letter Amino Acid Abbreviation

Abbreviations		Name	Molecular Weight
A	Ala	Alanine	71
D	Asp	Aspartic acid	115
F	Phe	Phenylalanine	147
H	His	Histidine	137
K	Lys	Lysine	128
M	Met	Methionine	131
P	Pro	Proline	97
R	Arg	Arginine	156
T	Thr	Threonine	101
W	Trp	Tryptophan	186
C	Cys	Cysteine	103
E	Glu	Glutamic acid	129
G	Gly	Glycine	57
I	Ile	Isoleucine	113
L	Leu	Leucine	113
N	Asn	Asparagine	114
Q	Gln	Glutamine	128
S	Ser	Serine	87
V	Val	Valine	99
Y	Tyr	Tyrosine	163
B	Asx	Asp or Asn (not distinguished)	—
Z	Glx	Glu or Gln (not distinguished)	—
X	X	Undetermined or atypical amino acid	

even window size is specified, the scores are plotted at the position following the midpoint. The program utilizes eight different sets of amino acid scoring values or alternatively allows users to devise their own scoring values. All of the data sets except those of Chou and Fasman, however, are different measures of hydrophilicity.

PRPLOT produces a plot consisting entirely of ASCII characters and is totally device independent. Although the plot may not be terribly elegant, it is easily interpreted and the device independence makes it available to a wide range of users who lack extensive computer equipment. Optionally, the output data may be displayed as a table and as a graph (see Figure 193). This allows users with their own plotters and software to capture the output and display it on their own plotting devices.

We must bear in mind, however, that these approaches may not be directly applicable to DNA regions encoding different spliced messages which in turn give rise to distinct genes products. This occurs, for example, in the early region of polyoma genome where three tumor antigens called large T, middle t, and small t [Schaffhausen et al., 1978; Ito et al., 1977a, b; Deffert and Walter, 1976; Simmons and Martin, 1978; Prives et al., 1977] are generated from a combination of the three different reading frames. In such cases, it is necessary to get precise information about the structure of the different RNA spe-

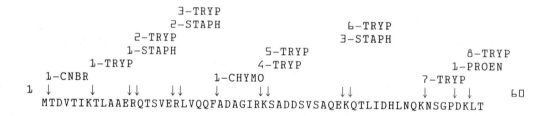

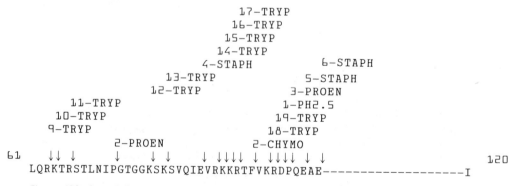

Figure 192. Search for positions of proteases and chemicals specific cleavage sites in proteins.

Table 74. Positions of Protease- and Chemical-Specific Cleavage Sites in Polypeptides

CNBR	NO: 1	POSITION: 1	STAPH	NO: 1	POSITION: 3
TRYP	NO: 1	POSITION: 8	PROEN	NO: 1	POSITION: 10
STAPH	NO: 2	POSITION: 11	TRYP	NO: 2	POSITION: 14
CNBR	NO: 2	POSITION: 16	CHYMO	NO: 1	POSITION: 24
PROEN	NO: 2	POSITION: 27	TRYP	NO: 3	POSITION: 30
TRYP	NO: 4	POSITION: 37	CHYMO	NO: 2	POSITION: 44
CNBR	NO: 3	POSITION: 45	STAPH	NO: 3	POSITION: 52
CNBR	NO: 4	POSITION: 55	PROEN	NO: 3	POSITION: 56
CHYMO	NO: 3	POSITION: 57	CHYMO	NO: 4	POSITION: 59
CHYMO	NO: 5	POSITION: 63	STAPH	NO: 4	POSITION: 64
STAPH	NO: 5	POSITION: 65	TRYP	NO: 5	POSITION: 67
STAPH	NO: 6	POSITION: 68	NTCB	NO: 1	POSITION: 70
STAPH	NO: 7	POSITION: 73	TRYP	NO: 6	POSITION: 76
TRYP	NO: 7	POSITION: 81	CHYMO	NO: 6	POSITION: 85
TRYP	NO: 8	POSITION: 86	TRYP	NO: 9	POSITION: 95
TRYP	NO: 10	POSITION: 96	TRYP	NO: 11	POSITION: 98

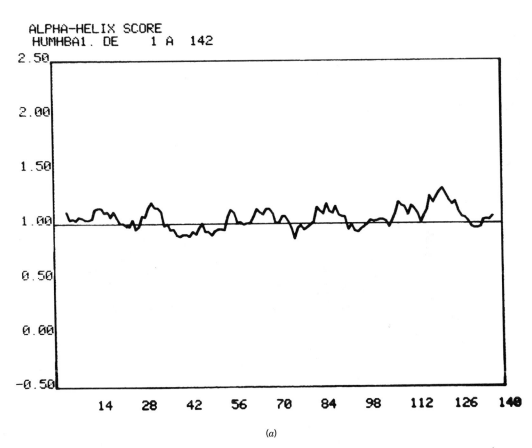

(a)

Figure 193. Search for secondary structures in proteins. (a) Graph for α helix portions in the 142 first amino acids of human hemoglobin (77% α helix total). (b) Graph for β sheet portions in the first 290 amino acids from concanavalin A (51% β sheet total). (c) Graph for hydrophobic regions (above 0.5 line) in the 300 first amino acids of bovine tyroglobulin. Note presence of the signal peptide at the N terminus. (d, e) Composite graph showing the positions of alpha helix (H) β sheet (E), turns (T), and coil (C) in human hemoglobin and concanavalin A, respectively (kindly provided by C. Mugnier).

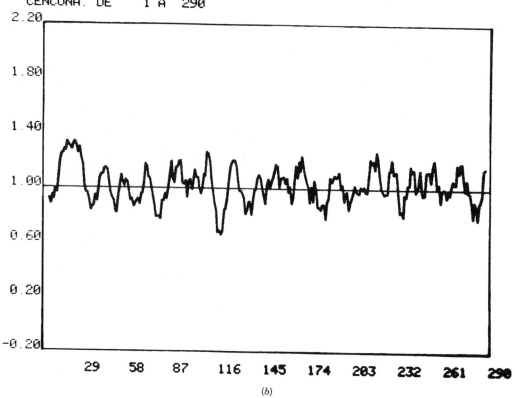

BETA-SHEET SCORE
CENCONA. DE   1 A   290

2.20
1.80
1.40
1.00
0.60
0.20
-0.20

29   58   87   116   145   174   203   232   261   290

(b)

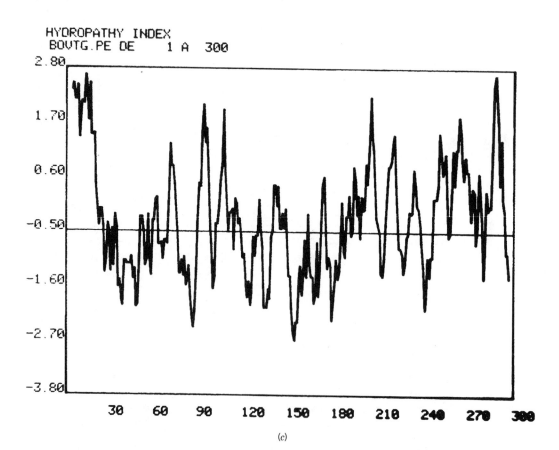

HYDROPATHY INDEX
BOVTG.PE DE   1 A   300

2.80
1.70
0.60
-0.50
-1.60
-2.70
-3.80

30   60   90   120   150   180   210   240   270   300

(c)

Figure 193 *(b, c)*

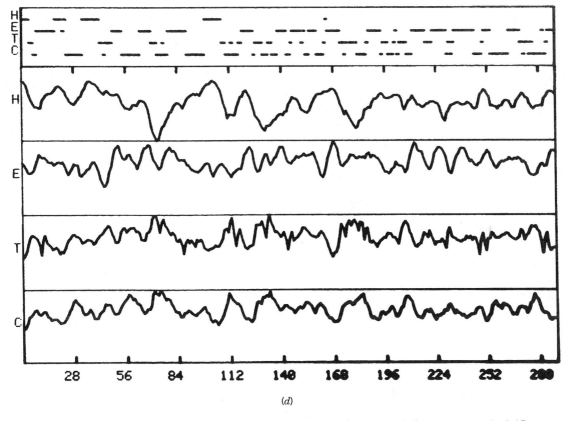

(d)

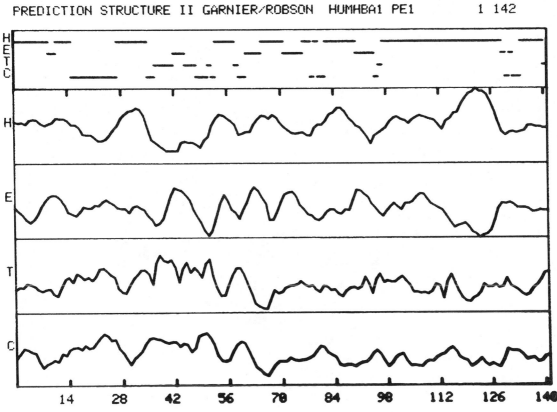

(e)

Figure 193 (d, e)

cies generated after transcription, or to know the position of potential splice donor and acceptor signals.

References

Chou, P. Y., and Fasman, G. D. (1978), *Ann. Rev. Biochem.,* **47**, 251.

Deppert, W., and Walter, G. (1976), *Proc. Natl. Acad. Sci. USA,* **73**, 2505.

Hopp, T. P., and Woods, K. R. (1981), *Proc. Nat. Acad. Sci. USA,* **78**, 3824.

Ito, Y., Brocklehurst, J. R., and Dulbecco, R. (1977a), *Proc. Natl. Acad. Sci. USA,* **74**, 4666.

Ito, Y., Spurr, N., and Dulbecco, R. (1977b), *Proc. Natl. Acad. Sci. USA,* **74**, 1259.

Kanehisa, M., and Goad, W. (1982), *Nucl. Acids Res.,* **10**, 265.

Prives, C., Gilboa, E., Revel, M., and Winocour, E. (1977), *Proc. Natl. Acad. Sci. USA,* **74**, 457.

Rose, G. D., and Roy, S. (1980), *Proc. Natl. Acad. Sci. USA,* **77**, 4643.

Schaffhausen, B. S., Silver, J. E., and Benjamin, T. L. (1978), *Proc. Natl. Acad. Sci. USA,* **75**, 79.

Simmons, D. T., and Martin, M. A. (1978), *Proc. Natl. Acad. Sci. USA,* **75**, 1131.

COMPARISON OF SEQUENCES*

There were many computer methods for comparing nucleic or amino acid sequences before the alignment programs were developed. The simplest way of proceeding consists of choosing a window of n bases (or amino acids) on the first sequence and moving it along the second sequence, with a fixed pitch value. The fraction of homologous bases is scored following each move on the second sequence. Any value below a predetermined minimal percentage of homology is not taken into account. The number of tests performed to compare complex sequences will vary with the length of the sequences (L_1 and L_2), the size of the window (w), and the pitch value (p). It can be calculated with the following formula

$$\text{number of tests} = L_1 \, (L_2/p) \, w$$

if L_1 is 2000 base pairs and L_2 is 1000 base pairs. With window size of 20 and a pitch value of 10 one gets a total of 4 million tests.

*With permission of Dr. Le Beux (CITI2, Paris).

Below is an example of data obtained with this program for the comparison of two sequences (208 and 1404 base pairs, respectively) from EMBL data bank.

Score Obtained for Different Position on the Sequences

Score	16	16	15	15	15	15	15
Sequence 1	70	179	13	60	91	130	170
Sequence 2	91	41	31	81	91	31	51

Quality of Homology at Each Position

```
7 0
    CATTACCACA  GGTAACGGTG  CGGGCTGACG
    * *   * **  * * * ** *       ** *
    TGTTTGCCCA  GTTGAAGGAG  CGCAAAGAAG
9 1
1 7 9
    ATGCGAGTGT  TGAAGTTCGG  CGGTACATCA
    * *   * * *   *** **   ** * ***
    ACGATATTTT  GAAAGCACGA  GGGGAAATCT
4 1
1 3
    TACAGAGTAC  ACAACATCCA  TGAAACGCAT
    * **  ** *  ** * **     ** * *
    TTCACCGTTC  ACGATATTTT  GAAAGCACGA
3 1
```

Another way of comparing sequences is provided by the DIAGON of Staden (1984) which is an interactive program for comparing and aligning nucleic acid or amino acid sequences (Figure 194).

The basic principle of the method involves producing a diagram that contains a representation of all the matches between a pair of sequences. The diagram consists of a two-dimensional plot in which the x axis represents one sequence (A) and the y axis the other (B). Every point (i, j) on the plane x, y is assigned a score which corresponds to the level of similarity between sequence characters $A(i)$ and $B(j)$. In the simplest use of the method a score of 1 could be assigned to every point (i, j) where $A(i) = B(j)$ and a score of 0 to every other point. If a plot of the points in the plane was made in which all scores of 1 were marked with a dot and all those of 0 left blank, then regions of identity would appear as diagonal lines. With the comparison displayed in this form, the human eye is very good at detecting regions of homology even if they are imperfect. The effects of mismatches, insertions, or deletions can be seen: matches interrupted by insertions or deletions

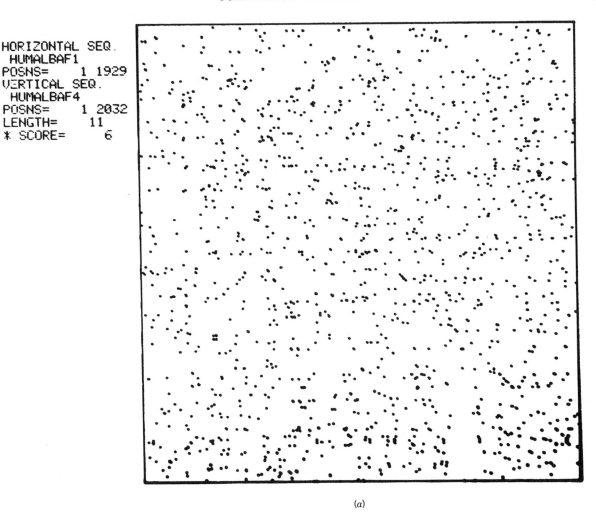

```
HORIZONTAL SEQ.
  HUMALBAF1
POSNS=     1 1929
VERTICAL SEQ.
  HUMALBAF4
POSNS=     1 2032
LENGTH=      11
* SCORE=      6
```

(a)

Figure 194. DIAGON graph for homologies between proteins. (a) Comparison of nucleic acid sequences from human serum albumin (F1) and alpha feto protein (F4). (b) Same comparison at a different scale. (c) Comparison of protein sequences from human serum albumin and human alpha feto protein (AFP). Homology result as an alignment of dots. This is particularly clear in (c) (kindly provided by C. Mugnier).

will appear as parallel diagonals, and matches interrupted by the odd mismatching pair of characters will appear as broken collinear diagonal lines. This diagram is a very useful representation but simply placing a dot for every identity is of limited value for the following reasons.

For nucleic acid sequences, around 25% of the plot will contain points and it will often be very difficult to distinguish significant homologies from chance matches. For proteins many significant alignments of sequences contain almost no identities but are formed from chemically and structurally similar amino acids so that simply looking for identity would be insufficient. What is required is to first find those points that correspond to fairly strong lo-

cal similarities and then to use the diagram of these points so that the human eye can be used to look for larger scale homologies. DIAGON uses two different algorithms to calculate the score for each point and the user defines a minimum score so that only those points in the diagram for which the score is at least this value will be marked with a dot. The first scoring method is to find the longest uninterrupted sections of perfect identity, that is, those that contain no mismatches, insertions, or deletions. The second method looks for sections where a proportion of the characters in the sequence are similar, again allowing no insertions or deletions. These two scoring methods, respectively referred to as the perfect and proportional algorithms, are described below.

```
HORIZONTAL SEQ.
  HUMALBAF1
POSNS=     1 1929
VERTICAL SEQ.
  HUMALBAF4
POSNS=     1 2032
LENGTH=    11
* SCORE=    5
```

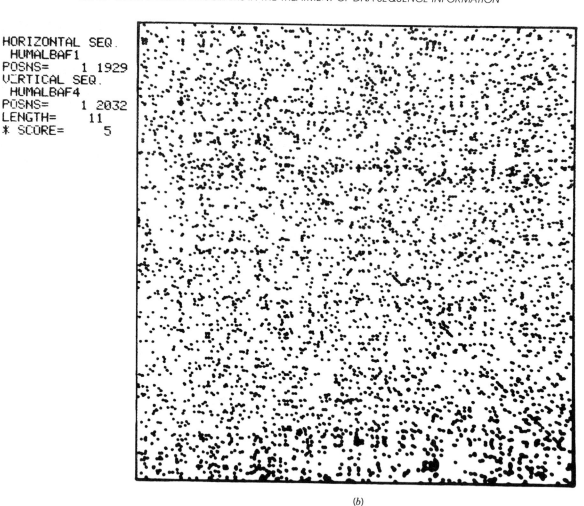

(b)

Figure 194 *(b)*

Perfect Matching. When looking for perfect matches the program stores the accumulated score for each diagonal at the current position, that is, the score at any point is the number of identities found looking back along the diagonal. At each point that a mismatch occurs the score is set back to zero so that, for example, a match of length six will have on its diagonal scores of 1, 2, 3, 4, 5, 6 and the next score will be zero. An advantage of this method for calculating the scores is that the user can see the heights of the peaks. For example, if the user plots all of peaks of 5 and above he knows that a diagonal of 2 points represents a match of 6 and a diagonal of 8 a match of 12.

Proportional Matching. This method, generally the most useful, involves calculating a score for each position in the matrix by summing points found when looking forward and backward along a diagonal line of a given length. This length, called the span,

should be an odd number so that the score for any point is correctly positioned at the center of the span. The algorithm does not simply look for identity but uses a score matrix that contains scores for every possible pair of characters. This matrix MDM78 was calculated by looking at accepted point mutations in 71 families of closely related proteins, and of those tested by Dayhoff, was found to be the most powerful score matrix for finding distant relationships between amino acid sequences.

The program is used on a simple graphics terminal, that is, a keyboard with a screen on which points and lines can be drawn. The user works at the terminal and produces plots for various combinations of values for the span length and minimum scores. The plots appear with the top left corner of the screen corresponding to the left ends of the two sections of sequence being compared. However large or small a region the user elects to compare, the program expands or contracts the diagram so that the plot always fills the screen. This allows the user to

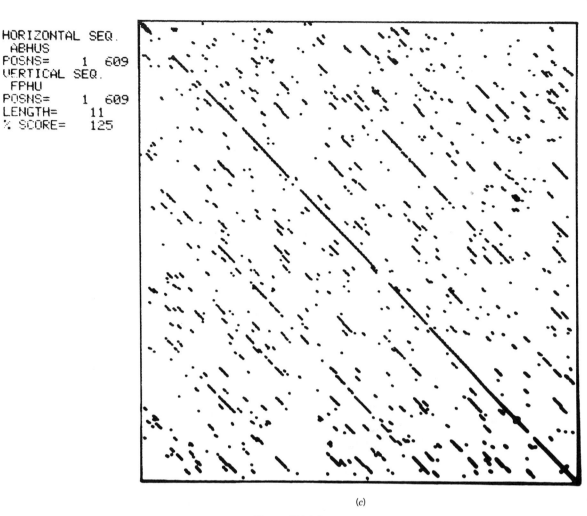

HORIZONTAL SEQ.
 ABHUS
POSNS=   1 609
VERTICAL SEQ.
 FPHU
POSNS=   1 609
LENGTH=   11
% SCORE= 125

(c)

Figure 194 (c)

gain an overall impression or to "home-in" on particular regions and examine them in more detail. Having found a region that looks interesting the user can determine its coordinates in terms of sequence positions by use of a crosshair facility. When this option is selected, a large cross appears on the screen. The user can move the cross around the screen by use of special directional keys, and the writing of characters and plotting of points are entirely independent so that the prompts and characters typed by the user elects to have it cleared. Giving the user control of screen clearing gives him the ability to overlay different plots.

The program has two statistical options to help the user choose score levels for plotting and to assess the significance of any similarity found. It can produce a cumulative histogram of observed scores for the current span length and region and it can calculate the "double matching probability" of McLachlan. The "double matching probability" is the probability of finding particular scores given two infinitely long sequences of the composition of those being compared, with the current span length and score matrix. By using these options the user can choose to plot all the matches for which the score exceeds a given significance level (such as 1%), using either empirical or theoretical probability values. Generally it is best to begin at a low level to avoid an overcrowded diagram.

If the user finds that the two sequences do contain stretches of homology, he will often want to align the sequences by inserting padding characters at deletion points. The program has a selection of options for this purpose: it can display on the screen the two sequences, one above the other, with asterisks marking identities, it has inbuilt editing functions, and it can save the aligned sequences on disk files. These options, combined with the diagonal display and crosshair facility, make the task of producing aligned sequences relatively easy.

A histogram of all of the scores is accumulated and printed, based on a user-specified scoring interval. The mean and standard deviation of the scores are also calculated. The statistical significance of the relationship between two sequences is evaluated by considering a numerical property of the segment scores. This property is calculated for the real sequences. The sequences are then randomly permuted and the same property is calculated for the randomized sequences. A statistical population of such comparisons is accumulated; typically, 100 random runs are performed. The score, expressed in SD units, is the difference between the value determined from the real sequences and the average value determined from the randomized sequences divided by the standard deviation of the values from the randomized sequences. The probability of finding a particular score by chance can be estimated from a table of standardized scores for the normal distribution.

Program RELATE* is designed to detect unusual similarity between sequences by comparing all possible segments of a given length from one sequence with all segments of the same length from the second sequence. A sequence can also be compared with all nonidentical segments from the same sequence (Dayhoff, 1979; Barker, 1979). A matrix of residue pair scores is supplied by the user. A segment score is accumulated from the pair scores of the residues occupying corresponding positions within the two segments. If the length of the segments used is 15 and the total lengths of the two sequences compared are 90 and 110 residues, then, 7296 scores will be tabulated. At most, 76 of these will come from comparisons of corresponding segments if the sequences had a common evolutionary origin.

The program saves a user-specified number of the highest segment scores. The values of these scores, the position of the first resdue of the segment from each sequence, and the differences between the first residue numbers of the two segments (the displacements) are displayed. For these highest scoring segment pairs, the number of occurrences of each displacement and the average scores of comparisons with the same displacement are also printed.

The segment scores are useful in finding small segments that have similar structures and so could have functional similarity. Any small repeated sequence can be discovered by examining this list. In-

verted repeats in nucleic acid sequences can be found by comparing a sequence with its (reverse) complement. In sequences with many tandem repeated segments, the displacements of the highest scoring segment comparisons tend to be multiples of the repeat length.

Two different kinds of numerical property are calculated by the program. The segment comparison score is based on the average magnitude of a predetermined number of the highest segment comparison scores, usually the number of scores to be expected if the sequences are related (i.e., 76 in the example above). The program also calculates a spectrum of count scores, $SC(N)$, in SD units. For each of these, the numerical property is the number of segment comparison scores within or above the Nth scoring interval. One can examine the spectrum to find the maximum $SC(N)$ and the corresponding number of segments from the real sequences with higher scores. For related sequences, sometimes there are only a few unexpectedly high scoring segments, which are the result of preferential conservation of a small active site. At times, in the case of extensively conserved sequences, all of the corresponding segments get scores above the rest of the distribution. On occasion, the maximum value corresponds to a surplus of hundreds of scores above expected, but all well below the maximum score. This situation would be expected when the sequences have evolved through many repetitions of similar segments. No single comparison may be outstanding, but the numerous intercomparisons of the repeated segments yield a bulge in the upper tail of the distribution.

This program is particularly useful for the comparison of a sequence with itself (omitting the comparison of identical segments) to show the existence and identity of single or multiple repeated segments.

The mutation data matrix (MD) has been found to be the most satisfactory scoring matrix for detecting distant relationships between protein sequences (Schwartz and Dayhoff, 1979). This matrix is based on amino acid replacements between present-day sequences and those inferred as common ancestors on evolutionary trees. The residues that did not change and the relative exposure of the sequences to mutational change were taken into account.

Two other scoring matrices are distributed with this program for use with protein sequences. The simplest of these, the unitary matrix (UP), assigns a value of 1 to residues that could be identical and 0 to all the others. A slightly more complicated scoring system reflects the maximum number of possible identities in the nucleotides of the genes coding for

*Information provided by National Research Foundation, with permission of H. P. Le Beux (CITI2, Paris).

```
QUERY SEQUENCE  ABHUS              FROM      1  TO    609  TOTAL   609
DEFINITION  Serum albumin precursor - Human.
QUERY SEQUENCE  FPHU              FROM      1  TO    609  TOTAL   609
DEFINITION  Alpha-fetoprotein precursor - Human.
           10        20        30        40        50        60
MKWVTFISLLFLFSSAYSRGVFRRDAHKSEVAHRFKDLGEENFKALVLIAFAQYLQQCPFEDHVKLVNE-
====  = =-==   - ==-- = -   - --   -   -=   =  = ===--=-  - - =-= --
MKWVESIFLIFLLNFTESRTLHRNEYGIASILDSYQCTAEISLADLATIFFAQFVQEATYKEVSKMVKDA
           10        20        30        40        50        60        70
        80        90       100       110       120       130
VTEFAKTCVADESAENC-DKSLHTLFGDKLCTVATLRETYGEMADCCAKQEPERNECFLQHKDDNP-NLP
-=   =   -== - = - = - = - ==  = = == ===- = - === ==   - -=
LTAIEKP-TGDEQSSGCLENQLPA-FLEELCHEKEILEKYGH-SDCCSQSEEGRHNCFLAHKKPTPASIP
        80        90       100       110       120       130
140       150       160       170       180       190       200
RLVRPEVDVMCTAFHDNEETFLKKYLYEIARRHPYFYAPELLFFAKRYKAAFTECCQAADKAACLLPKLD
  ==   = =- -- -- === =---======= == -= -= =   == = -  =
LFQVPEPVTSCEAYEEDRETFMNKFIYEIARRHPFLYAPTILLWAARYDKIIPSCCKAENAVECFQTKAA
140       150       160       170       180       190       200
210       220       230       240       250       260       270
ELRDEGKASSAKQRLKCASLQKFGERAFKAWAVARLSQRFPKAEFAEVSKLVTDLTKVHTECCHGDLLEC
 - = - == ==  == == =-= = -=--===-= - =-=- === =-- == == ==-=-=
TVTKELRESSLLNQHACAVMKNFGTRTFQAITVTKLSQKFTKVNFTEIQKLVLDVAHVHEHCCRGDVLDC
210       220       230       240       250       260       270
280       290       300       310       320       330       340
ADDRADLAKYICENQDSISSKLKECCEKPLLEKSHCIAEVENDEMPADLPSLAADFVESKDVCKNYAEA-
-=   -   === -==--= =- === ==-- == ==== = =   =- -= --- --
LQDGEKIMSYICSQQDTLSNKITECCKLTTLERGQCIIHAENDEKPEGLSPNLNRFLGDRDF-NQFSSGE
280       290       300       310       320       330       340
350       360       370       380       390       400       410
KDVFLGMFLYEYARRHPDYSVVLLLRLAKTYETTLEKCCAAADPHECYAKVFDEFKPLVEEPQNLIKQNC
=--==- =- ==-==== -= --==-==-=- ==== - -= == = -= --= = = ==
KNIFLASFVHEYSRRHPQLAVSVILRVAKGYQELLEKCFQTENPLECQDKGEEELQKYIQESQALAKRSC
350       360       370       380       390       400       410
420       430       440       450       460       470       480
ELFKQLGEYKFQNALLVRYTKKVPQVSTPTLVEVSRNLGKVGSKCCKHPEAKRMPCAEDYLSVVLNQLCV
== ==== ===  == == ==--- =- --= -- == -- == =
GLFQKLGEYYLQNAFLVAYTKKAPQLTSSELMAITRKMAATAATCCQLSEDKLLACGEGAADIIIGHLCI
420       430       440       450       460       470       480
490       500       510       520       530       540       550
LHEKTPVSDRVTKCCTESLVNRRPCFSALEVDETYVPKEFNAETFTFHADICTLSEKERQIKKQTALVEL
== === =- === = =======-= ====== = - = == =-= = == =--=
RHEMTPVNPGVGQCCTSSYANRRPCFSSLVVDETYVPPAFSDDKFIFHKDLCQAQGVALQTMKQEFLINL
490       500       510       520       530       540       550
560       570       580       590       600
VKHKPKATKEQLKAVMDDFAAFVEKCCKADDKETCFAEEGKKLVAASQAALGL
== == = === ==- ==-- -==== --- = ===== ==-- - ====-
VKQKPQITEEQLEAVIADFSGLLEKCCQGQEQEVCFAEEGQKLISKTRAALGV
560       570       580       590       600
```

(a)

Figure 195. Alignment of protein or nucleic acid sequences. (*a*) Comparison of protein sequences from human serum albumin and human alpha feto protein (AFP). (*b*) Comparison of mRNA sequences from human serum albumin and human alpha feto protein (AFP) (kindly provided by C. Mugnier).

the proteins. Identical amino acids obtain a score of 3; those for which two nucleotides could be identical, 2; one nucleotide, 1; and 0 if no nucleotides are ever shared in the codons for the amino acids. We refer to this as the genetic code matrix (GC). A unitary matrix for use with nucleic acids (UN) has also been supplied. This matrix assigns a score of 1 to identical nucleotides, to the pair *U/T*, and to any pairs of ambiguous nucleotides that could possibly be identical.

Program ALIGN* determines a best alignment of two protein or two nucleic acid sequences by computing the maximum match score using a version of the Needleman and Wunsch algorithm (1970). See Figure 195. An alignment score in standard

*Information provided by National Research Foundation, with permission of H. P. Le Beux (CITI2, Paris).

```
QUERY SEQUENCE  HUMALBAF1         FROM      1  TO   1929  TOTAL  1929
DEFINITION  Human serum albumin mRNA, complete cds.
==========================================
QUERY SEQUENCE  HUMALBAF4         FROM      1  TO   2032  TOTAL  2032
DEFINITION  Human alpha-fetoprotein (AFP) mRNA, complete cds.
  260       270       280           290       300          310
GAAGATCATGTAAAATTAGTGAATGAAGTA---ACTG-AATTTGCAAAAAC--A-TGTGTAGCTGATGAG
:::: : : : :::::  ::::: :: : :   :::: :::: : : ::::  : :: : :: :::  ::
GAAG-TAA-GCAAAATG-GTGAAAGATGCATTGACTGCAATT-G-AGAAACCCACTG-G-AGATGAACAG
         240        250       260        270        280         290
         330        340        350       360       370          380
TCAGCTGAAAATTGT-GACAAATCA-CTTCATACCCTTTTTGGAGACAAATTATGC-AC-AGTT-GCAAC
::  : :     :: : :: :: :: :: : : :: ::: :::: : :: :: ::: :  ::   : ::
TCTTCAGGGTGTT-TAGA-AAACCAGCTACCTGCC-TTTCTGGA-AGAACTT-TGCCATGAGAAAGAAAT
    300        310        320        330        340        350
    390        400       410       420          430       440
TCTTCGTGA-AACCTATGGTGAAATGGCTGACTGCTGT-GCAAAAC--AAGAACCTGAGAGAAATGAA-T
: :: : :: ::  :: ::  : ::  : :::::::  : :::   ::    :: ::: :: :: :: :
T-TT-G-GAGAAG-TACGG--ACATT-CAGACTGCTGCAGCCAAAGTGAAGAGG--GA-AGACAT-AACT
             370          380       390       400          410
          460       470         480         490        500         510
GCTT-CTTGCAACACAAAGATGACAAC-CCAA-ACC--TCCCCCGATTGGTGAGA-CCAGAGGTTG--AT
: :: :::::: :::::: : : : :: :::  : :  :::: :  ::    ::  :::::   ::  :
GTTTTCTTGCA-CACAAAAA-GCCCACTCCAGCATCGATCCCACTTTTCCA-AGTTCCAGAACCTGTCAC
420        430        440       450       460        470       480
       520       530         540         550       560        570
GTGATGTGCACTGCTTTTCATGAC-AATGA-AG--AGACATTTTTGAAAAAATAC-TTATATGAAATTGC
  : :::: :  :: :   :::  :: :: ::  :::::   :::: :::: : :: : :::: :: ::
AAGCTGTGAA--GCAT---ATGAAGAA-GACAGGGAGACATTCATGAACAAATTCATT-TATGAGATAGC
   490            500        510       520       530        540
  580        590       600        610       620       630        640
CAGAAGACATCC-TTACTTTTATGCCCCGG-AACTCCTTTTCTTTGCTAAAAGGTATAAAGCT-GCTTTT
 ::::: :::: :: :: :: : :::: ::   :: :: ::  :::::      ::  :   ::: ::: :
AAGAAGGCATCCCTTCCTGT-ATGCACCTACAATTC-TT--CTTTG------GGC-T---GCTCGCTATG
550       560        570       580                590           600
     650        660        670          680            690
ACAGAATG-TTGCCAAGCT-GCTGATAAAGCTGCC--TGCC-T-GT-TGC--C-AAAGCTC-G--ATG-A
::: :::  :: :::  :: ::::  :::::::    :::  : :  :::  : ::: :   :  : : :
ACAAAATAATT-CCAT-CTTGCTGC-AAAGCTGAAAATGCAGTTGAATGCTTCCAAA-CAAAGGCA-GCA
    610         620        630       640       650        660
               710          720       730        740        750
AC--TT-CGG--GA-TGAAGGGAA-G--GCTTCGTCTGCCAAA-CAGAGACT-CAAATGTGCCAGTC-TC
::  :: :    :: : ::: ::: :  :::: :: :   ::: :: : : : ::  ::::: :::  :
ACAGTTACAAAAGAATTAAGAGAAAGCAGCTT-GT-T---AAATCA-ACA-TGCA--TGTGC-AGTAATG
         680       690       700             710          720
 760        770        780        790       800       810        820
CAAAAATTT-GGAGAAA-GAGCTTTCAAAGC-ATGGGCAGTGGCTCGCCTGAGCCAGA-GATTTCCCAAA
 :::::::: :: ::   :: ::::: :::: ::   : ::  ::   ::::: :::: : ::: :::::
-AAAAATTTTGG-GACCCGAACTTTCCAAGCCATAA-CTGTTACTAAACTGAGTCAGAAG-TTTACCAAA
           740       750       760        770       780        790
     830       840         850        860       870         880
GCTGAGTTTGCAGAAGTTTCCA-AGT-TAGTGACAG-ATCTTACCAAAGTCCACACG-GAA-TGCTGCC-
: : : ::: : ::: :  :: :   ::::  : : :: :  :: : ::  ::: : : : :: :::
GTTAATTTTACTGAAAT--CCAGAAACTAGTC-CTGGATGTGGCCCATGT--ACATGAGCACTGTTGCAG
   800       810         820        830       840         850
 890       900       910        920          930       940       950
ATGGAGATCTGCTTGAATGTGCTG-ATGACAGGGCGGACCT--TG-CCAAGTATATCTGTGAAAATCAGG
: :::::: :::: :: ::: ::: : ::  :::   :  :  :: :: :  :::: : :   :: :: :
A-GGAGATGTGCTGGATTGT-CTGCAGGATGGGGAAAAAATCATGTCCTAC-ATATGT-TCTCAA-CAAG
         870        880       890       900        910        920
      960         970        980         990      1000       1010
ATTCGA-TCTC-CAGT-AAACTGA-AGGAATGCTGTGAAAA--ACCTCTGTTGGAAAAATCC-CACTGCA
:  : : :::  ::   ::: : : :: :::::::  :::   ::: : : :::::   :   :: :: :
A--C-ACTCTGTCAAACAAAATAACAG-AATGCTGC-AAACTGACCAC-GCTGGAACG-TGGTCAATGTA
      930       940        950        960        970        980
   1020      1030      1040       1050       1060        1070
T--TGCCGAAGTGGAAAATGATGAGATGCCTGCTGA-CT-TGCCTTCATTAGCTG--CTGATTTTGTTGA
:  : :  : :  :::::::::::: :  ::::  :  :: : : : ::  : ::   : : :::: : :
TAATTC--ATGCAGAAAATGATGAAAAACCTGAAGGTCTAT-C-TCCAA-ATCTAAACAGGTTTT-TAGG
990        1000      1010      1020        1030       1040       1050
         1090       1100       1110        1120      1130       1140
```

Figure 195 *(b)*

```
A-AGTA-AGGATGTTTGCAAA-AACTATGCTG-AGGC-AAAGGA-TGTCTTCCTGGGCATGTTTT-TGT-
: : :: :: ::: :: :: :: ::: ::: : : ::::: ::: :: ::::: : :
AGA-TAGAG-AT-TTT--AACCAATTTT-CTTCAGGGGAAAAAAATATCTTCTTGG-CAAGTTTTGT-TC
      1060         1070        1080        1090        1100        1110
   1150         1160        1170        1180        1190        1200
ATGAATATGCAAGAAGGCATCCTGA--TTACTCTGTC-GTGCTGCTGCTGAGACTTGCCAA-GACATATG
::::::: ::::::: :::::: : ::: :: :: ::: :: : :::
ATGAATATTCAAGAAGACATCCTCAGCTTGCTGTCTCAGTAAT--T-CTAAGAGTTGCTAAAGG-ATACC
      1120         1130        1140        1150        1160        1170
         1220        1230        1240        1250        1260        1270
AA-ACC-ACTCTAGAGAAGTGCTGTGCC-G-CTGCAGATCCTCATGAATGCTATGCCAAAGT-GTTCGAT
: : : : : : :::: : : :::: ::::::::: : : :::: : ::
AGGAGTTA-T-TGGAGAAGTGTT-T-CCAGACTGAAAACCCTCTTGAATGCCAAGATAAAGGAGAA-GAA
         1190        1200        1210        1220        1230        1240
   1280         1290        1300        1310        1320        1330
GAATTT-A-AACCT-CTTGT-GGAAGAGCCTCAGAATTTAATCAAA-CAAAACTGTGAGCTTTTTA-AGC
::::: : :: : : : ::: ::::: :: ::: :::: : :: ::: : :: : :: ::
GAATTACAGAAA-TACATCCAGGA-GAGCCA-AGCATTGG--CAAAGCGAAGCTGCG-GCCTCTTCCAGA
   1250         1260        1270        1280        1290        1300
1340         1350        1360        1370        1380        1390        1400
AGCTTGGAGAGTACAAATTCCAGAATGCGCTAT-TAGTT-CGTTACACCAAGAAAGTACCCCAAG-TGTC
: :: ::::: :: : :: :: :::: : :: ::: :: : :: ::::: ::::: : ::: :
AACTAGGAGAATATTACTTACAAAATGCGTT-TCTCGTTGC-TTACACAAAGAAAGCCCCCCA-GCTGAC
   1320         1330        1340        1350        1360        1370
1410         1420        1430        1440        1450        1460
AAC-TCCAACTCTTGTAGAGGTCTCAAGAAACCTAGGAAAAGTGG--GC-A--GCAA--A-T-GTTGTAA
: : :: : :: :: ::: :: :: : ::: : : ::: : ::: :
CTCGTCGGAG-CT-G-AT-GG-C-CATCA--CC-AGAAAAA-TGGCAGCCACAGCAGCCACTTGTTGCCA
1380         1390        1400        1410        1420        1430
   1470         1480        1490        1500        1510        1520
ACATCC-TGAAG-CAAAAAGAATGCCCTGTG-C-AGA-AGA--CT-A--TCTATCCGTGGTCCTGAACCA
:: :: : :: :::::: :: :: :: : ::: : : :: :: :
AC-TCAGTGAGGACAAACT-ATTGGCCTGTGGCGAGGGAGCGGCTGACAT-TAT---TA-TCG-GA-C-A
1440         1450        1460        1470        1480        1490
   1530         1540        1550        1560        1570        1580
GTTATGTGTGTTG-CATGAGAAA-ACGCCAGTAAGTGACAGA-GTCA-CAAAATGCTGCACAGAGTCCT-
::::::: : : ::::: :: :: ::::::: : : :: : : :::::::: :: :
CTTATGTATCA-GACATGA-AATGACTCCAGTAAACC-CTGGTGTTGGCCAG-TGCTGCACTTCTTCATA
1500         1510        1520        1530        1540        1550        1560
   1600         1610        1620        1630        1640        1650
TGGTGAACAGGCGACCATGCTTTT-CAGCTCTGGAAGTCGATGAAACATACGTTCC-CAAAG-AGTT-TA
:: ::::::: ::::: ::: :: ::::::::::: :: :: : : :: :
TGCC-AACAGGAGGCCATGCTTCAGCAGCT-TGGTGGTGGATGAAACATATGTCCCTCCT-GCA-TTCTC
   1570         1580        1590        1600        1610        1620
   1660         1670        1680        1690        1700        1710
ATGCTGAAACA-TTCACCTTCCATGCAG-ATATATGC-ACA-CTTTCTGAGAAGGAGAGA--CAAATCAA
:: ::: : :: :: :: :: ::::::::: :: : : : : :::: :
-TGATGACA-AGTTCATTTTCCATA-AGGATCTGTGCCA-AGCT--CAGGGT-GTAGCGCTGCAAA-CGA
         1640        1650        1660        1670        1680
         1730        1740        1750        1760        1770        1780
-GAAACAA-ACTGCACTTGTTGAGCTTGTGAAACACAAGCC-CAAGGCAACAAAAGAGCAACT-GAAAGC
::: ::: : :: :: : :::::::: :: ::::: ::: :::: : :: ::::: :: ::
TGAAGCAAGAGTTT-CTCATTAACCTTGTGAAGCAAAAGCCACAAAT-AACAGAGGAACACTTGAG-GC
   1790         1800        1810        1820        1830        1840
 1790         1800        1810        1820        1830        1840
TGTTATGG-ATGATTTCGCAG-CTT-TTGTAGAGAAGTGCTGC-AAGGCTGACGATA-AGGA-GACCTGC
::: :: : : ::::::: ::: : : ::: ::: :: ::::::: ::::: : : :::: : : :::
TGTCATTGCA-GATTTCTCAGGCCTGTTGGAGA-AA-TGCTGCCAAGGCC-AGGA-ACAGGAAGTC-TGC
1760         1770        1780        1790        1800        1810        1820
1860         1870        1880        1890        1900        1910
TTTGCCGAGGAGGGTA-AAAAACTTG-TTGCTGCAAGTCAAGCTGCCTTAGGC-TTATAACATCTACATT
::::: :: ::::: : ::::::: : :: :: ::::: :: ::: :: ::: ::
TTTGCTGAAGAGGG-ACAAAAACT-GATTTCAAAAACTCGTGCTGCTTTGGGAGTT-TAA-AT-TAC-TT
   1830         1840        1850        1860        1870        1880

            HOMOLOGY SCORE =   -452     (  -45 )
            PERCENT MATCH  =  62.4%    (1136 /1820 )
```

Figure 195 *(b continued)*

deviation units is calculated by taking the difference between this maximum match score and the average maximum match score for random permutations of the two sequences and dividing by the standard deviation of the random scores.

The best alignment between any pair of sequences is based on a computed numerical value. The contribution of each match of a residue or gap in one sequence with a residue or gap in the other sequence is accumulated to give a total score for the alignment. A scoring matrix defines the contribution assigned to each of the possible pairings between residues. A break penalty parameter, NPEN, can be assigned. A string of one or more consecutive residues in one sequence matched with a string of consecutive gaps (a break) within the other sequence contributes a score of $-$NPEN (independent of the length of the string). Residues matching a string of gaps at either end of a sequence are assigned a score of 0. A gap cannot be matched with a gap. Considering all possible alignments of the residues and any number of gaps, the basic algorithm of program ALIGN determines the maximum score possible and an alignment with that score. The scoring matrix is constructed from an input matrix and a matrix bias parameter, B, that is added to all terms of the input matrix. The net effect of adding B is that the score for any given alignment is increased by B times the number of positions where a residue matches another residue. Increasing B will often produce maximum scoring alignments with shorter overall lengths.

The maximum score, R, that is achieved by an alignment of a pair of real sequences is compared with the distribution of maximum scores for a large number (usually 100) of random permutations of the two sequences. The mean and standard deviation of this approximately normal distribution are M and D. The alignment score, A, is the number of standard deviations by which the maximum score for the real sequences exceeds the average maximum score for the random permutations:

$$A = (R - M)/D \text{ (in SD units)}$$

The probability that a score as high as that from the real sequences could have been obtained in a comparison of randomized sequences can be determined from a table for the cumulative standardized normal distribution.

The mutation data matrix (MD) has been found to be the most satisfactory scoring matrix for detecting distant relationships between protein sequences (Schwartz and Dayhoff, 1979; Dayhoff et al., 1979). A bias between 2 and 20 is added depending on the evolutionary distance of the pair of proteins, so that the alignment approximates the length adjustments that actually occurred during evolution. Judicious choice of the break penalty parameter produces alignment with a reasonable number of breaks. A value of 2 is appropriate for very distant pairs where there have been many changes in length, whereas 12 or more is appropriate for comparison of a short segment that closely matches a portion of a longer sequence. The values of these two parameters also affect the average maximum score and the standard deviation obtained from the randomized sequences. When these parameters are varied, the alignment score exhibits a maximum, presumably near a position where the gaps reflect the actual genetic events.

References

Barker, W. C., Ketcham, L. K., and Dayhoff, M. O. (1979), "Duplications in Protein Sequences," in *Atlas of Protein Sequence and Structure,* M. O. Dayhoff, ed., Vol. 5, suppl. 3, p. 353, National Biomedical Research Foundation, Washington, D.C.

Dayhoff, M. O. (1979), "Survey of New Data and Computer Methods of Analysis," in *Atlas of Protein Sequence and Structure,* M. O. Dayhoff, ed., Vol. 5, suppl. 3, p. 353, National Biomedical Research Foundation, Washington, D.C.

Dayhoff, M. O., Schwartz, R. M., and Orcutt, B. C. (1979), "A Model of Evolutionary Change in Proteins," in *Atlas of Protein Sequence and Structure,* M. O. Dayhoff, ed., Vol. 5, suppl. 3, p. 345, National Biomedical Research Foundation, Washington, D.C.

Needleman, S. B., and Wunsch, D. D. (1970), *J. Mol. Biol.,* **48**, 443.

Orcutt, B. C., and Dayhoff, M. O. (1982), *Nucleic Acid Sequence Database: Sequence File Format,* NBR Report 820530-08710, National Biomedical Research Foundation, Washington, D.C.

Orcutt, B. C., and Dayhoff, M. O. (1982), *Protein Sequence Database: Sequence File Format,* NBR Report 820535-08710, National Biomedical Research Foundation, Washington, D.C.

Orcutt, B. C., and Dayhoff, M. O. (1982), *Scoring Matrices,* NBR Report 820541-08710, National Biomedical Research Foundation, Washington, D.C.

Schwartz, R. M., and Dayhoff, M. O. (1979), "Matrices for Detecting Distant Relationships," in *Atlas of Protein Sequence and Structure,* M. O. Dayhoff, ed., Vol. 5, suppl. 3, p. 353, National Biomedical Research Foundation, Washington, D.C.

COMPARISON OF SEQUENCES WITH THE WHOLE CONTENT OF A DATA BANK

One of the most useful programs for this kind of approach has been developed in 1984 by Orcutt, Dayhoff, and Barker.

Program SEARCH* compares a user-specified segment with all segments of the same length (and with the shorter end segments) of every sequence in a database. The protein or nucleic acid sequence database or part of either database and/or a user-specified file of sequences can be searched. A matrix of residue pair scores is read into the program. A segment score is accumulated by adding the pair scores of the corresponding residues.

For detecting distant relationships among proteins we recommend the mutation data matrix. The distribution of scores from unrelated sequences is approximately normal. Typically, for a 25-residue piece, all corresponding sequences of the same family (sequences less than 50% different) appear above the distribution of scores of unrelated segments. About half of the more distantly related sequences in the same superfamily are also above this distribution, and the rest are in the upper tail. If a search produces no scores above the distribution, in all probability there are no other sequences from the same family in the data collection.

The SEARCH program using the unitary matrix is useful for locating short pieces that are identical with or very similar to a test piece, as in the problem of identifying the source or a peptide believed to originate by chemical degradation of a larger protein. Because fewer than 0.3% of the possible sequences of six residues actually occur in the collection of sequenced proteins, a search for a hexapeptide will usually turn up the source protein (and its relatives) or nothing at all.

By a judicious choice of segments and matrices many problems of sequence identification can be approached. For nucleic acid searches, a unitary matrix that gives a score of 1 to all residues that match (or could match, in the case of ambiguous residues) and 0 to other pairs is useful.

An accelerated way of aligning two sequences (nucleid acid or protein) has been described by Kaneisha et al. (1984). These programs proceed for alignment only in regions where a given number of consecutive bases (or amino acids) is homologous in the two sequences (see figure 195). The length of the treated sequences is limited to 3000 nucleotides (or 2000 amino acids).

Reference

Orcutt, B. C., and Dayhoff, M. O. (1982), *Scoring Matrices,* NBR Report 820541-08710, National Biomedical Research Foundation, Washington, D.C.

Orcutt, B. C., and Dayhoff, M. O. (1982), *Nucleic Acid Sequence Database: Sequence File Format,* NBR Report 820530-08710, National Biomedical Research Foundation, Washington, D.C.

Orcutt, B. C., and Dayhoff, M. O. (1982), *Protein Sequence Database: Sequence File Format,* NBR Report 820535-08710, National Biomedical Research Foundation, Washington, D.C.

DATA SOURCES

GENBANK

The Computer and Information Sciences Division
BBN Laboratories Inc.
10 Moulton Street
Cambridge, Massachusetts 02238 USA
Telephone: (617) 497 2742

EMBL Data Library

European Molecular Biology Laboratory
Meierhoff Str. 1
6900 Heidelberg FRG
Telephone: 6221-387409

Protein Identification Resource

National Biomedical Research Foundation
Georgetown University Medical Center
3900 Reservoir Road, N.W.
Washington, D.C. 20007 USA
Telephone: (202) 625 2121

C.I.T.I. 2

Centre Inter-universitaire de Traitement de
l'Information
Université René Descartes
45 rue des Saints Pères
75270 Paris Cedex 06, FRANCE
Telephone: (1) 42 96 24 89

*Information provided by National Biomedical Research Foundation, with permission of Dr. P. Le Beux (CITI2, Paris).

BIONET

Intelligenetics, 124 University Avenue
Palo Alto, California 94301 USA
Telephone: (415) 965 5576

C.I.T.I. 2 and BIONET are computer resources for molecular biology providing access to sophisticated programs for sequence analysis and communication networks.

GENERAL BIBLIOGRAPHY

Adrian, T., and Heinrich, W. (1986), COMAP: a comigrating analysis program for estimating the relationship of adenoviruses on the genome level, *Nucl. Acids Res.*, **14**, 559–565.

Akeroyd, R., Lenstra, A., Westerman, J., Vriend, G., Wirtz, K. W., and VanDeenen, L. M. (1982), Prediction of secondary structural elements in the phosphatidylcholine transfer protein from bovine liver, *Eur. J. Biochem.*, **121**, 391–394.

Arbanabel, R. M., Wieneke, P. R., Mansfield, E., Jaffe, D. A., and Brutlag, D. L. (1984), Rapid searches for complex patterns in biological molecules, *Nucl. Acids Res.*, **12**, 263–280.

Arentzen, R., and Ripka, W. C. (1984), Introduction of restriction enzyme sites in protein-coding DNA sequences by site specific mutagenesis not affecting the amino acid sequence: a computer program, *Nucl. Acids Res.*, **12**, 777–787.

Argos, P., and Siezen, J. (1983), Structural homology of lens crystallins. A method to detect structural homology from primary sequences, *Eur. J. Biochem.*, **131**, 143–148.

Arnold, J., Eckenrode, V. K., Lemke, K., Phillips, G. J., and Schaeffer, S. W. (1986), A comprehensive package for DNA sequence analysis in FORTRAN IV for the PDP–11, *Nucl. Acids Res.*, **14**, 239–254.

Atilgan, T., Nicholas, H. B., Jr., and McClain, W. H. (1986), A statistical method for correlating tRNA sequence with amino acid specificity, *Nucl. Acids Res.*, **14**, 375–380.

Auron, P. E., Rindone, W. P., Vary, C. P. H., Celentano, J. J., and Vournakis, J. N. (1982), Computer aided prediction of RNA secondary structures, *Nucl. Acids Res.*, **10**, 403–419.

Bach, R., Ywasaki, Y., and Friedland, P. (1984), Intelligent computational assistance for experiment design, *Nucl. Acids Res.*, **12**, 11–30.

Bach, R., Friedland, P., Brutlag, D. L., and Kedes, L. (1982), MAXIMIZE: a DNA sequencing strategy advisor, *Nucl. Acids Res.*, **10**, 295–304.

Bains, W. (1986), MULTAN: a program to align multiple DNA sequences, *Nucl. Acids Res.*, **14**, 159–177.

Bibb, M. J., Findlay, P. R., and Johson, M. W. (1984), The relationship between base composition and codon usage in bacterial genes and its use for the simple and reliable identification of protein coding genes, *Gene*, **30**, 157–166.

Bilofsky, H. S., Burks, C., Fickett, J. W., Goad, W. B., Lewitter, F. I., Rindone, W. P., Swindell, C. D., and Tung, C. S. (1986), The GenBank genetic sequence data bank, *Nucl. Acids Res.*, **14**, 1–4.

Biro, P. A. (1984), DNA sequence handling programs in Basic for home computers, *Nucl. Acids. Res.*, **12**, 627–631.

Bishop, J. M., and Rawlings, C. J. (1987), *Nucleic Acid and Protein Sequence Analysis, A Practical Approach,* IRL Press, Oxford.

Blanz, P. A., Kleindienst, S. (1986), A computer program for comparative analysis of nucleic acid sequences, *Nucl. Acids Res.*, **14**, 537–541.

Blattner, F. R., and Schroeder, J. L. (1984), A computer package for DNA sequence analysis, *Nucl. Acids Res.*, **12**, 615–617.

Blumenthal, R. M., Rice, P. J., and Roberts, R. J. (1982), Computer programs for nucleic acid sequence manipulation, *Nucl. Acids Res.*, **10**, 91–101.

Borer, P. N., Dengler, D., Tinoco, I., Jr., and Uhlenbeck, O. C. (1974), *J. Mol. Biol.*, **86**, 843–853.

Boswell, D. R., and McLachlan, A. D. (1984), Sequence comparison by exponentially damped alignment, *Nucl. Acids Res.*, **12**, 457–464.

Brendel, V., and Trifonov, E. N. (1984), A computer algorithm for testing potential prokaryotic terminators, *Nucl. Acids Res.*, **12**, 4411–4427.

Brutlag, D. L., Clayton, J., Friedland, P., and Kedes, L. H. (1982), A nucleotide sequence analysis and recombinaison system, *Nucl. Acids Res.*, **10**, 279–294.

Bucholtz, C. A., and Reisner, A. H. (1986), MBIS- an integrated system for the retrieval and analyses of sequence data from nucleic acids and proteins, *Nucl. Acids Res.*, **14**, 265–272.

Burnett, L. (1986), Development of a superior strategy for computer assisted nucleotide sequence analysis, *Nucl. Acids Res.*, **14**, 47–55.

Burnett, L., Basten, A., and Hensley, W. J. (1986), Should nucleotide sequence analysing computer algorithms always extend homologies by extending homologies, *Nucl. Acids Res.*, **14**, 425–430.

Busetta, B. (1982), Improving the residual representation of proteins, *J. Theor. Biol.*, **98**, 621–635.

Busetta, B., and Barrans, Y. (1982), The prediction of protein topologies, *Biochim. Biophys. Acta*, **709**, 73–83.

Busetta, B., and Hospital, M. (1982), An analysis of the prediction of the secondary structures, *Biochim. Biophys. Acta*, **701**, 111–118.

Busetta, B., and Barrans, Y. (1984), The prediction of proteins domains, *Biochim. Biophys. Acta,* **790**, 117–124.

Busetta, B., Tickle, I. J., and Blundell, T. L. (1983), Docker, an interactive program for simulating protein receptor DNA substrate interactions, *J. Appl. Cryst.,* **16**, 432–437.

Campione-Piccardo, J., and Ruben, M. (1986), An integrated software system for microcomputer management of recombinant DNA data, *Nucl. Acids Res.,* **14**, 571–574.

Caron, P. R. (1984), KLONER: A computer program to simulate recombinant DNA strategies by restriction map manipulations, *Nucl. Acids Res.,* **12**, 731–737.

Cech, T. R., Tanner, N. K., Tinoco, I., Jr., Weir, B. R., Zucker, M., and Perlman, P. S. (1983), Secondary structure of the Tetrahymena ribosomal RNA intervening sequence: Structural homology with fungal mitochondrial intervening sequences, *J. Appl. Cryst.,* **16**, 432–437.

Chou, P. Y., and Fasman, D. (1978), Empirical predictions of protein conformation, *Ann. Rev. Biochem.,* **47**, 251–276.

Cid, H., Bunster, M., Arriagada, E., and Campos, M. (1982), Prediction of secondary structure of proteins by means of hydrophobicity profiles, *FEBS Lett.,* **150**, 247–254.

Claverie, J. M., Sauvaget, I., and Bougueleret, L. (1985), Computer generation and statistical analysis of a data bank of protein sequences translated from Genbank, *Biochimie,* **67**, 437–444.

Claverie, J. M. (1984). A common philosophy and Fortran 77 software package for implementing and searching sequence databases, *Nucl. Acids Res.,* **12**, 397–408.

Claverie, J. M., and Bougueleret, L. (1986), Heuristic informational analysis of sequences, *Nucl. Acids Res.,* **14**, 179–196.

Clayton, J., and Kedes, L. (1982), GEL, a DNA sequencing project management system, *Nucl. Acids Res.,* **10**, 305–321.

Clift, B., Haussler, D., McConnell, R., Schneider, T. F., and Stormo, G. D. (1986), Sequence landscapes, *Nucl. Acids Res.,* **14**, 141–158.

Cohen, F. E., Abarbanel, R. M., Kuntz, I. D., and Fletterik, F. J. (1983), Secondary structure assignment for alpha/beta proteins by a combinatorial approach, *Biochemistry,* **22**, 4894–4904.

Cohen, F. E., Sternberg, M. J. E., and Taylor, W. (1981), Analysis of the tertiary structure of protein beta-sheet sandwiches, *J. Mol. Biol.,* **148**, 253–272.

Collins, J. F., and Coulson, A. F. W. (1984), Applications of parallel processing algorithms for DNA sequence analysis, *Nucl. Acids Res.,* **12**, 181–192.

Comay, E., Nussinov, R., and Comay, O. (1984), An accelerated algorithm for calculating the secondary structure of single stranded RNAs, *Nucl. Acids Res.,* **12**, 53–66.

Conrad, B., and Mount, D. W. (1982), Microcomputer programs for DNA sequence analysis, *Nucl. Acids Res.,* **10**, 31–37.

Cowin, J. E., Jellis, C. H., and Rickwood, D. (1986), A new method of representing DNA sequences which combines ease of visual analysis with machine readability, *Nucl. Acids Res.,* **14**, 509–515.

Dabrowiak, J. C., Skorobogaty, A., Rich, N., Vary, C. P. H., and Vournakis, J. N. (1986), Computer assisted microdensitometric analysis of footprinting autoradiographic data, *Nucl. Acids Res.,* **14**, 489–499.

Dardel, F. (1985), PEGASE: a machine language program for DNA sequence analysis on APPLE II microcomputer using a binary coding of nucleotides, *CABIOS,* **1**, 19–22.

Day, G. R., and Blake, D. (1982), Statistical significance of symmetrical and repetitive segments in DNA, *Nucl. Acids Res.,* **10**, 8323–8339.

Dayhoff, M. O., Hunt, L. T., McLaughlin, P. J., and Barker, W. C., Chromosomal Proteins, in *Atlas of Protein Sequence and Structure 1972, Vol. 5, Handbook and Reference Data Collection,* M. O. Dayhoff, ed., National Biomedical Research Foundation.

DeBanzie, J. S., Steeg, E. W., and Lis, J. T. (1984), Update for users of the Cornell sequence analysis package, *Nucl. Acids Res.,* **12**, 619–625.

Delaney, A. D. (1982), A sequence handling program, *Nucl. Acids Res.,* **10**, 61–67.

Devereux, J., Haerbeli, P., and Smithies, O. (1984), A comprehensive set of sequence analysis programs for the VAX, *Nucl. Acids Res.,* **12**, 387–395.

Doggett, P. E., and Blattner, F. R. (1986), Personal access to sequence databases on personal computers, *Nucl. Acids Res.,* **14**, 611–619.

Doolittle, R. F. (1981), Similar aminoacid sequences: Chance of common ancestry?, *Science,* **214**, 149–158.

Doolittle, R. F., Feng, D. F., and Johnson, M. S. (1984), Computer based characterization of epidermal growth factor precursor, *Nature,* **307**, 558–550.

Doolittle, R. F., Hunkapiller, M. W., Hood, L. E., Devare, S. G., Robbins, K. C., Aaronson, S. A., and Antoniades, H. M. (1983), Simian sarcoma virus Oncogene v-sis is derived from the gene (or genes) encoding a platelet-derived growth factor, *Science,* **221**, 275–276.

Douthart, R. J., Thomas, J. J., Rosier, S. D., Schmaltz, J. E., West, J. W. (1986), Cloning simulation in the cage environment, *Nucl. Acids Res.,* **14**, 285–297.

Duggleby, R. G., Kinns, H., and Rood, J. I. (1981), A computer program for determining the size of DNA restriction fragments, *Analytical Biochemistry,* **110**, 49–51.

Dumas, J. P., and Ninio, J. (1982), Efficient algorithm for folding and comparing nucleic acids sequences, *Nucl. Acids Res.,* **10**, 197–206.

Durand, R., and Bregegere, F. (1984), An efficient program to construct restriction maps from experimental data with realistic error levels, *Nucl. Acids Res.*, **12**, 703–716.

Elder, J. K., Green, D. K., and Southern, E. M. (1986), Automatic reading of DNA sequencing gel autoradiographs using a large format digital scanner, *Nucl. Acids Res.*, **14**, 417–424.

Fickett, J. W. (1984), Fast alignment algorithm, *Nucl. Acids Res.*, **12**, 175–179.

Finney, J. L. (1978), Volume, occupation, environment and accessibility in proteins. Environment and molecular area of RNase S, *J. Mol. Biol.*, **119**, 415–441.

Fitch, W. M. (1969), Locating gaps in aminoacid sequences to optimize the homology between two proteins, *Biochem. Genet.*, **3**, 99.

Fitch, W. M., and Margoliash, E. (1967), Construction of polygenetic trees, *Sciences*, **155**, 279–284.

Fitch, W. M., and Smith, T. F. (1983), Optimal sequence alignments, *Proc. Natl. Acad. Sci. USA*, **80**, 1382–1386.

Fondrat, C., Dessen, P., Le Beux, P. (1986), Principle of codification for quick comparisons with the entire biomolecule databanks and associated programs in FORTRAN 77, *Nucl. Acids Res.*, **14**, 197–204.

Friedland, P., Kedes, L. H. (1985), Discovery the secrets of DNA, *Communications of the ACM 28*, **11**, 1164–1186.

Friedland, P., Kedes, L., Brutlag, D., Iwasaki, Y., and Bachk, R. (1982), A knowledge based genetic engineering simulation system for representation of genetic data and experiment planning, *Nucl. Acids Res.*, **10**, 323–340.

Fritensky, B., Lis, J., and Wu, R. (1982), Portable microcomputer software for nucleotide sequence analysis, *Nucl. Acids Res.*, **10**, 6451–6463.

Fristensky, B. (1986), Improving the efficiency of dot matrix similarity searches through use of an oligomer table, *Nucl. Acids Res.*, **14**, 597–610.

Fuchs, C., Rosenwald, E. C., Honigman, A., and Szybalski, W. (1980), *Gene*, **10**, 357– .

Galas, D. J., Eggert, M., and Watterman, M. S. (1985), Rigorous pattern-recognition methods for DNA sequences. Analysis of promoter sequences of *Escherichia coli*, *J. Mol. Biol.*, **186**, 117–128.

Gariepy, C. E., Lomax, M. R., and Grossman, L. I. (1986), SPLINT: a cubic spline interpolation program for the analysis of fragment sizes in one dimensional, *Nucl. Acids Res.*, **14**, 575–581.

Garnier, J., Osguthorpe, D. J., and Robson, B. (1978), Analysis of the accuracy and implications of simple methods for predicting the secondary structure of globular proteins, *J. Mol. Biol.*, **120**, 97–120.

Gascuel, O. (1985), Structural descriptions. Learning and discrimination of these descriptions, *Biochimie*, **67**, 499–508.

Gates, M. A. (1985), Simpler DNA sequences representations, *Nature*, **316**, 219.

Gautier, C., Gouy, M., and Louail, S. (1985), Non-para-metrical statistics for nucleic acid sequence studies, *Biochimie*, **67**, 449–454.

George, D. G., Baker, W. C., and Hunt, L. T. (1986), The protein identification resource (PIR), *Nucl. Acids Res.*, **14**, 11–15.

Gibbs, A. J., McIntyre, G. A. (1970), The diagram, a method for comparing sequences, *Eur. J. Biochem.*, **16**, 1–11.

Gingeras, T. R., Milazzo, J. P., and Roberts, R. J. (1978), A computer assisted method for the determination of restriction enzyme recognition sites, *Nucl. Acids Res.*, **5**, 4105–4127.

Gingeras, T. R., Milazzo, J. P., Sciatky, D., and Roberts, R. J. (1979), Computer programs for the assembly of DNA sequences, *Nucl. Acids Res.*, **7**, 529–545.

Gingeras, T. R., Rice, P., and Roberts, R. J. (1982), A semi-automated method for the reading of nucleic acid sequencing gels, *Nucl. Acids Res.*, **10**, 103–114.

Gingeras, T. R., and Roberts, R. J. (1980), *Science*, **209**, 1322–1328.

Goad, W., and Kanehisa, M. I. (1982), Pattern recognition in nucleic acids sequences: I. A general method for finding local homologies and symmetries, *Nucl. Acids Res.*, **10**, 247–264.

Göll, D., and Roberts, R. J. (1984), The applications of computers to research on nucleic acids II. Part I and Part II, *Nucl. Acids Res.*, **12**, I, 1–430, II, 1–870.

Gordon, A. D. (1973), A sequence comparison statistic and algorithm, *Biometrika*, **60**, 197–200.

Gotoh, O. (1982), An improved algorithm for matching biological sequences, *J. Mol. Biol.*, **162**, 705–708.

Gotoh, O., and Tagashira, Y. (1986), Sequence search on a supercomputer, *Nucl. Acids Res.*, **14**, 57–64.

Gough, E. J., and Gough, N. M. (1984), Direct calculation of the sizes of DNA fragments separated by gel electrophoresis using programmes written for a pocket calculator, *Nucl. Acids Res.*, **12**, 845–853.

Gouy, M., Gautier, C., and Milleret, F. (1985), Systems analysis and nucleic acid sequence banks, *Biochimie*, **67**, 433–436.

Gouy, M., Marlière, P., Papanicolaou, C., and Ninio, J. (1985), Prediction of secondary structures in nucleic acids: algorithmic and physical aspects, *Biochimie*, **67**, 523–532.

Gouy, M., Milleret, F., Mugnier, C., Jacobzone, M., and Gautier, C. (1984), ACNUC: a nucleic acid sequence data base and analysis system, *Nucl. Acids Res.*, **12**, 121–128.

Gralla, J., and Crothers, D. M. (1973), *J. Mol. Biol.*, **73**, 497– .

Gray, A. J., Beecher, D. E., and Olson, M. V. (1984), Computer-based image analysis of one dimensional electrophoretic gels used for the separation of DNA restriction fragments, *Nucl. Acids Res.*, **12**, 473–491.

Gribskov, M., Devereux, J., and Burgess, R. R. (1984), The

codon preference plot: graphic analysis of protein coding sequences and prediction of gene expression, *Nucl. Acids Res.,* **12**, 539–549.

Gribskov, M., Burgess, R. R., and Devereux, J. (1986), PEPPLOT, a protein secondary structure analysis program for the UWGCG sequence analysis software package, *Nucl. Acids Res.,* **14**, 327–334.

Gross, R. H. (1986), A DNA sequence analysis program for the Apple Macintosh, *Nucl. Acids Res.,* **14**, 591–596.

Grymes, R. A., Travers, P., and Engelberg, A. (1986), A computer tool for DNA sequencing projects, *Nucl. Acids Res.,* **14**, 87–98.

Haber, J. E., and Koshlandjr, D. E. (1970), An evaluation of the relateness of protein based on comparison of aminoacid sequences, *J. Mol. Biol.,* **50**, 617–639.

Haiech, J., and Sallantin, J. (1985), Computer search of calcium binding sites in a gene data bank: use of learning techniques to build an expert system, *Biochimie,* **67**, 555–560.

Hamm, G. H., and Cameron, G. N. (1986), The EMBL data library, *Nucl. Acids Res.,* **14**, 5–9.

Hamori, E. (1985), A novel DNA sequence representations, *Nature,* **314**, 585–586.

Harley, E. H. (1986), A general DNA analysis program for the Hewlett-Packard Model 86/87 microcomputer, *Nucl. Acids Res.,* **14**, 467–477.

Harr, R., Haglbom, P., and Gustafsson, P. (1982), Two-dimensional graphic-analysis of DNA sequence homologies, *Nucl. Acids Res.,* **10**, 365–374.

Harr, R., Häggström, and Gustafsson, P. (1983), Search algorithm for pattern analysis of nucleic sequences, *Nucl. Acids Res.,* **11**, 2943–2957.

Harr, R., Fallman, P., Haggstrom, M., Wahlstrom, L., and Gustafsson, P. (1986), GENEUS, a computer system for DNA and protein sequence analysis containing an information retrieval system for the EMBL data library, *Nucl. Acids Res.,* **14**, 273–284.

Henaut, A., Limaiem, J., and Vigier, P. (1985), The origins of the strategy of codon use, *Biochimie,* **67**, 475–484.

Hider, R. C., and Hodges, S. J. (1984), Protein secondary structure: analysis and prediction, *Biochem. Educ.,* **12**, 9–18.

Jagadeeswaran, P., and McGuire, P. M. (1982), Interactive computer programs in sequence data analysis, *Nucl. Acids Res.,* **10**, 433–447.

Johnsen, M. (1984), JINN, an integrated software package for molecular geneticists, *Nucl. Acids Res.,* **12**, 657–664.

Johnston, R. E., Mackenzie, J. M., Jr., and Dougherty, W. G. (1986), Assembly of overlapping DNA sequences by a program written in BASIC for 64K CP/M and MS, DOS IBM compatible, *Nucl. Acids Res.,* **14**, 517–527.

Kabsch, W., and Sander, C. (1984), On the use of sequence homologies to predict protein structure: Identical pentapeptides can have completely different conformations, *Proc. Natl. Acad. Sci. USA,* **81**, 1075–1078.

Kanehisa, M. I. (1982), Los Alamos sequence analysis package for nucleic acids and proteins, *Nucl. Acids Res.,* **10**, 183–196.

Kanehisa, M. (1984), Use of statistical criteria for screening potential homologies in nucleic acids sequences, *Nucl. Acids Res.,* **12**, 203–213.

Kanehisha, M. I., and Goad, W. (1982), Pattern recognition in nucleid acids sequences: II. An efficient method for finding local stable secondary structure, *Nucl. Acids Res.,* **10**, 265–278.

Kanehisa, M., Fickett, J. W., and Goad, W. B. (1984), A relational database system for the maintenance and verification of the Los Alamos sequence library, *Nucl. Acids Res.,* **12**, 149–158.

Kanehisa, M., Klein, P., Greif, P., and Delisi, C. (1984), Computer analysis and structure prediction of nucleic acids and proteins, *Nucl. Acids Res.,* **12**, 409–416.

Kanehisa, M., Klein, P., Greif, P., and Delisi, C. (1984), Computer analysis and structure prediction of nucleic acids and proteins, *Nucl. Acids Res.,* **12**, 417–428.

Karlin, S. G., Ghandour, D. E., Ost, E., Tavare, S., and Korn, L. J. (1983), New approach for computer analysis of nucleic acid sequences, *Proc. Natl. Acad. Sci. USA,* **80**, 5660–5664.

Karp, R. M., Miller, R. E., and Rosenberg, A. L. (1972), *Proc. 4th annual ACM Symposium on Theory Computing,* 125–136.

Keller, C., Corcoran, M., and Roberts, R. J. (1984), Computer programs for handling nucleic acid sequences, *Nucl. Acids Res.,* **12**, 379–386.

Kieser, T. (1984), DNAGEL: a computer program for determining DNA fragment sizes using a small computer equipped with a graphics tablet, *Nucl. Acids Res.,* **12**, 679–689.

Komaromy, M., and Govan, H. (1984), An inexpensive semi automated sequence reader for the Apple II computer, *Nucl. Acids Res.,* **12**, 675–678.

Konopka, A. K. (1985), Theory of degenerate coding and informational parameters of protein coding genes, *Biochimie,* **67**, 445–468.

Konopka, A. K., and Brendel, V. (1985), The missense errors in protein can be controlled by suractived synonymous codon usage at the level of transcription, *Biochimie,* **67**, 469–474.

Korn, L. J., and Queen, C. (1984), Analysis of biological sequences on small computers, *DNA,* **3**, 421–436.

Korn, L. J., Queen, C. L., and Wegmen, M. N. (1977), Computer analysis of nucleic acid regulatory sequences, *Proc. Natl. Acad. Sci. USA,* **75**, 4401–4405.

Krishnan, G., Kaul, R. K., and Jadadeeswaran, P. (1986), DNA sequence analysis: a procedure to find homologies among many sequences, *Nucl. Acids Res.,* **14**, 543–550.

Kyte, J., and Doolittle, R. F. (1982), A simple method for displaying the hydropatic character of a protein, *J. Mol. Biol.,* **157**, 105–132.

Kyte, J., and Doolittle, R. F. (1982), A simple model for displaying the hydropathic character of a protein, *J. Mol. Biol.*, **157**, 105–132.

Lagrimini, L. M., Brentano, S. T., and Donelson, J. E. (1984), A DNA sequence analysis package for the IBM personal computer, *Nucl. Acids Res.*, **12**, 539–549.

Landau, G. M., Vishkin, U., and Nussinov, R. (1986), An efficient string matching algorithm with kappa differences for nucleotide and amino acid sequences, *Nucl. Acids Res.*, **14**, 31–46.

Lang, B. F., and Burger, G. (1986), A collection of programs for nucleic acid and protein analysis, written in FORTRAN 77 for IBM-PC compatible microcomputers, *Nucl. Acids Res.*, **14**, 445–465.

Lapalme, G., Cedergren, R. J., and Sankoff, D. (1982), An algorithm for the display of nucleic acid secondary structure, *Nucl. Acids. Res.*, **10**, 8351–8356.

Larson, R., and Messing, J. (1982), Apple II software for M13 shotgun DNA sequencing, *Nucl. Acids Res.*, **10**, 39–49.

Larson, R., and Messing, J. (1983), Apple II computer software for DNA and protein sequence data, *DNA*, **2**, 31–35.

Lautenberger, J. A. (1982), A program for reading DNA sequence gels using a small computer equipped with a graphic tablet, *Nucl. Acids Res.*, **10**, 27–30.

Lawrence, C. B. (1986), Data structures of DNA sequence manipulation, *Nucl. Acids Res.*, **14**, 205–216.

Lenstra, J. A., Hofsteenge, J., and Beintema, J. J. (1977), Invariant features of the structure of pancreatic ribonuclease, *J. Mol. Biol.*, **109**, 185–293.

Levitt, M., and Greer, J. (1977), Automatic identification of secondary structure in globular proteins, *J. Mol. Biol.*, **114**, 181–293.

Lewis, R. M. (1986), PROBFIND: a computer program for selecting oligonucleotide probes from peptide sequences, *Nucl. Acids Res.*, **14**, 567–570.

Lipmann, D. J., Wilbur, W. J., Smith, T. F., and Waterman, M. S. (1984), On the statistical significance of nucleic acid similarities, *Nucl. Acids Res.*, **12**, 215–226.

Lombardi, S., Seidell, H., Pulford, S., Dutton, W., and Parekh, S. (1984), Computer programs in nucleic acid synthesis: synthetic strategy development using solid phase chemical techniques with data storage, retrieval and analysis capabilities, *Nucl. Acids Res.*, **12**, 437–446.

Lonsdale, D. M., Hodge, T. P., and Stoehr, P. J. (1984), A computer program for the management of small cosmid banks, *Nucl. Acids Res.*, **12**, 429–436.

Luckow, V. A., Littlewood, R. K., and Rownd, R. H. (1984), Interactive computer program for the graphic analysis of nucleotide sequence data, *Nucl. Acids Res.*, **12**, 665–673.

Lyall, A., Hammond, P., Brough, D., and Glover, D. (1984), A DNA sequence analysis system in Prolog, *Nucl. Acids Res.*, **12**, 633–642.

Macmurray, A. (1986), DIGICALC: a restriction fragment analysis program, *Nucl. Acids Res.*, **14**, 529–536.

Maina, C. V., Nolan, G. P., and Szalay, A. A. (1984), Molecular weight determination program, *Nucl. Acids Res.*, **12**, 695–702.

Maizel, J. A., and Lenk, R. P. (1981), Enhanced graphic matrix analysis of nucleic acid and protein sequences, *Proc. Natl. Acad. Sci. USA*, **78**, 7665–7669.

Marcaud, H., Gabarro-Arpa, J., Ehrlich, R., and Reiss, C. (1986), An algorithm for studying cooperative transitions in DNA, *Nucl. Acids Res.*, **14**, 551–558.

Marck, C. (1986), Fast analysis of DNA and protein sequence on apple IIe: restriction sites search, alignment of short sequence and dot matrix analysis, *Nucl. Acids Res.*, **14**, 583–590.

Malthierry, B., Bellon, B., Giorgi, D., and Jacq, B. (1984), Apple II Pascal programs for molecular biologists, *Nucl. Acids Res.*, **12**, 569–580.

Marlière, P. (1983), Computer building and folding to fictious transfer-RNA sequences, *Biochimie*, **65**, 267–273.

Martinez, H. M. (1983), An efficient method for finding repeats in molecular sequences, *Nucl. Acids Res.*, **11**, 4629–4634.

Martinez, H. M. (1984), An RNA folding rule, *Nucl. Acids Res.*, **12**, 323–334.

Marvel, C. C. (1986), A program for the identification of tRNA like structures in DNA sequence data, *Nucl. Acids Res.*, **14**, 431–435.

McCormick, D. (1984), Microcomputer tools for Biotechnology, *Biotechnology*, **2**, 1002–1027.

McLachlan, A. D., and Boswell, D. R. (1985), Confidence limits for homology in protein or gene sequences. The c-myc oncogene and adenovirus Ela protein, *J. Mol. Biol.*, **185**, 39–49.

McLachlan, A. D., Staden, R., and Boswell, D. B. (1985), A method for measuring the non-random bias of a codon usage table, *Nucl. Acids Res.*, **12**, 9567–9576.

McLachlan, A. D. (1971), Tests for comparing related amino-acid sequences. Cytochrome c and cytochrome c551, *J. Mol. Biol.*, **61**, 409–424.

Meyers, S., and Friedland, P. (1984), Knowledge based simulation of genetic regulation in bacteriophage lambda, *Nucl. Acids Res.*, **12**, 1–9.

Michel, F., and Dujon, B. (1983), Conservation of RNA secondary structures in two intron families including mitochondrial-, chloroplastic- and nuclear-encoded members, *EMBO Journal*, **2**, 33–38.

Michel, F., Jacquier, A., and Dujon, B. (1982), Comparison of fungal mitochondrial introns reveals intensive homologies in RNA secondary structure, *Biochimie*, **64**, 867–881.

Minsky, M., and Papert, S. (1969), in *Perceptrons*, MIT Press, Cambridge.

Mizraji, E., and Ninio, J. (1985), Graphical coding of nucleic acid sequences, *Biochimie*, **67**, 445–448.

Modlevsky, J. L., Norris, F., and Griesinger, G. (1986), GENEVIEW and the DNACE data bus: computational tools for analysis, display and exchange of genetic information, *Nucl. Acids Res.*, **14**, 397–403.

Mount, D. W., and Conrad, B. (1984), Microcomputer programs for backtranslation of protein to DNA sequences and analysis of ambiguous DNA sequences, *Nucl. Acids Res.*, **12**, 819–823.

Mount, D. W., and Conrad, B. (1984), Microcomputer programs of graphic analysis of nucleic acid and protein sequences, *Nucl. Acids Res.*, **12**, 811–817.

Mount, D. W., and Conrad, B. (1986), Improved programs for NDA and protein sequence analysis on the IBM personal computer and other standard computer system, *Nucl. Acids Res.*, **14**, 443–454.

Mulligan, M. E., and McClure, W. R. (1986), Analysis of the occurrence of promoter sites in DNA, *Nucl. Acids Res.*, **14**, 109–126.

Mulligan, M. E., Hawley, D. R., Entricken, R., and McClure, W. R. (1984), *Escherichia coli* promoter sequences predict in vitro RNA polymerase selectivity, *Nucl. Acids Res.*, **12**, 789–800.

Murata, M., Richardson, J. S., and Sussman, J. L. (1985), Simultaneous comparison of three protein sequences, *Proc. Natl. Acad. Sci. USA*, **82**, 3073–3077.

Myers, E. W., and Mount, D. W. (1986), Computer program for the IBM personal computer which searches for approximate matches to short oligonucleotide sequences in long target DNA sequences, *Nucl. Acids Res.*, **14**, 501–508.

Nanard, M., and Nanard, J. (1985), A user-friend biological workstation, *Biochimie*, **67**, 429–432.

Needleman, S. B., and Wunsch, C. D. (1970), A general method applicable to the search, *J. Mol. Biol.*, **48**, 443–453.

Neumaier, P. S. (1986), A program package applicable to the detection of overlaps between restriction maps, *Nucl. Acids Res.*, **14**, 351–362.

Ninio, J. (1979), Prediction of pairing schemes in RNA molecules-loop contributions and energy of wobble and non-wobble pairs, *Biochimie*, **61**, 1133–1150.

Nolan, G. P., Maina, C. V., and Szalay, A. A. (1984), Plasmid mapping computer program, *Nucl. Acids Res.*, **12**, 717–729.

Noller, H. F., Kop, J. A., Wheaton, V., Brosius, J., Gutell, R. R., Kopilov, A., Doehme, F., and Herr, W. (1981), Secondary structure model for 23S ribosomal RNA, *Nucl. Acids Res.*, **9**, 6167–6189.

Novotny, J., and Auffray, C. (1984), A program for prediction of protein secondary structure from nucleotide sequence data: application to histocompatibility antigens, *Nucl. Acids Res.*, **12**, 243–356.

Novotny, J. (1982), Matrix program to analyse primary structure homology, *Nucl. Acids Res.*, **10**, 127–131.

Nussinov, R., and Jacobson, A. B. (1980), Fast algorithm for predicting the secondary structure of single stranded RNA, *Proc. Natl. Acad. Sci. USA*, **77**, 6309–6313.

Nussinov, R., Pieczenik, G., Griggs, J. G., and Kleitman, D. J. (1978), Algorithms for loop matchings, *SIAM J. Appl. Math*, **35**, 68–81.

Ninio, J. (1983), L'explosion des séquences: les années folles 1980–1990, *Biochem. System Ecology*, **11**, 305–313.

Ooi, T., and Takanami, M. (1981), A computer method for construction of secondary structure from polynucleotide sequence, *Biochim. Biophys. Acta*, **655**, 221–229.

Orcutt, B. C., George, D. G., and Dayhoff, M. O. (1983), Protein and nucleic acid sequence database system, *Ann. Rev. Biophys. Bioeng.*, **12**, 419–441.

Papanicolaou, C., Gouy, M., and Ninio, J. (1984), An energy model that predicts the correct folding of both the tRNA and the 5S RNA, *Nucl. Acids Res.*, **12**, 31–44.

Pavlov, M. Y., and Fedorov, B. A. (1983), Improved technique for calculating X-ray scattering intensity of biopolymers in solution: evaluation of the form, volume and surface of a particle, *Biopolymers*, **22**, 1507–1522.

Pearson, W. R. (1982), Automatic construction of restriction site maps, *Nucl. Acids Res.*, **10**, 217–227.

Peltola, H., Soderlund, H., and Ukkonen, E. (1986), Algorithms for the search of amino acid patterns in nucleic acid sequences, *Nucl. Acids Res.*, **14**, 99–107.

Pincus, M. R., VanRenswoude, J., Harford, J. B., Chang, E. H., Carty, R. P., and Klausner, R. D. (1983), Prediction of the three dimensional structure of the transforming region of the EJ/T24 human bladder oncogene product and its normal cellular homologue, *Proc. Natl. Acad. Sci. USA*, **80**, 5253–5257.

Pipas, J. M., and McMahon, J. E. (1975), Method for predicting RNA secondary structure, *Proc. Natl. Acad. Sci. USA*, **72**, 2017–2021.

Polner, G., and Orosz, L. (1984), PMAP, PMAPS: DNA physical map constructing programs, *Nucl. Acids Res.*, **12**, 227–236.

Ponnuswamy, P. K. (1978), Prediction of tertiary structures in globular proteins: somme essential parameters, in *Biomolecular Structure, Conformation, Function and Evolution*, R. Srinivasan, ed., Vol. 2, pp. 151–167, Pergamon Press.

Prophet: A national computing resource for life science research, (1986), *Nucl. Acids Res.*, **14**, 21–24.

Ptitsyn, O. B., and Finkekstein, A. V. (1978), Mechanism of self-organization of the tertiary structure of globular proteins, in *Biomolecular Structure, Conformation, Function, and Evolution*, R. Srinivasan, ed., Vol. 2, pp. 119–132, Pergamon Press.

Pustell, J., and Kafatos, F. C. (1982), A convenient and adaptable package of DNA sequence analysis programs for microcomputers, *Nucl. Acids Res.*, **10**, 51–59.

Pustell, J., and Kafatos, F. C. (1982), A high speed high capacity homology matrix: zooming through SV40, *Nucl. Acids Res.*, **10**, 4765–4782.

Pustell, J., and Kafatos, F. C. (1984), A convenient and adaptable package of computer programs for DNA and protein sequence management, analysis and homology determination, *Nucl. Acids Res.,* **12**, 643–655.

Pustell, J., and Kafatos, F. C. (1986), A convenient and adaptable microcomputer environment for DNA and protein sequence manipulation and analysis, *Nucl. Acids Res.,* **14**, 479–488.

Quenn, C., and Korn, L. J. (1980), Computer analysis of nucleic acids and proteins, *Meth. Enzymol.,* **65**, 595–609.

Queen, C., and Korn, L. J. (1984), A comprehensive sequence analysis program for the IBM personal computer, *Nucl. Acids Res.,* **12**, 581–600.

Queen, C., Wegman, M. N., and Korn, L. J. (1982), Improvements to a program for DNA analysis: a procedure to find homologies among many sequences, *Nucl. Acids Res.,* **10**, 449–456.

Quigley, G. J., Gehrke, L., Roth, D. A., and Auron, P. E. (1984), Computer aided nucleic acid secondary structure modeling incorporating enzymatic digestion data, *Nucl. Acids Res.,* **12**, 347–366.

Quinqueton, J. (1985), OURCIN: a tool to build expert systems, *Biochimie,* **67**, 485–492.

Raupach, R. E. (1984), Computer programs used to aid in the selection of DNA hybridization probes, *Nucl. Acids Res.,* **12**, 833–836.

Rawlings, C. J. (1986), *Software Directory for Molecular Biologists,* Globe (Macmillan Publisher Ltd).

Reisner, A. H., and Bucholtz, C. A. (1984), Utilization of sequence libraries on a 16-bit mini computer with particular reference to high speed searching, *Nucl. Acids Res.,* **12**, 409–416.

Reisner, A. H., and Bucholtz, C. A. (1986), The MTX package of computer programs for the comparison of sequences of nucleotides and amino acid residues, *Nucl. Acids Res.,* **14**, 233–238.

Richmond, T. J., and Richards, F. M. (1978), Packing of alpha-helices: geometrical constraints and contact areas, *J. Mol. Biol.,* **119**, 537–555.

Robson, B., and Osguthorpe, D. J. (1979), Refined models for computer simulation of protein folding. Applications to the study of conserved secondary structure and flexible hinge points during the folding of pancreatic trypsin inhibitor, *J. Mol. Biol.,* **132**, 19–51.

Rodier, F., and Sallantin, J. (1985), Localization of the initiation of translation in messenger RNAs of prokaryotes by learning techniques, *Biochimie,* **67**, 533–540.

Rood, J. I., and Gawthorne, J. M. (1984), Apple software for analysis of the size of restriction fragments, *Nucl. Acids Res.,* **12**, 689–694.

Rose, G. D. (1978), Prediction of chain turns in globular proteins on a hydrophobic basis, *Nature,* **272**, 586–590.

Rowe, G. W. (1985), A three-dimensional representation for base composition of protein-coding DNA sequences, *J. Theor. Biol.,* **112**, 433–444.

Russel, P. J., Crandall, R. E., and Feinbaum, R. (1984), GELYSIS: Pascal implemented analysis of one-dimensional electrophoresis gels, *Nucl. Acids Res.,* **12**, 493–498.

Sallantin, J., Haiech, J., and Rodier, F. (1985), Search for promoter sites of prokaryotic DNA using learning techniques, *Biochimie,* **67**, 549–554.

Salser, W. (1977), Globin mRNA sequences: analysis of base pairing and evolutionary implications, *Cold Spring Harbor Quantitative Symposia,* **42**, 985–1002.

Sankoff, D. (1972), Matching sequences under deletion/insertion constraints, *P.N.A.S.,* **69**, 4–6.

Sankoff, D. (1973), Shortcuts, diversions and maximal chains in partially ordered sets, *Discrete Math,* **4**, 287–293.

Sankoff, D. (1975), Minimal mutation trees of sequences, *SIAM J. Appl. Math,* **78**, 35–42.

Sankoff, D., and Krustal, J. B. (1983), *Time Warps, String Edits, and Macromolecules: The Theory and Practice of Sequence Comparison,* Addison and Wesley.

Saurin, W., and Marlière, P. (1985), Extraction of symbolic patterns common to a family of biosequences, *Biochimie,* **67**, 517–522.

Schneider, T. D., Stormo, G. D., and Gold, L. (1982), A design for computer nucleic-acid sequence storage, retrieval and manipulation, *Nucl. Acids Res.,* **10**, 3013–3024.

Schroeder, J. L., and Blattner, F. R. (1982), Formal description of a DNA oriented computer language, *Nucl. Acids Res.,* **10**, 69–84.

Schwindinger, W. F., and Warner, J. R. (1984), DNA sequence analysis on the IBM-PC, *Nucl. Acids Res.,* **12**, 601–604.

Sellers, P. H. (1974), An algorithm for the distance between two finite sequences, *Comb. Theory,* **16**, 253–258.

Sellers, P. H. (1974), On the theory and computation of evolutionary distances, *SIAM J. Appl. Math,* **26**, 787–793.

Sellers, P. H. (1979), Pattern recognition in genetic sequences, *Proc. Natl. Acad. Sci. USA,* **76**, 3041.

Sellers, P. H. (1980), The theory and computation of evolutionary distances. Pattern recognition, *J. Algorithms,* **1**, 359–373.

Shalloway, D., and Deering, N. R. (1984), Recombinant DNA data management at the restriction and function site level, *Nucl. Acids Res.,* **12**, 739–750.

Shapiro, M. B., and Senapathy, P. (1986), Automated preparation of DNA sequences for publication, *Nucl. Acids Res.,* **14**, 65–73.

Shapiro, B. A., Nussinov, R., Lipkin, L. E., and Maizel, J. V., Jr. (1986), A sequence analysis system encompassing rules for DNA helical distortion, *Nucl. Acids Res.,* **14**, 75–86.

Shepard, R. N. (1980), Multidimentioning scaling, tree-fitting and clustering, *Science, 210*, 390–398.

Sheridan, R. P., Dixon, J. S., Venkatataghavan, R., Kuntz, I. D., and Scott, K. P. (1985), Amino acid composition and hydrophobicity patterns of protein domains correlate with their structures, *Biopolymers, 24*, 1995–2023.

Singhal, R. P., Ray, R. C., and Dobbs, L. (1982), Computer program for storage and retrieval of the nucleic acid structures. Storing and updating of transfer RNA sequences. Drawing of the secondary structure for transfer RNA by computer. *Computer Prog. Biomed., 14*, 277–282.

Smith, T. F., and Waterman, M. S. (1981), Comparison of biosequences, *Adv. Appl. Math., 2*, 482–489.

Smith, T. F., Waterman, M. S., and Fitch, W. M. (1981), Comparative biosequence metrics, *J. Mol. Biol., 18*, 38–46.

Smith, T. F., and Waterman, M. S. (1981), Identification of common molecular subsequences, *J. Mol. Biol., 147*, 195–197.

Smith, D. H., Brutlag, D., Friedland, P., and Kedes, L. H. (1986), BIONET-(tm): national computer resource for molecular biology, *Nucl. Acids Res., 14*, 17–20.

Smith, T. F., Gruskin, K., Tolman, S., and Faulkner, D. (1986), The molecular biology computer research resource, *Nucl. Acids Res., 14*, 25–29.

Sobel, E., and Martinez, H. M. (1986), A multiple sequence alignment program, *Nucl. Acids Res., 14*, 363–374.

Soldano, H., and Moisy, J. L. (1985), Statistico-syntactic learning techniques, *Biochimie, 67*, 493–498.

Söll, D., and Roberts, R. J. (1982), The applications of computers to research on nucleic acids, *Nucl. Acids Res., 10*, 1–458.

Spouge, J. L. (1985), Improving sequence-matching algorithms by working from both ends, *J. Mol. Biol., 181*, 137–138.

Staden, R. (1977), Sequence data handling by computer, *Nucl. Acids Res., 4*, 4037–4051.

Staden, R. (1978), Further procedures for sequence analysis by computer, *Nucl. Acids Res., 5*, 1013–1015.

Staden, R. (1979), A strategy of DNA sequencing employing computer program, *Nucl. Acids Res., 6*, 2601–2610.

Staden, R. (1980), A new computer method for the storage and manipulation of gel DNA reading data, *Nucl. Acids Res., 8*, 817–825.

Staden, R. (1982), An interactive program for comparing and aligning nucleic acid and aminoacid sequences, *Nucl. Acids Res., 10*, 2951–2961.

Staden, R. (1982), Aromatisation of the computer handling of gel reading data produced by the shotgun method of DNA sequencing, *Nucl. Acids Res., 10*, 4731–4751.

Staden, R. (1983), Computer methods for DNA sequencer, in *Laboratory Techniques in Biochemistry and Molecular Biology*, T. S. Work and R. H. Burdon, eds., Vol. 10, pp. 311–368, Elsevier Biomedical Press.

Staden, R. (1984), Computer methods to locate signals in nucleic acid sequences, *Nucl. Acids. Res., 12*, 505–520.

Staden, R. (1984), A computer program to enter DNA gel reading data into a computer, *Nucl. Acids Res., 12*, 499–504.

Staden, R. (1984), Graphics methods to determine the function of nucleic acid sequences, *Nucl. Acids Res., 12*, 521–538.

Staden, R. (1984), Measurements of the effects that coding for a protein has on a DNA sequence and their use for finding genes, *Nucl. Acids Res., 12*, 551–568.

Staden, R. (1986), The current status and portability of our sequence handling software, *Nucl. Acids Res., 14*, 217–231.

Staden, R., and McLachlan, A. D. (1982), Codon preference and its use in identifying protein coding region in long DNA sequences, *Nucl. Acids Res., 10*, 141–156.

Stone, T. W., and Potter, K. N. (1984), A DNA analysis program designed for computer novices working in an industrial research environment, *Nucl. Acids Res., 12*, 367–378.

Stone, T. W., and Potter, K. N. (1986), Methylation blockage and other improvements to a comprehensive DNA analysis program, *Nucl. Acids Res., 14*, 255–264.

Stormo, G. D., Schneider, T. D., and Gold, L. (1982), Characterization of translational initiation sites in *E. coli, Nucl. Acids Res., 10*, 141–156.

Stormo, G. D., Schneider, T. D., Gold, L., and Ehrenfeucht, A. (1982), Use of the "Perceptron" algorithm to distinguish translational sites in *E. coli, Nucl. Acids Res., 10*, 2997–3011.

Stuber, K. (1986), Nucleic acid secondary structure prediction and display, *Nucl. Acids Res., 14*, 317–326.

Studnika, G. M., Rahn, G. M., Cummings, I. W., and Salse, W. A. (1978), Computer method for predicting the secondary structure of single-stranded RNA, *Nucl. Acids Res., 5*, 3365–3387.

Taylor, P. (1984), A fast homology program for aligning biological sequences, *Nucl. Acids Res., 12*, 447–455.

Taylor, P. (1986), A computer program for translating DNA sequences into protein, *Nucl. Acids. Res., 14*, 437–441.

Taylor, W. R., Thornton, J. M. (1983), Prediction of super-secondary structure in proteins, *Nature, 301*, 540–541.

Taylor, W. R., and Thornton, J. M. (1984), Recognition of super-secondary structure in proteins, *J. Mol. Biol., 173*, 487–514.

Tolstoshev, C., and Blakesley, R. W. (1982), RSITE: A computer program to predict the recognition sequence of a restriction enzyme, *Nucl. Acids Res., 10*, 1–17.

Tolstoshev, C. M., Matthes, H. W. D., and Oudet, P. (1986), Computer management of oligonucleotide synthesis on cellulose filters, *Nucl. Acids Res., 14*, 405–415.

Tolstoshev, C., Jeltsch, J. M., Fritz, R., and Oudet, P. (1983), A DNA recombinant data base management system, *Nucl. Acids Res.,* **11**, 4611–4627.

Tramontano, A., and Macchiato, M. F. (1986), Probability of coding of a DNA sequence: an algorithm to predict translated reading frames from their thermodynamic characteristics, *Nucl. Acids Res.,* **14**, 127–135.

Tung, C. S. (1986), Computer graphics programs to reveal the dependence of the gross three dimensional structure of the B-DNA double helix on primary structure, *Nucl. Acids Res.,* **14**, 381–387.

Ukkonen, E. (1983), On approximate string matching, *Proc. Int. Found. Comp. Theor. Lect. Notes Comp. Sci.,* **158**, 487–496.

Valiquette, G., Zimmerman, E. A., and Roberts, J. L. (1985), mRNA sequence predictions from homologous protein sequences, *J. Mol. Biol.,* **181**, 137–138.

Valiquette, G., Zimmerman, E. A., and Roberts, J. L. (1985), mRNA sequence predictions from homologous protein sequences, *J. Theor. Biol.,* **112**, 445–458.

VanBockstaele, F. (1985), Sequence representation, *Biochimie,* **67**, 509–516.

Van Den Berg, J., and Osinga, M. (1986), A peptide to DNA conversion program, *Nucl. Acids Res.,* **14**, 137–140.

Vass, J. K., and Wilson, R. H. (1984), "ZSTATS" a statistical analysis for potential Z-DNA sequences, *Nucl. Acids Res.,* **12**, 825–833.

Watanabe, K., Yasukawa, K., and Iso, K. (1984), Graphic display of nucleic acid structure by a microcomputer, *Nucl. Acids Res.,* **12**, 801–809.

Waterman, M. S. (1983), Frequencies of restriction sites, *Nucl. Acids Res.,* **11**, 8951–8956.

Waterman, M. S. (1983), Sequence alignement in the neighborhood of the optimum with general applications two the dynamic programming, *Proc. Natl. Acad. Sci. USA,* **80**, 3123–3124.

Waterman, M. S. (1984), General methods of sequence comparison, *Bull. Math. Biol.,* **46**, 473–500.

Waterman, M. S., Arratia, R., and Galas, D. J. (1984), Pattern recognition in several sequences: consensus and alignement, *Bull. Math. Biol.,* **46**, 515–527.

Waterman, M. S., Smith, T. F., and Beyer, W. A. (1976), Some biological sequences metrics, *Adv. Math,* **20**, 367–387.

Watterman, M. S. (1976), Secondary structure of single stranded nucleic acids, *Adv. Math. Suppl. Stud.,* **1**, 167–212.

White, C. T., Hardies, S. C., Hutchinson, C. A., and Edgell, M. H. (1984), The diagonal traverse homology search algorithm for locating similarities between two sequences, *Nucl. Acids. Res.,* **12**, 751–766.

Wilbur, W. J., and Lipman, D. J. (1983), Rapid similarity searches of nucleic acid and protein data banks, *Proc. Natl. Acad. Sci. USA,* **80**, 726–730.

Williams, A. L., Jr., and Tinocco, I., Jr. (1986), A dynamic programming algorithm for finding alternative RNA secondary structures, *Nucl. Acids Res.,* **14**, 299–315.

Wilson, I. A., Haft, D. H., Getzoff, E. D., Tainer, J. A., Lerner, A., and Brenner, S. (1985), Identical short peptide sequences in unrelated proteins can have different conformations: a testing ground for theories of immune recognition, *Proc. Natl. Acad. Sci. USA,* **82**, 5255–5259.

Wong, A. K. C., Reichert, T. A., Cohen, D. N., and Aygun, B. O. (1974), A generalized method for matching informational macromolecular code sequences, *Comp. Biol. Med.,* **4**, 43–57.

Yamamoto, K., Kitamura, Y., and Yoshikura, H. (1984), Computation of statistical secondary structures of nucleic acids, *Nucl. Acids Res.,* **12**, 335–346.

Yang, J., Ye, J., and Wallace, D. C. (1984), Computation selection of oligonucleotide probes from aminoacid sequences for use in gene library screening, *Nucl. Acids Res.,* **12**, 837–843.

Zehetner, G., and Lehrach, H. (1986), A computer program package for restriction map analysis and manipulation, *Nucl. Acids Res.,* **14**, 335–349.

Zweig, S. E. (1984), Analysis of large nucleic acid dot matrices on small computers, *Nucl. Acids Res.,* **12**, 767–776.

23

LOCALIZED MUTAGENESIS

The isolation of mutants altered in essential functions has proved to be an invaluable tool for the analysis of the biological regulatory processes governing gene expression. The burst of molecular cloning techniques has offered new ways of studying the effects of gene alterations on the expression of a particular function, therefore leading to the identification of functional domains in biologically important proteins and to the characterization of the DNA regions involved in regulatory mechanisms of cell biology.

OLIGONUCLEOTIDE-DIRECTED MUTAGENESIS

Initial attempts to isolate well-characterized mutations in cloned DNA were based on oligonucleotide-directed in vitro mutagenesis. The basic principles of this method involved the enzymatic extension of an oligonucleotide primer hybridized to a single stranded circular template (Figure 196). It had been shown for several years that DNA duplexes containing mismatches were fairly stable (Astell and Smith, 1971, 1972; Astell et al., 1973; Gillam et al., 1975). Therefore, synthetic oligonucleotides bearing a single base modification with respect to the original complementary sequence, can be annealed to the parental single stranded DNA and elongated by DNA polymerase to generate a mutant genome carrying the precise modification corresponding to that introduced in the oligonucleotide.

Once the complete new genome molecule is synthesized, it can be introduced in a host cell where it is replicated and give rise to a progeny which can be distinguished from the parental molecules by screening procedures.

There were examples of oligonucleotide-directed mutagenesis based on the use of φX174 (Hutchison et al., 1978; Gillam and Smith, 1979; Razin et al., 1978), filamentous fd bacteriophage (Wasylyk et al., 1980), and pBR322 (Wallace et al., 1980, 1981). Soon, the M13-derived vectors (see Chapter 6) appeared to represent a particularly suitable system for oligonucleotide-directed mutagenesis because they allow the easy isolation of single stranded circular templates.

When the partial extension product synthesized from the annealed synthetic oligonucleotide is transfected to competent *E. coli* for generation of the mutant progeny, the 5'-end of the primer is exposed. Thus, the region containing the mismatch can be edited by 5'-3' exonucleases in vivo, and modifications of the original technique were devised so as to include a protection of the 5'-end by ligation (Kramer et al., 1982; Zoller and Smith, 1982; Norris et al., 1983; Zoller and Smith, 1984). A typical protocol (Sor, personal communication) is described below.

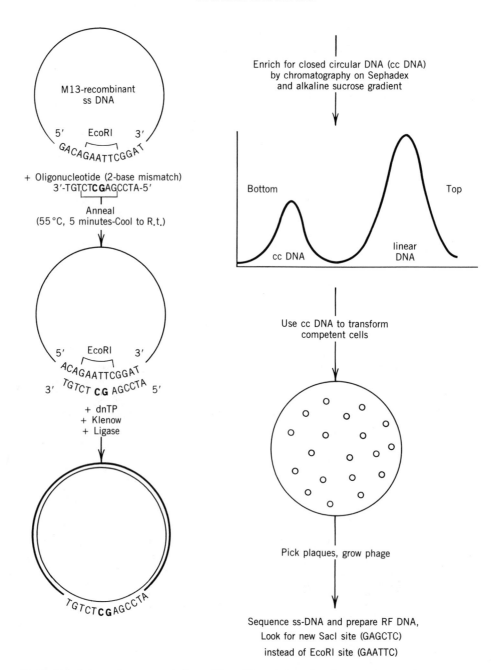

Figure 196. Schematic representation of the different steps for oligonucleotide-directed muta-genesis.

PROTOCOL

OLIGONUCLEOTIDE-DIRECTED MUTAGENESIS

Purification of the Oligonucleotide

Starting from the oligomer detritylated, in NH_4OH.

1. Dry half of the volume of oligomer in a Speed Vac. Store the other half at -20 or $-80°C$. When the oligomer is dry (overnight to 24 hours), you may sometimes get a viscous residue. Disregard it.

2. Resuspend the oligomer in 150 µl 1X TE (pH 8.0)

3. To 50 µl of oligomer, add 10 µl of 10X TBE and 20 µl of Ficoll/dye mix. Store the remaining oligomer at -20 or $-80°C$.

4. Prepare an acrylamide–urea gel.

	16%	20%
Acrylamide/bis 40%	32 ml	40 ml
Urea	40 g	40 g
10X TBE	8 ml	8 ml
H_2O	10 ml	0
APS 10%	0.56 ml	0.56 ml
Temed	25 µl	25 µl

Use either a 20 cm $\times$ 20 cm gel apparatus, or a Hoefer apparatus. Use a comb with 3.5 cm wide wells (3 lanes/gel). You can use a 20% gel for up to 16–18 mer. For longer oligomers (16–18 mer and up), use a 16% gel.

5. Prerun the gel for 30–60 minutes at 20 watts (constant power). Use 1X TBE as running buffer.

6. Run the gel in the same conditions until the Bromophenol blue gets somewhere between 2/3 and the bottom of the gel. The migration requires about 1.5–2 hours. The position of the oligomer depends on its length. On a 16% gel, a 18 mer migrates at about equal distance between xylene cyanol blue and Bromophenol blue. On a 20% gel, a 18 mer migrates one-third of the way between both dyes.

7. View band(s) by placing gel on Saran wrap on top of a TLC plate with fluorescent indicator. View under short-wave UV (hand-held UV lamp). Any type of chromatography plate with a fluorescent indicator works. If you want to take a picture of the gel, use only Eastman type 6065 (or equivalent), and the orange (DNA) filter (the color of the fluorescent indicator of other plates makes the bands disappear in the

background on the picture). The exposure is in the range 30 seconds to 1 minute at $f4.5$. You have to hold the UV lamp above the gel at a 40° angle during the exposure. It is even better if you can use two lamps. Be careful not to confuse the oligomer and the dyes. They also appear as dark bands on the plate. Depending on the purity of the oligomer you received from the synthesis facility, you will see anything between only one well-defined band, and a ladder of bands of heterogenous intensity. In the latter case, whatever it looks like, the upper band is the oligomer you are interested in.

8. Cut out the bands. Crush in a 1.5-ml microfuge tube with a round tipped glass rod.

9. Add 0.8 ml elution buffer. Mix well (vortex). Seal the cap with Parafilm.

10. Incubate overnight at 37°C in a rotating drum.

11. Filter gel bands through a 5-ml syringe in which 2 Whatman #1 filters have been placed. Gravity flow.

12. Wash with 3×1 ml H_2O. Rinse tube and gel pieces with the washes. Pool eluate and washes.

13. Run over DE52 column. Prepare the column in a Pasteur pipette: shorten the tip to about 1 inch. Place a bit of glass wool in the bottom, tight enough to prevent the DE52 flowing through, loose enough to allow the buffers to flow through. Put 1-in. Tygon tubing (1/16–1/8–1/32) at the end of the tip. Pour a 0.5-cm bed of DE52 (aspirate through the tubing with the help of a Pasteur pipette to get rid of the air bubbles trapped in glass wool). Allow the bed to stabilize for a few minutes (use any kind of clamp to close the column). Equilibrate the column with 5 ml ammonium acetate 0.2 M. Add the oligomer to the column. Reequilibrate the column with 5 ml ammonium acetate 0.2 M. Elute the oligomer off with 0.5 ml ammonium acetate 2 M (collect this elution).

14. Freeze and lyophilize (Speed Vac overnight).

15. Resuspend in 1 ml 1X TE and read optical density at 260 and 280 nm.

Mutagenesis

PHOSPHORYLATION OF PRIMER AND PREPARATION OF A ³²P-LABELED PROBE

1. Mix in tube C (cold):
   1 µg oligonucleotide (as prepared above)
   2 µl 1 mM ATP

4μl 5X kinase buffer
4 units of T4-polynucleotide kinase (sequencing grade)
sterile distilled water up to 20 μl

2. Mix in tube H (hot):
   1 μg of oligonucleotide (as prepared above)
   100–200 μCi of $\gamma[^{32}P]$-ATP
   4 μl of 5X kinase buffer
   4 units of T4-polynucleotide kinase (sequencing grade)
   sterile distilled water up to 20 μl

3. Incubate at 37°C for 60 min, add 2 μl 0.5 EDTA, heat at 70°C for 5–10 minutes (you can spike cold reaction with 1 μl of hot reaction to help keep track of oligonucleotide if desired).

4. Add 1 ml $TES_{0.2}$, load onto a 0.1-ml DEAE-cellulose (DE-52) column equilibrated with $TES_{0.2}$.

5. Wash column with two column volumes of $TES_{0.2}$ (10 ml) (you can check eluate with a Geiger counter to ensure all unbound material is washed off).

6. Wash column with 1.2 ml of $TES_{0.5}$ to elute oligonucleotide. Add 3 μl of 20 μg/μl glycogen (carrier), 133 μl 3 M sodium acetate, and mix well. Divide into three microfuge tubes, add 1 ml 95% ethanol to each, mix, and freeze. Spin 10 minutes in a microfuge, wash with 70% ethanol, and dry in Speed Vac for 5 minutes.

7. Resuspend pellets in 20 μl TE (total) for primer, and 300 μl TE for probe. Store frozen.

OLIGONUCLEOTIDE-PRIMED SECOND STRAND SYNTHESIS

1. Mix in a microfuge tube:
   1 μg M13 single stranded DNA carrying the sequence to be altered (see below for details)
   1.5 μl 10X sequencing buffer
   5 μl phosphorylated primer
   sterile distilled water up to 12 μl total

2. Incubate in 95°–100°C heating block for 1–2 minutes, then remove block from heating unit and allow to cool to 30°C (1 hour).

3. Add 3 μl dNTP mix, 2.5 units Klenow fragment (0.5 μl), and incubate at 15°C for 2–4 hours.

4. Add 1 μl 20 mM rATP, 1 unit T4 DNA ligase (1 μl), and incubate at 15°C for 2–12 hours (or overnight).

5. Run 4 μl of the reaction on a 1% agarose minigel next to a single stranded DNA control to check conversion of single strand to RF (generally, most of the single stranded DNA disappear and a dsDNA band appears).

IDENTIFICATION OF MUTANTS

1. Transform JM101/103/105 with 2 μl of the extension reaction and plate a range of dilutions onto TYE or LB plates in a soft agar overlay. Incubate overnight.

2. Remove plates from incubator; place in refrigerator 30–60 minutes to harden top agar.

3. Carefully place an 82-mm Schleicher & Schuell (BA85) nitrocellulose filter onto a plate with well-isolated plaques (100–500/plate). This is easily done by bowing the filter between your fingers (wear gloves), touching the middle of the filter to the plate, and then letting the filter go. Capillary action will quickly wet the filter and make a clean contact between the filter and the surface of the plate. Next, use an 18 gauge needle to place several stabs around the perimeter of the filter (and into the agar) to aid in identifying positive candidates later on. Leave the filter on the plate for 1–2 minutes and then carefully remove it with forceps.

4. Bake the filter at 80°C for 2 hours in a vacuum oven.

5. Incubate the filter for 2 hours at 65°C in prehybridization buffer, which contains:
   10X Denhardt's
   6X SSC
   0.2% SDS
   1 mM EDTA

6. Pour off the prehybridization solution, and add the hybridization solution, which is:
   5X Denhardts
   6X SSC
   ^{32}P-labeled probe (use entire reaction from step 6; preparation of probe).

7. Hybridize at 25°C for 2–12 hours (or overnight).

8. Carefully remove the hybridization solution and save at −20°C (can be reused). Wash the filter at 25°C in 6X SSC (3 washes × 5 minutes each × 100 ml each).

9. Wrap the filter in Saran wrap and autoradiograph for 1–2 hours (with intensifying screen). All plaques should light up at this stage.

10. Wash the filter in 6X SSC at successively higher temperatures (5–10°C increments) until only mutant spots are seen on autoradiograph.

11. Go back to the 25°C film and mark the positive plaques. Also mark the stab marks made earlier using the 18 gauge needle.

12. Align this film with the original plate and mark the positive plaques on the plate.

13. Prepare single stranded DNA from the mutants and sequence (see Chapter 20).

PREPARATION OF CLEAN M13 ssDNA FOR MUTAGENESIS AND SEQUENCING

1. Grow 5–10 ml of M13 phage in TYE for 6–8 hours.

2. Centrifuge at 9000 rpm for 5 minutes.

3. Precipitate phage by adding 2.5 *M* NaCl/20% PEG-8000 to a final concentration of 0.5 *M* NaCl/4% PEG. Let sit at 25°C for 30 minutes.

4. Centrifuge at 5000 rpm for 10 minutes to pellet phage. Resuspend pellet in 1/10 original volume of TE–0.5% sarcosyl.

5. Reprecipitate with same NaCl/PEG concentration. Centrifuge at 5000 rpm for 10 minutes to repellet phage.

6. Resuspend the pellet in 1/50 original volume in TE. Extract single stranded DNA by 2-3 phenol/chloroform extractions and precipitate DNA with ethanol. Wash the pellet with 70% ethanol and resuspend in 1/200 original culture volume.

Buffers and Solutions

Elution buffer
500 m*M* Ammonium acetate
10 m*M* MgCl$_2$
1 m*M* EDTA (pH 7.5)

1 liter
250 ml 2 *M* Ammonium acetate
10 ml 1 *M* MgCl$_2$
2 ml 0.5 M EDTA (pH 7.5)
738 ml H$_2$O

10X TBE
500 m*M* Tris–base
500 m*M* Boric acid
1 m*M* EDTA (pH 7.5)

1 liter
60.5 g Tris–base
30.9 g Boric acid
2 ml 0.5 *M* EDTA (pH 7.5)

DNA loading buffer
10% Ficoll
0.1% Xylene cyanol blue
0.1% Bromophenol blue
50 m*M* EDTA (pH 8.0)
Keep at room temperature.

TE buffer
10 m*M* Tris–HCl (pH 8.0)
0.1 m*M* EDTA (pH 8.0)

1 liter
10 ml 1 *M* Tris–HCl (pH 8.0)
0.2 ml 0.5 *M* EDTA (pH 8.0)
989.8 ml H$_2$O

5X Kinase buffer   *100 μl*
350 mM Tris–HCl (pH 7.6)   35 μl 1 M Tris–HCl (pH 7.6)
50 mM MgCl$_2$ 5 μl 1 M MgCl$_2$
25 mM DTT   2.5 μl 1 M DTT
     57.5 μl H$_2$O

10X sequencing buffer   *1 liter*
100 mM Tris–HCl (pH 8.0)   100 ml 1 M Tris–HCl (pH 8.0)
50 mM MgCl$_2$ 50 ml 1 M MgCl$_2$
     850 ml H$_2$O

50X Denhardt's solution   *200 ml*
1% Bovine serum albumin   2 g BSA (Pentex)
1% Ficoll   2 g Ficoll
1% Polyvinylpyrrolidone   2 g Polyvinylpyrrolidone
     H$_2$O to 200 ml

dNTP mix   *50 μl*
2.5 mM of each dNTP   6.25 μl 20 mM dATP
20 mM DTT   6.25 μl 20 mM dCTP
     6.25 μl 20 mM dGTP
     6.25 μl 20 mM TTP
     10.0 μl 100 mM DTT
     15 μl H$_2$O

TES 0.2   *100 ml*
10 mM Tris–HCl (pH 7.6)   1 ml 1 M Tris–HCl (pH 7.6)
1 mM EDTA (pH 7.5)   0.2 ml 0.5 M EDTA (pH 7.5)
200 mM NaCl   4 ml 5 M NaCl
     94.8 ml H$_2$O

TES 0.5   *100 ml*
10 mM Tris–HCl (pH 7.6)   1 ml 1 M Tris–HCl (pH 7.6)
1 mM EDTA (pH 7.5)   0.2 ml 0.5 M EDTA (pH 7.5)
500 mM NaCl   10 ml 5 M NaCl
     88.8 ml H$_2$O

To increase the yield of mutant phages in the progeny, and to allow the sequential addition of mutations in the same DNA strand, Carter et al. (1985) developed a coupled priming method in which the DNA is first cloned in a new M13 vector carrying a genetic marker (such as Eco K, Eco B, or amber) that can be selected against in the appropriate host. In this method, two primers are used simultaneously: one of them is designed to construct the silent mutation required while the other is used to eliminate the chosen selectable marker on the minus strand. Several rounds of mutagenesis can therefore be performed to generate multiple-mutant genomes.

More recently, two methods based on the use of thionucleotides were designed for high-efficiency in vitro mutagenesis (Taylor et al., 1985). They rely on the observation that several restriction enzymes are unable to linearize phosphorothioate DNA but, instead generate a nicked circular molecule from supercoiled DNA. These enzymes only cut the DNA which does not contain thionucleotides.

Although the oligonucleotide-directed mutagenesis has proved to be the most powerful technique of the last decade to dissect the regulatory mechanisms of gene expression in both procaryotes and eucaryotes, its use has unfortunately remained limited, for some time, to the laboratories equipped with oligonucleotide synthesizers, and, later, to the laboratories that could afford to purchase high-quality oligonucleotides from commercial firms offering custom synthesis programs (see, for example, Pharmacia, New England Laboratories, or International Biotechnologies Inc.).

LINKERS MUTAGENESIS

Linker-Scanning Mutagenesis

Linker-scanning mutagenesis has first been described by McKnight and Kingsbury (1982) in their study of the transcriptional signals governing the herpes simplex virus (HSV) thymidine kinase (TK) gene. The method is based on the substitution of small clusters of nucleotides residue by molecular linkers, without addition or deletion of other sequences.

In their original study, McKnight and Kingsbury generated different sets of 5' and 3' deletions of the TK gene by sequential treatment of linearized DNA with exonuclease III and S1 nuclease, under standard conditions (Sakonju et al., 1980). Sequencing of 43 different 5'-deletion mutants and 42 different 3'-deletion mutants allowed selection of "matching" mutants whose deletion termini were separated by 10 nucleotides (Figure 197). Addition of a 10 mer-linker at each of the mutant end allows to replace the 10 nucleotides that formerly separated the two deletion termini by the linker sequence (Figure 197) without changing anything else in the primary structure of the corresponding product. Linker scanning has been used with synthetic oligonucleotides of different length in a wide variety of biological systems to correlate the phenotypic effect of mutations with the corresponding nucleotide substitution created after linker insertion. Several modifications have been brought to the linker mutagenesis technique in order to facilitate the generation of deletions, insertions, and clustered point mutations.

These include the use of linker cassettes that can be used to fuse a synthetic restriction site sequence to the endpoints of deletions via intramolecular ligation (Haltiner et al., 1985). However, both the original and the modified procedures require the construction and the sequence analysis of a large number of 5' and 3' deletions to obtain finally a rather limited number of linker scanning mutants. For example, McKnight and Kingsbury (1982) had to sequence 43 5'- and 42 3'-deletions to obtain 15 matching pairs, Haltiner et al. (1985) analyzed 35 5'- and 45 3'-deletions to obtain 7 mutants, while Buetti and Kühnel (1986) had to sequence 60 5'- and 49 3'-deletions to construct 9 mutants.

To overcome the need of large-scale screenings, Luckow et al. (1987) developed a method relying on the use of formic acid, exonuclease III, and nuclease S1. First, the plasmid DNA containing the sequence to be mutated is treated with 0.2% formic acid to induce a partial apurination. Then, a specific nicking at the apurinic sites is performed by incubation in the presence of exonuclease III and the second strand of the DNA molecule is cleaved at a position opposite to a gap (or a nick) by incubation with S1 nuclease. These steps ensure a linearization and a slight shortening of the original DNA molecule. After addition of a linker sequence to the new generated ends, the recombinant plasmids are enriched under standard conditions.

The mutant plasmids are screened by comparing their topoisomer patterns with that of the parental plasmid, after relaxation with topoisomerase I (see Chapter 4). Since this technique can be used to detect differences in length as small as 1 base pair (Wang, 1979), it allows selection of DNA molecules which show the wild-type pattern and which therefore are likely to contain a linker molecule replacing an equivalent number of bases, or corresponding to parental DNA. In their analysis of 26 mutants displaying topoisomer patterns identical to that of wild type, Luclow et al. (1987) found that 21 were correct linker scanning mutants.

Although the linker scanning methods have proved to be very powerful in the functional study of small DNA regions involved in interactions with regulatory proteins, they have not been extensively used to elucidate the role of specific domains recognized in a considerable number of different proteins. Among the possible reasons for the little usage of this method are the disadvantages resulting from the possibility of creating frame shift mutations and those due to the limitation of the different codons that can be inserted.

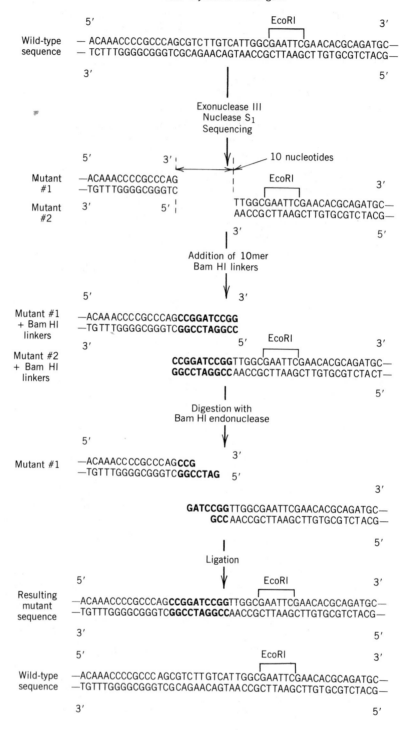

Figure 197. Schematic representation of linker scanning mutagenesis.

TAB Linker Mutagenesis

Mutagenesis using TAB linkers overcomes these problems by being more versatile and easier to use. A TAB linker is a single stranded hexameric oligonucleotide which contains a sequence complementary to either a two-base or a four-base restriction site overhang, the remaining bases being palindromic. Therefore, insertion of a TAB linker into a plasmid creates a new restriction site from the combination of the palindromic sequences of the TAB linker and the adjacent nucleotides of the insertion site (Figure 198).

In vitro mutagenesis using hexameric linkers has numerous technical advantages compared to other procedures. For example:

TAB linker insertion involves a simple ligation reaction resulting in two (or four) codon insertion which never destroys preexisting codons or causes frameshifts.

TAB linker insertion can be selected and assayed using either biological or biochemical selection.

TAB linker insertion creates a new restriction site easily assayed by conventional restriction enzyme analysis.

Thus, TAB linker mutagenesis provides a rapid and simple method for the introduction of numerous and easily detected mutations over a large region of DNA and for doing the initial mutagenesis studies on a cloned gene. The insertion of two amino acids causes relatively minor perturbation of protein structure and has been shown to generate both temperature sensitivity and modified enzyme-substrate specificity (Barany, 1985a, 1985b).

TAB linker insertion strategy is illustrated in Figure 199.

Biochemical Selection.* Biochemical selection involves plasmid recircularization followed by redigestion with the restriction endonuclease used for the initial cleavage. This process enriches for plasmids which contain TAB linkers and thus lack the original restriction site. Such selection is not possible in cases where the plasmid contains multiple cleavage sites for the enzyme. However, in many cases, it is possible to circumvent this problem. Having linearized the plasmid by partial digestion, additional restriction enzyme sites are methylated and thereby

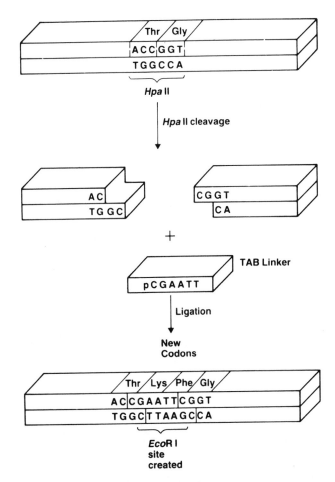

Figure 198. Schematic representation of TAB linker mutagenesis.

(1) Circular plasmid DNA is linearized using Hpa II, a restriction enzyme that recognizes multiple sites within most plasmids, leaving 5'-CG overhanging ends. Conditions for partial digestion are used to generate linearized molecules cleaved at random sites.

(2) Single stranded TAB linkers (5'-pCGAATT in the example shown) can now be ligated to the Hpa II ends since the two nucleotides at the 5'-end of the TAB linker are complementary to the overhang and thereby permit annealing and subsequent ligation with T4-DNA ligase. The ligation reaction converts the 5'-CG Hpa II cohesive end into a AATT-3' cohesive end and permits ligation of a second TAB linker to the complementary strand.

(3) The addition of TAB linker generates an Eco RI site and, since more than one linker may be added to each end, the plasmid ends are trimmed with Eco RI or Asu II (insertion of two linkers generates an Asu II site, TTCGAA). This removes excess linkers and ensures that the overhanging ends will be compatible with each other (for biochemical selection) or with the ends of a kanamycin resistance gene cartridge cassette (for biological selection). By inserting a TAB linker at a 3'-overhanging end, an unphosphorylated TAB linker can be ligated. As the insertion is limited to a single linker addition, the trimming step is not needed (step 3). TAB linker-containing plasmids are then selected and isolated using either biochemical (enzymatic) or biological (kanamycin resistance) selection (courtesy of Pharmacia).

*Courtesy of Pharmacia

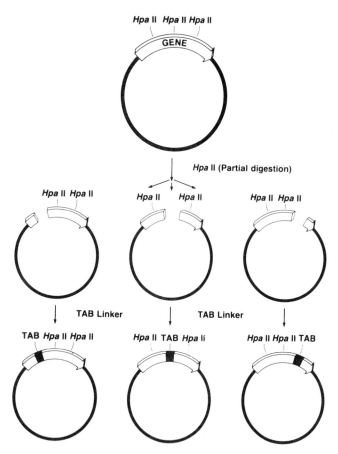

Figure 199. Schematic representation of TAB linker mutagenesis. Insertion of a TAB linker into a gene may create a mutation that is phenotypically different from the wild-type protein (e.g., mutant enzyme with temperature sensitivity, altered substrate specificity, decreased or increased activity), or silent (courtesy of Pharmacia).

Biological Selection.* When the inserted restriction site is unique in the TAB linker modified plasmid, biological selection is possible. Excess linkers are trimmed by digestion with the enzyme whose recognition site is being inserted. The plasmid DNA is then ligated to a kanamycin resistance (Kan[r]) gene cartridge bearing the same recognition site. Transformants are selected for growth in kanamycin containing media. DNA is then isolated from Kan[r] colonies and the Kan[r] gene cartridge is excised by digestion at the original site. TAB linker modified plasmids are religated and recircularized following transformation into *E. coli*. The precise location of insertion of the TAB linker is finally determined by simple restriction endonuclease mapping (Figure 200).

Choosing a TAB Linker.* TAB linker mutagenesis can be performed on any cloned gene or DNA fragment which has been mapped, in detail, with restriction endonucleases. Availability of the nucleotide sequence around TAB linker insertion sites makes it possible to determine which dipeptide sequences will be inserted.

The appropriate TAB linker can then be chosen according to one of the criteria listed below.

1. *TAB linker "shotgun" mutagenesis.* This method is used for creating numerous mutations in a single experiment, spread throughout a gene. Select the restriction site at which the mutation is to be created. Locate the chosen restriction site and corresponding TAB group in Table 75 and locate the TAB group in Table 76 (Table 76 lists the TAB linkers that can be inserted into the chosen site). Choose the most appropriate TAB linker according to whether you want to insert a specific restriction endonuclease site (under "New Site Conversion") or a specific amino acid sequence (under "Amino Acids Inserted"). Remember that the reading frame into which the TAB linker will be inserted is dependent on the cutting site of the restriction enzyme used.

2. *Restriction endonuclease site insertion.* Choose the restriction endonuclease site to be inserted into the cloned DNA fragment from the alphabetical listing in Table 75. In column 4, locate the corresponding TAB group, as listed in Table 76. Read across

rendered resistant to cleavage (Kessler et al., 1985). In the example shown in Figure 199, methylation of the linearized plasmid by *Hpa* II methylase prior to TAB linker ligation renders the uncleaved *Hpa* II recognition sites resistant to *Hpa* II digestion. After TAB linker insertion, the plasmid is recircularized by ligation under dilute conditions and plasmids lacking a TAB insertion regenerate an unmethylated *Hpa* II site. Unmodified plasmids are linearized by *Hpa* II digestion to greatly reduce their transformation efficiency compared to the circular, TAB linker-modified plasmids. After transformation, plasmid DNA is prepared and screened for the presence and location of the new restriction site created by TAB linker insertion.

*Courtesy of Pharmacia

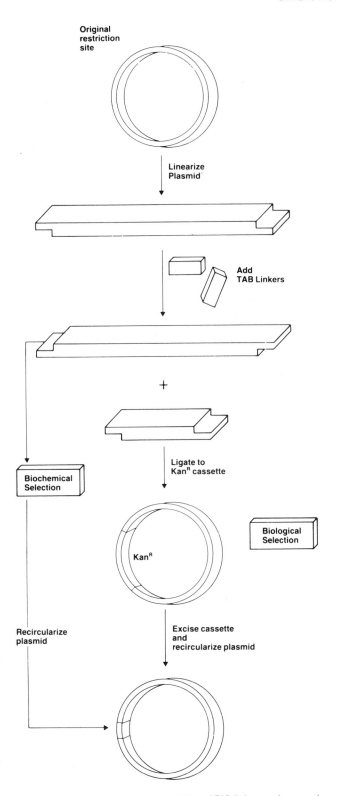

Figure 200. Schematic representation of TAB linker mutagenesis and selection (courtesy of Pharmacia).

Table 76 and find the desired restriction endonuclease site in column 5 (+1 linker) or column 6 (+2 linker). Single TAB linker addition will generate only the restriction site listed in column 5 (+1 linker). Multiple linker additions will generate the restriction sites listed in both columns 5 and 6 (+1 *and* +2 linkers). Choose the appropriate TAB linker based on the restriction sites available for TAB insertion into the gene (under "Insertion Restriction Site").

3. *Amino acid insertion.* Choose the amino acids to be inserted into the gene. Locate the desired dipeptide sequence in columns 7, 8, and 9 of Table 76 (under "Amino Acids Inserted"). Read across the columns and choose the appropriate TAB linker, listed under the column "Sequence," depending on the restriction endonuclease site(s) available in the gene (under "Original Restriction Site"). Consideration of the reading frame into which the TAB linker will be inserted is essential.

The choice of specific procedures for insertion and detection of a TAB linker is dependent on:

1. Whether the restriction enzyme used to linearize the plasmid leaves a 5'- or a 3'-overhanging end. Ligation to a 5'-overhanging end requires a 5'-phosphorylated TAB linker and results in multiple linker addition. Ligation of TAB linkers to a 3'-overhanging end utilizes unphosphorylated (3'-OH) linkers and results in single linker additions.
2. Whether the original restriction site is a unique site in the gene and the plasmid vector.
3. Whether the TAB linker-generated site will be unique in the gene *and* the plasmid vector.
4. Whether cleavage of the insertion restriction site is inhibited by methylation, and whether cleavage of the TAB linker-generated site is inhibited by methylation.

If the original restriction site occurs more than once within the plasmid, biochemical selection may not be possible unless cleavage at these additional sites can be inhibited. If the TAB linker-generated restriction site occurs at other locations within the plasmid, trimming excess linkers may not be possible unless cleavage at internal sites can be inhibited.

The protocol needed for successful use of TAB linkers follows.

Table 75. Alphabetical List of Restriction Enzymes Compatible with TAB Linker Mutagenesis

Enzyme[a]	Recognition Sequence	TAB Group of Insertion Site[b]	TAB Group of Generated Site[c]
Enzymes Leaving Two-Base or Four-Base Overhanging Ends			
Aat II	GACGT'C	10	2, 6, 10, 38
Acc I	GT'ATAC	5	1, 2, 5, 6, 10
	GT'CGAC	6	See Sal I
Afl II	C'TTAAG	40	3, 8, 11, 27, 40
Afl III	A'CATGT	29	8, 20, 36, Nsp HI
	A'CGCGT	31	See Mlu I
Aha II	GPu'CGPyC	6	See Aat II, Nar I
Asp 718	G'GTACC	36	See Kpn I
Asu II	TT'CGAA	6	1, 2, 5, 6, 25, 38
Ava I	C'CCGGG	30	See Xcy I
	C'TCGAG	38	See Xho I
Avr II	C'CTAGG	32	3, 19, 35
Bam HI	G'GATCC	33	2, 6, 30
Ban I	G'GCGCC	34	See Nar I
	G'GTACC	36	See Kpn I
Ban II	GAGCT'C	11	See Sac I
	GGGCC'C	19	See Apa I
Bcl I	T'GATCA	33	1, 5, 13, 29
Bbe AI	GGCGC'C	18	See Nar I
Bgl II	A'GATCT	33	1, 5, 8, 32, 33
Bss HII	G'CGCGC	31	2, 6, 18, 34
Cla I	AT'CGAT	6	2, 6, 8, 38
Eco RI	G'AATTC	25	1, 2, 5, 6, 25, 38
Hae II	PuGCGC'Py	18	2, 6, 18, 32, 34, Nar I
Hgi AI	GAGCT'C	11	See Sac I
	GTGCA'C	23	2, 6, 10
Hha I	GCG'C	2	—
Hind III	A'AGCTT	27	3, 8, 11, 27, 40
Hin PI	G'CGC	6	—
Hpa II	C'CGG	6	—
Kpn I	GGTAC'C	20	2, 6, 30
Mae I	C'TAG	8	—
Mae II	A'CGT	6	—
Mbo I	'GATC	33	—
Mlu I	A'CGCGT	31	8, 20, 36
Nar I	GG'CGCC	6	2, 6, 18, 30, 34
Nco I	C'CATGG	29	1, 3, 5, 13, 29, 35
Nde I	CA'TATG	8	3, 8, 23
Nhe I	G'CTAGC	32	2, 6, 18, 34
Nla III	CATG'	13	—
Not I	GC'GGCCGC	35	—
Nsi I	ATGCA'T	23	3, 8, 23
Nsp HI	PuCATG'Py	13	8, 20, 36, Sph I
Nsp II	CAGCT'G	11	See Pvu II
	CGGCC'G	19	See Xma III
	CTGCA'G	23	See Pst I

Table 75.   Alphabetical List of Restriction Enzymes Compatible with TAB Linker Mutagenesis (*continued*)

Enzyme[a]	Recognition Sequence	TAB Group of Insertion Site[b]	TAB Group of Generated Site[c]
Pst I	CTGCA'G	23	3, 11, 23, 27
Pvu I	CGAT'CG	1	1, 3, 5, 31, 33
Sac I	GAGCT'C	11	2, 3, 6, 11, 27, 38
Sac II	CCGC'GG	3	3, 19, 31, 35
Sal I	G'TCGAC	38	2, 6, 10, 38
Spe I	A'CTAGT	32	8, 20, 32, 36 (continued)
Sph I	GCATG'C	13	1, 2, 5, 6, 13, 18, 29, 34
Taq I	T'CGA	6	—
Xba I	T'CTAGA	32	1, 5, 8, 32, 33
Xcy I (Xma I)	C'CCGGG	30	2, 3, 6, 19, 30, 35
Xho I	C'TCGAG	38	2, 3, 6, 11, 27, 38
Xho II	Pu'GATCpy	33	See Bam HI, Bgl II
Xma III	C'GGCCG	35	3, 19, 31, 35

Enzymes That Leave Blunt Ends[d]

Enzyme	Recognition Sequence		TAB Group of Generated Site
Aha III	TTT'AAA		1, 5, 25
Bal I	TGG'CCA		1, 3, 5, 13, 19, 29, 35
Eco RV	GAT'ATC		2, 6, 8, 38
Fsp I	TGC'GCA		1, 2, 5, 6, 13, 18, 29, 34
Hpa I	GTT'AAC		2, 6, 10
Nae I	GCC'GGC		2, 6, 18, 30, 34
Nru I	TCG'CGA		1, 3, 5, 31, 33
Pvu II	CAG'CTG		3, 11, 23, 27
Sca I	AGT'ACT		8, 20, 32, 36
Sna BI	TAC'GTA		1, 2, 5
Ssp I	AAT'ATT		8, 40
Stu I	AGG'CCT		3, 18, 19, 32, 35

Isoschizomers

Enzyme	Listed Under	Enzyme	Listed Under
Aoc II	Nsp II	Nde II	Mbo I
Aos II	Aha II	Nsp III	Ava I
Ban III	Cla I	Nun II	Nar I
Bsp 1286	Nsp II	Pae R7I	Xho I
Bst I	Bam HI	Sau 3AI	Mbo I
Cfo I	Hha I	Sex I	Xho I
Dpn I	Mbo I	Sst I	Sac I
Dra I	Aha III	Sst II	Sac II
Msp I	Hpa II	Xma I	Xcy I
Mst I	Fsp I	Xor II	Pvu I

[a]Restriction endonucleases that recognize 4-base or 6-base palindromic sequences and generate 2-base or 4-base overhanging ends.

[b]See Table 2, "Insertion Restriction Site."

[c]See Table 2, "TAB Generated Restriction Site."

[d]TAB linkers cannot be ligated to blunt-end restriction sites. However, blunt-end restriction endonuclease recognition sites can be generated by TAB linker insertion.

Table 76. Correlation Between TAB-Generated Restriction Sites and Amino Acids Inserted in the Target Sequence

TAB Group[a]	Insertion Restriction Site[b] (R.E.1)	TAB Linker Product Number	Sequence (5'–3')	TAB-Generated Restriction Site (R.E.2)[c]		Amino Acids Inserted		
				+1 Linker	+2 Linker	RF 1	RF 2	RF 3
1	Pvu I		ACGTAT	SnaB I[f]	Acc I[g]	Ile Arg	Tyr Val	Thr Tyr
			CGCGAT	Nru I[f]	Pvu I	Ile Ala	Ser Arg	Arg Asp
			CTAGAT	Xba I	Bgl II	Ile ***	Ser Arg	Leu Asp
			GATCAT	Bcl I	Bcl I	Met Ile	*** Ser	Asp His
			GCGCAT	Fsp I[f]	Sph I	Met Arg	Cys Ala	Ala His
			GGCCAT	Bal I[f]	Nco I	Met Ala	Trp Pro	Gly His
		27-8548[h]	TCGAAT	Asu II	EcoR I	Ile Arg	Phe Glu	Ser Asn
			TTAAAT	Aha III[f]	Aha III	Ile ***	Phe Lys	Leu Asn
2	Hha I	27-8432	AATTCG	EcoR I	Asu II	Arg Ile	Glu Phe	Asn Ser
			ACGTCG	Aat II	Sal I	Arg Arg	Asp Val	Thr Ser
		27-8556[h]	AGCTCG	Sac I	Xho I	Arg Ala	Glu Leu	Ser Ser
			ATATCG	EcoR V	Cla I	Arg Tyr	Asp Ile	Ile Ser
		27-8562[h]	CATGCG	Sph I	Fsp I	Arg Met	Ala Cys	His Ala
			CCGGCG	Nae I[f]	Nar I	Arg Arg	Ala Gly	Pro Ala
			CGCGCG	BssH II	BssH II	Arg Ala	Ala Arg	Arg Ala
			CTAGCG	Nhe I	Hae II[g]	Arg ***	Ala Ser	Leu Ala
		27-8420[h]	GATCCG	BamH I	BamH I	Arg Ile	Gly Ser	Asp Pro
			GCGCCG	Nar I	Nae I	Arg Arg	Gly Ala	Ala Pro
		27-8542[h]	GGCCCG	Apa I	Xcy I	Arg Ala	Gly Pro	Gly Pro
		27-8564[h]	GTACCG	Kpn I	Kpn I	Arg Tyr	Gly Thr	Val Pro
			TATACG	Acc I[g]	SnaB I[f]	Arg Ile	Val Tyr	Tyr Thr
		27-8422[h]	TCGACG	Sal I	Ant II	Arg Arg	Val Asp	Ser Thr
			TGCACG	Hgi I[g]	Hgi I	Arg Ala	Val His	Cys Thr
			TTAACG	Hpa I[f]	Hpa I	Arg ***	Val Asn	Leu Thr
3	Sac II	27-8560[h]	AGCTGC	Pvu II[f]	Pst I	Ala Ala	Glu Leu	Ser Cys
			ATATGC	Nde I	Nsi I	Ala Tyr	His Met	Ile Cys
			CATGGC	Nco I	Bal I[f]	Ala Met	Pro Trp	His Gly
		27-8428	CCGGCC	Xcy I	Apa I	Ala Arg	Pro Gly	Pro Gly
			CGCGGC	Sac II	Xma III	Ala Ala	Pro Arg	Arg Gly
			CTAGGC	Avr II	Stu I[f]	Ala ***	Pro Arg	Leu Gly
			GATCGC	Pvu I	Nru I	Ala Ile	Arg Ser	Asp Arg
			GGCCGC	Xma III	Sac II	Ala Ala	Arg Pro	Gly Arg
		27-8424	TCGAGC	Xho I	Sac I	Ala Arg	Leu Glu	Ser Ser
		27-8566	TGCAGC	Pst I	Pvu II[f]	Ala Ala	Leu Gln	Cys Ser
			TTAAGC	Afl II	Hind III	Ala ***	Leu Lys	Leu Ser
4								
5	Acc I[e]		pATACGT	SnaB I[f]	Acc I[g]	Ile Arg	Tyr Val	Thr Tyr
			pATCGCG	Nru I	Pvu I	Ile Ala	Ser Arg	Arg Asp
			pATCTAG	Xho II	Bgl I	Ile ***	Ser Arg	Leu Asp
			pATGATC	Bcl I	Bcl I	Met Ile	*** Ser	Asp His
			pATGCGC	Fsp I[f]	Sph I	Met Arg	Cys Ala	Ala His
			pATGGCC	Bal I[f]	Nco I	Met Ala	Trp Pro	Gly His
			pATTCGA	Asu II	EcoR I	Ile Arg	Phe Glu	Ser Asn
			pATTTAA	Aha III[f]	Aha III	Ile ***	Phe Lys	Leu Asn
6	Acc I[e]	27-8534	pCGAATT	EcoR I	Asu II	Arg Ile	Glu Phe	Asn Ser
	Aha II		pCGACGT	Aat II	Sal I	Arg Arg	Asp Val	Thr Ser
	Asu II	27-8536	pCGAGCT	Sac I	Xho I	Ser Ser	Glu Leu	Arg Ala
	Cla I		pCGATAT	EcoR V[f]	Cla I	Arg Tyr	Asp Ile	Ile Ser
	HinP I	27-8544	pCGCATG	Sph I	Fsp I[f]	Arg Met	Ala Cys	His Ala
	Hpa II		pCGCCGG	Nae I	Nar I	Arg Arg	Ala Gly	Pro Ala
	Mae II		pCGCGCG	BssH II	BssH II	Arg Ala	Ala Arg	Arg Ala
	Nar I		pCGCTAG	Nhe I	Hae II[g]	Arg ***	Ala Ser	Leu Ala

Table 76. Correlation Between TAB-Generated Restriction Sites and Amino Acids Inserted in the Target Sequence (*continued*)

TAB Group[a]	Insertion Restriction Site[b] (R.E.1)	TAB Linker Product Number	Sequence (5'–3')	TAB-Generated Restriction Site (R.E.2)[c]		Amino Acids Inserted		
				+1 Linker	+2 Linker	RF 1	RF 2	RF 3
	Taq I	27-8538	pCGGATC	*Bam*H I	*Bam*H I	Arg Ile	Gly Ser	Asp Pro
			pCGGCGC	*Nar* I	*Nae* I[f]	Arg Arg	Gly Ala	Ala Pro
		27-8540	pCGGGCC	*Apa* I	*Xcy* I	Arg Ala	Gly Pro	Gly Pro
		27-8558	pCGGTAC	*Kpn* I	*Kpn* I	Arg Tyr	Gly Thr	Val Pro
			pCGTATA	*Acc* I	*Acc* I	Arg Ile	Val Tyr	Tyr Thr
		27-8546	pCGTCGA	*Sal* I	*Aat* II	Arg Arg	Val Asn	Ser Thr
			pCGTGCA	*Hgi*A 3[g]	*Hgi*A 3	Arg Alg	Val His	Cys Thr
			pCGTTAA	*Hpa* I	*Hpa* I	Arg ***	Val Asn	Leu Thr
8	*Mae* I		pTAAGCT	*Hind* III	*Afl* II	*** Ala	Lys Leu	Ser Leu
	Nde I		pTAATAT	*Ssp* I[f]	*Ssp* I	*** Tyr	Asn Ile	Ile Leu
			pTACATG	*Nsp*H I[g]	*Nsp*H I	Tyr Met	Thr Cys	His Val
			pTACGCG	*Mlu* I	*Mlu* I	Tyr Ala	Thr Arg	Arg Val
			pTACTAG	*Spe* I	*Sca* I[f]	Tyr ***	Thr Ser	Leu Val
			pTAGATC	*Bgl* II	*Xba* I	*** Ile	Arg Ser	Asp Leu
			pTAGCGC	*Hae* II[g]	*Hae* II	*** Arg	Ser Ala	Ala Leu
			pTAGGCC	*Stu* I[f]	*Avr* II	*** Ala	Arg Pro	Gly Leu
			pTAGTAC	*Sca* I[f]	*Spe* I	*** Tyr	Ser Thr	Val Leu
			pTATCGA	*Cla* I	*EcoR* V[f]	Tyr Arg	Ile Asp	Ser Ile
			pTATGCA	*Nsi* I	*Nde* I	Tyr Ala	Met His	Cys Ile
9								
10	*Aat* II		ATACGT	*Acc* I	*Acc* I	Val Tyr	Tyr *Thr*	Ile *Arg*
			CGACGT	*Sal* I	*Aat* III	Val Asp	Ser *Thr*	Arg *Arg*
			GCACGT	*Hgi*A I[g]	*Hgi*A I	Val His	Cys *Thr*	Ala *Arg*
			TAACGT	*Hpa* I[f]	*Hpa* I	Val Asn	Leu *Thr*	*** *Arg*
11	*Hgi*A I[e]	27-8536[i]	CGAGCT	*Xho* I	*Sac* I	Leu Glu	Ser *Ser*	Arg *Ala*
	Nsp II[e]	27-8550[i]	GCAGCT	*Pst* I	*Pvu* II[f]	Leu Gln	Cys *Ser*	Ala *Ala*
	Sac I		TAAGCT	*Afl* II	*Hind* III	Leu Lys	Leu *Ser*	*** *Ala*
12								
13	*Nla* II		ATCATG	*Bcl* I	*Bcl* I	*** Ser	Asp *His*	Ile *Met*
	Nsp I	27-8544[i]	CGCATG	*Fsp* I[f]	*Sph* I	Cys Ala	Ala *His*	Arg *Met*
	Sph I		GCCATG	*Bal* I[f]	*Nco* I	Try Pro	Gly *His*	Ala *Met*
14								
15								
16								
17								
18	*Bbe* I		ATGCGC	*Sph* I	*Fsp* I[f]	Ala Cys	His *Ala*	Met *Arg*
	Hae II		CGGCGC	*Nae* I[f]	*Nae* I	Ala Gly	Pro *Ala*	Arg *Arg*
			GCGCGC	*Bss*H II	*Bss*H II	Ala Arg	Arg *Ala*	Ala *Arg*
			TAGCGC	*Nhe* I	*Hae* II[g]	Ala Ser	Leu *Ala*	*** *Arg*
19	*Apa* I		ATGGCC	*Nco* I	*Bal* I[f]	Pro Trp	His *Gly*	Met *Ala*
	Nsp II[e]	27-8540[i]	CGGGCC	*Xcy* I	*Xcy* I	Pro Gly	Pro *Gly*	Arg *Ala*
			GCGGCC	*Sac* II	*Xma* III	Pro Arg	Arg *Gly*	Ala *Ala*
			TAGGCC	*Avr* II	*Avr* II	Pro Arg	Leu *Gly*	*** *Ala*
20	*Kpn* I		ATGTAC	*Nsp*H I[g]	*Nsp*H I	Thr Cys	His *Val*	met *Tyr*
			GCGTAC	*Mfu* I	*Mfu* I	Thr Arg	Arg *Val*	Ala *Tyr*
			TAGTAC	*Spe* I	*Sca* I	Thr Ser	Leu *Val*	*** *Tyr*
21								
22								
23	*Hgi*A I[e]		GCTGCA	*Pvu* II[f]	*Pst* I	Gln Leu	Ser *Cys*	Ala *Ala*
	Nsp II[e]		TATGCA	*Nde* I	*Nsi* I	His Met	Ile *Cys*	Tyr *Ala*
	Pst I							
	Nsi I							

Table 76. Correlation Between TAB-Generated Restriction Sites and Amino Acids Inserted in the Target Sequence (*continued*)

TAB Group[a]	Insertion Restriction Site[b] (R.E.1)	TAB Linker Product Number	Sequence (5'–3')	TAB-Generated Restriction Site (R.E.2)[c]		Amino Acids Inserted		
				+1 Linker	+2 Linker	RF 1	RF 2	RF 3
24								
25	*Eco*R I		pAATTGG	*Asu* II	*Eco*R I	Phe Glu	Ser *Asn*	Arg *Ile*
			pAATTTA	*Aha* III	*Aha* III	Phe Lys	Leu *Asn*	*** *Ile*
26								
27	*Afl* III[e]		pAGCTCG	*Xho* I	*Sac* I	Leu Glu	Ser *Ser*	Arg *Ala*
	Ban II[e]		pAGCTGC	*Pst* I	*Pvu* II[f]	Leu Gln	Cys *Ser*	Ala *Ala*
	*Hin*d III		pAGCTTA	*Afl* II	*Hin*d III	Leu Lys	Leu *Ser*	*** *Ala*
28								
29	*Afl* III[e]		pCATGAT	*Bcl* I	*Bcl* I	*** Ser	Asp *His*	Ile *Met*
	Nco I	27-8562	pCATGCG	*Fsp* I[f]	*Sph* I	Cys Ala	Ala *His*	Arg *Met*
			pCATGGC	*Bal* I[f]	*Nco* I	Trp Pro	Gly *His*	Ala *Met*
30	*Avr* I[e]	27-8552	pCCGGAT	*Bam*H I	*Bam*H I	Gly Ser	Asp *Pro*	Ile *Arg*
	Xma I/*Xcy* I	27-8428	pCCGGCG	*Nar* I	*Nde* I[f]	Gly Ala	Ala *Pro*	Arg *Arg*
			pCCGGGC	*Apa* I	*Xcy* I	Gly Pro	Gly *Pro*	Ala *Arg*
			pCCGGTA	*Kpn* I	*Kpn* I	Gly Thr	Val *Pro*	Tyr *Arg*
31	*Afl* III[e]		pCGCGAT	*Pvu* I	*Nru* I	Arg Ser	Asp *Arg*	Ile *Ala*
	*Bss*H II		pCGCGGC	*Xma* III	*Sac* II	Arg Pro	Gly *Arg*	Ala *Ala*
	Mlu I							
32	*Nhe* I		pCTAGAT	*Bgl* II	*Xba* I	Arg Ser	Asp *Leu*	Ile ***
	Spe I		pCTAGCG	*Hae* II[g]	*Nhe* I	Ser Ala	Ala *Leu*	Arg ***
	Xba I		pCTAGGC	*Stu* I[f]	*Avr* II	Arg Pro	Gly *Leu*	Ala ***
			pCTAGTA	*Sca* I[f]	*Spe* I	Ser Thr	Val *Leu*	Tyr ***
33	*Bam*H I		pGATCGC	*Nru* I	*Pvu* I	Ser Arg	Arg *Asp*	Ala *Ile*
	Bcl I	27-8554	pGATCTA	*Xba* I	*Bgl* II	Ser Arg	Leu *Asp*	*** *Ile*
	Bgl II							
	Mbo I							
	*Sau*3A I							
	Xho I							
34	*Ban* I[e]		pGCGCAT	*Sph* I	*Fsp* I	Ala Cys	His *Ala*	Met *Arg*
			pGCGCCG	*Nae* I[f]	*Nar* I	Ala Gly	Pro *Ala*	Arg *Arg*
			pGCGCGC	*Bss*H II	*Bss*H II	Ala Arg	Arg *Ala*	Ala *Arg*
			pGCGCTA	*Nhe* I	*Hae* II	Ala Ser	Leu *Ala*	*** *Arg*
35	*Ban* I[e]		pGGCCAT	*Nco* I	*Bal* I	Pro Trp	His *Gly*	Met *Ala*
	Cfr I	27-8542	pGGCCCG	*Xcy* I	*Apa* I	Pro Gly	Pro *Gly*	Arg *Ala*
	Not I		pGGCCGC	*Sac* II	*Xma* III	Pro Arg	Arg *Gly*	Ala *Ala*
			pGGCCTA	*Avr* II	*Stu* I	Pro Arg	Leu *Gly*	*** *Ala*
36	*Asp* 718		pGTACAT	*Nsp*H I[g]	*Nsp*H I	Thr Cys	His *Val*	Met *Tyr*
	Ban I[e]		pGTACGC	*Mlu* I	*Mlu* I	Thr Arg	Arg *Val*	Ala *Tyr*
			pGTACTA	*Spe* I	*Sca* I[f]	Thr Ser	Leu *Val*	*** *Tyr*
37								
38	*Ava* I[e]	27-8548	pTCGAAT	*Eco*R I	*Asu* II	Glu Phe	Asn *Ser*	Ile *Arg*
	Sal I		pTCGACG	*Aat* II	*Sal* I	Asp Val	Thr *Ser*	Arg *Arg*
	Xho I		pTCGAGC	*Sac* I	*Xho* I	Glu Leu	Ser *Ser*	Ala *Arg*
				*Eco*R V[f]	*Cla* I	Asp Ile	Ile *Ser*	Tyr *Arg*
39								
40	*Afl* II		pTTAAGC	*Hin*d III	*Afl* II	Lys Leu	Ser *Leu*	Ala ***
			pTTAATA	*Ssp* I[f]	*Ssp* I	Asn Ile	Ile *Leu*	Tyr ***

Source: Courtesy of Pharmacia.

***—Denotes translational STOP codon.

[a] A TAB group comprises those TAB linkers that can be inserted into a restriction site that contains specific two-base sequence or four-sequence overhanging ends.

Table 76.   Correlation Between TAB-Generated Restriction Sites and Amino Acids Inserted in the Target Sequence (continued)

[b]Those restriction enzymes that leave the same two-base or four-base overhanging ends and are therefore all compatible with those linkers in the designated TAB group.

[c]Restriction site generated by insertion of either 1 (+ 1 linker) or 2 (+ 2 linkers) TAB linkers.

[d]The two amino acids added to a protein as a result of TAB linker insertion can be predicted for all three reading frames. Reading frame 1 (RF1) is defined as the reading frame containing the first codon altered, reading 5′-3′, as a result of insertion of the TAB linker. Reading frames 2 (RF2) and 3 (RF3) are one and two bases, respectively, 3′ to reading frame 1.

The TAB linkers in the following examples contain numbers corresponding to the bases of the restriction site overlap (and are therefore the same for all linkers within a TAB group) and letters corresponding to the variable bases within a linkers of a TAB group.

TAB Group

1–4	(2 base 3′-overhang)	Restriction overhang: 12 TAB linker: *WXYZ12* Mutated sequence: 12*WXYZ12* Reading frames: 123
5–8	(2 base 5′-overhang)	Restriction overhang: 12 TAB linker: *12WXYZ* Mutated sequence: *12WXYZ*12 Reading frames: 123
9–23	(4 base 3′-overhang)	Restriction overhang: 12 TAB linker: *WXYZ12* Mutated sequence: 1234*YZ1234* Reading frames: 123
24–40	(4 base 5′-overhang)	Restriction overhang: 1234 TAB linker: *1234YZ* Mutated sequence: *1234YZ*1234 Reading frames: 123

[e]Because of variability in the recognition sequence of these restriction enzymes, these enzymes are listed in two or more TAB groups.

[f]These restriction enzymes recognize six-base palindromic sequences and leave blunt ends after cleavage.

[g]These restriction enzymes recognize degenerate sequences.

[h]Phosphorylated linker.

[i]Unphosphorylated linker.

PROTOCOL

TAB-LINKER MUTAGENESIS

Methylation of DNA

Proceed as described in Chapter 19.

Ligation of TAB Linkers to Linearized Plasmid DNA

For each TAB linker there is ratio of linker : plasmid that results in optimal ligation of TAB linker to DNA. When the concentration of linker is too low or too high, a low percentage of transformants contain linker insertions, as assayed by biochemical (enzymatic) selection. While it is desirable to optimize the linker concentration to yield the maximum percentage of transformants for each linker insertion, it is time consuming to do so. We therefore recommend the following conditions, which will result in linker insertion in at least 60% of the transformants isolated:

1. In a tube containing 0.5 μg of linearized DNA (as dried pellet), add successively:
    5 μl of sterile distilled water
    1 μl of 10X TAB buffer
    1 μl of 10 m*M* ATP
    0.1 unit (A$_{260}$) of the selected linker
2. Add sterile distilled water to obtain a final volume of 10 μl.
3. Add 3–4 units of T4 DNA ligase.
4. Incubate at 4°C for 1 hour.
5. Heat-denature T4 DNA ligase at 70°C for 10 minutes.

Digestion of Excess Linkers

When 5′-phosphorylated TAB linkers are used, multiple linker ligations can occur. Removal of those multimers is achieved by incubation with the restriction enzyme that recognizes the restriction site generated by TAB linker insertion. If the TAB linker restriction site is *not* unique in the plasmid, the plasmid should be methylated (provided the restriction enzyme is inhibited by methylation) before this step is attempted (see Chapter 5). Recommended conditions for digestion of excess linkers are:

Ligated DNA (50 μg/ml)	4 μl
10X Restriction enzyme buffer	1 μl
Restriction enzyme	25–50 units
H$_2$O	to 10 μl

Incubate for 1 hour at the appropriate temperature. The reaction is terminated either by incubating 5 minutes at 65°C, or by phenol extraction (Chapter 3).

Recircularization of Plasmid DNA

This procedure should be carried out only when biochemical selection is to be used. Plasmid DNA, to which TAB linkers have been ligated, can be recircularized even in the presence of large numbers of free linkers, provided that the DNA is sufficiently dilute (note that if nonphosphorylated TAB linkers have been inserted, they have to be phosphorylated before religation).

1. In the tube containing 5 μl of DNA (20 μg/ml) from step C add successively:
   35 μl of sterile distilled water
   5 μl of 10X ligase buffer
   5 μl of 1 m*M* ATP
   5 units of T4 DNA ligase
2. Incubate at 14°C for 16 hours.

Biochemical (Enzymatic) Selection

Plasmids that have been recircularized without incorporating a TAB linker can be linearized by restriction endonuclease digestion. The closed circular TAB-linker modified plasmids will have a much higher transformation efficiency than the unmodified linears produced by the second restriction enzyme digestion. The reaction can be carried out by adding the appropriate buffer and restriction enzyme to the 50-μl reaction mixture of the previous section, unless the enzyme is sensitive to high salt concentrations (see Table 45). In this case, the NaCl has to be removed, prior to restriction endonuclease digestion, by ethanol precipitation followed by resuspension of the DNA in water. After restriction endonuclease digestion, bacterial cells are transformed with plasmid DNA and linker-containing clones are selected (see Chapter 14).

Recircularization with a Kanamycin Resistance Gene Cartridge

This step should be followed only when biological selection is to be used. Choose the Kan^r gene cartridge cassette with the appropriate restriction enzyme cohesive ends from the available pUC4 derivatives shown in Figure 201. Digest the plasmid with the appropriate restriction endonuclease and purify the Kan^r-containing restriction fragment by gel electrophoresis (see Chapter 9). For ligation of a kanamycin cassette, add 0.2 μg of purified Kan^r DNA to the reaction mixture described under "recircularization of plasmid DNA" above and reduce the total volume by half.

Stop the reaction by heating for 5 minutes at 65°C. Transform competent *E. coli* and select for TAB-modified plasmids by plating on kanamycin-containing medium.

Excision of the Kanamycin Resistance Gene Cartridge

If the TAB linker insertion has been selected by biological selection, the kanamycin resistance cassette has to be removed to generate the complete TAB linker mutant. The cassette is removed by diges-

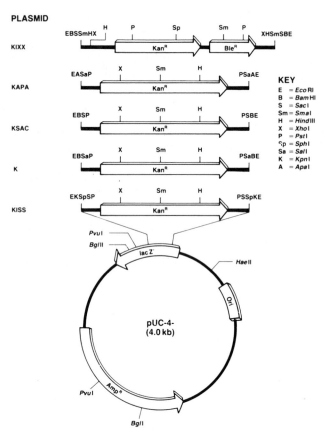

Figure 201. The pUC-4-kanamycin-resistance gene cartridge plasmids (courtesy of Pharmacia).

tion with the same restriction enzyme as was used when inserting it. Following religation the reaction mixture is diluted by the addition of T4 DNA ligase. This mixture can then be used for transformation of competent *E. coli* cells.

Excision

DNA	2–5 μl
10X Restriction enzyme buffer	1 μl
Restriction enzyme	5–20 units
H$_2$O	to 10 μl

Stop reaction by heating 5 minutes at 65°C or phenol extract (see Chapter 3).

Recircularization

10 μl of recleaved DNA (from above), add 5 μl of 10X DNA ligase buffer, 5 μl of ATP 1 mM, 4–5 units of T4 DNA ligase, and sterile distilled water to 50 μl. Incubate at 16°C for 16 hours.

References for Oligonucleotide-Directed Mutagenesis and Linkers Mutagenesis

Astell, C. R., and Smith, M. (1971), *J. Biol. Chem.*, **246**, 1944.

Astell, C. R., and Smith, M. (1972), *Biochemistry*, **11**, 4114.

Astell, C. R., Doel, M. R., Jahnke, P. A., and Smith, M. (1973), *Biochemistry*, **12**, 5068.

Carter, P., Bedouelle, H., and Winter, C. (1985), *Nucl. Acids Res.*, **13**, 4431.

Gillam, S., Waterman, K., and Smith, M. (1975), *Nucl. Acids Res.*, **2**, 625.

Gillam, S., and Smith, M. (1979), *Gene*, **8**, 81.

Haltiner, M., Kempe, T., and Tjian, R. (1985), *Nucl. Acids Res.*, **13**, 1015.

Hutchinson, III, C. A., Phillips, S., Eggell, M. H., Gillam, S., Jahnke, P. A., and Smith, M. (1978), *J. Biol. Chem.*, **253**, 6551.

Kramer, W., Schuggart, K., and Fritz, M.-J. (1982), *Nucl. Acids Res.*, **10**, 6475.

Luckow, B., Renkawitz, R., and Schütz, G. (1987), *Nucl. Acids Res.* **15,** 417.

McKnight, S. L., and Kingsbury, R. (1982), *Science,* **217**, 316.

Norris, K., Norris, F., Christiansen, L., and Fiil, N. (1983), *Nucl. Acids Res.,* **11**, 5103.

Razin, A., Hirose, T., Itakura, K., and Riggs, A. (1978), *Proc. Natl. Acad. Sci. USA,* **75**, 4268.

Sakonju, S., Bogenhagen, D. F., and Brown, D. D. (1980), *Cell,* **19**, 13.

Taylor, J. W., Schmidt, W., Cosstick, R., Okruszek, A., and Eckstein, F. (1985), *Nucl. Acids Res.,* **13**, 8749.

Wallace, R. B., Johnson, P. F., Tanakas, S., Schold, M., Itakura, K., and Abelson, J. (1980), *Science,* **209**, 1396.

Wang, J. C. (1979), *Proc. Natl. Acad. Sci. USA,* **76**, 200.

Wasylyk, B., Derbyshire, R., Guy, A., Molko, D., Roget, A., Teoule, R., and Chambon, P. (1980), *Proc. Natl. Acad. Sci. USA,* **77**, 7024.

Zoller, M. J., and Smith, M. (1982), *Nucl. Acids Res.,* **10**, 6487.

Zoller, M. J., and Smith, M. (1984), *DNA,* **3**, 479.

References for TAB Mutagenesis

Barany, F. (1985a), *Proc. Natl. Acad. Sci. USA,* **82**, 4202.

Barany, F. (1985b), *Gene,* **37**, 111.

Boeke, J. D., (1981), *Mol. Gen. Genet.,* **181**, 288.

Dalbadie-McFarland, G., Cohen, L. W., Riggs, A. D., Morin, C., Itakura, K., and Richards, J. H. (1982), *Proc. Natl. Acad. Sci. USA,* **79**, 6409.

Heffron, F., So, M., and McCarthy, B. J. (1978), *Proc. Natl. Acad. Sci. USA,* **75**, 6012.

Hunkapiller, M. W., Lujan, E., Ostrander, F., and Hood, L. E. (1983), *Methods in Enzymology,* C. H. W. Hirs and S. N. Timasheff, eds., Vol. **91**, Part 1, p. 227, Academic Press.

Hutchinson, C. A., Phillips, S., Edgell, M., Gillam, S., Jahnke, P., and Smith, M. (1978), *J. Biol. Chem.,* **253**, 6551.

Kadonaga, J. T., and Knowles, J. R. (1985), *Nucl. Acids Res.,* **13**, 1733.

Kessler, C., Neumaier, P. S., and Wolf, W. (1985), *Gene,* **33**, 1.

Kunkel, R. A. (1985), *Proc. Natl. Acad. Sci. USA,* **82**, 488.

Lobel, L. I., and Goff, S. P. (1984), *Proc. Natl. Acad. Sci. USA,* **81**, 4149.

McDonnel, M. W., Simon, M. N., and Studier, F. W., *J. Mol. Biol.,* **110**, 119.

McKnight, S., and Kingsbury, R. (1982), *Science,* **217**, 316.

Mandel, M., and Higa, A. (1970), *J. Mol. Biol.,* **53**, 159.

Osterlund, M., Luthman, H., Nilsson, S. V., and Magnusson, G. (1982), *Gene,* **20**, 121.

Parker, R. C., (1977), *Proc. Natl. Acad. Sci. USA,* **74**, 851.

Prentki, P., and Krisch, H. M. (1984), *Gene,* **29**, 303.

Shortle, D., and Nathans, D. (1978), *Proc. Natl. Acad. Sci. USA,* **75**, 2170.

Shortle, D., and Nathans, D. (1979), *J. Mol. Biol.,* **131**, 801.

Shortle, D., Koshland, D., Weinstock, G., and Botstein, D. (1980), *Proc. Natl. Acad. Sci. USA,* **77**, 5375.

Smith, H. O., Grisafi, P., Benkovic, S. J., and Botstein, D. (1980), *Methods in Enzymology,* L. Grossman and K. Moldave, eds., vol. **65**, p 371.

Stone, J. C., Atkinson, T., Smith, M., and Pawson, T., (1984), *Cell,* **37**, 549.

Tabak, H. L., and Flavell, R. A. (1978), *Nucl. Acids Res.,* **5**, 2321.

Templeton, H. F., and Exkart, W. (1982), *J. Virol.,* **41**, 1014.

Vieira, J., and Messing, J. (1982), *Gene,* **19**, 259.

Weislander, L. (1979), *Anal. Biochem.,* **98**, 305.

RANDOM INTRODUCTION OF SINGLE BASE MUTATIONS IN A DEFINED DNA FRAGMENT

This method is based on the in vitro synthesis of an oligonucleotide whose sequence corresponds to the DNA region in which point mutations are to be introduced. At each step of the synthesis, a mixture of the four bases is used so that there is a significant probability of incorporating another base in lieu of the wild-type base at almost any position in the synthetic DNA strand (McNeil and Smith, 1985). Theoretically, the procedure allows one to obtain a nest of DNA fragments containing point mutations at all possible positions. Following enzymatic synthesis of the complementary strand, the population of mutagenized fragments is cloned in an appropriate vector for isolation and characterization prior to use in biological assays. This procedure has been used by Hill et al. (1986) to study the DNA binding site of a yeast regulatory protein. The experimental protocol which is described in detail below is outlined in Figure 202.

Specific Points to Consider for the Synthesis of the Mutagenized Oligonucleotide

The size of the oligonucleotide is dictated by that of the corresponding DNA fragment under study. It is important that a few additional nucleotides be added at the 5' end of the oligonucleotide in order to allow efficient digestion by restriction endonuclease.

Choice of the Restriction Sites Located at the 3' and 5' End of the Oligonucleotide. In order to favor the annealing of 3' ends (in step 2) the corresponding restriction site should be palindromic, contain more bases than the 5'-proximal site, and be G/C rich. In the example described here, the 3' and 5' sites were specific for Bam HI and Mbo I, respectively.

site A   5'-A-G-C-**G-A-T-C**-C-oligonucleotide-3'
                    Mbo I

site B   5'-oligonucleotide-C-**G-G-A-T-C**-C-C-G-3'
                              Bam HI

In this case, annealing of Bam HI sites will be favored because sticky ends contain eight bases (over five for Mbo I), six of them being G and C. The same enzyme (Mbo I; see Table 25, page 148) will be used to cut both ends of the duplex. The addition of AGC at the 5' end of the oligonucleotide will allow efficient digestion by the enzyme without interfering with annealing. The addition of a C downstream to the Mbo I site will allow subcloning of the final fragments at the Bam HI site in vector DNA.

Level of Deoxynucleotide Degeneracy for Mutagenesis. It is possible to calculate the proper fraction (p) of wild-type phosphoramidite to be employed at each step of the synthesis reaction (McNeil and Smith, 1985). This value varies with the number (n) of nucleotides in the DNA fragment to mutagenize and with the number (r) of point mutations to be introduced in this fragment. The fraction $P(p)$ of oligo-

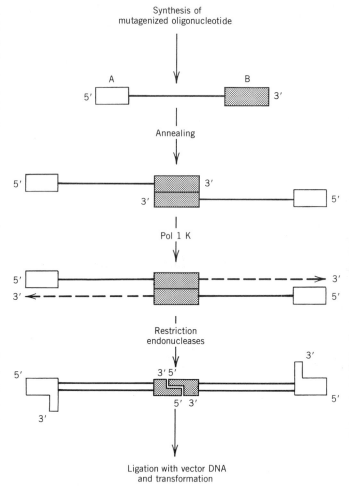

Figure 202. Saturation–mutagenesis of an oligonucleotide sequence. The first step consists in the synthesis of a mutagenized oligonucleotide containing at both ends recognition sequences for two different restriction endonucleases. Following annealing of the 3' proximal palindromic sequences (step 2) the complementary strands are enzymatically synthesized with Pol IK (step 3). Restriction endonuclease digestion of the duplex molecules generates clonable DNA fragments with cohesives ends (step 4).

nucleotide species containing (r) mutations is given by the binomial distribution equation

$$P_r(p) = C_n^r\, p^{n-r}\, (1 - p)^r$$
$$= \frac{n!}{(n - r)!\, r!}\, p^{n-r}\, (1 - p)^r \quad (1)$$

where $n! = 1 \times 2 \times 3 \times \cdots (n - 1) \times n$.

Let us consider, for example, a 14-nucleotide sequence in which one 1-point mutation is to be introduced.

$n = 14$ and $r = 1$

$P_1\ (p) = 14p^{13}\ (1 - p)$

To determine the fraction of wild-type phosphoramidite (p) to be employed for the synthesis of each residue in the mutagenized oligonucleotide, one calculates the maximum value of $P(p)$. This is done by setting to zero the first derivative of $P(p)$ with respect to p (designated as $P'(p)$).

In our example,

$P'(p) = 14p^2(13 - 14p)$

$P'(p) = 0$ when $p = \dfrac{13}{14} = 0.928 = 92.8\%$

Thus, in each step of the reaction, the relative proportion of wild-type phosphoramidite over the three other phosphoramidites must be 92.8% of wild type, 2.4% of each of the three others.

From equation (1) it is possible to calculate $P_r(p)$ for different values of r. The results obtained for a few examples are reported in Figure 203.

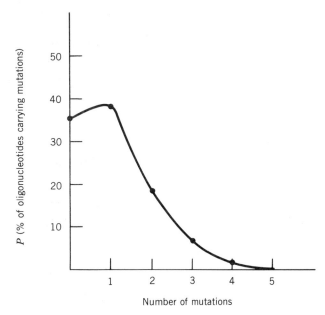

Figure 203. Fraction of oligonucleotides carrying a given number of mutations (see text for details).

PROTOCOL

PROCEDURE FOR CLONING THE RANDOMLY MUTAGENIZED OLIGONUCLEOTIDE*

Synthesis of the Oligonucleotide

In our example, the following 29-mer oligonucleotide was purchased from Genofit. The internal 14-mer portion was mutagenized by performing the synthesis steps in the presence of 92.8% wild-type base and 2.4% of the three others.

14 mer

5'-A-G-C-**G-A-T-C**-C-oligonucleotide-C-**G-G-A-T-C-C**-G-3'

Mbo I   Bam HI

*This protocol has been kindly provided by M. L. Vignais.

Annealing of the Oligonucleotide Strands

1. In a microfuge tube, mix:
   8 μl of sterile distilled water
   2 μl of oligonucleotide (50 pmol/μl)
2. Heat to 70°C in a water bath for 10 minutes.
3. Add 3 μl of 10X Pol buffer.
4. Mix and spin in a microfuge for a few seconds to collect droplets.
5. Incubate at 37°C for 1 hour.
6. Place the microfuge tube in a 5-ml tube containing prewarmed (37°C) water and let equilibrate slowly to room temperature.

Enzymatic Synthesis of the Complementary Strand

Due to the very small amounts of DNA being manipulated in this kind of experiment, it is advisable to use a radioactive tracer. The choice of the tracer is a function of the mean nucleotide composition of the mutagenized DNA.

1. To the 13-μl sample from step 6 above, add:
   8 μl of sterile distilled water
   1.5 μl of 5 mM dCTP
   1.5 μl of 5 mM dGTP
   1.5 μl of 5 mM dTTP
   3 μl of α[^{32}P]dATP (10 μCi/μl; 3.3 pmol/μl)
2. Mix gently by tilting the tube and spin in a microfuge for a few seconds.
3. Add 1 μl of Pol1K (5 units/μl).
4. Incubate at 16°C for 1 hour.

Yield of Second Strand Synthesis

The efficiency of synthesis is checked during the 16°C incubation period.

1. After 15 to 20 minutes incubation at 16°C, take 1 μl of the incubation mixture, and return tube to 16°C.
2. Add the 1-μl aliquot to 9 μl of sterile distilled water. Use 4 μl of this diluted mixture to measure specific radioactivity of the DNA and 4 μl to measure the total radioactivity in the reaction mixture.

ESTIMATION OF SPECIFIC RADIOACTIVITY
In a 5-ml plastic tube containing 2 ml of prechilled 5% TCA, add 5 μl of 2 mg/ml carrier tRNA and 4 μl of the diluted mixture. Let

stand in ice for 5 minutes and filter the precipitated DNA on a 0.45-μm pore size nitrocellulose filter. Wash 5 to 10 times with cold 5% TCA. Dry the filter and count.

ESTIMATION OF THE TOTAL RADIOACTIVITY IN THE REACTION MIXTURE

Spot 4 μl of the diluted mixture directly on a nitrocellulose filter. Dry without washing and count. Incorporated radiolabeled nucleotide should represent about 80% of input.

Addition of Cold dATP

The measurement of radioactivity requires about 10 minutes. Thus, incubation at 16°C has now proceeded for 25–30 minutes. Add 1.5 μl of 5 m*M* dATP to the DNA sample and return to 16°C until the 1-hour incubation time is over. At this stage, the sample can be frozen at −20°C.

Size of the Newly Synthesized DNA Fragment

The length of the DNA fragment synthesized by Pol1K is determined by electrophoresis in a denaturing polyacrymide gel. The conditions below apply to the example described above. Adjust to your own system. Prepare a medium-sized gel (for example, 200 × 150 × 0.4 mm) (20% polyacrylamide, 8 *M* urea, 1X TBE).

1. Mix an aliquot (equivalent to about 500 cpm) of the sample from step 5 with sample buffer. Use a mixture of xylene cyanol and bromophenol blue as migration marker. Also run as size controls G/A and T/C tracks from sequencing reactions performed according to Maxam and Gilbert (see Chapter 20).
2. Prerun gel at 25 watts for 15 minutes.
3. Heat the samples at 90°C for 3 minutes.
4. Load gel immediately and run the electrophoresis at 25–30 watts for about 1 hour (until xylene cyanol reaches 2/3 of the gel length).
5. Expose the gel after transfer to an old autoradiogram (see Chapter 20).

Interpretation of the Results

1. In our example (see Figure 204) the expected fragment size is [(3′-proximal site + 14-mer) × 2] + 5′-proximal site, i.e., [(7 + 14) × 2] + 8 = 50 nucleotides.
2. If only one band with the expected size is obtained, proceed to step 7.
3. If several intense bands with intermediate sizes are obtained,

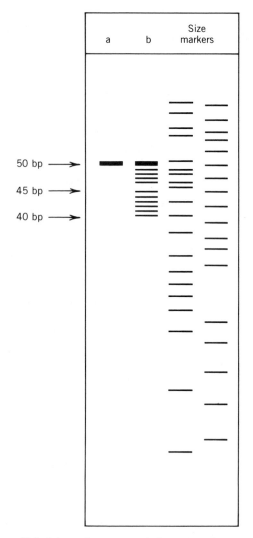

Figure 204. Schematic representation of a denaturing poly-acrylamide gel showing the results expected in the case of a complete copy of the oligonucleotide by Pol 1 k (a) and in the case of a partial copy of the template by Pol 1 k (b). Size markers (c) are provided by specific base cleavage of standard radio-labeled DNA.

phenol extract the sample as described in step 7, precipitate, and resuspend the pellet in 13 μl of sterile distilled water. Go back and repeat step 3.

Notes

1. In a 20% denaturing polyacrylamide gel xylene cyanol and bromophenol blue migrate as 28 and 8 nucleotides, respectively.

2. Since the oligonucleotide strands are not phosphorylated at

their 5′ end, a slight difference may be observed between the migrations of the oligonucleotide and the same fragment resolved in the sequencing G/A and T/C tracks. This difference is only obvious with small DNA fragments. For example, a nonphosphorylated 20-nucleotides fragment migrates as a phosphorylated 21-22 nucleotides fragment.

Phenol Extraction and Precipitation of the Duplex DNA Fragment

1. To the sample from step 5, add:
   70 μl of sterile distilled water
   200 μl of a 1/1 mixture of phenol/chloroform:isoamyl alcohol (24:1)
2. Vortex for 30 seconds, and centrifuge at 12,000 rpm for 30 seconds.
3. Take the upper phase (100 μl) and transfer in a tube containing 200 μl of a 1/1 mixture of phenol/chloroform:isoamyl alcohol (24:1).
4. Repeat step 2.
5. Transfer upper phase to a tube containing 11 μl of 1 *M* NaCl.
6. Add 300 μl of cold absolute ethanol.
7. Mix and let the DNA precipitate at −70°C for 1 hour.
8. Centrifuge at 12,000 rpm and 4°C for 30 minutes.
9. Aspirate carefully the ethanol with a drawn out Pasteur pipette.
10. Add 200 μl of 70% ethanol very carefully on the side of the tube. *Do not vortex.*
11. Centrifuge at 12,000 rpm and 4°C for 10 minutes.
12. Aspirate the ethanol carefully with a drawn out Pasteur pipette.
13. Dry the pellet in a vacuum centrifuge (Speed Vac, Savant).

Notes

Precipitation of small oligonucleotides without addition of carrier requires some caution.

1. Precipitation efficiency will vary with the nature of the salt being added. Precipitation is more efficient in the presence of 0.1 *M* NaCl than in the presence of 2 *M* ammonium acetate.
2. A better recovery is obtained if three volumes of ethanol are added to the sample for precipitation.

3. Be extremely careful not to disturb the pellet (sometimes visible) when aspirating the ethanol.

4. Refer to Chapter 3 for other tips.

Preparative Electrophoresis of the Duplex DNA Fragment

This step is carried out to eliminate the residual single stranded DNA molecules that did not anneal in step 2 of the procedure.

1. Resuspend the DNA pellet from step 7 in 36 µl of sterile distilled water.

2. Add 4 µl of 10X Ficoll buffer.

3. Load on a 6% polyacrylamide gel and run at 150 volts. (Migration time will depend upon the size of the DNA fragment.)

Note

In order to evaluate the amount of single stranded DNA present in the mixture, it is possible to incubate an aliquot in the presence of [^{32}P] ATP and polynucleotide kinase (see Chapter 16). The labeled DNA fragments are then separated by electrophoresis in a polyacrylamide gel under denaturing conditions to estimate the relative amounts of single stranded and annealed DNA fragments at this stage.

Autoradiography and Electroelution of the DNA Fragment

1. Following to electrophoresis, cover the gel with Saran wrap and expose for autoradiography at room temperature for 10–30 minutes (Figure 205).

2. Cut out the portion of the gel containing the DNA fragments and perform an electroelution in a small volume of 0.75X TBE at 4°C in the cold room [the use of a Schleicher and Schuell Biotrap (see Chapter 7) has proved to be very convenient at this stage].

3. Reduce the volume to about 100 µl by lyophilization.

4. Perform extraction and precipitation as described above.

Generation of DNA Fragments with Cohesive Ends

Digestion of the DNA fragments with appropriate restriction endonucleases is performed as described in Chapter 8. Make sure that complete digestion is obtained prior to performing the ligation step (Figure 206).

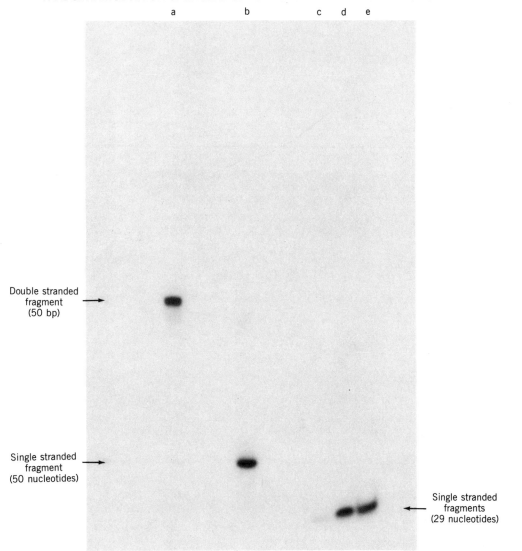

Figure 205. Preparative electrophoresis of double stranded labeled oligonucleotide (kindly provided by M.L. Vignais).

Ligation of the DNA Fragments with Vector DNA and Transformation

These steps are performed as described in Chapters 13 and 14. The following technique allows rapid determination of the number of inserted fragments in a given recombinant vector.

1. Digest recombinant DNA at a restriction site located at one boundary of the inserted fragment in the vector and generating a 5′ protruding end (for example, Eco RI in the polylinker sequence of pUC plasmids).
2. Fill in the protruding ends with Pol1K in the presence of a labeled nucleotide.

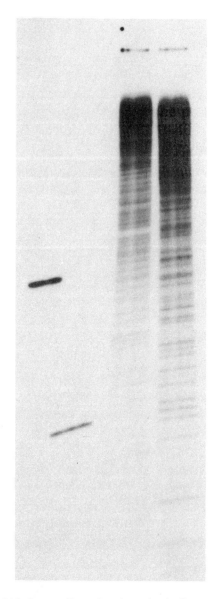

Figure 206. Preparation of mutagenized oligonucleotide with cohesive ends for cloning. The double stranded DNA fragment (50 bp, lane a) purified from gel shown in Figure 205 has been digested with Mbo I restriction endonuclease to generate the 20-bp DNA fragment shown in lane (b). (TC)(GA): Internal size markers provided by base-specific cleavage of pUC19 DNA radio-labeled at the Bam HI site with $\alpha[^{32}P]$ dGTP and $\alpha[^{32}P]$ dATP (kindly provided by M.L. Vignais).

3. Digest the linearized labeled DNA with a restriction enzyme whose recognition site is located on the other side of the insert in the vector.

4. Run a polyacrylamide gel under denaturing conditions to determine the size of the labeled fragments and establish whether one or several fragments have been inserted in the vector DNA (Figure 207).

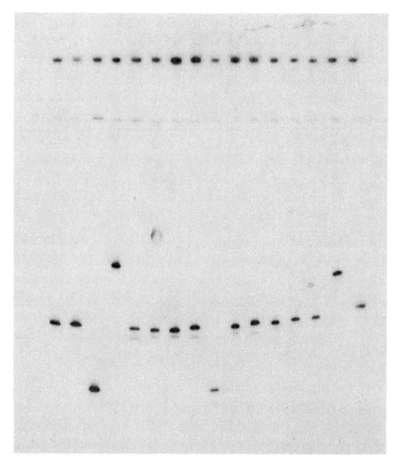

Figure 207. Number of DNA fragments inserted in the recombinant clones. Recombinant DNA molecules have been digested with Eco RI, treated with Pol 1 K in the presence of α[³²P] dATP, and then digested with Hind III to excise cloned DNA fragments. Samples were analyzed by poly-acrylamide gel electrophoresis as described in text (kindly provided by M.L. Vignais).

Sequencing of the Cloned DNA Fragments

Sequencing is performed directly on the double stranded cloned DNA as described in Chapter 20.

Buffer

10X Pol Buffer	100 μl
0.5 M Tris HCl (pH 7.2)	50 μl IM Tris HCl (pH 7.2)
0.1 M MgSO₄	10 μl 1 M MgSO₄
1 mM DTT	1 μl 100 mM DTT
500 μg/ml Bovine serum albumin	10 μl Bovine serum albumin (5 mg/ml)
	29 μl sterile distilled water

References

Hill, D. E., Hope, I. A., Macke, J. P., and Struhl, K. (1986), *Science,* **234**, 451.

McNeil, J. B., and Smith, M. (1985), *Mol. Cel. Biol.,* **5**, 3545.

24

METHODS FOR STUDYING DNA–PROTEIN INTERACTIONS

Several different techniques are now available for studying the interactions between potential regulatory proteins and corresponding target sites on DNA molecules at the molecular level. These techniques include: band shift or gel retardation measurement, analysis of dimethylsulfate contact points, DNAse footprinting, digestion with exonuclease III to characterize boundaries of the complexes, and retention on nitrocellulose filters. It is frequent that two such approaches are used in combination so as to obtain more information on the nature of the complex (for example, gel retardation and DNAse footprinting, or gel retardation and dimethylsulfate contact points). Recent improvements to this technology have emerged from the progress made in the fields of molecular cloning and molecular biology.

Since all these techniques rely on the formation of a protein–DNA complex, it is an essential requirement that optimal conditions favoring the formation of the complex be determined. The different parameters to consider are the following:

1. Ionic strength of the solution (concentration of NaCl, KCl, or $(NH_4)_2 SO_4$).

2. Presence of divalent cations such as Mg^{2+}.

3. Temperature and time of incubation.

4. Nature and quantity of the competing DNA, which is added to the reaction mixture in order to trap proteins that can bind nonspecifically to nucleic acids (this will lead to a reduction of background). Salmon sperm or pBR322 DNA and poly d(I–C) have proved to be satisfactory. The quantity of competing DNA employed will vary with the purity of the protein fraction used. Generally, equal amounts (w/w) of DNA and proteins are added when working with a crude proteic fraction. The optimal conditions may vary with the nature of both protein and DNA involved in the complex of interest.

ANALYSIS OF PROTEIN–DNA COMPLEXES BY POLYACRYLAMIDE GEL ELECTROPHORESIS (GEL RETARDATION)*

Refer to Chapter 3 for the general principles of electrophoresis. The following protocol uses a 5% polyacrylamide gel.

PROTOCOL

GEL RETARDATION FOR ANALYSIS OF DNA-PROTEIN COMPLEXES

Preparation of the Gel

Use siliconized plates of medium size (for example, SE 600 from Hoefer or homemade 13 × 11 cm) and 1 mm thick spacers.

1. In a 50 ml sidearm flask, mix:
   12.5 ml of distilled water
   0.4 ml of 1 *M* Tris–HCl pH 8.0
   5 ml of acrylamide solution (19.76% acrylamide, 0.24% bis acrylamide)
   2 ml of 50% glycerol
   200 μl of 0.1 *M* EDTA pH 8.0
2. Add 100 μl of 10% ammonium persulfate, and degas under vacuum.
3. Add 10 μl of TEMED and swirl to mix.
4. Pour the mixture between the two plates and let polymerize.

Prerun

Once the gel is polymerized, perform two successive preruns at 200 volts (35 mA) and room temperature. Change the buffer each time. After the second prerun, the pH of anode buffer should be stable (you will note a decrease of the pH during the prerun). Transfer the gel to the cold room, or refrigerate if using a SE 600 apparatus. It is possible to prepare the gel one day before it is used. In that case, perform only one prerun once the gel is polymerized and store it at room temperature overnight. The next day, perform the second prerun, just before use in the cold.

Formation of the Complex

As stated above, the conditions given below have been found to be optimal for a given DNA–protein couple and might require slight modification to meet your own needs.

1. In a microfuge tube, mix successively:
   5 μl of 4X "complex buffer"
   3 μl of 75% bidistilled glycerol
   0.5 μg of pBR322 DNA

2 to 10 × 10^3 cpm of labeled target DNA (468 bp in this example)

0.5 μg of the protein fraction

x mM KCl (or other salt), to be adjusted to each particular case (usually 10 to 100 mM)

H_2O to 20 μl

2. Incubate for 10 minutes at 25°C and load directly on the gel prepared as described above.

Notes

1. The glycerol concentration in the sample should not exceed 10% to avoid density problems.
2. The size of the labeled DNA fragment should be in the range of 50 to 1000 base pairs (preferably 200 to 600), and can be labeled at both ends.
3. When the protein sample is supposed to contain DNAses, it is advisable to use a "complex buffer" deprived of magnesium and calcium ions (activators of DNAses). Most complexes will form efficiently in the absence of divalent cations.
4. The KCl concentration (10 to 100 mM) should be adjusted for each particular case.

Electrophoresis

Electrophoresis is performed at 4°C. Use fresh buffer. Loading of the samples is eased by the presence of glycerol. A mixture of xylene cyanol and bromophenol blue is used as tracking dye. Mixing the dye mixture with the samples is not recommended. Apply 200 volts for 5 minutes in order to obtain a rapid penetration of the sample in the gel. Perform the run at 160 volts. It is important to change the buffer periodically (every hour) to avoid variations of pH. Let the electrophoresis proceed until the xylene cyanol dye is about 2 cm from the bottom of the gel. Under these conditions, the xylene cyanol migrates as a 480-base pair DNA fragment. An estimation of the degree of complexation will be possible only if the free target labeled DNA remains in the gel under the electrophoretic conditions used. If the target DNA fragment is in the range of 150 base-pair long, stop the electrophoresis when the bromophenol blue reaches the bottom of the gel.

Autoradiography

Following electrophoresis, treat the gel for autoradiography (Figure 208.)

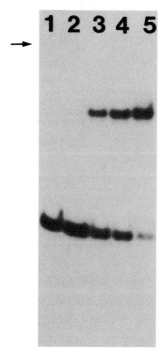

TEF1

Figure 208. Analysis of protein–DNA complexes by gel electrophoresis. Protein–DNA complexes were formed with TEF1 (486-bp) labeled probe in the presence of 0.54 μg of pBR322 DNA and varying amounts of proteins. The figure shows the autoradiogram of the slab gel. In lanes 1–5 the amount of proteins was 0, 0.07, 0.175, 0.35, and 0.7 μg respectively. The arrows indicate the origin (kindly provided by A. Sentenac).

Buffers and Solutions

Electrophoresis buffer
20 mM Tris–HCl (pH 8.0)
1 mM EDTA (pH 8.0)
5 mM 2-Mercaptoethanol

3 liters
60 ml 1 M Tris–HCl (pH 8.0)
30 ml 0.1 M EDTA (pH 8.0)
1 ml 14.26 M
   2-Mercaptoethanol
H_2O to 3 liters

4X "Complex buffer"
80 mM Tris–HCl (pH 8.0)
20 mM $MgCl_2$
2 mM DTT
2 mM $CaCl_2$
0.4 mM EDTA

1 ml
80 μl 1 M Tris–HCl (pH 8.0)
20 μl 1 M $MgCl_2$
20 μl 0.1 M DTT
20 μl 0.1 M $CaCl_2$
4 μl 0.1 M EDTA
H_2O to 1 ml

References

Fried, M., and Crothers, D. M. (1981), *Nucl. Acids Res.*, **9,** 6505.

Garner, M. M., and Revzin, A. (1981), *Nucl. Acids Res.*, **9,** 3047.

DNASE I FOOTPRINTING*

Originally described by Galas and Schmitz (1978), this technique (illustrated in the following Protocol) allows localization of the DNA regions protected from DNAse I digestion following binding of proteins.

PROTOCOL

DNASE I FOOTPRINTING

Formation of the Complex

1. In a microfuge tube, mix successively:

5 µl of 4X "complex buffer"

1 µl of bovine serum albumin (0.1 mg/ml)

2.5 µl 50% bidistilled glycerol

x mM KCl (or other salt), to be adjusted to each particular case (usually 10 to 100 mM)

y ng of competing DNA (usually same amount as protein)

5×10^{-15} mol of labeled target DNA (about 10 to 20,000 cpm total). This DNA must be labeled at one end only, and on one strand (see Chapter 16 for labeling DNA).

y ng of protein (quantity depends upon purity of the fraction, see note 3 above if sample contains DNAses)

H_2O to 20 µl

2. Incubate at 25°C for 5 to 30 minutes.

DNAse I Digestion

1. To each 20-µl sample, add 2 µl of DNAse I solution prepared according to the following chart. The quantity of DNAse to be added will vary with the ionic strength of the incubation mixture and with the amount of impurities in the protein sample (due to non-specific DNA binding proteins).

mM KCl

	30	55	80	105	130	155	180	205	230
ng DNAse	3	3	5	5	10	10	15	15	30

2. Incubate at 20°C for 30 seconds.

3. Stop the reaction by adding:

10 µl of stop buffer I if less than 0.2 µg of protein per sample

or

80 µl of stop buffer II if more than 0.2 µg of protein per sample

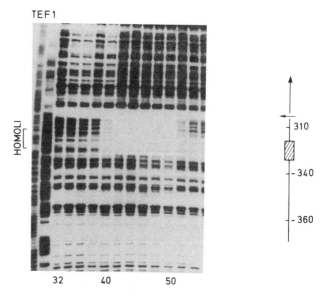

Figure 209. A DNA-binding activity interacts with TEF1 5′-flanking regions. Protein fractions from a salt gradient were assayed for proteins and for RNA polymerase B or C activity on 10 μl, and every second fraction (14 μl) was used in the footprint assay. For footprint analysis the protein fractions were incubated with the TEF1 probe 3′ end-labeled at the Bam HI site. The complexes were digested lightly with DNase I, and the DNA fragments were analyzed on an 8% polyacrylamide gel as described in text (kindly provided by A. Sentenac).

4. When working with low concentrations of proteins, precipitate the DNA directly by adding to the tube containing buffer I:
220 μl of 2.5 *M* ammonium acetate
750 μl of cold absolute ethanol

When working with high concentrations of proteins, incubate the tube containing proteinase K (buffer II) for 15 minutes at 37°C, and then extract DNA according to standard procedure (see Chapter 3).

Analysis of Protected DNA Sequences

1. Determine the radioactivity of the treated sample by Cerenkov procedure and resuspend in 5 μl of sample buffer.
2. Heat the samples for 3 minutes at 90°C and load on a sequencing gel as described in Chapter 20 (Figure 209).

Buffers and Solutions

4X Complex buffer (same as above)

DNAse I (BDH 39101, RNAse free)
1 mg/ml in 20 m*M* Tris–HCl (pH 8.0)

1 m*M* EDTA
25% Glycerol
5 m*M* CaCl$_2$
Store at −20°C. Do not keep diluted solutions.

Bovine serum albumin (New England Biolabs, nuclease free)
0.1 mg/ml in sterile distilled water

50% Glycerol
Double distilled glycerol, sterilized for 20 minutes at 110°C

Stop buffer I
95 m*M* EDTA
0.8% SDS
1.6 *M* ammonium acetate
0.3 mg/ml sheared calf thymus DNA

Stop buffer II
12.5 m*M* EDTA
12.5 µg/ml Proteinase K
125 µg/ml yeast tRNA
0.1% v/v SDS

Sample buffer
Formamide 90%
TBE 1X
0.05% w/v Xylene cyanol
0.05% w/v Bromophenol blue

Reference

Galas, D. J., and Schmitz, A. (1978), *Nucl. Acids Res.,* **5,** 3157.

LOCALIZATION OF CONTACT POINTS BY DIMETHYLSULFATE RESISTANCE*

Dimethylsulfate (DMS) is a reactive product for the N7 position of guanine residue in the DNA major groove and for the N3 position of adenine in the minor groove (see Chapter 20 for use in sequencing). When a DNA–protein complex is treated with DMS, the protected DNA sequences do not react and "contact points" can be visualized (Sickerlist and Gilbert, 1980).

PROTOCOL

LOCALIZATION OF CONTACT POINTS BY DIMETHYLSULFATE RESISTANCE

Formation of the Complex

1. In a microfuge tube, mix successively:
   7 μl of Hepes buffer
   9 μl of 50% bidistilled glycerol
   3.5 μl of bovine serum albumin (0.1 mg/ml)
   x mM salt (see protocol on DNAse footprinting)
   y ng of pBR322 DNA (see protocol on DNAse footprinting)
   5×10^{-15} mol of end-labeled DNA (10 to 20,000 cpm; see protocol on DNAse footprinting)
   y ng of protein sample (the amount of be used is larger than that needed for complete protection in the DNAse I footprinting technique)
   H$_2$O to 70 μl

2. Incubate at 25°C for 5 to 30 minutes.

Methylation by DMS

1. Add 1 μl of DMS (Aldrich) to each 70-μl sample.

2. Incubate at 20°C for increasing periods of time (30 seconds, 1 minute, 2 minutes, etc.), to statistically obtain one methylation per DNA molecule. The longer the DNA fragment, the shorter the time of incubation. For example, 3 minutes of incubation are satisfactory for a 200-base pair fragments, while 1 minute at 0°C is satisfactory for a 660 base pair fragment.

3. Add 35 μl of DMS stop buffer.

4. Add 350 μl of absolute ethanol and incubate 10 minutes in dry ice.

Treatment of the Methylated DNA

ELIMINATION OF DMS

1. Centrifuge the treated sample at 12,000 rpm for 2 minutes at 20°C.

2. Discard the supernatant and resuspend the pellet in 250 μl of 0.3 M sodium acetate.

3. Add 750 μl of absolute ethanol (at room temperature).

4. Centrifuge at 12,000 rpm for 2 minutes at 20°C.

5. Repeat steps 2, 3, and 4.

6. Resuspend the pellet in 1 ml of 80% ethanol.

7. Centrifuge at 12,000 rpm for 1 minute at 20°C.
8. Discard the supernatant and dry the pellet in a vacuum centrifuge (Speed Vac, Savant).

REACTION WITH PIPERIDINE (CUT AT THE LEVEL OF GUANINE RESIDUES)

1. Resuspend the dry samples in 100 μl of 1 M piperidine (Merck).
2. Heat at 90°C for 30 minutes.
3. Add 1.2 ml of 1-butanol.
4. Vortex vigorously until only one phase is obtained.
5. Centrifuge at 12,000 rpm for 2 minutes.
6. Discard the supernatant.
7. Add 150 μl of 1% SDS, vortex to mix.
8. Repeat steps 3 to 6.
9. Rinse the pellet twice with 80% ethanol.
10. Centrifuge at 12,000 rpm for 1 minute.
11. Discard supernatant and dry the pellet under vacuum.
12. Resuspend the sample in 5 μl of sample buffer, heat for 3 minutes at 90°C, and load on a sequencing gel (see Chapter 20).

REACTION WITH SODIUM HYDROXYDE (CUT AT THE LEVEL OF ADENINE AND GUANINE RESIDUES)

1. Resuspend the dry pellets from step 8 (after elimination of DMS) in 20 μl of 10 mM sodium phosphate buffer, pH 7.0, 1 mM EDTA.
2. Heat for 5 minutes at 90°C.
3. Add 2 μl of 1 M NaOH.
4. Heat for 30 minutes at 90°C.
5. Add 78 μl of sterile distilled water.
6. Precipitate DNA twice according to standard method (see Chapter 3).
7. Resuspend the dry DNA in 5 μl of sample buffer, heat for 3 minutes at 90°C, and load a sequencing gel (see Chapter 20).

Notes

1. DMS is dangerous. Use gloves, and work in a fume hood that has been checked recently (see Chapter 2).
2. Inactivate DMS with 5 M NaOH. All tips should be immersed in 5 M NaOH before being discarded.
3. Stock solution of piperidine Merck is 10 M. Prepare 1M fresh solution just before use.

Buffers

Hepes buffer
250 mM Hepes (pH 8.0)
10 mM DTT

DMS stop buffer
1 M 2-mercaptoethanol
1.5 M sodium acetate
0.5 M Tris HCl (pH 7.5)
5 mM MgCl$_2$

Reference

Siebenlist, U., and Gilbert, W. (1980), *Proc. Natl. Acad. Sci.*, **77**, 122.

ISOLATION OF DNA–PROTEIN COMPLEXES FOR THE ANALYSIS OF FOOTPRINTS FOLLOWING DNAse I DIGESTION OR PROTECTION FROM DMS MODIFICATION*

The binding of a protein to a specific DNA region will result in a protection of the corresponding bases from enzymatic or chemical reactions (DNAse I and DMS, respectively). The footprint generated under such conditions is clearly observed only if all DNA molecules are involved in complexes. Unfortunately, this is not always easy to achieve, especially when the concentration and (or) purity of the specific binding protein are not satisfactory. In order to alleviate this kind of problem, the protein and DNA molecules are incubated together, and once the complex is formed, the mixture is incubated in the presence of the chosen reagent. The subsequent electrophoretic isolation of the DNA–protein complex and analysis of the protected sequences in this complex allow the visualization of zones of specific binding, even if a low percentage of the initial DNA molecules has been complexed.

In some instances, when the complex is not stable, the bound protein can detach from the protected DNA fragment and reassociate to other fragments that have already been degraded by DNAse I, or modified by DMS. Then, it may be difficult to observe a footprint.

PROTOCOL

ISOLATION OF DNA-PROTEIN COMPLEXES

Isolation of the Complex Following DNAse I Footprinting

1. The protein is incubated with the labeled DNA fragment and the complex treated with DNAse I, as described above. (The specific radioactivity of the DNA fragments should be about 6000 cpm/fmol.)

2. The DNAse digestion is stopped by addition of 2.5 μl 0.2 *M* EDTA and the sample electrophoresed as described above.

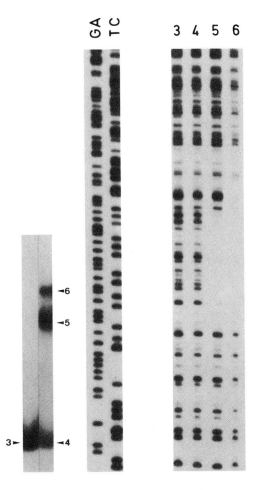

Figure 210. The footprinting activity is responsible for altered migration of DNA complexes on polyacrylamide gels. A ^{32}P-labeled TEF2 probe (10 fmol; 24,000 cpm) was incubated with a suboptimal concentration of proteins (0.7 µg; fraction 45) to leave some DNA uncomplexed. The mixture was treated with DNase I and subjected to electrophoresis. The gel was autoradiographed (inset) and the bands corresponding to free DNA and complexes were excised and DNA fragments analyzed on a sequencing gel as described in text, alongside degradation products of the G + A and C + T reactions (lanes GA and CT). Lane 3 DNA probe digested with nuclease, subjected to electrophoresis and excised from the gel; lanes 4, 5, and 6 correspond to free DNA, complex 1, and complex 2 bands, respectively, excised from the gel (kindly provided by A. Sentenac).

3. Following electrophoresis, the gel is left on one plate and covered with Saran wrap. Orientation marks are performed with radioactive ink and autoradiography is performed at 4°C.

4. Bands corresponding to both free and complexed DNA are cut out with a scalpel and each slice is transferred to a microfuge tube and incubated in 0.4 ml of elution buffer.

5. Elution is performed at 37°C for 15 hours. The buffer is renewed twice during this period of time. Do not crush the gel.

6. Combine the three solutions and filter through a 0.45-μm Schleicher and Schuell FP 030/2 filter.

7. The eluted DNA fragments are precipitated according to the standard method (see Chapter 3), and a minimum of 5000 cpm of material is then loaded onto a sequencing gel (see Chapter 20 and Figure 210).

Buffer

Elution buffer	*1 ml*
10 mM Tris–HCl (pH 8.0)	10 μl 1 M Tris–HCl (pH 8.0)
1 mM EDTA	10 μl 0.1 M EDTA
0.2% w/v SDS	10 μl 20% w/v SDS
0.3 M NaCl	100 μl 3 M NaCl
1 μg/ml Yeast tRNA	1 μl 1 mg/ml Yeast tRNA
	869 μl Sterile distilled water

Isolation of the Complex Following Methylation by DMS

1. The protein sample is incubated with the labeled DNA fragment in a final volume of 50 μl. The specific radioactivity of the DNA must be about 5 to 10 $\times$ 10^3 cpm/fmol.

2. Following to the complex formation at 25°C (see page 726), the mixture is cooled to 0°C.

3. Add cold DMS to the mixture and incubate at 0°C (see page 726). Addition of 1 μl DMS for 1 minute at 0°C has proved to be satisfactory for a 650-base pair fragment.

4. Add 5 μl of 2 M 2-mercaptoethanol to stop the reaction and isolate complex after gel electrophoresis as described above.

---

*These protocols have been kindly provided by A. Sentenac, J. Huet, and S. Camier.

Reference

Huet, J., Cotrelle, P., Cool, M., Vignais, M. L., Thiele, D., Marck, C., Buhler, J. M., Sentenac, A., and Fromageot, P. (1985), *EMBO J.*, **4**, 3539.

25

EXPRESSION OF CLONED DNA SEQUENCES IN VITRO OR IN PROCARYOTIC AND EUCARYOTIC CELLS

Gene expression in living cells proceeds through two basic processes known as transcription and translation. The considerable amount of work that has been performed over the past years to elucidate the salient features of the regulatory mechanisms governing gene expression has provide us with a rather fascinating picture of the complex interactions taking place at this level. Several excellent reviews and books have dealt with this topic in great detail. We present here a brief summary of our knowledge in this field because accurate expression of cloned genes, either in vitro with acellular systems, or in cultured cells and in vivo, in transgenic animals, is submitted to the same fundamental requirements.

The first step of genome expression is the transcription of genes into messenger RNAs, which in eucaryotic cells are processed by splicing from larger nuclear precursors. The accuracy and level of transcription are extremely well regulated in normal cells. Several consensus sequences in the DNA of both procaryotes and eucaryotes have been identified as potential regulatory signals for transcription. For example, in procaryotes, the startpoint of transcription is usually a purine and is often the central base of a CAT sequence. Nucleotide sequencing of more than 100 procaryotic promoters has led to recognition of a consensus TATAAT sequence (Pribnow box) located just upstream to the startpoint of transcription, while a TTGACA sequence has also been shown to be centered 35 base pairs upstream to the startpoint. In eucaryotes, most mRNA are polyadenylated at their 3′ end and all of them have a meth-

ylated cap at their 5′ end. The presence of these two particular structures is also dictated by specific signals (polyadenylation signal: AATAAA; poly A acceptor CACA; GCCA capping site; and CCATT upstream to the 5′-cap).

No extensive homology has been found at the startpoint of transcription in eucaryotes. However, the first base of mRNA is generally an A, flanked by pyrimidines on either side. Most of the eucaryotic promoters contain a TATA consensus sequence (known as TATA box, or Hogness box), and a CAAT box is also frequently found with a GC box (GGGCGG) in many eucaryotic promoters. Other sequences, designated enhancers and dehancers have also been found to play pivotal roles in the regulation of transcription in eucaryotes.

The second step in the expression of the genome results in the synthesis of the genes products. Again, this process, described as translation of the genetic message, involves a complex interplay of regulatory elements.

Protein synthesis is achieved through the ribosomes progression on mRNA molecules. Three stages are recognized in this process: initiation, elongation, and termination.

Initiation requires the binding of the ribosome to the mRNA at a specific site, and corresponds to the reactions preceding formation of the peptide bond between the first two amino acids of the polypeptide. Then, *elongation* proceeds. Amino acids are added to the growing chain, held by the ribosome. When the amino acids assembly is completed, the polypeptidic chain is released. This step is known as *termination* of protein synthesis.

The faithful translation of messenger RNAs involves the participation of several regulatory elements. These include cytoplasmic factors such as initiation factors (IF1–3, in bacteria, and IF1–6, CBPI, CBPII in eukaryotes), elongation factors (EF-Tu, EF-Ts in bacteria; EF1, EF2 in eucaryotes), and termination, or release factors (RF1, RF2 in bacteria, eRF in eucaryotes).

The synthesis of the correct product indeed requires appropriate starting and stopping of the mRNA translation. This is controlled by the presence of specific signals at the beginning and at the end of the mRNA molecule.

The signal that indicates the beginning of the reading frame is the initiation codon. In bacteria, AUG, GUG, and sometimes UUG codons may serve as initiation codon for protein synthesis while in eucaryotes, only the AUG codon is used. When the AUG is used as an initiation codon it codes for a formyl-methionine whereas it is encoding a methionine when located within the coding region.

Termination of protein synthesis is also governed by the presence of specific codons in the mRNA species. Three codons are used as termination signals.

The presence of either a UAG (amber), a UAA (ochre), or a UGA (opal) codon, is sufficient to terminate protein synthesis within a reading frame.

Therefore, expressing a cloned gene in procaryotic or eukaryotic cells will require compilation with the precise rules governing transcription and translation.

Several different vectors have been engineered to allow an efficient expression of cloned genes in both procaryotes and eucaryotes. One can schematically distinguish three classes of expression vectors:

1. Vectors that allow in vitro synthesis of mRNA from cloned genes.

2. Vectors harboring signals for proper transcription in either procarotic or eucaryotic cells but deprived of the signals needed for accurate translation.

3. Vectors harboring all regulatory signals for efficient transcription and translation of cloned genes in either procaryotic or eucaryotic cells.

A selection of commercial expression vectors of the three categories is described in Table 77.

Table 77.   Commercial Expression Vectors

Vectors	Selective Marker	Main Features	Suppliers
		A Selection of Commercial Procaryotic Vectors	
pRIT2T (4,250 bp)	Ampicillin resistance	Vector for expression of protein A–fusion proteins in *E. coli*. Contains the λ P_R promoter for high-level expression, multiple cloning site, IgG binding domain of protein A, protein A gene transcription termination sequence (I) downstream to cloning sites, *E. coli* origin or replication. Need for stop codon in translated cloned sequences.	PHA
pRIT5 (6,894 bp)	Ampicillin resistance Chloramphenicol resistance	Vector for expression of protein A–fusion proteins in either *E. coli* or *S. aureus*. Contains protein A promoter and IgG binding site, multiple cloning site, *E. coli,* and *S. aureus* origins of replication. Need for stop codon in translated cloned sequences.	PHA
pPL-λ (5,200 bp)	Ampicillin resistance	Contains P_L promoter and N gene of lambda, on a 1215 bp fragment inserted between EcoRI and BamHI site of pBR322. Unique Hpa I cloning site in the N gene, 321 bp downstream from P_L promoter. Expression of cloned sequences under the control of P_L-λcI857 thermoinducible system. No ATG initiation codon or translation termination codon provided.	PHA

Table 77. Commercial Expression Vectors (*continued*)

Vectors	Selective Marker	Main Features	Suppliers
pYEJ001 (4,100 bp)	Amipicillin resistance Chloramphenicol resistance Tetracycline resistance	Derivative of pBR327. Contains a synthetic consensus promoter downstream to a tandem of synthetic lactose operators, and upstream to chloramphenicol and tetracycline resistance genes. Expression of foreign sequences cloned at the promoter-proximal Hind III site can be monitored by a CAT assay. Need for both ATG and stop codon in cloned sequences.	PHA
pKK223–3 (4,585 bp)	Ampicillin resistance	Contains the strong *tac* promoter made of the − 35 region of the *trp* promoter and the − 10 region, operator, and ribosome binding site of the *lac* UV–5 promoter. The *tac* promoter located upstream to a multiple cloning site region (from pUC8) allows expression of cloned foreign sequences upon induction with IPTG. Ribosomal RNA transcription terminator sequence (rrn^B) adjacent to the cloning region stabilizes the host–vector system. Need for translation termination codon in the cloned sequences.	PHA
pKK233–2 (4,602 bp)	Ampicillin resistance	Contains the *trc* promoter made of the *trp–lac* promoter with the consensus 17 bp spacing between the *trp*–35 region and the –10 region of *lac* UV–5 promoter. Also contains the *lac* Z appropriate NcoI linkers it is possible to align the reading rame of the cloned fragment with the initiation codon. Pst I and Hind III cloning sites are conveniently placed downstream to the Nco I site. Need for translation termination codon in the cloned sequences.	PHA
pEX series (5,775 bp)	Ampicillin resistance	Vectors for efficient expression of β-galactosidase fusion proteins. Three vectors have different reading frames through the cloning linker. Unique cloning sites: Bam HI, Pst I, Sal I, and Sma I. The PEX$_2$ vector also contains a unique EcoRI cloning site. Stop codons in the 3 translational reading frames are provided on the 3′ side of the cloning region. Upon induction, the hybrid proteins accumulated may represent up to 30% of the bacterial proteins.	GEN
pUC series	Ampicillin resistance Lack of β-galactosidase activity	Vectors for β-galactosidase fusion proteins. Contain multiple cloning sites. Allow cloning in either orientation, therefore leading to expression of the correct product from in-phase cloned inserts. Need for translation stop codons in the cloned sequences.	AME, BOE, BRL, GEN, IBI, PHA

A Selection of Commercial Eucaryotic Expression Vectors

Vectors	Selective Marker	Main Features	Suppliers
pSVL (4,800 bp)	Ampicillin resistance	Vector for *transient* expression of cloned genes in mammalian cells. Hybrid of pBR322 and SV40 sequences. Contains the SV40 origin of replication, and the late-promoter sequences with the VP1 intron upstream to a multiple cloning sites region. Unique cloning sites: Xho I, Xba I, Sma I, Xcy I, Sac I, and Bam HI. Presence of late polyadenylation signals allow synthesis of authentic eucaryotic mRNAs from cloned sequences. Also contains pBR322 origin of replication and ampicillin resistance gene for selection in *E. coli* hosts. The plasmid "poison sequences" have been deleted to allow increased replication and efficient transient expression in *Cos* monkey cells. Does not provide ATG initiation codon or termination codon.	PHA

Table 77. **Commercial Expression Vectors (continued)**

Vectors	Selective Marker	Main Features	Suppliers
pMSG (7,635 bp)	Ampicillin resistance Ability to grow in HAT medium	Vector for *stable* expression of cloned genes in mammalian cells. Does not provide ATG initiation codon or termination codon. Contains the dexamethasone inducible transcription promoter from the mouse mammary tumor virus LTR, the *E. coli* gpt (xanthine–guanine phosphoribosyltransferase) gene for selection in HAT medium, a multiple cloning site downstream to the MMTV LTR, and the SV40 regulatory signals for efficient expression of the cloned sequences. Also contains pBR322 origin of replication and ampicillin resistance gene for convenient amplification of recombinant DNA.	PHA
pKSV–10	Ampicillin resistance	Vector for *stable* expression of cloned genes in mammalian cells. Analogous to the pSV2 plasmid. Contains the SV40 early promoter region, large T antigen intervening sequence, and early region polyadenylation site. Cloning at the Bgl II site located downstream to the early promoter will allow efficient expression in mammalian cells. Can replicate in both *E. coli* and *Cos* monkey cells. Does not provide ATG initiation codon or translation stop codon.	PHA
pSVN9	Ampicillin resistance Neomycin resistance	Eucaryotic expression vector carrying the neo^r gene downstream to SV40 early region. Contains SV40 early promoter/enhancer, origin of replication 3′ end of large T antigen intervening sequence and early region polyadenylation site. Does not provide signals for initiation and termination of protein synthesis. Poison sequences are not present in this vector which can also be propagated in *E. coli*. Insertion of DNA at the unique Xho I, Sma I, and Xma I sites inactivate the neo^r gene.	STR

Key for abbreviations of suppliers names: (AME) Amersham International; (BOE) Boehringer Mannheim; (BRL) Bethesda Research Laboratories; (GEN) Genofit; (IBI) International Biotechnologies Inc.; (PHA) Pharmacia; (STR) Stratagene; See Appendix 1 for addresses.

IN VITRO TRANSCRIPTION OF CLONED GENES

Several vectors containing two different viral transcription promoters on either side of a multiple cloning site region have been developed to provide means for high-level transcription of both strands from any inserted DNA fragment (see Figure 211). These systems can be used in conjunction with purified viral RNA polymerases, to synthesize: labeled probes (see Chapter 16); RNA molecular weight markers (see Chapter 18); antisense RNA species from the transcription of the "noncoding" strand of the cloned fragment; and high levels of specific transcripts which can be used for molecular studies of

RNA synthesis and processing and used to direct polypeptide synthesis with the in vitro translation systems or after injection in oocytes. The levels of transcription in bacteria harboring the recombinant plasmid can be increased notably after infection of the cells by the virus encoding the appropriate RNA polymerase.

In these systems, the mostly used viral promoters are those recognized by RNA polymerases from bacteriophages T3, T7, and SP6.

The T7 gene 1, coding for RNA polymerase has been cloned (Davanloo et al., 1984; Tabor and Richardson, 1985) and can be expressed to very high levels in *E. coli*. The T3 RNA polymerase gene has been

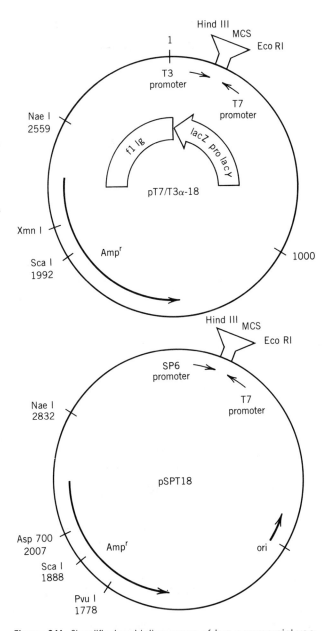

Figure 211. Simplified restriction maps of two commercial vectors for in vitro transcription of cloned DNA sequences.

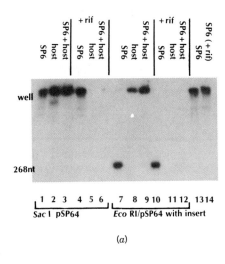

(a)

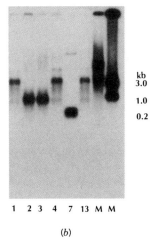

(b)

Figure 212. (a) Effect of rifampicin on transcription of two different templates by SP6 and *Salmonella typhimurium* RNA polymerases. The indicated template DNAs (0.2 μg) were transcribed with 2.5 units of *S. typhimurium* RNA polymerase, 4 units of SP6 RNA polymerase, or a mixture of these amounts, in a total volume of 20 μl in the presence and absence of 20 μ/ml rifampicin. Transcriptions and electrophoresis of samples (5% gel) were performed as described in Figure 143. The enzyme and rifampicin treatment are indicated for each lane. (b) Size analysis of large transcripts on an agarose gel containing formaldehyde. Samples of the same reaction mixture used for the urea–polyacrylamide gel in (a) were applied to a 1.4% agarose gel containing 1.1 *M* formaldehyde. Following electrophoresis the gel was blotted onto nitrocellulose in 20 × SSC and the transcripts bound to the membrane were visualized by autoradiography. The numbers under the lanes correspond to the samples in (a) that were run on this gel.

cloned in an expression vector in which it has been placed under the control of the inducible *lac* promoter (McAllister, unpublished). The SP6 RNA polymerase is accumulated in *Salmonella typhimurium* infected with the SP6 bacteriophage. The specificity of these RNA polymerases for the corresponding viral promoters is high enough to allow fairly efficient transcription of the DNA sequences inserted downstream to these promoters in the cloning vectors (Figures 212*a*, *b*).

PROTOCOL

IN VITRO SYNTHESIS OF LARGE AMOUNTS OF UNLABELED RNA WITH SP6-RNA POLYMERASE

1. In a microfuge tube prepare a 2.5 mM stock of ribonucleotide triphosphates (rXTP) by mixing:
   10 μl of 10 mM rATP
   10 μl of 10 mM rCTP
   10 μl of 10 mM UTP
   10 μl of 10 mM rGTP

2. Linearize the vector-insert recombinant DNA at a restriction site located immediately downstream to the inserted sequence in the multiple cloning site, with respect to the promoter that is going to be used. This step is performed as the standard restriction digestions (see Chapter 8). Clean DNA fragment is required.

3. In a microfuge tube mix successively:
   44 μl of sterile distilled water
   20 μl of 5X SP6-transcription buffer
   10 μl of 100 mM DTT
   4 μl of RNasin (25 units/μl)
   20 μl of the 2.5 mM rXTP mixture prepared in step 1
   2 μl (1 μg/μl) of linearized DNA plasmid prepared in step 2

4. Add 20 units of SP6 RNA polymerase; mix.

5. Incubate for 2 hours at 37°C.

6. Add 2 μl of DNase I (0.25 μg/μl) directly to the reaction mixture.

7. Incubate 10 minutes at 37°C.

8. Add 50 μl of buffer-saturated phenol and 50 μl of chloroform:isoamyl alcohol mixture (24:1).

9. Extract and recover RNA as described in Chapter 3.

Notes

1. All reactions must be performed in sterile polypropylene tubes with sterile solutions. Treat sterile distilled water with diethylpyrocarbonate.

2. The components of the reaction mixture should be added in order.

3. Do not keep the reaction mixture on ice. DNA would precipite because of spermidine.

4. Under the conditions used, transcription does not stop pre-

Table 78. The Effect of DNase I on the Size Distribution of RNA Transcripts[a]

Size range (kb)	Control (%)	5 μg/ml DNase (%)	220 μg/ml DNase (%)
7.5–10	37	34	23
6–7.5	20	20	17
4–6.0	20	21	25
2–4	16	17	24
1–2	6	7	9
<1	1	1	2

Source: Courtesy of Bethesda Research Laboratories, Life Technologies, Inc.

[a]SP6 RNA polymerase reactions were performed as described in text. DNase I was diluted, if necessary, in 10 mM sodium acetate (pH 6.5), 5 mM CaCl$_2$, and added directly to the SP6 RNA polymerase reactions. The DNase I digestions were incubated at 37°C for 10 minutes. Reactions were terminated, extracted, glyoxylated, and analyzed on a 1.2% agarose gel.

maturely. Templates up to 6 kb in length can be transcribed efficiently if the concentration of nucleotides triphosphate is kept above 400 μM in the reaction mixture (Table 62 and Figure 141).

5. It is not recommended to exceed 10 minutes of incubation when removing the template DNA with the DNase I. The effects of DNase digestion on the size distribution of RNA transcripts is shown in Table 78.

6. Concentrations of DNase as low as 1 μg/ml are generally sufficient for removal of the template.

7. Gross degradation of RNA is not observed after treatment with DNase (if the satisfactory grade is used). This does not mean that discrete alterations of the RNA do not occur. Such nondetectable RNA degradation may turn to be a real problem for further biological studies (such as in vitro translation).

8. We recommend addition of RNasin (Promega) to avoid degradation of RNA. The RNAguard (Pharmacia) compound and placenta RNase inhibitor from BRL have also been used.

9. Under the conditions described here, up to 10 μg of RNA can be obtained for 2 μg of plasmid DNA.

10. Linearization of the recombinant vector leads the enzyme run off the template at the terminus of linear DNA and therefore avoids transcription sequences of plasmid (Figure 213).

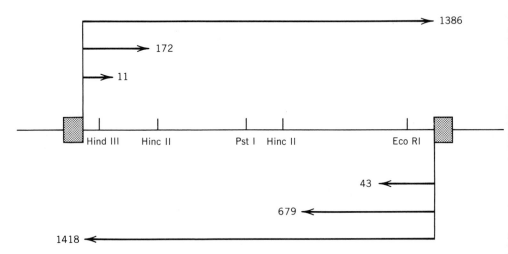

Figure 213. Schematic representation of the promoter/insert region of a transcription vector, showing the position of restriction sites which can be used to linearize plasmid DNA. For example, digestion with Pst I produces a template yielding transcripts of 557 and 863 bases with SP6 and T7 RNA polymerases, respectively. The sizes of other transcripts initiated from both promoters are also indicated (not drawn to scale). The RNA species synthesized after in vitro transcription of linearized plasmid DNA with SP6 and T7 RNA polymerases are shown in Figure 143 (redrawn with permission of Promega Biotec).

Buffers and Solutions

5X SP6 Transcription buffer   *1 ml*
200 mM Tris–HCl (pH 7.5 at   200 µl 1 M Tris–HCl (pH 7.5)
   37°C)
30 mM MgCl$_2$ 30 µl 1 M MgCl$_2$
10 mM Spermidine   100 µl 0.1 M Spermidine
50 mM NaCl   10 µl 5 M NaCl
                                       660 µl H$_2$O

Store as 100 µl aliquots at -20°C.

rXTP Stock solutions
10 mM rATP
10 mM rCTP
10 mM rGTP
10 mM UTP

The ribonucleotides mixtures are adjusted to the correct concentration on the basis of their UV absorption (see Chapter 3). The mixtures should be at pH 7.0. Neutralized if necessary.

100 mM DTT.
Dilute 1/10 a 1 M DTT stock solution stored at -20°C. It is possible to add 0.5 mM EDTA to increase the stability of DTT in the stock solution.

PROTOCOL

IN VITRO SYNTHESIS OF LARGE AMOUNTS OF UNLABELED RNA WITH T7-RNA POLYMERASE

1. Linearize the vector-insert recombinant DNA at a restriction site located immediately downstream to the cloned sequence, with respect to the promoter being used. This is achieved under standard conditions described in Chapter 8.

2. In a microfuge tube mix successively (at room temperature):
   10 µl of 5X T7-transcription buffer
   8 µl of the 2.5 mM rNTP mixture (see step 1, previous protocol)
   2.5 µl of 0.1 M DTT
   1 µl of linearized DNA (1 µg/µl)
   sterile distilled water 27.5 µl

3. Add 1 µl T7-RNA polymerase (10–20 units).

4. Incubate for one hour at 37°C.

5. Treat sample as in steps 6–9 of previous protocol to recover the RNA.

Notes

1. See notes for SP6 transcription.

2. Aberrant transcription may occur from the ends of the DNA digested with restriction endonucleases. This is particularly the case when 3′ protruding ends have been generated. Raising the NaCl from 25 mM to 100 mM results in a decrease of nonspecific transcripts (0.1 to 0.5%) but inhibits the polymerase activity by about 50%.

3. When using vectors carrying T3 and T7 transcription promoters on either side of the multiple cloning site region care should be taken that both polymerases can recognize the heterologous promoter. Transcripts initiated from the T3 promoter by T7 RNA polymerase represent about 0.5 to 1% of the correct transcripts, while transcripts initiated from the T7 promoter by T3 RNA polymerase can represent as much as 3% of the correct transcripts. These nonspecific initiations of transcription occur 50% less frequently in the presence of high salt concentrations (100 mM NaCl).

Buffer

5X T7-transcription buffer   *1 ml*
200 mM Tris–HCl (pH 8.0 at   200 µl 1 M Tris–HCl (pH 8.0)
  37°C)

$$40 \text{ m}M \text{ MgCl}_2$$
$$10 \text{ m}M \text{ Spermidine}$$
$$125 \text{ m}M \text{ NaCl}$$

$$40 \text{ }\mu\text{l } 1 \text{ }M \text{ MgCl}_2$$
$$100 \text{ }\mu\text{l } 0.1 \text{ }M \text{ spermidine}$$
$$25 \text{ }\mu\text{l } 5 \text{ }M \text{ NaCl}$$
$$635 \text{ }\mu\text{l H}_2\text{O}$$

Store as 100-μl aliquots at $-20°$C.

References

Davanloo, P., Rosenberg, A. H., Dunn, J. J., and Studier, F. W. (1984), *Proc. Natl. Acad. Sci. USA*, **81**, 2035.

Tabor, S., and Richardson, C. C. (1985), *Proc. Natl. Acad. Sci. USA*, **82**, 1074.

IN VITRO TRANSLATION OF RNA EXPRESSED FROM CLONED GENES

Until the recent development of efficient expression vectors (see below), the characterization of the protein products encoded by cloned DNA fragments of either procaryotic or eucaryotic origins was generally achieved by the means of the maxi cell technique (Sancar et al., 1979; Roberts et al., 1979); phage infection of irradiated cells (Jaskunas et al., 1975; Ptashne, 1967; Murialdo and Siminovitch, 1972); or selective expression of cloned genes following release of chloramphenicol inhibition (Neidhardt et al., 1980).

These procedures were based on the selective labeling of proteins in vivo. The characterization of the radioactive polypeptides was generally performed by electrophoresis and autoradiography or by immune precipitation with specific antisera.

The maxi cell technique relies on the use of *E. coli* mutants (*uvrA* and *recA*) which degrade chromosomal host DNA but not plasmid DNA, following a UV irradiation at low doses. The maxi cells go on synthesizing plasmid-specific products up to 24 hours after irradiation.

Similarly, UV irradiated *uvrA* mutants of *E. coli* can selectively express phage-coded proteins after infection. This method has been used either with wild-type λ phages or with derived vectors containing foreign DNA inserts. If lysogen cells are irradiated, the majority of the λ proteins are repressed. Depending on the presence of adequate control sig-

nals in the cloned fragment, expression may occur in these cells. Since read through from the phage λ promoters is possible, one can even obtain expression of cloned DNA at high levels in the absence of regulatory sequences in the insert.

The third method was based on the observation that ColE1-derived plasmids are synthesized at a high level in host cells treated with chloramphenicol (see Chapter 7). Since only a few dozen of *E. coli* proteins are expressed in these cells, a bidirectional SDS gel electrophoresis of labeled cell extracts may allow the characterization of gene products encoded by the cloned sequences.

Although these different systems may be useful for the expression of simple procaryotic or eucaryotic sequences cloned in plasmid or phage vectors, they do not permit the synthesis of more complex eucaryotic messenger RNAs which need splicing and capping for their accurate expression. With the development of vectors allowing highly efficient in vitro synthesis of specific RNA species, it is now possible to test directly for the coding capacity of a given gene by using the in vitro translation of purified mRNAs transcribed from cloned sequences by the SP6, T3, or T7 RNA polymerases.

In vitro translation of eucaryotic mRNA species has played a fundamental role in the characterization of the regulatory mechanisms governing gene expression. For example, its use has led directly to the discovery of precursor forms of viral and cellular proteins, and to the concept of "signal peptide" for the secretion of extracellular proteins.

Mainly two mRNA-dependent in vitro translation cell-free systems have been developed in the past years: (1) reticulocyte lysates and (2) wheat germ extracts. The reticulocyte lysates (Woodward and Herbert, 1974; Schimke et al., 1974; Berns and Bloementhal, 1974) are prepared from reticulocytes of animals made anemic by repeated injections of phenyl-hydrazine. The removal of endogenous globin mRNA is achieved by a brief treatment with the calcium-dependent micrococcal nuclease and the nuclease activity is subsequently inactivated by chelation of the divalent cations by EGTA (Pelham and Jackson, 1976: Jackson and Hunt, 1983). This treatment does not alter the protein synthetic activity of the extract which is then mainly dependent upon the addition of exogenous mRNA species.

Wheat germ extracts have very low endogenous mRNA activity, and have successfully been used for several years (Roberts and Paterson, 1973; Marcus et al., 1974; Marcu and Dudock, 1974). However, because this system had been reported to be inefficient for the synthesis of high molecular weight translation products (Davies, 1976; Shih and Kaesberg, 1976) it has not been widely used. Modifications to the previously published procedures (Zagorski, 1978; Anderson et al., 1983) have led to the constitution of a wheat-germ-derived system (Morch et al., 1986) which has proven very efficient for the in vitro synthesis of high molecular weight proteins (at least 200 kilodaltons) encoded by plant viral mRNAs. Since this system can be supplemented by beef liver tRNAs, it is expected to be satisfactory for the synthesis of mammalian proteins as well.

It should be pointed out that several other cell-free systems have been developed for specific in vitro translation of proteins from different species such as yeast (see review by Moldave and Gasior, 1983) drosophila (Scott et al., 1979), or bacteria (see review by Chambliss et al., 1983) and that Erlich ascite tumor cell extracts have also been used as a general system for in vitro translation (see review by Henshaw and Panniers, 1983). It is also important to note that aside from the use of cell-free systems, very efficient translation of purified mRNA species is also achieved after microinjection of *Xenopus oocytes* (Gurdon et al., 1971, 1973). In addition to its high sensitivity, this system offers the possibility of performing coupled transcription/translation of DNA sequences injected in the nucleus of the oocytes.

Only the use of the reticulocyte lysates and wheat germ extracts will be described in this manual.

Early attempts to use in vitro translation for the characterization of cloned gene products were based on the fact that translation of a specific mRNA species is inhibited when it is hybridized to complementary DNA. In this approach, known as "hybrid-arrested translation" (Paterson et al., 1977; Hastie and Held, 1978), the translation products corresponding to the hybridized RNA only appear when the DNA–RNA hybrids are heat-denatured prior to translation. Variations of the procedure included separation of the duplex from nonhybridized molecules (Woolford and Rosbach, 1979) or purification of specific mRNA species after hybridization with DNA sequences bound onto nitrocellulose filters, or paper filters (see review by Miller et al., 1983).

Unfortunately, the yield and quality of mRNA recovered by these methods are not always satisfactory and they should be used only when subcloning in transcription or expression vectors is not feasible. In most cases, in vitro transcription of cloned sequences is the method of choice to obtain large amounts of translatable RNA species corresponding to the cloned sequences of interest.

Synthetic eucaryotic mRNAs are effective templates for in vitro translation (in oocytes as well as in reticulocyte and wheat germ systems) and in vitro splicing, when they possess a 5'-capped end (see above). In vitro capping of synthetic eucaryotic RNA species can be achieved by transfer of the 5' terminal cap structure by guanylyltransferase in vitro (see Chapter 4). Alternatively, methylated nucleotides can be used in combination with the SP6 or T7 RNA polymerases as initiator of the in vitro synthesis of capped RNA.

PROTOCOL

ADDITION OF A CAP STRUCTURE TO RNA TRANSCRIPTS BY INCUBATION WITH GUANYLYLTRANSFERASE

1. In a microfuge tube containing 1.0 µg of RNA as a dried pellet, add successively:
   13.6 µl of sterile distilled water (mix by vortexing)
   1.5 µl of 1 *M* Tris–HCl (pH 7.9)

3.7 μl of 10 mM MgCl$_2$
1.8 μl of 100 mM DTT
3 μl of BSA (1 mg/ml)
3 μl of 1 mM S-adenosyl-methionine (SAM)
1.2 μl of RNasin (Promega, 25 units/μl)
1.2 μl of 1 mM GTP

2. Mix; centrifuge at 12,000 rpm a few seconds.
3. Add 1 μl of guanylyltransferase (BRL, 1–5 units/μl).
4. Incubate for 45 minutes at 37°C.
5. Add 70 μl of NTE (pH 7.5), and 100 μl of phenol/chloroform isoamyl.
6. Vortex; centrifuge at 12,000 rpm for 15 seconds.
7. Take the supernatant and repeat steps 5–6.
8. Add 50 μl of 7.5 M ammonium acetate, and 300 μl of cold absolute ethanol.
9. Let precipitate, and recover the RNA as described in Chapter 3.

Note. The GTP concentration may have to be empirically adjusted to meet your own requirements.

PROTOCOL

IN VITRO SYNTHESIS OF LARGE AMOUNTS OF CAPPED RNA WITH SP6 RNA POLYMERASE

In this protocol, the in vitro RNA synthesis is primed with a cap dinucleotide such as P^1-5′-(7-methyl)-guanosine-P^3-5′-adenosine triphosphate [m^{7}G(5′)ppp(5′)A] or P^1-5′-(7-methyl)-guanosine-P^3-5′-guanosine-triphosphate [m^7-G(5′)ppp(5′)G].

1. In a microfuge tube, mix ribonucleotides according to the following chart:

Cap Dinucleotide	Ribonucleotides (20 mM)				H$_2$O
	rATP	rCTP	rGTP	UTP	
m^7-G(5′)ppp(5′)A	0.5 μl	5 μl	5 μl	5 μl	4.5 μl
m^7-G(5′)ppp(5′)G	5 μl	5 μl	0.5 μl	5 μl	4.5 μl

2. Linearize vector DNA by restriction endonuclease as described in Chapter 8.
3. In a microfuge tube, mix successively at room temperature: 7.5 μl of sterile distilled water

    10 μl of 5X transcription buffer
    5 μl of 100 m*M* DTT
    5 μl of BSA (1 mg/ml)
    2.5 μl of RNasin (Promega, 25 units/μl)
    5 μl of the ribonucleotide mix prepared in step 1
    5 μl of the cap dinucleotide used (5 m*M* solution of [m⁷-G(5′)ppp(5′)G] or [m⁷-G(5′)ppp(5′)A]
    5 μl of the linearized template prepared in step 2
    2 μl of SP6 RNA polymerase (5 units/ml) to obtain a final concentration of about 100 units/ml

4. Incubate for 60 minutes at 40°C.
5. Add 2 μl of SP6 RNA polymerase.
6. Incubate for 30 minutes.
7. Purify RNA species after DNase I treatment, phenol extraction, and chromatography on Sephadex G100, as described in Chapter 3.

Buffer

5X Transcription buffer
200 m*M* Tris–HCl (pH 7.5 at 40°C)
30 m*M* MgCl₂
10 m*M* Spermidine
50 m*M* NaCl

Note. Cap dinucleotides are commercially available (for example, New England Biolabs or Boehringer).

PROTOCOL

IN VITRO TRANSLATION WITH WHEAT GERM EXTRACT

Preparation of the Wheat Germ Extract (Morch et al., 1986)

Start from 30 g of wheat germ obtained from industrial flour mills (make sure that it has not been previously heat treated).

1. Float twice with a mixture of cyclohexane–carbon tetrachloride (usually 50–70 ml of cyclohexane for 250 ml of carbon tetrachloride) so as to obtain about 50% of material floating on the surface.

2. Collect the floating wheat germ with a sieve and let dry on a filter overnight in a fume hood.

3. Remove remaining dark particles with tweezers.

All operations from the next step onward are performed in the cold room.

4. Prepare a Sephadex G25 column (2 × 15 cm) equilibrated with the extraction buffer.

5. Grind 3 g of floated wheat germ with 3 g of sterile broken Pasteur pipettes (use precooled mortar and pestle) for 2 minutes.

6. Add 5 ml of extraction buffer, mix.

7. Add 5 ml of extraction buffer and grind for about 30 seconds to obtain a smooth paste.

8. Transfer the mixture in a 15-ml Corex tube and centrifuge at 23,000 × g for 10 minutes (rpm in a SS34 Sorval rotor, or equivalent) at 4°C.

9. Transfer the supernatant in a clean 15-ml Corex tube and repeat step 8.

10. Pipette the supernatant and load immediately 4 ml onto the G25 column prepared at step 4.

11. Collect 1.25-ml fractions in microfuge tubes, as soon as the extract has penetrated into the Sephadex.

12. Combine the most opaque fractions (usually fractions 11–14 or 12–15) in a 15-ml Corex tube.

13. Leave on ice for 5 minutes to allow aggregation of high molecular weight inhibitor(s).

14. Centrifuge for 10 minutes at 23,000 × g in precooled rotor and tubes, and at 4°C.

15. Aliquots (100–200 μl) of the supernatant are immediately frozen in small tubes and stored in liquid nitrogen.

Notes

1. It may be necessary to adjust the proportions of the cyclohexane/carbone tetrachloride mixture.

2. The nonfloating material corresponds to endosperm-rich fragments which are not suitable for extract preparation.

3. The fresh wheat germs and the floated wheat germs can be stored at 0–4°C for several months in sealed bags.

4. It is important that centrifugation in step 14 be performed at a temperature close to 0°C because the aggregate dissolves above 6°C.

5. Since a low molecular weight inhibitor and a yellow pigment are retarded on the G25 column, it is necessary to wash it thoroughly with at least 500 ml of extraction buffer prior to use for a second preparation.

6. Extract samples stored at $-20°C$ for 3 weeks lost 70% of activity while no decrease was observed for samples kept at $-70°C$.

7. Loss of activity was observed upon thawing–freezing. A sample thawed and refrozen once in liquid nitrogen lost 5% of its activity.

In Vitro Translation Assay

1. In a microfuge tube mix successively:
   5 μl of 5X energy mix
   2.5 μl of 10X salt mix
   1 μl (10 μCi) of ^{35}S methionine (> 1000 Ci/mmol, 37 TBq/mmol)
   1 μl (3 μg) of tRNA (from plant, liver, or other adequate source)
   1 μl of mRNA (4 μg/μl)
   1.5 μl of sterile distilled water
   1 μl of RNasin (Promega 25 units/μl)
   12 μl of wheat germ extract

2. Incubate at 25°C for 2 hours.

3. Analyze translation products by direct electrophoresis, immunoprecipitation, immunotransfer, or any other adapted technique.

Notes

1. The volume of incubation can be reduced to 12.5 μl.

2. Optimal translation is obtained without volumes of extract ranging between 8 and 12 μl for a 25-μl incubation, depending on the intrinsic concentration of the preparation.

3. The use of Na^+, $NH4^+$, and Cl^- ions must be avoided during extraction or incubation since they are detrimental to the system.

Buffers and Solutions

Extraction buffer	*1 liter*
10 mM Tris–acetate (pH 7.5)	10 ml 1 M Tris–acetate (pH 7.5)
3 mM Magnesium acetate	3 ml 1 M Mg acetate
5 mM Potassium acetate	50 ml 1 M K acetate
	937 ml H_2O

5X Energy mix
2.5 mM ATP (pH 7.0)
1.5 mM GTP (pH 7.0)
100 mM Creatine phosphate
Sigma (pH 7.0)
250 μg/ml creatine
Phosphokinase (Sigma)

25 μl 10 mM ATP (diptassium salt)
15 μl 10 mM GTP (trisodium salt)
10 μl 1 M Creatine phosphate (dipotassium salt)
25 μl Creatine phosphokinase 1 mg/ml
25 μl H$_2$O

Prepare the 5X mix freshly from stock solutions kept at $-20°$C.

10X Salt mix
200 mM Hepes–KOH (pH 7.5)
1 M K acetate
6 mM Spermidine
2.4 mM Each amino acid except methionine
80 μM Methionine
Can be stored for weeks at $-20°$C.

PROTOCOL

IN VITRO TRANSLATION WITH RETICULOCYTE LYSATES

Preparation of the Lysates

1. Obtain an immature healthy rabbit (New Zealand, white rabbit) and perform five successive subcutaneous injections (at one-day intervals) with 1 ml of 1.25% acetyl phenyl hydrazine (Sigma) between the two scapula of the animal.

2. Three days after the fifth injection (day 8), perform an intracardiac punction to bleed the rabbit.

3. Collect the blood in 50-ml Falcon tubes containing 5 ml of cold "wash saline" and 1000 units of heparin. One usually obtains one full tube per animal. All subsequent steps are performed in the cold room or with the tubes and solutions in ice buckets.

4. Pellet the reticulocytes by 10-minute centrifugation at 2000 × g (4000 rpm in SS34 rotor) and 4°C.

5. Discard the plasma, add 10 ml of wash saline, and resuspend gently the cells by aspiration in a pipette.

6. Centrifuge as in step 4. Discard the supernatant and add 10 ml of wash saline to the cell pellet.

7. Repeat steps 5–6 twice.

8. Add 40 ml of gradient buffer in the tube and resuspend *gently* the cell pellet. Transfer to a 15-ml Corex.

9. Centrifuge for 15 minutes at 15,000 rpm in a SS34 rotor (or equivalent) at 4°C.

10. With a pipette, aspirate the top layer of white cells.

11. Change pipette, and aspirate the reticulocyte layer.

12. Transfer to a tube containing cold wash saline and wash the cells three times as described in steps 4–6.

13. The reticulocytes are lysed by addition of chilled sterile distilled water to the cell pellet (1 volume H_2O containing 2 mM DTT for 1 volume of cells).

14. Vortex for 1 minute to maximized osmotic lysis.

15. Centrifuge at 30,000 × g for 15 minutes (17,000 rpm in SS34 rotor) at 4°C.

16. The supernatant is distributed as 3-ml fractions in 50-ml Falcon tubes.

17. At this step it is possible to check the quality of the extract by performing a few control tests before initiating the nuclease-treatment steps (see note 2).

18. To each 3-ml fraction, add:
    14.3 μl of hemin 6.3 mM (30 μM final)
    60 μl of $CaCl_2$ 50 mM (1 mM final)
    23.5 μl (80 units/ml final) of Micrococcus nuclease (Boehringer, 10, 200 units/ml)

19. Mix gently and incubate for 15 minutes at 20°C.

20. Add 60 μl of 100 mM EGTA to stop the nuclease activity.

21. Add 12 μl of 40 mg/ml creatine phosphokinase (Sigma) to obtain a final concentration of 0.16 μg/ml.

22. Mix all treated samples and store in small tubes (1 ml) in liquid nitrogen.

Notes

1. Count the number of reticulocytes per milliliter of blood at day 1 and day 8 to check anemia.

2. The quality of reticulocyte lysates prior to micrococcal nuclease treatment can be checked as follows:
   a. The absorbance at 415 nm should be about 800–1000 (corresponding to an hemoglobin concentration of 200–250 mg/ml).

b. 20 to 25 A_{260} units of polysomes should be obtained after high-speed centrifugation (100,000 $\times$ g) of 1-ml extract sample.

c. Protein synthesis measured by incorporation of a labeled amino acid into TCA precipitable material, should be linear for about 30 minutes, and 50 μl of the preparation should allow the incorporation of about 300–400 pmol of labeled leucine into the newly synthesized polypeptides.

3. Overdigestion of endogenous mRNAs by micrococcal nuclease may result in a reduction of efficiency upon addition of exogenous RNA species.

In Vitro Translation Assay

Prepare a reaction mix as described below. The indicated volumes allow us to perform 4 tests and a control.

1. Thaw a 100-μl aliquot of nuclease-treated lysate on ice for 5–10 minutes.

2. Transfer 60 μl of lysate in a microfuge tube and add successively:
5 μl of master mix
5 μl of 1 M potassium acetate (50 mM final)
2.5 μl of 2 M potassium chloride (50 mM final)
1.6 μl of 50 mM magnesium acetate (0.8 mM final)
10 μl of 2.5 mg/ml calf liver tRNA (250 μg/ml final)
10 μl of ^{35}S methionine (>1000 Ci/mmol, 37 TBq/mmol, 10 mCi/ml)
15.9 μl of 20 mM Hepes (pH 7.5).

3. Resuspend 100 ng-samples of specific in vitro transcribed capped mRNAs (or 20 μg of total polyadenylated RNAs for other source) in 4 μl of sterile distilled water.

4. Heat at 68°C for 5 minutes and cool quickly in iced water.

5. Add 20 μl of the reaction-mixture to the RNA samples and incubate at 30°C for 2 hours.

6. Run a control experiment with 20 μl of the reaction mixture and 4 μl of H_2O instead of RNA.

7. To each tube, add:
0.7 μl of 0.1 M EDTA (pH 7.5)
0.7 μl of 0.1 M PMSF (phenylmethane sulfonyl fluoride, Fluka)

It is wise at this stage to determine the extent of incorporation obtained.

8. Spot 1-μl samples of the incubation mixtures on Whatmann no. 1 filters and dip in 100 ml of cold 10% TCA (in a beaker).
9. Boil the filters in 50 ml of 5% TCA for 2 minutes.
10. Immerse the filters in a fresh solution (100 ml) of cold 5% TCA for 2 minutes.
11. Transfer the filters for 2 minutes in a beaker containing a 50:50 mixture of ethanol:ethyl ether.
12. Dry the filters.
13. Incubate for 2 minutes in H_2O (50%).
14. Repeat step 9.
15. Dry the filters and count in scintillation mixture (use ^{14}C channel of the counter).
16. Treat the remainder of the sample for immunoprecipitation or immunotransfer (see pages 789, 792), or direct electrophoresis on polyacrylamide gels (see Chapter 3).

Notes

1. The concentrations of spermidine magnesium acetate, potassium acetate, and tRNA should be titrated to optimize the reaction efficiency. For example, spermidine should be assayed at 0, 0.5, 1.0, 1.55 mM, magnesium acetate at 0.1, 0.5, 1.0, 1.5, 2.0 mM, and potassium acetate at 50, 100, 150, 200 mM.
2. The presence of high concentrations of magnesium ions may be inhibitory to the in vitro translation of some mRNA species.
3. The master mix should not contain the amino acid that is used to label the synthesized polypeptides (most frequently ^{35}S-methionine, or ^{3}H-leucine).
4. The in vitro synthesis of an intact globin chain (15,000 daltons) requires 2 minutes at 30°C. One can estimate therefore that 20–30 minutes will be needed for the synthesis of a 150,000-dalton polypeptide.

Notes (for Reticulocyte Lysate)

1. It is advised to equilibrate the lysate at 20°C before adding the micrococcal nuclease, since the rate of warming up will depend on the thickness of the flask and the surface of lysate exposed, and so forth. You can leave the lysate at 20°C for this short period of time without notably loosing the activity.

2. There is no point digesting more than 20 minutes.

3. Boehringer calf liver tRNA is a suitable source for in vitro translation of mammalian mRNAs.

4. Hemin is difficult to dissolve, unless you know how to proceed. Weight out enough to make 10–50 ml. Dissolve in a small volume of 1 N NaOH (about 1/20th of the final volume) and add an equal volume of 1 M Tris–HCl pH 7.5 (A_{610} hemin = 4.6/1 mM in 0.01 NaOH). Add about 0.8 final volume of ethylene glycol. Adjust pH to 8.2 with 1 N HCl on a pH meter while stirring as quickly as possible. *Proceed slowly* because the pH meter responds very sluggishly in the glycol and if you overshoot, the hemin may come out of solution. Store at −20°C; it keeps forever.

Buffers and Solutions

Master mix
10 mM Magnesium acetate
10 mM Spermidine
200 mM Creatine phosphate
20 mM DTT
20 mM ATP (disodium) (pH 7.0)
1 mM GTP (disodium) (pH 7.0)
400 mM Hepes (pH 7.5)
1 mM Amino acid mix
1 mM Cystein (in 1 mM DTT)
The GTP, ATP, disodium solutions, and Hepes buffer are adjusted to pH 7.0 with NaOH.

References

Anderson, C. W., Straus, J. W., and Dudock, B. S. (1983), *Meth. Enzymology*, **101**, 635.

Berns, A. J. M., and Bloementhal, H. (1974), *Meth. Enzymology*, **30**, 675.

Chambliss, G. H., Henkin, T. M., and Leventhal, J. M. (1983), *Meth. Enzymology*, **101**, 598.

Davies, J. W. (1976), *Ann. Microbiol. (Paris)*, **127A**, 131.

Gurdon, J. B., Lane, C. D., Woodland, H. R., and Marbaix, G. (1971), *Nature*, **223**, 177.

Gurdon, J. B., Lingrel, J. B., and Marbaix, G. (1973), *J. Mol. Biol.* **80**, 539.

Hastie N. D. and Held, W. A. (1978), *Proc. Natl. Acad. Sci. USA*, **75**, 1217.

Henshaw, E. C., and Panniers, R. (1983), *Meth. Enzymology*, **101**, 616.

Jackson, R. J., and Hunt, T. (1983), *Meth. Enzymology*, **96**, 50.

Jaskunas, S. R., Lindahl, L., Nomura, M., and Burgess, R. R. (1975), *Nature*, **257**, 458.

Marcu, K., and Dudock, B. (1974), *Nucl. Acids Res.*, **1**, 1385.

Marcus, A., Efron, D., and Weeks, D. P. (1974), *Meth. Enzymology*, **30**, 749.

Miller, J. S., Paterson, B. M., Ricciardi, R. P., Cohen, L., and Roberts, B. E. (1983), *Meth. Enzymology*, **101**, 650.

Moldave, K., and Gasior, E. (1983), *Meth. Enzymology*, **101**, 644.

Morch, M. D., Drugeon, G., Zagorski, W., and Haenni, A. L. (1986), *Meth. Enzymology*, **118**, 154.

Murialdo, H., and Siminovitch, L. (1972), *Virology*, **48**, 785.

Neidhardt, F. C., Wirth, R., Smith, M. W., and Boglen, R. V. (1980), *J. Bacteriol.*, **143**, 535.

Paterson, B. M., Roberts, B. E., and Kuff, E. L. (1977), *Proc. Natl. Acad. Sci. USA*, **74**, 4370.

Pelham, H. R. B., and Jackson, R. J. (1976), *Eur. J. Biochem.*, **67**, 247.

Ptashne, M. (1967), *Proc. Natl. Acad. Sci. USA,* **57**, 306.

Roberts, B. E., and Paterson, B. M. (1973), *Proc. Natl. Acad. Sci. USA,* **70**, 2330.

Roberts, T. M., Bikel, I., Yocum, R. R., Livingston, D. M., and Ptashne, M. (1979), *Proc. Natl. Acad. Sci. USA,* **76**, 5596.

Sancar, A., Hack, A. M., and Rupp, W. D. (1979), *J. Bacteriol.*, **137**, 692.

Schimke, R. T., Rhoades, R. E., and McNight, G. S. (1974), *Meth. Enzymology*, **30**, 694.

Scott, M. P., Storti, R. V., Pardue, M. L., and Rich, A. (1979), *Biochemistry,* **18**, 1588.

Shih, D., and Kaesberg, P. (1976), *J. Mol. Biol.*, **103**, 77.

Woodward, W. R., and Herbert, E. (1974), *Meth. Enzymology,* **30**, 746.

Woolford, J. L., and Rosbach, M. (1979), *Nucl. Acids Res.*, **6**, 2483.

Zagorski, W. (1978), *Anal. Biochem.*, **87**, 316.

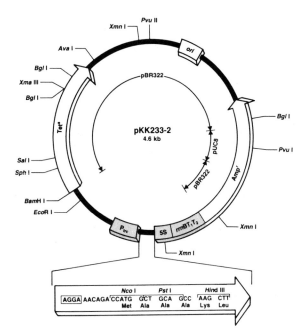

Figure 214. Features of the pKK233-2 expression vector eukaryotic gene fragments without a prokaryotic ribosome binding site and/or an ATG start codon can be inserted in the correct reading frame with one of three Nco I linkers. When the 8 or 10 base linkers are used to align the reading frame of the gene coding fragment with respect to the initiation codon, an alanine is added to the N-terminus of the protein being expressed. If the 12-mer Nco I linker is used, there is a 75% chance that a second neutral amino acid will be added. It is also possible to blunt-end ligate the DNA fragment of interest to the filled-in Nco I site after the reading frame has been adjusted by mild exonuclease digestion. Located just downstream from the Nco I site are unique Pst I and Hind III recognition sequences. Once the DNA is properly inserted, the plasmid is grown in a lacI host such as RB791 or JM105 and expression is induced from the *trc* promoter by adding IPTG (courtesy of Pharmacia).

EXPRESSION VECTORS FOR THE CHARACTERIZATION OF CLONED GENE PRODUCTS

A considerable number of highly efficient expression vectors have been engineered in the past few years to obtain high yields of proteins encoded by cloned genes of procaryotic and eucaryotic origins.

Schematically, there are two groups of vectors among the procaryotic and the eucaryotic expression vectors: (1) those in which specific signals for transcription and translation are represented and (2) those which carry only some or not any of the signals required for efficient expression of foreign proteins in host cells. For example, expression vectors derived from the popular cloning plasmid pBR322 have been listed in Table 28, while a selection of commercial expression vectors is presented in Table 77 and Figures 214–216.

Expression in Procaryotic Systems

Early efficient procaryotic cloning vectors were designed so as to contain strong promoters from (1) bacteriophage lambda (pL), (2) *E. coli lac* and *trp* operons, or (3) a construction in which the $P_{trp}-35$ and the P_{lac} (UV-5)-10 regions had been fused and gave rise to the so-called *tac* promoter (under the control of *lac* repressor).

Both the *trp, tac,* and P_L promoters can be purchased now from Pharmacia in the form of purified DNA fragments which can be easily inserted in a recipient vector with suitable cloning sites (see Figure 217) and several other regulatory sequences are also commercially available in the same form. These include universal translation terminator, ribosome site binding, translation initiation sites, transcription terminator, and T7 RNA polymerase promoter (see Figure 217). The introduction of such oligonucleotides in cloning vectors with multiple cloning sites and carrying genes of interest should allow anyone to construct the unique system especially tailored for efficient expression and characterization of a particular cloned gene product.

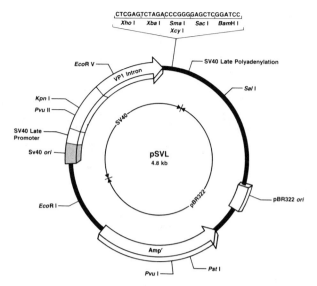

Figure 215. Simplified restriction map of pSVL expression vector (courtesy of Pharmacia).

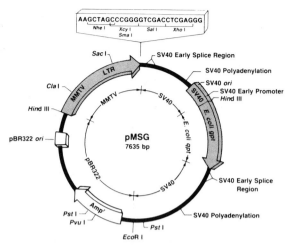

Figure 216. Schematic map of pMSG inducible Mammalian expression vector (courtesy of Pharmacia).

The reader interested in recent advances in the construction of plasmid-derived expression vectors should consult the book by Pouwels et al. (1986) and the review by Balbas et al. (1986). Several bacterio-phage-derived vectors have also been designed to reach a high-level expression of cloned gene products in infected cells, therefore allowing direct char-acterization of proteins encoded by cellular se-quences represented in cDNA libraries. The most popular system of this kind is the λgt11 vector (see Chapter 19).

The preparation of a cDNA library has been de-scribed in Chapter 19; we present a protocol below for the immunological characterization of the ex-pressed cloned sequences, and we also discuss the possibilities of obtaining purified polypeptides in a native form, after cleavage of fusion proteins.

PROTOCOL

IMMUNOLOGICAL SCREENING OF POLYPEPTIDES EXPRESSED IN λ gt11 INFECTED CELLS

1. Seed an overnight culture of *E. coli* Y1090 (r⁻) in 50 ml of TYE medium supplemented with 0.2% maltose and 100 µg/ml of ampicillin.

2. The next day, spin down the cells at 4°C for 10 minutes (4000 rpm in low-speed centrifuge).

3. Resuspend one volume of the cell pellet in half a volume of CaMg mixture.

4. Dechloroform 0.1 ml of phage (10^5–10^6 pfu) at 37°C for 10 minutes (see Chapter 7).

5. In a 5-ml tube mix 0.1 ml of bacteria with 0.1 ml of phage. Let the phage particles adsorb to the bacteria for 15 minutes at room temperature.

Transcription Promoters

tac

```
                          .                .       -35       .               .        -10    .
5'-AGCTTACTCCCCATCCCCCTG TTGACA ATTAATCATCGGCTCG TATAA TGTGTGG
   ‾‾‾‾‾‾                      ‾‾‾‾‾‾‾
   Alu I                        Hinc II
   Hind III

    +1        .                    .                  .     S/D        .S/D
     A ATTGTGAGCGGATAACAATTTCACAC AGGA AAC AG(GA TCC)-3'
                                                    ‾‾‾‾‾‾
                                                    BamH I
```

trp

```
                          .                .                 .     -35   .
5'-AGCTTACTCCCCATCCCCCCAGTGAATTCCCCTG TTGACA ATTAATCATCGAACT
   Hind III                        EcoR I                                 Taq I

          .   -10       .  +1          .                 .
    AGTT AACTA GTACGC A GCTTGG CTGCAG GTCGAC G(GATCC)-3'
         ‾‾‾‾‾      ‾‾‾‾      ‾‾‾‾‾‾ ‾‾‾‾‾‾ ‾‾‾‾‾‾
    Hpa I    Rsa I        Pst I   Sal I  BamH I
```

T7 RNA polymerase

```
5'-d[TAATACGACTCACTATAGGG]-3'
3'-d[ATTATGCTGAGTGATATCCC]-5'
```

Transcription terminator

```
5'-d[AGCCCGCCTAATGAGCGGGCTTTTTTTT]-3'
3'-d[TCGGGCGGATTACTCGCCCGAAAAAAAA]-5'
```

Ribosome binding site

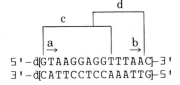

```
5'-d[GTAAGGAGGTTTAAC]-3'
3'-d[CATTCCTCCAAATTG]-5'
```

Translation terminator

```
5'-d[GCTTAATTAATTAAGC]-3'
3'-d[CGAATTAATTAATTCG]-5'
```

Figure 217. Some examples of commercial DNA cartridges containing regulatory signals for expression of cloned genes.

6. Add 4 ml of 1% melted agar (45°C) per tube.

7. Pour evenly on a dry TYE plate containing 100 µg/ml of ampicillin.

8. Incubate at 42°C for 3–4 hours until the lysis plaques appear in the overlay.

9. Meanwhile, prepare IPTG-saturated nitrocellulose filters (0.45-µm pore size) by immersing prewet filters in a solution of 10 m*M* IPTG for 1–2 hours.

10. Blot filters onto Whatman paper, and let dry at room temperature for 1 hour.

11. Remove the plates from the incubator. Carefully layer an IPTG-treated filter on the surface of each plate.

12. Immediately incubate for 3 hours at 37°C.

13. Remove the filters and wash them in TBS for 5 minutes to eliminate pieces of agar which came off with the filter. Repeat steps 11–12 with another filter on each plate.

14. Do not allow the filters to dry during the subsequent steps.

15. Incubate the filters for 30 minutes at room temperature in saturation buffer (5 ml/filter).

16. Incubate the filters for 1 hour at 37°C (or overnight at 4°C) in the antibody solution (5 ml/filter) (see note 1).

17. Wash the filters twice for 10 minutes each time in TBS at room temperature.

18. Wash the filters twice in saturation buffer for 10 minutes each time.

19. Incubate for 1 hour at 37°C with an anti-Ig peroxidase conjugate (diluted 1:250 to 1:500 in saturation solution; 5 ml per filter).

20. Wash the filters twice for 10 minutes each time in saturation buffer.

21. Wash the filters twice for 10 minutes each time in TBS.

22. Incubate in chloronaphtol reagent for revelation of peroxidase activity, until coloration is sufficient (10 to 60 minutes).

23. Stop the reaction by rinsing the filter in 0.1 *N* HCl (or 5% TCA), and wash the filter thoroughly with distilled water.

24. Localize on the masterplate the plaques that give a positive signal on the filter.

25. Aspirate a few positive plaques with Pasteur pipettes and prepare ministocks according to the procedure described in Chapter 7.

26. Repeat the screening procedure by plating 2–3 10³ phages of each ministock.

27. Pick positive plaques and perform a third screening with 4–

$5 \ 10^2$ phages plated on bacteria. All plaques should be positive at this step.

Buffers and Solutions

TBS	*100 ml*
50 mM Tris–HCl (pH 8.0)	5 ml 1 M Tris–HCl (pH 8.0)
150 mM NaCl	3 ml 5 M NaCl
	92 ml H_2O

Saturation buffer	*100 ml*
50 mM Tris–HCl (pH 8.0)	5 ml 1 M Tris–HCl (pH 8.0)
150 mM NaCl	3 ml 5 M NaCl
10% Fetal calf serum	10 ml Fetal calf serum
0.2% Triton X-100	0.2 ml Triton X-100
	81.8 ml H_2O

Antibody solution	*50 ml*
50 mM Tris–HCl (pH 8.0)	2.5 ml 1 M Tris–HCl (pH 8.0)
150 mM NaCl	1.5 ml 5 M NaCl
10% Fetal calf serum	5 ml Fetal calf serum
Specific antibody diluted 1/100	500 µl antiserum
	40.5 ml H_2O

Chloronaphtol reagent (120 ml)
20 ml Chloronaphtol solution (0.3 g/100 ml methanol)
100 ml PBS containing 60 ml of 35% H_2O_2

Notes

1. Since antibody preparations often react with *E. coli* proteins, it is advisable to perform a preincubation of the antiserum with a lysate of the bacterial strain used for the production of preparative amounts of the recombinant proteins. For this purpose, IPTG-induced *E. coli* 1089 (r⁻) are sonicated and the extract boiled at 100°C for 4 minutes. Two filters are incubated for 1 hour at room temperature in the bacterial lysate. Shake occasionally. Wash the filters four times with saturation buffer. Incubate one filter in an antibody solution (5 ml per filter) containing 1:10 diluted antiserum for 1 hour at room temperature. Remove filter, repeat process with the other filter. When the second filter is removed, store the adsorbed antibody solution for screening of the plaques. The antibody preparation obtained in this way is stable enough to allow its use for subsequent steps in screening (M. Vellard, personal communication).

2. The diluted antibody solution may be used repeatedly. Up to

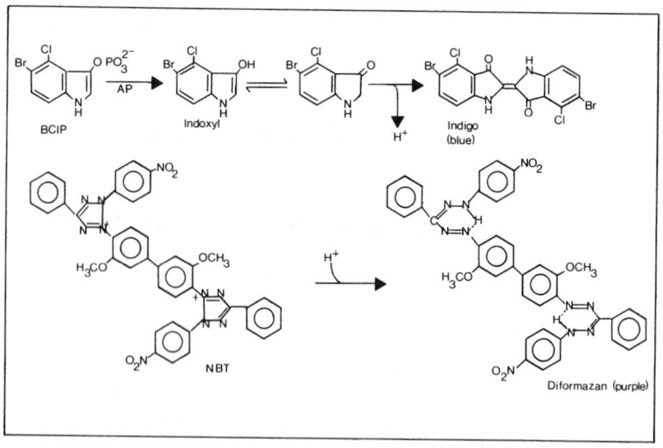

Figure 218. Summary of color development reaction catalyzed by alkaline phosphatase with BCIP as substrate combined with NBT (courtesy of Promega Biotec).

7–10 independent screenings can be performed with the same preparation.

3. 4-chloro-1-naphtol is being used in most laboratories, since it is as sensitive as the original benzidine derivatives, without being carcinogenic.

4. It is also possible to use anti-Ig-alkaline phosphatase conjugates. In these cases it is recommended to use 5-bromo-4-chloro-3-indolyl phosphate (BCIP) in combination with nitro blue tetrazolium (NBT) (see Figure 218) to improve the sensitivity of the enzyme detection.

5. The use of biotinylated second antibodies has been reported to provide a two- to three-fold enhancement in sensitivity usually accompanied by increased background.

6. A very good quality methanol is required for the preparation of the chloronaphtol solution.

7. Since drying the filters after coloration with chloronaphtol results in a considerable loss of sensitivity, it is recommended to photograph the *wet* blot.

Production of Preparative Amounts of Recombinant Proteins. Once a positive recombinant phage has been isolated from a cDNA library, it is often desirable to obtain larger quantities of the recombinant protein in order to perform further biological studies. Preparative amounts of recombinant proteins are generally obtained in the lysates of thermally induced λ gt11-recombinant lysogens. Since the Y1089 strain of *E. coli* is *lon* protease deficient and carries the high-frequency lysogenization mutation *hfl*A150, it is the host of choice to obtain high levels of fusion proteins with an increased stability.

PROTOCOL

PRODUCTION OF PREPARATIVE AMOUNTS OF RECOMBINANT PROTEINS

Preparation of Lysogenic E. coli Y1089

1. Seed 10 ml of TYE medium (supplemented with 0.2% maltose) with *E. coli* 1089 and let grow with agitation overnight.

2. Use 0.2 ml of the overnight culture to seed 20 ml of TYE medium supplemented with 0.2% maltose and 50 μg/ml ampicillin.

3. Let the bacteria grow at 37°C with agitation until the absorbance at 650 nm of the culture reaches 0.5. This requires about 2 hours.

4. Centrifuge the cells for 10 minutes at 4000 rpm (40°C).

5. Resuspend the cell pellet in 2 ml of CaMg solution.

6. Incubate the resuspended cells at 37°C for 1 hour with agitation.

7. Add 0.1 ml of recombinant phage suspension (about 10^5 pfu in TMG) to the cells (so as to have a m.o.i. of 1–2).

8. Incubate at 0°C for 30 minutes, and then at 30°C for 15 minutes.

9. During this incubation time, spread 0.1 ml of a λ im^{80} phage suspension (10^8 pfu/plate) on an ampicillin plate. Let dry a few minutes.

10. With a platinium loop, streak the phage–bacteria mixture from step 8 on this plate.

11. Incubate overnight at 30°C.

12. Take a few colonies that have grown on the plate and resuspend them in 0.2 ml of TYE medium.

13. With a platinium loop, streak a small volume of each resuspended colony on two sets of ampicillin plates.

14. Incubate one set of plates at 42°C and the other set at 30°C.

15. Select the colonies which grow at 30°C and do not grow at 42°C (thermal denaturation of CI repressor induces lysis of lysogenic bacteria).

Expression of Cloned DNA Fragments

1. Seed each colony in 5 ml of TYE medium containing 50 μg/ml of ampicillin and incubate overnight at 30°C.

2. Seed tubes containing 5 ml of TYE medium supplemented with 50 μg/ml of ampicillin, with 5 μl of each overnight cul-

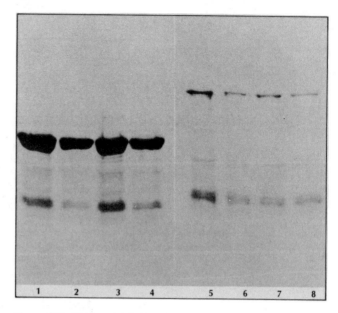

Figure 219. Western blot of proteins expressed in λ-gt11. Lanes 1–4 contain extracts prepared from an induced λ-gt11 lysogen that produces β-galactosidase (MW 116,000) and lanes 5–8 contain similar extracts of a λ-gt11 recombinant that produces an ovalbumin/β-galactosidase fusion protein (MW 144,000). The primary antibody was Promega's monoclonal anti-β-galactosidase, which was detected using the ProtoBlot immunoblotting system. The minor bands are proteolytic degradation products (courtesy of Promega Biotec).

tures. Freeze the remainder of the culture at $-70°C$ (in 30% glycerol).

3. Incubate at 30°C with agitation until the absorbance of the culture at 600 nm is about 0.4 to 0.5.

4. Shift the cell suspension rapidly to 42°C in a waterbath for 15 minutes with agitation.

5. Add 500 μl of IPTG 100 mM in the culture.

6. Incubate at 37°C for 2 hours with agitation.

7. Centrifuge 1.5-ml aliquots of the culture in microfuge tubes for 15 seconds at 12,000 rpm (room temperature).

8. Discard the supernatants, and resuspend the pellets in 150 μl of electrophoresis buffer by vortexing.

9. Proceed for detection of proteins by immunoprecipitation or immunoblotting (Figure 219).

Vectors for Mammalian Cells

For a long time, the wide range of plasmids and phage vectors available for cloning DNA fragments in *E. coli* has contrasted with the small number of cloning vehicles available for cloning in eucaryotic cells. This situation has now changed and several eucaryotic vectors have been used in the past years to express eucaryotic genes in mammalian cells

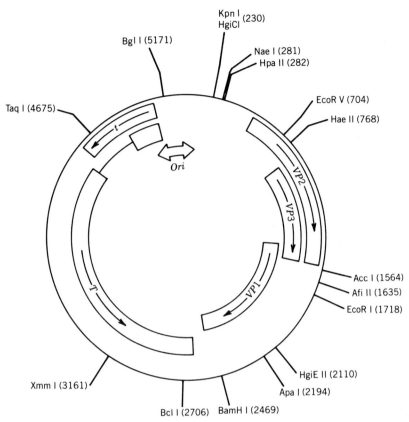

Figure 220. Restriction map of the SV40 genome. The position of restriction sites occurring only once is shown. Other sites are listed in Table 79. Also represented are positions and orientations of the origin of replication (*Ori*), the viral structural protein genes *VP1*, *VP2*, and *VP3*, and the tumor antigen genes *T* and *t* (courtesy of New England Biolabs).

(Gluzman, 1982). The properties of a few commercial vectors for expression of cloned sequences in eucaryotic cells are described in Table 77. We present a brief outline below of the progress made with different viral systems.

SV40 Vectors. One of the most widely used vectors for introducing cloned DNA into mammalian cells has been simian virus 40 (SV40), whose biology has now been studied at a very fine level for several years. The main advantages of using SV40 as a cloning vector can be summarized as follows:

1. SV40 has a small genome consisting of a single circular DNA molecule of known sequence.
2. Large quantities of SV40 DNA can be purified easily, free of contaminating cellular DNA.
3. The viral genome can become integrated into cellular DNA and then replicate as an integral part of the cellular genome.

4. The genes involved in the lytic cycle and transformation induced by SV40 have been mapped on the viral genome. Mutants have been obtained for both functions.
5. In vitro-manipulated SV40 DNA can be reintroduced easily by transfection methods into eucaryotic cells. Expression of the recombinant genomes in transfected cells can be detected at the level of transcription and translation. The resulting recombinant virus stocks can be used for further infection under standard conditions.

The structure and organization of the SV40 genome is shown in Figure 220, and Table 79 lists the restriction sites in SV40 DNA.

Insertion of foreign DNA has been possible in both the early and the late regions of the SV40 genome.

Table 79. Number and Location of Cutting Sites for Commercial Restriction Endonucleases in SV40 DNA

0 Site

AatII	BbiII	Eco81I	NruI	SnaBI
AccIII	BcnI	FnudII	Nsp7524III	SpeI
AflIII	BslII	FspI	NunII	SplI
AhaII	BspMI	FspII	OxaNI	SstI
AocI	BspMII	HgaI	PaeR7	SstII
AosI	BssHII	HgiAI	PvuI	ThaI
AosII	BstEII	MaeII	RsrII	Tth111I
ApaLI	CfrI	MluI	SacI	XbaI
AquI	ClaI	MstI	SacII	XcyI
Asp700	CvnI	MstII	SalI	XhoI
AsuII	DraIII	NarI	SauI	XmaI
AvaI	EaeI	NciI	ScaI	XmaIII
BalI	EagI	NheI	SexI	XmnI
BanIII	Eco0109	NotI	SmaI	XorII
BbeI	Eco52I			

	Location	Fragment Ends	Fragment Length
1 Site			
AccI	1629	1630–1629	5243
AflII	1699	1700–1699	5243
ApaI	2262	2263–2262	5243
Asp718	294	295–294	5243
BamHI	2533	2534–2533	5243
BanI	294	295–294	5243
BclI	2770	2771–2770	5243
BglI	5241	5242–5241	5243
BstI	2533	2534–2533	5243
BstXI	4766	4767–4766	5243
Cfr10I	345	346–345	5243
EcoRI	1782	1783–1782	5243
EcoRV	770	771–770	5243
Eco47III	834	835–834	5243
EspI	1711	1712–1711	5243
HaeII	836	837–836	5243
HapII	346	347–346	5243
HpaII	346	347–346	5243
KpnI	298	299–298	5243
MspI	346	347–346	5243
NaeI	347	348–347	5243
PpuMI	588	589–588	5243
SfiI	5241	5242–5241	5243
TaqI	4739	4740–4739	5243
Tth HB8I	4739	4740–4739	5243

	Location	Fragment Ends	Fragment Length
2 Sites			
AlwNI	640	641–3679	3039
	3679	3680– 640	2204
AvrII	1078	1079–5187	4109
	5187	5188–1078	1134
BanII	2262	2263–780	3761
	780	781–2262	1482
BspHI	2773	2774–1912	4382
	1912	1913–2773	861
CfoI	835	836–345	4753
	345	346–835	490
HhaI	835	836–345	4753
	345	346–835	490
HinpI	833	834–343	4753
	343	344–833	490
NdeI	4827	4828–3809	4225
	3809	3810–4827	1018
Nsp7524I	204	205–132	5171
	132	133–204	72
NspHI	204	205–132	5171
	132	133–204	72
PflMI	1013	1014–4564	3551
	4564	4565–1013	1692
PleI	4462	4463–2851	3632
	2851	2852–4462	1611
PstI	3208	3209–1992	4027
	1992	1993–3208	1216
SdnI	2262	2263–1292	4273
	1292	1293–2262	970
SphI	204	205–132	5171
	132	133–204	72
3 Sites			
AvaIII	200	201–3585	3385
	3585	3586–128	1786
	128	129–200	72
DraII	2798	2799–588	3033
	588	589–2259	1671
	2259	2260–2798	539
MflI	4769	4770–2533	3007
	2533	2534–4099	1566
	4099	4100–4769	670
NcoI	560	561–37	4720
	37	38–333	296
	333	334–560	227
NsiI	202	203–3587	3385
	3587	3588–130	1786
	130	131–202	72
PssI	2801	2802–591	3033
	591	592–2262	1671
	2262	2263–2801	539

Table 79. Number and Location of Cutting Sites for Commercial Restriction Endonucleases in SV40 DNA (continued)

Enzyme	Location	Fragment Ends	Fragment Length
PvuII	3508	3509–272	2007
	1718	1719–3508	1790
	272	273–1718	1446
XhoII	4769	4770–2533	3007
	2533	2534–4099	1566
	4099	4100–4769	670

4 Sites

Enzyme	Location	Fragment Ends	Fragment Length
AlwI	876	877–2536	1660
	2536	2537–4102	1566
	4772	4773– 876	1347
	4102	4103–4772	670
AocII	3389	3390–780	2634
	2262	2263–3389	1127
	1292	1293–2262	970
	780	781–1292	512
BsmI	4529	4530–1892	2606
	2683	2684–4529	1846
	1892	1893–2589	697
	2589	2590–2683	94
Bsp1286	3389	3390–780	2634
	2262	2263–3389	1127
	1292	1293–2262	970
	780	781–1292	512
HpaI	521	522–2668	2147
	3735	3736–501	2009
	2668	2669–3735	1067
	501	502–521	20
HphI	3249	3250–1078	3072
	1393	1394–3249	1856
	1078	1079–1257	179
	1257	1258–1393	136
NspBII	3508	3509–272	2007
	1718	1719–3508	1790
	664	665–1718	1054
	272	273–664	392
Nsp7524II	3389	3390–780	2634
	2262	2263–3389	1127
	1292	1293–2262	970
	780	781–1292	512

6 Sites

Enzyme	Location	Fragment Ends	Fragment Length
AflI	3538	3539–5118	1580
	2013	2014–3538	1525
	1018	1019–2013	995
	5118	5119–557	682
	588	589–1018	430
	557	558–588	31
AvaII	3538	3539–5118	1580
	2013	2014–3538	1525
	1018	1019–2013	995
	5118	5119–557	682
	588	589–1018	430
	557	558–588	31
ClaII	3538	3539–5118	1580
	2013	2014–3538	1525
	1018	1019–2013	995
	5118	5119–557	682
	588	589–1018	430
	557	558–588	31
Eco47I	3537	3538–5117	1580
	2012	2013–3537	1525
	1017	1018–2012	995
	5117	5118–556	682
	587	588–1017	430
	556	557–587	31
HindIII	1708	1709–3476	1768
	4002	4003–5171	1169
	5171	5172–1046	1118
	3476	3477–4002	526
	1046	1047–1493	447
	1493	1494–1708	215
SfaNI	2612	2613–4892	2280
	697	698–2612	1915
	188	189–697	509
	4934	4935–116	425
	116	117–188	72
	4892	4893–4934	42
SinI	3538	3539–5118	1580
	2013	2014–3538	1525
	1018	1019–2013	995
	5118	5119–557	682
	588	589–1018	430
	557	558–588	31
SspI	5085	5086–1305	1463
	1825	1826–3258	1433
	3258	3259–4129	871
	4282	4283–5085	803
	1305	1306–1825	520
	4129	4130–4282	153

7 Sites

Enzyme	Location	Fragment Ends	Fragment Length
AatI	1463	1464–5192	3729
	757	758–1236	479
	5192	5193–362	413
	362	363–710	348
	1236	1237–1463	227
	710	711–743	33
	743	744–757	14
HincII	3735	3736–472	1980
	521	522–2059	1538
	2668	2669–3735	1067
	2299	2300–2668	369
	2059	2060–2299	240
	472	473–501	29

Table 79.   Number and Location of Cutting Sites for Commercial Restriction Endonucleases in SV40 DNA (continued)

	Location	Fragment Ends	Fragment Length		Location	Fragment Ends	Fragment Length
	501	502–521	20		2770	2771–3715	945
HindII	3735	3736–472	1980		4099	4100–4709	610
	521	522–2059	1538		2137	2138–2533	396
	2668	2669–3735	1067		3715	3716–4099	384
	2299	2300–2668	369		2533	2534–2770	237
	2059	2060–2299	240		4709	4710–4769	60
	472	473–501	29	StyI	1078	1079–2812	1734
	501	502–521	20		3455	3456–4409	954
StuI	1463	1464–5192	3729		4409	4410–5187	778
	757	758–1236	479		2812	2813–3455	643
	5192	5193–362	413		560	561–1078	518
	362	363–710	348		37	38–333	296
	1236	1237–1463	227		333	334–560	227
	710	711–743	33		5187	5188–37	93
	743	744–757	14				
					10 Sites		
	8 Sites			HinfI	5135	5136–1739	1847
DpnI	4771	4772–875	1347		1739	1740–2824	1085
	875	876–2139	1264		3610	3611–4376	766
	2772	2773–3717	945		4592	4593–5135	543
	4101	4102–4711	610		2848	2849–3373	525
	2139	2140–2535	396		3373	3374–3610	237
	3717	3718–4101	384		4459	4460–4568	109
	2535	2536–2772	237		4376	4377–4459	83
	4711	4712–4771	60		4568	4569–4592	24
EcoT14I	1078	1079–2812	1734		2824	2825–2848	24
	3455	3456–4409	954				
	4409	4410–5187	778		*11 Sites*		
	2812	2813–3455	643	AsuI	3538	3539–5118	1580
	560	561–1078	518		5118	5119–557	682
	37	38–333	296		2259	2260–2798	539
	333	334–560	227		1507	1508–2013	506
	5187	5188–37	93		1018	1019–1507	489
MboI	4769	4770–873	1347		588	589–1018	430
	873	874–2137	1264		2798	2799–3171	373
	2770	2771–3715	945		3171	3172–3538	367
	4099	4100–4709	610		2013	2014–2258	245
	2137	2138–2533	396		557	558–588	31
	3715	3716–4099	384		2258	2259–2259	1
	2533	2534–2770	237	Cfr13I	3538	3539–5118	1580
	4709	4710–4769	60		5118	5119–557	682
NdeII	4769	4770–873	1347		2259	2260–2798	539
	873	874–2137	1264		1507	1508–2013	506
	2770	2771–3715	945		1018	1019–1507	489
	4099	4100–4709	610		588	589–1018	430
	2137	2138–2533	396		2798	2799–3171	373
	3715	3716–4099	384		3171	3172–3538	367
	2533	2534–2770	237		2013	2014–2258	245
	4709	4710–4769	60		557	558–588	31
Sau3A	4769	4770–873	1347		2258	2259–2259	1
	873	874–2137	1264	FokI	93	94–1455	1362

Table 79. Number and Location of Cutting Sites for Commercial Restriction Endonucleases in SV40 DNA (*continued*)

	Location	Fragment Ends	Fragment Length
	1455	1456–2468	1013
	3932	3933–4791	859
	2468	2469–3298	830
	5049	5050–93	287
	3382	3383–3641	259
	3689	3690–3932	243
	4913	4914–5049	136
	4791	4792–4913	122
	3298	3299–3382	84
	3641	3642–3689	48
Nsp7524IV	3538	3539–5118	1580
	5118	5119–557	682
	2259	2260–2798	539
	1507	1508–2013	506
	1018	1019–1507	489
	588	589–1018	430
	2798	2799–3171	373
	3171	3172–3538	367
	2013	2014–2258	245
	557	558–588	31
	2258	2259–2259	1
Nsp7524V	3538	3539–5118	1580
	5118	5119–557	682
	2259	2260–2798	539
	1507	1508–2013	506
	1018	1019–1507	489
	588	589–1018	430
	2798	2799–3171	373
	3171	3172–3538	367
	2013	2014–2258	245
	557	558–588	31
	2258	2259–2259	1
Sau96I	3538	3539–5118	1580
	5118	5119–557	682
	2259	2260–2798	539
	1507	1508–2013	506
	1018	1019–1507	489
	588	589–1018	430
	2798	2799–3171	373
	3171	3172–3538	367
	2013	2014–2258	245
	557	558–588	31
	2258	2259–2259	1
Tth111II	439	440–3488	3049
	4884	4885–439	798
	3941	3942–4336	395
	4336	4337–4598	262
	4598	4599–4813	215
	3744	3745–3941	197
	3488	3489–3631	143
	4813	4814–4884	71
	3643	3644–3705	62

	Location	Fragment Ends	Fragment Length
	3705	3706–3744	39
	3631	3632–3643	12

12 Sites

	Location	Fragment Ends	Fragment Length
AhaIII	5149	5150–1659	1753
	3909	3910–4648	739
	1800	1801–2365	565
	2729	2730–3159	430
	4648	4649–5059	411
	2365	2366–2729	364
	3591	3592–3909	318
	3159	3160–3474	315
	1659	1660–1800	141
	5059	5060–5149	90
	3474	3475–3545	71
	3545	3546–3591	46
DraI	5149	5150–1659	1753
	3909	3910–4648	739
	1800	1801–2365	565
	2729	2730–3159	430
	4648	4649–5059	411
	2365	2366–2729	364
	3591	3592–3909	318
	3159	3160–3474	315
	1659	1660–1800	141
	5059	5060–5149	90
	3474	3475–3545	71
	3545	3546–3591	46
MaeI	5188	5189–1079	1134
	2580	2581–3434	854
	4483	4484–5112	629
	2160	2161–2580	420
	1779	1780–2160	381
	1079	1080–1459	380
	4105	4106–4483	378
	3748	3749–4105	357
	3434	3435–3748	314
	1610	1611–1779	169
	1459	1460–1610	151
	5112	5113–5188	76
RsaI	1468	1469–3073	1605
	4169	4170–4877	708
	296	297–971	675
	4988	4989–296	551
	971	972–1468	497
	3226	3227–3577	351
	3577	3578–3871	294
	3943	3944–4169	226
	3073	3074–3226	153
	4877	4878–4988	111
	3871	3872–3928	57
	3928	3929–3943	15

Table 79. Number and Location of Cutting Sites for Commercial Restriction Endonucleases in SV40 DNA (continued)

	Location	Fragment Ends	Fragment Length		Location	Fragment Ends	Fragment Length
	14 Sites				3336	3337–4009	673
MaeIII	4677	4678–351	917		359	360–911	552
	3237	3238–3996	759		965	966–1409	444
	2638	2639–3237	599		2529	2530–2898	369
	351	352–819	468		5093	5094–161	311
	865	866–1332	467		2898	2899–3147	249
	4272	4273–4677	405		4893	4894–5093	200
	2035	2036–2423	388		1409	1410–1536	127
	1476	1477–1843	367		233	234–359	126
	3996	3997–4272	276		3147	3148–3248	101
	2423	2424–2638	215		3248	3249–3336	88
	1870	1871–2035	165		4009	4010–4070	61
	1332	1333–1476	144		178	179–233	55
	819	820–865	46		911	912–965	54
	1843	1844–1870	27		161	162–178	17
				BspNI	1536	1537–2529	993
	16 Sites				4070	4071–4893	823
MboII	1513	1514–2844	1331		3336	3337–4009	673
	2898	2899–3654	756		359	360–911	552
	5008	5009–473	708		965	966–1409	444
	473	474–1118	645		2529	2530–2898	369
	4488	4489–4916	428		5093	5094–161	311
	1118	1119–1513	395		2898	2899–3147	249
	4029	4030–4410	381		4893	4894–5093	200
	3654	3655–4029	375		1409	1410–1536	127
	4423	4424–4488	65		233	234–359	126
	4916	4917–4966	50		3147	3148–3248	101
	4966	4967–4997	31		3248	3249–3336	88
	2868	2869–2898	30		4009	4010–4070	61
	2854	2855–2868	14		178	179–233	55
	4410	4411–4423	13		911	912–965	54
	4997	4998–5008	11		161	162–178	17
	2844	2845–2854	10	BstNI	1536	1537–2529	993
NlaIV	2799	2800–4698	1899		4070	4071–4893	823
	4698	4699–158	703		3336	3337–4009	673
	1563	1564–1977	414		359	360–911	552
	1019	1020–1406	387		965	966–1409	444
	296	297–589	293		2529	2530–2898	369
	2260	2261–2535	275		5093	5094–161	311
	2535	2536–2799	264		2898	2899–3147	249
	779	780–1019	240		4893	4894–5093	200
	589	590–779	190		1409	1410–1536	127
	1977	1978–2133	156		233	234–359	126
	1406	1407–1509	103		3147	3148–3248	101
	2173	2174–2260	87		3248	3249–3336	88
	158	159–230	72		4009	4010–4070	61
	230	231–296	66		178	179–233	55
	1509	1510–1563	54		911	912–965	54
	2133	2134–2173	40		161	162–178	17
				EcoRII	1534	1535–2527	993
	17 Sites				4068	4069–4891	823
ApyI	1536	1537–2529	993		3334	3335–4007	673
	4070	4071–4893	823		357	358–909	552
					963	964–1407	444

Table 79. Number and Location of Cutting Sites for Commercial Restriction Endonucleases in SV40 DNA (*continued*)

	Location	Fragment Ends	Fragment Length		Location	Fragment Ends	Fragment Length
	2527	2528–2896	369		4893	4894–5093	200
	5091	5092–159	311		1409	1410–1536	127
	2896	2897–3145	249		233	234–359	126
	4891	4892–5091	200		3147	3148–3248	101
	1407	1408–1534	127		3248	3249–3336	88
	231	232–357	126		4009	4010–4070	61
	3145	3146–3246	101		178	179–233	55
	3246	3247–3334	88		911	912–965	54
	4007	4008–4068	61		161	162–178	17
	176	177–231	55				
	909	910–963	54				
	159	160–176	17			*19 Sites*	
MvaI	1536	1537–2529	993	BspRI	3202	3203–4863	1661
	4070	4071–4893	823		1508	1509–2260	752
	3336	3337–4009	673		2260	2261–2800	540
	359	360–911	552		2800	2801–3173	373
	965	966–1409	444		4863	4864–5192	329
	2529	2530–2898	369		7	8–332	325
	5093	5094–161	311		936	937–1236	300
	2898	2899–3147	249		362	363–661	299
	4893	4894–5093	200		1236	1237–1463	227
	1409	1410–1536	127		757	758–936	179
	233	234–359	126		661	662–710	49
	3147	3148–3248	101		1463	1464–1508	45
	3248	3249–3336	88		5192	5193–5235	43
	4009	4010–4070	61		710	711–743	33
	178	179–233	55		332	333–362	30
	911	912–965	54		3173	3174–3202	29
	161	162–178	17		743	744–757	14
NlaIII	564	565–1916	1352		5235	5236–1	9
	1916	1917–2545	629		1	2–7	6
	4800	4801–41	484	HaeIII	3202	3203–4863	1661
	4376	4377–4800	424		1508	1509–2260	752
	2859	2860–3237	378		2260	2261–2800	540
	3237	3238–3585	348		2800	2801–3173	373
	4055	4056–4376	321		4863	4864–5192	329
	3585	3586–3869	284		7	8–332	325
	2545	2546–2777	232		936	937–1236	300
	3869	3870–4055	186		362	363–661	299
	387	388–564	177		1236	1237–1463	227
	204	205–337	133		757	758–936	179
	41	42–132	91		661	662–710	49
	132	133–204	72		1463	1464–1508	45
	2777	2778–2843	66		5192	5193–5235	43
	337	338–387	50		710	711–743	33
	2843	2844–2859	16		332	333–362	30
ScrFI	1536	1537–2529	993		3173	3174–3202	29
	4070	4071–4893	823		743	744–757	14
	3336	3337–4009	673		5235	5236–1	9
	359	360–911	552		1	2–7	6
	965	966–1409	444	PalI	3202	3203–4863	1661
	2529	2530–2898	369		1508	1509–2260	752
	5093	5094–161	311		2260	2261–2800	540
	2898	2899–3147	249		2800	2801–3173	373

Table 79. Number and Location of Cutting Sites for Commercial Restriction Endonucleases in SV40 DNA (*continued*)

	Location	Fragment Ends	Fragment Length		Location	Fragment Ends	Fragment Length
	4863	4864–5192	329		556	557–601	45
	7	8–332	325		772	773–814	42
	936	937–1236	300		607	608–637	30
	362	363–661	299		3496	3497–3517	21
	1236	1237–1463	227		637	638–652	15
	757	758–936	179		652	653–655	3
	661	662–710	49		604	605–607	3
	1463	1464–1508	45		601	602–604	3
	5192	5193–5235	43				
	710	711–743	33			*24 Sites*	
	332	333–362	30	Fnu4HI	2653	2654–3506	853
	3173	3174–3202	29		4313	4314–5128	815
	743	744–757	14		803	804–1542	739
	5235	5236–1	9		1988	1989–2462	474
	1	2–7	6		3845	3846–4313	468
					5242	5243–341	342
		20 Sites			3509	3510–3845	336
DdeI	3290	3291–4387	1097		1716	1717–1988	272
	722	723–1337	615		2462	2463–2653	191
	1857	1858–2375	518		1542	1543–1716	174
	287	288–722	435		443	444–569	126
	2375	2376–2793	418		5128	5129–5242	114
	4918	4919–5228	310		341	342–443	102
	5228	5229–287	302		668	669–734	66
	3057	3058–3290	233		734	735–785	51
	1436	1437–1641	205		569	570–614	45
	4638	4639–4808	170		620	621–650	30
	1711	1712–1857	146		785	786–803	18
	4499	4500–4638	139		650	651–662	12
	2929	2930–3057	128		3506	3507–3509	3
	4387	4388–4499	112		665	666–668	3
	1337	1338–1436	99		662	663–665	3
	2846	2847–2929	83		617	618–620	3
	1641	1642–1711	70		614	615–617	3
	4858	4859–4918	60				
	2793	2794–2846	53			*34 Sites*	
	4808	4809–4858	50	AluI	2652	2653–3427	775
					4644	4645–5127	483
		22 Sites			4315	4316–4644	329
BbvI	2640	2641–3496	856		5227	5228–272	288
	4324	4325–5115	791		1987	1988–2273	286
	814	815–1553	739		3508	3509–3781	273
	1975	1976–2473	498		1098	1099–1351	253
	3856	3857–4324	468		2280	2281–2524	244
	5115	5116–328	456		805	806–1048	243
	3517	3518–3856	339		3781	3782–4004	223
	1727	1728–1975	248		436	437–613	177
	1553	1554–1727	174		4004	4005–4161	157
	2473	2474–2640	167		4161	4162–4315	154
	430	431–556	126		640	641–793	153
	328	329–430	102		1841	1842–1987	146
	655	656–721	66		1351	1352–1495	144
	721	722–772	51		272	273–407	135

Table 79. Number and Location of Cutting Sites for Commercial Restriction Endonucleases in SV40 DNA (*continued*)

	Location	Fragment Ends	Fragment Length		Location	Fragment Ends	Fragment Length
	2524	2525–2652	128		5072	5073–5149	77
	1718	1719–1841	123		651	652–717	66
	1635	1636–1710	75		1334	1335–1395	61
	5173	5174–5227	54		5149	5150–5203	54
	1582	1583–1635	53		4038	4039–4092	54
	1048	1049–1098	50		1798	1799–1852	54
	1495	1496–1544	49		320	321–373	53
	5127	5128–5173	46		1395	1396–1445	50
	3437	3438–3478	41		4435	4436–4482	47
	1544	1545–1582	38		1589	1590–1636	47
	3478	3479–3508	30		4482	4483–4527	45
	407	408–436	29		1098	1099–1143	45
	613	614–640	27		2442	2443–2485	43
	793	794–805	12		2713	2714–2753	40
	3427	3428–3437	10		5034	5035–5072	38
	1710	1711–1718	8		1636	1637–1674	38
	2273	2274–2280	7		2753	2754–2781	28
					729	730–754	25
					5231	5232–12	24
					4927	4928–4947	20
		51 Sites			3389	3390–3408	19
MnlII	3408	3409–4038	630		296	297–312	16
	2788	2789–3389	601		5203	5204–5215	12
	1852	1853–2208	356		2491	2492–2503	12
	754	755–1098	344		717	718–729	12
	4092	4093–4435	343		2704	2705–2713	9
	4527	4528–4817	290		2503	2504–2512	9
	373	374–651	278		2433	2434–2442	9
	18	19–296	278		312	313–320	8
	2208	2209–2433	225		5218	5219–5225	7
	2512	2513–2704	192		2781	2782–2788	7
	1143	1144–1334	191		5225	5226–5231	6
	1445	1446–1589	144		12	13–18	6
	1674	1675–1798	124		5215	5216–5218	3
	4817	4818–4927	110		2488	2489–2491	3
	4947	4948–5034	87		2485	2486–2488	3

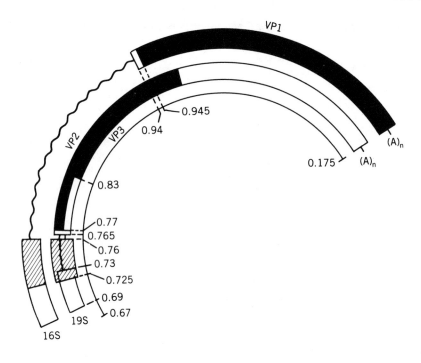

Figure 221. The composite structure of SV40 late mRNAs. The map of SV40 DNA from the origin of DNA replication (0.67) to the end of the late region (0.175) is shown on the inner circle. The regions coding for the structure of VP1, VP2, and VP3 appear as shaded regions within bars that define the "bodies" of the 19S and 16S mRNAs. The "leader" segments are shown as bars spanning map coordinates 0.69–0.76. The cross-hatched and stippled regions of the 16S mRNA leader, for example, are intended as symbolic, rather than literal representations of more than one type of leader segment; one leader segment spans the region from 0.725 to 0.76 (hatching) and another the region from 0.69 to 0.76 (stippling). For the 19S mRNA species, one class of leaders spans map coordinates 0.69–0.73 (stippling) and is joined to the body at 0.76; another class of leaders extends from 0.725 to 0.76 (hatching).

Insertion in the Late Region. The methods used for insertion in the late region (Figure 221) are based in most cases on: (a) replacement of the late SV40 sequences by insertion of foreign DNA between the Bam HI site and either the Kpn I, Hae II, or Hpa II sites; or (b) use of viable deletion mutants of the late region.

Some examples of late-region insertion are illustrated in Figures 222–227. In all of these cases, the recombinant viruses generated are defective. Their propagation is achieved in the presence of a helper virus which provides the late functions necessary for the production of infectious particles. Generally, the helper virus used is also defective because it has been altered in the early viral functions (*tsA* mutants).

Insertion in the Early Region. DNA corresponding to the 3′ end of the rat preproinsulin, to the mouse β globin, and to the mouse dihydrofolate reductase have been introduced in the SV40 early region between the Taq I and Bcl I sites. The hemagglutinin gene of influenza has been introduced in place of the early sequences located between the Hind III and Bcl I sites.

In these cases, propagation of the recombinant DNAs can be achieved by infection of COS cells which express T-antigen functions or by coinfection of CV1 monkey cells with a helper virus altered in the late viral functions.

Other Vectors. Although cloning of eucaryotic genes in SV40 vectors had made possible a great deal of progress in the understanding of the regulatory mechanisms governing gene expression in mammalian cells, one of the limitations of the system is the approximately 5-kb size limit for DNA that can be accommodated within the SV40 virion. To overcome

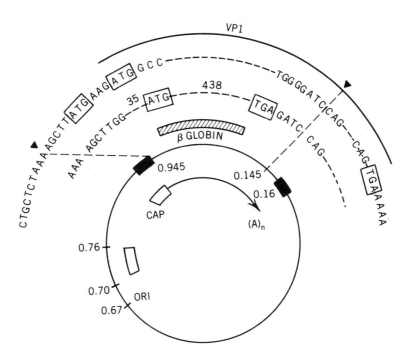

Figure 222. SV40 and rabbit β-globin cDNA nucleotide sequences relevant to the construction of SVGT5–RaβG. The origin of SV40 DNA replication at map position 0.67 is denoted as ORI. SV40 16S late mRNA, with its "capped" 5′ leader sequence derived from 0.72 to 0.76 map units (stippled box) and 3′ poly A, is represented inside the circle. The region of the map coding for VP1 is represented by the heavy black line labeled VP1. The regions of SV40 DNA implicated in splicing and polyadenylation of the 16S mRNA are shown as black rectangles on the circular map. The dashed line extending leftward from map position 0.945 indicates the position of a Hind III endonuclease cleavage site in SV40 DNA (the outer sequence) and in pBR322–βG2 DNA (the inner sequence). The nucleotide sequences between these dashed lines show the β-globin cDNA sequences inserted into SVGT5 and the SV40 sequences excised to construct SVGT5. The triplets enclosed in boxes show the initiation and termination signals for translation of VP1 and β-globin. The β-globin cDNA sequence in pBR322–βG2 contains 35 nucleotides proximal to the initiator AUG in β-globin mRNA. The small hatched bar spans the region containing the initiation β-globin codon, the 438 nucleotides specifying the entire β-globin protein amino acid sequence, and the translation terminator codon. [With permission of P. Berg. Copyright (1979) Macmillan Press.] Reprinted by permission from *Nature*, **277**, 108. Copyright 1979 Macmillan Journals Limited.

this problem, several other eucaryotic vectors have been developed in the past few years. They include retroviruses, adenoviruses, and herpes simplex virus.

All retroviruses have single stranded RNA genomes which are replicated through a double stranded DNA intermediate, known as proviral DNA. The double stranded form of the retroviral genome is a necessary intermediate in the life cycle of the virus, one or more copy of the proviral DNA becoming integrated in the host cell genome after its synthesis by reverse transcriptase.

The mechanism of reverse transcriptase is such that it generates, at each end of the integrated viral genome, a particular structure designated LTR (Long Terminal Repeat) which contains both the 5′ and 3′ unique viral sequences and the repeated sequences (R) from each end of the viral genome.

The two LTR of a given proviral DNA are identical. However, since the U3 and R region vary notably in different virus strains, the resulting LTR appear to differ significantly from one proviral DNA to the other (for a review, see Coffin, 1982).

Transcription of the proviral genome is promoted from the efficient retroviral promoter located in the U3 region of the 5′-LTR, whereas RNA synthesis ter-

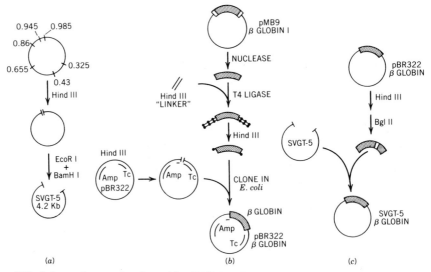

Figure 223. Scheme for construction of the SV40-RaβG recombinant genome. (a) Construction of SVGT5 vector. (b) Subcloning of β-globin cDNA. (c) Cloning of β-globin cDNA into SVGT5. Reprinted by permission from *Nature,* **277,** 109. Copyright 1979 Macmillan Journals Limited. [With permission of P. Berg. Copyright (1979) Macmillan Press.]

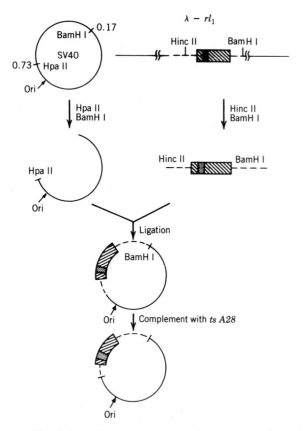

Figure 224. Schematic representation of the construction of SV40-rat preproinsulin gene *I* (*SVL₁–rl₁*) recombinants. ——, SV40 sequences; ---, *rl₁*, flanking sequences. The diagonally striped rectangular box indicates the location of the *rl₁* gene. The black square inside the box symbolizes the intron. Ori, functional replication origin (with permission of G. Khoury).

minates in the U5-LTR region which contains the polyadenylation signals.

The retroviral vectors make use of the strong retroviral transcription signals to promote the expression of both dominant selectable maker genes and nonselectable DNA or cDNA sequences.

Among the most efficient expression system of this type is the murine retrovirus shuttle vector derived from Moloney murine leukemia virus by Cepko et al. (1984). The viral sequences retained in this vector (pZIP-Neo SV(x)1) include the LTR sequences required for transcription initiation and termination, LTR sequences required for reverse transcription and encapsidation of RNAs, and the 5' and 3' splicing signals involved in the synthesis of the *env* mRNA. The foreign sequences can be inserted in this vector at the level of two unique restriction sites (Bam HI and Xho I), in place of the viral sequences coding for the *gag-pol* and *env* polypeptides. In addition to these viral sequences the pZIP-Neo SV(x)1 vector also contains the transposon Tn5 conferring G418 resistance (Davies and Jimenez, 1980; Colbère-Garapin et al., 1981) and kanamycin resistance (Jogersen et al., 1979) and the sequences corresponding to the pBR322 and SV40 origins of replication.

The generation of highly transmissible virus carrying the recombinant genome is performed through transfection of mouse cells (ψ2) which provide all the *trans* functions necessary for the encapsidation of the defective genome (Mann et al., 1983). This sys-

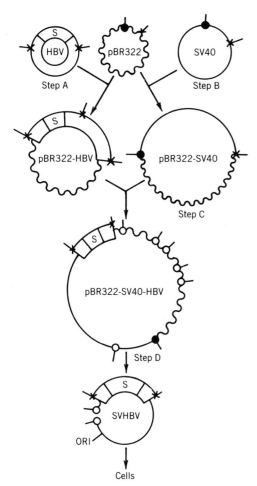

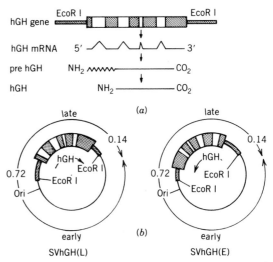

Figure 226. SV40–hGH (human growth hormone) recombinants. (*a*) Both hGH genes were cloned on 2.7-kb Eco RI fragments containing five structural sequences (solid bars) and four intervening sequences (hollow bars) together with 500-bp of 5′ flanking sequences and 550 bp of 3′ flanking sequences (thick lines). Translation of hGH mRNA yields pre-hGH containing an amino-terminal signal sequence of 23 amino acids. This is processed to generate mature hGH. (*b*) The two hGH genes were inserted, in both possible orientations, into an SV40*dl*2005 vector (thin lines) extending clockwise from the Bam HI site at 0.14 map units to the Hpa II site at 0.72 map units. Both vector sites were converted to Eco RI sites by using oligonucleotide linkers (unpublished results). Ori, origin of replication (with permission of D. H. Hamer).

Figure 225. Construction and analysis of the SV40–HBV (hepatitis B virus) recombinant. *Step A*: A 120-μl reaction mixture containing 200 ng HBV DNA and 1.4 μg plasmid pBR322 DNA was digested with 9 units Bam HI for 2 hours at 37°C, heat-inactivated at 70°C for 10 minutes, extracted with CHCl₃, precipitated with ethanol, resuspended in 20 μl ligation assay buffer, and treated with 4 units T4 DNA ligase for 5 hours at 14°C. This mixture was used to transform *E. coli* HB101, and a pBR322–HBV clone was identified by colony hybridization. DNA from this clone was cleaved with Bam HI and the 1350-bp fragment containing the HBsAG coding region was purified by preparative electrophoresis through a 0.8% agarose gel, subcloned in pBR322, and reisolated by electrophoresis. *Step B*: pBR322–SV40 was prepared by cleavage with Bam HI–Eco RI, treatment with ligase, and cloning in *E. coli*. *Step C*: pBR322–HBV was then ligated to Bam HI-cleaved pBR322–SV40 DNA and cloned in *E. coli* as described above. *Step D*: Hae II cleavaged of the resulting pBR322–SV40–HBV "double recombinant" DNA generated the 4950-bp linear SVHBV DNA fragment that was used to infect monkey kidney cells. Cleavage sites: X, Bam HI; ●, Eco RI; ○, Hae II (with permission of D. H. Hamer).

tem has proven useful for isolating cDNA copies of several genomic inserts such as adenovirus and SV40 early antigens (cited in Cepko et al., 1984).

In addition, the expression of cloned genes in such vectors can be modulated by the enhancer sequences located in the LTR and which are responsible for the tissue tropism and type of disease induced by murine leukemia virus (Desgroseillers and Jolicœur, 1984; Lenz et al., 1984; Chatis et al., 1983; Desgroseillers et al., 1983; Chatis et al., 1984).

Recent retroviral vectors have been designed to allow the insertion of both replication-competent and defective retroviral recombinants into preimplantation mouse embryos, resulting in germ-line insertion of foreign genes (Huszar et al., 1985; Van Der Putten et al., 1985). Such transgenic animals transmit proviral DNA genomes efficiently to subsequent generations, therefore allowing a modulated expression of single copy genes in the offspring via the enhancer elements located in the LTR.

However, a potential drawback of these systems comes from the fact that insertion of retroviral-derived genome in the host cell DNA may result in the loss of cloned sequences, and activation of silent cel-

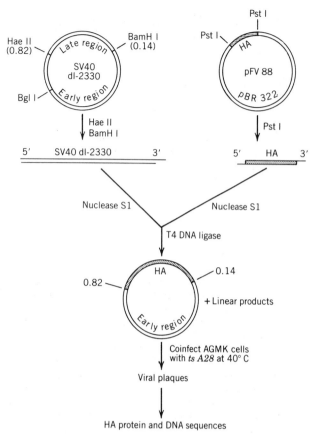

Figure 227. Construction and isolation of a hemagglutinin (HA)–SV40 hybrid virus (with permission of C. J. Lai).

lular genes by the enhancer/promoter sequences located in the proviral LTR. In such a case, cellular transformation may occur, followed by tumorization (Hayward et al., 1981).

Several adenovirus-derived vectors have been developed for cloning and expression of eucaryotic genes (see Chapter 6). Among them, an interesting system uses adenovirus-associated virus as a mammalian DNA cloning vector (Hermonat and Muzyczka, 1984). This virus (AAV)-2 is a parvovirus which is propagated either in a proviral form, or as a lytic virus. The viral genome is a 4675 nucleotides single stranded DNA flanked by inverted terminal repeats (145 bases each). Interestingly, this virus itself is not pathogenic to humans or animals. Lytic growth requires the presence of a helper virus (adenovirus and herpes virus work as well) and recovery of proviral integrated AAV genomes occurs readily upon superinfection with a helper.

It has been shown that AAV can be used to introduce and express foreign DNA into mammalian cells. Since the limit cloning size (5–10 kb) is in the same range of that offered by SV40-derived vectors, its only advantage lies in the fact that AAV-transduced sequences can be easily recovered from infected cells following superinfection.

The herpes simplex system developed by Spaete and Frenkel (1982) and Frenkel et al. (1983) appears to be very powerful and promising. It allows the cloning of long pieces of eucaryotic DNA (the HSV virion can accommodate up to 150 kb DNA) for research into their expression in eucaryotic cells. Defective HSV genomes have been shown to contain only limited subsets of the standard HSV DNA sequences and to utilize HSV-specific replication functions (provided in *trans* by a helper virus).

The defective DNAs are indistinguishable in size from their standard virus DNA counterparts. They consist of multiple-repeated units arranged in a head-to-tail tandem array. The size of the repeated units varies from 3 to 30 kb. Also, full-length defective genomes can be regenerated from individual monomeric repeats after cotransfection of cells with helper virus DNA. Subsets of HSV DNA sequences containing the viral replication functions have been linked to plasmid DNA and cloned into bacteria. When such amplicons are used to transfect eucaryotic cells (in the presence of helper virus DNA) full-length chimeric defective genomes of approximately 150 kb are generated. They consist of multiple head-to-tail reiterations of the inserted cloned amplicon, which can be introduced back into bacteria, providing a very convenient shuttle cloning system (Figure 228).

A 12-kb DNA fragment containing the chicken ovalbumin gene with its 5' and 3' flanking sequences and an 11.7-kb fragment containing the gene coding for the α subunit of human chorionic gonadotropin have been cloned in the HSV amplicon and stably propagated in virus stocks (Figure 229).

It is not yet clear if the promoters for eucaryotic genes inserted into defective genomes will be efficiently active in infected cells. However, there is strong evidence suggesting that repeat-unit DNA is preferentially transcribed in infected cells and that more abundant expression of the corresponding gene takes place under these conditions.

Because a high level of expression is in most cases a primary requirement, this field has been leading in the search for efficient promoters and enhancers. Two efficient systems described recently are based

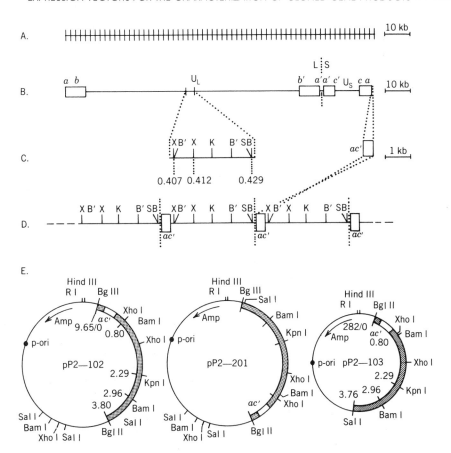

Figure 228. The structure of the HSV-1 (Patton) defective genomes and derivative recombinant plasmids. A. Schematic representation of the defective Patton genomes containing approximately 38 reiterations of 3.9-kb repeat units. The defective genomes terminate at one end with sequences corresponding to the *ac* terminus of standard virus DNA. B. Schematic representation of the standard HSV DNA displaying the arrangement of the unique and inverted repeat sequences of the S and L components. C. Arrangement of the Xho I (X), Bam I (B′), Kpn I (K), Sal I (S), and Bgl II (B) restriction enzyme sites in the U_L segment of standard HSV DNA bounded within the map coordinates shown. *ac** represents the portion of the *S* inverted repeat sequences present in the defective genome repeat units. This region has a maximum size of 500 bp. The asterisk denotes uncertainty in the number of c sequences present in the 3.9-kb Patton defective genome repeat units. D. The arrangement of the U_L and *ac** sequences within the Patton defective genomes. The junctions between adjacent repeats are represented by the dotted vertical lines; the *a* sequences begin approximately 190 bp from the Bgl II site. Arrows denote the Bgl II cleavage site used for the introduction of the repeat unit into pKC7. E. Representation of the pP2-102, pP2-201, and pP2-103 recombinant clones. p-ori denotes the replication origin of the plasmid, Amp denotes the β-lactamase gene (with permission of N. Frenkel).

upon a powerful enhancer sequence from the human cytomegalovirus (hCMV), and a recombinant vaccinia virus that synthesizes T7 RNA polymerase.

The enhancer sequence from the major immediate–early gene of hCMV has been shown to function in a wide variety of tissues (Boshart et al., 1985). The strength of this enhancer–promoter has been measured by placing the chloramphenicol acetylase

(cat) gene under the control of this sequence (Foecking and Hofstetter, 1986). The results obtained reveal that the hCMV enhancer–promoter is considerably stronger than that of SV40 and the Rous sarcoma virus LTR for driving expression of foreign genes in transfected cells, and therefore suggest that this system can be conveniently employed for transient or stable expression of cloned genes.

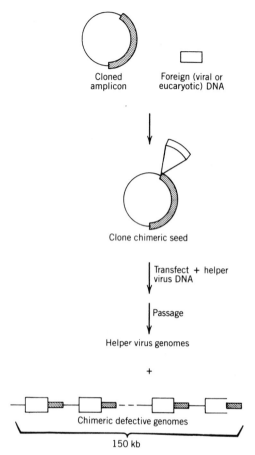

Figure 229. Steps in the introduction of foreign DNA into the reiterated defective genomes of herpes simplex virus (with permission of N. Frenkel).

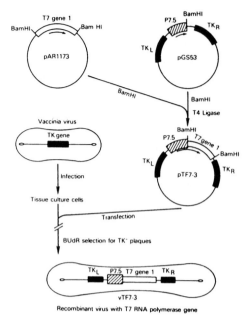

Figure 230. Insertion of bacteriophage T7 gene 1 into the genome of vaccinia virus. A 2.65-kbp Bam HI fragment containing T7 gene 1 was excised from pAR1173 and inserted into the unique Bam HI site of pGS53 to form pTF7-3. In the latter plasmid, the coding sequence for T7 RNA polymerase is downstream of the vaccinia P7-5 promoter and the chimeric gene is flanked by the left (TK$_L$) and right (TK$_R$) vaccinia TK gene sequences. DNA segments are not drawn to scale. CV-1 cells were infected with vaccinia virus and transfected with pTF7-3. After 48 hours, the cells were harvested and the virus was plaqued on TK-cells in the presence of BrdUrd ("BUdR"). Virus plaques were amplified and screened by dot blot hybridization to T7 gene 1 DNA (with permission of Dr. Morse).

Another potentially powerful system is represented by an infectious recombinant virus synthesizing T7 RNA polymerase in eucaryotic infected cells (Fuerst et al., 1986).

The choice of both a single subunit enzyme exhibiting a rather stringent specificity and a cytoplasmic DNA virus that encodes its own RNA-modifying enzymes might have been the key point that allowed the success of the construction (Fuerst et al., 1986).

This system is based on a plasmid containing the vaccinia P7.5 promoter inserted upstream to the T7 gene 1 coding for T7 RNA polymerase, these two elements being flanked by the left and right vaccinia TK gene sequences (TK$_L$ and TK$_R$, respectively) in order to direct recombination into the TK locus of the genome. The *E. coli* β-galactosidase and CAT gene derived from transposon Tn9 were used as target sequences to determine whether the T7 RNA polymerase expressed under the control of vaccinia virus could function efficiently in mammalian cells. This system was found to be very efficient (Fuerst et al., 1986) (Figures 230–232).

The development of a vaccinia/T7 hybrid virus system opens the road for efficient expression of cloned genes under control of the T7-operator sequence. It has already been observed that elevated expression of cloned sequences is achieved when the T7 RNA polymerase gene and the target gene are carried by the vaccinia-vector (Fuerst and Moss, unpublished).

Among the expression vectors that have been developed recently, the insect baculovirus *Autographa californica* nuclear polyhedrosis virus (Ac NPV) has been shown to be suitable as a helper-independent, viral expression vector for the efficient production of cloned gene products in cultured insect cells. This

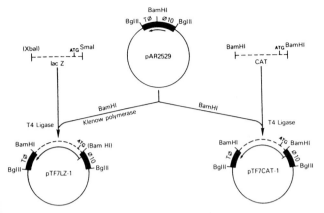

Figure 231. Construction of plasmids containing target genes flanked by T7 promoter and terminator sequences. A 3.2-kbp DNA segment containing the lacZ gene with translation and termination codons was obtained by cleavage of pWS61 (provided by A. Majmdar, National Institutes of Health) with Xba I, filling in the staggered end with the Klenow fragment of DNA polymerase I and dNTPs, and cleaving with Sma I. The fragment was then blunt-end-ligated to pAR2529 that had been cleaved with Bam HI and treated with Klenow fragment. The resulting plasmid, pTF7LZ-1, has the β-galactosidase coding sequence flanked by the T7 O10 promoter and TO terminator. Similarly, a 0.7-kbp Bam HI fragment from pGS30 containing the CAT gene was ligated to Bam HI-cleaved pAR2529 to form pTF7CAT-1 (with permission of Dr. Morse).

system has been used for the high-level production of human β-interferon (Smith et al., 1983), *E. coli* β-galactosidase (Pennock et al., 1984), human *c-myc* protein (Miyamoto et al., 1985), human interleukin (Smith et al., 1985), and influenza virus haemagglutinin (Kuroda et al., 1986).

This system has proved to be very valuable because of the very efficient promoter of Ac NPV and because biologically active products have been obtained with it. For example, it has been shown that under these conditions, β-interferon was glycosylated by an N-linked glycan, the preinterferon signal polypeptide was removed, and that the protein was secreted efficiently. Also, the c-*myc* protein was found to be phosphorylated, removal of the signal peptide from interleukin 2 was performed accurately, and hemaglutinin was processed in insect cells so as to exhibit a full biological activity.

All these examples show that baculovirus vectors are among the most promising host/vector systems for high-level expression and efficient accurate processing of cloned gene products.

References

Coffin, J. (1982), in *RNA Tumor Viruses*. p261, Weiss, R., Teich N., Varmus, H. and Coffin, J., Eds. Cold Spring Harbor Laboratory, Cold Spring Harbor, NY.

Boshart, M., Weber, F., Jahn, G., Dorsch-Häsler, K., Fleckenstein, B., and Schaffner, W. (1985), *Cell*, **41**, 521.

Cepko, C. L., Roberts, B. E., and Mulligan, R. C. (1984), *Cell*, **37**, 1053.

Colbère-Garapin, F., Horodniceaunu, F., Kourilsky, P., and Garapin, A. (1981), *J. Mol. Biol.*, **150**, 1.

Chatis, P. A., Holland, C. A., Hartley, J. W., Rowe, W. P., and Hopkins, N. (1983), *Proc. Natl. Acad. Sci. USA*, **80**, 4408.

Chatis, P. A., Holland, C. A., Silver, J. E., Frederickson, T. N., Hopkins, N., and Hartley, J. W. (1984), *J. Virol.*, **52**, 248.

Davies, J., and Jimenez, A. (1980), *Am. J. Trop. Med. Hyg. (Suppl.)*, **29**(5), 1089.

Desgroseillers, L., Rassart, E., and Jolicœur, P. (1983), *Proc. Natl. Acad. Sci. USA*, **80**, 4203.

Desgroseillers, L., and Jolicœur, P. (1984), *J. Virol.*, **52**, 945.

Foecking, M. K., and Hofstetter, H. (1986), *Gene*, **45**, 101.

Frenkel, N., Deiss, L. P., Kwong, A. D., and Spaete, R. R. (1983), in *Gene Transfer and Cancer*, M. L. Sternberg and M. L. Pearson, Eds., Raven Press, New York.

Fuerst, T. R., Niles, E. G., Studier, F. W., and Moss, B. (1986), *Proc. Natl. Acad. Sci. USA*, **83**, 8122.

Hayward, W. S., Neel, B. G., and Astrin, S. (1981), *Nature*, **290**, 475.

Hermonat, P. L., and Muzyczka, N. (1984), *Proc. Natl. Acad. Sci. USA*, **81**, 6466.

Huszar, D., Balling, R., Kothary, R., Magli, M. C., Hozumi, N., Rossan, J., and Bernstein, A. (1985), *Proc. Natl. Acad. Sci. USA*, **82**, 8587.

Jogersen, R. A., Rothstein, S. J., and Reznikoff, W. S. (1979), *Mol. Gen. Genet.*, **177**, 65.

Kuroda, K., Hauser, C., Rott, R., Klenk, H. D., and Doerfler, W. (1986), *EMBO J.*, **5**, 1359.

Lenz, J., Celander, D., Crowther, R. L., Patarca, R., Perkins, D. W., and Hazeltine, W. A. (1984), *Nature*, **308**, 467.

Mann, R., Mulligan, R. C., and Baltimore, D. B. (1983), *Cell*, **33**, 153.

Miyamoto, C., Smith, G. E., Garrell-Towt, J., Chizzonite, R., Summers, M. D., and Ju, G. (1985), *Mol. Cell. Biol.*, **5**, 2860.

Pennock, G. D., Shoemaker, C., and Miller, L. K. (1984), *Mol. Cell. Biol.*, **4**, 399.

Smith, G. E., Summers, M. D., and Fraser, M. J. (1983), *Mol. Cell Biol.*, **3**, 2156.

Smith, G. E., Ju, G., Ericson, B. L., Moschera, J., Lahm,

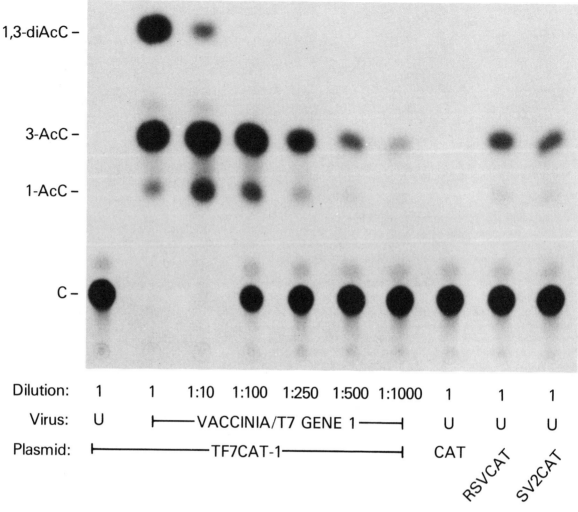

Figure 232. Comparison of transient-expression systems. Cell lysates were prepared at 48 hours after infection with vTF7-3 (vaccinia/T7 gene 1) and/or transfection with the indicated plasmid and assayed for CAT. (Virus "U" indicates uninfected cells). Samples were spotted on a silica gel plate and chromatographed. An autoradiogram is shown, with the positions of chloramphenicol (C) and acetylated forms of chloramphenicol (AcC) indicated (with permission of Dr. Morse).

H. W., Chizzonite, R., and Summers, M. D. (1985), *Proc. Natl. Acad. Sci. USA,* **82**, 8404.

Spaete, R. R., and Frenkel, N. (1982), *Cell,* **30**, 295.

Van Der Putten, H., Botteri, F. M., Miller, A. D., Rosenfeld, M. G., Fan, H., Evans, R. M., and Verma, I. M. (1985), *Proc. Natl. Acad. Sci. USA,* **82**, 6148.

RECOVERY OF INTACT GENE PRODUCTS FROM FUSION PROTEINS

The translation products of DNA sequences cloned in vectors such as λgt11 will be expressed as fusion proteins in which the β-galactosidase portion ac-counts for 114,000 daltons (see Chapter 6). Since only a few bacterial proteins are larger than β-ga-lactosidase, the fusion proteins can be easily de-tected, either by direct coloration in a SDS poly-acrylamide gel, or after electrophoretic transfer and immunodetection. When the fusion polypeptide is characterized, it can be purified by preparative elec-trophoresis and elution, to be used in most cases, as a source of antigenic material. Alternatively, the de-tection of fusion products is a confirmation that the cloned cDNA sequences of interest have been ob-tained and can be subcloned for further studies. This system has proved to be useful in many instances when DNA sequences corresponding to a given prod-uct were to be identified.

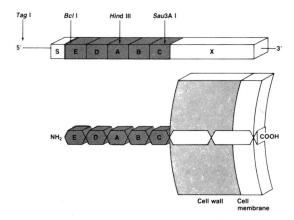

Figure 233. The upper part shows the organization of the protein A structural gene. Three regions can be distinguished, the signal sequence (S), five IgG-binding domains (E–C) and a cell wall/membrane associated region (X). The lower part shows the corresponding protein A protein and location within the cell (courtesy of Pharrnacia).

An interesting improvement in the purification of cloned proteins has been brought by the introduction of protein A gene fusion vectors.* In these systems, the coding sequence for a protein of interest is fused with a portion of the staphylococcal protein A gene, therefore allowing its direct purification by affinity chromatography.

The protein A gene from *Staphylococcus aureus* has been cloned and sequenced (Uhlén et al., 1984a), revealing three genetically distinct parts; a signal sequence (S), five highly homologous IgG-binding domains (E, D, A, B, C), and a cell wall/membrane associated region X (Figure 233). The promoter and sequence are functional in a wide range of bacterial species including *E. coli, B. subtilis* (Eliasson, unpublished), *S. aureus, S. xylosus, S. epidermidis, S. capitis,* and *Streptomyces lividans* (Emas, unpublished). Any construction can thus be expressed in a variety of hosts, allowing comparisons of yields and degradation (Uhlén et al., 1984b; Nilsson et al., 1985a,b; Abrahamsén et al., 1985).

The use of staphylococcal protein A as an affinity tail is based on the "pseudoimmune" interaction between the constant region of immunoglobulins and protein A. The dissociation constant for this interaction is 2×10^{-8} (Nilsson et al., 1985a), allowing fusion proteins containing a protein A moiety to be rapidly recovered from crude lysates under nondenaturing conditions. Furthermore, the IgG-binding

*Information courtesy of Pharmacia

region is composed of independently folded globular domains which interfere minimaly with the tertiary structure of the fusion protein.

A fusion protein containing a protein A tail is well suited for direct immunization because the repetitive structure of the protein A moiety may enhance the immune response. Also, the presence of the affinity tail may stabilize labile proteins from proteolytic degradation.

After expression of a gene product by the mean of genes fusion, sequence-specific cleavage of the hybrid protein may be used to release a pure gene product. One can distinguish two basic methods for generating specific cleavage of proteins: (1) chemical and (2) enzymatic methods.

The use of the chemical methods has been limited, not only because short stretches of amino acids are recognized by the compounds employed (Table 80), but essentially because they are restricted to the release of polypeptides lacking the corresponding amino acids or dipeptides. Until recently, the dominating chemical cleavage method for hybrid proteins relied on the use of cyanogen bromide which cleaves the peptide bond C-terminal of the methionine residues (Gross and Witkop, 1962).

Because the recognition sequences for some endopeptidase may involve as much as 4 to 8 amino acid residues (Table 80), they make proteolytic cleavage more attractive for general applications. However, several of them, such as chymotrypsin, V8 protease, elastase, trypsin, and clostripain, are not specific enough to be universally used. In addition, it has been shown (Friberger, 1982) that many of the primary sequences found in natural substrates do not function well using peptide substrates. Instead other specificities can be found using peptide substrates, often involving not naturally occurring amino acid residues in the optimal cleavage structure. The reason may be that the primary structure found in the natural substrate is involved in a complex tertiary structure, that is, the surface "seen" by the protease is not the primary sequence. These observations are of course relevant for the cleavage of hybrid proteins by any specific protease. Contamination of endopeptidase preparations with nonspecific proteases may be another disadvantage of the method.

As the cleavage of most of the presented proteases in Table 80 are on the C-terminal side of the recognition sequence, they are suitable for fusion vectors where the product gene is the 3' gene. Thus, a cleavage by any such protease would potentially release mature product.

Table 80. Examples of Reagents That Cleave Gene Fusion Products

Chemical

	Amino	Carboxyl side
Cyanogen bromide	Met ↓	
Formic acid	Asp ↓	Pro
Hydroxyl amine	Asn ↓	Gly
Oxidative reagents	Trp ↓	
2-Nitro-5-thiocyanobenzoate		↓ Cys

Enzymatic

Enterokinase	Asp Asp Asp Lys ↓
Factor Xa	Ile Glu Gly Arg ↓
Collagenase	Pro X ↓ Gly Pro Y ↓ (Pro Val ↓ Gly Pro)
Trypsin	Lys ↓ and Arg ↓
Clostridiopeptidase B	Arg ↓ (Lys) ↓
Staphylococcal protease	Asp ↓ and Glu ↓
Thermolysin	↓ Leu, ↓ Ile, ↓ Phe and ↓ Val (and others)
Chymotrypsin	Phe, ↓ Trp ↓ and Tyr ↓
Pepsin	Phe, ↓ Trp ↓ and Tyr ↓ (and others)

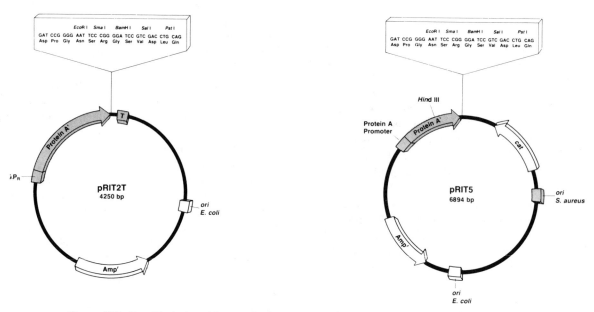

Figure 234. Simplified plasmid maps for two commercial protein A fusion vectors (courtesy of Pharmacia).

The introduction of a synthetic specific sequence at the fusion point between the two proteins has been a considerable improvement, and has allowed the recovery of biologically active products from cloned genes. For example, a synthetic IGF-1 (insuline-like growth factor) sequence has been fused to the protein A gene after addition of an acid labile Asp-Pro cleavage site by in vitro mutagenesis, at the fusion point between the two proteins. The modified DNA was inserted in the gene fusion vector pRIT5 (see Figure 234) and the resulting recombinant DNA was used to transform *S. aureus*. After temperature-induced expression and secretion of the protein A-IGF-1 hybrid protein into the growth medium, the

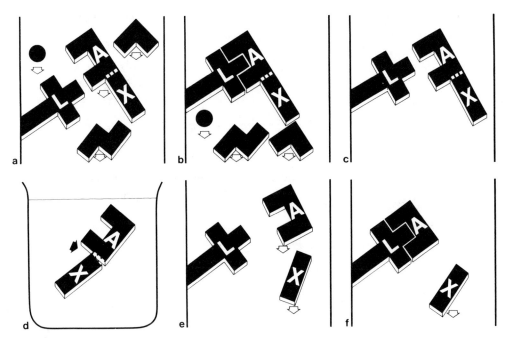

Figure 235. Schematic illustration of a downstream processing system (X = desired protein; A = affinity tail; L = ligand) (courtesy of Pharmacia).

product was purified by passing the medium through an IgG sepharose 6FF affinity column. Following elution, the purified hybrid protein was treated with 70% formic acid in order to cleave the Asp-Pro peptide bond. Biologically active IGF-1 was released showing identical specific activity to the native hormone purified from human serum (Nilsson et al., 1985b).

The basic concept of downstream processing that utilizes genetic engineering is outlined in Figure 235. A cell lysate containing a hybrid protein comprised of the desired protein (X) fused to the protein A affinity tail (A) is passed through an IgG sepharose 6FF affinity column containing the specific IgG ligand (L) (a,b). After elution (c), performed by lowering the pH, the purified fusion protein may be used directly for immunization. Alternatively, a specific chemical or enzymatic cleavage may be performed, cleaving the fusion protein at the junction between the protein A affinity tail and the desired product (d). The mixture is again passed through the column (e), the affinity tail (A) and noncleavaged material adsorbs to the IgG ligand (L), while the desired protein (X) is collected in the effluent (f).

The column is regenerated by desorbing the bound protein A moiety. This purification scheme offers unique advantages over traditional biochemical

methods, providing both a general and a specific scheme for purification of high yields of proteins.

The choice of a protein A fusion vector for gene expression is dependent upon the origin and properties of the gene product to which protein A is to be fused. Proteins that are secreted in their natural host are often unstable in the intracellular matrix; this is partly because stabilizing disulfide bridges cannot be formed in the reducing environment found inside most bacteria. In the case of these proteins, pRIT5 is the recommended host vector; it can be used either in *E. coli* or *S. aureus*. Alternatively, proteins that are found intracellularly in their natural host may be impossible to translocate through the bacterial membrane and should therefore be grown in pRIT2T, which expresses the fusion protein intracellularly. pRIT2T can only be utilized in *E. coli*, and disruption of the cells is required prior to purification.

Both pRIT2T and pRIT5 protein A fusion vector systems allow rapid and efficient purification of the expressed fusion proteins by IgG sepharose 6FF affinity (vectors commercialized by Pharmacia).

Another very elegant approach for the introduction of cleavage-site-specific sequences consists of introducing at the fusion point specially designed linkers which encode for well-defined amino acid residues. These synthetic oligonucleotides are known as MuRFI (Perlman and Halvorson, 1986) and TAB (Barany, 1985a,b) linkers. Both types are commercially available (for example, New England Biolabs, and Pharmacia) and are described in Chapters 12 and 23.

References

Abrahamsén, L., Moks, T., Nilsson, B., Hellman, U., and Uhlén, M. (1985), *EMBO J.,* **4**, 3901.

Barany, F. (1985), *Proc. Natl. Acad. Sci. USA,* **82**, 4202.

Barany, F. (1985), *Gene,* **37**, 111.

Friberger, M. (1982), *Scand. J. Clin. Lab. Invest. Suppl.,* 162.

Gross and Wikop, (1962), *J. Biol. Chem.,* **237**, 1856.

Nilsson, B., Abrahamsen, L., and Uhlén, M. (1985a), *EMBO J.,* **4**, 1075.

Nilsson, B., Holmgqren, E., Josephson, S., Gatenbeck, S., Philipson, L., and Uhlén, M. (1985b), Nucl. Acids. Res. **13**, 1151.

Perlman, D. and Halvorson, H. O. (1986), *Nucl. Acids Res.,* **14**, 2139.

Uhlén, M., Guss, B., Nilsson, B., Gatenbeck, S., Philipson, L., and Lindberg, M. (1984a), *J. Biol. Chem.,* **259**, 1695.

Uhlén, M., Guss, B., Nilsson, B., Gotz, F., and Lindberg, M. (1984b), *J. Bact.,* **159**, 713.

TRANSFECTION OF EUCARYOTIC CELLS WITH PURIFIED DNA

Manipulated DNA cannot always be reintroduced into eucaryotic cells by means of nonlytic or non-transforming viral vectors. Techniques have been developed that allow direct transfer of recombinant DNA molecules into recipient cells maintained in culture and therefore represent an unique tool for the study of gene expression.

Unfortunately, transfection procedures are very inefficient. In the best cases, no more than 15% of the cells acquire and express DNA fragments, while in other cases a maximum of 6% has been obtained. We present below a brief outline of the different procedures used along with their advantages and disadvantages.

Transfection of cells following DEAE Dextran treatment (Sompayrac and Danna, 1981)

In this technique, monolayers of cells are first treated with a non-lethal dose of DEAE Dextran (Di-ethylaminoethyl dextran, polysaccharide derivative) in order to alter the outer membrane and permit uptake of foreign DNA molecules.

PROTOCOL

USE OF DIETHYLAMINOETHYL DEXTRAN (DEAE DEXTRAN) ON CELL MONOLAYERS

1. Grow the cells on 35-mm Petri dishes under conditions such that the cultures will be subconfluent at the time of transfection.

2. Dissolve DEAE Dextran (molecular weight 5×10^5 or 2×10^6 daltons) in Dulbecco minimal medium buffered with 0.05 M Tris–HCl (pH 7.3). Since the percentage of surviving cells after DEAE Dextran treatment will be dependent on the origin of the cells and the concentration of DEAE Dextran used, you should run a trial experiment to optimize the concentration. Generally 100 µg/ml of DEAE Dextran is satisfactory.

3. Withdraw the culture medium and wash the cells two times with serum-free medium.

4. Withdraw the medium and replace with 1 ml of medium containing 1 μg of the DNA to be transfected and DEAE Dextran at the desired concentration. Incubate either under a laminar-flow hood or at 37°C for 30 minutes.

5. Wash the cells once with serum-free culture medium and incubate the cells in complete medium until performing biological or biochemical assays for transfection. Always run control plates transfected with calf thymus or salmon sperm DNA.

DNA–Calcium Phosphate Coprecipitation Technique with Cell Monolayers (Graham and Van der Eb, 1973)

One can distinguish two major steps in this method: (a) precipitation of the DNA in the presence of divalent cations (Ca^{2+}) and phosphate; and (b) uptake of the DNA–calcium phosphate precipitate by the cells.

PROTOCOL

DNA-CALCIUM PHOSPHATE COPRECIPITATION TECHNIQUE WITH CELL MONOLAYERS

1. Plate cells at a density of about 5×10^5 cells/60-mm dish in regular culture medium.

2. Dilute the DNA solution in HeBS* to a concentration of about 20–40 μg/ml (plan on using 0.5 ml/plate).

3. Add a one-tenth volume of 1.25 M $CaCl_2$ to the DNA solution. Mix and let stand on the bench for 20 minutes. A fine precipitate should be visible.

4. Mix the precipitate by inverting the tube several times.

5. Add 0.5 ml of the DNA suspension to each cell plate. Allow the liquid to cover all the cells.

6. Incubate the cells + DNA at 37°C for 4 hours.

7. Aspirate the solution of each plate and add 5 ml culture medium to the plate. Swirl gently. Aspirate the medium.

*HeBS is: Hepes, 5 g/liter; NaCl, 8g/liter; KCl, 0.37 g/liter; $Na_2HPO_4 \cdot H_2O$, 0.13 g/liter; dextrose 1 g/liter; pH 7.05.

8. Add 1 ml 20% DMSO in culture medium per plate. Incubate for 5 minutes. Aspirate the medium.

9. Add 5 ml culture medium per plate to wash the cells. Aspirate the medium.

10. Add 4 ml fresh culture medium.

11. Incubate the cells at 37°C. Change the medium every 3 days until performing a test for successful transfection (growth of transformed cells in agar, selection on restrictive medium, etc.).

Transfection of Cells in Suspension with DNA–Calcium Phosphate Precipitates (Chu and Sharp, 1981)

This protocol, a modification of the method presented above, consists of exposing cells in suspension to the DNA precipitate. A comparison of the up-take yields obtained under different conditions reveals that up to 15% of CV1 cells can be transfected by using the suspension method as compared with 6–7% for transfection on monolayers with calcium phosphate precipitate and 2% for DEAE Dextran.

PROTOCOL

TRANSFECTION OF CELLS IN SUSPENSION WITH DNA-CALCIUM PHOSPHATE PRECIPITATES

1. Grow cells up to confluency on 15-cm diameter plates.

2. Trypsinize cells until about 95% of the cells have become round.

3. Detach trypsinized cells by firmly tapping the edge of the dish.

4. Add culture medium and calf serum (to a final concentration of 10%) to suspend the cells and stop trypsinization.

5. Pellet the cells by centrifugation at 800–1000 rpm for 10 minutes.

6. Resuspend the cells in 2.5 ml of the transfection cocktail (1X HeBS, 0.125 M CaCl$_2$) containing 25 μg/ml DNA.

7. Incubate resuspended cells at room temperature for 15 minutes.

8. Break up the cell clumps by repeatedly pipetting the transfected cells. Transfer the cells to 15-cm dishes (at about two-thirds of their original density). Incubate at 37°C.

9. Once cells are attached to the plate, and 5 hours after transfection, aspirate the medium and replace with 2.5 ml 25% glycerol (or 25% DMSO depending on the sensitivity of the cells) for 1 minute.

10. Wash the cells with culture medium to remove glycerol (or DMSO) and incubate cells in 8 ml culture medium containing serum until performing tests for transfection.

Liposome-Mediated DNA Transfection (Schaffer-Ridder et al., 1982)

This protocol is based on the entrapment of DNA into liposome vesicles which are then used to transfect cells. The efficiencies of gene transfer reported by different groups are in the same range as those reported for calcium phosphate precipitation, and are therefore higher than those obtained with the DEAE Dextran procedure.

The preparation and use of DNA-containing liposomes are as follows.

PROTOCOL

LIPOSOME-MEDIATED DNA TRANSFECTION

1. Dilute the DNA to be used in PBS to a concentration of approximately 50–500 μg/ml.

2. In a test tube, mix L-α-phosphatidyl-L-serine and cholesterol (Sigma) in a 1:1 molar ratio. Evaporate the solvents under reduced pressure by rotary evaporation.

3. Dissolve the lipids in diethyl ether to a concentration of 10 μmol/ml.

4. Mix 1 ml of the lipid solution with 0.33 ml of the DNA solution.

5. Sonicate the two-phase mixture for 10 seconds (Branson sonifier B15, equipped with microtip, 500-millisecond pulse frequency). This will form a one-phase dispersion.

6. Evaporate the ether under reduced pressure by rotary evaporation. This step enables formation of liposomes from the inverted micelles generated by sonication.

7. Pellet the liposomes by centrifugation at 40,000 rpm for 30 minutes.

8. The supernatant contains nonentrapped DNA, which can be extracted to evaluate the efficiency of liposome trapping.

9. Resuspend the liposome pellet in 1 ml PBS and add 50 µg DNase I to the suspension.

10. Incubate for 1 hour at room temperature. This step allows better separation of liposomes and free DNA.

11. Apply the liposome suspension to a Sepharose 4B column equilibrated with S buffer. Collect 1-ml fractions. Monitor separation of digested DNA and liposomes by absorption at 260 nm.

12. The DNA-loaded liposomes can be kept at 4°C for up to a month.

13. Recipient cells grown in the regular culture medium should be plated 24 hours before transformation at a density of 5×10^5 cells/10-cm Petri dish.

14. Before transfecting, remove the culture medium and rinse the cells once with culture medium deprived of serum.

15. Mix the DNA-loaded liposomes with culture medium without serum to obtain approximately 10–500 nmol phospholipid/ml of medium.

16. Add 1 ml of the culture medium containing liposomes directly to the cells and incubate for 30 minutes at 37°C.

17. Aspirate the medium and replace with culture medium containing 25% glycerol for 1 minute or 25% DMSO for 4 minutes.

18. Rinse the cells with culture medium and incubate in the presence of serum-supplemented medium until performing tests for transfection.

Direct Transfer of Plasmid DNA from Bacterial to Mammalian Cells (Schaffner, 1980)

This protocol relies on a membrane fusion between *E. coli* protoplasts and mammalian cells allowing transfer of genetic material from the bacteria to the eucaryotic cells. Bacteria containing the suitable plasmid are grown and amplified as described elsewhere (Chapter 7) and protoplasts are prepared as follows.

PROTOCOL

DIRECT TRANSFER OF PLASMID DNA FROM BACTERIAL TO MAMMALIAN CELLS

1. Spin down amplified bacteria from a 100-ml culture at 5000 rpm for 10 minutes and resuspend in 1 ml protoplast buffer.

2. Add 200 μl lysozyme solution (10 mg/ml) and incubate at 4°C for 10 minutes.

3. Dilute little by little the bacterial protoplasts with 3.8 ml culture medium without serum containing 7% wt/vol sucrose, 10 mM MgCl$_2$, and 1 μg DNase I (0.2 μg/ml).

4. Incubate for 10 minutes at room temperature to digest any DNA that may have been released from burst spheroplasts.

5. For transfection, use plates containing subconfluent cultures (2×10^5 cells/35-mm dish, 8×10^5 cells/60-mm dish). Aspirate the medium and wash the cells once with 2–5 ml prewarmed PBS. Aspirate.

6. Add 0.2–0.5 ml spheroplast suspension per plate.

7. Put the Petri dishes onto the flat bottoms of swing-out centrifuge buckets and centrifuge the protoplasts on the cell monolayer.

8. Carefully aspirate the clear supernatant and immediately replace with 2.5 ml 48% polyethylene glycol (PEG) 1000/60-mm dish (0.8 ml/35-mm dish).

9. Distribute the PEG on the cells by gentle tilting and let incubate at room temperature for 90 seconds.

10. Wash the cell monolayer three times with 4 ml PBS to eliminate the residual PEG. Add culture medium supplemented with serum and incubate at 37°C until testing for DNA transfer.

Selection of Cells Expressing Newly Introduced DNA

Above were presented different methods for introducing foreign DNA into eucaryotic cells. In most cases, the transfection efficiencies reported were based on the expression of a selective marker. Since it is not always possible to select cells expressing a particular gene (e.g., globin gene) and, more generally, any gene whose function is unknown, it is necessary to perform cotransfections with DNA coding for a selectable dominant marker. In such cases, the DNA to be used is coprecipitated with or physically linked to another DNA fragment (e.g., a plasmid) coding for a selectable function such as an enzymatic activity or a drug resistance. The most commonly used selective markers are the herpes simplex virus thymidine kinase (HSV-*tk*), and the cellular adenine phosphoribosyl transferase or dihydrofolate reductase (Wigler et al., 1979; Mantai et al., 1979; Wold et al., 1979; Mulligan and Berg, 1980; Jimenez and Davies, 1980).

Clones have been constructed which carry genes for HSV thymidine kinase and procaryotic dihydrofolate reductase. More recently the bacterial gene coding for the enzyme xanthine:guanine phosphoribosyl transferase (*Eco gpt*) has been cloned in an SV40 vector and used as a transfection marker. Mammalian cells do not efficiently use xanthine for purine nucleotide synthesis; on the other hand, cells transfected with DNA carrying the *Eco gpt* gene can synthesize guanosine monophosphate from xanthine via xanthine monophosphate. This property is used to select transfected cells on a medium containing aminopterin and mycophenolic acid (which block *de novo* purine nucleotide synthesis) and xanthine as the sole precursor for guanine nucleotide synthesis. This system has been used successfully as a selective marker for cotransfection of nonselectable genes. However, it necessitates testing for sensitivity to both inhibitors (aminopterin and mycophenolic acid) and adapting the concentration of xanthine for each cell line intended for transfection (Mulligan and Berg, 1980).

Another dominant selective marker suitable for cotransfection is represented by the enzyme aminoglycoside 3' phosphotransferase (coded by the *Tn*-5 transposon). Mammalian cells are usually sensitive to the aminoglycoside G 418 (a 2-deoxystreptamine antibiotic). On transfection with a bacterial plasmid coding for the aminoglycoside 3' phosphotransferase, the mammalian cells become resistant to the action of G 418 and can thereby be selected (Colbere-Garapin et al., 1980).

Once transfectants have been determined to be positive, they can then be assayed for expression of other genes that may have been introduced.

Notes

1. A variety of agents have been used to enhance DNA uptake by eucaryotic cells. Of them, DMSO (25%, for 4 minutes), PEG (44%, for 90 seconds) and glycerol (25%, for 1 minute) appear to be the most successful. It should be noted, however, that these treatments may have dramatic effects (detachment or killing) on some particularly sensitive cell lines, and it is strongly recommended that you test the cells before transfecting.

2. In the liposome-mediated gene transfer technique, phosphatidyl serine is used because a high percentage of negatively charged phospholipids has been reported to increase the binding capacity of liposomes to cells and their cellular uptake. The choice of cholesterol was dictated by its ability to increase vesicle size.

3. Electroporation of cells has been used successfully to introduce foreign DNA in fragile cells. Several commercial devices are available.

References

Chu, G., and Sharp, P. A. (1981), *Gene,* **13**, 197.

Colbere-Garapin, F., Horodniceaunu, F., Kourilsky, P., and Garapin, A. C. (1980), *J. Mol. Biol.,* **150**, 1.

Graham, F. L., and Van der Eb, A. J. (1973), *Virology,* **52**, 456.

Jimenez, A., and Davies, J. (1980), *Nature,* **287**, 869.

Mantai, N., Boll, W., and Weissmann, C. (1979), *Nature,* **281**, 40.

Mulligan, R., and Berg, P. (1980), *Science,* **209**, 1422.

Mulligan, R., and Berg, P. (1981), *Proc. Natl. Acad. Sci. USA,* **78**, 2072.

Schaffer-Ridder, M., Wang, Y., and Hofschneider, P. H. (1982), *Science,* **215**, 166.

Schaffner, W. (1980), *Proc. Natl. Acad. Sci. USA,* **77**, 2163.

Sompayrac, L. M., and Danna, K. J. (1981), *Proc. Natl. Acad. Sci. USA,* **78**, 7575.

Wigler, M., Sweet, R., Sim, C. K., Wold, B., Pellicer, A., Lacy, E., Maniatis, T., Silverstein, S., and Axel, R. (1979), *Cell,* **16**, 777.

Wold, B., Wigler, M., Lacy, E., Maniatis, T., Silverstein, S., and Axel, R. (1979), *Proc. Natl. Acad. Sci. USA,* **76**, 5684.

See also for a general review: Baserga, R., Croce, C., and Rovera, G., eds., *Introduction of Macromolecules into Viable Mammalian Cells,* Alan R. Liss, New York (1980).

CONSTRUCTION OF COSMID LIBRARIES USED TO TRANSFORM EUCARYOTIC CELLS

Introduction of DNA by transfection or microinjection is not always sufficient to study gene expression in eucaryotic cells. Methods have been developed, based on the properties of cosmid vectors and on the availability of dominant selective markers for gene transfer, which allow the construction of cosmid vectors that can be used to generate libraries of any DNA and that can subsequently be directly employed for the study of gene expression in mammalian cells.

The structures of some expression plasmid vectors with selectable markers, are shown in Figure 236.

Plasmid pRT1 contains the HSV thymidine kinase (*tk*) gene as a selective marker for introduction in *TK⁻* cells and the ampicillin resistance marker from pBR322. Insertion of the replication origin of SV40 led to the construction of pOPF. The vectors pHEP and pSAE are similar to pRT1 and pOPF except that the cosmid sequences between the Cla I and Pvu I sites were replaced with the plasmid pBR327 sequences (this plasmid is free of sequences thought to inhibit replication in eucaryotic cells). Replacement of the central part of the HSV-*tk* gene by the dominant marker aminoglycosyl-3'-phosphotransferase (*agpt*) led to the construction of pTM and pMCS.

Placing the Bam HI site of these vectors at differ-

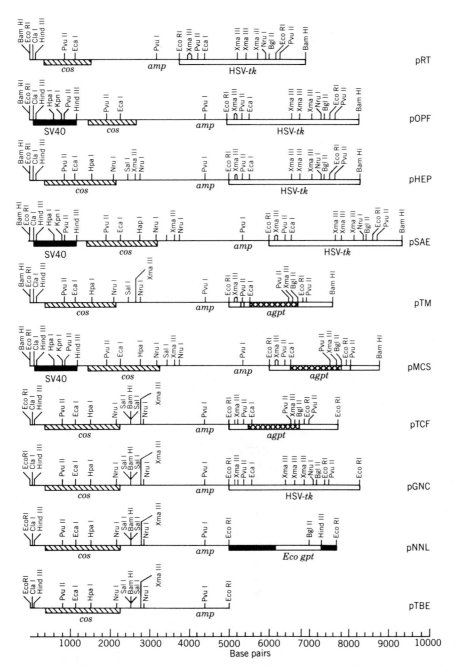

Figure 236. Structure of plasmid vectors. *cos* reters to the phage λ cohesive ends, *amp* to the β-lactamase gene of pBR322, *tk* to the thymidine kinase gene of herpes simplex virus, *gpt* to the guanine phosphoribosyl transferase gene of *E. coli*, and *agpt* to the aminoglycosyl-3'-phospho-transferase gene from *Tn* 5 (with permission of R. A. Flavell).

ent locations has enabled the construction of recombinant plasmids for eucaryotic marker exchange. These cosmids are pTBE, pTCF, pGNC, and pNNL. One of them (pNNL) contains the guanine phosphoribosyl transferase (*Eco gpt*) gene as a selectable marker.

Marker exchange has two advantages: (a) The marker is usually chosen in accordance with the phenotype of the cells used; and (b) eucaryotic cells can be transformed several times by different recombinant cosmids, using a different marker for each transformation.

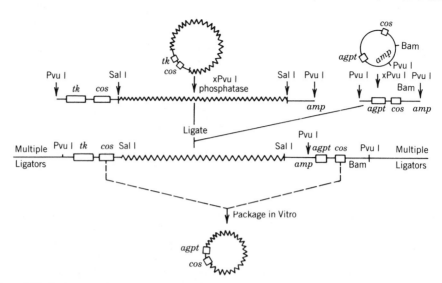

Figure 237. Exchange of eucaryotic markers in a recombinant cosmid; symbols as in Figure 236 (with permission of R. A. Flavell).

The procedure is based on (a) the position of the *cos* site, (b) the position of the cloning site (Bam HI), (c) the position of the eucaryotic markers with respect to each other and with respect to three restriction enzyme sites which contain two CG base pairs in their recognition sequences (i.e., Xma III, Pvu I, Nru I), and (d) the low occurrence of CG base pairs in eucaryotic DNAs.

The probability of any of the three restriction sites occurring in a random piece of 40-kb human DNA is 19%. The chance that all three sites will occur in the same stretch of DNA is less than 1%. As a consequence, more than 99% of the recombinant cosmids obtained from a library of 40-kb fragments will not contain one of these three sites in the inserted DNA.

The recombinant can therefore be cleaved with one of these enzymes in the vector only (e.g., Pvu I). After treatment with alkaline phosphatase to prevent self-ligation, the linear cosmid can be ligated to another cosmid also cleaved with the same enzyme and carrying another marker. Packaging of these ligated molecules will generate a new recombinant cosmid in which the eucaryotic marker has been exchanged. This process is schematically illustrated in Figure 237.

Libraries prepared in this way have been screened for several genes and allowed in some cases to study directly the expression of particular genes in eucaryotic cells.

Reference

Grosveld, F. G., Lund, T., Murray, E. J., Mellor, A. L., Dahl, H. H. M., and Flavell, R. A. (1982), *Nucleic Acids Res.*, **10**, 6715.

DETECTION OF POLYPEPTIDE SYNTHESIS BY IMMUNE PRECIPITATION

In many cases, it is possible to obtain antibodies against polypeptides corresponding to cloned genes, allowing one to detect the expression of the cloned sequences when introduced into procaryotic or eucaryotic cells. The procedure described below is based on the observation that some strains of *Staphylococcus aureus* have a protein component (protein A) on their cell walls that exhibits a strong affinity for mammalian immunoglobulins. This protein specifically and strongly binds to the *Fc* region of immunoglobulin G (IgG) molecules, thereby providing a means for isolation of antigen–antibody complexes. Protein A is available commercially in a soluble form or immobilized on agarose or sepharose (e.g., from Sigma). We present here a protocol using protein A–sepharose or a protein A solution.

DETECTION OF POLYPEPTIDE SYNTHESIS BY IMMUNOPRECIPITATION

Labeling Proteins with [³⁵S] Methionine

Although the radioactive amino acid most commonly used to label cellular proteins in culture is methionine, any other amino acid can be employed if the methionine content of the polypeptide of interest is too low.

1. Rinse cell monolayers twice with PBS prewarmed to 37°C.
2. Add culture medium deprived of methionine to monolayers (0.4 ml/35-mm plate; 1 ml/60-min plate) and let incubate at 37°C for 1 hour, in the presence of 2% dialyzed calf serum.
3. Aspirate the medium off the plates and replace with the same medium containing [³⁵S]-methionine (100–150 µCi/ml). Incubate at 37°C for 4 hours to label the proteins.

Preparation of Cellular Extracts

1. Wash the cell monolayers twice with cold PBS (2 × 2 ml/60-mm plate). Keep plates on ice.
2. Add 500 µl cold lysis buffer/60-mm plate and scrape cells with a rubber policeman.
3. Transfer extracts into microfuge tubes and vortex three times for 30 seconds.
4. Spin at 12,000 rpm for 10 minutes at 4°C.
5. Save the supernatants. Keep the lysates frozen at −70°C or −170°C.
6. Perform a Lowry determination of protein content on an aliquot for each sample (see Chapter 3), and determine radioactivity incorporated by TCA precipitation.

Immunoprecipitation with Protein A–Sepharose

1. In a microfuge tube, mix x µl cellular extract (containing 10–50 µg protein) with NET/NP40 buffer, to obtain a 50–100-µl final volume.
2. Add 1 µl of PMSF 100 mM. Vortex.

3. Add 5 μl of antitubulin serum diluted 1/250.

4. Incubate on wet ice for 30 minutes.

5. Add 20 μl of protein A–sepharose solution (50% v/v in NET/NP40 buffer).

6. Let stand on wet ice for 15 minutes.

7. Centrifuge at 12,000 rpm in a microfuge for 5 minutes at 4°C.

8. Transfer the supernatant in a microfuge tube, and add 5 μl of sample antiserum at a predetermined dilution.

9. Incubate at 4°C for 30 minutes (on wet ice).

10. Add 20 μl of protein A–sepharose solution and let stand on wet ice for 30 minutes.

11. Add 1 ml of NET/NP40 solution and vortex.

12. Centrifuge at 12,000 rpm for 2 minutes at 4°C.

13. Discard the supernatant and resuspend the pellet in 50 μl of NET 0.5 *M*. Vortex.

14. Add 1 ml of NET 0.5 *M*.

15. Centrifuge at 12,000 rpm for 2 minutes at 4°C.

16. Repeat steps 13 to 15 once with NET 0.5 *M* and twice with NET 0.25 *M*.

17. Resuspend the precipitate in 50 μl of NET 0.15 *M* by vortexing.

18. Add 1 ml of NET 0.15 *M*, and transfer to a new microfuge tube.

19. Centrifuge at 12,000 rpm for 2 minutes at 4°C.

20. Pour off the supernatant, spin again for 30 seconds, and aspirate the remaining liquid.

21. Resuspend the precipitate in 50 μl sample buffer containing 0.1 *M* DTT.

22. Heat the samples for 3 minutes in boiling water.

23. Load samples on polyacrylamide gels.

Immunoprecipitation with Protein A Solution

PREPARATION OF STAPH-A SOLUTION

1. Spin the *Staphylococcus* cells for 20 minutes at 3000 rpm.

2. Resuspend in 0.5% NP40 to obtain a 10% staph-A solution.

3. Let incubate at room temperature for 15 minutes.

4. Spin at 3000 rpm for 20 minutes. Remove the supernatant.

5. Resuspend in 0.05% NP40.

6. Spin at 3000 rpm for 20 minutes. Remove the supernatant.

7. Resuspend in 0.05% NP40 and albumin at a final concentration of 1 mg/ml.

PREPARATION OF THE LYSATES

1. Thaw the lysates on wet ice.
2. Add 0.2 ml of the 10% staph-A solution prepared above.
3. Let stand on ice for 30–60 minutes.
4. Centrifuge at 35,000 rpm for 1 hour at 4°C in a Beckman Ti 50 rotor. Save the cleared supernatant.

IMMUNE PRECIPITATION

1. In a microfuge tube mix 250 μl lysate and 5 μl serum (containing the antibody).
2. Incubate overnight at 4°C (3 hours of incubation is sufficient).
3. Add 250 μl of the 10% staph-A suspension previously prepared.
4. Incubate on ice for 3–4 hours.
5. Spin at 3000 rpm for 15 minutes.
6. Discard the supernatant and resuspend the pellet by vortexing in 1 ml cold RIPA buffer.
7. Spin at 3000 rpm for 15 minutes.
8. Repeat steps 6 and 7 three times.
9. Resuspend the pellet in 20 μl protein sample buffer for polyacrylamide gel electrophoresis.
10. Boil the samples for 3 minutes, centrifuge at 12,000 rpm for 5 minutes, and load on polyacrylamide gels.

Buffers and Solutions for Immunoprecipitation

NET/NP40 Solution   *10 ml*
50 mM Tris–HCl (pH 7.5)   0.5 ml 1 M Tris–HCl (pH 7.5)
150 mM NaCl   0.3 ml 5 M NaCl
5 mM EDTA   0.1 ml 500 mM EDTA
0.5% NP 40   50 μl NP40
1 mg/ml BSA   1 ml BSA 10 mg/ml
    9 ml H$_2$O

NET 0.5 M and 0.25 M. Same as NET/NP40 except that NaCl is 0.5 and 0.25 M, respectively.

NET 0.15 M   *10 ml*
50 mM Tris–HCl (pH 7.5)   0.5 ml 1 M Tris–HCl (pH 7.5)
150 mM NaCl   0.3 ml 5 M NaCl
5 mM EDTA   0.1 ml 500 mM EDTA

Sample buffer
65 mM Tris–HCl (pH 6.8)
3% SDS
10% Glycerol
100 mM DTT
0.004% Bromophenol blue

10 ml
0.65 ml 1 M Tris–HCl (pH 6.8)
1.5 ml 20% SDS
1.0 ml Glycerol
1.0 ml 1 M DTT
0.4 ml 0.1% Bromophenol blue
5.45 ml H_2O

RIPA solution
50 mM Tris–HCl (pH 7.4)
150 mM NaCl
1% Triton X-100
1% Sodium deoxycholate
0.1% SDS

100 ml
5 ml 1 mM Tris–HCl (pH 7.4)
3 ml 5 M NaCl
1 ml Triton X-100
1 g Deoxycholate
0.5 ml 20% SDS
H_2O to 100 ml

Lysis buffer (RIPA)
90% RIPA
22 mM EDTA
1% Trasylol

10 ml
9 ml RIPA solution
0.9 ml 250 mM EDTA (pH 7.5)
0.1 ml Trasylol

DETECTION OF POLYPEPTIDE SYNTHESIS BY IMMUNOBLOTTING

In this protocol, polypeptides are separated on a polyacrylamide gel and transferred onto a nitrocellulose sheet which is subsequently incubated in the presence of specific antibodies. The considerable advantage of the immunoblotting over immunoprecipitation is that almost no background will be observed if working with clean monospecific antibodies because nonspecific precipitation does not occur. The main disadvantage of immunoblotting is that larger quantities of antiserum are needed to perform the incubation with nitrocellulose strips.

PROTOCOL

DETECTION OF POLYPEPTIDE SYNTHESIS BY IMMUNOBLOTTING

1. Run the protein extract on a denaturing polyacrylamide gel, as described in Chapter 3.
2. Do not fix or stain.

3. Set up a sandwich for transfer, as described in Chapter 3.

4. Immerse the whole sandwich in electrophoresis buffer, and place in a suitable system for transfer (for example, BioRad or Hoefer). Always place the nitrocellulose sheet toward the anode (+).

5. Fill up the tank with prechilled buffer (we keep the buffer in the cold room at least 12 hours before use).

6. Transfer is performed in the cold room at 50 volts overnight, or for 3 hours at a constant current of 300 mA.

7. Check that transfer worked by staining the gel with Coomassie blue (see Chapter 3).

8. Cut off from the blot the lane containing the molecular weight markers and stain for 10 minutes with Ponceau red (0.2% in 3% TCA). Rinse with bidistilled water to localize markers. If you use radioactive markers you don't need to stain them.

9. Rinse the nitrocellulose sheet in PBS, and place it in saturation solution for at least 90 minutes at 37°C with agitation (preferably overnight).

10. Rinse the nitrocellulose membrane three times for 10 minutes each in saturation solution.

11. Dilute antiserum (accurate dilution must be predetermined) in incubation buffer.

12. Place the nitrocellulose blot in a plastic bag (seal-a-meal type) and add diluted antiserum.

13. Incubate for 90 minutes at room temperature with agitation.

14. Remove the blot from the bag and wash it four times for 15 minutes each in saturation solution.

15. Place the blot in a new bag and add anti IGg peroxidase conjugate (diluted 1:500 in incubation solution).

16. Incubate for 1 hour at room temperature with agitation.

17. Remove the blot and wash it three times for 15 minutes each with saturation solution and once with PBS.

18. Transfer the blot in a tray and add freshly prepared staining mixture.

19. Incubate with agitation for 10 to 60 minutes at room temperature.

20. Take a picture when the blot is still wet to increase visualization.

21. You can store the dried blot between two pieces of 3MM Whatman paper.

Buffers and Solutions

Transfer buffer
20 mM Tris–Base
50 mM Glycine
20% Methanol
Adjust to pH 8.5.
Store at 4°C.

4 liters
9.68 g Tris–base
45.04 g Glycine
1 liter Methanol
H_2O to 4 liters

Saturation solution
5% Nonfat dried milk
0.2% Triton X100 in PBS
  (pH 7.4)

Incubation solution
10% v/v Fetal calf serum
0.2% Triton X100 in PBS
  (pH 7.4)

Staining mixture
Solution 1: 60 mg of HRP
  Biorad color develop-
  ment reagent in 20 ml
  of methanol
Solution 2: 60 μl of 35%
  v/v H_2O_2 in 100 ml of
  PBS
Make solutions fresh each
  time. Mix just before
  use.

EPILOGUE

If You want Ava I to meet Bam HI . . .

Once you have succeeded in cloning, sequencing, expressing a gene of interest, or in anything else, it is often a good time for celebration. Since you might wish to share your success with numerous friends from surrounding laboratories it might be a good idea to try the following protocol.

Reproduced with permission of New England Biolabs

A few weeks before celebration, buy a nice, big, ripe pineapple to the nearest market. Back to your place:

1. Clean an empty glass bottle (usually 75 cm^3) and let it dry.
2. Cut the pineapple in small cubes (about 5 mm) to fill half the bottle.
3. Add 500 ml of rum (from Martinique).
4. Close the bottle with a cork, invert a few times to mix, and leave the mixture in a cool place (5–10°C) for at least three weeks. Go back to work, and be patient . . .

Two days before the celebration, prepare the following base mixture:

1. In a 1-liter sterile glass beaker, add successively:
   500 ml of plain water
   500 g of crystalized sugar
   the outer skin of half a lemon (you may save remainder of lemon)
   a pinch of cinnamon
2. Cook slowly on a Bunsen burner with constant stirring, until the solution becomes a syrup with an oily consistency.
3. Add a pinch of ground nutmeg to the syrup and let it cool on a bench, until it reaches room temperature.
4. Filter the syrup to remove solid particles and save syrup as a source of sweetener.

Depending on the desired strength of the final mixture,

1. Mix varying amounts of syrup and rum solution from steps 4 above. A 1:4 ratio of syrup to rum will generate a rather strong drink, while a 1:1 ratio of both components will give rise to a softer mix.
2. Adjust the amounts of solutions and put the final mixture in a refrigerator for at least 18 hours.
3. Drink slowly . . .

Notes

1. Read Chapter 2 carefully before initiating this experiment in your laboratory.
2. Do not throw away the remainder of the mixture in the sink if you prepared too much. You may keep it for another celebration, or drink it at home.

3. You may prefer to use strawberries, limes, lichees, or other exotic fruits instead of pineapple, to prepare the rum in step 2.

Reference

Crochet, J., (1987), personnal communication.

APPENDIX

ADDRESSES OF MAIN OFFICES FOR SUPPLIERS CITED IN THIS MANUAL*

AMERSHAM INTERNATIONAL plc (AME)
Amersham Place, Little Chalfont
Buckinghamshire
England, HP7 9NA
Telephone: (24 04) 4444
Telex: 838818 ACTIVL G

APPLIGENE (APL)
74 Route du Rhin
B.P. 72
Illkirch Cedex
France
Telephone: 88 67 22 67
Telex: 880 179 F

BECKMAN (BEC)
P.O. Box 10200
Palo Alto
California 94304
USA
Telephone: (415) 857 1150
Telex: 0678413

BETHESDA RESEARCH LABORATORIES (BRL)
GIBCO-BRL
P.O. Box 6009
Gaithersburg
Maryland 20877
USA
Telephone: (301) 840 8000 or (800) 638 4045
Telex: 64210 BRL GARG UW

BIO-RAD LABORATORIES
2200 Wright Avenue
Richmond
California 94804
USA
Telephone: (415) 234 4130
Telex: 335 358

BOEHRINGER MANNHEIM (BOE)
Biochemica
P.O. Box 310 120
D-6800 Mannheim 31
West Germany
Telephone: (49) 621 759 1
Telex: 041 462 420

BIOTECHNOLOGY RESEARCH ENTREPRISES S.A. Pty. Ltd.
G.P.O. Box 498, Adelaide,
South Australia, 5001
Telephone: (08) 228 5361
Telex: AA89141

*Abbreviations in brackets are those used in Tables 27, 39, and 77. See for local representatives.

CALBIOCHEM (CAL)
P.O. Box 12087
San Diego
California 92112-4180
USA
Telephone: (800) 854 9256 or (619) 450 9600
Telex: 697934

GENOFIT (GEN)
110 Ch. du Pont du Centenaire
C.P. 239
1212 Grand Lancy 1
Switzerland
Telephone: (22) 71 44 44
Telex: 423 082 gfit ch

GEN-PROBE
9620 Cheasapeake Drive,
San Diego
California 92123
USA
Telephone: (619) 268 8400

HOEFER SCIENTIFIC INSTRUMENTS
P.O. Box 77387
654 Minnesota Street
San Francisco
California 94107
USA
Telephone: (800) 227 4750 or (415) 282 2307
Telex: 470778

**INTERNATIONAL BIOTECHNOLOGIES INC.
(IBI)**
P.O. Box 9558
275 Winchester Avenue
New Haven
Connecticut 06535
USA
Telephone: (800) 243 2555 or (203) 562 3878
Telex: 643993

MILES LABORATORIES (MIL)
Elkhart
Indiana 46514
USA
Telephone: (219) 264 8111
Telex: 258450 MILES LAB EKR

MOLECULAR BIOSYSTEMS INC.
11180-A Roselle Street
San Diego
California 92121
USA
Telephone: (714) 452 0681

NEW ENGLAND BIOLABS (NEB)
32 Tozer Road
Beverly
Massachusetts 01915-9990
USA
Telephone: (617) 927 5054
Telex: 6817316

NEW ENGLAND NUCLEAR (NEN)
549 Albany Street
Boston
Massachusetts 02118
USA
Telephone: (800) 225 1572 or (617) 482 9595
Telex: 94 0996

PBS-ORGENICS
47 rue Charles-Heller
B.P. 107
94402 Vitry sur Seine Cedex
France
Telephone: (1) 46 81 73 07

PHARMACIA (PHA)
S-751 82 UPSALA
Sweden
Telephone: (018) 16 30 00

PROMEGA BIOTEC (PRB)
2800 S. Fish Hatchery Road
Madison
Wisconsin 53711
USA
Telephone: (800) 356 9526 or (608) 274 4330
Telex: 910-286-2738

SCHLEICHER AND SCHUELL GmbH
P.O. Box 4
D 3354 Dassel
West Germany
Telephone: 05561-791-0
Telex: 965632

STRATAGENE (STR)
3770 Tansy Street
San Diego
California 92121
USA
Telephone: (800) 424 5444 or (619) 458 9151
Telex: 910380 9841

SIGMA CHEMICAL COMPANY (SIG)
P.O. Box 14508
St. Louis
Missouri 63178
USA
Telephone: (800) 325 3010 or (314) 771 5750
Telex: 434475

STEHELIN/ANGLIAN BIOTECHNOLOGY (ANB)
Spalentorweg 62
4003 Basel
Switzerland
Telephone: (61) 23 39 24/23
Telex: 962 317 PHSTE CH

TAKARA SHUZO Co Ltd (TAS)
Kyoto
Japan
Telex: 5422127 TAKARA J

UNITED STATES BIOCHEMICAL CORPORATION (USB)
P.O. Box 22400
Cleveland
Ohio 44122
USA
Telephone: (800) 321 9322 or (216) 765 5000
Telex: 980718

INDEX